# AUTOMOTIVE ENGINE PERFORMANCE

## TUNEUP, TESTING, AND SERVICE

## VOLUME 1: TEXT

**KEN LAYNE** CMAT, SAE

# AUTOMOTIVE ENGINE PERFORMANCE

## TUNEUP, TESTING, AND SERVICE

## VOLUME 1: TEXT

**Second Edition**

PRENTICE HALL CAREER & TECHNOLOGY
ENGLEWOOD CLIFFS, NEW JERSEY 07632

*Library of Congress Cataloging-in-Publication Data*

Layne, Ken.
Automotive engine performance: tuneup, testing, and service / Ken Layne.—2nd ed.
p. cm.
Contents: v. 1. Text
Includes index.
ISBN 0-13-059775-9
1. Automobiles—Motors. 2. Automobiles—Motors—Maintenance and repair. I. Title.
TL210.L38 1993 92-24357
629.25'04—dc20 CIP

Acquisitions Editor: Robert Koehler
Cover Designer: Marianne Frasco
Production Editor: Ruth Cottrell
Manufacturing Buyer: Ed O'Dougherty
Prepress Buyer: Ilene Levy


**Prentice-Hall, Inc.**
**A Simon & Schuster Company**
**Englewood Cliffs, New Jersey 07632**



Printed in the United States of America.

10 9 8 7 6 5

ISBN 0-13-059775-9

Prentice-Hall International (UK) Limited, *London*
Prentice-Hall of Australia Pty. Limited, *Sydney*
Prentice-Hall Canada, Inc., *Toronto*
Prentice-Hall Hispanoamericana, S.A., *Mexico*
Prentice-Hall of India Private Limited, *New Delhi*
Prentice-Hall of Japan, Inc., *Tokyo*
Simon & Schuster Asia Pte. Ltd., *Singapore*
Editora Prentice-Hall do Brasil, Ltda., *Rio de Janeiro*

# DEDICATION

A lot of writers dedicate books to their wives, husbands, or other family members whose forbearance has helped them through the trials and tribulations of getting a book into print. I think Roseann would find that patronizing. In fact, my wife is a far better teacher than I. She is also a skilled librarian. Working in both capacities, she contributed many of the chapter review questions to this text and catalogued several thousand illustrations. If you find value in these areas, you may thank the lady.

The acknowledgments page of this book lists the contributions of many people and companies who helped make this work possible. I want to extend special thanks to two people. Bob Mahaffay, former engineering director of the Gousha/Chek-Chart company, has done more in his career to aid the automobile service profession than any other individual that I can think of. Don Nilson, retired head of the automotive technology department at Las Positas College in Livermore, California, is a "teacher's teacher." His career defines "professionalism" in automobile service. Both of these gentlemen have been looking over my shoulder and keeping me straight in automobile service publications for more years than I care to recount. Without their assistance, this book would not exist.

Additionally, I must acknowledge and thank one of the outstanding college automotive technology programs in the country: the Auto Tech Department at DeAnza College in Cupertino, California. This school has been a leader in the automotive service education field for well over 20 years even though it provides me a classroom forum to hold forth on the subject of automotive electronics.

Any textbook, any educational program, is less than half complete without skilled teachers. The best information in print and the best training plans are enhanced twofold by a teacher who can bring a subject to life and demonstrate its importance. Without teachers, there is no learning; so this book is dedicated to:

*Bill Abbott*
*Paul "X" Derry*
*Paul Hebert*
*Hank Nesel*
*Karl Pape*
and especially,
*Jerry Talbert . . . teachers.*

# BRIEF CONTENTS

# EXPANDED CONTENTS

## PART TWO



## PART THREE

# PREFACE

No technical service profession demands more knowledge and higher skills from its members than the automobile service profession. In the early days of the automobile, a person could do a lot of service jobs with a good measure of horse sense and a box full of blacksmith's tools. Those days are long gone.

We cannot underestimate the knowledge and skills of an old-time mechanic, however. Professional mechanics of past generations laid the foundation for modern automobile service. A professional service technician today, though, must have equal measures of:

**1.** Knowledge of automobile system operation

**2.** The service skills necessary to keep those systems operating correctly.

## TWO VOLUMES—SYSTEM DESCRIPTION AND SERVICE

Because a professional technician must know how systems operate and how to service them, we have presented this body of information in two volumes: AUTOMOTIVE ENGINE PERFORMANCE: TUNEUP, TESTING, AND SERVICE, the first volume, will help you learn how engine combustion control systems operate. The second volume, AUTOMOTIVE ENGINE PERFORMANCE: TUNEUP, TESTING, AND SERVICE PRACTICE MANUAL, will help you learn the service skills that are the true measure of a master mechanic.

AUTOMOTIVE ENGINE PERFORMANCE explains the operation of fuel, ignition, emission control, and electrical systems on late-model vehicles. It progresses from the fundamentals of combustion control through specific examples of these systems from major carmakers. These descriptions emphasize the similarities among various systems and the basic combustion control principles used by all manufacturers.

The Practice Manual provides the testing and repair instructions that will be the foundation of your professional service skills. This volume contains step-by-step procedures similar to those you will find in service manuals from carmakers and service guide publishers. These instructions will help you to understand professional reference manuals and use them as basic tools in your service work. Moreover, the Practice Manual explains the *reasons* for different service procedures and the common features of all carmakers' service methods.

## SERVICE INSTRUCTIONS COMPLEMENT INDUSTRY REFERENCE MANUALS

We do not have the room in this book to include every manufacturer's specific test and service procedures. No textbook can do that. The service procedures in the Practice Manual are basic guidelines for engine performance testing and service. They will familiarize you with the principles and common points of engine maintenance. These service instructions will also acquaint you with the ways in which carmakers and independent manual publishers produce service procedures and specifications.

Reference manuals are essential tools for service of late-model cars. The second chapter of the Practice Manual explains how to use various service publications. For every job that you do, you will need the following information from factory shop manuals or independent service guides:

- Specifications for parts and adjustments
- Step-by-step test and service procedures
- Carmakers' preventive maintenance schedules
- Test and service equipment operating instructions
- Flat-rate labor time standards.

The basic service skills that you learn as you study engine performance service will lay the foundation for your career as a professional automobile service technician.

## SYSTEM ENGINEERING AND SYSTEM SERVICE

This text applies a system-engineering and system-service viewpoint to engine performance maintenance. The first chapter of AUTOMOTIVE ENGINE PERFORMANCE lays this foundation. On today's cars, fuel delivery, ignition, and emission control are fully integrated combustion control systems, regulated by on-board computers. You can't think of emission controls as items that are separated from fuel and ignition components. Also, when you work on fuel and ignition components, you must know how a change in one part or system will affect other systems. In addition, you must know how the operation of any one part will affect overall performance, economy, and emission control.

Don't get the idea that a good mechanic ever ignored any fuel, ignition, or emission system component when servicing an engine for maximum efficiency. The point is that on today's cars, you can't think about them separately for a moment. You have to understand total engine operation. Engine performance service, or "tune-up" has always been an interesting and profitable automotive service field. Whether you specialize in this area or in general automobile service, this book will help you understand modern engine service.

Each volume is divided into six major parts that contain chapters on related subjects and engine systems:

- An introduction to the principles of engine performance and to common shop practices, tools, and materials
- Basic engine lubrication, ventilation, cooling, and exhaust system operation and maintenance
- Engine electrical system operation and maintenance
- Ignition system operation and service
- Fuel systems and emission controls
- Electronic engine control systems

These divisions follow the traditional viewpoint toward comprehensive engine performance service. This arrangement also highlights the fact that all engine systems on today's cars are fully integrated through electronic controls.

## INDUSTRY TESTING AND TRAINING STANDARDS

The arrangement of the subjects in these volumes covers the skill requirements outlined in the specifications for the Automotive Service Excellence (ASE) engine performance certification test. Review questions at the end of the chapters are written in the style of ASE test questions. These will not only help you review the chapter contents, they will help to prepare you for the actual certification tests.

Additionally, the organization and contents of these volumes are based on automotive industry training standards contained in American National Standards Institute (ANSI) publication D18.1-1980. This is your assurance that this text covers the knowledge and skills that the service industry requires of its members.

## FULLY ILLUSTRATED TEXT

The two volumes contain almost 2,000 illustrations. Many are original drawings and photographs, prepared expressly for these volumes. Others have been selected from carmakers' late-model service manuals and adapted to illustrate the contents of these books. The result is a fully integrated combination of text and illustrations that *shows and tells* you how vehicle systems operate and how to service them.

The Practice manual contains several step-by-step photographic overhaul procedures for alternators, starter motors, carburetors, and ignition distributors. These procedures provide detailed illustrations and instructions for typical methods required to service these vital engine performance components.

Major sections of the chapters in the Practice Manual also explain the use of electrical diagrams, troubleshooting charts, and carmakers' illustrated service procedures.

## OTHER FEATURES OF THESE VOLUMES

In addition to the fully illustrated text, these volumes contain the following features that will help you learn and practice professional engine performance service:

- *Safety*—Safety is good business, and following all necessary precautions will ensure the safety of yourself, your shop, and your customers. The first chapter of the Practical Manual summarizes the general safety practices for engine performance service. Specific CAUTIONS and WARNINGS appear in the procedures of other chapters to help you do various repair jobs safely.
- *Learning Objectives*—Each chapter begins with a list of basic knowledge or skills that you should acquire from studying the chapter. Again, these objectives follow the general requirements of ASE certification testing and industry training standards.
- *Technical Terms*—Key words and phrases important to a service technician are printed in boldface type. These terms are explained in the text and collected in the glossary at the end of each volume. Studying and understanding these will help you build the vocabulary of a professional mechanic.
- *Essays on Related Topics*—Short essays on topics related to engine performance service are included throughout the first volume. These short subjects deal with specific service tips, system designs, and the historical development of automobile engine systems. They will add depth to your understanding of the automobile industry and help you in the service profession.
- *Measurement Data*—An appendix at the end of each volume contains important information on threaded fasteners and on customary and metric measurement standards that you will use in your service work.

## NEW FEATURES FOR THE SECOND EDITION

Additions and changes to these volumes for the second edition consist basically of bringing the technology up to date. Engine control systems and engine performance service have evolved rapidly during the 1980's. Fuel injection became the "conventional" fuel delivery system for most automobiles, and direct ignition systems (DIS) are showing the same trend in the 1990's. The self-diagnostic capabilities of on-board computer systems also have improved dramatically. The new on-board diagnostic II (OBD II) systems of 1994 models promise standardization of basic test methods among all carmakers.

The second edition of these volumes offers:

• Increased coverage of DIS installations on domestic engines.

• Improved explanations of computer principles shared by all carmakers.

• Expanded overviews of fuel injection systems and basic service procedures for domestic and imported vehicles.

• Comprehensive insight into the self-diagnostic methods used on GM, Ford, and Chrysler vehicles.

# ACKNOWLEDGMENTS

The author and the publisher wish to thank the vehicle manufacturers and other companies in the automotive service industry that contributed information and illustrations for this book. Automobile service is a highly technical, rapidly advancing profession. This book would not have been possible without the assistance of the following companies:

American Isuzu Motors Inc.

Atlas Supply Company

Automotive Technician Associates, Inc. (An Organization Dedicated to Better Mechanics)

Balco, Inc.

B&M Automotive Products

Champion Spark Plug Company

Chilton Book Company

Chrysler Corporation

Clayton Industries

Federal-Mogul Corporation

FEL-PRO Incorporated

Ford Motor Company (Ford Parts and Service Division)

Garrett AiResearch Industrial Division

The Gates Rubber Company

General Motors Corporation:
- AC-Delco Division
- Buick Motor Division
- Cadillac Motor Car Division
- Chevrolet Motor Division
- Delco-Remy Division
- Oldsmobile Division
- Pontiac Motor Division
- Rochester Products Division
- Saginaw Steering Gear Division

Hennessy Industries, Inc.: Coats Diagnostic Division

Industrial Indemnity Company

K-D Manufacturing Company

Kent-Moore Tool Division

Mazda Motor Corporation

McCord Gasket Corporation

Mitchell Manuals, Inc.

Mitsubishi Motor Sales of America, Inc.

Motor Publications (The Hearst Corporation)

National Institute for Occupational Safety and Health

Nissan Motor Corporation in U.S.A.

Porsche + Audi

Prestolite Division of Bunker Ramo-Eltra Corporation (a subsidiary of Allied Corporation)

Snap-on Tools Corporation

Stanley Proto Industrial Tools

Society of Automotive Engineers (SAE)

The L. S. Starrett Company

Sun Electric Corporation

Robert Bosch Sales Corporation (Robert Bosch GmbH)

Volkswagen Of America, Inc.

Volvo of America Corporation

Weatherhead Division, Dana Corporation

Additionally, the following individuals contributed information and research materials or reviewed selected parts of the manuscript. The author wishes to thank:

*Robert "Charlie" Chapman*—Software Engineering, Tujunga, CA

*Michael DeMiniconi*

*Peter Felice*—Red Line Synthetic Oil Corporation, Martinez, CA

*James Geddes*

*William J. Hanley*

*Carlton Hardy*—Ford Motor Company Training Center, Milpitas, CA

*Bob Kruze*—Saturn Corp.

*Russ Ostler*—Chrysler Corporation, Livermore, CA

*Ben Sims*—California State Automobile Association

*Thomas M. Terrell*—Garrett Air-Research Industrial Division, Torrance, CA

*Chris Wood*—Grant & El Camino Chevron, Mountain View, CA

Photographs for the component overhaul sequences and many other photos are by *Kalton C. Lahue*.

Finally, this text wouldn't exist at all if Ruth Cottrell hadn't put it all together.

# PART ONE
# INTRODUCTION TO ENGINE PERFORMANCE

# INTRODUCTION

This text is about automobile engine systems (combustion control systems). This text has two main themes:

• The best way to understand how any single part of an automobile works is to know how it operates as part of a complete system. The best way to test and service a part is to treat it as part of a larger system. The best way to understand any system is to understand its overall purpose and how each part works to help the system do its job.

• All systems of an automobile operate on common scientific principles. If you understand some of the basic facts of science, you can recognize them in different systems. This, in turn, will help you understand how the systems and their parts work.

In Part One of this text, you will learn about the "system concept" for automobile design and service. You also will review many of the basic facts of science that operate in all automobile systems. Even though these first four chapters deal with general principles, we use examples from various automobile systems to illustrate the basic ideas.

Part One also includes an explanation of the measurement systems used for all parts of a car. Your ability to understand and use measurement units in the U.S. customary and metric systems will make you a better professional mechanic—an automotive technician.

# 1 INTRODUCTION TO ENGINE AND COMBUSTION CONTROL SYSTEMS

## WHAT

This book is about automobile engine and combustion control systems, how they work, and how to service them. This book approaches engine performance service from a *systems viewpoint*. Therefore, it makes sense to begin with a chapter that introduces the nature of systems in general and automobile engine systems in particular. This chapter will help you to see the automobile powerplant as a complete system. Then you will recognize how other systems work together so that the engine can do its job.

## WHY

It's easy enough to look at an engine and its combustion controls and see the separate pieces one at a time: engine block, crankshaft, pistons, connecting rods, carburetor, distributor, emission control devices, and so on. People drive cars very well every day without completely understanding how they work. A person can even do simple repair jobs like replacing spark plugs or an air filter or changing the engine oil without knowing why spark plugs and filters are built as they are or why engines even need oil. But a professional mechanic—an automotive technician—is not a parts swapper. A professional needs to know more.

If a professional mechanic is going to be a *master* of the profession, he or she needs to understand how and why an engine and its combustion control systems are built as they are and how all of the parts and systems work together. Through this understanding, you, as a professional mechanic, will be able to do a better job providing complete engine performance service.

## GOALS

After studying this chapter, you should be able to:

1. List the common features of all systems.
2. Identify and list the major systems of an automobile.
3. List the engine-related systems of an automobile and the basic function of each.
4. Explain how these major systems work together and relate to each other.
5. Develop a system viewpoint toward complete engine performance service.

## SYSTEM ENGINEERING

In this book, you will study the engine and the fuel, ignition, electrical, and emission systems as a *single complete system* that powers an automobile. That seems like a simple idea, but you will see that the automobile powerplant is a complex system, made up of several other equally complex subsystems. Each is distinct from the others, but they all share some common features and goals that allow them to work together. Moreover, the powerplant is part of the larger, complete vehicle system. As you develop the system viewpoint, it will be easier for you to understand how all the parts of a modern automobile work together and how you can provide complete system service.

Engineers use the term "system engineering" to describe the process of designing all parts of a system to do their individual jobs and to accomplish the purpose of the complete system. "System integration" describes the operations that tie together the functions of several systems so that each can work properly, or work better, in relation to the others.

Electronic control systems provide the system integration on modern automobiles, and you will study them thoroughly in this book. Because electronic systems integrate the operations of the engine and its combustion controls, you must practice "service integration" when you work on a late-model vehicle. You can't service *just* the carburetor, the distributor, or a single emission device without un-

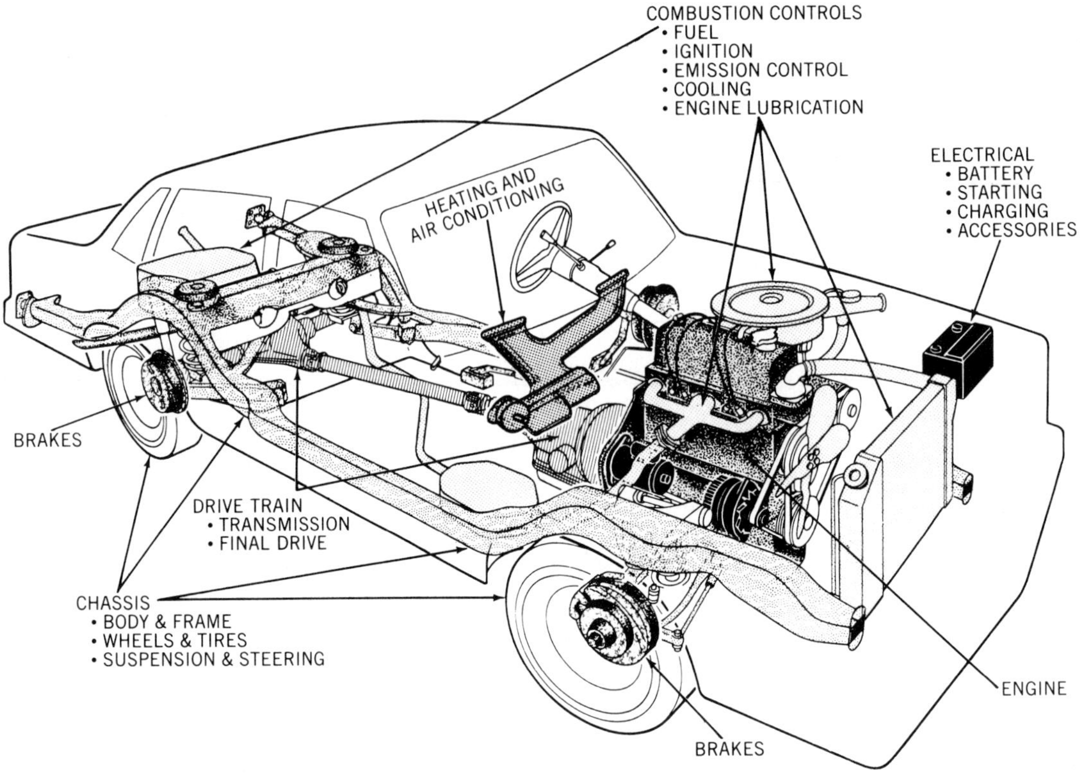


**Figure 1-1.** Engineers and professional service personnel recognize these seven major systems in an automobile.

derstanding how its operation affects the complete combustion process. One problem in a single electronic sensor can drastically upset ignition, fuel delivery, and emission control.

The system viewpoint is not new. It developed in the automobile engineering and service professions as various vehicle systems developed and became more complex. Professional mechanics have always had this viewpoint toward automobile service. Today, it is essential. If you learn system integration and service integration as you study engine performance service, you will be a truly professional service technician.

## SYSTEM FEATURES

We apply the name "system" to many different things: school systems, computer systems, transportation systems, and so on. The system viewpoint is based on the idea that all systems have common features. The purpose of this book is not to explain the philosophy of system engineering. But, if you understand the characteristics of any system, you can use this knowledge for a clearer view of your service work. Here are some features of all systems in general that you can see in the systems of an automobile:

**1.** Every system has one final purpose, or goal.

**2.** Every system has several parts, or smaller systems, each with its own purpose. The final purpose of the larger system, however, is different from that of each of its parts.

**3.** Every system has interaction among its parts, all working toward the final purpose.

**4.** The final goal of the system cannot be reached by anything less than *all* of the parts working together. If you remove one part or subsystem, the total system stops working.

**5.** Every automobile system works on common principles of chemistry, physics, and electronics.

At this point, you may be wondering about the use of the words "system" and "subsystem." When we talk about the fuel "system" is that different from the fuel "subsystem" of the total engine "system"? No, it isn't. It really doesn't matter whether you use the word "system" or "subsystem." What does matter is that you get the idea of a system as a group of related parts that work together for a common purpose. You also must understand that two or more smaller systems can be part of a larger one. Each of the smaller systems has its own purpose, but they depend on each other to achieve the larger purpose of the larger system.

## AUTOMOBILE SYSTEMS

The automobile is a complete transportation system, but engineers and professional service people generally recognize seven major systems, or groups of systems, within the automobile, figure 1-1:

**1.** The engine

**2.** Combustion control systems

**3.** Electrical systems

**4.** Heating, air conditioning, and other

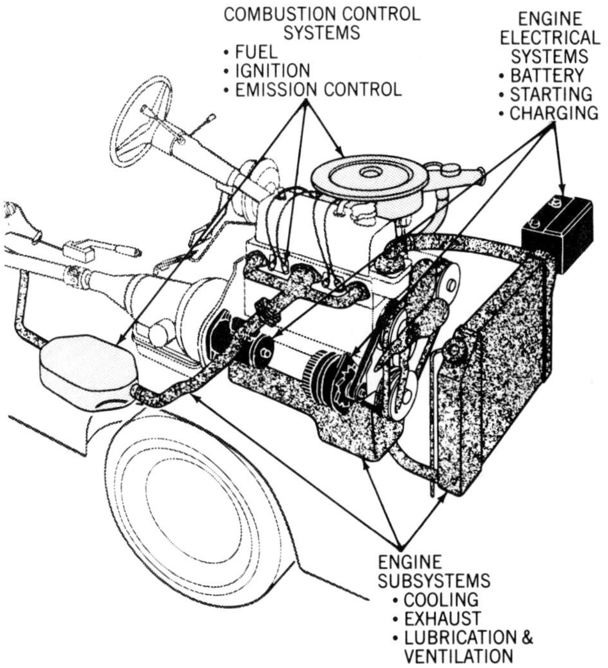


**Figure 1-2.** The three major engine systems have nine related subsystems.

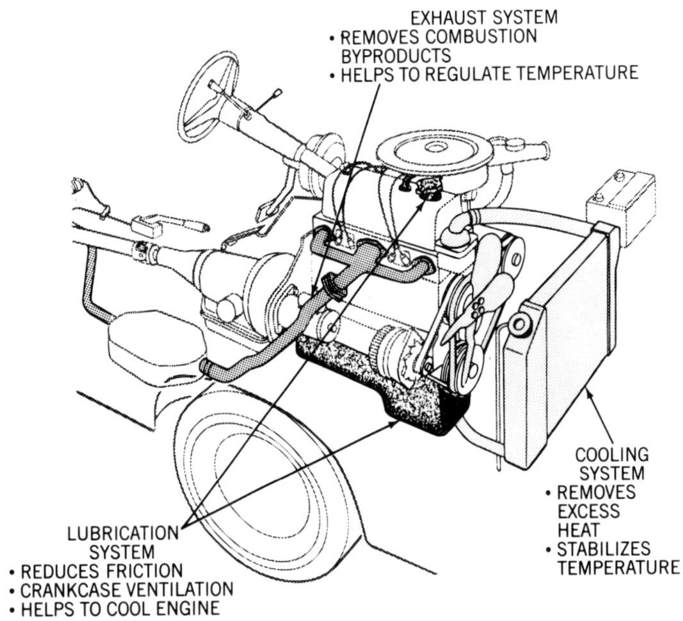


**Figure 1-3.** The engine lubrication, cooling, and exhaust systems contain these principal parts.

accessory systems

**5.** Drive trains (power transmission systems)

**6.** Chassis systems (wheels, tires, suspension, and steering)

**7.** Braking systems

These vehicle systems are not separate and independent. You don't need combustion controls if you don't have an internal-combustion engine. You can't operate air conditioning and electrical accessories without a basic electrical system. A braking system needs wheels, tires, and suspension to operate effectively.

### Engine Systems

This book deals with the first three major vehicle systems, figure 1-2:

**1.** The engine and its related subsystems:

- Engine lubrication and ventilation systems
- The cooling system
- The exhaust system

**2.** Combustion control systems:

- The fuel system
- The ignition system
- Emission controls

**3.** Engine electrical systems:

- The battery
- The starting system
- The charging system

This is the organization that you will follow in this book.

## SYSTEM FUNCTIONS

We said that every system has a final purpose or an overall function. We also said that the functions of several small systems combine to achieve the purpose of a larger system. Understanding the basic functions of engine systems is an important step toward complete engine performance service.

### Engine Functions

Obviously the engine's function is to power the automobile, and the engines in all modern automobiles are internal-combustion engines. Electric motors and steam engines can be (and have been) used to power automobiles, but for 100 years, the internal-combustion engine has been the most practical, reliable, and efficient automobile powerplant.

An internal-combustion engine changes chemical energy of fuel into heat energy through combustion. The engine then converts heat energy to mechanical power. The engine is an air pump that draws in oxygen in the air and compresses it for the combustion reaction. The combustion process is the same whether the engine burns gasoline, propane, or diesel fuel. An engine develops power through one of several common mechanical cycles. To develop combustion and mechanical power, an engine needs the support of other systems. Here are the basic functions of three major engine-related systems, figure 1-3:

- *The lubrication system* reduces friction between moving parts and helps to cool the engine.
- *The cooling system* stabilizes engine temperatures for the most efficient combustion and removes excess heat.
- *The exhaust system* removes combustion byproducts and also helps to regulate combustion temperatures.

### Combustion Control Functions

On older automobiles, it often was possible to think of fuel, ignition, and

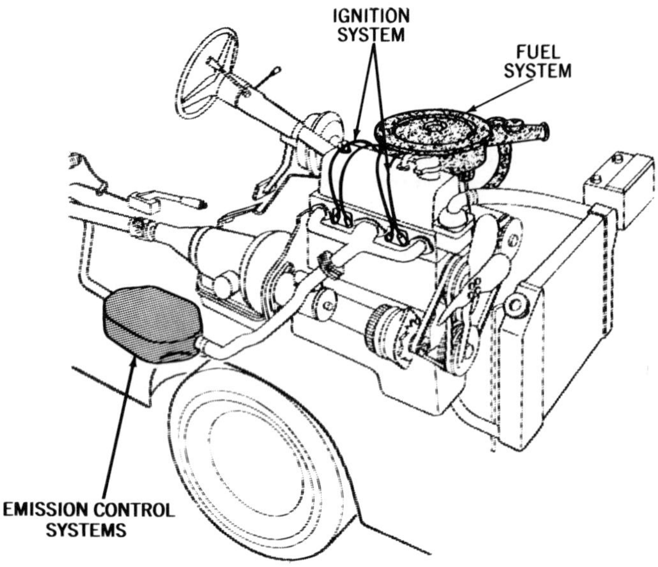


**Figure 1-4.** The fuel, ignition, and emission control components are the complete combustion control systems.

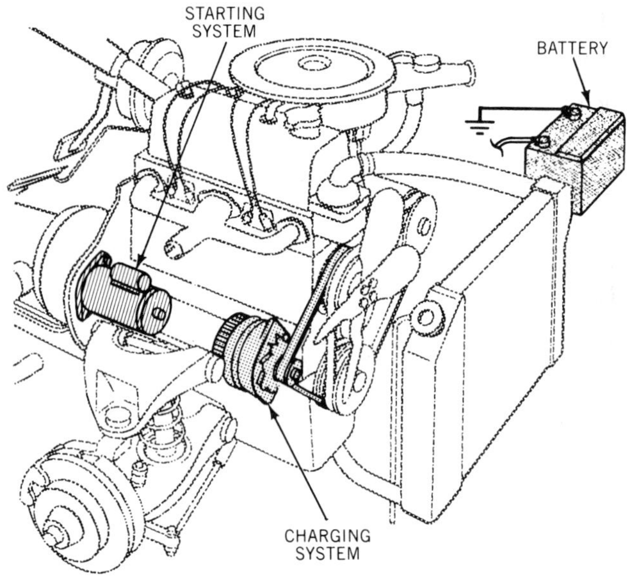


**Figure 1-5.** The battery, the starting system, and the charging system are electrical systems that affect overall engine performance.

emission control functions separately. These systems really don't operate independently; and on late-model cars, the system functions are fully integrated by electronic controls. We refer to these systems as combustion controls because they work together to deliver fuel to the engine, mix it with air, ignite it for efficient combustion, and reduce the emission of harmful combustion byproducts. However, each combustion control system, figure 1-4, has its own individual functions:

- *The fuel system* stores fuel, delivers it to the engine, and mixes it proportionally with the engine intake airflow for maximum combustion efficiency. These purposes are the same whether the fuel is gasoline, diesel fuel, or propane.
- *The ignition system* controls combustion timing to develop maximum power. In a gasoline engine, the ignition system causes and controls combustion with a timed electric spark. In a diesel engine, ignition is controlled by fuel injection timing.
- *The emission controls* work with the fuel and ignition systems to reduce air pollutants emitted by the engine. The fuel and ignition systems, themselves, are major emission control systems. Other emission devices reduce pollutants in engine exhaust after combustion or limit their formation in the first place.

### Engine Electrical System Functions

The earliest automobiles did not have electrical systems. The driver had to crank the engine with muscle power to start it, and many cars had *no* electrical accessories. Ignition was provided on early engines by something called a flywheel magneto, or even earlier by nonelectric, "hot-tube" ignition.

Internal-combustion engines could not have advanced to their current state of development if it were not for battery ignition systems, electric starter motors, and alternators to provide electric current. The fuel, ignition, and electronic control systems on modern engines rely on the functions of these electrical systems, figure 1-5:

- *The battery* provides electric power to the starter and a reserve electrical capacity to operate other electrical systems.
- *The starting system* cranks the engine to start it.
- *The charging system* recharges the battery and provides electric current to all other electrical devices on the automobile.

Complete engine performance service must consider the functions of these systems and their effects on overall engine operation.

### System Integration

The preceding sections provide just thumbnail sketches of system functions. You will learn more about them as you work through this book. If you understand the basic *purpose* of a vehicle system, you can understand more easily *how* and *why* it works as it does. Then you can understand completely how to service and repair it. Also, if you understand the basic purpose of a system, you can recognize how the system has been developed and improved in today's automobiles. More importantly, you will recognize how it will continue to develop and improve in tomorrow's cars.

## ENGINE PERFORMANCE

Because this is a book about engine performance service, we can't end this chapter without defining performance. This means different things to different people, but for today's cars, it means the best combination of:

- Power

- Fuel economy
- Emission control

for any given engine and vehicle. Your goal as a professional service person is to maintain optimum engine performance.

## SUMMARY

We said at the start of this chapter that the system concept is an interesting way to look at an automobile and that it can help you be a better service technician—a master mechanic. Using this concept, we have seen that the automobile engine system is made up of other systems. Each has its own purpose, but all systems are inseparable if the engine is to accomplish its overall performance function.

A system is a group of related parts that work together for a common purpose. Several small systems can be part of a larger one. Each of the smaller systems has its own purpose, but they depend on each other to achieve the larger purpose of the larger system. Furthermore, all systems work on common principles.

The next few chapters outline the engineering principles of automobile engine and combustion control systems. With a basic knowledge of engineering fundamentals, you can see how common principles operate in different systems. The important thing to do with this knowledge is to understand the development of automobile systems: what they were in last year's cars, and how they will develop next year.

### EARN YOUR STRIPES

Every group of professional people has a professional association. Doctors have the American Medical Association (AMA); lawyers have the American Bar Association (ABA). Engineers have various engineering societies, and accountants have accounting organizations. Among the features of any professional association are the goals of helping its members improve their professional abilities, setting high standards of competence, and gaining public recognition for skills of its members.

Professional auto mechanics—automotive technicians—have a professional society that is the equal of any for advancing and certifying the skills of its members. That is the National Institute for Automotive Service Excellence (NIASE). Founded in 1972 by leaders in the automobile manufacturing and service industries, consumer groups, and educators, the NIASE is an independent, nonprofit corporation that provides a program to test and certify the skills of professional mechanics. Today, the NIASE offers a series of eight competency tests for automobile mechanics, six for heavy-duty truck mechanics, and two for body repair and painting specialists.

You can earn NIASE certification in any or all of these automotive service fields by taking the tests and working in the service profession for at least two years. Mechanics who pass the tests receive shoulder patches and stripes with the name of each test passed, plus certificates and credentials that inform the public of their skills. Moreover, NIASE requires certified mechanics to take recertification tests every five years to ensure that their skills are up to date. Few other professional societies have this requirement.

In 1983, the NIASE changed its certification symbol from NIASE to simply ASE for "Automotive Service Excellence," while retaining the full corporate NIASE name. In its two decades of service to the automobile industry, NIASE has administered about 2.5 million separate tests and certified almost 250,000 individual mechanics in one or more service specialties.. Perhaps you already have taken the NIASE tests. If not, you probably will if you work as a professional mechanic. They are a great opportunity for you to improve (and prove) your skills and to earn your stripes.

ASE CERTIFIED ®

# 2 THE AUTOMOBILE AND BASIC SCIENCE

## WHAT

Chapter 1 introduced the idea of looking at an automobile as a complete system made up of other related systems. It also encouraged you to look for the principles of how and why things work so that you can understand them completely. We will start by reviewing the principles of science that went into the design of a part or a system. If you understand these principles, you can recognize the same ones in several different systems. This, in turn, will help you understand more completely and easily how the systems work. This chapter is a brief review of principles you have learned in earlier studies. They are the foundation of engine and combustion control system operation.

## WHY

This chapter will not help you to be an engineer. That's not its purpose. It will help you understand the principles that engineers have applied to the design of several automobile systems. The automobile is one of the best examples of applied science that we have. If you understand the simple starting points of chemistry, physics, and mechanics that are shared by engines, electrical systems, and fuel systems, your job of servicing and repairing them will be easier.

## GOALS

After studying this chapter, you should be able to:

**1.** Explain physical states of matter and define elements, molecules, compounds, and mixtures.

**2.** Define the three parts of a simple atom and explain the structure of all atoms.

**3.** Explain positive and negative charges.

**4.** Define oxidation and reduction reactions.

**5.** Define the following terms: energy, work, force, mass and weight, torque, power, inertia, momentum, and friction.

**6.** Explain pressure, temperature, and volume relationships in gases and liquids and the effects of pressure on gases and liquids.

**7.** Define atmospheric pressure and vacuum.

**8.** Define gear ratios and explain torque, speed, and gear ratio relationships.

**9.** Explain the operation of belt drives and chain drives in engine systems.

## MATTER AND ATOMS—CHEMICAL BUILDING BLOCKS

One of the simplest scientific principles is that everything in the universe consists of **matter**, and the simplest way to view matter is in its physical form. Matter exists in three physical forms: solid, liquid, and gas. Metals in an engine are solid. Gasoline and engine coolant are liquids. The air that an engine takes in and the exhaust it gives off are gases.

Matter can change its form, of course. Water is normally liquid, but it can freeze to solid ice or **vaporize** to gaseous steam. The actions by which matter changes its physical state are important in the operation of automobile engine systems. An important job of the fuel system is to vaporize liquid gasoline, for example.

Later in this chapter you will study the way energy is used to change the physical form of matter. First, let's review the chemical nature of matter.

### Atoms and Elements

Whether it exists as a solid, a liquid, or a gas, all matter begins with simple building blocks called **atoms**. There are only a few more than 100 different kinds of atoms. Of these, 92 exist in nature, and another dozen or so have been created in laboratories.

Each different atom is called an **element**. Although the atoms of one element are different from the atoms of all other elements, all have the same basic parts. The atomic parts are the things that let atoms of one element join with each other or combine with atoms of other elements. The actions of atomic parts also are the basis for electricity and combustion.

### Electrons, Protons, and Neutrons

As simple as an atom is, its parts are simpler. They are the **electron**, the **proton**, and in most atoms the **neutron**. Protons and neutrons are at the center, or **nucleus**, of every atom. The electrons of an atom circle, or orbit, the nucleus, figure 2-1. The differences between atoms of different elements are the numbers of electrons and

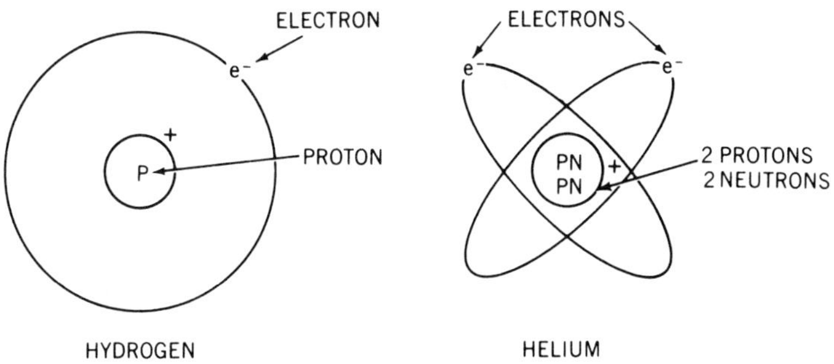


**Figure 2-1.** Hydrogen is the simplest element. It has one electron and one proton. Helium has two electrons, two protons, and two neutrons.

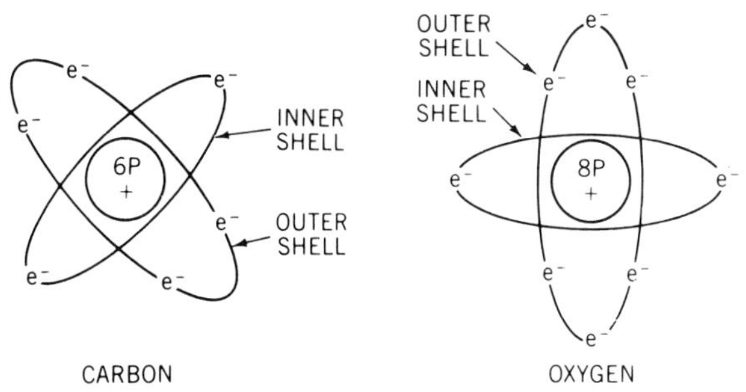


**Figure 2-2.** A carbon atom has six protons and electrons. An oxygen atom has eight of each. Carbon and oxygen both have two electron shells.

protons each element has. The simplest element is hydrogen, represented by the letter "H." It has one electron and one proton. Helium (He) is the second simplest element. It has a nucleus of two protons and two neutrons, with two electrons in its orbit, figure 2-1.

### Positive and Negative Charges

Electrons and protons have opposing electrical charges. This is a basic atomic theory that is the foundation for electrical theory and combustion. An electron has a negative charge, shown by a minus sign (−). A proton has a positive charge, shown by a plus (+) sign. A neutron has no charge; it is neutral.

In a balanced atom, the number of electrons equals the number of protons. The balance of the opposing negative and positive charges holds the atom together. The principle of charges is:

- Like charges (+ and +, or − and −) repel each other.
- Opposite charges (+ and −) attract each other.

The positive proton in the nucleus attracts the negative electron and keeps it from flying out of the atom. Because the electron is moving in orbit, **centrifugal force** keeps it from being drawn into the nucleus by the proton. The positive charges on protons in the nucleus don't force the protons apart because the neutrons neutralize the effects of the positive charges on each other. That's why all atoms (except hydrogen with only one proton) have neutrons.

### Simple and Complex Atoms

More complex atoms have more electrons, protons, and neutrons. Carbon (C) and oxygen (O) are common elements, figure 2-2, but are more complex than hydrogen and helium. A carbon atom has six protons in its nucleus and six electrons in its orbit. An oxygen atom has eight of each. These atoms are still balanced because they have equal numbers of electrons and protons. Atoms can become unbalanced by giving up or taking on electrons. An atom cannot give up or take on protons, however. Only electrons can be lost or gained.

An unbalanced atom with more or fewer electrons than protons is an **ion**. An atom with more electrons than protons is a negative ion. One with fewer electrons is a positive ion. Electron movement is the basis for electricity and for chemical combinations. Ions and ionization are important when you study engine ignition and combustion. Basic chemistry also will help you understand the formation and control of automobile emissions.

**Electron Movement.** Atoms of different elements can link together chemically by sharing electrons with each other. Look again at the carbon and oxygen atoms in figure 2-2. The electrons are not all in the same orbit. The orbits are arranged in shells, and each shell is progressively farther from the nucleus. Carbon and oxygen have two electron shells. More complex atoms have more shells. Each shell of any atom can hold certain maximum numbers of electrons. From the inner shell outward, there are seven possible shells. The maximum numbers of electrons in each shell are: 2, 8, 18, 32, 32, 9, and 2.

The *outer* shell, whether it is the second or the seventh, never has more than eight electrons. The next higher element forms another shell or adds an electron to a partially filled inner shell. The outer electron shell, or **valence shell**, is the most important for understanding chemistry and electricity.

Scientists say that the electron shells of an atom are energy levels. Atoms use this energy to try to complete their outer electron shells. An atom with few electrons in its outer, valence, shell combines easily with other atoms. This allows each atom in the combination to complete its outer shell by sharing electrons. Atoms with full outer shells don't combine with other atoms.

Atoms with only one or two valence electrons in their outer shells require little energy to cause these electrons to move to the shells of other atoms. Most such elements are metals: iron, copper, aluminum, and so on. Because it takes little energy to move the valence electrons of these atoms into other orbits, these elements combine easily with other elements, and they are good conductors of electric current.

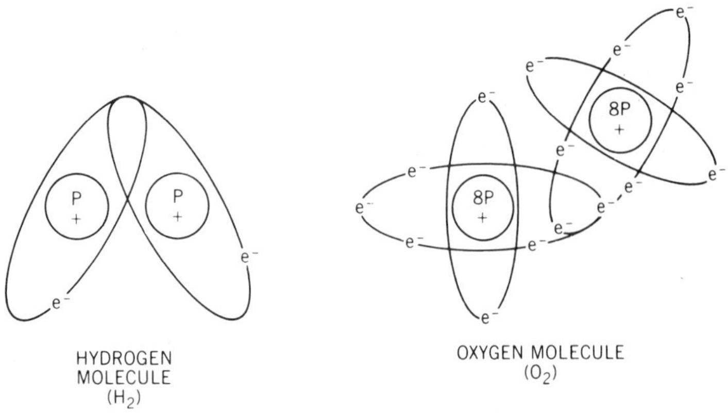


**Figure 2-3.** Hydrogen and oxygen in their natural states both exist as molecules of two atoms: $H_2$ and $O_2$.

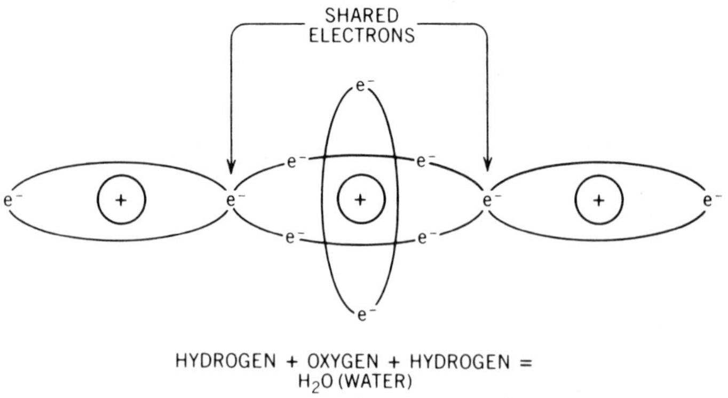


**Figure 2-4.** Water is a compound of two hydrogen atoms and one oxygen atom: $H_2O$.

Elements with six or seven valence electrons combine selectively with elements that have one or two valence electrons. But, atoms with many outer electrons need more energy to move the electrons into other orbits. Many of these elements are gases, and they are poor conductors of electricity because they need so much energy to move their electrons. You will learn more about electron energy and electricity later. These examples, here, show that electron energy is the basis for chemical reactions and for electric current.

## CHEMICAL REACTIONS

Atoms share valence electrons in two ways. They combine with another atom of the same element or they combine with an atom of another element. When two or more atoms join together, they form a **molecule**. Some elements in their natural state exist as molecules of two identical atoms. Hydrogen and oxygen exist naturally as molecules of two identical atoms, $H_2$ and $O_2$, figure 2-3.

### Compounds and Mixtures

Atoms of one element can combine with those of another in a **compound**. Two hydrogen atoms, each with one electron, can combine with one oxygen atom lacking two electrons in its outer shell. In this way the atoms are chemically bound together in a molecule of the compound $H_2O$, figure 2-4. There is another form of matter besides elements and compounds. That is a **mixture**. Mixtures are formed when molecules of elements or compounds mix together but *do not* share electrons.

Water ($H_2O$) is a simple compound. Ethylene glycol in engine antifreeze is a complex compound. When we put water and ethylene glycol together in a radiator, we do not have another compound. We have a coolant *mixture*.

The important things to remember about elements, compounds, and mixtures are:

- *Elements* are matter in its purest form. You can't break an atom of an element into smaller pieces and still have recognizable matter.
- *Molecules* are two or more atoms joined together by sharing their electrons in their outer shells. They can be atoms of the same or different elements.
- *Compounds* are molecules of atoms of two or more different elements joined by sharing the electrons in their outer shells.
- *Mixtures* are two or more elements or compounds that join together physically but do not share electrons in a chemical bond.

### Two Important Kinds of Reactions

Earlier, we used oxygen as an example of a common element that combines easily with other elements. Water ($H_2O$) is the most obvious oxygen compound. When oxygen combines with iron (Fe), we have iron oxide ($FeO_2$), or rust. When oxygen combines with silicon, we have $SiO_2$, or sand. Because reactions that form oxygen compounds are so common, they are called **oxidation reactions**. Oxidation reactions are basic to the operation of your automobile. Gasoline, a very complex mixture of compounds, is mixed with air that contains oxygen and drawn into the engine cylinders. There it is ignited and burns. The heat causes the oxygen to recombine with some of the gasoline molecules and form new compounds. Thus, combustion in an engine is an oxidation reaction.

If oxygen can be combined easily with other elements and compounds, it also can be separated from them. This is the opposite of an oxidation reaction. It is called a **reduction reaction**, which also is important in your automobile. You know that some of the new compounds formed when oxygen reacts with gasoline are harmful air pollutants. If some harmful pollutants are formed by oxidation, it stands to reason that it will take a reduction reaction to get rid of them. Other pollutants are formed by incomplete oxidation, and it takes further oxidation to eliminate them. You will learn more about oxidation and reduction when you study emission controls, but if you understand these two chemical principles now, that later study will be easier.

In discussing oxidation, we started

to get ahead of ourselves by mixing physics with chemistry when we talked about the heat of combustion. Actually, chemistry and physics can't be separated in an engine.

## SOME BASIC PHYSICS

As we have seen how atoms join to form molecules of elements and compounds, we have mentioned the **energy** of atoms, particularly of their electrons. Energy is not matter, however. Atoms and molecules are matter. Energy is a feature, or an ability, of matter. Here, physics joins chemistry.

Seventeenth-century scientists believed that the total amount of matter and its energy in the universe were absolutely fixed. The quantities never changed; only the form changed. Atomic scientists of the 20th century proved that isn't always true, but we needn't concern ourselves with atomic physics. The older, simpler theory that matter and energy do not increase or decrease, they only change, still applies in our cars.

If you ever have trouble understanding what happens when an alternator charges a battery, remember that an alternator only changes mechanical energy into electrical energy, figure 2-5. You can change it back again when you use the battery to power the starter motor and crank the engine. If you wonder what happens to matter in the form of air and gasoline during combustion, remember that the same *amount* of each taken into the engine goes out the exhaust pipe in new forms.

### Energy

Energy is the ability to do work, and we can view it in several ways. Let's go back to the electrons of an atom for a minute to look at their energy.

**Electron Energy.** We said earlier that all atoms have electrical energy in the form of attraction and repulsion among their positive and negative charges. We also noted that electrons are in constant motion. Does this mean that

**IT'S NOT JUST A GOOD IDEA, IT'S THE LAW**

You can't study any area of science for long without learning about one or more "laws." These laws are just simple statements of fundamental principles that govern how matter and energy relate to each other. We take these laws for granted today because scientists proved them centuries ago. Usually, we know these laws by the names of the people who discovered and explained them. There are Ohm's and Kirchhoff's laws that explain some of the principles that help us understand electricity. There is Pascal's law that explains hydraulic pressure and Boyle's laws about the pressure and volume of gases. One of the most important groups of physical laws are Newton's three laws of motion, and you can see all of them at work in your automobile:

*First Law.* A body at rest remains at rest, and a body in motion remains in motion unless some outside force acts on it. Newton's first law explains our understanding of inertia and the combined effects of momentum and friction.

*Second Law.* A body's acceleration is directly proportional to the force applied to it and moves in a straight line away from the force. When you apply two forces to a body at angles to each other, the combined result is centrifugal force. Also, the greater the force applied to an object, the harder it is to change its direction.

*Third Law.* For every action, there is an equal and opposite reaction. This is the best known of Newton's Laws of Motion. Among other things, it is a simple explanation of rocket propulsion. It works in your car, too. As the driving wheels apply force on the pavement in one direction, the car moves in the other direction.

Sir Isaac Newton (1642-1727) didn't invent these laws, just as he didn't invent gravity. He simply provided a basic explanation for natural phenomena. When later scientists understood these principles, they used that understanding to develop new technology that eventually made the automobile possible. These were more than good ideas. They were laws.

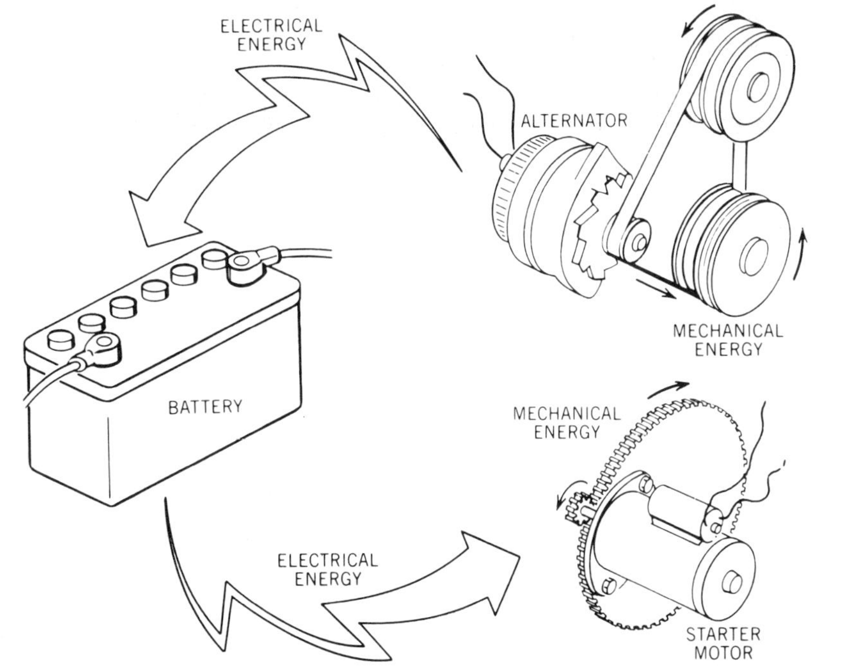


**Figure 2-5.** An alternator changes mechanical energy of rotation into electrical energy when it charges a battery. A starter motor changes electrical energy to mechanical energy when it cranks the engine.

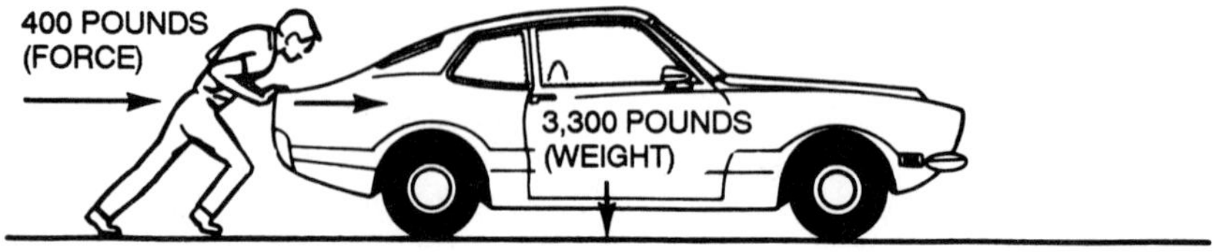


**Figure 2-6.** When you push against this automobile, you don't do any work until you move the car.

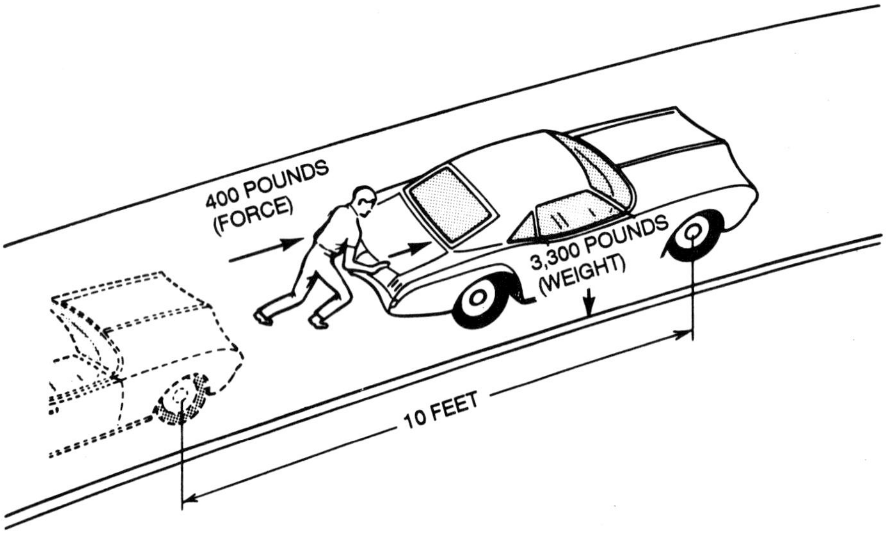


**Figure 2-7.** If you apply 400 pounds of force to move the car 10 feet, you do 4,000 foot-pounds of work.

electricity exists in all things? The answer is yes and no. The same electron energy that binds atoms together in different molecules is the basis for electricity, but with an important difference. The normal electron movement in most molecules is a random drift. It has no direction. When an outside force is imposed on this random electron drift, the electron movement takes on a definite direction. Then we have electric current. The same electron energy that binds atoms together becomes useful electricity only when a direction of current flow is established. This idea will be the starting point for your study of electricity in a later chapter.

**Potential and Specific Energy.** The difference between electron energy and useful electricity illustrates one way that energy is classified: whether it is released for work or whether it is stored. Unreleased, stored energy is **potential energy**. When it is released to do work, it becomes **specific energy** and takes other forms. When we apply one kind of force to potential electron energy, we create specific electrical energy that can do work.

The energy in an atom is also potential chemical energy and can be released in other ways. If we apply heat to some elements, compounds, or mixtures, we can release potential chemical energy as specific **thermal energy**. That *is* the energy of heat, and it can be used to do mechanical work.

Energy can change its form as different forces are applied to it. Chemical energy can become electrical energy or thermal energy. Either of these can become mechanical energy used to move an object. Whenever energy changes its form, some kind of force is involved. Force and energy combine to produce work.

## Work

If energy is the ability to do work, **work**, itself, is the result of using the ability. When energy is released to move a car, work is produced as motion. When energy is released to light an electric lamp, that's another form of work. Mechanical work is the most visible form of released energy in an automobile. The energy of mechanical work is **kinetic energy**, which causes things to move. Whenever kinetic energy is applied to a solid object in an automobile, it is through mechanical force.

## Mechanical Force

Mechanical force acts on an object to start, stop, or change the direction of motion. You can, however, exert force on an object and not move it; but you don't produce mechanical work unless you start, stop, or redirect motion. If you push against a 3,300-pound automobile, you exert force, but you don't produce any work unless you move the automobile forward or keep it from rolling backward, figure 2-6.

Mechanical work is force multiplied by distance. For example, if you apply 400 pounds of force to push that car 10 feet, you do 400 pounds times 10 feet of work, or 4,000 foot-pounds of work, figure 2-7. You converted the chemical energy in your body (from the food that you ate) into kinetic energy to move the weight of the automobile. In doing so, you applied force and did work.

### Inertia

To continue developing the idea of mechanical work, you should know about **inertia**. Inertia is the tendency of an object at rest to remain at rest or the tendency of an object in motion to remain in motion. Further, inertia of an object at rest is **static inertia**. Inertia of an object in motion is **dynamic inertia**. Inertia exists in all forms of matter: gas, liquid, and solid. Inertia affects fuel and airflow in an engine. It is easiest to see in solids, however, so we will concentrate on solid objects.

When you pushed that car from a standstill, you overcame the static inertia of the vehicle. When someone steps on the brakes to stop the motion, the brakes overcome dynamic inertia, figure 2-8.

### Momentum and Friction

When force overcomes static inertia of an object, the object gains **momentum**. This is a kind of movement. Momentum is the mathematical product of the object's weight times its speed. When another force overcomes the dynamic inertia of the moving object, the object loses momentum. Momentum and inertia are related kinds of mechanical energy.

If you pushed that car in a straight line on level ground, it would keep moving in the same direction at the same speed if no other forces were applied to it. This would mean that its momentum perfectly equals its weight. Other forces do act on any object in motion, though. The most common is **friction**. Friction is a force that resists motion between the surfaces of two objects. The objects can be solids, liquids, or gases. A car moving through the air develops friction between its surface and the surfaces of gas molecules in the air. Its tires also develop friction with the road. Friction would eventually stop the car if you didn't keep pushing on it, figure 2-9.

When friction opposes the mechanical energy of inertia and momentum, some of the mechanical energy is changed to heat energy. The amount of mechanical energy available to move the object is reduced by the energy converted to heat. This is how the brakes on a car work. Friction of the brakes against the wheels changes the mechanical energy (momentum) of the moving wheels into heat energy. Much of the mechanical energy developed in an engine is changed to heat energy by friction of moving parts.

### Mass, Weight, and Volume

The amounts of energy and force needed to overcome momentum and either kind of inertia are related to the **mass** and **weight** of an object. We use these terms interchangeably to talk about objects on the surface of the earth. They are not the same, however. Mass really is the measurement of an object's inertia. Weight is the measurement of the earth's gravitational pull on the object. For example, a space satellite has the same mass on earth as it has 4,000 miles out in orbit,

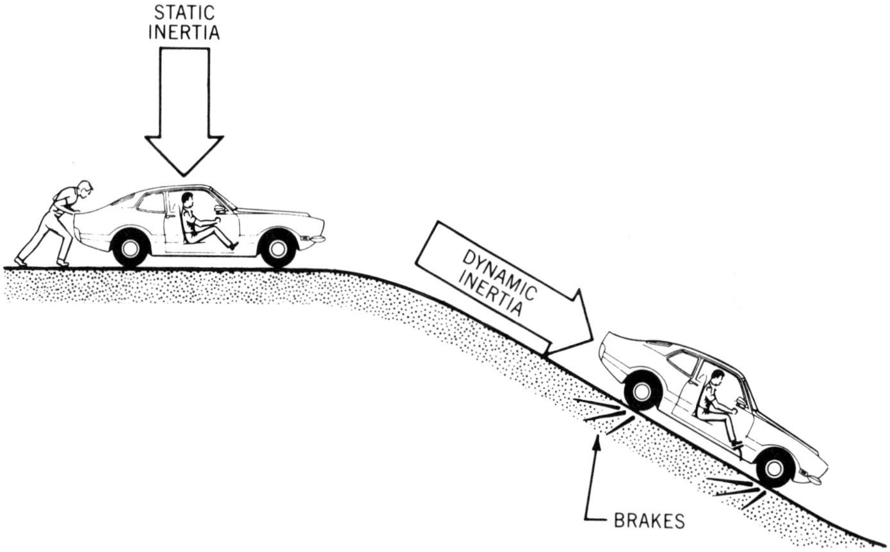


**Figure 2-8.** You overcome static inertia to get the car moving. Once it is moving, another person applies the brakes to overcome dynamic inertia and stop it.

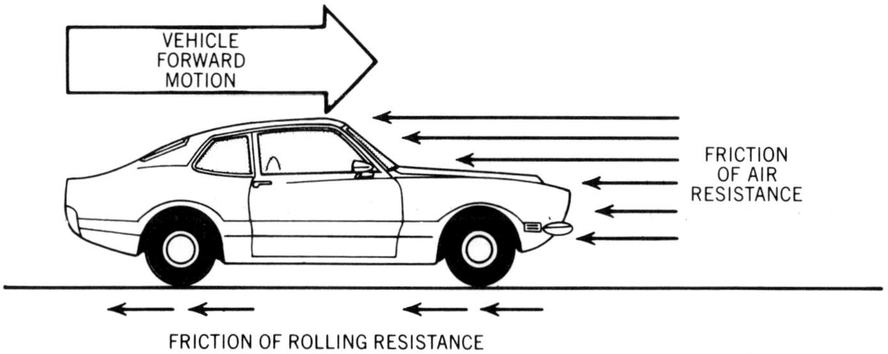


**Figure 2-9.** The friction of the moving car against the air and of its tires against the ground would eventually stop its momentum if no other force keeps it moving.

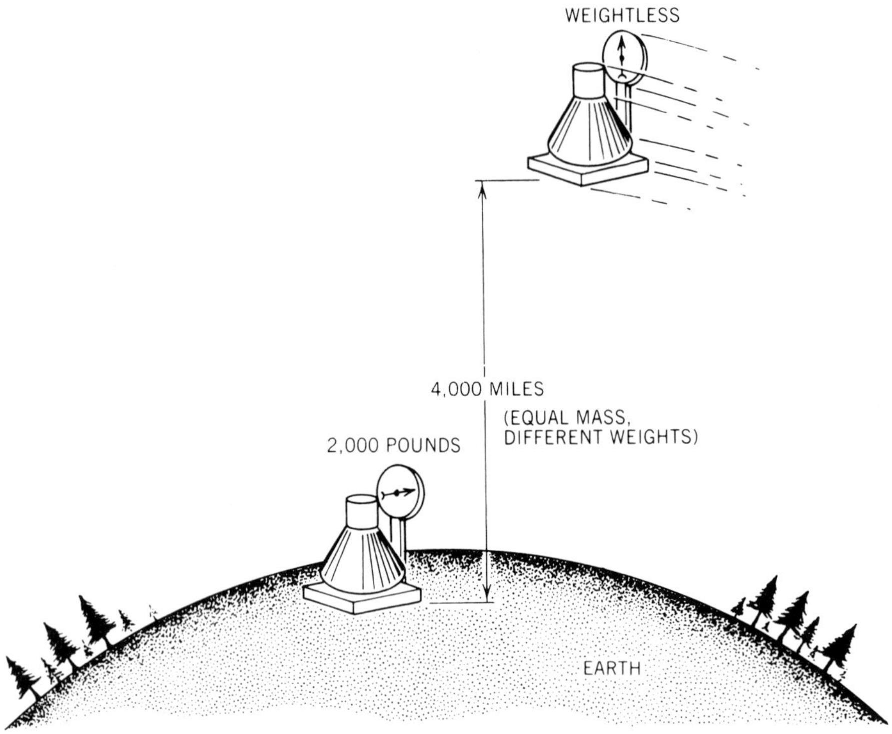


**Figure 2-10.** This satellite has equal mass but different weights when it is on Earth and when it is in orbit.

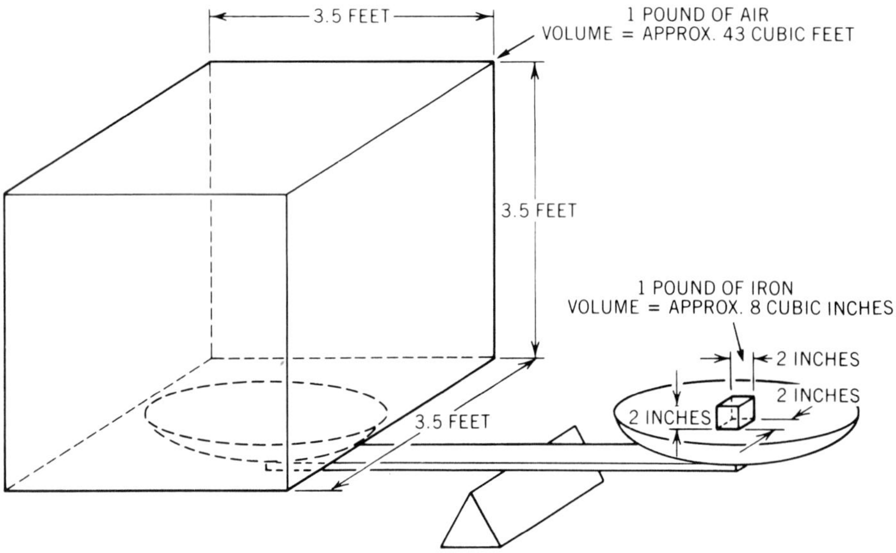


**Figure 2-11.** A pound of air and a pound of iron are the same weight, but air occupies more than 100 times as much volume as iron.

figure 2-10. On earth, though, it weighs several thousand pounds and needs a lot of force to move it. In space, it may be almost weightless and need little force to move it.

To measure mass, weight, and force on the surface of the earth, we use pounds and ounces in the customary English system or kilograms in the metric system. It doesn't confuse or mislead us to think of mass and weight interchangeably, nor does it confuse us to use pounds, ounces, or kilograms to measure force.

Volume is related to mass and weight, but it is a measurement of size. **Volume** is the space that a particular form of matter occupies in length, width, and height. It is measured in cubic inches, cubic feet, cubic centimeters, quarts, gallons, or liters or similar units. You can measure the weight of iron and oxygen in pounds, for example, but a pound of iron occupies much less volume than a pound of air, figure 2-11.

### Torque

Another term you will use when discussing mechanical work is **torque**. It is related to force just as work is. You know that work is force multiplied by distance. You also know that mechanical work, or motion, isn't always in a straight line. Torque is force that twists or turns an object to do work. Like work, torque can be measured in a combination of distance and force measurement.

In the customary English system, we measure torque in a combination of pounds or ounces of force and feet or inches of distance. Some engineering books express torque measurements in pound-feet or pound-inches. However, the expressions foot-pounds and inch-pounds are more common in the automobile service industry. These are the expressions for torque measurements in the customary system that we use in this book. When you study the metric system, you will learn another unit for torque measurement. No matter how you express torque, the important thing to remember is that it is a measurement of force times distance.

To study auto mechanics, we can use the same formula for torque and for work. Torque, and mechanical work, however, are not always the same. Mechanical work always involves motion, but you can apply torque without creating motion. When you put a 1-foot wrench on a bolt and apply 20 pounds of force, you have applied 20 foot-pounds of torque. The bolt may or may not turn, figure 2-12. If it turns, you have created motion and done work with your torque.

You have probably read torque ratings for various automobile engines or for electric motors. An engine or a motor can be rated at 200 foot-pounds of torque at 3,000 revolutions per minute, or 10 foot-pounds of torque at 1,750 revolutions per minute, or some other rating depending on its size and design. You may have asked yourself, "Is torque a measure of an engine's power?" No, it isn't, but torque and power are both related to doing mechanical work.

## Power

Torque measures how much work an engine *can* do. **Power** measures the rate, or speed, of doing that work. The power of an engine or an electric motor is commonly measured in horsepower or in metric kilowatts. The customary English system measures power in horsepower because a Scotsman, James Watt, was the first person to scientifically calculate power. He based his calculations on the amount of work that a horse could do in a specific amount of time. Watt determined that in one minute, one horse could move 200 pounds 165 feet.

When you multiply 165 feet times 200 pounds, you get 33,000 foot-pounds of work. If one horse can do that work in one minute, then one **horsepower** equals 33,000 foot-pounds of work per minute or 550 foot-pounds per second, figure 2-13. Two horsepower can do the same work in one-half minute (twice as fast). Four horsepower can do it in one-quarter minute (four times as fast).

If you apply 400 pounds of force to move that car 10 feet, you are doing 4,000 foot-pounds of work. If it takes 8 seconds, you develop about 1 horsepower. If you move it 10 feet in 4 seconds, you develop 1.8 (almost 2 horsepower, figure 2-14.

Now you can see how torque, speed, and power relate to each other. As power increases, speed increases (time decreases) for a constant amount of torque. However, if speed and power increase together, the torque that is applied decreases. We will go into these torque and power relationships in more detail later when we study engine torque, speed, and power. For now, remember that torque is the amount of work; power is the work times the speed.

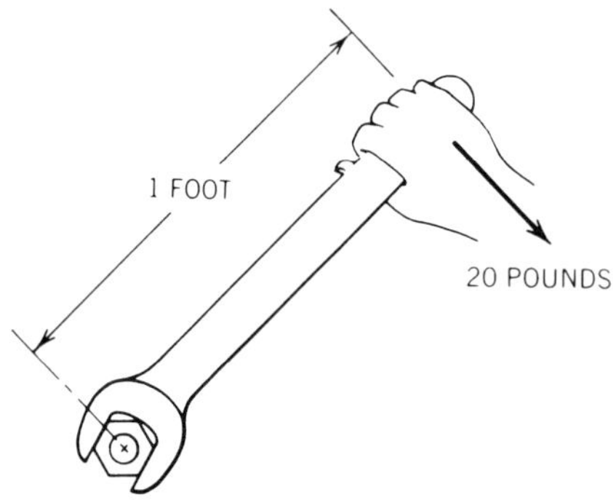


**Figure 2-12.** A 20-pound force on the end of this wrench applies 20 foot-pounds of torque to the bolt. Torque is a form of work.

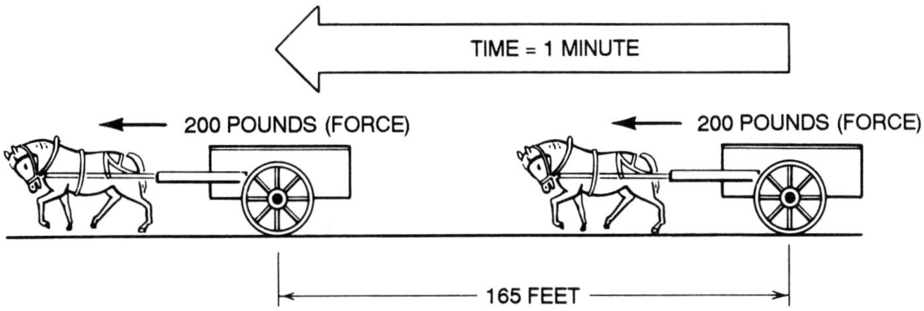


**Figure 2-13.** A force of 200 pounds that moves an object 165 feet in one minute produces one horsepower.

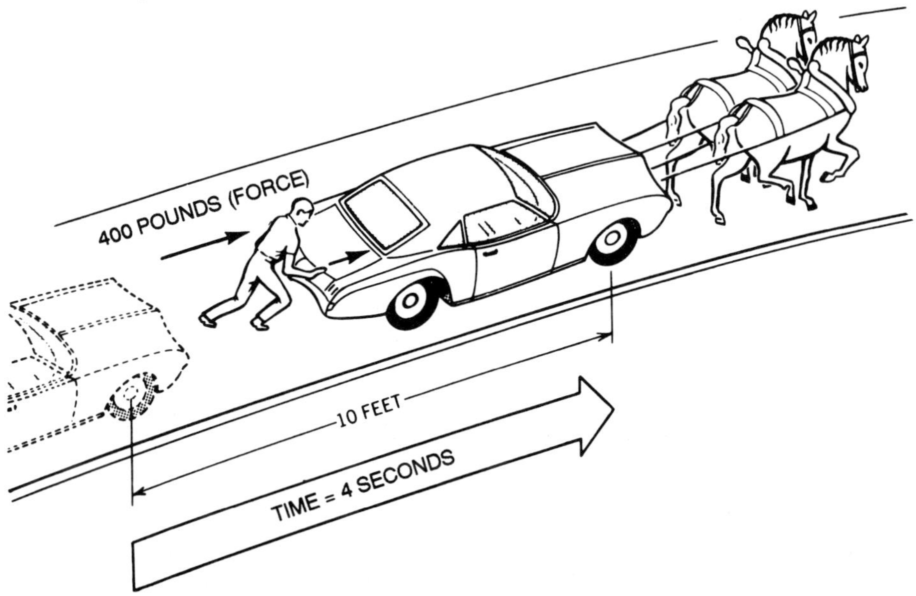


**Figure 2-14.** If 400 pounds of force moves the car 10 feet in 4 seconds, it produces almost two horsepower.

## WORKING WITH GASES AND LIQUIDS

Now that you understand how energy produces mechanical force, work, torque, and power, let's back up for a moment and review energy in gas and liquid systems. The next few paragraphs summarize how the principles of force, mass, weight, and volume apply to gases and liquids in engine systems. When most people think of automobile hydraulic systems, they think of brakes, power steering, and automatic transmissions. But, hydraulic principles also are the basis for many fuel system operations. In fact, a fuel injection system has all the features of a power steering or transmission hydraulic system.

Similarly, most people think of **pneumatic** systems in terms of shop air systems or compressed air systems on large trucks. Remember, an engine *is* an air pump. It is able to operate because it draws in air and compresses it for combustion. It operates on the principles of differences in air pressure. We call low air pressure in an engine **vacuum**. The differences between atmospheric air pressure and low air pressure (vacuum) are fundamental to fuel system operation and to vacuum controls for all engine systems. If you review these principles now, you can apply your understanding quickly and easily to fuel, emission, and engine control systems. Also, if you recognize that air has mass, weight, and volume and is subject to inertia, momentum, and friction, you will understand the operation of air systems better.

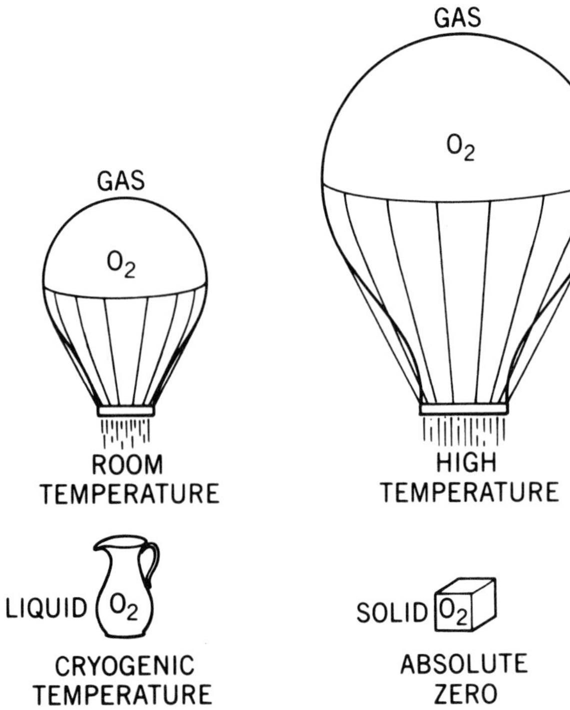


**Figure 2-15.** The volume of a gas for any given mass will increase or decrease as temperature increases or decreases.

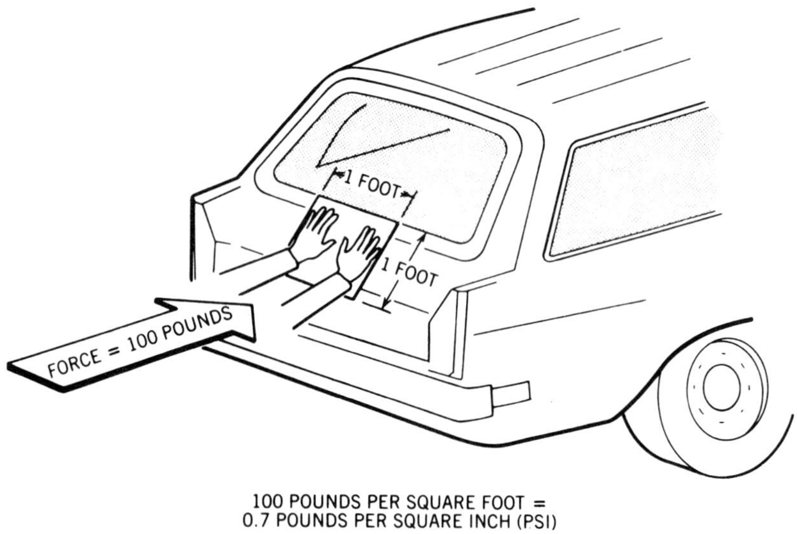


**Figure 2-16.** Pushing with 100 pounds of force over an area of 1 square foot creates a pressure of 100 pounds per square foot or 0.7 pounds per square inch (psi).

### Molecular Energy

Remember that kinetic energy is the energy of motion and that the electrons of atoms and molecules are constantly in motion around the nucleus. Therefore, all atoms and molecules must have kinetic energy. A basic principle about the behavior of all molecules—gas, liquid, or solid—is the kinetic theory of matter. It says:

**1.** Molecules are in constant motion because of their kinetic energy.

**2.** Kinetic energy increases as temperature increases and decreases as temperature decreases. The higher the temperature, the faster the molecules move.

**3.** Gas molecules move faster than liquid molecules, and liquid molecules move faster than solid molecules.

**4.** Molecules of solids vibrate around fixed points. Molecules of liquids slide over each other freely. Gas molecules are in constant motion and expand to fill the space available to them.

The kinetic theory explains why gases do not have constant volume or shape. The kinetic theory says that gas molecules spread out to fill all the space available. It also says that if you in-

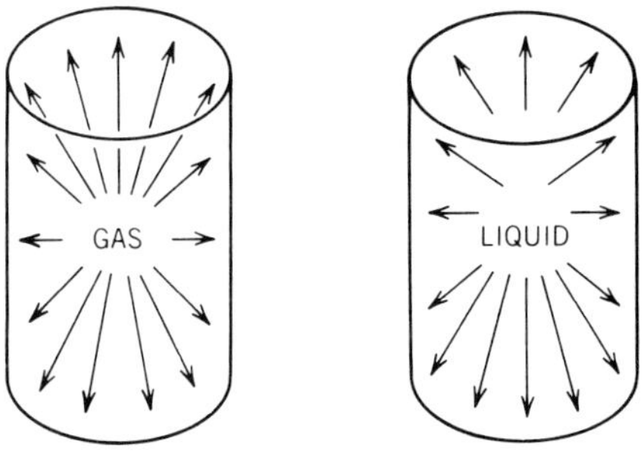


**Figure 2-17.** Gases and liquids exert equal pressure in all directions against the insides of their containers.

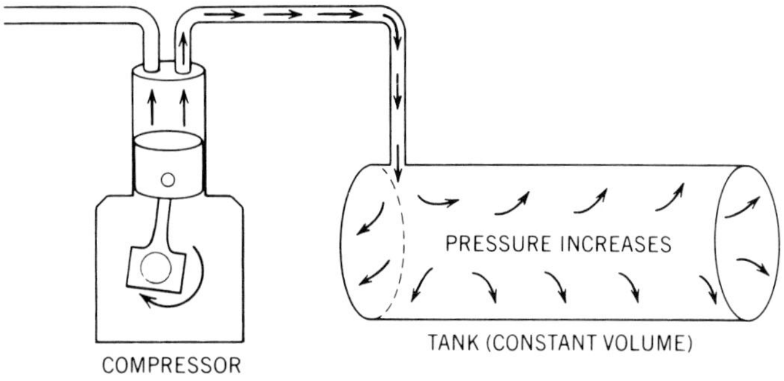


**Figure 2-18.** As more air is compressed into a tank of fixed volume, the pressure rises.

crease the temperature, the molecules spread out more. If you decrease the temperature, the molecules move closer together. This is the principle that accounts for matter changing from solid to liquid to gas as temperature rises.

## Temperature

Temperature affects all objects, but its effects on gases are most important for engine systems. Oxygen and nitrogen are gases at normal atmospheric temperatures. If temperature increases, they are still gases, but they expand to fill more space (volume), figure 2-15. If temperature decreases, these gases contract to fill less space until their molecular motion decreases to that of a liquid. At that point, you have liquid oxygen and liquid nitrogen. If you could lower the temperature still further to the point where molecular motion stops (or almost stops) you would have *solid* oxygen and *solid* nitrogen. This temperature is called **absolute zero**. It is theoretically possible to reach this point, but it doesn't apply in automobile physics. What does apply, though, is the principle that the volume of any matter increases as temperature increases and decreases as temperature decreases. Among other things, this principle is important for its effects on engine air-fuel ratios.

## Pressure

Pressure is based on molecular energy and related to temperature and volume. You can't study gases and liquids without understanding how temperature, pressure, and volume relate to each other. When we talked about mechanical energy and work, we talked about force. We said it is the amount of push or pull exerted on an object needed to cause motion. We also said it is measured in the same units as mass and weight: pounds or kilograms, for example.

**Pressure** is a kind of force. It is the force exerted on a given unit of surface area. You can apply pressure as a force when you push against an automobile. If you push with 100 pounds of force, and your hands cover a surface area of 1 square foot, you have applied pressure of 100 pounds per square foot, figure 2-16. This is a simple example of mechanical pressure, solid against solid. Whether it is applied by a solid, a liquid, or a gas, *pressure equals the force divided by the surface area*.

The pressure of a gas or liquid is the force exerted against the inner surface area of its container. In customary English units, pressure is measured in pounds per square inch (psi). An important fact about pressure applied by gases and liquids is that *gases and liquids exert equal pressure in all directions against their containers*, figure 2-17. The container can be the inside of a fuel line, a turbocharger housing, or an intake manifold; the principle is the same. Two important relationships are:

- Temperature and pressure are *directly* related. If you increase one, you increase the other.
- Pressure and volume are *inversely* related. If you increase one, you decrease the other.

**Pressure on Gases.** To understand these principles in an automobile, you need to know one more thing about the pressure and volume of gases and liquids. *Gases can be compressed; liquids cannot*. That is, if you apply physical force or pressure to a volume of gas, you can squeeze it into a smaller volume. Or you can squeeze more gas into the same volume that you started with. In either case, the pressure that the gas exerts on its container goes up. This happens when an air compressor pumps air into a tank, figure 2-18. The tank limits the volume that the air can fill. As the compressor squeezes in more air, the pressure inside the tank rises. The same principle applies in an engine, but as the piston compresses air in the cylinder, the volume decreases and pressure rises, figure 2-19. In either case, the temperature of the compressed gas also rises. Here we have combined forms of potential kinetic and thermal energy that can be released to do work.

This is an important point that works in several automobile systems. *When any gas is compressed, its temperature rises, and it gives off heat to anything that surrounds it. When pressure of a gas is released, it cools and takes on heat from anything that surrounds it.* This is the principle that allows diesel engines to operate. The kinetic theory explains why temperature increases with pressure. As pressure forces more gas into a given space, each atom has less space in which its molecules can

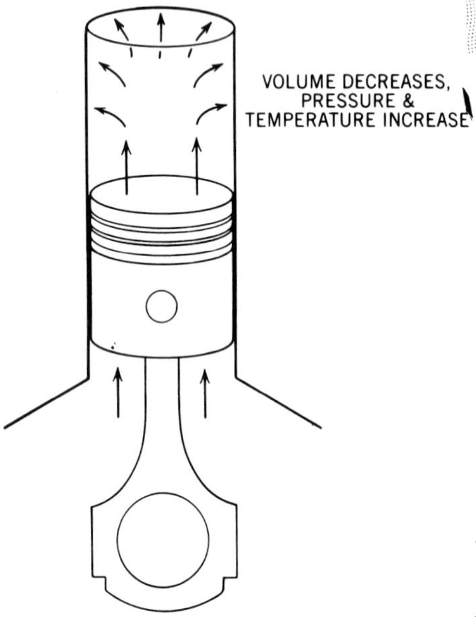


**Figure 2-19.** As the piston rises and decreases the volume of the cylinder, gas pressure rises.

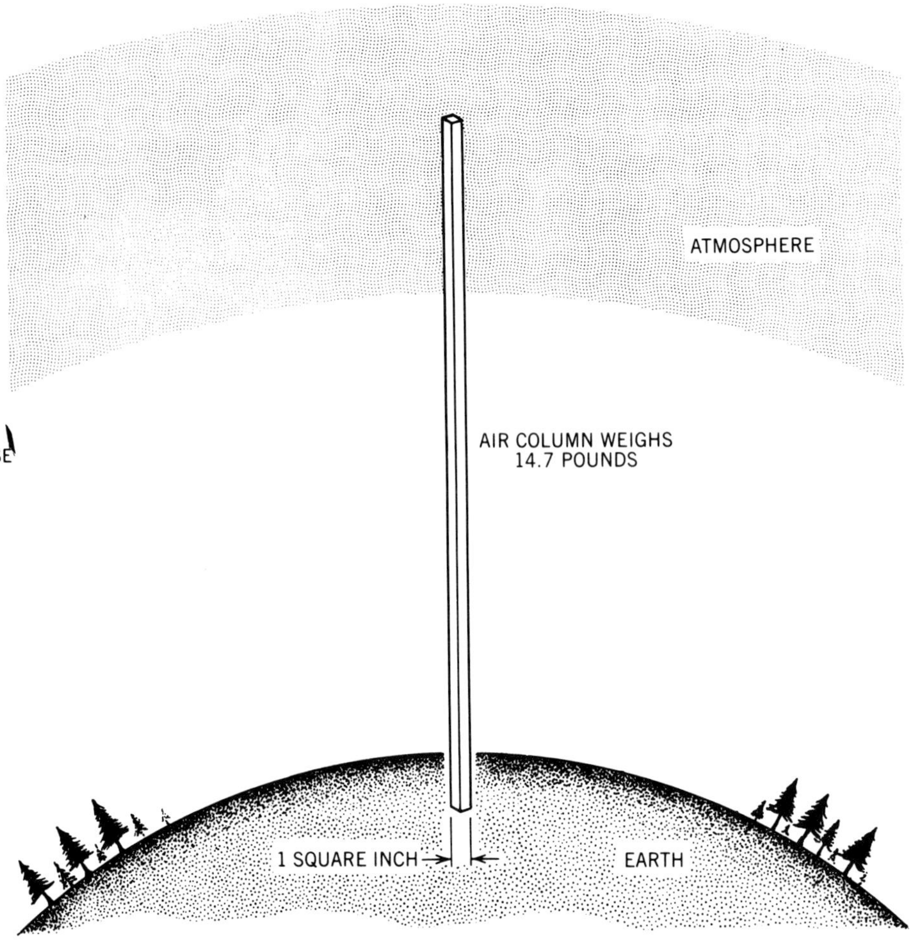


**Figure 2-20.** A column of air (gaseous matter) with a 1-square-inch area and extending from the surface of the earth to the edge of the atmosphere weighs 14.7 pounds.

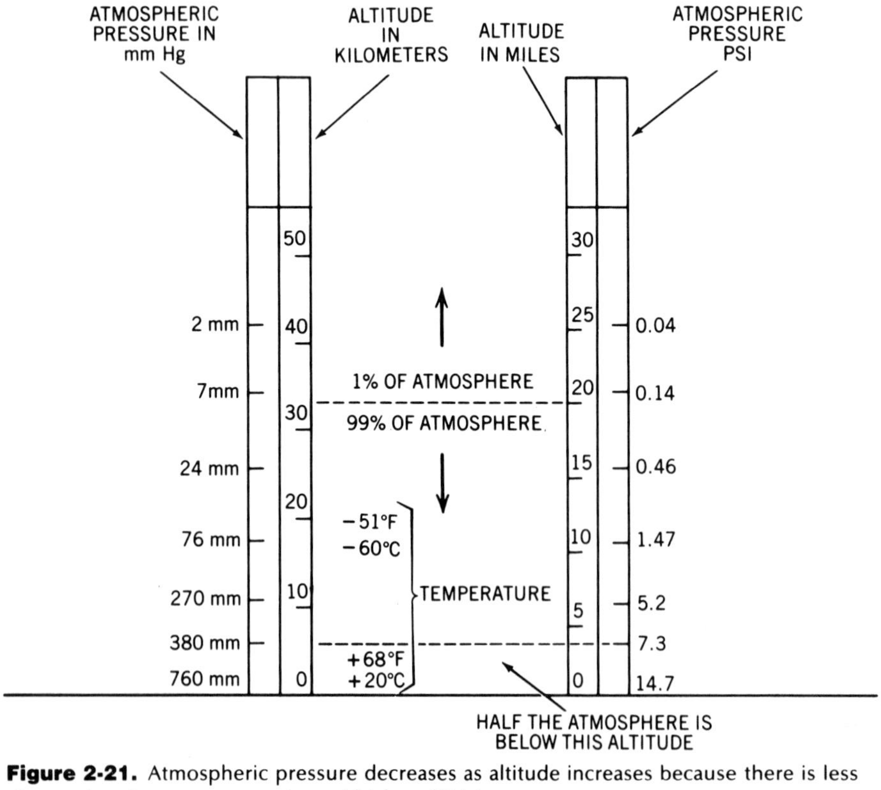


**Figure 2-21.** Atmospheric pressure decreases as altitude increases because there is less air pressing down on any surface at higher altitudes.

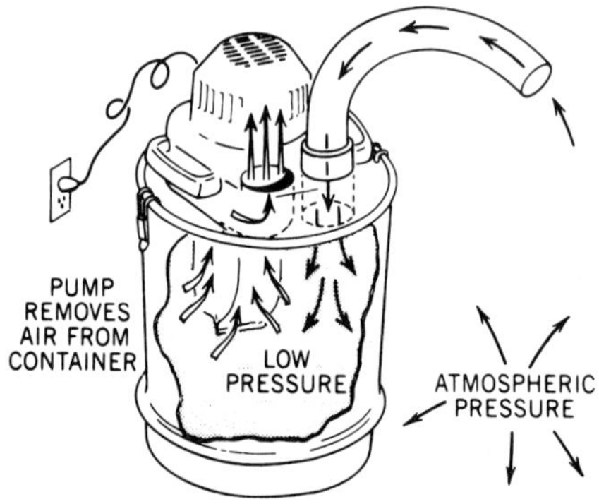


**Figure 2-22.** A vacuum cleaner works with an air pump, creating a low-pressure area which is then filled with more air forced in by atmospheric pressure.

move. Therefore, they create more friction against each other and hence more heat.

**Pressure on Liquids.** When you apply pressure to a liquid in a container, you cannot compress it as you can a gas. The pressure of the liquid pushes back with the same amount of force. This is why you cannot put 20 gallons of gasoline into a 15-gallon tank. You still are working with energy that can be used to do work, however. If force cannot compress liquid, force can be used to move liquid and create motion and do work just as solid objects can. This principle is the foundation for the science of hydraulics.

**Pressure and Resistance.** If we want to move gases and liquids or use them to do work, we must develop pressure within their containers. We can't develop pressure without resistance to molecular motion. That's why air and hydraulic systems have valves to open and close the systems and control pressure and resistance. An engine or an air compressor can't create air pressure unless there is a valve to let air in and out of a cylinder or chamber.

## Air Pressure

Air is gaseous matter. It has mass and weight, and it exerts pressure. Among other things, air exerts **atmospheric pressure** on the earth. We can think of atmospheric pressure as the total weight of the earth's atmosphere. Air exerts a pressure of about 14.7 pounds per square inch on the earth at sea level. If you could see a volume of air as a column of matter extending to the outer edge of the atmosphere, and having a surface area on its end of one square inch, it would weigh 14.7 pounds, figure 2-20.

**Air Pressure and Temperature.** As temperature rises, atmospheric air expands, becomes lighter for any volume, and exerts less pressure. As temperature falls, atmospheric air contracts, becomes heavier for any constant volume, and exerts more pressure.

**Air Pressure and Height.** As you climb above sea level, the volume of air pressing down on you decreases. Since there is less volume, the amount of air weighs less and exerts less pressure. Air pressure continues to decrease as the distance above sea level increases, figure 2-21. At 30,000 to 40,000 feet above sea level, air pressure is only about 5 pounds per square inch (psi). A few hundred miles above sea level, the air runs out and there is only vacuum.

**Vacuum.** We can talk about vacuum as an absence of air pressure. As vacuum works in most machines—including automobiles—it is a condition where air pressure in one area is lower than atmospheric pressure. Consider a simple vacuum cleaner for a minute. We think of a vacuum cleaner sucking dirt up from our rugs. What really happens is that an electric motor runs a pump that draws air out of a canister. The end of the canister opposite the pump is open to the atmosphere. Atmospheric pressure forces more air through this opening and carries dirt along with it into the canister, figure 2-22.

The same principle works in an automobile engine, which also is an air pump and the source of most vacuum in an automobile. You will study this fully when you get to engine operation, but for now, think of a piston in an engine cylinder moving downward to create a larger air volume. This decreases the pressure in the cylinder below atmospheric pressure. Air pressure outside the engine forces more air into the cylinder to fill the larger volume, figure 2-23. Vacuum is just a difference in air pressure between two areas or containers. Air pressure is another form of kinetic energy that we put to work in our automobiles.

## Hydraulic Principles

**Hydraulics** is the science of using liquids to do work. Because liquids can't be compressed, they can be used like levers or gears to transmit motion and

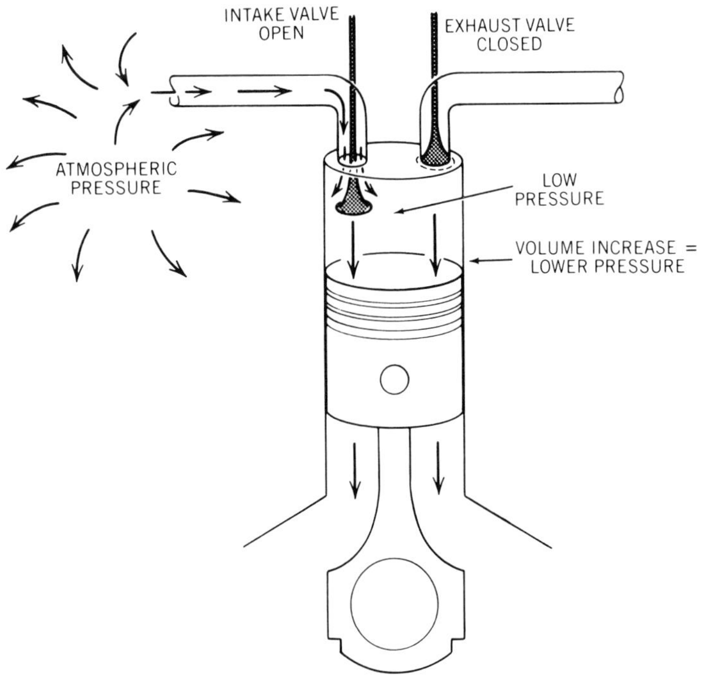


**Figure 2-23.** An engine (or an air compressor) is really an air pump. As a piston moves down in a cylinder, it increases the cylinder volume and creates a low-pressure area. Atmospheric pressure then fills the area with more air through an open valve.

force and to increase one or the other. We use hydraulic systems to do mechanical work in automobile braking, steering, and transmission systems. Engine fuel, lubrication, and cooling systems also are hydraulic systems. These systems use hydraulic principles to move liquid fuel and to control heat energy. The same relationships that exist in mechanical systems between torque and motion exist in hydraulic systems between pressure and motion.

**Hydraulic Force, Pressure, and Motion.** Every hydraulic system has an input device to develop pressure. Oil pumps, water pumps, and fuel pumps all develop hydraulic pressure by applying force to a liquid. Remember that pressure equals force divided by area. In this case, it is the mechanical force of the pump divided by the area of the pump chamber.

These pumps work like any other pump—an air compressor or an engine. They draw liquid into a system and force it through lines by creating a difference in pressure between the pump inlet and outlet. Since liquid cannot be compressed, a hydraulic system forces liquid through the system to do work. The work may be simply the movement of liquid fuel from a tank to an engine. The work also may be the control of heat energy through heat transfer (the cooling system) or friction reduction (the lubrication system).

### Pressure Control in Gas and Liquid Systems

Earlier, we said that gas and liquid systems cannot do any work without controlled resistance in the system to develop pressure and control its application. Hydraulic, air, and vacuum systems use valves to control and apply pressure. You will be most concerned with **pressure-regulating valves** in fuel, cooling, lubrication, and emission systems.

**Pressure-Regulating Valves.** Pressure-regulating valves control the amount of pressure developed in a system and the release of that pressure. Three of the simplest pressure-regulating valves are **orifices**, **relief valves**, and **check valves**. You can find examples in all systems. An orifice is a restriction in a fluid or air line. When gas or liquid under pressure hits a restriction, it meets more resistance and increases its pressure on the inlet side of the orifice. Pressure on the outlet side of the orifice drops, figure 2-24. This **pressure differential** exists as long as fluid flows. When fluid flow stops, pressures equalize on both sides of the orifice. Engine vacuum systems use orifices to control the time of an air pressure change on a diaphragm. A carburetor jet meters fuel accurately because of a controlled pressure differential across its orifice.

A relief valve releases system pressure when it gets above a desired level. A simple relief valve is a ball, a piston, or a disc with spring pressure on one side and gas or fluid pressure on the other. When gas or fluid pressure exceeds spring pressure (the pressure-limiting force), the ball or piston moves and releases the system pressure. Oil pumps have this kind of relief valve. A radiator cap also contains pressure relief valves, figure 2-25, and air injection systems use relief valves to limit system pressure.

A check valve allows gas or fluid flow in one direction but prevents flow in the opposite direction. Thus, pressure builds up on one side of the valve. A check valve can be a simple flapper or a ball in a passage or line, or it can be a spring-loaded disc or ball, figure 2-26.

Pressure-regulating valves also maintain system pressure at a specific, required operating level. You will learn in later chapters that all fuel injection systems must have pressure-regulating valves to control fuel delivery. The simple inlet valve in a carburetor, too, is a pressure-regulating valve.

The principles of force, pressure, and resistance operate in the air and fluid systems that support combustion in an engine. As you study specific systems, you will see their specific applications.

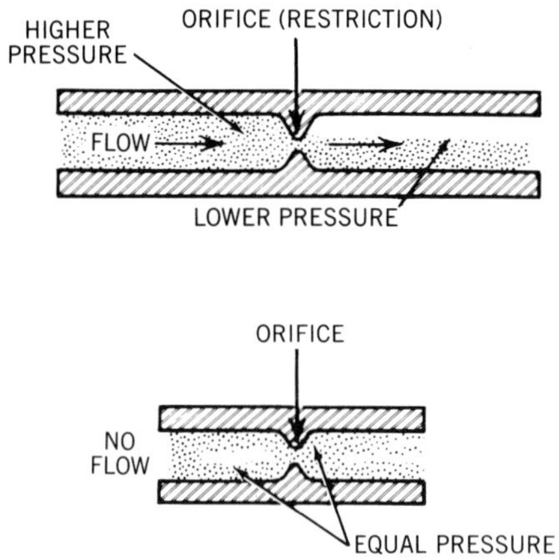


**Figure 2-24.** Orifices are used in hydraulic and vacuum systems to regulate pressure.

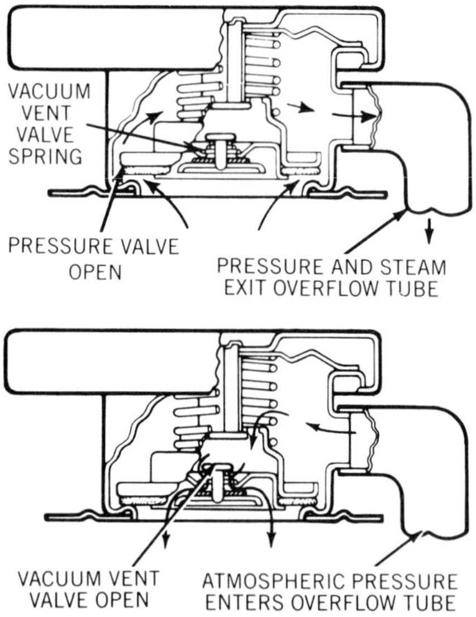


**Figure 2-25.** Pressure and vacuum relief valves in the radiator cap control cooling system pressure.

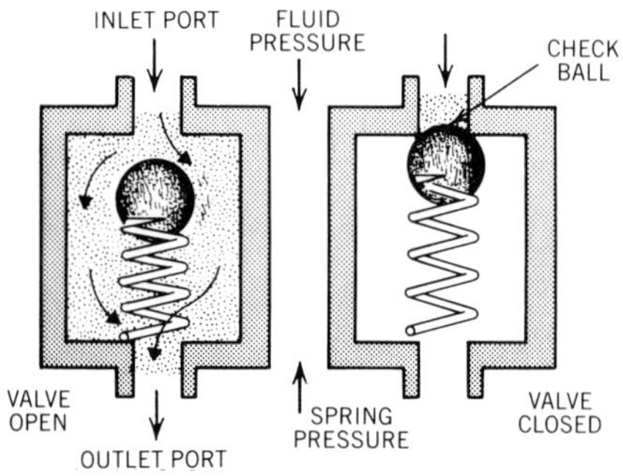


**Figure 2-26.** A check valve allows fluid flow in one direction but restricts it in the other.

## TORQUE, SPEED, AND POWER IN GEARS AND MECHANICAL DRIVES

An engine wouldn't do any useful work without some way to transmit its power and torque. Automobiles use gears to transmit power and torque to the driving wheels. Gears and other mechanical drives (belts and chains) transmit power and torque within the engine and to engine accessories. A complete study of engine performance should include a review of the mechanical drive devices used by engine systems. These devices are simply ways to apply the principles of torque and power to do work.

When you applied 20 pounds of force to a 1-foot wrench in figure 2-12 you applied 20 foot-pounds of torque. That wrench is a simple lever. A gear is a circular lever and applies force just as a wrench does. Therefore, it applies torque. A gear with a 1-foot radius and a 20-pound force applied to its teeth applies 20 foot-pounds of torque on a shaft at its center, figure 2-27. It works just like the wrench in figure 2-12. In figure 2-27, a drive gear applies the 20 pounds of force to the teeth of the driven gear.

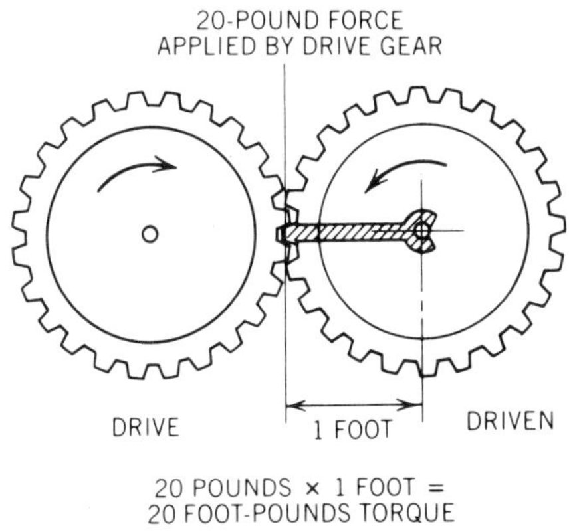


**Figure 2-27.** A gear is a circular lever. The lever length is the radius of the gear. Thus, 20 pounds of force on the teeth of this gear creates 20 foot-pounds of torque on its shaft.

### Gear Ratios

In all gear combinations, one gear drives another. The speed at which each turns in relation to the other is the **gear ratio**. It is determined by the diameter and the number of teeth on each gear and represents the number of turns the drive gear makes to rotate the driven gear one revolution. Figure 2-28 shows the three basic gear ratio combinations explained in the following sections.

**Direct Drive.** With two gears of the same size and the same number of teeth, figure 2-28*A*, the drive gear revolves once to turn the driven gear one revolution. Both gears turn at the same speed. The gear ratio is 1 to 1 (usually written as 1:1) and is called **direct drive**. You can find direct-drive gear combinations used to drive many engine accessories.

**Gear Reduction.** If one gear drives another that is twice as large with twice as many teeth, figure 2-28*B*, the drive gear turns twice to turn the driven gear once. This is a ratio of 2:1 and is called **gear reduction**. Gear reduction refers to any gear ratio where the driven gear turns at *reduced speed*. It could be 3:1, 16:1, 1.5:1, or any similar combination.

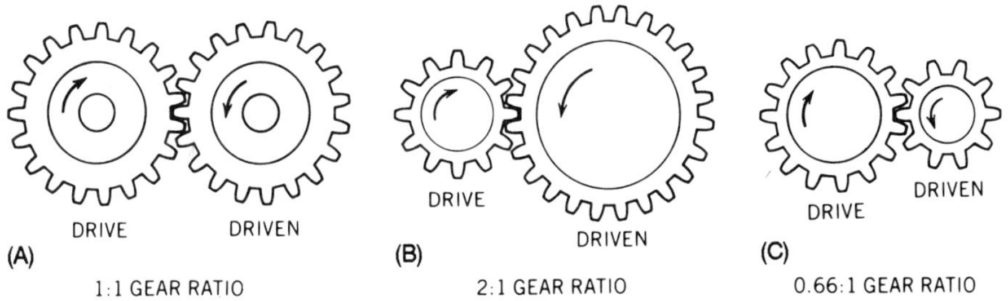


**Figure 2-28.** Direct drive, reduction, and overdrive gear ratios indicate torque and speed relationships in gears.

Sometimes reduction ratios are used just to change the speed of rotating parts and keep the speeds synchronized at a given ratio. This is the case with engine timing gears. In other cases, reduction ratios are used to increase the torque available from the driving gear.

**Overdrive.** When a drive gear has more teeth than a driven gear, the driven gear turns faster. This is **overdrive** and is the opposite of gear reduction. In figure 2-28*C*, the drive gear turns once for every one and one-half turns of the driven gear. This could be written as a 1:1.5 ratio, but gear ratios are usually given with the driven speed as 1. Therefore, this ratio is written 0.66:1. Many engine accessories, such as alternators, are overdriven because they need to turn faster than the engine.

**Calculating Gear Ratios.** To calculate the ratio between two gears, you count the teeth on each and divide the number on the driven gear by the number on the drive gear, thus:

$$\frac{15 \text{ (driven)}}{10 \text{ (drive)}} = 1.5:1 \text{ ratio}$$

You calculate the overall ratio for any combination of more than two gears in two ways. First, if each gear drives the next in order, you simply divide the number of teeth on the last (driven) gear by the number of teeth on the first (drive) gear and ignore all the rest. Figure 2-29 illustrates this with a set of four gears. You will find timing gear arrangements like this in some engines to drive oil pumps, distributor drive shafts, and overhead camshafts.

The first way to calculate multiple gear ratios is really a shortcut. What you are really doing is calculating the ratio of each pair of gears in sequence and multiplying that ratio by the next ratio. This is important if you need to calculate the overall ratio of sets of gears that *do not mesh* with each other.

### Torque Multiplication

Gears also can multiply torque. If you want to increase the 20 foot-pounds

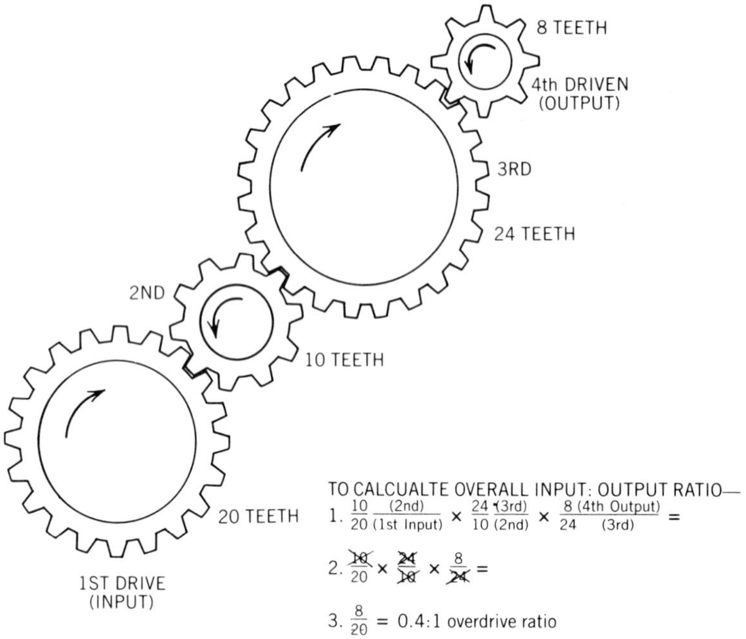


**Figure 2-29.** To calculate the overall ratio from the first drive (input) gear to the last driven (output) gear, divide the number of teeth on the last gear by the number on the first gear. The illustration shows the complete math for the calculation.

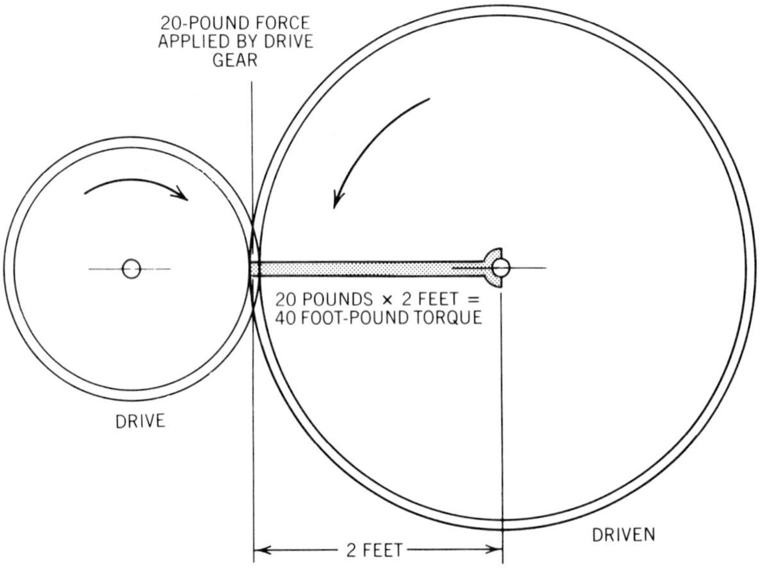


**Figure 2-30.** With a 2:1 gear ratio, the driven gear develops 40 foot-pounds of torque on its shaft for 20 pounds of force on its teeth. Gears can multiply torque just as levers can.

of torque on the driven gear shaft in figure 2-27, you can install a larger driven gear or a smaller drive gear. That is, you can change the gear ratio. If you double the size of the driven gear so that it has a 2-foot radius (a longer lever), and twice as many teeth, you can apply 40 foot-pounds of torque to its shaft with the same 20 pounds of force from the drive gear, figure 2-30.

When you change a gear ratio, you change the relative speeds of the gears. The gears in figure 2-27 have a 1:1 ratio and turn at the same speed. The gears in figure 2-30 have a 2:1 ratio; the driven gear turns at one-half the speed of the drive gear. And, we just showed that increasing the size of the driven gear increases the torque that it delivers to its shaft. Therefore, as driven (or output) *torque increases*, output *speed decreases*.

Here you have an important fact about any gearset. When gears increase output torque, they reduce output speed. When they increase output speed, they reduce output torque. This also shows that the gear ratio indicates both a speed and a torque relationship. If you assume a *constant input (drive) speed*, here are how some different gear ratios work:

- A 1:1 ratio gives equal input (drive) and output (driven) speed and torque.
- A 2:1 ratio doubles the output torque and cuts the output speed in half.
- A 0.66:1 ratio gives only 66 percent as much output torque but 150 percent as much output speed.

## Gear Types

Most gears in engine systems are simple **spur gears** and **helical gears**, figure 2-31. The teeth of a spur gear are parallel to the **axis**, or centerline, of the gear. The teeth of a helical gear are at an angle to its axis. The spur and helical gears in figure 2-31 are external gears. The teeth are on the outside. Gears with teeth on the inside circumference are **internal ring gears**, figure 2-32. Just as external gears can have spur or helical teeth, so can internal gears. There are few internal gears in engine systems, except in some oil pumps. But the principles of internal and external gear rotation apply to pulleys and belts and sprockets and chains, as you will see.

## Gear Rotation

When two external gears mesh and turn, they rotate in opposite directions, figure 2-33. If you want two shafts connected by external gears to turn in the same direction, you must put a third gear, called an **idler**, between the first two, figure 2-34.

In any *even* combination of gears (2, 4, 6, and so on), the first and last gears in the set always turn in *opposite* directions. In any *uneven* combination (3, 5, 7, and so on), the first and last gear in the set always turn in the *same* direction. You can use this fact to figure out camshaft and distributor rotation in any engine. When an internal gear and an external gear mesh and turn, both gears turn in the same direction.

When any combination of external and internal gears mesh and turn, they transmit power and motion along the *same or parallel axes of rotation*, figure 2-35. Bevel gears and worm-and-roller gears are used to change the axis of rotation, usually 90 degrees, figure 2-36. When we look inside an engine in later chapters, you will see distributor drive gears that are based on the worm-and-roller design.

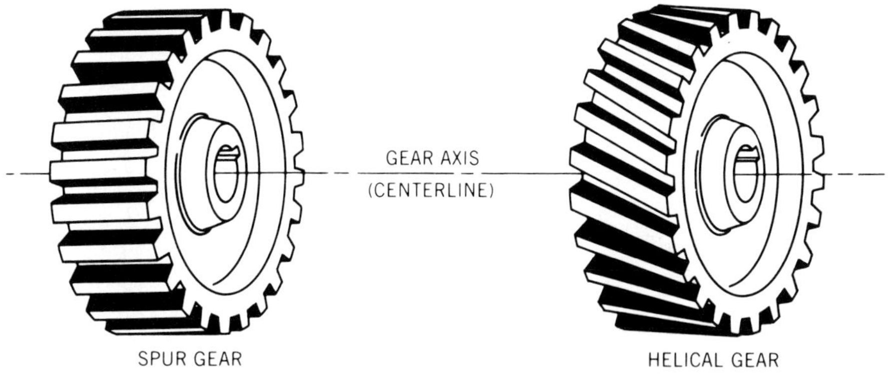


**Figure 2-31.** On simple spur gears, the teeth are parallel to the gear axis, or centerline. On helical gears, the teeth are at an angle to the gear axis

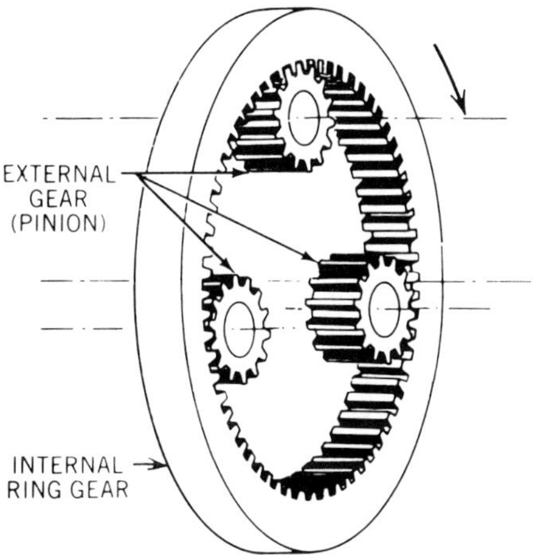


**Figure 2-32.** An internal ring gear has teeth on the inner circumference of an open gear wheel.

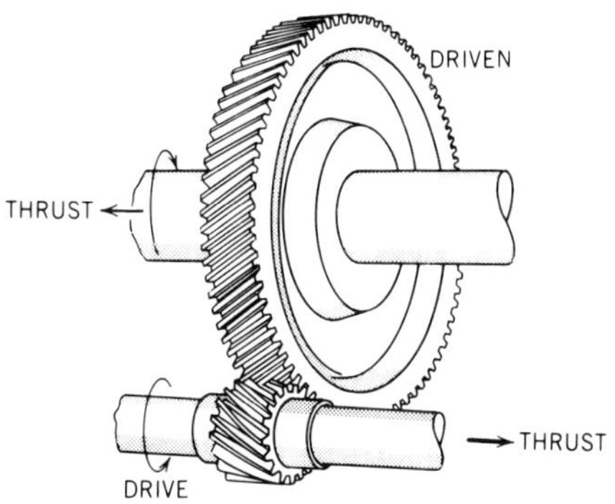


**Figure 2-33.** When two external gears mesh and turn, they rotate in opposite directions.

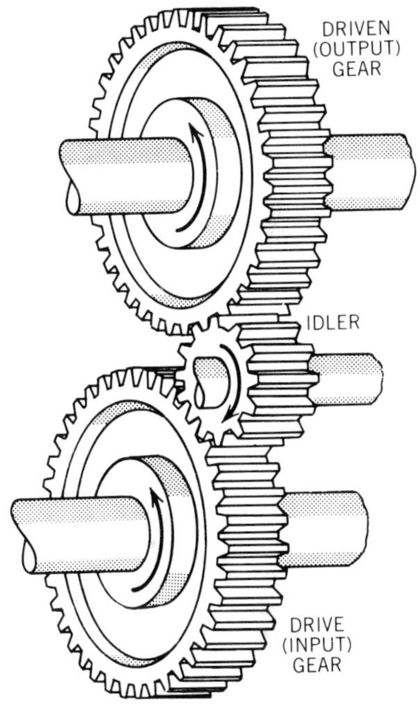


**Figure 2-34.** When three external gears mesh and turn, the first (drive) and last (driven) turn in the same direction. The idler reverses the direction of rotation.

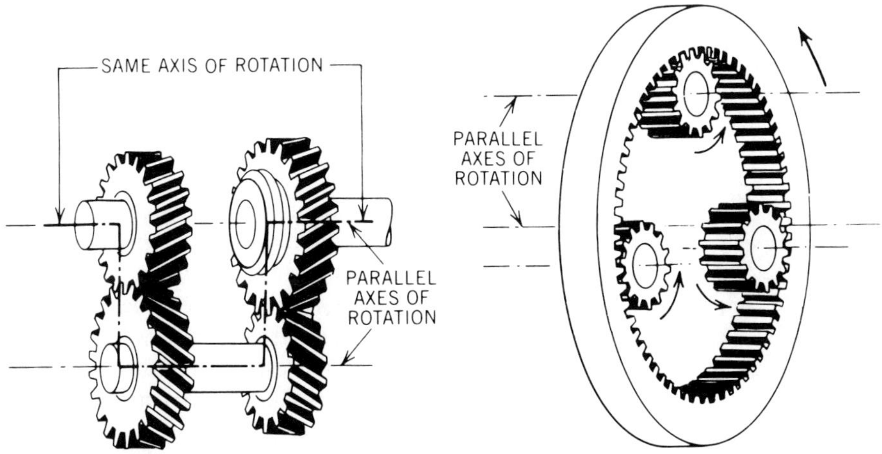


**Figure 2-35.** Any combination of internal and external gears, no matter how complex, transmits motion on the same axis, or parallel axes, of rotation.

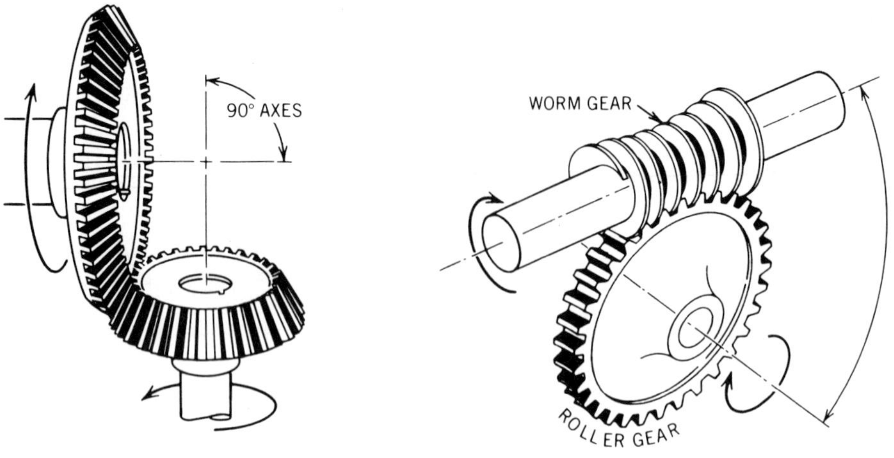


A: STRAIGHT BEVEL GEARS B: WORM AND ROLLER GEARS

**Figure 2-36.** Bevel gears and worm-and-roller gears change the axis of rotation.

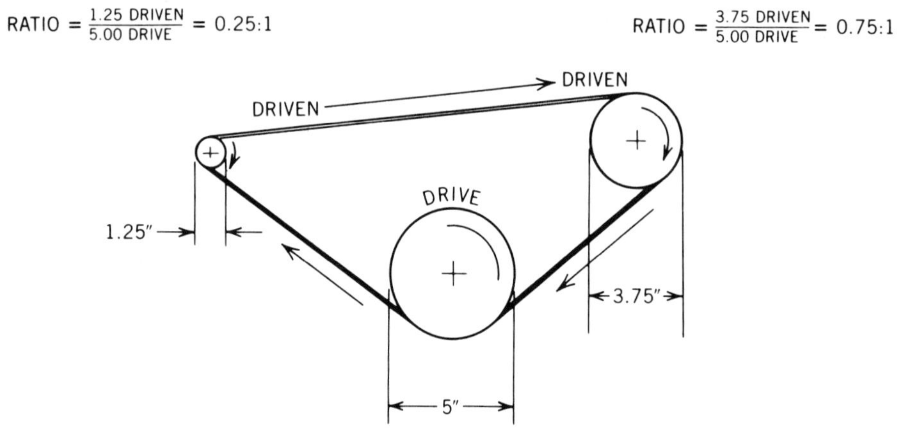


**Figure 2-37.** With a single belt, one drive pulley can drive several driven pulleys at different speeds.

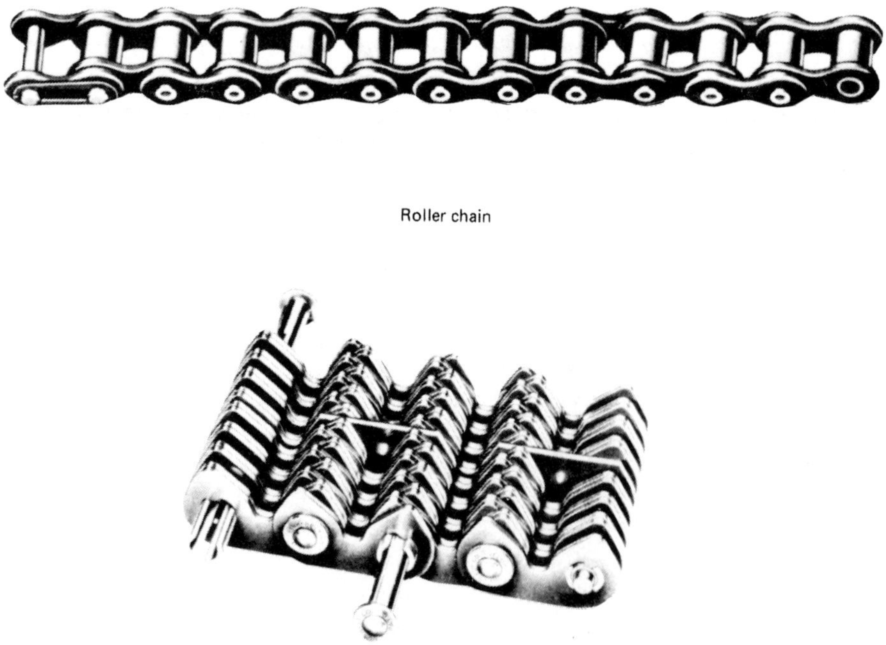


**Figure 2-38.** Roller chains and silent chains are used in many automotive applications.

Bevel gears and worm-and-roller gears are external gears, which change the direction of rotation as well as the axis of rotation. This is easiest to see in a pair of bevel gears. If one shaft rotates clockwise (viewed from the face of the gear), the other shaft rotates counterclockwise, figure 2-36.

## Other Power Transmission Devices

There are other ways to change rotating direction, speed, and torque in an automobile engine. Belt drives and chain drives are common on engines, and they work on the same principles as gear drives.

You can think of a belt or a chain as a flexible internal ring gear. If you use one **sprocket** or **pulley** as a drive member, you can turn several other sprockets or pulleys at different speeds with one belt or chain. The ratio between the drive member and each of the driven members will be a simple ratio of the drive pulley or sprocket to one other driven pulley or sprocket. Figure 2-37 shows this with an arrangement where a 5-inch drive pulley drives a belt that, in turn, drives several other pulleys of different diameters.

To calculate the ratios of a belt-and-pulley system, use the diameters of the drive and driven pulleys. To calculate the ratios of a chain-and-sprocket system, count the teeth on the drive and driven sprockets and use these numbers to figure the ratios.

**Chain Drive Systems.** The chain drives commonly used on automobiles are **roller chains**, and **silent chains**, figure 2-38. Each link of a roller chain is a small roller bearing that engages the teeth of the sprockets. Most roller chains in automobiles are single-stranded types.

Silent chains are made from many flat metal links pinned together. The links are toothed at each end, and when the chain is assembled, the links form a series of flexible gear teeth that engage the sprockets.

Roller chains are used in some en-

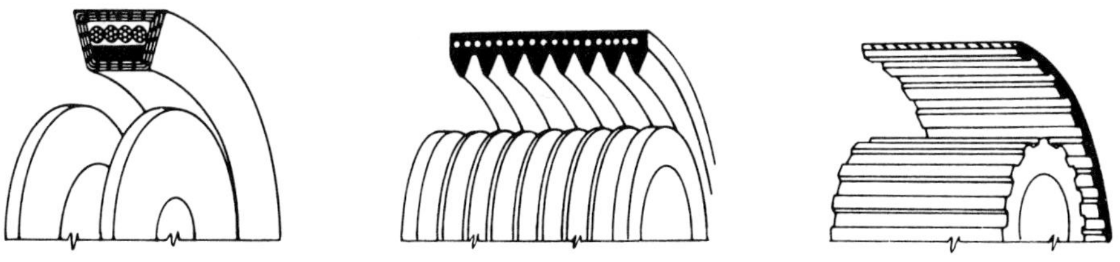


**Figure 2-39.** V-belts, V-ribbed belts, and cogged timing belts are used in many automotive installations.

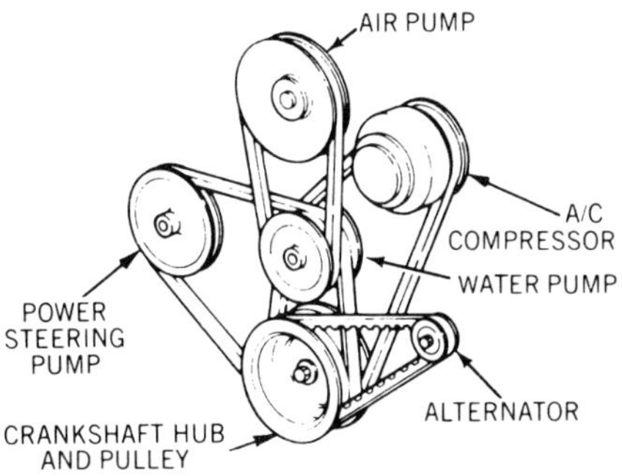


**Figure 2-40.** The most common use of V-belts on an automobile is to drive several accessories from a pulley on the front of the engine crankshaft. (Chevrolet)

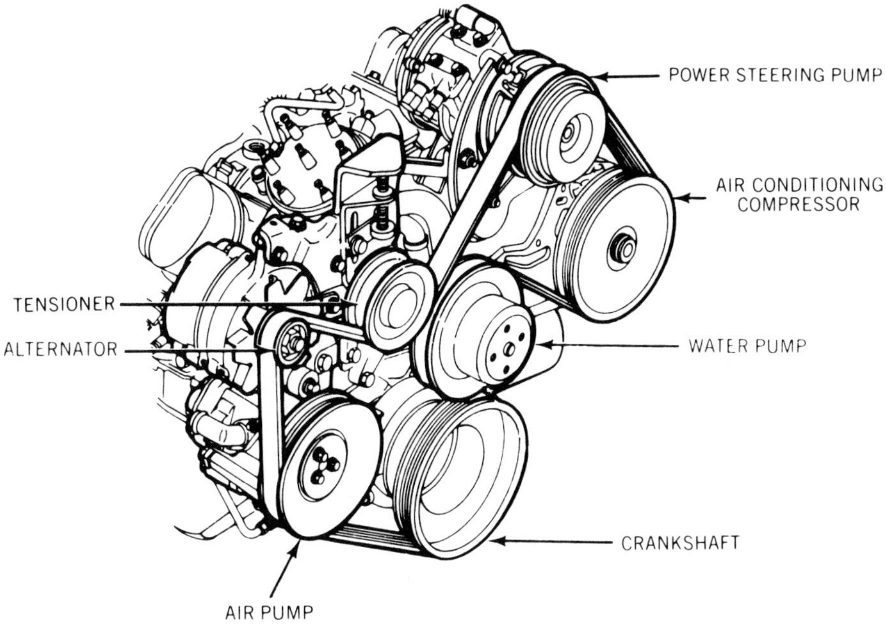


**Figure 2-41.** A single V-ribbed, or "serpentine," belt can drive several accessories. Some may be driven by the back, or outside, of the belt. (Ford)

gines to drive camshafts and other shafts. Silent chains also are used inside engines to drive various shafts. Roller chains are somewhat more flexible than silent chains and can drive several driven sprockets from a single drive sprocket. Silent chains can operate at higher speeds with less looseness than roller chains. Silent chains require complete enclosure and constant lubrication. Roller chains work best when enclosed and continually lubricated, but they can be used for exposed drives with intermittent lubrication.

**Belt Drives.** The most common automobile belt drive systems use V-belts, V-ribbed belts, or cogged timing belts, figure 2-39. The most visible use of belts and pulleys is to drive engine accessories, such as water pumps, alternators, and air conditioning compressors, figure 2-40. V-belts are used singly or in groups of two or more. In some complicated installations, more than one pulley is attached to a driven shaft. One of the pulleys is driven by one belt, while another pulley drives another belt to turn another shaft. This arrangement allows more flexibility in belt drive installations and in selecting speed ratios.

Flat, V-ribbed belts provide more flexibility and driving power than individual V-belts. These features allow a single V-ribbed belt to drive several accessories in an installation that might require several individual V-belts, figure 2-41. This reduces the cost, weight, and space needed for the belt drive. These installations often are called **serpentine drive belts**. Notice in figure 2-41 that some of the driven pulleys are turned by the outside, or back, of the drive belt. In this case, the driven pulleys rotate in the opposite direction from the drive pulley, and the belt works like an external gear to change the rotation direction.

V-belts and V-ribbed belts slip slightly on their pulleys, particularly when starting rotation. Therefore, V-belts and V-ribbed belts can't be used when the speeds and rotation of drive and driven pulleys must be synchronized. Chains and sprockets do not have this natural slippage and can be used for synchronized rotation, such as in engine timing drives. A cogged timing belt, however, provides the light weight and flexibility of a belt, as well

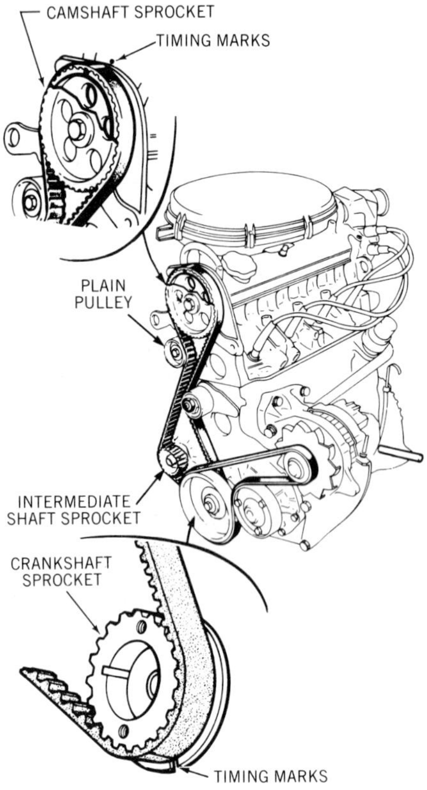


**Figure 2-42.** Cogged timing belts are used where rotating speeds must be perfectly synchronized. (AMC-Renault)

as the positive nonslip drive of a chain, figure 2-42.

A timing belt has cogs like gear teeth formed on its inside surface. Sprockets used with timing belts are shaped like gears. Their teeth engage the teeth of the belt to give a nonslip drive. Timing belts are used instead of chains to drive some engine camshafts. Timing belts eliminate slippage and work like silent chains, but they are lighter and do not need lubrication.

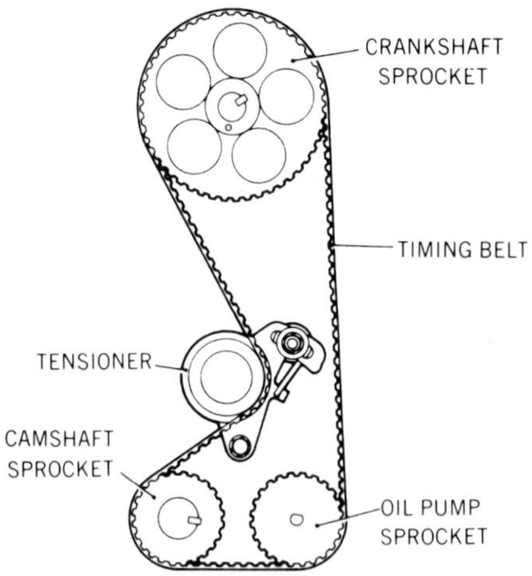


**Figure 2-43.** Long or complicated belt or chain installations often have idlers or tensioners to maintain proper tension and keep the chain or belt from slipping. (Chrysler)

Even the most precise timing belts and chains have some looseness. Proper tension is calculated by engineers when they design belt and chain drives. If a belt or chain is too tight, it will overheat and eventually break. If a belt or chain is too loose, it can slip on the pulleys or jump off the sprockets.

If several sprockets or pulleys are driven by one chain or belt, the chain or belt must wrap around enough of each one to drive it smoothly, without slippage. To achieve this, idlers or **tensioners** are used to remove slack and locate the chain or belt properly on the sprocket or pulley, figure 2-43. Every chain or belt has a tension side and a slack side. The tension side is the side of the belt that the drive sprocket or pulley pulls against. Idlers or tensioners are always installed on the slack side of the chain or belt.

## SUMMARY Looking Ahead to More Automobile Science

We have covered a lot of ground very fast in this chapter, from the nucleus of an atom to the mechanical drives that transmit torque and power. At this point, some of these principles may still seem abstract. You may wonder what they have to do with repairing an automobile. The answer is, "Quite a bit." If you understand *how* and *why* a particular system operates, you will understand *what* to do to make it operate correctly.

The important things to remember from this chapter are terms such as energy, work, force, pressure, volume, torque, power, and others that we have explained here. Remember what each means and how they relate to each other. We will put them to work in the following chapters as we study how common principles operate in different systems.

# REVIEW QUESTIONS

### Multiple Choice

**1.** Student A says that good electrical conductors are materials with few electrons in the outer shell. Student B says that good conductors have full valence shells. Who is right?

- **a.** A only
- **b.** B only
- **c.** Both A and B
- **d.** Neither A nor B

**2.** Student A says that two hydrogen atoms joined with one oxygen atom form a molecule. Student B says that two hydrogen atoms joined with one oxygen atom form a compound. Who is right?

- **a.** A only
- **b.** B only
- **c.** Both A and B
- **d.** Neither A nor B

**3.** Student A says that an oxidation reaction adds oxygen to other compounds or elements. Student B says that a reduction reaction removes oxygen from compounds. Who is right?

- **a.** A only
- **b.** B only
- **c.** Both A and B
- **d.** Neither A nor B

**4.** A reduction reaction occurs when:

- **a.** Rust forms on sheet metal
- **b.** Combustion occurs in an engine
- **c.** Emission devices remove oxygen from exhaust gases
- **d.** All of the above

**5.** Electron energy becomes useful electricity when:

- **a.** The negative charge is high enough
- **b.** The positive charge is high enough
- **c.** Atoms share many valence electrons
- **d.** All free electrons are forced to move in the same direction from point to point

**6.** Gravitational pull on an object is measured by its:

**a.** Mass
**b.** Density
**c.** Volume
**d.** Weight

**7.** Student A says that torque is a measurement of force times distance. Student B says that torque does not always produce motion. Who is right?

**a.** A only
**b.** B only
**c.** Both A and B
**d.** Neither A nor B

**8.** Student A says that power is a measurement of the distance that force moves an object. Student B says that power is a measurement of the rate, or speed, at which work is done. Who is right?

**a.** A only
**b.** B only
**c.** Both A and B
**d.** Neither A nor B

**9.** Pressure equals force divided by:

**a.** Temperature
**b.** Volume
**c.** Surface area
**d.** Distance

**10.** Student A says that pressure increases when temperature increases. Student B says that pressure increases when volume decreases. Who is right?

**a.** A only
**b.** B only
**c.** Both A and B
**d.** Neither A nor B

**11.** Student A says that liquids are compressed when pressure is applied to them. Student B says that temperature rises when any gas is compressed. Who is right?

**a.** A only
**b.** B only
**c.** Both A and B
**d.** Neither A nor B

**12.** Student A says that vacuum is air pressure lower than atmospheric pressure. Student B says that temperature drops as pressure drops (vacuum increases). Who is right?

**a.** A only
**b.** B only
**c.** Both A and B
**d.** Neither A nor B

**13.** If you increase output torque in a gear system, you also:

**a.** Increase output speed
**b.** Increase output resistance
**c.** Decrease output speed
**d.** Decrease output resistance

**14.** A gear ratio of 2.5:1 is:

**a.** A direct-drive ratio
**b.** An overdrive ratio
**c.** A reduction ratio
**d.** A compound ratio

**15.** When 5 external gears are used in combination, the first and last gears always turn:

**a.** In opposite directions
**b.** Clockwise
**c.** In the same direction
**d.** Counterclockwise

**16.** All of the following kinds of drive belts are used on automobile engines, *except:*

**a.** V-belts
**b.** V-ribbed belts
**c.** Flat belts
**d.** Cogged belts

### Fill in the Blank

**17.** When a liquid becomes a gas, it is __________.

**18.** A balanced atom has equal numbers of __________ and __________.

**19.** An unbalanced atom is an __________.

**20.** The outer electron orbit, or shell, of an atom is the __________ shell.

**21.** Electrons that move easily from their shells to the shells of other atoms are __________ electrons.

**22.** Atoms or molecules with many bound electrons are good electrical __________.

**23.** Chemical reactions that form oxygen compounds are __________ reactions.

**24.** Chemical reactions that separate oxygen from a compound are __________ reactions.

**25.** Force multiplied by distance equals __________.

**26.** A rate of 550 foot-pounds of work per second equals one __________.

**27.** To use gases or liquids to do work, you must develop pressure and __________ in the system.

**28.** The speed at which one gear turns in relation to another is the gear __________.

**29.** A gear ratio of 0.8:1 is an __________ ratio.

**30.** As output speed increases, output torque __________.

**31.** If you want two external gears to turn in the same direction, you must put an __________ gear between them.

### JAMES WATT—METRIC SCIENTIST?

The Scotsman, James Watt (1736–1819), was one of history's most prolific inventors and engineers. Although he didn't invent the steam engine, his improved engine with a separate condensor (which he patented in 1769) improved the engine's efficiency more than 70 percent. Watt also developed a furnace that consumed its own smoke (a leader in air pollution control) and an early copying machine. (He didn't invent Xerox, however.) A little known fact is that he was the first scientist to prove that water is a compound of oxygen and hydrogen, not an element.

Remember that Watt worked out the calculations that gave us the horsepower as a standard unit of power measurement. His formula for computing horsepower was based on the fact that a horse could move 200 pounds 165 feet in one minute. Now, 165 feet may seem like an odd unit of distance measurement, but was it? Watt started with a traditional English unit, the rod, which was used in surveying and mining engineering of his time. One rod equals 16.5 feet, and a convenient distance measurement for his calculations was 10 rods, or 165 feet. Do you suppose he was thinking metric and actually invented a standard unit we could call "one dekarod?"

# 3 WORKING WITH MEASUREMENT SYSTEMS AND PRECISION MEASURING TOOLS

## WHAT

A measurement system is a group of related, standard units that indicate length, area, volume, mass, weight, time, power, pressure, force, and anything else we want to quantify. Measurement is the act of expressing the properties of something in quantities that people can understand. To be useful, the units of any measurement system must be easy to work with, understandable, and have some practical value to the people who use them.

The two measurement systems widely used in the automobile industry are the U.S. customary system and the international metric system. While the United States still uses U.S. customary units for many measurements, the metric system is the most widely used system in the world. This chapter explains the relationships between U.S. customary units and metric units and the use of precision measuring tools, such as micrometers, graduated in metric and in customary inch units.

## WHY

Some of the cars you work on are engineered in metric units; others are engineered in customary units. Many late-model cars have parts designed in both systems and specifications in both customary and metric units.

The auto service profession today requires that professional mechanics be equally proficient in both measuring systems. This should be no problem if you understand both systems. When you work on a car built to metric standards, use metric tools and think in metric units. The same goes for a car built to U.S. customary standards. Occasionally, you may have to convert a metric measurement to customary units, or vice versa. When you do, it's no problem if you understand the equivalent units in each system and the simple **conversion factors**. This chapter summarizes equivalent customary and metric units that you will use in engine and combustion control system service.

## GOALS

After studying this chapter, you should be able to:

**1.** List base, supplementary, and derived SI metric units and symbols. Identify comparable customary units and symbols.

**3.** Do simple calculations using the powers of 10 and metric prefixes.

**4.** Convert measurements between customary and metric units for the following: length, area, mass or weight, volume, force, pressure, torque, power, and temperature.

**5.** Take measurements with inside and outside micrometers and vernier calipers graduated in inch and metric units.

## THE SI METRIC SYSTEM

All of the mathematical operations for the metric system—multiplying, dividing, adding, and subtracting—are based on **powers of 10**, or the decimal system. There is nothing unique about the decimal system. We use decimal-inch fractions in the customary system for precise measurements. We can say that the diameter of a cylinder is 2-5/8 inches or 2.625 inches.

This kind of conversion from fractions to decimals isn't complicated because we have learned to do it instinctively. The complicated thing about the customary system is that we use different factors for different conversions, or different increasing and decreasing units. The fraction we used above was 5/8 of an inch. The decimal equivalent of 1/8 is 0.125, so to get a decimal equivalent of 5/8, we can multiply $0.125 \times 5 = 0.625$. Or, we can divide 5 by 8 to get 0.625. Either way, we work with different factors for whole numbers and fractions.

The easy feature of the metric system is that *all* multiplying and dividing to get from large to small units is done by 10 or powers of 10. For example, the metric base unit of length is the meter (which is roughly 1 U.S. yard). We can convert the meter to larger or smaller metric units by multiplying or dividing successively by 10 and only 10.

## SI METRIC UNITS

Although the metric system was conceived in the 17th century, it was not until 1960 that an international conference of scientific leaders adopted the modern international system, known as **SI metric** units. The name comes from the French term, *Le Système International d'Unités*. SI metric units are the metric system used by the automobile industry today.

SI metric units as standardized in 1960 were not a revolutionary new system, but many older combination units were discarded and new ones adopted. Figure 3-1 lists the seven SI metric **base units** (foundation units). The first five base units are the foundation for most metric specifications used in engine performance service.

| *Quantity* | *Unit* | *Symbol* |
|---|---|---|
| length | meter | m |
| mass | kilogram | kg |
| time | second | s |
| electric current | ampere | A |
| thermodynamic temperature | kelvin | K |
| amount of substance | mole | mol |
| luminous intensity | candela | cd |

**Figure 3-1.** These are the seven SI metric base units from which all other measurements are derived.

| *Quantity* | *Unit* | *Symbol* | *Formula* |
|---|---|---|---|
| frequency (cycles per second) | hertz | Hz | 1/s |
| force | newton | N | kg · m/s² |
| pressure, stress | pascal | Pa | N/m² |
| energy, work, quantity of heat | joule | J | N · m |
| power | watt | W | J/s |
| torque | Newton meter | N · m | (N · m/rad) · rad |
| electric potential, potential difference, electromotive force | volt | V | W/A |
| capacitance | farad | F | C/V |
| electric resistance | ohm | Ω | V/A |
| volume | cubic meter | m³ | |
| | liter | l | 0.1 m³ or 1,000 cm³ |

**Figure 3-2.** These are the derived metric units used most often in the automobile service industry.

The SI metric system also has two **supplementary units** for angular measurements: the radian for plane 2-dimensional angles and the steradian for solid 3-dimensional angles. You will not find these units used for automobile engine service, however. Degree specifications for such things as ignition timing and valve timing are normally given in customary degrees for all automobiles.

### Derived Units

Besides base and supplementary units, the metric system uses many **derived units** for common measurements. Derived units are compound units formed by combining base units, supplementary units, and other derived units by multiplying and dividing. Many derived units are formed by squaring or cubing base units, such as a common derived unit for volume: the cubic centimeter ($cm^3$, or cc).

Still other derived standard units are formed by combining two or more units through multiplication or division, but the combinations have special names. For example, the metric unit for force is kilograms times meters per second squared ($kg \cdot m/s^2$). That is a complicated symbol and formula, so the unit of force is simply called the newton (N) to make further combinations simpler. The metric derived unit for pressure is the pascal (Pa), or newtons per square meter ($N/m^2$). Figure 3-2 is a table that lists the common derived units that you will see most often in engine performance service.

## METRIC MATHEMATICS

Whether a metric unit is a base, supplementary, or derived unit, you can always express larger or smaller quantities by multiplying or dividing by 10. The powers of 10 are a mathematical shorthand that we express by assigning **exponents** to the number 10 to show how many times it is multiplied or divided by itself. You know that when you "square" any number, such as 10, you multiply the number by itself (10 × 10) and write the formula with the exponent 2: $10^2$. Ten squared is 100. Similarly, 10 cubed is 10 × 10 × 10, or $10^3$, which is 1,000.

When you divide 10 by itself (10/10), you get 1. When you divide by 10 again (1/10), you get 0.1. You also can write this as 10 with a negative exponent ($10^{-1}$). You can multiply or divide 10 by itself as many times as you like and keep assigning successive exponents. Exponents that raise 10 to a larger number are **multiples**. Exponents that reduce 10 to smaller decimal fractions are **submultiples**. This becomes useful shorthand when you have another large or small number that you want to show as a multiple or submultiple of 10 without writing a lot of extra digits. Thus:

$$2{,}500 = 2.5 \times 10^3$$
$$0.025 = 2.5 \times 10^{-2}$$
$$2{,}500{,}000 = 2.5 \times 10^6$$
$$0.000\,025 = 2.5 \times 10^{-5}$$

The metric system might be awkward if you had to write 0.025 meter as $2.5 \times 10^{-2}$ meter, or 2,500 meters as $2.5 \times 10^3$ meters. So, the metric system has a set of prefixes that you add to any unit to show the power of 10 by which you are multiplying or dividing. Thus $2.5 \times 10^{-2}$ meter becomes 2.5 *centi*meters, and $2.5 \times 10^3$ meters becomes 2.5 *kilo*meters. Figure 3-3 lists all of the metric prefixes and their powers of 10 from $10^{18}$ to $10^{-18}$. You won't have to deal with all of these regularly as an auto mechanic, so the ones most often used for automobile specifications are shown in boldface type.

| | *Multiplication Factor* | *Prefix* | *Prefix Symbol* |
|---|---|---|---|
| Multiples | 1 000 000 000 000 000 000 = $10^{18}$ | exa | E |
| | 1 000 000 000 000 000 = $10^{15}$ | peta | P |
| | 1 000 000 000 000 = $10^{12}$ | tera | T |
| | 1 000 000 000 = $10^{9}$ | giga | G |
| | **1 000 000 = $\mathbf{10^{6}}$** | **mega** | **M** |
| | **1 000 = $\mathbf{10^{3}}$** | **kilo** | **k** |
| | 100 = $10^{2}$ | hecto | h |
| | 10 = $10^{1}$ | deka | da |
| | 1.0 | | |
| Submultiples | 0.1 = $10^{-1}$ | deci | d |
| | **0.01 = $\mathbf{10^{-2}}$** | **centi** | **c** |
| | **0.001 = $\mathbf{10^{-3}}$** | **milli** | **m** |
| | **0.000 001 = $\mathbf{10^{-6}}$** | **micro** | **$\mu$** |
| | 0.000 000 001 = $10^{-9}$ | nano | n |
| | 0.000 000 000 001 = $10^{-12}$ | pico | p |
| | 0.000 000 000 000 001 = $10^{-15}$ | femto | f |
| | 0.000 000 000 000 000 001 = $10^{-18}$ | atto | a |

**Figure 3-3.** These are the metric prefixes used to increase or decrease a measurement by a power of 10. Those used most often in the auto industry are listed in boldface type.

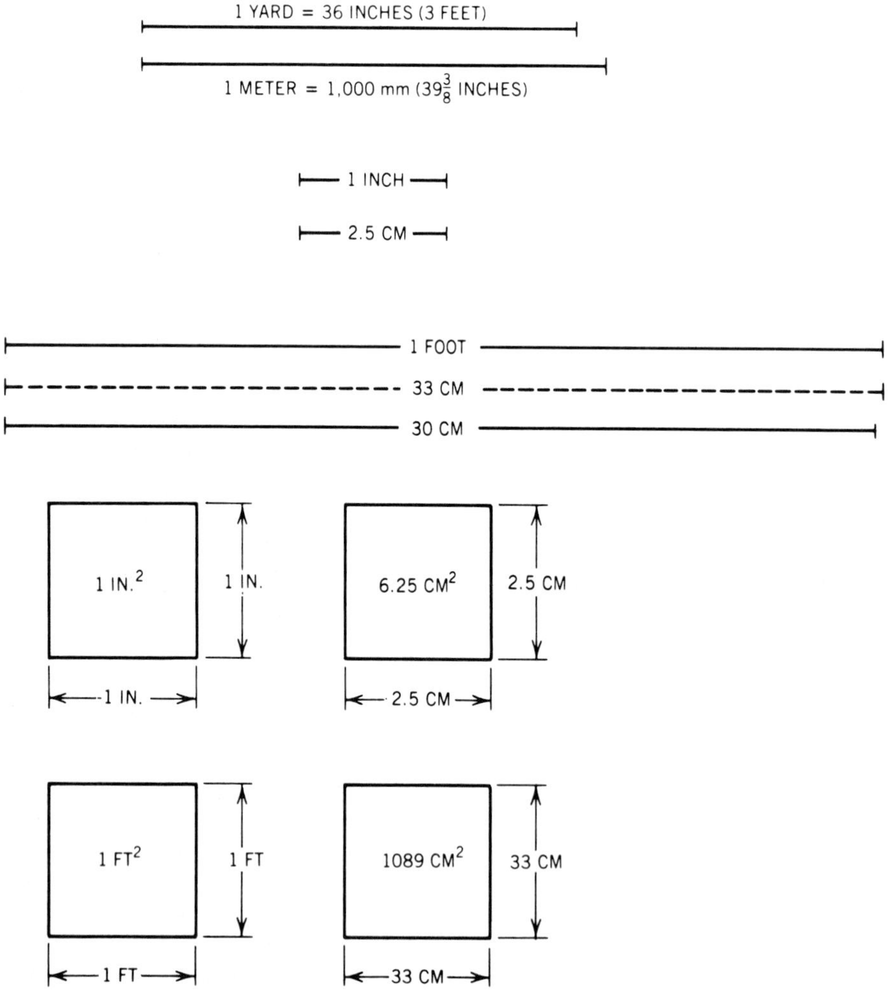


**Figure 3-4.** Comparable metric and customary units for length and area measurements.

## METRIC UNITS FOR AUTO MECHANICS

The following sections explain the metric units—base, supplementary, and derived—that are used most often in automobile design and service. Comparable U.S. customary units also are given for comparison, and simple conversion formulas are explained.

### Length and Area

The metric unit for measuring length is the meter (m). The multiples and submultiples you will use most often are the:

kilometer (km)—1,000 meters

centimeter (cm)—0.01 meter

millimeter (mm)—0.001 meter

As an engineering convention, specifications for any part of an automobile are always given in millimeters. You can square any of these to get an area measurement, such as the square centimeter ($cm^2$). The comparable customary units are, of course, the inch and fractions of an inch, the foot, the yard, and the mile. Metric units and customary units are not exact multiples of each other, but there are some close approximations that are useful for estimating equivalent measurements.

A meter is approximately 39 3/8 inches, a little more than the 36 inches of a yard. A kilometer is approximately 0.6 mile. There is no single metric unit formed from a prefix and a base that is approximately equivalent to 1 foot, but 30 centimeters (30 cm) or 300 millimeters (300 mm) are close. One inch is approximately 2.5 centimeters (2.5 cm) or 25 millimeters (25 mm). To estimate fractional equivalents, 10 mm are approximately 3/8 inch, or 0.375 inch. One millimeter is about 1/25 inch, or 0.039 inch. One square inch is approximately 6.5 square centimeters ($cm^2$). Figure 3-4 shows these approximately equivalent metric and customary units, and the exact conversion factors are given below.

millimeters (mm) × 0.03937 = inches

millimeters (mm) = 25.400 × inches

centimeters (cm) × 0.3937 = inches
centimeters (cm) = 2.540 × inches
meters (m) × 3.2809 = feet
meters (m) = 0.3048 × feet
kilometers (km) × 0.621377 = miles
kilometers (km) = 1.6093 × miles
square centimeters ($cm^2$) × 0.1550 = square inches
square centimeters ($cm^2$) = 6.4516 × square inches

## Volume

The metric unit for volume is the cubic meter ($m^3$) and its multiples and submultiples. They are based on the base unit of length, the meter. We use these to measure the volume, or capacity, of things like fuel tanks, radiators, cylinders, and oil pans.

The customary equivalents for metric volume measurements are complicated because in the customary system, we use volume measurements based on length units cubed (cubic inches and cubic feet) and volume measurements based on liquid measure (ounces, pints, quarts, and gallons). Rather than worry about 3-way conversions, the best way to understand these volume relationships is to remember the customary and metric units used for particular kinds of measurements.

In the customary system, we use cubic inches to measure engine volume, or displacement, and the corresponding metric unit is the cubic centimeter ($cm^3$). The "official" symbol is $cm^3$ for consistency with other metric symbols, but the auto industry has used cc for cubic centimeter so long that its use continues. (Either $cm^3$ or cc is both singular and plural; "cc's" is not a correct symbol.) One cubic inch equals approximately 16.4 cc, or 1 cc equals approximately 0.061 cubic inch.

In the customary system, we use fluid ounces, pints, quarts, and gallons to measure the volume of liquids. The metric system still uses units based on the cubic meter, but it allows a special unit that we use often in the auto industry: the liter. The liter is not an "official" metric unit, but it is a recognized name for a volume of one cubic decimeter for liquids and gases. Thus, one liter equals 0.1 cubic meter or 1,000 cubic centimeters. The only prefix used with liter is milli, so one milliliter (1 ml) equals one cubic centimeter (1 cc).

One liter is a little more than one quart, and four liters are a little more than a gallon. The exact conversion factors for metric and customary volume measurements are:

cubic centimeter (cc) × 0.061025 = cubic inch
cubic centimeter (cc) = 16.3866 × cubic inch
1 liter = 1,000 cubic centimeters (1,000 cc)
liter (l) × 61.023 = cubic inches
liter (l) = 0.01639 × cubic inches
liter (l) × 0.946 = quart
liter (l) = 1.057 × quart
liter (l) × 0.26418 = gallon
liter (l) = 3.7854 × gallon

Figure 3-5 shows comparisons between common metric and customary measurements. Although 1 liter equals 1,000 cubic centimeters, it is common to round off engine displacement and other volume measurements in cubic centimeters to the nearest liter or tenth of a liter. Thus a 1,943-cc engine might be labeled either a 1.9- or a 2-liter engine. A 15-gallon fuel tank that holds exactly 56,781 cc (56.781 liters) would be considered a 57-liter tank.

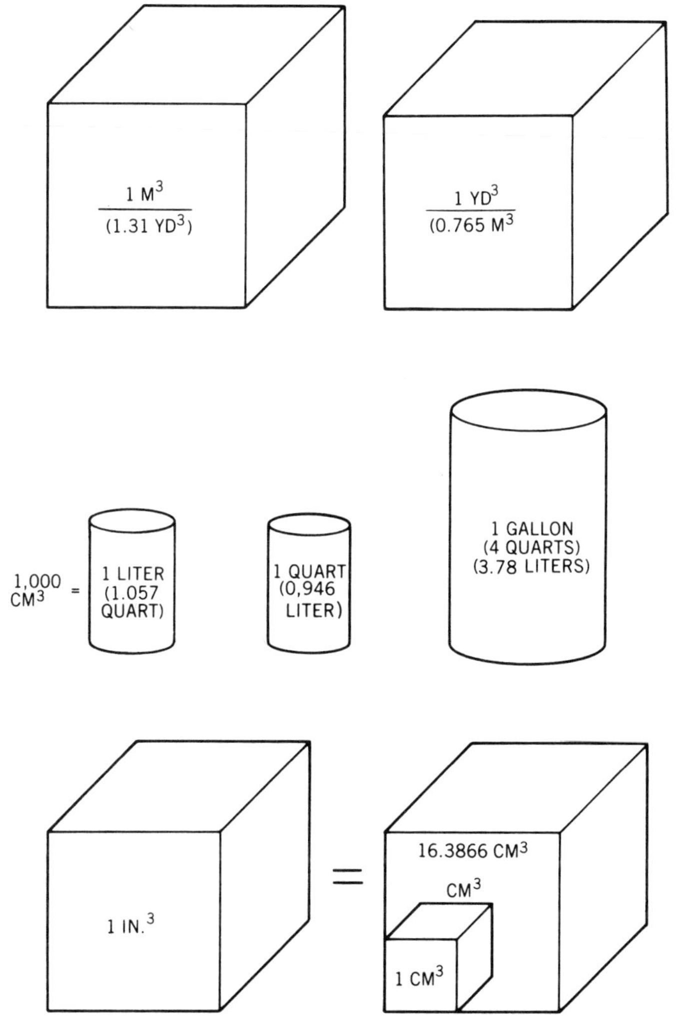


**Figure 3-5.** Comparable metric and customary units for volume measurements.

### Mass and Weight

Weight is a measure of the earth's gravity pulling on an object's mass. In the customary system, we use ounces, pounds, and tons to measure mass and weight. The metric base unit for mass is the kilogram, but the metric system doesn't define weight. Therefore, we use kilograms as we use pounds to measure both.

The kilogram is a base unit, but it is a multiple of the gram **stem unit**. One kilogram equals 1,000 grams. Metric units of weight or mass are expressed by adding prefixes to the gram. One kilogram is approximately 2.2 pounds, or 0.45 kilogram is approximately 1 pound. Figure 3-6 shows these and other mass comparisons. The exact conversion factors for grams, ounces, kilograms, and pounds are:

grams (g) × 0.03527 = ounces
grams (g) = 28.3495 × ounces
kilograms (kg) × 2.2046 = pounds
kilograms (kg) = 0.4536 × pounds

### Force

In the customary system, we measure force in the same units we use for mass or weight: pounds and ounces. The SI metric system has a special unit for force: the newton (N). One newton is a force of one kilogram applied at a rate of one meter per second squared: $1\ N = 1\ kg \cdot m/s^2$. You can see why it was easier to create a single symbol, N, for that equation. The newton, then, is used to derive the metric standard unit for energy or work, the joule, which is newtons times meters. The newton also is the basis for the SI metric units for pressure and torque.

### Pressure

The metric unit for pressure is the pascal (Pa). The corresponding customary unit is pounds per square inch (psi). A single pascal is a small amount of pressure, so the automobile industry gives pressure specifications in kilopascals (kPa). One psi equals approximately 6.9 or 7 kilopascals (Kpa). The exact conversion factors are:

kilopascals (kPa) × 0.145 = psi
kilopascals (kPa) = 6.895 × psi

### Torque

The customary units of torque measurement are compound units of distance times force, such as foot-pounds or inch-pounds. The SI metric unit for torque is the newton-meter. The symbol is N·m, or sometimes simply Nm. The conversion factors are:

newton-meters (N·m) × 0.737 = foot-pounds
newton-meters (N·m) = 1.356 × foot-pounds
newton-meters (N·m) × 0.089 = inch-pounds
newton-meters (N·m) = 0.113 × inch-pounds

The newton-meter has been the SI unit for torque measurement since about 1971. Previously, metric torque measurements were given in the compound units of force and distance, kilogram-meters (kg·m) or kilogram-centimeters (kg·cm). You may find torque specifications for older European cars that are given in kilogram-meters or kilogram- centimeters, and you can buy torque wrenches graduated either in newton-meters or the older metric units.

### Power

Power is the rate of doing work, and the horsepower (hp) is the customary unit for measuring mechanical power. Electric power is measured in watts.

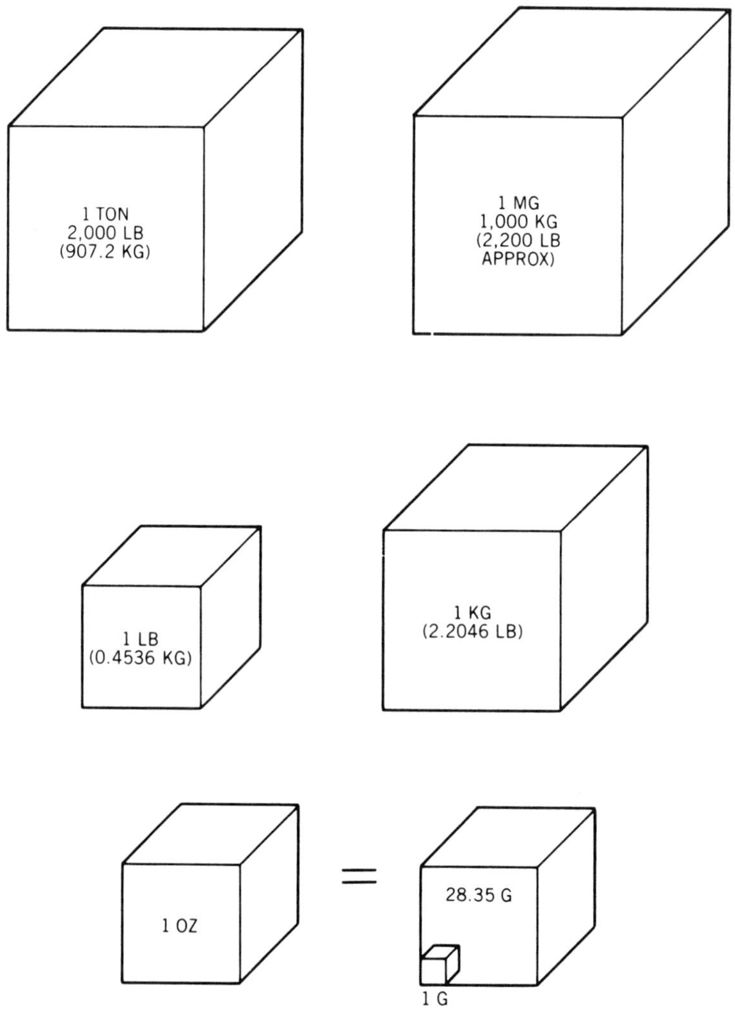


**Figure 3-6.** Comparable metric and customary units for mass and weight measurements.

Both are measurements of the amount of work per unit of time, usually per second. Horsepower equals foot-pounds per second; watts equal joules per second. You needn't worry about the mathematics of converting foot-pounds to newton-meters to joules because there are direct conversion factors for horsepower and watts:

watts (W) × 746 = horsepower

watts (W) = 0.0013 × horsepower

### Temperature

The SI base unit for temperature is the kelvin (K), but we use the Celsius scale for metric temperature measurements. The zero-point of the Celsius scale is the freezing point of water, while kelvin units begin at absolute zero, which is −273.15°C. One degree Celsius (°C) equals one kelvin (1 K; the Celsius scale simply begins at a higher point.

In the customary Fahrenheit scale, the freezing point of water is 32° F. Therefore, 0° C = 32° F. The boiling point of water is 100° C or 212° F. The conversion factors for Celsius and Fahrenheit temperatures are:

- Celsius (°C) = °F − 32 × $\frac{5}{9}$ or
- Celsius (°C) = °F − 32 /1.8
- Fahrenheit (°F) = (1.8 × °C) + 32

1° Fahrenheit (1° F) = 0.56° C

1° Celsius (1° C) = 1.8° F

Figure 3-7 shows a comparison of Celsius and Fahrenheit temperature scales. Notice that the two scales are equal at −40°.

### Electrical Units

Electrical units such as the ampere, volt, ohm, farad, and others are used interchangeably in the metric and customary systems. Therefore, you don't have to worry about converting metric volts to customary volts. A volt is a volt. You will use metric prefixes with electrical units, however, for easier measurements. Therefore, 1,500,000 ohms is expressed as 1.5 megohms (1.5 MΩ), which is 1.5 × $10^6$ ohms. Also, you will work with measurements such as 15 milliamperes (15 mA), which is 0.015 ampere, and 25 kilovolts (25 kV), which is 25,000 volts. Most electrical test equipment is built with switches that do the multiplying and dividing for you.

## GUIDELINES FOR CONVERSION

We convert measurements in one system to measurements in another by either hard conversion or soft conversion. **Hard conversion** is what engineers do when they redesign a part that had been built to customary dimensions so that it is built to metric dimensions. Thus, a car with a 103-inch wheelbase (2,616.2 mm) might be redesigned so that its wheelbase is an even metric unit—say 2,610 or 2,620 millimeters.

**Soft conversion** involves no physical change in a part. It is simply the process of restating customary measurements in metric units. Thus, you could identify a 10-inch iron bar as a 254-mm bar without making any physical change to the bar. Measurements such as pressure and temperature that aren't dimensions for physical objects can be given in either customary or metric units without redesigning anything. These are still soft conversions.

Fortunately, you won't have to do any hard conversion as a professional mechanic, but you occasionally may have to do some soft conversion. In this text, we list equivalent customary and metric measurements when soft conversion is relevant. We list many linear, pressure, temperature, power, and torque values in both metric and customary units because it is possible for you to measure them in both ways on the job. In other cases, we give measurements in only one value because they are hard values, and you can't do anything to change them. For example, a 3/8-inch diameter bolt is always that size in customary units. The fact that 3/8-inch is approximately 9.5 millimeters is irrelevant. There is no such thing as a 9.5-mm diameter bolt. There are 8-mm and 10-mm bolts,

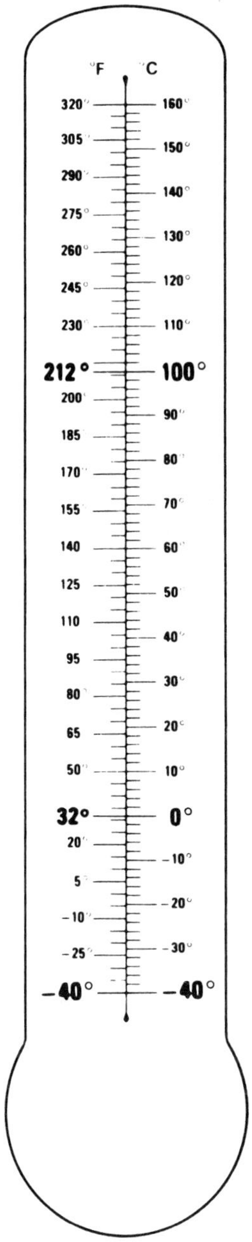


**Figure 3-7.** The Fahrenheit and Celsius scales are equal at −40°.

and the facts that these are approximately 0.32 inch and 0.39 inch diameters are equally irrelevant. The point of all this is to remind you that the best way to work with any measurement system is to *think* and *work* in its units.

Let's look at a couple examples of the kinds of soft conversion you might have to do on the job. Suppose a spark plug gap is listed as 1.4 mm, but you have only a decimal-inch plug gapping tool. Simply divide 1.4 by 25.4

millimeters per inch to find an equivalent decimal-inch gap measurement:

$$1.4/25.4 = 0.055118$$

Your 0.055-inch plug gapping gauge will give you an accurate adjustment for these plugs.

As another example, suppose the bolt torque specification for an intake manifold is 34 Nm and you have a foot-pound torque wrench. Multiply 34 by 0.737 Nm per foot-pound:

$$34 \times 0.737 = 25.058$$

Use your foot-pound torque wrench to tighten those bolts to 25 foot-pounds.

These are just two short examples of how you can use soft conversion on the job. Of course, the jobs would be faster and easier if you had metric measuring equipment to work with. That's another good reason to think and work in the system by which the part was designed.

Notice in these examples that we did any rounding off of numbers *after* the conversion. Don't round off either an original specification or measurement or the conversion factor. After the measurement is converted, round off to the accuracy of your measuring equipment or to the accuracy of the original specification, whichever is more precise.

## PRECISION MEASURING TOOLS

Many parts of a car's engine and other systems are made to precise specifications that require precise measurements for accurate service. As a professional mechanic, you will use many precision measuring tools, graduated in both customary and metric units. Among these are simple scales, rulers, straightedges, and thickness (feeler) gauges. You also will use a variety of calipers, micrometers, and dial-indicating gauges. Micrometers and precision calipers have **vernier scales** graduated in thousandths or ten-thousandths of an inch or hundredths of a millimeter. A vernier scale is an auxiliary, or fine, scale that indicates fractional parts of a larger scale. Whether an instrument is an inside or outside micrometer, a micrometer depth gauge, or inside or outside calipers, you read the scales according to the principles in the following sections.

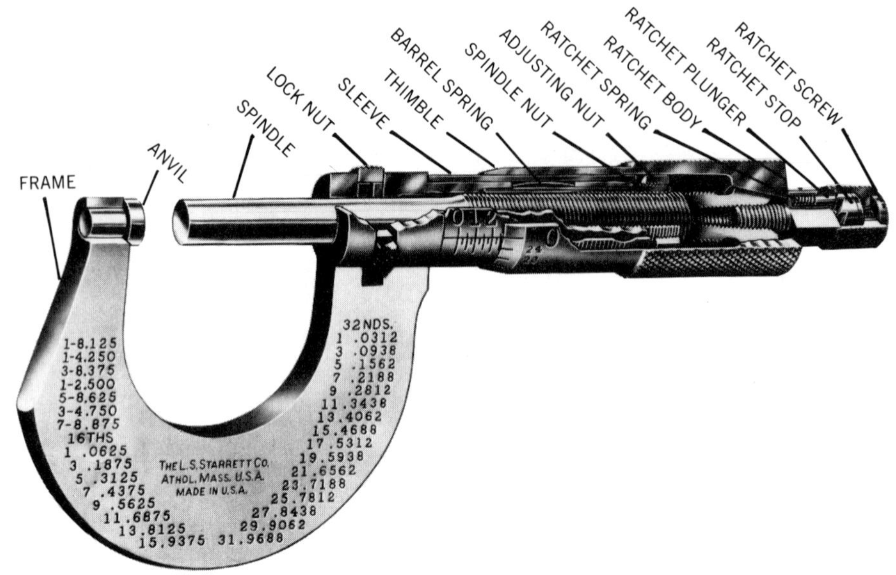


**Figure 3-8.** Regardless of the frame opening, a typical outside micrometer has these basic parts. (Starrett)

### Micrometers

A micrometer is an accurately ground screw thread that is rotated in a fixed nut to change the distance between two measuring surfaces. A typical outside micrometer has a fixed anvil and a movable spindle in a C-shaped frame, figure 3-8. Measurement graduations are on the fixed sleeve and movable thimble. Turning the thimble moves the spindle in and out toward the anvil. The screw thread of a decimal-inch micrometer has 40 threads per inch so that one revolution of the thimble moves the spindle 1/40, or 0.025, inch.

The lines on the fixed sleeve correspond to the threads of the micrometer, and each equals 1/40, or 0.025, inch on a decimal-inch micrometer. Every fourth line is numbered to indicate hundreds of thousandths, figure 3-9. Line 1 equals 0.100; 2 equals 0.200; 3 equals 0.300, and so on. The beveled edge of the movable thimble has 25 lines, each of which equals 0.001. Rotating the thimble from one line to the next moves the spindle longitudinally (in or out) 0.001 inch.

To read a micrometer graduated in thousandths of an inch, multiply the number of vertical divisions on the spindle by 0.025 and then add the number of thousandths shown by the line on the thimble that matches with the horizontal line on the spindle. The example in figure 3-9 shows the following measurement:

*Line 1 on the spindle equals 0.100*
*Three more lines on the spindle equal 3 × 0.025 = 0.075*
*Line 3 on the thimble coincides with the horizontal line on the spindle, which equals 3 × 0.001 = 0.003*
*The micrometer reading is 0.178*

Some micrometers have additional sleeve graduations to provide readings in ten-thousandths of an inch. These graduations are a 10-division vernier scale on the sleeve, which occupies the same space as 9 divisions on the thimble, figure 3-10. The difference between one space on the fixed vernier and one space on the movable thimble is 0.1 of a thimble division, or 0.0001

To read a micrometer graduated in ten-thousandths of an inch, first obtain the reading in thousandths as described above. Then see which line on the vernier matches a line on the thimble. If it is line 1, add 0.0001 to the thimble reading; if it is line 2, add

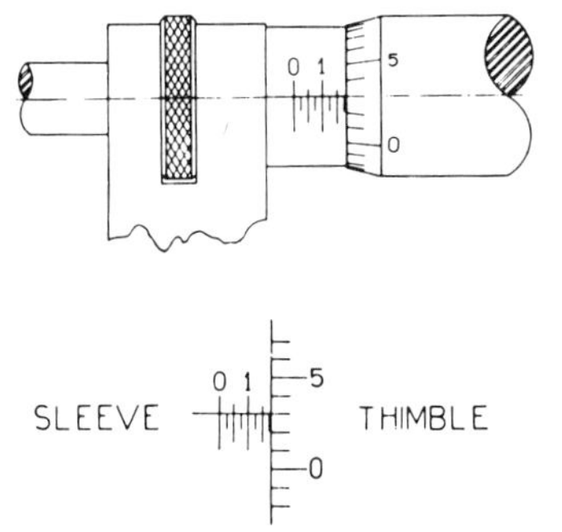


READING .178"

**Figure 3-9.** Typical reading on a micrometer graduated in thousandths of an inch. (Starrett)

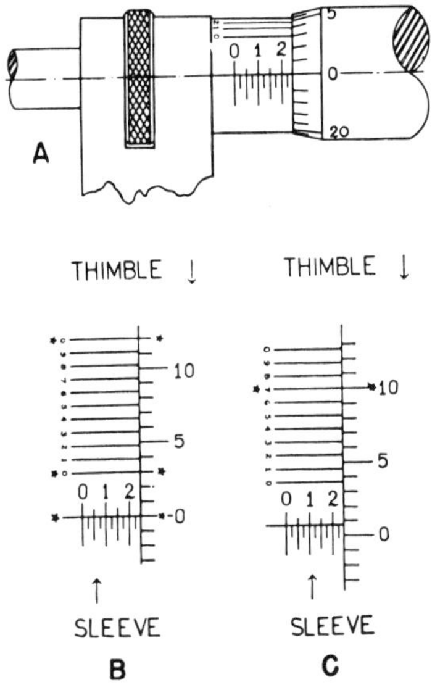


**Figure 3-10.** Typical reading on a micrometer graduated in ten-thousandths of an inch. (Starrett)

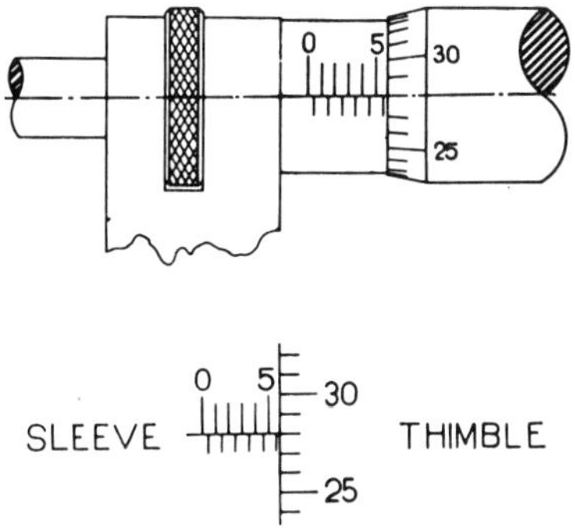


READING 5.78MM

**Figure 3-11.** Typical reading on a micrometer graduated in hundredths of a millimeter. (Starrett)

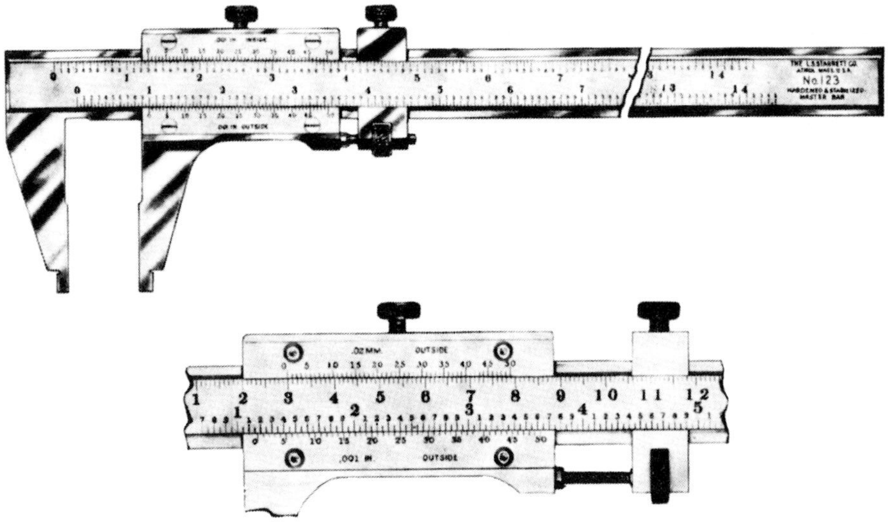

**Figure 3-12.** This vernier caliper is graduated for both inside and outside measurements. (Starrett)

0.0002, and so on. The example in figure 3-10 shows the following measurement:

*Line 2 on the sleeve equals 0.200*

*Two more lines on the spindle equal 2 × 0.025 = 0.050*

*The longitudinal line on the sleeve is between the 0 and 1 on the thimble, indicating that a ten-thousandth measurement must be read from the vernier.*

*Line 7 on the vernier coincides with a line on the thimble: 7 × 0.0001 = 0.0007*

*The micrometer reading is 0.2507*

Most metric micrometers have a spindle screw thread pitch of 0.5 mm so that each full turn of the thimble moves the spindle 0.5 mm. Two turns move it 1.0 mm. The longitudinal line on the sleeve is graduated from 0 to 25 mm and marked in 1.0- and 0.5-mm increments, figure 3-11. The beveled edge of the thimble has 50 divisions, and every fifth line is numbered. Since a full turn of the thimble moves the spindle 0.5 mm, each graduation on the thimble equals 1/50 of 0.5, or 0.01 mm.

To read a metric micrometer, add the millimeter reading from the sleeve to the reading in hundredths of a millimeter shown on the thimble. The example in figure 3-11 shows the following measurement

*Line 5 on the sleeve equals 5.0 mm*

*One more half-millimeter line on spindle equals 0.05 mm*

*Line 28 on the thimble is on the longitudinal line of the spindle (each line is 0.01 mm): 28 × 0.01 = 0.28 mm*

*The micrometer reading is 5.78 mm*

Outside micrometers come in many different sizes, but the measurement range of the spindle is usually just one inch, or 25 millimeters. That is, a 4-inch micrometer measures from 3 to 4 inches.

**CAUTION.** A micrometer is a precision instrument. Be sure the anvil is clean before measuring any item. Always hold the part being measured squarely between the anvil and the spindle. Do not overtighten the thimble, which could damage the micrometer, the part being measured, or both. If the micrometer has a torque-limiting ratchet, use it to tighten the thimble.

### Vernier Calipers

Vernier calipers, figure 3-12, are available for inside and outside measurements, and depth and height gauges are made with vernier scales for measurements to 0.001 inch, or 0.02 millimeter.

On a 25-division vernier scale that measures thousandths of an inch, the bar is graduated in 40 units of 0.025 inch, figure 3-13. Each fourth graduation is numbered to show tenths of

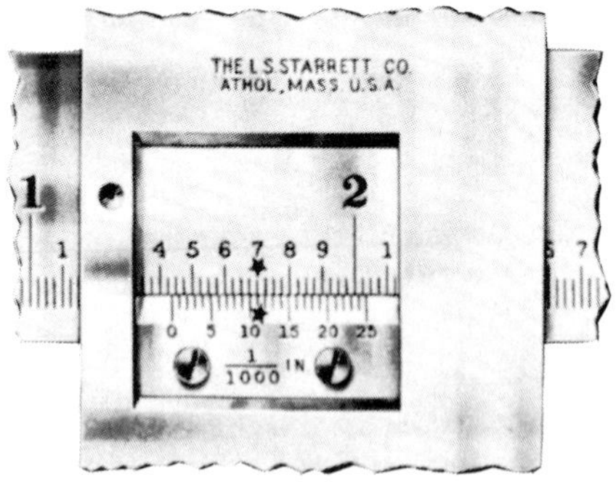


ENGLISH MEASURE—READING 1.436 INCHES

**Figure 3-13.** Typical decimal-inch measurement on a 25-division vernier scale. (Starrett)

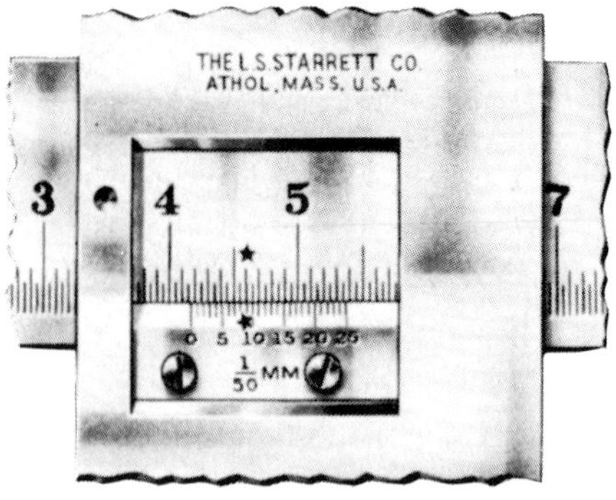


METRIC MEASURE—READING 41.68 MM

**Figure 3-14.** Typical metric measurement on a 25-division vernier scale. (Starrett)

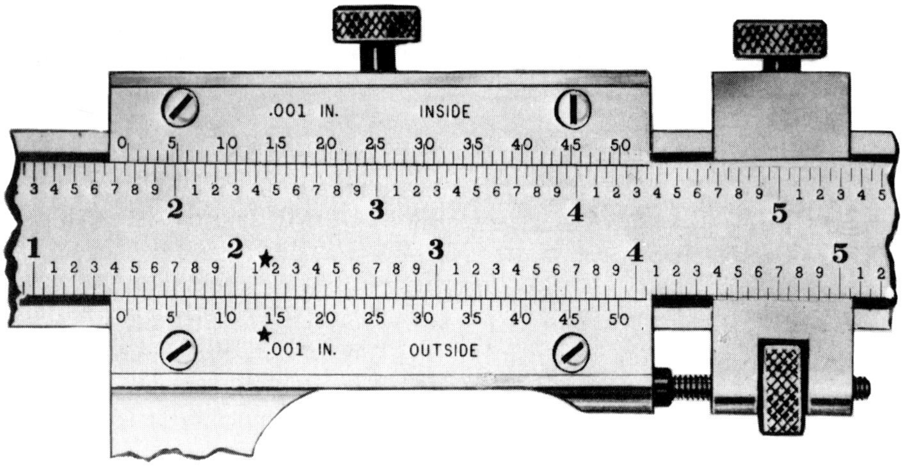


**Figure 3-15.** You read this 50-division vernier scale similarly to a 25-division scale. (Starrett)

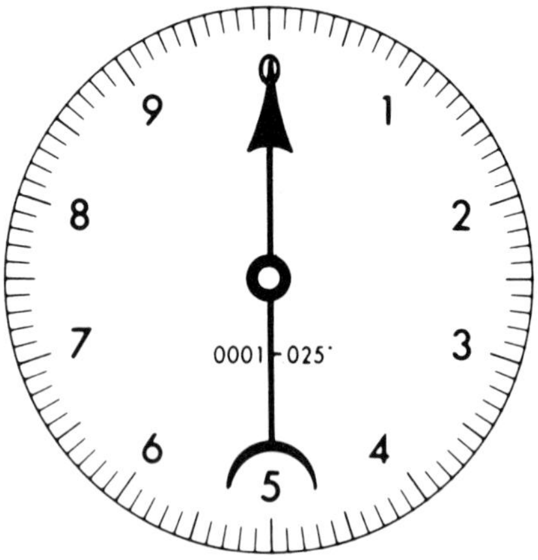


**Figure 3-16.** This continuous-reading dial indicator reads in just one direction from zero.

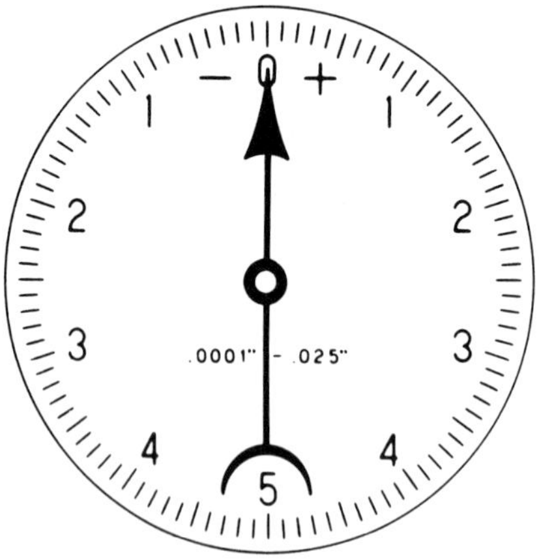


**Figure 3-17.** This balanced dial indicator reads in both directions from zero.

an inch. The 25 divisions on the movable vernier occupy the same space as 24 divisions on the bar. Since one division on the bar equals 0.025 inch, 24 divisions equal 0.600 inch (24 × 0.025 = 0.600). The 25 divisions on the movable plate also equal 0.600 inch, so each division equals 1/25 × 0.600, or 0.024 inch. Therefore, the difference between each mark on the bar and on the plate is 0.001 inch.

To read a 25-division vernier scale, note how many inches, tenths (0.100), and fortieths (0.025) the 0 mark on the plate is from the 0 mark on the bar. Then add the number of thousandths shown by the line on the plate that exactly matches a line on the bar. The example in figure 3-13 shows a reading of 1.000 inch, plus 0.400, plus 0.025, or 1.425 inches. The eleventh line on the plate matches a line on the bar, as shown by the stars, so add 0.011 to 1.425 for an exact reading of 1.436 inches.

On a 25-division metric vernier, the graduations on the movable plate are 1/50, or 0.02 mm. The bar is graduated in centimeters, millimeters, one-half (0.5) millimeters. The example in figure 3-14 shows a reading of 4.00 cm (40.00 mm), plus 1.00 mm, plus 0.50 mm, or 41.50 mm. The ninth line on the plate matches a line on the bar, as shown by the stars, so add 9 × 0.02 mm, or 0.18 mm to the measurement for a final reading of 41.68 mm

Some vernier scales have 50 divisions on the movable plate that correspond to 49 divisions on the bar, figure 3-15. You read them in the same general way as 25-division scales.

## Dial Indicators

You probably will use various kinds of dial indicators to measure hole depth, surface smoothness, and out-of-round and runout of cylinders and rotating parts. You will usually use them to check measurement variations rather than single linear measurements.

Dial indicators are measuring instruments with a plunger and precision gear mechanism to turn a pointer on a dial. Most are graduated in thousandths (0.001) or ten-thousandths

(0.0001) of an inch or hundredths (0.01) or thousandths (0.001) of a millimeter. Continuous dial indicators read in one direction from the zero setting, figure 3-16. Balanced dial indicators read in both directions, plus and minus, from zero, figure 3-17. Dial indicators come with a variety of mounting fixtures to hold them for different jobs. On most indicators, the dial is turned to align the 0 mark with the pointer and then locked in place before starting a measurement operation. Pointer movement then indicates the variation from zero.

**U.S. METRIC STANDARDS**

Did you know that the metric system is the only measurement system "officially" sanctioned by the U.S. government? Article 1, Section 8, of the U.S. Constitution gave Congress the power to "fix the Standard of Weights and Measures," but the U.S. customary system developed without an act of Congress. In 1866, Congress did pass a law that legalized the metric system as the only primary standard for weights and measures. In 1893, the international meter and kilogram became the primary standards of length and mass in the United States, both for metric and customary weights and measures. Thus, an inch is not officially defined as $\frac{1}{12}$ of a foot or $\frac{1}{36}$ of a yard. An inch is *officially* 2.54 centimeters. A foot is not officially 12 inches; rather it is *officially* 0.3048 meter—by Act of Congress.

## SUMMARY Think Metric, Think Customary

Occasionally, you may have to convert from one measuring system to another in your job. The things you have learned in this chapter will help you do this. Conversion is simple if you remember the comparable units in each system. Sometimes it's easier to look up corresponding measurements in tables such as those at the end of this book. If so, fine; conversion tables often can help you do your job fast and accurately. The easiest and most accurate way to work with any measuring system, however, is to *think* in the system to which the automobile was designed and use tools and measuring equipment graduated in units of that system.

If you grew up in the United States, you probably learned to "think" instinctively in inches, pounds, and gallons. If you grew up in Canada, Europe, or some other part of the world, you probably learned the metric system equally as well. Perhaps you were fortunate enough to learn *both* the customary and the metric systems when you were younger. If so, you have the advantage of a person who can speak and "think" in more than one language. The key to speaking any language fluently is to be able to think in its terms. The key to using any measurement system is to be able to think in its units.

# REVIEW QUESTIONS

### Multiple Choice

**1.** All multiplying and dividing to get from large to small metric units is done by:
- **a.** Submultiples
- **b.** Prefixes
- **c.** Powers of 10
- **d.** Base units

**2.** A common metric measurement of volume is the:
- **a.** Kilogram
- **b.** Pascal
- **c.** Millimeter
- **d.** Cubic centimeter

**3.** Ten millimeters times $10^3$ equals one:
- **a.** Kilometer
- **b.** Decimeter
- **c.** Centimeter
- **d.** Meter

**4.** One centimeter times $10^{-1}$ equals one:
- **a.** Centimeter
- **b.** Kilometer
- **c.** Millimeter
- **d.** Cubic centimeter

**5.** One inch equals:
- **a.** 0.254 centimeter
- **b.** 2.54 centimeters
- **c.** 25.4 centimeters
- **d.** 254 centimeters

**6.** A meter is approximately:
- **a.** 30 inches
- **b.** 36 inches
- **c.** 39 inches
- **d.** 42 inches

**7.** A kilometer is approximately:
- **a.** 0.6 mile
- **b.** 0.8 mile
- **c.** 1.2 miles
- **d.** 1.6 miles

**8.** One cubic centimeter (cc or $cm^3$) equals approximately:
- **a.** 0.61 cubic inch
- **b.** 0.061 cubic inch
- **c.** 6.1 cubic inches
- **d.** 61 cubic inches

**9.** One cubic inch equals approximately:
- **a.** 1.64 cc
- **b.** 16.4 cc
- **c.** 164 cc
- **d.** 32.6 cc

**10.** One quart equals approximately:
- **a.** 0.95 liter
- **b.** 1.2 liters
- **c.** 1.5 liters
- **d.** 0.84 liter

**11.** One gallon equals approximately:
- **a.** 37 liters
- **b.** 16 liters
- **c.** 3.8 liters
- **d.** 8.4 liters

**12.** Pressure measurements for automobile service are usually expressed in metric:

**a.** Kilograms
**b.** Newton-meters
**c.** Liters
**d.** Kilopascals

**13.** Torque measurements for automobile service are usually expressed in metric:

**a.** Kilograms
**b.** Newton-meters
**c.** Radians
**d.** Kilopascals

**14.** One psi equals approximately:

**a.** 1.45 kPa
**b.** 14.5 kPa
**c.** 0.69 kPa
**d.** 6.9 kPa

**15.** Rotating the thimble of decimal-inch micrometer exactly one turn moves the spindle in or out exactly:

**a.** 0.1 inch
**b.** 0.01 inch
**c.** 0.001 inch
**d.** 0.0001 inch

**16.** Rotating the thimble of metric micrometer exactly one turn moves the spindle in or out exactly:

**a.** 0.5 centimeter
**b.** 0.5 millimeter
**c.** 5.0 millimeters
**d.** 0.01 centimeter

### Fill in the Blank

**17.** One thousand cubic centimeters (1,000 cc) equal one ________.

**18.** A pressure of 30 psi equals approximately ________ kPa.

**19.** The metric temperature scale is the ________ scale.

**20.** At ________ the Fahrenheit and Celsius scales are the same.

**21.** The metric base unit to measure mass (or weight) is the ________.

**22.** The metric unit for electric current measurement is the ________.

**23.** One inch equals ________ mm.

**24.** One newton-meter equals approximately ________ foot-pound.

**25.** A number used to multiply or divide a measurement to change it to units of another system is called a ________.

**26.** Redesigning a part to convert its dimensions from one measurement system to another is called ________.

**27.** Restating the dimensions or other specifications of a part in the units of a different measurement system is called ________.

### METRIC HISTORY

Decimal measurement systems developed in the 16th century as people tried to establish order from the confusion of different weights and measures. Even as these measurement systems developed, they were based on old English units, such as the foot, yard, and bushel. And these units required nonuniform conversion factors, such as 16 × 2 × 4 to get from small to large measures, or vice versa. While the U.S. customary system has its origins in the British system, they are not identical. For example, the British Imperial pint is larger than the U.S. pint; it has 20 ounces. Therefore, the Imperial gallon is larger than the U.S. gallon even though both have 8 pints or 4 quarts.

The metric system was proposed in France in 1670, but it was not until 1790 that the French government (in the wake of the French Revolution) asked the Academy of Sciences to work out a uniform measurement system. The metric system has grown and changed in the past 200 years. Originally, it provided a set of units to measure length, area, volume, capacity, and mass based on two fundamental units: the meter and the kilogram. Science and commerce required measurements of other quantities, so other fundamental and derived units were added to the system.

In the late 1800's, the second was adopted as the base metric unit for time. In the early 1900's, the metric system was known as the centimeter-gram-second (cgs) system or the meter-kilogram-second (mks) system. In 1935, the ampere was recognized as the metric standard for measuring electric current, and the system became known as the meter-kilogram-second-ampere (mksa) metric system. This system was used by most countries through the 1950's, but it continued to grow and become more precise.

In 1960, the modern SI metric units became the official metric system, even though some older units, such as the *torr* for pressure measurements and kilogram-meters for torque measurements continue in use.

# 4 ENGINE DESIGN AND OPERATION

## WHAT

This chapter summarizes the design and operation of automobile engines. It begins with engine operating cycles and the structure of an engine from its block and crankshaft to its manifolds and accessory covers. This chapter also explains engine specifications and introduces the relationships among the fuel, ignition, and electrical systems. Understanding basic engine operation is the starting point for understanding combustion control systems.

## WHY

To understand engine combustion controls, you have to begin with engine operation. If you know about the 4-stroke cycle, basic engine design, fuel delivery, and ignition, you then can study various systems in detail and understand how they work together.

Efficient engine operation depends on the individual parts of seven separate but interrelated systems:

1. The engine mechanical parts
2. The engine lubrication system
3. The engine cooling system
4. The fuel system
5. The ignition system
6. Related emission controls
7. The engine electrical system (the battery and the starting and charging systems).

A breakdown in any one of these systems can reduce overall engine performance. A professional mechanic must know how one part in any of these systems can affect other systems.

## GOALS

After studying this chapter, you should be able to:

1. Explain 4-stroke operation of a reciprocating engine.
2. Explain the ignition methods and differences between spark-ignition (otto-cycle) and compression-ignition (diesel) engines.
3. Explain ignition intervals and firing orders.
4. Illustrate various multicylinder engine designs.
5. Explain the operating relationships of engine blocks, heads, and reciprocating and rotating parts.
6. Define the following terms: bore, stroke, displacement, compression ratio, torque, horsepower and kilowatts, and engine pressure and efficiency.
7. Explain basic fuel system operation for spark-ignition and diesel engines.

## ENGINE OPERATION

All engines burn fuel to change chemical energy to heat energy to kinetic energy and develop mechanical power. An engine that burns its fuel in a closed chamber and uses the heat energy directly to produce mechanical power is an **internal-combustion engine**. All automobile engines are internal-combustion engines. An engine, such as a steam engine, that burns its fuel outside of the working chamber where mechanical power is developed is an external combustion engine.

A mixture of fuel and air is compressed, ignited, and burned to develop pressure and force in an internal-combustion engine. The pressure and force produce mechanical power. To get the most energy from a burning liquid fuel, it must be **atomized**, vaporized, and compressed into a small area before it is burned. Without compression, the burning air-fuel mixture would not produce any force, pressure, and power. The kinetic theory of matter and the principles of compressed gases that you learned in chapter 2 are the principles that govern engine operation.

Other than the Wankel rotary engine, automobile engines are **reciprocating engines** in which a piston moves up and down or back and forth in a cylinder and compresses the air-fuel mixture in the cylinder. When this mixture ignites and burns, it produces the high pressure that moves the piston. To drive an automobile, the inline reciprocating, or **linear**, power of the moving piston must be changed to rotary power by the crankshaft, figure 4-1.

### Basic Engine Parts

Most gasoline and diesel engines are built on an engine block, or cylinder block, that contains the cylinders. The block also has coolant passages, or water jackets, for cooling, figure 4-2. For an overhead-valve (ohv) engine, a cylinder head with more water jackets

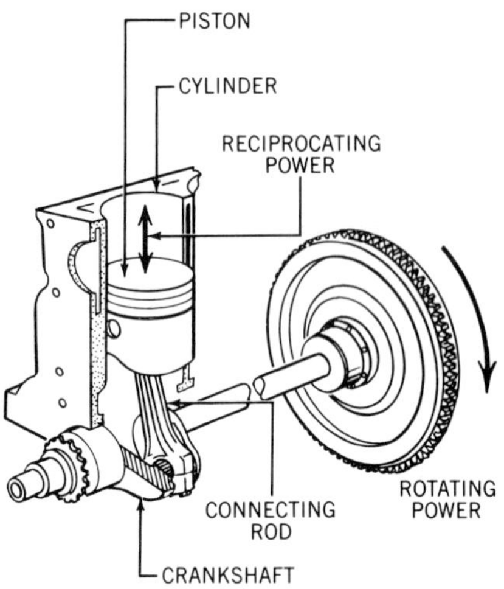


**Figure 4-1.** Power from the reciprocating motion of the piston is changed to rotary motion by the crankshaft. This rotating power ultimately drives the wheels of the car.

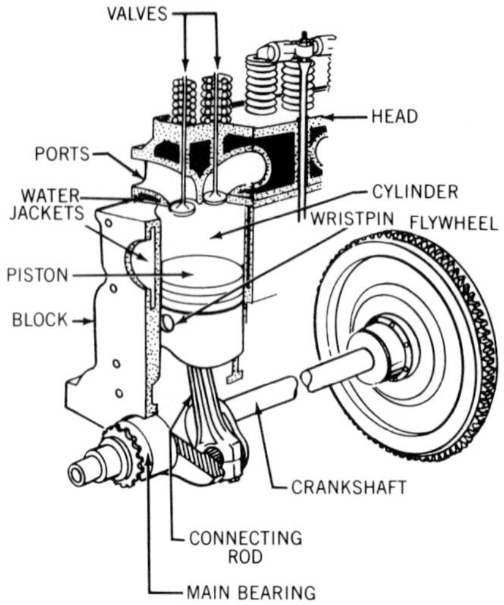


**Figure 4-2.** Most gasoline and diesel engines have the major engine parts shown here.

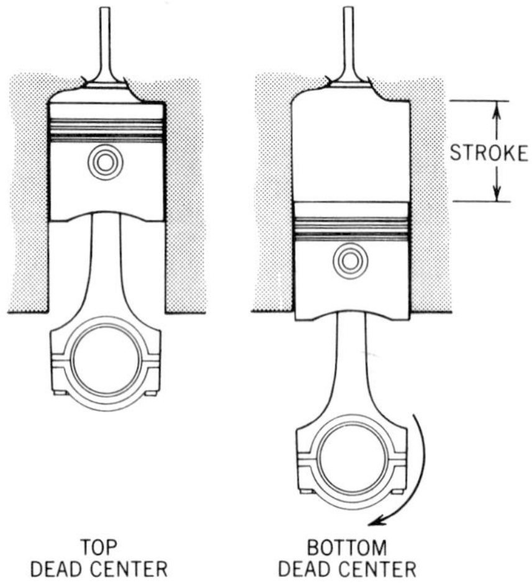


**Figure 4-3.** One piston stroke from top to bottom or from bottom to top of the cylinder equals one-half crankshaft revolution, or 180 degrees.

covers the top of the block and forms the combustion chamber. The ohv head also contains the fuel intake and exhaust passages (ports) and the intake and exhaust valves.

A connecting rod attaches the piston to the crankshaft. At its top, or small end, the connecting rod has a piston pin, or wristpin, that allows the rod to swing sideways as the piston moves up and down. The rod is mounted to the crankshaft with bearings that allow the crankshaft journal to rotate inside the big end of the rod. Similar bearings, called main bearings, support the crankshaft in the crankcase of the block and allow it to rotate. One end of the crankshaft has a large flywheel that provides rotating inertia for the engine.

The movement of the piston from the top of the cylinder to the bottom, or from bottom to top, is called a **stroke**. The top of a stroke is **top dead center (tdc)**. The bottom of a stroke is **bottom dead center (bdc)**. Each stroke of the piston turns the crankshaft one-half revolution, or 180 degrees, figure 4-3. Two piston strokes turn the crankshaft 360 degrees through one complete revolution. The term **revolutions per minute (rpm)** indicates the number of revolutions that the crankshaft makes in one minute.

## ENGINE OPERATING CYCLES

An engine operating cycle is the process of drawing air and fuel into a cylinder, compressing it, burning it to develop power, and exhausting the burned mixture. In a reciprocating engine, an operating cycle is measured in the number of piston strokes needed for one complete cycle. To understand engine-operating cycles, you must remember the principles of air pressure, temperature, and vacuum explained in chapter 2. Most gasoline and diesel engines operate on a **four-stroke cycle**.

### The 4-Stroke Cycle

The 4-stroke cycle consists of two piston strokes up and two down. Each stroke is named after its principal action: intake, compression, power, and exhaust. Here is how they work in a 4-stroke gasoline engine, figure 4-4:

**1.** *Intake stroke*—Starting at the top of its stroke (tdc), the piston moves down. This increases the available cylinder volume and creates a low-pressure area, or vacuum, in the cylinder. The vacuum allows atmospheric pressure to force air into the engine. This draws a vaporized air-fuel mixture into the cylinder through an open intake valve.

**2.** *Compression stroke*—As the piston reaches the bottom of the intake stroke, the intake valve closes. Then the piston moves up to compress the air-fuel mixture into the combustion chamber. Compression develops high potential energy in a small space.

**3.** *Power stroke*—As the piston nears top dead center of the compression stroke, the spark plug fires to ignite the air-fuel mixture. As the mixture burns in the confined space of the combustion chamber, it releases its energy and drives the piston downward on the power stroke. It is only during this stroke that the engine develops mechanical power.

**4.** *Exhaust stroke*—Near the bottom of the power stroke, the exhaust valve opens. The exhaust valve opens slightly before bdc so that pressure in the cylinder will help to expell exhaust gases. At bottom dead center, the piston starts back up on the exhaust stroke to force the burned air-fuel mixture out of the cylinder. As the piston nears the top of the exhaust stroke, the exhaust valve closes and the cycle starts over with another downward intake stroke.

The 4-stroke cycle is repeated continuously in every cylinder of an engine

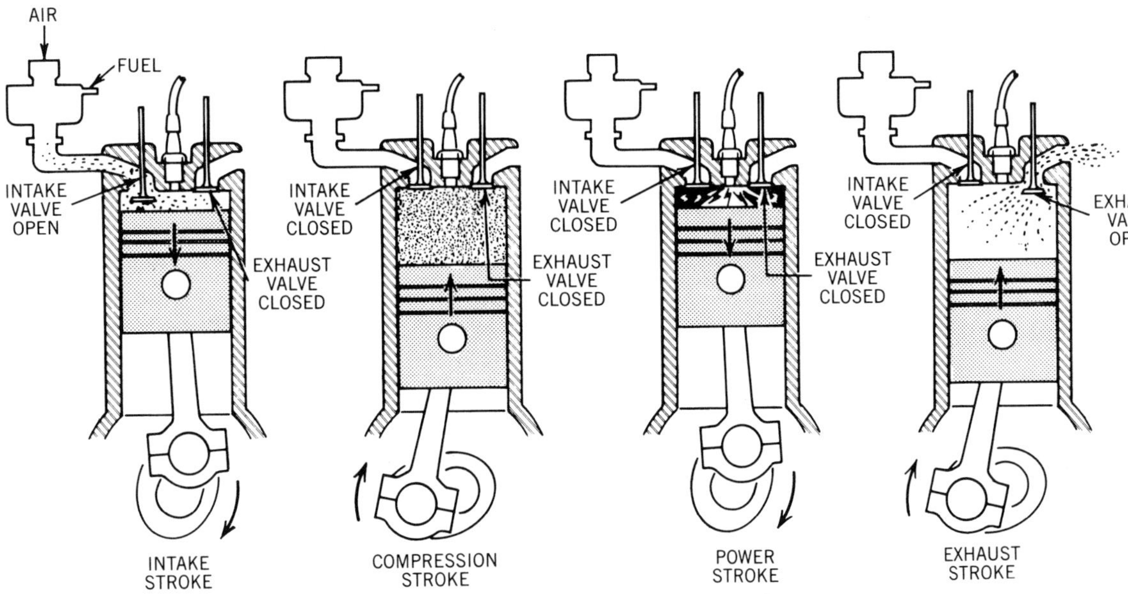


**Figure 4-4.** The 4-stroke cycle of intake, compression, power, and exhaust is repeated continuously in every cylinder as long as the engine is running.

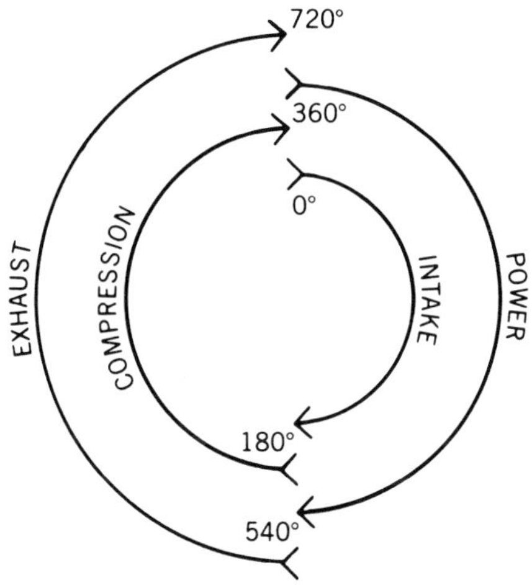


**Figure 4-5.** Each piston stroke—intake, compression, power, and exhaust—requires 180 degrees of crankshaft rotation. Four strokes equal 720 degrees of rotation.

as long as it is running. Figure 4-5 shows that each stroke for one piston requires 180 degrees of crankshaft rotation. Therefore, four strokes require 720 degrees, or two revolutions of the crankshaft. Four-stroke-cycle engines often are called simply "4-stroke" engines. They also are called "Ottocycle" engines, after Dr. Nikolaus Otto, the German inventor who perfected their operation in 1876.

## The 2-Stroke Cycle

Gasoline and diesel engines also can be built to operate on a **two-stroke cycle** in which intake and compression occur on one stroke, and power and exhaust occur on a second stroke. A 2-stroke engine requires only one crankshaft revolution (two strokes) for a complete operating cycle. Because 2-stroke engines have poorer fuel economy and higher exhaust emissions than 4-stroke engines, they are not used in late-model automobiles.

## Air-Fuel Mixture Ignition

The principal difference between gasoline and diesel engines is the method used to ignite the air-fuel mixture. Gasoline engines use a timed electric spark. Diesel engines rely on the heat of highly compressed air and control ignition by timing the injection of fuel into the combustion chamber. Ignition timing must be controlled in either type of engine so that combustion occurs when the piston is in a position to develop maximum power.

In a gasoline engine, air and fuel are mixed in the carburetor or in the intake manifold by the fuel injection system. The air-fuel mixture is compressed in the cylinder to a pressure of 100 to 200 psi (700 to 1,400 kPa) before ignition. This raises the temperature of the mixture to 750° to 1,100° F (400° to 600° C), which is not high enough to ignite it spontaneously. Ignition is provided by the spark plug in the combustion chamber, figure 4-6. Speed and power of a gasoline engine are controlled by regulating the amount of air-fuel mixture that enters the cylinders. This is done by **throttling** the mixture. A throttle valve in the carburetor or manifold inlet varies the size of the opening through which air enters the engine. Fuel is mixed with the incoming air, and the throttling process regulates the volume of the combined air-fuel mixture.

In a diesel engine, only air is drawn into the cylinder through an intake valve. The piston compresses the air volume in the cylinder to a pressure of 450 to 800 psi (3,000 to 5,600 kPa). This raises the air temperature to 1,300° to 1,650° F (700° to 900° C), which is high enough to ignite any fuel that might be present. Fuel must be kept out of this hot compressed air until the time when combustion will develop the most useful power. If fuel were present in the highly compressed air, the mixture would ignite before the piston reached the desired point on the power stroke. Therefore, ignition timing in a diesel is controlled by injecting fuel into the combustion chamber at the exact time of maximum combustion efficiency, figure 4-7.

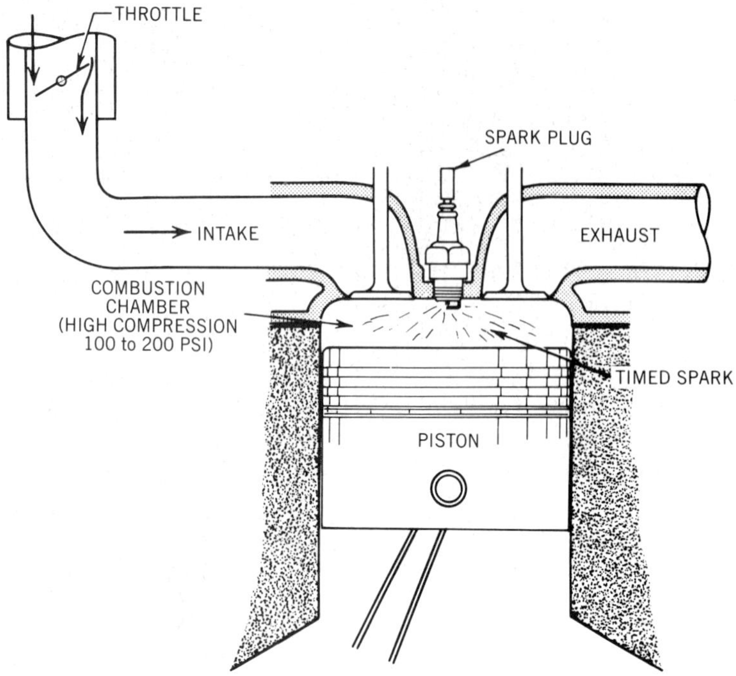


**Figure 4-6.** Ignition in a gasoline engine occurs when a timed spark ignites the air-fuel mixture.

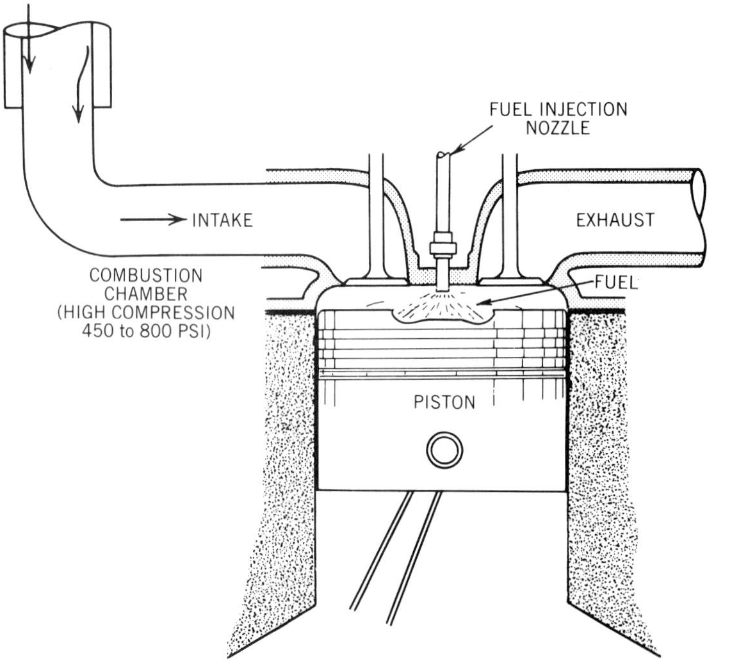


**Figure 4-7.** Ignition in a diesel engine occurs with the timed injection of fuel into highly compressed air in the cylinder.

Diesel engines run with an unthrottled air intake. Air volume in a diesel cylinder on the intake stroke is relatively constant at any engine speed. It must be to maintain uniform combustion temperatures at all speeds because only air volume and compression pressure cause ignition. Diesel engine speed and power are regulated by varying the amount of fuel injected into the compressed air in the cylinder.

Gasoline engines operate with a ratio of air to fuel by weight that can vary from 8:1 at the richest to 18.5:1 at the leanest. However, engine throttling varies the *volume* of the mixture greatly from idle to full power. Diesel engines operate with a constant volume of air but air-fuel ratios that vary from as rich as 20:1 at full power to as lean as 85:1 at idle.

## Engine Cooling

Not all of the chemical energy of fuel is changed to mechanical energy during combustion. As much as 40 percent of the fuel energy is lost as heat energy that does no mechanical work. Peak combustion temperatures range from 3,500° to 6,000° F (2,000° to 3,330° C), and engine temperatures average from 1,200° to 1,700° F (650° to 925° C) during a complete operating cycle. Excess heat must be removed from the engine or it would literally melt down. This is the job of the cooling system.

Actually, "temperature regulation system" would be a better name because an engine cannot operate efficiently if it is too cold. Engine temperature must be high enough for complete combustion but not so high that it will damage the engine. The cooling system carries excess heat away from the engine *and* maintains uniform temperatures throughout the engine. Two methods are used for engine cooling, or temperature regulation.

The most common engine cooling system is a liquid cooling system. Liquid is circulated through passages in the engine called water jackets, and heat is transferred from the metal to the liquid, figure 4-8. The liquid then is circulated through a radiator where the heat is transferred to the surrounding air. Although parts of the cooling system are called the "water pump" and "water jackets," the liquid coolant really is a mixture of approximately half water and half ethylene glycol compound, with oxidation inhibitors.

The other method used to cool an engine is air cooling, in which air is circulated directly around the surfaces of the engine block, cylinders, and heads. The cylinders and heads of air-cooled engines have cooling fins that increase the surface area exposed to the circulating air and aid heat transfer from the metal to the air.

Air-cooled engines have been used in some automobiles since the earliest days but not as much as liquid-cooled engines have been used. The best known air-cooled engines are those of the Type 1 Volkswagen "Beetle" and many Porsches. Air cooling does not provide the uniform heat transfer and temperature control that liquid cooling does. Air-cooled engines rely on the incoming air-fuel mixture for a large part of combustion chamber cooling and therefore run with richer mixtures than liquid-cooled engines. This is the main reason that air-cooled engines are impractical for the emission control and fuel economy requirements of late-model automobiles.

This discussion of engine air-fuel ignition and cooling has been a preview of subjects that you will study in more detail later in this and other chapters. This introduction will help you understand the principles of fuel induction, ignition, cooling, and lubrication. Before we get to those subjects, however, let's look at the nuts and bolts of engine construction in detail.

## ENGINE DESIGN

Our discussion of engine operation so far has been based on a single piston in a single cylinder, but automobile engines have more than one cylinder. One basic way to classify engines is by the number of cylinders and the arrangement of the cylinders in the block, such as V-6, V-8, inline-4, or flat-4. Most automobile engines have 4, 6, or 8 cylinders, although there are engines with 2, 3, 5, 12, and 16 cylinders. Regardless of the number of cylinders, all pistons are connected to a single crankshaft. Engines with more cylinders produce more power strokes for each turn of the crankshaft and run more smoothly than engines with few cylinders. For example, an 8-cylinder engine has twice as many power strokes as a 4-cylinder engine for each crankshaft revolution. The 8-cylinder engine runs more smoothly because its power strokes are closer together in time and in degrees of crankshaft revolution. An 8-cylinder engine produces one power stroke for every 90 degrees of crankshaft rotation. A 4-cylinder engine produces a power stroke every 180 degrees of crankshaft rotation.

Remember, there are 720 degrees (two crankshaft revolutions) for every 4-stroke cycle. Divide 720 by the number of cylinders to determine the number of degrees between power strokes, or power overlap:

$$\frac{720}{8} = 90$$

$$\frac{720}{4} = 180$$

This rule works for any *even-firing* engine. As you will learn later, some V-6 engines fire unevenly and have different numbers of degrees between power strokes.

### Multicylinder Engines

Figure 4-9 shows the most common arrangements for the cylinders of automobile engines:

**1.** *Inline engine*—All cylinders are in a straight line, in a single bank. Most 4- and 6-cylinder engines are built like this, as are 3- and 5-cylinder engines.

**2.** *V-type*—Cylinders are in two equal banks, separated by an angle of 60 or 90 degrees. Most V-type engines have 6 or 8 cylinders, but V-4 engines also have been built. Engines with 10, 12, or 16 cylinders also are usually V-type engines.

**3.** *Horizontally opposed*—Cylinders are in two equal banks 180 degrees apart. Volkswagen and Porsche 4- and 6-cylinder engines are the best known "flat" or "pancake" engines. Subaru also builds a flat 4-cylinder engine.

### Engine Block and Crankcase

Most engines are built with a 1-piece block and crankcase made of cast iron or aluminum, figure 4-10. The block

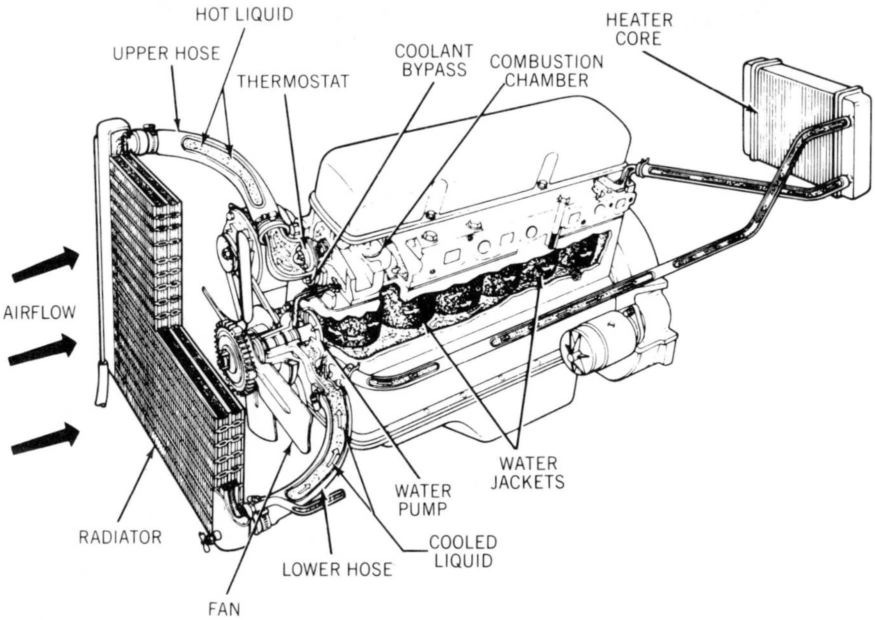


**Figure 4-8.** Liquid cooling systems are the most common cooling systems for automobile engines. (Chrysler)

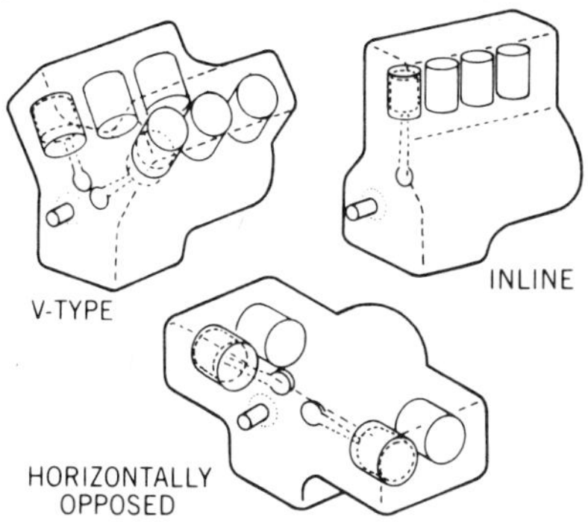


**Figure 4-9.** Most automobile engines are built with inline, V-type, or horizontally opposed cylinder arrangements.

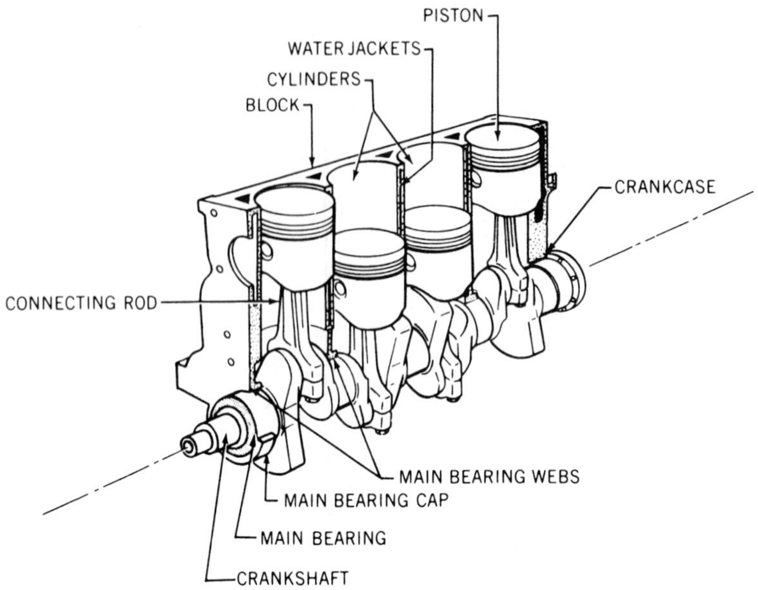


**Figure 4-10.** Most engines are built from a cast, 1-piece block that contains the cylinders, water jackets, and crankcase.

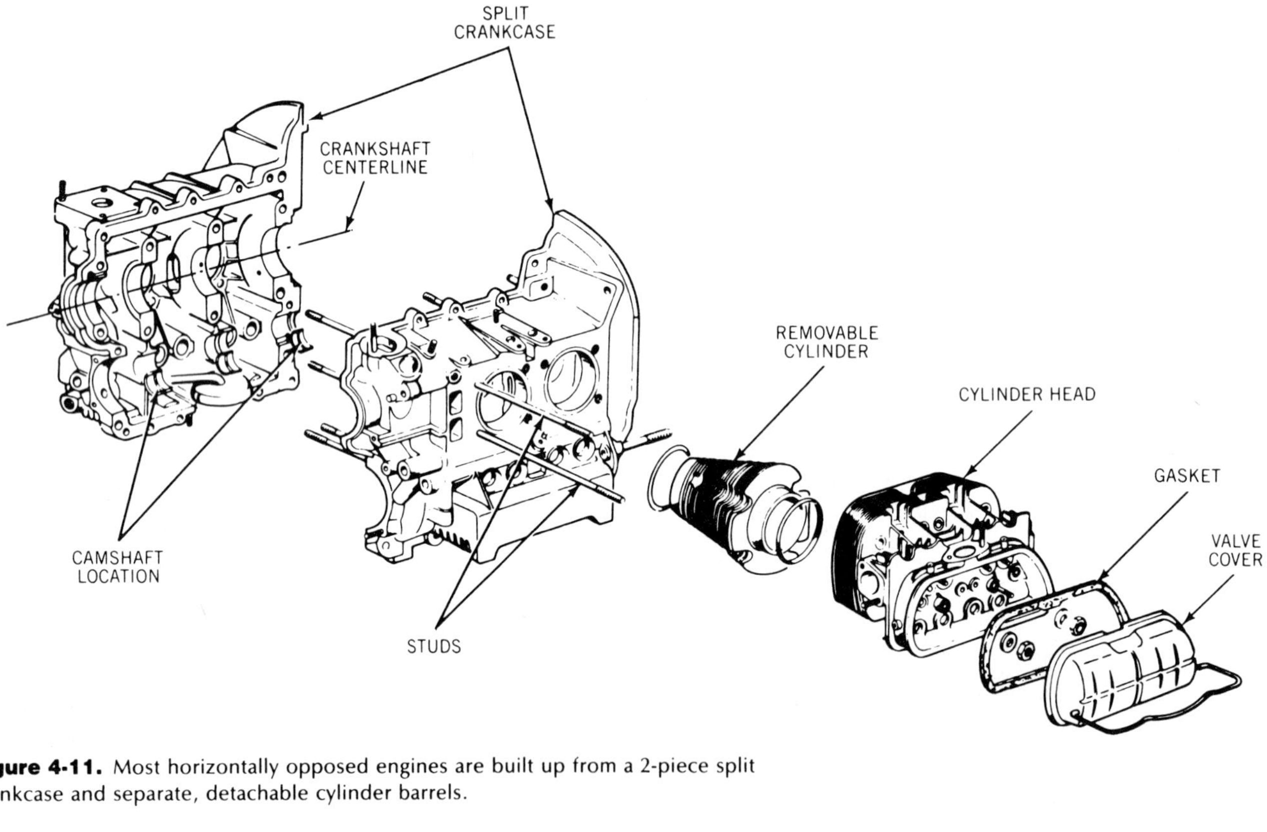


**Figure 4-11.** Most horizontally opposed engines are built up from a 2-piece split crankcase and separate, detachable cylinder barrels.

casting forms the cylinders and water jackets. The lower part of the block forms the upper portion of the crankcase, which supports the crankshaft in main bearings. The main bearings are in two pieces. One-half of each is installed in a **bearing web or saddle** that is part of the block. The other half is in a removable cap that allows the crankshaft to be installed and removed. The block has main bearings at both ends and usually one between each crankshaft throw. The oil pan forms the lower portion of the crankcase.

Horizontally opposed engines usually are built with a crankcase in two pieces that join at the centerline of the crankshaft, figure 4-11. Half of each main bearing is in each half of the crankcase. The cylinders, particularly of flat air-cooled engines, often are separate detachable castings that are bolted to the crankcase.

## TWO STROKES = FOUR

Like a 4-stroke engine, a 2-stroke engine can be either a spark-ignition or a compression-ignition type. A two-stroke engine produces a power stroke for each revolution of the crankshaft, while a 4-stroke engine produces one power stroke for every two revolutions. Gasoline 2-stroke, spark-ignition engines don't have poppet valves as 4-stroke engines do. Instead, the piston uncovers intake and exhaust ports in the side of the cylinder as it nears the bottom of its stroke. Here is how a 2-stroke engine combines intake and compression into one single stroke, and power and exhaust into a second stroke.

1. *Intake and compression stroke*—As the piston moves up in the cylinder, it creates a low-pressure area in the crankcase. A reed valve or rotary valve opens in the side of the crankcase, and the air-fuel mixture is drawn in. At the same time, the piston compresses a previous air-fuel charge in the cylinder. As the piston nears top dead center, the spark plug fires to ignite this mixture.

2. *Power and exhaust stroke*—As the piston is forced down by the burning air-fuel mixture on the power stroke, it uncovers exhaust ports in the side of the cylinder, toward the bottom. By the time the ports are uncovered, the mixture is almost completely burned, and the exhaust flows out through the ports.

As the piston continues down, it compresses the air-fuel charge in the crankcase. The reed valve or rotary valve closes to hold the charge in the crankcase. Crankcase compression forces this mixture through a transfer passage to intake ports in the side of the cylinder. The force of the incoming mixture helps drive the remaining exhaust out the exhaust ports on the opposite side of the cylinder, but a ridge on the piston top deflects the incoming mixture upward so that it doesn't flow out the exhaust ports. As the piston moves upward, it covers both the intake and the exhaust ports and compresses the new air-fuel charge in the cylinder to repeat the cycle.

Because the crankcase of a 2-stroke gasoline engine is used as a "precompression chamber" for the air-fuel mixture, it can't serve as a lubricating oil reservoir. Therefore, oil must be mixed with the fuel or injected intermittently into the engine.

The most common 2-stroke diesel engines are truck engines built by the Detroit Diesel Allison Division of GM. Detroit Diesels have cam-operated exhaust valves rather than exhaust ports in the cylinder walls. A Roots blower supplies the intake air charge through cylinder-wall intake ports. Timed diesel injection controls fuel delivery and ignition timing. Because the crankcase does not have to serve as an intake precompression chamber, Detroit Diesels use a conventional oil pan and lubricating oil supply.

The accompanying diagrams show the differences—and similarities—beween 4-stroke and 2-stroke operating cycles.

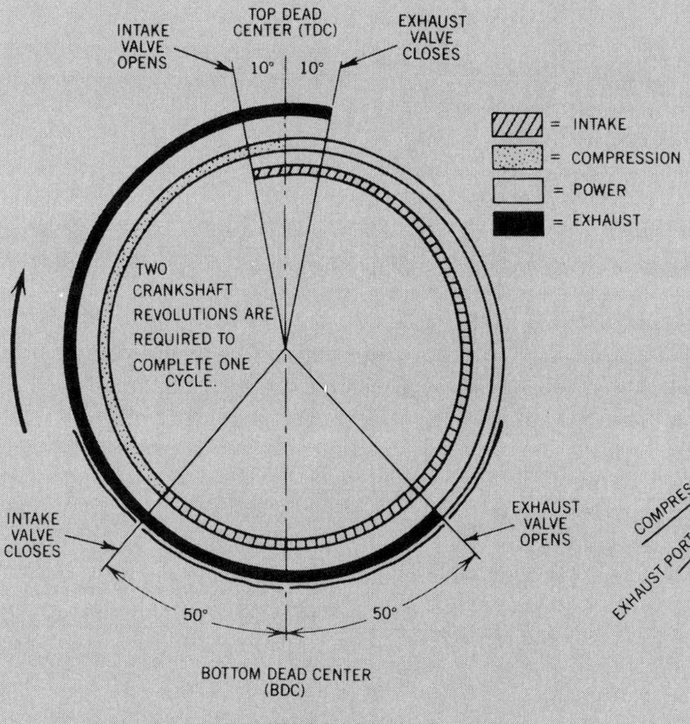


TYPICAL 4-STROKE CYCLE OPERATION

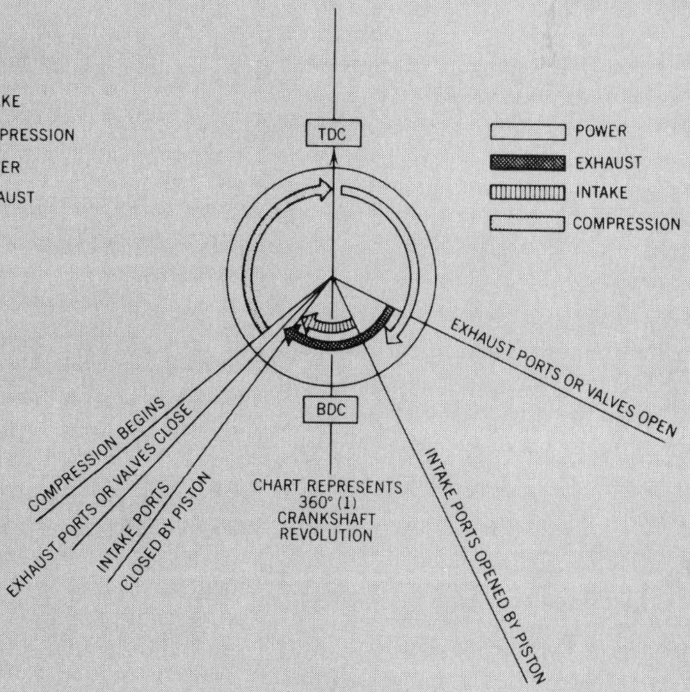


TYPICAL 2-STROKE CYCLE OPERATION

## Crankshaft and Flywheel

The main bearing journals of the crankshaft are arranged in a straight line on the center of the shaft. The crankshaft rotates around this centerline in the block, figure 4-12. The crankshaft centerline is the **datum line**, or foundation point, for the entire engine. This means that all dimensions and measurements of the engine are made with the crankshaft centerline as the primary reference point.

The **crankshaft throws** are offset from the centerline. Each throw has a journal to which one or two connecting rods are bolted. The crankshaft throws provide the leverage for the force of the moving piston to develop rotating power. In most multicylinder engines, the rod journals are offset equally around the crankshaft centerline so that the firing impulses of the power strokes are evenly spaced. The degree spacing of the crank throws depends on the number of cylinders and the block arrangement. You will learn more about crankshaft rod journal spacing and firing intervals when you study ignition intervals later in this chapter.

All engines have a flywheel on one end of the crankshaft, figure 4-12. The weight of the flywheel provides rotating inertia to the crankshaft that carries it smoothly through the periods between firing impulses. The flywheel also is a convenient point to locate a transmission to transmit engine power to the drive train of the car. Most engines also have a smaller rotating weight, called a vibration damper or harmonic balancer, at the opposite end of the crankshaft, figure 4-12. The vibration damper further smooths out the engine's firing impulses and other vibrations. The crankshafts of most engines rotate clockwise when viewed from the front, or the end opposite the flywheel.

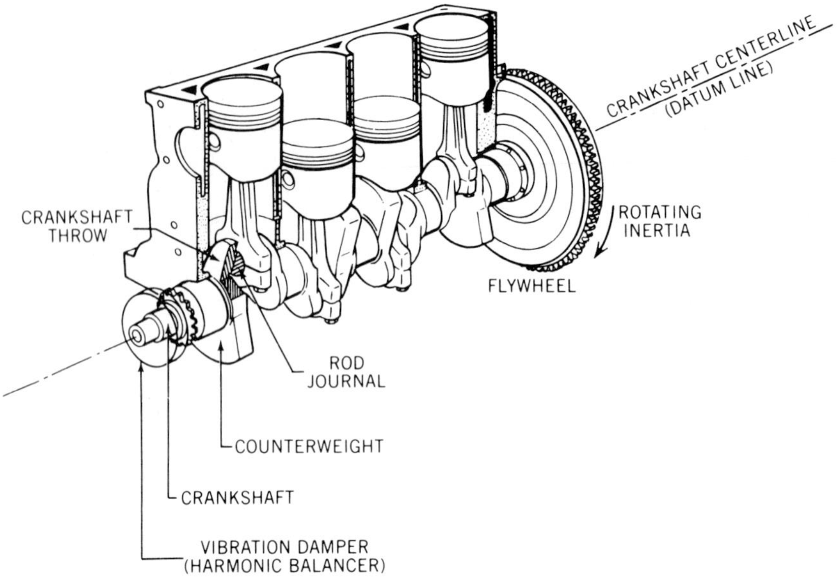


**Figure 4-12.** The crankshaft rotates in main bearings in the block. A flywheel at one end of the crankshaft provides rotating inertia.

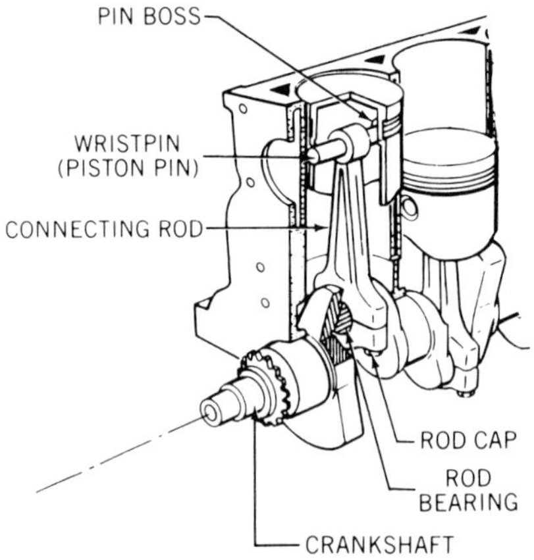


**Figure 4-13.** Connecting rods attach the pistons to the crankshaft.

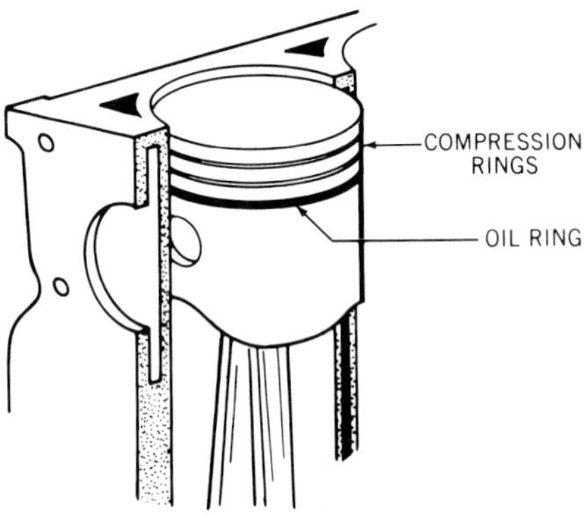


**Figure 4-14.** Piston rings seal the piston in the cylinder and control the amount of oil on the cylinder walls.

## Pistons and Connecting Rods

Most crankshafts are solid, 1-piece units. The rods that connect the pistons to the crankshaft have removable caps on their big ends. The bottom of each connecting rod and its cap are bolted around a crankshaft journal and contain a split bearing called a rod bearing. The bearings allow the crankshaft rod journals to rotate within the rods as the crankshaft revolves, figure 4-13. In an inline engine and most flat engines, one rod is bolted to each crank throw. In most V-type engines, two rods are bolted to each throw. The opposite, or small, end of the rod holds a piston pin, or wristpin, to which the piston is attached. The pin passes through **bosses** in opposite sides of the piston and through the opening in the rod. The pin can be pressed solidly into the piston or the rod, or it can float in both and be held in place with locking rings. Regardless of how it is mounted, the wristpin has to allow free movement of either the rod, the piston, or both so that the rod can swing back and forth as the piston moves up and down in the cylinder.

The piston has three or four grooves around its circumference that hold piston rings, figure 4-14. The rings seal the piston in the cylinder so that compression is not lost through the clearance between the piston and the cylinder. The rings also control the amount of oil on the cylinder walls for piston lubrication.

## Cylinder Heads and Valves

The cylinder head covers the top of the block and cylinders and contains the combustion chambers. Modern engines have the intake and exhaust valves in the head and are called **overhead-valve (ohv) engines**, figure 4-15. Intake passages, or ports, carry the incoming air-fuel mixture (or just air in a diesel engine) to the intake valve. Similar exhaust ports carry the exhaust away from the exhaust valve and out of the engine.

The valves in automobile engines are **poppet valves**. They open and close in linear motion against fixed seats. When closed, the valves are sealed against their seats by the compression of coil springs. The valves are opened by **cams**, which are irregularly shaped lobes on a rotating shaft. The cam for each valve must have the right shape and position on the camshaft to open its intake or exhaust valve at the desired point in the 4-stroke cycle. Similarly, each cam must allow its valve to close at another precise point in the cycle. Each valve opens and closes once for each revolution of its cam.

**Valve Timing.** The valves do not open exactly at the top and bottom of a piston stoke. The intake valve opens several degrees before tdc at the end of the exhaust stroke, figure 4-16. This ensures that the valve is fully open when the piston starts downward on the intake stroke. The intake valve closes after bdc on the compression stroke to ensure that the cylinder receives as much air for the intake charge as possible, figure 4-17.

Similarly, the exhaust valve opens slightly before bdc on the power stroke so that remaining pressure in the cylinder will help expel exhaust gases, figure 4-18. The exhaust valve closes slightly after tdc on the compression stroke, figure 4-19, so that the exhaust is completely removed from the cylinder. This also allows the inertia of the exhaust gas to help draw a fresh intake charge into the cylinder.

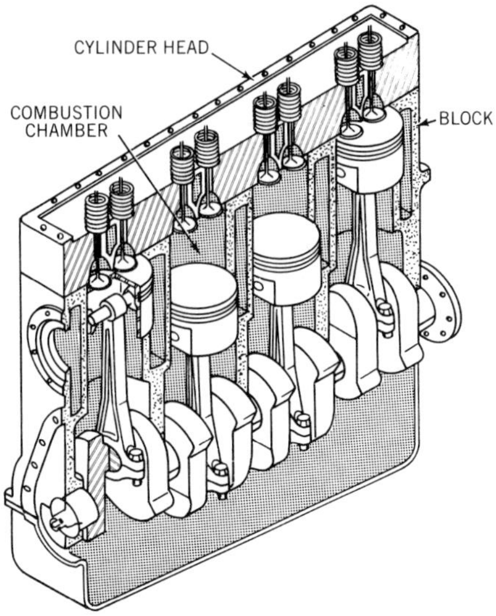


**Figure 4-15.** The cylinder head covers the top of the cylinders and contains the combustion chamber and valves.

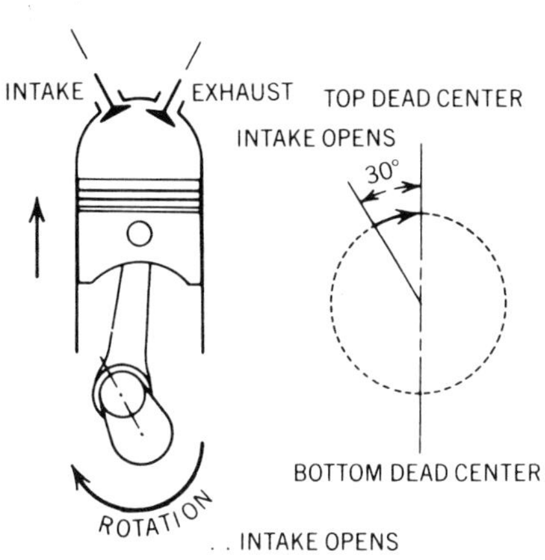


**Figure 4-16.** The intake valve opens before top dead center.

The relationship between the crankshaft position and the opening and closing positions of the intake and exhaust valves is the **valve timing**. Valve timing has a major effect on power, emission control, and fuel economy. It varies for different engines but always follows the principles outlined here. The diagram in figure 4-19 shows the valve timing for a complete 4-stroke cycle. Here are some other important terms you will hear in relation to valve timing:

- **Valve lift** is the maximum distance that a valve opens, expressed in a decimal-inch fraction or millimeters.
- **Valve duration** is the number of crankshaft degrees during which a valve is open. A valve is always open for 180° of the intake or exhaust stroke plus the number of degrees it is open before tdc and after bdc. In figure 4-19, the duration for both the intake and the exhaust valve is 30° + 180° + 70° = 280° duration.
- **Valve overlap** is the period during which both the intake and the exhaust valves are open before and after tdc between the exhaust and intake strokes. In figure 4-19, valve overlap is 60°.

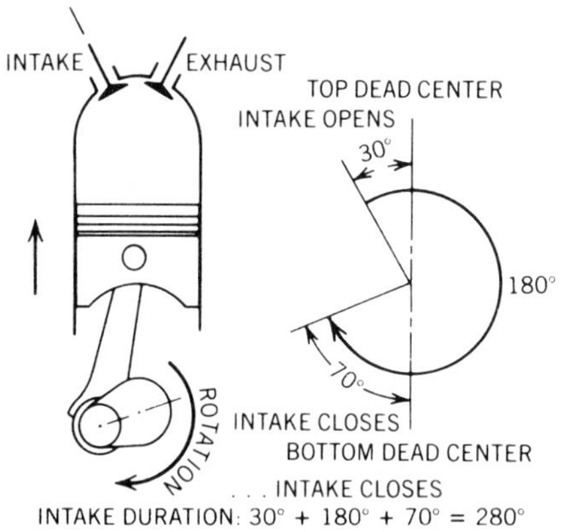


**Figure 4-17.** The intake valve closes after bottom dead center.

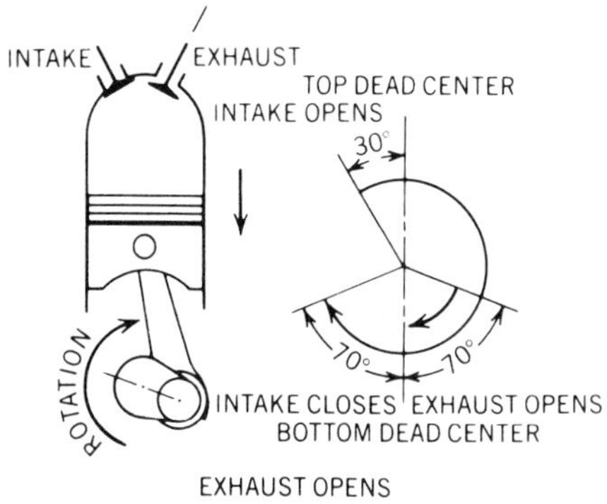


**Figure 4-18.** The exhaust valve opens before bottom dead center.

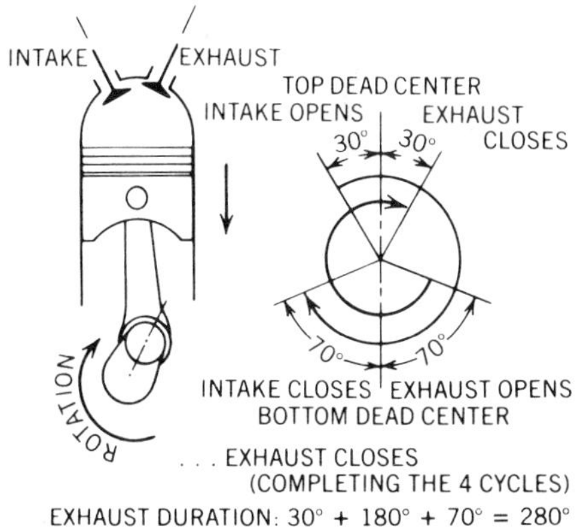


**Figure 4-19.** The exhaust valve closes after top dead center.

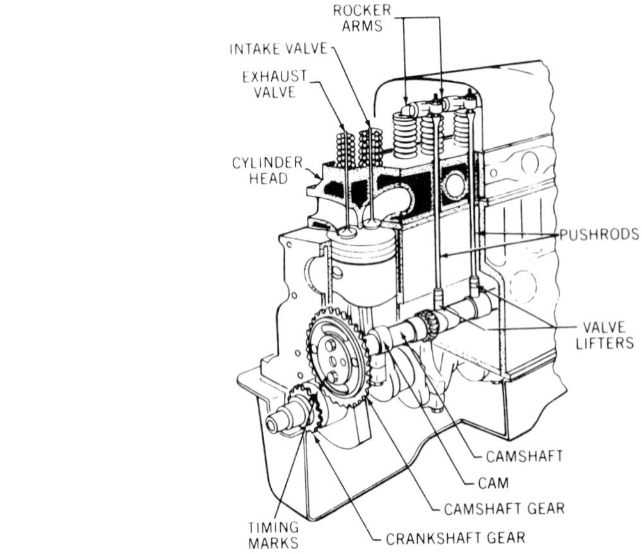


**Figure 4-20.** The camshaft is driven by the crankshaft through gears as shown here or by a chain or belt. Timing marks synchronize camshaft and crankshaft rotation.

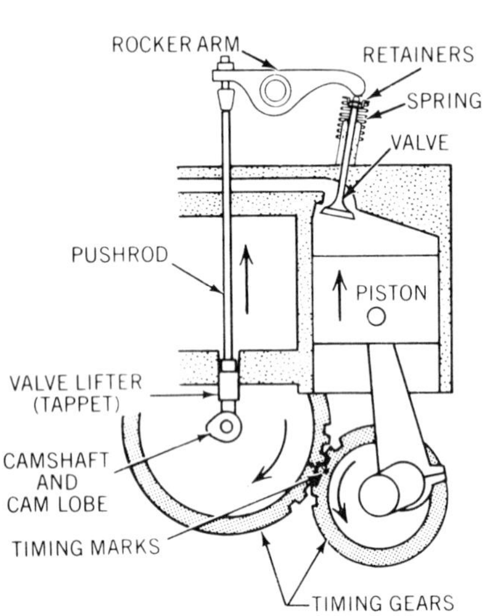


**Figure 4-21.** These are the typical parts of the valve train of an overhead-valve engine with the cam in the block.

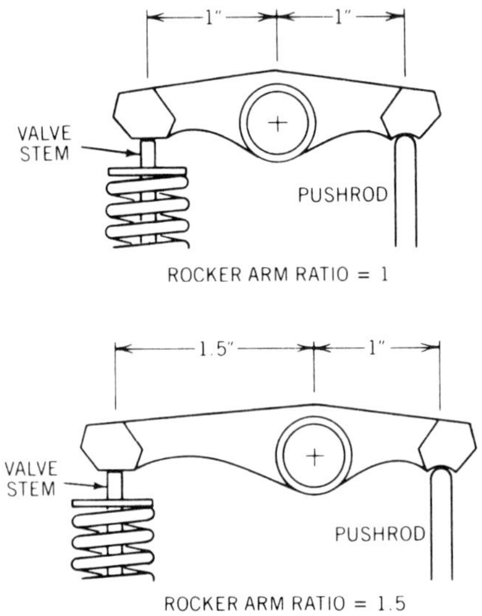


**Figure 4-22.** The rocker arm ratio multiplies cam lift to determine actual valve lift.

## Camshafts and Valve Trains

Because each valve opens and closes once in each 4-stroke cycle for a cylinder, the camshaft makes one revolution for every two revolutions of the crankshaft. The crankshaft drives the camshaft through timing gears, chains, or belts to synchronize valve opening with piston position during the 4-stroke cycle, figure 4-20. Timing marks on the gears or sprockets are used to match the relative positions of the crankshaft and camshaft so that their rotations are synchronized. The camshaft rotates at a 2:1 reduction ratio, or at one-half crankshaft speed.

**Overhead-Valve Engines.** Although the valves are in the head of the engine shown in figure 4-20, the camshaft is in the block. This is one common arrangement for an overhead-valve engine. It is cheap and easy to cast and machine the block to hold the camshaft, and the cam drive mechanism is relatively simple. However, the valve-opening mechanism is more complicated. The up-and-down motion created by the rotating cam must be transferred to the head and then reversed to open the valves. All of the parts that open and close the valves, and the valves themselves, are called the **valve train**. Here are the typical valve train parts in an ohv engine with the camshaft in the block, figure 4-21:

**1.** *The valve lifter, or tappet,* provides a smooth bearing surface as it rides against the cam lobe and transmits linear motion to the rest of the valve train.

**2.** *The pushrod* connects the valve lifter to the rocker arm in the head.

**3.** *The rocker arm* reverses the motion of the lifter and pushrod and bears against the valve to open it. The cam, lifter, pushrod, and rocker arm all work against the compression of the valve spring. The rocker arm also determines the final amount of valve lift by its leverage ratio. The **rocker arm ratio** is the ratio of the distance from the pivot point to the pushrod center on one end and the valve stem center on the other. If the distances are equal, the ratio is 1, figure 4-22. Most rocker arms, however, have a ratio greater than 1. If the rocker arm ratio is 1.5, for example, 0.200-inch lift at the camshaft produces 0.300-inch lift at the valve:

$$= 0.200 \times 1.5 = 0.300$$

**4.** *The valve spring* holds the valve face closed against the valve seat when the rest of the valve train is not forcing it open.

**5.** *The retainers* are the parts that lock the valve spring to the top of the valve stem and maintain the compression of the spring.

**Overhead-Cam Engines.** Every engine design is a compromise. While it is simple to locate the camshaft in the block, the valve train necessary to get the cam motion to the valve is complicated. Also, the weight of the lifters,

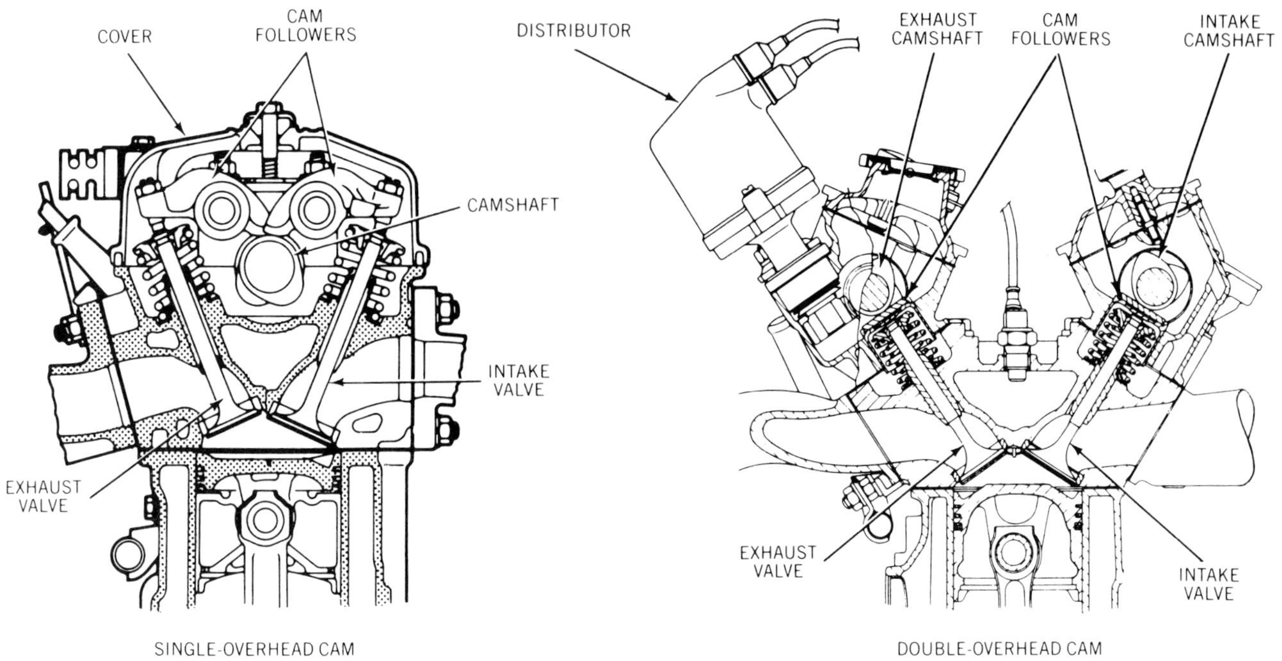


**Figure 4-23.** Typical single- and double-overhead-cam engines. (Chrysler) (Clymer Publications)

pushrods, and rocker arms creates a lot of reciprocating inertia that limits engine speeds. The valve train could be much simpler if the camshaft were in the head, closer to the valves. Of course, then the drive mechanism would be more complicated, requiring longer chains or belts and probably one or more tensioners or idler sprockets.

Many engines are built, however, with the camshaft in the head because this design allows higher engine speeds and placement of the valves for more efficient engine breathing. When an **overhead-cam (ohc) engine** has a single camshaft for intake and exhaust valves, it is a **single-overhead-cam (sohc) engine**, figure 4-23. If an ohc engine has separate camshafts for intake and exhaust valves, it is a **double-overhead-cam (dohc) engine**.

Gears, chains, or belts can be used to drive overhead camshafts and synchronize their rotation with crankshaft rotation. Most passenger car engines use chains or a timing belt to drive overhead camshafts. Depending on the location of the camshaft and the design of the head, various kinds of cam followers are used to transmit cam motion to the valves. Some are rocker arms; others are inverted cups placed directly between the cam lobes and the valve stems.

**Multiple Valve Engines.** During the 1980's, many carmakers developed ohc engines with more than one intake and one exhaust valve per cylinder. Most of these multiple valve engines have two intake and two exhaust valves per cylinder and often are called 4-valve engines. Honda and Toyota also have built 3-valve engines with two intake valves and one exhaust valve per cylinder.

Although four valves per cylinder have been common in racing engine design since the early 1920's, they were popularized in production car use by Japanese manufacturers in the last decade. Domestic carmakers also now have several versions, such as GM's Quad 4 and LT5 V-8, Ford's SHO V-6 and Modular V-8's, and Chrysler's R/T series.

The smaller and lighter valve train parts in 4-valve engines allow higher engine speeds for increased power. Additionally, combustion chamber shapes usually allow two intake and exhaust valves to provide greater gas flow area than single intake and exhaust valves can provide. This, in turn, leads to improved volumetric efficiency. Many manufacturers also design the two intake or two exhaust valves with different timing to improve intake and exhaust tuning and gas flow. More complicated valve trains and increased components, however, make 4-valve engines more expensive and limit their use to higher-priced cars.

## Intake and Exhaust Manifolds

Every multicylinder engine (unless it has a separate carburetor or fuel injector bolted to each intake port) needs an intake manifold. The intake manifold is a metal casting with passages that route the intake air or air-fuel mixture from a central inlet to the individual cylinders. Similarly, every engine needs an exhaust manifold to route the exhaust away from the cylinders to a central exhaust system.

Most V-type engines have the intake manifold between the cylinder banks and a separate exhaust manifold on the outboard side of each head, figure 4-24. Inline engines have the intake manifold on one side of the head, figure 4-25. The exhaust manifold can be on the same side or the opposite side, depending on valve and port design. A cylinder head with intake and exhaust manifolds on opposite sides is called a **crossflow head**. Exhaust manifolds collect the exhaust from each cylinder and route it to a central exhaust system.

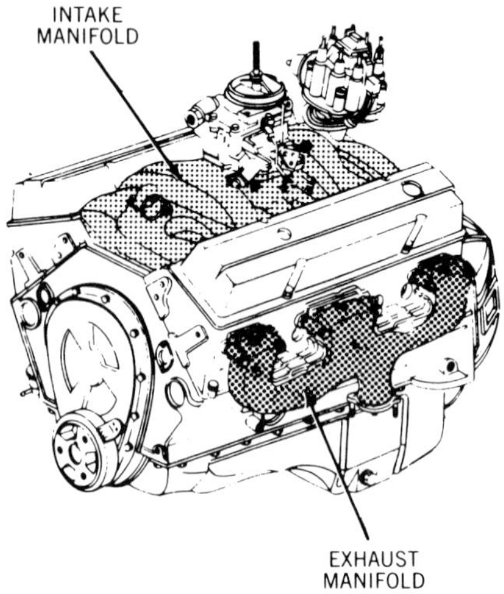


**Figure 4-24.** V-type engines have the intake manifold between the cylinder banks and exhaust manifolds bolted to the outside of each head.

### Engine Covers

The largest cover on most engines is the oil pan that closes the bottom of the crankcase and acts as a reservoir for the engine's lubricating oil, figure 4-26. The top of each cylinder head above the valves is enclosed by a valve cover or cam cover, figure 4-26. Different kinds of covers also enclose the timing gears or camshaft drive mechanism on the front of the engine, figure 4-27. The size and shape of these covers vary according to the valve train design of the engine. The engines in many front-wheel-drive cars often have other covers over the flywheel end of the engine and transaxle assembly, figure 4-28. All of these covers are usually stamped from light-gauge sheet metal or cast from aluminum, although on some late-model engines they are made from plastic materials.

## ENGINE DISPLACEMENT AND COMPRESSION RATIO

Everything you have read in the last few pages has dealt with the mechanical parts of an engine. There are some basic specifications for all engines that you must know about to understand engine operation completely. Two of these are engine **displacement** and **compression ratio**.

### Engine Displacement

Engine displacement is often referred to as "engine size." Displacement is a measurement of the total volume of all cylinders. The number of cylinders partly determines displacement, but the arrangement of the cylinders does not.

To calculate total engine displacement, you start with one cylinder. You know that the distance each piston travels from the bottom to the top of a cylinder is its stroke, figure 4-29, which is measured in inches or millimeters. The displacement of a single cylinder is the volume of gas or air displaced by the top surface of the piston as it travels from bottom dead center (maximum cylinder volume) to top dead center (minimum cylinder volume) during its stroke.

The diameter of the cylinder is the **bore**. To calculate cylinder displacement, use the following formula with a bore of 3.000 inches and a stroke of 3.000 inches, figure 4-30:

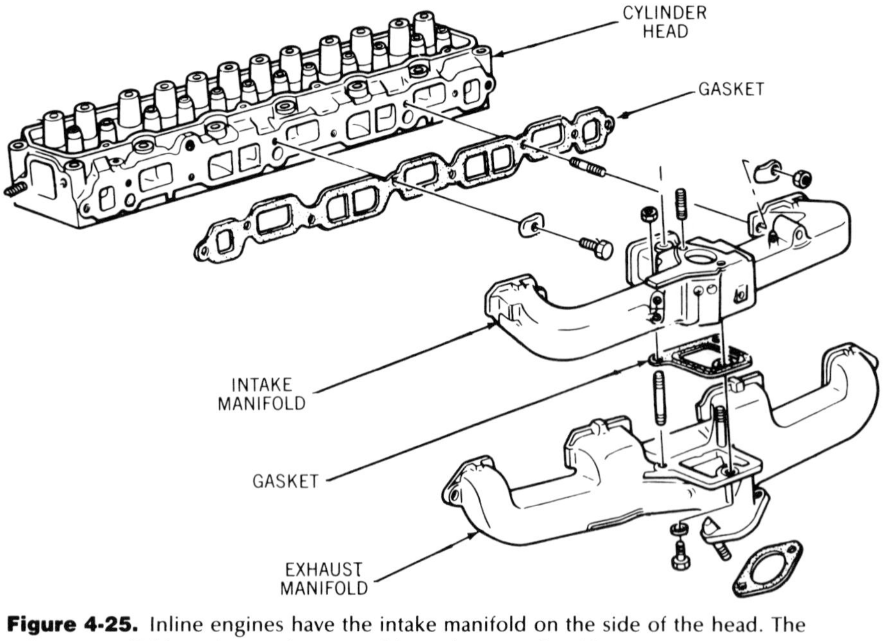


**Figure 4-25.** Inline engines have the intake manifold on the side of the head. The exhaust manifold can be on the same side or the opposite side.

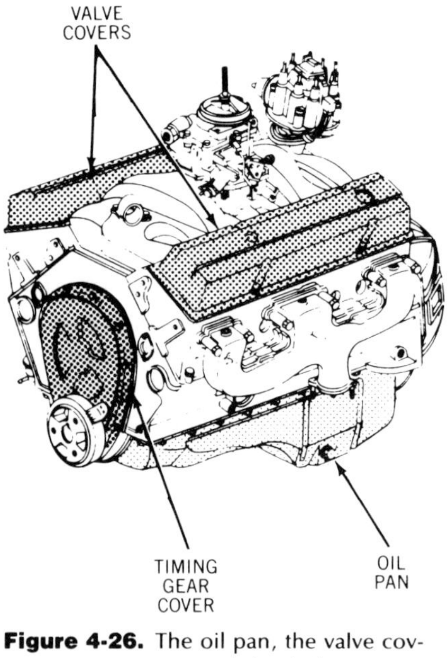


**Figure 4-26.** The oil pan, the valve covers or cam covers, and the timing cover are the most visible of an engine's covers.

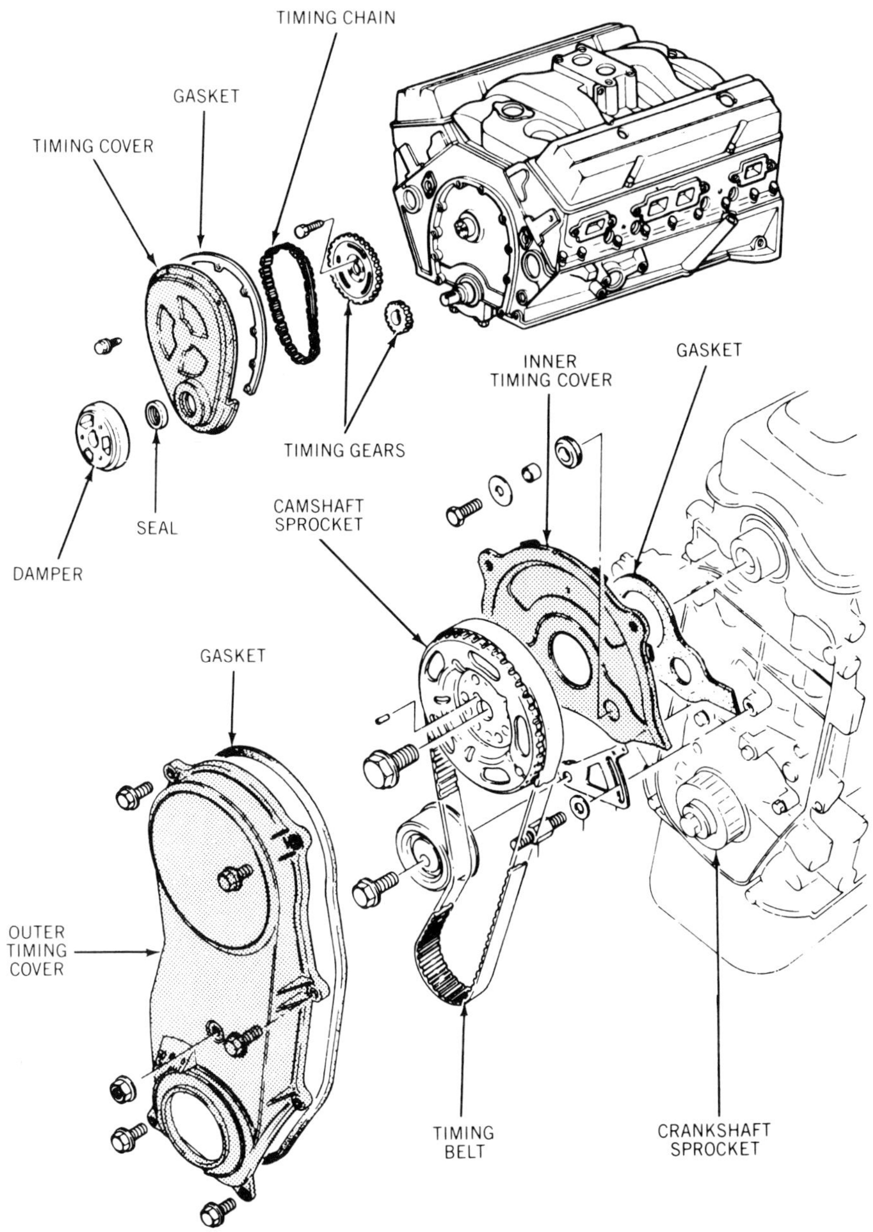


**Figure 4-27.** The sizes and shapes of timing covers vary according to the valve train design of an engine.

**1.** Divide the bore (cylinder diameter) by 2 to determine the radius of the cylinder: 3/2 = 1.5

**2.** Square the radius (multiply it by itself): 1.5 × 1.5 = 2.25.

**3.** Multiply the square of the radius, 2.25, by pi ($\pi$), or 3.1416, to find the area of the cylinder cross section: 2.25 × 3.1416 = 7.069.

**4.** Multiply the area of the cylinder cross section by the stroke (3.000 inches) to determine the displacement of one cylinder: 7.069 × 3.000 = 21.207 cubic inches

**5.** Multiply the displacement of one cylinder by the total number of engine cylinders to determine engine displacement. If the engine has four cylinders, its displacement is: 4 × 21.207 = 84.828 cubic inches. In practice, this would be rounded off to an even 85-cubic- inch displacement. This often is abbreviated "85 cid."

If an engine with the same bore and stroke dimensions had six cylinders, its displacement would be 6 × 21.207 = 127.242 cubic inches, or simply 127 cubic inches. An 8-cylinder engine with the same bore and stroke would have a displacement of 8 × 21.207 = 169.656 cubic inches. Rounded off, we would say it was a 170-cid engine.

You can use a simplified formula, based on the same calculations:

$\text{bore}^2 \times 0.7854 \times \text{stroke} \times \text{cylinders}$

Here's how it works for the same 3-inch bore and stroke measurements:

$3^2 \times 0.7854 \times 3 \times 8$ cylinders
$9 \times 0.7854 \times 3 \times 8$
$= 169.6$ cubic inches.

**Metric Displacement Specifications.** In the example above, we used measurements in customary inch units. The same formula works with bore and stroke measurements in millimeters. Let's say that the bore is 75 millimeters, and the stroke is 70 millimeters, figure 4-31. Displacement is expressed in cubic centimeters, so change the millimeter dimensions to centimeters by moving the decimal one place to the left (75 mm = 7.5 cm and 70 mm = 7.0 cm). Then proceed as follows:

**1.** The bore, 7.5 cm, divided by 2, equals a radius of 3.75 cm.

**2.** 3.75 × 3.75 = 14.0625

**3.** 14.0625 × 3.1416 ($\pi$) = 44.1788 $mm^2$ cylinder cross-sectional area.

**4.** 44.1788 × 7.0 cm (stroke) = 309.2516 $cm^3$, or cc, cylinder displacement.

If the engine has four cylinders, the total displacement is 4 × 309.3516 cc = 1,237.406 cc. We would normally round this off to 1,237 cc or 1.2 liters. (Actually, if you were a carmaker, you would probably advertise it as a 1.3-liter engine.)

Using the simplified formula:

$\text{Bore}^2 \times 0.7854 \times \text{stroke} \times \text{cylinders}$
$7.5^2 \times 0.7854 \times 7.0 \times 4$ cylinders
$56.25 \times 0.7854 \times 7.0 \times 4 = 1{,}237$ cc

Remember that you can convert metric to customary displacement measurements with the following factors:

- To change cubic centimeters to cubic inches, multiply by 0.061 (1,237 cc × 0.061 = 75.457 cubic inches).

- To change liters to cubic inches, multiply by 61.02 (1.2 liters × 61.02 = 73.224 cubic inches).
- To change cubic inches to cubic centimeters, multiply by 16.39 (75 cubic inches × 16.39 = 1,229.25 cc).

You can see that conversion accuracy depends on where you round off the numbers. To maintain best accuracy, work in one system, customary or metric, and use conversions only for approximate measurements.

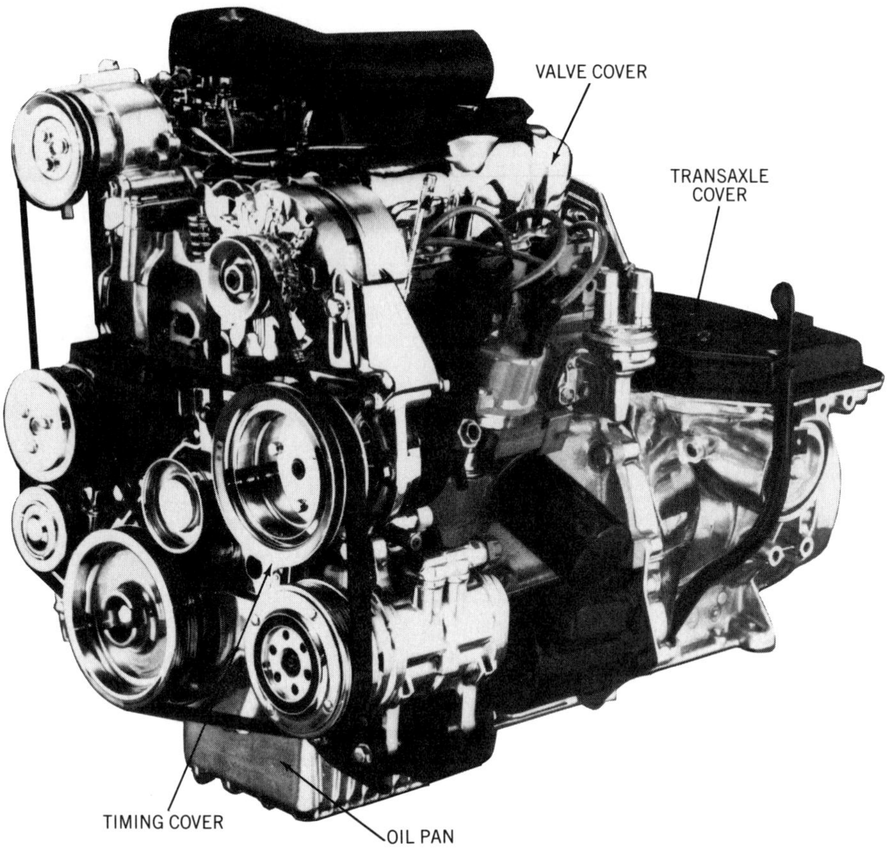


**Figure 4-28.** Many front-wheel-drive engines have covers to enclose the flywheel end of the engine.

### Compression Ratio

Engine compression ratio is the ratio of total cylinder volume (piston at bdc) to combustion chamber volume (piston at tdc), figure 4-32. Total cylinder volume is piston displacement *plus* combustion chamber volume. Combustion chamber volume also is called **clearance volume**. Therefore, the compression ratio is total cylinder volume divided by clearance volume. If the clearance volume is 1/8 of the total, the compression ratio is 8:1. Here is the simple formula:

$$\frac{\text{total cylinder volume}}{\text{compression ratio}} = \text{clearance volume}$$

Here is an example of a compression ratio calculation. First, calculate cylinder displacement from the bore and stroke dimensions. Let's say it is 21.207 cubic inches as in our earlier example. Let's also say that the combustion chamber volume is 3.0 cubic inches. Now, add the two volumes:

$$21.207 + 3.0 = 24.207 \text{ (total cylinder volume)}$$

Then, divide total volume by clearance volume:

$$\frac{24.207}{3.0} = 8.069$$

The compression ratio is 8.069:1, which would be rounded off and listed as an 8:1 compression ratio.

Theoretically, a high compression ratio improves the efficiency of an engine and allows it to develop more power from a given amount of fuel. Combustion occurs faster with higher compression because fuel molecules are packed more tightly and the combustion flame travels fast.

Most gasoline engines, however, are limited to compression ratios of about 11.5:1 or less. If compression is too high, the air-fuel mixture will overheat and ignite before the spark plug fires, which can damage an engine. Compression ratios as high as 10:1 or 11:1 were common in the 1960's, but the higher octane fuel available then burned more evenly than today's gasoline. Higher compression also raises combustion temperatures and increases emission of oxides of nitrogen ($NO_x$) In the 1970's, carmakers lowered compression, and oil companies introduced lower octane unleaded fuel to meet stricter emission limits. Compression ratios on today's gasoline engines range from 8:1 to 9:1.

Diesel engines have higher compression ratios than gasoline engines because they rely on high compression to create heat for fuel ignition. Diesel compression ratios range from about 16:1 to 22:1.

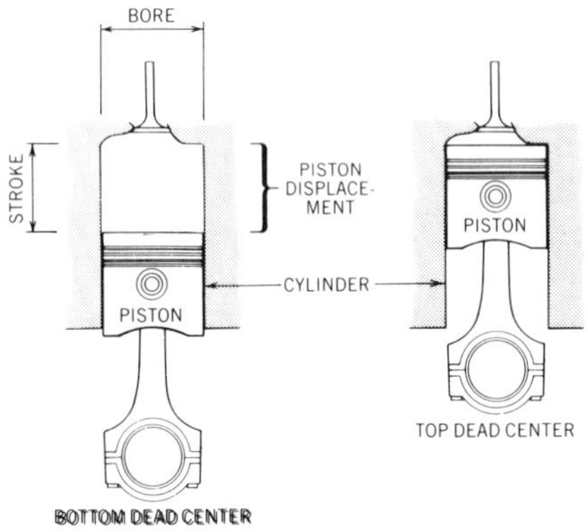


**Figure 4-29.** The bore is the diameter of the cylinder, and the stroke is the distance that the top of the piston travels from bottom dead center to top dead center.

## ENGINE TORQUE AND POWER

You know that torque is rotating work measured in foot-pounds, inch-pounds, or newton-meters. You also know that power is the rate, or speed, of doing work and can be measured in horsepower or watts. Engines produce both torque and power. They are related to each other and they are common specifications that are given for almost all automobiles.

## Engine Torque

The engine crankshaft changes reciprocating work into rotary torque, so engine torque is measured at the end of the crankshaft. To measure torque with the engine running, we can install a friction clutch, or adjustable clamp, on the end of the crankshaft. We can connect an arm to this clutch, and we can connect the other end of the arm to a scale, figure 4-33. When we tighten the clutch, the rotating crankshaft starts to exert force on the scale. Let's say the force is 200 pounds.

To know what this represents in foot-pounds (ft-lb) of torque, we must consider the length of the arm. If it is 2 feet, the 200 pounds of force represent 2 feet times 200 pounds, or 400 foot-pounds of torque, figure 4-33. If the engine is running at 2,000 rpm and we tighten the clutch so that the engine almost stalls at full throttle as it develops 400 foot-pounds of torque, we can say that this is its maximum torque at 2,000 rpm.

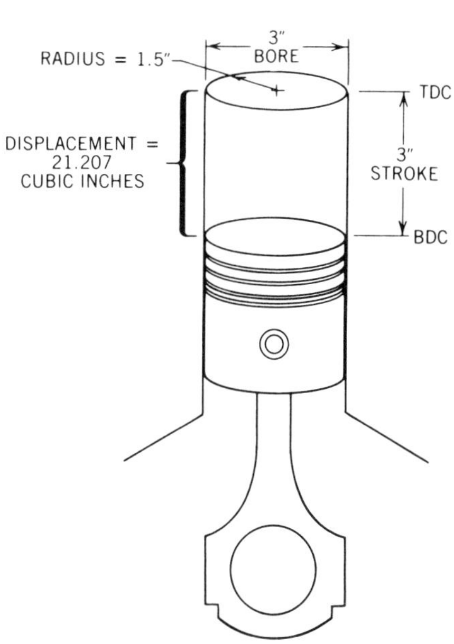


**Figure 4-30.** Calculate total cylinder displacement by multiplying the surface area of the cylinder cross section, or piston top, times the stroke.

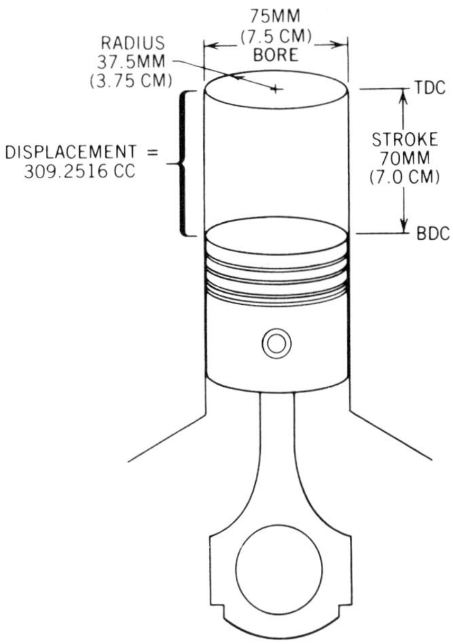


**Figure 4-31.** Use the same formula to calculate cylinder displacement with either metric or customary dimensions.

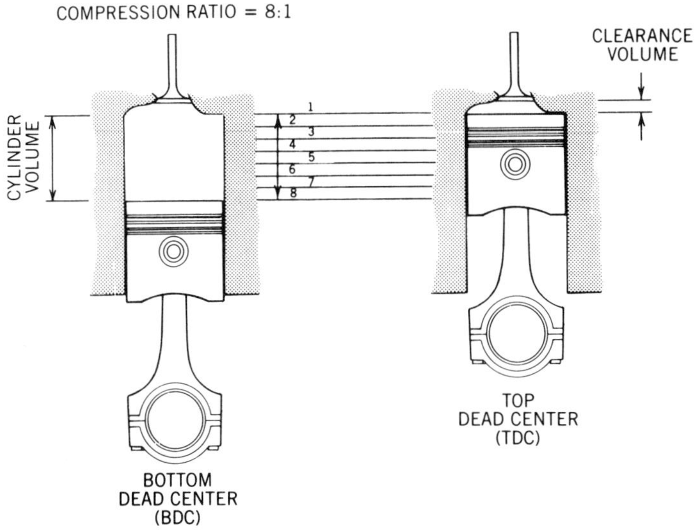


**Figure 4-32.** The compression ratio is total cylinder volume divided by combustion chamber, or clearance, volume.

## Engine Horsepower

To develop power, an engine must move some weight, not simply exert force on a scale. To determine what the power is, we have to know the weight, the distance, and the speed of movement. Let's allow our engine to rotate a 200-pound weight on the end of the 2-foot arm at a speed of 2,000 rpm. To calculate torque now, we must consider the distance through which the weight moves, not just the force on the end of a 2-foot arm. The distance is the circumference of a circle formed by the arm as it turns through a complete revolution, figure 4-34. That distance is

$$2 \times \text{radius (2 feet)} \times 3.1416 = 12.57 \text{ feet}$$

Now, multiply 12.57 feet by 200 pounds to determine the work done during one engine revolution:

$$12.57 \text{ feet} \times 200 \text{ pounds} = 2{,}514 \text{ foot-pounds}$$

The engine is running at 2,000 rpm, so multiply 2,514 by 2,000 to determine the amount of work done per minute:

$$2{,}000 \text{ rpm} \times 2{,}514 \text{ ft-lb} = 5{,}028{,}000 \text{ ft-lb/min}$$

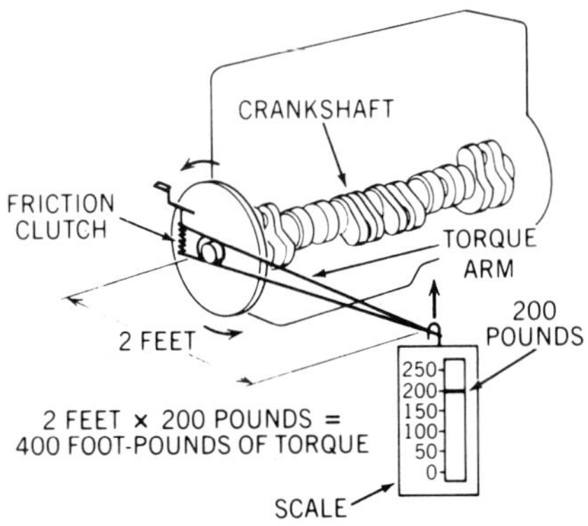


**Figure 4-33.** This engine applies 200 pounds of force through a 2-foot arm and therefore develops 2 × 200, or 400 foot-pounds of torque.

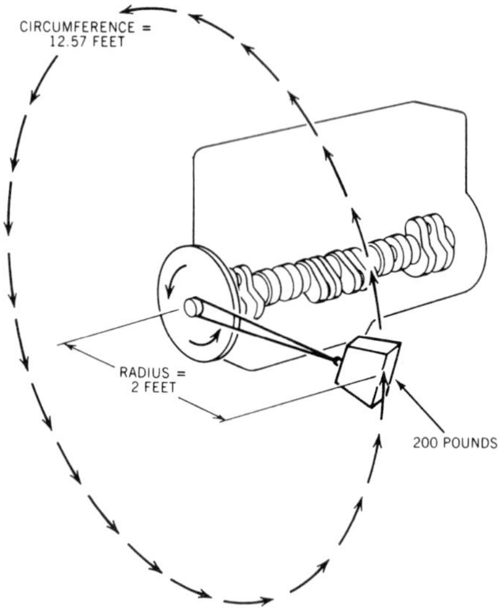


**Figure 4-34.** If the engine moves 200 pounds through a 12.57-foot-diameter circle, it does 2,514 foot-pounds of work.

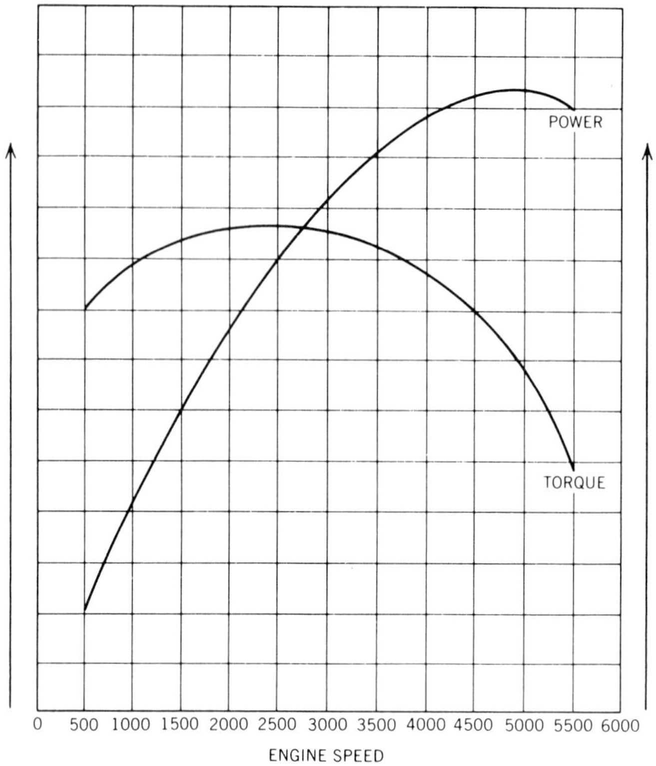


**Figure 4-35.** Most engines develop maximum torque at a lower speed than maximum power.

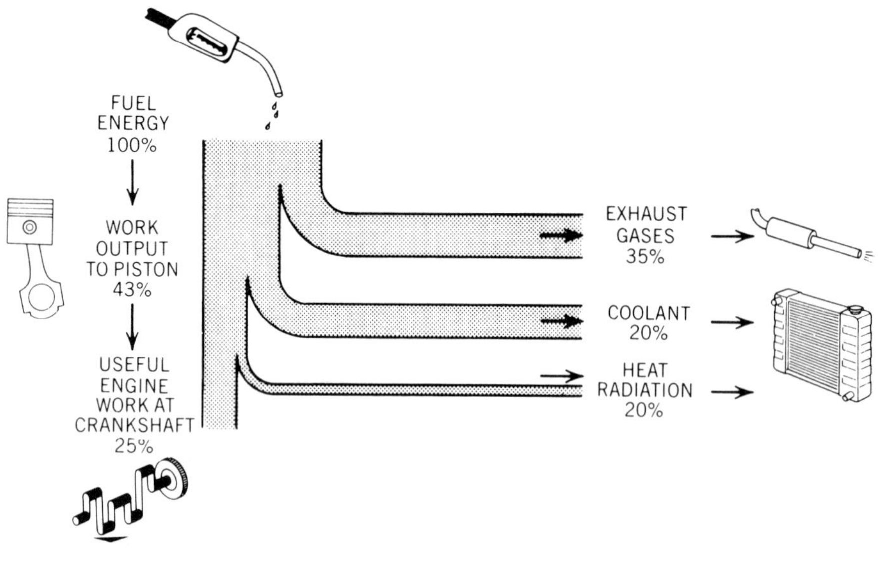


**Figure 4-36.** Only about 25 percent of an engine's potential fuel energy is available as mechanical power at the crankshaft.

Remember that 1 horsepower equals 33,000 foot-pounds per minute, so divide 5,028,000 by 33,000 to determine horsepower:

$$\frac{5{,}028{,}000}{33{,}000} = 152.36 \text{ horsepower}$$

If you know engine torque at any rpm, you can calculate horsepower with this formula:

$$\text{horsepower} = \frac{\text{torque} \times 2\pi \times \text{rpm}}{3{,}000}$$

If you divide 33,000 by $2\pi$ (6.2832), you have the shorter formula:

$$\text{horsepower} = \frac{\text{torque} \times \text{rpm}}{5{,}252}$$

**Metric Torque And Power.** The previous examples of torque and power calculations used customary units of foot-pounds and horsepower. Most carmakers now list the torque and power of their engines in both customary units and the metric units of newton-meters (Nm) and kilowatts (kW). Remember, there are direct relationships between foot-pounds and newton-meters and between horsepower and kilowatts. For review, here are the conversion factors you can use to change from one specification to another:

foot-pounds × 0.7376 = newton-meters

newton-meters × 1.356 = foot-pounds

horsepower × 0.746 = kilowatts

kilowatts × 1.3405 = horsepower

### Torque and Power Relationships

The arm and scale that we used to measure torque and power are a simple **dynamometer**. All dynamometers work by applying a load to an engine and measuring torque at different speeds.

Torque can be measured at any rpm, usually at wide-open throttle to determine maximum torque at a given engine speed. Most automobile gasoline engines develop maximum torque at about 2,500 rpm and maximum power at 3,500 to 5,000 rpm. Above 2,500 rpm, torque decreases because an engine takes in less air at higher speeds and combustion efficiency drops. There is less compression pressure and therefore less force on the piston to develop torque. Power, however, continues to increase as speed increases because speed goes up faster than torque drops off. As long as the torque decrease is less than the speed increase, power continues to rise. In the upper part of an engine's speed range, torque drops faster, and power similarly drops off. Figure 4-35 shows typical power and torque curves for an automobile gasoline engine. Most diesel engines develop maximum torque and power at lower speeds than gasoline engines.

### Brake Horsepower

The horsepower measured at an engine's crankshaft is called **brake horsepower (bhp)** because early engine dynamometers used a friction brake, called a prony brake, to apply a load to the engine. Brake horsepower is the horsepower available at the flywheel. **Gross brake horsepower** is the maximum power output with only the fuel, oil, and water pumps and built-in emission controls as accessory loads on the engine. **Net brake horsepower** is the power output of a fully equipped engine with all accessories such as the cooling system, exhaust system, air filter, charging system, and add-on emission controls installed. Before 1970, carmakers advertised the gross bhp ratings of engines. Since the early '70's, they have advertised net bhp ratings, which are lower.

### Indicated Horsepower and Friction Horsepower

The difference between gross and net bhp shows that not all of an engine's power is available to drive the car. Some is used to power various accessories. Energy that otherwise would be mechanical power also is lost as heat energy in the cooling and exhaust systems.

Remember, too, that friction is heat energy, and a lot of engine power is used to overcome the friction of its own moving parts. This lost horsepower is **friction horsepower**. We can't measure friction horsepower with simple equipment, but we can calculate it if we know the **indicated horsepower** of an engine.

Indicated horsepower is the theoretical maximum power that could be produced inside the cylinders of a particular engine. Engineers measure indicated horsepower with instruments that measure gas pressures in the cylinders through all four strokes of an operating cycle. The simple engineering comparisons of indicated friction, and brake horsepower are:

brake hp = indicated hp − friction hp

friction hp = indicated hp − brake hp

Brake horsepower (the power an engine *does* produce) divided by indicated horsepower (the power it *could* produce) is a measure of an engine's mechanical efficiency. For example, if brake hp is 100 and indicated hp is 125, the mechanical efficiency is: 100/125 = 0.8, or 80% efficiency.

By now, you can see that all of the potential energy in an engine's gasoline or diesel fuel is not changed to useful mechanical energy or power. Figure 4-36 shows that only about 25 percent of the potential energy in a gallon of fuel is available as mechanical power at the engine crankshaft.

## ENGINE PRESSURE

An engine is basically an air pump that draws in air, compresses it, and then exhausts it. Engine cylinder pressure varies during the 4-stroke cycle, and it varies according to the compression ratio and combustion efficiency of an engine.

The simplest pressure measurement is the **compression pressure**. This is the pressure that the piston creates as it moves from bottom dead center to top dead center on the compression stroke *without having the air-fuel mixture ignited*. In a gasoline engine it ranges from 100 to 200 psi (700 to 1,400 kPa). In a diesel engine, it ranges 450 to 800 psi (3,000 to 5,600 kPa), figure 4-37. Compression pressure is a useful tuneup specification that you can test with a pressure gauge to see if the cylinder is properly sealed. It is not, however, the operating pressure of the engine.

Operating pressures in a gasoline engine are as high as 700 psi during the power stroke because the heat of combustion raises the pressure of the compressed air and fuel. This pressure increase is what develops force on the piston and torque and power in an

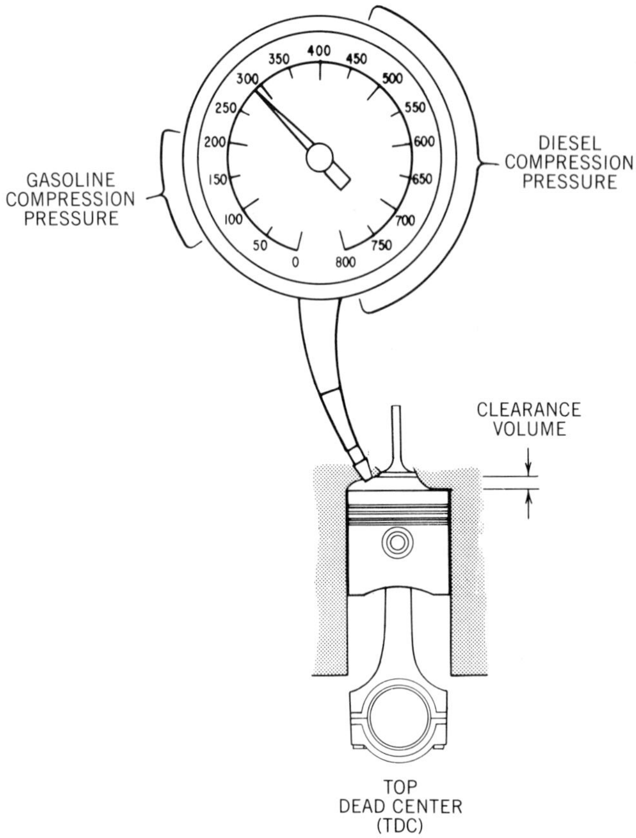


**Figure 4-37.** Compression pressure is the pressure created by the piston on the compression stroke without having the air-fuel mixture ignited.

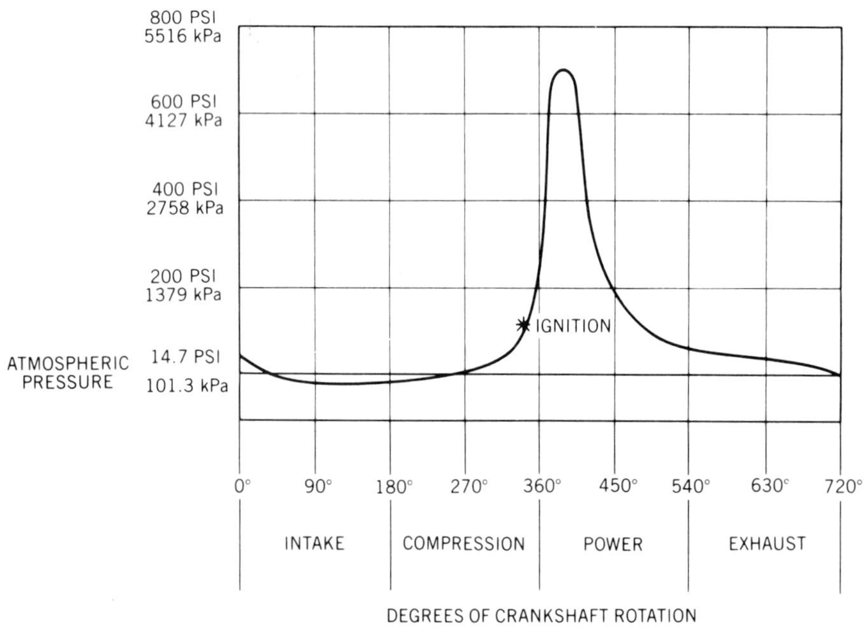


**Figure 4-38.** Cylinder pressures vary widely during the 4-stroke cycle.

engine. If the mixture doesn't burn on one power stroke (the cylinder misfires), there is no pressure rise, and that power stroke produces no power.

During the 4-stroke cycle, cylinder pressure varies from less than atmospheric pressure (14.7 psi) during the intake stroke to about 700 psi during combustion, figure 4-38. Engineers use the average pressure during all four strokes to calculate indicated horsepower. This is the **mean effective pressure**. Engineers also use **brake mean effective pressure (bmep)** to indicate engine efficiency. This is calculated from the brake horsepower measured on a dynamometer as follows for a 4-stroke gasoline engine:

$$\text{bmep} \frac{792{,}000 \times \text{bhp}}{\text{rpm} \times \text{cid}}$$

## FUEL SYSTEM PREVIEW

Now that you have a fundamental understanding of engine design and operation, we can look at the basics of fuel delivery. The fuel system, which is a major combustion control system, has the following major parts, figure 4-39:

- A fuel storage tank
- Fuel lines
- A fuel pump
- Fuel filters
- Air inlet and air filter
- Fuel delivery components (carburetor or fuel injectors)
- Evaporative emission controls
- Other related emission controls

The fuel tank stores the gasoline or diesel fuel. Steel lines and rubber hoses connect the tank to the fuel delivery parts on the engine. One or more fuel pumps move fuel from the tank to the carburetor or fuel injection system. One or more fuel filters remove water, dirt, and other impurities from the fuel before it enters the engine. Another filter at the engine air inlet removes dirt from the incoming air. Evaporative emission controls keep fuel vapors from escaping into the atmosphere from the tank, the carburetor, and other fuel system parts.

The carburetor or the fuel injection

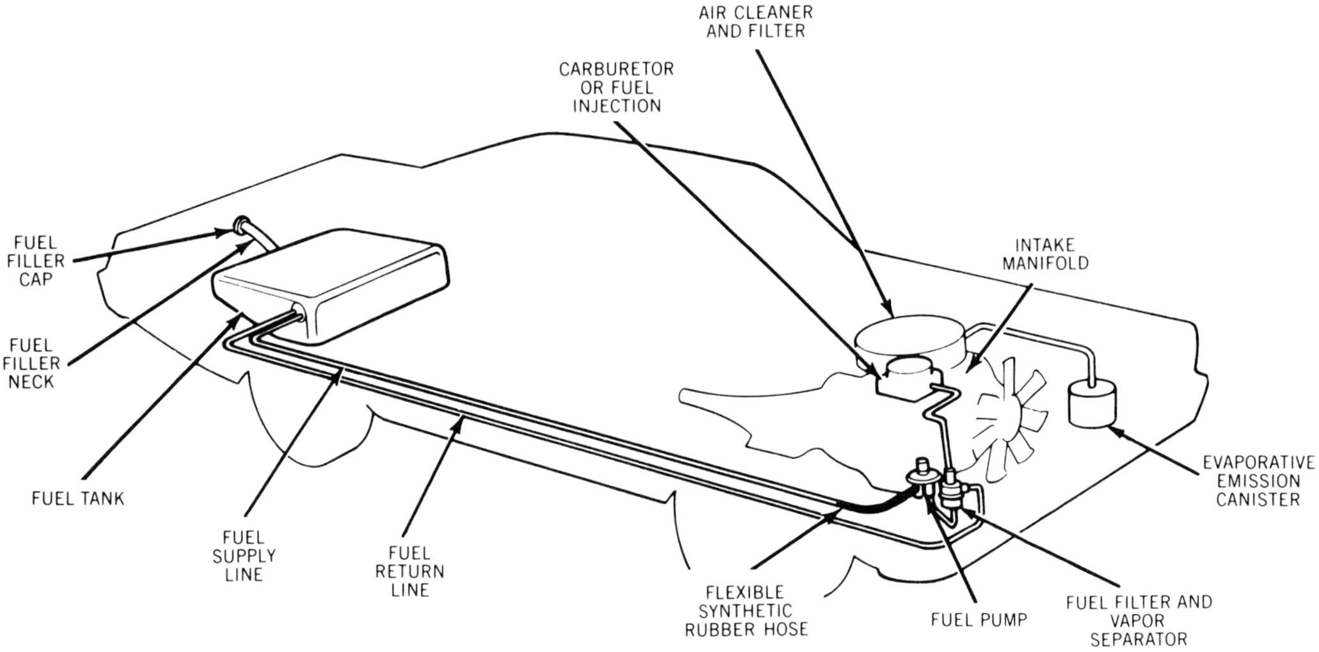


**Figure 4-39.** These are the major parts of a gasoline fuel system.

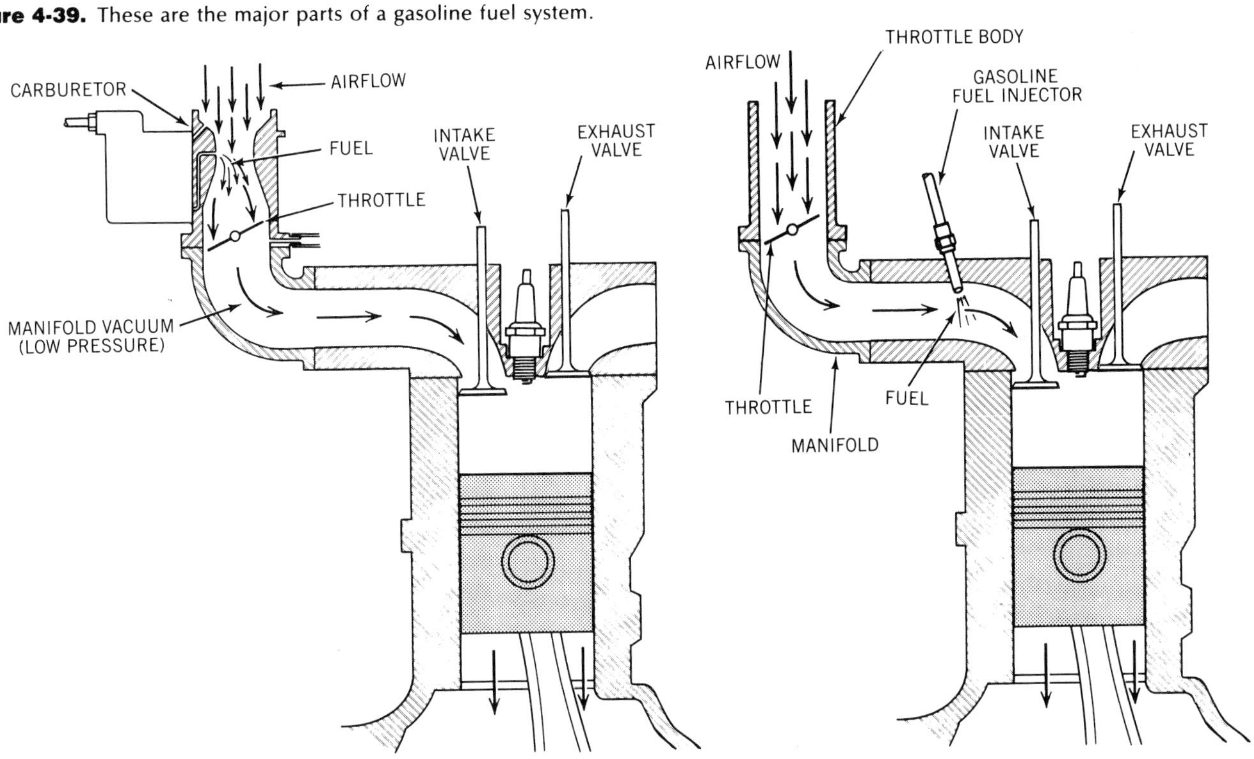


**Figure 4-40.** In a gasoline engine, fuel is vaporized and mixed with air outside of the combustion chamber.

system atomizes and vaporizes the fuel and mixes it with air in ratios that satisfy an engine's operating needs from light load to full load.

## Air and Fuel Mixing

Traditionally, the carburetor has been the standard device to mix air and fuel for a gasoline engine, but fuel injection systems are now as common as carburetors. With either a carburetor or fuel injection, gasoline is atomized, vaporized, and mixed with air outside the combustion chamber, figure 4-40.

When an intake valve opens, engine vacuum draws air into the engine's air intake and a mixture of air and fuel into the combustion chamber. Actually, atmospheric pressure forces air into the low-pressure area in the manifold and cylinders that is created as the pistons move down on their intake strokes. Air passing through a carburetor draws gasoline from a fuel bowl, or float bowl. As gasoline enters the intake airstream, it is atomized and then vaporized. Gasoline flow is precisely metered through jets so that it mixes with air in controlled ratios.

## Air-Fuel Ratios

Air-fuel ratios are measurements of the amount of air, *by weight*, mixed with an amount of fuel, *by weight*. An air-fuel ratio of 15:1 means that the mixture contains 15 pounds of air for every 1 pound of gasoline. In volume measurements, rather than weight, that equals approximately 9,000 gallons of air for every 1 gallon of gasoline, figure 4-41.

The air-fuel ratios for a gasoline engine range from 8:1 at the richest (most fuel) to 18.5:1 at the leanest (least fuel), figure 4- 42. With more or less fuel, the mixture would not be combustible. The best ratios for maximum power are approximately 12:1 to 13.5:1. The best ratios for fuel economy are about 15:1 to 16:1. The most efficient ratio that provides almost complete combustion is approximately 14.7:1. This is called a **stoichiometric ratio** and is important for emission control.

## Fuel Distribution

Carburetors, fuel injectors, and intake manifolds do not provide perfect air-fuel distribution. For example, if one carburetor mixes air and fuel for several cylinders, it is hard to locate the carburetor an equal distance from all cylinders so that each gets an equal mixture. Long passages and sharp turns in the manifold slow down the airflow and cause liquid fuel to drop out of the mixture. Multiple carburetors and fuel injection eliminate some of these problems.

Electronic fuel injection (EFI) on late-model cars provides precise air-fuel ratio control with the best combination of power, economy, and emission control. EFI systems are either throttle body injection systems or direct port or manifold injection systems.

Throttle body injection systems have one or two electronically controlled injection nozzles on a throttle body that is mounted on a manifold similarly to a carburetor, figure 4-43. Gasoline is injected into the airflow before it passes through the throttle opening. Fuel vaporization starts in the throttle body, and air-fuel mixing continues through the manifold to the intake valves. Injecting the fuel under pressure provides better atomization and vaporization than a carburetor does, and electronic control provides faster and more precise fuel metering changes.

Direct port, or manifold, injection systems have separate injectors in each manifold passage, near the intake ports of the cylinder head, figure 4-44. Only air passes through the intake throttle and the manifold up to where the injectors are located. Air-fuel mixing takes place immediately before the mixture enters the cylinder.

## Volumetric Efficiency

This discussion of the volume and ratios of air and fuel taken in by an engine leads us to the principle of **volumetric efficiency**. This is a concept that compares the airflow volume actually entering an engine to maximum volume that *could* enter, which is the same as the displacement. Volumetric efficiency is stated as a percentage, and it changes with engine speed.

To determine volumetric efficiency, engineers measure airflow volume in units such as cubic feet per minute (cfm) at standard temperature and pressure. Standard pressure is atmospheric pressure of 14.7 psi (101.28 kPa) or 760 millimeters of mercury (mm Hg). Standard temperature is 32° F (0° C). If intake airflow is equivalent to engine displacement, volumetric efficiency is 100 percent. We often determine carburetor airflow capacity for an engine based on 100 percent volumetric efficiency, but few engines ever reach complete efficiency. Volumetric efficiency for a stock passenger car engine also will vary under different operating conditions. At 2,000 rpm, an engine might be 85-percent efficient, At 1,000 rpm, it might be rated at 75 percent, and at 4,000 rpm, it might have a volumetric efficiency of only 60 percent.

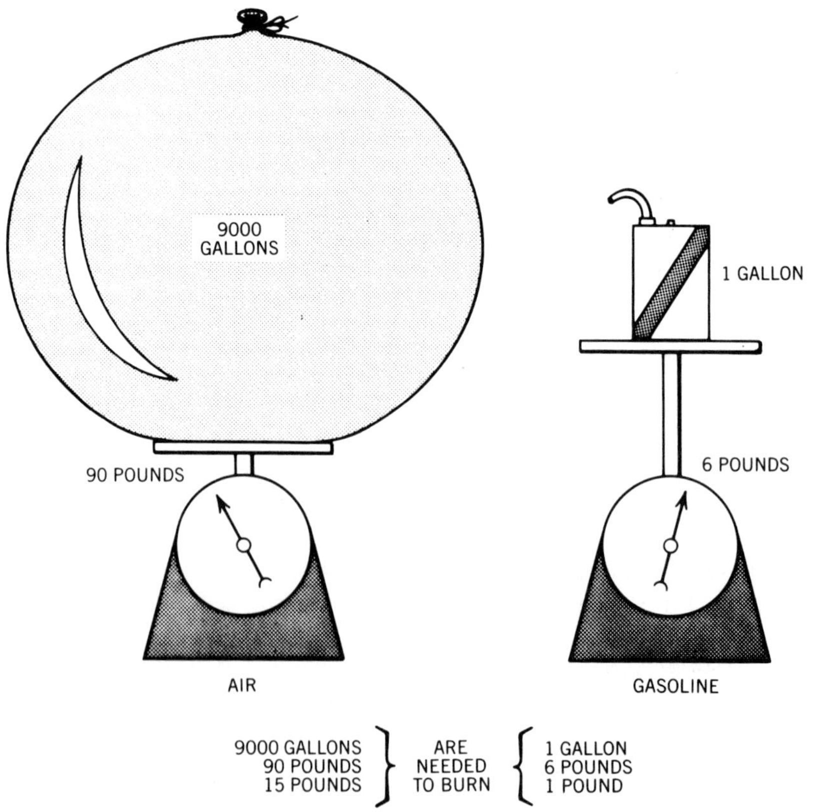


**Figure 4-41.** An air-fuel ratio of 15:1 means that 15 pounds of air are mixed with every 1 pound of gasoline. That equals 9,000 gallons of air for every one gallon of gasoline.

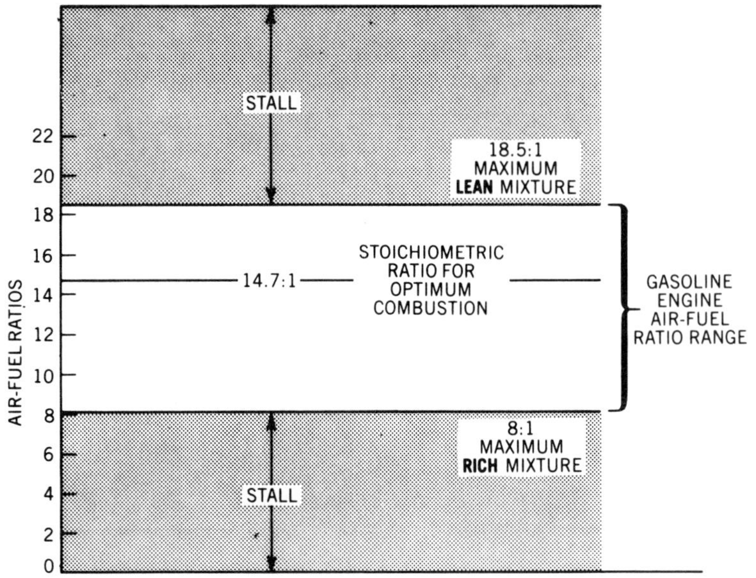


**Figure 4-42.** Air-fuel ratios for a 4-stroke gasoline engine can range from 8:1 at the richest to about 18.5:1 at the leanest.

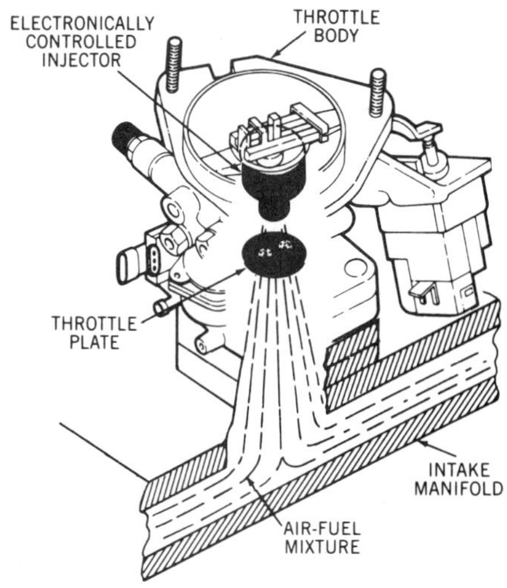


**Figure 4-43.** The injectors of a throttle body fuel injection system are mounted on the throttle body.

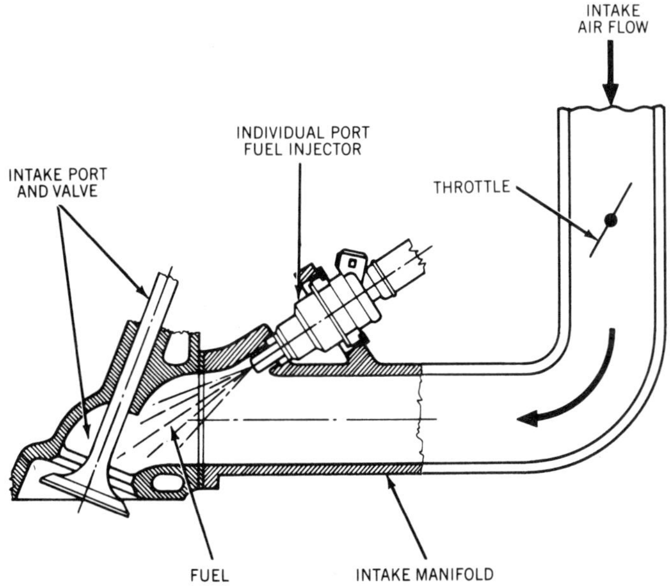


**Figure 4-44.** This direct port injection system has separate injectors for each cylinder, near the cylinder head intake ports.

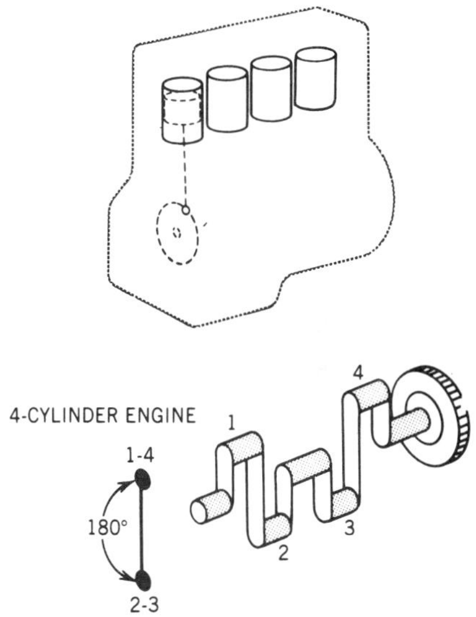


**Figure 4-45.** Inline 4-cylinder engines fire 1-3-2-4 or 1-2-4-3.

## IGNITION SYSTEM PREVIEW

After the air-fuel mixture is compressed in the cylinders, it must be ignited. For either a gasoline or a diesel engine, the timing of ignition—by spark or by fuel injection—is critical to engine operation. Ignition must occur near the start of the power stroke. If it is too early or too late, the engine won't develop full power and may even be damaged. The principles of ignition timing are the same for gasoline and diesel engines, but since gasoline engines are more common in passenger cars, we will concentrate on spark ignition in the following sections.

### Ignition Intervals

Remember that every piston stroke rotates the crankshaft 180 degrees, and there are 720 degrees in a complete 4-stroke cycle. During each cycle, the spark plug in each cylinder fires once. Therefore, ignition occurs in a 1-cylinder engine once every 720 degrees. The 720 degrees are the **ignition interval** for a 1-cylinder engine. The ignition interval is the number of degrees of crankshaft rotation between ignition, or power strokes, for each cylinder of an engine.

A 4-cylinder engine has four power strokes in every 720 degrees of crankshaft rotation, so the ignition interval of a 4-cylinder engine is 180 degrees (720/4 = 180), figure 4-45. An inline 6-cylinder engine has six power strokes

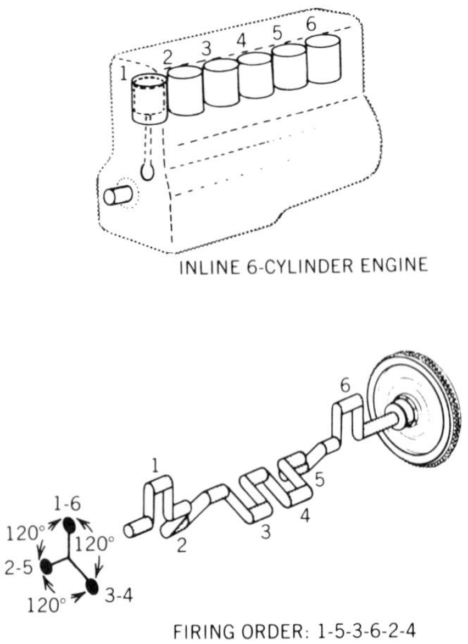


**Figure 4-46.** All inline 6-cylinder engines fire 1-5-3-6-2-4.

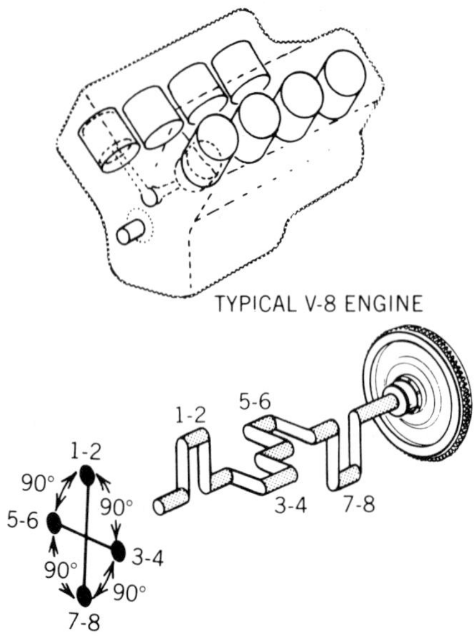


**Figure 4-47.** V-8 firing orders vary according to the cylinder numbering sequence.

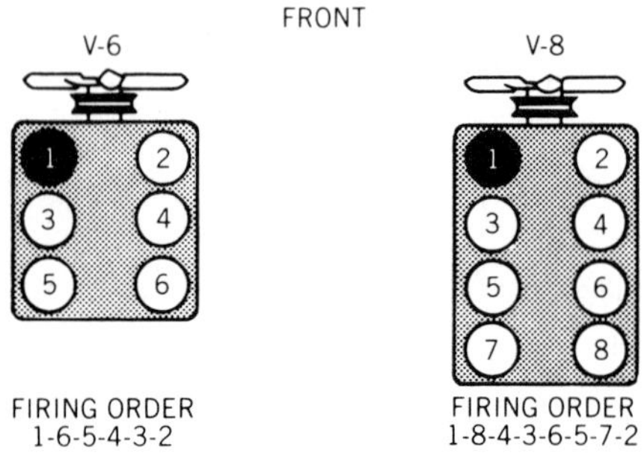


**Figure 4-48.** These are the most common cylinder numbering patterns and firing orders for V-type engines.

every 720 degrees, so its ignition interval is 120 degrees (720/6 = 120), figure 4-46. An 8-cylinder engine has eight power strokes in 720 degrees of rotation for an ignition interval of 90 degrees (720/8 = 90), figure 4-47. Three-cylinder engines, such as the Isuzu-Chevrolet Sprint engine, have 240-degree ignition intervals; 5-cylinder VW, Audi, and Mercedes-Benz engines have 144-degree intervals.

Most 4-, 6-, and 8-cylinder engines have the ignition intervals listed above. V-6 engines can have several different intervals. A V-6 engine with a 60-degree angle between cylinder banks has a 120-degree ignition interval, the same as an inline 6-cylinder engine. A 90-degree V-6 can have 120-degree ignition intervals if the crankshaft rod journals are split, such as on 1978 and later Buick-built V-6's. Buick 90-degree V-6's built before 1978 did not have split journals and fired alternately at 90- and 150-degree intervals. Chevrolet 90-degree V-6's have split rod journals that create alternate 108- and 132-degree intervals. V-12 engines usually have 60-degree blocks and 60-degree ignition intervals.

### Firing Order and Cylinder Numbering

Ignition interval is the amount of crankshaft rotation between ignition in each cylinder. **Firing order** is the sequence in which ignition occurs in the various cylinders. So that the cylinders fire at regular intervals and each cylinder fires once every 720 degrees, the crankshaft throws are arranged in a particular order. This crank throw arrangement creates the firing order, and it varies depending on the number of cylinders and the block design.

Engine cylinders are numbered for identification, but they usually don't fire in cylinder number order. Inline engine cylinders are numbered from front, or the end opposite the flywheel, to rear. For an inline 4-cylinder engine, the crankshaft rod journals are offset and the cylinders are numbered as shown in figure 4-45. The cylinders must fire at 180-degree intervals, so the firing order must be 1-3-4-2 (most common) or 1-2-4-3. For an inline 6-cylinder engine, the rod journals are arranged and the cylinders are numbered as shown in figure 4-46. All inline 6-cylinder engines fire 1-5-3-6-2-4.

Except for Ford products and a few older GM engines, most domestic V-type engines have the cylinder number patterns and firing orders shown in figure 4-48. Odd-numbered cylinders are on the left bank viewed from the flywheel end; even-numbered cylinders are on the right. The one major exception to the firing order shown for V-6 engines is the GM-Chevrolet 60-degree, 2.8-liter (170-cid) V-6. It fires 1-2-3-4-5-6.

Ford V-type engines have the number 1 cylinder at the right front. Cylinder numbers follow in order down the right bank and then continue from front to rear on the left. Figure 4-49 shows Ford V-6 and V-8 cylinder numbering and firing orders. Fortunately, most carmakers cast the firing order of an engine on the intake manifold to avoid confusion during engine service.

### Cylinder Running Mates

Notice in figure 4-45 that the crankshaft throws for cylinders 1 and 4 of a 4-cylinder engine are in the same position, as are the throws for cylinders 2 and 3. In an inline 6-cylinder engine, figure 4-46, the throws for cylinders 1 and 6, 3 and 4, and 2 and 5 share common positions. Cylinders with crank throws in the same positions in relation to firing order are called **running mates**. Although the pistons are at tdc simultaneously, they are 360° apart in firing order. When piston 1 of a 4-cylinder engine is at tdc at the start of the power stroke, piston 4 is at tdc at the start of its intake stroke.

Visualizing cylinders that are running mates in a V-type engine is a bit more difficult because you must account for the angle between cylinder banks. In the V-8 engine in figure 4-47, cylinders 1 and 6, 8 and 5, 4 and 7, and 3 and 2 are running mates because you must add the 90° cylinder angle to the crankshaft angle.

You can determine cylinder running mates in any engine from its firing order. Simply take the first half of the firing order and write it above the second half of the firing order, like this:

$$\frac{1-8-4-3}{6-5-7-2}$$

This establishes the running mates for the V-8 engine in figure 4-47. It works the same way for an inline 6-cylinder engine:

$$\frac{1-5-3}{6-2-4}$$

and for an inline-4:

$$\frac{1-3}{4-2} \text{ or } \frac{1-2}{4-3}$$

Knowing the cylinder running mates is useful when you adjust mechanical valve lifters and for ignition timing on some engines, as you will learn when you study these subjects.

## Engine and Ignition Synchronization

During each 4-stroke cycle, the intake and exhaust valves of each cylinder open and close at precise times in relation to piston and crankshaft position. Ignition must be similarly synchronized with piston and crankshaft position at the end of the compression stroke and the start of the power stroke. The ignition distributor delivers one spark to each cylinder in each 4-stroke cycle. Distributors are built so that one revolution of the distributor shaft delivers ignition voltage once to each cylinder; that is, every 360 degrees of distributor rotation. But, each cylinder fires once for every *two* revolutions of the crankshaft: 720 degrees. Therefore, the distributor turns at one-half crankshaft speed. Because the camshaft also turns at one-half crankshaft speed, most distributors are driven by the camshaft. In this way, crankshaft position, valve timing, and ignition timing are all synchronized, figure 4-50.

Diesel fuel injection must be timed similarly to the spark ignition of a gasoline engine, and each cylinder must receive fuel once for every two crankshaft revolutions. Diesel fuel injection pumps distribute fuel to each cylinder in firing-order sequence and time the injection the way a distributor delivers and times spark ignition. In fact, some diesel pumps are called "distributor pumps." Because diesel fuel injection pumps operate at one-half engine speed, some are driven and synchronized by the camshaft. In other diesel engines, the pumps are driven by separate gears or belts but synchronized to time injection with crankshaft and valve position, figure 4-51.

## Crankshaft Position and Burn Time

Combustion is not an instantaneous explosion, but a controlled burning, although the **burn time** is extremely short. It takes about 3 milliseconds (0.003 second) from the time of ignition until combustion is complete in a gasoline engine. In a diesel engine, a similar, but shorter, amount of **ignition lag** occurs from the time fuel is injected until it mixes with the hot compressed air and burns.

Burn time or ignition lag is strictly

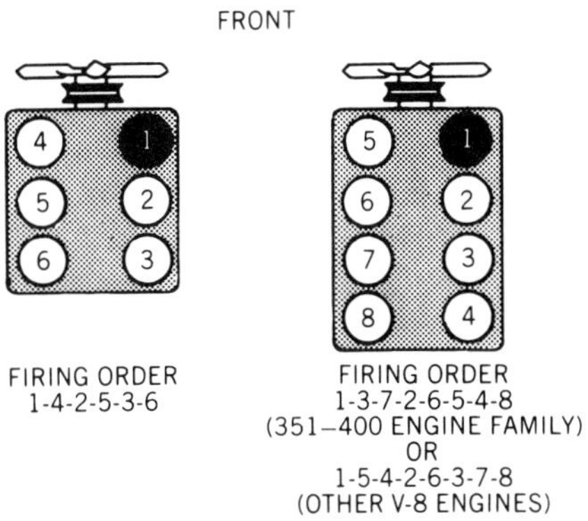


**Figure 4-49.** Ford numbers the cylinders of its V-type engines like this.

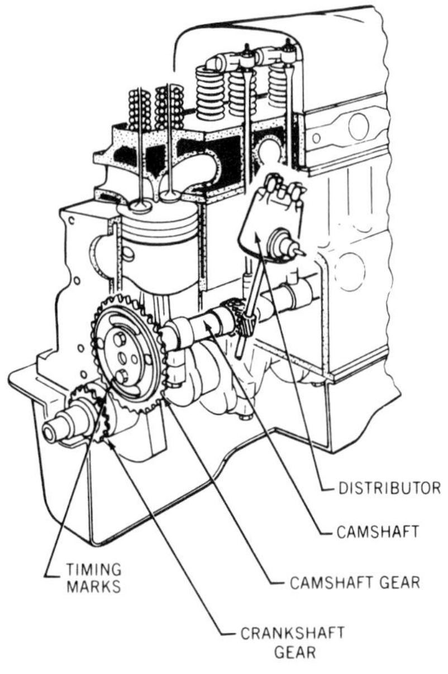


**Figure 4-50.** The camshaft and the ignition distributor rotate at one-half engine speed, so that crankshaft position, valve timing, and ignition timing are synchronized.

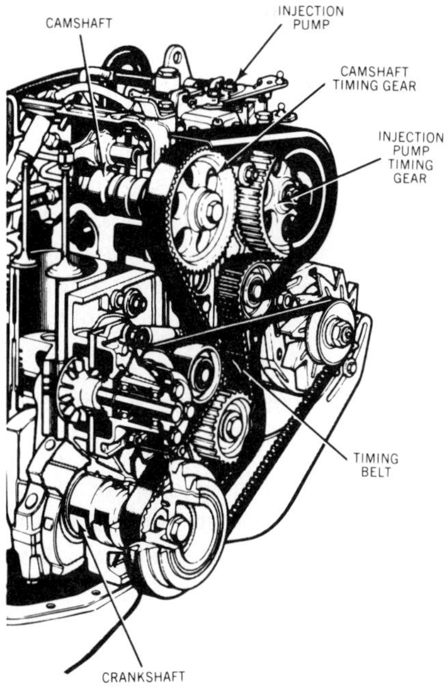


**Figure 4-51.** Diesel injection pumps operate at one-half engine speed to time fuel injection similarly to spark ignition in a gasoline engine. (AMC-Jeep)

a function of time, not of piston or crankshaft position. Ignition must occur early enough so that maximum combustion pressure develops when the piston is starting down on its power stroke. Combustion should be finished about 10 degrees **after top dead center (atdc)**. If ignition occurs too early when the piston is **before top dead center (btdc)** on the compression stroke, the rising piston is opposed by combustion pressure. If ignition occurs too late atdc, combustion force is reduced as the piston is farther down the power stroke. In either case, the combustion process doesn't develop full power, and the engine may be damaged.

As engine speed increases, the piston and crankshaft travel farther (rotate through more degrees) in a given amount of time. Burn time or ignition lag, however, remains relatively constant. Therefore, ignition timing must vary with engine speed so that maximum combustion pressure occurs at the correct piston position.

Consider, for example, an engine that needs a 3-millisecond burn time and is most efficient with combustion ending at 10° atdc. Let's say that at 800 rpm, the crankshaft rotates 12 degrees in 3 milliseconds. So, ignition must occur 2° btdc for proper burn time in relation to piston and crankshaft position, figure 4-52. At 1,200 rpm, the crankshaft rotates 20 degrees in 3 milliseconds, so ignition must occur 10° btdc for combustion to end 10° atdc. At 2,400 rpm, the crankshaft rotates 40 degrees in 3 milliseconds. Now ignition must be at 30° btdc for combustion to end at 10° atdc, figure 4-52.

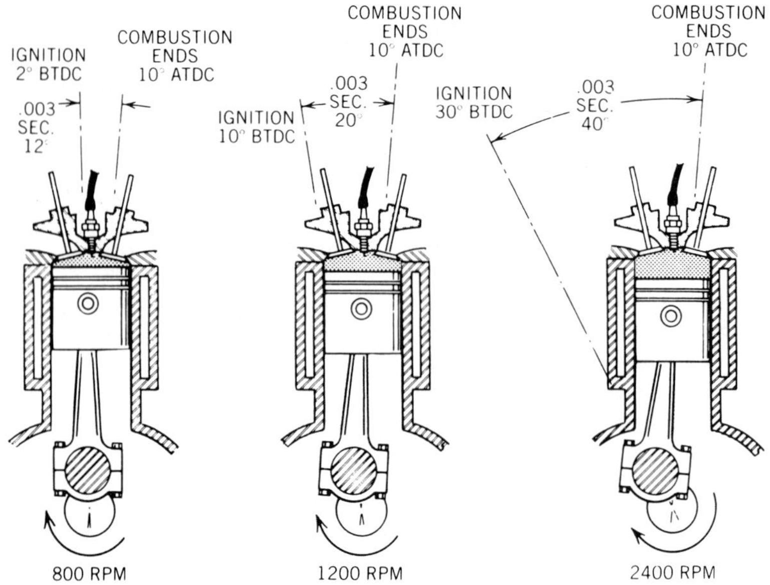


**Figure 4-52.** Ignition timing must be advanced as engine speed increases.

### Ignition Timing and Advance

The point where ignition occurs is called ignition timing, or spark timing, and it is measured in terms of crankshaft degrees before or after top dead center. The ignition timing of an engine at idle speed is called **basic, or initial, timing**. It is usually a few degrees before or after tdc. As engine speed increases, ignition must occur sooner before top dead center. This is called **timing, or spark, advance**. As engine speed decreases from high rpm to idle, timing must be retarded from its advanced condition back to its basic setting. In some operating conditions (principally for some emission control needs), it is necessary to retard the timing to several degrees atdc.

Spark ignition systems have devices to advance and retard timing automatically as engine speed changes. Older distributors used centrifugal flyweights and vacuum diaphragms connected to intake vacuum to control timing for varying engine speeds and loads. Late-model electronic ignitions control timing electronically by sensing engine speed and other conditions and processing the sensor signals in a computer.

Diesel injection pumps have devices that advance and retard injection timing as engine speed and load change.

## ELECTRICAL SYSTEM PREVIEW

We sometimes don't think of the battery, the charging system, and the starting system as combustion control systems, but an engine can't operate at maximum efficiency if these systems aren't working correctly. You will study them in detail later in this text, but a short preview now will help you understand them in relation to overall engine operation.

### The Battery and Charging System

The battery is the heart of a car's electrical system. It supplies voltage and current for every other electrical device on the automobile. We often call it a "storage battery," and it is a voltage and current reservoir for the rest of the electrical system. The battery does more than store electricity, however. It creates a voltage potential between its positive and negative terminals through electrochemical action of the different materials on its positive and negative plates and the electrolyte solution it contains. When the battery is connected to a circuit, the voltage it has created causes current to flow through the circuit.

An automobile battery delivers current at about 12 volts. Therefore, all lamps, motors, and relays on an automobile are 12-volt devices. However, 12 volts are not enough to create a spark between the electrodes of a spark plug. Spark ignition needs thousands of volts to overcome the resistance of the spark plug gap and the compressed air-fuel mixture. The ignition system uses an induction coil to raise ignition voltage to tens of thou-

sands of volts (10 to 40 kilovolts) to ignite the air-fuel mixture in the engine cylinders.

As a battery delivers current to the electrical devices on an automobile, it cycles from a charged to a discharged condition. If left to itself, a battery will eventually become completely discharged, but automobile batteries can be recharged. This is the job of the charging system.

The principal part of an automobile charging system is the alternator, which generates current through the induction of an electromagnetic rotor rotating within a coiled conductor. The alternator delivers current at 12 to about 14.5 volts to the battery to reverse its discharge cycle.

### The Starting System

The earliest automobiles were started by the driver turning a crank that engaged the engine crankshaft. The driver turned the crank fast enough to compress the air-fuel mixture enough for ignition. That was fine for low-power, low-compression engines of 75 years ago, but not for modern automobile engines.

For 60 or 70 years, automobiles have used electric motors to crank the engine. The starter motor draws current from the battery when you turn the ignition switch to start. The motor has a small drive gear, or pinion, that engages a larger driven gear, or ring gear, on the engine flywheel. Starter motors are fractional horsepower motors, but they have to overcome the friction and compression resistance of the automobile engine to crank it. Therefore the reduction ratio of the starter pinion to the flywheel ring gear is in the range of 15:1 to 25:1, and the starter motor draws 200 to 400 amperes of current. The extreme reduction ratio and high current draw allow the starter to crank the engine at 200 to 400 rpm. This is fast enough to draw air and fuel into the cylinders, compress the mixture, and create ignition. The battery, however, must supply this high current for a few seconds until the engine starts.

## SUMMARY

All gasoline and diesel automobile engines are internal-combustion engines that operate on the 4-stroke cycle of intake, compression, power, and exhaust. Engines commonly are identified by the number of cylinders and cylinder arrangement. Engine size is measured by displacement, which is the total volume of all cylinders.

Engine torque is the amount of rotating work that an engine can do, measured in foot-pounds or newton-meters. Power is the amount of torque delivered in a given amount of time. An engine develops power and torque as pressures rise in the cylinders during combustion.

The fuel system delivers a measured amount of fuel to the engine, where it is mixed with air before combustion. In a gasoline engine, air and fuel are mixed outside of the cylinders. In a diesel engine, fuel is mixed with air when it is injected into the combustion chamber at the time of ignition.

Ignition in a gasoline engine is provided by a timed electric spark from an electrical ignition system. Ignition in a diesel is controlled by timing fuel injection to the cylinders. Ignition timing for any engine must be advanced as engine speed increases to allow adequate burn time and develop full power.

## REVIEW QUESTIONS

### Multiple Choice

**1.** Mechanic A says that gasoline and diesel engines are both internal-combustion engines. Mechanic B says that steam engines also are internal-combustion engines. Who is right?

**a.** A only
**b.** B only
**c.** Both A and B
**d.** Neither A nor B

**2.** Piston travel from the top to the bottom of a cylinder is called a:

**a.** Cycle
**b.** Revolution
**c.** Rotation
**d.** Stroke

**3.** An engine block casting contains all of the following, *except* the:

**a.** Cylinders
**b.** Flywheel
**c.** Water jackets
**d.** Main bearing webs

**4.** Mechanic A says ignition in a gasoline engine is controlled by a timed electric spark. Mechanic B says ignition in a diesel engine is controlled by air-fuel ratio. Who is right?

**a.** A only
**b.** B only
**c.** Both A and B
**d.** Neither A nor B

**5.** Speed and power of a diesel engine are controlled by:

**a.** Varying the amount of fuel mixed with a constant volume of air
**b.** Varying the amount of air mixed with a constant volume of fuel
**c.** Advancing or retarding ignition timing
**d.** Throttling the air-fuel mixture

**6.** The 4-stroke cycle for each cylinder requires which of the following number of degrees of crankshaft rotation:

**a.** 180
**b.** 270
**c.** 360
**d.** 720

**7.** Most 8-cylinder engines produce a power stroke every:

**a.** 60 degrees of crankshaft rotation
**b.** 90 degrees of crankshaft rotation
**c.** 120 degrees of crankshaft rotation
**d.** 180 degrees of crankshaft rotation

**8.** Even-firing 6-cylinder engines produce a power stroke every:

**a.** 60 degrees of crankshaft rotation
**b.** 90 degrees of crankshaft rotation
**c.** 120 degrees of crankshaft rotation
**d.** 180 degrees of crankshaft rotation

**9.** Four-cylinder engines produce a power stroke every:

**a.** 60 degrees of crankshaft rotation
**b.** 90 degrees of crankshaft rotation
**c.** 120 degrees of crankshaft rotation
**d.** 180 degrees of crankshaft rotation

**10.** The intake valve usually opens:

**a.** At bottom dead center on the exhaust stroke
**b.** At top dead center on the intake stroke
**c.** After top dead center of the intake stroke
**d.** Before top dead center of the intake stroke

**11.** The exhaust valve usually closes:

**a.** After top dead center on the compression stroke
**b.** At top dead center on the exhaust stroke
**c.** After bottom dead center of the power stroke
**d.** Before bottom dead center of the power stroke

**12.** The number of crankshaft degrees during which any valve is open is:

**a.** Valve lift
**b.** Valve overlap
**c.** Valve duration
**d.** Valve timing

**13.** A simplified formula to calculate engine displacement is:

**a.** Bore $\times$ 0.7854 $\times$ stroke $\times$ no. cylinders
**b.** Bore $\times$ 0.7854 $\times$ stroke$^2$ $\times$ no. cylinders
**c.** Bore$^2$ $\times$ 0.7854 $\times$ 1/2 stroke $\times$ no. cylinders
**d.** Bore 2 $\times$ 0.7854 $\times$ stroke $\times$ no. cylinders

**14.** To change cubic inches to cubic centimeters, multiply by:

**a.** 0.016
**b.** 61.39
**c.** 0.061
**d.** 16.39

**15.** The total cylinder volume of one engine cylinder is 27 cubic inches. The clearance volume is 3 cubic inches. The compression ratio, therefore, is:

**a.** 8.5:1
**b.** 8:1
**c.** 9:1
**d.** 9.2:1

**16.** $\frac{\text{Torque (in foot-pounds)} \times \text{rpm}}{5252}$ is the formula for:

**a.** Engine kilowatts
**b.** Torque in newton-meters
**c.** Engine horsepower
**d.** Torque rise in relation to engine speed

**17.** The stoichiometric gasoline air-fuel ratio for maximum combustion efficiency is approximately:

**a.** 13.5:1
**b.** 14.0:1
**c.** 14.7:1
**d.** 15.2:1

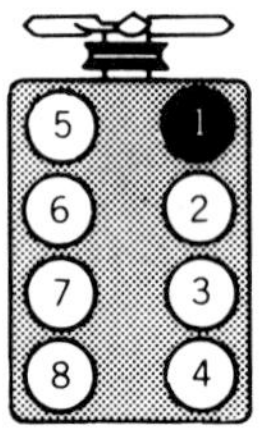


FIRING ORDER
1-3-7-2-6-5-4-8
(351–400 ENGINE FAMILY)
OR
1-5-4-2-6-3-7-8
(OTHER V-8 ENGINES)

**18.** The figure above illustrates cylinder numbering for V-8 engines built by:

**a.** Chrysler Corporation
**b.** Mercedes-Benz
**c.** Ford Motor Company
**d.** General Motors

**19.** The firing order for all inline 6-cylinder engines is:

**a.** 1-6-5-4-3-2
**b.** 1-2-3-4-5-6
**c.** 1-4-6-3-5-2
**d.** 1-5-3-6-2-4

**20.** The firing order for *most* V-8 engines except Ford products is:

**a.** 1-3-7-2-6-5-4-8
**b.** 1-5-4-2-6-3-7-8
**c.** 1-8-7-2-6-5-4-3
**d.** 1-8-4-3-6-5-7-2

### Fill in the Blank

**21.** The exact top of a piston stroke is called __________.

**22.** Diesel ignition air temperatures are approximately __________ to __________ F.

**23.** Diesel air-fuel ratios vary from as rich as __________ at full power to as lean as __________ at idle.

**24.** The range of gasoline air-fuel ratios varies from approximately __________ at the richest to approximately __________ at the leanest.

**25.** The number of crankshaft degrees when the intake and exhaust valves of one cylinder are open simultaneously is __________.

**26.** To change liters to cubic inches, multiply by __________.

**27.** Compression ratio is the ratio of total cylinder volume to __________ volume.

**28.** Camshaft lift is multiplied by the __________ to determine final valve lift.

**29.** The firing order of *most* inline 4-cylinder engines is __________.

**30.** In a gasoline engine, an air-fuel ratio of approximately 14.7:1 is called a __________ ratio.

# PART TWO
# BASIC ENGINE SYSTEMS

# INTRODUCTION

The two chapters in this part deal with the engine lubrication, ventilation, cooling, and exhaust systems. Many motorists do not think of these systems when they consider a traditional engine "tuneup." No engine can perform efficiently, however, unless these systems operate properly.

Lubrication minimizes the friction between moving parts and helps to cool the engine. The lubrication system and the cooling system thus work hand in hand to ensure proper engine performance.

The cooling system does not simply "cool" the engine. Its most important job is to stabilize engine temperature for efficient combustion. If an engine runs too hot or too cold, air-fuel ratios cannot be maintained for the best combination of performance, economy, and emission control.

The exhaust system removes combustion byproducts (exhaust) from the engine. That's obvious. It is essential for engine "breathing" and thus affects the intake air-fuel mixture. The exhaust system also removes excess heat, helps to maintain proper engine temperature, heats the intake air-fuel mixture of a cold engine, and provides exhaust gas recirculation to reduce $NO_x$ emissions. Exhaust emission control has been a critical part of automobile engineering for more than 20 years. Catalytic converters are important parts of the exhaust systems on all late-model cars.

Positive crankcase ventilation (PCV) systems were the first emission control systems installed on automobiles over 20 years ago, but proper crankcase ventilation has been a critical part of engine design since the invention of the Otto-cycle engine. Crankcase ventilation is essential for proper lubrication and engine breathing. PCV systems, thus, are important emission controls and an integral part of basic engine design.

Complete engine performance service includes testing, adjustment, and repair of the lubrication, ventilation, cooling, and exhaust systems.

# 5 ENGINE LUBRICATION AND VENTILATION

## WHAT

The last chapter introduced the principles of engine design and explained the basic operation of fuel and ignition systems. It did not, however, explain one of the most important engine systems: the lubrication system. Without proper lubrication, an engine could not run. This chapter explains the lubrication system and the properties and classifications of motor oil. It also introduces the equally important subject of engine crankcase ventilation.

## WHY

The close tolerances, high temperatures, and high speeds of an automobile engine require a sophisticated lubrication system. Whenever you service an engine, you must understand and check the operation of its oiling system. Even a basic tuneup won't provide the necessary performance and economy if the engine is not getting proper lubrication. Regular oil and filter changes also are a required emission control service. Additionally, proper crankcase ventilation goes hand in hand with lubrication to control emissions and prolong engine life. Your professional ability to provide complete engine care includes your understanding of correct lubrication and crankcase ventilation.

## GOALS

After studying this chapter, you should be able to:

**1.** Explain engine lubrication and the conditions under which oil operates.

**2.** Explain motor oil service classifications and viscosity ratings.

**3.** Identify the parts of an engine lubricated by splash and by pressure.

**4.** Explain the operation and relationships of engine lubrication system parts.

**5.** Explain the reasons for crankcase ventilation and the operation of a closed positive crankcase ventilation (PCV) system. Identify PCV system parts.

## ENGINE LUBRICATION REQUIREMENTS

It would be hard to design any machine that puts greater demands on its lubrication system than an automobile engine. Many engine parts, even those away from the direct heat of combustion, operate at temperatures of many hundreds of degrees Fahrenheit. Also, many rotating and reciprocating parts operate at close clearances with surface speeds of several hundred feet per minute. These conditions place severe demands on the lubricating ability of motor oil.

Motor oil does six important jobs in an engine:

**1.** It reduces friction, which in turn reduces wear and heat, between moving parts.

**2.** It serves as a coolant by removing heat from lubricated parts of the engine.

**3.** It helps to keep the engine clean of sludge, varnish, and combustion chamber deposits. It also removes dirt particles from moving parts.

**4.** It prevents rust and corrosion of precision parts.

**5.** It forms an oil film between piston rings and cylinder walls and thus helps to seal the combustion chamber.

**6.** It serves as a shock absorber to cushion reciprocating and rotating parts and protect them from combustion forces.

For proper lubrication and protection of modern engines, oil also must flow readily to permit easy starting, and it must dissipate foam to maintain a noncompressible lubricating film.

## MOTOR OIL COMPOSITION

Most motor oils are petroleum hydrocarbon compounds or mixtures of compounds. Natural petroleum oil, however, does not have all the qualities needed to lubricate an engine. To improve the properties of motor oil, refiners add many additives. Additives have the following features:

**1.** They may add a property that the oil did not originally have.

**2.** They may strengthen a natural quality of the oil.

**3.** They counteract an undesirable natural property of the oil, such as a tendency to foam.

The following sections describe some of the general classes of oil additives.

### Antioxidants

The heat of an engine tends to oxidize oil. The oil literally cooks and combines with oxygen to form **varnish** and carbon deposits on engine parts, figure 5-1. **Antioxidant** additives keep the oil compounds stable at high temperatures and reduce this problem.

**Figure 5-1.** Antioxidant additives reduce the formation of varnish and carbon deposits like these.

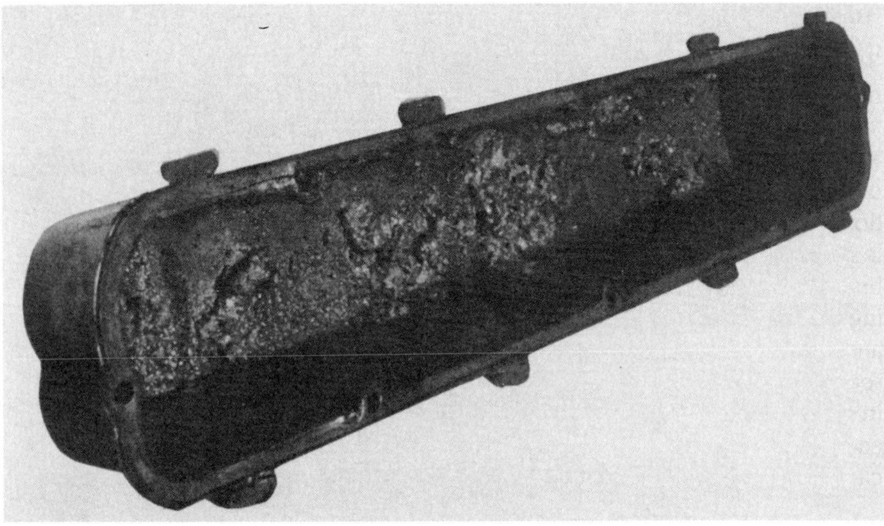

**Figure 5-2.** Sludge forms when blowby gases mix with cold oil in the engine.

### Rust and Corrosion Preventives

Even the best engine has some **blowby**, which is the collection of combustion gases and byproducts that slip past the piston rings and into the crankcase. The crankcase ventilation system removes most of these combustion byproducts when they are vapors, but some water and acids condense from the vapors and settle in the oil in the crankcase. Rust and corrosion preventives counteract the harmful effects of these acids and water.

### Detergent-Dispersant Additives

Water, acids, and some unburned fuel from blowby gases will mix with cold oil in the engine and form **sludge**, figure 5-2. **Detergent-dispersant** additives reduce sludge and varnish formation by keeping sludge particles suspended in the oil so that they will be removed when the oil and filter are changed.

### Foam Inhibitors

Oil in an engine is continuously churned by moving parts, particularly the crankshaft counterweights and lower ends of the connecting rods. Like any liquid, oil will mix with air when it is churned and create foam. Foam, however, does not lubricate moving parts. Motor oil, therefore, contains foam inhibitors to reduce its formation and to dissipate any foam that does form.

### Viscosity Index Improvers

The tendency of a liquid to resist flowing is its **viscosity**. High-viscosity liquids are thick and flow slowly. Low-viscosity liquids are thin and flow easily. The viscosity of a motor oil is critical because it must flow easily under widely varying temperature and pressure conditions. But it must not flow so easily that it does not maintain a lubricating film on moving parts. Viscosity index (VI) improvers are oil additives that maintain uniform viscosity under all operating conditions. VI improvers also extend the viscosity range of a motor oil. A 10W-40 multigrade oil usually has more viscosity improvers than a 20W-40 oil. Oil with a lower viscosity base stock requires more VI improvers than oil with a higher viscosity base. You will learn more about viscosity ratings later in this section.

### Pour Point Depressants

The **pour point** is the lowest temperature at which an oil will pour, or flow readily. All oils contain some paraffin, or wax. At low temperature, the paraffin solidifies into wax crystals, which stick together and reduce the oil's ability to flow. Pour point depressants are additives that coat the wax crystals and keep them from sticking together. This lowers the pour point so that the oil will flow at lower temperatures.

### Antiwear and Extreme Pressure Additives

These additives improve an oil's ability to form a film on close-fitting parts under high pressure. They reduce wear and scuffing on parts such as cam lobes, valve lifters, and gears.

### Wetting Agents

These additives help oil to maintain a lubricating film on high-temperature surfaces.

### Friction-Reducing Additives

To improve fuel economy, many oils have friction-reducing additives to lower friction between moving parts. These additives may be solid materials, such as graphite or molybdenum compounds, or they may be oil-soluble organic compounds.

### Additive Precautions

Because motor oil chemistry is a sophisticated science, many oil manufacturers advise against the use of oil additives that are sold separately from oil and mixed in the engine. Additionally, many mechanics and petroleum engineers prefer not to mix oil brands within an engine. While mixing brands of oil of the same viscosity and service classification will not cause serious engine damage, the additives in different brands may not blend with each other to provide the best lubrication and engine protection.

## MOTOR OIL CLASSIFICATIONS

While all engines have the same basic lubrication needs, a diesel truck engine operates under different conditions than a gasoline passenger car engine, for example. Therefore, oil companies make different kinds of oil for the *specific* needs of different engines. The two classification systems for motor oil are:

- API service classifications
- SAE viscosity ratings

Additionally, many oil companies are rating modern oils according to their energy-conserving abilities, and some oil refiners are producing blends of "synthetic" oils.

### API Service Classification

The American Petroleum Institute (API) has standardized a series of **API Service Classifications** for motor oils. The oil is used in a series of laboratory engines to determine its service classification; its resistance to wear, oxidation, corrosion, and deposit formation is monitored. The API Service Classifications and related test methods are developed, published, and monitored in cooperation with the Society of Automotive Engineers (SAE) and the American Society for Testing and Materials (ASTM).

Oils for gasoline engines have service classifications beginning with an "S" (for *Service*) and followed by a letter "A" through "G" to indicate increasing performance levels. Oils for diesel engines have classifications beginning with "C" (for *Commercial*) and followed by a letter "A" through "E." Since the current service classification categories were introduced in 1970, gasoline engine oils designated SA through SG and diesel engine oils designated CA through CE have been defined.

**Gasoline Engine Service Classifications.** Service classifications SA and SB never were suitable for use in modern automobile engines and should not be used in any engine unless specifically recommended by the manufacturer. SA oils have no specific performance requirements. SB oils only moderately resist oxidation, scuffing, and corrosion.

From 1964 to 1988, API Service Classifications SC through SF were developed for increasingly more demanding engine protection requirements. Each succeeding classification had specific performance requirements and contained additives to protect against high- and low-temperature deposits, corrosion, wear, and rust. The test techniques for these oils are now considered obsolete, but some oil companies still market SE and SF oils. Oils that meet service classification SF can be used where SC, SD, and SE oils were originally recommended at the time of vehicle manufacture.

In 1988, service classification SG was developed for 1989 and later gasoline engines. It is defined as suitable for use in current gasoline engines in cars, vans, and light trucks operated under the manufacturers' recommended maintenance schedules. SG oils also include the performance properties of diesel service classification CC oils. (Some gasoline engine manufacturers also require diesel CD oil properties for their engines.) SG-quality oils provide better control of engine deposits, oil oxidation, rust, corrosion, and engine wear than previous categories. SG oils can be used in engines for which SE, SF, SE/CC or SF/CC oils were originally recommended.

### METRIC EXCHANGE

It's always best to use the "right" tool for any job, and this includes using metric wrenches on metric nuts and bolts. However, if a situation arises where your metric wrenches aren't available, there are some inch-sized wrenches that will fit metric fasteners closely enough to use in an emergency. The interchangeability works the other way around, too—some metric wrenches will fit inch-sized nuts and bolts. In most cases, though, the interchange works only in one direction. A 3/8-inch wrench will work on a 9-mm bolt, but it takes a 10-mm wrench to fit a 3/8-inch bolt head. Here are the wrench swaps that you can use to get out of a bind on some occasions.

**Customary to Metric**

| *Wrench* | *Nut or Bolt Head* |
|---|---|
| 3/8 | 9 mm |
| 7/16 | 11 mm |
| 9/16 | 14 mm |
| 11/16 | 17 mm |
| 3/4 | 19 mm |

**Metric to Customary**

| | |
|---|---|
| 10 mm | 3/8 |
| 11 mm | 7/16 |
| 13 mm | 1/2 |
| 17 mm | 5/8 |
| 19 mm | 3/4 |
| 21 mm | 13/16 |

These interchanges usually fit within 1 to 0.5 millimeter or 0.040 to 0.020 inch, depending on the precision of the wrenches and fasteners. Again, it's better to use the correct wrench for a specific metric or customary fastener, but these interchanges will work safely for most low-torque applications. And, they beat relying on the universal adjustable speed handle.

Today, SG oil is the only S-category service classification for which there is an active test method. Earlier oils for which test methods are obsolete should only be used in engines when explicitly recommended by the manufacturer.

**Diesel Engine Service Classifications.** Diesel oils CA, CB, and CC were in common use from the 1940's through the 1980's. CA-quality oils were suitable for light-duty, nonsupercharged diesels with high-quality fuels. They provided moderate protection against bearing corrosion and high-temperature deposits. CB-quality oils were suitable for moderate-duty, nonsupercharged diesels with lower-quality fuels and provided more protection than CA oils. Both CA and CB oils are not recommended for late-model automobile engines.

CC-quality oils were introduced in 1961 and were suitable for use in naturally aspirated, supercharged, and turbocharged diesel engines in moderate to severe use. CC oils protect against high- and low-temperature deposits, rust, and corrosion in diesel and some gasoline engines. The current gasoline engine SG service classification includes the requirements of diesel CC-quality oils. The test methods for CA, CB, and CC oils are now obsolete.

Oils that meet service classification CD were first used in 1955. They are suitable for most supercharged and naturally aspirated diesel engines in high-speed, high-output use. These oils protect against wear, rust, corrosion, and deposits when using fuels of a wide quality range.

Service classification CD-II defines an oil for 2-stroke diesel engines that require highly effective control of wear and deposits. CD-II oils also meet all performance requirements of CD oils.

CE oils were developed for certain supercharged and turbocharged diesel engines built since 1983 and used in low-speed, high-load and high-speed, high-load conditions. CE oils also meet the performance requirements of CC and CD oils. The test methods for CD, CD-II, and CE oils are currently active.

The API developed the current service classification system in 1971 to replace the older service designation system. Some oils that met or meet "S" and "C" classifications were introduced before 1971. You may see original recommendations for older engines that refer to oils such as "MS" and "DG." Figure 5-3 compares the older service designations with the current service classifications. For practical purposes, you can use SG oils satisfactorily in older gasoline engines.

| *Service Classifications* | *Old Service Designations* |
|---|---|
| SA | ML |
| SB | MM |
| SC | MS (1964) |
| SD | MS (1968) |
| SE | None |
| SF | None |
| SG | None |
| CA | DG |
| CB | DM |
| CC | DM |
| CD | DS |
| CD-II | None |
| CE | None |

**Figure 5-3.** Current API oil service classifications have more stringent performance requirements than the older service designations.

**Multiple Service Classifications.** Most oil companies sell oils with multiple service classifications, such as SF/SG or SG, SF/CC, CD or SG/CD. These classifications indicate that the oil meets the requirements of all listed service classifications and is suitable for different kinds of engines operating under varying conditions. Some carmakers are very specific about service classification recommendations, particularly for diesel and turbocharged engines. Many carmakers specify oils that meet *both* designated "S" and "C" classifications. If a manufacturer calls for an oil with a dual service classification, do not use oil that meets only a single classification. To do so may void the engine warranty.

You may hear these multiple classification oils called "fleet oils." This refers to their use by large fleets that choose to operate different kinds of gasoline and diesel vehicles on a single motor oil.

**API Service Classification Symbol.** To provide uniform identification and all consumer information in one location, the API developed the service classification symbol, figure 5-4. The appropriate service classifications are shown in the upper part of the symbol. SAE viscosity grades are listed in the center. The words "Energy Conserving" may be printed at the bottom if the oil meets those requirements. You will see this symbol on many oil containers, but an oil company can use the symbol only with specific licensing approval of the API.

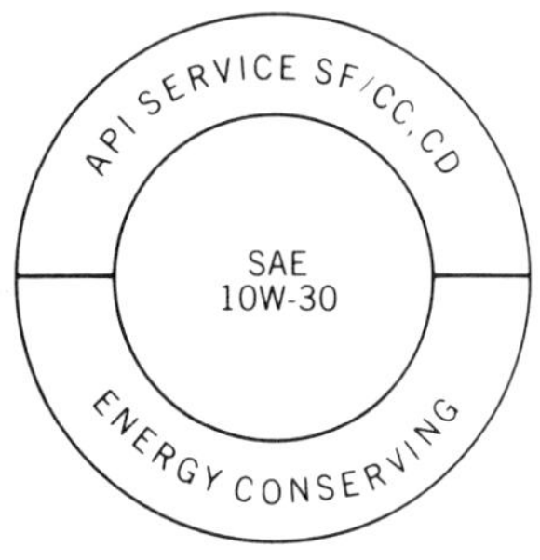


**Figure 5-4.** The API service classification symbol lists the oil service classification, viscosity, and energy-conserving features in one location.

## SAE Viscosity Grades

The Society of Automotive Engineers (SAE) established the viscosity grading system that indicates how easily an oil flows at a specific temperature. The **SAE viscosity grade** numbers for motor oil are 0W, 5W, 10W, 15W, 20W, 25W, 20, 30, 40, 50, and 60. Low-viscosity (thin) oils have low numbers.

High-viscosity (thick) oils have high numbers.

Oils tested for viscosity and pumpability at various temperatures from 23° F (−5° C) down to −31° F (−35° C) have a "W" suffix on their viscosity numbers to indicate their cold viscosity. The SAE cold-viscosity grades are 0W, 5W, 10W, 15W, 20W, and 25W. Oils tested at 210° F (99° C) have grades of 20, 30, 40, 50, or 60 to indicate their hot viscosity.

**Multiviscosity Oils.** Oils tested at both high and low temperatures have two or more viscosity numbers. An SAE 20W-50 oil has a viscosity of 20W at low temperature and a viscosity of 50 at high temperature. These oils are **multigrade** or **multiviscosity oils**. Oils with only one viscosity grade are **single-grade** or **straight-grade oils.**

**Carmakers' Viscosity Recommendations.** Most carmakers recommend several oil viscosity grades for their engines, depending on the air temperature range in which the vehicle is operated. These recommendations often overlap so that for a particular temperature range, several single-viscosity or multiviscosity oils may be suitable. Figure 5-5 shows recent oil viscosity charts from two major carmakers.

For many late-model engines, some carmakers advise *against* certain viscosities. For example, no carmaker recommends 10W-40 oil for diesel engines. GM also does not list 10W-40 as a recommended viscosity for gasoline engines, although other carmakers do. Ford advises against low-viscosity 5W-30 and 10W oils for sustained high-speed driving, even in very cold temperatures. For proper engine service, always check the carmaker's viscosity and service classifications thoroughly.

## Synthetic and Energy-Saving Oils

Since the mid-1970's **synthetic motor oils** have attracted attention and con-

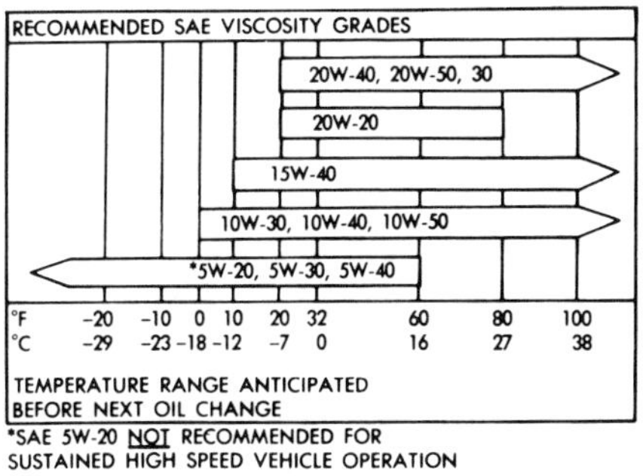


CHRYSLER

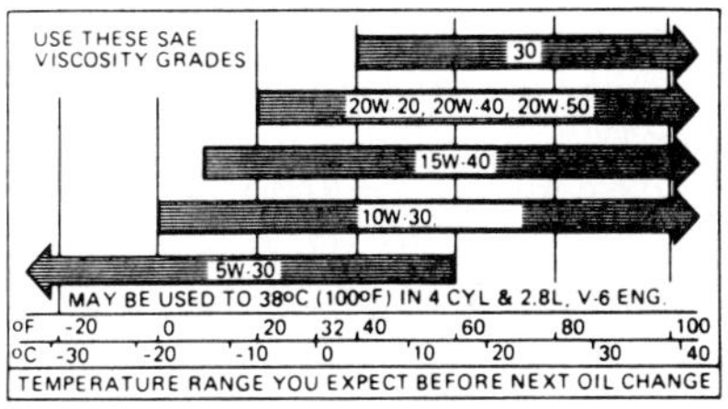


GM GASOLINE

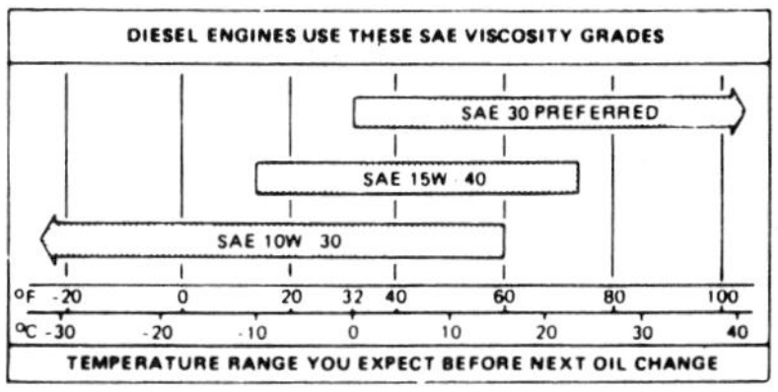


GM DIESEL

**Figure 5-5.** These oil viscosity recommendation charts show that several single or multiviscosity oils are acceptable for a given temperature range. (Chevrolet) (Chrysler)

troversy, but synthetic-base lubricants have been used in aircraft and other special engines for 40 years or more.

The term, "synthetic oil," is not precise because all motor oils have some nonpetroleum additives. To this extent, they are all slightly synthetic. Also, many oils labeled as synthetic are based on crude oil petroleum stocks. The differences between conventional and synthetic oils are in the refining process.

When conventional oil is refined, the molecular structure of the crude oil is not changed after the initial refinery cracking process. When synthetic oil is formulated, small molecules from crude oil and from other sources are combined into larger molecules with the lubricating properties needed for motor oil. After the synthetic stock is prepared, additives are blended with it as with conventional oils. These additives are specifically designed for antiwear and seal protection characteristics. Synthetic oils

### WHO INVENTED SMOG?

Back in the 1940's smog was something that Bob Hope joked about on the radio. It was funny then. On a clear day in Hollywood, you could see your hand in front of your face. Actually, nobody is certain who coined the word, "smog," by combining smoke and fog. We are certain, however, that smog had been around for almost a hundred years before scientists figured out its chemical nature and how it was formed.

In the 1920's, Donora, Pennsylvania, suffered an air pollution attack that killed 17 people and made thousands ill. In 1952, London suffered an industrial air pollution incident that killed 4,000 people. Neither of these air pollution incidents could be blamed on automobile pollution, however.

It wasn't until the late 1940's and early 1950's that the pioneering research of Dr. Arie Haagen-Smit proved the link between automobiles and smog. Dr. Haagen-Smit was a leader in the early research on air pollution and the first chairman of the California Air Resources Board. His work established that the smog that plagued Los Angeles was caused by a photochemical reaction between sunlight and hydrocarbons (HC) and oxides of nitrogen (NOx) in the atmosphere. He further proved that the automobile was a major source of HC and NOx emissions.

We are not sure if Dr. Haagen-Smit should be given the dubious recognition as the man who invented smog, but it was because of his research that engineers were able to attack the sources of air pollution and give us the clean-running, almost pollution-free cars we have today.

usually have different additive combinations and percentages than conventional oils.

The API and the SAE have not yet set standards for synthetic oils, but many manufacturers rate their synthetic oils by service classifications and viscosity grades. Among some of the performance claims that manufacturers make for synthetic oils are the following:

- They last up to 50,000 miles or more, compared to the 6,000- to 12,000-mile change intervals that carmakers recommend for conventional oils.
- They protect an engine over extreme temperature ranges. This is usually because of a lower oxidation rate at high temperatures and their higher viscosity index.
- They reduce friction and wear of moving parts and improve fuel mileage.

Synthetic oils tend to have lower volatility and less viscosity change with temperature changes than conventional oils do, but engine wear tests have yielded conflicting results. Not all makers claim an extended drain interval for synthetic oils, and carmakers have been reluctant to accept any extended oil change recommendations.

To help save energy and improve fuel mileage, oil companies have developed oils that reduce friction between lubricated surfaces. Oil refiners do this by using a lower viscosity base oil with friction-reducing additives such as graphite or molybdenum disulfide, or they add oil-soluble organic compounds to the oil.

These oils usually have the words ENERGY CONSERVING or ENERGY CONSERVING II in the API service symbol on the container, figures 5-4 and 5-6. A standard issued in 1988 by SAE, API, and ASTM allows oil companies to label an oil as ENERGY CONSERVING if it increases fuel economy by 1.5 percent compared to a standard test oil of the same viscosity and service classification. An oil can be labeled ENERGY CONSERVING II if it improves fuel economy by 2.7 percent or more. These energy-conserving oils are rated by the same service classifications and viscosity grading systems as other oils.

## LUBRICATION SYSTEM PARTS AND OPERATION

The engine lubrication system has all the parts of a basic hydraulic system that you have already studied. It has a reservoir to hold the oil supply, a pump to develop pressure, and valves to control oil flow and pressure.

To lubricate, clean, and cool an engine, oil is circulated by pressure and splash methods to the moving parts. These are often considered as two separate circulation methods, but modern engines use both, figure 5-7. The oil pump develops system pressure and delivers oil to the crankshaft and camshaft bearings and to valve train parts in the top of the engine. Oil passes through the bearing surfaces of the crankshaft, camshaft, and rocker arms and is splashed onto cam lobes, valve lifters, cylinder walls, and other moving parts.

The principal parts of the lubrication system, figure 5-8, are:

- The reservoir, or oil pan
- The pump and oil pickup
- The pressure relief valve
- The filter
- Galleries and lines to circulate oil through the engine
- Oil level and pressure indicators
- The crankcase ventilation system

### Oil Reservoir

Most automobile engines have a **wet-sump oil system** in which oil is stored in a reservoir—the oil pan—at the bottom of the engine. The lowest part of the pan is the sump, figure 5-9, which holds the oil supply. The pump draws oil through the pickup, located in the sump, and circulates it through the engine. After the oil passes through the engine, it drains back to the sump where it cools.

Oil is added to the engine through an oil filler at the top of the engine. Oil capacity varies according to engine size and design. Most small 4-cylinder engines hold 3 or 4 quarts (or liters) of oil. Most V-6 and V-8 engines hold 4 or 5 quarts. Some diesel engines require 6 or 7 quarts because of their high operating temperatures and pressures.

Some high-performance engines have **dry-sump oil systems** in which the engine oil supply is stored in a tank separate from the engine. After it circulates through the engine, oil collects in a shallow sump at the bottom of the engine. An oil return pump, or scavenger pump, pumps the oil from the sump to the separate storage tank. The oil delivery (pressure) pump draws oil from the tank and circulates it

**Figure 5-6.** Many oil companies sell energy-saving oils that help to improve fuel mileage.

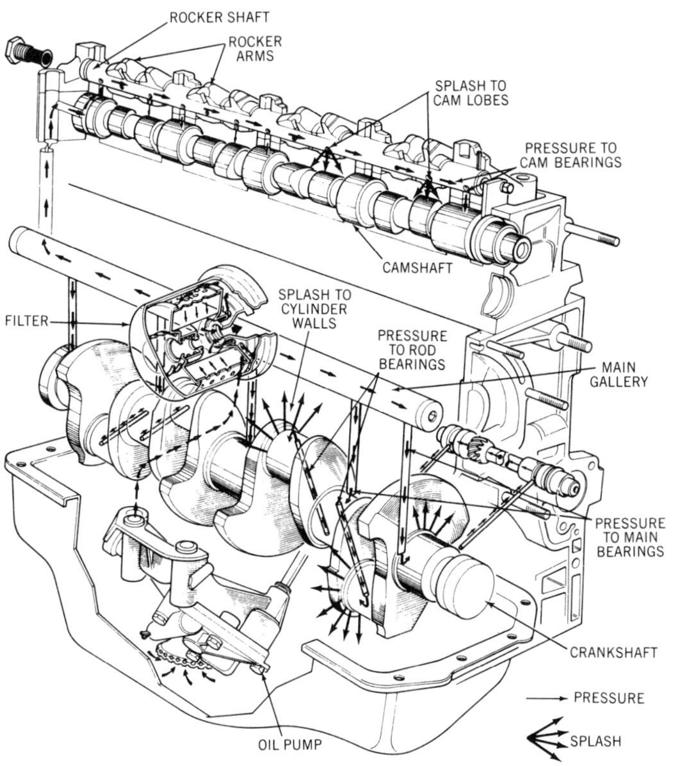


**Figure 5-7.** Modern engines use both pressure and splash lubrication.

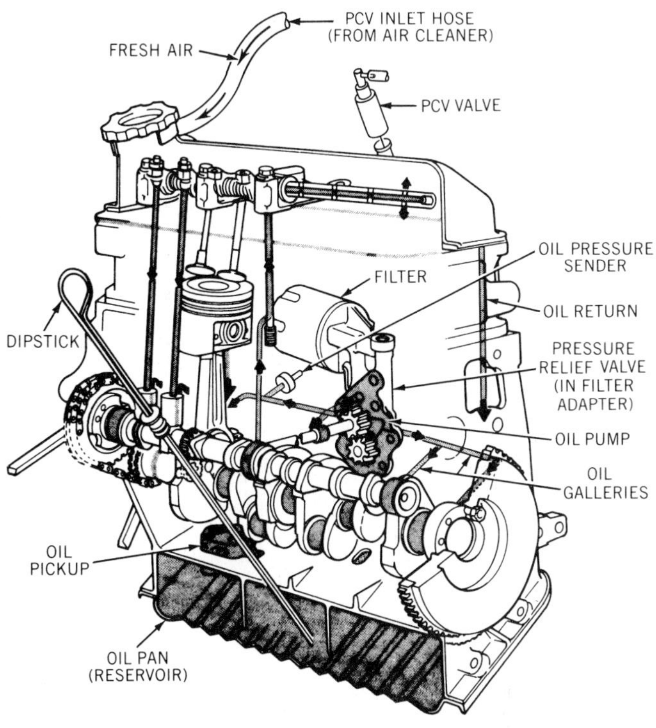


**Figure 5-8.** All engine lubrication systems have these principal parts. (Chrysler)

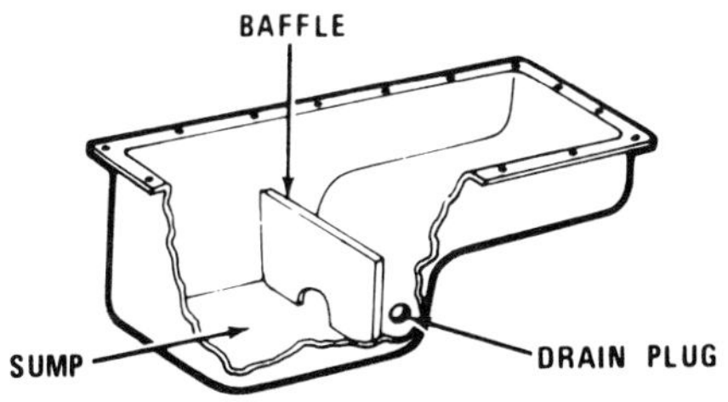


**Figure 5-9.** The low part of the oil pan is the sump. Many oil pans have baffles to prevent oil sloshing. (Chrysler)

through the engine. Porsche 6-cylinder engines, figure 5-10, are some of the most common examples of dry-sump oiling systems.

Oil in the pan of a wet-sump system will slosh from end to end or side to side under hard braking, acceleration, or cornering. This can momentarily uncover the oil pump pickup and starve the pump, figure 5-11. Most oil pans have baffles to minimize this sloshing. Some high-performance engines have special plates, called windage trays, below the crankshaft to keep the crankshaft from churning the oil and creating foam.

Dry-sump systems provide a constant supply of oil to the engine under all driving conditions. Dry-sump systems also allow somewhat better filtering and cooling of the oil than wet-sump systems do, but they are more complicated and expensive. Wet-sump systems are simple and quite satisfactory for most passenger car engines.

## Oil Pump and Pickup

The oil pickup is located in the sump, below the minimum oil level, but not so low that it picks up sludge and other contamination from the bottom of the pan, figure 5-12. The pickup has a screen over the inlet to strain out heavy sludge and dirt, but the screen is not the primary filter for the oil system. Most pickups have a bypass valve that opens if the screen gets completely clogged. This allows unstrained oil into the pump but still provides lubrication for the engine. A tube connects the pickup to the pump.

The oil pumps of most inline and V-type overhead valve engines are driven by the camshaft, either directly by gears or by an extension of the distributor shaft, figure 5- 13. The oil pumps of most overhead cam engines are driven by a crankshaft gear or by an auxiliary shaft that is driven by the camshaft timing chain or belt, figure 5-14. The auxiliary shaft also may drive the distributor.

Most engines have gear-type or rotor-type oil pumps. However, some late-model 4-cylinder Ford engines have gear-and-crescent oil pumps.

A gear-type pump has two spur gears in a closed body, figure 5-15. One gear is driven by a shaft; the other gear meshes with the first as an idler. As the gears rotate, they create a low pressure at the pump inlet and draw oil through the pickup. The gears rotate away from the inlet, and the space between each gear tooth carries oil around the outside of the pump housing. As the teeth mesh near the pump outlet, they develop pressure on the oil and force it through the outlet.

A rotor-type pump has an inner and an outer rotor that look like a pair of

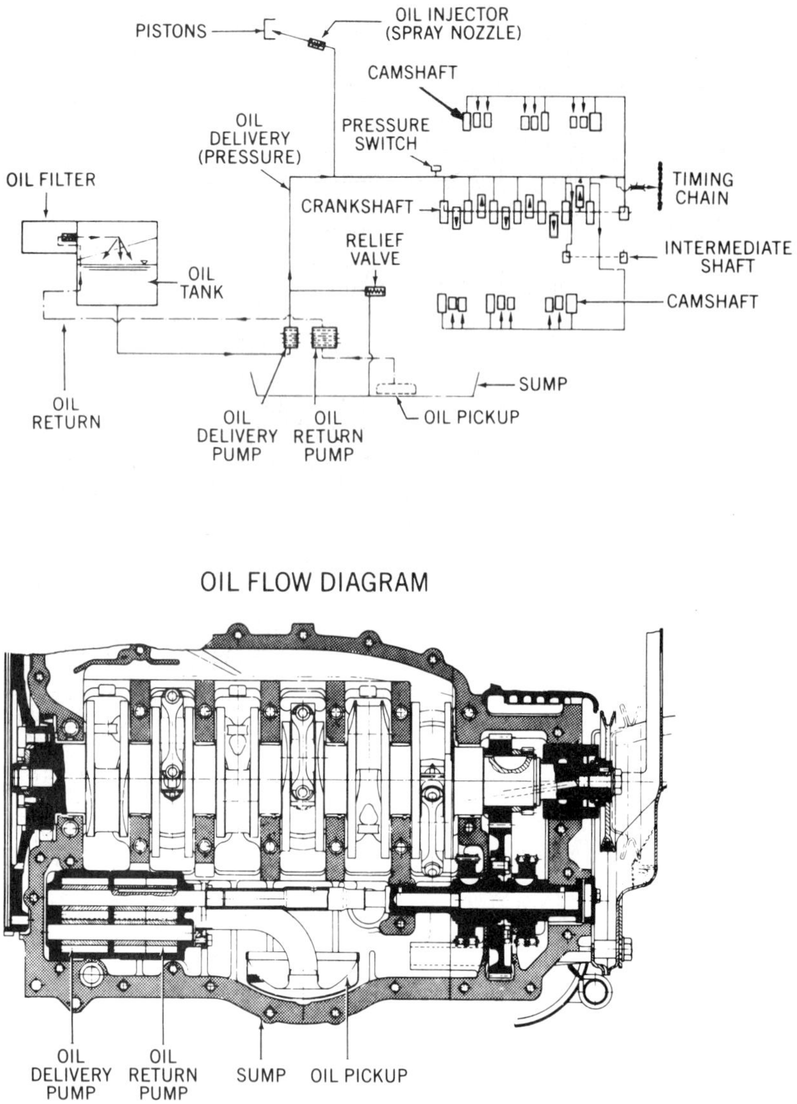


**Figure 5-10.** In a dry-sump oil system, the oil reservoir is a tank, separate from the engine. (Porsche)

internal and external gears, figure 5-16. The inner rotor is off center and driven by the crankshaft or camshaft. On one side of the pump body, the inner and outer rotors mesh closely. On the other side, they have more clearance. As the rotors turn, they open a pocket at the pump inlet. This expanding clearance creates low pressure that draws oil from the pickup. As the rotors continue to turn, they hold oil in the pocket and carry it toward the outlet. At the pump outlet, clearance between the rotors decreases to develop pressure on the oil and force it out the outlet.

Gear-and-crescent oil pumps are used on Ford 4-cylinder, 1.6-liter gasoline engines and some small 4-cylinder Ford and Toyo-Kogyo diesels. The inner drive gear is driven by the front of the crankshaft, figure 5-17. At one side of the pump body, the inner gear meshes with an outer gear. At the other side, the gears are separated by a crescent-shaped area of the pump body. The pumping action is similar to that of a rotor pump or a gear pump. As the gears rotate, they create low pressure at the point where they separate at the pump inlet. The rotating gears carry oil through the space around the crescent and then to the pump outlet. At the outlet, clearance between the gears decreases to develop pressure on the oil and force it out of the pump.

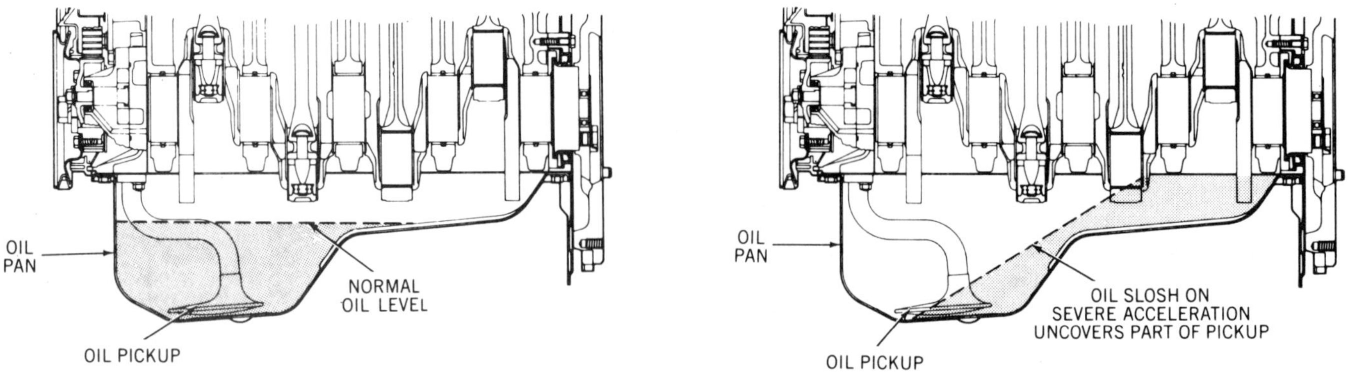


**Figure 5-11.** Hard acceleration, stopping, and cornering can cause the oil to slosh away from the pickup. Most oil pans have baffles to prevent this. (Chrysler)

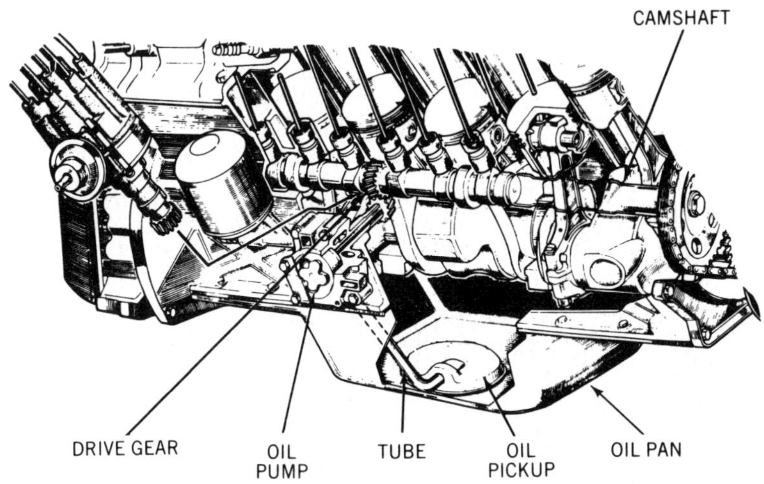


**Figure 5-12.** The oil pickup is located in the bottom of the sump and connected to the pump by a tube. (Chrysler)

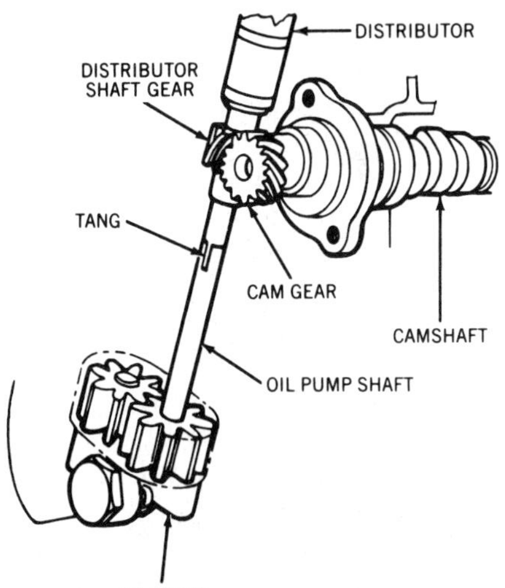


**Figure 5-13.** Driving the oil pump with an extension of the distributor shaft is the most common method used in overhead-valve engines. (Chrysler)

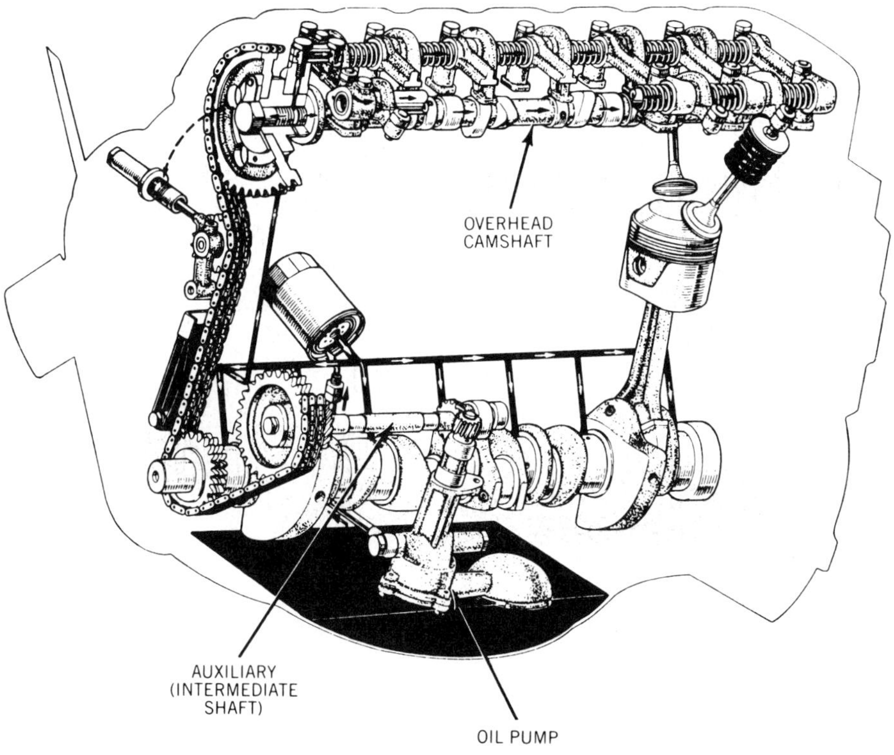


**Figure 5-14.** The oil pump in this overhead-cam engine is driven by an auxiliary engine shaft. (Toyota)

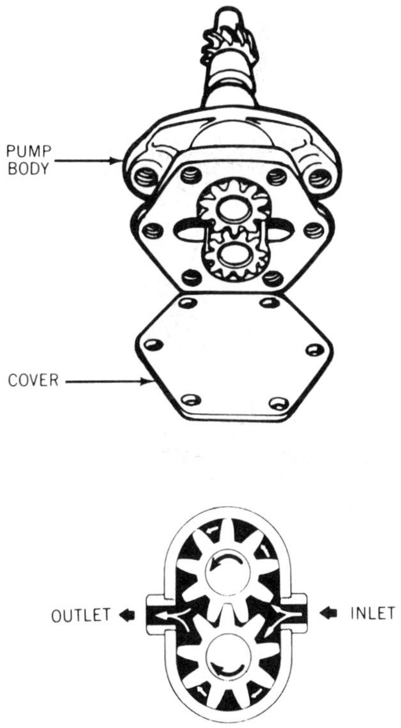


**Figure 5-15.** Typical gear-type oil pump.

Rotor, gear, and gear-and-crescent pumps are all **positive displacement pumps**, which means that each revolution delivers the same volume of oil regardless of speed. As pump speed increases, however, pressure increases in the system. System pressure must be high enough to deliver enough oil to all bearings and other parts, but it cannot be too high or it will damage bearings and blow out oil seals.

## System Pressure and Pressure Relief Valve

Three factors determine oil system pressure:

**1.** The speed of the pump

**2.** The viscosity of the oil (Cold oil is more viscous than hot oil.)

**3.** The controlled leakage out of bearings, oil nozzles, and metering orifices in the oil galleries and lines.

Oil pumps deliver 12- to 15-psi (83- to 103-kPa) pressure to the system at idle with a warm engine. However, when the engine and the oil are cold, the oil is more viscous, and pressure is higher. As engine speed increases, so does oil pressure. But, as the oil warms up, it thins (becomes less viscous), and pressure drops. Most engines operate with average oil system pressure of 40 to 60 psi (275 to 415 kPa) at cruising and high speeds when warm.

Pressure within this range ensures full lubrication to all moving parts, but if pressure rises above this range, it can blow out oil seals or gallery plugs and actually gouge out soft bearing metal. Two methods are used to keep oil pressure within correct limits.

All pumps have a pressure relief valve, figure 5-18, which is a simple spring-loaded poppet or ball valve of the kind you have studied in chapter 2. The spring is calibrated so that as pump pressure rises above the desired limit, it forces the ball or poppet off its seat and allows some of the oil to return to the inlet side of the pump or to the sump. Enough oil flows through the system, however, to maintain a constant pressure.

Much of the time, the relief valve is closed, and pressure is regulated by the controlled leakage from bearings and oil nozzles or metering orifices. The oil system has these controlled leaks so that fresh oil circulates continuously to lubricate, cool, and clean moving parts. The pump maintains pressure because the volume of all the leaks is less than the output volume of the pump. Most pumps deliver 3 to 6 gallons of oil per minute, depending on engine speed.

Engine bearing clearances are critical factors for maintaining correct oil pressure. Crankshaft main and rod bearing clearances are determined by the diameter and width of the journal and the kind of bearing material. Clearances typically range from 0.001 to 0.003 inch (0.025 to 0.075 mm). These clearances allow a continuous oil film between the shaft journal and the bearing when the engine is running. This supports the shaft and centers it in the bearing. The oil film is not formed by oil pressure, but by the turning shaft rolling up on a wedge of oil. The shaft almost acts like a pump to distribute the oil evenly around the bearing. Bearing clearance allows oil to pass through the bearing area and drain out the edge. If clearance is excessive, too much oil passes through the bearing, and pressure drops. If clearance is too tight, too little oil will circulate, and a complete oil film may

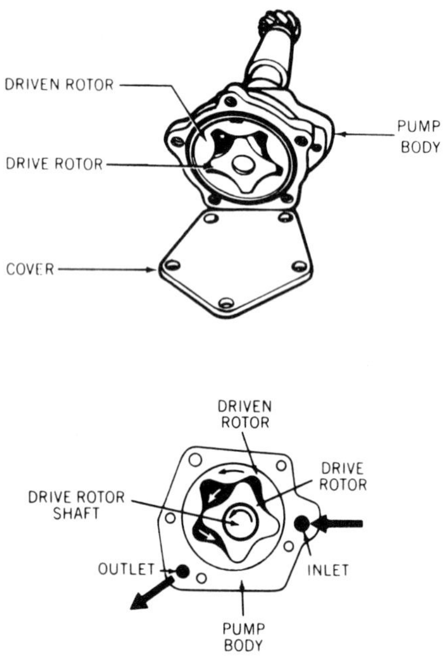


**Figure 5-16.** Typical rotor-type oil pump.

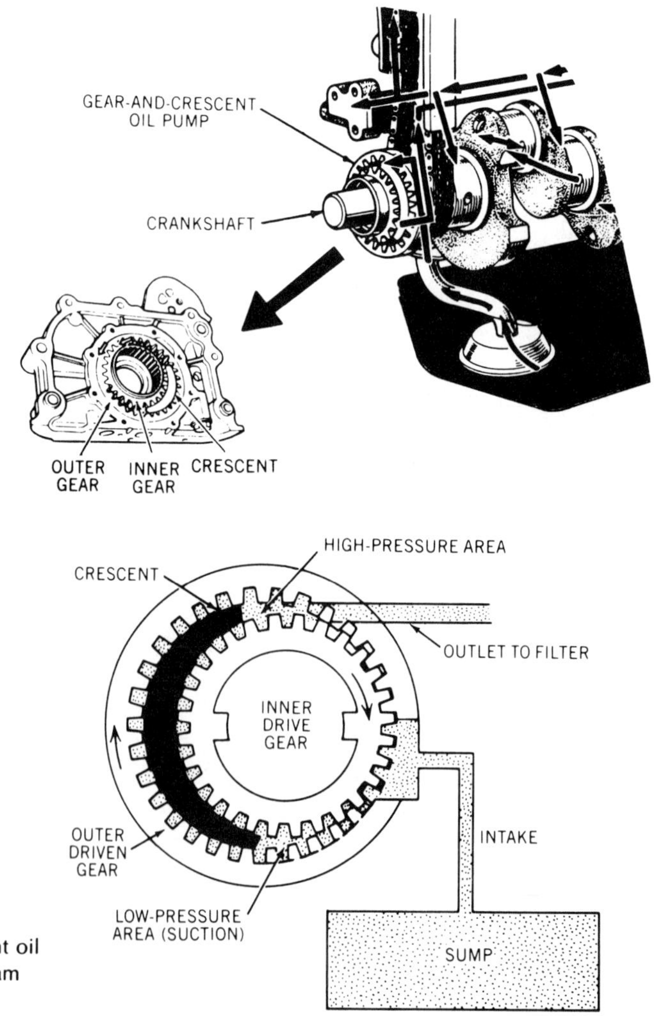


**Figure 5-17.** This gear-and-crescent oil pump is used in several overhead-cam engines.

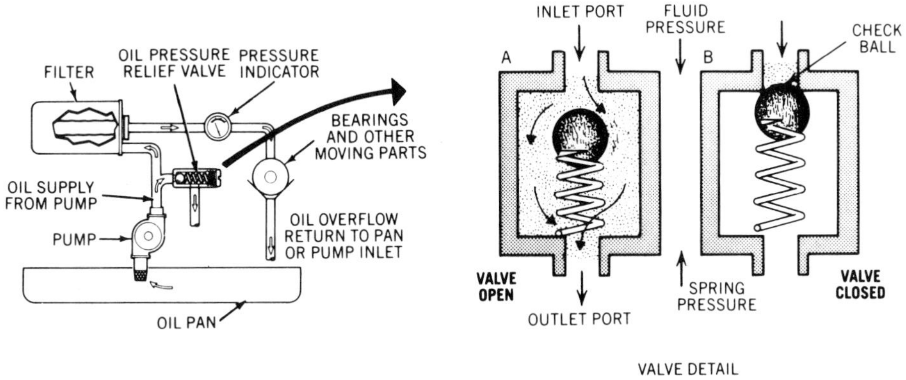


**Figure 5-18.** The oil pump relief valve opens when pressure rises above the desired limit. (Federal-Mogul)

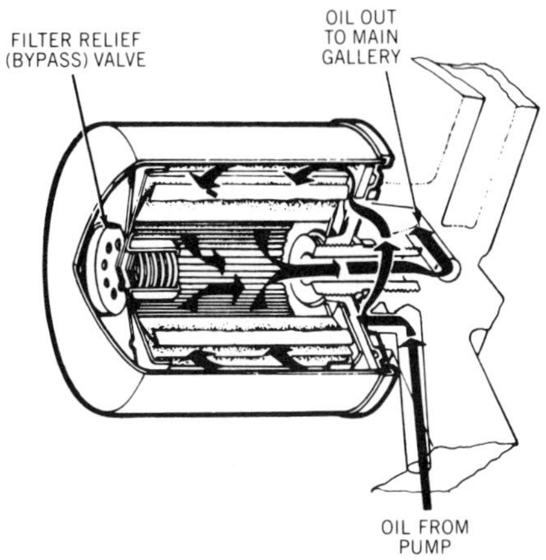


**Figure 5-19.** With a full-flow oil filter, all oil from the pump circulates through the filter before going to the engine. The filter relief valve opens if the filter gets clogged. (Chrysler)

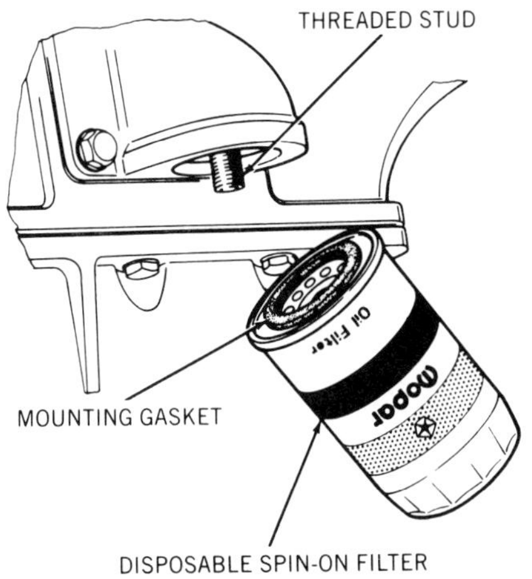


**Figure 5-20.** Most full-flow oil filters are screw-on, disposable cartridges. (Chrysler)

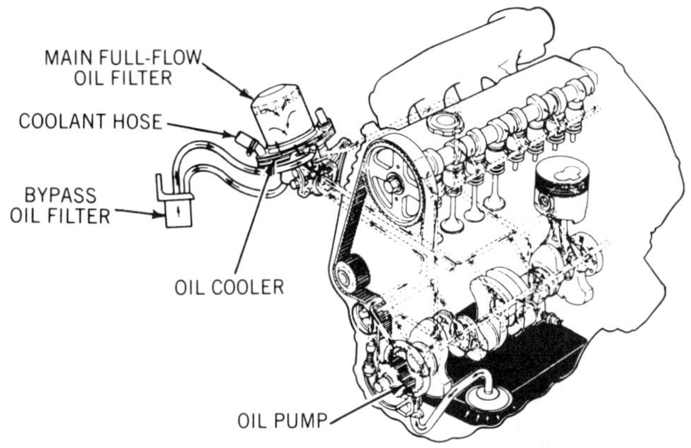


**Figure 5-21.** Bypass filters are used as auxiliary filters on some Ford diesel engines. (Ford)

not spread over the entire bearing. Bearings will overheat as a result. Oil pressure will rise, but the relief valve will open to limit system pressure.

## Oil Filter

All modern engines have **full-flow oil filters**, figure 5-19, which means that all oil flow from the pump goes to the filter before it circulates through the engine. The filter is made of paper or cloth fibers that trap dirt particles but allow liquid to pass through. Most filters are self-contained disposable cartridges that screw onto a fitting on the outside of the engine, figure 5-20.

As a filter collects dirt, oil flow through it can get restricted. To keep a clogged filter from blocking oil flow and harming the engine, a bypass valve is built into the filter or the filter housing on the engine, figure 5-19. The bypass valve opens when the pressure drop across the filter exceeds 5 to 15 psi (34 to 104 kPa). This allows unfiltered oil to circulate through the engine. The engine will not lack lubrication, but it may be damaged by dirt and other contaminants in the oil.

Before the 1960's, many engines had **bypass oil filters**, in which only part of the oil flow circulated through the filter. An oil line delivers oil from the main engine oil gallery to the bypass filter, and another line returns oil from the filter to the engine sump. Some late-model diesel engines have auxiliary bypass oil filters, along with the main full-flow filter, figure 5-21. The bypass filter provides additional oil filtration; and if the main filter gets clogged, the bypass filter ensures that some of the oil is filtered.

## Oil Galleries and Lines

After passing through the filter, oil circulates through passages, called oil galleries. Oil galleries are cast or drilled through the block, the head, and the crankshaft. Most inline and some V-type engines have one main gallery; other V-type engines have two. Smaller passages carry oil from the main gallery to the crankshaft main bearings and to the cam bearings if the cam is in the block. In an overhead cam engine, a vertical gallery or oil line carries oil to another gallery in the head

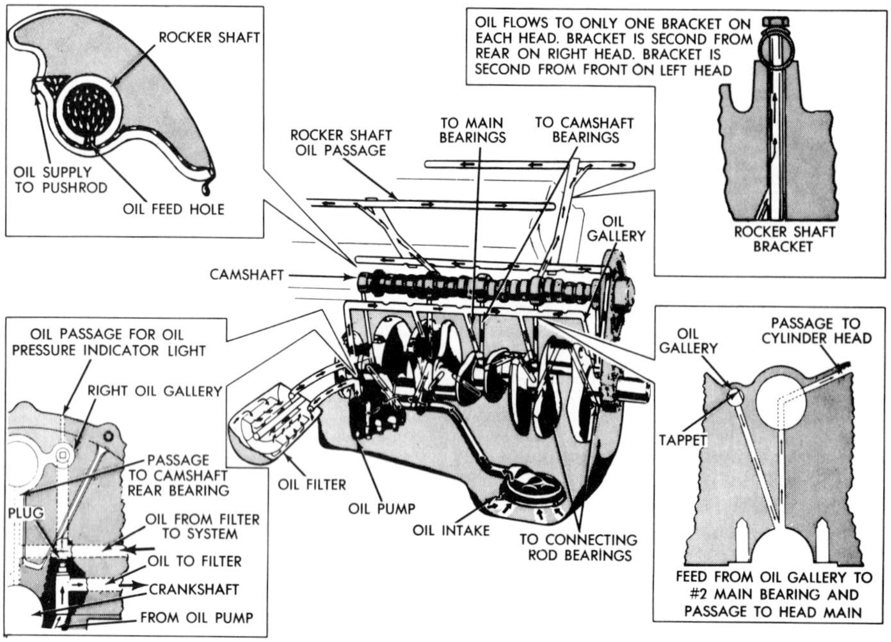


**Figure 5-22.** Typical lubrication system for a V-type overhead-valve engine. (Chrysler)

that distributes it to the cam bearings. Some engines also have separate oil lines made of tubing to carry oil to certain areas.

Figure 5-22 shows the oil circulation system of a typical V-type overhead-valve engine. Figure 5-7 shows the lubrication system for a 4-cylinder, overhead-cam engine. Figure 5-23 is a chart that traces the oil flow through a typical engine. Refer to these illustrations as you study the following paragraphs.

Oil enters the crankshaft main bearings through holes in the bearing webs and flows through grooves in the bearings. Holes in the main journals carry oil to passages drilled in the crank-

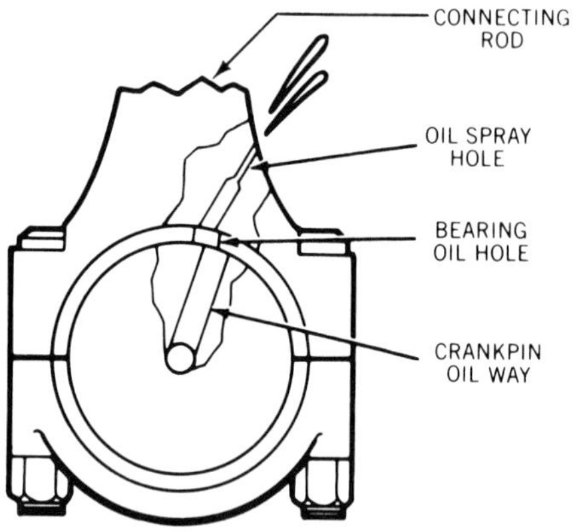


**Figure 5-24.** This hole in the connecting rod sprays oil from the crankshaft onto the cylinder walls. (Federal-Mogul)

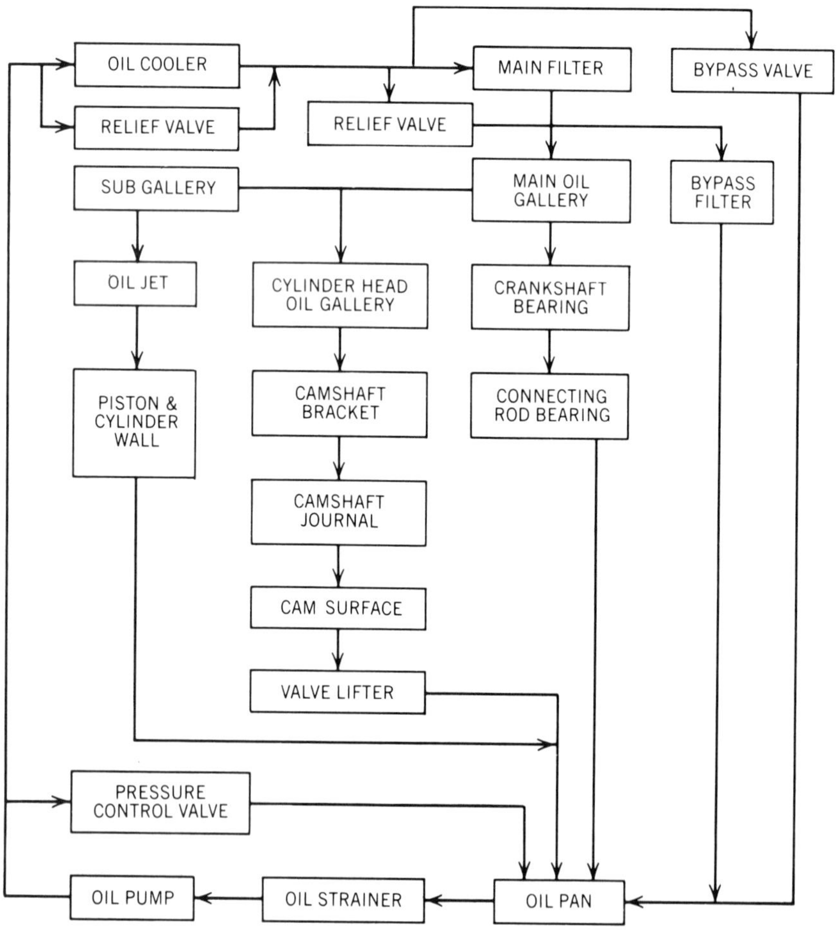


**Figure 5-23.** This oil circulation chart traces oil flow through an engine with a complex lubrication system. (Ford)

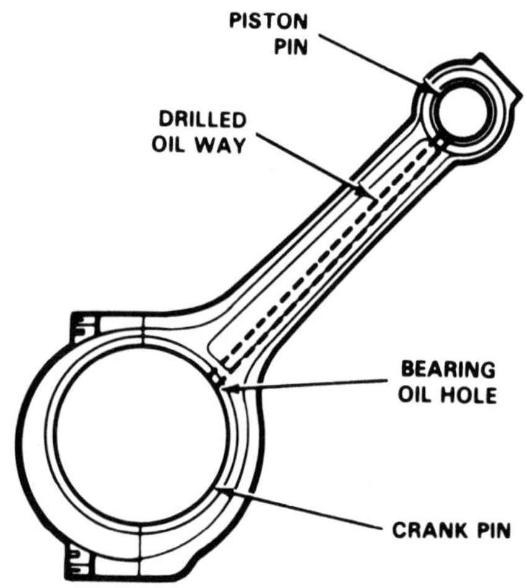


**Figure 5-25.** A drilled passage in this connecting rod carries oil from the crankshaft to the piston pin and upper cylinder area. (Federal-Mogul)

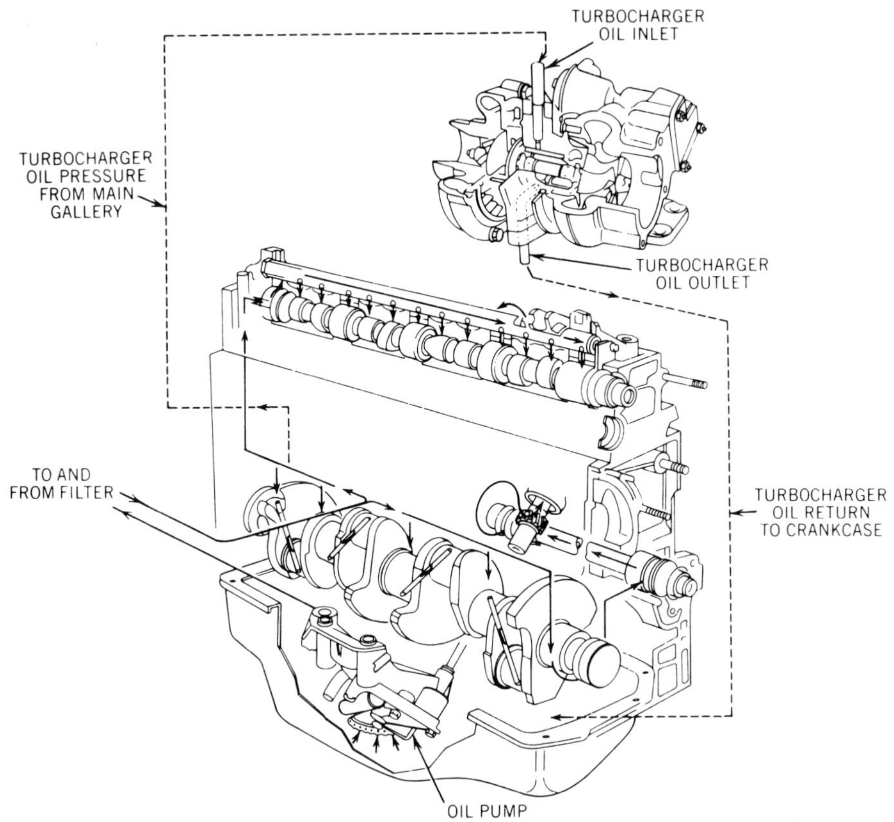


**Figure 5-26.** Turbochargers require their own oil supply for shaft bearing lubrication.

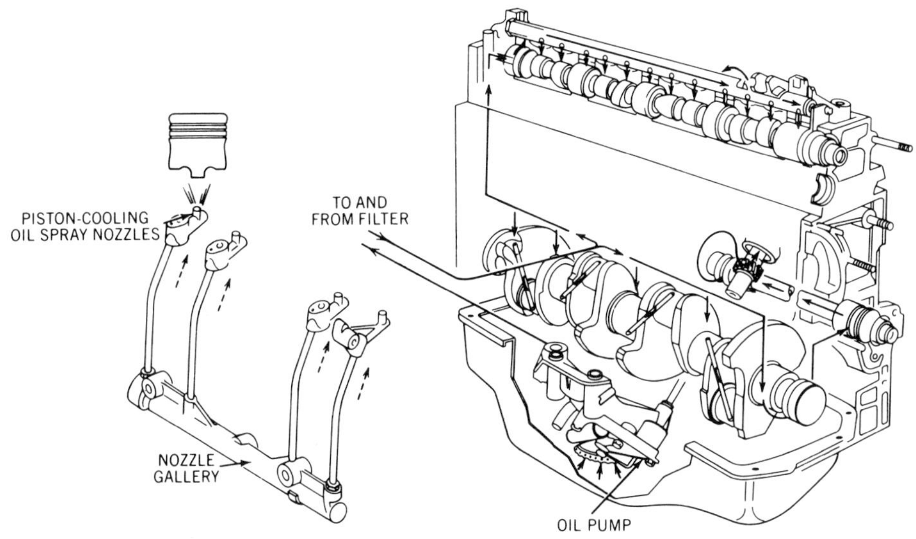


**Figure 5-27.** Many diesel engines have nozzles that spray oil into the bottoms of the pistons to aid piston cooling.

shaft. These passages route oil to the rod bearings. Some connecting rods have spray holes in the upper half of the big end to spray oil on the cylinder walls for piston and ring lubrication, figure 5-24. Other engines rely on spray from the edges of the crankshaft journals to do this. Some connecting rods are drilled from the big end to the small end to carry oil to the piston pin and upper cylinder area, figure 5-25.

Oil also flows from the main gallery through passages to lubricate the valve lifters and cam lobes if the cam is in the block, figure 5-22. In an ohc engine, cam lobe and cam follower lubrication is provided by passages from a gallery or line in the head, figure 5-7. If an engine has hydraulic valve lifters or cam followers, the camshaft oil supply also keeps them pressurized. Some overhead-valve engines have hollow pushrods that carry oil to the rocker arms. Other engines have galleries or passages in the block and head that carry oil to hollow rocker arm shafts, figure 5-22. These passages sometimes run through an oversize bolt hole for a head or rocker shaft capscrew. Lubrication for timing gears and distributor drive gears often comes from a nozzle or orifice that sprays oil from a gallery onto the gears.

Turbochargers have critical lubrication requirements because turbocharger shafts turn at speeds above 10,000 rpm, and the bearings are exposed to the high temperatures of exhaust gases. Therefore, turbocharged engines have an oil line from the main oil gallery to the turbocharger housing, figure 5-26, and a return line to the engine sump. This ensures an adequate supply of oil to the turbocharger shaft at all times.

Most diesel engines have oil nozzles that spray oil from a gallery onto the bottoms of the pistons or into oil ports in the pistons, figure 5-27. This extra oiling improves upper cylinder lubrication and helps to cool the pistons, which run much hotter than those of a gasoline engine.

When we think of proper engine lubrication, the crankshaft rod and main bearings often are the first things that come to mind. Actually, the parts with the most critical lubrication needs are the cam lobes and valve lifters. High valve train pressures and the reciprocating action of cam lobes and lifters cause these surfaces to scuff and wear faster than others without proper lubrication. Moreover, the lobes and lifters are not surrounded by a bearing surface that will hold some oil when the engine is off. Cam lobes are not lubricated by positive pressure but rely on oil splash from connecting rods or from gallery orifices.

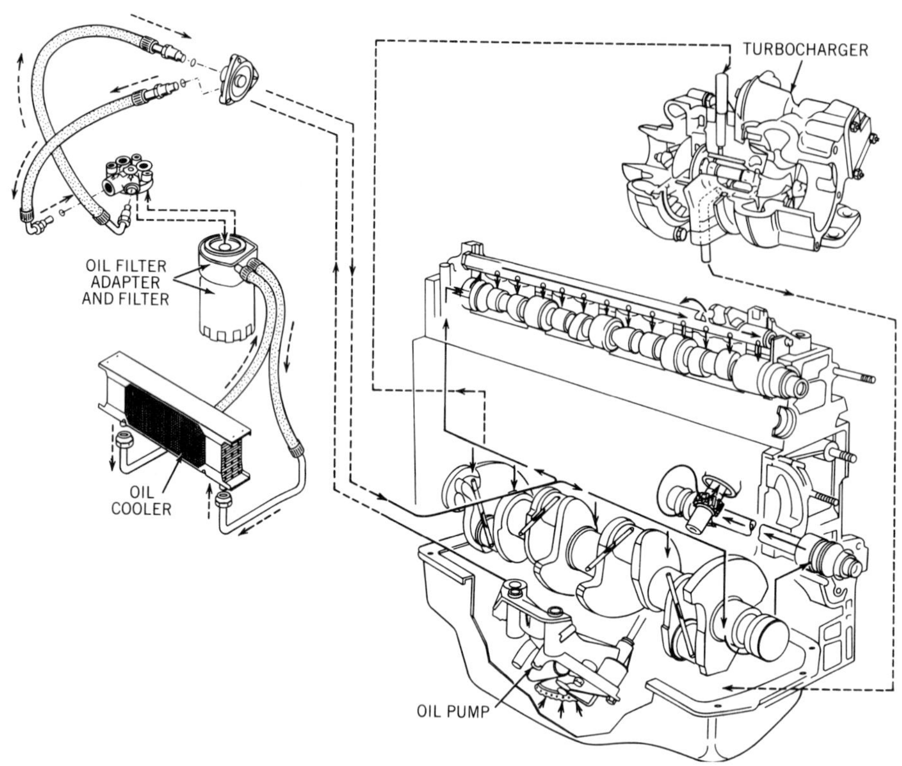


**Figure 5-28.** Oil coolers use air circulation or engine coolant to cool the oil.

oil warning lamp or controls an indicator gauge.

**Oil Pressure Gauge.** Some cars have oil pressure gauges that are operated electrically or directly by oil pressure. A mechanical gauge is a bourdon tube gauge connected to the main oil gallery by small-diameter tubing. Oil pressure in the gauge tube moves the gauge needle to indicate pressure on the dial, figure 5-31.

An electric oil pressure gauge uses a diaphragm pressure sensor, or sending unit, that is tapped into the main oil gallery. The sending unit is connected by wire to the gauge. Changing pressure against the diaphragm moves a variable resistor that changes current flow through electromagnetic coils in the gauge, figure 5-32. Changing electromagnetism of the coils moves the gauge needle in proportion to oil pressure.

### Oil Coolers

Many diesel engines, turbocharged engines (gasoline and diesel), and air-cooled engines operate at temperatures high enough to overheat the oil. Overheated oil cannot properly lubricate and cool an engine and is subject to accelerated oxidation and varnish buildup. Therefore, many of these engines have oil coolers. The oil cooler usually is connected to the system near the pump outlet so that oil is under full system pressure as it flows through the cooler. Oil coolers use either airflow or engine coolant circulation to cool the oil, figure 5-28.

### Oil Level and Pressure Indicators

To ensure that an engine is properly lubricated, car owners and mechanics must be able to check the oil level, or capacity, and the oil pressure.

**Oil Level Indicators.** The dipstick is the oldest, simplest, and most universal oil level indicator. It is simply a metal measuring rod that extends through a tube from outside the engine into the pan, figure 5-29. The end of the dipstick is graduated with marks that show the minimum and maximum oil levels for a particular engine. The difference between the two marks is usually about one quart or liter.

The dipstick gives you the most accurate oil level indication when the car is on level ground. If the car is not level, oil will flow to one side or end of the pan, and the dipstick may indicate more or less oil than is actually in the engine. If the engine is running or has just been shut off, some oil will be in the galleries and the top of the engine, and the dipstick will indicate less oil in the pan. For accurate measurement, the engine should be warm but shut off for 3 to 5 minutes to allow oil to drain from the lines into the pan.

Some late-model cars, particularly some Renault and Ford products, have electronic oil level indicators on the instrument panel. A float-type sensor in the oil pan, figure 5-30, or a sensor on the end of the dipstick lights a low-

**Oil Pressure Warning Lamp.** On many cars, an instrument panel warning lamp lights when oil pressure drops below minimum. A warning lamp uses a sending unit similar to the kind used with an electric gauge. This sending unit, however, opens and closes electrical contacts rather than varying current flow to the gauge, figure 5-33. The contacts close to light the lamp when oil pressure is low.

## CRANKCASE VENTILATION

The crankcase ventilation system is an integral part of the engine lubrication system, but it also is a major emission control system. Two conditions of engine operation require crankcase ventilation.

You know that as pistons move up in the cylinders, they compress air in the combustion chambers. As they move down, they also compress air in the crankcase. If pressure were allowed to build up in the crankcase, it would work against the pistons on their downward strokes. It also would blow out oil pan gaskets and seals. There-

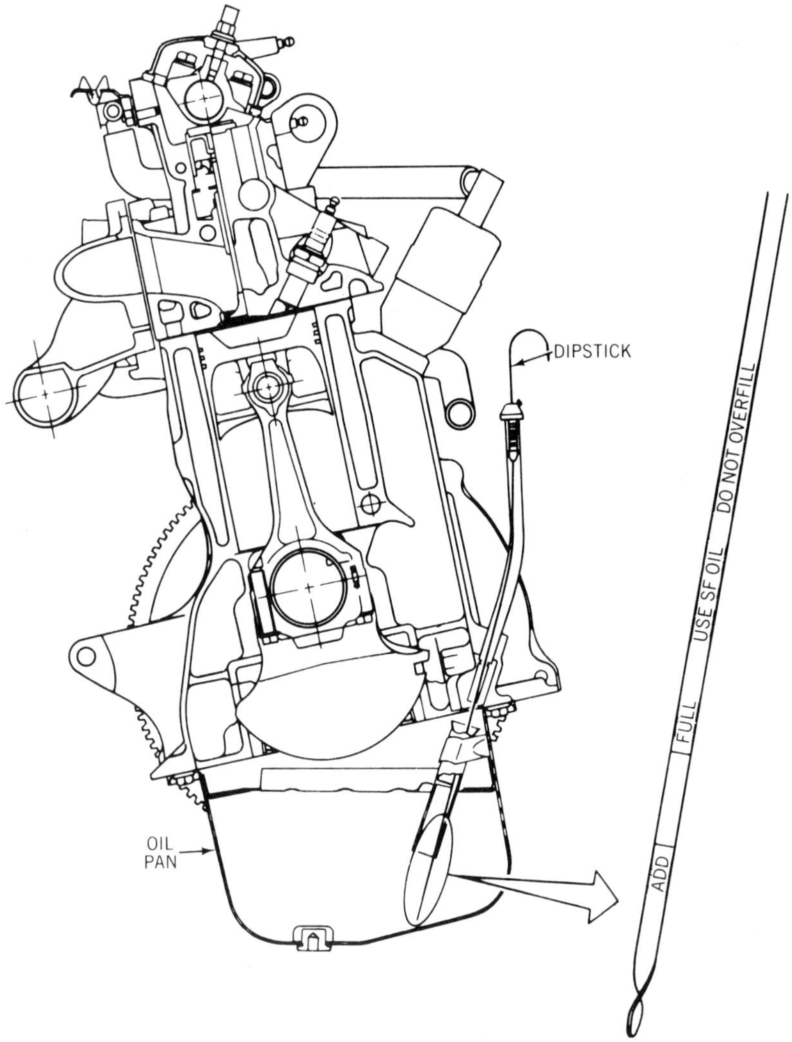


**Figure 5-29.** Typical dipstick installation and dipstick markings. (Chrysler)

fore, crankcase pressure must be relieved.

Even more importantly, some combustion blowby gases always leak past the piston rings and carry some exhaust, unburned fuel, and other combustion byproducts into the crankcase. These blowby gases not only increase crankcase pressure but they also contaminate the oil. Unburned fuel dilutes the oil and reduces its lubricating ability, and blowby gases also carry water vapor and acids into the crankcase. So, the crankcase must have a ventilation system that not only relieves excess pressure, but circulates fresh air to carry away blowby gases.

## Road Draft Ventilation

Before the 1960's, most cars had road draft systems to ventilate the crankcase. These consisted of a simple tube attached to the crankcase and bent so that its open end extended into the airstream under the moving car, figure 5-34. Airflow past the end of the tube created a vacuum that drew vapors out of the crankcase. Fresh air entered the crankcase through a vented oil filler cap or crankcase air filter at the top of the engine. Road draft ventilation, however, works only when a pressure difference exists between the crankcase and the end of the tube. At speeds below 25 mph, there is not enough airflow to create low pressure at the end of the tube to draw all of the blowby vapor out of the crankcase. At high

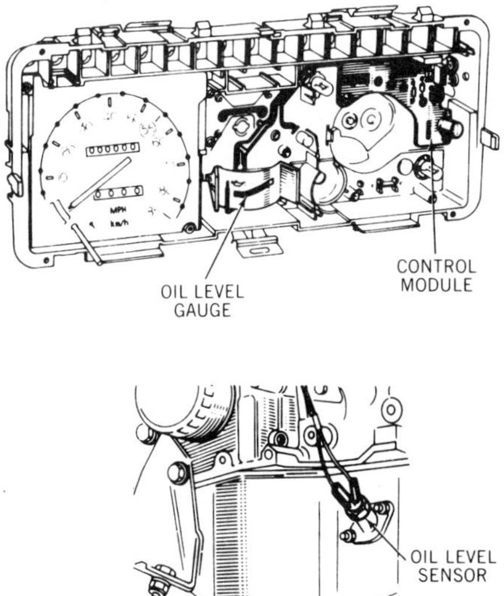


**Figure 5-30.** Some cars have oil level indicators on the instrument panel, controlled by a sensor in the oil pan.

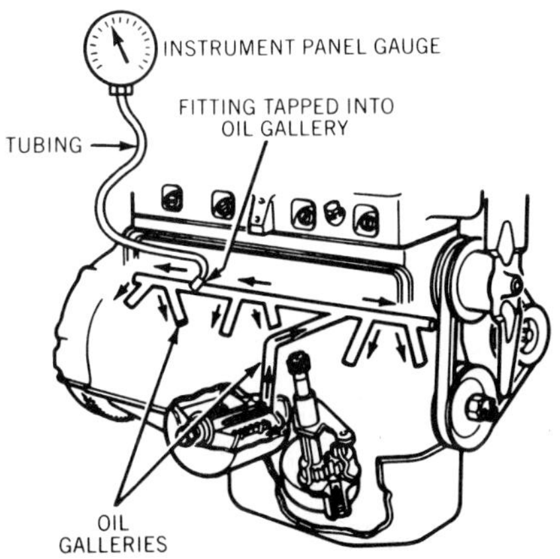


**Figure 5-31.** A direct-reading oil pressure gauge is connected by tubing to the main oil gallery of the engine. (Federal-Mogul)

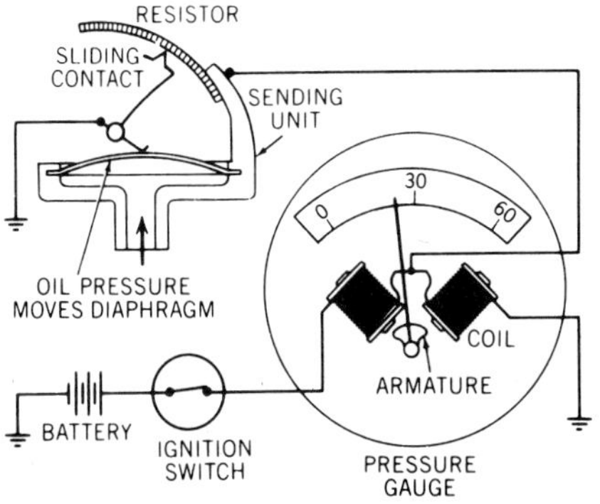


**Figure 5-32.** The sending unit for this oil pressure gauge is a variable resistor that changes current flow through the gauge coils in proportion to oil pressure.

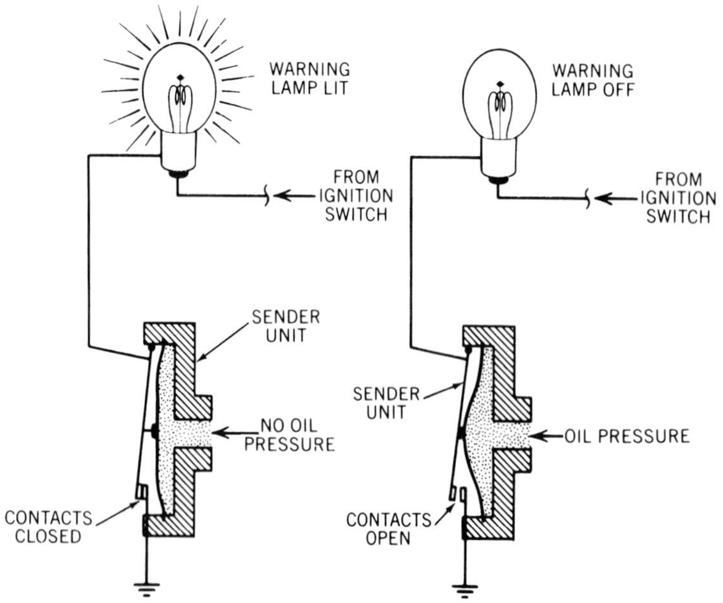


**Figure 5-33.** The sender unit for this warning lamp simply closes the lamp circuit when oil pressure drops below the minimum.

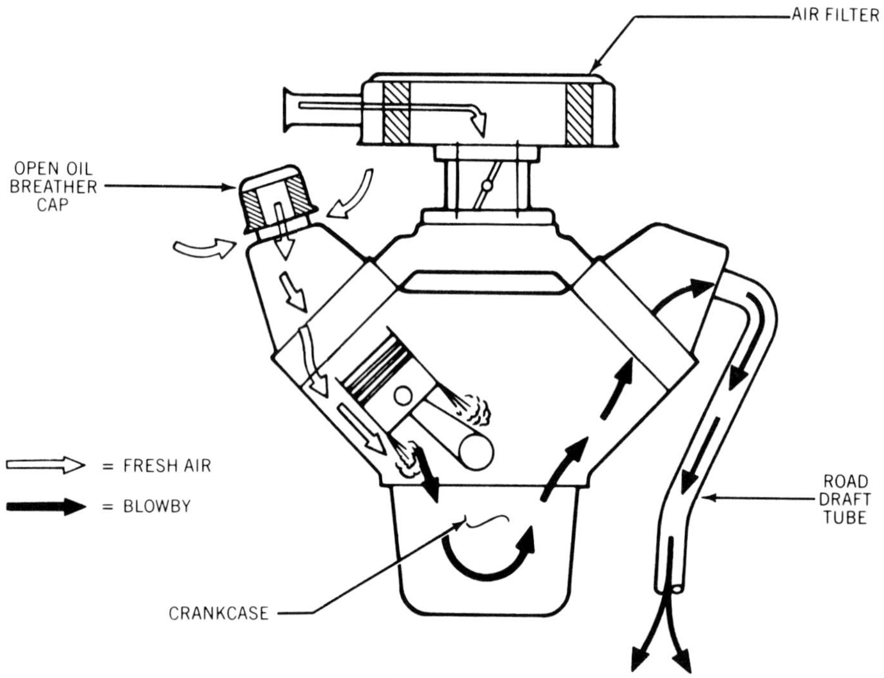


**Figure 5-34.** Road draft crankcase ventilation releases HC vapors to the atmosphere and contributes to air pollution.

vehicle speeds, the combination of high crankcase pressure and low pressure at the end of the tube may draw oil out of the engine.

Most importantly, blowby and crankcase vapors are principally hydrocarbons (HC), and the road draft tube releases them to the atmosphere where they are a major cause of air pollution.

## Open Positive Crankcase Ventilation (PCV) Systems

PCV systems overcome the problems of road draft ventilation by using intake manifold vacuum to draw vapors from the crankcase. This creates uniform, positive airflow through the crankcase whenever the engine is running. The vapors mix with the intake air and return to the combustion chamber.

PCV systems are classified as open or closed, according to their design. Open PCV systems appeared in large scale on 1961 cars in California and nationwide in 1963. They are divided into three types.

**Type 1 Open PCV System.** In a type 1 system, a hose with a PCV valve runs from the crankcase to the manifold. Fresh air enters the crankcase through a vented oil filler cap, figure 5-35. In normal operation, crankcase vapors are metered through the PCV valve and drawn into the manifold. If blowby volume exceeds the PCV valve flow capacity (such as with a worn engine), some blowby passes out through the oil filler cap to the atmosphere.

**Type 2 Open PCV System.** A type 2 system has a diaphragm PCV valve that is controlled by crankcase vacuum, figure 5-36. When vacuum (low pressure) exists in the crankcase, the valve is closed. As crankcase pressure rises, the valve opens and allows vapors to recirculate through a hose to the manifold. The oil filler cap has an orifice that allows air into the crankcase for ventilation but restricts this airflow to maintain crankcase vacuum at low engine speeds. Type 2 systems are most common on older imports, particularly British cars.

**Type 3 Open PCV System.** A type 3 system has no PCV valve. A large-diameter hose connects the crankcase to the air cleaner intake, figure 5-37. Suction at the air intake draws vapors from the crankcase, through the carburetor, and into the manifold. Fresh air enters the crankcase through a vented oil filler cap.

Open PCV systems were used on many cars in the early and mid-1960's but were replaced by closed systems on most engines in the late sixties and early seventies.

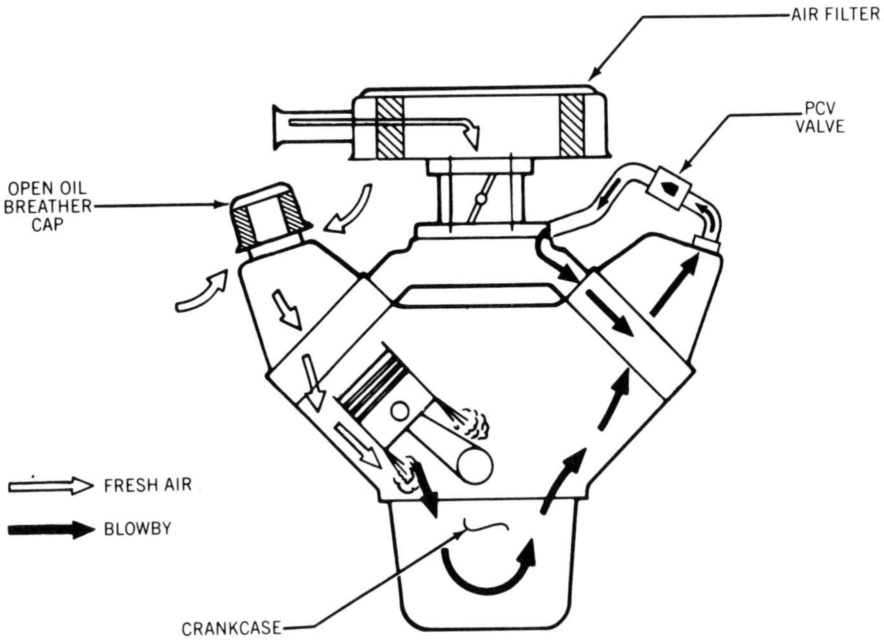


**Figure 5-35.** A type 1 PCV system has a vented oil filler cap and a PCV valve controlled by manifold vacuum.

## Closed PCV Systems

Closed PCV systems were required on all 1964 California cars and became common nationwide in 1968. Most closed systems are called type 4 systems. They are similar to an open type 1 system, except that the oil filler cap is sealed, not vented to the atmosphere. Fresh air enters the crankcase from the engine intake air cleaner and passes through a hose to a valve cover or crankcase inlet, figure 5-38. If air comes from the dirty side (outside) of the air cleaner, the inlet line has a separate PCV air filter, figure 5-39. If fresh air comes from the clean side (inside) of the air cleaner, the intake air filter removes dirt from the incoming air. In this case, the air inlet line also has a flame arrester to prevent a crankcase explosion if the engine backfires.

Because the crankcase fresh air inlet is connected to the air cleaner, any excess blowby beyond the capacity of the PCV valve is recycled to the intake air and drawn back into the engine. This keeps the blowby from adding to air pollution and is why the system is called a "closed" PCV system.

One major exception to the type 4 closed PCV system design is on Ford 1.6- and 1.9-liter engines. These engines have a fixed orifice rather than a PCV valve to meter crankcase vapors

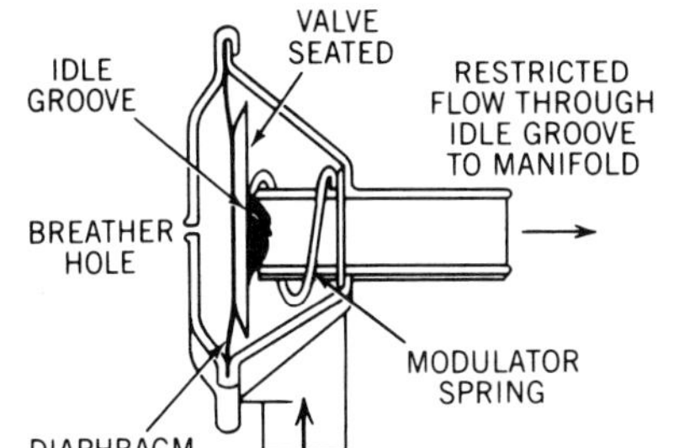


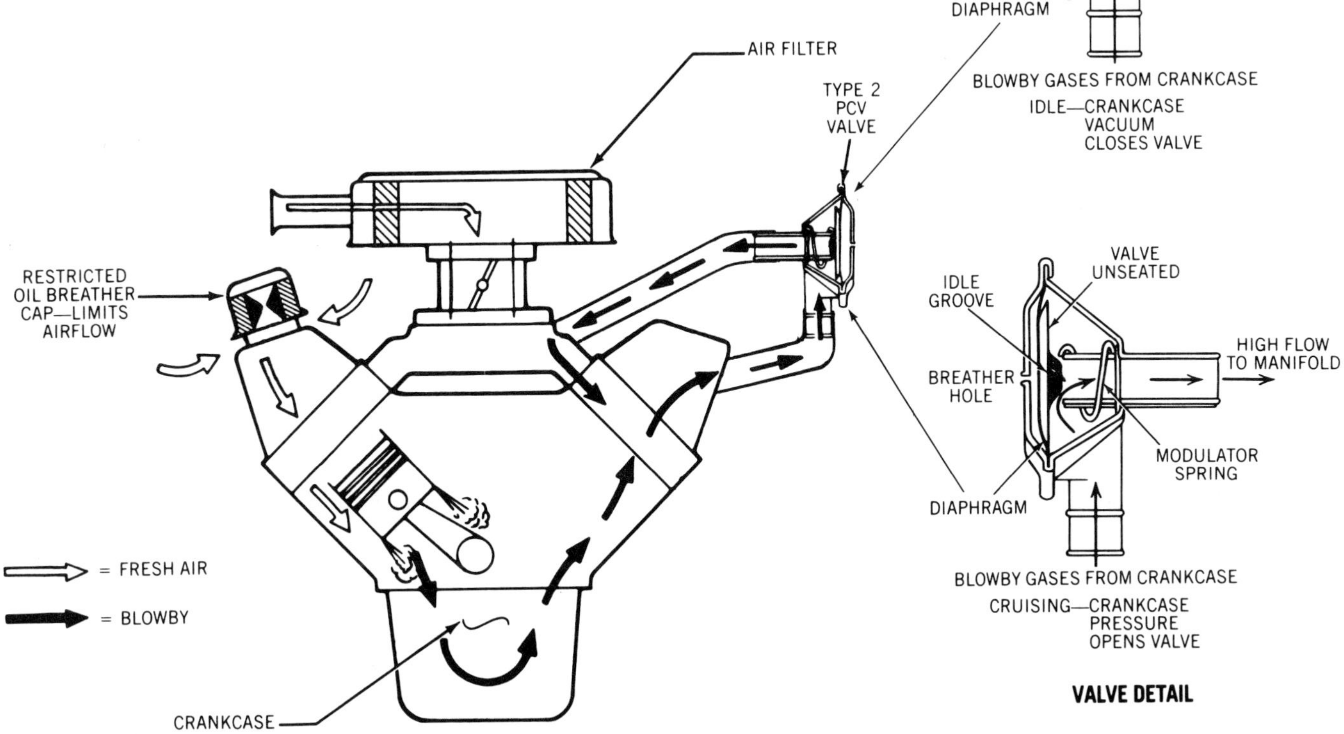


**Figure 5-36.** A type 2 open PCV system has a valve controlled by crankcase vacuum.

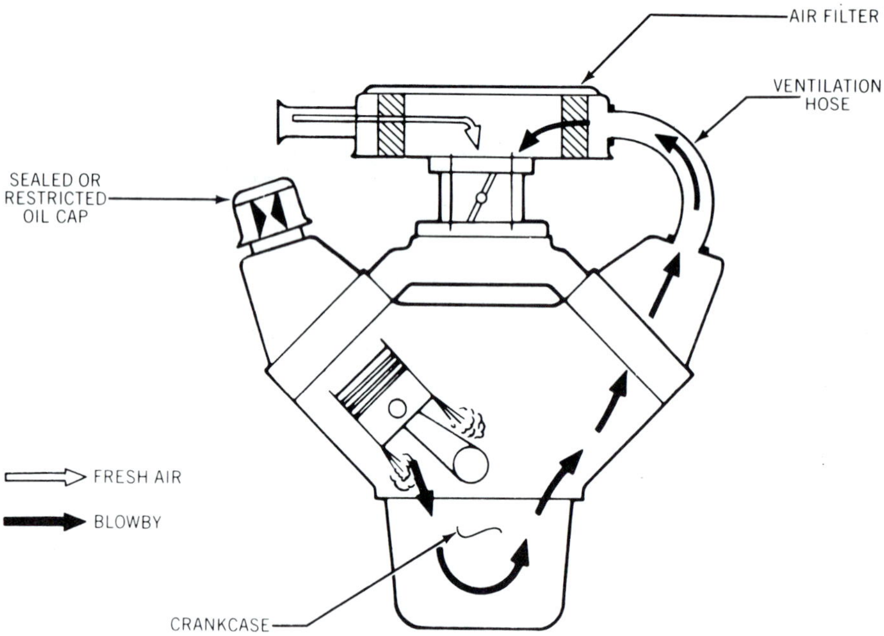


**Figure 5-37.** A type 3 open PCV system has no PCV metering valve.

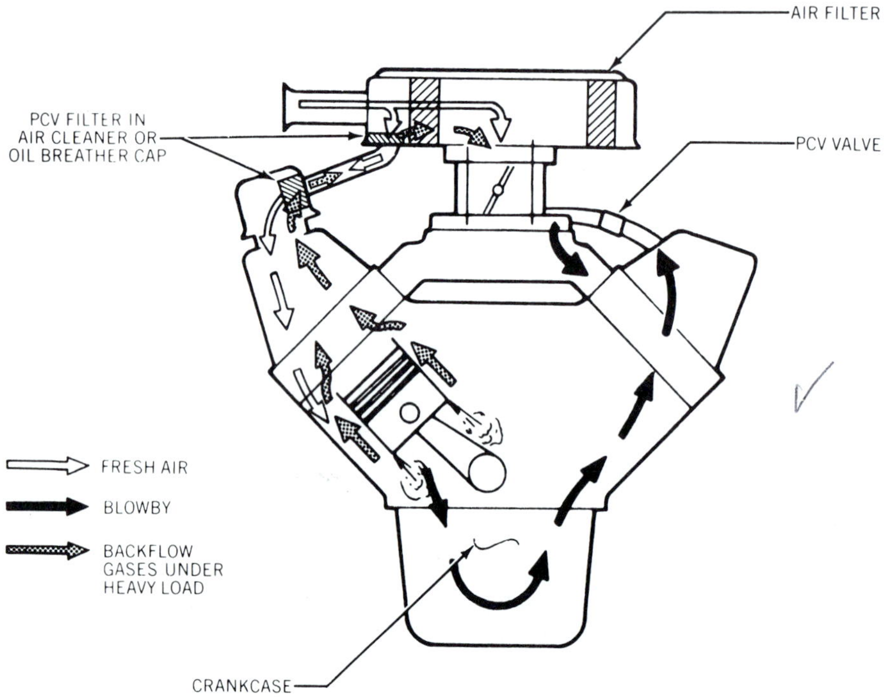


**Figure 5-38.** Type 4 closed PCV systems are the most common crankcase ventilation systems used for the past 20 years.

to the intake manifold, figure 5-40. The orifice also is connected to a fresh air inlet on the clean side of the air filter. Excess blowby at high speed and load flows through the orifice to the intake airflow. This system is thus closed to the atmosphere as all other late-model systems are. Airflow and vapor flow through the orifice are the critical factors of system operation.

### PCV Valve Operation

A simple spring-loaded valve regulates crankcase airflow in all type 1 and type 4 PCV systems, figure 5-41.

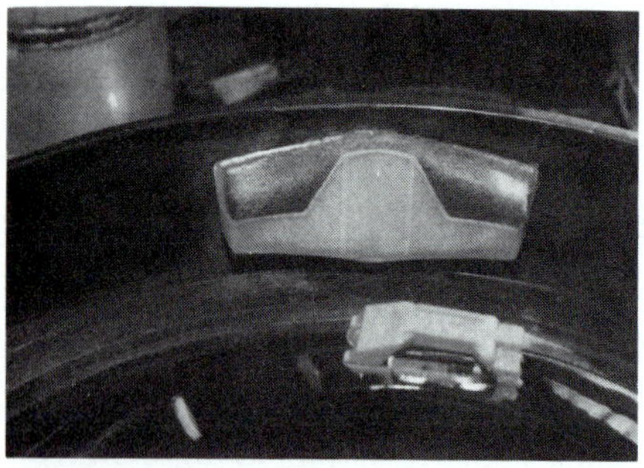

**Figure 5-39.** If PCV inlet air comes from the outside of the engine air filter, the system has a separate PCV air filter.

The forces of crankcase pressure and manifold vacuum tend to move the valve plunger to a closed position at the valve outlet. These forces are opposed by spring force that tends to move the plunger in the opposite direction. The combination of these opposing forces controls the valve opening under different driving conditions.

At idle, or whenever manifold vacuum is high, the PCV valve plunger moves forward to restrict airflow to about 1 to 3 cubic feet per minute, figure 5-41*A*. This keeps crankcase vapors from richening the air-fuel mixture, and it keeps manifold vacuum from drawing oil out of the crankcase.

When manifold vacuum decreases at high speed or heavy load, the valve plunger moves back to increase the PCV airflow to about 3 to 6 cubic feet per minute, figure 5-41*B*.

If the engine backfires, manifold pressure forces the valve plunger back against the rear seat, figure 5-41*C*, to prevent a crankcase explosion.

The modulating action of crankcase pressure, manifold vacuum, and PCV valve spring force provide almost uniform crankcase ventilation at all times. The flow rates of different valves are designed for engines of various sizes, and a valve should not be used on an engine for which it was not designed.

### Diesel PCV Systems

Because a diesel engine runs with an unthrottled air intake, there is no manifold vacuum to control a PCV valve. Most diesel PCV systems have a large

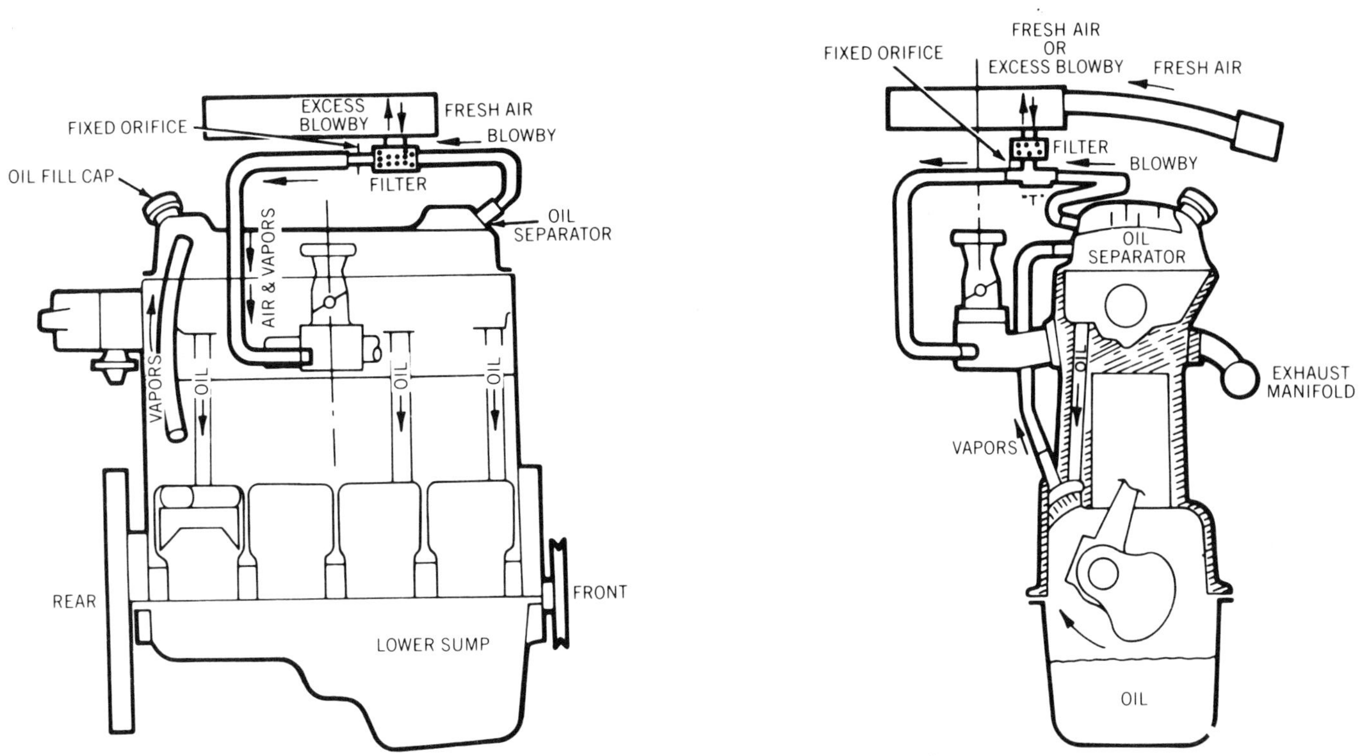


**Figure 5-40.** The PCV system on Ford's 1.6-liter engine uses a fixed orifice, rather than a valve, to control crankcase vapor flow. (Ford)

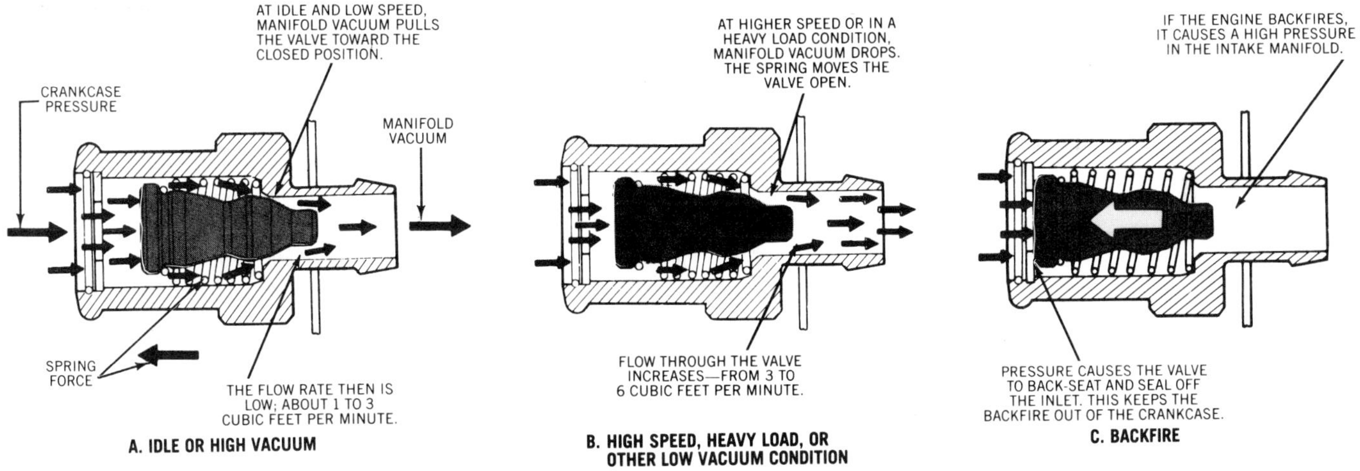


**Figure 5-41.** Crankcase pressure, manifold vacuum, and spring force control PCV valve operation. (Ford)

hose or tube that runs from the crankcase or valve cover to the engine air intake, figure 5-42. Fresh air enters the crankcase through the oil filler cap or an inlet air filter in the air intake system. Crankcase ventilation airflow is controlled by crankcase pressure and airflow past the PCV line in the air intake. Some diesel PCV systems have a flow control valve, similar to the valve used with a type 2 system, regulated by crankcase pressure and vacuum.

## SUMMARY

Motor oil lubricates, cleans, and cools an engine. The lubrication system and oils designed for gasoline or diesel engines minimize engine wear and maximize engine performance and economy. Oils are classified by API service classifications and SAE viscosity grades. Oil of the recommended API service classification reduces engine wear because it is blended to match an engine's operating requirements. Changing the oil when recommended, or sooner, reduces engine wear by replacing depleted additives and removing dirt from the engine. Using high-quality oil is an economical investment to ensure long engine life and top performance.

Oil viscosity is measured at high temperature (210° F, 99° C), low temperature (0° F, − 18° C), or both. If oil viscosity is too high for low-speed or

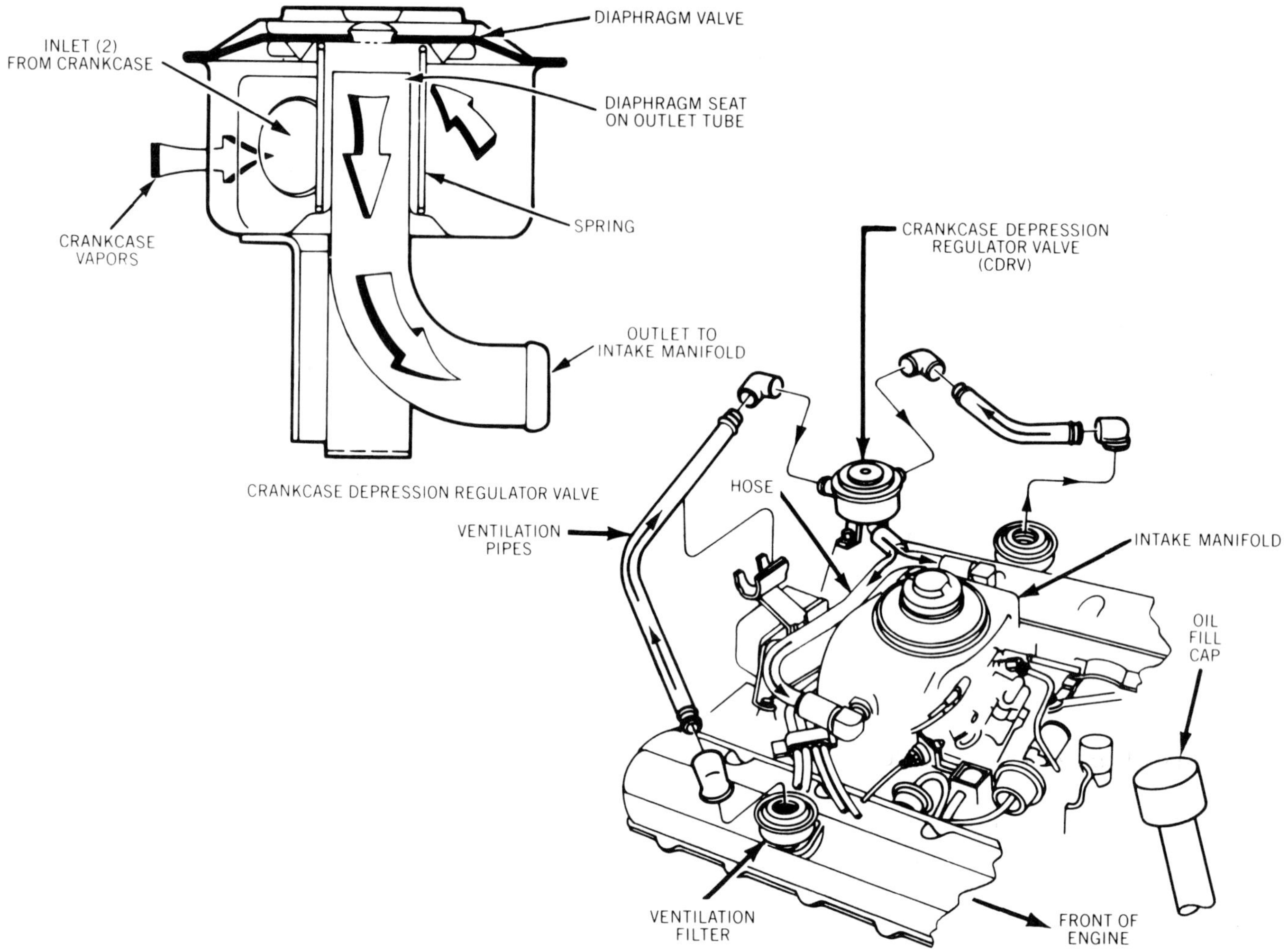


**Figure 5-42.** Typical diesel crankcase ventilation system. (Cadillac)

low-temperature operation, the engine wastes power to overcome the resistance of thick oil. If oil viscosity is too low for high-speed or high-temperature operation, the oil film on moving parts will break down and allow metal-to-metal contact. Also, if oil viscosity is too low, oil vapors add to crankcase blowby vapors and increase oil consumption. When oil is old or diluted by blowby, or when wear has increased clearances between engine parts, excess oil enters the combustion chamber past piston rings and valve guides and adds to HC exhaust emissions.

All engine oiling systems consist of an oil reservoir or oil pan, a pump, a pressure- relief valve, a filter, oil galleries and lines, pressure indicators, and volume indicators (dipsticks). All modern engines have full-pressure lubrication systems with full-flow oil filters, but many engine parts are lubricated by splash methods. The crankcase ventilation system relieves crankcase pressure and removes harmful blowby gases from the crankcase. PCV systems provide constant ventilation under all conditions and prevent HC emissions from crankcase blowby from adding to air pollution.

## REVIEW QUESTIONS

### Multiple Choice

**1.** Motor oil does all of the following jobs in an engine EXCEPT:

**a.** cushions moving parts from the shock of combustion

**b.** coats valve seats to help seal the combustion chamber

**c.** serves as a coolant

**d.** reduces friction between moving parts.

**2.** The oil additives which keep oil stable at high temperature (reducing varnish formation) are called:

**a.** foam inhibitors

**b.** pour point depressants

**c.** detergent-dispersants
**d.** antioxidants.

**3.** A motor oil won't pour easily. Mechanic A says it could be a low-viscosity oil. Mechanic B says the oil may be colder than its pour point. Who is right?

**a.** A only
**b.** B only
**c.** Both A and B
**d.** Neither A nor B

**4.** Which of the following types of oil additive is used to improve fuel economy?

**a.** Rust and corrosion preventives
**b.** Friction-reducing additives
**c.** Antioxidants
**d.** Extreme pressure additives

**5.** Which of the following statements about API SF-rated oil is true?

**a.** It is not suitable for modern engines.
**b.** It was developed specifically for turbodiesel engines
**c.** It can be used in place of SC, SD, and SE oils.
**d.** It is usually made from a synthetic base.

**6.** Mechanic A says that SAE 5W-30 oil has been tested at both high and low temperature. Mechanic B says "5W" indicates the oil's viscosity when it is warm. Who is right?

**a.** A only
**b.** B only
**c.** Both A and B
**d.** Neither A nor B

**7.** Which of the following statements about synthetic oils is true?

**a.** Synthetic oils were first developed during the 1970s.
**b.** Synthetic oils contain no petroleum products.
**c.** The API and SAE have not yet set standards for synthetic oils.
**d.** Synthetic oil stock does not require additional additives.

**8.** Which of the following is NOT a part of the engine lubrication system?

**a.** Crankcase ventilation system
**b.** Oil level indicator
**c.** Pressure relief valve
**d.** Thermostat

**9.** Mechanic A says that dry-sump oil systems help prevent oil starvation during hard cornering. Mechanic B says windage trays are used for the same purpose. Who is right?

**a.** A only
**b.** B only
**c.** Both A and B
**d.** Neither A nor B

**10.** Which of the following factors does NOT affect oil system pressure?

**a.** Engine mechanical wear
**b.** The oil's SAE viscosity rating
**c.** The oil's API Service Classification
**d.** Oil pump speed

**11.** Mechanic A says oil system pressure created an oil film between crankshaft journals and bearing surfaces. Mechanic B says that if bearing clearance is too small, the film will not form and the bearing will overheat. Who is right?

**a.** A only
**b.** B only
**c.** Both A and B
**d.** Neither A nor B

**12.** Mechanic A says if the oil filter clogs, a bypass valve opens to keep the engine lubricated. Mechanic B says any oil filter with a bypass valve is called a bypass oil filter. Who is right?

**a.** A only
**b.** B only
**c.** Both A and B
**d.** Neither A nor B

**13.** All of the following statements about oil flow to the cylinder walls are true EXCEPT:

**a.** Some connecting rods are drilled from end to end to carry oil to the upper cylinder area.
**b.** Spray from the edges of the crankshaft journals may lubricate the cylinder walls.
**c.** Oil can be sprayed onto the cylinder walls by an orifice or nozzle in the cylinder head.
**d.** Some connecting rods have spray holes in the upper end to help lubricate the cylinder walls.

**14.** Mechanic A says camshaft lobes and valve lifters have the most critical lubrication requirements. Mechanic B says they're lubricated only by splash. Who is right?

**a.** A only
**b.** B only
**c.** Both A and B
**d.** Neither A nor B

**15.** Crankcase ventilation systems are used to counteract:

**a.** the emission of fuel vapor from the carburetor
**b.** the leakage of combustion gases past the piston rings
**c.** the emission of pollutants from the crankcase oil
**d.** the vacuum created in the crankcase by piston action.

### Fill in the Blank

**16.** When water, acids, and unburned fuel combine with cold oil in the crankcase, __________ is formed.

**17.** Viscosity is the tendency of a liquid to resist __________.

**18.** The API Service Classifications for diesel engine oils begin with the letter __________.

**19.** Carmakers' viscosity recommendations are based on the __________ range in which the vehicle is operated.

**20.** The type of oil system in which oil is stored in the oil pan at the bottom of the engine, rather than in a separate reservoir, is called a __________.

**21.** An oil system in which all of the pump output must go to the oil filter before reaching the engine has a __________ oil filter.

**22.** For the most accurate measurement of engine oil level, the engine should be level and __________.

**23.** Positive crankcase ventilation systems use __________ vacuum to draw vapors from the crankcase.

**24.** The API Service Classifications for gasoline engine oils begin with the letter __________.

**25.** Rotor, gear, and gear-and-crescent oil pumps are all __________ type pumps.

**26.** At idle, or whenever manifold vacuum is high, the PCV valve plunger moves to __________ airflow.

**27.** The parts of an engine are lubricated by both __________ and __________ methods.

# 6 COOLING AND EXHAUST SYSTEMS

## WHAT

You have learned in earlier chapters that peak combustion temperatures can reach 6,000° F (3,300° C), and average combustion chamber temperatures are about 1,500° F (800° C) during the 4-stroke cycle in a gasoline engine, higher in a diesel. You also know that an engine converts only about one-third of this heat energy to mechanical energy. The cooling and exhaust systems remove the other two-thirds of the heat and regulate temperature so that the engine can operate efficiently. Therefore, the cooling and exhaust systems are important combustion control systems. This chapter explains these systems and how they affect performance, economy, and emission control.

## WHY

If an engine is too hot or too cold, several things happen:

- High temperatures cause oil to thin and engine parts to expand. Lubrication breaks down, and moving parts can be damaged.
- Low temperatures cause sludge formation and water condensation in the oil. Oil may not circulate well to all parts of the engine, and damage can occur from wear, rust, and corrosion.
- High temperatures expand the air-fuel charge and reduce volumetric efficiency. They also increase oxides of nitrogen ($NO_x$) emissions.
- Low temperatures cause incomplete combustion, reduce fuel economy, and increase hydrocarbon (HC) and carbon monoxide (CO) emissions.

To service an engine for proper operation, you must understand how the cooling and exhaust systems relate to other combustion control systems and affect overall engine performance.

## GOALS

After studying this chapter, you should be able to:

**1.** Explain engine temperature requirements under various operating conditions and how the cooling system meets these needs.

**2.** Trace coolant circulation through various cooling system parts under different operating conditions.

**3.** Explain the reasons for a pressurized cooling system.

**4.** Explain the reasons for using ethylene glycol coolant.

**5.** Identify the parts of a typical exhaust system and explain their operations.

## COOLING SYSTEM OPERATION

When we talk about engine temperature while testing and servicing a car, we talk about average coolant temperature in the engine. This is what the temperature gauge measures, not combustion chamber temperature. Average operating temperature of the coolant varies from engine to engine, but is usually about 200° F (94° C). Earlier engines without extensive emission controls may run about 180° F (82° C). Some late-model engines run at 212° F (100° C) or higher.

With few exceptions, such as opposed 6-cylinder Porsches, all late-model engines have liquid cooling systems. The coolant circulates through the engine to absorb heat. The coolant also circulates through a radiator outside the engine, where it is exposed to air circulation. The air absorbs heat from the coolant, which then returns to the engine to absorb more heat, figure 6-1. In this way, both liquid and airflow cool the engine. A large temperature difference between the coolant and the engine allows the coolant to absorb more heat. Similarly, a large temperature difference between the coolant and the airflow through the radiator allows the air to absorb more heat from the coolant.

### Coolant Circulation

Coolant circulates in two ways to maintain correct engine temperatures. Inside the engine, it circulates through water jackets. Outside the engine, it circulates through hoses and the radiator. The engine thermostat, a temperature-sensitive valve, opens and closes to control these two circulation paths, figure 6-2.

When an engine is cold, it must warm up quickly to operating temperature. If coolant circulated through the radiator with a cold engine, warm-

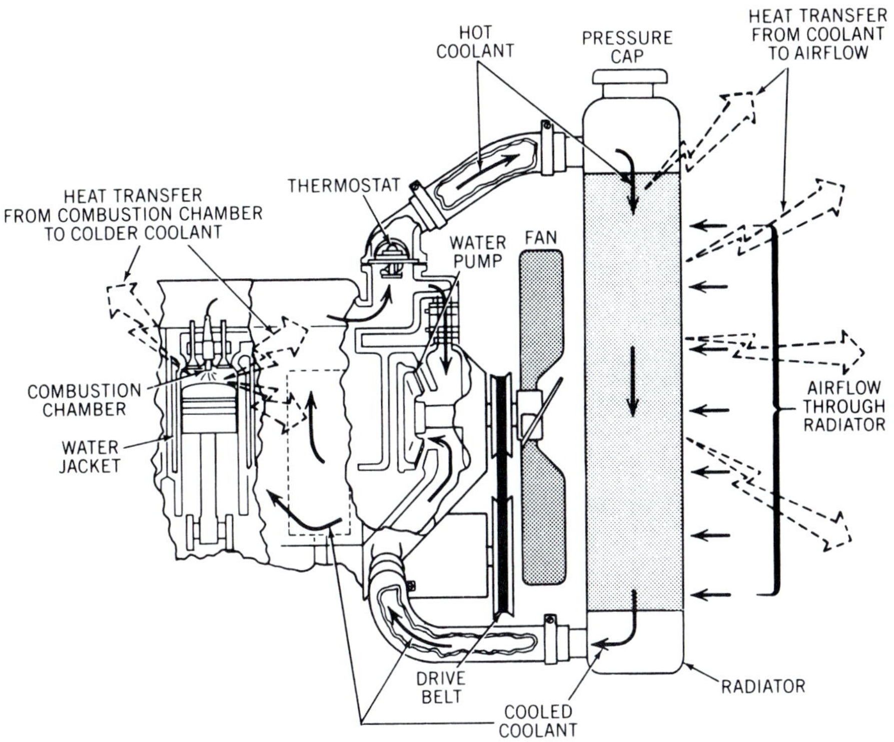


**Figure 6-1.** Heat always moves from a hot area to a cooler one. Therefore, heat moves from the engine to the colder coolant and from the coolant to the colder airflow through the radiator.

up would take too long, so the thermostat stays closed. This keeps coolant within the engine circulating through the engine, and it keeps colder coolant in the radiator from entering the engine. Because there is little or no heat transfer from the coolant to the outside airflow, the engine warms quickly.

As coolant in the engine warms, the thermostat opens to allow full circulation through the engine and the radiator. Engine temperature then is controlled by the amount of coolant in the system, the size of the radiator, and the opening temperature of the thermostat.

### Cooling System Pressure

All cooling systems are pressurized for two reasons:

1. To increase water pump efficiency
2. To raise the coolant boiling point

In an unpressurized system, the water pump does not get a complete supply of coolant at all times. Some air is present in the pump, and the pump churns the air with the coolant. This causes some amount of cooling system **aeration** and reduces the water pump coolant output to about 85 percent of its maximum. Air bubbles also reduce heat transfer from the engine to the coolant. Pressurizing the cooling system to about 14 psi (96 kPa) increases water pump output efficiency to about 100 percent and eliminates aeration.

At atmospheric pressure of 14.7 psi (101.4 kPa) at sea level, water boils at 212° F (100° C). A typical engine coolant solution of water and ethylene glycol boils at about 223° F (106° C). Even this higher temperature is close to the operating temperature of many engines.

Pressurizing the system raises the coolant boiling point. Each 1 psi of pressure increase raises the boiling point about 3° F (1.7° C). Most cooling systems operate at pressures of 12 to 17 psi (83 to 117 kPa) above atmospheric pressure when completely warm. This raises the coolant boiling point range to about 260° to 275° F (126° to 135° C). The spring-sealed radiator cap maintains system pressure.

## COOLING SYSTEM PARTS

The principal parts of a cooling system, figure 6-3, are:

- The coolant
- The water jackets and core plugs
- The thermostat and thermostat bypass
- The water pump
- The radiator
- The radiator cap
- The coolant recovery system
- The hoses
- The heater
- The fan
- The drive belts

### Coolant

Although some parts of the cooling system are called the "water" pump and the "water" jackets, engine coolant is not pure water. It is a mixture of water and **ethylene glycol**. Pure water transfers heat well, but it rusts and corrodes water jackets. Also, when not pressurized, water freezes at 32° F (0° C), which can damage engine and cooling system parts. Ethylene glycol does not transfer heat as well as water, but it freezes at lower temperatures and boils at higher temperatures. Ethylene glycol solutions for cooling systems also contain rust and corrosion inhibitors and water pump lubricant to protect cooling system parts. Together, water and ethylene glycol form an ideal engine coolant.

Most carmakers recommend about 50-percent water and 50-percent ethylene glycol as an engine coolant mixture. With 15-psi (103 kPa) system pressure, this protects against freezing to about −30° F (−34° C) and against boiling to about 264° F (129° C). Many engines need the ethylene glycol in the coolant to activate the high-temperature warning lamp. The switches for many of these indicators close at temperatures above the boiling point of water. If the system contained pure water, it could boil away and never light the warning lamp.

If a car is operated in cold climates,

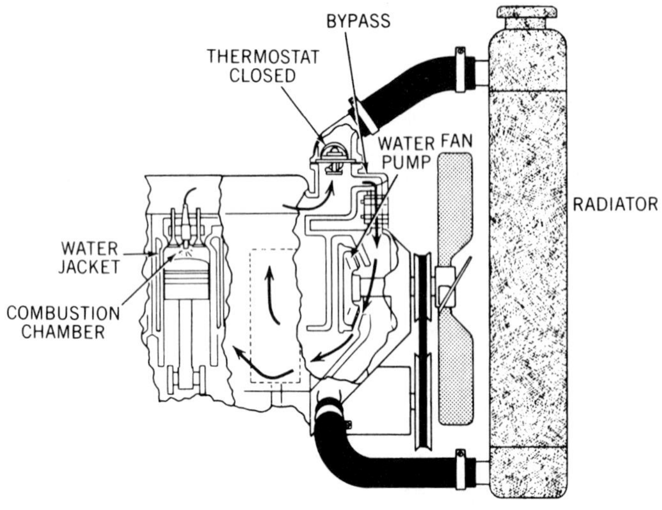


COLD ENGINE—BYPASS CIRCULATION

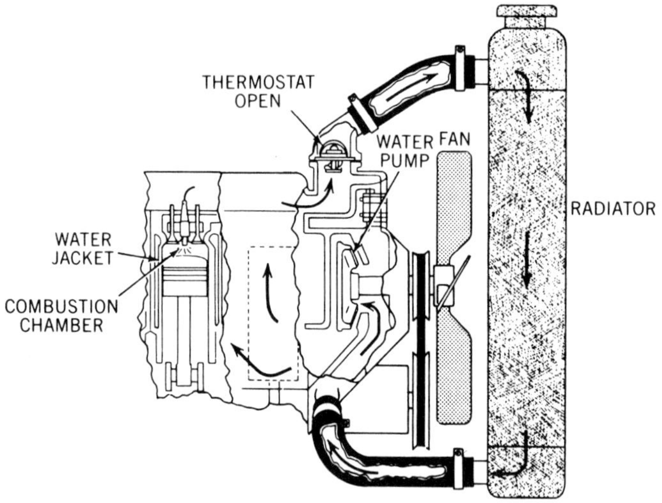


WARM ENGINE—CIRCULATION THROUGH RADIATOR

**Figure 6-2.** When the engine is cold, coolant circulates through the engine but not through the radiator. When the engine is warm, the thermostat opens to allow circulation through the radiator.

more ethylene glycol can be added to the mixture. Coolant with 68-percent ethylene glycol protects against freezing to −90° F (−62° C). If the coolant contains more than 68-percent ethylene glycol, the freezing point actually rises. Pure ethylene glycol freezes just below 0° F (−18° C). At the other temperature extreme, high concentrations of ethylene glycol do not remove engine heat well. The coolant may not boil, but the engine will not be properly cooled.

While half-and-half water and ethylene glycol is the industry norm, some carmakers recommend only 33-percent ethylene glycol in the coolant. This protects against freezing and boiling at moderately low and high temperatures and provides basic rust and corrosion protection. Additionally, some carmakers recommend ethylene glycol solutions with specific corrosion inhibitors and other additives because of special aluminum alloys used in their engines.

The additives in engine coolant wear out and lose their effectiveness just as the additives in motor oil do. Also, the cooling system accumulates dirt and other contaminants. For these reasons, engine coolant should be drained and replaced periodically, just as motor oil is. Carmakers' recommendations for coolant change intervals range from 12 to 48 months.

## Water Jackets and Core Plugs

The water jackets and coolant passages cast into engine blocks and heads are carefully designed so that coolant circulates evenly and does not collect in pockets. If coolant circulation is uneven, or fast in some parts of the engine and slow in others, it will not remove engine heat evenly. Some parts of the engine may be overcooled, and others may not be cooled enough.

Because coolant enters and leaves the front of the engine, designers use several methods to ensure uniform circulation. Some inline engines have water distribution tubes that receive coolant at the front of the engine and distribute it near each valve seat. Holes in the distribution tube are graduated so that all areas receive an equal coolant volume. Water jacket transfer holes between the block and the head often are graduated to force incoming coolant toward the rear of the engine. Many head gaskets have holes for coolant transfer that restrict coolant flow at the front of the engine and force it toward the rear.

Engine blocks and heads have core plugs pressed into holes left during the casting process, figure 6-4. These are often called freeze plugs because some people mistakenly believe they are in an engine to relieve pressure if the coolant freezes. These plugs actually fill the holes used to remove core sand in the foundry when blocks and heads are cast.

Core plugs can rust and leak if the coolant does not have enough rust inhibitors. This is another good reason to change coolant regularly. A plug also may pop out if the coolant freezes, but if freezing is severe enough to dislodge a plug, it probably also has damaged the block or head.

## Thermostat and Thermostat Bypass

The thermostat that regulates coolant flow is located at the point where

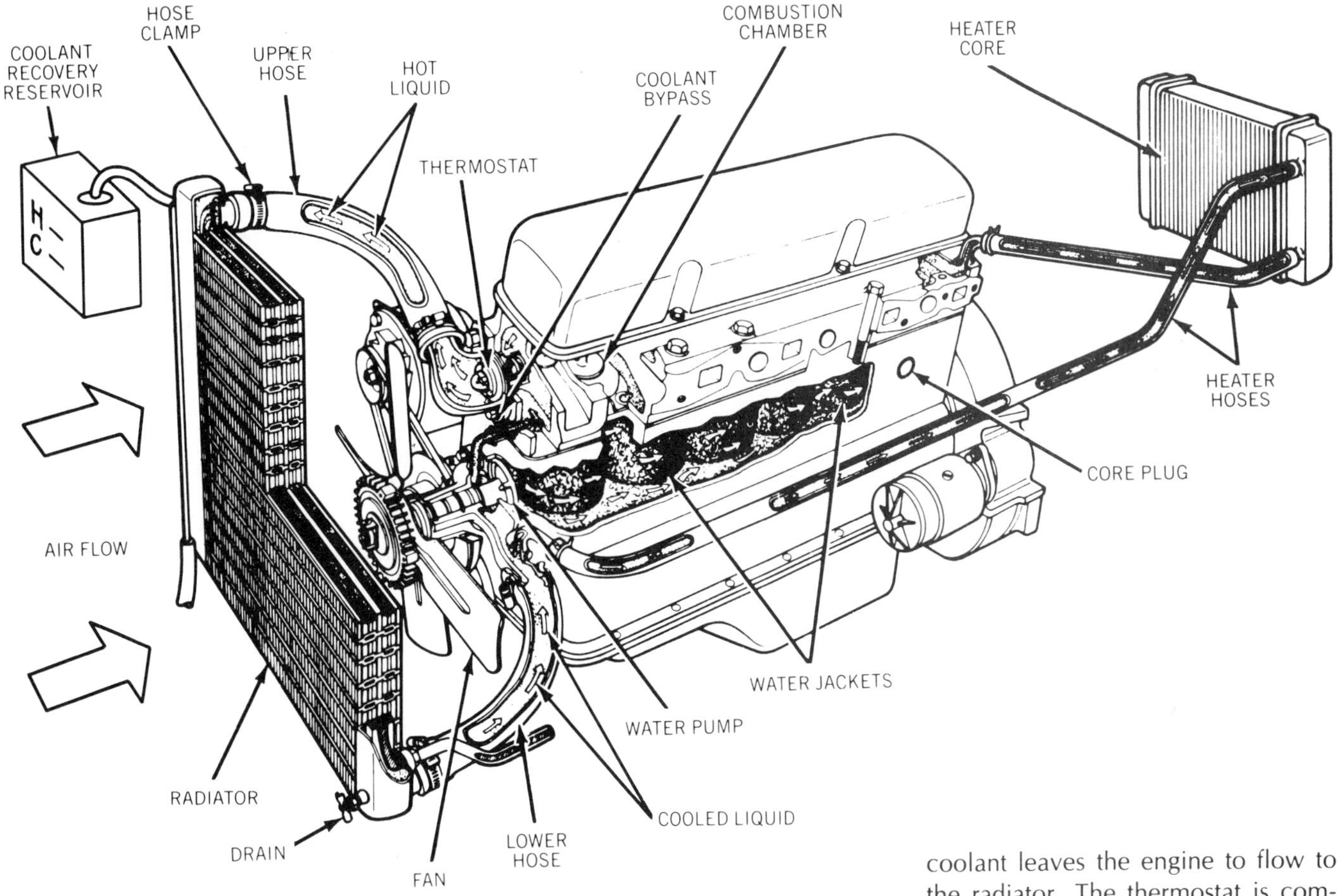


**Figure 6-3.** A typical cooling system contains these major parts. (Chrysler)

**Figure 6-4.** Core plugs fill holes used to remove sand when blocks and heads are cast.

coolant leaves the engine to flow to the radiator. The thermostat is completely closed when the engine is cold, but the water pump still circulates coolant within the engine. This is done through a bypass circuit. On some engines, the bypass is a short hose that connects the thermostat housing to the water pump housing, figure 6-5.

In some engines, the bypass is an internal passage at the bottom of the thermostat housing. There are two general types. A blocking-bypass thermostat has a small secondary valve, figure 6-6. When the primary thermostat valve is closed, the secondary valve is open for bypass circulation. When the primary valve is open, the secondary valve blocks the bypass hole. The second type of internal bypass has a separate spring-loaded bypass valve, figure 6-7. The spring is weaker than the thermostat spring and allows coolant circulation to open the bypass when the thermostat is closed. When the thermostat opens, the bypass closes. A few engines have an unrestricted internal bypass, and its flow is designed into total cooling system flow.

**Figure 6-5.** This engine has an external coolant bypass hose from the thermostat housing to the water pump. (Chrysler)

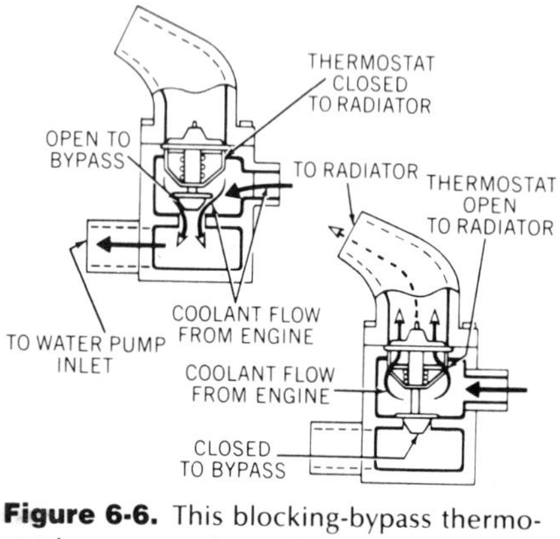


**Figure 6-6.** This blocking-bypass thermostat has a secondary valve to open and close the bypass opening. (Chrysler)

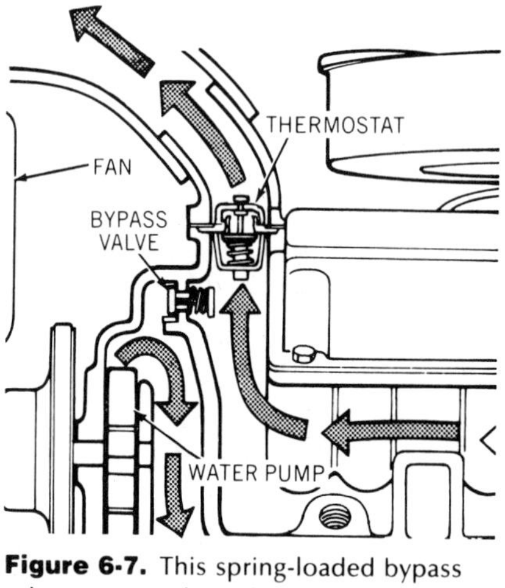


**Figure 6-7.** This spring-loaded bypass valve opens with coolant pressure. (Chrysler)

Thermostats are rated by the temperature at which they *start* to open, and this usually is stamped on the thermostat body. A 200-degree thermostat, for example, starts to open at 200° F (93° C). It may not be fully open until 220° F (104° C). The thermostat opening varies during normal operation, but it is rarely open fully unless the engine is working very hard or the weather is very hot. The thermostat modulates the coolant flow to keep the engine temperature within the desired range.

Thermostats in all late-model engines are operated by a wax pellet in a sealed chamber, figure 6-8. The wax is solid when cold, but it melts and expands when hot. When the wax is cold, a spring holds the thermostat valve closed. As the engine and the wax warm, the wax expands and forces the thermostat open against spring pressure. A wax-pellet thermostat works reliably throughout the system's normal pressure range.

Some people mistakenly believe that removing a thermostat will cure engine overheating. Obviously, if you remove a thermostat, the engine will take longer to warm up. Fuel economy, emission control, and performance suffer with a cold engine, and engine wear increases. Some engines may run hotter without a thermostat than with one. Coolant must flow at a controlled rate to remove heat properly. If coolant passes through the engine too fast, it does not have time to absorb enough heat. Similarly, fast coolant flow through the radiator doesn't allow enough time for heat to transfer to the airflow. Therefore, the engine retains more heat than it should, even though coolant temperature is low.

On late-model engines with computer-controlled fuel and ignition systems, thermostat temperature is critical to correct air-fuel ratios and ignition timing. Most of these systems have a coolant temperature sensor that tells the computer the exact engine temperature. If a thermostat is removed or a low-temperature thermostat is installed in one of these engines, the coolant sensor may never signal the computer that the engine has warmed up. Fuel delivery and ignition timing will remain at cold-engine settings.

## Water Pump

The water pump is a centrifugal pump consisting of a fan-shaped impeller inside a circular housing. The inlet and outlet passages are curved so that centrifugal force draws coolant through the inlet near the center of the chamber and pumps it out the outlet at the edge, figure 6-9. The impeller is turned by a belt from the engine crankshaft. The impeller and its shaft are supported in the housing by two bearings, which are permanently lubricated. A

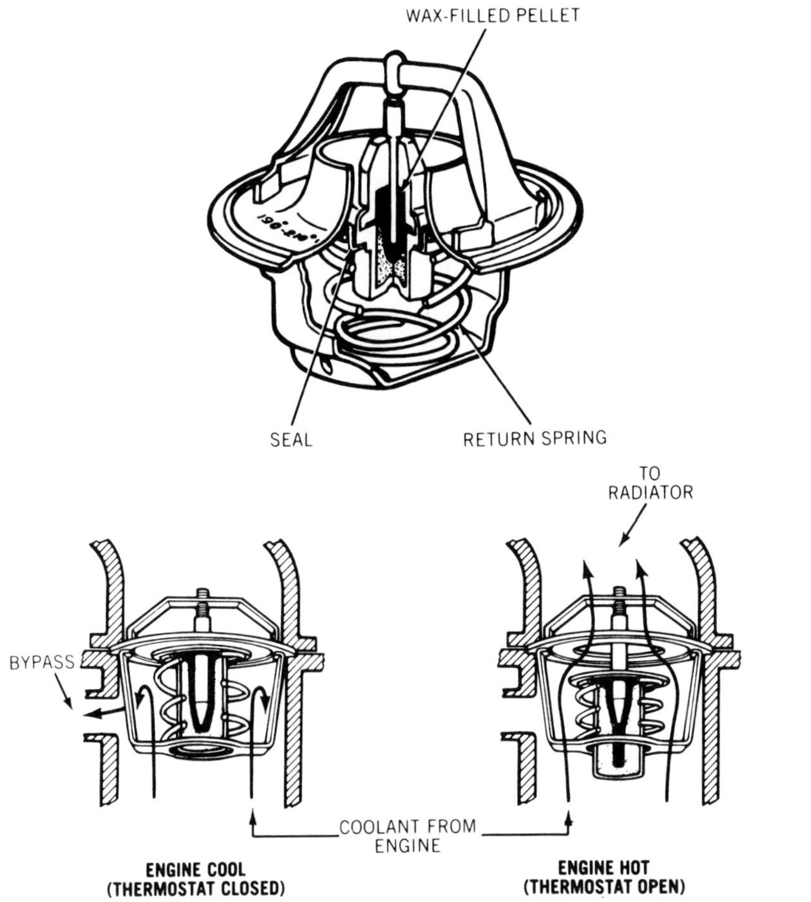


**Figure 6-8.** As the wax pellet warms up, it expands and forces the thermostat valve open. (Chrysler)

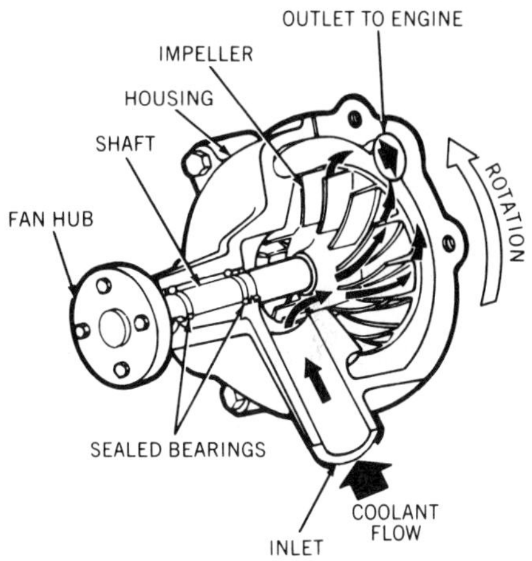


**Figure 6-9.** A water pump circulates coolant by centrifugal pumping action. (Chrysler)

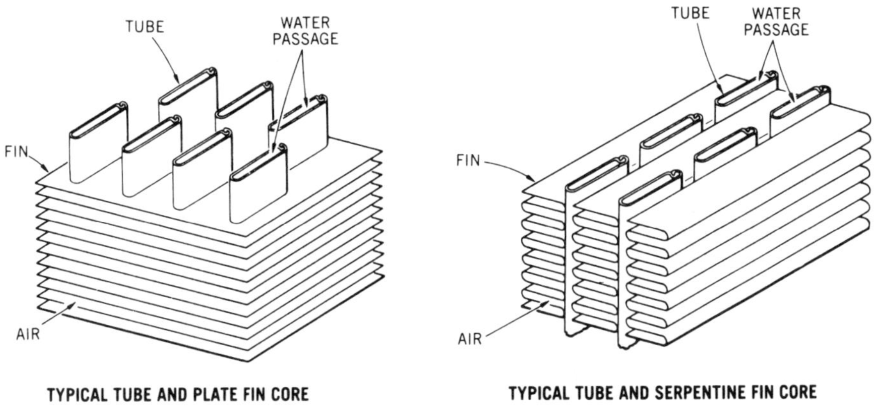


**Figure 6-10.** Most radiator cores are a series of tubes separated by flat or serpentine cooling fins. (SAE)

seal keeps coolant from leaking out of the pump along the shaft.

Because the water pump is driven by the crankshaft, it circulates coolant whenever the engine is running. Centrifugal pumps, however, are **variable-displacement pumps**. The output volume changes directly with speed. The water pump for any engine provides adequate circulation at all engine speeds, but a loose or slipping drive belt can reduce pump speed and circulation.

## Radiator

The coolant circulates through a radiator core that exposes a large surface area to airflow. This transfers the maximum amount of heat to the airflow.

Most radiator cores are a series of copper or aluminum tubes separated by cooling fins, figure 6-10. Heat transfers from the coolant to the metal of the tubes and fins and then to the airflow passing through the core. The number and size of tubes and fins determine the radiator surface and heat transfer ability.

Some radiators have a cellular, or honeycomb, core made by soldering pairs of metal ribbons along their edges to form coolant passages, figure 6-11. Fins separate the cellular passages to complete the radiator core. Cellular construction provides larger coolant passages for a given size core, but tube-and-fin cores are stronger and cheaper to build.

For uniform coolant flow, radiators have brass tanks at both ends of the core. The coolant inlet and outlet connections are attached to the tanks.

Downflow radiators have tanks at

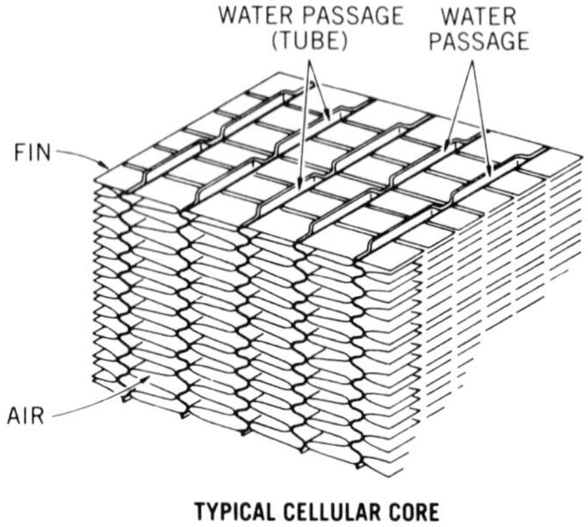


**Figure 6-11.** A cellular, or honeycomb, radiator core is made by soldering metal ribbons together to form coolant passages. (SAE)

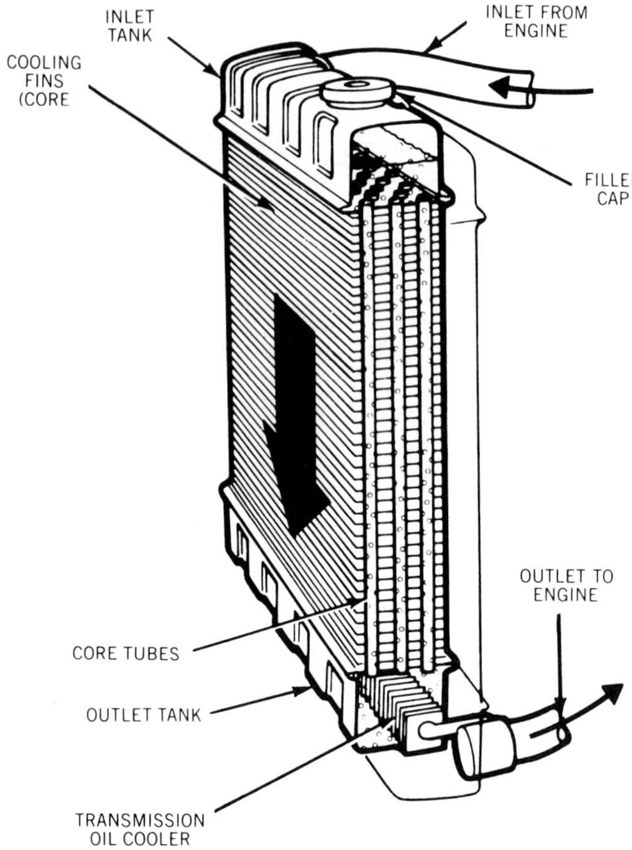


**Figure 6-12.** In a downflow radiator, hot coolant enters the top tank and flows downward through the core to the bottom tank. (Chrysler)

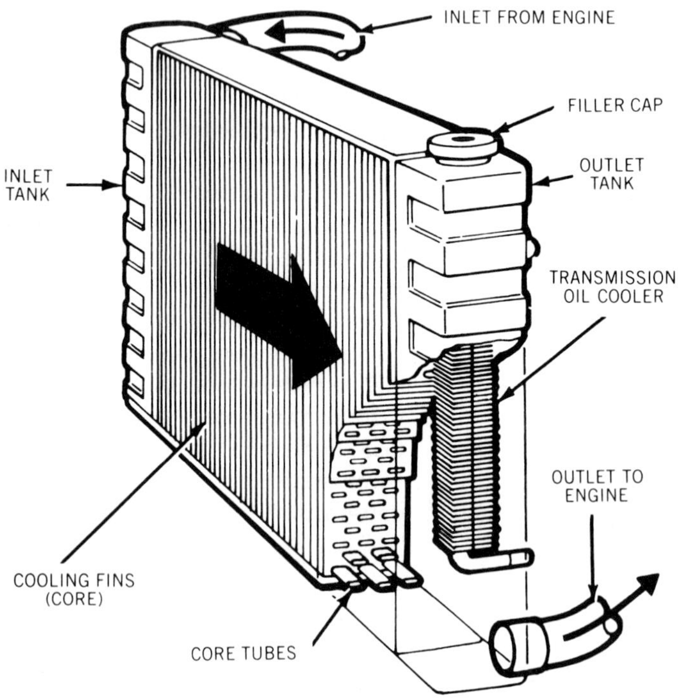


**Figure 6-13.** In a crossflow radiator, hot coolant enters the top of one side tank and leaves at the bottom of the other side tank. (Chrysler)

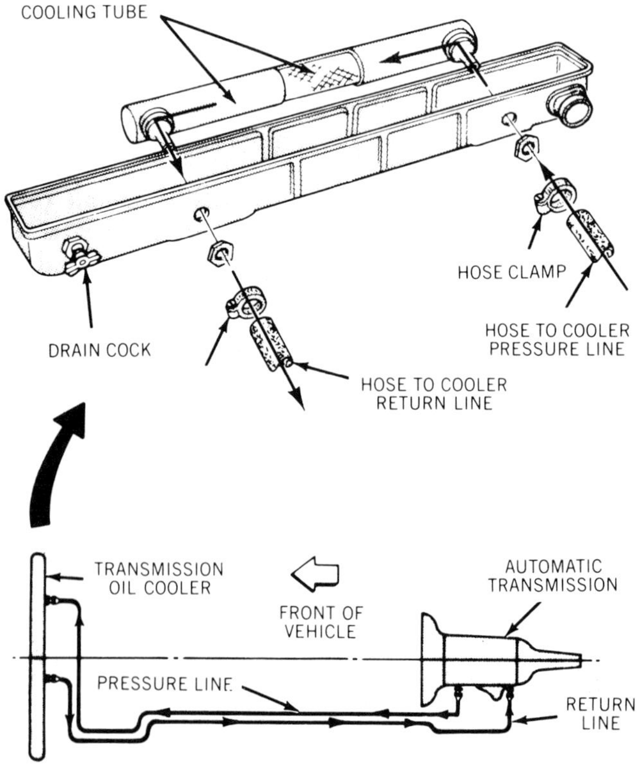


**Figure 6-14.** A car with an automatic transmission has a transmission fluid cooler in the radiator outlet tank. (Chrysler)

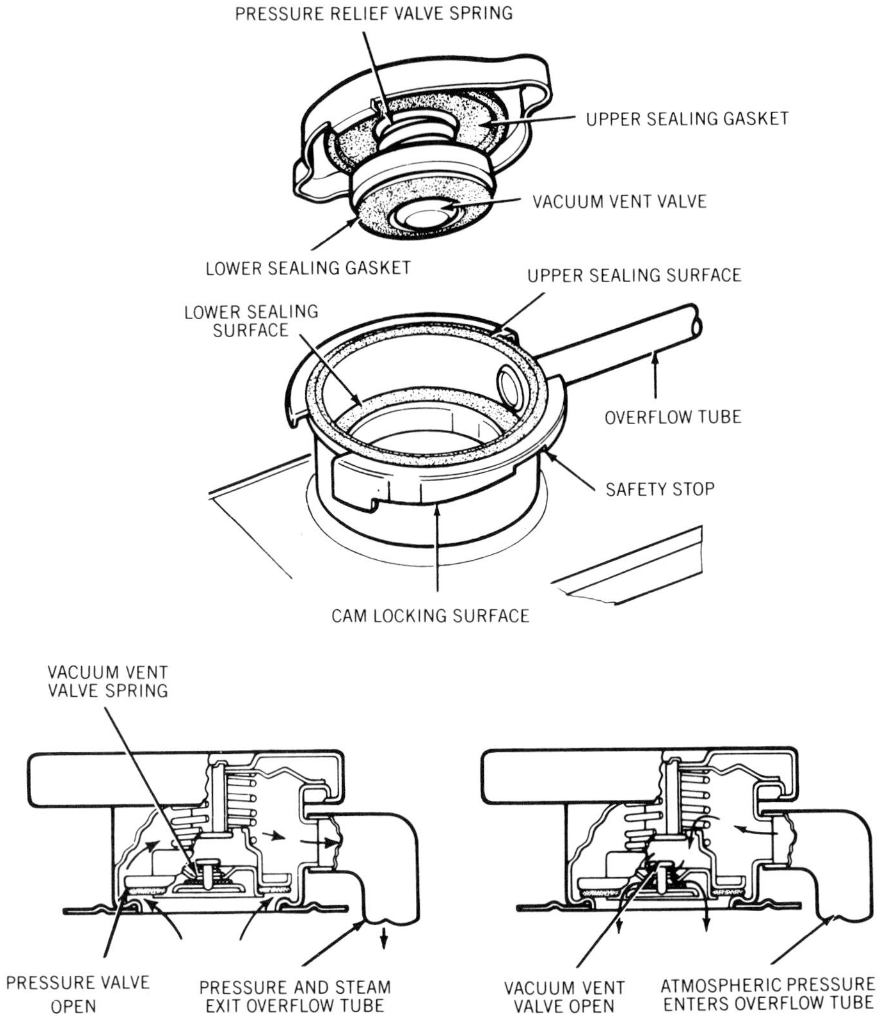


**Figure 6-15.** A constant-pressure radiator cap has a spring-loaded vacuum vent valve. (Chrysler)

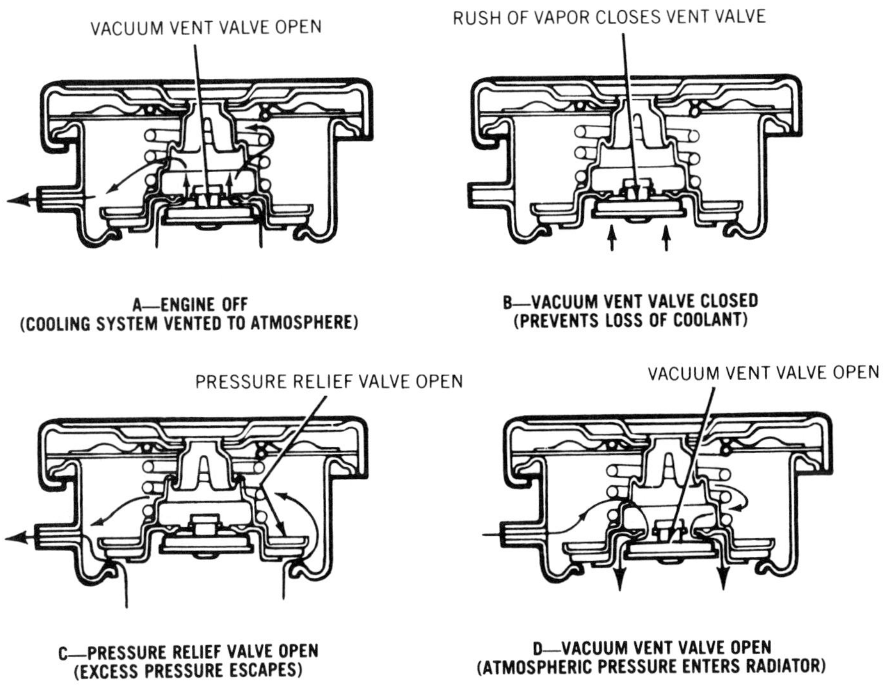


**Figure 6-16.** A pressure-vent cap has no spring on the vacuum valve. (Chrysler)

the top and bottom. Hot coolant enters the top tank and flows downward to the bottom as it cools, figure 6-12. From the bottom tank, coolant returns to the engine.

Crossflow radiators have a tank at each end. Hot coolant enters an inlet at the top of one tank, flows across the core, and returns to the engine through an outlet at the bottom of the other tank, figure 6-13.

The top tank, or one side tank of a crossflow radiator, has an open neck, sealed with a pressure cap. The bottom tank, or the outlet side tank, is the coolest part of the radiator. On a car with an automatic transmission, this tank contains a transmission fluid cooler, figure 6-14. Fluid from the transmission circulates through the cooler and transfers heat to the coolant just as the radiator transfers heat to the air. Because this adds to the work of the cooling system, radiators with transmission coolers usually are larger than those without.

## Radiator Cap

Cooling system pressure is developed by the action of the water pump and by expansion of the coolant as it warms. Correct pressure is maintained by the radiator cap. The radiator cap has two valves, figures 6-15 and 6-16. The pressure valve relieves excess pressure during engine operation and limits maximum system pressure. This pressure rating is marked on the cap. The vacuum valve allows air to enter the radiator as the coolant cools and contracts when the engine is off. Without these two valves, the cooling system would rupture from excess pressure when hot, and hoses or radiator tanks would collapse from the vacuum created in the system as the coolant cools.

Two slightly different kinds of caps are used on most automobiles:

**1.** A constant-pressure cap, figure 6-15, has a spring-loaded vacuum valve that opens only when radiator pressure is low enough that atmospheric pressure overcomes the light spring. This kind of cap seals the system to the atmosphere when air pressure inside and outside the radiator is equal. You will find these caps on most Ford and GM cars.

**2.** A pressure vent cap, figure 6-16, has no spring on the vacuum valve. The weight of the valve causes it to fall open whenever radiator pressure is equal to, or less than, atmospheric pressure. This kind of cap opens the system to the atmosphere whenever radiator pressure is less than air pressure. You will find these caps on many Chrysler cars.

The pressure valves on both kinds of caps work in the same way, but it is important that the right kind of cap be installed on a particular cooling system.

### Coolant Recovery System

If the radiator is overly full, coolant is forced past the cap pressure valve as it expands. To let this coolant out of the radiator and prevent damage from high pressure, every radiator has an overflow tube in the filler neck. On older cars, the overflow tube has a hose that lets the coolant drain out under the vehicle. This means that part of the coolant is lost.

All late-model vehicles have a coolant recovery system, in which the overflow tube is routed to an unpressurized reservoir, figure 6-17. The reservoir, or overflow tank, is one-third to one-half full of coolant at all times. The overflow tube extends to the bottom of the tank. As coolant is forced out of a hot, pressurized radiator, it flows to the overflow tank, and the coolant level in the tank rises. As cooling system pressure drops when the engine is shut off, atmospheric pressure in the tank forces coolant back to the radiator through the cap vacuum valve.

The cooling system must be airtight when pressurized for the recovery tank to operate correctly. If the system has a leak in a hose, a radiator tank, or elsewhere, air will be drawn into the radiator and coolant will not be withdrawn from the overflow tank. If the system is operating right, the recovery tank helps to keep air out of the system. As pressure drops in the cooling system, only coolant is drawn back through the overflow tube and cap relief valve.

When extra water is added to the system, it is usually added at the overflow tank. Tanks are marked with "hot" and "cold" or "min" and "max" levels. Coolant recovery systems improve overall cooling system operation up to 10 percent. But, you should check coolant level in the radiator as well, to ensure that air is not entering the system and keeping coolant from being withdrawn from the overflow tank.

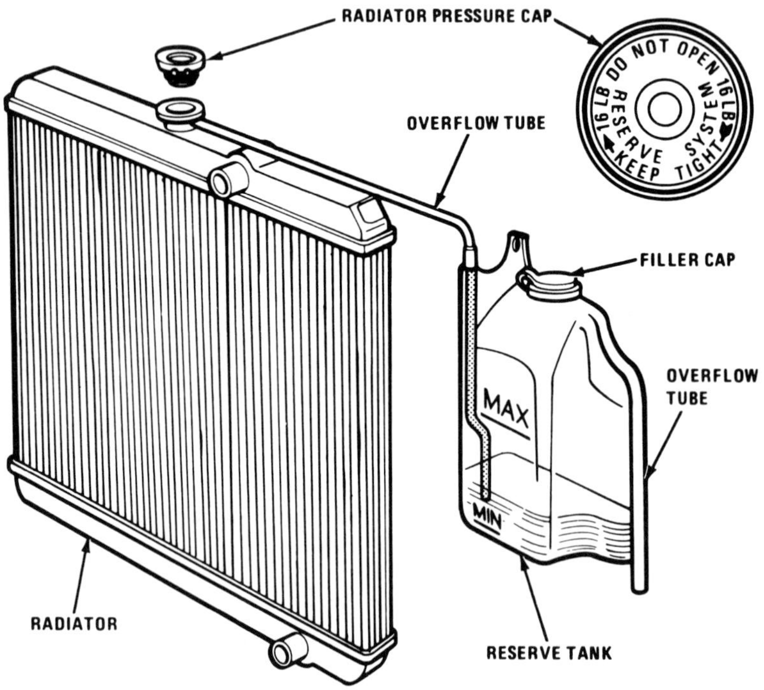


**Figure 6-17.** A coolant recovery system retains coolant overflow in an unpressurized reservoir and returns it to the engine as the engine cools. (Chrysler)

### Hoses

The hoses that connect the engine to the radiator are made of a rubber compound that is compatible with ethylene glycol and that can stand system pressure and high temperature. Most radiator hoses are reinforced with wire to reduce swelling under pressure and collapse as pressure drops when the system cools, figure 6-18. Many hoses are formed to fit particular engine and radiator combinations in specific cars. Others are flexible and made in varying lengths to fit several different vehicles.

Engine coolant is the heat source for the car's heater. Smaller diameter hoses carry coolant to and from the heater and the engine. Coolant flow is controlled by a water valve in the heater system. All cooling system hoses are secured by various kinds of hose clamps, figure 6-19.

### Heater

A vehicle's heating system is a complete accessory system with electric and vacuum controls and sophisticated air distribution. However, the heater core also is part of the cooling system. The heater core is a small radiator inside the passenger compartment, figure 6-3. Coolant flow through the heater core transfers heat to airflow through the heater just as it does to airflow through the car's radiator. Incorrect operation of a car's heater (too much heat or not enough heat) often indicates a problem in the cooling system.

### Fan

When a car travels at cruising speed or faster, there is enough airflow

**Figure 6-18.** Radiator hoses are reinforced with wire. They may be flexible or molded for a particular vehicle installation.

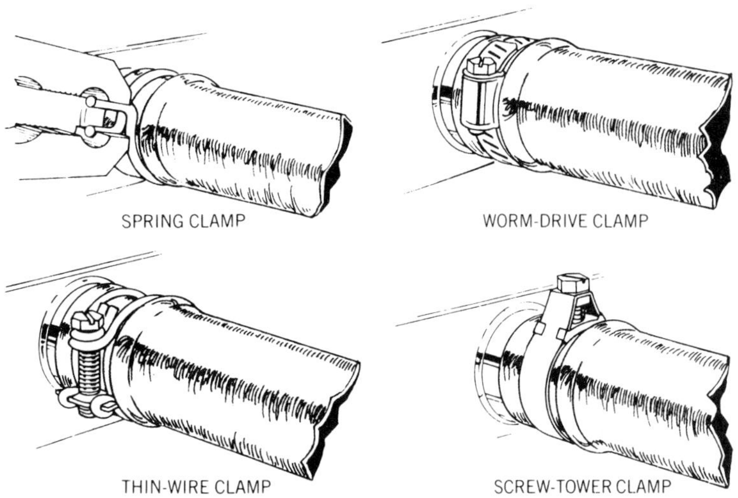


**Figure 6-19.** Most hose clamps are the spring type or the screw type. (Gates)

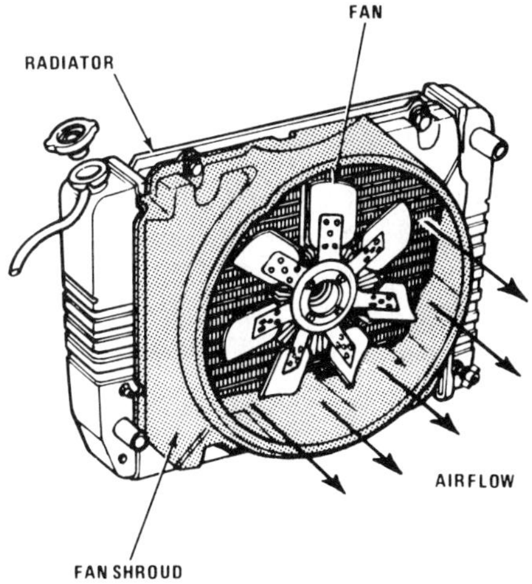


**Figure 6-20.** Many late-model cars have fan shrouds that improve airflow through the radiator. (Chrysler)

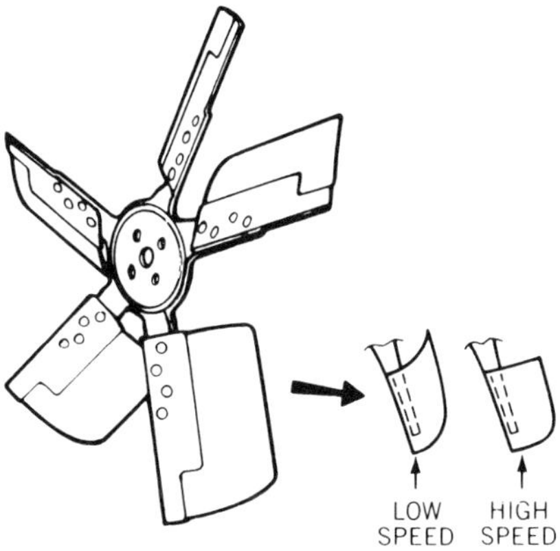


**Figure 6-21.** The pitch, or blade angle, of a flexible fan changes with engine speed. (Chrysler)

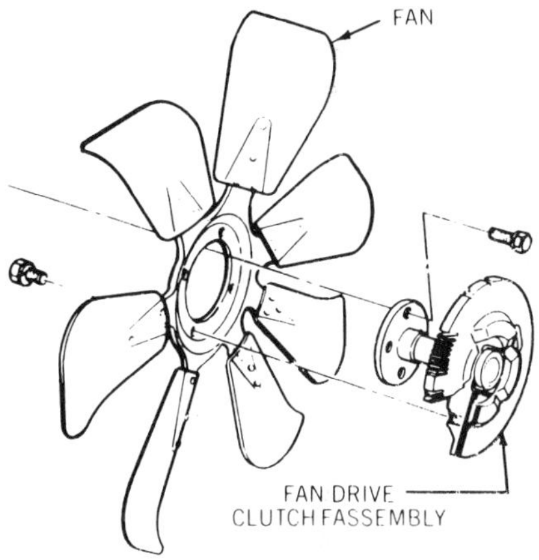


**Figure 6-22.** The viscous-drive clutch of this fan controls fan engagement at high and low temperatures. (Ford)

through the radiator to absorb all the heat the coolant can release. At low speed and idle, there is not enough airflow from vehicle movement. To provide enough airflow for heat transfer at low speed, the cooling system has a fan that draws air through the radiator. In the simplest fan installations, the fan is mounted on the water pump shaft and driven by the same belt that drives the pump.

Many cars have fan shrouds, figure 6-20, that channel the airflow through the radiator and create a venturi effect similar to that in a carburetor (which you will learn about later). To reduce noise, many fans have their blades arranged asymetrically

A fan driven by a belt from the crankshaft pulley wastes power at high speed when it is not needed. Three kinds of special fans eliminate this problem:

1. Flexible fans
2. Clutch fans
3. Electric fans

Flexible fans have blades made of thin, flexible steel or fiberglass that change their pitch, or angle, with engine speed, figure 6- 21. At low speed, the blades are curved sharply, which moves a lot of air but uses a lot of engine power. At high speed, centrifugal force straightens the blades so that they move less air but use less power.

A clutch fan, or fluid-drive fan, has a **viscous-drive** clutch in its hub, figure 6-22. The clutch is filled with silicone fluid that transmits drive belt pulley rotation to the fan. There are two kinds of fluid-drive fans:

1. In one type, the fluid transmits driving torque to the fan. But, at a certain speed, the torque load on on the fluid coupling causes the fan to slip. This limits top speed of the fan and reduces the engine load.

2. In the other kind of fluid-drive fan, the fluid is held in a small reservoir in the clutch hub and metered into the drive mechanism by a **bimetal temperature sensor**. The sensor is in the center of the hub and exposed to air-

flow through the radiator. When airflow is cool, the sensor meters little or no fluid into the clutch drive. The fan is disengaged or only partly engaged so that it idles or turns slowly. When airflow is hot, the sensor meters more fluid into the clutch drive to engage the fan completely.

Many late-model cars have fans driven by electric motors, figure 6-23. This completely removes the fan load from the engine and saves fuel. All front-wheel-drive (fwd) cars with transverse engines have electric fans because there is no practical way to drive the fan from the crankshaft pulley.

A temperature switch that senses coolant temperature turns the fan motor on and off. The switch is connected to the battery side of the ignition switch so that the fan will operate whether the engine is on or off. This allows the fan to circulate air through the radiator when the engine is first turned off and coolant temperature rises in the hot engine and radiator because it is no longer circulated by the water pump.

Whenever you work on a car with an electric fan, disconnect the motor power lead before getting near the fan because it can turn on with no warning if the temperature rises.

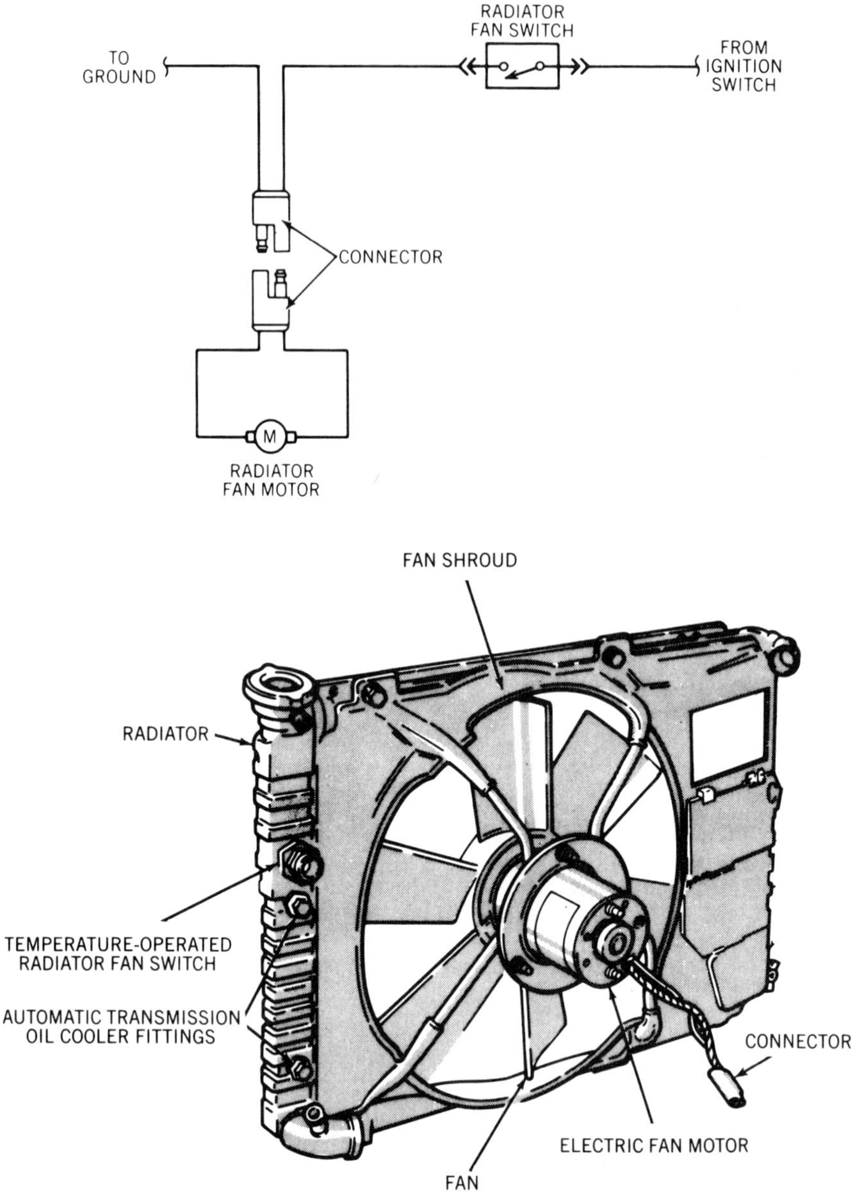


**Figure 6-23.** Electric fans are used on all front-wheel-drive cars with transverse engines, as well as on other cars. The electric motor removes the fan load from the engine completely. (Chrysler)

### Drive Belts

The simplest belt installation is a single V-belt that turns the water pump and alternator. Belt tension is adjusted by moving the alternator on its bracket. Most late-model engines have several drive belts to operate other accessories, such as the air pump, the power steering pump, and the air conditioning compressor, figure 6-24. Some engines use a pair of matched belts to drive several accessories, including the water pump. Other engines have a V-ribbed belt arranged in a serpentine installation to operate several accessories, figure 6-25. Regardless of the number and kind of belts on any engine, the belt that drives the water pump also usually drives the alternator.

When you install a new belt and run the engine for a minute or less, the belt stretches and tension changes. That is why manufacturers provide different tension specifications for new and used belts. That also is why matched pairs of belts must be replaced as pairs. If you install one new belt with a used belt, they will run at different tensions, which can damage the belts or cause slippage. You can check belt tension with a strand tension gauge or by measuring belt deflection, as you will learn in the service chapters of the shop manual that accompanies this text.

## THE EXHAUST SYSTEM

The exhaust system does four important jobs:

**1.** It collects engine exhaust and routes it to the rear of the car so that dangerous exhaust fumes do not enter the passenger area.

**2.** It quiets the exhaust noise.

**3.** It helps to remove air pollutants from the exhaust.

**4.** It helps to regulate engine temper-

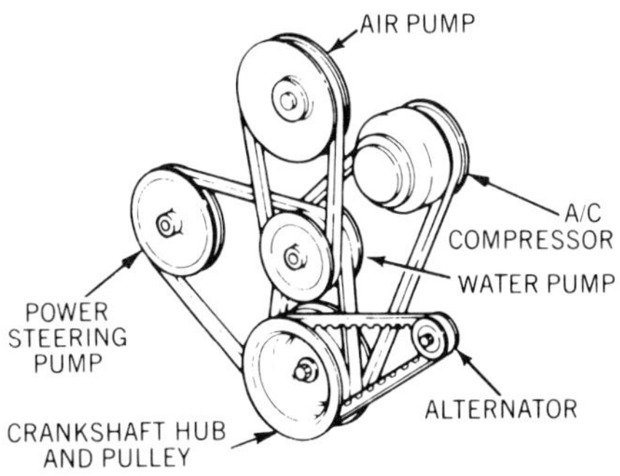


**Figure 6-24.** On this engine, several belts drive multiple accessories. (Chevrolet)

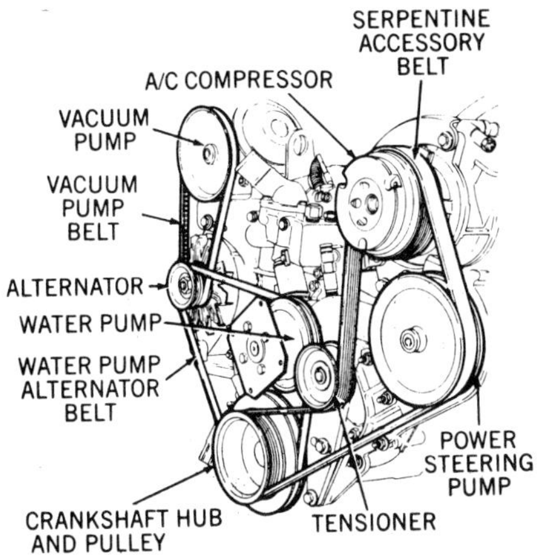


**Figure 6-25.** The serpentine installation of a single V-ribbed belt drives several engine accessories. (Ford)

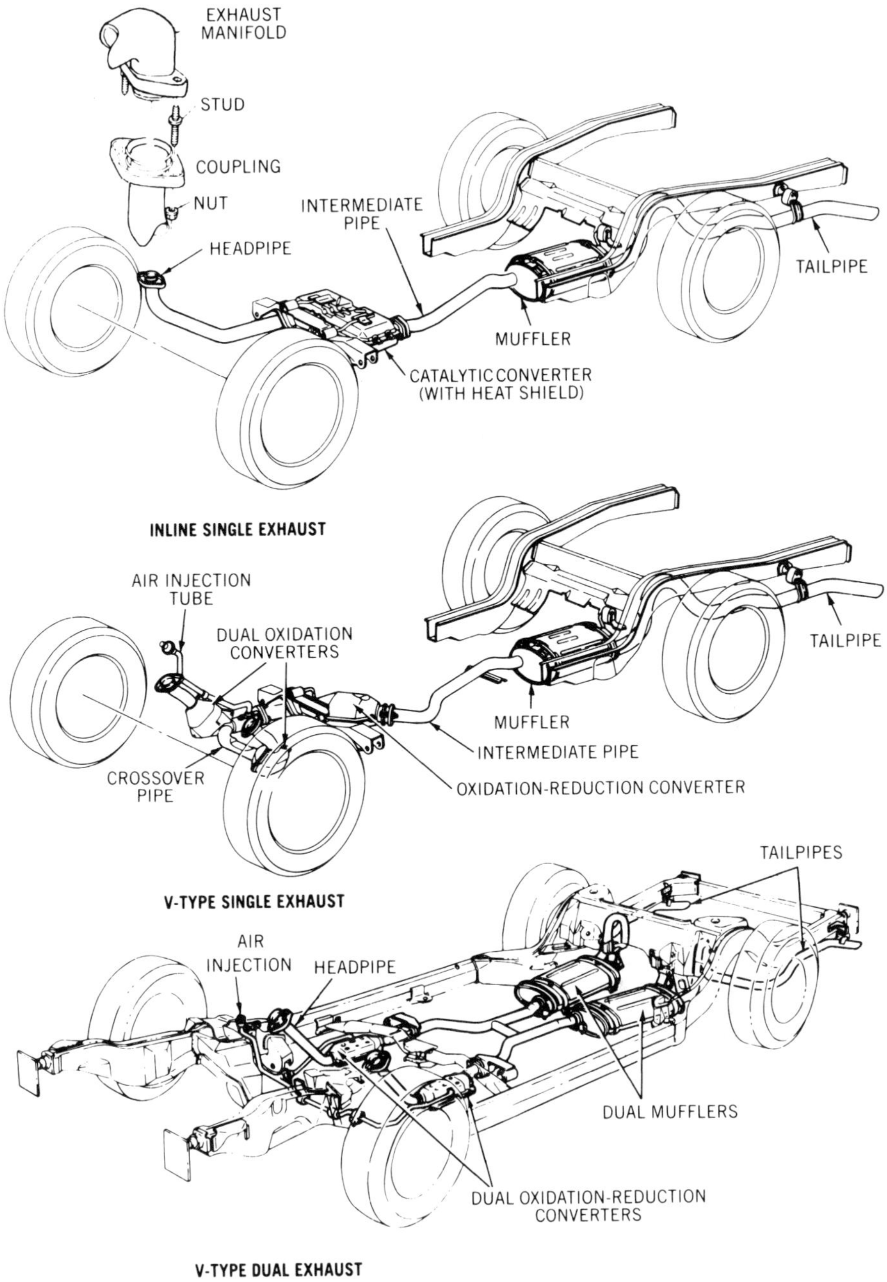


**Figure 6-26.** Typical passenger car exhaust systems. (Ford)

ature and warm the air-fuel mixture of a cold engine.

There are three basic exhaust system designs, figure 6-26:

**1.** *Inline*—Most inline engines have a single exhaust manifold, connected to a single system of pipes, muffler, and catalytic converter.

**2.** *Single V-type*—V-type engines have two exhaust manifolds, one on each head. Most such engines have a crossover pipe, or Y-pipe, beneath or behind the engine that connects the exhaust from both manifolds to a single system of pipes and other parts.

**3.** *Dual V-type*—Some V-type engines have a separate system of pipes and mufflers for each cylinder bank. Some cars have dual tailpipes that branch from a single exhaust pipe near the rear of the vehicle.

Regardless of design, all exhaust systems share the same basic parts:

- Exhaust manifolds
- Manifold heat control valves
- Exhaust pipes and tailpipes
- Mufflers and resonators
- Clamps and brackets
- Catalytic converters.

## Exhaust Manifolds

Like the engine block and head, the exhaust manifold operates at temperatures that range from very cold when the engine is started to over 1,000° F (538° C) when hot. Most exhaust manifolds are made of cast iron for strength and resistance to heat and temperature changes. The manifold is bolted to the head, and its passages match the exhaust ports in the head, figure 6-27. The manifold collects the exhaust from several cylinders and routes it through a single outlet to an exhaust pipe.

The manifold is designed for smooth, uniform exhaust flow from all cylin-

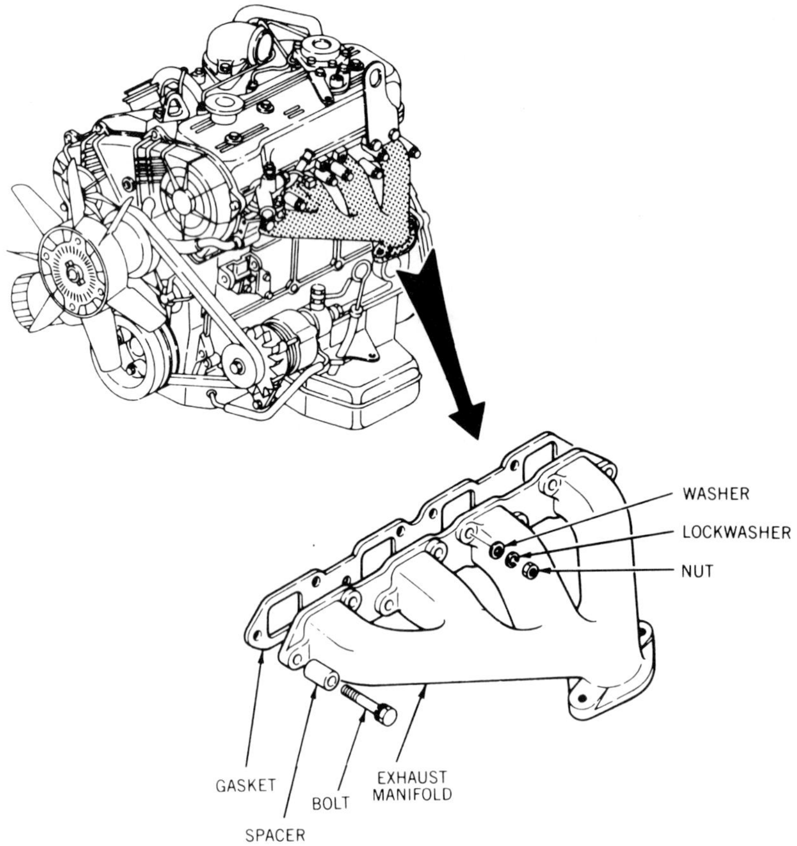


**Figure 6-27.** The exhaust manifold routes exhaust from several cylinders to the exhaust headpipe. (Chevrolet)

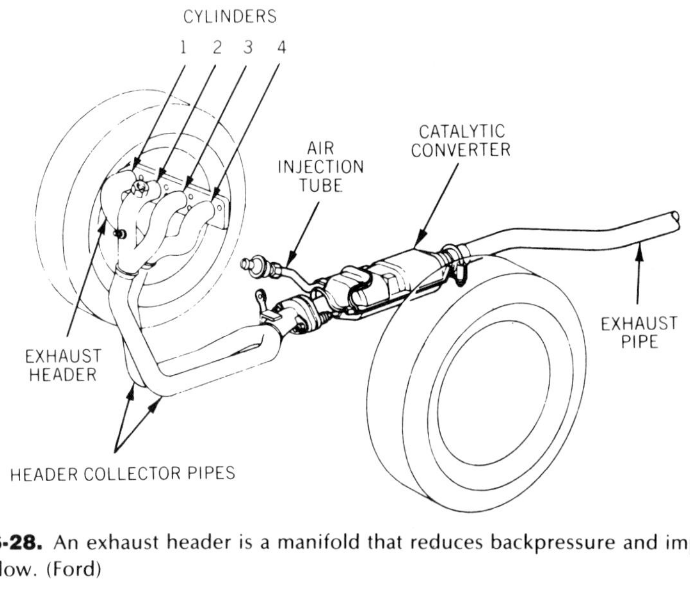


**Figure 6-28.** An exhaust header is a manifold that reduces backpressure and improves exhaust flow. (Ford)

ders to control **backpressure**. This is the pressure that tries to force the exhaust back into the engine. Too much backpressure can prevent all the exhaust from leaving the cylinder before the next intake charge enters. This can preheat and dilute the air-fuel mixture and reduce efficiency. In extreme cases, too much backpressure can damage the engine. Manifold location in the engine compartment, particularly on V-type engines, sometimes causes compromises in efficient exhaust flow. Manifolds must be designed to clear body and steering parts and to keep heat away from batteries, ignition wires, and other parts that could be damaged by high temperature.

Some high-performance engines have exhaust manifolds called **headers**, figure 6-28, formed from heavy steel tubing welded to flanges. Headers often are tuned to improve exhaust flow by scavenging exhaust from the cylinders. Depending on the individual header pipe lengths, low pressure occurs in the headers at different engine speeds. Short header pipes and large collectors improve exhaust flow at high speed. Longer header pipe lengths are tuned to improve torque at mid-range speeds.

Exhaust tuning is a specialized science and was once reserved for high-performance vehicles. However, exhaust backpressure has become a critical factor for the emission systems of electronically controlled engines. Engineers design modern exhaust systems to minimize backpressure that would harm engine volumetric efficiency. But, they provide a specific amount of backpressure to regulate exhaust gas recirculation (EGR).

You will learn more about EGR systems and backpressure control in later chapters. The important point to remember about manifold and exhaust system design is that any change to the system on a late-model car will probably upset the calculated backpressure. This will create incorrect EGR operation and may cause an engine computer to operate in a way that will reduce overall performance.

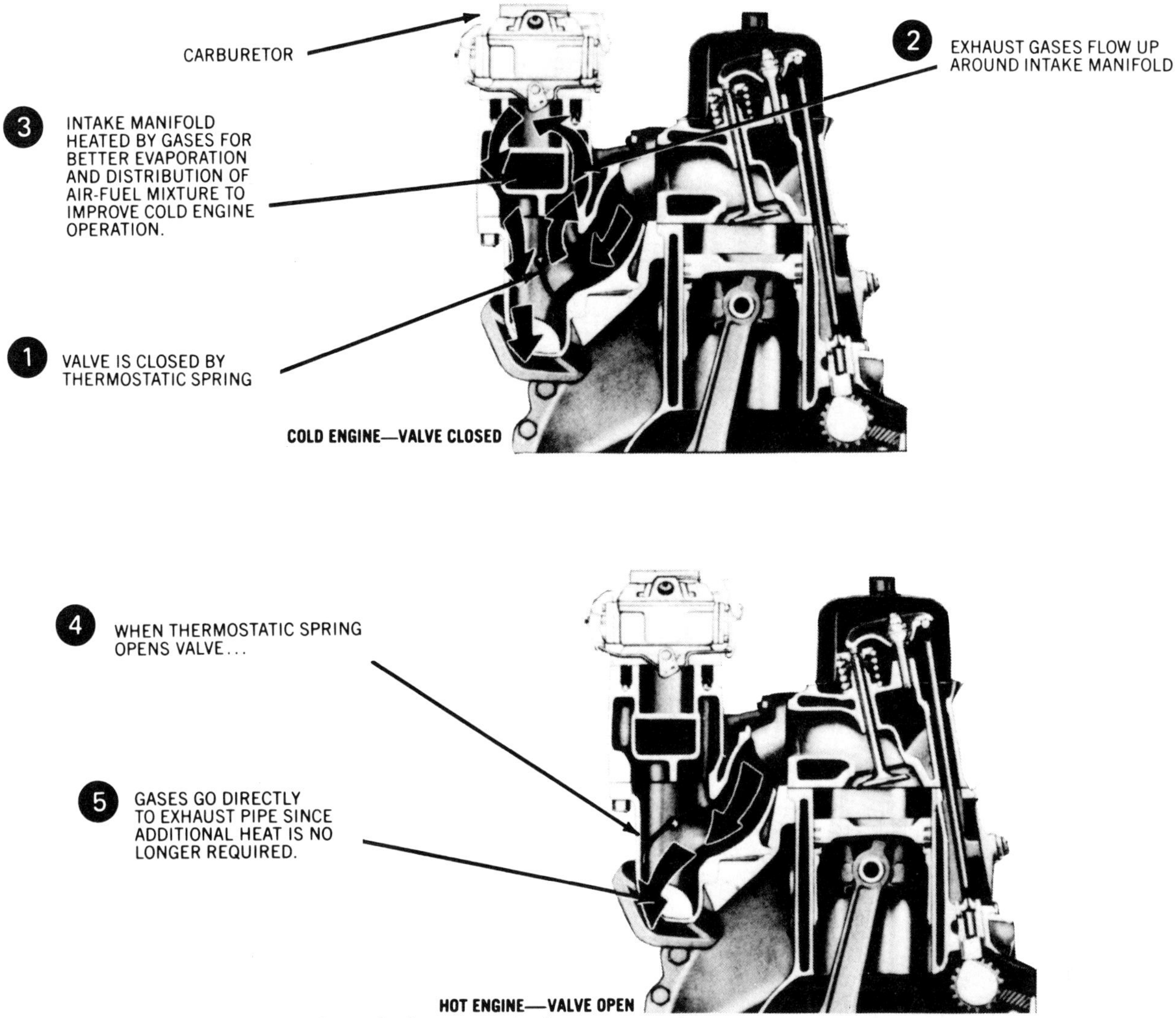


**Figure 6-29.** The thermostatic spring closes the heat control valve to route exhaust to the intake manifold hot spot when the engine is cold. (Ford)

## Manifold Heat Control Valves

When a gasoline engine is cold, fuel does not vaporize easily. Gasoline droplets will fall out of the air-fuel mixture in the intake manifold, and the correct mixture does not reach the cylinders. Gasoline can actually form puddles in the manifold, which cause uneven fuel distribution. For these reasons, the air-fuel mixture in a cold engine must be heated in the intake manifold as soon as the gasoline is mixed with air. In most engines, this is done by routing part of the hot exhaust through bypass passages from the exhaust to a "hot spot" in the intake manifold before it passes out the exhaust pipe. These passages are separate from the intake manifold runners, and the hot spot is below the carburetor or fuel injection throttle body.

Exhaust is directed to the intake manifold by a manifold heat control valve, or heat riser. There are two basic kinds of exhaust-type heat control valves:

1. A thermostatic spring-operated valve
2. A vacuum-operated valve

A thermostatic spring-operated valve is a flapper in the exhaust manifold, with a thermostatic spring and a counterweight. When the engine is cold, spring tension holds the valve closed to route part of the exhaust to the intake manifold hot spot, figure 6-29. As the engine warms, spring tension relaxes. The exhaust flow, along with the counterweight, then causes the flapper valve to open and allow the exhaust directly out of the manifold.

On inline engines with exhaust and intake manifolds on the same side, the two manifolds usually are bolted together near the exhaust outlet. The heat control valve and intake hot spot are in a chamber formed at this point, figure 6-29. On V-type engines, the heat control valve is in one exhaust man-

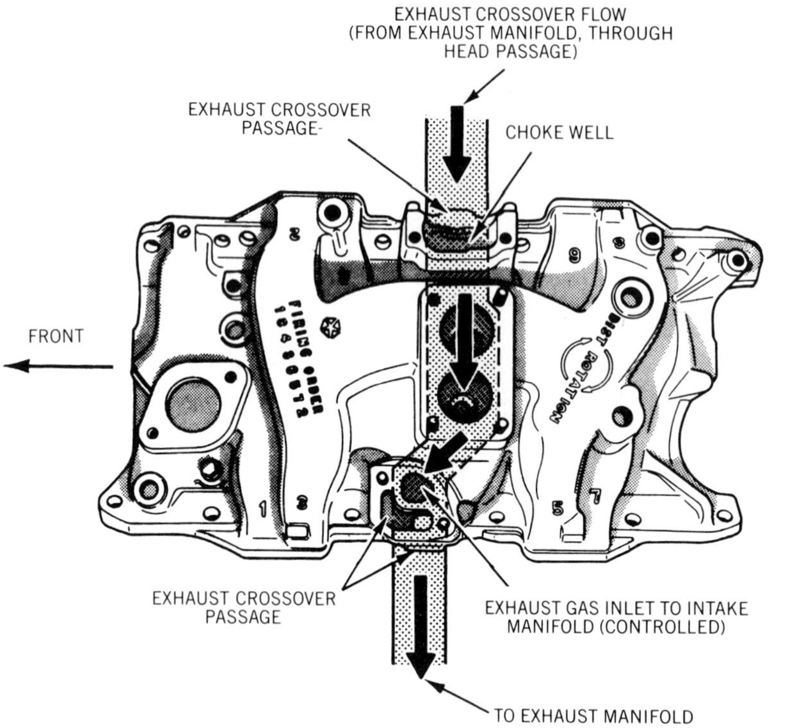


**Figure 6-30.** On this V-type engine, exhaust flows from one manifold through head and intake manifold passages to supply the EGR valve, as well as heat the air-fuel mixture. (Chrysler)

**WHY IS IT CELSIUS, NOT CENTIGRADE?**

The metric temperature scale originally was called the centigrade scale because it measured the temperature of water from freezing to boiling in 100 units of degrees. Centigrade means "one hundred graduations." However, in some European languages, "grad" or "grade" means $\frac{1}{100}$ of a right angle in scientific usage. Therefore, centigrade would mean 0.009 degree of an angle. To eliminate any possibility of confusion in any language, the International Conference on Weights and Measures changed the name of the temperature scale to Celsius in honor of the Swedish astronomer, Anders Celsius (1701-1744), who devised the scale in the first place.

ifold to divert part of the exhaust through passages in the head and intake manifold and out the other side of the engine, figure 6-30.

In 1975, some carmakers started to use vacuum-operated heat control valves for more precise control of intake heat. General Motors was among the first and called its systems early fuel evaporation (EFE) valves. Such a valve has a flapper in the exhaust manifold and exhaust bypass passages similar to a thermostatic spring-operated valve. Vacuum valves, however, are operated by manifold vacuum and spring pressure.

A large vacuum diaphragm is connected by linkage to the heat control valve flapper, figure 6-31. The vacuum line to the diaphragm is connected to a thermostatic vacuum valve or a vacuum solenoid. The valve or solenoid is connected to the intake manifold vacuum. When the engine is cold, vacuum is applied through the open vacuum valve or the solenoid to the diaphragm, which closes the heat valve flapper. A thermostatic vacuum valve is installed in an engine coolant passage. As the coolant warms, the vacuum valve closes and shuts off vacuum to the heat control valve diaphragm, figure 6-31. A solenoid is controlled by an engine computer that gets temperature signals from an engine sensor. When vacuum is shut off, a strong spring in the diaphragm housing opens the heat control valve to remove exhaust heat from the intake manifold.

Because all manifold heat control valves are exposed directly to hot, corrosive exhaust, they tend to collect carbon deposits, or they rust and stick. If the valve sticks open, it can cause hard starting, poor warmup, poor fuel mileage, and crankcase dilution. If the valve sticks closed, it can cause overheating, misfiring, and poor fuel mileage when hot. Heat control valves should be checked periodically and any rough operation corrected before the valve sticks completely.

**Other Manifold Heat Control Methods.** A few engines do not use exhaust to heat the intake manifold. Some engines circulate hot coolant through passages in the intake manifold to warm the air-fuel mixture. Coolant flow is controlled by a thermostat to remove manifold heat when the engine is warm. Other engines, particularly some late-model GM and Ford engines, have an electrically heated wire grid between the carburetor or fuel injection throttle body and the manifold, figure 6-32. The mixture flowing through this wire mesh is heated as it enters the manifold. Current to the heating grid is controlled to remove heat when the engine is warm.

Diesel engines do not use exhaust to heat the air-fuel charge because air and fuel are mixed in the cylinders, not in the manifold. On a diesel, glow plugs preheat the combustion chambers before the engine is started to ensure that the temperature is high enough for ignition.

## Exhaust Pipes And Tailpipes

Exhaust pipes carry the exhaust from the manifolds to the rear of the car. Exhaust pipes are made of large-diameter steel tubing and formed with gradual bends to minimize backpressure. The pipes that run from the manifolds to the catalytic converter or the first muffler are called exhaust pipes, or headpipes, figure 6-26. Because exhaust temperature is highest and water condensation and corrosion are

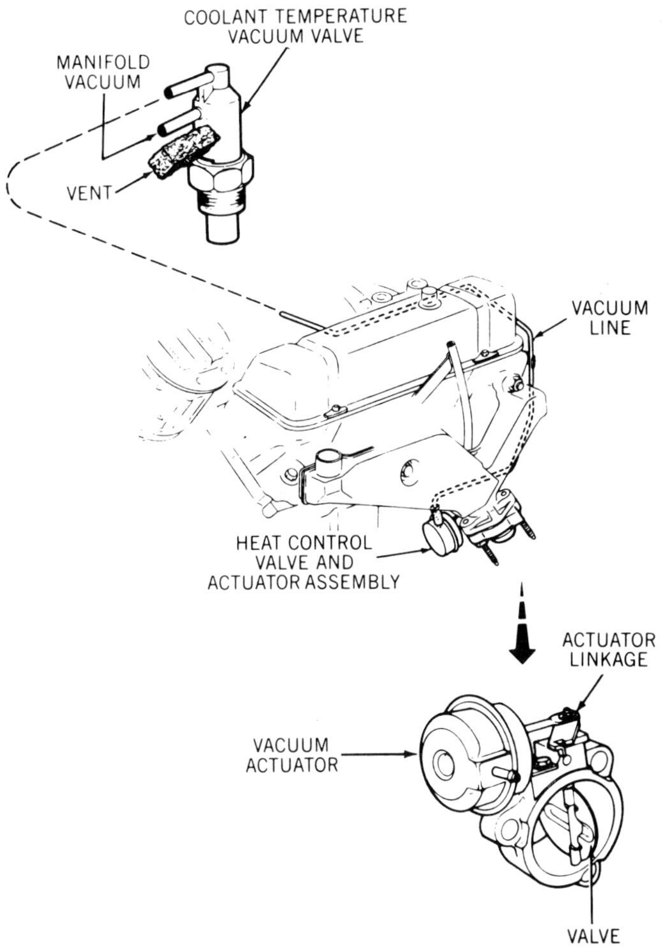


**Figure 6-31.** Manifold vacuum closes this heat control valve. A spring opens the valve when vacuum is cut off by the coolant temperature switch. (Buick)

greatest at the front of the system, these pipes often are made of double-wall tubing. Sometimes, the inner wall can collapse and restrict the system, even though the pipe looks normal on the outside.

The pipes that run from the muffler to the rear of the car are the tailpipes. These usually are single-wall tubing because they are not as susceptible to heat and corrosion as the headpipes. Many late-model cars have several mufflers or resonators in the exhaust system, along with a catalytic converter, connected by intermediate exhaust pipes, figure 6-26.

## Mufflers and Resonators

Mufflers reduce exhaust system noise. Many systems have one large main muffler, followed by one or two smaller resonators. These are carefully designed to reduce noise while developing minimum backpressure. Even though we speak of these parts as mufflers and resonators, depending on their size and location, they work the same way.

Two kinds of mufflers are used on passenger cars, figure 6-33. Most mufflers are the **resonator** type, which contain a series of baffles with the exhaust passages offset from each other.

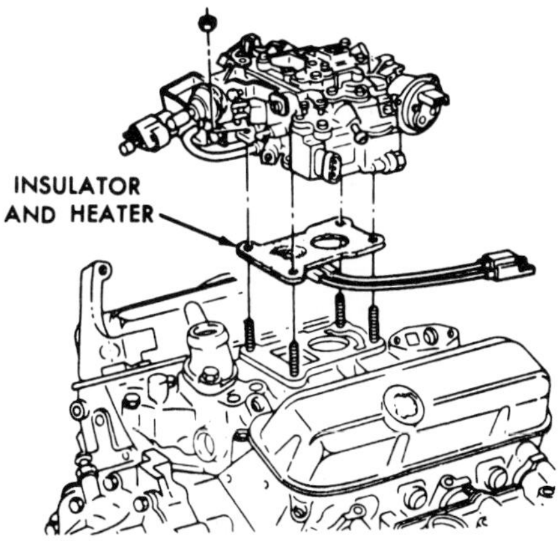


**Figure 6-32.** Some late-model engines have electric grid heaters between the carburetor and the intake manifold to warm the air-fuel mixture of a cold engine. (Chevrolet)

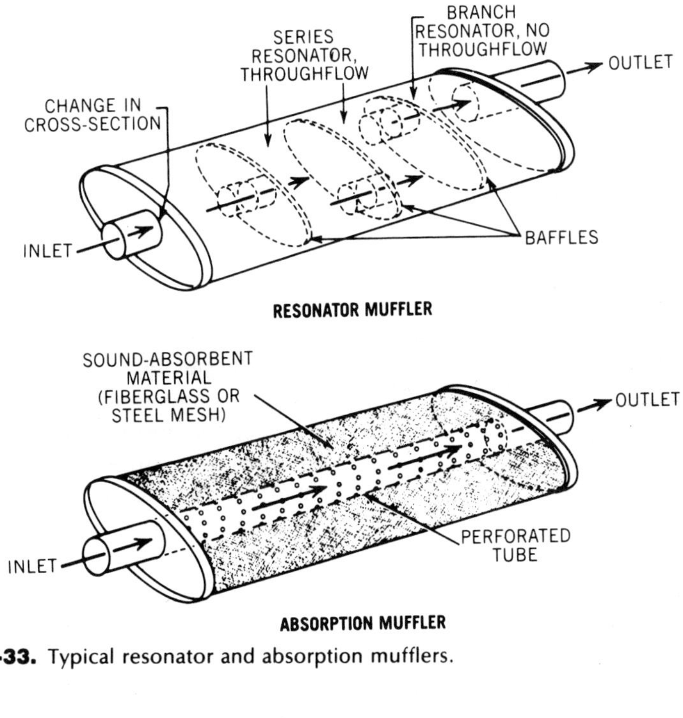


**Figure 6-33.** Typical resonator and absorption mufflers.

Exhaust sound waves reflect off the baffles back toward the noise source (the engine), while exhaust flows through the chambers and offset openings. Resonator mufflers have two to five chambers. Their lengths, diameters, and passage openings are designed so that sound frequencies cancel each other to reduce noise.

**Absorption mufflers** have a straight-through exhaust passage. The inner tube is perforated, and the muffler housing is lined with sound-absorbing material such as fiberglass. Noise penetrates the sound-absorbent material, and sound energy is consumed as friction energy.

## Clamps and Brackets

Exhaust pipes are attached to each other and to mufflers and resonators by U-bolt clamps. The outlet of a front pipe fits inside the inlet of a pipe or muffler to its rear. The U-bolt clamp compresses the joint for a leak-free seal. Exhaust pipes are attached to manifolds and some catalytic converters by a flanged coupling or by a circular compression-type gasket and a bolted collar (a ball joint connection). Figure 6-34 shows typical exhaust system connections. Exhaust system gaskets are made of hard, composition asbestos and metallic materials. All connections are designed to resist heat, vibration, and leakage. To further reduce vibration and noise, exhaust system components are hung from the chassis on rubber-insulated brackets or hangers, figure 6-34.

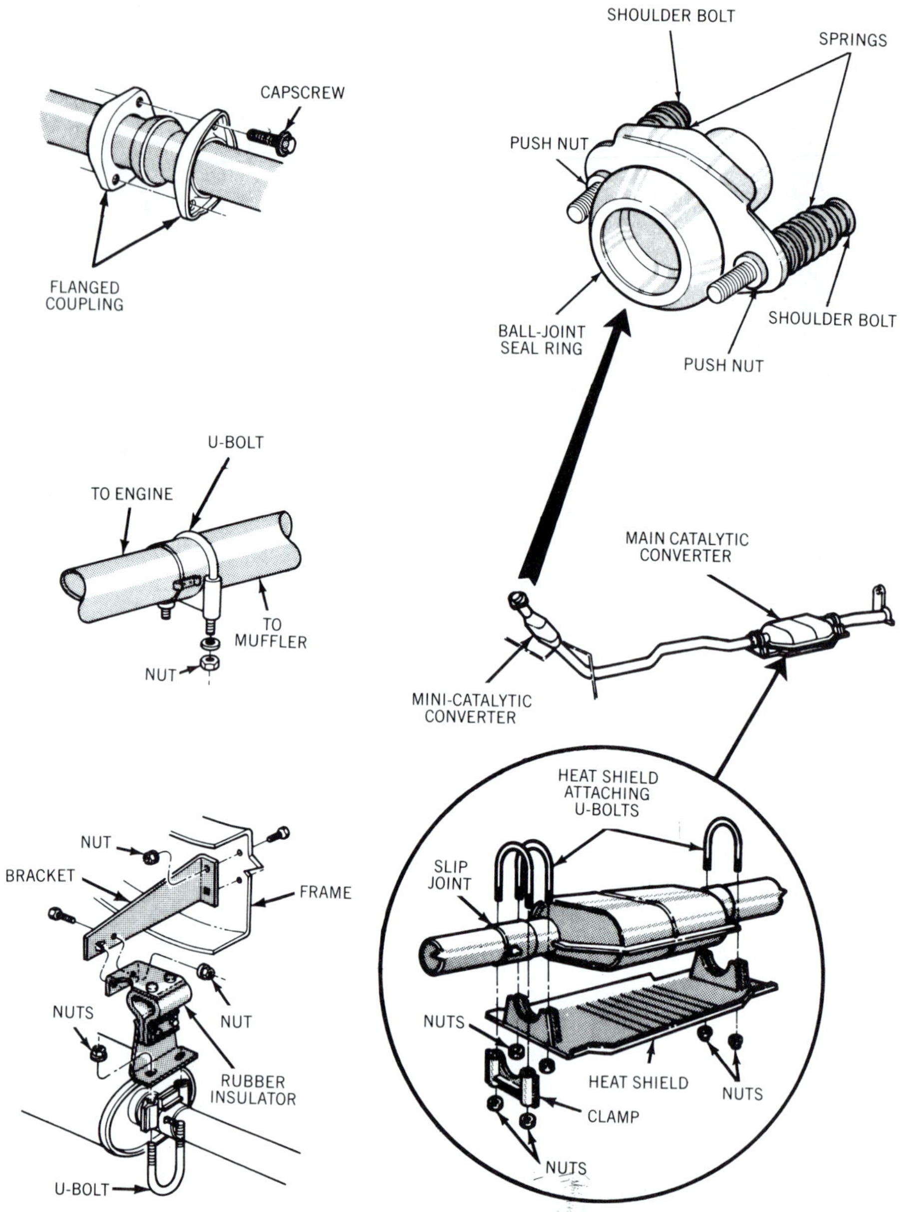


**Figure 6-34.** Typical exhaust system connections and mountings. (Chrysler)

## Catalytic Converters

The catalytic converters on late-model, gasoline-engined vehicles are emission control devices that are part of the exhaust system. Catalytic converters were first used on some 1975 and 1976 cars sold in the United States. The earliest converters were oxidation converters that controlled HC and CO emissions by oxidizing them to $CO_2$ and water vapor. In 1977 and 1978, some carmakers began using reduction converters to reduce $NO_x$ emissions.

An oxidation converter uses a **catalytic reaction** to add oxygen to HC and CO. A reduction converter uses a similar reaction to remove oxygen from $NO_x$. Converter systems that control HC, CO, and $NO_x$ emissions are called 2-stage or 3-way systems because they use both oxidation and reduction reactions to control the three major exhaust emissions. These systems became almost universal on all cars sold in the U.S. by the early 1980's. You will learn more about catalytic converter design and operation in later chapters. For now, we will examine their position in the exhaust system.

In the simplest catalytic converter installation, a single converter is installed between the exhaust manifold and the first muffler. Converters are always installed early in the exhaust system to treat the exhaust soon after it leaves the manifold, figure 6-35. This location also allows exhaust to heat the converter to the necessary reaction temperature. Some engines have a small converter bolted directly to the manifold outlet, figure 6-34. V-type engines with true dual-exhaust systems have separate converters for each

## STRATIFIED CHARGE ENGINES

The stratified charge principle is a way to control combustion. It is not a distinction between designs such as 2-stroke and 4-stroke engines or rotary and reciprocating engines are. In fact, any of these kinds of engines, gasoline or diesel, can be built to operate with a stratified charge.

The idea of the stratified charge is to develop the air-fuel mixture in the cylinders in layers, or strata. In a gasoline engine, the mixture may be as rich as 10:1 or 12:1 in one area, and as lean as 18:1 or 20:1 in another. With these variations in stratified charges, engineers try to create an overall air-fuel ratio for the best combination of power, economy, and emission control.

Honda is the only major carmaker to produce a stratified charge gasoline engine in large numbers. Honda's Compound Vortex Controlled Combustion (CVCC) engine has a small precombustion chamber in each cylinder with its own small intake valve. The carburetor delivers a rich air-fuel mixture through separate manifold passages to the prechamber. The carburetor also delivers a much leaner mixture to the main combustion chambers through other manifold passages. The spark plug is in the prechamber where it ignites the rich mixture for the first stage of combustion. As the mixture burns, the flame front moves through a passage to the main chamber, where it ignites the lean mixture in the second stage of combustion. Except for these two distinct stages of combustion and the prechamber and intake valve, the Honda CVCC engine is a conventional 4-stroke gasoline engine.

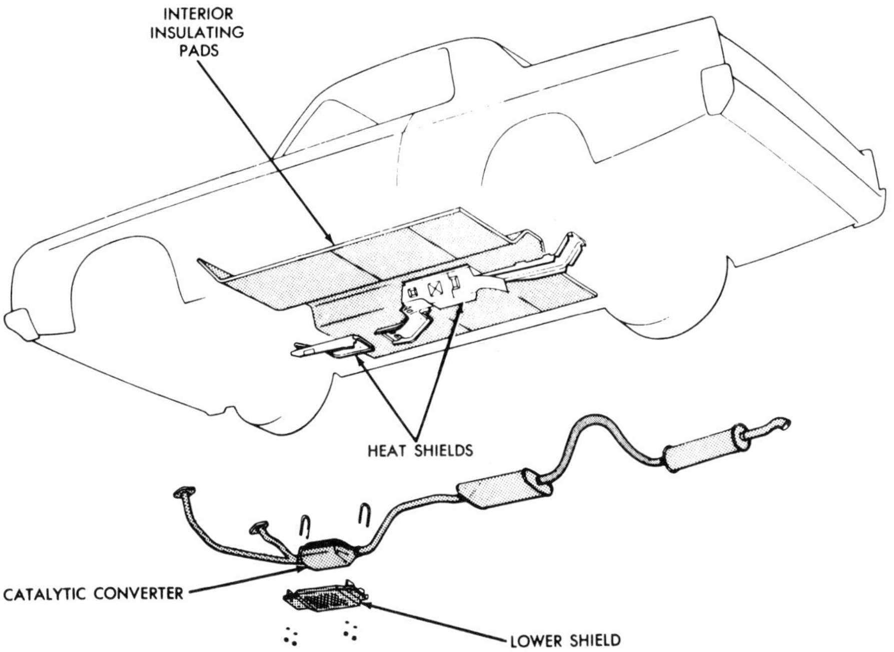


**Figure 6-35.** Typical catalytic converter and heat shield locations. (Chrysler)

engine bank, figure 6-26. Some single-exhaust systems have two converters in series.

Catalytic converters have no moving parts and need no adjustment. Any backpressure created by the converter is calculated into overall system design for efficient operation. Converters are warranted to operate effectively for 5 years or 50,000 miles if the engine is run on unleaded gasoline. Tetraethyl lead used to raise octane in leaded gasoline will coat the surface of catalyst material and reduce or eliminate the catalytic reaction. In extreme cases, "lead poisoning" may restrict exhaust flow through the converter.

The catalytic reaction in a converter creates temperatures as high as 1,500° F (816° C). Most converters have heat shields to reduce heat radiated to the passenger compartment and underneath the vehicle, figure 6-35. Because of these high temperatures, proper engine operation is critical to catalytic converter operation. If an engine misfires repeatedly or runs too rich or too lean, converter temperature can rise to the point where it will damage the catalyst material.

## SUMMARY

Cooling and exhaust systems remove about two-thirds of the heat energy produced in an engine and regulate engine temperature within the best operating range. The exhaust system also aids engine breathing by evacuating exhaust gases from the combustion chambers with minimum backpressure.

Most cooling systems use liquid coolant—half water and half ethylene glycol—that circulates through engine water jackets, hoses, and a radiator. The coolant absorbs heat from the engine and transfers it to cooler airflow through the radiator. A water pump circulates the coolant, and a thermostat controls circulation to maintain correct temperatures. Cooling systems are pressurized at 12 to 17 psi (83 to 117 kPa) to raise the boiling point of the coolant and maintain circulation efficiency. As coolant heats, it expands, and most vehicles have a coolant recovery system to retain coolant overflow and return it to the system as it cools. A radiator cap keeps the cooling system pressurized and relieves vacuum as the system cools. An engine-driven or electric fan draws airflow through the radiator to improve cooling at low speeds.

If an engine runs too cold, performance suffers because of poor fuel vaporization, which produces less power and more HC emissions. Liquid fuel entering the cylinders also washes oil off the cylinder walls, reduces lubrication, and dilutes oil in the crankcase. Cold oil forms sludge, which increases friction and wear and reduces power and efficiency.

If an engine runs too hot, it has lowered volumetric efficiency and performance. The air-fuel mixture may ignite prematurely as it is compressed in an overheated combustion chamber. The oil in a hot engine is overheated, which reduces its viscosity and increases engine wear. Hot oil also forms carbon and varnish deposits that reduce engine efficiency. Hot, thin oil also can be drawn into the combustion chamber where it adds to HC exhaust emissions.

The exhaust system removes engine heat and exhaust with specific, minimum backpressure. Catalytic converters in the exhaust systems of late-model vehicles remove HC, CO, and $NO_x$ from the exhaust through catalytic oxidation and reduction reactions.

## REVIEW QUESTIONS

### MULTIPLE CHOICE

**1.** Which of the following does NOT help determine normal engine operating temperature?

**a.** Radiator size
**b.** Water pump speed
**c.** Amount of coolant in system
**d.** Thermostat opening temperature

**2.** Mechanic A says pressurizing the cooling system decreases the coolant's boiling point. Mechanic B says pressurizing the cooling system increases water pump efficiency. Who is right?

**a.** A only
**b.** B only
**c.** Both A and B
**d.** Neither A nor B

**3.** Which of the following is NOT one of the advantages of an ethylene glycol coolant solution?

**a.** It transfers heat better than pure water.
**b.** It inhibits rust and corrosion.
**c.** It freezes at a lower temperature than pure water.
**d.** It boils at a higher temperature than pure water.

**4.** Head gasket coolant passage restrictions, graduated holes in coolant distribution tubes, and specially shaped water jacket transfer holes are all used to:

**a.** increase the speed of coolant flow through the engine
**b.** direct the coolant toward moving parts that need its lubricating quality
**c.** ensure equal coolant distribution between the front and rear of the engine
**d.** force the coolant toward the radiator inlet tank.

**5.** Mechanic A says a thermostat's temperature rating indicates the temperature at which it is fully open. Mechanic B says cooling system thermostats are either fully open or fully closed at all times. Who is right?

**a.** A only
**b.** B only
**c.** Both A and B
**d.** Neither A nor B

**6.** Which of the following statements is true?

**a.** Removing a thermostat will make an engine run cooler at all times.
**b.** The faster coolant flows through the engine, the better it does its job.
**c.** Removing the thermostat may affect computer-controlled engine systems.
**d.** If a computer-controlled engine overheats, you should install a lower temperature thermostat.

**7.** Mechanic A says the water pump, like an oil pump, may be gear-driven by the camshaft. Mechanic B says the water pump output volume is the same at all engine speeds. Who is right?

**a.** A only
**b.** B only
**c.** Both A and B
**d.** Neither A nor B

**8.** Which of the following materials is not usually used in auto engine radiators?

**a.** Brass
**b.** Iron
**c.** Copper
**d.** Aluminum

**9.** Mechanic A says tube-and-fin radiator cores are stronger than cellular construction cores. Mechanic B says cellular cores are cheaper to build than tube-and-fin. Who is right?

**a.** A only
**b.** B only
**c.** Both A and B
**d.** Neither A nor B

**10.** Automatic transmission fluid coolers:

**a.** transfer heat directly from the fluid to the air
**b.** are usually separate from the cooling system radiator
**c.** are only needed on high-performance models
**d.** are mounted in the coolest part of the radiator.

**11.** On an engine equipped with a coolant recovery system:

**a.** coolant level should be checked only at the radiator
**b.** the coolant level in the tank drops as engine temperature increases
**c.** coolant can flow both out of and into the radiator
**d.** the system will operate correctly even if there's an air leak.

**12.** Mechanic A says the blades on a flexible fan curve more sharply as fan speed increases. Mechanic B says the speed of a flexible fan has no effect on the amount of air it moves. Who is right?

**a.** A only
**b.** B only
**c.** Both A and B
**d.** Neither A nor B

**13.** Clutch (fluid-drive) fans:

**a.** are used on most front-wheel drive cars
**b.** always turn at the same speed as the engine
**c.** use engine oil to drive the fan blades
**d.** may be controlled by a bimetal temperature sensor.

14. Which of the following is not a common arrangement for exhaust systems?

a. Transverse
b. Dual V-type
c. Inline
d. Single V-type

15. Which of the following methods is NOT used to heat the incoming air-fuel mixture in a cold gasoline engine?

a. Coolant flow
b. Glow plugs
c. Exhaust gases
d. Electric heating

16. The type of muffler that uses baffles to cancel sound frequencies is called:

a. header
b. absorption
c. backpressure
d. resonator.

17. All of the following statements about catalytic converters are true EXCEPT:

a. converters are installed as far to the rear in an exhaust system as possible.
b. some engines have more than one catalytic converter
c. catalytic converter backpressure is figured into total system design
d. use of leaded gasoline can reduce the effectiveness of a catalytic converter.

## Fill in the Blank

18. Coolant flow between the internal engine passages and the external radiator is controlled by the ________.

19. The specific percentage of ethylene glycol in a coolant solution ranges from 33% to ________.

20. When the engine is cold, the thermostat ________ coolant flow between the engine and radiator.

21. The thermostat in late-model engines is operated by a ________.

22. Radiators with a tank at each end are called ________.

23. Cooling system pressure is caused by the ________ and by ________.

24. When exhaust cannot leave the cylinder before the next intake charge enters, the exhaust system has too much ________.

25. The exhaust system pipes connecting the manifold to the catalytic converter or the first muffler are called ________.

26. In an unpressurized cooling system, water pump efficiency can be reduced because of ________.

27. The plugs sometimes called "freeze plugs" are really ________ plugs.

28. The two valves in a radiator cap are the ________ and the ________.

29. The small radiator that heats the passenger compartment is called the ________.

30. Matched pairs of belts should always be replaced ________ when one or both break.

# PART THREE BASIC ELECTRICAL SYSTEMS

# INTRODUCTION

The earliest automobiles at the turn of the century had no electrical systems. Electrical systems became essential parts of the automobile, however, by the second decade of the industry's growth. The basic reasons are obvious. Electric power provides the best way to light a lamp, crank an engine, and operate dozens of accessories. Since Charles "Pop" Kettering invented the electric self-starter and the battery-powered inductive-discharge ignition, electrical systems have been integral parts of the combustion-control systems on automobiles.

Today, solid-state electronic systems are equally important parts of engine control systems. A skilled service technician must understand the basics of electricity and the principles of solid-state electronics to do his or her job professionally on late-model vehicles.

Chapters 7 and 8 in this part review the principles of automobile electrical and electronic systems. As you study the starting, charging, ignition, and electronic engine control systems in later chapters, you will see these fundamental principles repeated in various applications.

Chapters 9, 10, and 11 explain the roles of the battery, the starting system, and the charging system in overall engine performance.

# 7 WORKING WITH ELECTRICITY

## WHAT

Chapter 2 introduced you to some principles of science that operate in our automobiles. This chapter explains ways that energy, force, and power work electrically. Unlike the gears, belts, and hydraulic systems of our cars, electricity can't be touched, smelled, seen, or heard. We can, however, see, hear, touch, and measure its effects and actions. In this chapter, you will learn about the electrical principles that light the lamps, crank the engine, ignite the fuel, and do other jobs in all the systems of an automobile.

## WHY

People often consider the electrical system of a car as comprising the battery, the alternator, the starter motor, the ignition, and the vehicle lighting. These are not independent systems, however, particularly in the area of engine performance. On late-model cars, electrical and electronic controls integrate the operation of all combustion control systems. Although these systems may seem complex at first, they operate on the basic electrical principles explained in this chapter. If you understand the fundamentals of current, voltage, and resistance, you can recognize their operation in several vehicle systems and make your service work easy and accurate.

## GOALS

After studying this chapter, you should be able to:

**1.** Explain and give examples of materials that are conductors, insulators, and semiconductors.

**2.** Define current flow and explain the conventional and electron flow theories.

**3.** Define current, voltage, and resistance and the units for measuring each.

**4.** Explain Ohm's and Kirchhoff's laws.

**5.** Explain and illustrate series, parallel, and series-parallel circuits.

**6.** Explain how electricity is generated by friction, heat, light, pressure, chemistry, and magnetism.

**7.** Explain magnetic lines of force, polarity, permeability, and reluctance.

**8.** Explain electromagnetic principles and electromagnetic induction.

**9.** Explain capacitance and capacitor uses.

## ATOMIC REVIEW

Remember from chapter 2 that the outer, valence, electron shell of an atom never has more than eight electrons. If an atom has three or fewer valence electrons, they are loosely held **free electrons**. These electrons will move easily into the electron shells of other atoms.

If an atom has five or more valence electrons, they are held tightly in the valence shell. It takes more electron energy to dislodge them or to let another electron enter the shell. These are called **bound electrons**.

Remember also that electrons are constantly in motion, either around the nucleus of their own atom or among the valence shells of other atoms. This electron movement can be a random drift, or it can be a controlled movement from one point to another. If it is the latter, electron movement becomes **electric current**. Current flow produces electric power. You know that for any energy to produce useful work and power, some force must be applied to it. The **electromotive force** that produces current flow is **voltage**. Before we look at current and voltage more closely, however, we should examine a few facts about the materials that conduct and insulate current flow.

### Conductors, Insulators, and Semiconductors

Materials whose atoms have three or fewer valence electrons are good **conductors** of electric current because it takes little force to move the free electrons from one atom to another. Most metals, such as gold, copper, aluminum, and iron, are good conductors because of their valence electron structure, figure 7-1.

Materials with five or more valence electrons are poor conductors because it takes more force to move the bound outer electrons from one atom to another. These materials are **insulators**. Most gases, such as oxygen and nitrogen, have five or more valence electrons and are good insulators. Many good insulators used in electrical systems are not single elements but compounds that have full valence shells. Glass and rubber are examples of these. Many plastics are compounds created especially for use as insulators.

There are five elements that have just four electrons in their outer shells.

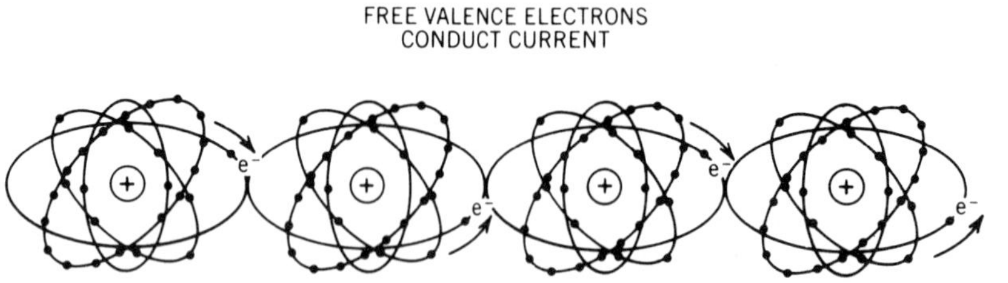


**Figure 7-1.** Copper is a good conductor because its free valence electrons move easily from atom to atom.

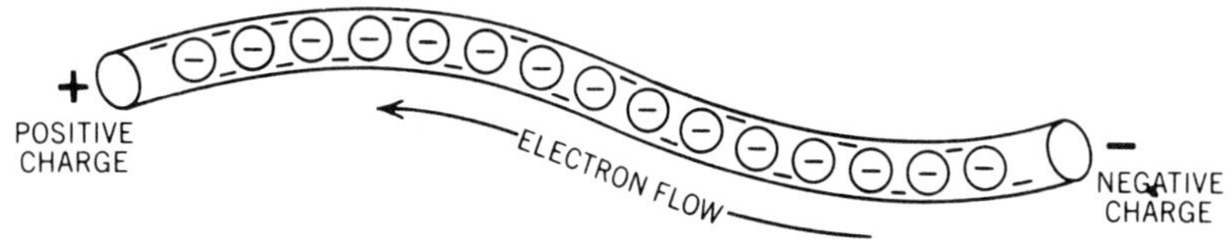


**Figure 7-2.** Electrons can be forced through a wire by placing a positive charge at one end and a negative charge at the other.

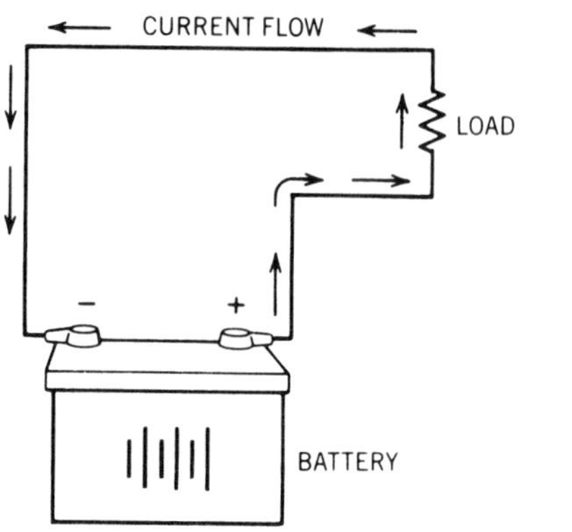


**Figure 7-3.** The conventional theory says that current flows from positive to negative.

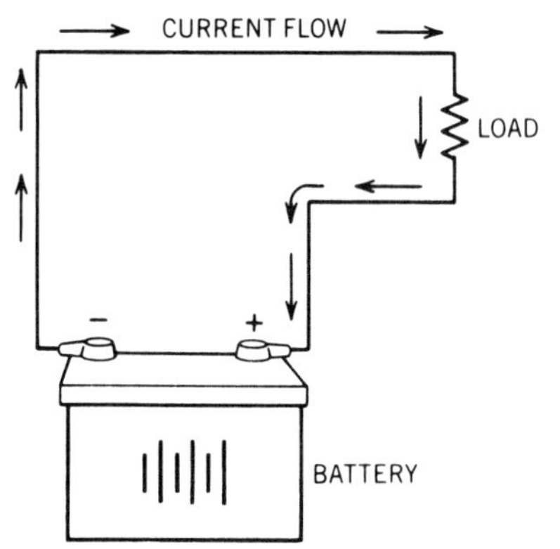


**Figure 7-4.** The electron theory says that current flows from negative to positive.

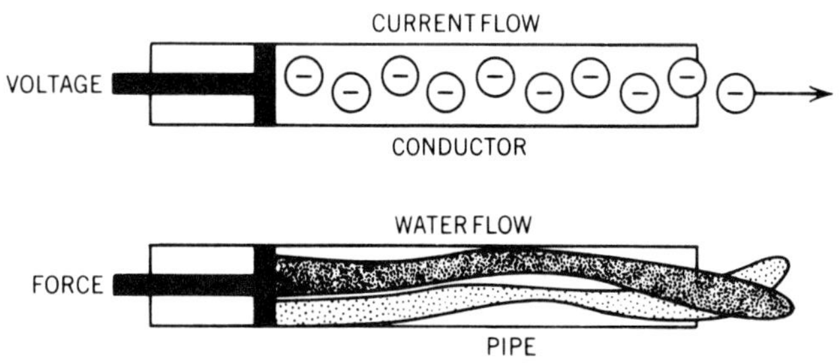


**Figure 7-5.** Voltage causes current to flow in a conductor just as force causes water to flow in a pipe.

These are neither good conductors nor good insulators. Electric current can flow in them under controlled conditions. These elements are **semiconductors**. The most important from an electrical viewpoint is silicon. Silicon-based semiconductors are the foundation of solid-state electronic devices, which you will study later.

## CURRENT FLOW

To understand current flow, you should remember the rules of positive and negative charges. Unlike charges (+ and −) attract each other. Like charges (+ and + or − and −) repel each other. The electrons of an atom have negative charges and are attracted to positive charges. The free electrons in a good conductor, like a piece of copper wire, are attracted by positive charges and will move easily from atom to atom. To create current flow in a copper wire, we need a positive charge at one end and a negative charge at the other.

Later in this chapter, we will explain how these charges develop at opposite ends of a wire. For now, just assume that they exist, figure 7-2. The free electron of an atom near the positive end of the wire is attracted by the positive charge. As the electron leaves its valence shell and moves toward the positive charge, the atom becomes unbalanced; it becomes positively charged. This atom then attracts the electron from another nearby atom. The electrons of neighboring atoms move toward the positive end of the wire. They are helped along by the negative charges at the other end of the wire. The result is controlled, directed current flow. Current will continue to flow as long as the positive and negative charges exist at opposite ends of the wire.

### Current Flow Theories

Historically, current flow has been described by two theories. These are the **conventional theory** and the **electron theory**. The conventional theory is the older of the two. It says that current flows from a positive charge to a negative charge, figure 7-3. The electron theory says that current flows from a negative charge to a positive charge, figure 7-4.

Traditionally, the conventional theory has been used to describe current flow in automobile systems and is still more common. Since the development of automotive electronic systems, such as on-board computers, the electron theory also has been used in automotive service. You can use either theory accurately as long as you use it consistently and don't mix the two. For some examples in this chapter, we will use both; but we will identify each theory as we use it.

### The Ampere

Current flow in a conductor is similar to water flow in a pipe, figure 7-5. When we measure water flow, we count ounces or gallons of water moving past one point in a given time: gallons per minute, for example. When we measure current flow, we count electrons moving past one point in a given time. Since electrons are so small and move so fast, we don't count individual electrons to measure current flow. We use the **ampere** as the unit to measure current. An ampere is to one electron what a gallon is to a single molecule of water. One ampere equals 6.28 billion billion electrons. When that many electrons pass one point in a conductor in one second, one ampere of current flows.

### Direct and Alternating Current

There are two kinds of current flow that we use for practical electric power: **direct current (dc)** and **alternating current (ac)**. Direct current is current that always flows in the same direction. It is the kind of current obtained from batteries. Alternating current changes directions between two points from positive to negative and back to positive. Each cycle of alternating current occurs in an equal amount of time. The electric power in our homes and shops is ac power and alternates at 60 cycles per second, or 60 **hertz (Hz)**.

Both ac and dc systems exist in an automobile. Most of the electrical devices in a car are 12-volt dc devices. They receive voltage and current from a 12-volt battery, which is a source of dc power. The battery, however, is charged by an ac generator, or alternator. The alternating current from the alternator is changed to direct current before it reaches the battery.

### Circuit—A Current Path

The single wire in figure 7-2 is not a complete electrical circuit. For current to flow, it must have a complete **circuit**, or circular path, from negative charge to positive and back to negative again. Figure 7-6 shows a simple circuit that consists of an energy source (a battery), conductors (wires), and a unit to do work (a lamp).

Following the electron flow theory, energy of the battery causes current to flow from the negative terminal, through the wires, to the lamp. The lamp changes electrical energy into light and heat energy. Current then flows back to the positive terminal of the battery. The battery, itself, completes the circuit as current flows from the positive terminal to the negative terminal. For current to flow, a circuit must have **continuity**, that is, it must be continuous. If the circuit is broken at any point, current will not flow, figure 7-7. To open and close a circuit, we use a switch, as shown in figure 7-7.

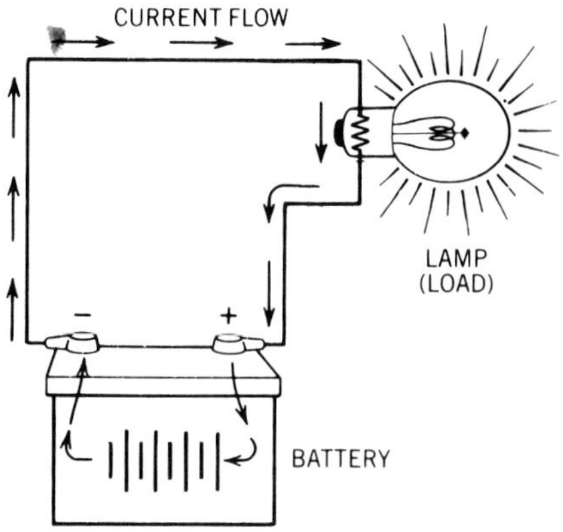


**Figure 7-6.** Current flow requires a complete circuit.

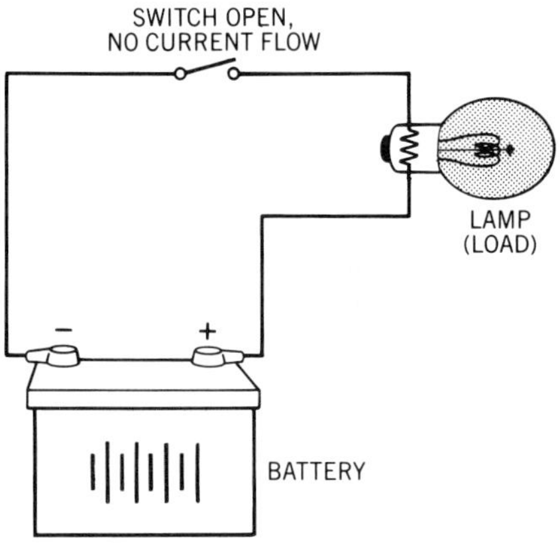


**Figure 7-7.** If a circuit is broken, no current will flow.

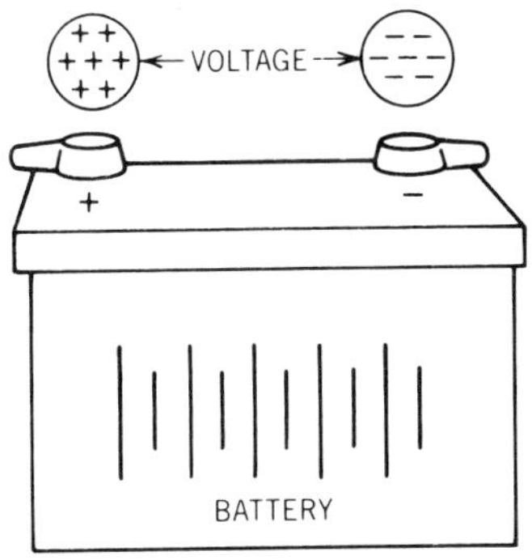


**Figure 7-8.** Voltage is a potential difference in force between two points.

## VOLTAGE

Voltage is the electromotive force that causes current to flow. Voltage is a form of energy. When a circuit is complete and current is flowing, voltage is released electrical energy. Voltage also is potential energy. When a battery is not connected to a circuit, it still contains potential voltage in the form of a negatively charged terminal at one end and a positively charged terminal at the other, figure 7-8. The strength of this energy, or the amount of force, depends on the strength of the positive and negative charges and the difference between them. Voltage is high if the charges are strong with a large difference between them. Voltage is low if the charges are weak and have little difference. A potential of 120 volts is 10 times as strong as a potential of 12 volts.

Voltage also exists between any two points in a circuit when current is

**JUMPING JUMPER CABLES**

You've probably used jumper cables several times to start a car with a dead battery, but did you know that you can use them as a quick troubleshooting tool? Remember that current flows in opposite directions through the two cables, to and from the dead battery, and a magnetic field forms around any current-carrying wire. The higher the current flow, the stronger the field. Also remember that the fields around two cables with current flowing in opposite directions will repel each other.

To use your jumper cables as troubleshooting tools, connect them in the usual way between the booster battery and the disabled vehicle, but be sure they are lying together, side by side. Now try to start the disabled car. If the starter is frozen or shorted, there will be an exceptionally high current draw. Watch the jumper cables. If they jump apart, current draw is high enough to create magnetic fields strong enough to force the cables away from each other. If they do jump, a shorted or jammed starter is a good bet. If they don't jump, but the car doesn't start, look for a problem in the starter control circuit or someplace other than the motor. Only a problem in the motor circuit will draw enough current to make your jumper cables jump.

flowing. The only time voltage does not exist is when the potential difference in force falls to zero. You will often hear voltage referred to simply as "potential" because of this.

### The Volt and Voltage Sources

Just as the ampere is the unit used to measure current, the **volt** is the unit used to measure voltage. One volt is the amount of energy, or force, needed to move one ampere through one ohm of resistance. You will learn about resistance and ohms in the next section.

In an automobile, the principal voltage sources are the battery and the alternator. In a battery, chemical energy creates the potential electrical energy between the two battery terminals. Mechanical energy creates the voltage potential between two terminals of the alternator.

## RESISTANCE

When you studied hydraulics, you learned that a hydraulic system must have some resistance to fluid flow in order to develop pressure and do useful work. So it is with electrical systems, too. An electric circuit must contain resistance to current flow to change electrical energy to heat, light, or movement. If a circuit contained no resistance, the flow of electrons would be like a flood of water from a broken dam. It would do more damage than good.

All conductors have some resistance to current flow, just as all pipes have some resistance to fluid flow. This resistance is the friction of the moving molecules. In electrical wire, it is almost negligible. The important resistance in an electric circuit is in the form of the working devices, such as lamps and motors. These are often called the **electrical load** of the circuit. The lamp in figures 7-6 and 7-7 is the load on the circuit. It has more resistance than the conductors or switches. Current flows *through* the lamp, however, and the same amount that enters one side leaves the other. Voltage, though, *decreases* from one side of the lamp to the other because the lamp is the point where electrical energy is changed to heat and light energy.

Resistance also exists at a point where a circuit is broken, or opened by a switch, figure 7-7. This is infinite resistance. It prevents current flow but does no useful work because electrical energy is not changed to any other form of energy. The important differences between the resistance of a load and the resistance of a broken, or open, circuit are:

- Current flows through a load, and voltage decreases from one side of the load to the other.
- Current does not flow through a break, and the same potential voltage exists on both sides of the break as exists at the voltage source.

### The Ohm and Factors of Resistance

The unit used to measure the amount of resistance in a circuit is the **ohm**. One ohm is the resistance present when one volt is required to move one ampere of current through a circuit. Five factors determine the amount of resistance in any part of a circuit. The following examples refer to conductors, but remember that loads are also conductors even though they have resistance.

1. *Atomic structure*—Materials with few free electrons offer high resistance to current.
2. *Conductor length*—Moving electrons constantly collide with the atoms of a conductor. The longer the conductor, the farther current must flow, and the more collisions occur.
3. *Conductor cross-sectional area*—A thin conductor has a narrow current flow path and more resistance than a thick conductor.
4. *Conductor temperature*—High temperature causes high resistance in most conductors.
5. *Conductor condition*—A completely broken conductor has infinite resistance. Anything that similarly reduces the area for current flow through a conductor also increases resistance. This includes corrosion, looseness, or a partially broken conductor.

## OHM'S AND KIRCHHOFF'S LAWS

The work of two 19th-century German scientists gave us some principles, or laws, that explain the relationships of current, voltage, and resistance in circuits. These laws have practical value in your automobile electrical service work.

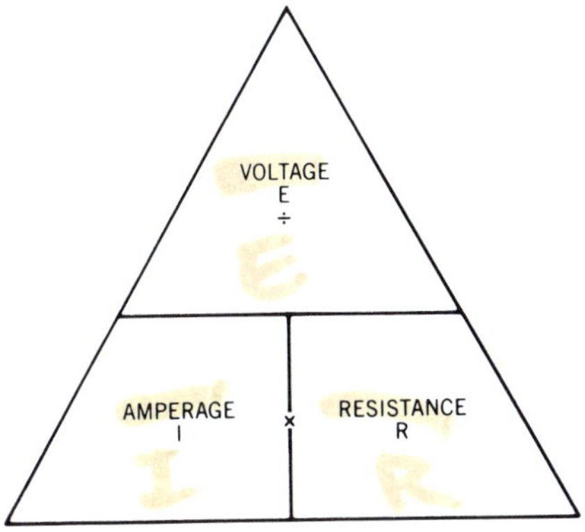


**Figure 7-9.** Use Ohm's law to calculate voltage, current, or resistance in a circuit if you know two of the values. Cover the unknown value and multiply or divide the other two.

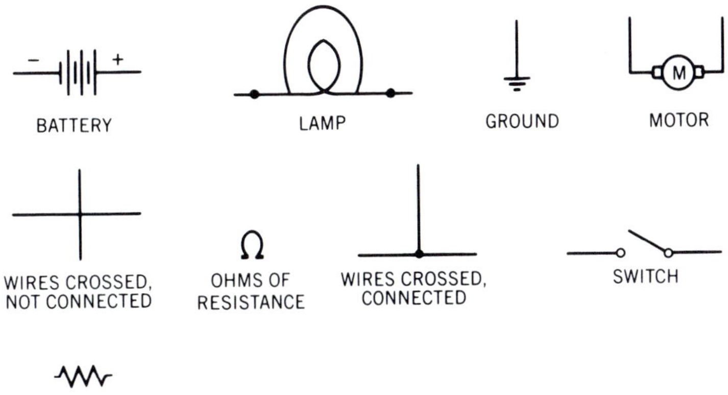


**Figure 7-10.** These are just a few of the electrical symbols you will learn as you study automobile electrical systems.

### Ohm's Law

The measurement of electrical resistance was named after Georg Simon Ohm, and we touched on his law when we said that one ohm is the amount of resistance present when one volt moves one ampere of current through a circuit. Ohm stated the relationship of voltage, current, and resistance as a simple equation that can be written three ways:

1. Voltage = current × resistance.
2. Current = voltage/resistance.
3. Resistance = voltage/current.

These equations usually are written with voltage abbreviated as *E* for electromotive force, current (amperes) abbreviated as *I* for intensity, and resistance abbreviated as *R*. The equations above can be written:

1. $E = IR$.
2. $I = E/R$.
3. $R = E/I$.

Figure 7-9 shows that if you know any two values for a circuit, you can find the third through simple multiplication or division.

### Kirchhoff's Laws

Another 19th-century German scientist, Gustav Robert Kirchhoff, discovered two more facts about voltage and current. His laws are really examples of the principles of conservation of matter and energy. Neither matter nor energy can be created nor destroyed, only changed. Kirchhoff's voltage law says:

> The sum of the voltage drops across all of the loads (items of resistance) in a circuit must equal the source voltage.

Kirchhoff's current law is:

> The sum of the currents flowing into any point in a circuit equals the sum of the currents flowing out of the same point.

Electrons do not collect at any load or other point in a circuit and stop flowing. This is one reason that a circuit must be unbroken for any current to flow at all.

## SIMPLE CIRCUITS

Ohm's and Kirchhoff's laws may seem a little abstract; but by looking at some simple circuits, you can see how they work. When you study and troubleshoot specific electrical systems, you will put these laws to work. To understand these laws better, we will look at three basic kinds of circuits: series, parallel, and series-parallel. To follow the diagrams for each circuit, you must recognize some symbols for electrical devices, such as batteries, loads, resistance, switches, and so on, figure 7-10.

### Series Circuit

In a **series circuit**, current flow has only one path, figure 7-11. Following conventional current flow theory (+ to −), you can see that current flows from the battery, through the resistor, and back to the battery. Figure 7-12 shows a circuit of a battery, two motors, and a switch in series. When the switch is closed and the circuit complete, current flows through both motors. Here are some applications of Ohm's and Kirchhoff's laws to a series circuit:

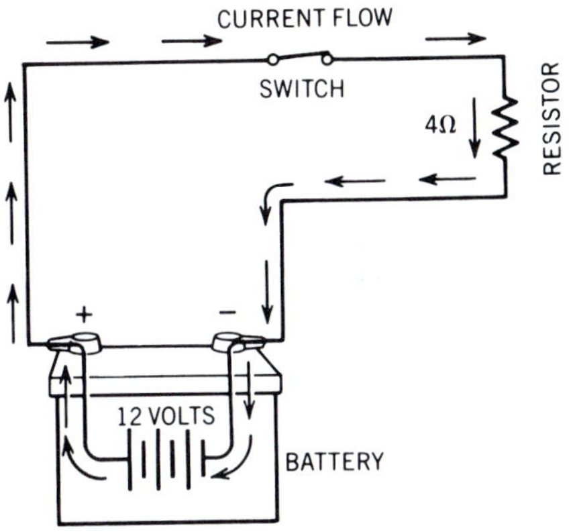


**Figure 7-11.** In a series circuit, all current flows equally through every part of the circuit. There is only one current path.

1. Because there is only one path for current flow, amperage is the same at every point in the circuit.
2. The total resistance of the circuit is the sum of the individual loads (resistors) in series. This can be one or any number of separate loads.
3. The total source voltage must be dropped across the total resistance load of the circuit. If the load is just one resistor, all voltage must be dropped across that one load. If there are several loads in the circuit, voltage drop

across each is proportional to its resistance. Total voltage drop equals total source voltage.

Using Ohm's law, you can figure out the voltage, current flow, and resistance of the circuits in figures 7-11 and 7-12. Assume that the switches have no resistance when closed. In figure 7-11, you have a 4-ohm resistor connected to a 12-volt battery. You can calculate the amperage using Ohm's law thus:

$E/R = I$

$12/4 = 3$ amperes

In figure 7-12, you have two 3-ohm resistors in the form of motors. Total series resistance equals 6 ohms. Voltage is again 12 volts, and again you calculate the circuit amperage:

$E/R = I$

$12/6 = 2$ amperes

If you know the amount of current and the individual resistance of each load, you can calculate the source voltage or the voltage drop across each load.

For figure 7-11:

$I = 3$ amperes, $R = 4$ ohms

$I \times R = E$

$3 \times 4 = 12$ volts

For figure 7-12:

$I = 2$ amperes

$R_1sb = 3$ ohms,

$R_2sb = 3$ ohms,

$R_t$ (total) $= 3 + 3 = 6$ ohms

$I \times R = E$

$2 \times 6 = 12$ volts

For figure 7-12, the voltage drop across each motor equals 2 amperes × 3 ohms, or 6 volts. The drop of 6 volts across each of two motors equals the source voltage of 12 volts, which proves Kirchhoff's voltage law.

If you know the current flow in a series circuit and the source voltage, you can calculate the total resistance as follows:

For figure 7-11:

$E/I = R$

$12/3 = 4$ ohms

For figure 7-12:

$12/2 = 6$ ohms total for both motors

You can use the same formula to calculate the resistance of each motor in series, but you must know the voltage drop across each. The 6 ohms of total resistance in figure 7-12 could be 4 ohms and 2 ohms. But to prove it, you would have to know that one motor dropped 8 volts and the other dropped 4 volts.

### Parallel Circuit

In a **parallel circuit**, current flow has two or more paths that it can follow. Current flow splits into parallel branches at junction points. The parallel branches are **shunt** circuits. Here are some applications of Ohm's and Kirchhoff's laws in parallel circuits:

**1.** The voltage across each branch, or applied to each parallel branch, is the same.

**2.** Total resistance of a parallel circuit is less than the lowest individual resistance because as you add resistors, or loads, in parallel, you are adding more conductors.

**3.** Total amperage in a parallel circuit equals the sum of the branch amperages.

Let's take the two motors from figure 7-12 and put them in parallel to see

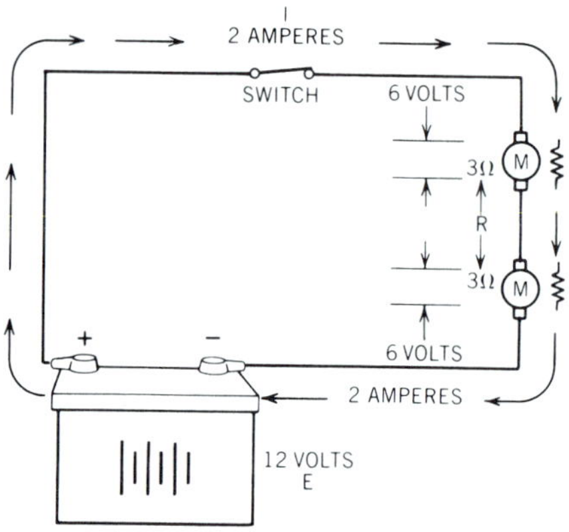


Total Resistance 3Ω + 3Ω = 6Ω
Amperage E ÷ R = I
12 ÷ 6 = 2
Voltage I × R = E
2 × 6 = 12
Resistance E ÷ I = R
12 ÷ 2 = 6

**Figure 7-12.** When the switch is closed and the circuit complete, current flows equally through both motors in series.

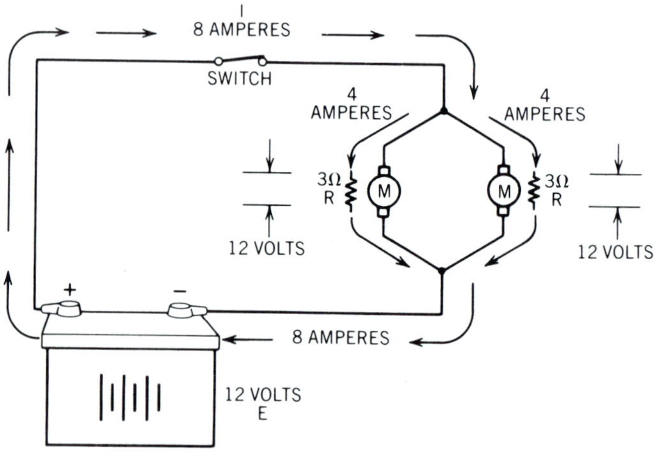


Total (Equivalent) Resistance $\frac{3\Omega \times 3\Omega}{3\Omega + 3\Omega}$ $\frac{9\Omega}{6\Omega}$ 1.5Ω

Branch Amperage E ÷ R = I
12 ÷ 3 = 4
Branch Voltage I × R = E
4 × 3 = 12
Branch Resistance E ÷ I = R
12 ÷ 4 = 3

Total Amperage E ÷ R = I
12 ÷ 1.5 = 8
Total Voltage I × R = E
8 × 1.5 = 12
Total Resistance E ÷ I = R
12 ÷ 8 = 1.5

**Figure 7-13.** In a parallel circuit, current flow has two or more alternate paths.

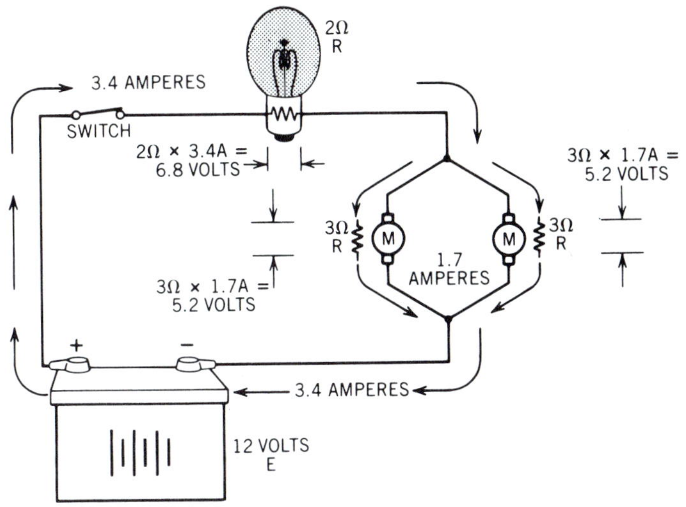


Total Series-Parallel Resistance 2Ω + 1.5Ω = 3.5Ω

Total Amperage $E \div R = I$
$12 \div 3.5 = 3.4$

Equivalent Resistance $\dfrac{3\Omega \times 3\Omega}{3\Omega + 3\Omega}$
$\dfrac{9\Omega}{6\Omega}$
1.5Ω

Branch Amperage $E \div R = I$
$5.2 \div 3 = 1.7$

**Figure 7-14.** In a series-parallel circuit, some loads are in series; others are in parallel.

how it affects current flow and total circuit resistance, figure 7-13. Current now flows independently through each motor. If you remove one, you still have a complete circuit through the other.

Total resistance of a parallel circuit is called **equivalent resistance** because it is equivalent to the resistance of a single load in series with the voltage source. You can calculate this equivalent resistance in two ways. If the circuit has only two loads, as in figure 7-13, you can use the product-over-the-sum method:

$$\frac{3 \times 3 \text{ ohms}}{3 + 3 \text{ ohms}} = \frac{9}{6} = 1.5 \text{ ohms}$$

If you want to use this method to calculate the equivalent resistance for a parallel circuit with more than two loads, you must repeat the formula for two resistances at a time until you determine one final resistance for the complete circuit. You also can use another formula that allows you to calculate the equivalent resistance for any number of parallel loads:

$$\frac{1}{R_t} = \frac{1}{R_1} + \frac{1}{R_2} + \frac{1}{R_3} + \cdots + \frac{1}{R_n}$$

To calculate amperage in a parallel circuit, you must treat each branch as a separate circuit because current in each branch is different. Voltage is equal across each branch, so divide the source voltage by the branch resistance to determine the branch amperage. For either of the motors in figure 7-13, do it like this:

$$E/R = I$$
$$12/3 = 4 \text{ amperes in each branch}$$

Total current flow for the complete circuit is 8 amperes because there are two identical branches. If you didn't know the resistance of each branch but did know the source voltage and the branch amperages, you could calculate each resistance as follows:

$$E/I = R$$
$$12/4 = 3 \text{ ohms}$$

You also can calculate the equivalent resistance for both motors in parallel as follows, using the source voltage and the total circuit amperage:

$$E/I = R$$
$$12/8 = 1.5 \text{ ohms}$$

That is the same value you got earlier, using the product-over-the-sum method with the two branch resistances.

If you know the resistance and current flow of any part of the circuit, or of the entire circuit, you can determine source voltage two ways. Using the resistance and amperage for one branch:

$$I \times R = E$$
$$4 \times 3 = 12 \text{ volts}$$

Using the total circuit amperage and equivalent resistance:

$$I \times R = E$$
$$8 \times 1.5 = 12 \text{ volts}$$

### Series-Parallel Circuits

To troubleshoot automobile circuits, you sometimes test one part of a large system as a simple series circuit. Other times, you will have to test loads in parallel as in the examples above. Most complete automobile circuits, however, are **series-parallel circuits**. Some loads are in series with others, and some are in parallel. Let's take the same two motors in parallel from figure 7-13 and add an indicator lamp in series with them, figure 7-14. The two 3-ohm motors remain in parallel with each other, but the 2-ohm indicator is in series with both.

All of the rules you have learned for series circuits and for parallel circuits apply in series-parallel circuits. To calculate the voltage, amperage, or resistance for any part, or all, of the circuit, begin by reducing the parallel branches to equivalent series loads. Then combine the equivalent values with any actual series loads.

The 3-ohm motors in parallel, figure 7-14, have an equivalent resistance of 1.5 ohms. Now, add this value to the 2-ohm resistance of the lamp to determine total circuit resistance:

$$1.5 + 2 = 3.5 \text{ ohms}$$

You know that the source voltage is 12 volts, so divide 12 volts by 3.5 ohms total resistance to determine total current:

$$12/3.5 = 3.4 \text{ amperes}$$

Total current in figure 7-14 is less than in figure 7-13 because total resistance is greater. Also, voltage across the parallel branches is less than 12 volts because part of the source voltage is dropped across the 2-ohm lamp. To calculate the voltage drop across the lamp, multiply the total current by the resistance of the lamp:

$$I \times R = E$$

$3.4 \times 2 = 6.8$ volts dropped across, indicator lamp

This leaves 5.2 volts across both parallel branches. Current in each branch is

$$E/R = I$$

$$5.2/3 = 1.7 \text{ amperes}$$

All values balance as they are supposed to. The sum of the voltage drops equals the source voltage ($6.8 + 5.2 = 12$). The amperage in each parallel branch adds up to the total circuit amperage ($1.7 + 1.7 = 3.4$).

This has been a brief introduction to basic electrical circuits. Testing and repairing electrical systems are not difficult if you understand the principles of circuits and take one troubleshooting step at a time.

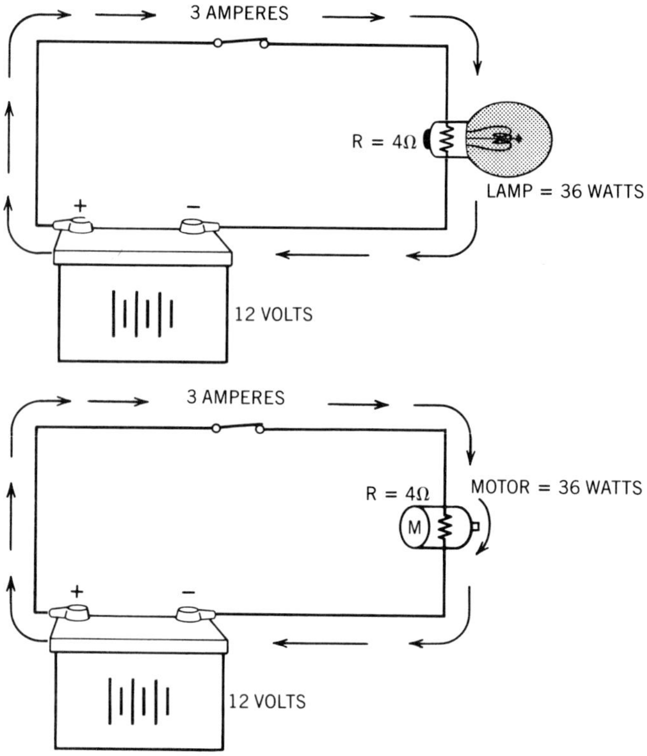


**Figure 7-15.** Watts, or wattage, is the measurement of electrical power. Watts = amperes × volts.

## ELECTRIC POWER—WATTS

You have learned that current flow is electricity at work and that voltage is the force, or energy, that causes the work to be done. You also know from chapter 2 that power is the rate at which work is done. The unit of electric power is the **watt**. You calculate watts by multiplying current times voltage. Using the symbol *P* for power, the formula is

$$P = E \times I$$

Thus, a circuit with 3 amperes of current flowing and a source voltage of 12 volts produces 36 watts of power. This power can be an electric motor, or a lamp, or some other device that does work, figure 7-15. Wattage ratings are often given for bulbs, motors, and other devices. Using this rating and the formulas of Ohm's law, you can calculate other circuit values. For example, a 36-watt lamp in a 12-volt circuit draws 3 amperes of current because

If $P = E \times I$, then $I = P/E$

36 watts/12 volts = 3 amperes

Therefore, the resistance of that 36-watt bulb must be 4 ohms because, using Ohm's law,

$$E/I = R$$

$$12/3 = 4 \text{ ohms}$$

The electric power formula ($P = E \times I$) relates directly to Ohm's law:

If $E = IR$, then $P = (IR)I$ or $P = I^2R$

If $I = E/R$, then $P = E(E/R)$ or $P = E^2/R$

These formulas can help your electrical troubleshooting. If you know the wattage rating of a bulb, you can calculate what the circuit resistance ought to be and compare that to the measured resistance of the circuit.

### Watts and Horsepower

Since the watt and the horsepower are both units for measuring the rate at which work is done, there is a relationship between them. Horsepower equals force times distance times time. One horsepower equals 200 pounds moved 165 feet in one minute, or 33,000 foot-pounds per minute. This equals 550 foot-pounds per second. Voltage is a measurement of force, similar to pounds. Amperage is a measurement of rate of movement, similar to feet per minute. (Remember that 1 ampere = 6.28 × 1018 electrons per second.) When one volt of force moves that many electrons per second (one ampere), we have one watt of power. To think of the difference between one watt and one horsepower very simply, one electron is much smaller than a foot. Therefore, one watt is smaller than one horsepower. In fact, 1 horsepower = 746 watts.

## SOURCES OF ELECTRICITY

Up to this point, we have talked about voltage and current simply as existing without much thought about how electricity is generated. Remember that electrical energy, like any energy, is generated by changing the form of some other energy. There are six common ways to do this. Examples of all six exist in your car, but chemistry and magnetism are the most important.

**1.** *Friction*—If you rub some materials together, friction transfers electrons from one surface to another. One surface becomes positively charged; the other becomes negatively charged. These charges are stationary, or static, so this kind of voltage is called **static electricity**. The surfaces remain positively and negatively charged until they are connected or brought close enough together so that the potential difference can be discharged through brief current flow. Tires rolling on pavement can build up a static charge on a vehicle frame that might be discharged if the frame is connected back to the ground. That's why gasoline trucks have straps connected to the frame that drag on the ground to eliminate the static charge as fast as it collects.

**2.** *Heat*—If you join two pieces of two different metals at both ends and heat one junction, a small amount of current will flow. The current might be only a millionth of an ampere, but it can be used in a temperature-sensing device called a **thermocouple**. This kind of electricity is **thermoelectricity**, and it is used in some engine temperature sensors for electronic control systems.

**3.** *Light*—If you expose certain metals to light, some of the light energy releases free electrons in the metal. If you connect one end of a conductor to the metal and the other end to a point with a different potential charge, the free electrons will flow as current. This is called **photoelectricity**, and it is used in automatic headlamp dimmers.

**4.** *Pressure*—If you apply pressure to some crystals, such as quartz or a diamond, voltage develops between the surfaces of the crystal. This is **piezoelectricity** and is how a phonograph needle works. Piezoelectricity also is used to create voltage or to vary resistance in some engine detonation sensors.

**5.** *Chemistry*—If you put two different materials (usually metals or metal compounds) in a conductive and reactive solution (such as an acid), you can create a voltage between the two materials. One becomes positively charged; the other becomes negatively charged. This happens because of the action of the valence electrons of the two metals and the conductive solution. The action is based on the principles you learned in chapter 2 and is called **electrochemistry**. This is the way an automobile battery develops voltage.

**6.** *Magnetism*—If you pass an electrical conductor through the lines of force that surround a magnet, you can cause current to flow in the conductor. **Electromagnetic** action is the most important source of electricity in an automobile. Magnetism generates electricity in an alternator and allows the electrical energy to do work in various electric motors. Electromagnetic action also provides the high voltage necessary to ignite the air-fuel mixture in an engine. The next section outlines the electromagnetic principles that operate in different automobile systems.

## PRINCIPLES OF MAGNETISM

Magnetism is caused by the movement of electrons in some materials and is recognized by the force it exerts on other materials. The properties of *magnetism* are similar to, but not the same as, the properties of electricity. All materials have electrical conductivity and resistance. All materials have the magnetic properties of permeability and reluctance. Although these properties are not the same, their relationships are similar. Also, electric current flow depends on the force of potential energy between opposite positive and negative terminals. Magnetic lines of force depend on the attraction and repulsion of opposite magnetic poles. Every electrical principle has a magnetic **analogy**.

### Lines of Force

Iron is the most common magnetic material. Some other metals have magnetic properties but not nearly as strong as those of iron. Still other materials—elements or compounds—such as aluminum, glass, wood, and all gases, cannot be magnetized at all. We recognize magnetism by the presence of magnetic lines of force around an object. These lines of force are a **magnetic field** caused by the alignment of atoms within the material. One theory says that the electrons of an atom have circles of force around them. When the electrons of an iron bar are aligned so that the circles of force add together, the iron is magnetized.

In a bar of iron that is magnetized, the lines of force in the magnetic field concentrate at the ends of the bar and form closed parallel loops around the bar, figure 7-16. The lines have direc-

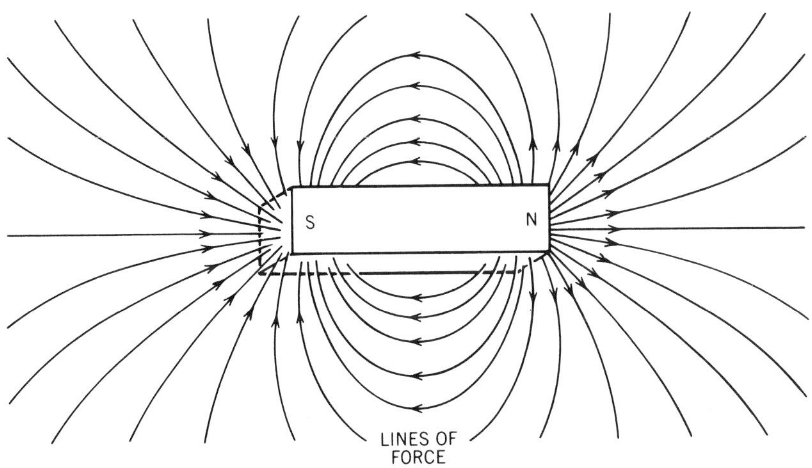


**Figure 7-16.** Magnetic lines of force form a magnetic field that surrounds a magnet.

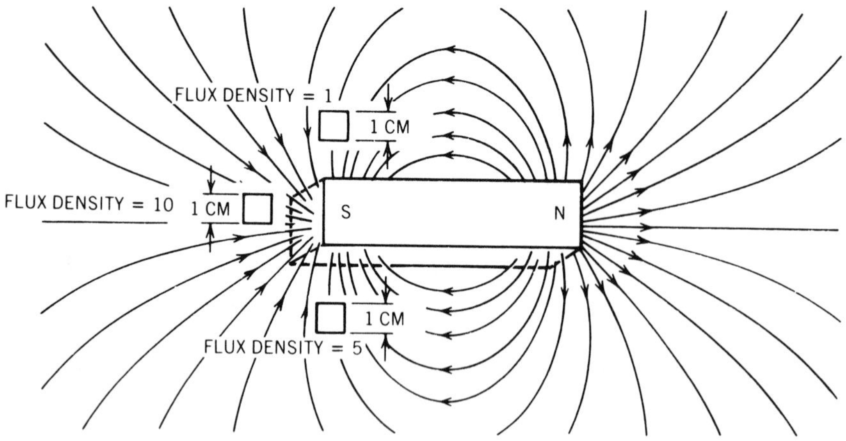


**Figure 7-17.** Flux density is the number of flux lines per square centimeter. It is strongest near the poles of a magnet.

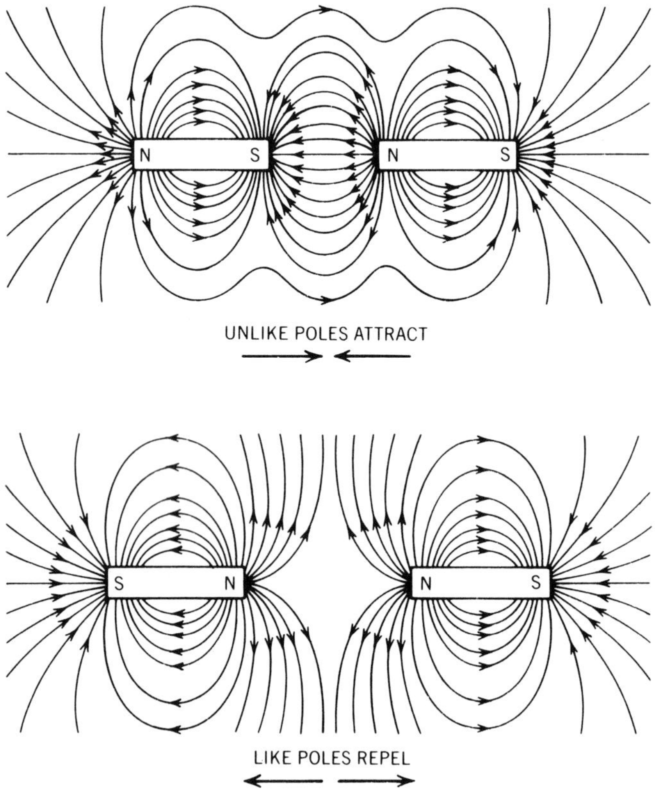


**Figure 7-18.** Magnetic poles attract and repel each other just as + and − electrical charges do.

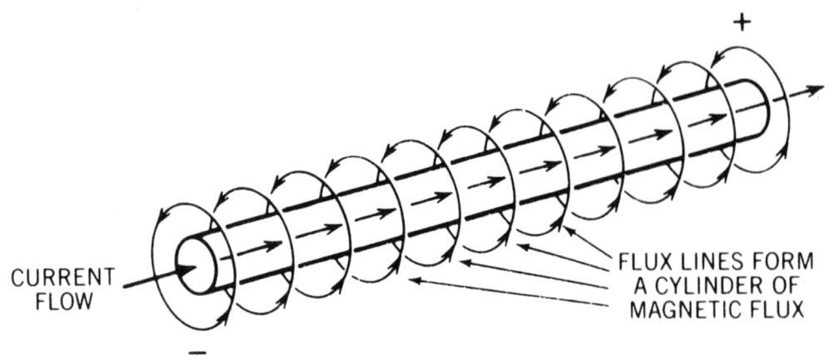


**Figure 7-19.** Current flow in a conductor forms cylinders of magnetic flux.

tion and exist between the opposite ends, or poles, of the magnet. The lines are called **flux lines**, and a magnetic field often is called **magnetic flux**. The **flux density** of a magnetic field indicates the number of flux lines per square centimeter of any area. If 100 flux lines pass through 10 square centimeters, the flux density of that area is 100 divided by 10, or simply 10, figure 7-17. Flux density is strongest near the poles of a magnet.

## Magnetic Polarity

Every magnet has a north (N) and a south (S) **pole**. The poles of a magnet relate to each other as positive and negative electrical charges relate to each other. Unlike poles (N and S) attract each other; like poles (N and N or S and S) repel each other, figure 7-18. We call this magnetic **polarity**. We also use the term polarity to describe the opposite + and − terminals of an electric circuit.

Flux lines exit from the north pole of a magnet and enter the south pole. Flux density is equal at each pole because an equal number of lines enter and exit. The flow of flux lines is what causes the poles to attract and repel each other. If you bring two south poles close together, flux lines try to enter both, and the flux density forces the poles apart. If you bring a north and a south pole close together, flux lines exit from one and enter the other, so their natural flow draws the poles together.

## Permeability and Reluctance

**Permeability** describes the ease with which flux lines pass through a material. Iron has high permeability because it allows flux lines to pass easily. Gases—including air—have low permeability because they do not allow easy passage of magnetic flux. **Reluctance** is the opposite of permeability. A material has high reluctance if it resists the passage of flux lines. Iron has low reluctance because it allows easy flow of flux lines.

Magnetic permeability and reluctance relate to each other as electrical conductivity and resistance relate to each other. They are not the same, however. Aluminum and iron are both good electrical conductors. Iron has high permeability; aluminum has very low permeability.

## ELECTROMAGNETISM

An important relationship between electricity and magnetism provides the major source of electric power in an automobile. As current flows through a conductor, a magnetic field forms around the conductor. There is a direct relationship between the amount of current in amperes and the strength (flux density) of the field. Moreover, there is a relationship between the direction of current flow and the polarity of the field. Magnetism that develops because of current flow is called **electromagnetism**.

### Electromagnetic Field

The magnetic field around a current-carrying wire is a series of concentric cylinders of flux lines, figure 7-19. The greater the current flow, the greater the flux density. The flux lines of the cylinders have direction like the flux lines of a bar magnet. The direction of current flow in the wire determines the direction of the flux lines.

We use arrows to show the direction of current flow, which you can see easily in the side view. If you look at the end of a wire in which current is flowing toward you, you see the head of an arrow, shown by a dot, figure 7-20. If you look at the end of a wire with current flowing away from you, you see the tail of an arrow, shown by a cross or a +.

If you know the direction of current, you can figure out the direction of the flux lines by using the right-hand rule or the left-hand rule. Using the conventional current theory from + to −, if you grasp a wire with your right hand so that your thumb points in the direction of current flow, your fingers wrap around the wire in the direction of the flux lines. This is the **right-hand rule**, figure 7- 21.

Using the electron theory of current flow from negative to positive, if you grasp a wire with your left hand so that your thumb points toward current flow, you fingers again wrap around the wire in the direction of the flux lines. This is the **left-hand rule**, figure 7-22. You can use either the right-hand or the left-hand rule for current flow and magnetic field relationships as long as you use the correct rule for each current flow theory and don't mix them.

### Field Interaction

The flux cylinders around conductors react with each other just as the fields around bar magnets do because all flux lines have direction and set up magnetic poles. If you bring two wires with current flowing in opposite directions close together, their fields oppose each other and force the wires apart, figure 7-23. If you bring two wires with current flowing in the same direction close together, their fields attract each other and draw the wires closer. You can do the same things with the electromagnetic fields of conductors and the

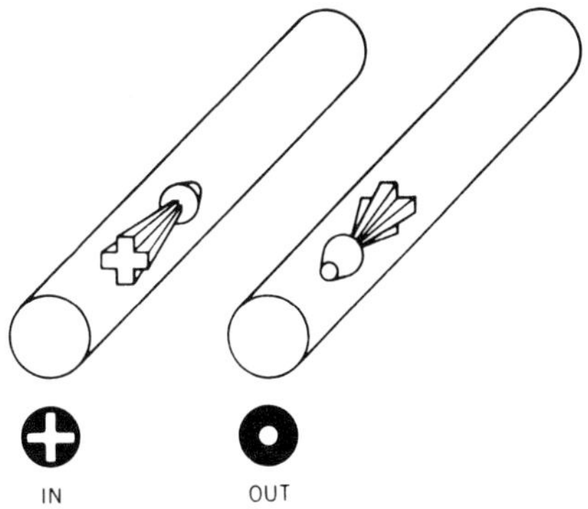


**Figure 7-20.** We use these current flow direction symbols in electrical and electromagnetic diagrams.

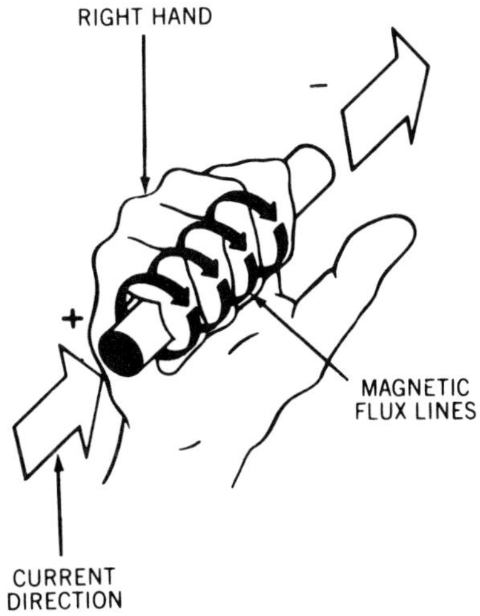


**Figure 7-21.** The right-hand rule of current flow and field direction is based on the conventional theory of current flow. (Delco-Remy)

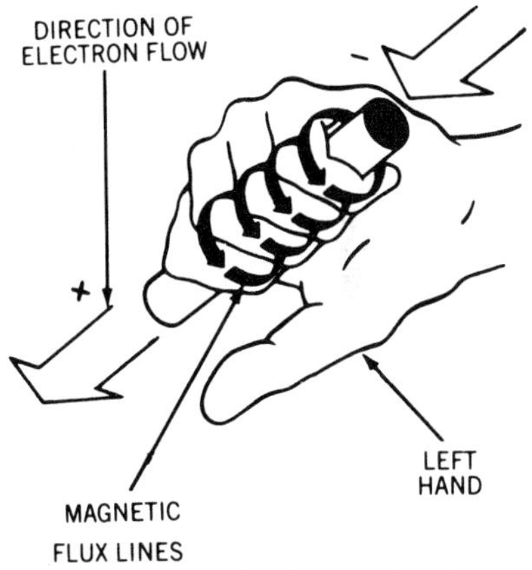


**Figure 7-22.** The left-hand rule of current flow and field direction is based on the electron current flow theory. (Delco-Remy)

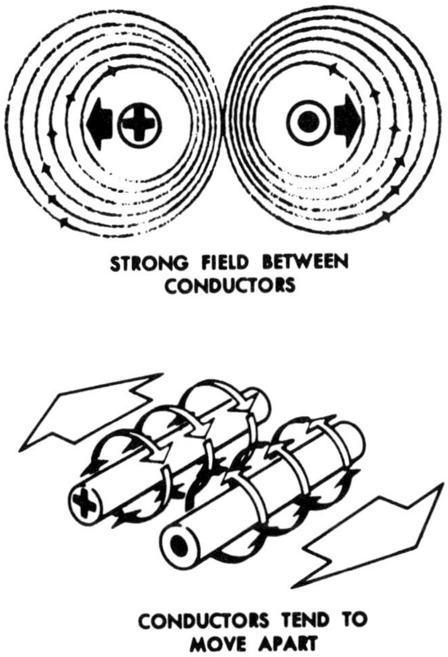


**Figure 7-23.** When current flows in opposite directions, the resulting magnetic fields oppose each other and force the conductors apart. (Delco-Remy)

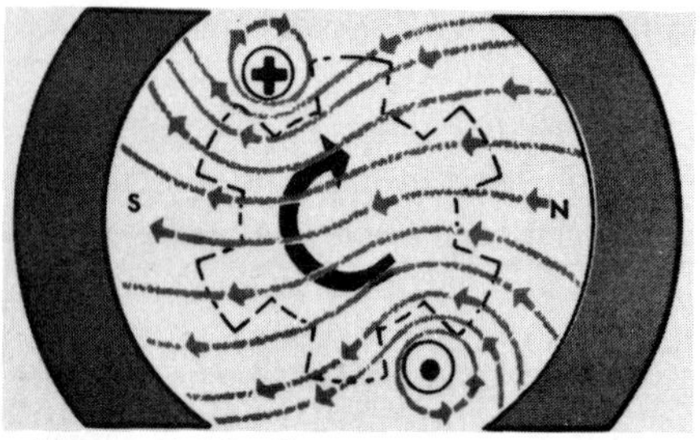


**Figure 7-24.** The interaction of magnetic fields causes electric motors to operate. (Delco-Remy)

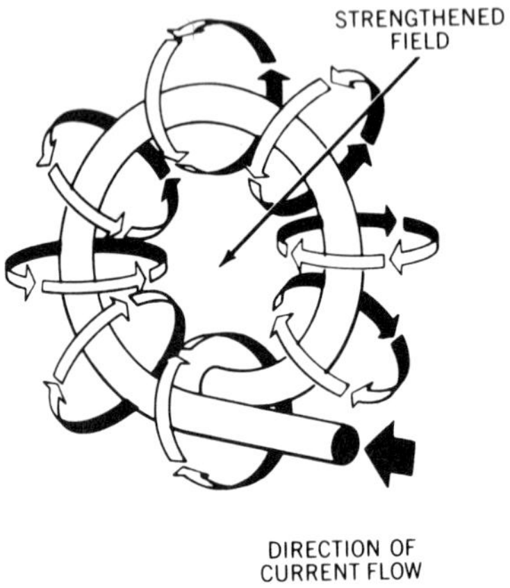


**Figure 7-25.** The magnetic field in the center of a loop is strengthened because the flux lines combine their strength. (Delco-Remy)

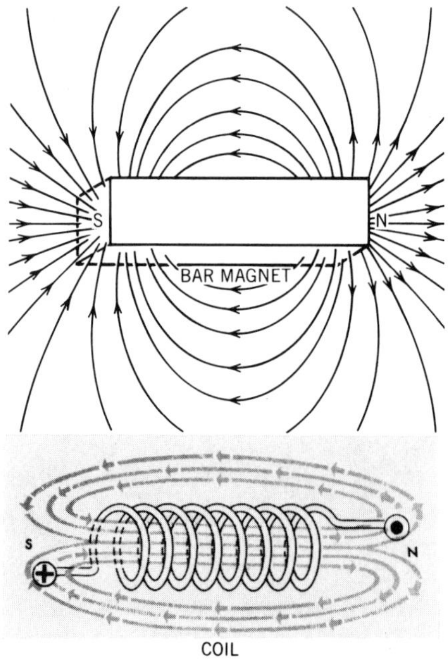


**Figure 7-26.** The magnetic field around a coil has north and south poles similar to those of a bar magnet. (Delco-Remy)

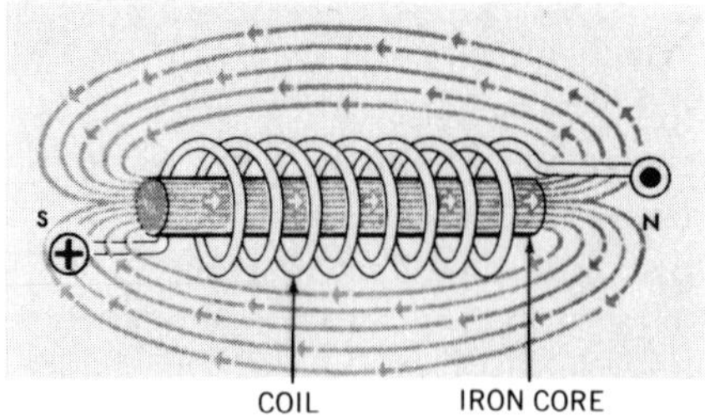


**Figure 7-27.** An iron bar placed in a current-carrying coil becomes an electromagnet. (Delco-Remy)

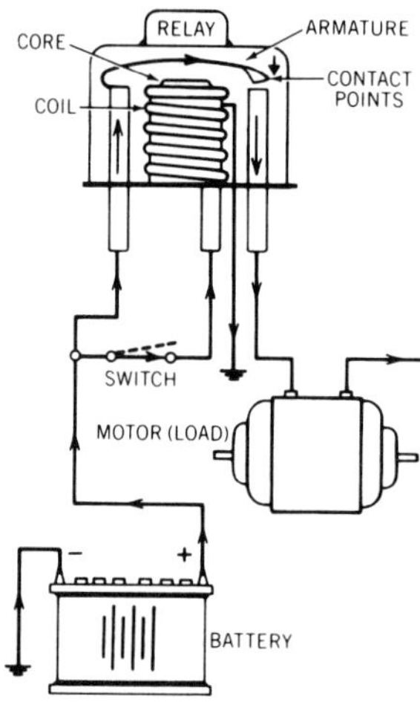


1. SWITCH CLOSES, CURRENT FLOWS THROUGH COIL AND CREATES ELECTROMAGNETIC FIELD IN RELAY CORE.
2. CORE ATTRACTS MOVABLE ARMATURE DOWNWARD AGAINST SPRING TENSION TO CLOSE CONTACT POINTS.
3. ARMATURE CONTACT POINTS CLOSE CIRCUIT TO MOTOR SO THAT CURRENT FLOWS THROUGH MOTOR.
4. WHEN SWITCH OPENS, CURRENT STOPS THROUGH RELAY COIL AND FIELD COLLAPSES.
5. SPRING TENSION OPENS ARMATURE CONTACT POINTS AND CURRENT TO MOTOR STOPS.

**Figure 7-28.** This relay is a remote control switch that opens and closes a motor circuit.

fields of permanent magnets, figure 7-24. These principles of field interaction are what cause electric motors to operate.

## Conductor Shape and Field Strength

You can strengthen the field around a conductor by bending it into a loop, figure 7-25. This causes the fields that meet in the center of the loop to attract each other or combine their strengths. You can strengthen the field even more by continuing to wind the conductor into a coil. When you do this, the field around the coil takes the shape of a field around a bar magnet, figure 7-26. The coil forms a north and a south pole from which the flux lines exit and enter. The strength of this field is determined by the number of turns in the coil and the amount of current flowing in it.

## Electromagnets

You can further strengthen the field of a coil by placing an iron bar inside it. Because the iron is more permeable than air, flux lines concentrate in the iron. When you do this, you create an **electromagnet**, figure 7-27. Electromagnets are used in **relays** and **solenoids** in many automobile systems. Relays are used as remote switches that allow a small amount of current in one circuit to open or close a switch in a circuit with more current. Solenoids are used to create mechanical motion. Figure 7-28 shows the operation of a relay. You should be able to follow the circuit diagram, current flow, and electromagnetic action from the principles you have learned in this chapter. Notice that the control circuit through the relay coil and the motor circuit are in parallel.

A final point to remember about electromagnets is that simple relays and solenoids operate on direct current. Alternating current that constantly reverses directions would cause the field of an electromagnet to reverse. You can understand why this will happen from what you have learned here about the relationships of current flow and flux direction.

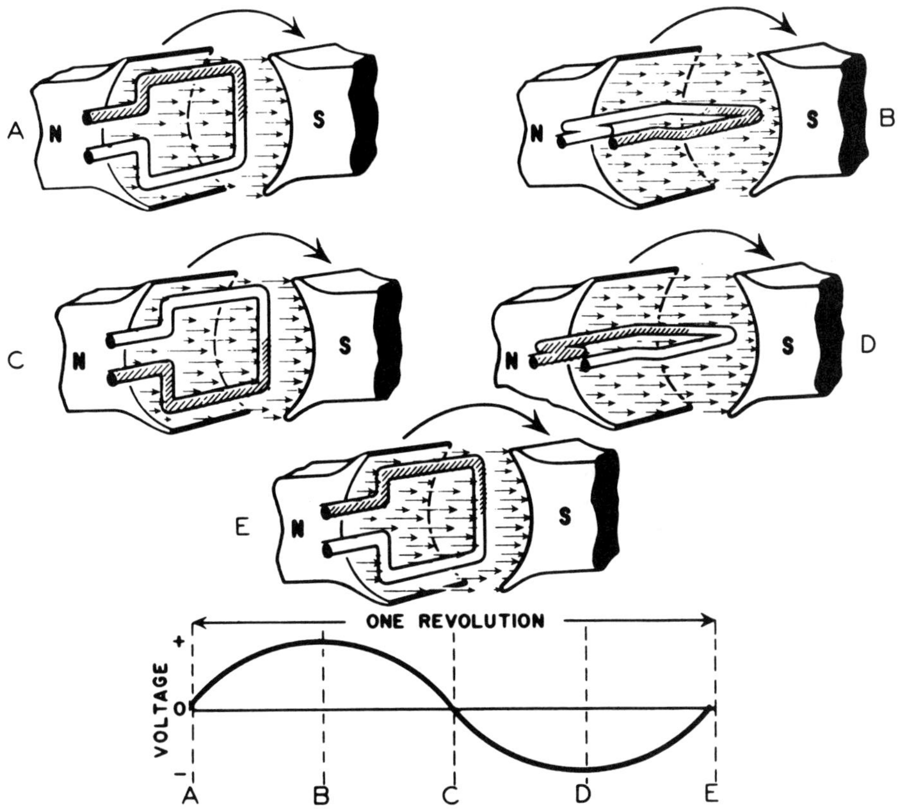


**Figure 7-29.** An ac voltage is induced as a rotating conductor cuts through the flux lines of a stationary magnetic field. (Prestolite)

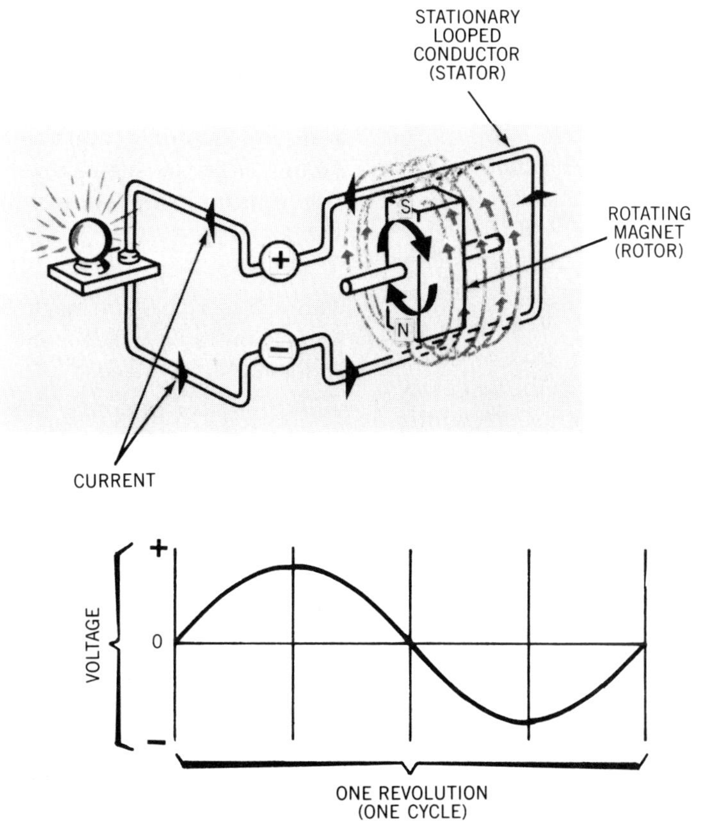


**Figure 7-30.** An ac voltage is induced as the flux lines of a rotating magnet cut through a stationary conductor. (Delco-Remy)

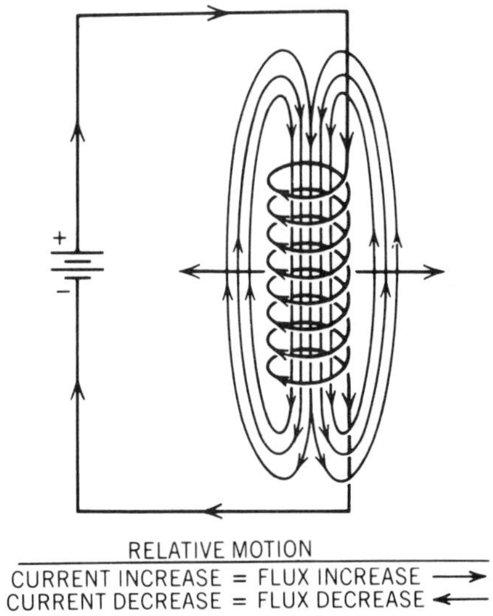


**Figure 7-31.** As a magnetic field expands and contracts, the relative motion of the changing flux density induces voltage in a conductor.

## Electromagnetic Induction

Magnetic flux lines generate a voltage in a conductor if either the field or the conductor is moving. This is called an **induced voltage**, and the process is called **electromagnetic induction**. Voltage is induced by **relative motion** between the conductor and the field. This can happen in three ways:

**1.** The conductor can move through a stationary field, figure 7-29.

**2.** The field can move past or around a stationary conductor, figure 7-30.

**3.** The flux lines of a magnetic field expand or collapse as current flow increases or decreases through a conductor, figure 7-31.

All three methods are used to induce voltage in automobile electrical systems. The first two ways depend on mechanical movement of a conductor or a magnet and are easy to visualize. The third way depends on the motion of increasing or decreasing magnetic energy and is not as easy to visualize.

If you put a loop conductor in the field between two stationary magnets and rotate the conductor, you will induce a voltage in the conductor, figure 7-29. The polarity of the voltage changes every half revolution because the rotating conductor cuts the flux lines in opposite directions. Therefore, cur-

**DO YOU KNOW YOUR WEIGHTS AND NUMBERS?**

The three principal parts of an atom—electrons, protons, and neutrons—are the things that distinguish one element from another and allow elements to react chemically and conduct electricity. This atomic structure has other features that allow scientists to identify other features of elements.

Because atoms never give up protons and the number of protons in an atom's nucleus remains constant, it is a good way to identify atoms of different elements. This is called the *atomic number*. Hydrogen has one proton; its atomic number is 1. Oxygen has eight protons; its atomic number is 8. Uranium has 92 protons; its atomic number is 92.

Remember that a balanced atom has equal amounts of protons and electrons, but an atom can become electrically unbalanced. An unbalanced atom with more or fewer electrons than protons is an *ion*. An atom with more electrons that protons is a *negative ion*. One with fewer electrons than protons is a positive ion.

Atoms of the same element can differ in another way. They can have different numbers of neutrons, and this is the basis for two other important terms: *atomic weight* and *isotope*. Protons and neutrons are the heaviest parts of an atom, by far. Therefore, scientists have said that the total number of protons and neutrons equals the atomic weight of an element. Helium has two of each; its atomic weight is 4. Oxygen has eight protons and eight neutrons; its atomic weight is 16.

Isotopes are atoms that have the same number of protons but different amounts of neutrons. Carbon, for example, in its most common form has six protons and six neutrons. Its atomic number is 6, and its weight is 12. Carbon also can exist with eight neutrons. Its number is still 6, but its weight is 14. The two atoms are chemically identical (same number of positive and negative charges), but they are physically different. Carbon with six neutrons is the isotope, carbon-12. With eight neutrons, it is the isotope, carbon-14.

rent that results from the induced voltage is alternating current. This is the method used in older generators for automobiles.

If you put a magnet inside a stationary conductor loop and rotate the magnet, you also create relative motion between the field and the conductor, figure 7-30. Again, the induced voltage changes polarity with every half revolution because the direction of the flux lines passing through the conductor changes. Current flow from this voltage also is alternating current. This method is used in the alternator of your car, and you will learn more about it when you study charging systems.

As current flow increases and decreases through a conductor, the resulting magnetic field expands and contracts. That is, it becomes more and less dense. If you put another conductor within the field created by current flow in the first and then vary the current flow in the first conductor, you create relative motion between the magnetic field and the second conductor, figure 7-31. This is the kind of relative motion that exists in ignition coils and transformers.

**Self-Induction.** As current flows through a coil, the magnetic field expands and induces a voltage in the coil that opposes the direction of current that created the field in the first place. This countervoltage is called **counter-electromotive force (cemf)**. As original current increases, cemf opposes its increase. As original current decreases, cemf opposes the decrease. If current is steady, there is no relative motion between the coil and the field, and there is *no* induced voltage. Induction occurs only when the field expands or collapses.

**Mutual Induction.** If you place two coils close together, you can transfer energy from one to the other through **mutual induction**. As current flows through the first coil, its increase or decrease causes the field around the first coil to increase or decrease. This changing field will induce a voltage in a second coil, figure 7-32. In practice, the two coils are wound around each other and around an iron core to increase the field strength, figure 7-33. One coil is connected to a current and voltage source and is called the **primary winding**. This is the coil winding that induces the voltage. The second coil is connected to another circuit. This is the coil winding in which the voltage is induced and is called the **secondary winding**.

Changing current in the primary winding induces a countervoltage in both the primary and secondary windings. Figure 7-32 shows how this works as current starts and stops in the primary winding. There is a direct relationship between the number of turns in a coil and the voltage that can be induced. Therefore, induction coils are made with secondary windings that have many more turns than the primary windings. In this way, low voltage in the primary can induce much higher voltage in the secondary. An ignition coil is an excellent example of this.

**Induced Voltage Strength.** Four factors affect the strength of the voltage induced in any of the three ways explained here:

**1.** The amount of current in a coil and thus the strength of the magnetic field. High current flow and a strong field induce a high voltage.

**2.** The number of conductors cutting the flux lines. Many turns in a coil secondary winding or many conductors in an alternator will cause a higher induced voltage.

**3.** The speed of relative motion between the conductor and the flux lines. A rapidly expanding or collapsing field or a fast moving conductor will cause a higher induced voltage.

**4.** The angle at which the conductor and flux lines intersect. If the conductor and the flux lines are parallel, the conductor does not cut any of the lines, and no voltage is induced, figure 7-

34. If the conductor and the flux lines are at 90 degrees to each other, the conductor cuts the most possible lines, and the greatest voltage is induced, figure 7-35.

## CAPACITANCE

**Capacitance** is the ability, or capacity, of two conducting surfaces to store an electric charge, or voltage, when they are separated by an insulator. A device that has this ability is a capacitor. A capacitor also is called a **condenser** because electric charges collect, or condense, on the plates as water vapor condenses on a cold glass.

Capacitance is an electrical feature that operates in many areas of an automobile. The principle that opposite + and − charges attract each other leads to the idea that there is field of electrical energy, or potential voltage, between any two oppositely charged points. This is called an **electrostatic field** because the charges do not move. They are stored on the two points as a form of static electricity. If you can cause opposite charges to collect on two surfaces, you can build a simple capacitor to store voltage.

A simple capacitor consists of two conductive plates, usually metal such as aluminum, zinc, steel, or copper. A conductor is attached to each plate, and an insulator is placed between them, figure 7-36. The insulator might be nothing more than the air in the space between the plates, or it might be some nonconductive material such as ceramic, glass, paper, or plastic. This insulator is called a **dielectric**.

If you install this capacitor in a circuit and start current flow by closing the switch, figure 7-37, electrons begin to flow from the battery around the circuit toward one plate of the capacitor. Because electrons can't flow through the dielectric, they collect on one plate, which becomes negatively charged. Simultaneously, electrons also flow from the opposite plate of the capacitor toward the positive terminal of the battery. This plate then becomes positively charged.

Current flows only as long as it takes for the positive and negative voltage difference across the capacitor plates to equal the positive and negative difference of the source voltage. At that point, the negative terminal of the battery and the negative plate of the capacitor are at the same negative potential. The positive plate and the positive battery terminal are at equal

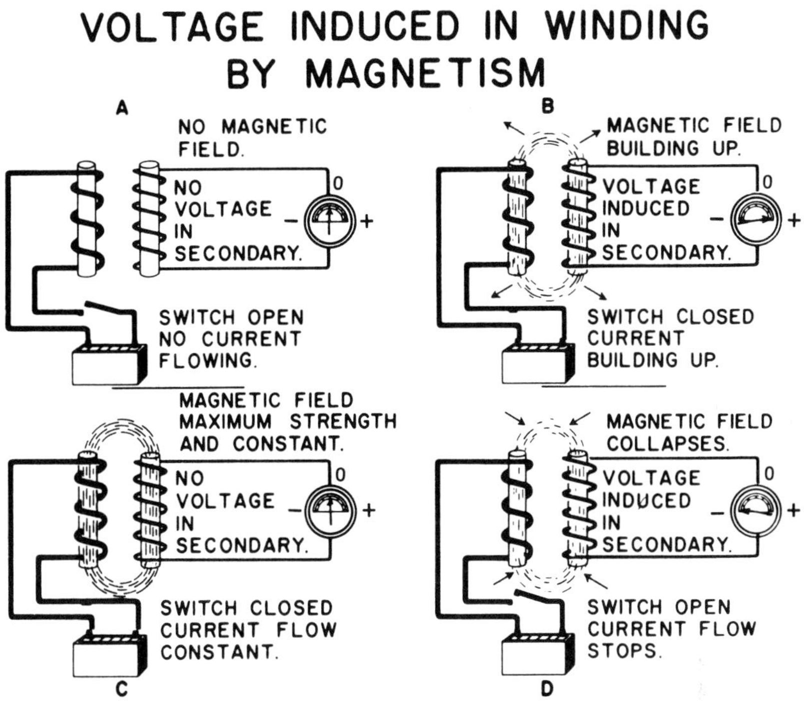


**Figure 7-32.** Mutual induction results as a changing field, caused by current flow in one coil, induces voltage in a second coil. (Prestolite)

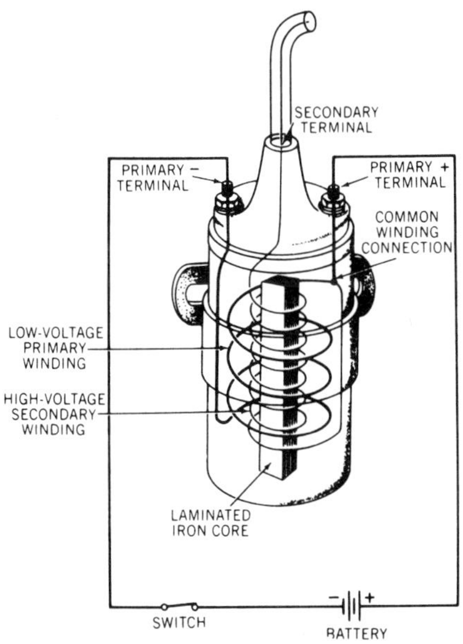


**Figure 7-33.** Induction coils have two concentric windings around an iron core. (Robert Bosch)

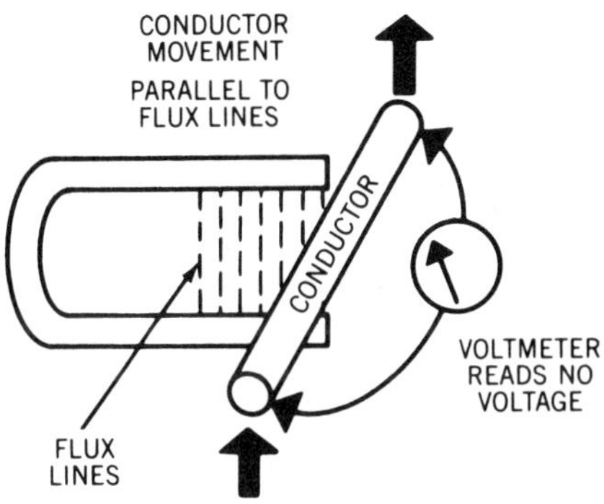


**Figure 7-34.** No voltage is induced if the conductor and the flux lines are parallel and no flux lines are cut. (Delco-Remy)

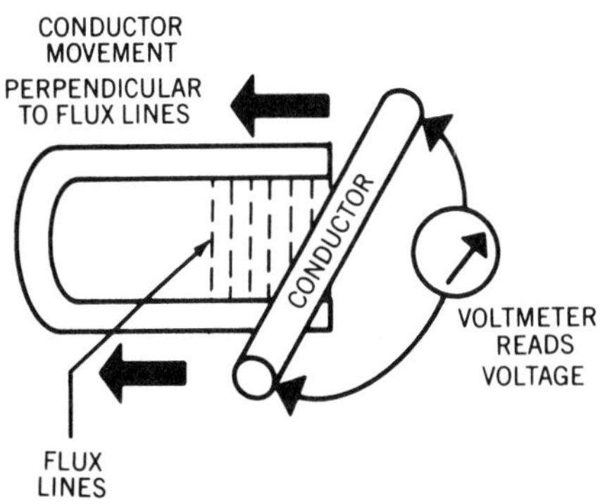


**Figure 7-35.** Maximum voltage is induced when the conductor and the flux lines intersect at 90 degrees. (Delco-Remy)

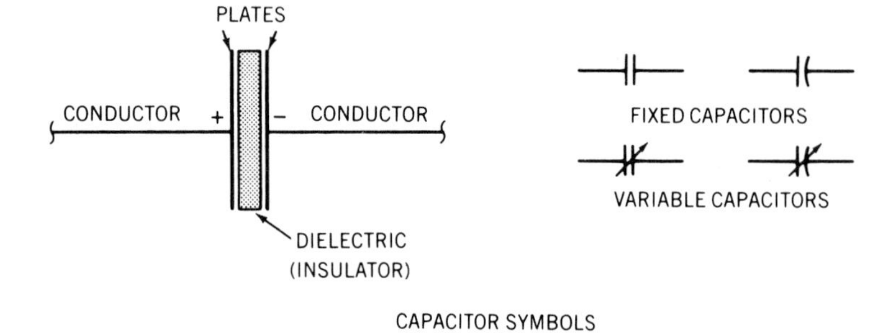


**Figure 7-36.** A capacitor is two conductive surfaces, or plates, separated by an insulator, or dielectric.

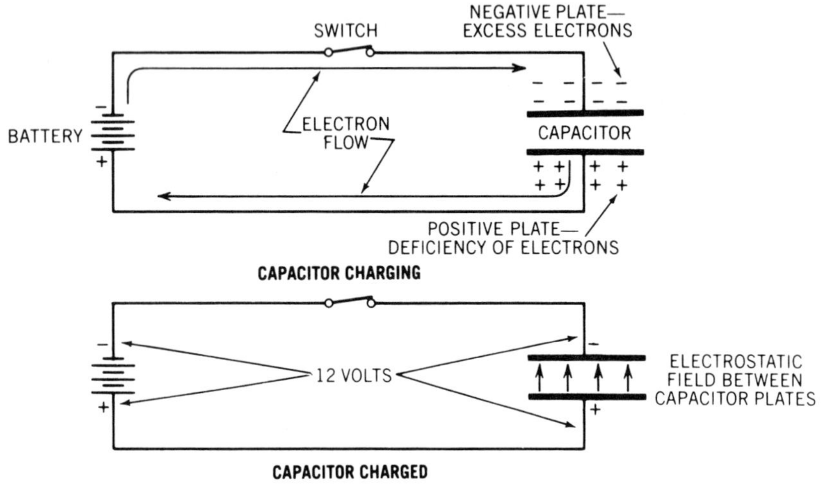


**Figure 7-37.** Current flow charges the capacitor to the same voltage as the battery, then current flow stops.

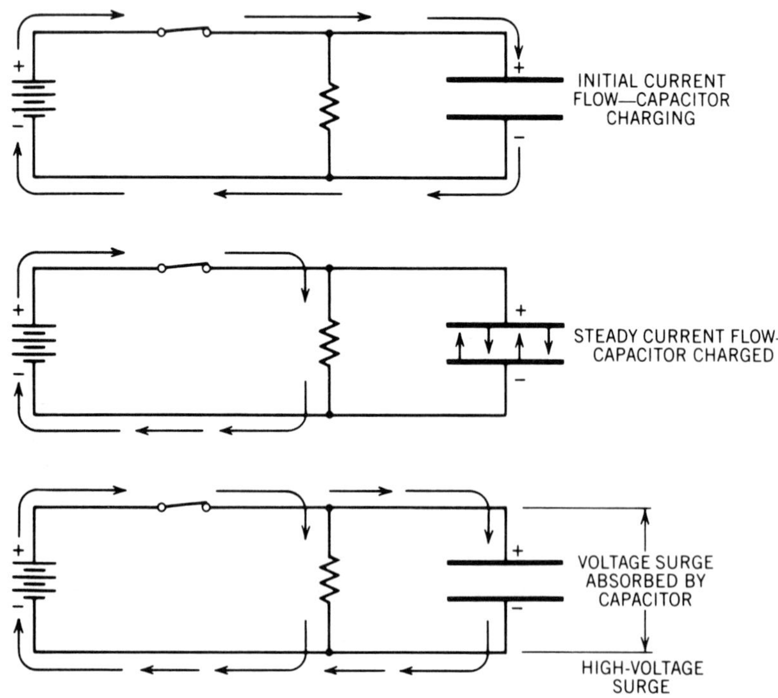


**Figure 7-38.** The capacitor absorbs voltage changes in this circuit and acts as a voltage shock absorber.

positive potential. The voltage charges across the capacitor and across the battery are equal, and no more current will flow because the dielectric of the capacitor creates an open circuit. The charges on the capacitor will not change as long as the plates are not connected to each other or to some other point with a different potential.

Capacitors are rated in **farads** (F). One farad is a very large charge, so the capacitors in automobiles and other electronic systems have capacitance of microfarads ($\mu$F), which is one-millionth (0.000001) of a farad. Three factors affect capacitance:

**1.** The larger the plate areas, the larger the charge the capacitor can hold because there is a larger area on which the charges can collect.

**2.** The closer the plates, the larger the charge because a strong electrostatic field exists between opposite charges that are close together.

**3.** The better the insulating qualities of the dielectric, the higher the capacitance.

Unlike a resistor, a capacitor does not consume any power or change any energy in a circuit. All of the voltage stored on a capacitor is returned to the circuit when it is discharged. The capacitance of a capacitor can be fixed or variable.

Because capacitors will store voltage, they will slow down any voltage change in a circuit. They are often used to absorb voltage changes in a circuit and are thus "shock absorbers." A capacitor has low resistance to current flow until it is charged. Then its resistance is infinite.

If you connect a capacitor in parallel across a 12-volt circuit, initial current flow charges the capacitor and then flows through other devices in the circuit, figure 7-38. If a high-voltage pulse hits the circuit (as can happen in an automobile), the capacitor absorbs the extra voltage as an additional charge before the current can damage any other part of the circuit. Capacitors also are used as short-circuit shunts to cause current to stop quickly when a circuit is opened. This is one of the jobs of an ignition condenser. Also, a capacitor can store a high-voltage charge and then discharge it quickly when connected across a circuit that needs the voltage. These are a few of the general uses for capacitors in automobile systems that you will discover in this text.

## ELECTRICAL SYSTEM POLARITY

Earlier in this chapter, you learned the two theories of current flow:

- The conventional theory says that current flows from positive (+) to negative (−).
- The electron theory says that current flows as electrons from negative (−) to positive (+).

We almost always use the conventional current flow theory for automotive electrical systems. This will help you understand system polarity.

Electrical polarity indicates the + and − connections of the voltage source to the system. The primary voltage source on a vehicle is the battery, and the battery connections establish system polarity. Since 1956, all domestic passenger cars and light trucks have used positive-polarity systems. The battery + terminal is connected to the insulated, or "hot," side of the system. This is the point of highest voltage in the system. The battery − terminal is connected to the ground side of the system, which is the lowest voltage point. Current flows from the + terminal of the battery, through the system to the − terminal. Voltage is highest at the + terminal, drops through the circuit, and is lowest at the − terminal. A positive-polarity system is also called a negative-ground system.

Before 1956, 6-volt Ford and Chrysler vehicles had negative-polarity, positive-ground systems. As late as 1969, some imported cars also had positive-ground systems.

System polarity affects the operation of electrical components. An alternator must be installed so that its output terminal connects to the battery + terminal, or the insulated side of the system. If alternator and battery connections are reversed, the alternator can be damaged. Ignition coils and other electromagnetic parts are also sensitive to system polarity. Some motors may run backwards, and some relays and solenoids may not work properly if their connections are reversed in relation to system polarity. Expensive electronic modules can be destroyed if their polarity is reversed. Electrical test equipment must be connected to the vehicle system in relation to system polarity. The positive lead from a test instrument must be connected to the + side of the system and the negative lead to the − side of the system.

You will learn more about specific effects of electrical polarity in later chapters of this book. Although all late-model vehicles have positive-polarity, negative-ground electrical systems, you should always double-check all connections to be sure that they are correct in relation to system polarity. Understanding the principle of electrical polarity also will help you follow the explanations of circuit conductors and electrical parts in the rest of this volume.

## SUMMARY

This chapter has summarized a basic course in dc electrical systems in a few pages. Its purpose is to help you understand the basics of all automobile electrical devices. Whether the next subject you study is batteries, alternators, starter motors, or ignition systems, remember these basic facts:

**1.** Current is the flow of electrons measured in amperes. It flows *through* a complete path called a circuit.

**2.** Current is either direct or alternating. The conventional current flow theory says that current flows from + to −; the electron flow theory says it flows from − to +.

**3.** Voltage is force *applied to or across a circuit* that causes current to flow.

**4.** Resistance opposes current flow and is measured in ohms. It is *contained within* a circuit.

**5.** Circuits are either series, parallel, or series-parallel. A series circuit has only one path for current flow. A parallel circuit has several paths for current flow. A series-parallel circuit has some loads in series and others in parallel.

**6.** Ohm's law says, "One ohm of resistance is present when one volt forces one ampere to flow in a circuit." Mathematically, Ohm's law is written:

Voltage $E$ = current $I$ × resistance $R$
Current $I$ = voltage $E$/resistance $R$
Resistance $R$ = voltage $E$/current $I$

**7.** The unit of electric power is the watt. Wattage equals current $I$ times voltage $E$.

**8.** Electricity can be generated by magnetism, which is energy caused by the alignment of atoms in some materials.

**9.** Magnetic flux lines form a magnetic field with lines of force concentrating at the poles. Unlike poles attract; like poles repel.

**10.** A magnetic field forms around a current-carrying conductor. The field can be used to operate a motor or to generate voltage through induction.

**11.** Capacitance is the ability of two surfaces to store voltage as an electrostatic charge.

## REVIEW QUESTIONS

### Multiple Choice

**1.** Good electrical conductors have how many free electrons in the outer shells of their atoms or molecules?

**a.** 3
**b.** 4
**c.** 6
**d.** 8

**2.** Controlled movement of electrons from one point to another is:

**a.** voltage
**b.** current

c. resistance
d. capacitance

3. Electromotive force that moves electrons from point to point is:
a. voltage
b. current
c. resistance
d. capacitance

4. Student A says that conventional current moves from positive (+) to negative (−). Student B says the electron theory states that electrons flow from negative (−) to positive (+). Who is right?
a. A only
b. B only
c. both A and B
d. neither A nor B

5. Student A says that an alternator generates direct current. Student B says that automobile electrical systems operate on alternating current. Who is right?
a. A only
b. B only
c. both A and B
d. neither A nor B

6. For current to flow and produce useful work, every circuit must have:
a. capacitance
b. inductance
c. resistance
d. reluctance

7. An open circuit has no:
a. resistance
b. reluctance
c. voltage
d. continuity

8. Student A says that voltage across a circuit equals current times resistance of the circuit. Student B says that power consumed by a circuit equals voltage times current. Who is right?
a. A only
b. B only
c. both A and B
d. neither A nor B

9. If voltage across a series circuit is 12 volts and current is 3 amperes, resistance must be:
a. 36 ohms
b. 4 ohms
c. 0.25 ohm
d. 6 ohms

10. If current in a circuit is 10 amperes, and resistance is 1.4 ohms, voltage must be:
a. 7 volts
b. 12 volts
c. 14 volts
d. 10 volts

11. Student A says that a parallel circuit has several current paths. Student B says that a series-parallel circuit has several current paths. Who is right?
a. A only
b. B only
c. both A and B
d. neither A nor B

12. Total current in a parallel circuit equals the:
a. sum of the resistance in each branch
b. voltage minus the resistance of each branch
c. sum of the amperage in each branch
d. sum of the voltage across each branch

13. If total circuit resistance is 5 ohms, 9 volts will deliver how much current?
a. 1.8 amperes
b. 2.4 amperes
c. 0.55 ampere
d. 0.41 ampere

14. A 12-volt circuit with 6 amperes of current consumes how much power?
a. 2 watts
b. 20 watts
c. 48 watts
d. 72 watts

15. Which of the following actions will cause a magnetic field to induce voltage in a conductor?
a. The conductor moves through the stationary magnetic field.
b. The magnetic field moves past the stationary conductor.
c. The magnetic field expands and collapses around the conductor.
d. All of the above.

16. Student A says that an induction coil has a primary and a secondary winding. Student B says that a coil produces voltage by rotating a conductor within a magnetic field. Who is right?
a. A only
b. B only
c. both A and B
d. neither A nor B

17. The field of electrical energy between two oppositely charged points (+ and −) is:
a. magnetic flux
b. a dielectric
c. an electrostatic field
d. permeability

### Fill in the Blank

18. Loosely held valence electrons that move easily into the electron shells of other atoms are called __________ electrons.

19. Tightly held valence electrons that take quite a bit of energy to dislodge them are called __________ electrons.

20. The electromotive force that produces current flow is called __________.

21. Elements with four valence electrons that are neither good conductors nor good insulators are called __________.

22. To conduct electricity, a copper wire must have a positive charge at one end and a __________ charge at the other end.

23. The unit of current measurement is the __________.

24. Current that flows, at equal intervals, from positive to negative and back to positive is called __________ current.

25. The principal voltage sources in an automobile are the battery and the __________.

26. The unit used to measure the amount of resistance in a circuit is the __________.

27. To calculate the voltage across a circuit, you must know the current and the __________.

28. Total resistance of a parallel circuit is called __________ resistance.

29. To calculate electric power (watts), you need to know current and __________.

30. The lines of force in a magnetic field are called __________ lines.

31. The unit used to rate capacitors is the __________.

32. Magnetism that develops because of current flow is called __________.

# 8 SEMICONDUCTOR ELECTRONIC PRINCIPLES

## WHAT

A little more than 100 years ago, the Industrial Revolution of the 19th century brought us the technology that made the automobile possible. For the past 35 years, all industries—and all of us—have been going through an electronic revolution that is equally as important as was the Industrial Revolution. The automobile has been the target of the electronic revolutionaries because it offers the opportunity to put electronics to work in many ways for the comfort, convenience, and safety of motorists.

The preceding chapters summarized the principles and sources of electricity, as well as basic automotive electrical systems. This chapter introduces the principles of solid-state electronics that operate in modern automobiles. As with other systems, all of our automobile electronic devices operate similarly, regardless of where they are used. The electronic fundamentals are simply an extension of the electrical fundamentals you have learned.

## WHY

You can't service any system on a modern car without dealing with some electronic device. That device might be a simple sensor that warns of low oil level, or it might be a sophisticated computer that performs complicated system integration for fuel delivery, ignition, emission control, and transmission shifting. Just as you should understand the fundamentals of electricity before working with switches, circuits, and electric power, you should understand the principles of electronics before dealing with them in different systems.

## GOALS

After studying this chapter, you should be able to:

**1.** Explain solid-state electronics, semiconductor current flow, and the difference between N and P material.

**2.** Explain the construction, operation, and voltage-current relationships in diodes, bipolar transistors, field-effect transistors, and other electronic devices.

**3.** Explain the differences between discrete devices, integrated circuits, and hybrid circuits.

## SOLID-STATE ELECTRONICS

Both electricity and electronics deal with the controlled movement of electrons to do work. We usually use the word, "electrical" to describe systems in which electric current flows through wires and electromechanical devices such as switches, motors, lamps, and other units to produce light, heat, or motion. Also, an electrical circuit usually does not increase, or amplify, the current or voltage delivered to the circuit by a battery, a generator, or an induction coil.

"Electronic" systems usually involve the transmission of electrons or voltage signals through vacuum or gases or semiconductor materials in at least part of a circuit. Also, many electronic components have no moving electromechanical parts. Electronic devices often amplify voltage or current and can perform logical decision-making functions.

For over 60 years, electronic systems used vacuum tubes, figure 8-1, to transmit electrons between two conductors and to amplify current or voltage. With the invention of the transistor in 1948, scientists had a practical way to do the same thing with solid, crystal materials. These crystal materials are the group of elements called semiconductors. Because they have all of their electrical properties in solid, one-piece forms, the science in which they are used came to be called **solid-state electronics**.

Today, it is difficult to separate electrical from electronic systems. For decades, automotive electrical systems have used induction coils for many purposes. Today, these traditional systems are controlled by solid-state electronic devices. Similarly, electrical generating systems have solid-state rectifiers and electronic voltage regulators. To test electromechanical systems for automatic door locks and other accessories, for example, you will have to recognize the purpose of electronic diodes in the circuits.

Traditionally, the science of electronics began with the study of electron flow in vacuum tubes. Today, it begins with semiconductor materials.

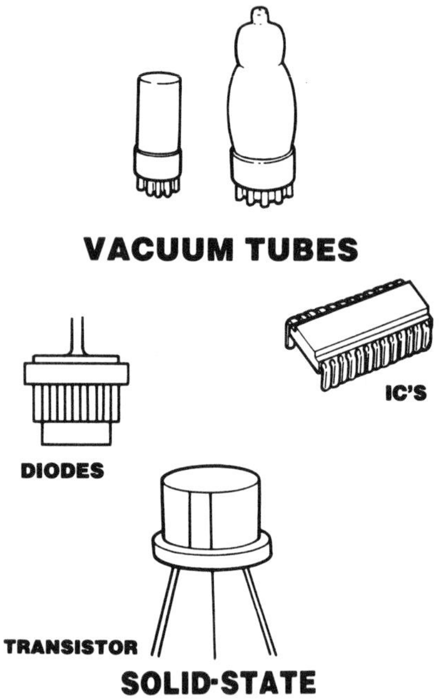


**Figure 8-1.** Vacuum tubes were the earliest method of transmitting electrons between conductors and amplifying current or voltage. Transistors do the same job much faster, require far less space, and do not overheat. (Chrysler)

## SEMICONDUCTORS

Five elements have just four electrons in their valence shells: carbon, silicon, germanium, tin, and lead. They are neither good conductors (three or fewer valence electrons) nor good insulators (five or more valence electrons). They are semiconductors. While all of these elements are chemically semiconductors, the two most important for solid-state electronics are silicon and germanium. Of these two, only silicon is used for modern integrated circuits, and we will concentrate on silicon devices.

As a pure element, silicon is a crystal solid. Each of its atoms shares its four valence electrons with four adjoining atoms, figure 8-2. In this way, every atom has a full valence shell, and elemental silicon is a relatively good insulator. The uniform crystal structure of silicon can be upset, however, by adding impurities to a silicon crystal in the form of atoms of other elements. This process is called **doping**. The amount of other material added to pure silicon is quite small, usually in a ratio of one part **dopant** to 10,000,000 parts silicon.

### N-Material

If an element such as phosphorus or arsenic with five free electrons is added to silicon, the resulting mixture has a negative charge because it has extra free electrons not permanently attached to any valence shell. The extra electrons continuously bump from shell to shell and dislodge other electrons in a random drift, figure 8-3. This kind of doped semiconductor material is called **N-material** because it has a negative charge.

### P-Material

If an element such as boron or gallium with three free electrons is added to silicon, the resulting mixture has a shortage of free electrons and, therefore, a positive charge. This is called **P-material**. The places in the incomplete valence shells that would normally be filled by electrons are called **holes,** figure 8-4. Free electrons from adjoining atoms continuously try to fill the holes; and, as they do, they leave other holes behind. Random electron drift also exists in P-material, but as electrons drift, so do holes. Semiconductor electrical theory defines holes as positive charge carriers and electrons as negative charge carriers.

### Hole Flow

You can think of electrons in semiconductors as moving in one direction and holes as moving in the other, figure 8-5. Moreover, if an electron has a negative charge (as you have learned), you can think of a hole as having a positive charge. Therefore,

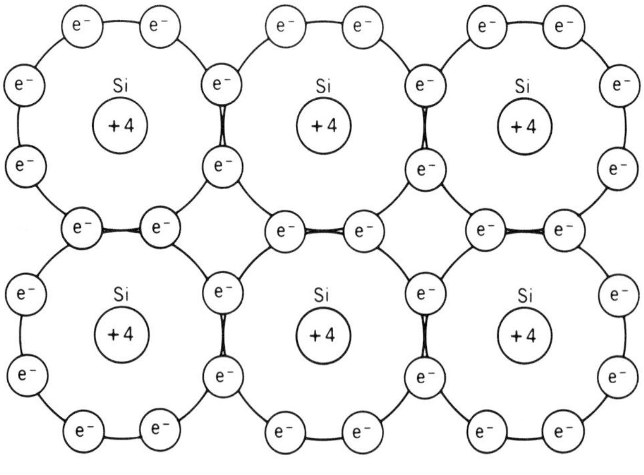


**Figure 8-2.** Each atom of silicon in its pure form shares its four valence electrons with four adjoining atoms so that each completes its valence shell.

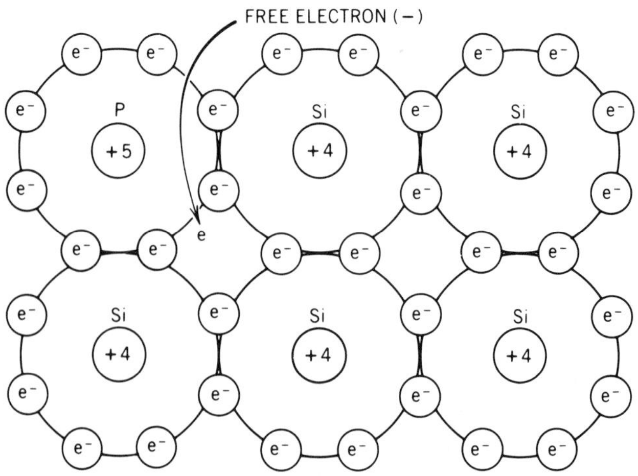


**Figure 8-3.** N-material has extra free electrons not permanently bound to any valence shells.

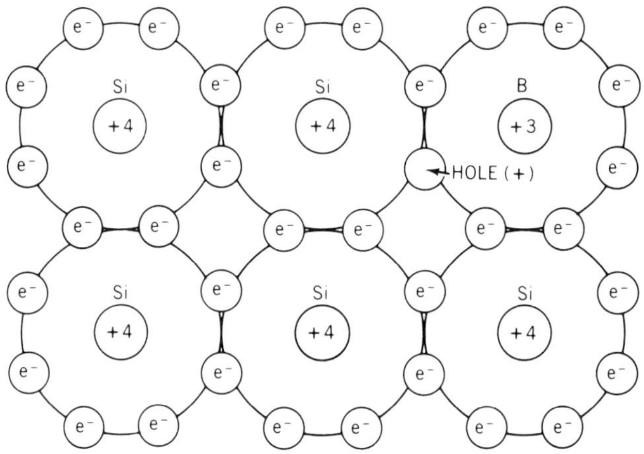


**Figure 8-4.** P-material has holes in some valence shells that attract electrons from adjoining atoms.

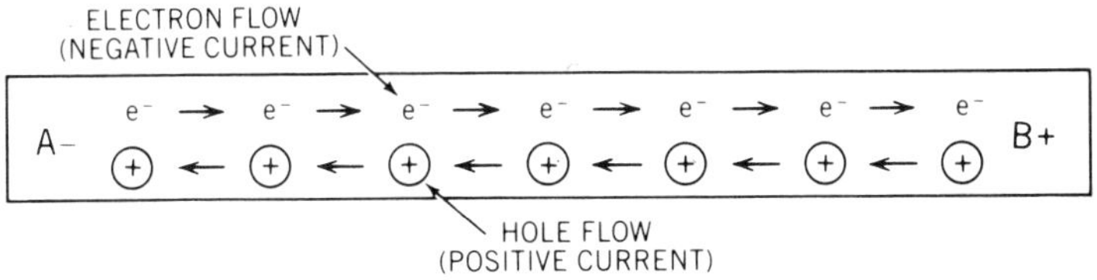


**Figure 8-5.** Electron flow and hole flow move in opposite directions through a semiconductor.

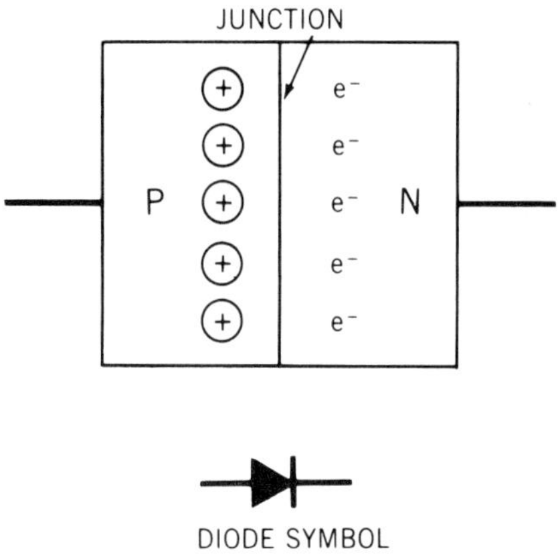


**Figure 8-6.** A diode is formed by the junction of P-material and N-material.

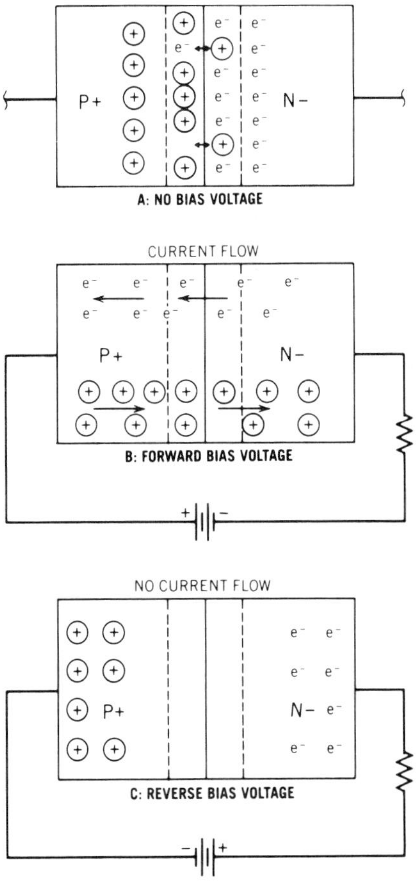


**Figure 8-7.** Forward bias voltage causes current flow through a diode; reverse bias prevents current flow.

as electron flow carries a negative charge, or negative current, from point A to point B, hole flow carries a positive charge, or positive current, from point B to point A, figure 8-5. Hole flow is simply another way to look at the conventional current flow theory, which says that current flows from positive to negative. To study semiconductors, we use both the electron flow theory (− to +) and the conventional, or hole flow, theory (+ to −).

## DIODES

Let's take a piece of semiconductor N-material and a piece of P-material and put them together to see what happens electrically. When we join the two pieces, we form a **PN junction**. Figure 8-6 shows such a junction and the symbol for the device called a **diode**.

The arrowhead in the diode symbol indicates the P-material and the direction of hole flow, or conventional current flow. The same arrowhead indication is used in transistor symbols, as you will see later. In a diode, the P-material is the anode, with + current carriers (holes); and the N-material is the cathode, with − charge carriers (electrons).

With no voltage applied to the diode, some electrons from the N-material cross the junction to fill holes in the P-material. This forces a few holes in the opposite direction across the junction to be filled by electrons in the N-material. This random hole and electron movement soon reaches a point where the junction stabilizes and no more holes and electrons drift from one side to the other, figure 8-7A.

The stabilized junction is also called the depletion layer because mobile charge carriers (holes and electrons) are not sufficient to overcome the fixed charge density of P- and N-material. In other words, conductivity is depleted, or reduced, at the PN junction. Without voltage applied to the diode, it will not do anything useful electrically. But if we apply voltage across the diode (positive to one side, negative to the other), we change its electrical characteristics. This is called **bias voltage**.

## Forward Bias

If we apply positive voltage to the P-material and negative voltage to the N-material, we have **forward bias**, which causes current to flow across the junction and through the diode. This is because the positive voltage repels positively charged holes in the P-material forward and across the junction into the N-material. At the same time, the negative voltage on the N-material attracts the holes and repels electrons toward the junction, where they move into the P-material. Thus, we have current flow: holes in one direction, electrons in the other, figure 8-7B.

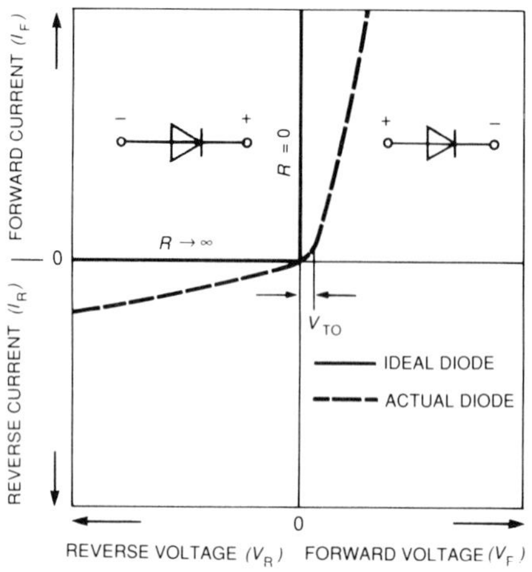


**Figure 8-8.** The ideal resistance curve of a PN diode.

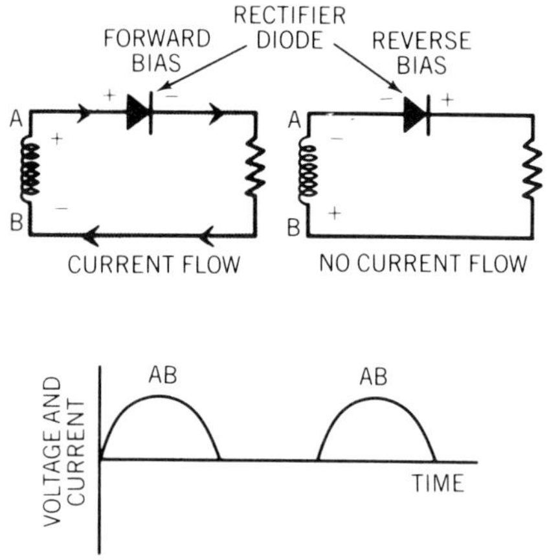


**Figure 8-10.** One diode allows current to flow in one direction but blocks it in the other. (Delco-Remy)

## Reverse Bias

If we reverse the voltage and apply positive to the N-material and negative to the P-material, we have **reverse bias**, figure 8-7C. The reverse voltage potential, or polarity, attracts holes and free electrons away from the PN junction, and no current flows through the diode.

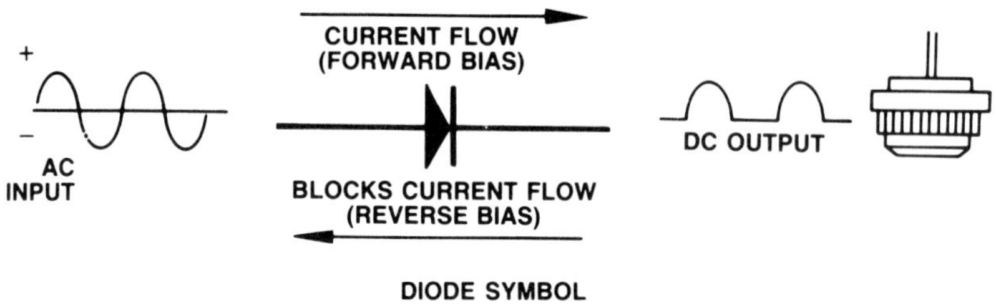


**Figure 8-9.** Forward bias allows current to flow in one direction in a diode; reverse bias blocks current flow. (Chrysler)

## Junction Diode Operation

The resistance of the PN junction is very low in the forward direction. Ideally, it would be almost zero. Therefore, it takes very low forward-bias voltage to make a diode conductive. The minimum conduction voltage is the **threshold voltage**. For a germanium diode, threshold voltage is about 0.3 volt; for a silicon diode, it is about 0.7 volt. Any further increase in voltage increases the current. With forward bias, a diode is not a circuit load and produces little or no voltage drop in the circuit. Current through a diode must be limited by a resistor or a circuit load. Otherwise, the diode would be operating in a short-circuit condition and would burn up from excessive current.

With reverse bias voltage, the resistance of the PN junction is very high, almost infinite. The diode acts like an open circuit with reverse bias. All of the reverse-bias voltage appears across the PN junction, just as it would across an open switch. Figure 8-8 shows the ideal resistance curve of a PN diode as a 90-degree angle. In the forward direction, resistance drops to zero. The diode conducts full current with any forward bias voltage. Similarly, resistance is infinite with any reverse bias voltage, and no current flows. No diode has the operating characteristics of the ideal curve, but most diodes used in automotive systems operate according to the typical curve shown in figure 8-8.

## Junction Diode Uses

Now you see a practical use for diodes. They can be used to control the direction of current flow. In this way, a diode acts in an electrical system as a check valve acts in a hydraulic system. Remember that alternating current changes its direction in regular cycles (+ to − and − to +). The alternator on your car generates alternating current, but the electrical devices need direct current. The first major use of diodes in automotive electrical systems was to rectify alternating current to direct current.

With forward bias applied to a diode, it allows current to flow in one direction, figure 8-9. With reverse bias, the diode blocks, or checks, current flow, figure 8-9. If we connect a diode to one end of an alternator conductor, figure 8-10, and then connect an external circuit, the diode allows current in one direction and blocks it in the other. The result is pulsating dc current.

The single diode in figure 8-10 rectifies only the positive (+) half of the voltage cycle. Adding more diodes and making the final output connection in parallel allows us to rectify the negative (−) half of the cycle as well, figure 8-11. A series of alternator conductor windings and an arrangement of six diodes produces a rectified dc output with only a trace of ac fluctuation, figure 8-12.

We can also use a diode in a warn-

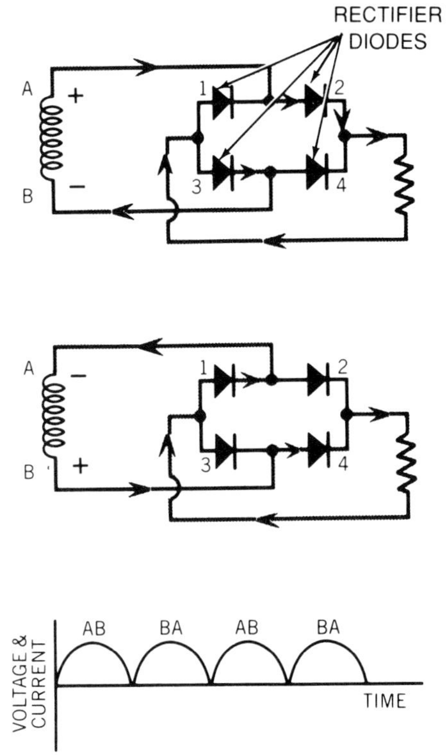


**Figure 8-11.** Four diodes provide full-wave rectification from one output winding. (Delco-Remy)

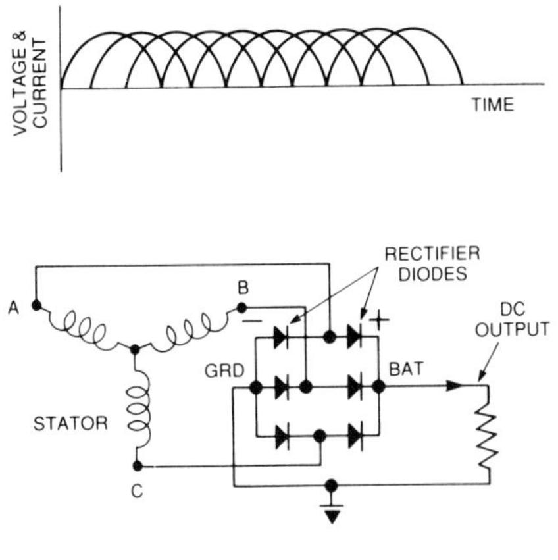


**Figure 8-12.** Six diodes provide full-wave rectification from three output windings (three-phase rectification).

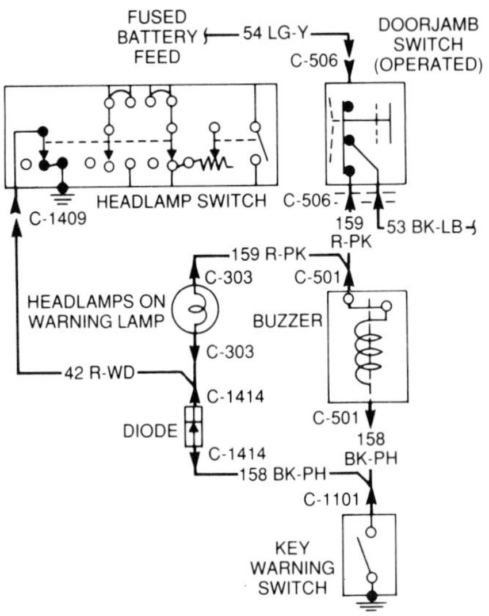


**Figure 8-13.** This warning circuit uses a diode to light an indicator lamp and sound a buzzer. (Ford)

ing circuit to light an indicator lamp or sound a buzzer, or both. Figure 8-13 shows a circuit that will light a warning lamp if the headlamps are left on or sound a buzzer if the ignition key is left in the lock. Power is supplied through the doorjamb switch. The circuit is grounded through the headlamp switch or through the key-warning switch. Here is how the diode works in the circuit.

• The headlamps are off; the ignition key is left in the lock to close the key switch; the door is opened to close the door switch. Current flows through the door switch, the buzzer, and the key-warning switch to ground. The diode blocks current through the warning lamp so it cannot reach ground at the key switch. The headlamp switch is open, so the warning lamp is not grounded there, either. The buzzer sounds, but the warning lamp does not light.

• The headlamps are on; the door is opened. Current flows through the warning lamp to ground at the headlamp switch. The lamp lights. Current also flows through the buzzer, through the diode in the forward direction, to ground at the headlamp switch. The buzzer also sounds.

The diode allows one buzzer to warn of the key left in the ignition and the headlamps left on. The diode allows the warning lamp to light only when the headlamps are left on. In this way, the diode acts as a simple electronic logic device. You will learn more about electronic logic circuits in chapter 20.

The current control of a PN diode is not perfect. Even with reverse bias, there is a small amount of reverse current flow across the junction because holes and electrons attract each other. Normally, this current is so small that it does not matter. There may be occasions, however, when diode leakage current leads to a diode being diagnosed as faulty. The very small test current used by a digital voltmeter may pass through a good diode in each direction. The meter reading then may lead you to reject it as being defective.

If reverse bias is too high, however, the voltage force can overcome the natural resistance of the diode. The covalent bond structure of the diode then breaks down, allowing full reverse current flow. If the magnitude and duration of the reverse current through the diode is great enough, the excessive heat will damage the diode.

## Zener Diodes

The **breakdown voltage** depends on the thickness of the kind of semiconductor material, the amount of doping, and the thickness of the PN junction. The breakdown voltage is called the zener breakdown voltage or the avalanche breakdown voltage, depending on the amount and rate of reverse current. Zener breakdown provides a more gradual reverse current than avalanche breakdown, figure 8-14.

Every diode has three operating regions, figure 8-15:

**1.** The conducting region with forward bias voltage. Resistance is low, and forward current is high.

**2.** The blocking region with low reverse bias voltage. The resistance is high, and current is low.

**3.** The breakdown region with high reverse bias voltage. Resistance breaks down, and reverse current is high.

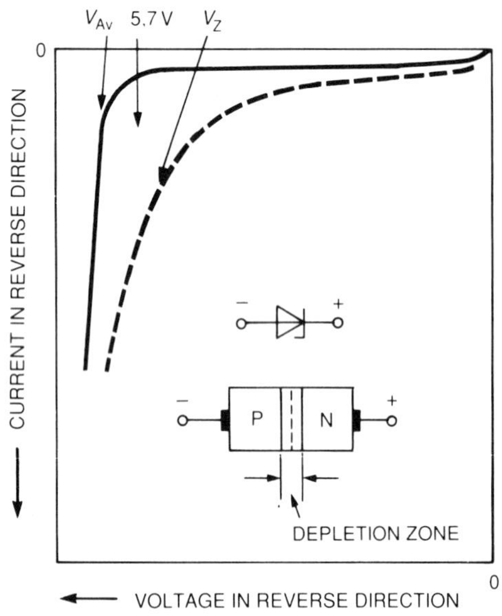


**Figure 8-14.** Avalanche breakdown ($V_{AV}$) is more abrupt than zener breakdown ($V_Z$). (Bosch)

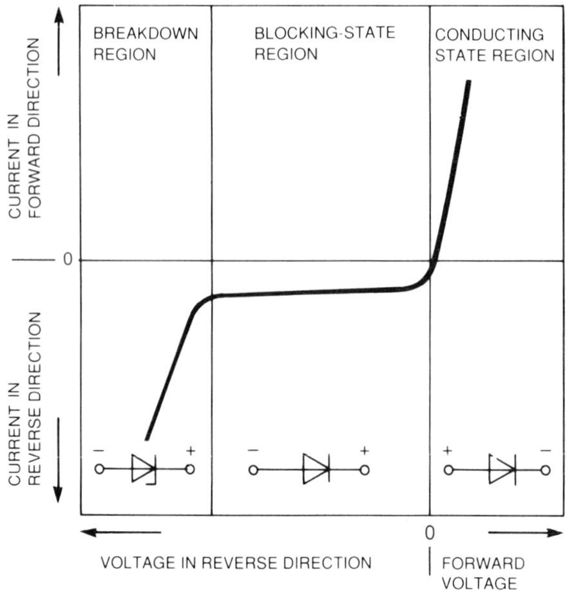


**Figure 8-15.** The three operating regions of a diode.

## ELECTROSTATIC DISCHARGE AND ELECTRONIC CIRCUITS

Walk into the service department of topflight new car dealerships these days and you're likely to see a strange sight—the tuneup technician is wearing a special wrist strap and is plugged into a grounded mat while working on a vehicle. He's probably learned the hard way that you can't be too careful about electrostatic discharge when working around electronic components.

Electrostatic discharge is generated when two dissimilar materials (especially synthetics) rub together or separate. The amount of charge that builds up depends on a variety of factors: the type of materials, their proximity to each other, the speed with which they are separated, and the amount of humidity present.

The static discharge we've all seen and felt has to be greater than about 7,000 volts before we even notice it. Static discharges of less voltage may pass unnoticed by humans, but not by a computer chip. A surge of less than 100 volts can put the computer chip on a junkpile. Under the right conditions, static discharges of 40,000 to 50,000 volts can occur. These give a human a good sting, but are the death knell for electronic circuits and components.

A static-related failure can take one of two forms:

**1.** It immediately blows a hole in a microcircuit, resulting in a complete failure.

**2.** It causes only minor damage in the microcircuit at first; but as time passes, the circuit gradually deteriorates and eventually results in a complete failure.

Technicians who slide across the car seat to unplug the computer module first noticed the results of static discharge. Anyone who has blown a computer in that manner has learned his lesson the hard way. But static charges and discharges can also affect solid-state components right from the parts department.

Electronic components are wrapped in static-free packing material and should not be removed from their containers in the usual manner. To prevent any static charge or discharge problem, ground the wrapping with an alligator clip lead before removing the component. It's also a good idea not to touch the component connections if you're not grounded. Since synthetic materials are especially prone to static charges or discharges, another good idea is to watch the kind of clothing you wear while working on the vehicle.

The technician described at the beginning has learned his lesson. Working on the grounded mat with the special wrist strap provided, he can handle the component in the vehicle, remove the component to be installed from its packing, and install it in the vehicle without fear of electrical reprisal. Of course, once he breaks the continuity that provides a path of equalization for a static charge by removing the wrist strap or stepping off the grounded mat, a static charge can start to build. For this reason, he must remain grounded during the entire repair procedure.

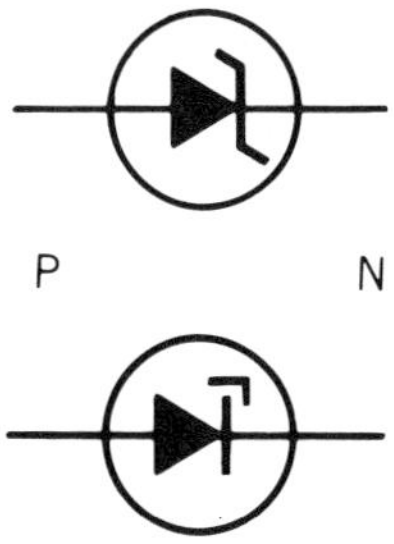


**Figure 8-16.** These symbols represent zener diodes in circuit diagrams.

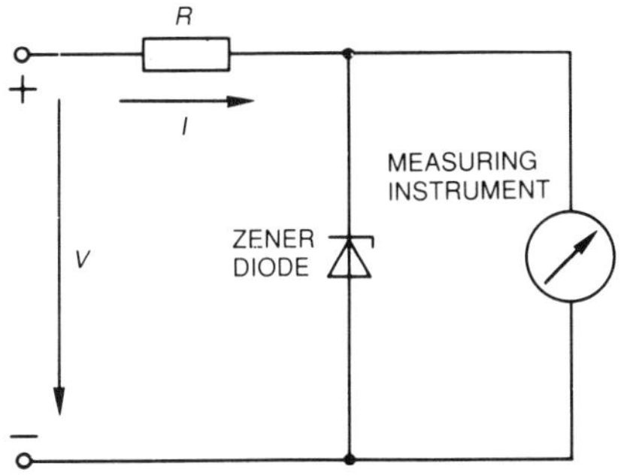


**Figure 8-17.** Overvoltage protection is provided for the measuring instrument (MI) by inserting a zener diode in parallel. (Bosch)

Reverse current can destroy a simple PN diode, but special diodes can take advantage of this characteristic.

If we build a diode of heavily doped materials that will withstand reverse current and that will allow reverse current only above a specific voltage, we have a **zener diode**. We might build a zener diode to conduct reverse current only when voltage is above 6 volts, for example. Below that voltage, the diode acts as a simple PN diode. We can use such a zener diode to open and close a circuit at a prescribed voltage. Figure 8-16 shows the symbols for a zener diode.

Figure 8-17 shows a zener diode in parallel with a measuring instrument to protect the instrument from high voltage surges. The diode breakdown voltage equals the maximum safe voltage that can be applied to the instrument. Let's assume that it is 10 volts. When system voltage reaches 10 volts, the zener diode breaks down and conducts reverse current. This causes an additional voltage drop across the resistor and limits voltage at the instrument to 10 volts. When system voltage drops below 10 volts, the zener diode again blocks reverse current. A silicon zener diode with a steep breakdown curve is ideal for precise voltage control in this kind of circuit. You will find such diodes in electronic voltage regulators, ignition control units, and instrumentation systems. Zener diodes also are used in Chrysler air conditioning systems to prevent compressor clutch circuit surges.

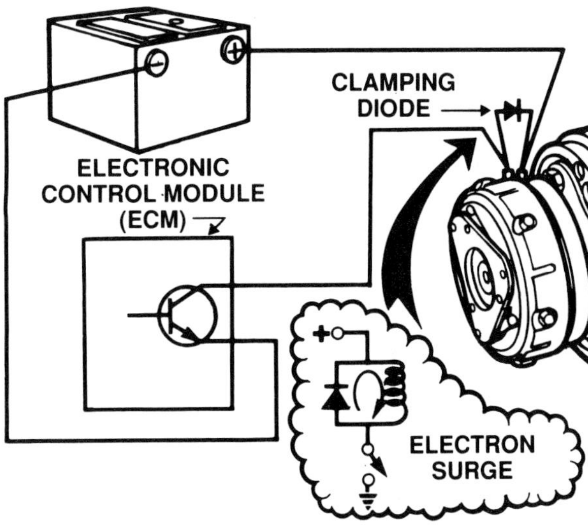


**Figure 8-18.** A clamping diode acts like a pressure relief valve. It protects a circuit from surge currents caused by abrupt interruption of the current in an electromagnetic device. (Delco-Remy)

## Clamping Diodes

A voltage surge results when current flowing through a solenoid or relay is interrupted. This surge is caused by electrons passing through the solenoid or relay coil that do not want to stop at the open in the circuit. As they run into the open, they attempt to jump it. If successful, a spark results. Installing a diode across the coil provides a bypass for the electrons during the period the circuit is open. In effect, this "clamping diode" acts as an electronic pressure relief valve and prevents the spark, figure 8-18.

One important application is the compressor clutch diode used in late-model air conditioning systems. Release of the electromagnetic compressor clutch coil produces voltage spikes that can damage the onboard computer. The clutch diode prevents this voltage surge from reaching the computer.

## Light-Emitting Diodes

When the two dopant materials, gallium and arsenic, form a gallium-arsenide compound in a diode, an interesting thing happens. When holes and electrons meet, they are neutralized as electrical charge carriers. When this happens, electrical energy is released as light. Such a diode is a **light-emitting diode (LED)**, figure 8-19.

The construction of an LED differs only slightly from that of a standard PN junction diode. A typical silicon diode requires 0.5 to 0.7 volt to turn on; an LED needs about 1.5 to 2.2

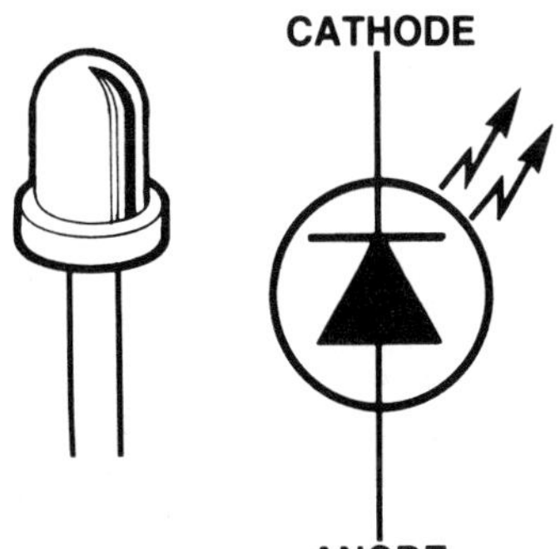


**Figure 8-19.** A light-emitting diode (LED) is represented by this symbol. (Delco-Remy)

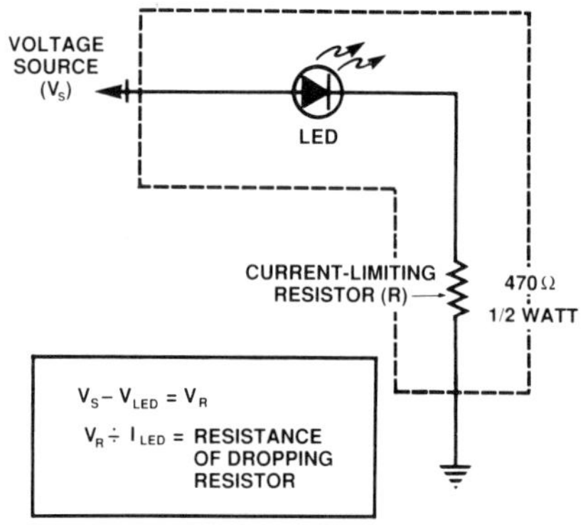


**Figure 8-20.** A current-limiting resistor is used in series to prevent an LED from failing. (Delco-Remy)

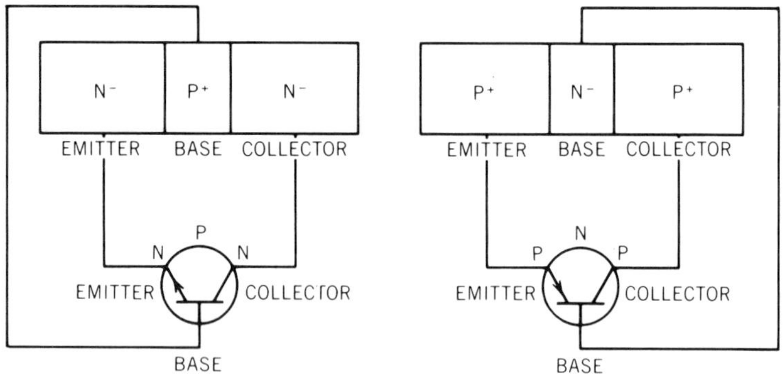


**Figure 8-21.** A transistor is a 3-element semiconductor made of alternate P and N materials (PNP or NPN).

volts to forward bias it. Although most LED's have no problem handling 20 to 200 mA, a current-limiting resistor is generally connected in series, figure 8-20.

If we arrange such diodes in geometrical shapes and turn current flow on and off through selected diodes in the pattern, we can create the lighted LED displays used in electronic calculators and some automobile instruments.

## TRANSISTORS

A diode can control the direction of current and the amount and polarity of voltage applied across all or part of a circuit. It cannot increase, or amplify, current or voltage, however. Electronic systems usually need current or voltage amplification to develop enough power to do useful work. A **transistor** is a semiconductor device that can provide this amplification. A transistor also can work like a relay to switch a high-current circuit on and off in response to a voltage signal across a low-current circuit.

We can build a transistor from a PN diode by adding a third layer of doped semiconductor material. We construct the transistor so that the outer layers are the same kind of material (both P or both N), and the center layer is the opposite. Thus, we can build either an NPN or a PNP transistor, figure 8-21. The name comes from transfer and resistor because a transistor transfers signals across the resistance of two PN junctions. The three parts of a transistor are the **base**, the **emitter**, and the **collector**, figure 8-21. Notice that the arrowhead on a transistor symbol is always on the emitter and, as on a diode symbol, it always points in the direction of hole flow (conventional current flow).

This kind of transistor is a **bipolar** transistor because it uses both electrons and holes as current carriers. Current flows with positive and negative polarity in different circuits. The emitter provides the majority current carriers (either holes or electrons), and the collector receives them when the transistor is switched on, or conducting. Let's look at an NPN transistor to see how it operates.

Because there are two PN junctions (between the emitter and the base, and between the base and the collector), there are two places where we can apply bias voltage. If we apply forward bias to the emitter-base junction (− potential on the N emitter and + potential on the P base), current flows as in a simple PN diode, figure 8-22. If we apply reverse bias to the base-collector junction, current does not flow. This junction again acts like a diode, but with voltage polarity reversed, figure 8-23.

The three parts of a transistor are doped and constructed differently to affect the amount of current through different parts. The base is thin and doped for just a few carriers—in this case, holes. The emitter is thicker and doped for maximum carriers, free electrons in an NPN transistor. The collector is larger (thicker) than the emitter but doped less. It has fewer majority carriers (electrons in an NPN transistor) and allows reverse current when bias voltage is applied.

To get our NPN transistor to conduct and amplify current, let's assume that the N-type emitter is at a potential of zero volts. If we apply a low forward bias voltage of 2 or 3 volts across the emitter-base junction, electrons flow from the emitter into the base, and holes flow in the opposite direction. Even at a low voltage, there are extra electrons that gather in the base. The base has fewer holes than the emitter has electrons because of the different doping ratios.

Now let's apply a higher positive voltage of 10 or 12 volts to the collector. This creates a reverse bias across the base-collector junction. High positive voltage on the collector attracts electrons away from the base-collector junction, so little or no current flows between the base and collector. However, the extra electrons that have gathered in the base due to forward bias on the emitter-base junction now pass through the base and into the collector toward the high positive voltage, figure 8-24.

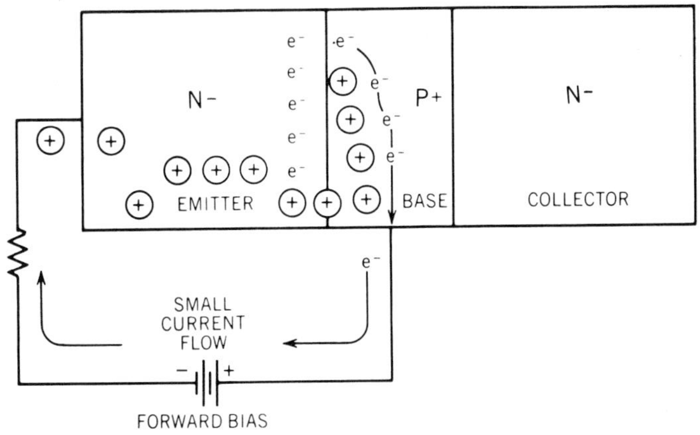


**Figure 8-22.** Forward bias on the emitter-base junction causes forward current flow, but excess free electrons accumulate in the base.

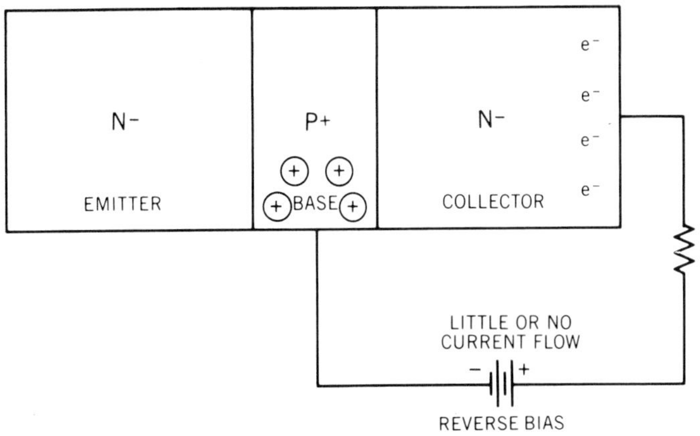


**Figure 8-23.** Reverse bias on the base-collector junction prevents current flow unless the emitter-base junction is forward biased.

The 10 or 12 volts across the emitter and collector would not be enough to cause electron flow from emitter to collector. (The voltage examples we are using are purely arbitrary; the actual voltages are a design consideration for each transistor.) However, the forward bias across the emitter-base junction boosts electron flow enough to cause the electrons to flow through the base, through the collector, to the higher voltage. The overall electron flow is forward, from low (or negative) potential to high (or positive) potential.

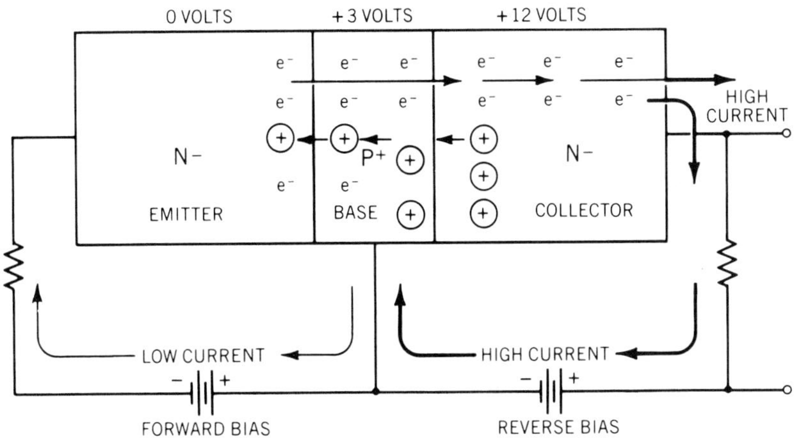


**Figure 8-24.** A small amount of current flow through the emitter-base junction controls a larger current flow through the emitter and collector.

## Transistors as Amplifiers

Figure 8-24 is a simple example of how a transistor can increase current flow through its parts and how a small voltage and current flow across one junction can control a larger current flow through the complete device. For a PNP transistor, the voltages and flows of electrons and holes are reversed, but the principle is the same.

Suppose that the forward bias across the emitter-base is supplied by the low voltage from a radio signal. This low voltage and current represent a fraction of a watt of power. The larger voltage and current flow between the emitter and collector represent several watts of power, enough to drive the electromagnet of a car radio speaker. Moreover, the ratio of emitter-base power and emitter-collector power remains constant. The ratio of amplification is called **gain** and is a design factor of the transistor.

Consider the forward bias voltage and current between the emitter and base as input power. If it fluctuates between one-tenth and five-tenths of a watt (100 to 500 milliwatts), and the gain of the transistor is 100 (100 to 1), then the output power between the emitter and collector will fluctuate between 1 and 5 watts. Such a transistor can increase power enough to drive a speaker or to switch ignition current on and off in response to low input voltage and current.

Amplifying transistors like this have been used for decades in automobile radios, charging systems, and ignition systems to handle large amounts of current in response to small signal voltages. They are often called power transistors, and figure 8-25 shows several examples.

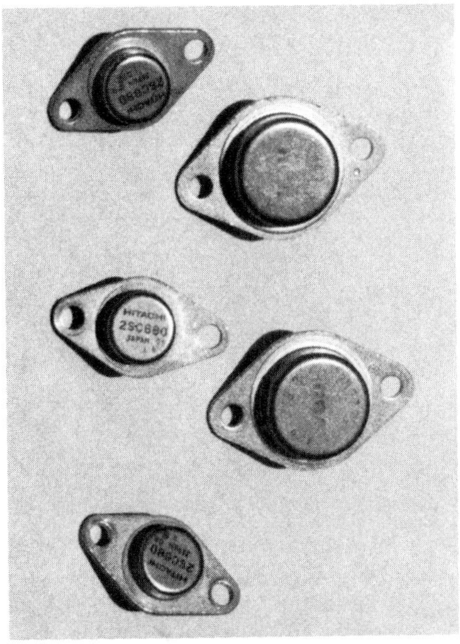

**Figure 8-25.** Power transistors are used in automotive charging and ignition systems.

The amount of forward bias voltage on the emitter-base junction at which the transistor starts to conduct is the switching point. When the transistor is on, or conducting, the ratio of emitter-base current to emitter-collector current stays relatively constant. This is called the linear conduction region because 0.1 ampere of emitter-base current may equal 1 ampere of emitter-collector current, for example. Eventually, however, the transistor will reach a saturation condition. At this point, large increases in emitter-base current produce only small increases in emitter-collector current. The transistor is simply conducting as much current as it can handle. Increasing the emitter-base bias voltage will not significantly increase output current. Voltage drop across the emitter-collector circuit is very small, and the transistor is operating as a simple relay.

Like a diode, a transistor has three operating conditions; however, they are not the same as those of a diode. In summary, the operating conditions of a transistor are:

1. *Cutoff*—Bias voltage on the base is below the switching point. The transistor is not conducting; it is blocking. It acts as an open switch.
2. *Conduction*—Bias voltage is high enough to switch the transistor on. It is conducting current, but output current changes in proportion to current through the base. This is the amplification stage.
3. *Saturation*—Bias voltage, base current, and output current are at the maximum. The transistor acts as a closed switch or relay.

## Transistors as Relays

You might ask what the difference is between a transistor as an amplifier and a transistor as a switch or relay. The answer is, not much electrically. The difference is in how the transistor is used: whether it switches from cutoff to saturation, or whether it operates in the conduction region. Generally, a given transistor is not used as both an amplifier and as a switch or relay. Specific transistors are designed to work better as one or the other.

A transistor can control collector output current or voltage in proportion to input current or signal voltage across the emitter-base junction. In this conduction region, the transistor amplifies the output in proportion to the input. Amplifying transistors are used in radio systems. The amplified output is an **analog** signal, which means that it changes in direct proportion to the input signal.

When a transistor acts as a switch or relay, it switches directly from cutoff (no conduction) to saturation (full conduction). Such a transistor can be used to switch current on and off in a power circuit. The switching action also is the foundation for digital computers. As you will learn later, a **digital** signal is one that is either on or off.

Remember the explanation of simple electromagnetic relays in chapter 7. Electromagnetism in the relay control coil moves the armature to close the contacts of the relay power circuit. The power circuit switches from open (no conduction) to closed (full conduction). A small amount of current in the control circuit can switch a large amount of current in the power circuit.

A transistor can act as a solid-state relay in a circuit, and this is one of the most important uses of transistors in automotive electrical systems. When maximum forward bias voltage and emitter-base current are applied, the collector current goes to complete saturation, and voltage drop across the emitter-collector circuit goes to a minimum. These are the same operating conditions that a relay has.

The emitter-collector circuit is the power circuit. The emitter-base circuit is the control circuit. Base current can be controlled by a simple single-throw switch, figure 8-26, or by a double-throw switch that changes the polarity on the base. When the switch closes, low emitter-base current biases the transistor to allow higher emitter-collector current to flow through a circuit load, figure 8-27. The emitter base circuit has a resistor to limit bias voltage and emitter-base current.

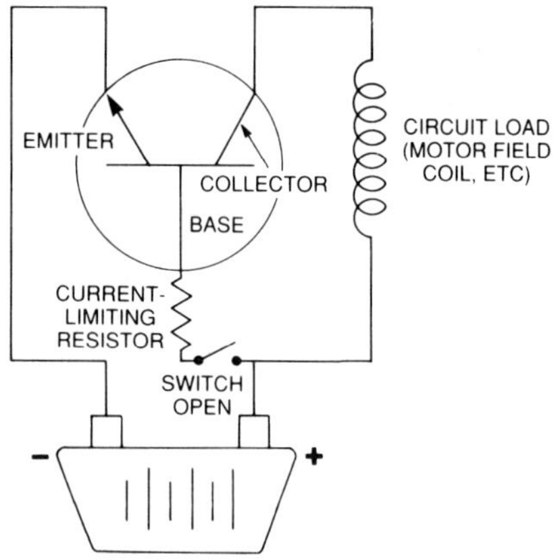


**Figure 8-26.** A simple diagram of a transistor used as a power relay.

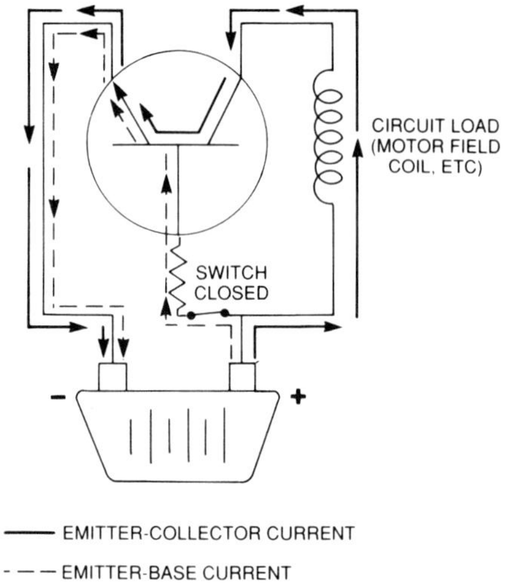


**Figure 8-27.** Low emitter-base current and bias voltage allow a higher emitter-collector current to flow in the power circuit. This transistor is a solid-state relay.

Although figures 8-26 and 8-27 show the symbol for a mechanical switch in the base circuit, combinations of diodes, resistors, and zener diodes can control voltage on the base. This allows a transistor to act as an electronic relay with no moving parts and very fast switching action.

## THYRISTORS

A **thyristor** is a semiconductor switching device made up of alternating P and N layers, or regions. Not only can it switch current on and off, but it also can rectify ac to dc. The most com-

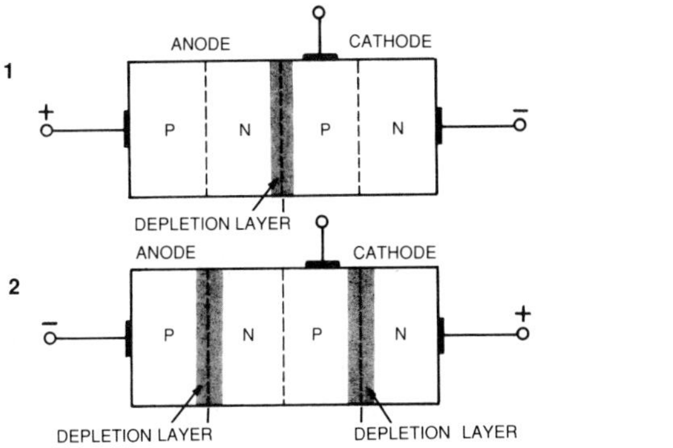


**Figure 8-28.** The center PN junction blocks current when the P-type anode is connected to the positive voltage side of the circuit and the N-type cathode is connected to the minus side. (Bosch)

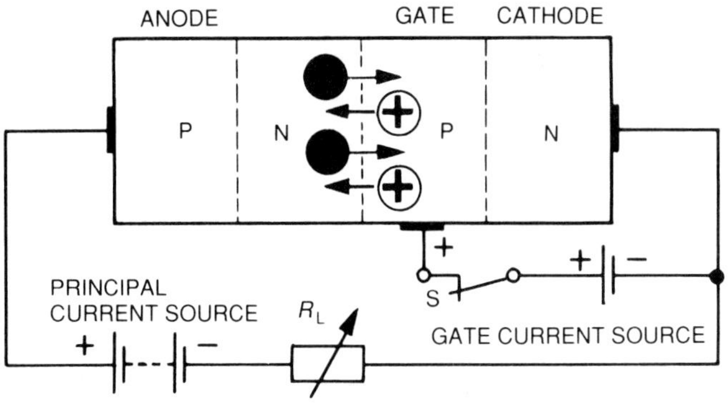


**Figure 8-29.** Applying a positive voltage to the P-type anode turns the transistor on. (Bosch)

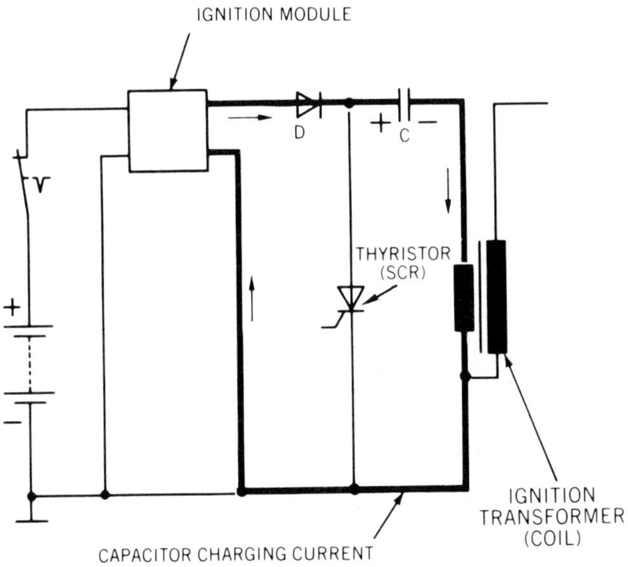


**Figure 8-30.** A CD ignition stores energy by charging the capacitor. The thyristor, or SCR, blocks the discharge circuit. (Bosch)

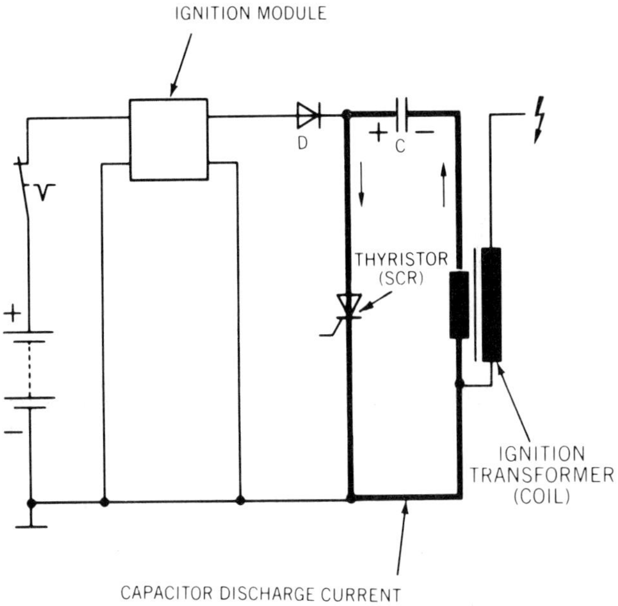


**Figure 8-31.** When the thyristor, or SCR, receives a trigger voltage from the distributor, it allows the capacitor to discharge energy to the coil primary winding. (Bosch)

mon kind of thyristor used in some automotive systems is the **silicon-controlled rectifier (SCR)**.

An SCR consists of four semiconductor regions, arranged PNPN. Like a transistor, an SCR has two terminals for working current (the power circuit) and one terminal for control current. These are:

- The anode (+), or P-terminal
- The cathode (−), or N-terminal
- The gate, one of the two inner regions

The SCR behaves as a PNP transistor joined to an NPN transistor. Unlike a transistor, however, an SCR does not need continuous control current. It needs only a trigger pulse applied to the gate to become conductive.

An SCR thyristor can be connected into a circuit in the forward or the reverse direction. The forward direction is the simplest example to show SCR operation, figure 8-28. In this condition, the P-type anode is connected to the + voltage side of the circuit, and the N-type cathode is connected to the − side. The center PN junction then blocks current. When a + voltage pulse is applied to the P-type gate, it acts as base current in an NPN transistor, figure 8-29. The transistor turns on, and its working current acts as base current on the PNP-transistor portion of the SCR. As long as voltage across the anode and cathode remains high enough, the SCR will continue to conduct. Current continues to flow even with the gate trigger voltage removed. Current stops if gate voltage is reversed or if system voltage falls below the breakdown voltage for the center PN junction.

This kind of thyristor effectively blocks reverse current from cathode (−) to anode (+) under all conditions. Thus it can rectify alternating current and gets the name "silicon-controlled rectifier." SCR's are used as high-power switches. You'll find them in some solid-state voltage regulators and ignition control units. Figures 8-30 and 8-31 show an SCR in a capacitive-discharge ignition circuit. Ignition primary current charges a large capacitor through diode D. The SCR acts as a diode to block current through the capacitor discharge circuit. When a signal voltage is applied to the SCR gate, it closes the capacitor discharge circuit through the ignition coil primary winding. Diode D then blocks reverse current to the ignition control unit and the rest of the system. When capacitor discharge voltage drops below the breakdown voltage of the SCR, the SCR stops conducting and opens the discharge circuit.

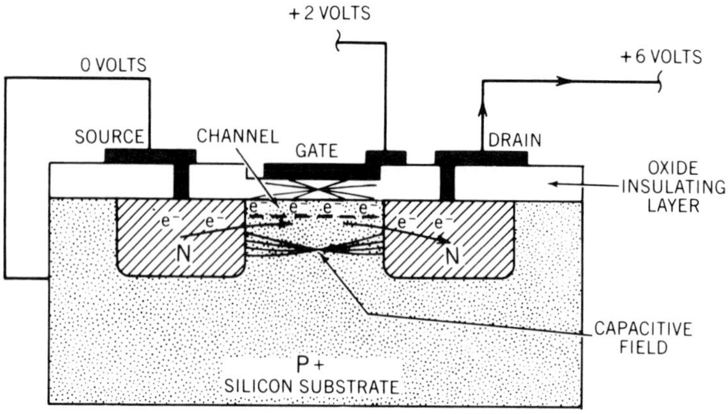


**Figure 8-32.** A field-effect transistor is controlled by gate voltage that creates a capacitive field and causes current to flow through the channel from source to drain. This is an N-channel enhancement-type FET.

## FIELD-EFFECT TRANSISTORS

A **field-effect transistor (FET)** is another 3-element semiconductor device that allows high voltage and large current flow to be controlled by a small signal voltage. The bipolar NPN transistor in the earlier examples is controlled by changing the current through the emitter-base junction. A field-effect transistor is controlled by changing voltage in a capacitive field. Its three parts are:

1. The **source**, which supplies the current-carrying electrons or holes and is similar to the emitter in a bipolar transistor.
2. The **drain**, which collects the current carriers and is similar to the collector.
3. The **gate**, which is similar to the base of a bipolar transistor and which creates the capacitive field that allows current to flow from the source to the drain.

A field-effect transistor is not a bipolar device because current involves only one group of charge carriers, either holes or electrons, and flows between similar materials, N to N or P to P. It does not require bias voltage or forward and reverse current as does a bipolar transistor. Briefly, here is one way in which a field-effect transistor works.

The source and the drain are the same kind of doped material, either N or P. Let's say both are N-material, figure 8-32. The source and drain are separated by a thin channel of either N- or P-material. If we hold the source at zero voltage and apply 6 volts to the drain, no current will flow between the two. Then we install the gate as a metal strip above the channel between the source and drain and apply a lower positive voltage to it. The gate forms a capacitive field between itself and the channel. The voltage of this field attracts electrons from the source just as the base does in a bipolar transistor. The electrons then flow across the channel to the higher positive voltage on the drain.

This kind of FET is called an enhancement-type because the field effect enhances, or improves, current flow from the source to the drain. Another kind of FET is the depletion-type, where the field effect depletes, or cuts off, current flow. In simple terms, you can think of enhancement-type and depletion-type FET's as working like normally open and normally closed switches. Remember that bipolar transistors are controlled by current flow between the emitter and base. FET's are controlled by voltage that forms a capacitive field between the source and the drain.

Like a bipolar transistor, a field-effect transistor can work as an amplifier or as a relay. Many radio and video amplifiers, as well as voltage regulators, use FET's to control output voltage. More importantly, field-effect transistors, built in miniature sizes, act as switching circuits for digital computers. Thousands of FET's can be built on a small silicon chip as an integrated circuit (IC). These IC's form the computing logic circuits of automotive computers, which you will study later.

## DARLINGTON PAIR

Some control circuits are too small or too sensitive to produce the required output. Increasing the operating current of the control circuit is not effective, because it can cause undesirable effects. To operate the circuit properly and produce the required output, a Darlington pair is used.

A Darlington pair consists of two transistors connected together as shown in figure 8-33. This is often referred to as a "piggyback" transistor. The first transistor acts as a preamplifier. It produces a large operating current for the second transistor, which is isolated from the control circuit. The second transistor functions as a final amplifier, boosting the current to produce the required output. Darlington pairs of transistors are the basis of most electronic ignition control modules, as you will see in chapter 13.

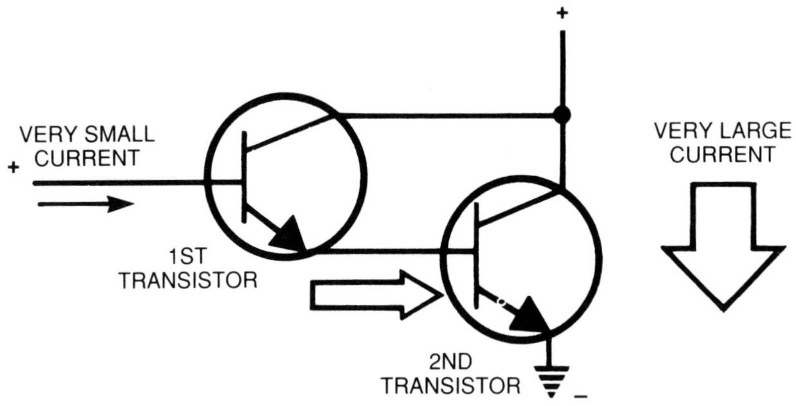


**Figure 8-33.** A Darlington pair consists of one transistor acting as a preamplifier to create a larger base current for the second transistor, or final amplifier. (Delco-Remy)

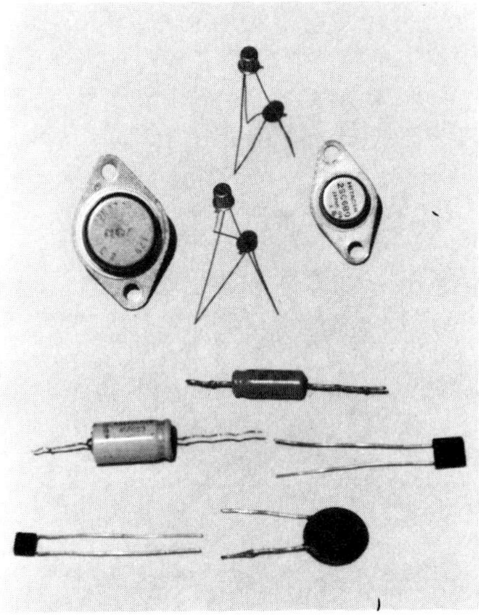

**Figure 8-34.** These discrete electronic devices are used in various automotive systems.

## DISCRETE DEVICES, INTEGRATED AND HYBRID CIRCUITS

All of the electronic devices we have seen so far can be manufactured and assembled into circuits in different ways. Device manufacture and circuit assembly is usually divided into three categories: discrete devices, integrated circuits, and hybrid circuits.

### Discrete Devices

Any of the electronic devices we have examined can be made of an individual part. Engineers call these **discrete devices**, which means that each one is made separately and has wire leads for connecting it into a complete circuit. Most early solid-state devices were discrete devices, such as the rectifying diodes on older alternators. You can still find discrete devices in many automotive systems, especially in charging, ignition, and headlamp circuits that handle large amounts of power. Figure 8-34 shows several discrete capacitors and transistors from automotive systems.

### Integrated Circuits

Although discrete devices are quite small compared to the vacuum tubes of older electronic circuits, they are enormous compared to a modern **integrated circuit (IC)**. An integrated circuit is a complex electronic circuit of perhaps several thousand transistors and other devices, all formed on a tiny chip of silicon.

Integrated circuits are made by photographically reproducing circuit patterns on a silicon slice or wafer and then depositing various conductive P and N materials and insulating materials. Conductive and insulating materials are formed through chemical reaction of gases and by depositing metals such as aluminum. The doping of alternating silicon layers is carried out during wafer fabrication.

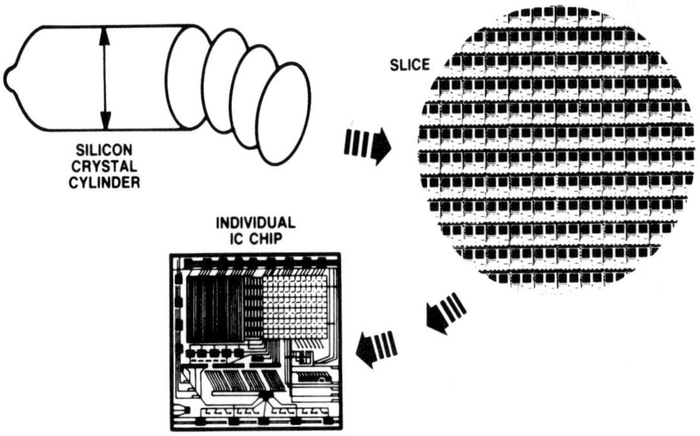


**Figure 8-35.** Hundreds of IC chips are made on a single slice or wafer of a silicon crystal cylinder. (Chrysler)

This process often requires more than 100 separate steps to build an IC. Several hundred identical IC's can be built on a single 4-, 5-, or 6-inch silicon slice or wafer, figure 8-35. Each is cut from the wafer as an individual "chip" and installed in a package device that can be plugged into a larger circuit. Figure 8-36 shows the main stages of IC manufacturing from silicon wafer to packaged circuit.

The field-effect transistors that you studied earlier are a major factor in the creation of integrated circuits. FET's can do most of the switching, processing, and amplifying jobs of discrete diodes, resistors, capacitors,

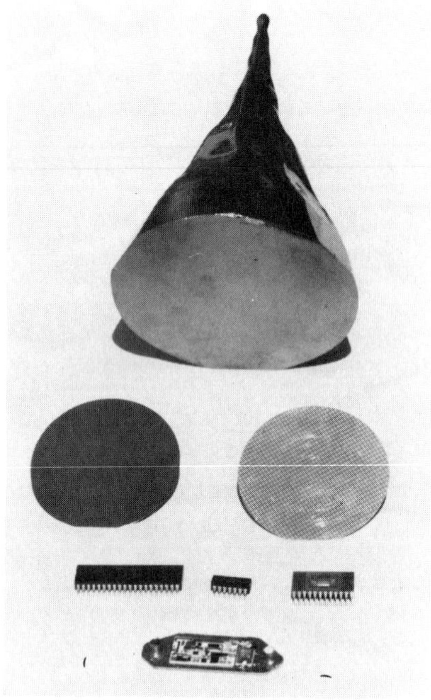

**Figure 8-36.** (Rear) a silicon ingot from which wafers are sliced for semiconductor manufacture. (Left center) a polished silicon wafer. (Right center) a finished device wafer. Each small square is a complete integrated circuit. (Foreground) each circuit chip is cut from a wafer and packaged as a microprocessor device or as a control chip in an ignition module.

**MOORE'S LAW**

There's a widely known law in the electronics industry that is more a historical observation than a statement of scientific fact. Nevertheless, Moore's Law seems true.

The modern integrated circuit—several electronic devices on a single silicon chip—was developed at Fairchild Semiconductor in 1959, where Gordon Moore was a research scientist. In 1964, Moore was Fairchild's research director. Looking back over the previous five years, he noted that the number of electronic devices that engineers were able to fabricate on a single chip had been doubling every year.

This observation gave rise to his prediction that integrated circuit complexity would continue to double annually for the foreseeable future—and he was right, for almost 20 years. This came to be known as "Moore's Law."

By 1969, 10 years after the invention of the integrated circuit, engineers were packing more than 1,000 diodes, resistors, capacitors, and transistors on a single chip. By 1974, the number was approaching 20,000 per chip; and by 1977, it had passed 250,000. Today, single IC chips that can do the jobs of one-half million separate transistors, diodes, and other devices are practical.

Gordon Moore, along with his partner, Robert Noyce, and others, is credited with developing the microprocessor—a digital computer on a chip. In 1968, Moore and Noyce founded Intel Corporation, a leading manufacturer of microprocessors and other IC devices.

and other transistors. Thousands of FET's can be built on a tiny chip called a metal-oxide semiconductor (MOS). This kind of IC is called a **MOSFET** and is the heart of modern computer systems—in automobiles and elsewhere.

Diodes, resistors, capacitors, bipolar transistors, and field-effect transistors all can be made as discrete devices, or as parts of integrated circuits, figure 8-37. The only electrical or electronic device that cannot be made as part of an IC is an inductor, or induction coil. To create a usable magnetic field, an induction coil is made by winding a conductor around a permeable core. No way of fabricating a usable induction coil in an integrated circuit has yet been invented.

As sophisticated and versatile as integrated circuits may be, many automotive systems have requirements that IC's cannot meet. These are chiefly systems with either, or both, of the following features:

**1.** Circuits that carry high current and high power that would overload the small conductors in IC's.

**2.** Circuits that employ electromagnetic induction.

For these reasons, many automotive systems use solid-state devices in combination with traditional electrical and electromagnetic components. Hybrid circuits also help to solve these problems by combining discrete devices with IC devices.

### Hybrid Circuits

A circuit built with several discrete devices and several IC's is called a **hybrid circuit**. Many automotive systems have used, and continue to use, hybrid circuits, figure 8-38. As more and more functions are packed into single IC's, the need for discrete diodes, resistors, capacitors, and transistors is decreasing. Many jobs done by hybrid circuits on cars 10 years ago are now done by one or two IC's. Hybrid circuits continue in use, however, for high-current circuits and for electromagnetic systems.

## SUMMARY Electronic System Integration

It's a long way from a PN diode to an engine management system computer; but if it were not for the simple idea of doping silicon atoms, neither one would be possible. You will be dealing with electronic devices regularly when you service late-model vehicles. Don't be surprised if you start out to adjust idle speed and wind up replacing an electronic control unit.

An important point to remember is that you can't repair any electronic control device on a modern car; you can only replace it. You can overhaul a starter motor or repair an alternator,

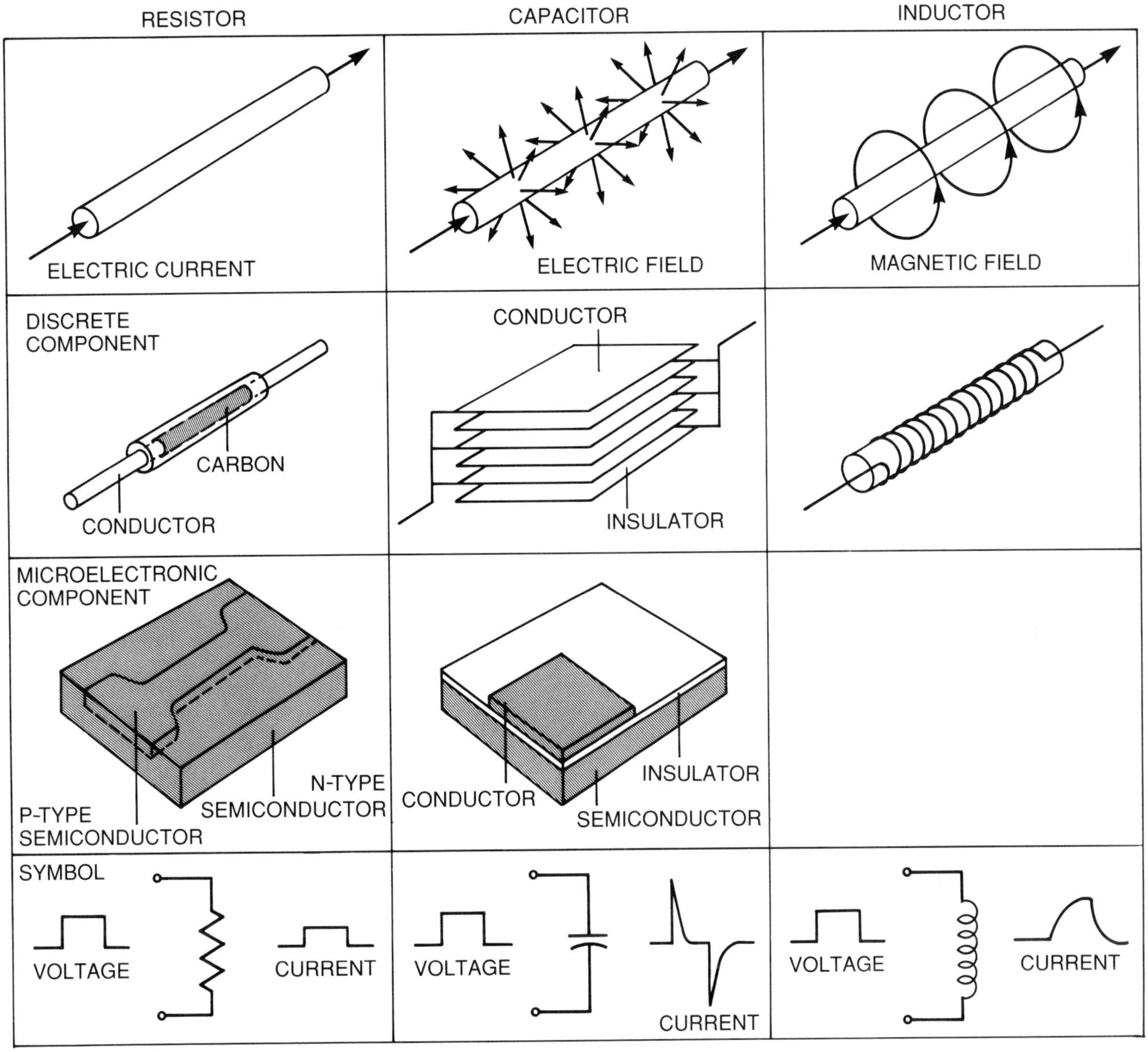


**Figure 8-37.** With the exception of inductors, discrete devices can be made as parts of an integrated circuit. There is no practical way to fabricate a usable induction coil in an integrated circuit.

but you can't repair or adjust the electronic chips in a computer. You can, however, test system operation, identify good and bad components, and replace only those that you can prove are defective. To do this professionally, you have to understand how they operate.

You have to understand what the input and output signals from sensors to computers to actuators mean. You have to know what they should be, and you have to identify those that are right and wrong. The next two chapters summarize the operation of automotive computers and the sensors and actuators of computer systems. All computer system operations are based on the principles of current flow and voltage levels in electronic semiconductor devices that you have learned here.

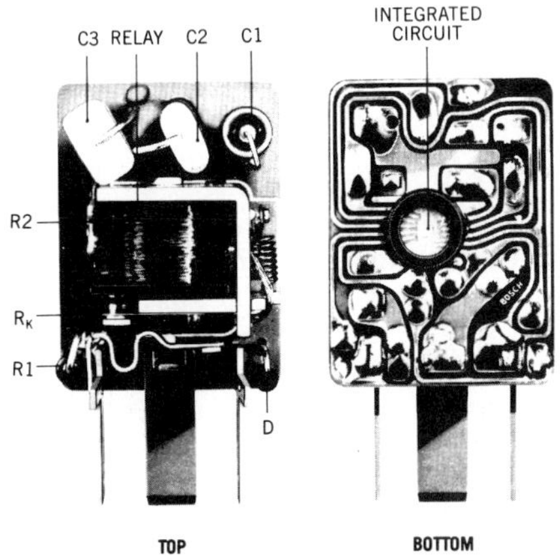


**Figure 8-38.** Hybrid circuits are made from discrete devices and integrated circuit chips assembled on a printed circuit board. (Bosch)

## REVIEW QUESTIONS

### Multiple Choice

**1.** Student A says that phosphorus and arsenic have five valence electrons and are P-dopant materials. Student B says that boron has three valence electrons and is an N-dopant. Who is right?

- **a.** A only
- **b.** B only
- **c.** both A and B
- **d.** neither A nor B

**2.** Student A says that hole flow is the opposite of electron flow. Student B says that hole flow is similar to conventional current flow theory. Who is right?

- **a.** A only
- **b.** B only
- **c.** both A and B
- **d.** neither A nor B

**3.** A zener diode allows reverse current flow at the point of:

- **a.** bias voltage
- **b.** breakdown voltage
- **c.** system voltage
- **d.** forward voltage

**4.** Student A says that a transistor can work like a relay to switch current. Student B says that a transistor can amplify current. Who is right?

- **a.** A only
- **b.** B only
- **c.** both A and B
- **d.** neither A nor B

**5.** All of the following are parts of a bipolar transistor, except the:

- **a.** base
- **b.** emitter
- **c.** collector
- **d.** gate

**6.** All of the following are parts of a field-effect transistor, except the:

- **a.** source
- **b.** collector
- **c.** drain
- **d.** gate

**7.** An individual transistor, diode, capacitor, or other electronic part is:

- **a.** an integrated circuit
- **b.** a hybrid circuit
- **c.** a discrete device
- **d.** an analog device

**8.** Student A says that an analog signal is either on or off. Student B says that an analog signal changes proportionally to the quantity measured. Who is right?

- **a.** A only
- **b.** B only
- **c.** both A and B
- **d.** neither A nor B

**9.** Student A says that temperature sensors are analog devices. Student B says that temperature sensors are digital devices. Who is right?

- **a.** A only
- **b.** B only
- **c.** both A and B
- **d.** neither A nor B

**10.** Student A says that a silicon-controlled rectifier (SCR) is a common type of thyristor that acts like a PNP transistor connected to an NPN transistor. Student B says that a hybrid circuit is built with two or more IC's and discrete devices. Who is right?

- **a.** A only
- **b.** B only
- **c.** both A and B
- **d.** neither A nor B

**11.** A major difference between a light-emitting diode (LED) and a standard silicon junction diode is that the LED:

- **a.** requires about three times as much voltage to turn it on
- **b.** allows current to pass in either direction
- **c.** is a specially doped zener diode
- **d.** none of the above

**12.** A piggybacked transistor is called a:

- **a.** bipolar transistor
- **b.** unipolar transistor
- **c.** field-effect transistor
- **d.** Darlington pair

### Fill in the Blank

**13.** The element commonly used as the basis for semiconductor devices is ________ .

**14.** The device created when N-material and P-material are joined is called a ________ .

**15.** The method used to induce electrical flow in a diode is called ________ voltage.

**16.** The process of adding an impurity to a silicon crystal is called ________ .

**17.** The ________ is used to increase, or amplify, current or voltage.

**18.** The ratio of current or voltage amplification in a transistor is the ________ .

**19.** The three parts of a transistor are the base, the ________ , and the collector.

**20.** In a field-effect transistor, the element that supplies the current-carrying electrons is called the ________ .

**21.** A circuit built on a single silicon chip without discrete transistors, resistors, diodes, and other devices, is called an ________ .

**22.** To measure changing temperature continuously across a range, a computer would use an ________ signal.

# 9 BATTERIES

## WHAT

Early automobiles did not have batteries. They also didn't have much in the way of electrical systems. Ignition was accomplished by a magneto of some kind, and it took muscle power to crank and start the engine. Headlamps (if a car had them) were fired by acetylene gas, and early cars had *no* electrical accessories. In these respects, our modern cars would be similar to those of 80 years ago if it were not for storage batteries. If our cars did not have batteries, they would not have self-starters; they would not have inductive ignition systems or safe lighting; and they would not have the reservoir of steady electrical power for accessories such as radios, air conditioners, and other systems.

This chapter explains the electrochemical principles of the batteries that make modern automobile electrical systems possible.

## WHY

None of the combustion control systems or accessories on a modern car can operate right if the battery is discharged or defective. These are the jobs of a battery:

1. To operate the starter motor—the primary function
2. To provide ignition current and power for the engine computer and combustion control systems during cranking
3. To act as a voltage stabilizer, or electrical reservoir, for the entire electrical system of the vehicle
4. To provide power for lights and accessories when the engine is off
5. To provide reserve power when the electrical demands of the car exceed the output of the charging system
6. To provide reserve electrical power in case of charging system failure

This list also outlines the vehicle systems that you *can't* test or service if the battery is defective. *Every* service procedure that you follow for ignition, starting, charging, lighting, accessory, and other electrical and electronic systems will begin with instructions to *check the battery voltage and state of charge.* Understanding the principles of battery construction and operation is the first step in automobile electrical service.

## GOALS

After studying this chapter, you should be able to:

1. Explain the electrochemical action of a lead-acid automotive battery.
2. Identify the parts of a battery and explain battery construction.
3. Explain battery charging voltage and the effects of temperature and the state of charge on charging voltage.
4. Define various battery rating methods.
5. Explain the factors that affect battery life.

## BATTERY TYPES

Every battery is either a **primary battery** or a **secondary battery**. In a primary battery, the electrochemical action is not reversible. Eventually, one or both of the **electrode** materials are destroyed. The chemical structure of the electrode changes so that it no longer reacts to create voltage and deliver current. Most dry-cell batteries in flashlights and portable radios are primary batteries. Rechargeable nickel-cadmium batteries, however, are secondary batteries.

The electrochemical action in a secondary battery is reversible. The chemical state of the electrolyte and the electronic state of the electrode materials in any battery (primary or secondary) change as the battery applies voltage and delivers current to a circuit. This is the discharging action of a battery, but it can be reversed. A secondary battery can be recharged, which is important for a car battery.

If we apply sufficiently high voltage and current from an outside source (an alternator) to the battery terminals, we can force current through the battery. This charging current flows in the direction opposite to the battery's discharge current, figure 9-1. In a secondary battery, this charging current reverses the electrochemical action and eventually restores the battery materials to their original conditions. In other words, the battery again stores chemical energy, which can be converted to electrical energy. The battery then can produce more current and once again supply power to an outside circuit. The electrochemical condition of battery electrolyte and plate materials at any given time is the battery's **state of charge**.

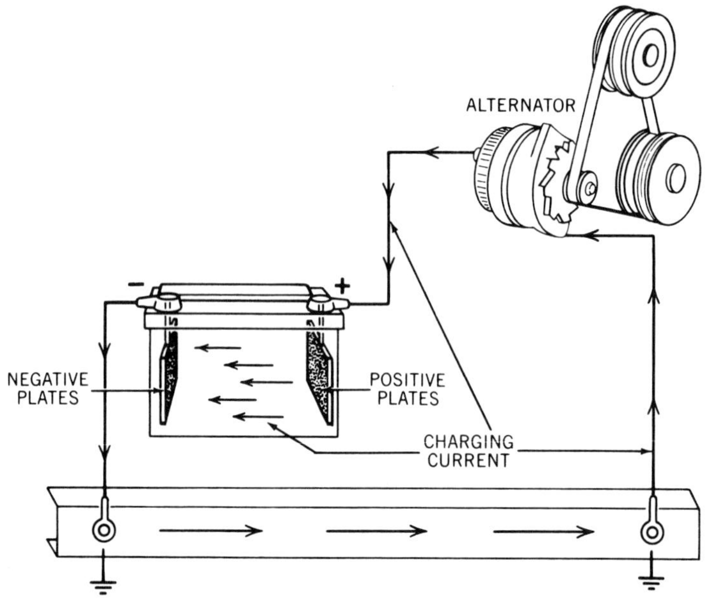


**Figure 9-1.** Charging current flows into the battery in the direction opposite discharge current flow.

## BATTERY ELECTROCHEMICAL ACTION

We call automobile batteries "storage" batteries, but they do more than store electricity. The **electrochemical** action of a battery:

- Changes chemical energy to electrical energy
- Changes electrical energy to chemical energy

The battery stores energy in chemical form until it is connected to an external circuit. The stored energy then changes to electrical energy that flows from one battery terminal, through the circuit, and back to the other terminal, figure 9-2. The electrochemistry of a battery works on the reaction that occurs when two dissimilar materials, or electrodes, are placed in a conductive and reactive solution called an **electrolyte**.

The chemical reaction causes the battery terminals to become oppositely charged—positive and negative. This creates a difference in potential, or voltage, between the two terminals of the battery.

The battery electrolyte is the medium through which the dissimilar materials react. These materials are lead (Pb) and lead dioxide ($PbO_2$) on the battery plates. Battery electrochemistry is another application of the oxidation and reduction reactions that you learned about in chapter 2.

In a fully charged battery, the lead (Pb) plates have surplus electrons and a negative (−) potential. When a battery is connected to an external circuit, oxidation occurs at the negative plates, which are joined to form a negative (−) electrode terminal, or cathode. Reduction occurs at the positive plates, which are joined to form a positive (+) electrode terminal, or anode, which has a deficiency of electrons. You can think of it in these terms:

- Oxidation at the negative (−) plates, or cathode, releases electrons and drives them through the external circuit.
- Reduction at the positive (+) plates, or anode, collects electrons that travel through the circuit.

As the negative plates give up electrons and the positive plates acquire electrons, their charges change. The negative plates become less negative, and the positive plates become less positive. The voltage difference between the battery terminals decreases, and the battery becomes discharged.

Remember that automobile electrical systems operate on the conventional current flow theory. Current flow is opposite to electron flow. So, if electrons flow from negative to positive during discharge, *current flows from the positive battery terminal, through the external circuit, to the negative terminal*. That is the way we view all electric current in an automobile.

During the discharge reaction, the chemical structures of the electrolyte and the plates change. The reaction is an example of the chemical principles of electron transfer and compound formation that you learned in chapter 2. As you will learn later, charging the battery reverses the electrochemical discharge reaction. The electrochemical properties of the battery materials allow a battery to do more than store energy. The battery creates voltage and provides current when connected to a circuit.

If we connect a conductor and a load (a motor, a lamp, or some other electrical device) across the battery terminals, we complete an external circuit. Current flows from the positive (+) battery terminal, through the circuit, and back to the negative (−) terminal, figure 9-2. Because the internal parts are part of the *total* circuit, conventional current theory also assumes that discharge current flows from negative to positive *inside* the battery.

### Electrochemical Discharge and Recharge

An automobile battery has negative plates of lead (Pb) in a spongy metallic form. The positive plates are lead dioxide ($PbO_2$). When the battery is fully charged, the electrolyte is a mixture of sulfuric acid ($H_2SO_4$) and water ($H_2O$).

As the battery discharges, figure 9-3, electrochemical action reduces the acid percentage in the electrolyte and increases the water percentage. Sulfate ($SO_4$) ions from the electrolyte combine with lead ions generated in both plates to form lead sulfate ($PbSO_4$). The plates then become similar, not dissimilar, materials, and the electrolyte contains less acid and more water. Two similar electrodes in an electrolyte cannot produce voltage (an electrical difference) or current between two terminals. The battery is then discharged. Because all the plates in a discharged battery contain sulfate, we often say that a completely discharged lead-acid battery is "sulfated."

A discharged battery, figure 9-3, has a weak acid solution for electrolyte, and the plates are mostly lead sulfate. Recharging reverses this condition. The lead sulfate in the positive plates changes back to lead dioxide ($PbO_2$), and the lead sulfate in the negative plates changes back to spongy lead (Pb). The sulfate ions recombine in the electrolyte, and hydrogen ions are generated, which increase the sulfuric acid ($H_2SO_4$) percentage. The battery then returns to a fully charged state, figure 9-3. The process of discharging and recharging a battery is called **cycling**.

During charging, a vented, lead-acid battery releases oxygen ($O_2$) and hydrogen ($H_2$) gases as part of the reaction. Hydrogen is very explosive, and any spark or open flame around a battery, particularly during charging, can cause an explosion.

Always keep sparks and open flame away from a battery. To be safe, treat a battery as if hydrogen were present at all times.

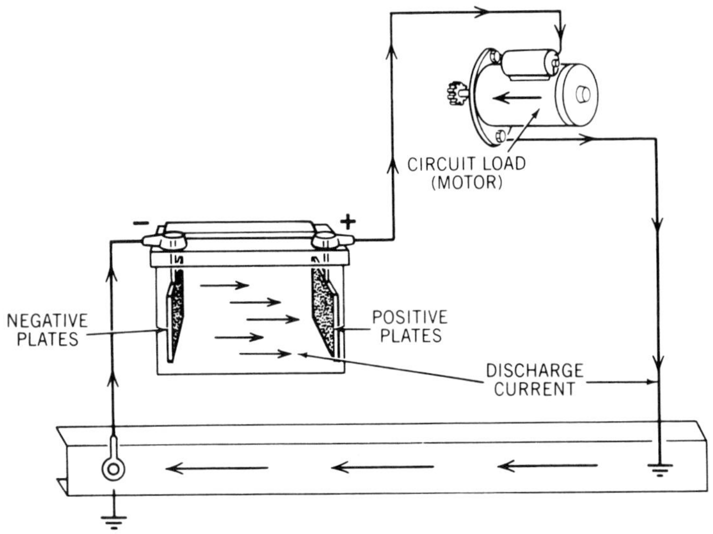


**Figure 9-2.** A storage battery does not simply store electricity. It stores chemical energy that it changes to electrical voltage and current, which it delivers to a circuit.

## BATTERY CONSTRUCTION

We have talked about battery plates, terminals, and electrolyte, which are a few parts of a battery. The following sections outline features of complete battery construction.

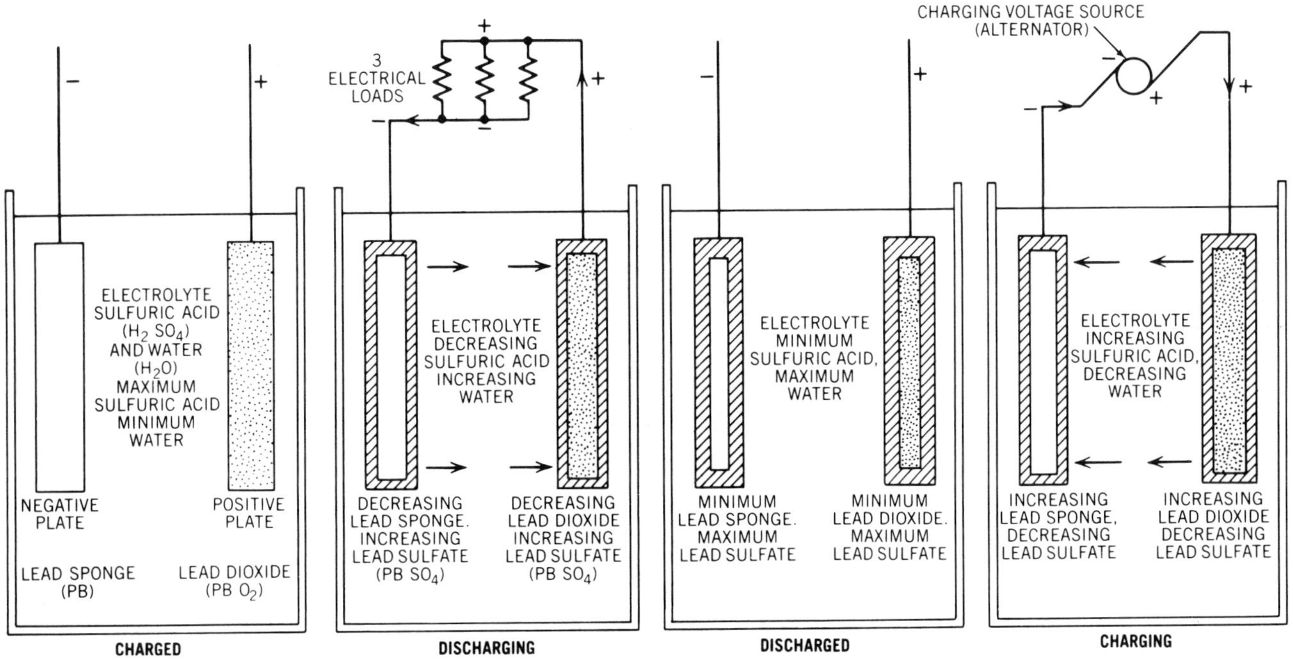


**Figure 9-3.** Battery electrochemical action is reversed as a battery changes from charged to discharged and back to charged.

### Plates and Grids

Plates are made of grids of conductive metal, figure 9-4, which are frames for the reactive lead and lead dioxide. The Pb and $PbO_2$ are spread uniformly on different grids in paste form. When the paste dries, it is solid but porous so the electrolyte can penetrate the plates. The grids are the conductors for the current generated by the plate materials.

Groups of similar all-positive or all-negative plates, figure 9-4, are welded to a plate strap by "lead burning." Each plate strap has a connector or post to join similar positive or negative plate groups together in the battery.

### Elements, Separators, and Cells

You will hear the active units of a battery called **battery cells**. Cell construction begins by placing a positive and a negative plate group together so their plates alternate. The negative group has one more plate than the positive group. Two plate groups form a battery element, not yet a complete cell. To finish the cell, we first must add separators, figure 9-5.

The separators are thin sheets of inert material that prevent short circuits between the positive and negative plates. Separators in older batteries were often wood, rubber, or coated paper. Modern batteries have plastic

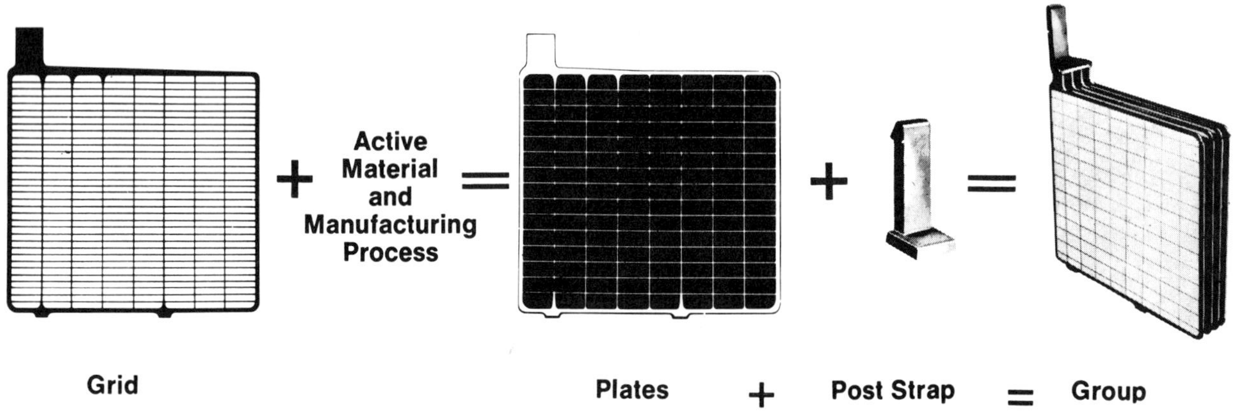


**Figure 9-4.** The grid is the framework for battery plate construction. Several plates, all + or all −, are joined in a plate group. (Atlas)

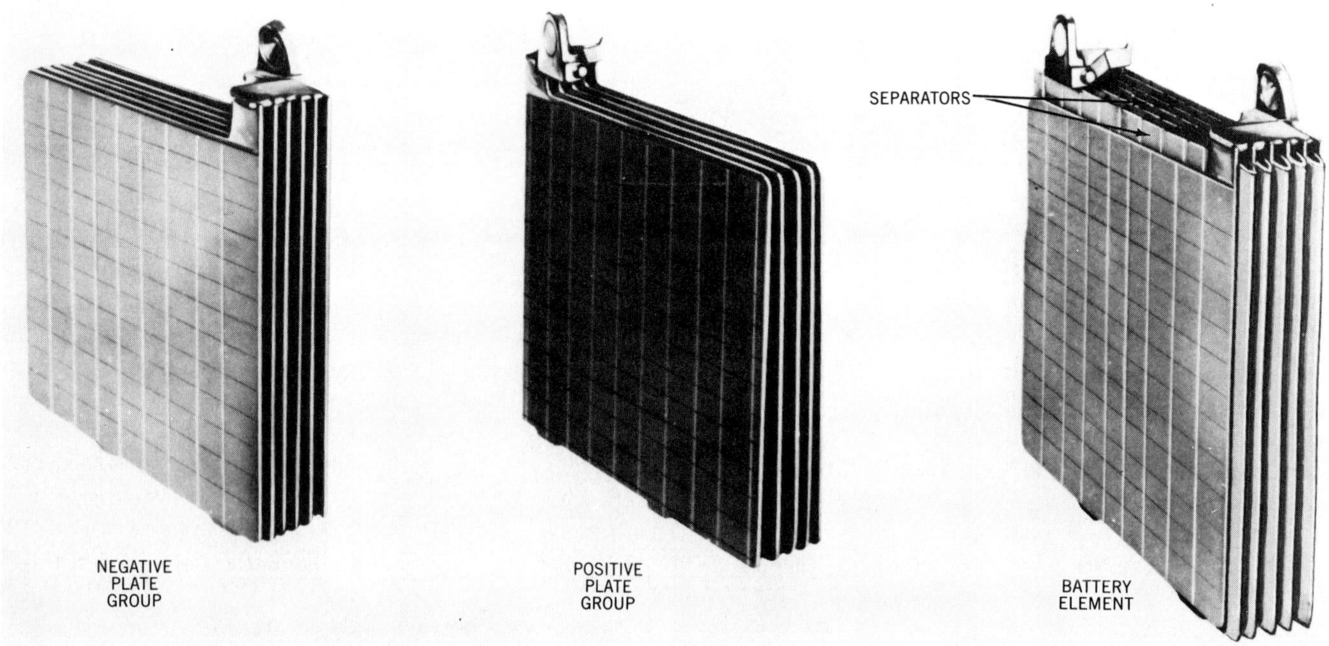


**Figure 9-5.** A complete battery element is a positive plate group, a negative plate group, and separators. When placed in a case with electrolyte, the element becomes a battery cell. (Atlas)

**Figure 9-6.** Many late-model batteries have envelope-type separators that retain active materials near the plates.

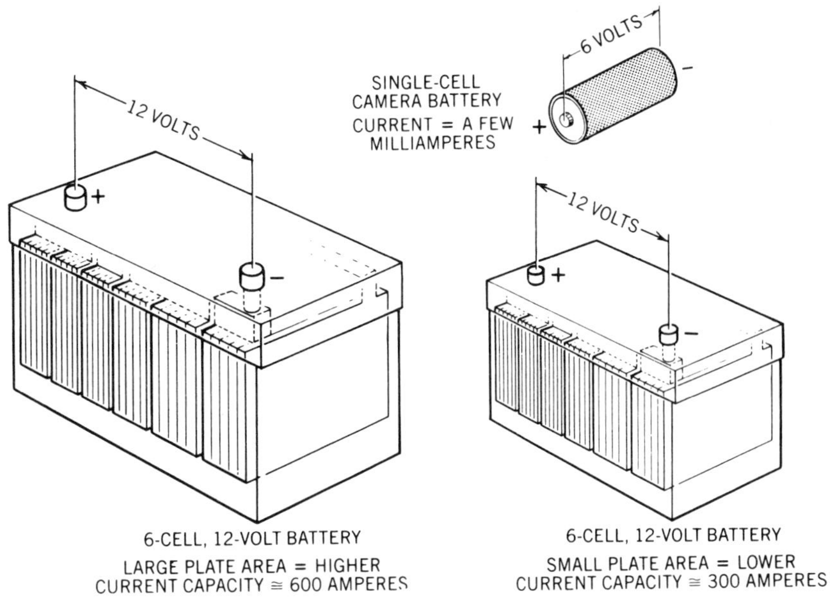


**Figure 9-7.** Total plate area of a battery determines current capacity. Different kinds of batteries can develop higher voltage per cell but very low current capacity.

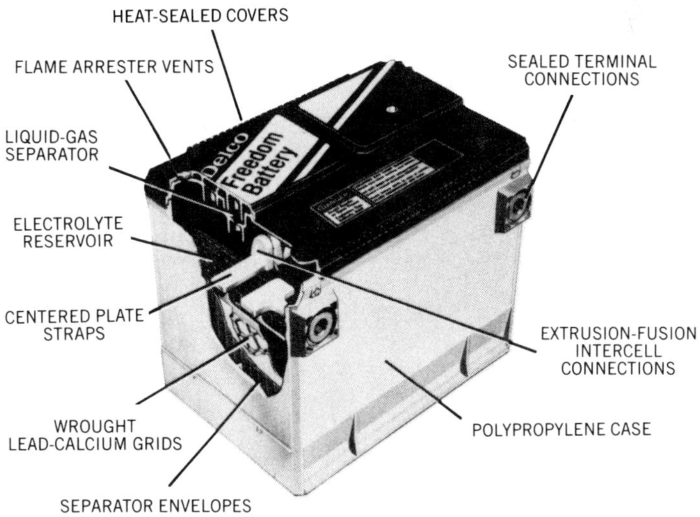


**Figure 9-8.** Modern batteries have 1-piece plastic cases and plastic tops. (Delco-Remy)

or fiberglass separators. Most separators have ribs on the sides near the positive plates to keep electrolyte near these plates for an efficient reaction. Many late-model batteries have porous plastic envelopes that surround the plates and act as separators, figure 9-6. The envelope construction catches active material that flakes off the positive plates during discharging. The envelopes hold the material closer to the plates so that it is redeposited more completely during charging.

When we put the assembled plate groups and separators into a battery case and add electrolyte, the battery cell is complete. Whether an automobile battery cell has 3 or 23 or any other number of plates, the lead and sulfuric acid compounds cause the cell to develop about 2.1 volts when fully charged. Compounds used in other kinds of batteries develop different voltages. An electronic camera battery, for comparison, is very small and has only one cell, yet it develops 6 volts. Its current capacity, however, is very small.

While the number and size of the plates in an automobile battery do not determine fully charged, open-circuit voltage, they do determine current capacity. The ampere rating of a battery is based on the total plate area in the battery. Batteries with high current capacity have more or larger plates than batteries with low capacity, figure 9-7.

Because each battery cell produces 2.1 volts, a 12-volt battery has six cells. The cells are connected in series inside the battery (+ to −) so that their voltages add together. The fully charged open-circuit voltage, then, of a 12-volt battery is actually 12.6 volts.

Although plate area and current capacity do not determine fully charged, open-circuit voltage, they do affect the rate of discharge. A large-capacity battery maintains high voltage longer than a small-capacity battery.

## Cases and Terminals

The cases of late-model batteries are made of lightweight, but strong, plastic, figure 9-8. Cells are separated from each other by plastic partitions. Cell-connecting straps pass over or through the partitions. Short connectors, passed through the partitions, reduce internal resistance. Plate groups in many batteries sit on raised supports in the case. The supports form chambers at the bottom of the battery where sediment that flakes off the plates can collect without shorting across the plates. In batteries with envelope separators, the

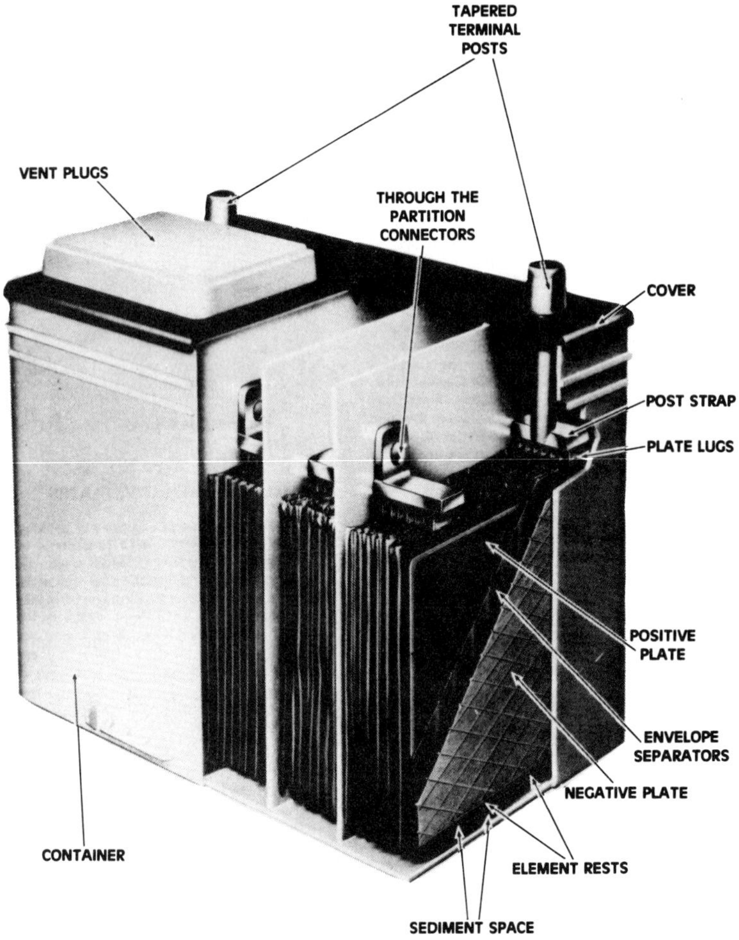


**Figure 9-9.** Automobile batteries have either top posts or side terminals.

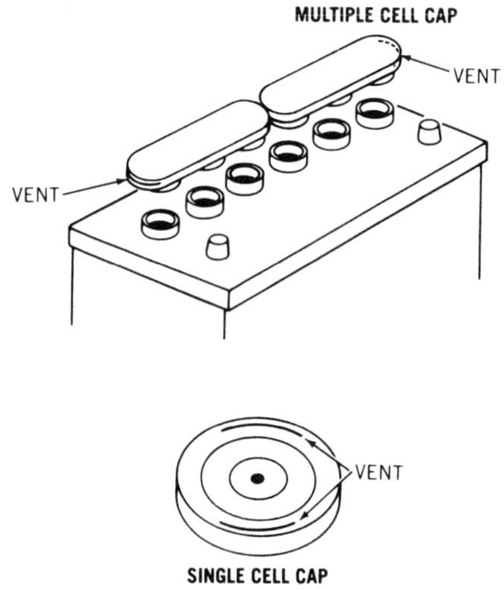


**Figure 9-10.** Vent-cap batteries have removable cell caps with gas vents.

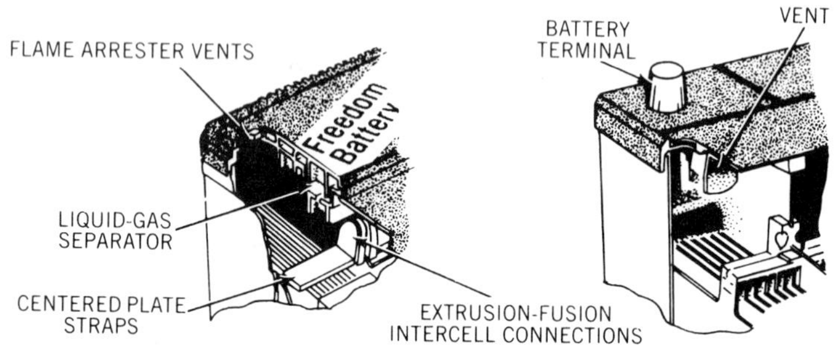


**Figure 9-11.** Maintenance-free batteries have closed covers with one or two small gas vents in the edges.

plates may rest on the bottom of the case because the envelopes retain sediment and prevent short circuits.

On modern batteries, the case is closed with a hard plastic cover to form a liquid-tight container. Older batteries had hard rubber cases, and the tops were sealed with a soft compound. Cell-connecting straps were exposed on the top of the battery. Some industrial batteries are still made this way, but few automobile batteries are.

Every battery has two terminals (one + and one −) to connect it to an external circuit. The terminals are connected to either end of the plate group connectors inside the battery, figure 9-9. Battery terminals are either tapered posts on top of the battery or side terminals with female (internal) threads. Some manufacturers say that side terminals develop less corrosion than top terminals because vapors from gassing and evaporation do not collect as easily on side terminals.

To ensure correct installation in a car's electrical system (to be sure battery polarity isn't reversed), the positive and negative terminals are slightly different. The positive post is slightly larger than the negative post. Battery cables have terminals that match either the positive or the negative terminal of a battery.

## Battery Vents and Cell Caps

Because a traditional battery releases hydrogen ($H_2$) gas during charging, the case cannot be completely sealed. Every such automobile battery has one or more vent openings. In a vent-cap battery (not maintenance free), the vents are in the cell caps, figure 9-10. In a maintenance-free battery (which you will learn more about later), the vent is a small opening in the edge of the cover, figure 9-11.

The cell caps of a vent-cap battery are removable so that you can inspect and test the electrolyte and add water to compensate for evaporation.

## Electrolyte

Electrolyte is a solution (a mixture) of water and sulfuric acid. The acid is the chemically active part of the solution that reacts with the lead compounds to generate voltage.

Acid is about 35 to 39 percent of the electrolyte solution in a fully charged battery. As a battery discharges, sulfate ions ($SO_4$) combine with lead ions formed at the plates. This reduces the $H_2SO_4$ percentage in the electrolyte. Some of the freed hydrogen ions combine with oxygen from the lead dioxide ($PbO_2$) to increase the water percentage. The percentages of water and acid in the electrolyte cycle back and forth as the battery cycles from charged to discharged. The volume of electrolyte stays relatively constant. A battery always loses some electrolyte volume, however, through water evaporation and hydrogen release during charging.

We won't go any farther into battery chemistry except to point out that 35- to 39-percent acid in the electrolyte is the proper molecular volume for a complete reaction with Pb and $PbO_2$. Any higher acid percentage would not increase voltage or current capacity. It would simply corrode the plate grid and connector metals. Over the life of a battery, the acid concentration decreases very slowly because of small, irreversible reactions. The ratio of acid to lead compounds stays relatively constant, however. That is why we add only water, not acid, to replenish electrolyte and compensate for water evaporation.

**Specific Gravity. Specific gravity** is the measurement of the density of any material to that of water. We usually use it to compare another liquid to water. Specific gravity also directly indicates the percentage of any liquid mixed in a solution with water. If we know what the specific gravity *should be* and then measure what it *is*, we can determine the percentage concentration of another liquid mixed with water. We use specific gravity measurements to determine acid concentration in battery electrolyte and antifreeze concentration in engine coolant.

The specific gravity of pure water is always 1.000. The specific gravity of electrolyte in a fully charged battery should be 1.260 to 1.280. This is equivalent to the 35- to 39-percent acid solution. As a battery discharges, the specific gravity decreases because the acid percentage decreases and the water percentage increases.

The electrolyte specific gravity can tell you the approximate state of charge of a battery:

| *Specific Gravity* | *Charge Percentage* |
|---|---|
| 1.265 | 100 |
| 1.225 | 75 |
| 1.190 | 50 |
| 1.155 | 25 |
| 1.120 or lower = discharged | |

Specific gravity measurements are based on a standard temperature of 80° F (26.7° C). Specific gravity is lower at higher temperature and higher at lower temperature. Every 10° F above or below 80° F changes the specific gravity by 4 points (0.004), as follows:

- Every 10° above 80° F—*add* 0.004 to the specific gravity measurement.
- Every 10° below 80° F—*subtract* 0.004 from the specific gravity measurement.

You can measure electrolyte specific gravity of a vent-cap battery with a **hydrometer**. When you study battery service, you will learn to do this and to correct specific gravity measurements for different temperatures.

## MAINTENANCE-FREE BATTERIES

Manufacturers call maintenance-free batteries such things as "sealed batteries" or "lifetime batteries." Most are not truly sealed; most have vents. While they usually last longer than older style vent-cap batteries in comparable service, they are not truly lifetime items. They are not even truly maintenance free because they do need occasional cleaning, testing, and possible recharging. Since their introduction in the mid-1970's, however, maintenance-free batteries have become standard original equipment on most late-model cars. They also are the most common kind sold for replacements.

Most maintenance-free batteries do not have removable cell caps to test

**BATTERY SERVICE TIPS**

Do the terminal clamps on the battery charger in your shop get corroded? Do you have to clean them regularly and perhaps replace them every year or so? You can reduce corrosion on the terminal clamps by always connecting them horizontally on batteries with removable vent caps. It's often easier to attach the clamps to the battery terminals in a vertical position, but this places them closer to the vents. Hydrogen gas that escapes from the vents during charging promotes the corrosion on the terminal clamps, so attach the charger clamps horizontally with the handles pointing away from the battery top.

Here's another point to remember about using a wire brush to clean battery terminals. When you are cleaning a battery in a car, you may have only a couple inches of clearance between the positive terminal and the body, frame, or engine. if your wire brush accidentally strikes the battery positive terminal and part of the car body or engine, a short-circuit spark can occur. This spark may burn you or, worse, ignite hydrogen fumes from the battery. If you can't remove the battery from the car to clean it, at least disconnect both cables from the terminals before going at them with your wire brush.

specific gravity and add water. Most also have a large electrolyte capacity to eliminate the need to add water during the battery lifetime. Large electrolyte volume alone does not make a maintenance-free battery, however.

The most important difference between maintenance-free batteries and older vent-cap batteries is in the plate grid material. Older batteries used 6- to 12-percent antimony alloyed with lead as the grid metal. Antimony is a good conductor, easy to manufacture, and supports the lead materials well. But, it also generates heat from the flow of charging current. Heat adds internal resistance to the battery and is the major cause of water evaporation.

Maintenance-free batteries have grids made of a calcium alloy, which has less resistance to charging current and generates less heat. Reducing heat reduces water loss so that electrolyte does not need water added to it. The heat reduction and better conductivity of calcium grid alloys also allow a maintenance-free battery to have a current capacity about 20 percent higher than an antimony-grid battery.

Battery heat and water loss, along with the charging reactions, are major causes of battery gassing. As hydrogen escapes from a battery and water evaporates, they carry a small amount of sulfuric acid vapors. These vapors condense on the outside of the battery and corrode the terminals, cables, and holddown fasteners. Heat, water loss, and gassing are all reduced in a maintenance-free battery. This, in turn, reduces corrosion.

Also, as an antimony-grid battery ages, the antimony moves from the grids to the active lead material of the negative plates. This causes two problems. Antimony reacting with the negative plates causes a slow self-discharge action. It also makes the battery less able to resist overcharging, which creates more heat. Calcium-grid batteries eliminate these problems.

While you can't add water or check the specific gravity of a maintenance-free battery, you can test it accurately with a load test. You will learn to do this when you study battery service.

Delco and some other maintenance-free batteries have built-in state-of-charge indicators that work like a simple hydrometer, figure 9-12. The indicator is a plastic rod inserted in the battery top. The rod has a small cage on the bottom that holds a small green ball. When the specific gravity is high (about 1.255 to 1.265), the ball floats in the electrolyte and touches the end of the rod. This makes the indicator appear green, which means that the battery is fully, or almost fully, charged. When the specific gravity is lower, the ball drops in the electrolyte and does not touch the rod. The indicator then appears black, which means that the battery should be tested and possibly charged. The bottom of the indicator rod is the minimum level for battery electrolyte. If electrolyte level falls below this point because of age or overcharging, the indicator appears clear. The battery then must be replaced.

Two problems can arise in using this state-of-charge indicator:

**1.** A battery that is left on charge too long or one that has been overcharged by the alternator may percolate, giving a false reading. The battery should be shaken gently or left to stand long enough for the bubbles to disperse before reading the indicator.

**2.** Stratification may take place if a battery has been deeply discharged, as in cranking the engine for a long period of time. Stratification is a condition in which the heavier sulfuric acid rests in a separate layer at the bottom of the case, with only the weak electrolyte surrounding the indicator. This condition can be corrected by shaking the battery gently or by tilting it a few degrees back and forth several times before reading the indicator.

### Second-Generation Maintenance-Free Batteries

In the mid-1980's several manufacturers introduced new maintenance-free batteries that are truly sealed. Envelope-type separators in these batteries retain free hydrogen ions near the plates for recombination with the electrolyte during charging. While hydrogen is generated during charging, none is released from the battery. Additionally, improved grid materials allow many such batteries to develop about 2.2 volts per cell for total open-circuit voltage of about 13.2 volts. Although these new batteries are an advancement in battery technology, test and service requirements are generally the same as for other maintenance-free batteries. However, fast charging at a high current rate may damage some of these batteries. Check the manufacturer's test and service instructions.

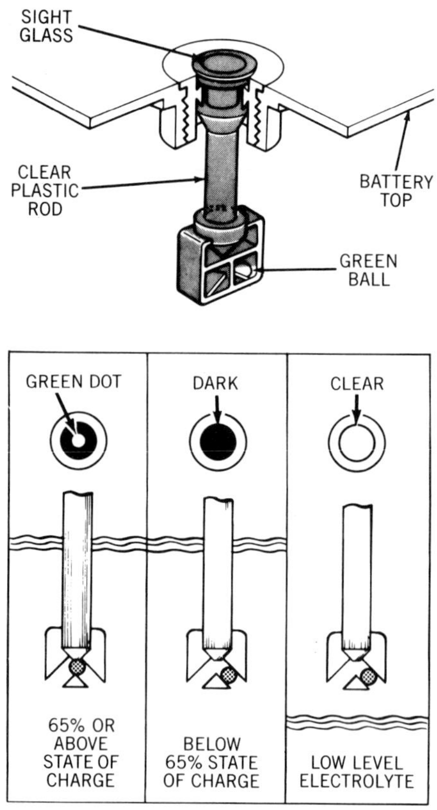


**Figure 9-12.** Delco and other maintenance-free batteries have small, built-in hydrometers to show the general state of charge. (Chrysler) (Delco-Remy)

## BATTERY CHARGING VOLTAGE

Charging current flows into a battery in the direction opposite to the battery discharge current, figures 9-1 and 9-2. When a battery supplies power to a circuit, current flows *out* of the battery + terminal, through the circuit, and back to the − terminal. Charging current flows *into* the battery + terminal and through the battery to the − terminal. To do this, charging current must overcome the battery's own resistance, which exists in two forms:

1. The physical resistance of the battery parts: the electrolyte, the plate metals, and the internal connectors
2. The countervoltage, or **counter-electromotive force (CEMF)**, of the battery.

Charging current enters the battery only when charging voltage is higher than the battery CEMF *plus* the voltage drop created by the battery internal resistance. When a battery is fully charged, CEMF is high; and very little charging current flows through the battery. When a battery is discharged, CEMF is low; and more charging current flows through the battery. Figure 9-13 is a graph that shows the relationships of charging voltage, charging current, and battery CEMF.

Understanding the relationship of CEMF to battery state of charge will help you understand battery charging methods. A discharged battery needs and will accept high charging current. A fully charged battery with high CEMF resists charging current. High charging current applied to a fully charged battery will overheat and damage the battery.

Battery temperature also affects battery resistance to charging current. A cold battery has higher resistance than a warm battery, largely because of the resistance of the electrolyte. When you learn to service batteries and charging systems, you will learn to consider the effects of temperature. Very high charging current applied to a cold battery, even if it is completely discharged, may damage the battery as the current tries to overcome the high resistance of cold electrolyte. Similarly, high charging system voltage in very hot weather may overheat a battery, and low charging system voltage in cold weather may not keep a battery fully charged.

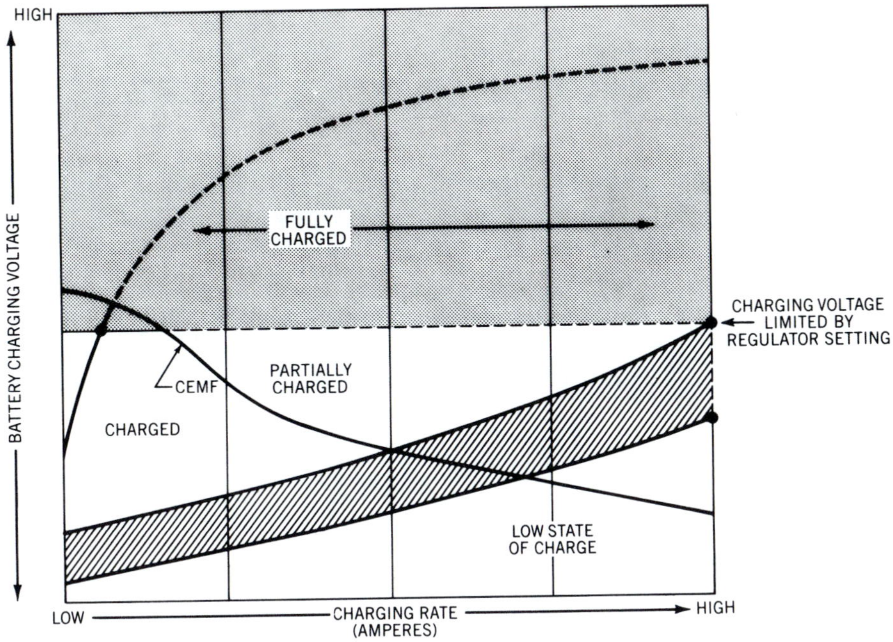


**Figure 9-13.** This graph shows the relationships of battery charging voltage, charging current, and CEMF.

## BATTERY SELECTION AND REPLACEMENT

Engineers select batteries as original equipment for a vehicle to supply the current requirements of a particular car or truck with specific electrical loads. As a professional mechanic, you must select a replacement on the same basis. The major factors that influence battery selection are:

1. Battery type
2. Battery current capacity rating
3. Battery size and group number

### Battery Type

All modern automobile batteries are 12-volt, lead-acid, wet-cell batteries. They come in different sizes and shapes and with different current ratings. Yet, battery companies also make 12-volt automotive-type batteries that are not intended for automobile use.

Batteries are designed for one or the other of two uses:

1. To start an automobile or truck engine
2. To be the power source for an electric vehicle or an electrical accessory system without continual recharging

The two basic types of batteries, therefore, are starting batteries and cycling batteries. Automobile batteries are starting batteries, designed principally to deliver large power output for a short time to start an engine. The vehicle charging system then supplies most of the current. The battery stabilizes the system and provides current when electrical loads exceed charging current output at any time or engine speed.

Cycling batteries deliver low, steady current for long periods of time. They are designed for electric vehicles such as industrial trucks (forklifts), golf carts, electric wheelchairs, and electric outboard motors. They also are used for marine and recreational vehicle (RV) electrical accessory systems. Most of the current capacity is depleted in each cycle before the battery is recharged. Figure 9-14 shows the differences between the current requirements of starting and cycling batteries.

The current capacity of a battery used to start a car engine is a small percent of its rated capacity, but actual current is very high for a few seconds. It may be as high as 600 amperes to crank an automotive diesel engine. This high current for a few seconds is obtained by using thinner plates in a starting battery than in a cycling battery. The battery can be subjected to high current drain many times, followed by recharging, and still maintain its rated capacity.

A cycling battery has thicker plates to sustain a steady current drain for several hours. If an automobile starting battery is subjected to such deep cycle use, the plates wear out rapidly. The positive plates get soft and shed active material. Conversely, a cycling battery used to start an automobile engine will get hot from the momentary high current draw and take longer to recharge. Its overall life probably will be shorter than that of a starting battery in this application.

An automobile battery is not the best choice to power an electric vehicle, and a cycling battery may not have the expected service life if used for automobile starting applications. Understanding these basic differences in battery types makes it easier to understand why there are different ways to rate battery current capacity. A single battery can be, and often is, rated by more than one method. To ensure adequate electrical power for any application, a replacement battery should have the same or higher current-capacity rating as the OEM battery, but never a lower rating.

**Starting Battery**
Typical starting current = 300 amperes
Typical starting time = 4 seconds
Typical current drain from the battery =
300 amperes × 4 seconds = 1200 ampere-seconds or 0.33 ampere-hour

Nominal ampere-hour capacity = 62 ampere-hours

**Cycling Battery**
Typical steady current drain of
an electric outboard motor = 10 amperes

Typical running (battery discharge)
time for a fishing trip = 3 hours
Typical current drain from the battery =
10 amperes × 3 hours = 30 ampere-hours

Nominal ampere-hour capacity = 80 ampere-hours

**Figure 9-14.** These comparison figures show the different requirements for a starting battery and a cycling battery.

## Battery Current-Capacity Ratings

A battery's current-capacity rating indicates its ability to provide cranking power for an engine, reserve power for an electrical system, or both. The following sections explain test standards and rating methods developed by the Battery Council International (BCI) and the Society of Automotive Engineers (SAE), as well as the types of batteries for which they are used.

**Ampere-Hour Rating.** Also called the 20-hour discharge rating, the **ampere-hour rating** is the oldest current-capacity rating method. It also is the best way to rate a cycling battery. This rating indicates the steady current flow that a battery will deliver for 20 hours at 80° F (26.7° C) before cell voltage drops below 1.75 volts. This equals 10.5 volts for a 12-volt battery. For example, a battery that continuously delivers 4 amperes for 20 hours is an 80-ampere-hour battery (4 amperes × 20 hours = 80 ampere-hours).

**Cold-Cranking Rating.** The **cold-cranking rating** is the newest and most common way to rate the current capacity of a modern automobile battery. This rating indicates the power that a battery can supply to start an engine at low temperature. Cold-cranking ratings are given in total amperes, such as 300, 375, 425, or 500. This represents the current a battery can deliver for 30 seconds at 0° F (−17.8° C) before cell voltage drops below 1.2 volts (7.2 volts, total, for a 12-volt battery).

**Reserve-Capacity Rating.** The **reserve-capacity rating** indicates the number of minutes that a fully charged battery can deliver 25 amperes before cell voltage drops below 1.75 volts (10.5 volts for a 12-volt battery). These ratings are listed in minutes, such as 60, 110, 120, and so on. The reserve capacity indicates the time that a vehicle can be driven if the charging system fails.

## Battery Size and Group Numbers

A battery must fit the space available for it in a vehicle and connect to the cables. Batteries of various current-capacity ratings and both types (starting and cycling) are built in different sizes and shapes. **BCI group numbers** indicate the length, width, height, terminal design, holddown locations, and other physical features of a battery, figure 9-15. The group number indicates only the size, shape, and voltage of a battery, not the current capacity.

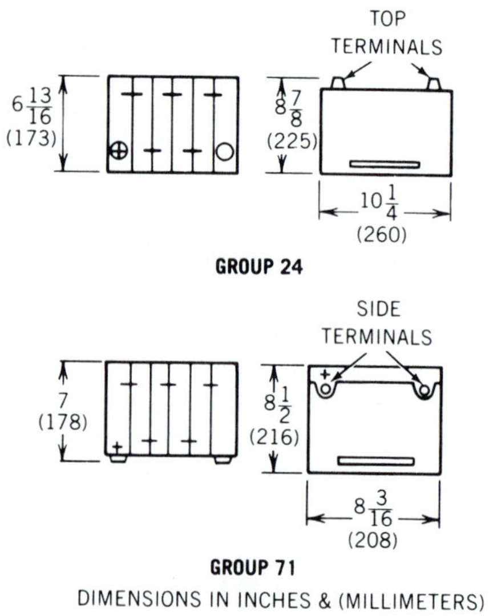


**Figure 9-15.** Battery group numbers indicate battery dimensions and terminal types and locations.

## BATTERY INSTALLATIONS

Most automobiles have a single 6-cell, 12-volt battery mounted in the engine compartment. Several factors determine battery location:

1. The battery should be close to the engine and the alternator so that long cables do not increase electrical system resistance.
2. The battery should be located away from exhaust system parts in a place where air can flow over it to cool it.
3. The battery should be mounted securely to protect it from vibration that can damage the plates.
4. The battery must be accessible for service and located so that its weight does not upset the balance of the vehicle.

The need to reduce weight and increase fuel mileage in late-model cars has led engineers to design lighter and smaller batteries with current capacity equal to the larger batteries of the past. Lightweight plastic case materials and improved grid and plate metals and designs have made this possible.

Some older cars and some race cars have the battery in the trunk or under the floor to keep it away from engine heat and put its weight over the rear driving axle. These installations require long cables, which usually are heavier gauge wire than cables used for engine compartment installations. The heavier cables reduce resistance but add cost and weight to vehicle design.

Many early-1980's GM diesel automobiles have two 12-volt batteries connected in parallel, figure 9-16. The + terminals of both batteries are connected to each other and to the + battery cable attached to the starter motor. Similarly, the − terminals are connected to each other and to the battery ground cable. This parallel connection doubles the available current for starting a high-compression diesel, but the two batteries still provide just 12 volts to the electrical system. The alternator charges both batteries simultaneously.

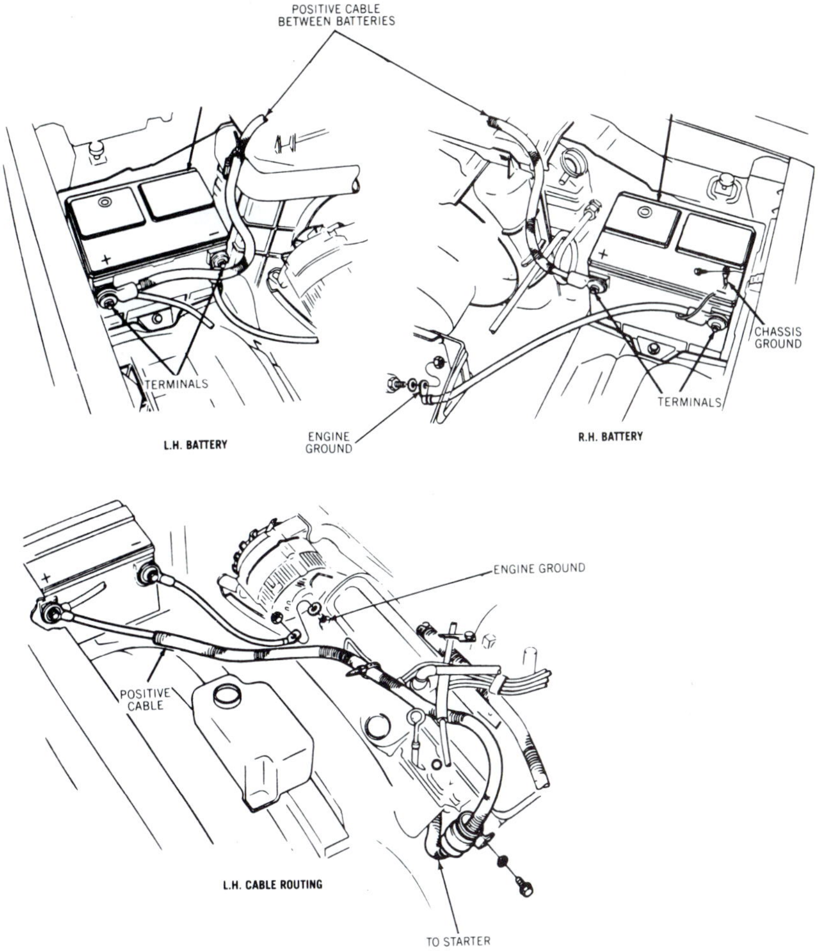


**Figure 9-16.** Many GM diesel automobiles have two 12-volt batteries for increased cranking current with a 12-volt starter. (Oldsmobile)

Remember that if two or more similar batteries are connected in series, the voltages add together. If two or more similar batteries are connected in parallel, the total system voltage is the same as with a single battery, but current capacity increases.

### Battery Installation Parts

Correct design and selection of battery installation parts and their proper maintenance are essential for good battery service life and operation.

**Battery Cables.** Battery cables are large-diameter, multistrand wire to carry the high current needed by the starter motor. Twelve-volt electrical systems usually have number 4 or number 6 gauge battery cables. You can install a larger gauge cable than the OEM cable on a vehicle, but not a smaller one. A smaller gauge cable will increase the voltage drop in the starter circuit and may cause poor starting operation.

The cables must have connectors that match the posts or the side terminals of the battery on which they are used, figure 9-17. Some cables have smaller gauge wire pigtails soldered into the battery connectors for specific circuits that get power directly from the battery on some cars. The ground cable on a battery installation may be a flat braided cable without insulation. These have the same resistance and electrical characteristics as insulated cables. Insulation is not required for the low-voltage ground side of the system.

The battery ground cable is connected to *either* the engine or the vehicle chassis. To ensure a low-resistance ground connection *between* the engine and chassis, some vehicles have a braided strap or a small cable between the engine and frame, figure 9-18. While not actually a battery cable, this ground "bonding" strap is necessary on some cars for a complete electrical system ground path.

**Battery Carriers and Holddown Parts.** A battery is usually mounted on a shelf or tray, called a carrier, in the engine compartment. The battery must be held securely to protect it from vibration and to keep it from spilling acid. Various kinds of holddown clamps and brackets are used on different vehicles, figure 9-19. Regardless of design, all holddown parts must be installed correctly and securely to ensure proper battery operation.

**Battery Heat Shields.** Underhood temperatures are high on late-model vehicles. Heat can shorten battery life and reduce battery performance. Therefore, most vehicles have shields to protect the battery from engine heat or to direct cooling airflow from the front of the car around the battery, figure 9-20. Most shields are made of heat-resistant plastic. Some are part of the holddown components; others are separate ducts to direct airflow around the battery. Like correctly installed holddown brackets, heat shields are necessary for proper battery performance and life. Heat shields must be reinstalled whenever a battery is removed and replaced.

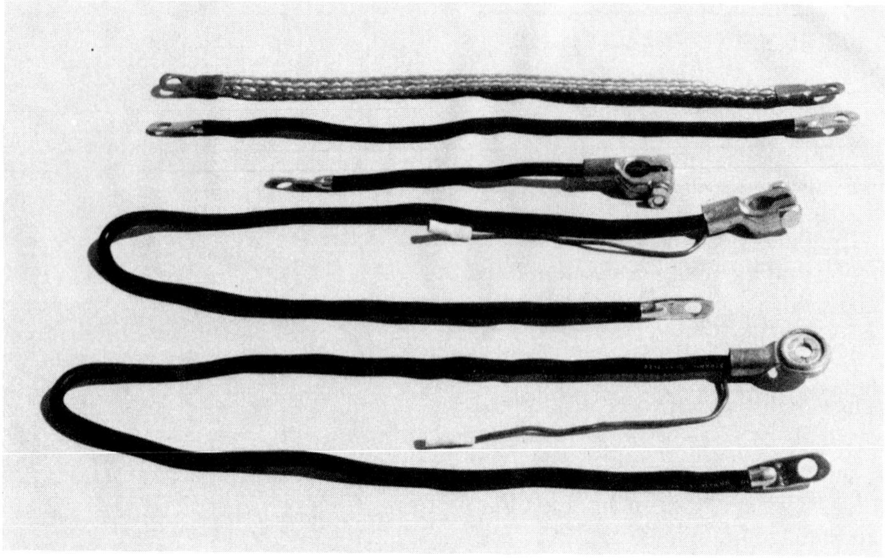

**Figure 9-17.** Typical battery cables for top-post and side-terminal batteries.

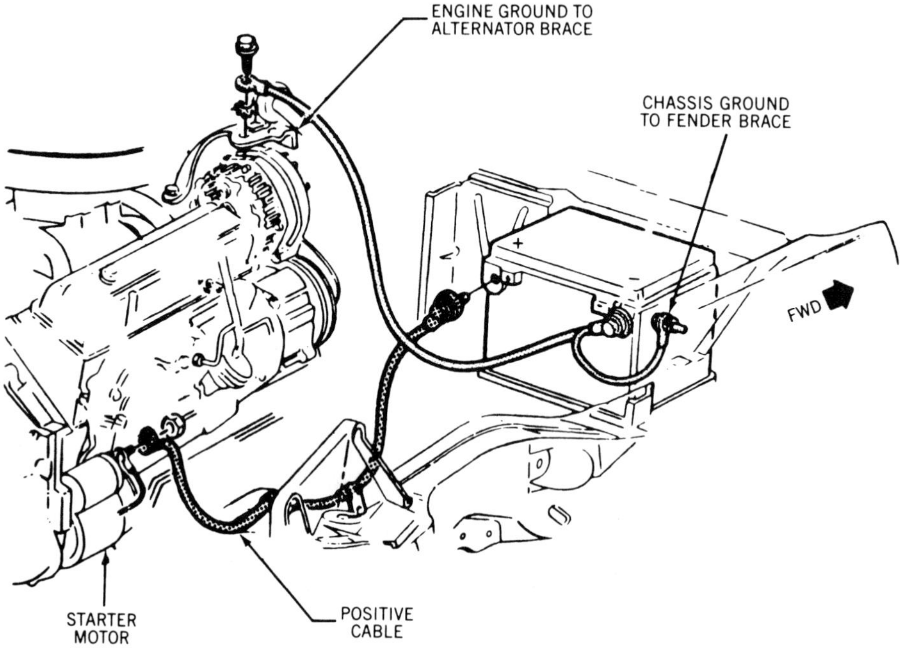


**Figure 9-18.** This bonding strap ensures a good ground between the battery, the engine, and the vehicle chassis.

## FACTORS OF BATTERY LIFE AND PERFORMANCE

A well-maintained modern battery, operating with a charging system in good condition, will last 4, 5, 6, or more years. Yet, even the finest maintenance-free "lifetime" batteries eventually wear out. Several factors (operating conditions) can extend or shorten battery life and affect performance.

### Overcharging

A battery can be overcharged by the vehicle charging system or by a shop battery charger. Overcharging results from continually high charging voltage or high charging current rates, or both. In any case, excess current is forced through the battery, which creates heat. Heat can buckle plates and oxidize the positive plate material. Overcharging also causes the electrolyte to break down and release hydrogen and oxygen gases. The released gases can wash active material off the plates, decrease electrolyte level, and increase battery corrosion.

### Undercharging and Sulfation

If a charging system does not deliver enough voltage and current to a battery, the battery will not be fully charged. Undercharging may be due to a fault in the charging system, continually high electrical loads from accessories, or a battery with lower current capacity than the vehicle needs.

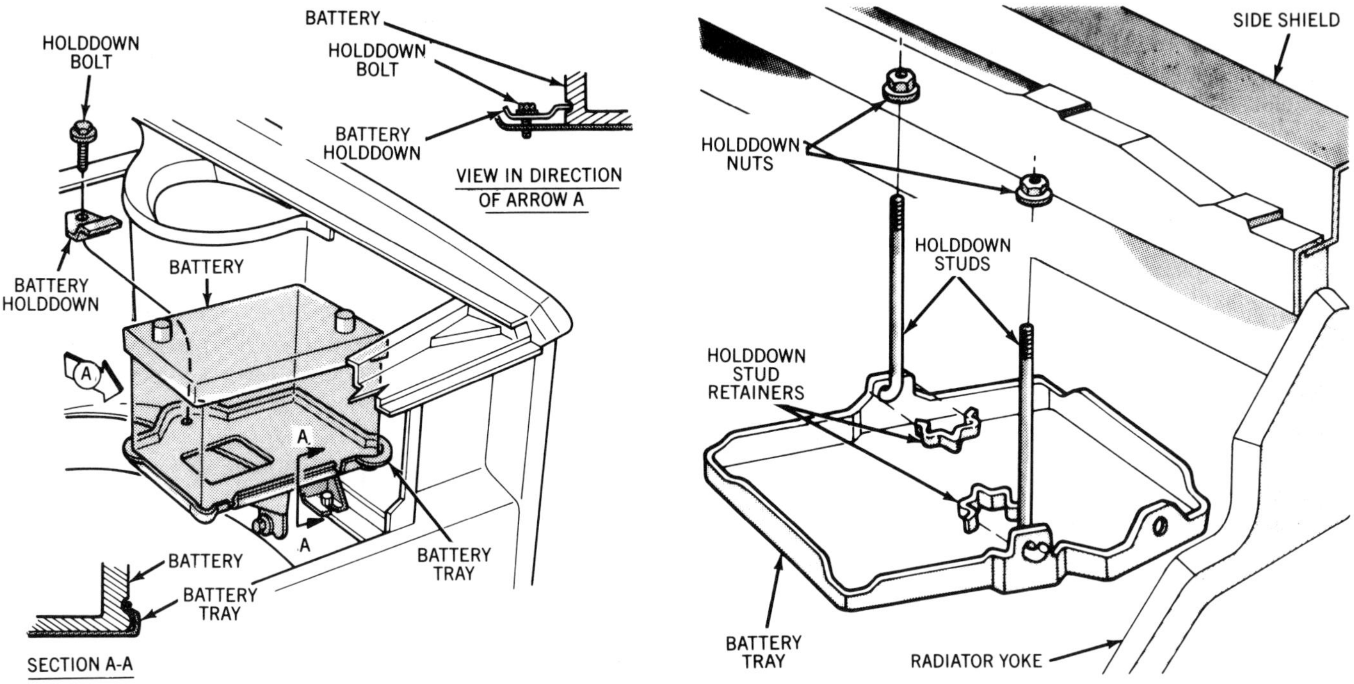


**Figure 9-19.** Battery holddown brackets and fasteners must be installed securely. (Chrysler)

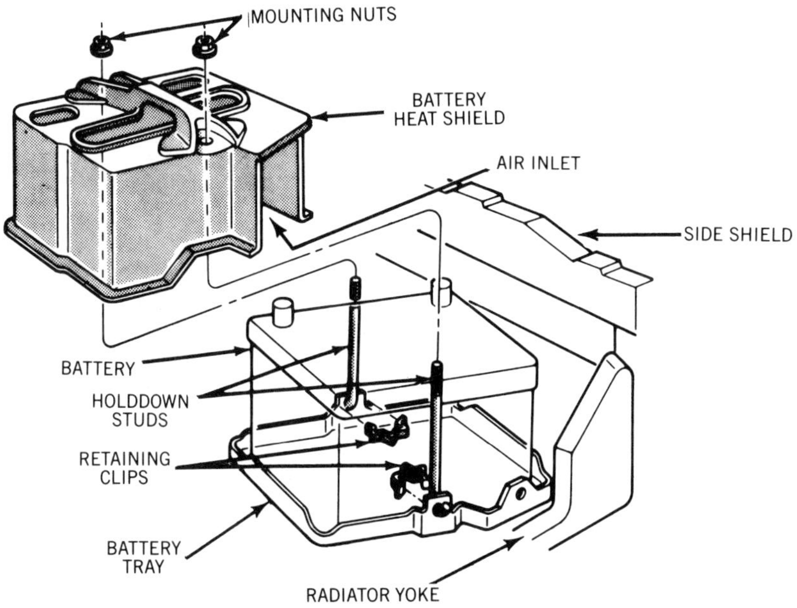


**Figure 9-20.** Many late-model cars have heat shields to protect batteries from underhood heat. (Chrysler)

The undercharged condition can range from 70 or 80 percent charged to fully discharged.

When a battery is less than fully charged, lead sulfate remains on the plates. The amount of sulfation varies with the state of charge. If a battery is constantly undercharged, the sulfate can crystalize and will not recombine with the electrolyte. Eventually, a constantly undercharged battery can never reach a state of full charge. A completely sulfated battery cannot be effectively recharged.

## Cycling

Automobile batteries are not designed for continual deep-cycle use. If a car battery is cycled repeatedly from fully charged to almost discharged, the positive plates will shed active material. The material cannot be restored to the plates, and battery current capacity and life will be reduced.

## Temperature

High and low temperatures affect battery life and performance in several ways. High temperature caused by engine heat, overcharging, or ambient air temperature will increase electrolyte loss and shorten battery life. Low temperature also can damage a battery.

If electrolyte freezes, it can expand and crack the battery case. The freezing point of electrolyte depends on specific gravity and, therefore, on the battery state of charge. A fully charged battery with specific gravity of 1.265 to 1.280 will not freeze until temperature is below −60° F (−51° C). A discharged battery with electrolyte specific gravity around 1.120 to 1.150 can freeze at 18° F (−7.7° C).

Low temperature also increases electrolyte resistance, which reduces battery cranking power. But, an engine needs *increased* cranking power to start at low temperatures. Figure 9-21 is a chart that compares the cranking power available from a bat-

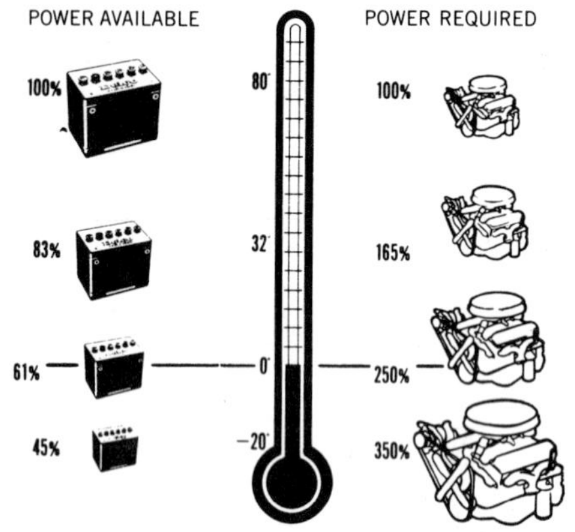


**Figure 9-21.** A battery delivers less cranking power at low temperature than at high temperature, but an engine needs more power to start. (Delco-Remy)

tery and the cranking power needed by an engine at 80°, 32°, and 0° F (26.7°, 0°, and −17.8° C).

Electrolyte resistance also means that a cold battery is harder to recharge than a warm battery.

### Electrolyte Level

Calcium-grid maintenance-free batteries have reduced electrolyte water loss to the point that battery cases can be closed. Additional water is not needed during the normal use and life of the battery. You know, however, that high temperature, overcharging, and deep cycling and recharging all create heat that can cause battery gassing and water loss. Even maintenance-free batteries lose some water. Electrolyte inspection, testing, and replenishment are parts of regular battery care for vent-cap batteries.

If electrolyte level drops below the tops of the plates, the plate materials are exposed to the air and harden. They do not take part in the electrochemical reaction. Even when water is added, hardened plate materials are not restored to a fully active condition. Additionally, electrolyte that has lost water has a higher acid concentration, which increases plate corrosion.

Overfilling a battery also is harmful. Too much water reduces the acid concentration and lowers specific gravity. It also raises electrolyte level closer to the vents, which increases battery corrosion due to gassing.

### Corrosion

Battery corrosion is caused by spilled electrolyte and by normal battery gassing. Sulfuric acid destroys metal connectors, terminals, holddown brackets, and carriers. Corroded terminals and connectors increase circuit resistance. Corrosion also weakens holddown brackets and damages sheet metal of the car body.

### Vibration

We have mentioned several times that vibration can shake active materials off battery plates. Vibration also can loosen internal connections between cells and plate groups, which increases resistance. Similarly, vibration can loosen external battery terminal connections. Severe vibration can even crack a battery case or shake a battery loose from its carrier.

## SUMMARY

Automobile batteries are lead-acid secondary batteries, which can be recharged after discharging. Batteries do more than store electrical power. They generate voltage and current through electrochemical action between the dissimilar plates (Pb and $PbO_2$) and sulfuric acid ($H_2SO_4$) in the electrolyte.

Each cell of a battery contains groups of positive (+) and negative (−) plates with separators between them. Each cell produces about 2.1 volts regardless of the number of plates. Cells are connected in series (+ to −) so that six cells produce about 12.6 volts in a fully charged 12-volt battery. Current capacity depends on the total surface area of all plates in all cells.

Electrolyte specific gravity indicates the battery state of charge. Electrolyte in a fully charged battery should have specific gravity of 1.265 to 1.280. Maintenance-free batteries have calcium-alloy grids that reduce battery heat and water loss. Electrolyte cannot be checked, and water cannot be added to a maintenance-free battery.

Automobile batteries are designed principally for engine-starting service, not for continual deep cycling. Cycling is the operation of a battery from fully charged to discharged and back to fully charged. Battery current capacity can be rated by several methods, such as the ampere-hour rating, the cold-cranking rating, and the reserve-capacity rating. Ampere-hour ratings are most appropriate for cycling batteries. Cold-cranking ratings are most appropriate for starting-type batteries.

Battery charging rate and charging current are affected by the battery state of charge and CEMF. Overcharging, undercharging, cycling, temperature, vibration, corrosion, and electrolyte level all affect battery service life and performance.

**WORD TO THE WISE**

When you are servicing a battery with removable cell caps, don't take the caps off and set them on any painted surface of a car. Moisture that collects on the caps as the battery gases contains sulfuric acid. If it contacts a car's paint, it will leave a mark and probably leave you with an angry customer. When you remove a cell cap to check or add water or test specific gravity, set the cap on the top of the battery case or away from the car on your workbench.

## REVIEW QUESTIONS

### Multiple Choice

1. An automobile battery:
   - a. stores energy in chemical form
   - b. changes chemical energy to electrical energy
   - c. changes electrical energy to chemical energy
   - d. all of the above

2. Student A says that the positive (+) plates of a battery are lead (Pb). Student B says that oxidation occurs at the positive plates when the battery delivers current to a circuit. Who is right?
   - a. A only
   - b. B only
   - c. both A and B
   - d. neither A nor B

3. Open-circuit voltage of a fully charged 12-volt battery is usually:
   - a. 11.8 volts
   - b. 12 volts
   - c. 12.6 volts
   - d. 13.3 volts

4. Student A says the plate area in a battery cell determines cell voltage. Student B says that the plate area in a battery cell determines current capacity. Who is right?
   - a. A only
   - b. B only
   - c. both A and B
   - d. neither A nor B

5. Battery electrolyte is approximately:
   - a. one-third sulfuric acid and two-thirds water
   - b. two-thirds sulfuric acid and one third water
   - c. half water and half sulfuric acid
   - d. one-quarter sulfuric acid and three-quarters water.

6. Specific gravity of the electrolyte in a fully charged battery should be approximately:
   - a. 1.235 to 1.255
   - b. 1.260 to 1.280
   - c. 1.275 to 1.290
   - d. 1.290 to 1.310

7. Specific gravity measurements must always be corrected to which of the following temperatures:
   - a. 60° F
   - b. 32° F
   - c. 80° F
   - d. 100° F

8. The grid material of a maintenance-free battery is alloyed with
   - a. antimony
   - b. silicon
   - c. cadmium
   - d. calcium

9. Student A says that maintenance-free batteries are completely sealed. Student B says that specific gravity cannot be checked in a maintenance-free battery. Who is right?
   - a. A only
   - b. B only
   - c. both A and B
   - d. neither A nor B

10. When selecting a replacement battery, you must consider all of the following, except:
    - a. charging voltage rating
    - b. battery type
    - c. current capacity rating
    - d. battery size and group number

11. A battery cold-cranking rating is based on:
    - a. steady current that the battery will deliver for 20 hours before cell voltage drops below 1.75 volts.
    - b. current that a battery can deliver for 30 seconds at 0° F before cell voltage drops below 1.2 volts.
    - c. current required to crank a 300-cid engine at 0° F for 6 seconds, or one-tenth of a minute.
    - d. the number of minutes that a battery can deliver 25 amperes before cell voltage drops below 1.75 volts

12. A battery ampere-hour rating is based on:
    - a. steady current that the battery will deliver for 20 hours before cell voltage drops below 1.75 volts.
    - b. current that a battery can deliver for 30 seconds at 0° F before cell voltage drops below 1.2 volts.
    - c. current required to crank a 300-cid engine at 0° F for 6 seconds, or one-tenth of a minute.
    - d. the number of minutes that a battery can deliver 25 amperes before cell votage drops below 1.75 volts

13. A battery reserve-capacity rating is based on:
    - a. steady current that the battery will deliver for 20 hours before cell voltage drops below 1.75 volts.
    - b. current that a battery can deliver for 30 seconds at 0° F before cell voltage drops below 1.2 volts.
    - c. current required to crank a 300-cid engine at 0° F for 6 seconds, or one-tenth of a minute.
    - d. the number of minutes that a battery can deliver 25 amperes before cell voltage drops below 1.75 volts

### Fill in the Blank

14. A battery in which one or both of the electrode materials are destroyed is called a ________ battery.

15. The electrochemical action in a secondary battery is ________.

16. A car battery is recharged by high voltage and current delivered by the ________.

17. In a battery, the medium through which the positive and negative terminals react is called an ________.

18. The negative terminal in a battery is the ________.

19. The positive terminal in a battery is the ________.

20. Current always flows from the ________ battery terminal.

21. Battery electrolyte is a mixture of ________ and water.

22. When both plates in a battery become similar, the battery is ________.

23. The process of discharging and recharging a battery is called ________.

24. During charging, a vented, lead-acid battery releases ________ which is explosive.

25. The active units of a battery are called ________.

26. To replenish electrolyte, we add ________ to a battery.

27. The measurement of the density of any material to that of water is called its ________.

28. The __________ is the instrument used to measure the specific gravity in a vent-cap battery.

29. Maintenance-free batteries use a calcium alloy because of its conductive properties and because it produces less __________.

30. The two basic types of batteries are charging batteries and __________ batteries.

# 10 STARTING SYSTEMS

## WHAT

In chapter 9, you learned that automobile batteries are designed primarily to provide high current for the starting system. The starting system is one of the simplest electrical systems on an automobile. It consists of one or two switches, a relay or a solenoid, and a high-torque motor. It also is one of the most important automobile systems.

Obviously, if the starting system cannot crank an engine, the engine cannot start and run. Problems in the starting system also can affect the battery state of charge and battery life. Additionally, starting system problems can reduce ignition voltage so that the engine may not fire and run even if it cranks. This chapter explains the design and operation of starter motors and other parts of this vital electrical system.

## WHY

Hard-starting complaints are among the most common engine performance problems. The cause of an engine starting problem may be in the parts of the starter system; it may be a high-resistance fault in the circuit; or it may be a discharged battery. Starting problems also may be due to the effects of the starting circuit on other vehicle systems. In any case, a thorough knowledge of starting system operation and how it relates to other vehicle systems is essential for complete engine performance service.

## GOALS

After studying this chapter, you should be able to:

1. Explain the operation of starting control circuits and motor circuits and identify the parts of each.
2. Explain the operation of various starter motors and identify principal motor parts.
3. Identify and explain the operation of various starter drives.

## STARTING SYSTEM CIRCUITS

The starting system, figure 10-1, consists of two related circuits: the motor circuit and the control circuit. The starter motor draws very high current for a few seconds to crank the engine. Cranking speed for most engines is about 200 rpm. If the starter cannot crank the engine at this speed, a hard-start or a no-start problem will probably result.

The amount of starter current varies from 150 or 200 amperes for small 4- and 6-cylinder engines to more than 300 amperes for large gasoline V-8 and diesel engines. Some diesels may require 500 or 600 amperes of cranking current to start in very cold weather. The starter motor draws this high current for only a few seconds. A well-running gasoline engine should start with 2 to 3 seconds of cranking time.

The high-current requirement of the starter motor is the reason that battery cables are made of heavy-gauge wire. Compare the several hundred amperes of starting current to the 10, 20, or 30 amperes needed by other circuits. If it were not for the starter motor, the battery could be connected to the rest of the electrical system with 10- or 12-gauge primary wiring.

The high-current requirement of the starter also is the reason that the starting system has two circuits. Heavy cables connect the motor to the battery as directly as possible. The motor circuit uses a relay, or the relay action of a solenoid, to connect the motor to the battery momentarily. The relay or solenoid is mounted on or near the motor and connected directly to the battery. The ignition switch energizes the relay or solenoid to operate the starter. If the battery were connected to the motor through the ignition switch, the cables would be excessively long and expensive. Moreover, the extra cable length would add resistance to the circuit, which could cause excessive voltage drop to the starter. The ignition switch also would have to be large and cumbersome to hold the heavy contacts needed to conduct high current to the starter.

Figure 10-2 shows the basic arrangement of the starting system motor and control circuits. The motor circuit carries high current from the battery, through the magnetic switch (relay or solenoid), to the motor. The control circuit energizes the relay or solenoid through the ignition switch and the starting safety switch. The battery ground connection is part of both the motor and the control circuits. You have already studied the battery, which is the power source for the starting system and all other electrical circuits. The following sections outline the op-

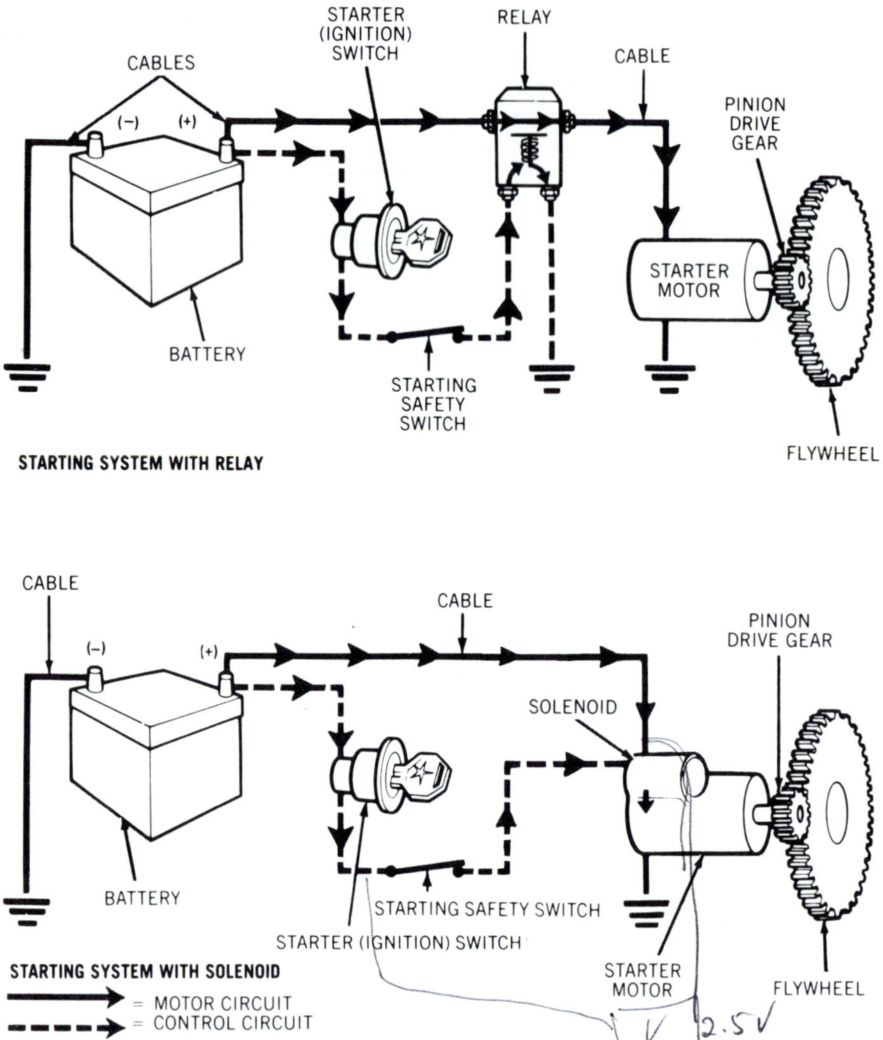


**Figure 10-1.** All starting systems have a motor circuit and a control circuit. (Chrysler)

eration of the starting circuits and their major parts.

## The Starter Control Circuit

The starter control circuit contains:

- The ignition switch
- The starting safety (neutral start) switch
- The control side of the relay or solenoid

These components are connected to the battery by primary wiring. The ignition switch usually receives battery voltage from the terminal on the relay or solenoid to which the positive (+) battery cable is connected. The ignition switch and the starting safety switch are in series. When both switches close, current flows through the relay or solenoid coil. Electromagnetic action closes the heavy contacts in the relay or solenoid and connects the motor to the battery through the heavy cables.

**The Ignition Switch.** Late-model ignition switches have multiple switch contacts and five positions.

1. Accessories
2. Lock
3. Off
4. On (Run)
5. Start

The ignition switch is the power distribution point for many circuits. In the accessories position, it supplies power to accessory circuits but not to ignition and engine control circuits. In the lock and the off positions, all circuits are open at the switch. Additionally, the lock position mechanically locks the steering and, on many cars, the transmission gear selector. In the run position, the switch provides power to the ignition and the engine control circuits, as well as other circuits. In the start position, the switch provides power only to the starter control circuit, the ignition, and other engine control circuits.

The switch start position is a **momentary contact** position. It is spring loaded so that when the driver releases the key, the switch automatically moves to the on (run) position. All other switch positions are **detented**. The switch stays in one selected position until turned to another.

With few exceptions, late-model ignition switches are mounted on the steering column. There are two basic designs:

1. Most switches are mounted remotely from the key and operated by linkage, figure 10-3.
2. Some switches are attached directly to the key at the upper end of the steering column, figure 10-4.

All ignition switches connect battery voltage to the ignition system in the start and the run positions. For ignition systems with a ballast resistor and a starting bypass circuit, the switch connects full battery voltage to the ignition when starting and lower voltage (through a resistor) when running. For ignitions without a ballast resistor, the switch delivers full battery or charging system voltage during both starting and running. The actual starting circuit to the ignition may be connected through the switch or through the starter relay. The ignition system chapters in Part Four of this book explain more about ignition ballast resistor and starting bypass circuits.

Figure 10-5 shows a typical ignition switch symbol that you will see on electrical diagrams.

**The Starting Safety Switch.** Once used only with automatic transmissions, the starting safety switch is now used on many cars with manual trans-

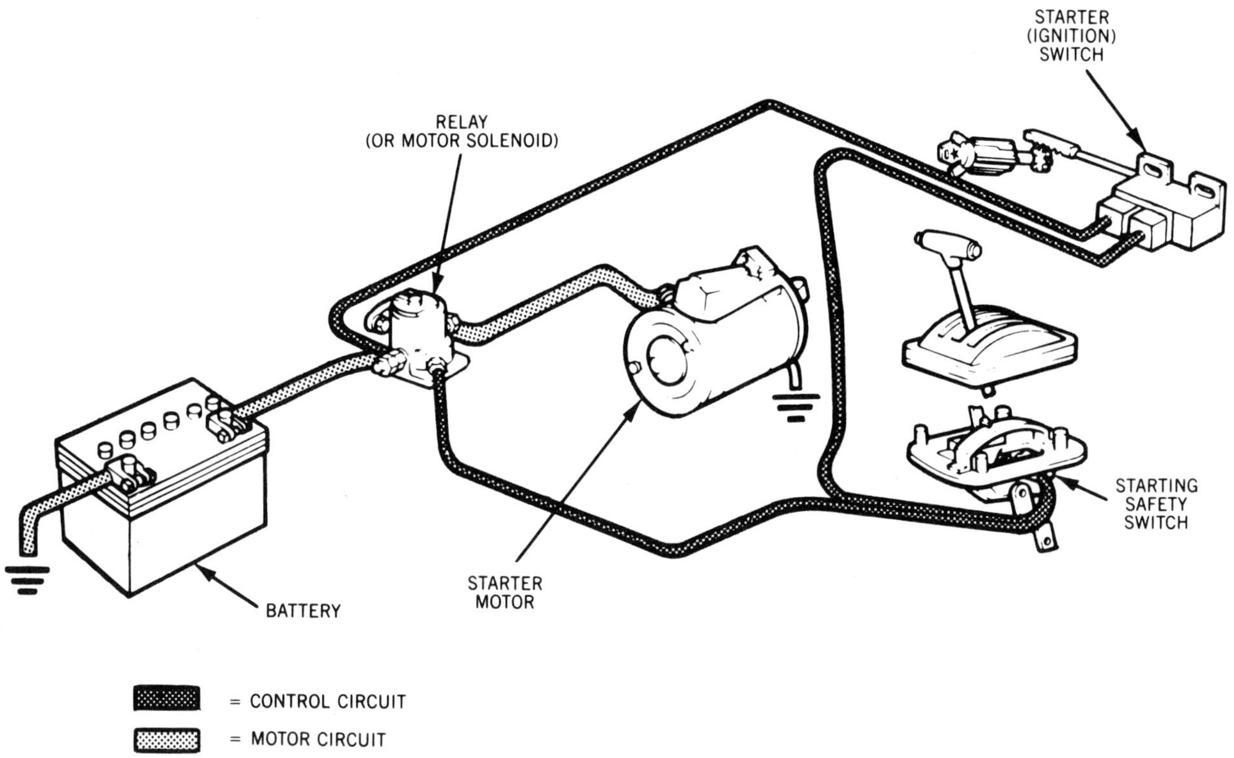


**Figure 10-2.** This is the basic arrangement of starting system motor circuits and control circuits. (Chrysler)

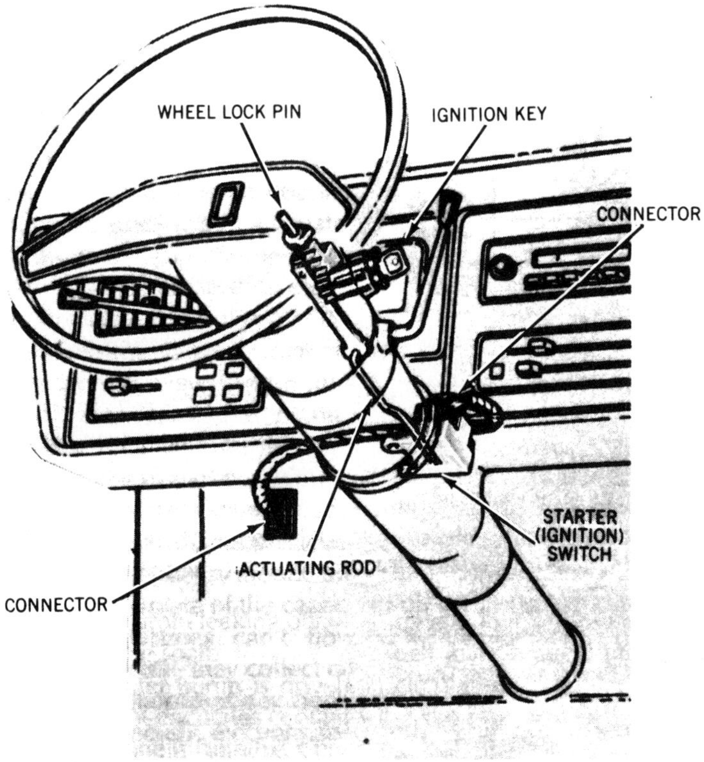


**Figure 10-3.** Most ignition switches are mounted away from the key lock and operated by linkage. (Ford)

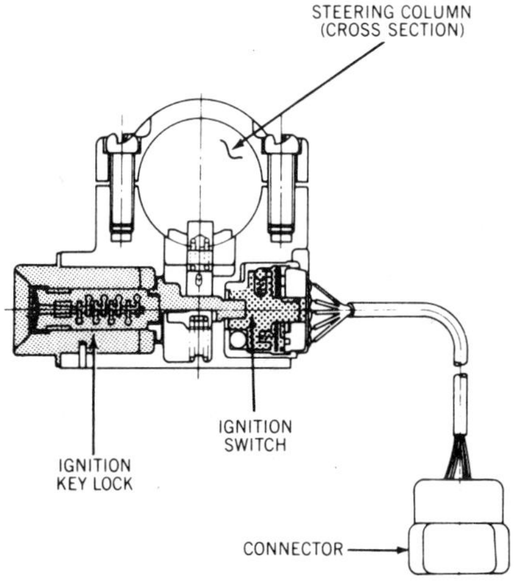


**Figure 10-4.** Some ignition switches are mounted in the steering column and attached directly to the key lock. (Chrysler)

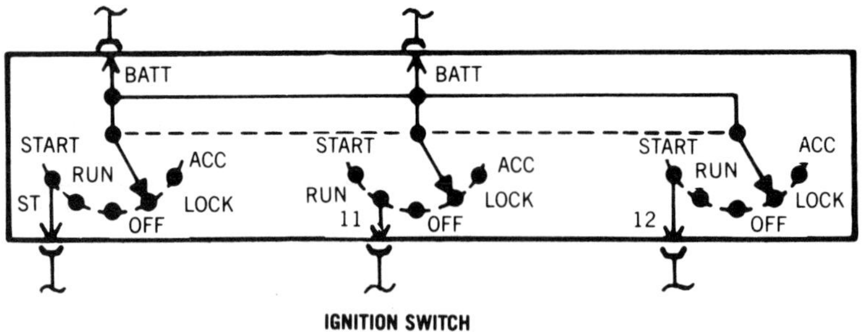


**Figure 10-5.** This is a typical ignition switch symbol you will see on wiring diagrams. (Ford)

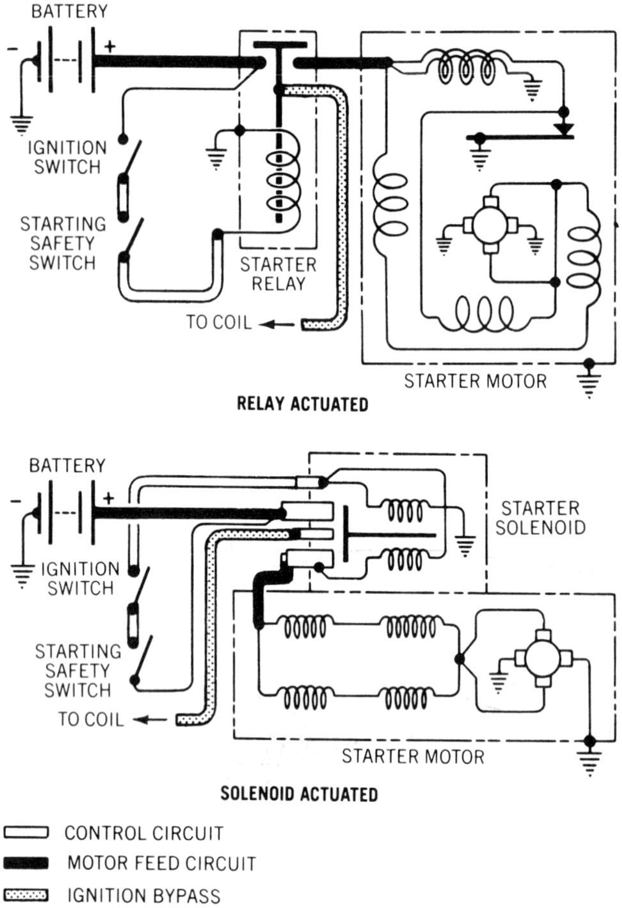


**Figure 10-6.** The starting safety switch can be in series between the ignition switch and the starter relay or solenoid. (Ford)

missions as well. The starting safety switch, or neutral start switch, closes the control circuit only when the automatic transmission is in park or neutral or the manual transmission is in neutral or the clutch is disengaged. This normally open switch is in series with the start position of the ignition switch but can be connected to the circuit in two ways:

**1.** The safety switch can be between the ignition switch and the relay or solenoid, figure 10-6.

**2.** The switch can be on the ground side of the relay or solenoid coil, figure 10-7.

In either case, the safety switch must be closed before current can flow through the relay or solenoid coil and close the motor circuit.

The physical location of the starting safety switch depends on the kind of transmission and the location of the shift lever. Switches for automatic transmissions can be on the steering column or on a floor-mounted shift

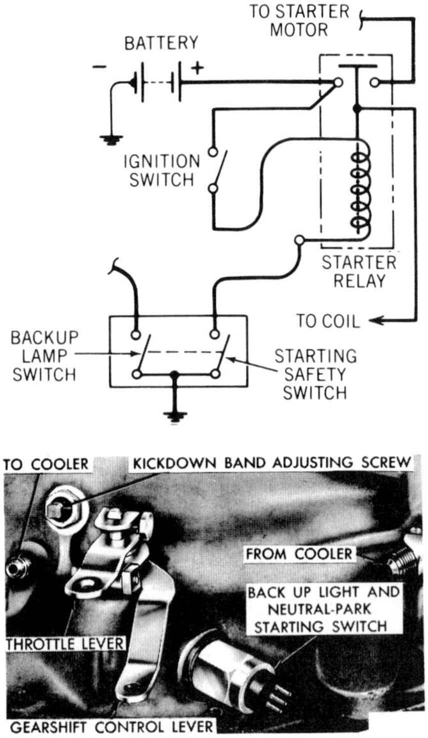


**Figure 10-7.** This Chrysler transmission-mounted safety switch is in series on the ground side of the relay. (Ford) (Chrysler)

console, figure 10-8, or on the transmission, figure 10-7.

Some installations use mechanical linkage to block the movement of the ignition key until the gear selector is in park or neutral, figure 10-9. These devices are not electrical switches, but they serve the same purpose. They disable the starting switch unless the transmission is disengaged.

Some cars with manual transmissions have safety switches operated by shift linkage. These are similar to automatic transmission installations. The switch closes only when the shift lever is in neutral. Most manual transmission cars with safety switches, however, have clutch-start switches. The clutch linkage closes the switch when the pedal is depressed, figure 10-10.

Regardless of the design and location of the starting safety switch, it is always in series with the start position of the ignition switch. The relay or solenoid coil cannot receive current unless *both* switches are closed. Many safety switches are adjustable to allow

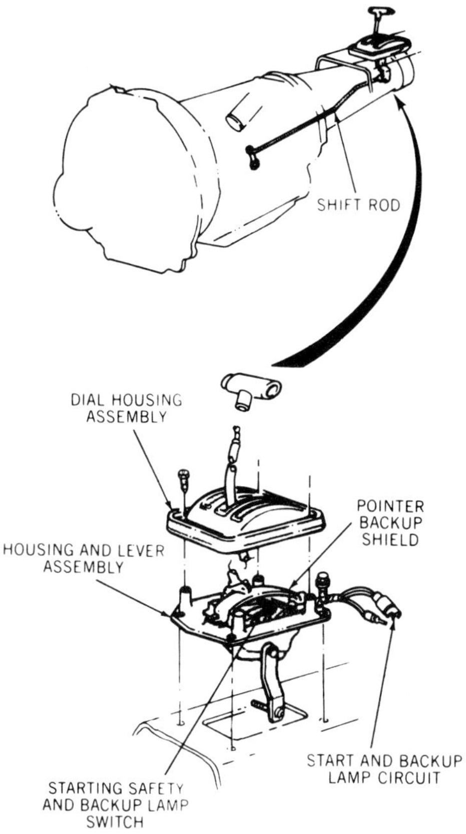


**Figure 10-8.** Most safety switches for floor- shift automatic transmissions are on the shift console. (Ford)

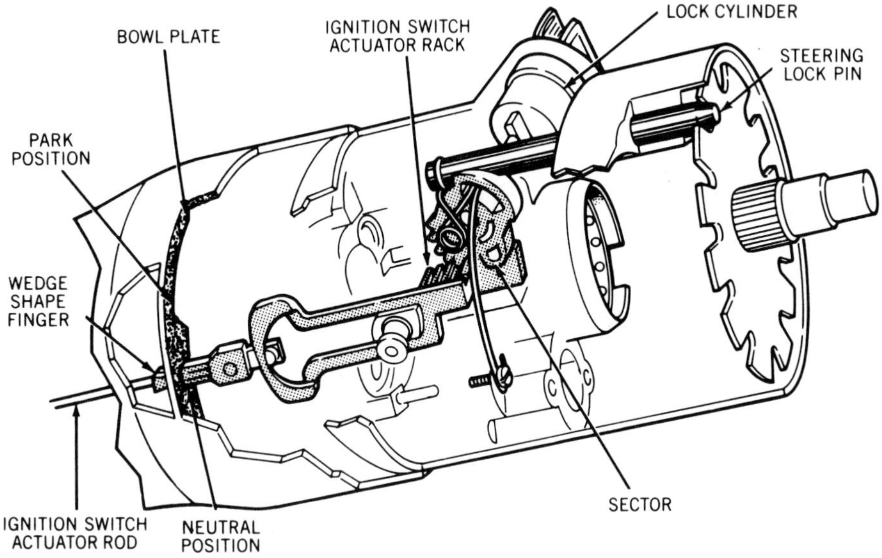


**Figure 10-9.** Many GM vehicles and other cars have mechanical linkage to disable the ignition key unless the transmission is in park or neutral. (Cadillac)

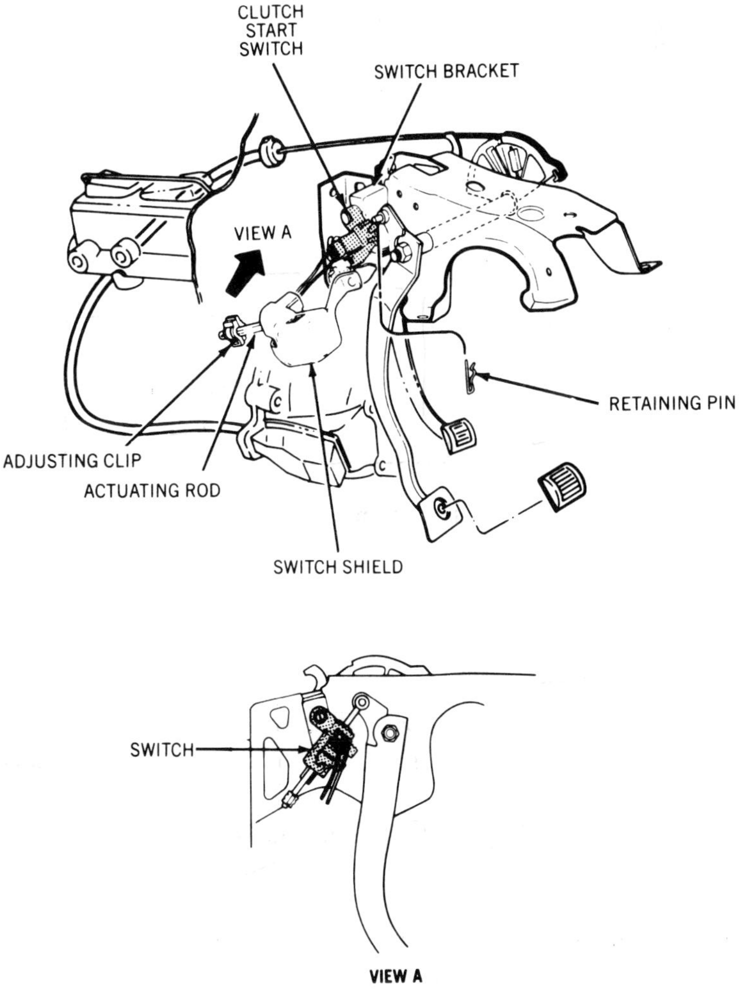


**Figure 10-10.** Typical clutch-start safety switch installation for a manual transmission. (Ford)

for shift linkage or clutch pedal adjustment. Often, you can fix a no-start problem by checking and adjusting the safety switch.

**Relays and Solenoids.** Chapter 7 outlined the basic electromagnetic action of relays and solenoids. Both kinds of devices use current through an electromagnetic coil to move an armature or a movable iron core. The ignition switch and the starting safety switch control current through the coil. When the control circuit is closed, current in the relay or the solenoid coil forms a magnetic field that moves the armature or the core. This movement closes the heavy contacts of the starter motor circuit. In this way, both relays and solenoids can act as magnetic switches for the starter motor, figure 10-11.

The solenoids used in many starting systems also have a second function. The movement of the core operates

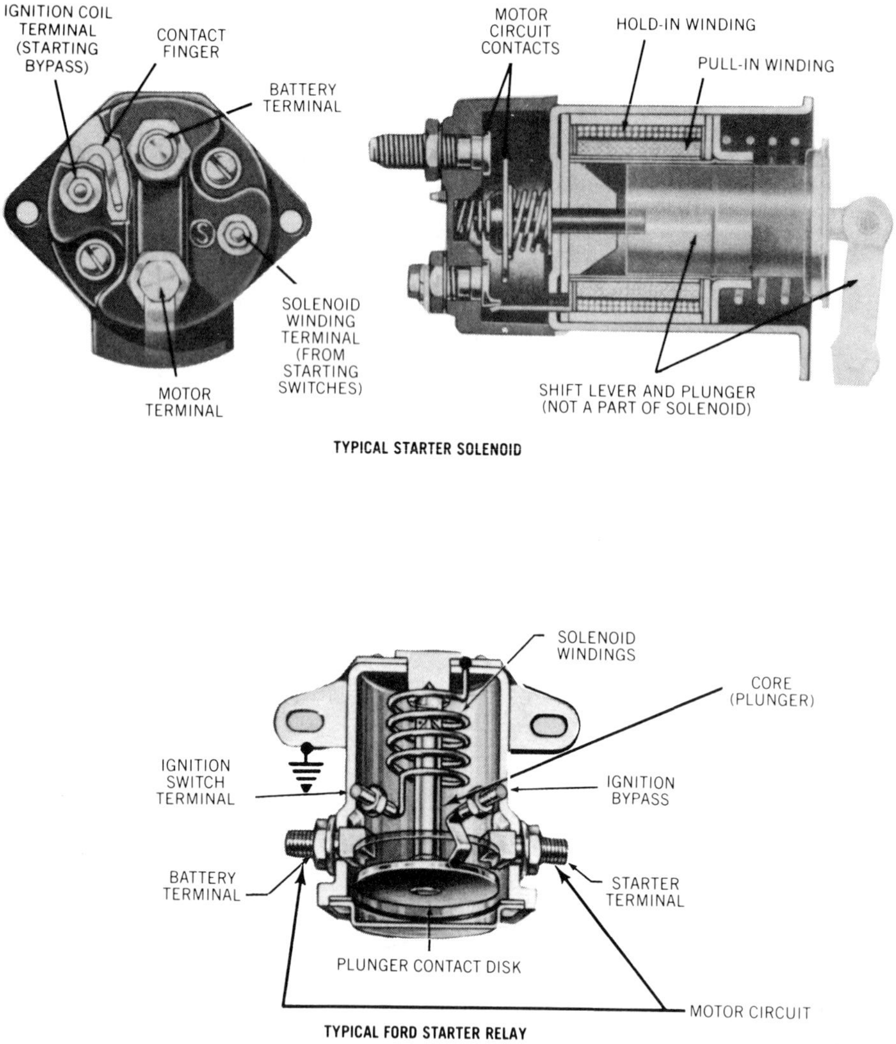


**Figure 10-11.** Relays and solenoids both can act as magnetic switches for the motor circuit. (Delco-Remy) (Ford)

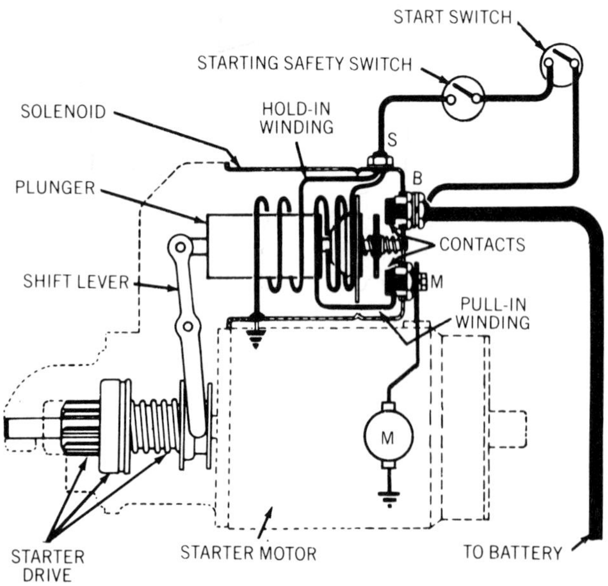


**Figure 10-12.** Typical solenoid-actuated starting system. (Delco-Remy)

linkage that engages the starter drive mechanism with the ring gear on the engine flywheel. The terms used to describe switching functions in starter motor circuits can be confusing. A simple solenoid-controlled circuit can use the solenoid as a relay to operate the magnetic switch for the motor and as a mechanical device to engage the starter drive. With slightly different connections, a separate relay can be added to the circuit. The solenoid may or may not switch the motor current. Ford starter relays, for example, are actually solenoid-type magnetic switches. They act as a relay, but they open and close the motor circuit. Chrysler relays switch the control circuit for a motor solenoid. We will examine starter motor circuits more fully in the next section.

## The Starter Motor Circuit

The motor circuit is a simple circuit that contains just the starter relay or solenoid and the motor. As explained previously, this circuit provides the most direct path possible for the momentary high-current needs of the motor.

**Solenoid-Actuated Circuit.** In a solenoid-actuated motor circuit, the solenoid engages the starter drive with the engine and closes the high-current motor circuit. Delco-Remy starting systems on GM vehicles are the most common examples, figure 10-12. Similar systems are used on most Japanese and European cars as well.

In a basic solenoid-actuated circuit, the battery positive (+) cable is connected to the solenoid battery terminal. The control circuit energizes the solenoid windings through the ignition switch and the safety switch. The electromagnetic field of the windings moves the solenoid core, or plunger, which does two things. The solenoid plunger moves the starter drive pinion gear to engage the flywheel ring gear. It also closes the heavy contacts for the motor circuit, figures 10-11 and 10-12.

Starter solenoids have separate pull-in windings and hold-in windings. The hold-in winding has many turns of light-gauge wire. The pull-in winding has fewer turns of heavier wire. Both windings are energized when the control circuit closes. Together they create the strong magnetic field needed to move the heavy plunger and starter drive mechanism.

The pull-in winding is grounded through the motor terminal and the armature and field windings, figure 10-12. When the plunger disc contacts the solenoid motor terminals, battery voltage is applied to both ends of the pull-in winding. This deactivates the pull-in winding. The lighter hold-in winding is grounded through the solenoid case and remains activated. Less current is needed to hold the plunger engaged than is needed to move it. The double-winding construction reduces solenoid current draw when the starter motor and the ignition are both drawing current during cranking.

The solenoid plunger disc completes the high-current motor circuit through the heavy terminals at the end of the solenoid. A large motor feed strap, or connector link, connects the solenoid motor terminal to the motor field coils and armature. We will look at the mechanical action of starter solenoids further in a later section of this chapter.

**Relay-Actuated Circuit.** There are two basic kinds of relay-actuated starter motor circuits. A Ford starter circuit uses a large solenoid-type relay to conduct high starting current to the motor, figure 10-11. Full motor current flows through the relay when it is energized.

Most Ford vehicles have movable pole shoe starter motors, which require this kind of relay for operation. The battery + cable is connected to the battery terminal of the relay, figure 10-13. A similar heavy cable connects the relay motor terminal to the motor. The control circuit from the ignition and safety switches energizes the relay. Some Ford relays have a fourth terminal that provides the starting bypass connection for the ignition. Systems without the fourth terminal on the relay provide the starting bypass connection through the ignition switch.

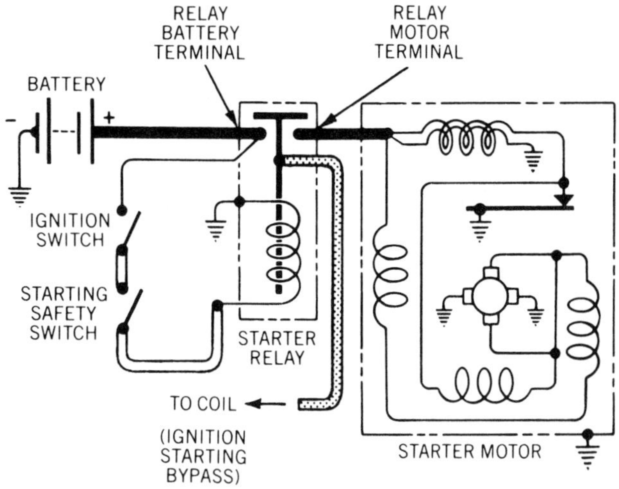


**Figure 10-13.** Typical Ford relay-actuated starter system with the relay in the motor circuit. (Ford)

**THE STARTERLESS ENGINE**

If it weren't for permanent-magnet starter motors, you might be learning to service engines without starter motors today. In the early 1980's, General Motors engineers were struggling with the problem of getting weight out of the new, smaller cars. Starter motors and high-capacity batteries account for 30 or 40 pounds of total car weight. Some engineers thought that if they could eliminate the starter completely, along with a large battery that provided several hundred amperes of cranking current, they could reach a substantial weight-reduction goal.

The method they tried required the use of a direct ignition system, electronic port fuel injection, and a sophisticated engine control computer. Direct ignition eliminated the mechanical distributor and allowed a spark plug in any given cylinder to be fired by computer control. Electronic port fuel injection allowed fuel to be directed to any cylinder by computer control. The idea was to apply spark and fuel to a cylinder that was in position to start a power stroke. The resulting combustion would crank the engine and allow the other cylinders to fire up.

GM engineers did have some success with the project. They got a few engines to start (and may have blown up a few others). To make the system truly reliable, however, would have required sensors to detect piston position in each cylinder and auxiliary startup injectors inside each cylinder.

The availability of powerful permanent magnets put the starterless engine on the shelf. Today's permanent magnet starters weigh much less than motors with field coils, and elimination of field current requirements allows the use of smaller, lighter batteries.

Some Ford products with large V-8 or diesel engines have solenoid-operated starter motors. For manufacturing simplicity, these vehicles have the same kind of starter relay used with movable pole shoe starters. The control circuit actuates the relay, rather than the solenoid. The solenoid engages the starter drive and completes the motor circuit. To energize the solenoid, a connector link is installed between the battery and switch terminals of the solenoid, figure 10-14

The second kind of relay-actuated motor circuit has a relay in the control circuit to actuate a solenoid-operated starter. Chrysler systems are the most common examples, figure 10-15, but this arrangement also is used on many imported vehicles. The relay does not conduct full starting current to the motor as in a Ford system. Rather, the control circuit through the ignition and safety switches energizes the relay, which completes the control circuit from the battery to the motor solenoid. The solenoid motor circuit is connected to the battery as in a Delco-Remy system. The relay simply provides a simple, low-current connection from the battery directly to the solenoid control windings. The solenoid engages the starter drive and completes the motor circuit as in other solenoid-operated motor circuits.

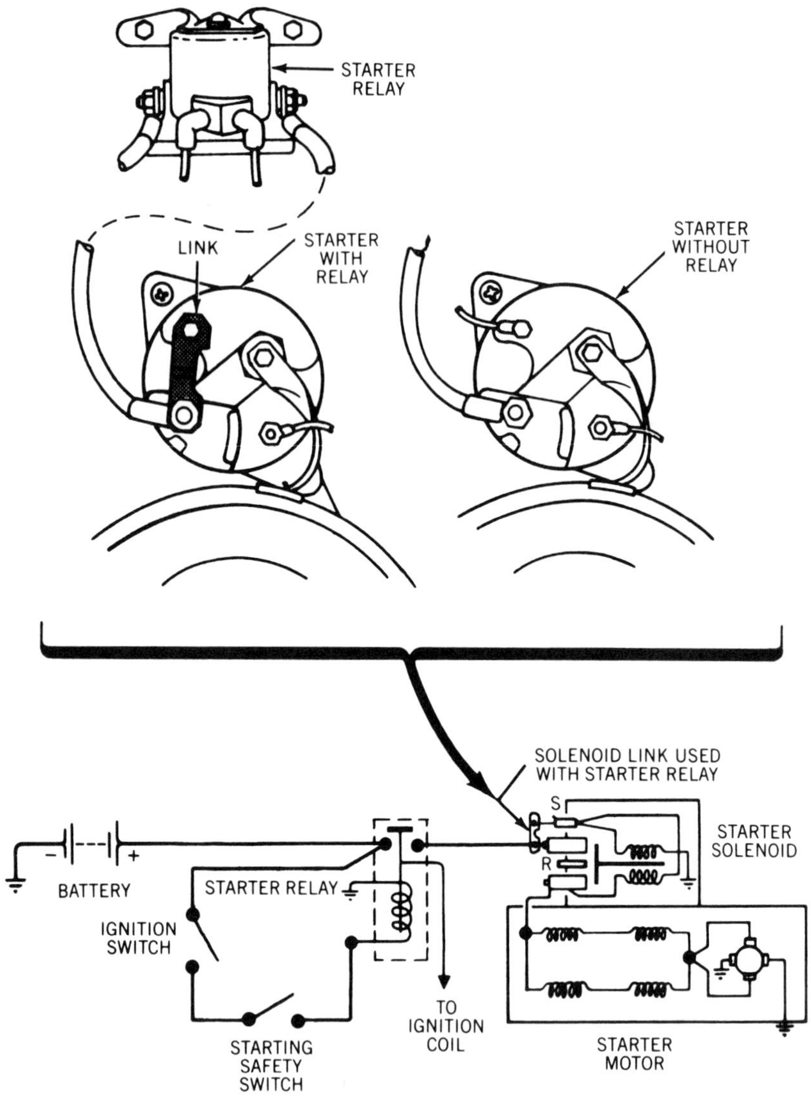


**Figure 10-14.** Typical Ford solenoid-actuated starter motor with a relay in the motor circuit. (Ford)

## STARTER MOTORS

All electric motors operate on the principles of electromagnetism that you learned in chapter 7. A current-carrying conductor develops a magnetic field around itself. The strength of the field depends on the amount of current, and the magnetic polarity depends on the direction of current flow. If you bring two current-carrying conductors close together, their magnetic fields will interact. Opposite poles will attract each other; like poles will repel each other, figure 10-16. This interaction of magnetic fields can change electrical energy to mechanical energy. This is the basis for all electric motors.

### The Motor Principle

If we place a current-carrying conductor in a strong magnetic field, the field created by conductor current tries to move the conductor out of the surrounding field. The conductor tries to move from the stronger field to a weaker field, figure 10-17. If we place two conductors with current in opposite directions between strong north and south poles, the conductor magnetic fields will interact with the surrounding field, figure 10-18. The clockwise field of the top conductor adds to the field of the poles. This creates a strong field beneath the conductor, which forces it upward. The counterclockwise field of the lower conductor also adds to the field of the poles and creates a stronger field above the conductor. This forces the lower conductor downward.

**THE CAR THAT HAS NO CRANK**

If you ask a car owner today about an engine starting crank, you will probably get a blank stare in response. Some people have heard about them; few people have ever seen one. Yet, in the early days of the automobile, a starting crank was standard equipment on any motor vehicle.

In 1912, Cadillac revolutionized the industry with the first electric self-starter. Invented by Charles F. Kettering and built by the "Delco Electrical Plant," this patented, combination starter and generator allowed Cadillac to advertise *"The Car That Has No Crank."* Here is how Cadillac introduced the subject in the 1912 owner's manual:

The past year was and is known in the annals of the motor car as the year of "self-starters" so called. It redounded to the credit of Cadillac Company, however, to provide the motoring world with **the** device which really performed the function for which it was designed.

When the Cadillac plans became known, a "wild scramble" ensued by many makers for almost anything that bore the name—"self-starter". We will not attempt to depict the resulting disappointment experienced by thousands of motor car purchasers who found that "self-starter" meant—much of the time—start it yourself. The disappointed ones will bear witness that the inference is not overdrawn.

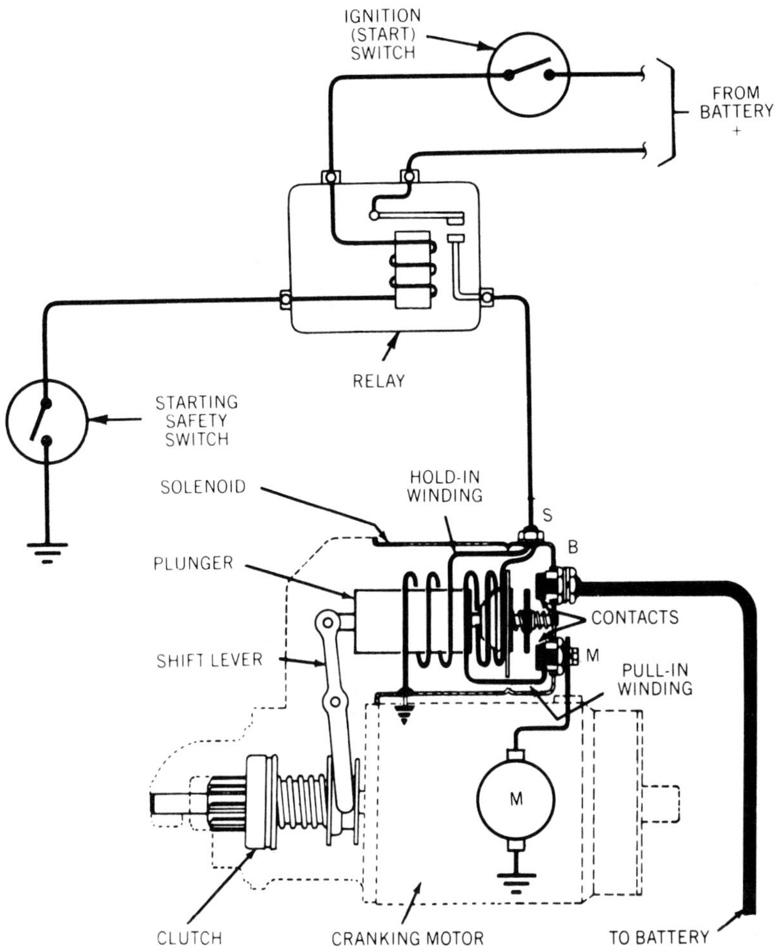


**Figure 10-15.** Typical relay-actuated starting system with the relay in the control circuit. (Delco-Remy)

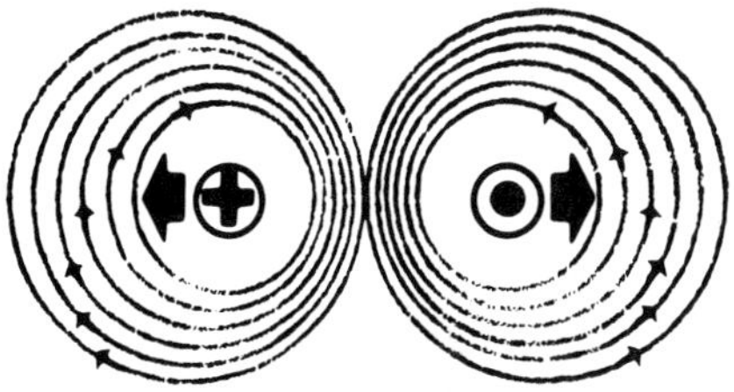

**Figure 10-16.** Two conductors with current in opposite directions form magnetic fields that repel each other. (Delco-Remy)

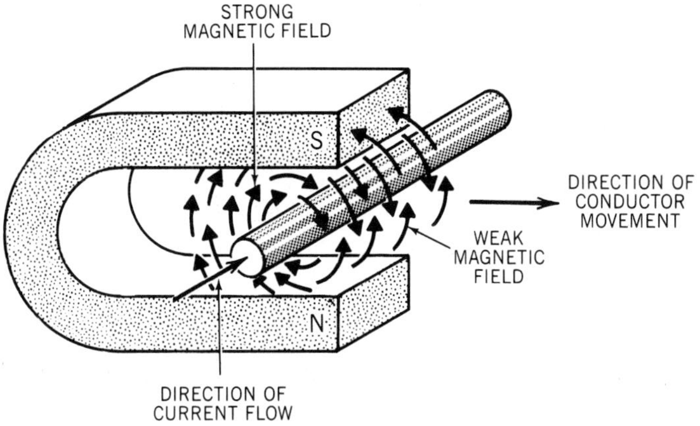


**Figure 10-17.** A current-carrying conductor in a magnetic field tries to move from the strong field to the weak field.

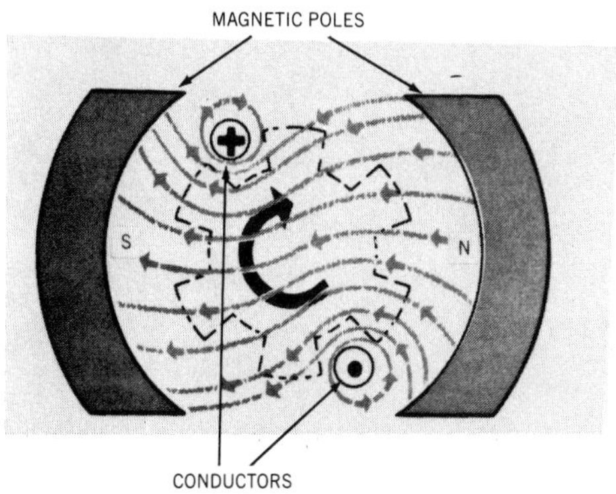


**Figure 10-18.** The interaction of magnetic fields causes electric motors to operate. (Delco-Remy)

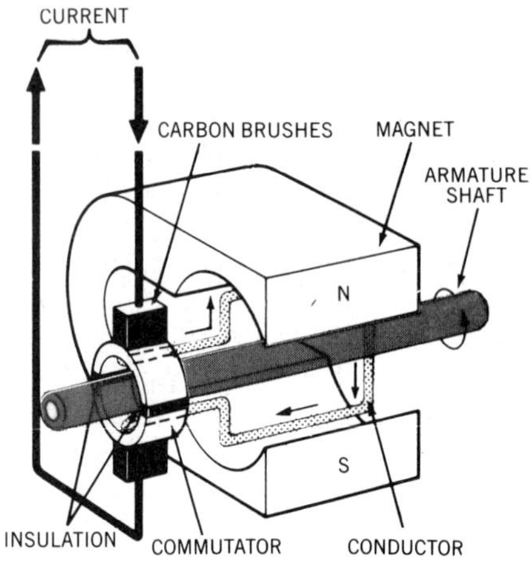


**Figure 10-19.** A loop conductor on a shaft within a magnetic field is a simple electric motor. (Bosch)

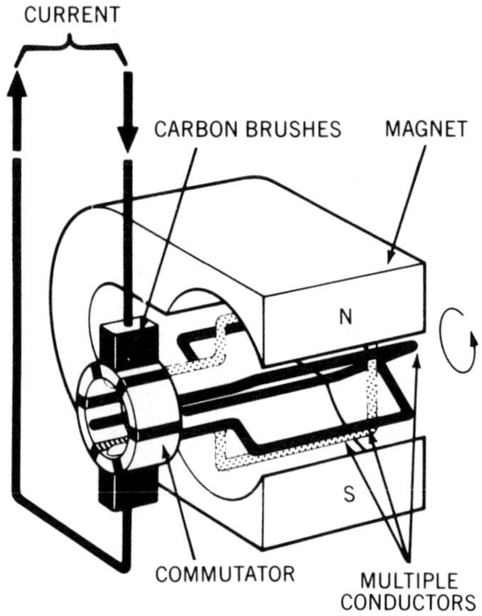


**Figure 10-20.** An actual motor has many loop conductors on the armature. (Bosch)

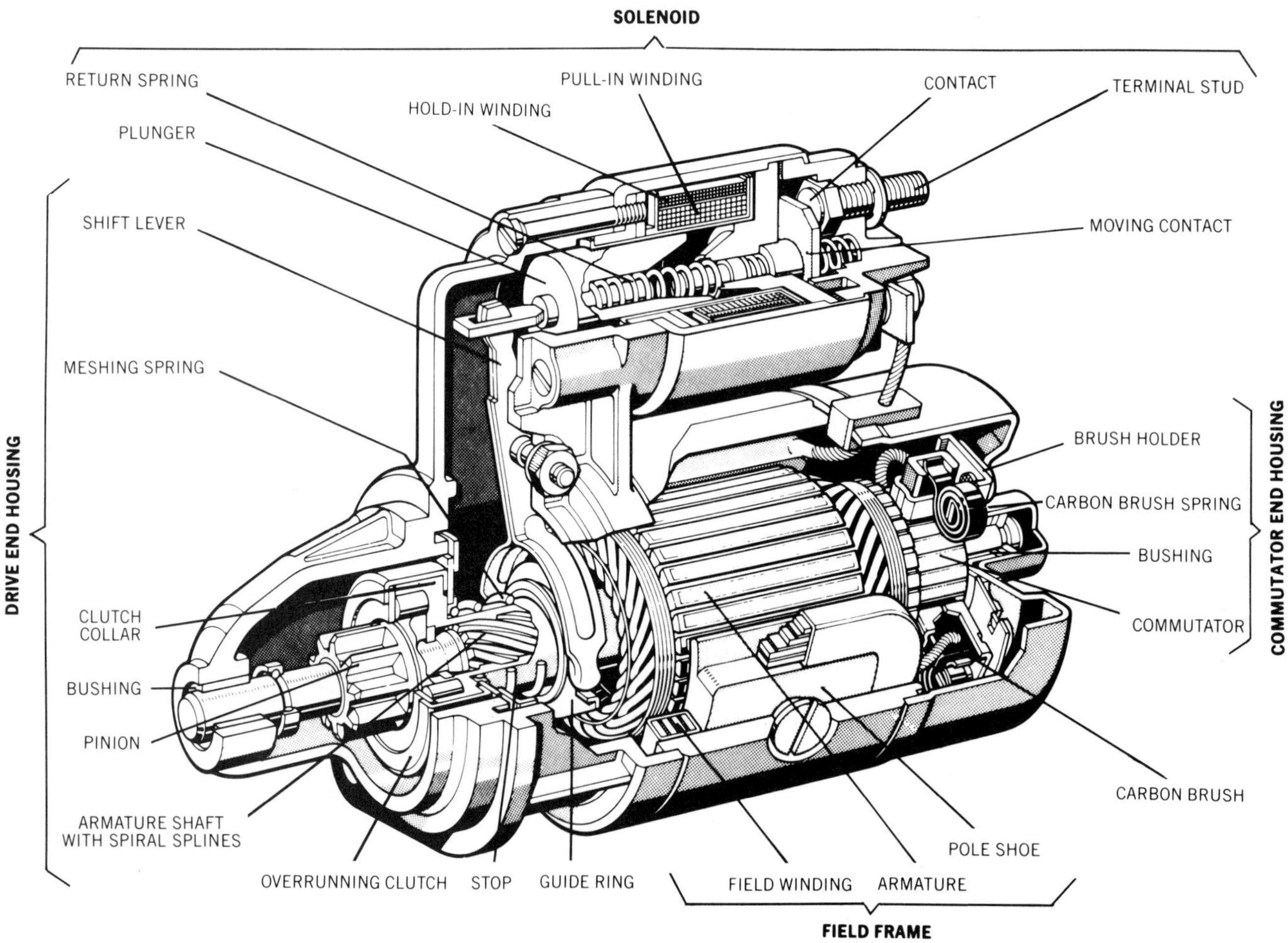


**Figure 10-21.** This Bosch starter motor is typical of most solenoid-operated starters. (Bosch)

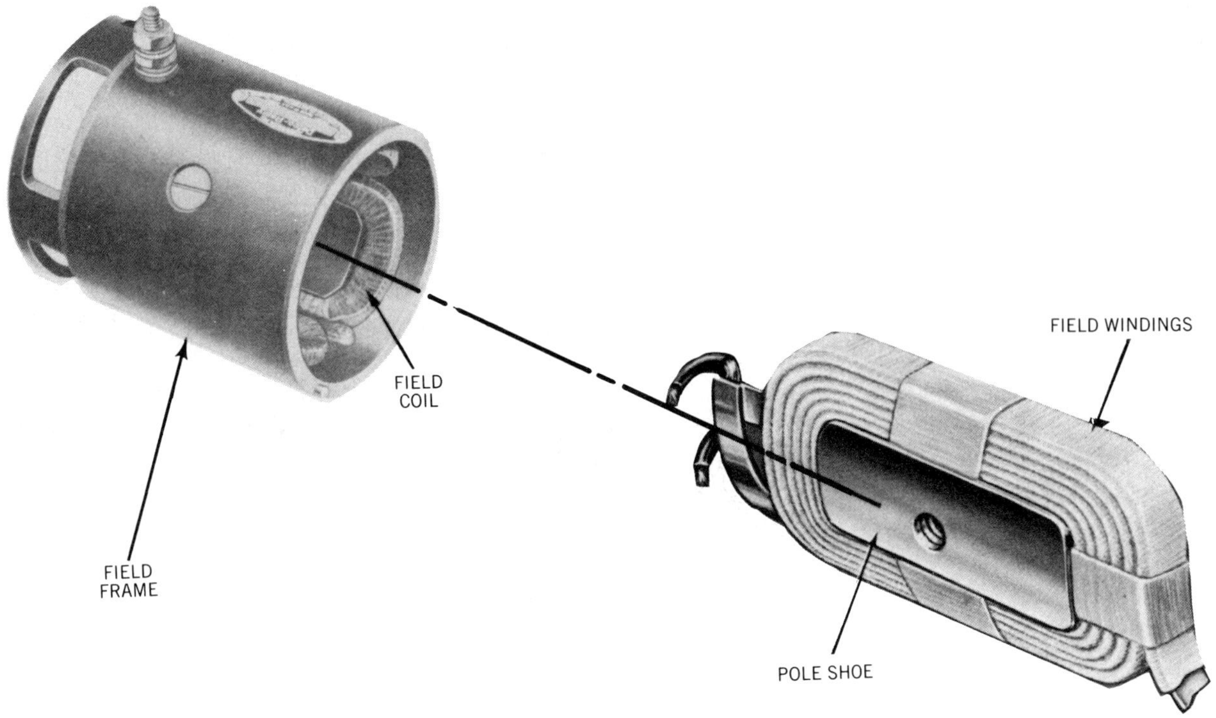


**Figure 10-22.** Typical field frame construction. (Ford) (Delco-Remy)

The two conductors in figure 10-18 can be opposite sides of a single loop. Each side carries current in an opposite direction. If the loop conductor is mounted on a shaft, figure 10-19, we have constructed a simple motor **armature**. An actual motor armature has many conductors to react with the stationary magnetic poles. Continuous interaction of the fields causes the armature shaft to rotate, figure 10-20.

The magnetic poles surrounding the current-carrying conductors can be permanent magnets or electromagnets. Many small motors are made with permanent magnet poles. Starter motors, however, have electromagnetic poles.

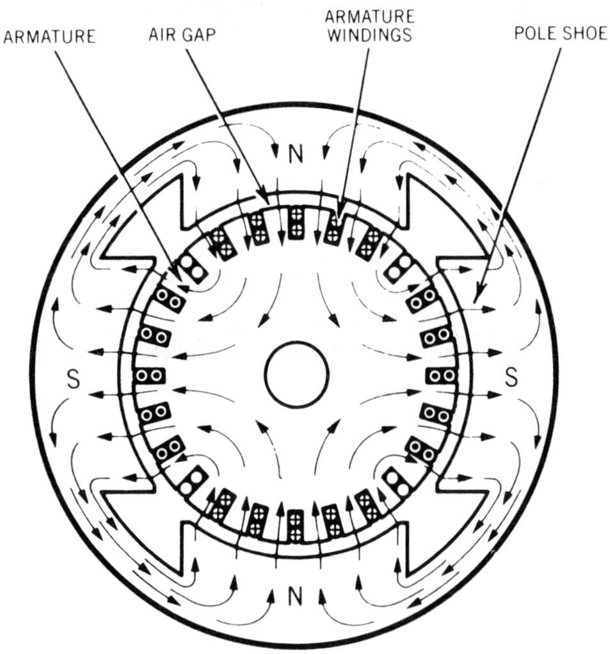


**Figure 10-23.** Typical magnetic fields of a 4-pole starter motor. (Bosch)

### Starter Motor Construction

A typical starter motor, figure 10-21, contains:

- A field frame with pole shoes and electromagetic field windings
- A rotating armature and shaft with current-carrying conductors
- A commutator end, or brush end, housing that holds the motor electrical terminals, and the brushes that conduct current to the armature
- A drive end housing that supports the starter drive gear, which engages the engine flywheel
- A solenoid, or a movable pole shoe, to close the motor circuit and engage the starter drive

Both end housings have bearings, or bushings, to support the armature shaft.

The motor field frame is an iron housing that holds the iron pole shoes and the field windings, figure 10-22. Iron of the frame and pole shoes increases and concentrates the field strength of the poles. Most starter motors have four pole shoes and four sets of field windings. The pole shoes are attached to the frame with large screws.

The field windings are heavy copper ribbon material. Current for the electromagnetic poles enters the motor through the motor terminal from the solenoid or the relay. Figure 10-23 shows the magnetic fields in a 4-pole motor.

The armature holds the armature windings (conductors), the **commutator**, and the armature core, figure 10-24. The armature core is made of thin laminated iron discs. Slots in the iron laminations hold the armature windings. The iron strengthens the armature field. Insulation between the laminations reduces **eddy currents** in the armature.

Eddy current is current caused by induced countervoltage in an armature core. If the core were solid, eddy current would be strong enough to overheat the armature. Counterelectromotive force (CEMF) would also be high enough to develop current and an opposing magnetic field that would resist motor rotation.

The commutator is a series of small copper bars mounted on the armature shaft. Each commutator bar, or segment, is insulated from the others and from the shaft by mica or plastic. A motor commutator acts as a rotating switch. One end of two different armature windings is soldered to each commutator segment, figures 10-18 and 10-20. The opposite end of each winding is connected to a different segment. This arrangement connects the windings in a long series of loop conductors.

Armature windings can be connected to the commutator in two ways:

1. In a lap winding, figure 10-25, the two ends of each winding are connected to adjacent commutator segments.
2. In a wave winding, figure 10-26, the two ends of each winding are connected to segments that are 90 or 180 degrees apart on the commutator.

Wave windings have less resistance than lap windings and are used in most starter motors.

Carbon brushes conduct current to the commutator segments and the armature windings, figure 10-27. Four-

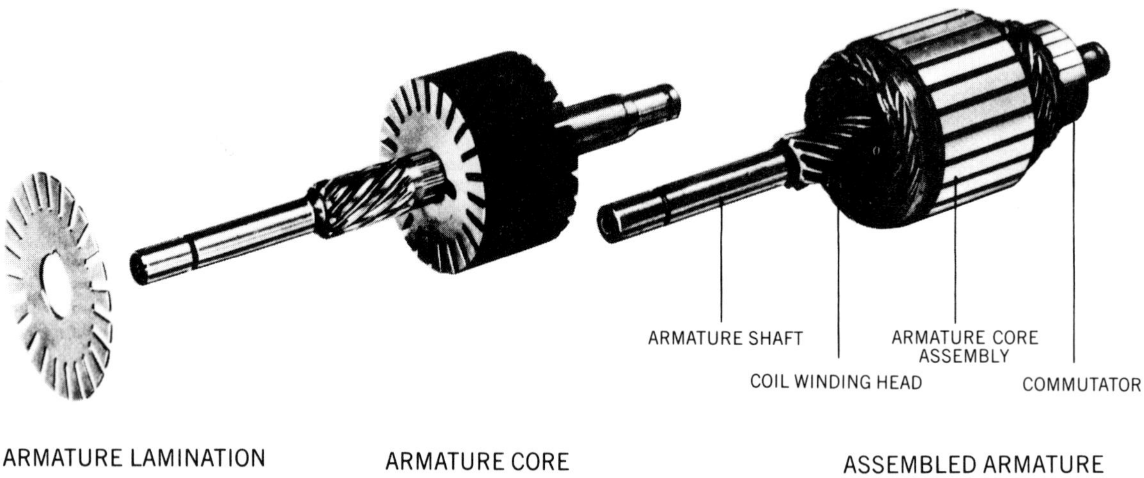


**Figure 10-24.** Typical motor armature construction. (Bosch)

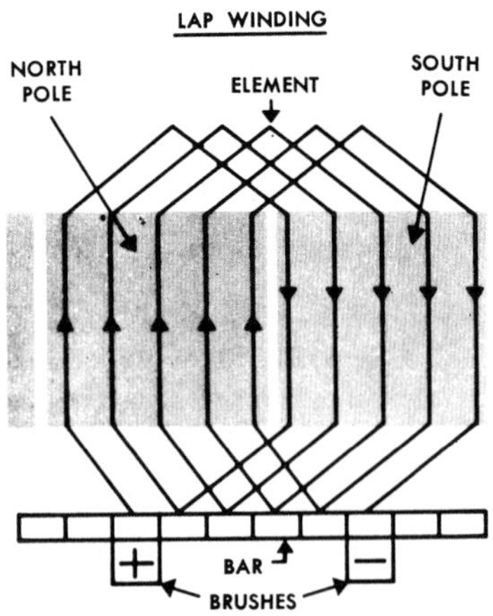


**Figure 10-25.** This diagram represents a lap-wound armature. (Delco-Remy)

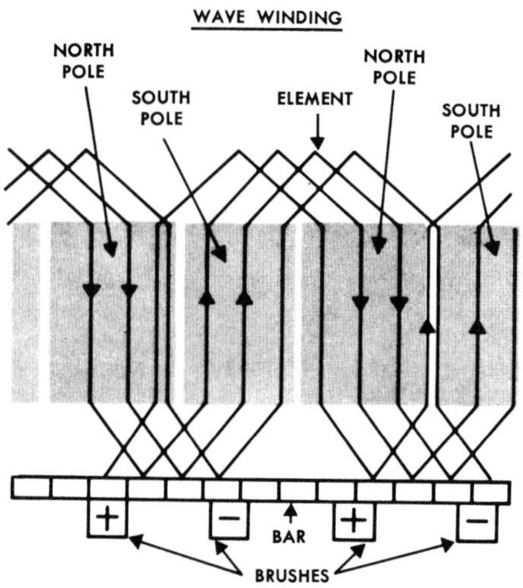


**Figure 10-26.** This diagram represents a wave-wound armature. (Delco-Remy)

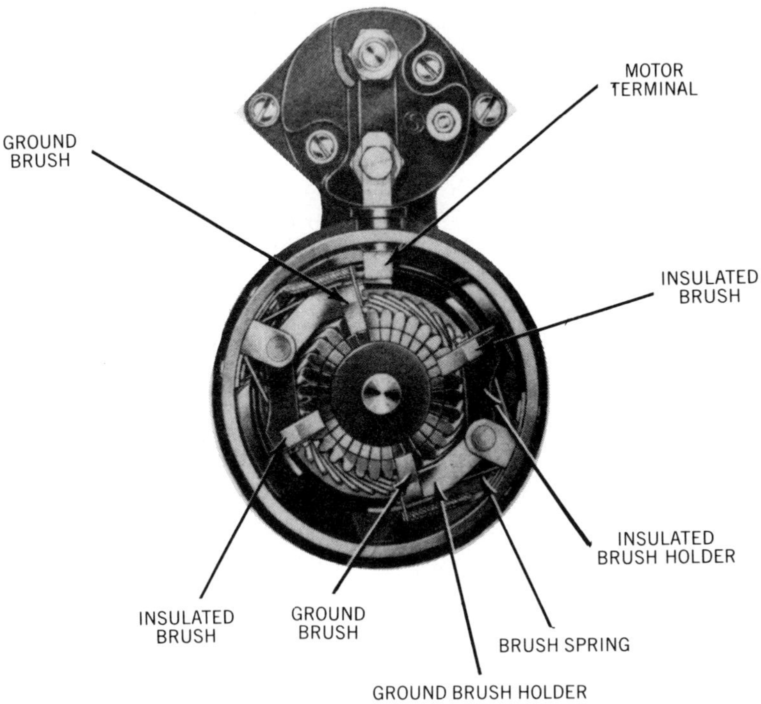


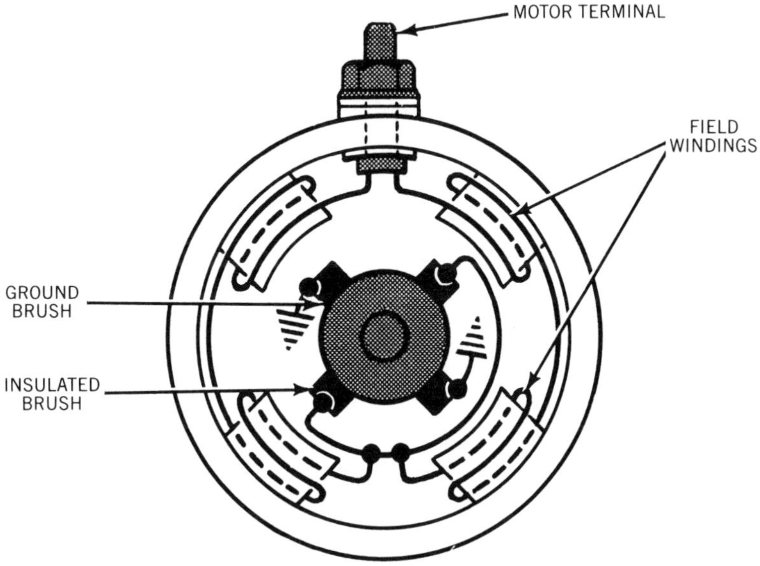


**Figure 10-27.** Typical 4-pole, 4-brush motor construction. (Ford) (Delco-Remy)

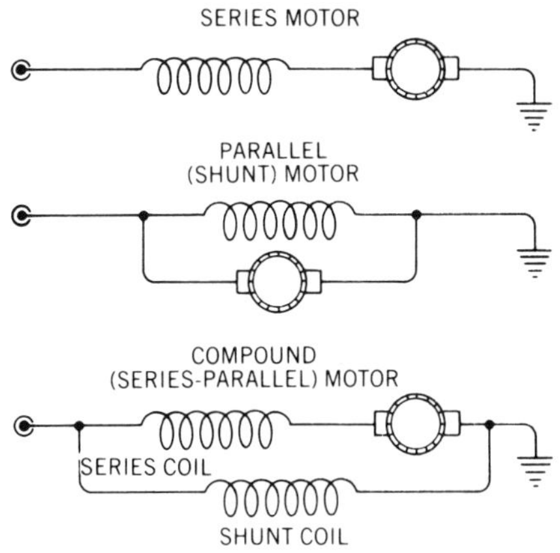


**Figure 10-28.** Three basic kinds of dc motor circuits. (Ford)

pole motors must have four brushes to carry current to the commutator. Each pair of brushes completes a parallel circuit through the armature. Each of the two armature circuits reacts with the field poles to rotate the shaft. One brush of each pair is connected in parallel to the motor battery terminal. The other brush of each pair is grounded to the motor frame.

The brushes are held against the commutator by springs in the brush holders of the commutator end frame. The commutator end frame of most motors also holds the terminals that connect the insulated brushes and the field windings to battery voltage. Commutator end frames on most motors are made of aluminum to reduce weight.

The drive end frame supports the armature shaft and the starter drive. It also provides the mounting flange where the motor is bolted to the engine.

## Motor Field and Armature Circuits

Field current and armature current enter the motor through a single terminal. Therefore, the field windings and the armature windings must be connected in a single complete circuit. There are three ways to do this in any dc electric motor, figure 10-28:

1. Series connection
2. Parallel (shunt) connection
3. Compound (series-parallel) connection

The field and armature connections affect the torque and speed relationship of the motor, figure 10-29.

**Series Motor.** A series-connected motor develops maximum torque at startup speed—zero rpm, figures 10-28 and 10-29. This is because induced CEMF in the armature and field windings is low when the motor is not rotating. CEMF generates current that opposes battery current and a magnetic field that opposes motor rotation. Because CEMF is low at low motor speed, however, it produces little opposition to armature and field current. A series motor, therefore, develops maximum torque as it starts to rotate. Torque then decreases as motor speed and CEMF increase. This high starting torque characteristic makes series motors ideal for many engine starting applications.

**Parallel (Shunt) Motors.** In this kind of motor, the field coils are in parallel with the armature, figure 10-28. Torque is low at startup speed and increases as motor speed increases, figure 10-29. Changing the load on the motor changes the CEMF in the armature but not in the field. The motor adjusts its torque output to the load and operates at a relatively constant speed. Parallel motors are ideal for applications such as power windows and power seats, but not for engine cranking. Combining the features of a series motor and a parallel motor produces a compound motor, which can be used to crank an engine.

**Compound (Series-Parallel) Motors.** A compound motor has one or more field coils in series with the armature and one or more coils in parallel, figure 10-28. A compound motor develops high torque soon after startup and maintains comparatively high torque as speed increases, figure 10-29. Many starter motors for large engines are the compound type.

**Typical Field and Armature Circuits.** We could go farther into the electrical theory of starter motor construction, but that isn't necessary for starter system service. Recognizing the different kinds of field and armature connections is important, however, when testing motors for open or short circuits. If you can recognize the connections for different series and compound motors, you can trace the circuits with an ohmmeter or self-powered test lamp.

Remember that most automobile starter motors are 4-pole, 4-brush motors. A simple series motor can be connected with all four field windings in series and then with parallel connections to the two insulated brushes, figure 10-30. A series motor also can be connected with two field windings in series with each insulated brush. These two circuits then can be connected in parallel with a common connection, figure 10-31. A compound motor typically will have three field coils in series with two parallel brushes and a shunt coil connected to a third insulated brush, figure 10-32.

Engineers select starter motors to meet the cranking speed and torque needs of different engines. Electrical diagrams for starting systems will show motor field and armature connections to help you troubleshoot motor problems.

## STARTER MOTOR AND DRIVE DESIGNS

We have examined starting motor circuits and the electrical design and construction of starter motors. The final major feature of starting systems is the kind of drive mechanism that engages the motor with the engine flywheel.

For any motor to apply enough torque to crank the engine at starting speed, the torque must be increased through gear reduction. The motor turns a pinion gear that engages a ring gear on the flywheel. A typical flywheel ring gear has approximately 150 to 200 teeth. A typical starter drive pinion has about 10 teeth. The ratio between the pinion and the flywheel gear, then, is approximately 15:1 to 20:1. The motor turns at 3,000 to 4,000 rpm to crank the engine at about 200 rpm.

Remember that the motor turns for only a few seconds. The starter drive

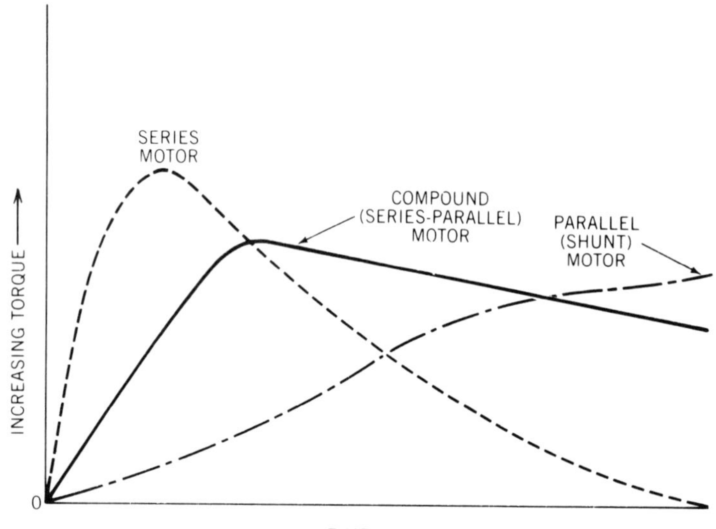


**Figure 10-29.** Typical torque and speed relationships of different dc electric motors.

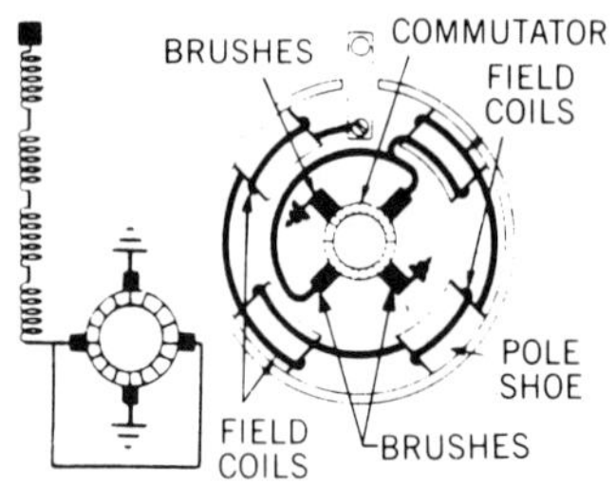


**Figure 10-30.** This motor has four field coils in series with two parallel brushes. (Oldsmobile)

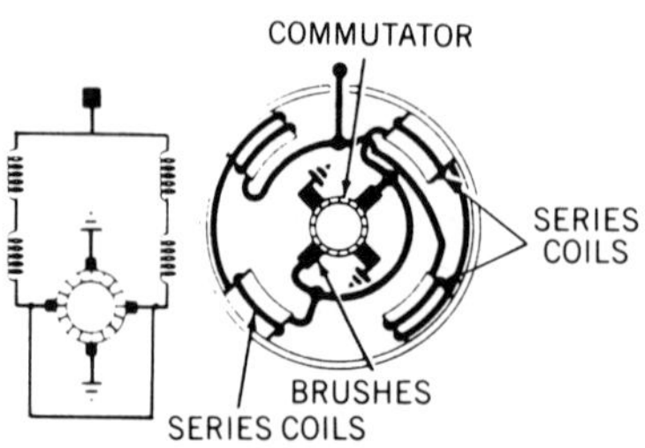


**Figure 10-31.** This motor has four field coils. Two are in series with each other and one brush. (Oldsmobile)

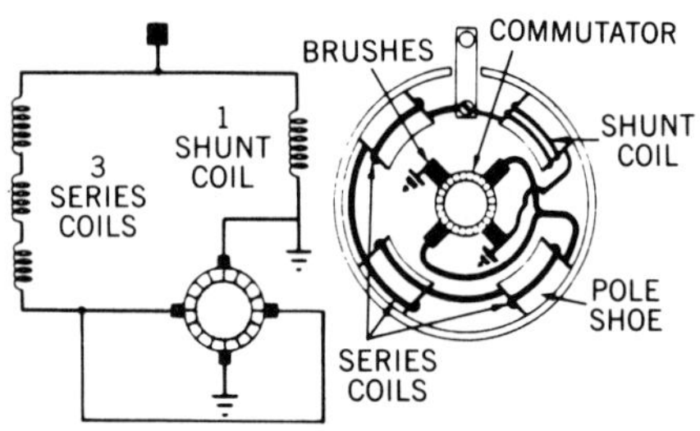


**Figure 10-32.** This compound motor has three field coils in series and one in parallel. (Oldsmobile)

is a mechanism that moves the motor pinion to engage the flywheel. Just as importantly, the starter drive disengages the pinion from the flywheel when the engine starts and turns faster than cranking speed. At an idle speed of 600 rpm, a starter motor would be turned at 9,000 to 12,000 rpm if it stayed engaged with the flywheel. At 3,000 engine rpm, the starter would turn at 45,000 to 60,000 rpm if it were engaged with the engine. These speeds would literally tear the motor apart. Occasionally, a starter drive may fail and leave the pinion engaged with the flywheel. The results will be a destroyed motor.

Starter motors on almost all late-model automobiles are the **positive-engagement** type. They use mechanical leverage provided by a solenoid or a movable pole shoe on the motor to engage the starter drive pinion with the engine flywheel. Older domestic cars and a few late-model European cars use **inertia-engagement** starters. These use what is commonly called a Bendix-drive mechanism. Positive-engagement starters are far more common on current vehicles than Bendix-drive starters. Positive-engagement starter drives are divided into three major groups:

1. Solenoid-operated, direct drive
2. Solenoid-operated, reduction drive
3. Movable pole shoe
4. Permanent magnet gear reduction (PMGR)

The following paragraphs explain their design and operation.

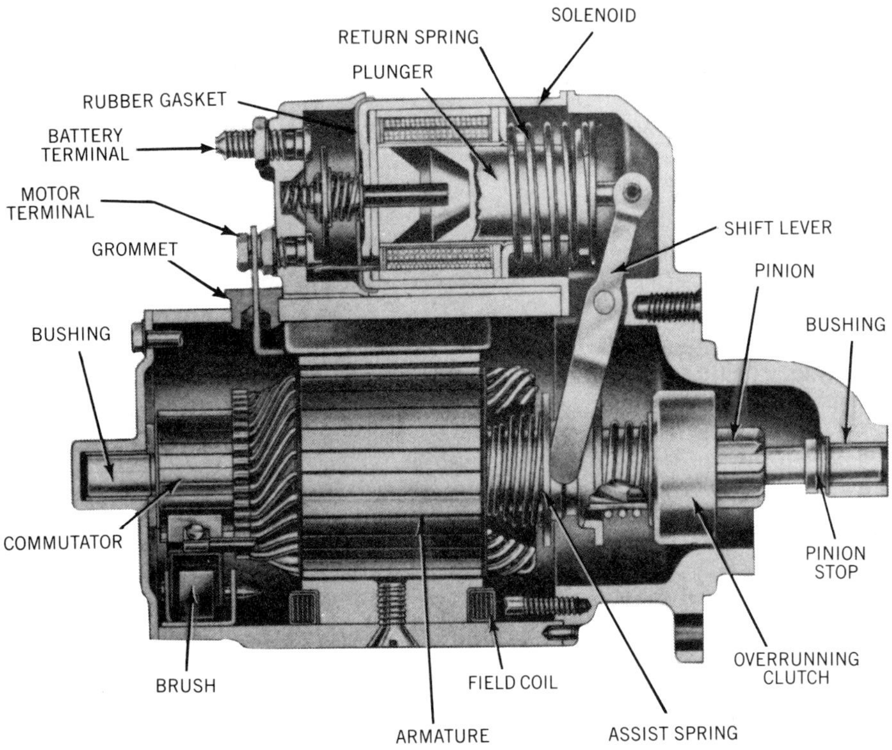


**Figure 10-33.** Delco MT series starter motors have been used on GM cars for 30 years. (Delco-Remy)

### Solenoid-Operated, Direct Drive

The most common solenoid-operated, direct-drive starter motors are made by the Delco-Remy Division of General Motors. The Delco-Remy MT series of starter motors, figure 10-33, has been used on almost all GM cars and light trucks for decades. These motors vary in size, as well as in field and armature connections, for different sizes of engines. They may be either series or compound motors. However, all have a solenoid-operated, positive-engagement drive mechanism. Although construction details vary, operating principles are the same for all of these motors.

Some large Ford engines also have solenoid-operated, direct-drive engagement. Most Bosch starter motors, figure 10-21, and Japanese motors from Nippondenso and other manufacturers work on the same principles. Late-model Chrysler engines also have similar direct-drive starters. We examined the solenoid action earlier as part of the starter motor electrical circuit. Now, let's look at the mechanical part of the solenoid-operated starter drive.

When the control circuit energizes the pull-in and hold-in windings of the solenoid, the plunger moves to close the high-current motor circuit. The plunger also moves a shift lever that slides the drive pinion assembly along the armature shaft, figure 10-34.

The pinion must engage the flywheel *before* the solenoid completes the motor circuit. To do this, the shift lever may have a mechanical leverage ratio that will engage the pinion before the solenoid closes the motor circuit. The solenoid plunger also may have two parts. The first part engages the drive pinion and then forces the second part to close the motor circuit. This is the common design used in Delco-Remy starters.

If the pinion teeth hit the flywheel teeth and do not engage fully, a spring on the starter drive compresses so that the solenoid plunger can complete its stroke. The spring forces the pinion to engage with the flywheel as soon as the armature shaft moves. This positive engagement prevents damage to the pinion or the flywheel ring gear when the starter motor applies cranking torque.

When the engine starts and runs faster than cranking speed, a 1-way, or overrunning, clutch disengages the pinion from the flywheel. The 1-way clutch has rollers held between a collar on the pinion and an outer clutch housing and shell. Motor torque is transferred through the shell to the rollers and the pinion. The shell has tapered slots in which the rollers can ride freely or wedge between the collar and the shell, figure 10-35. When the motor armature turns to crank the engine, the rollers are wedged against

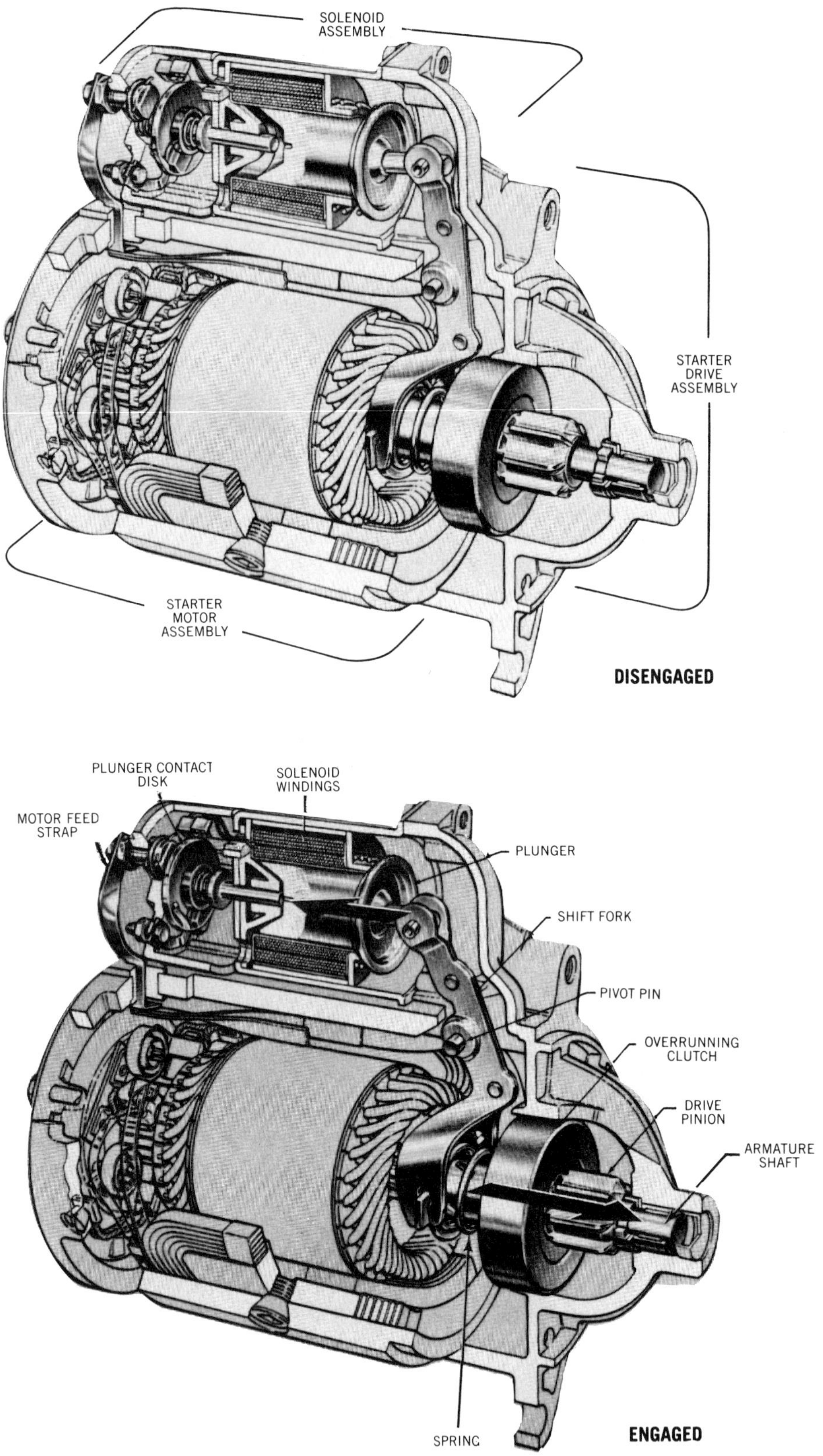


**Figure 10-34.** Solenoid action closes the motor circuit and mechanically engages the starter drive. (Ford)

spring pressure into the narrow ends of the slots. When the engine starts and turns faster than the motor, the armature accelerates and releases pressure on the rollers. The springs then force the rollers into the wider areas of the slots. This uncouples the motor from the pinion and lets the pinion spin freely.

When the driver releases the ignition switch from the start position, the solenoid is deenergized. A return spring in the solenoid moves the plunger and the shift lever to withdraw the starter drive from the flywheel.

### Solenoid-Operated, Reduction Drive

A reduction-drive starter also uses a solenoid to engage the pinion with the flywheel and to complete the motor circuit. The motor armature, however, does not drive the pinion directly. Instead, it turns a small drive gear that meshes with a larger gear, figure 10-36. This provides a reduction ratio of about 2:1 to 3.5:1. The exact ratio depends on motor design and engine cranking requirements.

The armature gear and the reduction gear are permanently meshed. The reduction gear turns the shaft for the starter drive pinion, which is engaged by the solenoid shift lever. The pinion drive mechanism works exactly like a solenoid-operated, direct-drive starter. A 1-way clutch releases the pinion when the engine starts. A solenoid return spring withdraws the drive from the flywheel when the solenoid is deenergized.

Reduction-drive starters allow a small, lightweight, high-speed motor to develop enough torque to crank a large or high-compression engine. For years, Chrysler was the major user of reduction-drive starters. Today, reduction-drive starters are used for small, high-compression engines and diesel engines in many imported and domestic vehicles. Reduction-drive starters are usually compound motors.

### Movable Pole Shoe

Ford Motor Company is the only maker of movable pole shoe starter motors,

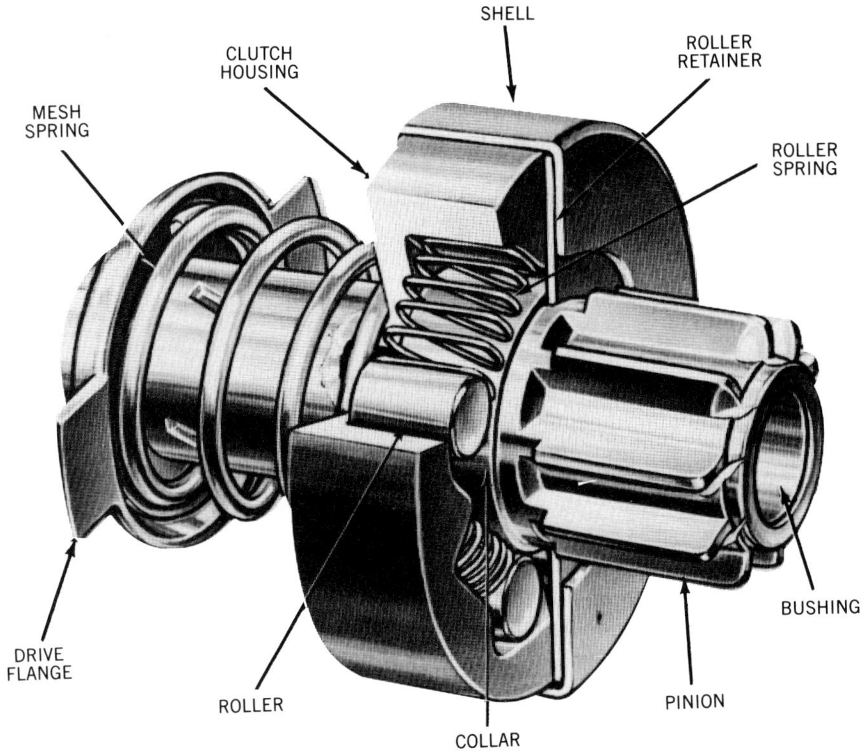


**Figure 10-35.** Most starter drives have a 1-way (overrunning) clutch. (Ford)

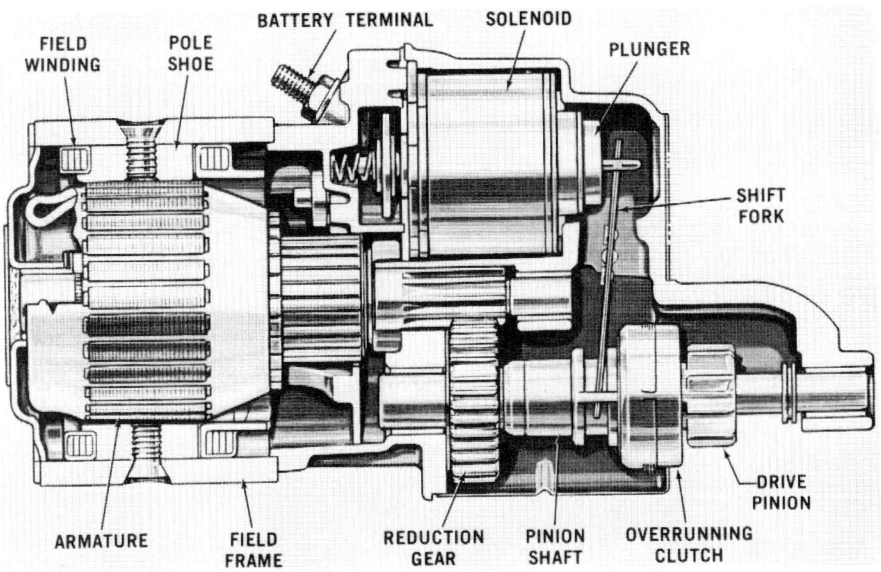


**Figure 10-36.** Typical Chrysler gear-reduction starter motor. (Ford)

figure 10-37. Ford has used these motors for over 25 years, and they are the most common Ford starters. Ford calls the movable pole shoe starter a "positive-engagement" starter. In fact, the drive mechanism is similar to the drive used on solenoid-operated, direct-drive starters. It is engaged, however, by movement of one of the pole shoes in the motor.

One of the pole shoes is mounted on a pivot in the drive end housing. A return spring holds the movable pole shoe upward when the relay is de-energized. The field winding for this pole shoe has a parallel holding coil that is independently grounded. All four field windings also have a normally closed ground contact.

When the Ford starter relay closes, current flows through all four field windings and the holding coil to ground, figure 10-38. This creates a strong magnetic field that pulls the pole shoe toward the motor armature. The movable pole shoe is linked to the starter drive by a shift lever, or yoke. Movement of the pole shoe shifts the drive pinion to engage the flywheel. When the pole shoe moves to operating position, it opens the normally closed grounding contacts. The motor field circuit is then completed through the armature and the ground brushes, and the motor rotates. The holding coil remains independently grounded to hold the movable pole shoe in operating position.

When the engine starts, a 1-way clutch in the starter drive allows the pinion to freewheel. When the driver releases the key, the motor circuit opens. The return spring then moves the pole shoe and shift lever to withdraw the drive from the flywheel.

The movable pole shoe starter requires a separate relay to provide motor current. These motors, made by Ford's Motorcraft Division, are used on many older AMC vehicles, as well as Ford products.

## Permanent Magnet Gear Reduction (PMGR) Starters

Permanent magnet (PM) starter motors were introduced on some 1986 Chrysler and AMC models and on 1988 Chevrolet and GMC light trucks. This type of starter motor has no electromagnetic field coils or pole shoes. The magnetic field is provided by four or six small permanent magnets, depending upon the size of the motor, figure 10-39. As a result, there is no motor field circuit and thus there are no potential field wire-to-frame shorts or other electrical problems related to the field. The motor uses only an armature circuit. The permanent magnets, figure 10-40, are made from an alloy of iron and rare-earth materials that delivers a magnetic field strong enough to operate the motor with the same cranking performance as a comparable starter motor with electromagnetic fields.

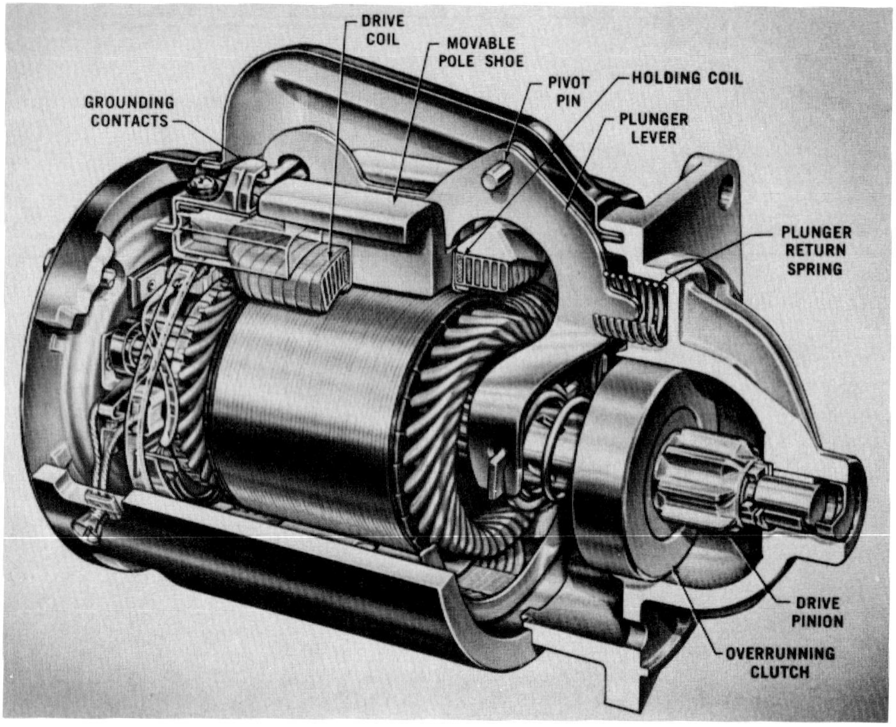


**Figure 10-37.** Ford calls the movable pole shoe starter a "positive engagement" starter. (Ford)

A planetary gear train transmits power between the armature and the pinion shaft, resulting in a high-speed, low-torque starter motor. (The armature revolves at speeds up to 7,000 rpm). Figure 10-41 shows the Bosch design used on Chrysler and AMC vehicles; figure 10-42 shows the Delco-Remy design used with Chevrolet and GMC vehicles. The Bosch and Delco-Remy starters both use a solenoid to operate the starter drive and close the motor armature circuit. While the drive mechanism is the same as that used on other solenoid-actuated starters, some models use lightweight plastic shift levers.

The planetary gearset installed between the armature and starter drive increases the torque at the drive pinion while reducing the speed. The gearset is quite compact, measuring only 1/2- to 3/4-inch deep. It is mounted inline with the drive pinion and armature. By keying the internal ring gear to the field frame, it remains stationary in the motor. The armature

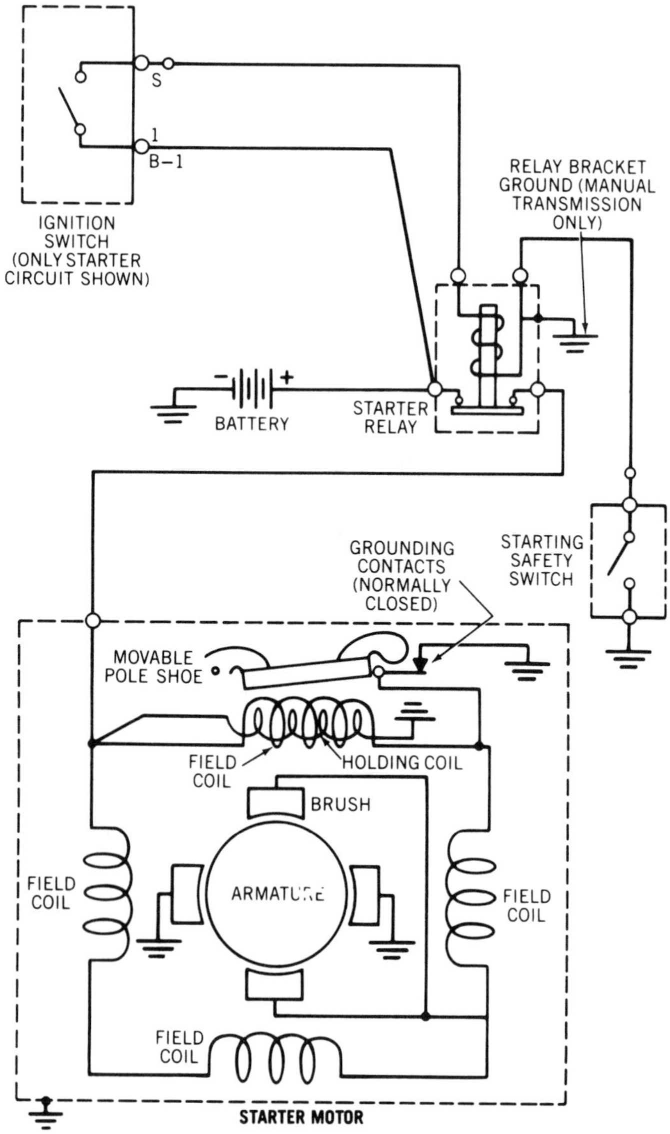


**Figure 10-38.** Typical circuit diagram for a Ford movable pole shoe starter system.

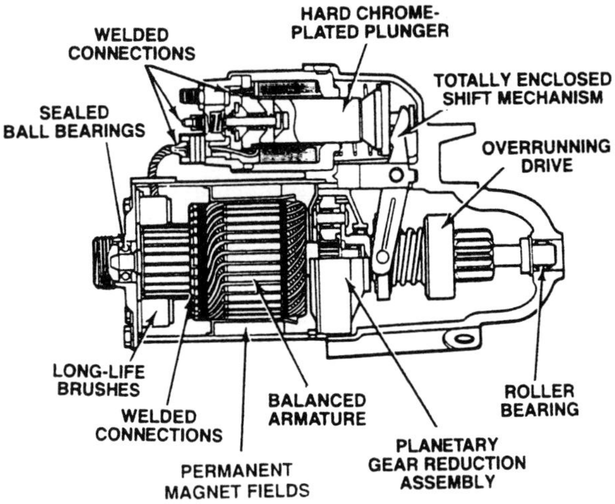


**Figure 10-39.** The PMGR starter motor uses permanent magnets to create a magnetic field instead of the field coils and pole pieces of a conventional starter design. (Delco-Remy)

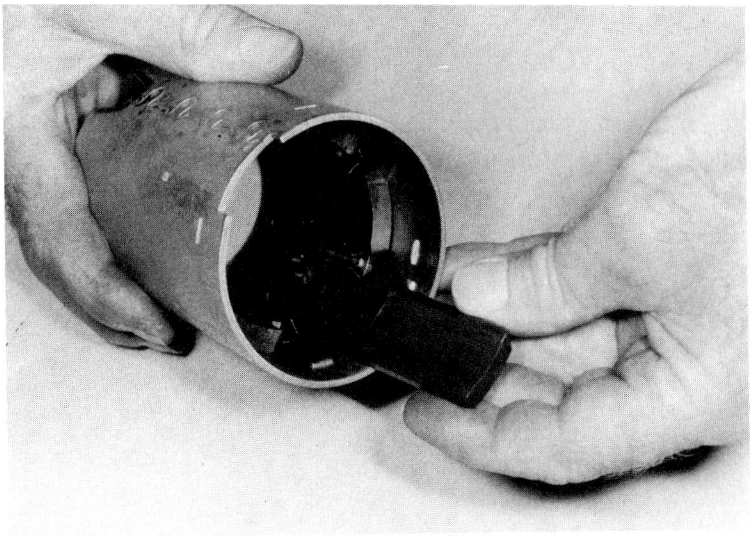

**Figure 10-40.** The magnets used in PMGR starter motors are made of a new alloy of iron and rare-earth materials, but are brittle and require careful handling.

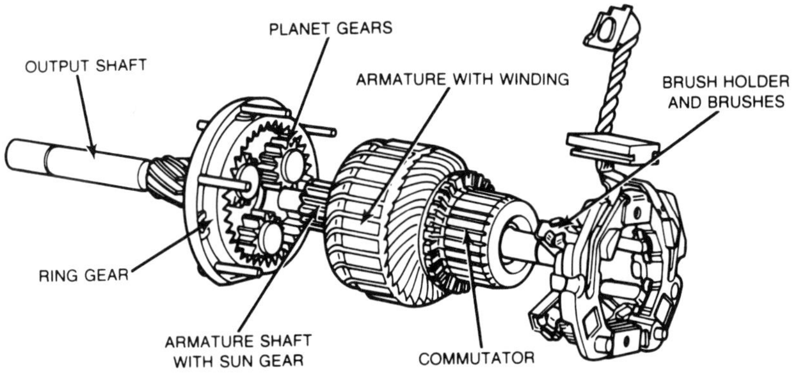


**Figure 10-41.** The Bosch planetary gearset design. (Chrysler)

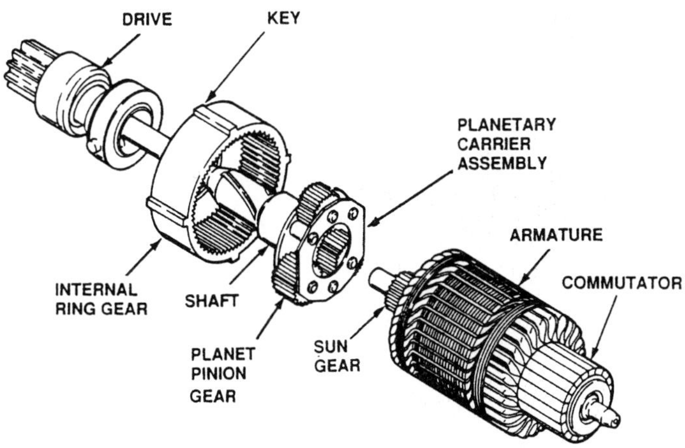


**Figure 10-42.** The Delco-Remy planetary gearset design. (Delco-Remy)

shaft drives the sun gear, which meshes with three planetary pinions. The pinions drive the pinion carrier in reduction during their rotation around the ring gear. Because the starter drive shaft is mounted on the carrier, it is driven at a reduced speed with increased torque.

Although the Bosch and Delco-Remy PMGR starters are mechanically different, their electrical operation is identical to that of a field coil design, figure 10-43. Field circuit testing and service are not required with PMGR starters, but other service such as brush, commutator, and armature testing are essentially the same as for a field coil motor.

Like any other permanent magnet motor, PMGR starters require care in handling. The permanent magnets are quite brittle, and the magnetic field can be destroyed by a sharp impact or by dropping the starter motor.

In 1990, Ford began using a permanent magnet starter on many engines, figure 10-44. This motor is similar in design and operation to the Delco-Remy and Bosch starters. Figure 10-45 is a comparison of the circuits for the Ford movable pole shoe starter and the Ford PMGR starter.

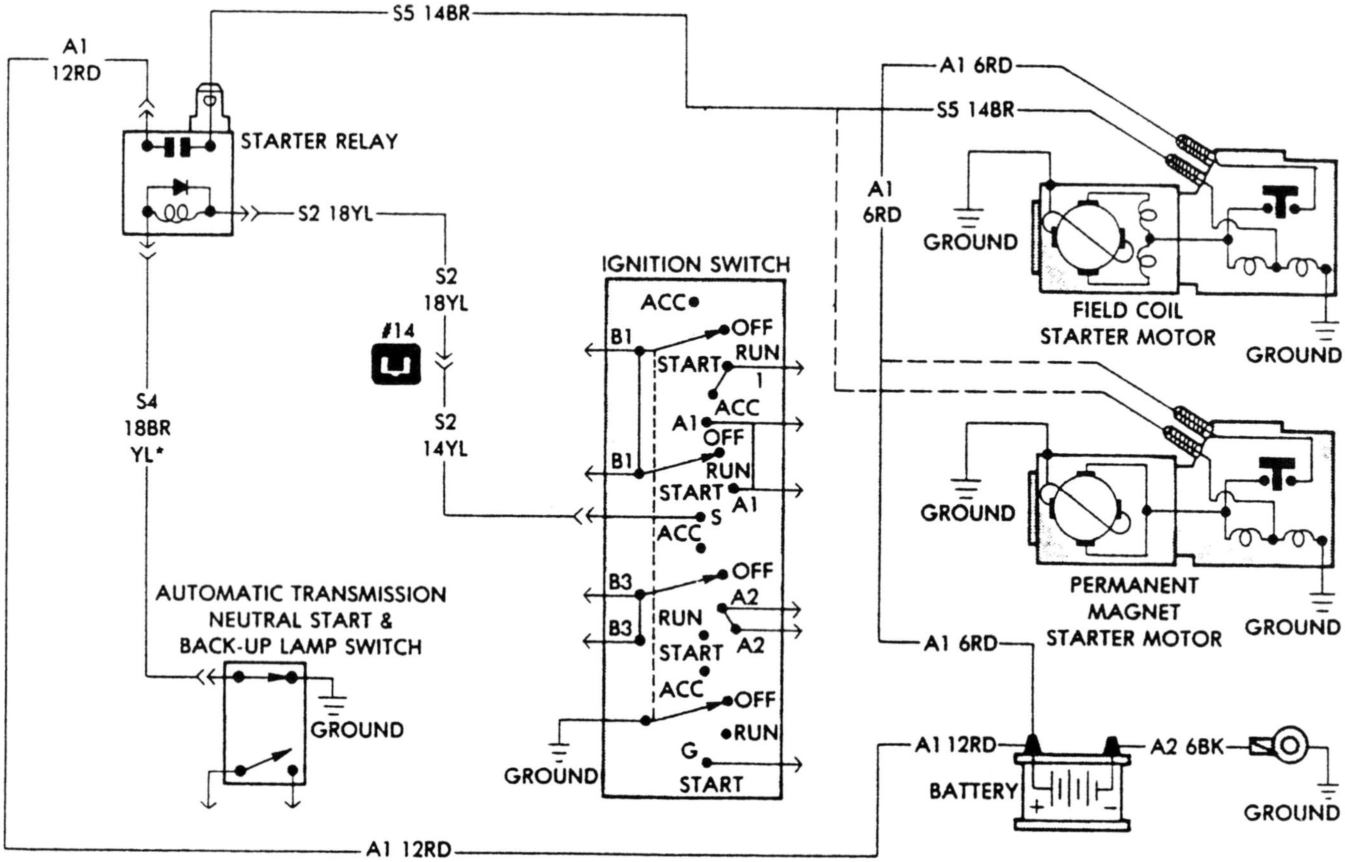


**Figure 10-43.** The electrical circuitry of PMGR and conventional starter motors is the same. (Chrysler)

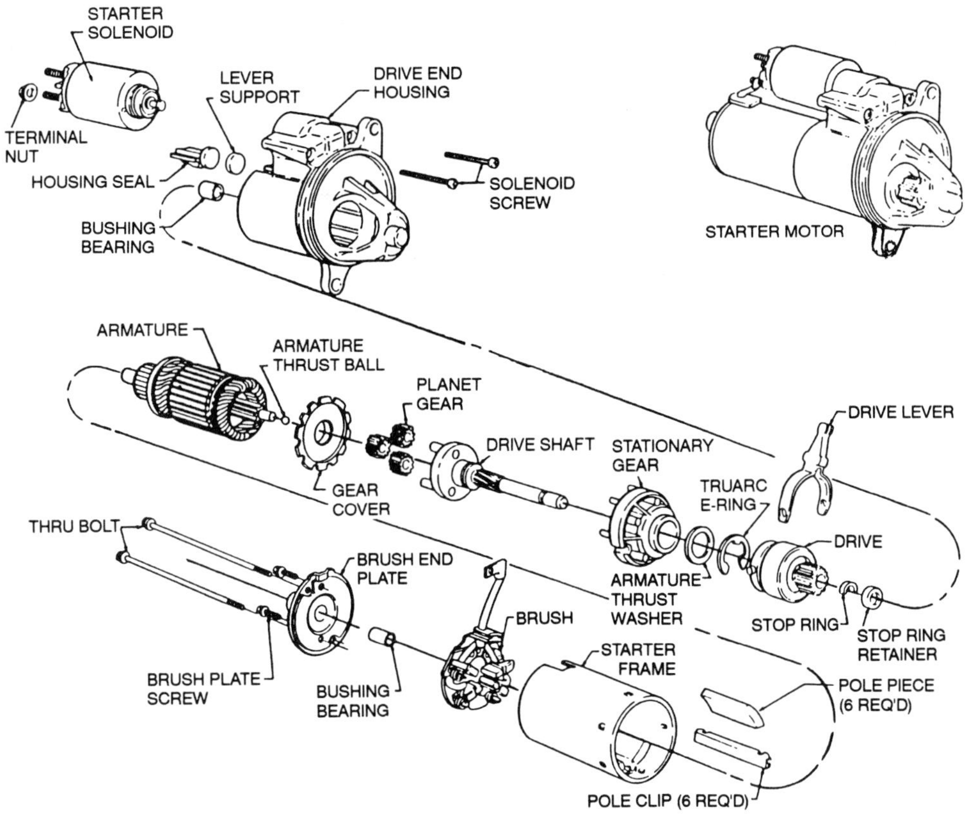


**Figure 10-44.** The Ford permanent magnet starter is a solenoid-operated unit, similar to the Bosch and Delco-Remy designs. (Ford)

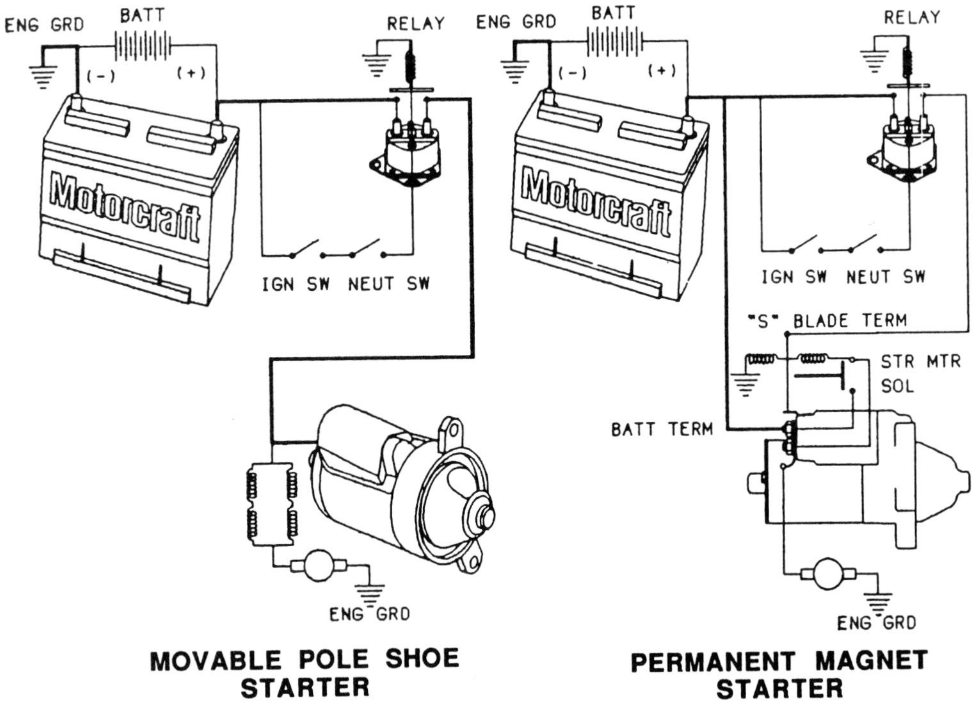


**Figure 10-45.** This diagram compares the installation of the Ford movable pole shoe starter and the PMGR starter. (Ford)

## SUMMARY

The starting system consists of two related circuits: the control circuit and the motor circuit. The control circuit contains the ignition switch, the starting safety switch, and the control side (coil) of the starter relay or solenoid. The motor circuit is a high-current circuit that conducts several hundred amperes of current to the high-torque starter motor for a few seconds. Starter control circuits can use either a relay or a solenoid to close the motor circuit. In either case, the relay or solenoid acts as a magnetic switch.

Starter motors, like all dc motors, operate on the electromagnetic action of current flowing through conductors. The electromagnetic interaction of current in the armature and the field windings causes the motor armature to rotate. Starter motor armatures and fields draw current from a common source—the battery. The field and armature windings can be connected in series or in a compound (series-parallel) arrangement.

Almost all late-model starters use a positive-engagement drive mechanism to engage the motor pinion with the flywheel ring gear. There are three common kinds of starter drives:

1. Solenoid-operated, direct drive
2. Solenoid-operated, reduction drive
3. Movable pole shoe
4. Permanent magnet gear reduction (PMGR)

All of these starter drives use a 1-way (overrunning) clutch to release the starter drive pinion from the flywheel when the engine starts.

Hard-starting complaints are common engine performance problems. The cause of an engine starting problem may be in the parts of the starting system, or the problem may be due to the effects of the starting circuits on other vehicle systems. Because of the high current draw of the starter motor, starting system problems also can affect battery performance and ignition operation.

## REVIEW QUESTIONS

### Multiple Choice

**1.** Automobile batteries are designed primarily to provide high current to:

- **a.** the distributor
- **b.** the ignition switch
- **c.** the starter
- **d.** the voltage regulator

**2.** The starting system consists of two related electrical circuits. They are:

- **a.** the motor circuit and the ignition circuit
- **b.** the motor circuit and the cranking circuit
- **c.** the control circuit and the ignition circuit
- **d.** the control circuit and the motor circuit

**3.** Cranking speed for most engines is approximately:

- **a.** 200 RPM
- **b.** 100 RPM
- **c.** 600 RPM
- **d.** 700 RPM

**4.** Battery cables are made of heavy guage wire because:

- **a.** they must withstand engine heat
- **b.** they must carry high electrical current
- **c.** they must provide a good electrical contact
- **d.** smaller gauge wire breaks too easily

**5.** The ignition switch usually receives battery voltage from a terminal on the:

- **a.** relay or solenoid
- **b.** voltage regulator
- **c.** distributor
- **d.** starter safety switch

**6.** The starter control circuit contains:

- **a.** the ignition switch
- **b.** the starting safety switch and the control side of a relay or solenoid
- **c.** the battery
- **d.** all of the above

**7.** The ignition switch start position is:

- **a.** a momentary contact position
- **b.** a detented position
- **c.** a spring-loaded switch position
- **d.** none of the above

**8.** Ballast resistors are used to:

- **a.** increase the voltage to the ignition after the engine has started
- **b.** decrease the voltage to the ignition after the engine has started
- **c.** connect the battery to a starter solenoid
- **d.** reduce the voltage supplied to the starting circuit

**9.** A starting safety switch:

- **a.** controls transmission operation
- **b.** is a neutral start switch
- **c.** is a normally open switch
- **d.** both b and c

**10.** A solenoid uses current through an electromagnetic field to move an armature or a movable iron core. In a starter control circuit it can be used to:

- **a.** cause the starter drive to engage the ring gear
- **b.** control safety bypass voltage
- **c.** close the contacts in a starter motor circuit
- **d.** all of the above

**11.** A basic starter control circuit energizes the solenoid windings through the ignition switch and the:

- **a.** solenoid core
- **b.** starting safety switch
- **c.** electromagnetic field
- **d.** pinion gear

**12.** Starter solenoids have separate:

- **a.** pull-in windings and pull-out windings
- **b.** pull-out windings and hold-in windings
- **c.** pull-in windings and hold-in windings
- **d.** pull-out windings and hold-out windings

**13.** A motor armature has many conductors designed to react with stationary magnetic poles. This interaction causes the armature shaft to:

a. remain stationary
b. move from one direction to another
c. turn in 90 degree increments
d. rotate

## Fill in the Blank

14. The starter motor field frame is an iron housing that holds the ________ and the ________.

15. Insulation between the laminations of the armature core reduces ________ in the armature.

16. CEMF means ________.

17. The ________ is a series of small copper bars mounted on the armature shaft.

18. ________ are used to conduct current to the commutator segments and the armature windings.

19. A series motor develops maximum ________ at startup speed.

20. ________ motors are ideal for use with power windows and power seats.

21. Most automobile starter motors are ________ motors.

22. Starter motors are mostly of the ________ type.

23. General Motors makes heavy use of the ________ starter.

24. Chrysler has been the major user of the ________ starter.

25. Ford Motor company is the only maker of the ________ starter.

26. All starter types use a ________ to release the starter drive pinion from the flywheel ring gear.

27. ________ is one the most common engine performance problems.

28. The starter control circuit contains the ________, the ________, and a ________.

29. All electric motors operate on the principle of ________.

30. Armature windings can be connected to the commutator by ________ or ________.

31. Series motors are ideal for engine starting applications because they develop ________ at startup speed.

# 11 CHARGING SYSTEMS

## WHAT

The battery is the source of electric power for all circuits on an automobile. But, a battery can't supply the vehicle's voltage and current needs indefinitely. As it delivers current, it cycles from a charged to a discharged condition. Eventually, it would discharge completely if the vehicle did not have a charging system.

The alternator is the principal part of the charging system. It generates current through electromagnetic induction and delivers current to the battery at about 12.6 to 14.5 volts. Whenever the alternator is operating, it also delivers current and applies charging voltage to the rest of the electrical system.

The entire electrical system depends on correct alternator operation. This chapter describes the construction and operation of alternators and other charging system parts.

## WHY

The battery and the alternator rely on each other for proper operation. Obviously, the alternator must maintain the battery state of charge. The alternator can't operate, however, without initial field current from the battery. The charging system also affects other vehicle circuits. The alternator must deliver enough output current and steady system voltage for lights, air conditioning, power accessories, ignition, and engine control systems. The charging voltage must be held within regulated limits. If charging voltage is lower than battery voltage, the alternator can't charge the battery or deliver adequate current to other circuits. If charging voltage is too high, it can damage the battery, ignition parts, and solid-state electronic components. These are a few basic reasons why charging system testing and service are part of total engine performance service.

## GOALS

After studying this chapter, you should be able to:

**1.** Identify the major parts of a charging system and their relationships: the battery, the alternator, the regulator, and indicators.

**2.** Explain current generation and rectification in an alternator. Describe the functions of the rotor, stator, brushes, sliprings, and diodes.

**3.** Explain the operation of solid-state and electromechanical voltage regulators. Explain field relay operation.

**4.** Explain the operation of vehicle ammeters, voltmeters, and indicator lamps.

## CHARGING VOLTAGE

Remember that each battery cell produces about 2.1 volts. The open-circuit voltage of a 12-volt, 6-cell battery is about 12.6 volts. Even though the vehicle electrical system is nominally a 12-volt system, the charging system must produce more than 12 volts. If the alternator were regulated at 12 volts, it could not charge the battery until system voltage fell below 12 volts. This would not allow an adequate voltage reserve for the rest of the electrical system. Charging voltage also must be above 12 volts to overcome the battery's internal resistance and CEMF.

Most charging systems are regulated to develop about 14.5 volts, maximum. This voltage will deliver adequate charging current to the battery as well as current to other electrical loads. Some systems may be regulated as low as 13.5 volts. Lower voltage slightly reduces heat from high current and may prolong the life of lamp bulbs and other parts. Some systems are regulated at 15 to 15.5 volts maximum. Higher voltage ensures adequate current under all operating conditions. Expensive electronic parts of some vehicles, however, can be damaged by excess current if voltage exceeds 15.5 or 16 volts.

If charging voltage is below the carmaker's specifications, the battery gradually will become sulfated. If charging voltage is above specifications, the battery may overheat with excessive gassing. In either case, battery life will be shortened.

All of these factors are reasons why the carmaker's maximum charging system voltage specification is important for engine performance service.

## CHARGING SYSTEM CIRCUITS

The charging system, figure 11-1, consists of:

- The battery, which is charged by the alternator and which provides initial field current to the alternator
- The alternator, which is turned by an engine drive belt and which is the charging voltage and current source
- The regulator, which limits maximum charging voltage
- An ammeter, a voltmeter, or an indicator lamp, which indicates charging system operation

Like the ignition, starting, and other electrical systems, the charging system has more than one major circuit. The two circuits of the charging system, figure 11-2 are:

- The output circuit, which delivers voltage and current to the battery and other electrical loads
- The field circuit, which delivers current to the alternator field.

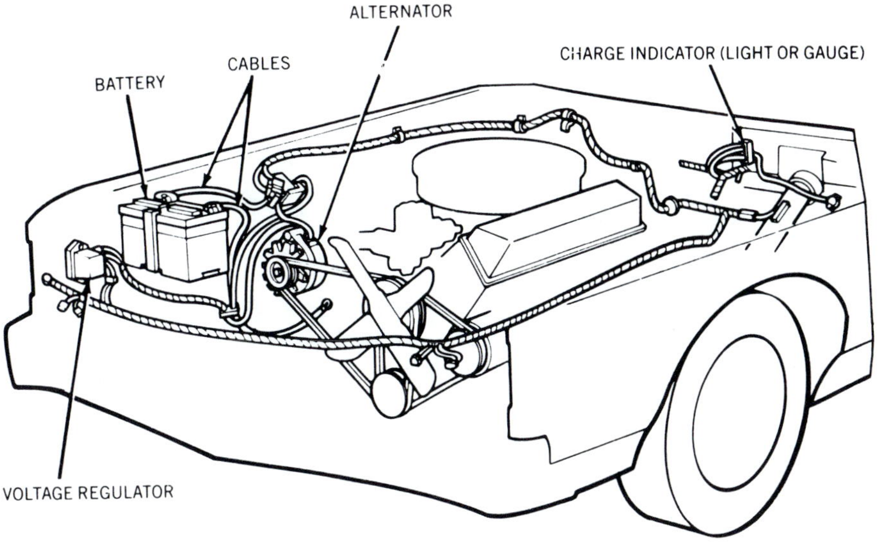


**Figure 11-1.** The charging system contains these major parts. (Chrysler)

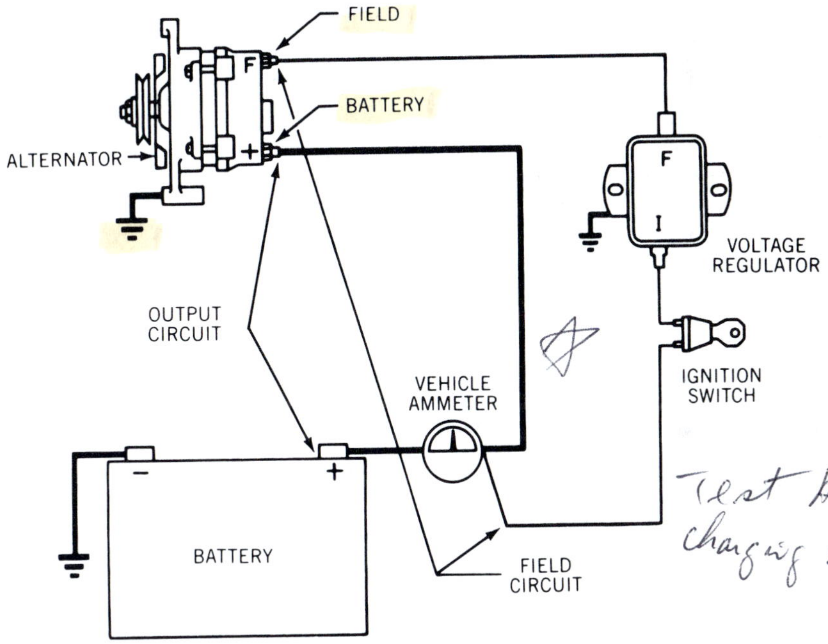


**Figure 11-2.** The charging system contains the output circuit and the field circuit. (Prestolite)

## ALTERNATORS

Alternators generate current and voltage by the principles of electromagnetic induction introduced in chapter 7. One way to induce voltage and generate current is to rotate a magnet inside a stationary looped conductor, figure 11-3. That is an alternator.

As the magnet rotates, its field induces a varying voltage and alternating current in the conductor. The amount of current and the voltage polarity (current direction) depend on:

1. The direction of magnetic polarity
2. The strength of the magnetic field
3. The number of conductors
4. The number of magnetic flux lines cut by each conductor
5. The speed of relative motion between the magnetic field and the conductors

### Alternating Current Generation

Figure 11-4 shows the voltage polarity and alternating current induced in the top half of a single looped conductor during one rotor revolution. Look at these points in the drawing:

**A.** The conductor is parallel with the magnetic field. The conductor is not cut by any flux lines. Voltage and current are at zero.

**B.** The rotor has turned 90 degrees, and the magnetic field is at a right angle to the conductor. The maximum number of flux lines cut the conductor at the north pole. Voltage and current are at maximum positive values.

**C.** The rotor has turned another 90 degrees, and the field is again parallel with the conductor. No flux lines cut the conductor, and voltage and current return to zero.

**D.** The rotor has turned another 90 degrees, but the magnetic field has reversed at the top conductor. The maximum number of flux lines cut the conductor at the south pole. Voltage and current increase to maximum negative values.

**E.** Voltage and current return to zero as the rotor completes one revolution and returns to its starting point.

One rotor revolution produces one ac cycle in a single conductor. The changing voltage polarity between positive and negative is called **sine wave voltage**, figure 11-5. The angle between the magnetic field and the conductor determines the sine wave shape. This is based on the trigonom-

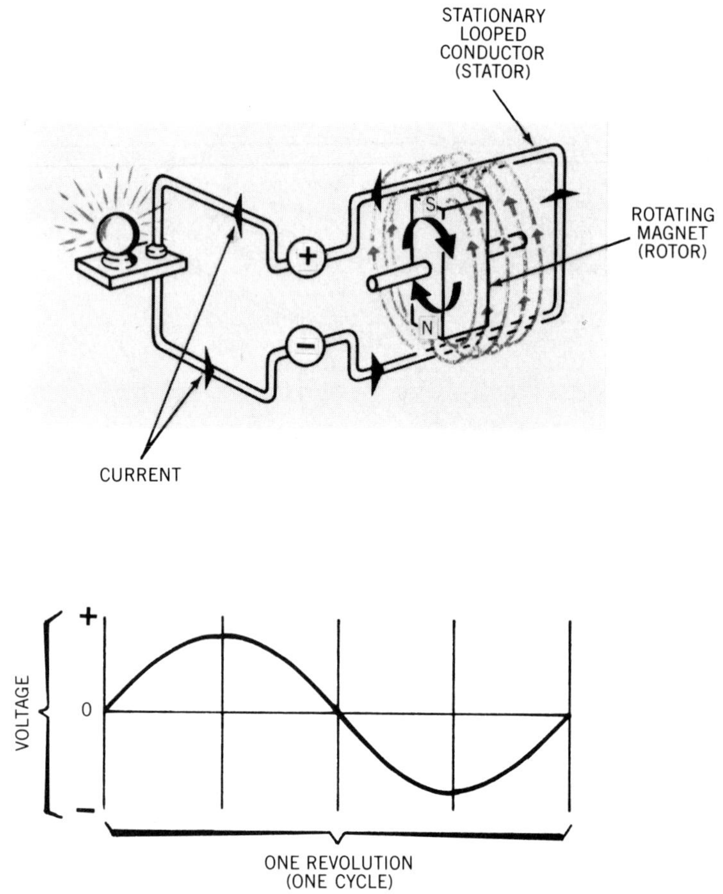


**Figure 11-3.** The field of a rotating magnet induces an ac voltage in a stationary conductor. (Delco-Remy)

etry sine function of angles. The sine wave voltage induced in one conductor during one rotor revolution is **single-phase voltage**.

We could build a simple alternator as described here, using a bar magnet and a length of wire. It wouldn't have any practical use in a charging system, however.

## Current Rectification

Automobile electrical devices are dc devices because the battery is a dc voltage source. Automobile devices require a steady supply of direct current at a relatively steady voltage of constant polarity. Most importantly, you can't charge a battery with alternating current. The ac output of an alternator must be rectified to direct current. There are several ways to **rectify** ac to dc, but the only practical way to do it in an alternator is with diodes.

Chapter 8 introduced a diode as a semiconductor electrical check valve, figure 11-6. It allows current flow in one direction but blocks it in the other. If we put a diode in the alternator output circuit, figure 11-7, it allows current to flow from A (+) to B (−) but not from B to A. This is **half-wave rectification**. A single diode can rectify current with positive or negative potential, but not both. A single diode produces intermittent pulsating dc. This is not useful to charge a battery or power other dc devices.

If we add more diodes to the circuit, we can rectify the negative half of the ac sine wave voltage, figure 11-8. Here, diodes 2 and 3 allow current to flow in one direction (positive polarity) from A, through the outside circuit, and back to B. When alternator polarity reverses, diodes 1 and 4 allow current to flow from B, through the circuit, and back to A. Polarity remains positive in the outside circuit because of diode action. The result is pulsating, full-wave direct current. Current fluctuates from zero to maximum positive and back to zero. This is steadier than half-wave rectified current but still not good enough for an automobile electrical system.

To get the steady direct current needed by an automobile dc system we must add more conductors to the alternator. Alternators have three output conductors, coiled in multiple loops, figure 11-9. The conductors are placed at different angles to the rotating field so that the ac voltage cycles overlap. These are the **phase angles** of the conductors. The result is **three-phase voltage**. Adding additional diodes, figure 11-9, allows us to rectify 3-phase voltage into steadier dc output.

A final requirement for an automobile alternator is to replace the rotating permanent magnet with an electromagnet. Alternator induced voltage varies with field strength and rotating speed. Because the alternator is driven by the engine, its rotating speed changes greatly. If the alternator field were a permanent magnet, induced voltage could not be controlled within the limits needed by the electrical system. Using an electromagnet for the field allows us to regulate field current and the strength of the field. This produces voltage within controlled limits and generates current for the load requirements of the electrical system. Also, an electromagnet can produce a stronger magnetic field, when needed, than a comparably sized permanent magnet.

We will return to alternator current generation and voltage regulation later in this chapter. To understand alternator operation completely, however, we should look at alternator construction.

## Alternator Construction

All alternators have these major parts, figure 11-10:

- A housing that supports the stationary output conductor windings and the rotating field

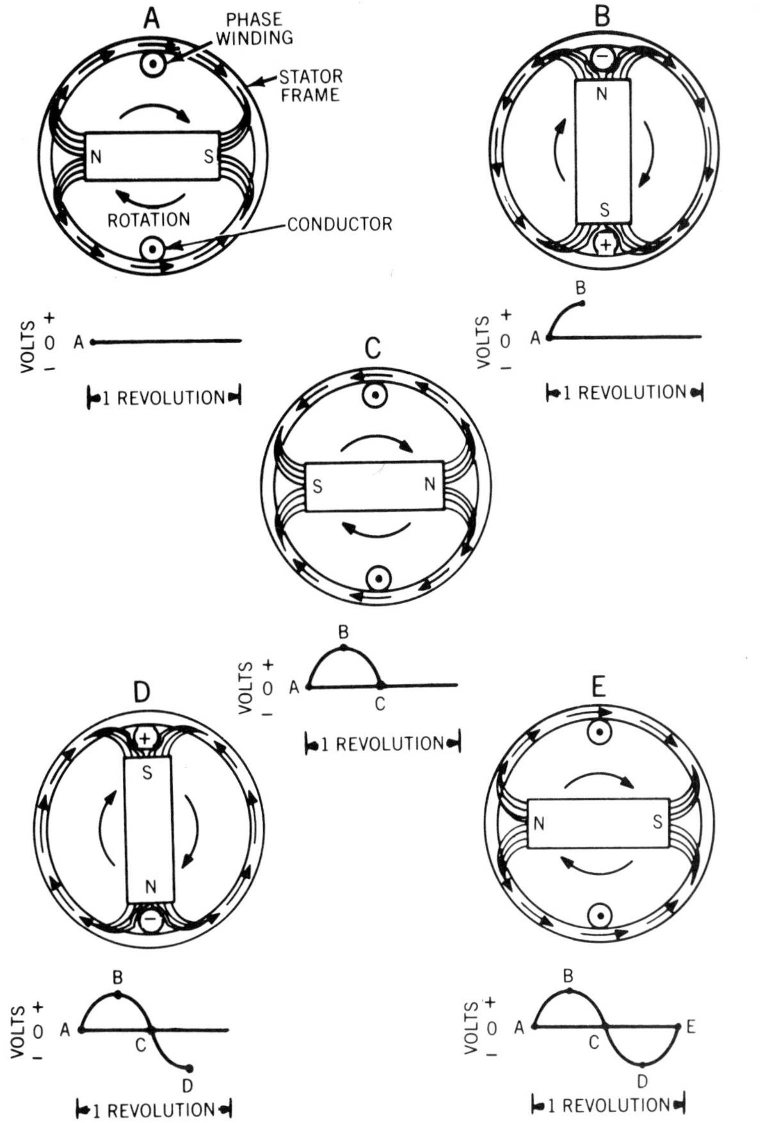


**Figure 11-4.** Positive and negative voltage induced in the upper half of one conductor during one rotor revolution. (Prestolite)

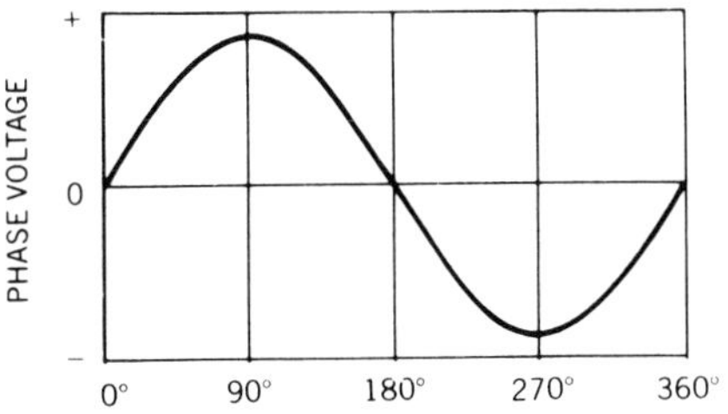


**Figure 11-5.** Sine wave voltage, + to − generated during one rotor revolution. (Bosch)

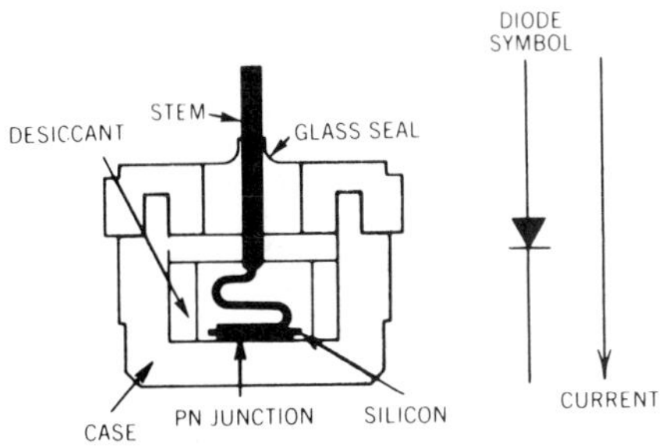


**Figure 11-6.** A diode is an electrical check valve. (Delco-Remy)

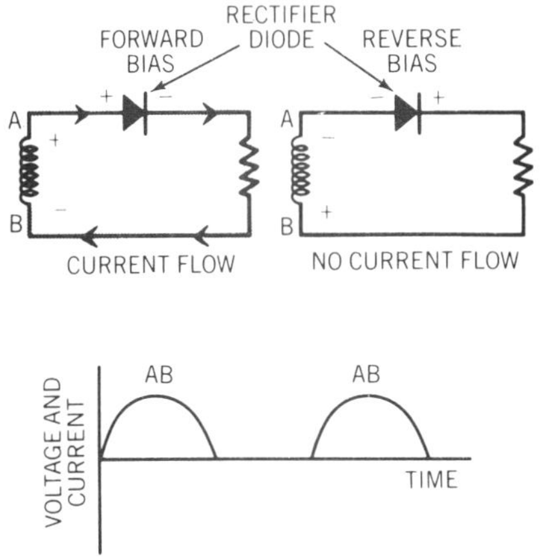


**Figure 11-7.** One diode provides half-wave rectification from one output winding. (Delco-Remy)

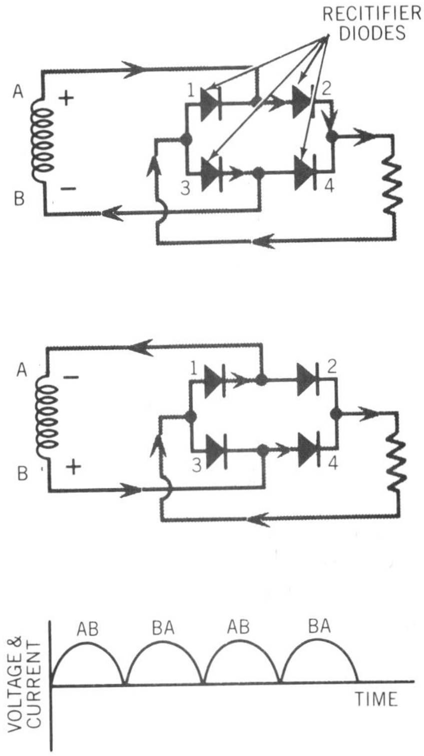


**Figure 11-8.** Four diodes provide full-wave rectification from one output winding. (Delco-Remy)

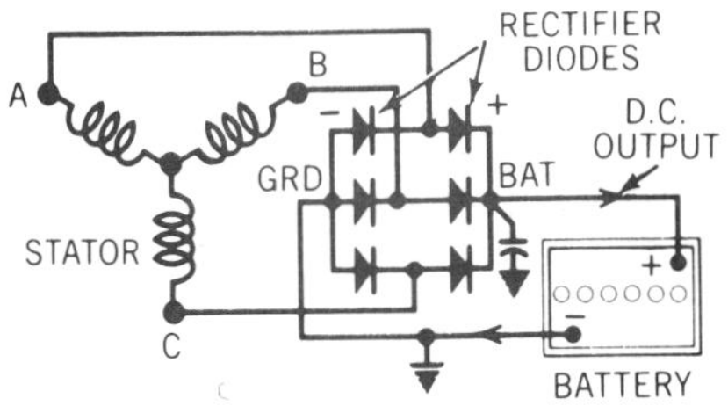


**Figure 11-9.** Six diodes provide full-wave rectification from three output windings (3-phase rectification). (Delco-Remy)

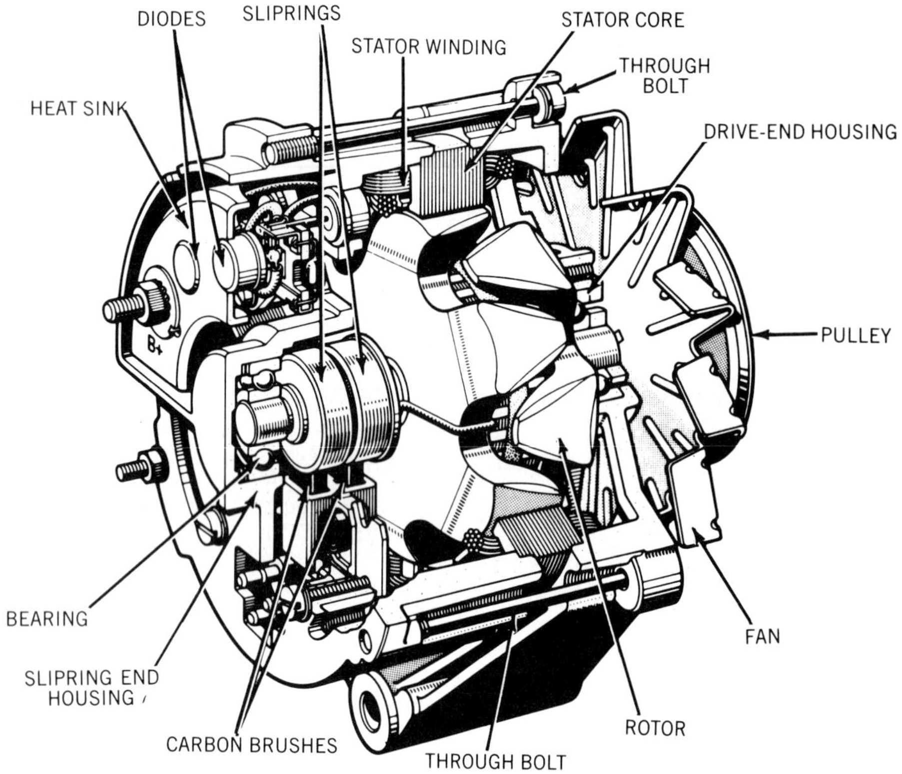


**Figure 11-10.** Typical automobile alternator construction. (Bosch)

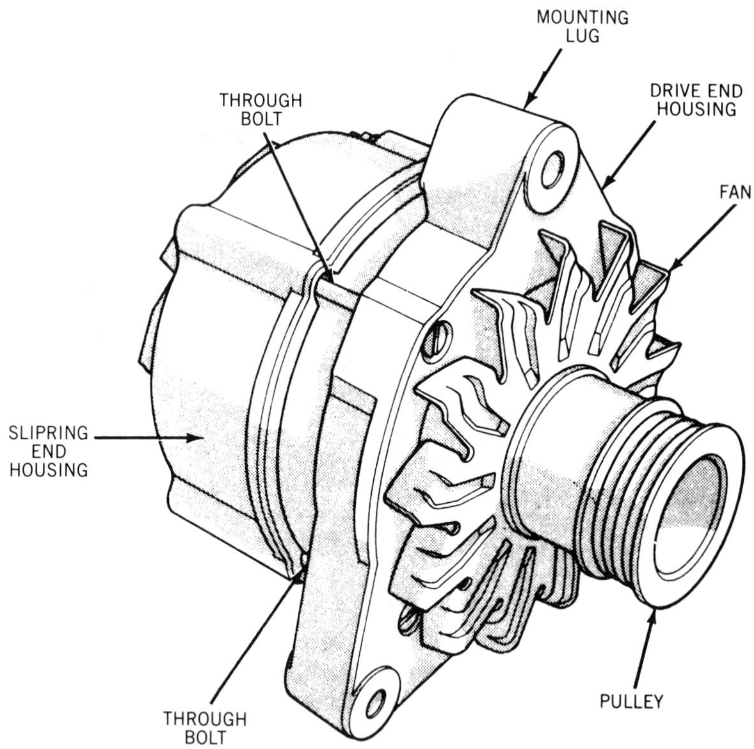


**Figure 11-11.** Automobile alternators have 2-piece housings. (Chrysler)

- A rotor, which is the rotating magnetic field
- Sliprings and brushes that conduct field current to the rotor
- A stator, which contains the output windings
- Rectifying diodes or 1-piece rectifier bridges to change ac to dc output.

Most late-model alternators have solid-state electronic voltage regulators. Many are mounted on, or inside, the alternator housing. Earlier alternators used electromechanical regulators, similar to those used with older dc generators. These are mounted away from the alternator. We will examine regulator operation later in this chapter.

**The Alternator Housing.** Most automobile alternators have 2-piece housings made of cast aluminum, figure 11-11. Heavy-duty alternators used on large trucks, buses, and industrial equipment often have separate center housings to hold larger stators and rotors.

The drive end housing, or end frame, holds a bearing for the rotor shaft. The shaft extends through the housing and holds the drive pulley and the cooling fan. The fan draws air through openings in the opposite end of the alternator. Cooling air leaves the alternator through openings behind the fan in the drive end housing.

The other housing is the slipring end housing (end frame), or rear end housing. It, too, holds a bearing for the shaft. It contains the brushes that carry field current to the rotor sliprings, and it has all electrical terminals for alternator connections. The slipring end housing also holds the regulator for an alternator with a built-in solid-state regulator.

Three or more through bolts hold the end housings together and provide a rigid mounting for the stator core. On some alternators, the stator core is exposed. On others, it is inside the housing.

The alternator housing is bolted directly to the engine so that it is part of the electrical ground path. Anything that is attached to the housing, and not insulated from it, is grounded.

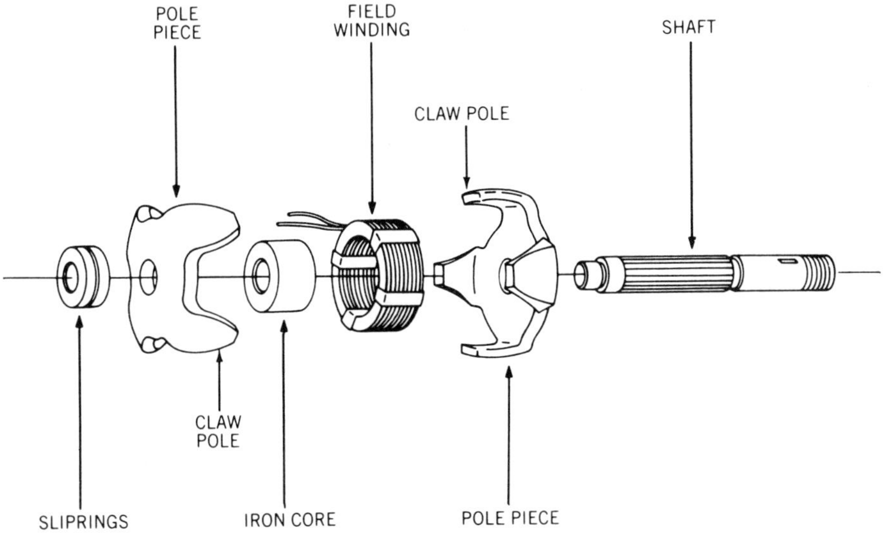


**Figure 11-12.** Typical alternator rotor construction. (Prestolite)

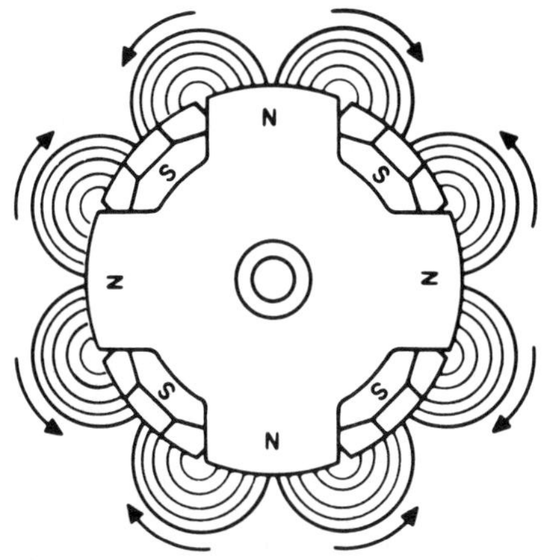


**Figure 11-13.** Rotor field north and south poles alternate. (Prestolite)

Checking ground connections through the housing is an important part of alternator troubleshooting. Some parts must be insulated; others specifically must be grounded.

**The Rotor.** The rotor is the alternator's magnetic field. It contains two magnetic pole pieces, a field winding, an iron core, and a pair of sliprings, figure 11-12. All of these are mounted on the rotor shaft. The opposite pole pieces have fingers, or **claw poles**, that fit together, or interlace with each other. The fingers are the magnetic poles of the field. One pole piece has all north poles; the other has all south poles. When the pole pieces are assembled on the shaft, the poles alternate: N–S, N–S, and so on, figure 11-13. Automobile alternators usually have rotors with 8, 10, 12, or 14 poles. Each of the pole pieces thus has 4, 5, 6, or 7 fingers. Most rotors have 12 or 14 poles.

Magnetic flux travels between adjacent north and south poles. Because of the alternating arrangement of the poles, the flux lines move in opposite directions between adjacent poles. Rotors are built this way so that several alternating magnetic fields intersect the stator conductors as the rotor turns, figure 11-14. The induced voltages alternate as in our example of a 1-conductor alternator. But, the voltages alternate at different, closely spaced phase angles in many conductors. This helps to produce uniform, rectified dc voltage.

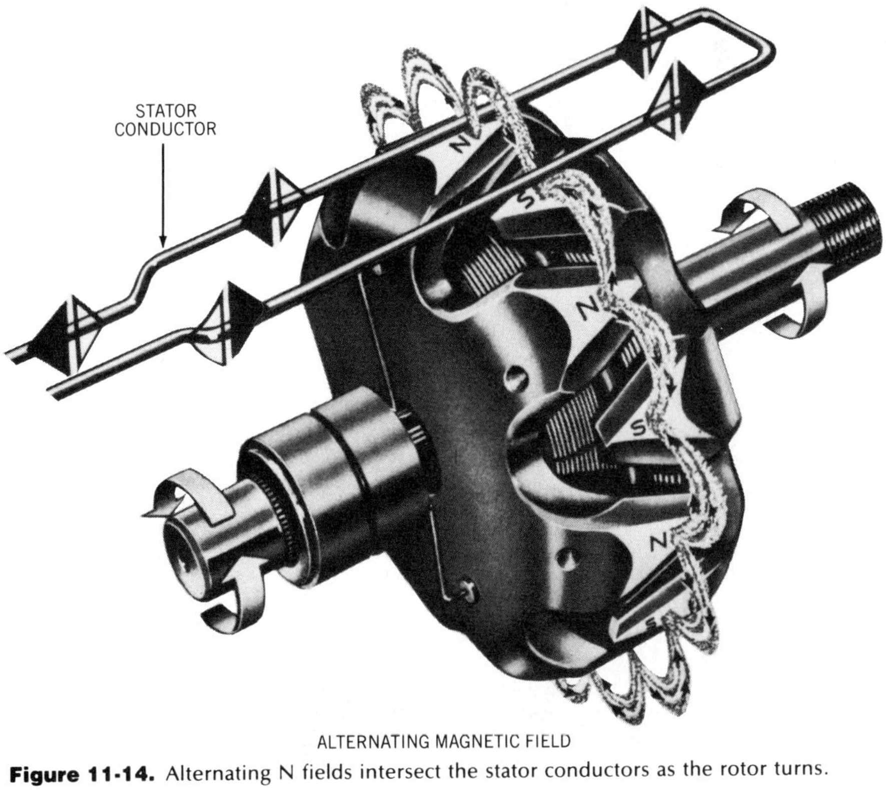


**Figure 11-14.** Alternating N fields intersect the stator conductors as the rotor turns. (Delco-Remy)

The rotor winding carries the current that develops magnetic fields between the rotor poles. The iron core inside the winding concentrates and strengthens the field. One pole piece is attached to either end of this core. When current flows through the winding, it magnetizes the core. Each pole piece develops the magnetic polarity of the end of the core to which it is mounted.

Changing the amount of current through the field winding changes the magnetic strength of the alternator field. When the alternator is not turning, the poles retain a little magnetism, but not enough to induce any voltage in the stator. Therefore, the field winding must receive field current and voltage from

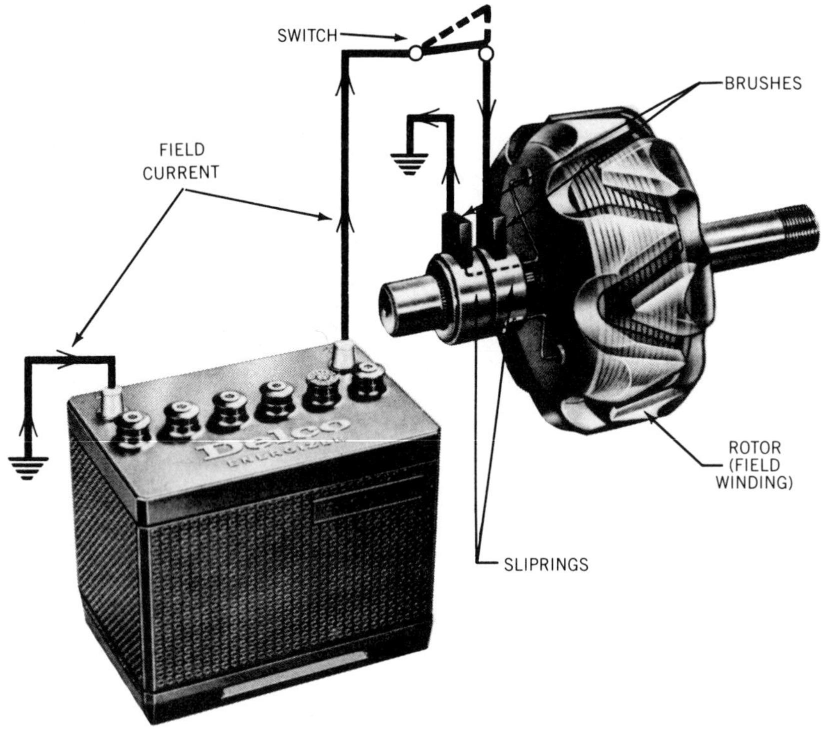


**Figure 11-15.** The field winding receives current from the battery when the alternator starts. (Delco-Remy)

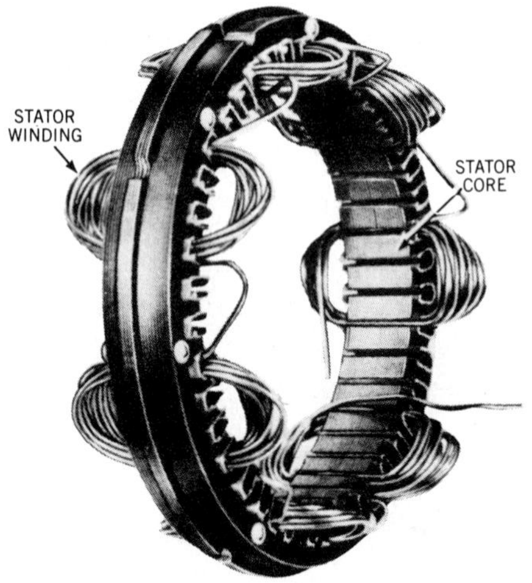


**Figure 11-16.** The coils of one stator winding are 120 degrees apart, electrically. (Delco-Remy)

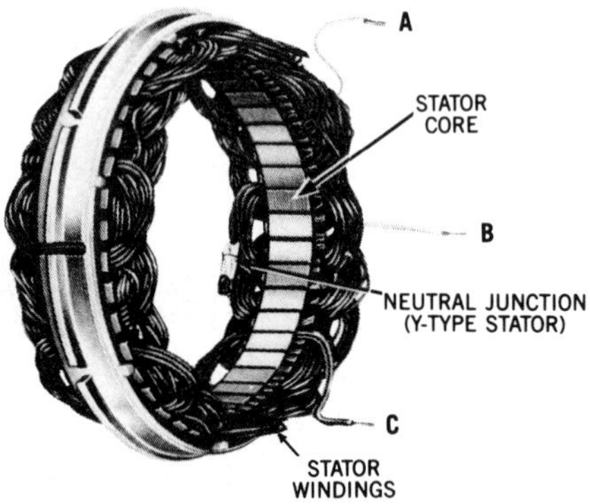


**Figure 11-17.** A complete stator has three overlapping windings. (Delco-Remy)

the battery when the alternator starts to turn, figure 11-15.

Field current often is called **excitation current**, and the field winding may be called the excitation winding. After the alternator starts to generate, field current usually comes from the output connections of the stator. Voltage that delivers field current, from the battery or from the stator output, is **excitation voltage**. We will examine this in more detail later.

**The Sliprings and Brushes.** The field winding receives current through a pair of sliprings on the rotor shaft, figure 11-15. One end of the winding is connected to each of the sliprings. The sliprings are insulated from each other and from the shaft.

The sliprings receive current through a pair of brushes. The insulated brush is connected to the current source—the battery or the stator. The other brush is grounded to the alternator housing or through the regulator to complete the field circuit.

The sliprings are smooth (not segmented like a commutator), and the brushes conduct field current of only 1.5 to 3.0 amperes. Therefore, alternator brushes do not suffer the wear and high-current arcing of motor or generator brushes.

**The Stator.** An automobile alternator has three separate output conductors in mutiple windings on the stator core. The core is made of laminated sections to reduce unwanted eddy currents in the stator. Eddy currents would oppose induced voltage and current.

Each of the three conductors has the same number of coils as the rotor has pairs of north and south poles. For example, an alternator with 7 pairs of N–S poles (a 14-pole rotor) has a stator with 7 coils in each winding. The coils of each conductor are evenly spaced around the core, figure 11-16. The three conductors, or sets of windings, alternate and overlap, figure 11-17, to produce the required phase angles.

The three stator windings are con-

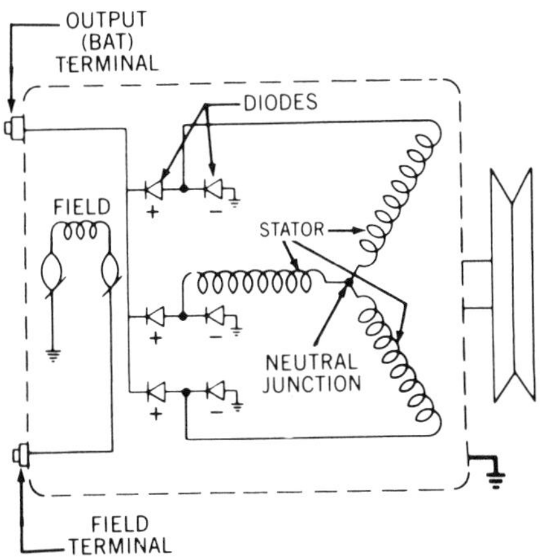


**Figure 11-18.** Y-type stator electrical schematic diagram. (Prestolite)

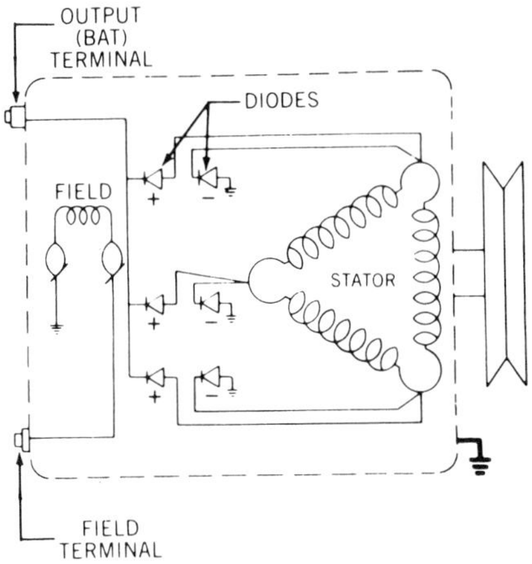


**Figure 11-19.** Delta stator electrical schematic diagram. (Prestolite)

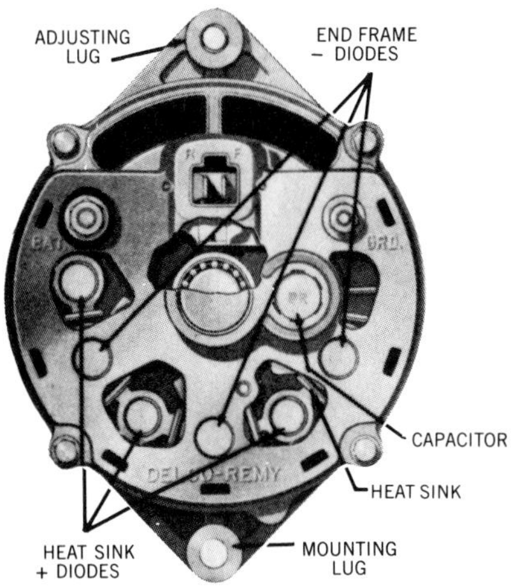


**Figure 11-20.** Early alternators had six diodes in the end housing and a heat sink. (Delco-Remy)

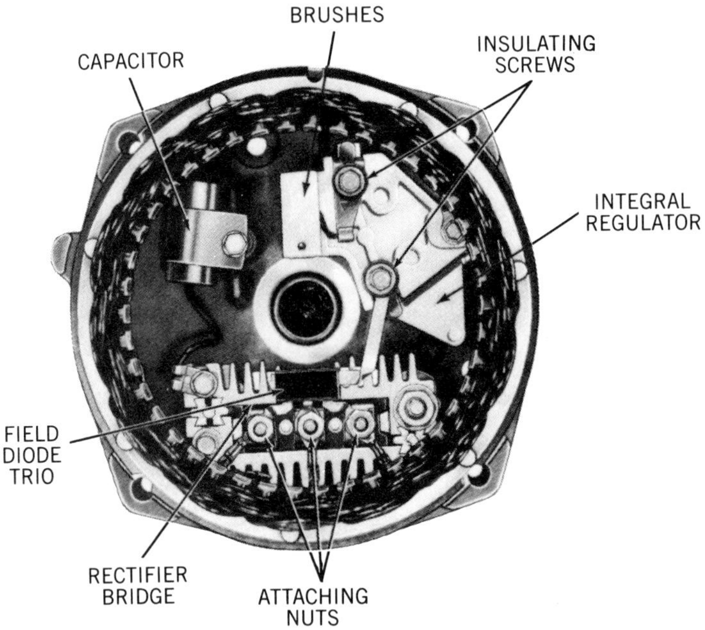


**Figure 11-21.** Most late-model alternators have the diodes in one or two rectifier bridge assemblies. (Delco-Remy)

nected to each other and to the rectifying diodes in one of two ways:

**1.** In a **Y-type stator**, one end of each winding is connected to a **neutral junction**, figure 11-18. The other end of each winding is connected between a positive and a negative diode.

**2.** In a **delta stator**, the three windings are connected end to end, figure 11-19. Each connection point is also connected to a pair of positive and negative diodes. There is no neutral junction. The circuit diagram for this arrangement looks like the Greek letter "delta," or a triangle.

Delta-type and Y-type alternators both produce 3-phase current, which is rectified for dc output. Both kinds of alternators are used in automotive applications, and the stator design affects voltage and current output, as follows:

- Y-type stators produce high voltage at lower alternator speed than delta stators. Most passenger cars have Y-type alternators.
- Delta stators produce higher maximum current than Y-type stators. They are used on large vehicles with high current loads. Most alternators rated at 100 amperes or more have delta stators.

**The Diodes or the Rectifier Bridge.** Three-phase alternators use three pairs of diodes to rectify output current. Figures 11-18 and 11-19 show the basic diode circuits for Y-type and delta-type stators. Each conductor is attached to one positive and one negative diode.

Early alternators had six separate diodes pressed into the slipring end housing and an insulated **heat sink**, figure 11-20. Because of improved semiconductor manufacturing, many late-model alternators have all six diodes in a small **rectifier bridge**, mounted in the end housing, figure 11-21. Either the end housing or the rectifier bridge housing, or both, act as a heat sink to keep the diodes from overheating due to high charging current.

Regardless of the physical construction of the diodes or the rectifier bridge, the circuit connections are basically as shown in figures 11-18 and 11-19. Three positive (+) diodes are insulated from the alternator housing and connected to the output terminal. The terminal is connected to the battery + terminal and the rest of the electrical system. The battery can't discharge through the alternator connection because diode bias blocks discharge current. The positive diodes conduct cur-

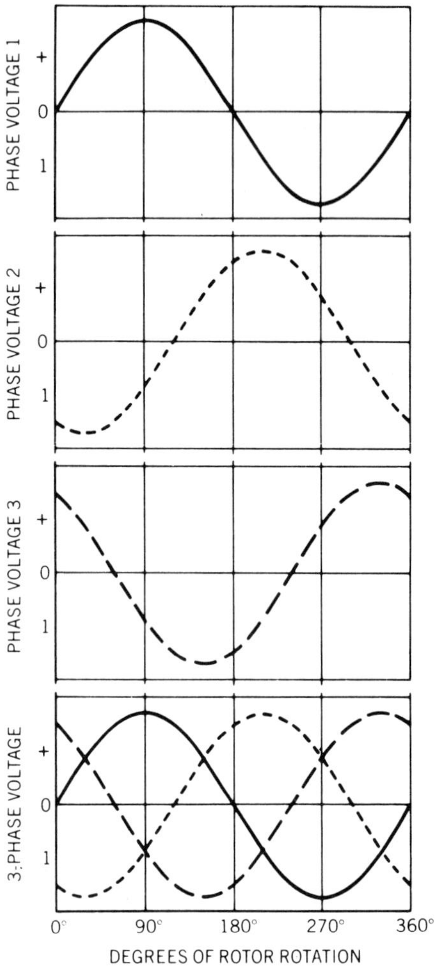


**Figure 11-22.** The sine wave voltages of the three stator windings are 120 degrees apart. (Bosch)

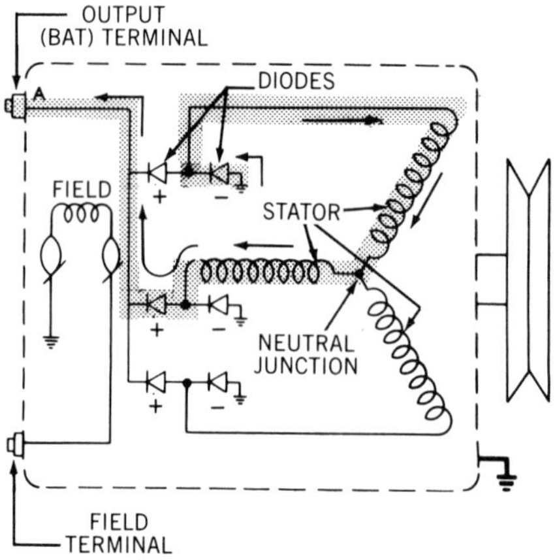


**Figure 11-23.** Current generation in a Y-type stator. (Prestolite)

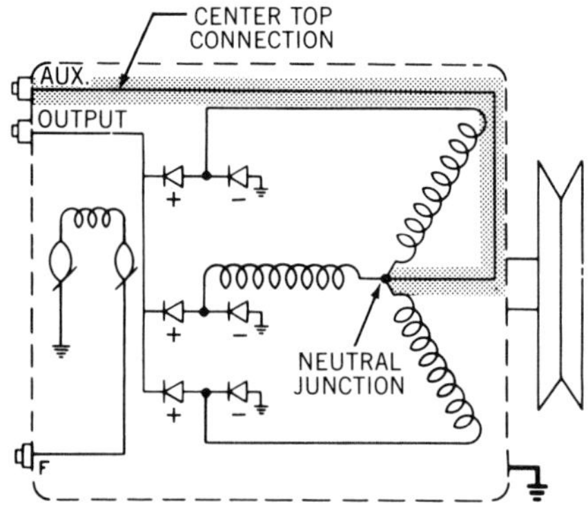


**Figure 11-24.** The neutral junction of a Y-type stator can be connected to a center tap. (Prestolite)

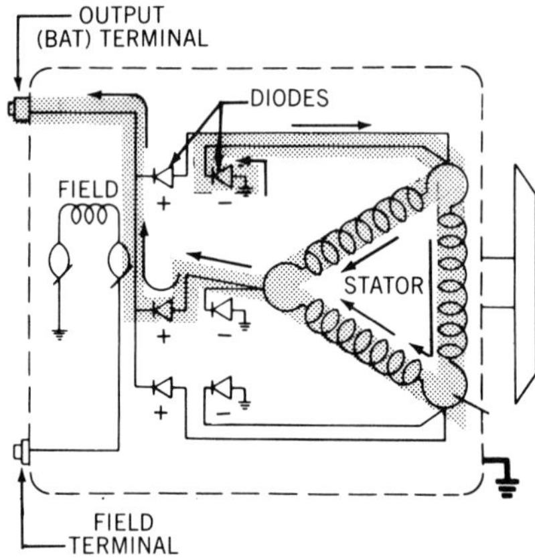


**Figure 11-25.** Current generation in a delta stator. (Prestolite)

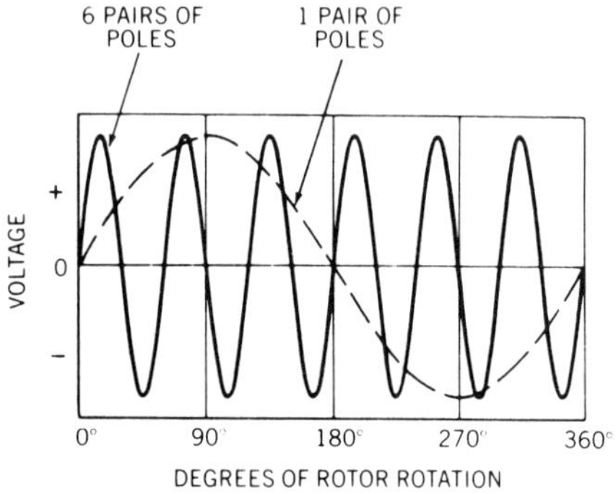


**Figure 11-26.** Sine wave voltage produced in one stator winding by one pair of poles and six pair of poles. (Bosch)

rent only from the alternator to the battery.

Three negative (−) diodes are grounded in one of three ways:

**1.** They may be pressed or threaded into the alternator housing.

**2.** They may be grounded to the housing through the rectifier bridge.

**3.** They may be grounded through the regulator.

The negative diodes are a ground connection for the alternator output circuit. Their bias allows them to conduct current only from the ground side of the alternator into the circuit.

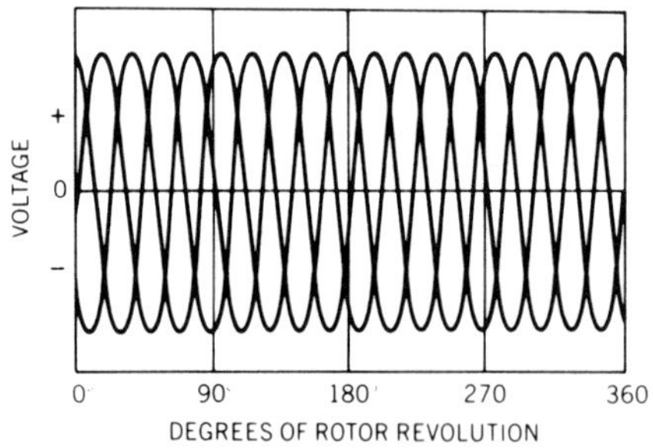


**Figure 11-27.** Overlapping sine wave voltages from three windings and six pairs of poles (12 poles). (Bosch)

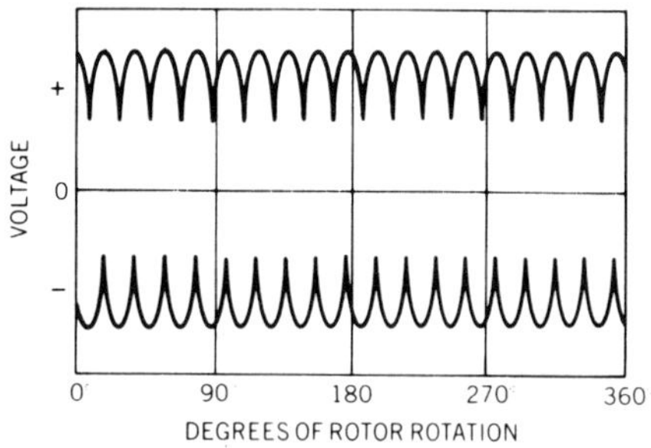


**Figure 11-28.** Rectifier diodes receive maximum positive and negative voltage. (Bosch)

## Three-Phase Current Generation

The coils of the three stator windings are spaced evenly around the stator core. Separate voltages are induced in each winding at different phase angles, figure 11-22. The alternator produces three overlapping, evenly spaced, single-phase sine waves. Because the windings are connected to each other, the result is **three-phase alternating current**. Here is how current is generated and rectified in the two kinds of alternators.

**Y-Type Stator Current.** In a Y-type stator, two windings are always in series between a + and − diode. The position of the rotor at any point determines current direction through the two windings. It also determines the voltage level in each winding. Current flows from ground (−), through the − diode, through the two windings, to the + diode, and to the output terminal, figure 11-23. Induced voltages across the two series windings add together to produce total output voltage. Because the series voltages add together, the Y-type stator can produce high regulated voltage at startup and low engine speeds.

The neutral junction of a Y-type stator is connected to a "center tap" terminal on some alternators, figure 11-24. The center tap voltage is lower than output voltage, but it can be used to control field current, light an indicator lamp, power an electric choke heater, or do other jobs.

**Delta-Type Stator Current.** The windings of a delta stator always form two parallel circuits between a + and a − diode, figure 11-25. Again, the rotor position at any instant determines current direction through the windings and voltage level in each winding. Because there are always two parallel circuits, however, more current can flow in a delta stator than in a Y-type stator.

### Three-Phase Rectification

The 3-phase ac voltage in figure 11-22 is the sine wave voltage of three stator windings, generated by a rotor with only one pole. Real rotors have several pairs of north and south poles. Each *pair* of poles produces one complete sine wave in each winding during each revolution. Figure 11-26 compares the single sine wave produced in one winding by one pair of poles with multiple sine waves produced by multiple poles. During one revolution, a 12-pole rotor (6 pairs of poles) produces 6 sine waves.

Remember that the stator has three windings, equally spaced at overlapping phase angles. The 12-pole rotor generates three overlapping sine wave voltage cycles in the stator. In one revolution, the 3-winding stator output from 6 pairs of rotor poles is 18 sine wave cycles, figure 11-27. The + and − voltage peaks of each cycle force current through the + and − diodes, figure 11-28. Diode rectification changes the − current to + output. The final rectified dc voltage appears at the alternator output terminal as shown in figure 11-29.

A common test of alternator operation is to observe the output voltage on an oscilloscope. The scope pattern of a good alternator is similar to figure 11-29. The small waves at the top are called an **ac ripple**. They are the peaks of the ac voltage sine waves. Any dc voltage from a rectified ac source always has a slight ac ripple. Dc voltage from a dc source, such as a battery, does not. You will learn to recognize the causes of abnormal alternator scope patterns when you study alternator testing.

### The Field Circuit

The magnetic field of the rotor is produced by current from a source outside the rotor. When the engine and the alternator start, the source is the battery, figure 11-15. The field circuit connects the battery to the rotor through the alternator output terminal. When the engine is off, the battery must be disconnected from the field circuit, or it would discharge through the rotor. Some systems use a field relay to do this. Others use diodes to block discharge current. The regulator or the ignition switch also can be connected to open the field circuit when the engine is off.

When the alternator starts, field current comes from the stator output. In some alternators, the field is connected to rectified dc output. In others it is connected to separately rectified voltage from the stator. In this latter design, three more diodes are connected to the three stator windings to provide direct current for the field, figure 11-30. These three diodes may be called "field diodes," "exciter diodes," or a "diode trio" if they are joined in a single assembly. Field current is fully rectified because the three negative output diodes also are in the circuit with the field diodes.

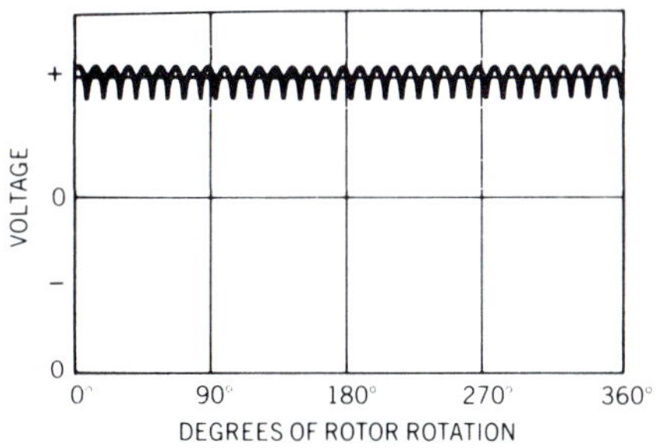


**Figure 11-29.** Rectified alternator output is pulsating dc with a slight ac ripple. (Bosch)

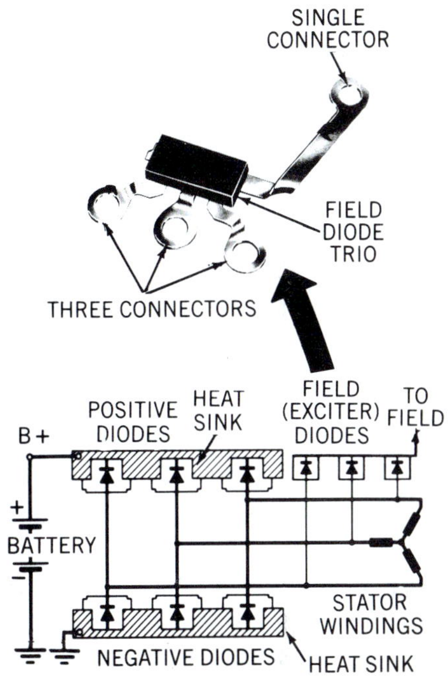


**Figure 11-30.** Many alternators have separate field diodes. (Bosch) (Delco-Remy)

Every alternator field circuit includes the regulator. As an introduction to regulator operation in the next section, we will look at three basic field circuit designs.

**A-Circuit.** An A-circuit alternator has an externally grounded field, figure 11-31. One insulated brush is connected to the output circuit inside the alternator to draw field current. The other brush also is insulated from the alternator housing and connected to the regulator through the alternator field terminal. The regulator is between the field and ground. The field is grounded through the regulator.

**B-Circuit.** A B-circuit alternator has an internally grounded field, figure 11-32. One insulated brush is connected to the output circuit *through the regulator* to draw field current. The other brush is grounded to the alternator housing. The regulator is between the current source and the field. The field is grounded inside the alternator housing.

A good way to remember the difference between these two circuit designs is:

- An A-circuit regulator is located *after* the stator.
- A B-circuit regulator is located *before* the stator.

**Isolated-Field Circuit.** The isolated-field circuit is a variation of the A-circuit. It is most commonly used in some Chrysler alternators. The field is grounded through the regulator, figure 11-33. Field current comes from the output circuit outside the alternator. It enters through a third terminal on the housing. Field current is switched by a field relay or a switch. Both brushes are insulated from the alternator housing.

## ALTERNATOR REGULATORS

An alternator regulator regulates field current and the magnetic strength of the field. This produces output voltage within controlled limits, while allowing the alternator to generate current for the load requirements of the electrical system.

### Current Regulation

Direct current generators used on most cars until the early 1960's needed separate current regulators. Alternators do not. Alternator design limits maximum current output. You learned in chapter 7 that induced voltage and current in a conductor create a countervoltage, or counterelectromotive force (CEMF). The CEMF produces current that opposes the induced current in the con-

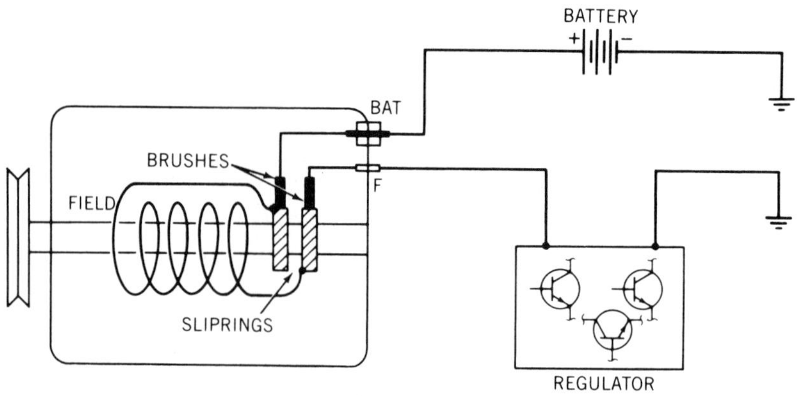


A CIRCUIT

**Figure 11-31.** Typical A-circuit regulator installation.

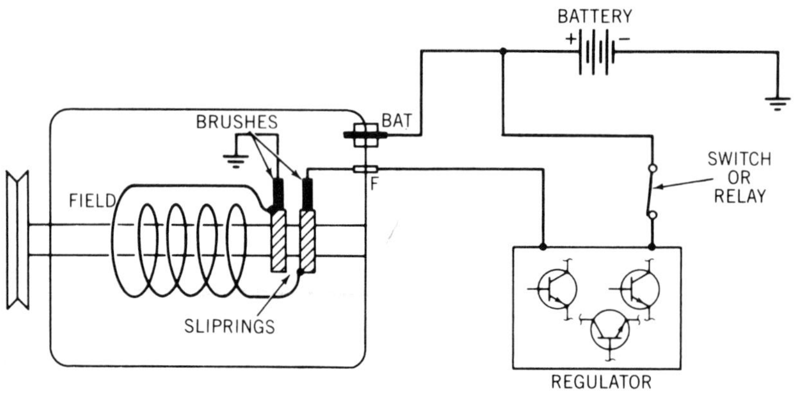


B CIRCUIT

**Figure 11-32.** Typical B-circuit regulator installation.

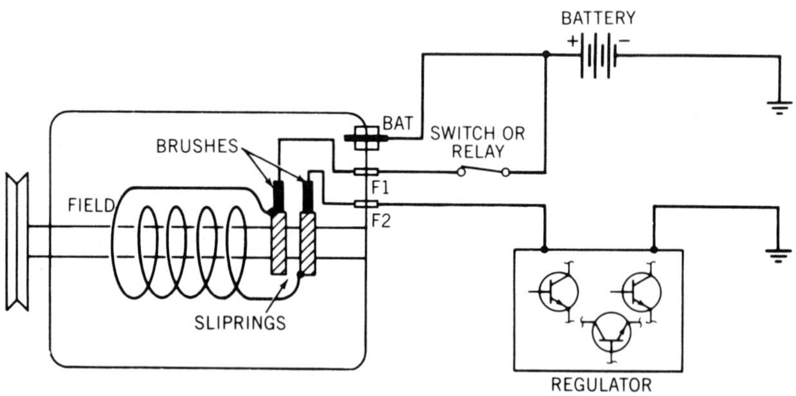


ISOLATED-FIELD CIRCUIT

**Figure 11-33.** Typical isolated-field regulator installation.

ductor. This is what happens in an alternator stator.

As induced current generated by the alternator increases, CEMF increases. At some point, current produced by the CEMF is high enough to block further output current. This point varies for different alternators according to the design of the stator and the rotor. In any alternator, however, the point where CEMF blocks further output current is the design limit of the alternator. It may occur at 30 amperes, 60 amperes, 100 amperes, or any other point.

Although alternator design limits maximum current, output voltage could continue to rise as alternator speed increases. Rotor speed is governed by engine speed and can't be changed to control alternator voltage. Field current can be controlled electrically to regulate output voltage.

### Voltage Regulation

Alternator voltage depends on field current and rotor speed. At low speed, the regulator allows full current to the field, and voltage rises quickly to its regulated level, figure 11-34. As speed increases, the regulator reduces field current to keep output voltage at a regulated maximum, even as rotor speed increases.

Raising the output current load on an alternator will cause output voltage to drop. This happens, for example, when you turn on the headlamps or the air conditioner. In this case, the regulator allows field current to increase, which raises output voltage to its regulated maximum, figure 11-35.

At low speeds, the regulator lets field current flow for relatively long periods and reduces it for short periods, figure 11-36. This produces high average field current. At high speeds, the regulator lets field current flow for shorter periods and reduces it for longer periods. This produces low average field current.

### Regulator Design

Early alternators used electromechanical, vibrating-contact regulators, similar to those used with dc generators.

**FORD'S FAMOUS FLIVVER**

The Ford Model T, symbol of low-cost reliable transportation, chugged into our lives and history over 75 years ago on October 1, 1908. Designed to be simple to operate, easy to repair, inexpensive, and durable, the Model T retired the horse to pasture. The first year's production of 10,660 units broke all industry records.

From the first, Ford conceived of his T as a car for everyone, but it wasn't until 1913 that Ford began building the T chassis on a moving assembly line. From that point on, Ford perfected his mass production methods that eventually brought the price for a basic T as low as $450.00. By 1914, Model T's were rolling out of Ford's Highland Park, Michigan, plant at the rate of one every 40 seconds.

The Model T not only was the first mass-produced affordable automobile, it was the first mass production car to have its engine block and crankcase cast as a single unit. It was the first production car to popularize the steering wheel on the left side and the first to have a removable cylinder head for easy engine service. But, it was a no-frills car. The fuel level was checked with a stick, and standard equipment did *not* include a spare tire, a water pump, an oil pump, a horn, windshield wipers, a radio, rear-view mirrors, a trunk, or bumpers. Its simplicity challenged inventors who developed more than 5,000 separate accessories for the Model T during its 19-year production run.

The official farewell for the Model T came on May 26, 1927 with the production of the 15-millionth car at Highland Park. That Ford plant then was retooled for the Model A, but many more T's were built during the summer of 1927 at other Ford factories. Official production records say that Ford built 15,458,781 Model T's from 1908 through 1927.

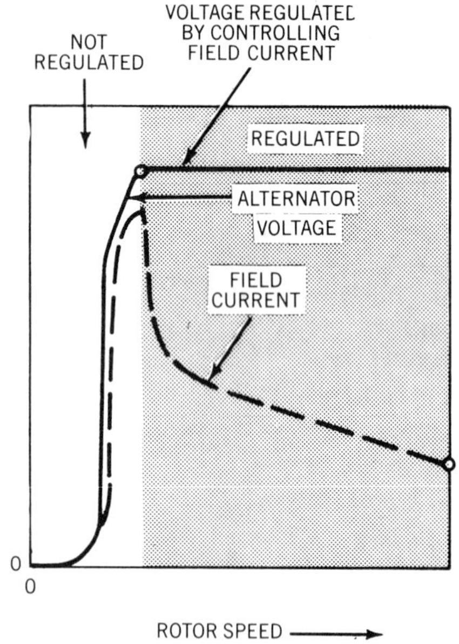


**Figure 11-34.** The regulator reduces field current as rotor speed increases to maintain a regulated voltage. (Bosch)

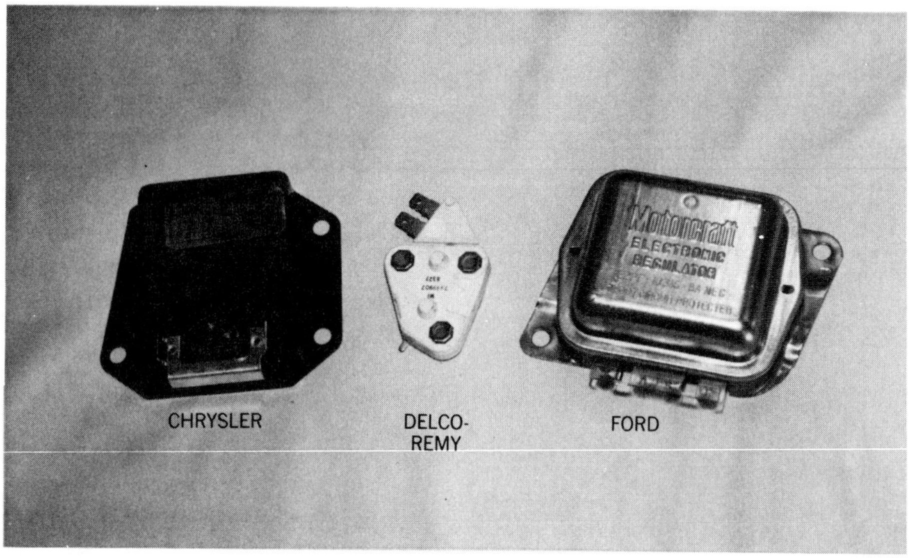


**Figure 11-37.** Solid-state regulators can be built into the alternator or mounted remotely.

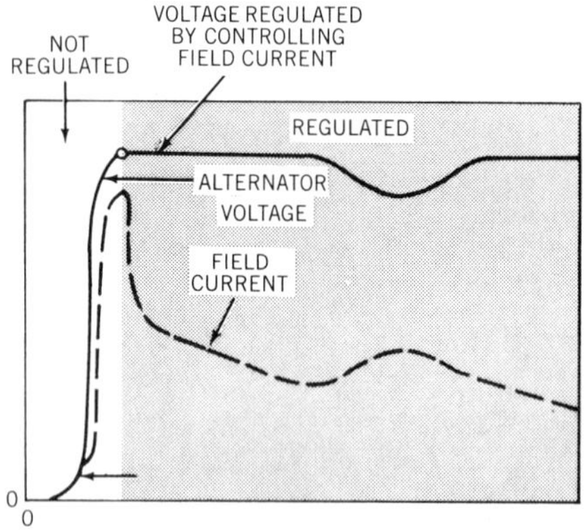


**Figure 11-35.** Increasing vehicle current loads cause field current to increase to maintain regulated voltage. (Bosch)

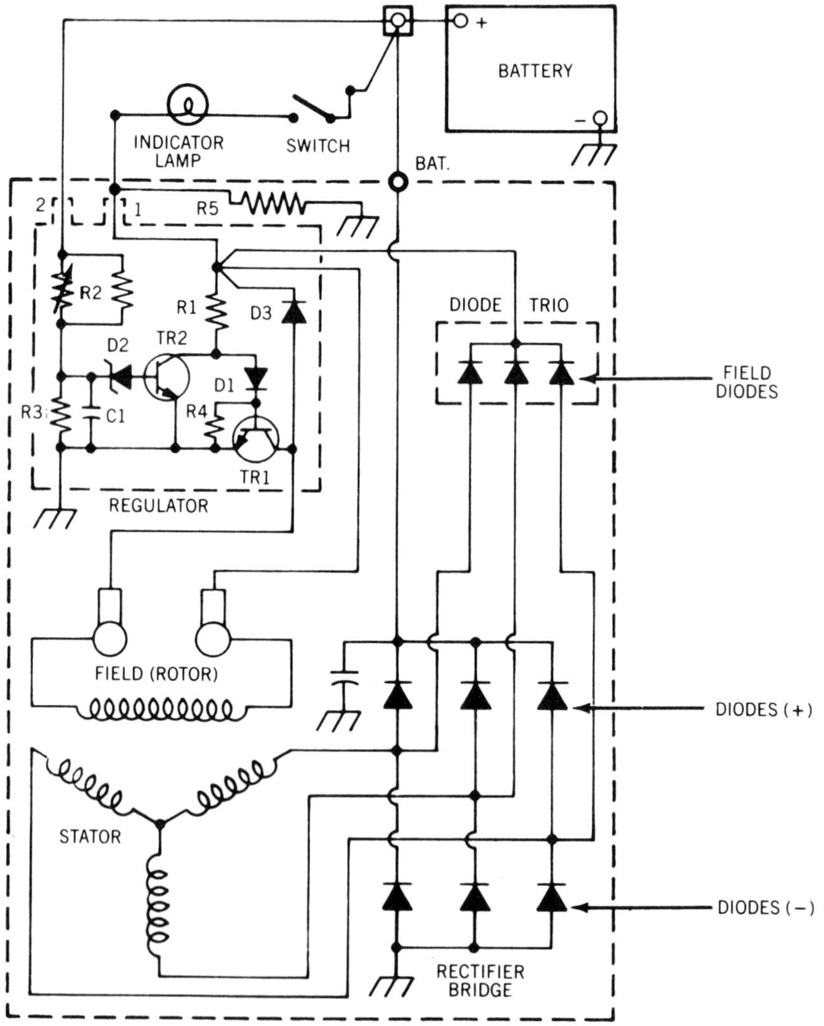


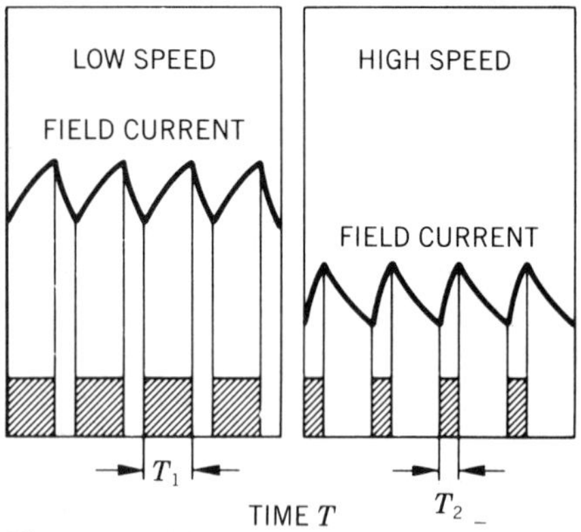


**Figure 11-36.** Field current flows for less time at low speed than at high speed to reduce average current. (Bosch)

**Figure 11-38.** Typical circuit for an alternator with a solid-state regulator. (Delco-Remy)

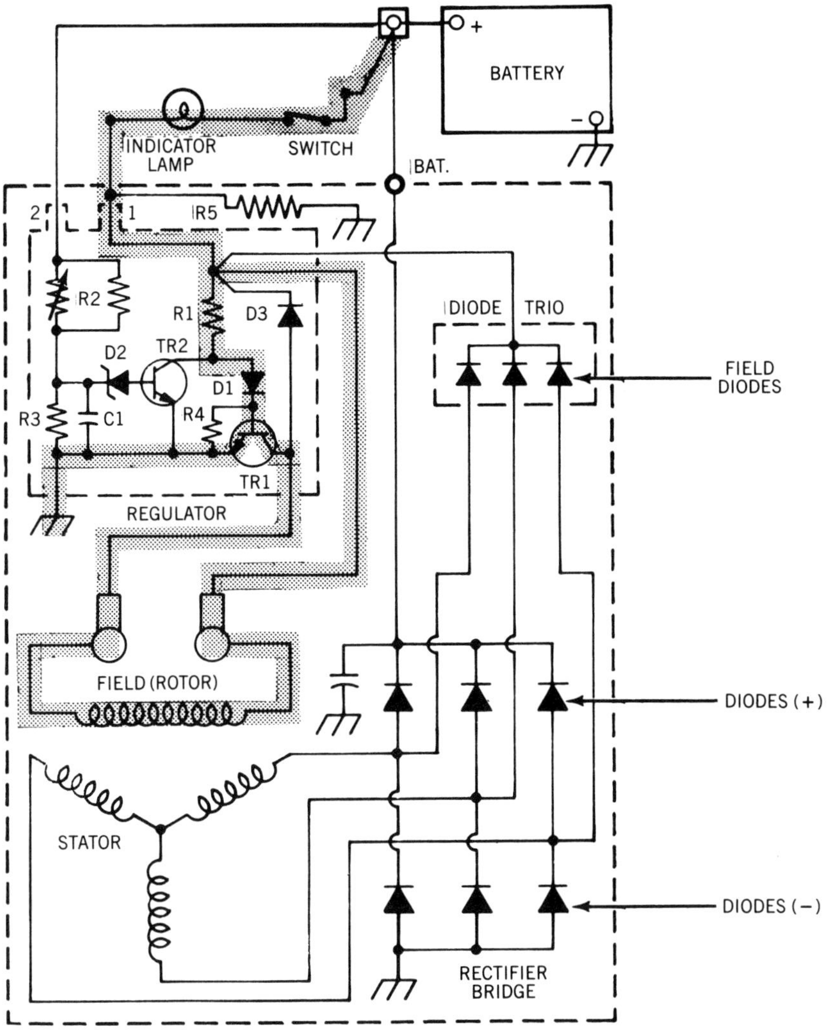


**Figure 11-39.** Field current through a solid-state regulator when starting. (Delco-Remy)

Since the early 1970's, most charging systems have used solid-state electronic voltage regulators. Solid-state regulators have no moving parts; they are compact and not affected much by extreme temperatures. These electronic regulators control field current and output voltage with diodes, transistors, capacitors, and other electronic devices you learned about in chapter 7.

Regulator shapes and sizes vary from manufacturer to manufacturer, figure 11-37. Some small units are installed in the alternator housing. Some regulators are attached to the outside of the alternator, and others are mounted away from the alternator and connected by a wiring harness.

**Electronic Regulator Circuit Operation.** We will use the integral regulator of a Delco-Remy 10-SI alternator to explain the general operation of electronic voltage regulators. These alternators are an A-circuit design (regulation between the field and ground) and have been standard equipment on GM vehicles for more than a decade.

The circuit diagram in figure 11-38 shows the alternator output, or BAT, terminal connected to the battery. The regulator number 2 terminal also is connected to the battery. The regulator number 1 terminal is connected to the dashboard indicator lamp and the ignition switch.

When the engine and the alternator are off, the battery can't discharge through the field because resistors R2 and R3 restrict current. The resistors (R) and diodes (D) in the regulator are arranged to let the regulator sense voltage of the battery and the charging system.

When the ignition switch closes, figure 11-39, current flows from the battery, through the indicator lamp, through terminal 1 of the regulator, and to transistor TR1. The base-emitter circuit of TR1 conducts current to ground. This biases TR1 so that the emitter-collector circuit conducts current through the alternator field. This current magnetizes the field to start generation. It also turns on the indicator lamp.

When the alternator starts to generate, field current is drawn from the stator and rectified by the diode trio (field diodes), figure 11-40. Stator

**FIRING ORDER CHANGES**

The service manuals that you use to find carmakers' specifications and repair procedures are essential tools for a professional mechanic. Your shop probably has an up-to-date library of current manuals, but even the most recent books may not have *all* the specs. Here are a couple of firing order changes you might want to remember. You never can tell when you may need them:

- Unlike most Ford V-type engines, the Lincoln Zephyr V-12 has odd-numbered cylinders on the left bank and even-numbered cylinders on the right, from front to rear. The firing order is 1-4-9-8-5-2-11-10-3-6-7-12.
- Like most other GM V-type engines, Cadillac V-16's have odd-numbered cylinders on the left bank, even-numbered on the right from front to rear. But watch out; Cadillac changed the firing order between 1937 and '38. The 1935-37 models fire 1-8-9-14-3-6-11-2-15-10-7-4-13-12-5-16. The 1938-40 models fire 1-4-9-12-3-16-11-8-15-14-7-6-13-2-5-10.

If your shop doesn't have a complete set of manuals, check the Detroit Public Library.

voltage is higher than battery voltage, so no current flows from the battery to the regulator. The indicator lamp goes out because system voltage is applied to both ends of the lamp circuit. Voltage across the lamp is essentially zero. Resistor R5 is a parallel connection to ground that prevents minor voltage differences and current from lighting the lamp. Field current flows from the diode trio through the same circuit to TR1 that the battery current followed before the alternator started.

When system voltage rises to the regulated maximum, current flows through terminal 2 of the regulator, figure 11-41. When voltage between resistors R2 and R3 is high enough, it forces zener diode D2 to conduct current to the base of transistor TR2. The base-emitter circuit of TR2 conducts current to ground and biases TR2 to complete the emitter-collector circuit. The emitter-collector circuit of TR2 is in parallel with TR1. It provides a shunt path to ground for field current. TR1 turns off so that no current reaches the field.

With TR1 off, system voltage drops. Zener diode D2 then blocks current to TR2. TR2 turns off, and TR1 turns back on. The two transistors TR1 and TR2 cycle on and off many times per second to regulate maximum charging voltage.

Among the other electronic parts of the regulator:

- Capacitor C1 absorbs voltage surges to maintain uniform voltage across R3 and TR2.
- Resistor R3 shunts high current past TR1 at high temperature.
- Diode D3 bypasses high induced voltage in the field when TR1 turns off.
- Resistor R2 in the regulation circuit is a thermistor that changes regulated voltage as temperature changes.

Since the mid-1970's many Delco-Remy 10-SI alternators have used regulators with three, instead of two, transistors. Circuit operation is a bit more complicated, but the principles of regulation are essentially the same as those explained above. Test procedures are basically similar for all 10-SI regulators and alternators.

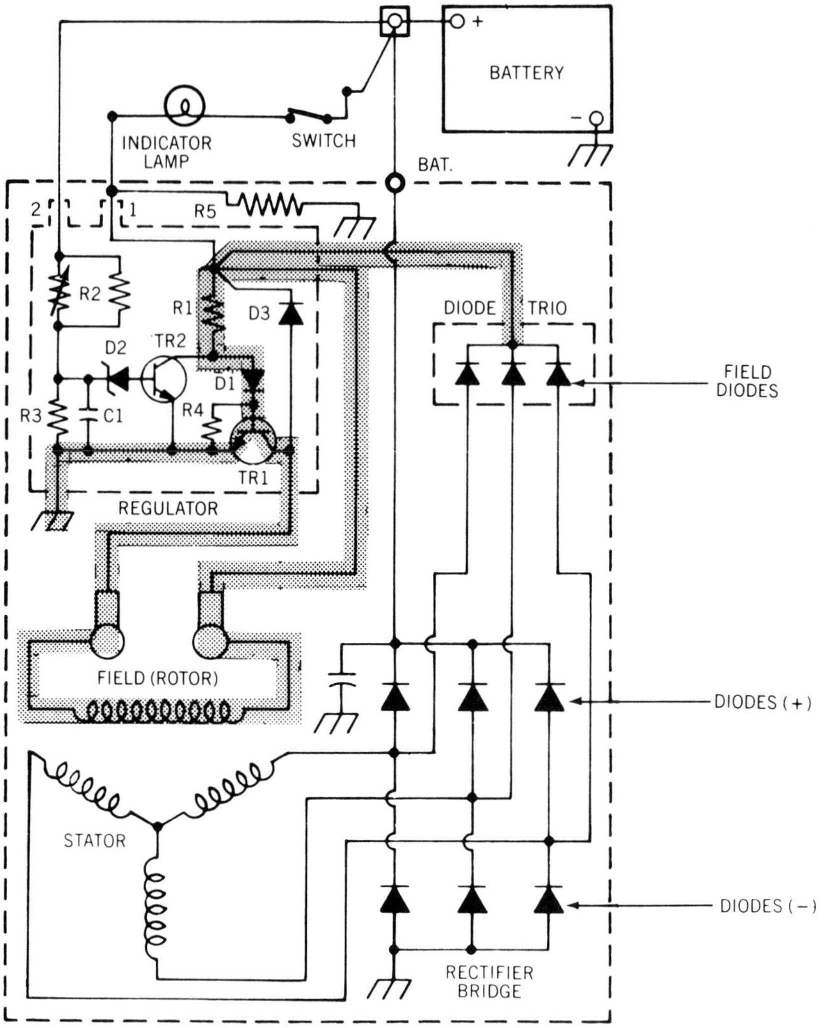


**Figure 11-40.** Field current through a solid-state regulator from the alternator output. (Delco-Remy)

**Electromechanical Regulator Operation.** Older electromechanical regulators are usually connected for a B-circuit field design. The regulator is between the current source and the field. It regulates field current by adding resistance to the field or by providing a shunt path to ground.

An electromechanical regulator is built like a relay. Current flow through a coil creates a magnetic field that moves the regulator armature. Figure 11-42 shows a simple electromechanical field circuit. The alternator field is connected to the battery through the regulator and the ignition switch. When the ignition switch closes, current flows through regulator terminal 3, through the closed lower contacts of the regulator, and to the alternator rotor. Current also flows through the regulator coil winding and a resistor to ground. The resistor limits coil current until system voltage rises to a regulated level.

When the alternator starts, field current continues through terminal 3, and system voltage starts to rise. When voltage rises high enough, current through the regulator coil and resistor produces a strong enough magnetic field to move the regulator armature. The lower contacts open as the armature moves toward the coil. Field current then flows through the resistor between terminals 3 and F. This reduces current to the alternator rotor.

At high engine and alternator speeds, field current may be high enough, even with the series resistor, to raise system voltage too high. However, current

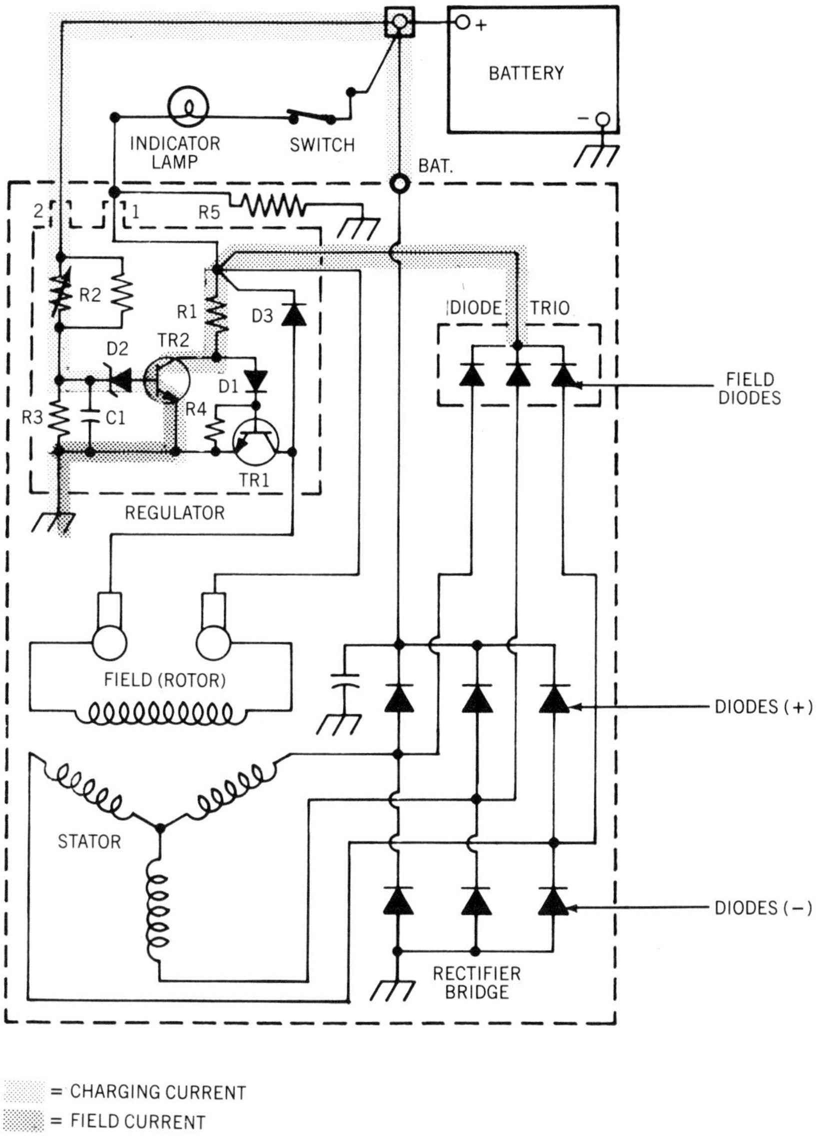


**Figure 11-41.** Field current through a solid-state regulator during regulation. (Delco-Remy)

through the regulator coil also increases and moves the regulator armature farther. This closes the upper set of contacts and grounds the field current, figure 11-43.

Electromagnetic regulators are more sensitive to temperature change than electronic regulators. To control field current through a changing temperature range, these older regulators include one or more of the following devices, figure 11-44:

- *A bimetallic hinge* for the armature. Spring tension is high when the hinge is cold and requires higher coil current to open the regulator contacts. At higher temperature, hinge spring tension decreases, and lower current will open the contacts.
- *A magnetic shunt* is a piece of nickel steel on top of the regulator coil. At low temperature, its permeability is high, and coil magnetism concentrates in the shunt. A stronger field is needed to open the regulator contacts. At higher temperature, shunt permeability decreases. The magnetic field of the coil increases with less current and opens the contacts.
- *A ballast resistor*, similar to an ignition resistor, can be placed in series with the regulator coil. The ballast resistor provides almost constant resistance at all temperatures. The regulator coil can then be designed with low resistance. Temperature changes

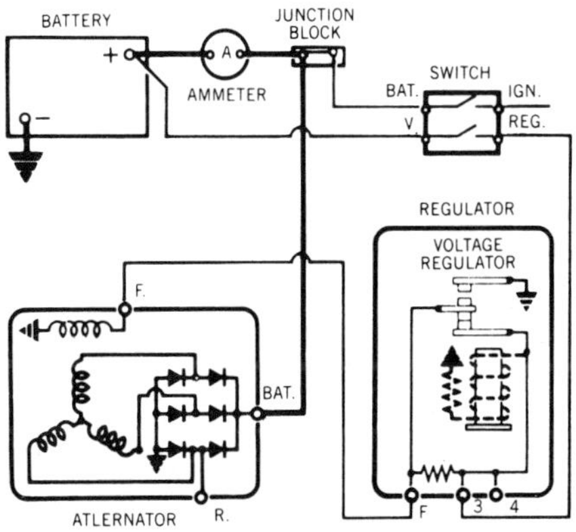


**Figure 11-42.** Typical wiring diagram for a single-unit electromechanical voltage regulator. (Delco-Remy)

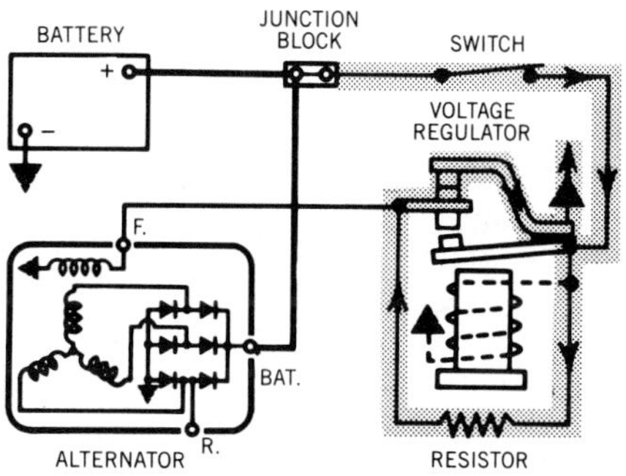


**Figure 11-43.** At high speed, with high field current, a double-contact regulator grounds the field. (Delco-Remy)

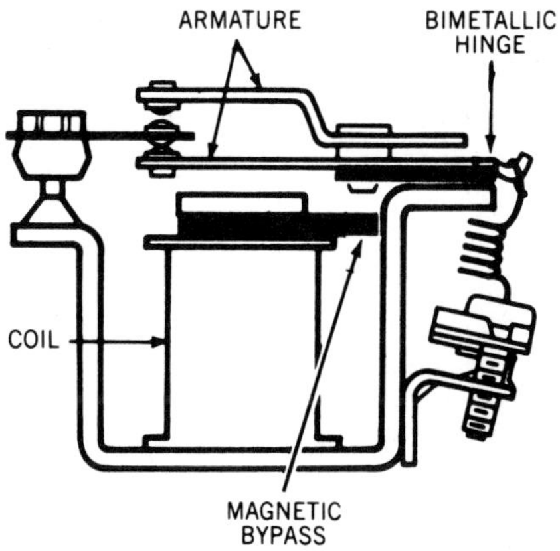


**Figure 11-44.** Electromechanical regulators use magnetic shunts and bimetallic hinges to compensate for temperature changes. (Delco-Remy)

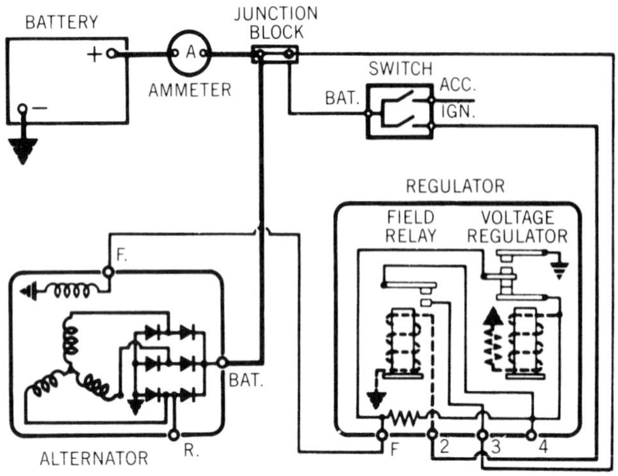


**Figure 11-45.** Typical wiring diagram for a 2-unit electromechanical voltage regulator with a field relay and ammeter. (Delco-Remy)

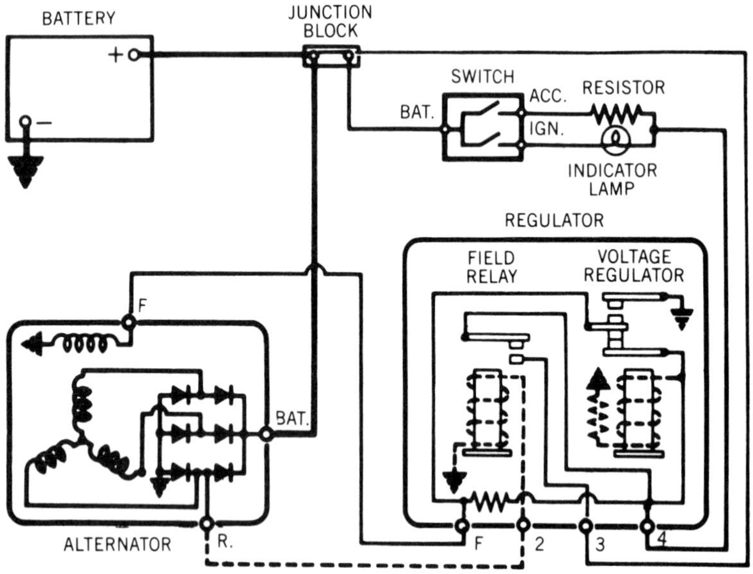


**Figure 11-46.** Typical wiring diagram for a 2-unit electromechanical voltage regulator with a field relay and indicator lamp. (Delco-Remy)

in the coil will change the total resistance very little so that field current remains relatively constant.

### Field Relays

Several charging systems use a field relay along with either an electronic or an electromechanical regulator. The relay reduces the wiring in the field circuit and the voltage drop across the field. In a system with an ammeter, the field relay opens and closes the field circuit. In a system with an indicator lamp, the relay also controls the lamp circuit.

In a system with an ammeter and a relay, field current flows from the battery, through the ammeter, through the ignition switch, to the relay coil, figure 11-45. The relay closes the contacts, and field current flows through the relay and the regulator to the field. The relay stays closed as long as the ignition switch is closed. Field current comes from the battery during starting and from the alternator output when running.

In a system with an indicator lamp and a relay, field current flows from the battery, through the lamp, through the regulator, to the field, figure 11-46. Current lights the lamp and excites the alternator field. When the alternator starts to supply voltage, this voltage is applied to the relay coil. The coil closes the relay contacts, which complete a circuit to supply voltage directly to the regulator. The relay also applies voltage to the opposite side of the indicator lamp. The voltage difference across the lamp is then zero, and the lamp goes out. The field relay stays closed as long as the alternator is operating.

## INDICATORS

Every vehicle charging system includes an indicator for the driver to monitor system operation. The indicator may be an ammeter, a voltmeter, a warning lamp, or a combination of these devices.

### Ammeters

An instrument panel ammeter measures charging system current into and out of the battery and other parts of the electrical system. As with all ammeters, it is in series with the vehicle electrical system, figures 11-42 and 11-45. When battery current flows to the alternator field or any other part of the electrical system, the ammeter shows a discharge (negative current). When the alternator generates current, the alternator shows a charge (positive current).

Although the ammeter may indicate charging current flow, the amount of current may not be high enough to recharge the battery if vehicle electrical loads are heavy and the battery is discharged

### Indicator Lamps

Most charging systems have an indicator lamp, or warning lamp, to indicate general charging system operation. The previous descriptions of regulator operation explained how battery voltage and system voltage are applied to the lamp circuit.

The lamp lights when field current flows through it from the battery. When the alternator produces output voltage, this voltage is applied to the opposite side of the lamp circuit. The lamp goes out because there is no voltage difference across it. The lamp will not warn the driver of an overcharged battery or high charging voltage, but it indicates low voltage from the alternator. Low charging voltage will lead to an undercharged battery.

An indicator lamp often has a parallel resistor, figure 11-46, to carry field current to the alternator if the lamp bulb fails.

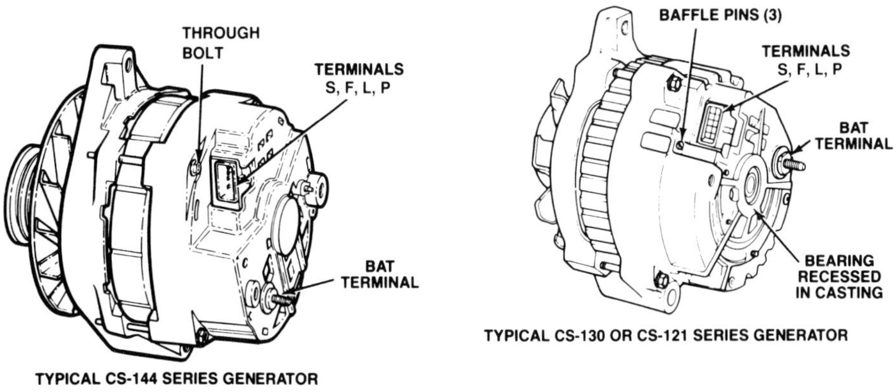


**Figure 11-47.** Delco-Remy CS series alternators are used on late-model GM vehicles. (GM)

### Voltmeters

Many late-model cars have instrument panel voltmeters to indicate charging system voltage. Most such systems also include an indicator lamp. Voltage is applied to the voltmeter through the ignition switch.

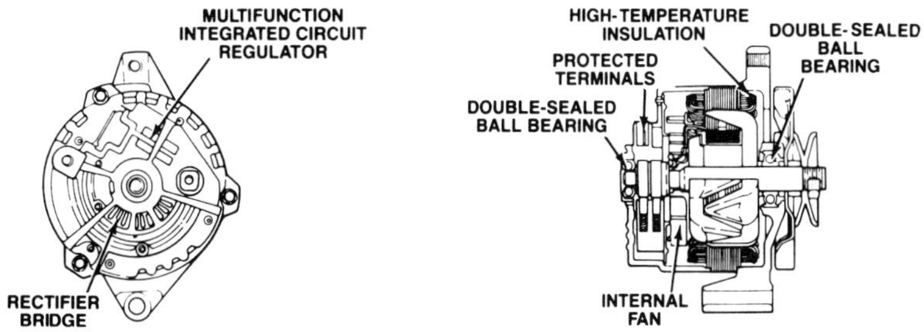


**Figure 11-48.** The Delco-Remy CS alternator uses a multifunction integrated circuit regulator and has an internal fan for cooling on some models. (GM)

## CARMAKERS' CHARGING SYSTEMS

Although all alternators and regulators work on the principles explained in this chapter, carmakers' charging systems vary for different vehicle applications. Different manufacturers also approach alternator and charging circuit design differently to achieve common goals. Recognizing the major alternator makers and their charging systems will help you to service charging systems efficiently.

### Delco-Remy

The Delco-Remy Division of General Motors supplies charging system components and parts for other electrical systems of GM cars and trucks. Delco-Remy refers to its alternators as Delcotron generators.

**CS Series.** The Delco CS alternator series, figure 11-47, was introduced on some car lines in 1986. These downsized alternators use a totally different integral regulator than other Delco alternators and produce a higher amperage output for their size than previous Delco SI series. The CS designation number indicates the outer diameter of the stator laminations in millimeters and can be used to compare relative size and ampere output of different CS models.

All CS series alternators take their field current directly from the delta-type stator. This eliminates the need for a diode trio. An internal cooling fan mounted on the rotor of CS-121 and CS-130 models, figure 11-48, is used to pull air through one end frame. After cooling the rectifier bridge and regulator assembly, the air passes out through openings in the other end frame. Sealed and lifetime-lubricated bearings are used in both end frames.

A conventional BAT output terminal is used on all CS series alternators, but the regulator may have either a 1-or 2-wire connector, depending upon its application. Figure 11-49 and

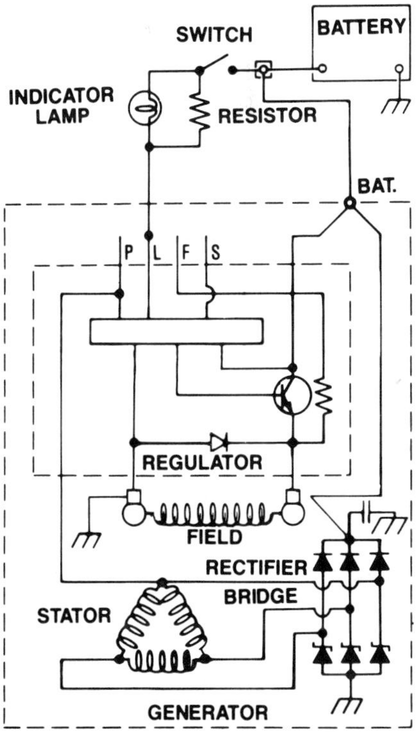


**Figure 11-49.** The basic circuit configuration for a CS series alternator using an L terminal. (GM)

figure 11-50 show the two basic circuits used with CS alternators, but the vehicle wiring diagram must be consulted for the complete charging system circuits. Use of the P, F, and S terminals, figure 11-49, is optional:

- The P terminal connects to the stator, and may be externally connected to a tachometer or comparable device.
- The F terminal is connected internally to field positive. It is used as a fault indicator on vehicles with a body computer module (BCM).
- The S terminal may be connected to battery voltage as a means of sensing the voltage to be controlled.
- The L terminal connects the regulator with an indicator lamp and battery voltage.

Full battery voltage should never be applied to the L terminal or the voltage regulator will be destroyed. Full battery voltage can be applied to the I terminal, however, because of the resistor in the regulator circuit, figure 11-50.

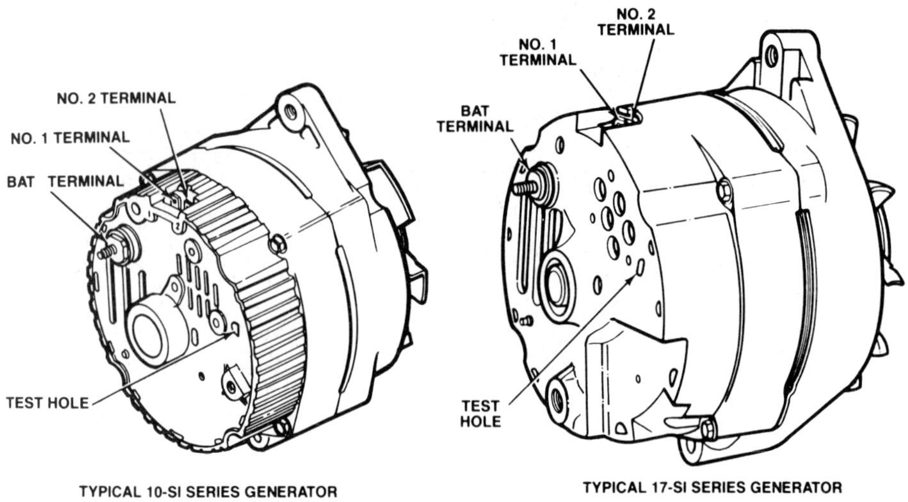


**Figure 11-51.** The Delco-Remy SI series alternators appeared in the early 1970's and are still used on some vehicles. (GM)

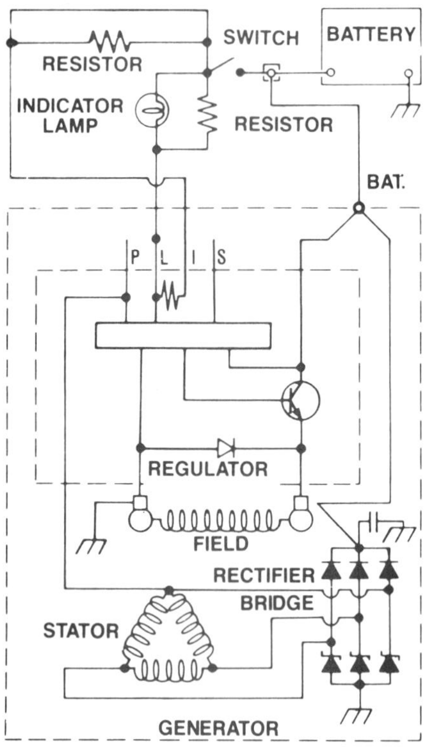


**Figure 11-50.** The basic circuit configuration for a CS series alternator using L and I terminals. (GM)

The I terminal on regulators so equipped, figure 11-50, turns the regulator on and supplies field current (in addition to that applied internally), either through a resistor or directly from the switch. Figure 11-50 shows the alternate circuit connections. The I circuit can be used alone or with the L circuit. On vehicles with a body computer module (BCM), such as the Cadillac Allanté, the BCM uses the I circuit to shut the alternator off during wide-open throttle conditions to improve performance, if battery voltage is above a calibrated value.

BCM-equipped vehicles also have a fault detection capability built into the alternator regulator. Whenever the ignition is on, the BCM sends 12-volts to the I terminal. This causes the alternator to start operation and regulate system voltage. The alternator sends a pulse-width-modulated (PWM) signal back to the BCM on the F terminal circuit. This signal keeps the BCM informed of the regulator duty cycle.

Vehicles with a BCM can set trouble codes, which light warning lamps in the instrument cluster. The trouble codes and their use are described in the test procedures in chapter 11 of Volume 2.

**SI Series.** The Delco SI alternator series, figure 11-51, was standard equipment on GM vehicles for over 15 years, and is still used on vehicles not yet equipped with CS series alternators. All SI series alternators have integral voltage regulators in the slipring end housing. The 10- and 12-SI alternators use a Y-type stator; 15-, 17-, and 27-SI models use delta-type stators. A 14-pole rotor is used in most SI alternators. The 27-SI alternator is primarily used on commercial vehicles and has an adjustable voltage regulator located under an adjustment cap on the rear end frame.

Early 10-SI alternators had six separate diodes in the slipring end housing and a separate ground (GRD) terminal. The 10-SI alternators used for the past 10 years have six diodes in a small rectifier bridge to replace the six large diodes of early models, figure 11-21. Regulator connections and circuit operation are explained earlier in this chapter. All Delco-Remy SI alternators have capacitors in the rear housing to protect the rectifier bridge from sudden voltage surges and to reduce radiofrequency interferences.

**DN Series.** GM vehicles built before about 1970 have Delco-Remy DN alternators with electromechanical regulators. The 10-DN is the most common model. It has six diodes in the rear end housing, figure 11-52. Like the 10-SI, the 10-DN has a 14-pole rotor and a Y-type stator.

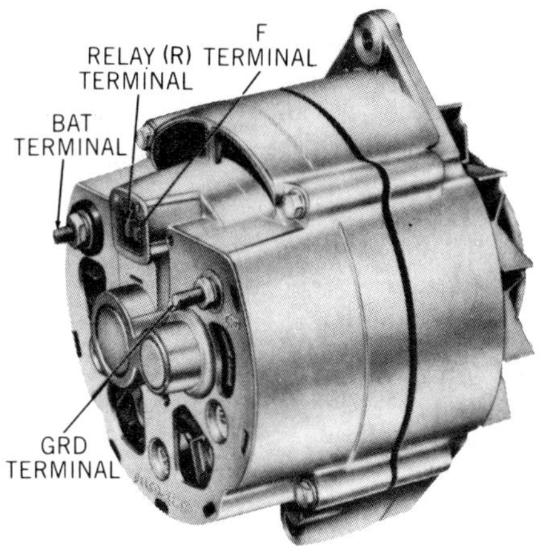


**Figure 11-52.** Typical Delco 10-DN alternator. (Delco-Remy)

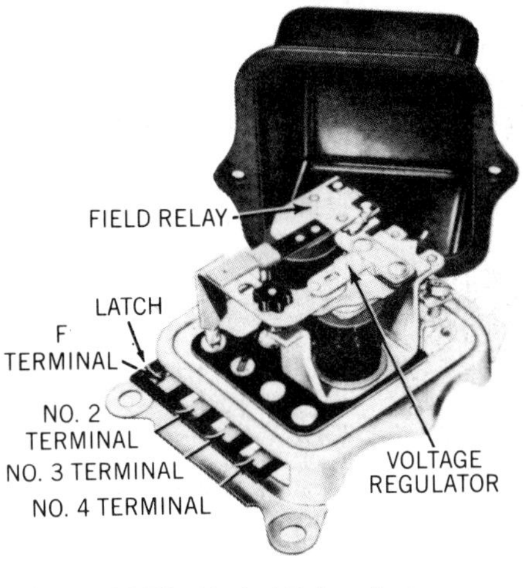


**Figure 11-53.** Typical Delco electromechanical regulator. Units without a field relay have no number 2 terminal. (Delco-Remy)

A 10-DN alternator has the following terminals:

- BAT is the output terminal, connected to the battery.
- GRD is an additional ground connection.
- R is connected to a field relay for an indicator lamp.
- F is connected to the B-circuit electromechanical regulator.

Delco-Remy electromechanical regulators are built with and without field relays. Regulator terminals are marked F, 2, 3, and 4, figure 11-53. Single-unit regulators, without a field relay, do not have a 2 terminal. Terminal connections are as follows:

- F connects the regulator to the alternator field.
- 2 connects the alternator output to the field relay, if a relay is installed.
- 3 connects the battery to the contacts of the regulator and the relay.
- 4 connects the ignition switch to the contacts of the regulator and the relay.

## Motorcraft

The Motorcraft Division of Ford Motor Company supplies parts for charging systems and other electrical systems used on Ford-produced vehicles and those of some other carmakers. Before the early 1970's, components were built under the Autolite name. Late-model Motorcraft alternators use a color code stamped on the front end frame to identify the model and current rating.

Before 1985, all Motorcraft charging systems, except one, used either an electromechanical voltage regulator (through 1978) or a remotely mounted, solid-state regulator (1978 and later). The 1969-71 charging system with a 55-ampere alternator had a solid-state regulator mounted on the rear end frame, figure 11-54. This Motorcraft alternator used an A-circuit; all others have a B-circuit.

**IAR Series.** The Motorcraft integral alternator regulator (IAR) charging system was introduced on some 1984 Ford Motor Company vehicles. Its name comes from the solid-state regulator assembly, figure 11-55, mounted on the alternator rear housing. The rectifier assembly is sealed and secured to the slipring end frame. Its BAT and STA terminals are enclosed in a plastic housing that extends through the side of the end frame, figure 11-56. The brushes are attached to the regulator, requiring that both

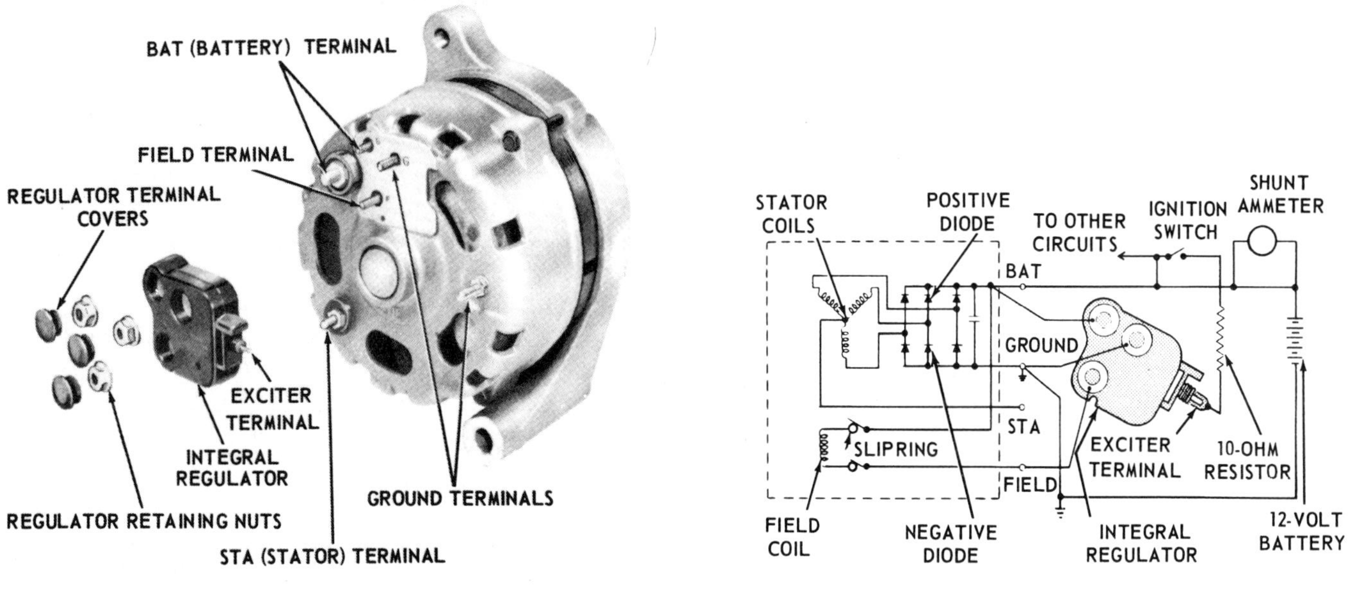


**Figure 11-54.** This Ford Motorcraft 55-ampere alternator has a solid-state regulator attached to the outside of the rear housing. (Ford)

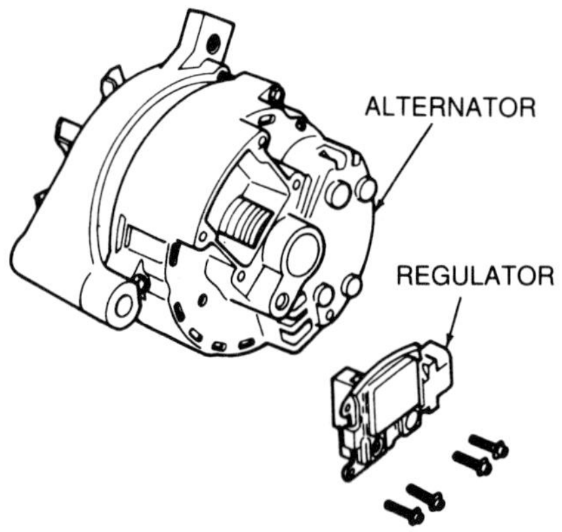


**Figure 11-55.** A solid-state regulator assembly is mounted on the rear housing of the Motorcraft IAR alternator. (Ford)

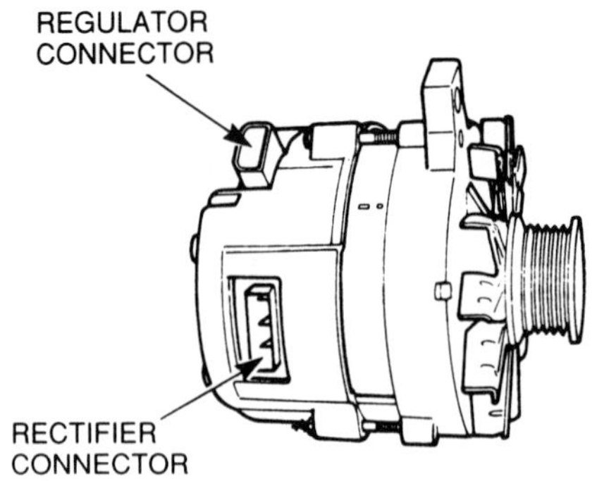


**Figure 11-56.** The Motorcraft IAR regulator connector protrudes from the rear of the unit; the rectifier connector extends through the end frame. (Ford)

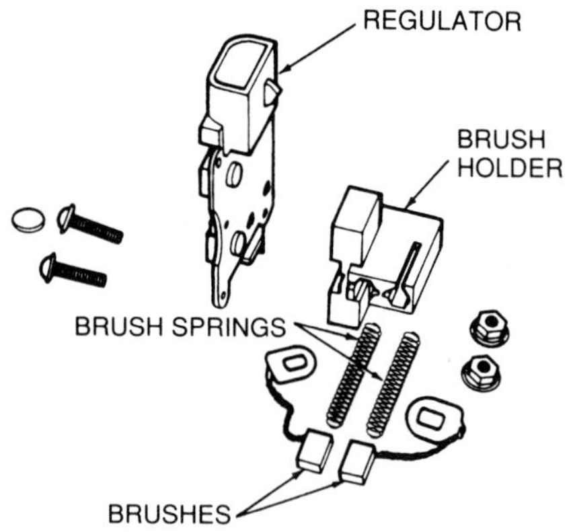


**Figure 11-57.** The IAR regulator and brushes form a single assembly. (Ford)

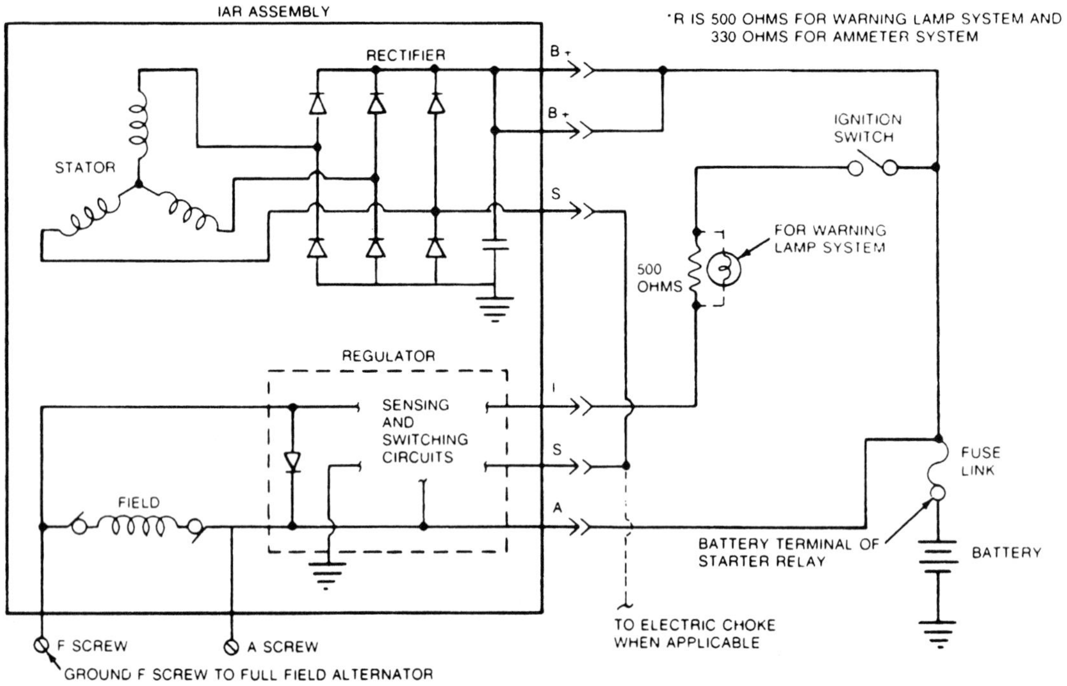


**Figure 11-58.** The basic IAR charging circuit diagram. (Ford)

assemblies be removed if either requires service, figure 11-57. A 12-pole rotor is used with the Y-type stator.

When the ignition is turned on, voltage is sent to the regulator I terminal through a resistor in the I circuit. As long as the ignition is on, system voltage is sensed and field current is drawn through the regulator A terminal. Field current is supplied from the regulator to the rotor through the brushes riding on the sliprings. When the ignition is turned off, the control circuit is shut down. Figure 11-58 shows the basic IAR alternator circuit.

**EVR Series.** Motorcraft charging systems using an external solid-state regulator are now known as external voltage regulator (EVR) systems. They may use either a side-terminal or a rear-terminal alternator, figure 11-59. The rear-terminal alternator is built in 40- to 65-ampere ratings. The same housing is used for all models, with the higher ampere ratings obtained by using more wire in the stator and rotor windings. The side-terminal alternator is built in 70- to 100-ampere ratings. Output at lower engine speeds is greater than the rear-terminal models. High-current 90- and 100-ampere models have delta stators. All others have Y-type stators. All use a 12-pole rotor.

Motorcraft alternators with external voltage regulators or electromechanical regulators have the following terminals:

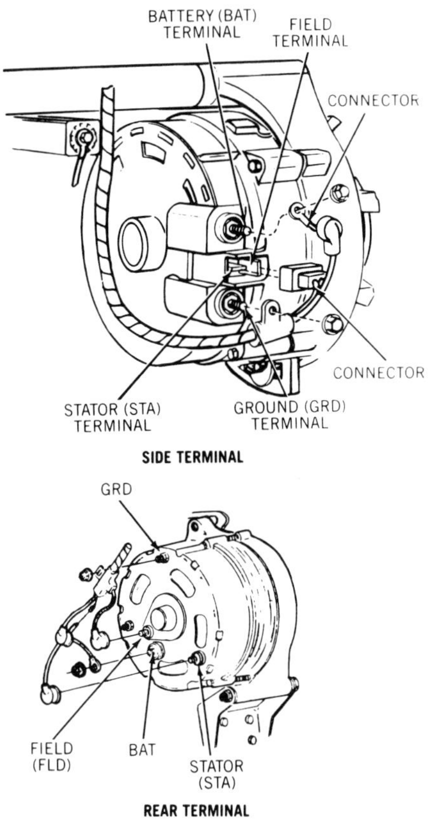


**Figure 11-59.** Motorcraft builds rear-terminal and side-terminal alternators of various current ratings. (Ford)

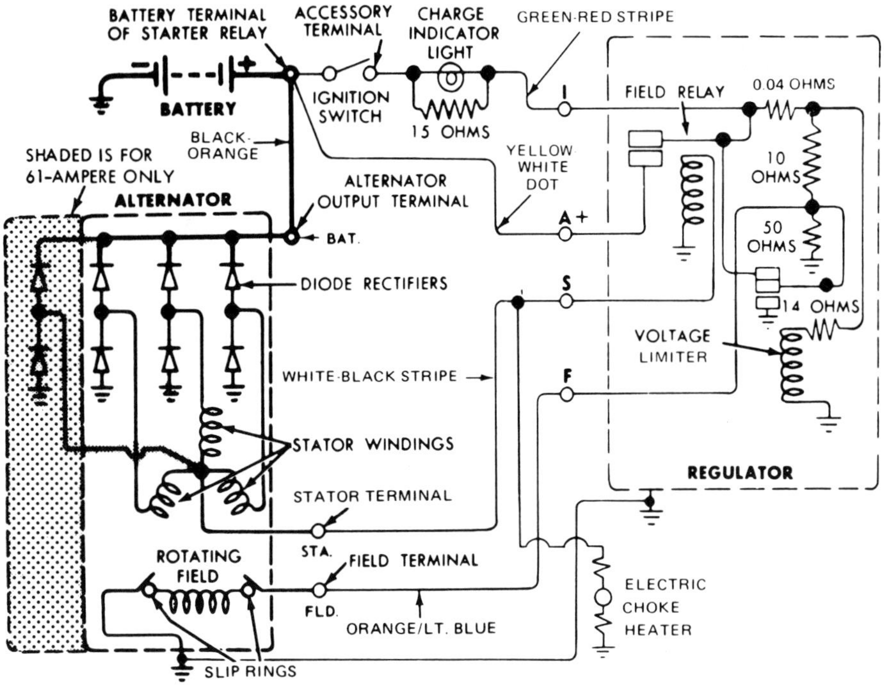


**Figure 11-60.** Typical Motorcraft charging system diagram. The center tap of the 61-ampere alternator (shaded area) is connected to another pair of rectifying diodes. (Ford)

- BAT is the output terminal, connected to the battery.
- GRD is a ground connection.
- STA connects the center tap (neutral junction) of a Y-type stator to different units, such as a field relay and lamp and the electric choke heater.
- FLD connects the rotor field winding to the regulator.

The 61-ampere Motorcraft rear-terminal alternator has the stator center tap connected to two more diodes, figure 11-60. This creates a parallel output circuit to increase current at low speed. The STA terminal on a 90-ampere, side-terminal alternator connects to rectified output because the delta stator has no center tap.

Motorcraft regulator terminals are marked as follows, figure 11-61:

- I is used only without an ammeter to connect the ignition switch to the field relay contacts and the regulator contacts.

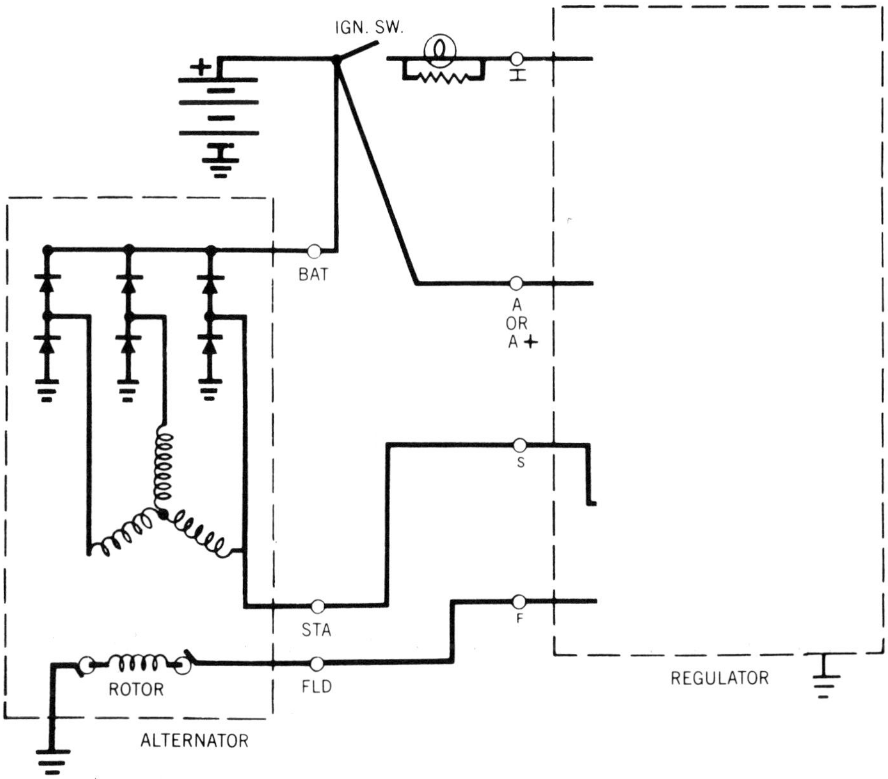


**Figure 11-61.** Ford regulator terminals are the same for solid-state and electromechanical units. (Ford)

• A+ or A connects the battery to the field relay contacts.

• S connects the alternator output to the field relay coil.

• F connects the field to the regulator contacts or to the regulator transistors.

## Chrysler

In the late 1950's, Chrysler was the first domestic carmaker to use ac charging systems as standard equipment on its passenger cars. In 1985, Chrysler introduced a dual-output 40- or 90-ampere alternator with computer-regulated voltage regulation on some models. Other models use dual-output Bosch or Mitsubishi alternators with external or integral solid-state regulators. The output of the new design exceeds the previous Chrysler alternator by 56 percent at idle and 35 percent during cruising. The Chrysler dual-output alternator has a delta-type stator; Bosch and Mitsubishi dual-output alternators use a Y-type stator.

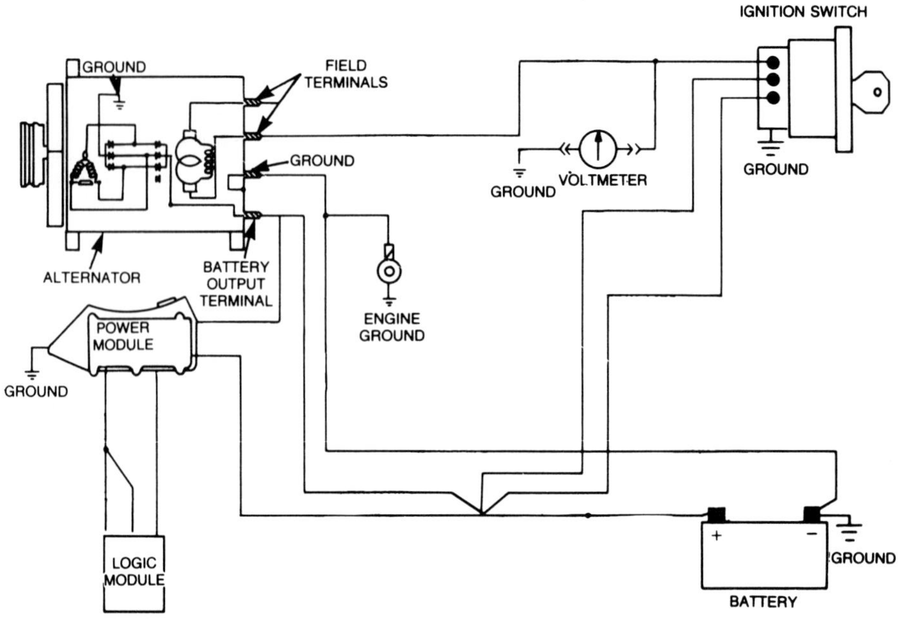


**Figure 11-62.** A typical Chrysler dual-output alternator charging circuit with computer-controlled voltage regulation. (Chrysler)

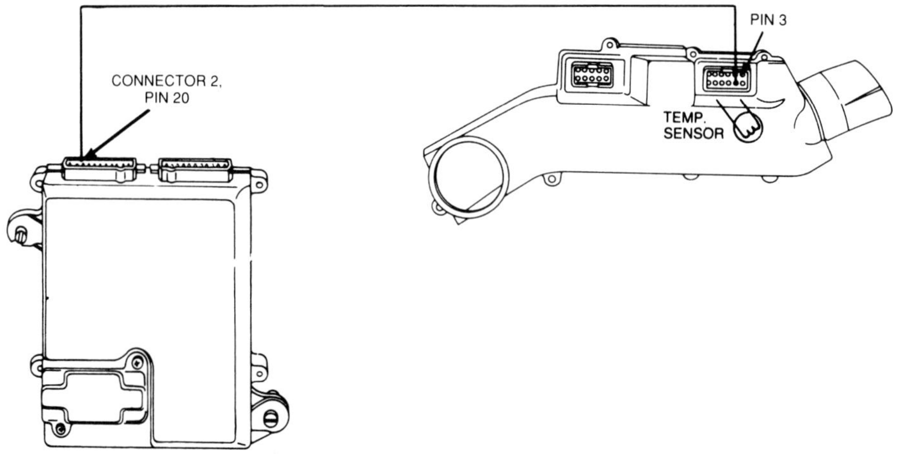


**Figure 11-63.** The Chrysler charging system with computer-controlled voltage regulation. (Chrysler)

**Computer-controlled Alternator.** The regulator circuitry used with the computer-controlled Chrysler alternator is an isolated-field type, but integrated circuitry in the logic and power modules controls field current, figure 11-62. The logic module senses battery temperature according to system resistance and switches field current on and off in a duty cycle to regulate charging voltage.

The Chrysler computer-controlled charging system has two unique capabilities:

• It can vary charging voltage according to ambient temperature, as well as in response to the system's voltage needs.

• The computer can detect problems in the charging system and set fault codes.

Whenever the ignition is turned on, the logic module checks battery temperature, figure 11-63. This determines the control voltage. A predriver transistor in the logic module signals the power module driver translator to turn on the alternator field current, figure 11-64. System voltage and battery temperature are continually monitored by the logic module, which uses this data to inform the power module driver when field current adjustment is required to maintain output voltage between 13.6 and 14.8 volts, ± 0.3 volt, figure 11-65.

On vehicles with computer-controlled voltage regulation, charging system problems can be detected by the computer, which records fault codes in the system memory. Some codes light a POWER LIMITED, POWER LOSS, or CHECK ENGINE lamp on the instrument cluster; others do not. The fault codes and how to use them are described in the test procedures in chapter 11 of Volume 2.

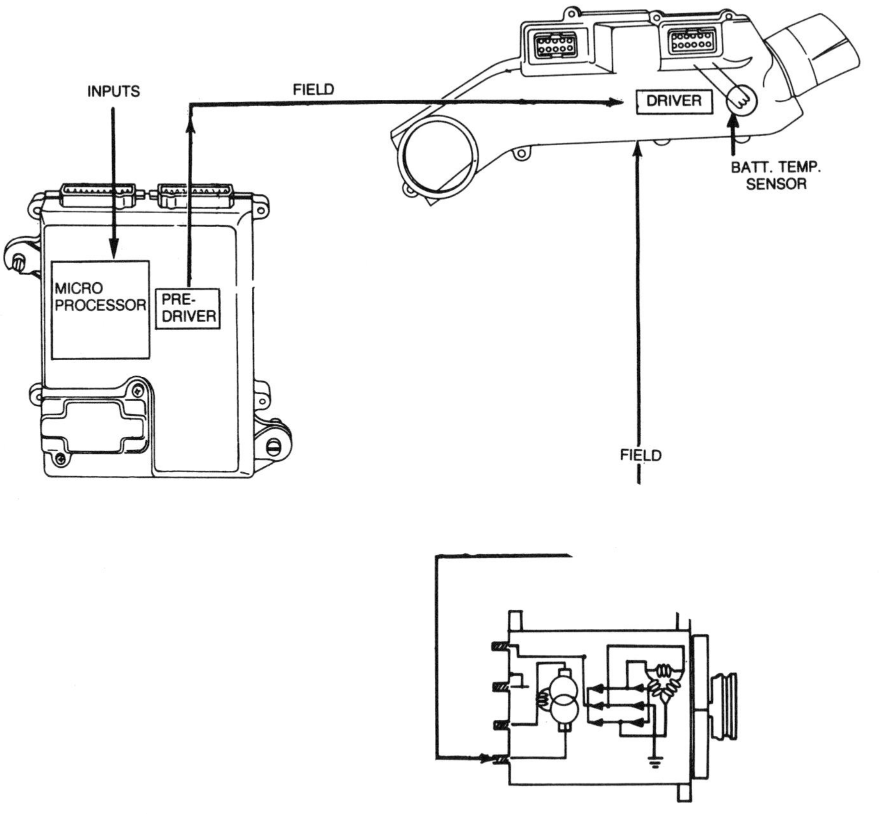


**Figure 11-64.** A predriver in the logic module activates a driver in the power module to turn on the alternator field. (Chrysler)

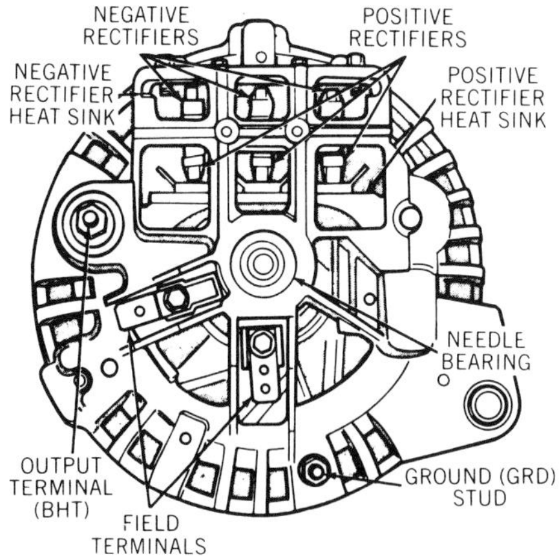


**Figure 11-66.** Typical late-model Chrysler alternator. (Chrysler)

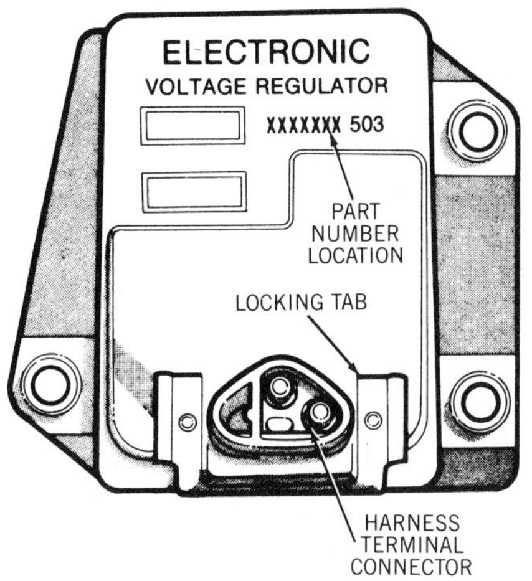


**Figure 11-67.** Typical Chrysler solid-state regulator. (Chrysler)

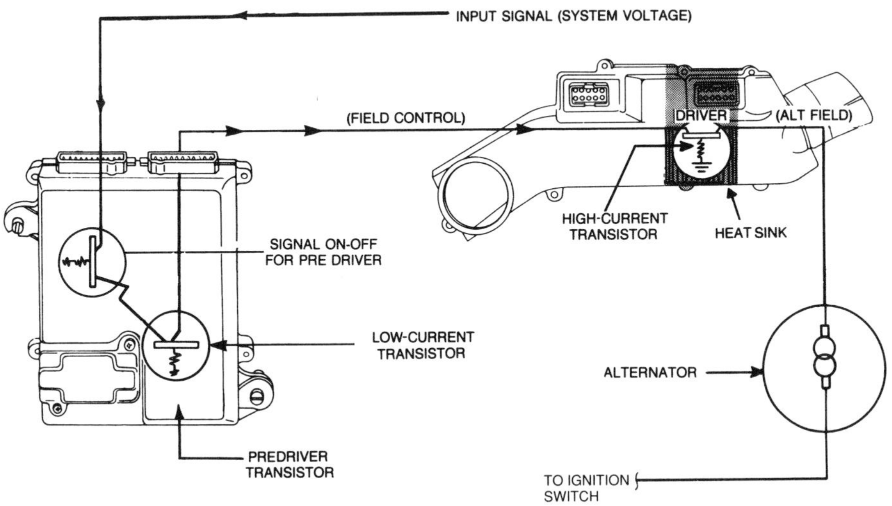


**Figure 11-65.** Internal field control of the Chrysler charging system with computer-controlled voltage regulation. (Chrysler)

**1972–1984 Systems.** Between 1972 and 1984, Chrysler built single-output alternators of two basic designs. Models rated at less than 100 amperes have an exposed stator core. Heavy-duty 100- and 117-ampere alternators have an enclosed stator. All Chrysler alternators have 12-pole rotors. All have Y-type stators, except the heavy-duty models, which have delta stators. Six diodes are mounted on positive and negative heat sinks in the rear housing of standard-duty alternators, figure 11-66. The heavy-duty 100-ampere model has each stator winding attached to two positive and two negative diodes. The 12 diodes provide extra rectification circuits for high-current output.

Chrysler alternator terminals are marked as follows:

- BAT is the output terminal.
- GRD is a ground connection.
- Two FLD terminals connect the regulator to the field brushes.

Except for the computer-controlled alternator, Chrysler products have used a remotely mounted, solid-state regulator, figure 11-67, since 1972.

Most systems have a field relay to open and close the field circuit. Before 1972, Chrysler used an alternator with six separate diodes pressed into a heat sink and into the rear housing. These alternators used a single-unit electromechanical regulator in a B-circuit arrangement.

## Imported Vehicle Charging Systems

Alternators and charging systems on imported vehicles follow the same design and operating principles as domestic equipment, even though physical construction may vary somewhat.

Robert Bosch is a major European maker of electrical equipment. Most Bosch alternators on passenger cars are the type G1 and K1 designs, figure 11-68. These have Y-type stators and are used with either solid-state or electromechanical regulators.

Japanese carmakers use alternators from several manufacturers, such as Nippondenso, Hitachi, Mitsubishi, and others. Some feature unique construction but operate on the principles explained here. One Mitsubishi alternator with an integral regulator, for example, has a double-Y stator and 12 diodes in two rectifier assemblies, figure 11-69. This design provides high current, along with high voltage at low speed.

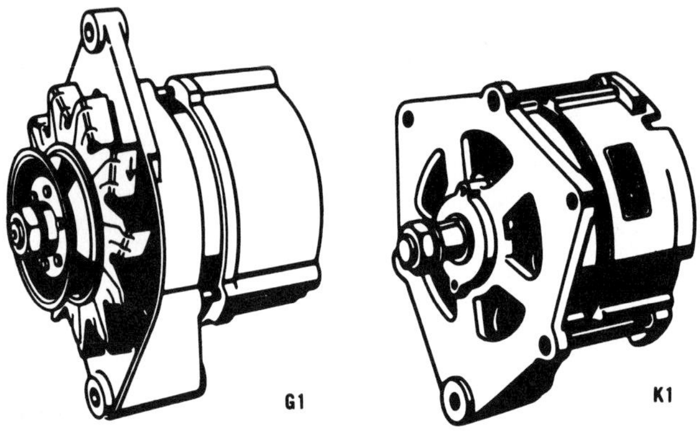


**Figure 11-68.** Bosch G1 and K1 alternators. (Bosch)

## SUMMARY

Charging systems contain the battery, the alternator, the regulator, and an indicator. The output circuit and the field circuit are the two major circuits of the charging system.

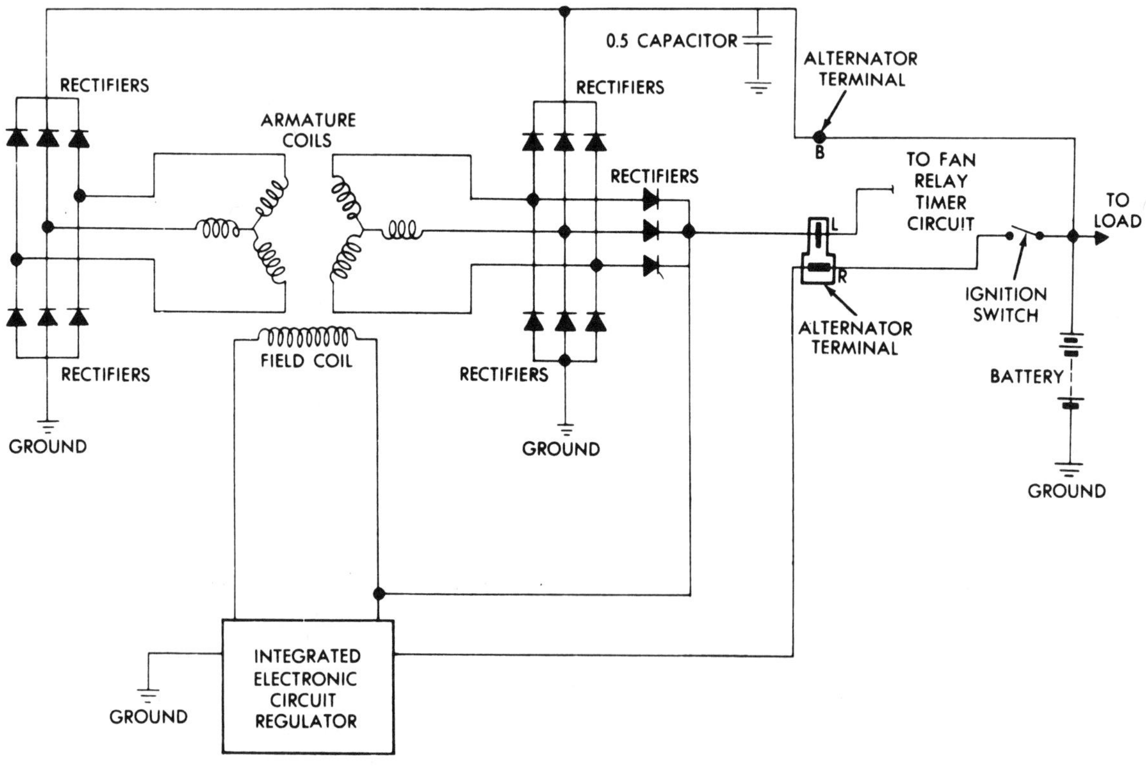


**Figure 11-69.** Mitsubishi alternator charging system. (Chrysler)

The alternator uses a rotating magnetic field (the rotor) to generate alternating current (ac) in stationary conductors (the stator). Diodes rectify the ac sine wave voltage to direct current (dc). Most alternators have 12- or 14-pole rotors (6 or 7 pairs of N–S magnetic poles) and three sets of windings in the stator. Stator windings are installed at overlapping phase angles to generate multiple sine wave voltage output. Stator windings are connected to each other in two ways:

1. A high-voltage Y-type connection
2. A high-current delta connection

Field current flows to the rotor through two brushes and sliprings on the rotor shaft. Alternator design regulates maximum current output, but a voltage regulator is placed in the field circuit to control maximum output voltage. The regulator controls field current and thus limits maximum voltage.

Although charging systems with computer-controlled voltage regulation are beginning to appear, most late-model vehicles have internal or external solid-state regulators that control field current through a pair of parallel transistors. Earlier systems used electromechanical regulators that controlled field current through a vibrating armature and an electromagnetic coil. Regulators can be connected to the field in three ways:

1. *A-circuit*—The field is externally grounded through the regulator.
2. *B-circuit*—The field is internally grounded, with the regulator between the current source and the field.
3. *Isolated field*—The field is isolated and externally grounded through the regulator.

The charging system indicator lets the driver monitor system operation. The indicator may be an ammeter, a voltmeter, or a warning lamp. Some systems use a combination of two, or all three, kinds of indicators.

# REVIEW QUESTIONS

### Multiple Choice

1. An alternator generates current through:
   a. voltage drop
   b. electromagnetic induction
   c. CEMF
   d. the field and relay circuits

2. A battery cell in a normal state of charge produces:
   a. about 3 volts
   b. about 2.5 volts
   c. about 2.1 volts
   d. about 2.9 volts

3. Most charging systems are regulated to develop:
   a. about 14.5 volts
   b. about 12 volts
   c. about 12.6 volts
   d. about 13 volts

4. Battery life can be shortened by:
   a. sulfation caused by charging voltage that is below specifications
   b. excessive gassing caused by charging voltage that is above specifications
   c. unregulated charging voltage
   d. all of the above

5. The two circuits of the charging system are:
   a. the motor circuit and the control circuit
   b. the input circuit and the output circuit
   c. the output circuit and the field circuit
   d. the field circuit and the input circuit

6. The simplest form of an alternator is:
   a. a rotating magnet in a stationary looped conductor
   b. a fixed magnet in a stationary looped conductor
   c. a stator in a stationary looped conductor
   d. a rotor

7. Voltage polarity means:
   a. magnetic polarity
   b. current direction
   c. the direction of sine wave voltage
   d. voltage cycle

8. When a rotor revolves in an alternator, and no flux lines cut the conductor:
   a. voltage and current increase
   b. voltage increases and current decreases
   c. voltage and current return to zero
   d. voltage increases and current returns to zero

9. When an alternator rotor makes one revolution and the voltage polarity changes from positive to negative, it is called:
   a. the sine function of angles
   b. sine wave voltage
   c. rectified voltage
   d. dc voltage

10. An alternator uses diodes to:
   a. increase voltage to the battery
   b. rectify alternating current to direct current
   c. allow current to flow in two directions
   d. reduce current flow

11. Alternators have three output conductors in order to produce:
   a. three-phase voltage
   b. two-phase voltage
   c. single-phase voltage
   d. sine-wave voltage

12. Field current is often called:
   a. flux current
   b. excitation current
   c. polarity current
   d. stator current

13. A voltage regulator controls output voltage by regulating:
   a. CEMF
   b. battery voltage

c. field current

d. output current

14. Student A says that the Chrysler charging system with computer-controlled voltage regulation can set fault codes. Student B says that its control voltage is determined by battery temperature. Who is right?

a. A only

b. B only

c. both A and B

d. neither A nor B

**Fill in the Blank**

15. Most passenger car alternators use ________ stators.

16. Most alternators rated at 100 amperes or more use ________ stators.

17. A rectifier bridge is used to keep diodes from ________ .

18. An ________ can be used to view alternator output voltage waveforms.

19. In an alternator, the field current connects the battery to the ________ through the alternator output terminal.

20. After an alternator starts working, field current comes from the ________ output.

21. An A-circuit alternator has an ________ grounded field.

22. A B-circuit alternator has an ________ grounded field.

23. Turning on the headlamps of a vehicle will cause alternator ________ to drop.

24. The charging system consists of the ________, the ________, the ________, and a ________ .

25. An alternator can't operate without initial ________ from the battery.

26. Field current in the Chrysler charging system with computer-controlled voltage regulation is adjusted according to voltage and ________ temperature.

# PART FOUR
# IGNITION SYSTEMS

# INTRODUCTION

The two chapters in this part explain the principles of electrical spark ignition systems for gasoline engines. The ignition system is essential to ignite the air-fuel mixture for combustion in a gasoline engine. It is equally important as an emission control system to reduce the formation of harmful combustion byproducts.

The principles of timed, inductive-discharge ignition systems have not changed for over 85 years, since Charles Kettering patented his first system in the first decade of the 20th century. The applications of the principles, however, have evolved and improved continuously.

For generations, engineers relied on mechanical breaker points to control ignition operation. Breaker-point ignitions have been obsolete in new-car design for about two decades. They are not yet obsolete, however, in automobile service or in the study of ignition operation. There are still many breaker-point ignitions in vehicles on the road that require regular service. Moreover understanding the simple mechanical switching action of breaker points makes understanding the electronic switching of modern ignitions much easier and clearer.

Chapter 12 explains the principles of breaker-point ignition operation. Most importantly, it also covers the functions of ignition secondary circuits, dwell, and timing that are common to breaker-point and electronic ignitions, alike. Understanding the electromechanical fundamentals of breaker-point ignitions and engine operation provides the background to understand electronic ignition control.

Chapter 13 continues the explanation of ignition systems by outlining the principles of electronic ignition control used by most carmakers. It also summarizes the ignition system development of major domestic and foreign manufacturers, including the latest direct (distributorless) ignition systems (DIS). This information will give you the foundation to understand and service all late-model ignition systems. It also provides an introduction to fully integrated electronic engine control systems.

# 12 BREAKER-POINT IGNITION SYSTEMS, SPARK TIMING, AND SPARK ADVANCE

## WHAT

Since Charles Kettering invented the inductive-discharge battery ignition system in 1908, it has been almost the only spark ignition system used on automobile engines. For about 70 years, ignition systems used breaker points as an electromechanical switch to open and close a low-voltage primary circuit and to time high-voltage ignition in the engine cylinders. Since the mid-1970's almost all ignition systems have used solid-state electronic switching devices to control the primary circuit and ignition timing. Whether an ignition system uses breaker points or electronic devices, however, the principles of producing high voltage through electromagnetic induction are exactly the same.

This chapter explains the components and operation of the low-voltage primary and the high-voltage secondary ignition circuits. It describes a simple breaker-point system and the functions of high-voltage induction, ignition dwell, spark timing, and advance. Understanding a breaker-point ignition system lays the foundation for understanding the features shared by breaker-point and electronic systems.

## WHY

A spark-ignition, gasoline engine requires an air-fuel mixture compressed and ignited in the cylinders at the exact moment when it will burn to produce the most power. If either fuel or spark is missing, or not timed properly, the engine will not run. The fuel and ignition systems are the most important combustion controls on any engine. One can't work without the other. To understand engine performance and to service an engine professionally, you must know how the fuel and ignition systems work separately and together.

Although breaker-point ignitions are not used on late-model automobiles, there are still millions of vehicles on the road that have these systems and that require service. There are other equally important reasons to study breaker-point ignitions as an introduction to electronic ignitions. All inductive-discharge ignitions operate on the principles of magnetic induction. The principles of ignition dwell, timing, and spark advance are the same, whether they are controlled mechanically or electronically. Ignition secondary circuits from the coil to the spark plugs operate the same way whether the primary circuit has an electronic switch or electromechanical breaker points. This chapter explains the basic operating requirements of ignition primary and secondary circuits. When you understand these principles in the context of simple breaker-point operation, you can understand electronic systems more easily.

## GOALS

After studying this chapter, you should be able to:

**1.** Explain how ignition secondary voltage is produced in an induction coil.

**2.** Illustrate a simple ignition circuit and identify the primary and secondary components.

**3.** Explain voltage and current action in primary and secondary circuits.

**4.** Explain ignition dwell angle and point gap.

**5.** Explain spark plug firing action and the following design features: reach, heat range, diameter and seat type, and gap.

**6.** Explain the operation of centrifugal and vacuum advance mechanisms and the principles of spark advance and initial timing.

## IGNITION VOLTAGE

Ignition in a gasoline engine requires that an electrical circuit be completed across an air gap to create a spark. This requires very high voltage because such a circuit contains very high resistance. Resistance exists because:

- The air gap between the spark plug electrodes is an open-circuit condition that creates several thousand ohms of resistance.
- High pressure and high temperature in the engine cylinders create high resistance.
- The air-fuel mixture is a high-resistance conductor.

To complete a circuit through these conditions and create a spark requires 5,000 to 25,000 volts, or more. However, the electrical system of an automobile receives voltage from a 12-volt battery and an alternator. System voltage ranges from 10 to 15 volts under various operating conditions, about one-thousandth of the voltage needed for ignition.

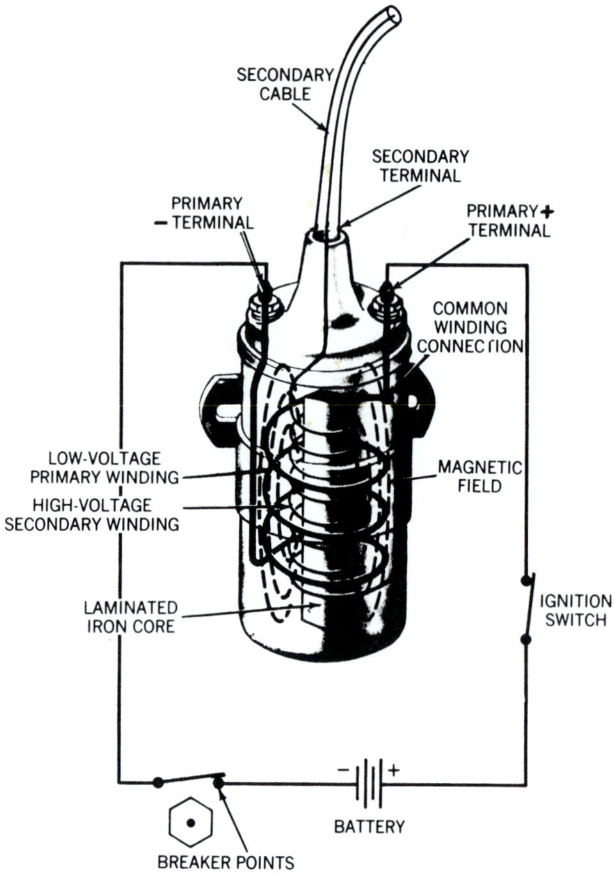


**Figure 12-1.** The coil primary winding receives low-voltage current from the primary circuit and induces high voltage in the secondary winding. (Bosch)

### High-Voltage Induction

High voltage for ignition is provided by an induction coil. To begin to understand ignition voltage, you should review what you learned about electromagnetic induction in chapter 7.

An ignition coil uses mutual induction to transform, or increase, low battery voltage to high ignition voltage. The coil contains two windings of copper wire around an iron core. The primary winding has a few hundred turns of heavy wire and is connected to the low-voltage primary circuit, figure 12-1. The **primary circuit** receives low-voltage current from the battery and uses it to induce higher voltage in the coil secondary winding.

The secondary winding has several thousand turns of fine wire and is connected to the high-voltage **secondary circuit** of the ignition system. We will look at the primary and secondary ignition circuits in detail later. First, we will review high-voltage induction in the ignition coil.

Remember from chapter 7 that one way to induce voltage in a coil is to place it near another primary coil, through which current flows. Current in the primary coil creates magnetic flux lines that cut the windings of both the primary and secondary coils. If you designed a circuit so that you could switch primary current on and off rapidly, the flux lines would expand and contract with each *change* in current flow. Each rapid expansion and contraction of the magnetic field would induce voltage in the primary and secondary coils. This is how an **inductive-discharge ignition system** works. An important point to remember is that induction occurs when primary current flow *changes* and the magnetic field expands and collapses. The changing current flow creates the relative motion between the field and the coil windings. When current flow is steady, the magnetic field does not move, and no voltage is induced.

Several factors determine the amount of voltage induced in the coil secondary winding.

1. The ratio of the number of turns in the secondary winding to the number of turns in the primary winding. This is fixed by the design and manufacture of the coil and is between 100:1 and 200:1. The **turns ratio** is a voltage multiplier. The voltage induced in the secondary winding is 100 to 200 times higher than the voltage applied to the primary winding.
2. The amount of current in the primary winding and the strength of its magnetic field. More current creates a stronger magnetic field with more flux lines to cut the secondary winding.
3. The speed with which primary current flow stops and the magnetic field collapses. Fast collapse of the primary field creates fast relative motion between the field and the turns of the secondary winding.

Under most conditions, 9 to 10 volts are applied to the coil primary winding. At high engine speeds, voltage may rise to 12 volts. This voltage produces 1 to 4 amperes of current in the primary winding. Primary current increases gradually and builds up a magnetic field around both the primary and secondary windings. When this field reaches full magnetic strength, it is called **magnetic saturation**.

When the primary circuit opens, current stops abruptly, and the magnetic field collapses rapidly. The fast collapse of the field induces a higher voltage of 250 to 400 volts in the primary winding through self-induction. The turns ratio of the coil multiplies this voltage to produce several thousand volts, or kilovolts (kV), in the secondary winding. For example if the self-induced voltage in the primary winding is 300 volts and the turns ratio is 150:1, the induced voltage in the secondary winding is 45,000 volts (45 kV). We now have voltage high enough to ignite the air-fuel mixture for all conditions.

Although a modern ignition coil can produce 30,000 to 50,000 volts (30 to 50 kV), it delivers only enough voltage to jump the air gap and create an arc (spark) at the spark plug. This voltage requirement changes with changing resistance conditions in the cylinder. At idle, only 5,000 volts may be needed to create an ignition spark. At high speed or heavy load, 25,000 volts or more may be needed. Electrical resistance increases in a cylinder with:

- A wider spark plug gap
- Higher temperatures and compression pressures
- A leaner air-fuel mixture.

As soon as the spark is created, the remaining coil energy is expended as current flow in the secondary circuit. A coil must have the capability to produce more than the voltage required for an ignition spark. You will learn more about required voltage, available voltage, and voltage reserve when you study coils in more detail as part of the ignition secondary circuit.

## IGNITION CIRCUITS AND CURRENT FLOW

We have reviewed the principles used to induce high ignition voltage and seen where the primary and secondary circuits come together in the coil. We can now examine the primary and secondary circuits and their major parts. The low-voltage primary circuit, figure 12-2, is basically a series circuit with a couple of parallel branches. It contains:

1. The battery
2. The ignition switch
3. The primary (ballast) resistor
4. The starting (ballast) bypass
5. The coil primary winding
6. The breaker points
7. The condenser (capacitor).

When the driver turns on (closes) the ignition switch, current flows:

- From the battery through the ignition switch
- Through the primary wiring and primary (ballast) resistor

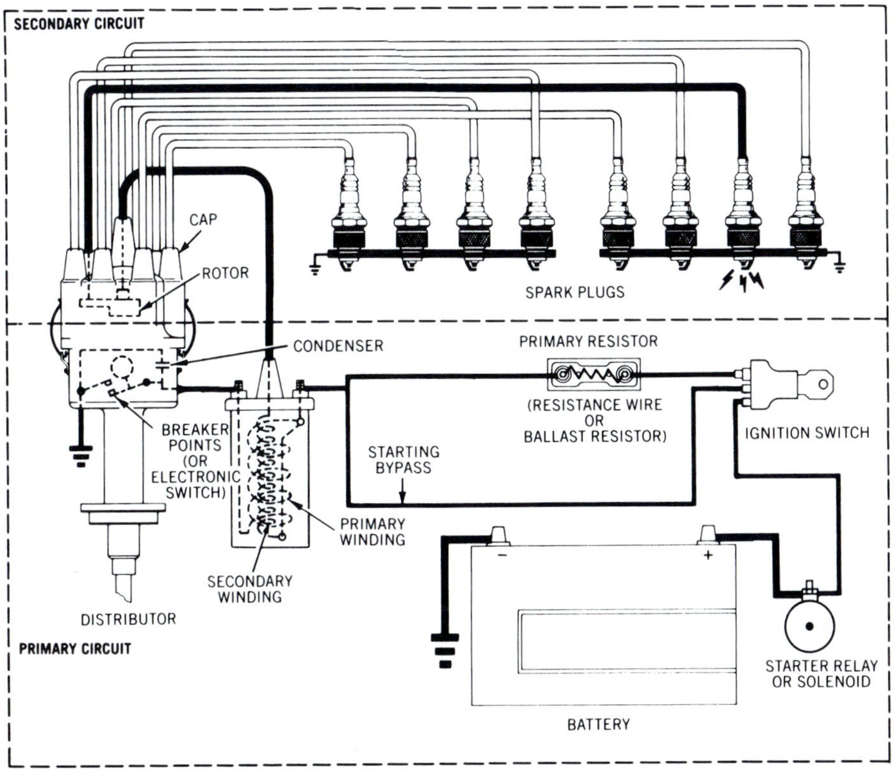


**Figure 12-2.** The ignition system consists of the low-voltage primary and the high-voltage secondary circuits. (Prestolite)

### HERE'S A 20-kV LEAK TESTER

When you have to test for high-voltage leakage in secondary ignition parts such as the distributor cap, plug cables, or spark plugs, it's helpful to have a high-voltage power supply as a test aid. If you don't have an independent power supply, you can use the ignition coil of a breaker-point ignition system. Take a length of metal-core spark plug cable and strip about an inch of insulation off one end. Attach a connector on the other end of the cable and install it in the distributor cap in place of one plug cable. Run the engine at idle, hold the stripped end of the test cable with insulated pliers, and pass it over suspected high-voltage leakage areas in the cap, cables, and elsewhere. If high voltage jumps from the stripped cable end to ground on the cap, cables, or plug insulators, you have found a point where normal ignition voltage also can short, or leak, to ground. Use this test *only* on breaker-point ignitions, *not* on electronic systems.

• Through the coil primary winding

• Through the breaker points, which are an automatic switch

• To ground and back to the grounded terminal of the battery.

The breaker points turn low-voltage primary current on and off. As current starts and stops in the coil primary winding, it creates the expanding and contracting magnetic field that induces high secondary voltage. The high voltage from the coil secondary causes high-voltage current to flow from the coil:

• Through a cable to the distributor cap

• To and through the distributor rotor, the rotor air gap, and another terminal in the distributor cap

• Through another cable to one spark plug

• Across a spark plug gap, where it creates an arc, and back to ground.

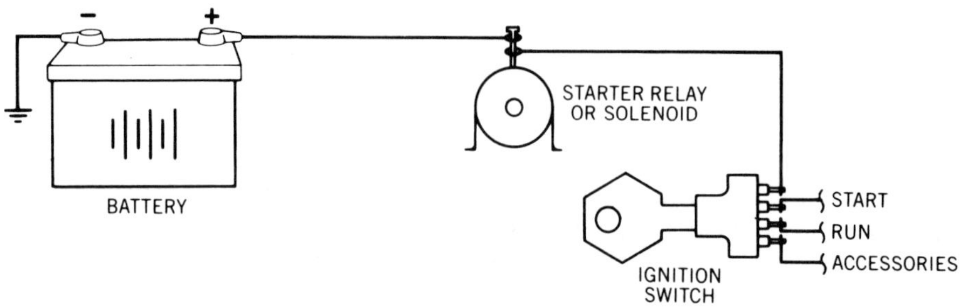


**Figure 12-3.** The ignition primary circuit starts and ends at the battery.

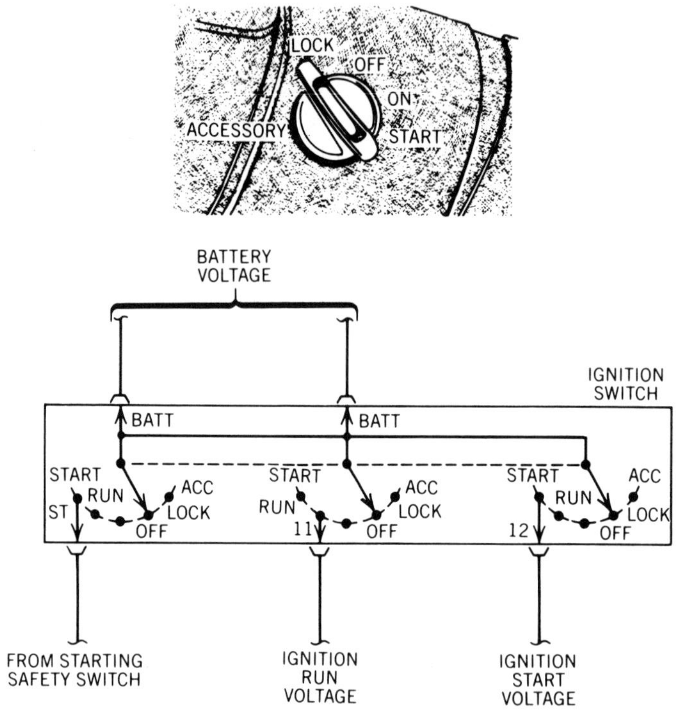


**Figure 12-4.** The ignition switch controls primary ignition current for starting and running, and it controls other electrical circuits on the car. (Ford)

## PRIMARY CIRCUIT COMPONENTS

The following sections explain the purposes of the major parts of the primary circuit.

### The Battery

The battery is the source of low-voltage current for all electrical systems on the car, including the ignition. Current flows to the ignition system, figure 12-3, when the ignition switch is turned to start or run.

### The Ignition Switch

The ignition switch, figure 12-4, is the driver's control switch for the ignition and other electrical circuits on the car. Current flows through the ignition primary circuit when the driver turns the switch to start or run. The switch also has other positions to deliver current to accessory circuits when the ignition circuit is open (off). On most cars, the ignition switch also mechanically locks the steering wheel, and it also may lock automatic transmission shift linkage.

All carmakers use basically similar switches. General Motors vehicles receive ignition current from a connection on the starter solenoid. Ford vehicles receive current from a connection on the starter relay. Chrysler vehicles receive ignition current from a connection between the battery and the alternator.

### The Primary (Ballast) Resistor

All 12-volt breaker-point ignition systems have a **primary, or ballast, resistor** to provide stable voltage and current flow to the coil primary winding, figure 12-2. For a coil to have consistent secondary voltage capabilities under all engine operating conditions, it must have uniform and complete saturation of its magnetic field under all conditions. Magnetic saturation depends on the amount of voltage applied to the coil and the amount and time of current flow through the primary winding.

However, the primary circuit does not have uniform voltage and current at all times. During cranking, the starter motor draws several hundred amperes of current, and battery voltage drops to about 10 or 11 volts. The ignition coil must produce enough secondary voltage to start the engine with this low primary voltage. When the engine starts, system voltage rises above 12 volts. At high speed, system voltage reaches 14 or 15 volts.

Additionally, the amount of time that

current flows through the primary circuit affects coil saturation. At high speed, there is less current because there is less time, and magnetic field strength decreases. At low speed there is more current because there is more time, and field strength increases. But, too much current can overheat the coil and burn the breaker points.

An ignition coil must produce satisfactory secondary voltage with varying primary voltage and current. Most coils are designed to operate with a voltage of 9 or 10 volts under most conditions. To provide this stable voltage, the primary circuit must have a device to compensate for changing system voltage. Moreover, this device must compensate for varying amounts of current at different engine speeds. These are the jobs of the ballast resistor.

At low engine speed, current flows through the resistor longer than at high speeds. This heats the resistor and increases its resistance, which decreases voltage applied to the coil. At high speeds, current flows through the resistor for less time, and it cools. This decreases its resistance and allows more voltage to the coil. The higher voltage but shorter current flow time produces about the same secondary voltage capabilities as the lower voltage and longer current flow that the resistor allows at low speeds. The ballast resistor simply *stabilizes* primary voltage and current for uniform secondary voltage capabilities at all speeds.

Most ballast resistors have about 1.5 ± 0.25 ohms of resistance at 75°F (24° C) with no current flow. However, Chrysler breaker-point ignitions have ballast resistors rated at about 0.5 ohm. The ballast resistor provides about half of the ignition primary circuit resistance. The coil primary winding provides the other half. In most GM, Ford, AMC, and other primary circuits, the ballast resistor is a length of resistance wire between the ignition switch and the coil, figure 12-5. In Chrysler primary circuits, and a few others, the ballast resistor is a coil of wire in a ceramic block, mounted on the firewall, figure 12-6. These resistors, too, are connected between the run contacts of the ignition switch and the coil.

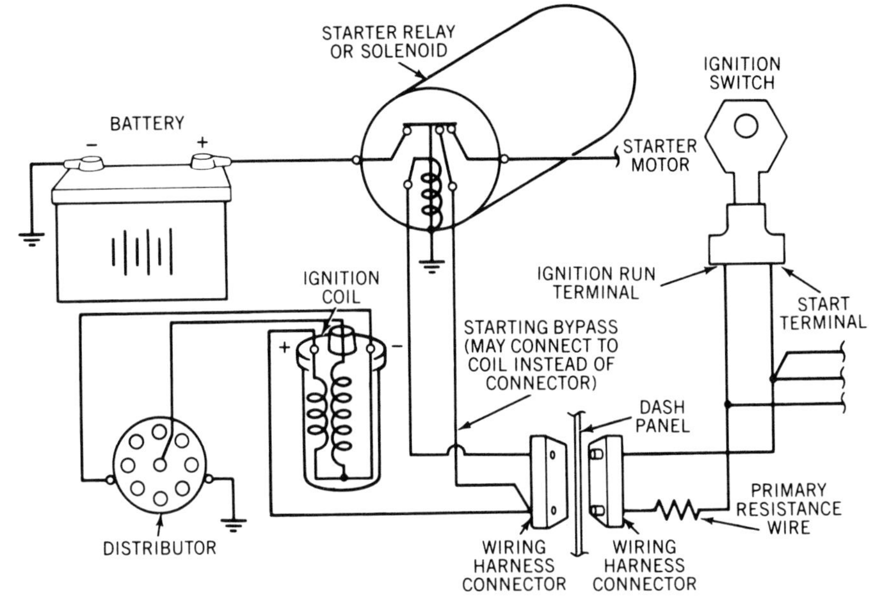


**Figure 12-5.** Most GM, Ford, and other breaker-point ignitions have a length of resistance wire as the ballast resistor.

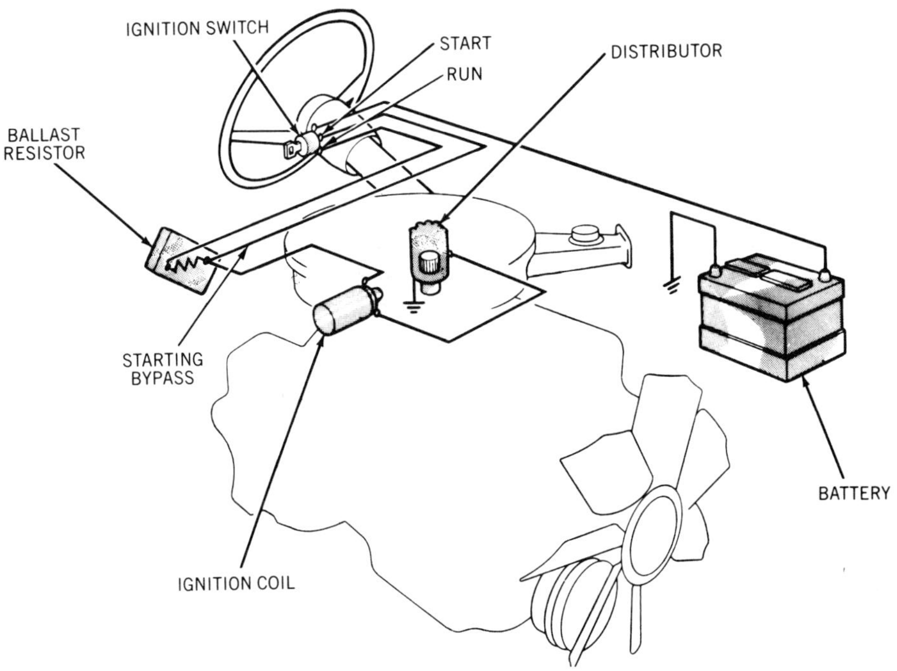


**Figure 12-6.** Chrysler and a few other carmakers use a ballast resistor in a ceramic block. It is connected to the circuit in the same way a resistance wire is. (Chrysler)

## The Starting (Ballast) Bypass

The current draw of the starter motor drops primary voltage to about 10 volts, so the ballast resistor is bypassed during starting. A low-resistance bypass circuit branch, in parallel with the ballast resistor, does this. The **starting bypass** is closed by the start contacts of the ignition switch. When the engine starts, and the ignition switch moves to the run position, the bypass opens, and primary current flows through the ballast resistor. Figure 12-5 shows a starting bypass between the starter relay or solenoid. When the ignition switch closes the relay or solenoid, the bypass applies battery voltage to the coil. Figure 12-6 shows a starting bypass connected from the start contacts of the ignition switch to the coil.

### The Coil Primary Winding

The coil primary winding contains 100 to 150 turns of heavy copper wire, figure 12-7, with a resistance of about 1.5 to 2.0 ohms. The turns of the winding are insulated from each other by a coating of enamel. If they were not insulated, the winding would develop a short circuit, and the primary magnetic field would not develop.

The two ends of the primary winding are connected to the coil positive (+) and negative (−) primary terminals, figure 12-7. In a **negative-ground electrical system**, the coil + terminal is connected to the battery through the ballast resistor and the ignition switch. The coil − terminal is connected to ground through the breaker points. These connections are reversed in a **positive-ground electrical system**.

### The Breaker Points

The coil primary magnetic field must collapse quickly and completely to induce high secondary voltage. After the secondary voltage has discharged to fire one cylinder, the primary magnetic field must build up again quickly to start inducing high secondary voltage for the next cylinder. Because an ignition system must supply as many as 15,000 to 25,000 timed sparks per minute, it must have a precise automatic switch to open and close the primary circuit synchronized in time with the engine firing order.

Until the mid-1970's ignition systems used breaker points as a mechanical switch to open and close the primary circuit. Even though breaker points have been replaced by electronic switching devices on late-model ignition systems, there are still millions of breaker-point-equipped automobiles on the road. Because of their simplicity, breaker points are a good starting point to learn the principles of ignition timing control.

A complete breaker-point assembly, figure 12-8, contains:

- A fixed contact point
- A movable contact point and arm
- A pivot point for the movable arm
- A spring
- A rubbing block
- A breaker plate.

Both contact points are made of tungsten for hardness and good conductivity. The point spring is the connector that joins the movable contact and its arm to the primary terminal of the distributor. The movable contact, the arm, the spring, and the primary terminal are insulated from the distributor housing and connected to the coil negative (−) terminal, which is at a higher voltage potential than the fixed contact point. The fixed contact is grounded through the breaker plate to the distributor housing and completes the primary circuit back to the engine ground and the battery, figure 12-9. When the points are closed, current flows from the movable contact at + potential to the fixed contact at − potential in a negative-ground system.

The distributor shaft has a cam with as many lobes as the engine has cylinders, figures 12-8 and 12-9. As the distributor shaft rotates, its cam pushes the rubbing block on the movable arm to open the points. As the cam lobe rotates away from the rubbing block, the spring closes the points. This is the automatic switching action for the primary circuit. Because opening and closing the primary circuit must be synchronized with engine speed and firing order, the distributor turns at en-

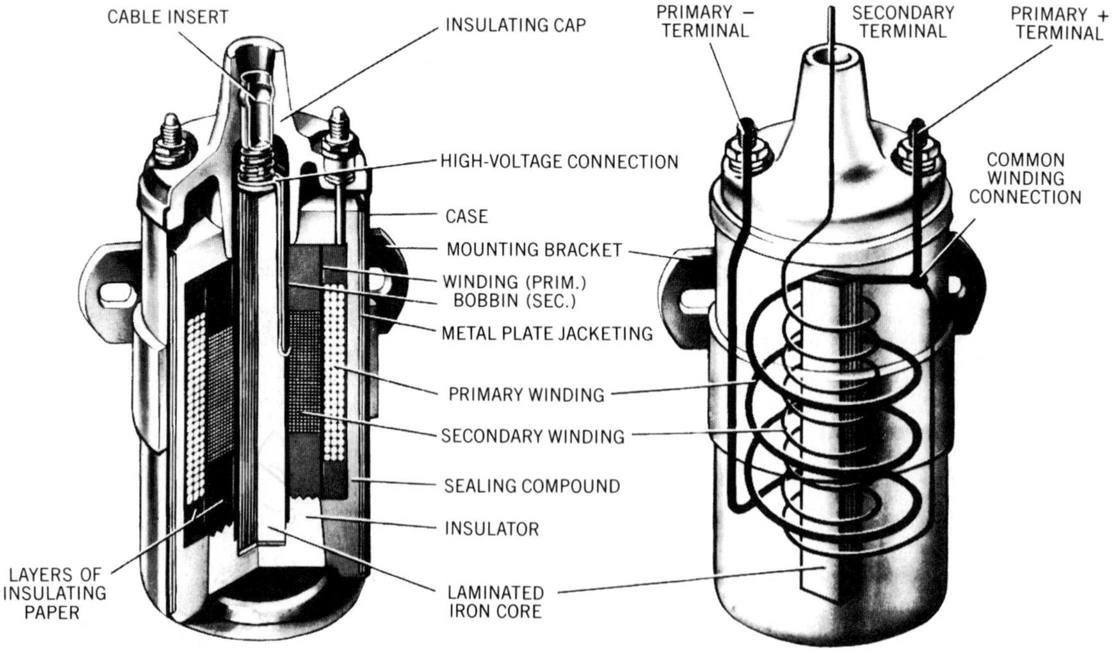


**Figure 12-7.** The coil primary and secondary windings are both wound around a soft iron core. (Bosch)

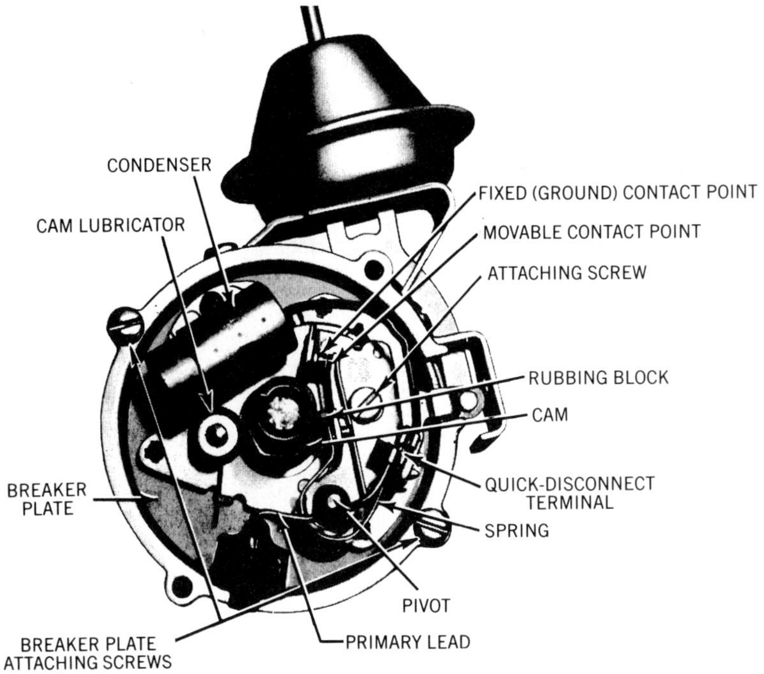


**Figure 12-8.** Individual designs may vary somewhat, but all breaker-point assemblies have these parts. (Delco-Remy)

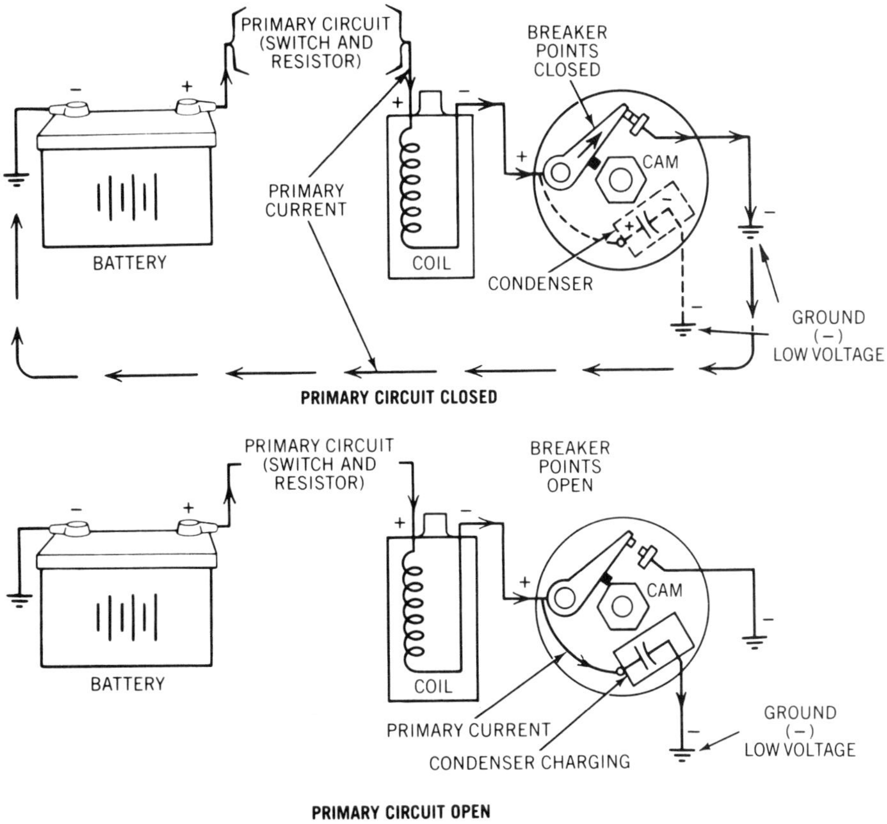


**Figure 12-9.** The breaker points are a switch on the ground side of the coil to open and close the primary circuit. The condenser absorbs a voltage charge when the points open.

## DUAL POINTS

In the days before electronic ignitions, short dwell time was a problem for high-performance engines at high rpm. A 32-degree dwell angle on a V-8 engine at 6,000 rpm, for example, simply did not allow enough time for complete coil saturation. The result was usually a high-speed misfire as available voltage fell below required voltage. If a mechanic tried to adjust the points for a longer dwell angle, they would arc and burn at low speeds because of the insufficient gap.

A common solution to the problem was to build a distributor with two sets of breaker points in parallel. The points are offset in relation to the distributor cam. One set opens the primary circuit; the other set closes the circuit. The dwell on each set of points can be 27° to 31°, for example, with sufficient gap to minimize arcing. The combined dwell of both sets, however, can be 36° to 40° to provide adequate coil saturation and voltage at high speed.

Because the point sets are in parallel, the primary circuit is closed whenever either or both of the point sets are closed. The first set to open is the "closing" point set because the other set maintains primary current flow. The second set to open is the "opening" point set. Dual-point distributors, like this, were common on high-performance engines before electronic ignitions became common.

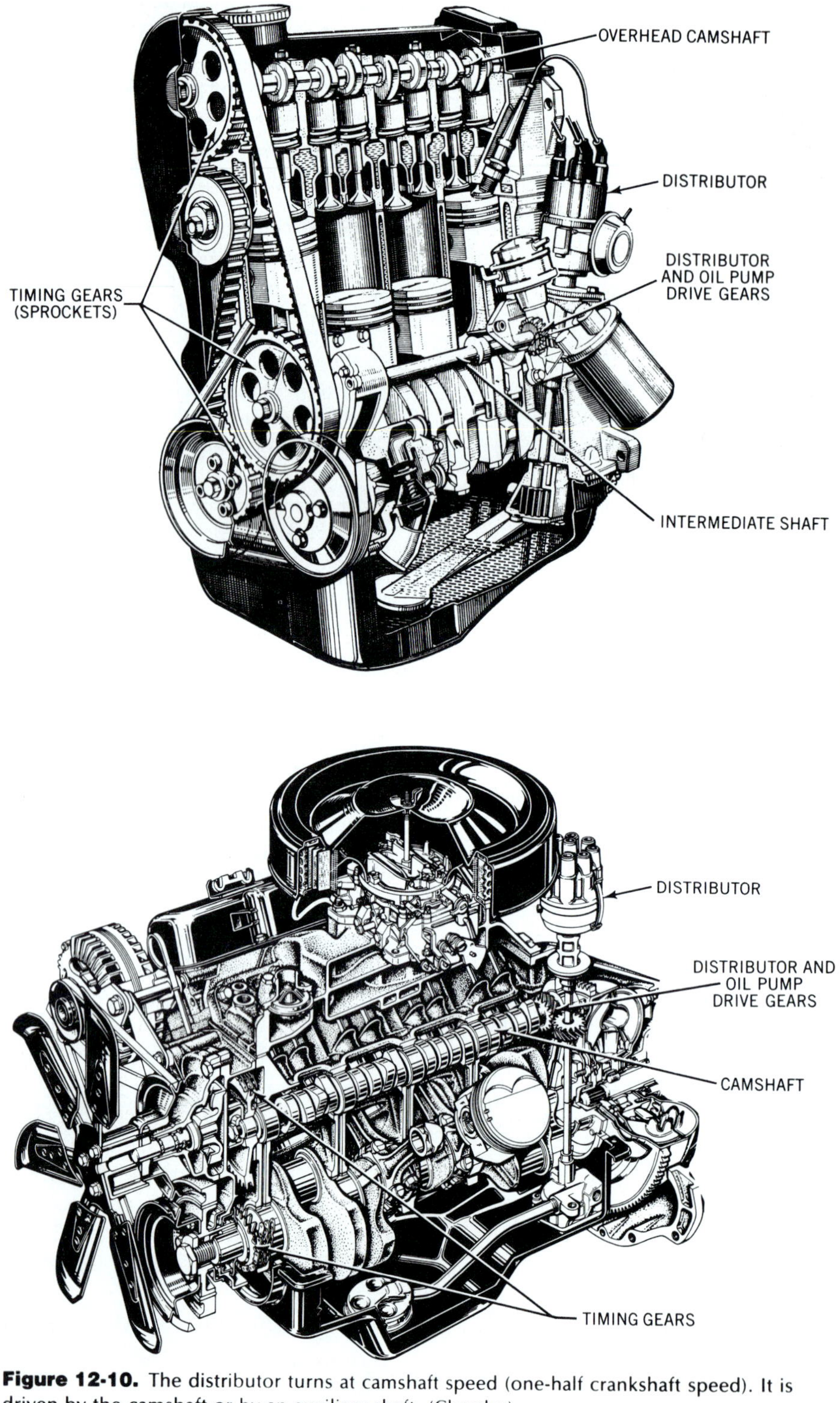


**Figure 12-10.** The distributor turns at camshaft speed (one-half crankshaft speed). It is driven by the camshaft or by an auxiliary shaft. (Chrysler)

gine camshaft speed (one-half crankshaft speed). Distributors on overhead-valve engines are driven by the camshaft. The distributor on an overhead-cam engine can be driven by the camshaft or by an auxiliary shaft that turns at camshaft speed, figure 12-10.

**Ignition Dwell Angle.** Each time the breaker points open, the coil secondary winding discharges high voltage to a spark plug. In this way, the points control ignition timing. When the points are closed, they control the amount of magnetic saturation that

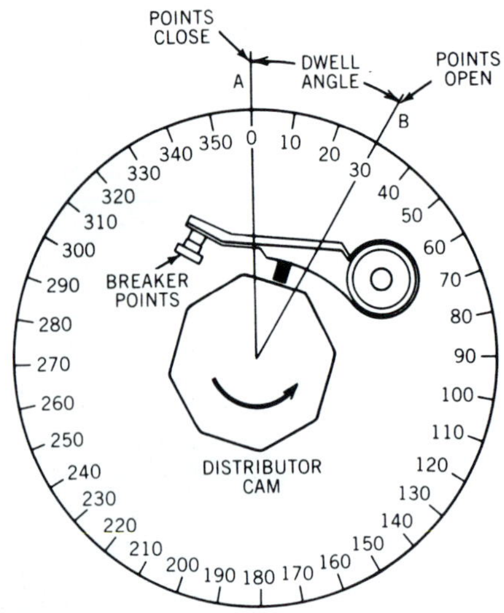


**Figure 12-11.** The dwell angle is the number of degrees during which the points are closed.

develops in the coil. The distributor, or ignition, **dwell angle** is the number of degrees of distributor rotation during which the points are closed.

In figure 12-11, the points close at line A and open again at line B. The number of degrees between A and B is the dwell angle. This is the period when current flows through the coil primary winding to build up the magnetic field.

Remember that magnetic field strength is proportional to the *time* that current flows. The dwell angle actually controls this time, but we measure it in degrees of distributor rotation because this is the easiest way to measure and adjust it. At a slow rotation speed, a dwell angle of 30 degrees allows more current time than the same dwell angle allows at high speed. However, at high speed, the ballast resistor allows a higher primary voltage than it does at low speed. Short current time at higher voltage produces about the same magnetic field strength as longer current time at lower voltage.

In chapter 7, you learned that current in any coil induces a countervoltage that opposes original current direction. This means that current flow and the magnetic field in a coil do not reach full strength instantly. It takes about 10 to 15 milliseconds (0.010 to 0.015 second) for an ignition coil to develop complete magnetic saturation from primary current, figure 12-12.

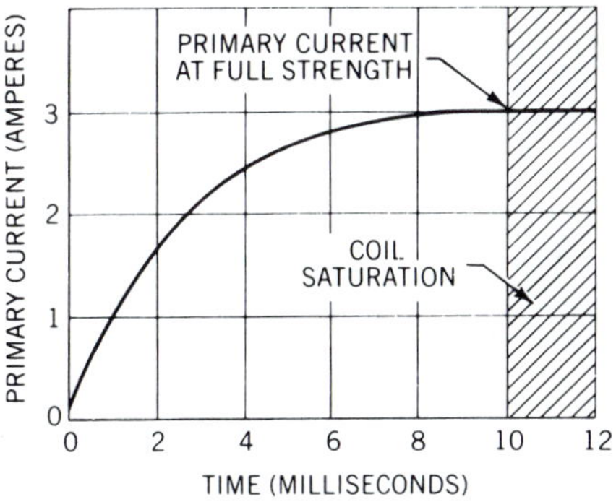


**Figure 12-12.** Primary current requires about 10 to 15 milliseconds to reach full strength. (Bosch)

At idle speed, an ignition dwell angle of 30 degrees allows about 15 milliseconds of primary current flow. At higher speeds, the dwell angle may not change, but the dwell *time* decreases. At 1,000 rpm, a 30-degree dwell angle allows about 9 milliseconds of current. At 2,000 rpm, the same dwell angle allows only 4.5 milliseconds of dwell time.

If current time were the only factor controlling coil secondary voltage, the voltage could be too low at high speed to fire the spark plugs. However, ignition systems are designed to balance the factors of current amperage, current time, primary voltage, and the turns ratios of the primary and secondary coil windings. Careful design ensures that the system will provide the secondary voltage needed for ignition under all conditions. As you understand all the factors of correct ignition operation, you can appreciate why correct maintenance of all parts of the system is essential for proper engine operation.

Now that you have learned about ignition dwell, you should have a better idea how the breaker points and the ballast resistor work together to control primary voltage and current. These combined actions allow the coil secondary winding to have relatively constant secondary voltage capabilities at all engine speeds. When you study electronic ignitions, you will see other interesting ways that electronic devices accomplish the same thing.

Ignition dwell angle is related to an engine's firing intervals and crankshaft and camshaft rotation. Remember that in an 8-cylinder engine, there are 90 degrees of crankshaft rotation between cylinder firing intervals. This is the same as 45 degrees of camshaft and distributor rotation. Theoretically, an 8-cylinder engine could have 45 degrees of ignition dwell. But, the points must open and close during several of those 45 degrees. Most 8-cylinder distributors rotate about 15 degrees from the time the points open (break contact) until they close again. Therefore, the dwell angle is about 30 degrees.

In an even-firing 6-cylinder engine, the crankshaft rotates 120 degrees between cylinder firing intervals (60 degrees of camshaft and distributor rotation). In a 4-cylinder engine, the crankshaft rotates 180 degrees, and the camshaft rotates 90 degrees between firing intervals. For a 6-cylinder engine, the dwell angle could be about 45 degrees, and it could be about 75 degrees for a 4-cylinder engine. However, a larger dwell angle would allow longer current flow time at low speed. This is not necessary for complete coil saturation and could overheat the primary winding. Therefore, 4- and 6-cylinder distributors have less dwell than they theoretically could have, about 45 to 50 degrees for a 4-cylinder distributor and about 35 to 40 degrees for a 6-cylinder distributor.

### VISUAL POLARITY TEST

Modern ignition system design makes the probability of reversed coil polarity unlikely. Moreover, modern test equipment makes the condition easy to detect when it does occur. An upside-down oscilloscope pattern, for example, is an immediate clue to possible reversed system polarity.

Old-time mechanics had simpler ways to check for engine problems, such as reversed polarity, however. Here's an easy test that you can use on any breaker-point ignition.

CAUTION

Do not use this test on a Delco HEI or similar high-voltage electronic ignition. A high-voltage arc may damage system components. Do not use a metal mechanical pencil or you will get shocked.

You can use a common lead pencil to check coil polarity on a breaker-point ignition, as follows:

**1.** Disconnect one spark plug cable from its plug.

**2.** Hold the cable terminal about 1/4 inch from the plug, with the tip of a wooden pencil about halfway between the cable and plug terminals, as shown in the illustration.

**3.** Crank the engine briefly and watch the spark flare at the pencil tip. If the flare goes toward the plug, polarity is correct. If the flare goes toward the cable, polarity is reversed.

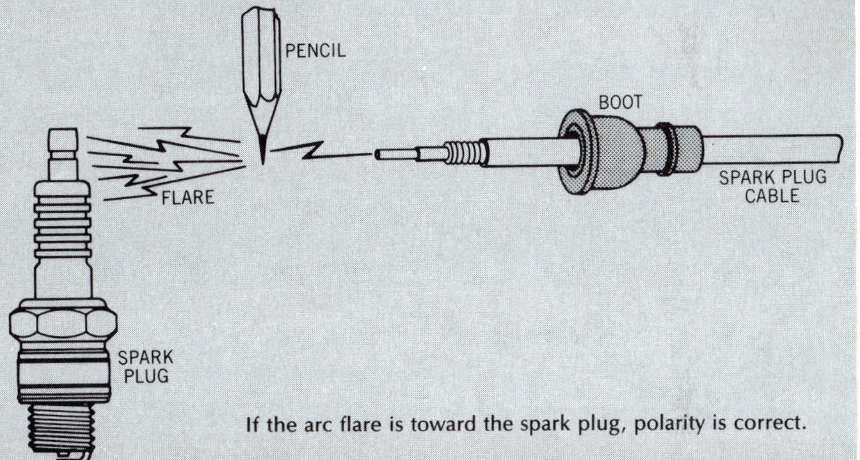


If the arc flare is toward the spark plug, polarity is correct.

**Point Gap.** The **point gap** is the maximum distance between the point contact surfaces when the rubbing block is on the high point of the distributor cam and the points are fully open. The points in any distributor must open far enough to avoid primary current arcing between them. Point gap is a common distributor specification and is inversely related to the dwell angle, figure 12-13. A large point gap provides a small dwell angle. A small point gap provides a large dwell angle. Point gap is measured in thousandths of an inch or hundredths of a millimeter. For an 8-cylinder engine, a gap of about 0.020 inch (0.50 mm) provides a dwell angle of about 30 degrees. A gap of 0.018 inch (0.46 mm) for the same set of points may provide a dwell angle of 34 degrees.

## The Condenser

As breaker points start to open, current tries to keep flowing across the gap. Moreover, the countervoltage induced in the primary winding increases the force that tries to move current across the point gap. This would create an arc between the points that would have two harmful effects. It would burn the points, and it would delay the collapse of the primary magnetic field.

All breaker-point ignitions have a capacitor, or condenser, to minimize point arcing and to help cut off primary current quickly. Recall from your study of electrical principles that a capacitor is a device that will store a voltage charge when it is connected between two points of different voltage potential. A capacitor has two conductive plates, or surfaces, separated by a dielectric insulating material. An ignition condenser is a capacitor made of two thin foil strips separated by insulating paper, figure 12-14. The two strips are installed in a small metal canister. One strip is connected to a small wire lead. The other is connected to the canister shell, which provides a ground connection for the capacitor.

The condenser wire lead is connected to a positive voltage source at the breaker points. The canister shell is grounded to a negative potential by connecting it to the distributor housing, figures 12-8 and 12-9. The condenser is in parallel with the breaker points. When the points are closed, they offer less resistance than the condenser, so current flows through the points to ground. As the points start to open, their resistance increases. At this time, the condenser is discharged, so

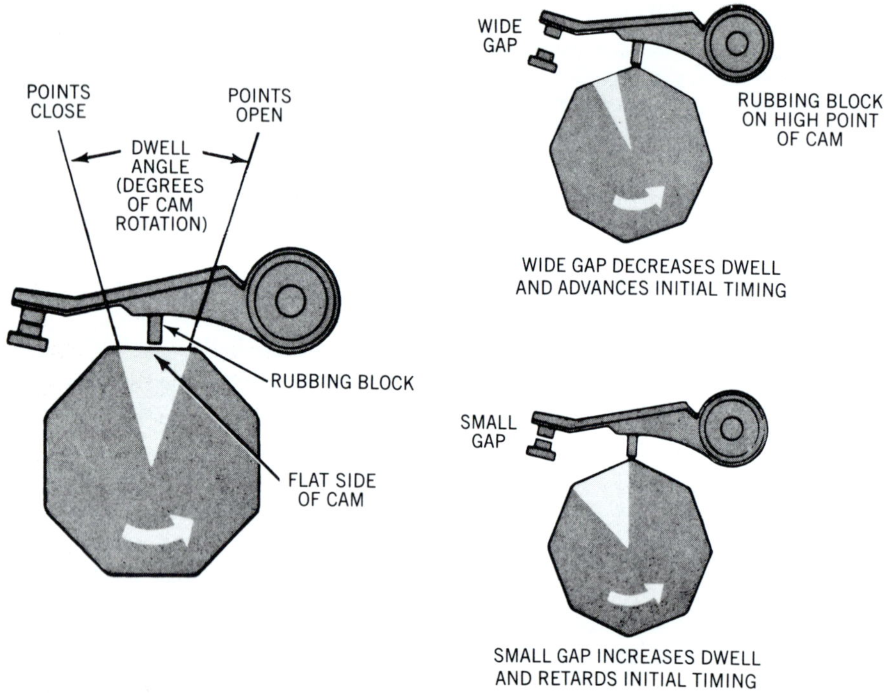


**Figure 12-13.** A close point gap creates a long dwell angle. A wide gap creates a short dwell angle. (Ford)

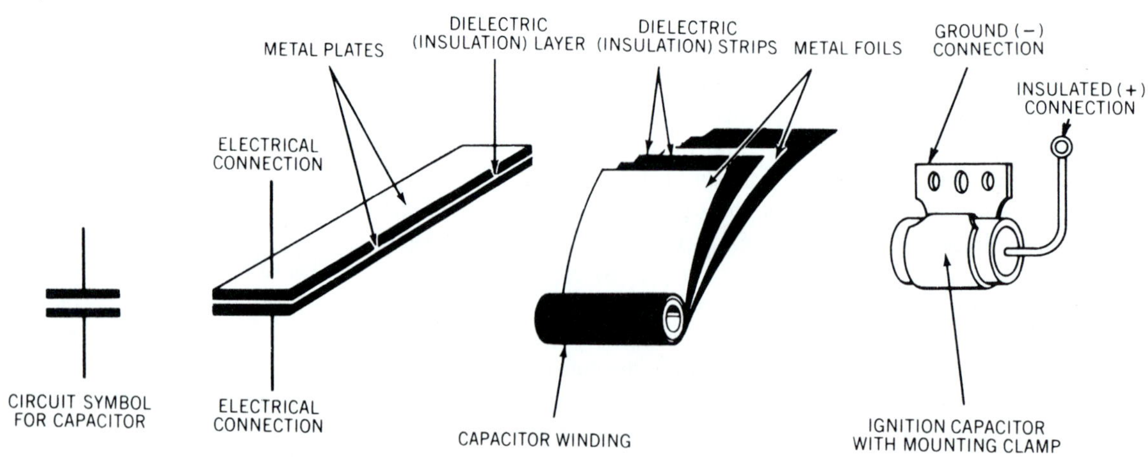


**Figure 12-14.** An ignition condenser is a capacitor made of two foil strips installed in a canister. (Bosch)

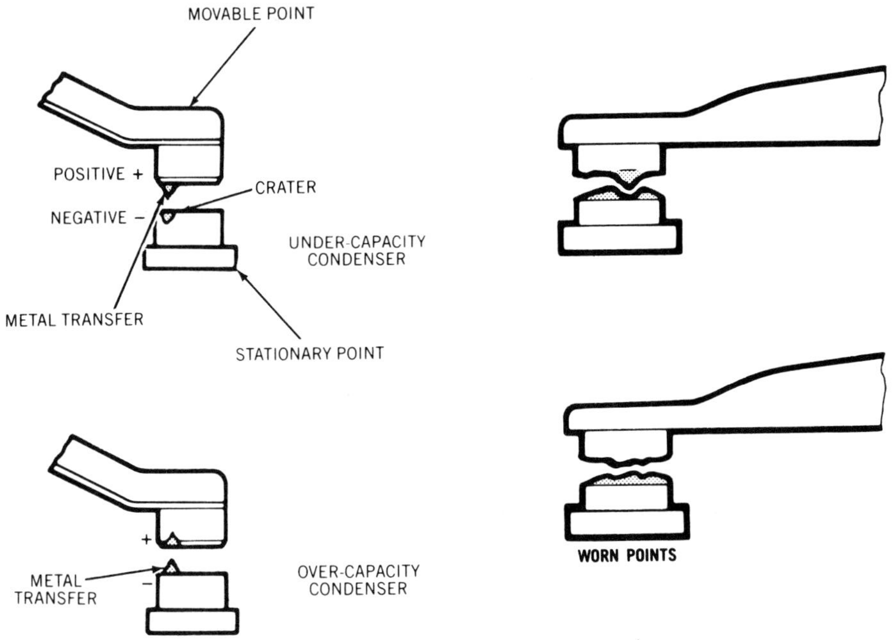


**Figure 12-15.** A low-capacity condenser causes pitting on the ground point. An over-capacity condenser causes opposite metal transfer. (Chrysler)

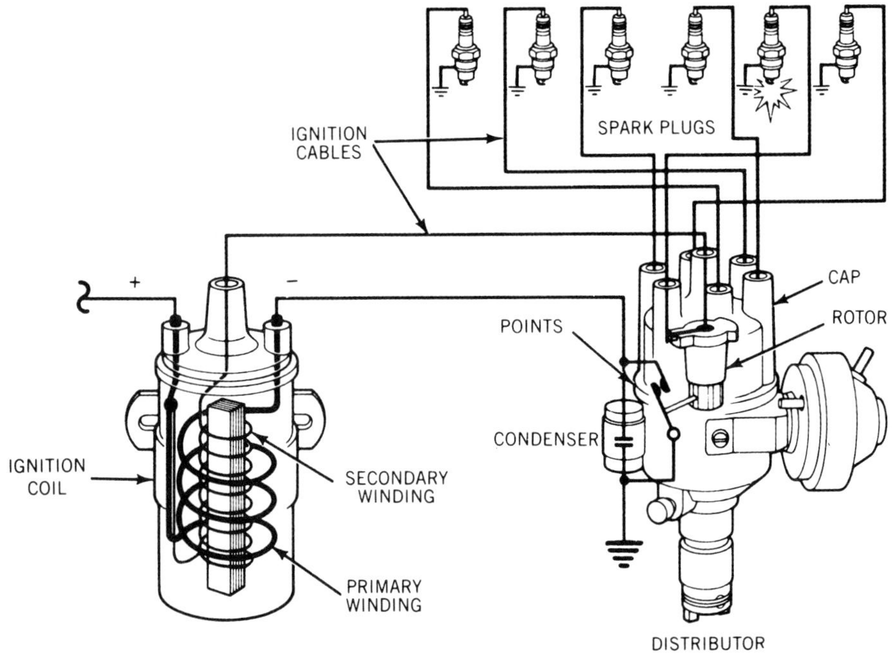


**Figure 12-16.** The high-voltage secondary circuit includes the coil secondary winding, the distributor cap and rotor, the spark plugs, and cables. (Bosch)

current flows to the positive plate of the condenser because it has less resistance than the point gap.

Current flows to the positive plate of the condenser until its voltage potential is the same as the positive voltage of the primary circuit. Current cannot pass through the dielectric of the condenser, so the condenser develops a voltage charge between its positive and negative plates. The condenser is charged by the self-induced countervoltage of the primary winding, so it actually stores several hundred volts.

Charging the condenser takes about 0.1 millisecond, during which time the points open almost completely. Primary current stops abruptly, and the magnetic field collapses quickly. The field collapses about 20 times faster with the condenser in the circuit than it would without it.

While the points are open, the condenser maintains the voltage charge between its plates. When the points close, primary current is at zero, and there is no magnetic field in the coil. The condenser releases its charge to the primary circuit and helps to start current flow. The condenser energy then **oscillates** between the condenser and the coil primary winding.

The charge capacity of any capacitor is rated in farads. A typical ignition condenser has a capacity of 0.18 to 0.32 microfarad. If a condenser with too little capacity is used in a circuit, primary current will charge the condenser and still be able to arc across the points. This causes burning and metal transfer from the – to the + contact point, figure 12-15. If a condenser capacity is too large, metal transfer occurs in the opposite direction. The **minus rule** will help you remember how condenser capacity affects point burning:

> Minus metal on the minus (–) point means a minus-capacity condenser.

Badly worn points often have metal transfer and erosion on both surfaces, figure 12-15.

## SECONDARY CIRCUIT COMPONENTS

Figure 12-2 shows the relationships of the primary and secondary circuit components. The ignition secondary circuit, figure 12-16, consists of:

- The coil secondary winding
- The distributor cap
- The distributor rotor
- The ignition cables
- The spark plugs.

### The Coil Secondary Winding

The ignition secondary circuit begins at the coil secondary winding. The secondary winding has 15,000 to 30,000 turns of fine copper wire, figure 12-8, insulated from each other by a coating of enamel. Remember that the turns ratio between the secondary and the primary windings is

100:1 to 200:1 and acts as a voltage multiplier. To further increase the coil's magnetic field strength, both windings are installed around a soft iron core. The primary and secondary windings are insulated from each other, and the coil case is filled with wax or oil for more insulation. Most coils are contained in a metal cylinder. The coil top is made of an insulating material and has terminals for primary and secondary connections.

**Coil Polarity.** The primary and secondary circuit connections to the coil determine coil polarity and the polarity of the secondary circuit. One end of the secondary winding is connected to the primary + terminal in a negative-ground electrical system. The other end of the secondary winding is connected to the high-tension terminal of the coil. In this way, the secondary circuit is grounded through the primary circuit. The ground side of the secondary circuit is electrically positive. The other side of the circuit—the center electrode of the spark plug—is negative, figure 12-17.

The polarity of the secondary circuit is opposite to that of the primary circuit in most vehicles. Regardless of whether the primary circuit has negative or positive polarity, the secondary circuit must have negative polarity (positive ground) because:

**1.** Electrons (current) flow more easily from a negative point to a positive point than they do in the opposite direction.

**2.** High temperature of the spark plug center electrode increases the rate of electron movement (current).

The center electrode of a spark plug is hotter than the side electrode because it can't transfer heat to the cylinder head as easily. If the center electrode is at a negative potential, electrons will move more easily to the side electrode. Less secondary voltage will be required to force current across the plug gap. If coil connections are reversed, figure 12-18, or the wrong coil is installed, the direction of the coil magnetic field is reversed. This reverses the secondary voltage polarity and forces electrons to move from the side electrode to the center electrode of the spark plug. This **reverse polarity** increases the required secondary voltage 20 to 40 percent, figure 12-19.

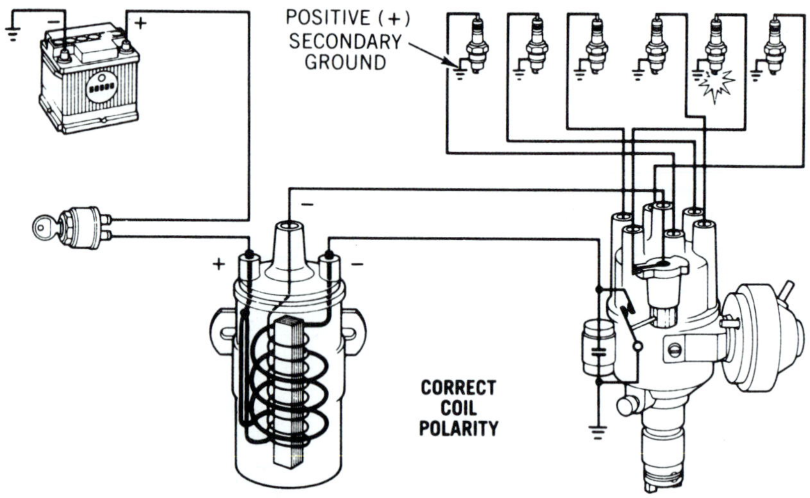


**Figure 12-17.** Correct (negative) secondary polarity exists when the coil is connected so that the spark plug center electrode is negative and the ground electrode is positive. (Bosch)

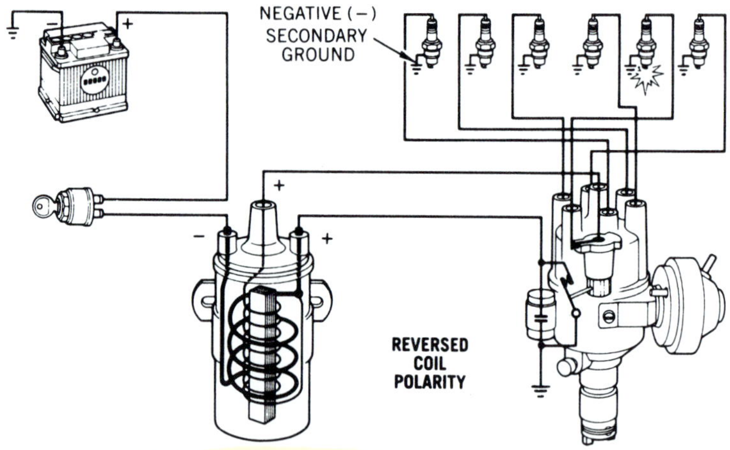


**Figure 12-18.** Reverse (positive) polarity exists when the coil is connected so that the spark plug center electrode is positive. (Bosch)

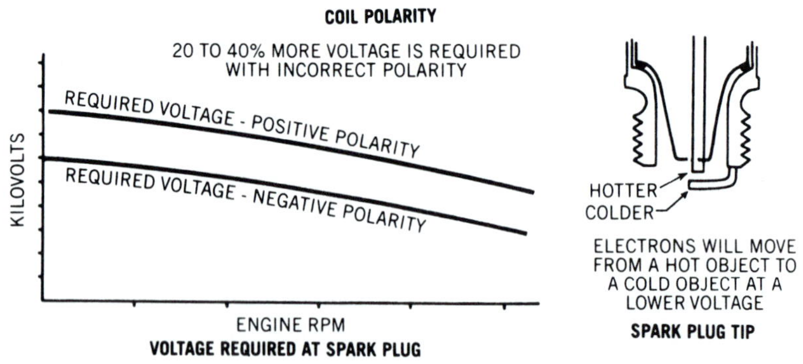


**Figure 12-19.** Reversed polarity requires 20 to 40 percent more secondary voltage to fire the plugs. (Prestolite)

**Coil Secondary Voltage.** You learned earlier that a coil delivers only the secondary voltage required to jump the spark plug gap at any given instant.

This is the **required voltage** of the ignition system, and it changes with varying engine conditions. You also learned that the coil must be *able* to deliver more secondary voltage than the engine requires so that it will always have enough voltage capability for *all* conditions. The maximum, or peak, voltage that a coil can deliver is its **available voltage**. The difference between the required voltage and the available voltage under any condition is the coil's **voltage reserve**.

Available voltage is determined by a coil's secondary-to-primary turns ratio and other factors of its design. Available voltage also is affected by primary circuit voltage, current, and resistance. Required voltage is determined by conditions in the engine cylinders and in the secondary circuit. You know that engine compression pressure and temperature and the spark plug gap create high resistance in the secondary circuit. The air-fuel mixture, too, is a high-resistance conductor. Here are some conditions that will *increase* the secondary circuit resistance and the required voltage:

1. A wider spark plug gap or an overheated plug
2. A leaner air-fuel mixture (fewer conductive gasoline molecules)
3. An overheated engine
4. Faster engine speeds and heavier engine loads, which increase temperature and pressure
5. Loose or corroded connections or other high-resistance conditions in the secondary cables from the coil to the distributor and the distributor to the spark plugs.

Here are a few conditions that will *decrease* secondary resistance and required voltage:

1. A narrower spark plug gap or a fouled plug that provides a low-resistance path for secondary current
2. A richer air-fuel mixture (more gasoline molecules)
3. Slow engine speeds and light loads, which decrease temperature and pressure.

These last three conditions can occur *inside* the cylinders. Secondary circuit resistance also can be reduced by a short circuit *outside* the cylinders, such as a conductive path on the outside of the coil, a spark plug, a cable, or in the distributor cap. This kind of lower resistance short circuit still requires secondary voltage from the coil, but the voltage jumps to ground outside the cylinder. It never gets near the spark plug gap, and the engine misfires.

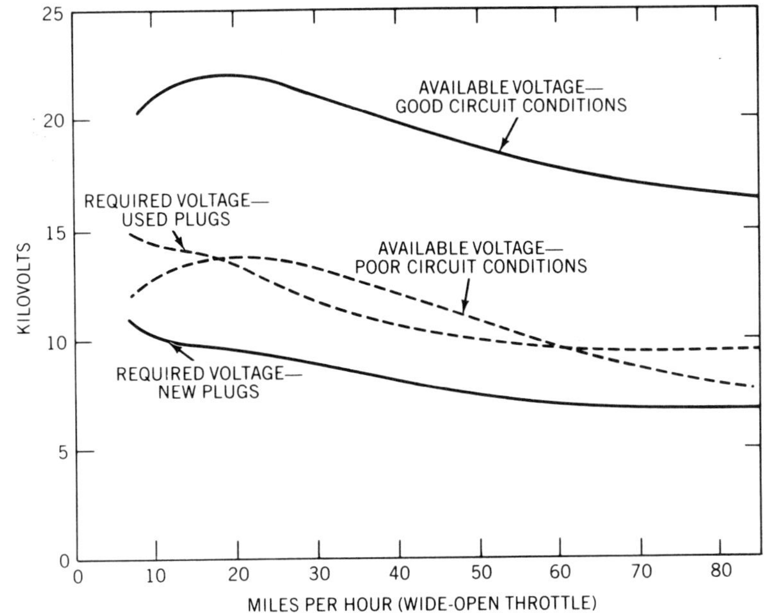


**Figure 12-20.** A well-tuned ignition system should maintain a 60-percent voltage reserve. (Prestolite)

Conditions in an engine and in an ignition system can affect both available voltage and required voltage. Low voltage or high resistance in the primary circuit decreases available secondary voltage because either will reduce primary field strength. If available voltage decreases when secondary voltage requirements are high, the coil may not have enough voltage reserve to fire the spark plugs. Figure 12-20 compares available and required voltage under good and bad circuit conditions. To ensure correct engine operation, a well-tuned ignition system should have a 60-percent voltage reserve under most conditions.

We can get a picture of the changing secondary voltage conditions by looking at a series of graphs that plot voltage levels over a period of time. These graphs are similar to voltage waveforms displayed on an **oscilloscope**, which you will use for ignition service.

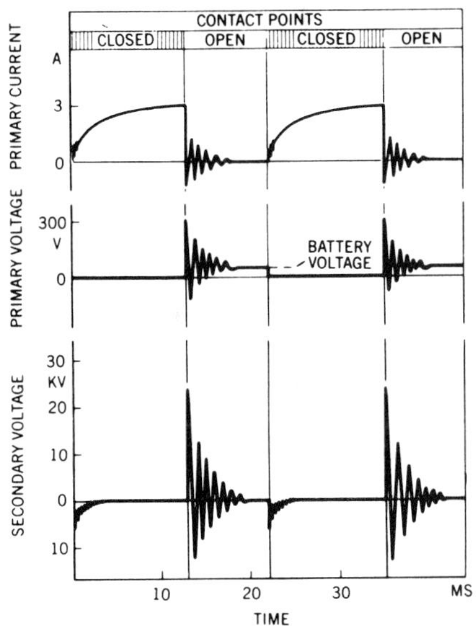


**Figure 12-21.** Open-circuit (no-load) voltage is the coil's maximum available voltage. (Bosch)

If the secondary circuit were open—that is, if there were no spark plug or cable between the coil and ground—the maximum available voltage of the coil could not complete a circuit. The voltage would oscillate through the coil windings, and between the coil and condenser, and dissipate its energy as heat. Figure 12-21 shows the pattern of this **open-circuit (no-load) voltage**. This is the same as the coil's available voltage and ranges from 20 kV to 50

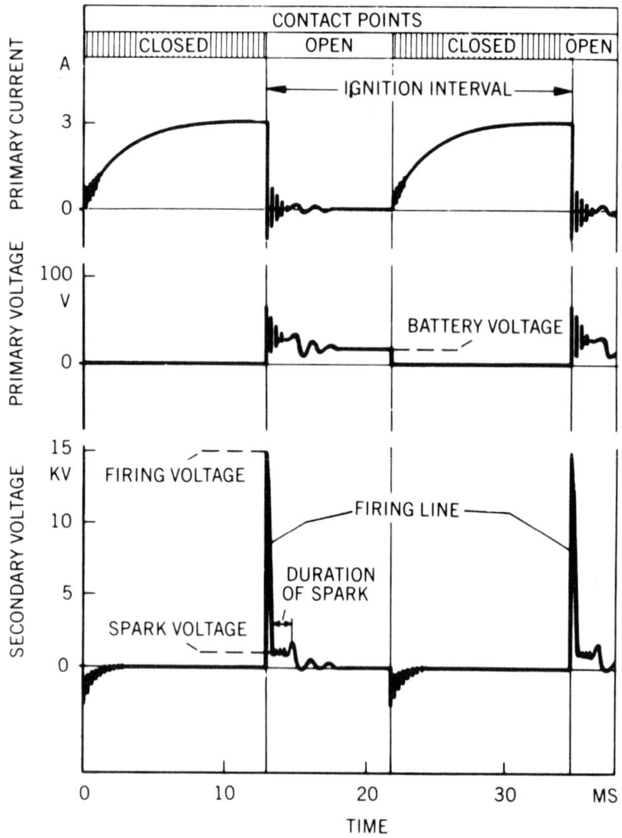


**Figure 12-22.** Firing voltage is the voltage required to create a spark at the plug. It is lower than open-circuit voltage. (Bosch)

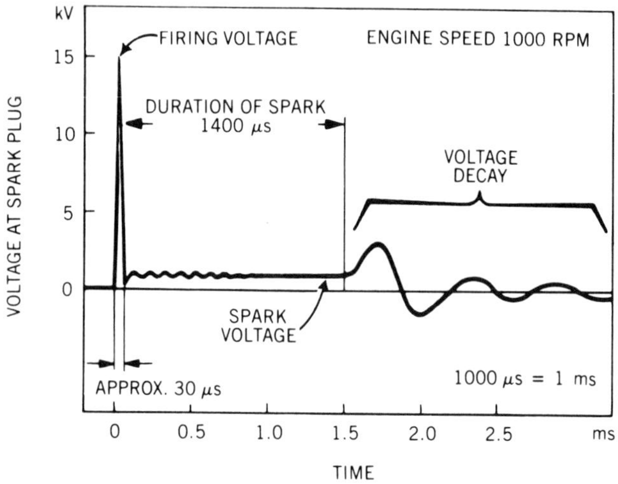


**Figure 12-23.** This complete secondary voltage pattern shows the lower spark voltage that maintains the spark after the plug fires. (Bosch)

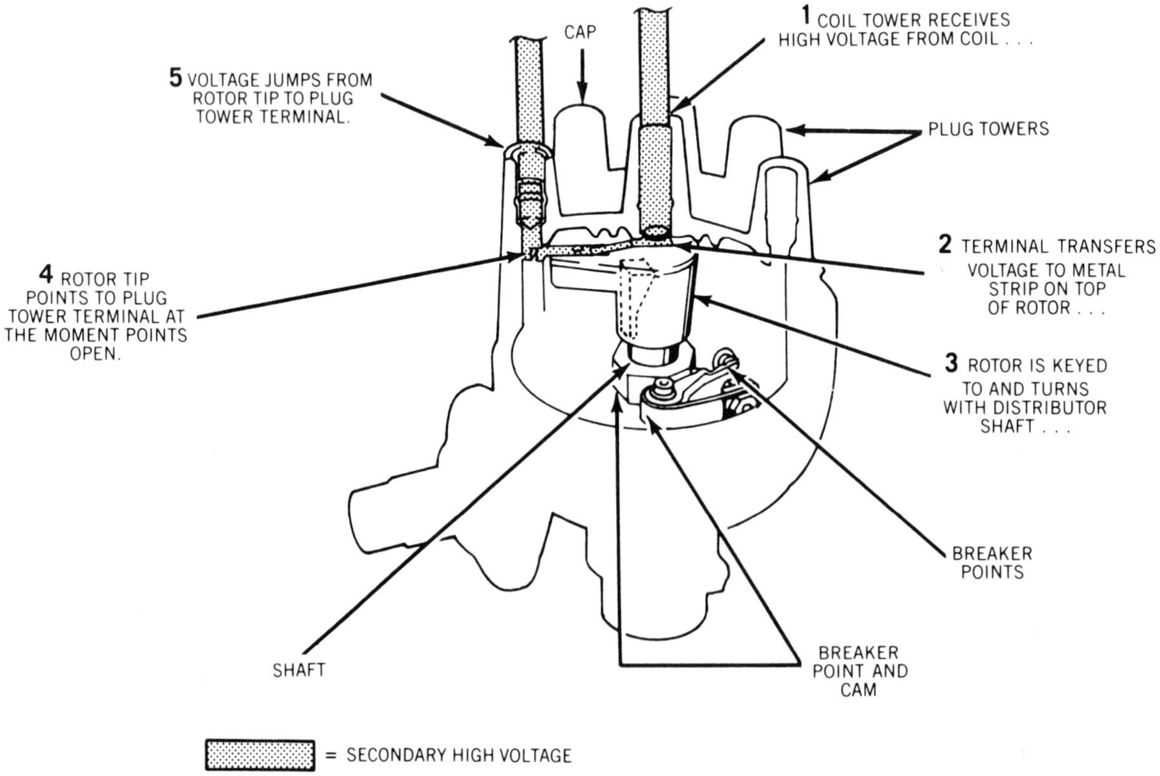


**Figure 12-24.** The distributor cap and rotor receive high-voltage current from the coil and distribute it to the spark plugs. (Ford)

kV for different coils. Figure 12-21 also shows primary current and voltage during the same time periods.

With a spark plug in the circuit, coil voltage completes the circuit by creating an arc across the gap. When the secondary winding discharges, voltage rises to the level needed to jump the gap. The high voltage ionizes the air and fuel molecules at the plug gap and makes them more conductive. When voltage overcomes resistance at the plug gap, the plug fires. Figure 12-22 shows this **firing (ionization) voltage**, which is the same as the required voltage. Engineers also call this the capacitive portion of the secondary voltage discharge because the spark gap acts as a capacitor in the circuit. Secondary voltage charges the capacitive terminals formed by the plug electrodes until the voltage overcomes the dielectric resistance of the gap.

Once the plug fires, less voltage is needed to sustain the spark across the gap. Voltage drops almost instantly to a much lower level, called the **spark voltage**, figure 12-22. This is the inductive portion of the secondary discharge because the coil now discharges the rest of its voltage as an induction coil in a closed circuit. Figure 12-23 shows an enlarged view of the secondary voltage trace from the point at which the plug fires until voltage drops too low to sustain the spark across the gap. This figure also shows the time in microseconds for each part of the plug-firing sequence. The final part of the secondary voltage oscillates in the coil windings and between the coil and condenser. It dissipates its energy as heat and is called the voltage decay.

We have taken quite a bit of time to examine ignition secondary voltage before looking at the other parts of the secondary circuit. We did this because it is easier to understand the design and operation of these parts if you know something about the voltage requirements of the circuit and how the high voltage originates.

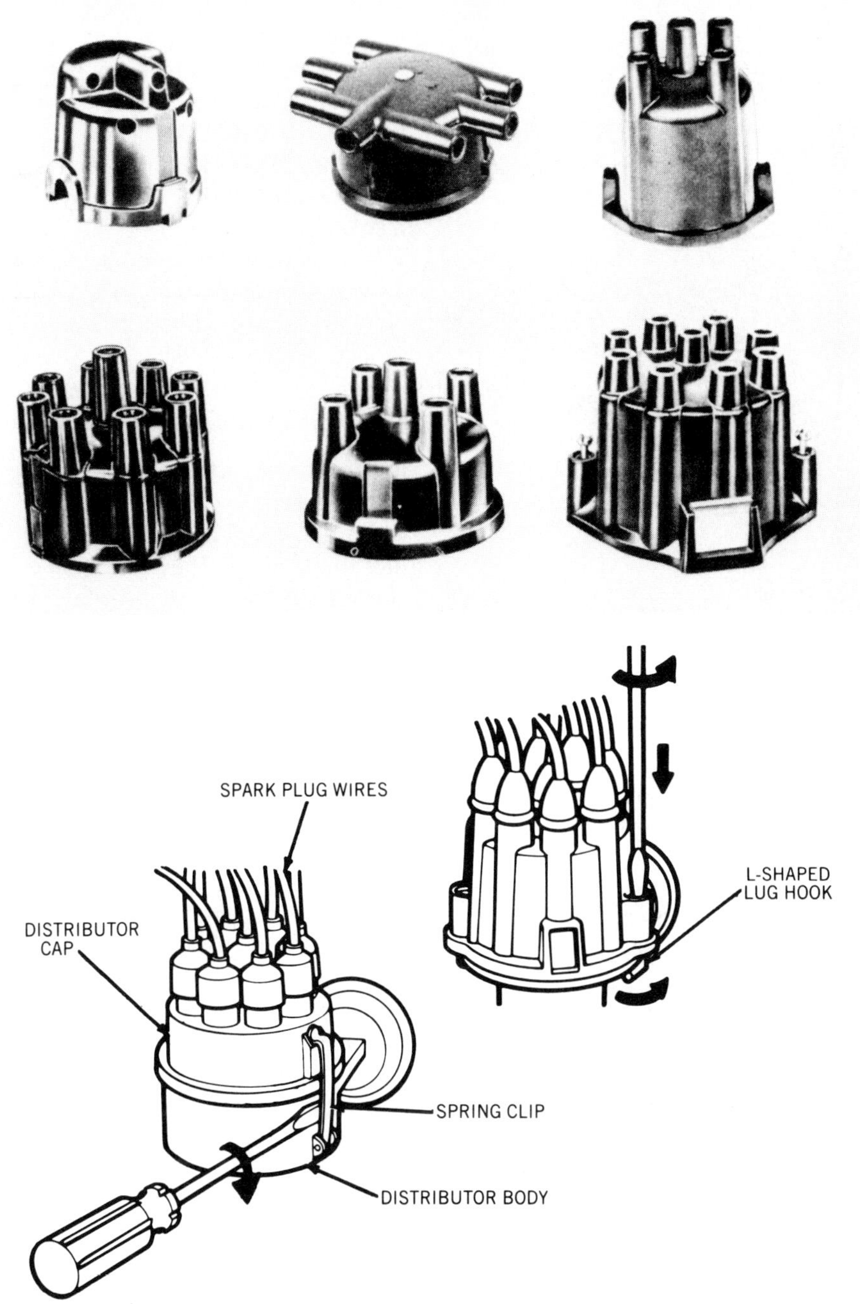


**Figure 12-25.** Regardless of the shape, every distributor cap serves the same purpose. (Chrysler)

### The Distributor Cap and Rotor

High-voltage current from the coil secondary winding travels to the distributor cap and rotor, figure 12-24, from where it is distributed to the spark plugs in firing order. Current enters the cap through a central terminal, passes through the rotor, crosses an air gap from the rotor tip to a terminal in the cap, and travels through another cable to a spark plug.

**Distributor Cap.** The distributor cap, figure 12-25, is made of bakelite, silicone plastic, or a similar insulating material. The central terminal for the coil cable and metal inserts in the spark plug towers are circuit connections for the cables and the rotor. The coil terminal inside the cap may be a carbon insert of a spring-loaded metal button that contacts the rotor. The cap is keyed to the distributor body and held in place by spring clips, spring-loaded latches, or screws.

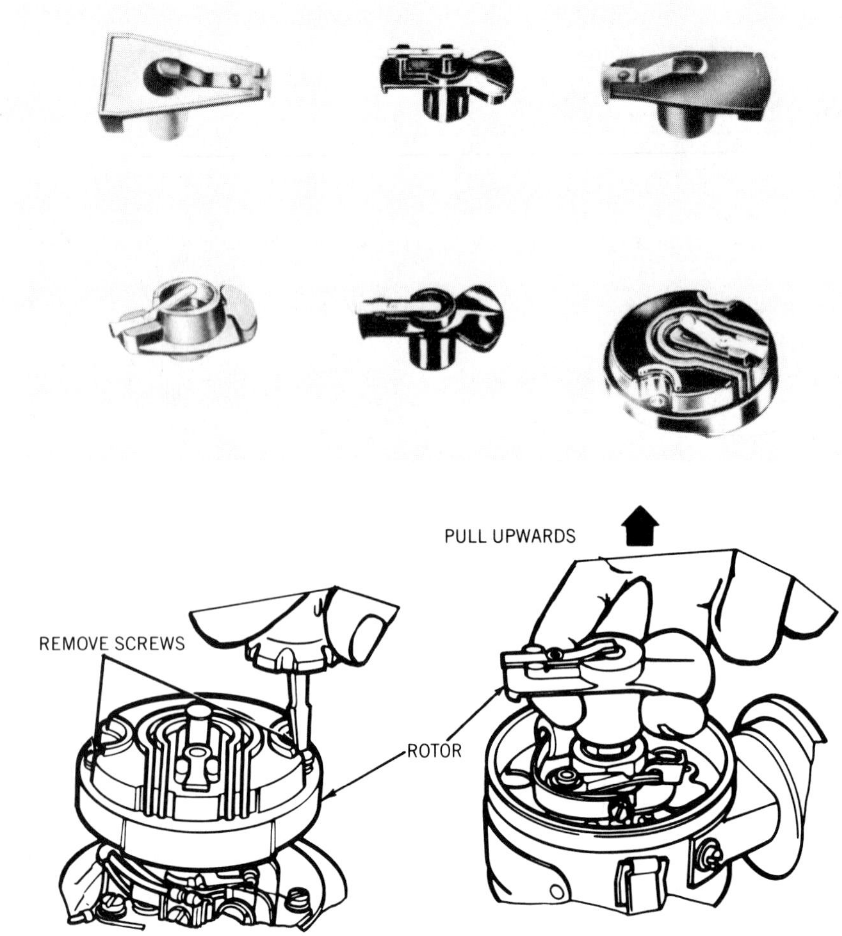


**Figure 12-26.** Most rotors press onto the top of the distributor shaft. Some rotors are secured by screws. (Chrysler)

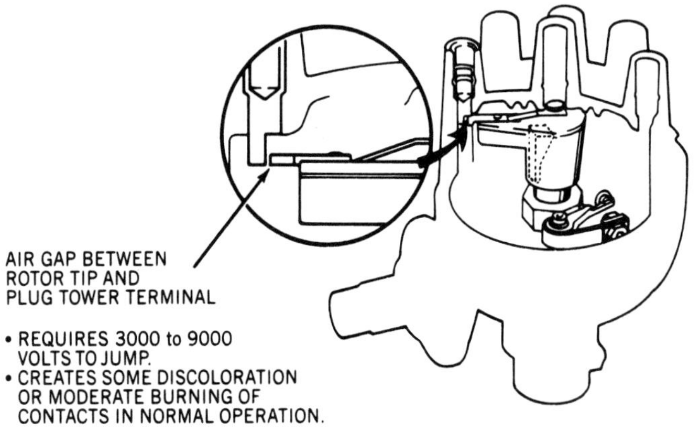


**Figure 12-27.** The rotor air gap adds resistance to the secondary circuit. (Ford)

**Distributor Rotor.** The rotor is attached to the distributor shaft and rotates with it to distribute secondary voltage to the spark plug terminals in the cap. The rotor may simply press onto the shaft and be located by a key, or it may be mounted with screws, figure 12-26. The rotor is made of an insulating plastic material, similar to the cap. The metal electrode on the rotor conducts current from the coil terminal of the cap to the spark plug terminals.

The rotor tip cannot touch the plug terminals in the cap, or the rotor and cap would be damaged. An air gap of 0.003 to 0.009 inch exists between the rotor tip and the plug terminals of the cap, figure 12-27. The air gap adds resistance to the secondary circuit that requires 3,000 to 9,000 volts to create a small arc and complete the circuit. The voltage needed to bridge the rotor air gap is an important service specification, as you will learn when you service ignition systems.

### Ignition Cables

Ignition cables, or spark plug cables, carry high-voltage current from the coil to the distributor and from the distributor to the plugs. They are heavily insulated so that the high-voltage current will not jump to ground before reaching the plugs. Cables for breaker-point ignitions have 7-mm-diameter insulation around a conductive core. Many electronic ignitions use 8-mm cables.

Years ago, most ignition cables were made of copper or steel wire. Although this kind of cable is still made, most cars for over 30 years have used cables with high-resistance, nonmetallic conductors. Figure 12-28 shows both kinds of cable construction. The nonmetallic conductor is made of carbon or from linen or fiberglass strands impregnated with graphite. High-resistance cables are used for two basic reasons:

**1.** High-voltage discharge from the coil causes **radiofrequency interference (RFI)**, which interferes with radio and television transmission. High resis-

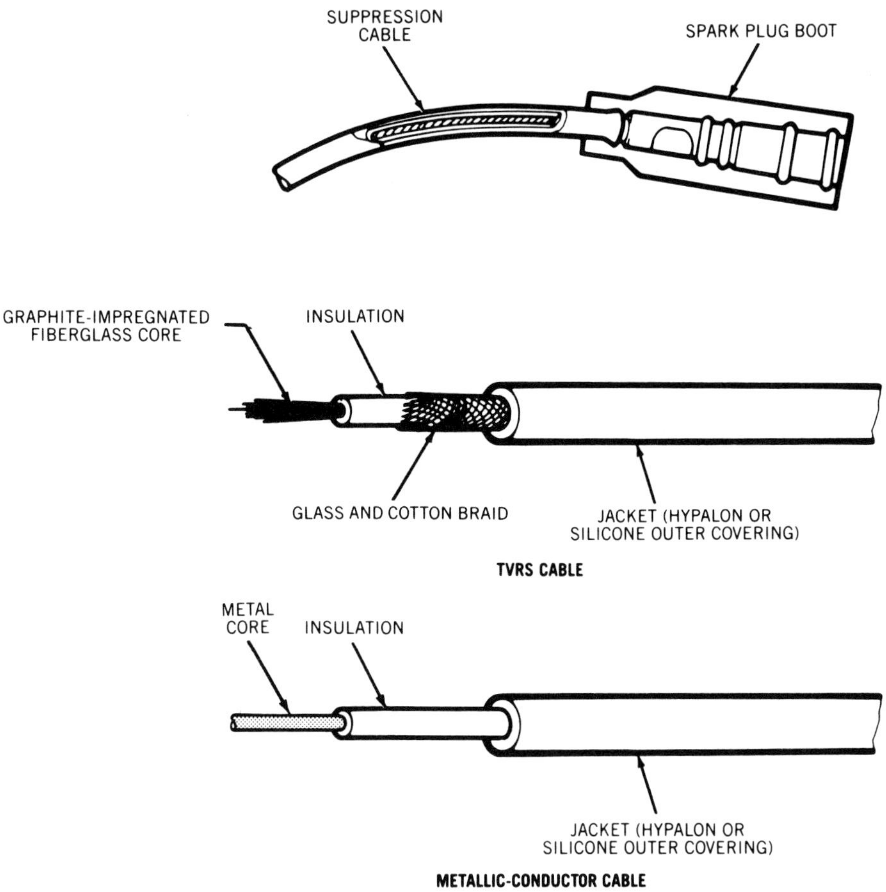


**Figure 12-28.** Modern automobile ignition systems use nonmetallic, high-resistance spark plug cables. Metallic cables are used for special applications. (Chrysler)

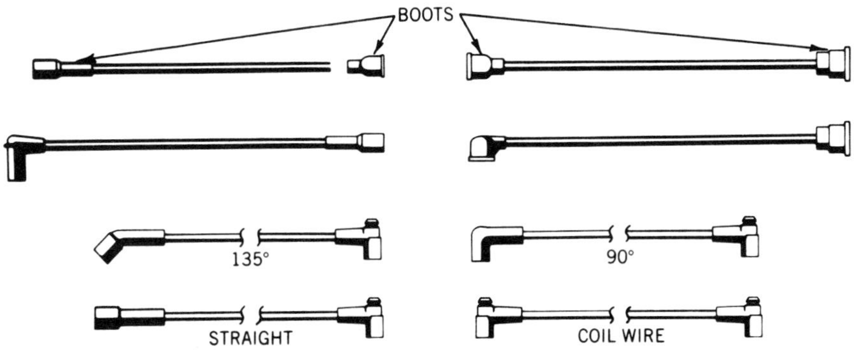


**Figure 12-29.** Insulating boots on the ends of spark plugs and coil cables keep out dirt and moisture and help keep high voltage from short circuiting to ground.

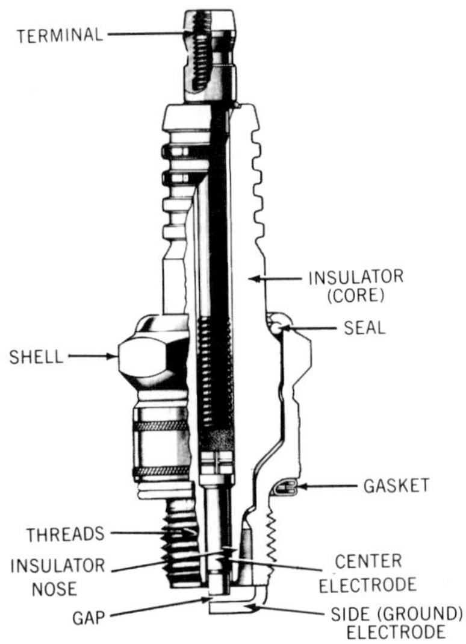


**Figure 12-30.** These are the basic parts of all spark plugs. (Bosch)

tance in the cables reduces this RFI, and the cables are often called **television-radio-suppression (TVRS) cables**.

**2.** High resistance in the cables reduces current, which would burn spark plug and distributor cap electrodes. This resistance also raises the voltage applied to the spark plug, which improves plug firing in the high-resistance conditions in the cylinder.

Resistance in the cables and in the spark plugs affects secondary voltage and current after the plug fires. Before that time, the circuit is not complete. TVRS cables and resistor spark plugs suppress radio interference, and they lower secondary current that would erode spark plug electrodes.

Ignition cables have insulating boots around the terminals that connect them to the distributor cap and the spark plugs, figure 12-29. The boots keep dirt and water away from the cable connections and help insulate the secondary circuit.

## Spark Plugs

The spark plugs complete the secondary ignition circuit. They cause high-voltage current to arc across a gap and ignite the air-fuel mixture. Spark plugs have always been critical parts of the ignition system. When you service ignition systems, you will learn that spark plugs are also good indicators of engine operating condition.

The major parts of a spark plug, figure 12-30, are:

**1.** The center electrode that carries high-voltage current to the plug gap and becomes the negative terminal of the plug

**2.** The ceramic insulator (core) that insulates the center electrode and its high voltage

**3.** The side, or ground electrode, which is attached to the shell and becomes the positive electrode

**4.** The steel shell that holds the other parts in a gas-tight assembly and has threads so that the plug can be installed in the cylinder head.

You already have learned something about the high-voltage spark when you read about coil voltage. Remember that the firing voltage develops because of a capacitive charge across the plug electrodes, negative at the center and positive at the side (ground) electrode. When voltage becomes high enough, it ionizes the gases in the plug gap, which makes them more conductive. The resistance of the gap breaks down, and the plug fires. Ignition systems are designed to have a high firing, or ionization, voltage at the plug gap. This is the voltage that ignites the mixture.

The current that continues to flow across the gap after ignition provides the visible part of the spark. This current contributes nothing to combustion, but it does create electrical interference and burns the plug electrodes. A high firing voltage reduces this current after the spark is established because more of the secondary voltage is discharged for initial firing. Modern ignition systems—including spark plugs—are designed to have high firing voltages, but not higher than the available voltage from the coil. These high voltages also improve ignition of lean air-fuel mixtures.

While all spark plugs have the same purposes, plugs have different design features for different engine requirements. The most important for spark plug selection and service are:

- Thread diameter and seat design
- Reach
- Heat range
- Gap.

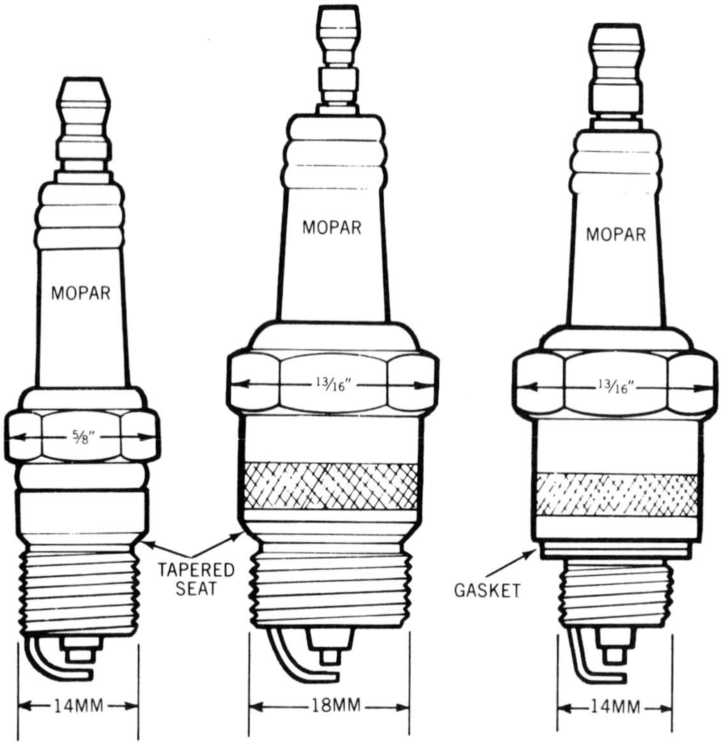


**Figure 12-31.** Common spark plug thread diameters are 14 and 18 millimeters, with gasketed or tapered seats. (Chrysler)

**Thread Diameter And Seat Design.** All modern spark plug threads are made to metric dimensions. Most plugs, today, have a thread diameter of 14 millimeters, but some older Ford engines use 18-mm plugs. Forty years ago, some GM engines used 10-mm plugs, and even older plugs had pipe threads.

The threads and shell of any spark plug must provide a leakproof seal when the plug is installed in the engine. To do this, the shell has a gasket or a tapered seat where it compresses against the cylinder head, figure 12-31. All 18-mm plugs, and many 14-mm plugs, have tapered seats. Most 14-mm plugs have gaskets that compress between the plug seat and the cylinder head.

The steel shell has a hexagonal section that fits a wrench used for plug installation and removal. Standard sizes for spark plugs and wrenches are:

- 14-mm gasketed plug—13/16-inch wrench
- 14-mm taper-seat plug—5/8-inch wrench
- 18-mm taper-seat plug—13/16-inch wrench.

**Reach.** The **spark plug reach** is the distance from the plug seat to the end of the threads, figure 12-32. This includes threaded and unthreaded portions because some plugs have threads for only part of their reach. The reach dimension must match the depth of the tapped hole in the cylinder head, figure 12-33. If the plug reach is too short for the cylinder head hole, the plug will be shrouded in a pocket, and the spark will not ignite the air-fuel mixture completely. If plug reach is too long, the plug will extend into the combustion chamber. In this case, several problems can occur:

**1.** The plug will overheat and ignite the mixture prematurely.

**2.** Carbon deposits will form on the exposed threads and make plug removal difficult.

**3.** The piston may strike the plug and damage both the plug and the piston.

Mixing up a 3/4-inch reach plug with a 3/8-inch reach plug doesn't happen too often because the dimensions are obviously different. However, it is not hard to interchange 3/8-, 7/16-, and 1/2-inch plugs because they look almost the same at first glance. Always check the carmaker's specifications for plug part numbers and reach before installing new plugs.

**Heat Range.** Spark plug **heat range** does not indicate ignition temperature, or the temperature of the spark. It indicates the ability of the plug to dissipate heat from its firing end, or insulator core. Heat range is deter-

**Figure 12-32.** Spark plug reach is the distance from the seat to the tip of the threads, or end of the shell. (Champion)

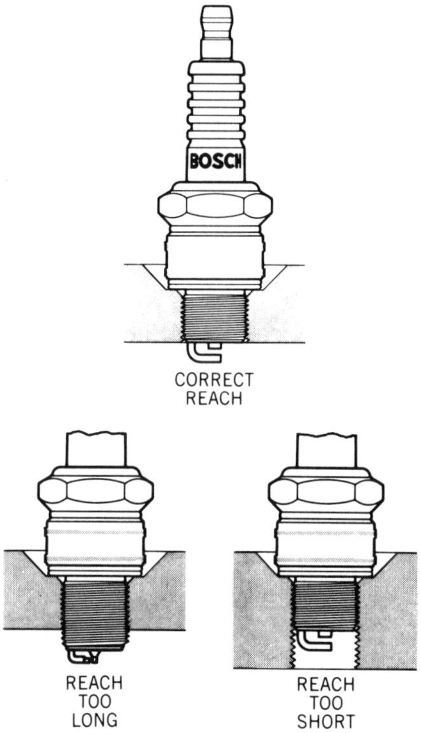


**Figure 12-33.** Spark plug reach must match the depth of the tapped hole in the cylinder head. (Bosch)

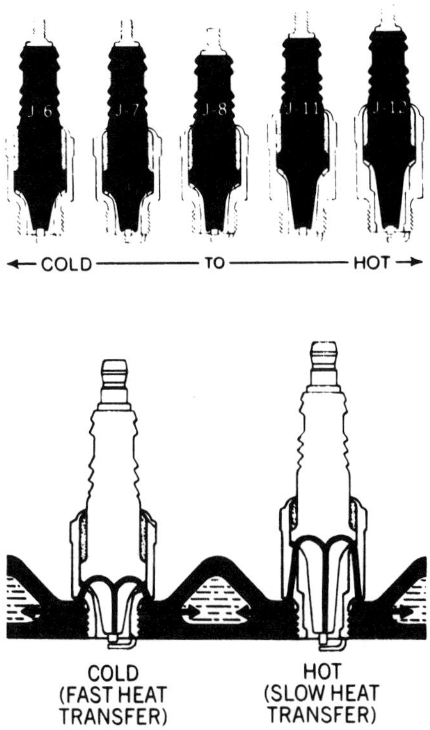


**Figure 12-34.** The distance of the heat path from the center electrode through the insulator to the shell determines the heat range.

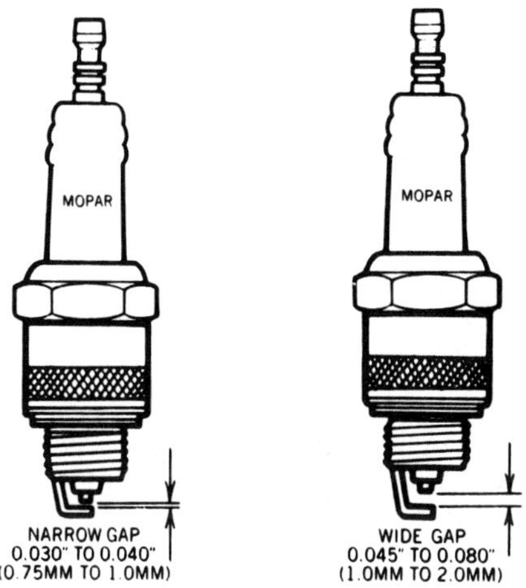


**Figure 12-35.** Narrow-gap and wide-gap plugs are made differently. Do not substitute one for the other. (Chrysler)

mined by the design of the insulator, figure 12-34. The insulator of a cold plug has a short length from the electrode to the shell. This promotes fast heat transfer and a cooler center electrode. The insulator of a hot plug has a longer length from electrode tip to shell, which promotes slower heat transfer and a hotter tip.

Heat range is related to expected engine operating temperature. A cold-running engine used for short trips or easy driving requires a hotter plug to burn deposits off the end of the plug. A hot-running engine requires a colder plug so that the electrodes will not overheat. An overheated plug can create high resistance and a misfire, or it can ignite the mixture prematurely.

**Spark Plug Gap.** The spark plug gap is one of the most important engine service specifications. It is the distance between the center and ground electrodes, measured in thousandths of an inch or hundredths of a millimeter, figure 12-35. A narrow gap requires less voltage to create a spark

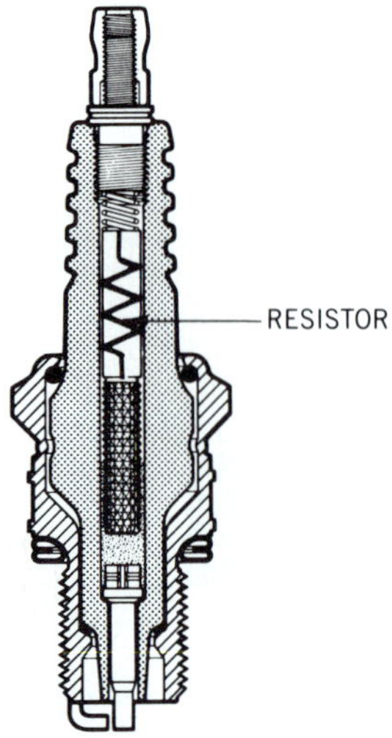


**Figure 12-36.** The resistor in the plug electrode reduces radio interference and plug errosion. (Bosch)

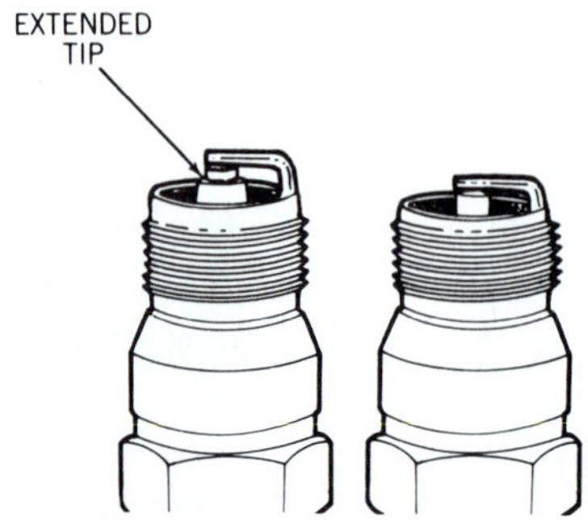


**Figure 12-37.** The insulators and electrodes of extended-core spark plugs extend farther into the combustion chamber. (Chrysler)

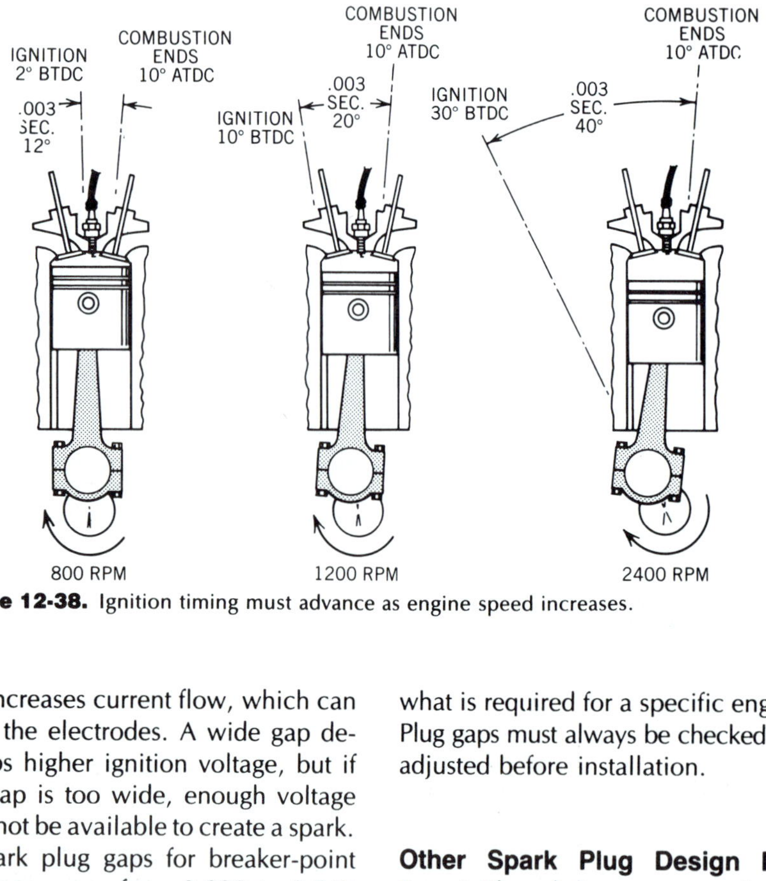


**Figure 12-38.** Ignition timing must advance as engine speed increases.

but increases current flow, which can burn the electrodes. A wide gap develops higher ignition voltage, but if the gap is too wide, enough voltage may not be available to create a spark.

Spark plug gaps for breaker-point ignitions range from 0.030 to 0.040 inch (0.75 to 1.0 mm). Some electronic ignitions require plug gaps of 0.045 to 0.080 inch (1.0 to 2.0 mm), figure 12-35. All plugs are made with a gap between the electrodes, but that does not mean that the gap is exactly what is required for a specific engine. Plug gaps must always be checked and adjusted before installation.

**Other Spark Plug Design Features.** Thread diameter, seat design, reach, heat range, and gap are the most common features you must consider when you install spark plugs. There are other features, or special-purpose spark plugs that you must be aware of, however.

*Resistor spark plugs* have a resistor of about 5,000 to 10,000 ohms in the center electrode, figure 12-36. This resistor reduces the inductive portion of the spark that causes radio interference and plug erosion.

*Extended-core spark plugs* have insulators and electrodes that extend farther into the combustion chamber, figure 12-37. Part numbers for many extended-core plugs indicate the same heat range as a standard plug. Most extended-core plugs, however, operate efficiently over a wider temperature range than standard plugs. They operate hotter at slow speeds to help burn off plug deposits. They also operate slightly colder at high speeds, which prevents overheating.

Other special-purpose plugs are made for 2-stroke engines, Wankel rotary engines, racing engines, and marine and motorcycle engines.

## IGNITION TIMING AND SPARK ADVANCE

In chapter 4, you learned about ignition intervals and how ignition is synchronized with crankshaft rotation and firing order. You also learned that ignition timing must change as engine speeds and loads change. This is called advancing and retarding the timing.

Remember that the air-fuel mixture has a relatively constant burn time of about 3 milliseconds, figure 12-38. Peak combustion pressure must occur when the piston, connecting rod, and crankshaft are in the position to develop the most power. At high engine speeds, combustion must start sooner to allow enough burn time. At low engine speeds, less spark advance is needed in relation to crankshaft speed and position.

Although burn time is *relatively* constant, it does vary slightly with engine load. The lean air-fuel mixtures provided by the carburetor under light load take longer to ignite and burn. Rich mixtures under heavy load ignite and burn slightly faster.

The first principle of ignition timing for any internal combustion engine is that *all changes in timing are related to engine speed and load.*

- Timing advances as speed increases and retards as speed decreases.
- Timing decreases as load increases and increases as load decreases.

Ignition timing at any moment, with any combination of speed and load, is a combination of these actions. Ideal ignition timing also varies for different engine designs. Ideal timing for any engine under any combination of speed and load produces optimum cylinder pressure to deliver the most power with best fuel economy and emission control.

If ignition occurs too early, combustion pressure acts as a brake on the piston. Increased combustion pressures with overly advanced timing also cause engine knocking and lower efficiency. If ignition is too late, peak combustion occurs when the piston is farther down on the power stroke. This reduces cylinder pressure and overall efficiency. Figure 12-39 shows different cylinder pressures with different amounts of spark advance.

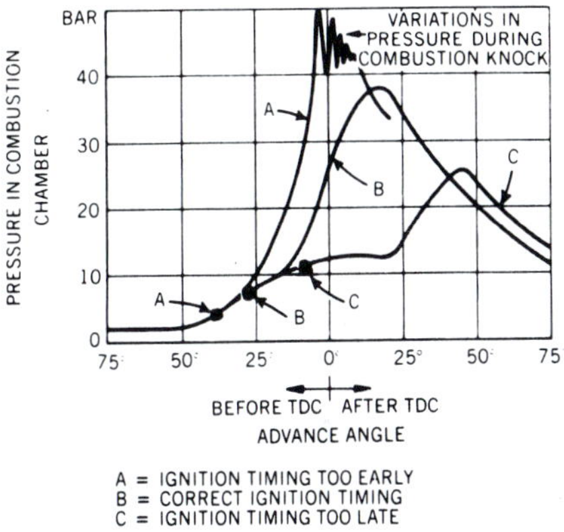


**Figure 12-39.** These graphs show the relationship of ignition timing and combustion pressure. (Bosch)

Breaker-point distributors use simple mechanical devices—centrifugal weights and vacuum diaphragms—to control timing. Many electronic distributors use the same devices, but late-model computer-controlled engine systems alter timing electronically. Whether timing is controlled by weights, vacuum diaphragms, or computers, it is controlled in relation to engine speed and load.

We will begin the study of ignition timing with the devices used on breaker-point distributors and simple electronic distributors. In later chapters, you will learn about variations on these systems that have been used to help control exhaust emissions. You also will learn about electronic systems that control timing more precisely than mechanical devices can.

## Speed Control—Centrifugal Advance

The distributor shaft rotates at camshaft speed, or one-half crankshaft speed. Its speed increases and decreases directly with engine speed. If we could rotate the distributor cam around the center of the shaft so that it opens the points sooner as speed increases and later as speed decreases, we could control timing in relation to engine speed. Centrifugal advance weights provide this kind of control.

The distributor cam and rotor are on a separate shaft that fits over the distributor main drive shaft and pivots around it. The **centrifugal (mechanical) advance** mechanism consists of two centrifugal weights that pivot at the edge of a weight base on the drive shaft. A spring connects each weight to the shaft of the distributor cam, figure 12-40.

As distributor speed increases, the weights swing outward through centrifugal force and rotate the cam and rotor ahead of the main drive shaft. Higher speed creates more centrifugal force, which advances the timing. The springs control the movement of the weights and pull them back inward as speed decreases. A lug or pin on the advance mechanism limits the movement for maximum advance and acts as a stop for the no-advance position. Figure 12-41 shows the positions of a centrifugal advance mechanism with no advance and with full advance.

The amount and rate of advance for any distributor are determined by the size and shape of the weights and the tension of the springs. Engineers select springs and weights to design distributor advance "curves" for different engines. Figure 12-42 shows centrifugal advance curves for different engines. Notice that advance does not

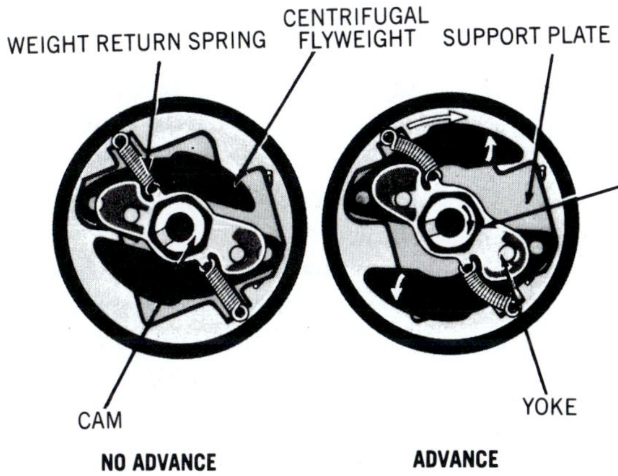


**Figure 12-40.** As the centrifugal weights move outward, they rotate the distributor cam ahead of shaft rotation. (Bosch)

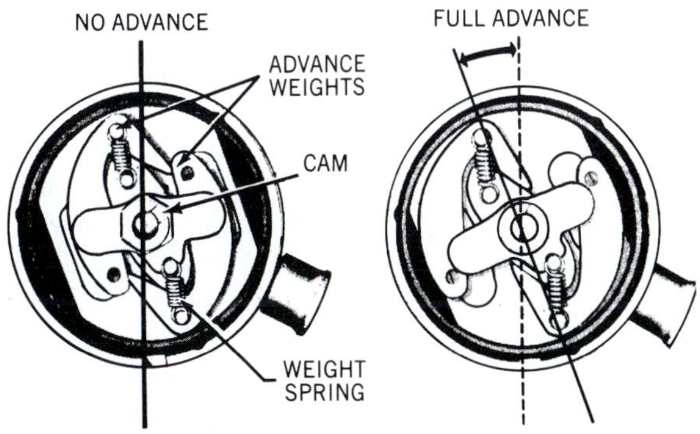


**Figure 12-41.** No-advance and full-advance positions of a typical centrifugal advance mechanism. (Delco-Remy)

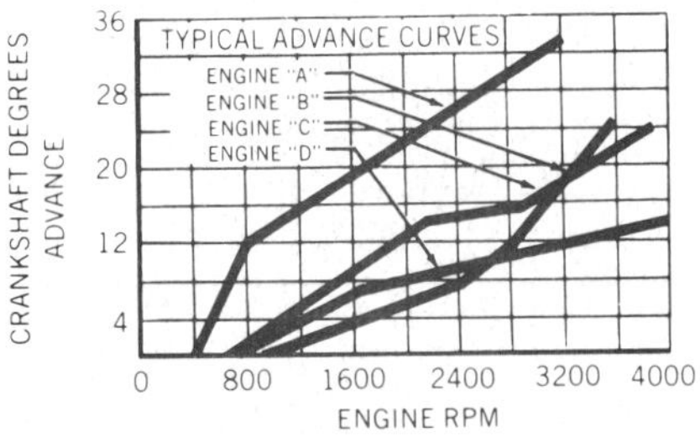


**Figure 12-42.** Centrifugal advance varies for different engines. It occurs gradually over a wide speed range. (Delco-Remy)

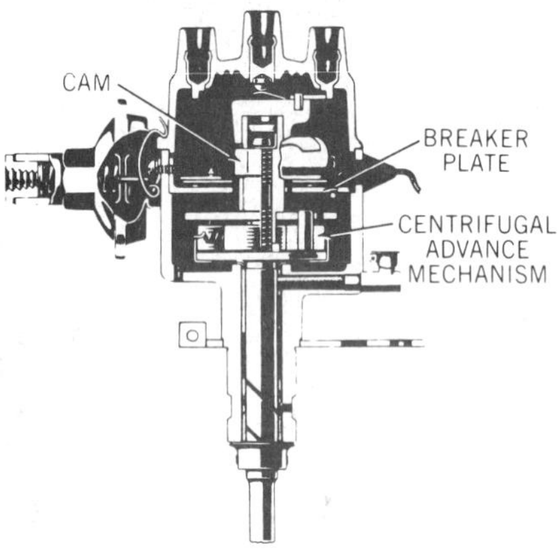


**Figure 12-43.** Most distributors have the centrifugal advance weights below the breaker plate. (Prestolite)

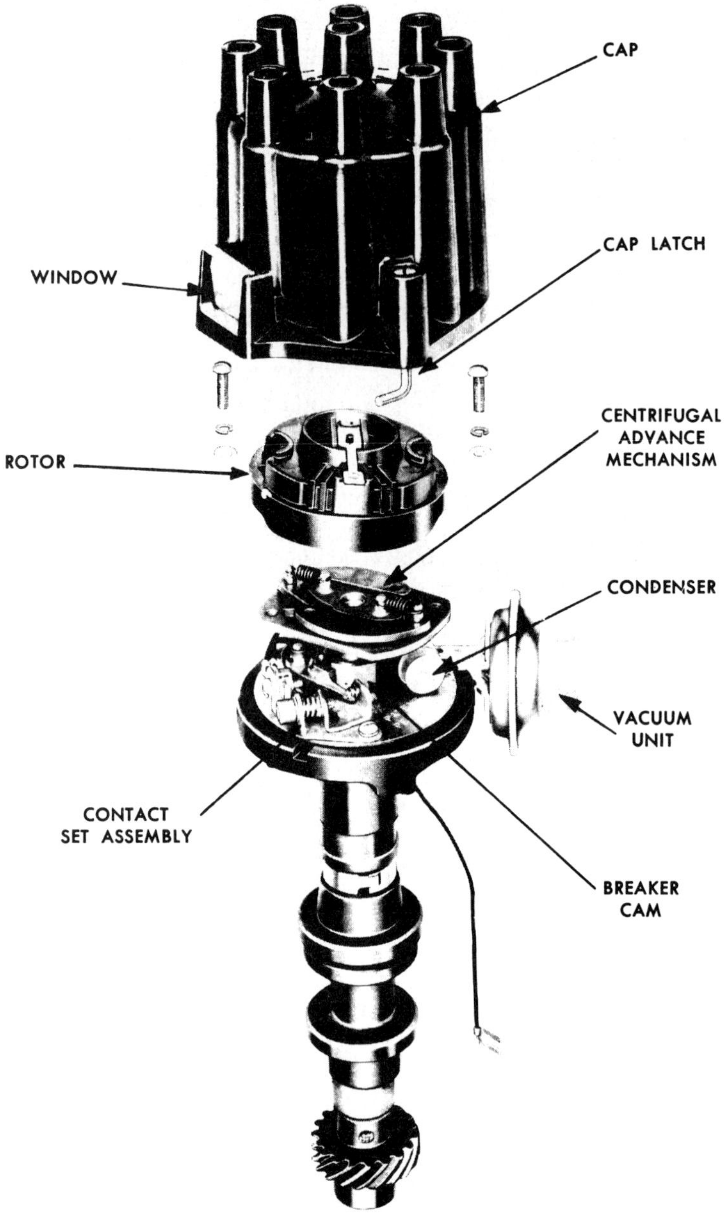


**Figure 12-44.** Delco-Remy V-8 and V-6 distributors have the centrifugal advance weights above the cam and breaker plate. (Delco-Remy)

occur suddenly, but increases gradually over a specific speed range.

The centrifugal weights in most distributors are below the cam and breaker plate, figure 12-43. Delco distributors for V-8 and V-6 engines, as well as some Japanese distributors, have the weights above the cam and breaker plate, figure 12-44.

Centrifugal advance curves are usually designed for full-load conditions. This assumes a constant air-fuel mixture and engine load so that timing can be regulated only in relation to increasing speed. In operation, however, engine loads can change independently of speed. Vacuum advance changes timing in relation to additional load changes.

## Load Control—Vacuum Advance

Centrifugal force varies directly with engine speed and provides a way to control timing in relation to speed. Engine vacuum varies directly with engine load and provides a way to control timing in relation to load. Under light load, vacuum is high. Under heavy load, vacuum is low.

If we connect a vacuum diaphragm to a ported vacuum outlet just above the carburetor throttle, we can get a vacuum signal that is proportional to engine load and throttle position:

- High vacuum at part throttle (light load)
- Low vacuum at full throttle (heavy load) or closed throttle.

Then, if we link that diaphragm to a movable breaker plate in the distributor, we can rotate the plate so that the points open sooner to ignite the lean mixtures at part throttle and light load.

The **vacuum advance** mechanism uses a vacuum diaphragm to rotate the breaker plate in the direction opposite distributor rotation, figure 12-45. That is, if the distributor rotates clockwise, the diaphragm moves the plate counterclockwise. This causes the point rubbing block to contact the cam lobes sooner, which advances the timing. A spring in the diaphragm housing forces

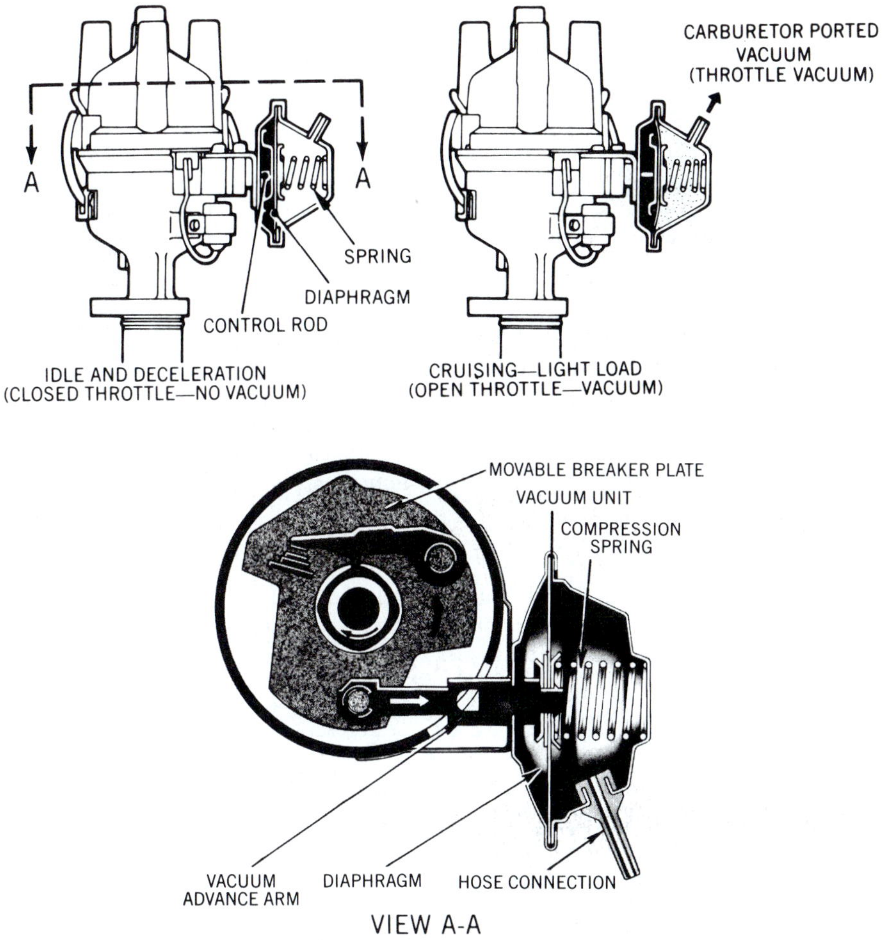


**Figure 12-45.** The vacuum advance mechanism moves the breaker plate opposite to distributor rotation to advance timing. (Chrysler) (Bosch)

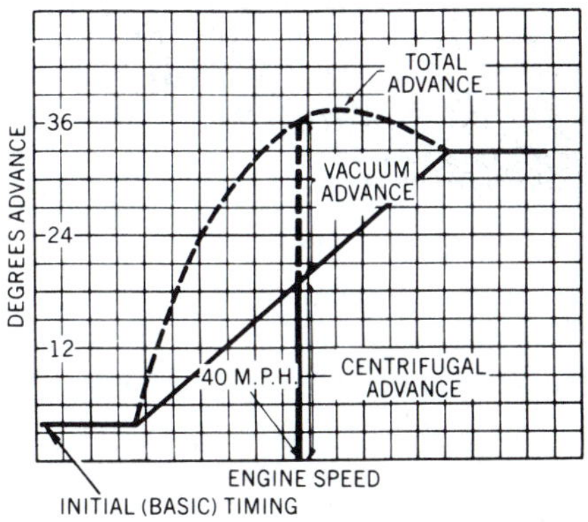


**Figure 12-46.** Total advance equals initial timing plus centrifugal and vacuum advance at any combination of speed and load. (Delco-Remy)

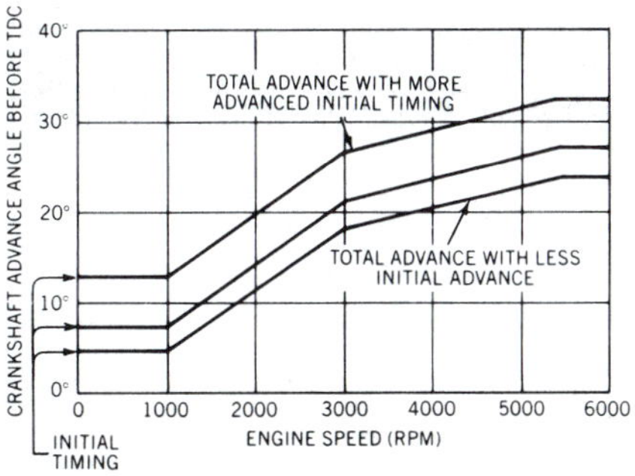


**Figure 12-47.** The same distributor can provide different amounts of spark advance, depending on initial timing.

the plate in the opposite direction as vacuum decreases. This retards the timing slightly for increased load and reduces the detonation ping that would result with too much advance.

## Ignition Advance and Initial Timing

Up to this point, we have looked at centrifugal and vacuum advance separately, but they don't operate separately. Under any combination of speed and load, total ignition advance is determined by combined centrifugal and vacuum action, figure 12-46. This graph also shows that total ignition advance for any engine is affected by *initial timing*.

Remember from chapter 4 that initial (basic) timing is usually set at engine idle speed. It could be set at top dead center (tdc) of the compression stroke, or it could be set a few degrees before or after tdc. The centrifugal and vacuum advance devices for most breaker-point ignitions do not operate at idle. They add nothing to initial timing, and the engine runs at its basic timing setting. When engine speed moves off idle and speed and load change, the advance mechanisms start to operate.

To determine total spark advance from this point, you must add initial timing to the combined centrifugal and vacuum advance. For example, if the centrifugal and vacuum combination provides 20 degrees of spark advance at 2,000 rpm with light load and initial timing is 8° btdc, total advance is 28° btdc. Suppose, on the other hand, that initial timing is 4° *atdc*. The same 20 degrees of centrifugal and vacuum advance then provides only 16 degrees of total spark advance. Figure 12-47 shows the spark advance curves for the same distributor with three different initial timing settings.

## Ignition Timing and Emission Control

When an engine is idling, or when the car is overrunning the engine on deceleration, the throttle is almost closed. This reduces the air intake and creates

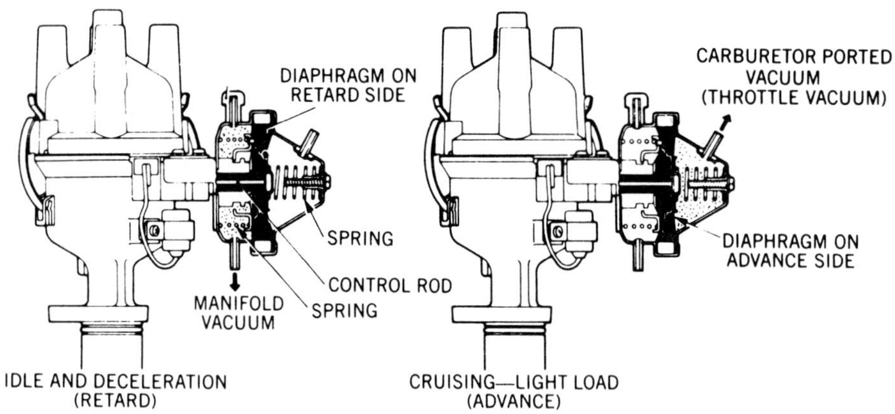


**Figure 12-48.** A dual-diaphragm vacuum unit can both advance and retard timing. (Chrysler)

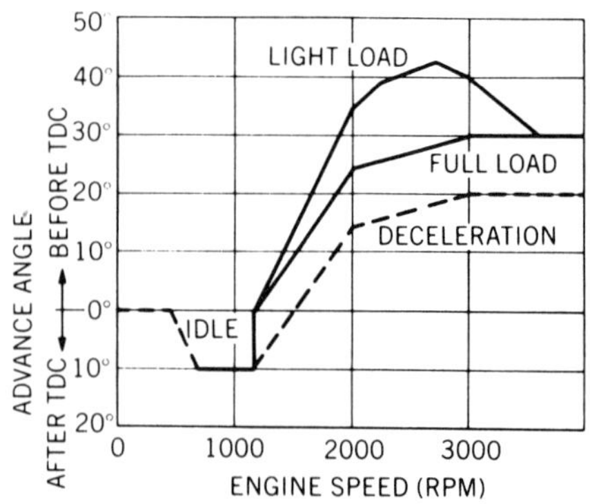


**Figure 12-49.** These graphs show the effect of vacuum advance and retard under different operating conditions. (Bosch)

rich air-fuel ratios. Combustion is incomplete, and HC and CO emissions increase. The engine can misfire, also. Emissions under these can be reduced if timing is retarded or advanced less. This produces more heat in the combustion chamber, which reduces emissions.

Many distributors have dual-diaphragm vacuum units, figure 12-48. Vacuum is switched from the advance diaphragm to the retard diaphragm during idle and deceleration. This moves the breaker plate in the direction of distributor rotation and delays ignition. At low engine speed, this action can produce truly retarded timing where ignition occurs later than initial timing or after tdc. At higher speeds, the retard action reduces the amount of advance. Figure 12-49 is a set of advance curves that show this vacuum retard action.

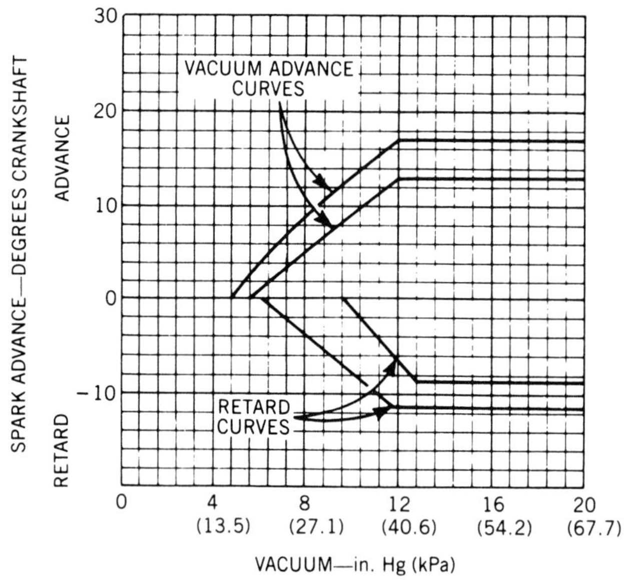


**Figure 12-50.** Typical vacuum advance and retard curves. (Chrysler)

## Spark Advance Specifications

When you service ignition systems, you will find spark advance specifications given in several different ways:

- Degrees of crankshaft advance at engine rpm
- Degrees of distributor advance at distributor rpm
- Degrees of vacuum advance with different amounts of vacuum

Remember that the distributor rotates at one-half engine speed. Therefore, the point where the primary circuit opens advances only half as far in distributor degrees as spark timing advances in crankshaft degrees. Remember also that vacuum advance is independent of distributor and engine speed. At any speed, you can have high or low vacuum with light or heavy load.

Centrifugal advance can be measured out of the engine in a distributor test machine. For this, you need advance specifications in distributor degrees at distributor speed. Centrifugal advance also can be measured with the distributor running in the engine. For this, you need advance specifications in crankshaft degrees at engine speed. *Both* of these are double the distributor specifications and must be added to the initial timing setting when testing a distributor in an engine.

Vacuum advance can be tested with the distributor in or out of the engine. Vacuum advance specifications are given in distributor degrees or crankshaft degrees at various vacuum levels, such as 4, 8, 12, 16, and 20 inches of mercury (in. Hg), figure 12-50. Some curves also include specifications for vacuum retard.

All of these specifications do relate to each other and to the specific ignition requirements of any engine. As you learn more about ignition system service, you will learn how to use these specifications for professional engine service.

## ELECTRONIC SPARK TIMING CONTROL

*Ignition timing is controlled in relation to engine speed and load.*

The preceding is a fundamental rule of engine operation. You have seen how traditional centrifugal and vacuum advance devices accomplish this in simple breaker-point and electronic distributors. On most late-model engines, however, the centrifugal weights and vacuum diaphragms have been removed from the distributor. The distributor has become simply that: a device to distribute secondary voltage to the spark plugs. It has little or no control over timing advance or retard after the engine is started and running. On many engines, the distributor has been eliminated completely by direct (distributorless) ignition systems (DIS).

Yet the fundamental requirement is still to control timing in relation to speed and load. On modern engines, the engine control computer does this job electronically. It controls the switching and timing of the coil charging and discharge far more accurately than simple weights and vacuum devices can.

Various sensors send the computer information about engine speed and load. By using these signals and its programs, the computer calculates when to turn on the coil primary circuit and when to turn it off to create the secondary discharge and fire the spark plug for each cylinder ignition cycle.

You will learn about these sensors and computers in detail in Part 6. For a preview, however, we can point out that engine load sensors, such as manifold absolute pressure (MAP) sensors, provide the computer with information about engine load. In this regard, a MAP sensor does the job of a vacuum advance diaphragm. Timing and crankshaft position sensors provide information about engine speed and piston position in relation to tdc on the compression stroke. The computer uses these signals to control timing according to engine speed and thus does the job of the traditional centrifugal advance weights.

## SUMMARY

Almost all ignition systems for gasoline engines are the inductive-discharge type. Electromagnetic induction in the ignition coil uses low-voltage current from the battery to develop high voltage needed for ignition. Low-voltage current in the coil primary winding creates a magnetic field. When primary current stops, the field collapses and induces high voltage in the secondary winding. Secondary available voltage can be as high as 30 to 50 kV, but the coil delivers only the required voltage needed to fire the plugs. The difference between required voltage and available voltage is the voltage reserve, which should be about 60 percent in a well-tuned engine.

The low-voltage primary circuit of the ignition consists of the battery, the ignition switch, the primary (ballast) resistor, the starting bypass, the breaker points, the condenser, and the coil primary winding. Since the mid-1970's breaker points have been replaced by electronic switching devices in ignition primary circuits. Whether the primary circuit is controlled by points or electronic devices, the principles of current flow and secondary voltage induction are the same.

The high-voltage secondary circuit contains the coil secondary winding, the distributor cap and rotor, the ignition (high-tension) cables, and the spark plugs. Secondary voltage always has negative potential (positive ground) for efficient firing action.

Important specifications of a spark plug are its thread diameter and seat design, the reach, the heat range, and the gap. Some special-purpose spark plugs have resistors in the center electrode. Others have extended tips.

All breaker-point and many electronic distributors use centrifugal weights to advance ignition timing in relation to engine speed. These distributors also have vacuum advance mechanisms to advance timing in relation to engine load.

# REVIEW QUESTIONS

### Multiple Choice

1. All of the following create resistance in the ignition primary or secondary circuit, EXCEPT:
   - **a.** the spark plug gap
   - **b.** the rotor air gap
   - **c.** the condenser
   - **d.** the coil primary winding.

2. Mechanic A says that ignition secondary voltage is created by electromagnetic induction. Mechanic B says that secondary voltage is created by primary current. Who is right?
   - **a.** A only
   - **b.** B only
   - **c.** both A and B
   - **d.** neither A nor B

3. All of the following determine ignition secondary voltage levels, EXCEPT:
   - **a.** the direction of primary current
   - **b.** the coil turns ratio
   - **c.** the amount of primary current
   - **d.** the speed at which primary current stops or changes.

4. Mechanic A says that high compression pressures increase ignition resistance. Mechanic B says that lean air-fuel mixtures decrease ignition resistance. Who is right?
   - **a.** A only
   - **b.** B only
   - **c.** both A and B
   - **d.** neither A nor B

5. The ignition primary circuit includes all of the following EXCEPT the:
   - **a.** ignition switch
   - **b.** rotor air gap
   - **c.** battery
   - **d.** starting bypass circuit.

6. An ignition primary (ballast) resistor limits maximum primary voltage to approximately:

a. 6 volts
c. 5 volts
b. 9 volts
d. 12 volts.

7. For a breaker-point ignition on any engine, the dwell angle is usually between:

a. 15 and 25 degrees
b. 20 and 30 degrees
c. 30 and 45 degrees
d. 45 and 60 degrees.

8. Mechanic A says that as breaker-point gap decreases, dwell increases. Mechanic B says that as breaker-point gap decreases, timing retards. Who is right?

a. A only
c. both A and B
b. B only
d. neither A nor B

9. A typical ignition condenser has a capacity of:

a. 0.02 to 0.03 microfarad
b. 0.18 to 0.32 microfarad
c. 1.8 to 3.2 microfarads
d. 2.0 to 3.0 microfarads

10. Mechanic A says that a spark plug center electrode should always have positive polarity. Mechanic B says that reverse polarity increases voltage reserve. Who is right?

a. A only
b. B only
c. both A and B
d. neither A nor B

11. All of the following are measurements of ignition voltage, EXCEPT:

a. reference voltage
b. required voltage
c. available voltage
d. reserve voltage.

12. Ignition secondary circuit resistance is increased by:

a. a narrow spark plug gap
b. a rich air-fuel mixture
c. low engine temperature
d. high engine speeds.

13. Coil open-circuit (no-load) secondary voltage is the same as:

a. reference voltage
b. required voltage
c. available voltage
d. ionization voltage.

14. Mechanic A says that spark voltage is higher than firing voltage. Mechanic B says that firing voltage is affected by resistance in the cylinder. Who is right?

a. A only
b. B only
c. both A and B
d. neither A nor B

15. The voltage required to jump the rotor air gap is approximately:

a. 12 to 14 volts
b. 500 to 1,000 volts
c. 10 kV to 20 kV
d. 3 kV to 9 kV.

16. Ignition (spark plug) cables are made with nonmetallic conductors primarily to:

a. reduce required voltage
b. suppress radiofrequency interference
c. increase spark voltage
d. reduce secondary resistance.

17. The thread diameter of most spark plugs for late-model engines is:

a. 10 mm
b. 12 mm
c. 14 mm
d. 18 mm

18. Mechanic A says that plugs with 3/8-inch and 1/2-inch reach are interchangeable. Mechanic B says that 14-mm plugs with gaskets and tapered seats are not interchangeable. Who is right?

a. A only
b. B only
c. both A and B
d. neither A nor B

19. Mechanic A says that ignition timing advances as speed increases to maintain a relatively constant burn time. Mechanic B says that timing advances under light load because lean mixtures take longer to burn. Who is right?

a. A only
b. B only
c. both A and B
d. neither A nor B

20. Centrifugal advance controls timing in relation to:

a. engine load
b. engine speed
c. air-fuel ratio
d. ignition lag.

21. Vacuum advance controls ignition timing in relation to:

a. ignition firing voltage
b. engine speed
c. required voltage
d. engine load.

22. Mechanic A says that timing should advance as speed increases. Mechanic B says that timing should advance as load increases. Who is right?

a. A only
b. B only
c. both A and B
d. neither A nor B

23. Mechanic A says that initial timing affects total centrifugal advance for any engine. Mechanic B says that initial timing affects total vacuum advance for any engine. Who is right?

a. A only
b. B only
c. both A and B
d. neither A nor B

### Fill in the Blank

24. The __________ circuit receives low voltage from the battery and uses it to induce higher voltage in the coil secondary winding.

25. The coil secondary winding has several thousand turns of fine wire and is connected to the high-voltage __________ circuit of the ignition system.

26. In an inductive-discharge ignition system, voltage is induced in the primary and secondary coils with each __________ in current flow.

27. If the turns ratio between the primary and secondary windings is 200:1, then the voltage in the secondary winding would be __________ times greater than the voltage in the primary winding.

28. When the magnetic field produced in the coil primary and secondary windings is at full strength, it is called __________.

29. An ignition coil produces only enough voltage to create an __________.

# 13 ELECTRONIC IGNITION SYSTEMS

## WHAT

This chapter continues with ignition systems by examining the solid-state electronic ignitions that have become the "standard," or "conventional" systems used by almost all carmakers. There are several reasons for the wholesale changeover from breaker-point ignitions that occurred in the mid-1970's:

- By the 1970's electronic technology had advanced to the point that economical and reliable solid-state devices could be mass produced by the millions.
- Increasingly more stringent emission control and fuel economy requirements demanded more accurate and uniform control of ignition and spark timing.
- Engineers found that electronic systems allowed them to control engine operation more precisely and with fewer compromises than electromechanical devices did.

Even though the 1970's are considered the decade when the "electronic revolution" started in the auto industry, electronic ignitions were not a radical change. A basic electronic ignition simply has the breaker points replaced by an electronic switch. Most original equipment electronic ignitions are inductive-discharge systems that generate and distribute high ignition voltage in the same way that breaker-point systems have done since 1912. You will find that all the principles of current flow, dwell, coils, distributors, spark plugs, and spark advance are present in both electronic ignitions and breaker-point systems.

## WHY

For the past decade, automobile service has been going through a transition. Nowhere is the transition more apparent than in ignition service. Modern ignition service has become a field of electronic testing and diagnosis. You will find it easier to master electronic skills if you understand how electronic principles are similar to electrical principles that have always been used in automobiles. Basic electronic ignition systems are a good place to learn these similarities. Today, a professional mechanic must be able to service both breaker-point and electronic ignitions. If you understand how the systems are similar, you can do your job faster and better.

## GOALS

After studying this chapter, you should be able to:

1. Explain the similarities between inductive-discharge electronic ignition and breaker-point ignition.
2. Explain the operation of magnetic-pulse, metal-detection, and Hall-effect ignition trigger circuits.
3. Explain dwell and current regulation in electronic ignitions.
4. Explain the differences between capacitive-discharge and inductive-discharge ignitions.
5. Identify various original equipment electronic ignition systems and the principal parts of each.
6. Recognize ignition timing and spark advance devices for electronic ignitions.

## PRIMARY CIRCUIT CONTROL

Breaker points are simple, cheap, and (within limits) reliable devices for switching primary current. They have disadvantages, however, that make them less than ideal for the precise ignition control needed on modern engines:

1. Mechanical wear of the point rubbing block on the distributor cam decreases the gap and increases the dwell angle. This retards basic ignition timing and reduces performance and fuel economy.
2. Force applied by the point spring to the rubbing block increases wear on the distributor cam and shaft bushings. This too causes timing and dwell to vary, not only over a period of time, but from cylinder to cylinder.
3. At high speeds, the points open and close so fast that the spring can't maintain uniform pressure on the movable arm. The points can actually bounce,

which upsets both dwell and timing and creates a high-speed misfire.

**4.** Even with a condenser of required capacitance and with perfect point alignment and spring tension, some arcing always occurs across the points as they open. This causes burning and pitting, and also upsets timing and dwell.

**5.** Because of their small size and an opening gap of a few thousandths of an inch, points can only switch current of 4 to 6 amperes, maximum. Even with a coil that could produce 50 kV of available voltage, breaker points cannot handle the current needed for full magnetic saturation.

**6.** Breaker points rely on physical movement to control primary current time, and this time changes with engine speed. Therefore coil saturation and available voltage are always a compromise between high and low engine speeds.

## Electronic Primary Circuit Advantages

Electronic ignitions do not rely on mechanical switching devices to control primary current.

**1.** Electronic ignitions have no moving parts that bear against each other to cause mechanical wear. Distributor shaft bushings can still wear from rotational forces, but there is no sideways force from a spring and rubbing block to accelerate the wear.

**2.** Dwell and timing do not change during several thousand miles of operation because there is no rubbing block to wear down.

**3.** Electronic switching devices cannot "bounce" at high speeds. Switching operation is uniform from idle to maximum rpm.

**4.** The triggering devices in electronic distributors do not carry primary current. Therefore, current is not limited by the physical capacity of breaker points. Engineers can provide any amount of primary current based on the design of the coil and the capacity of the transistors in the electronic system.

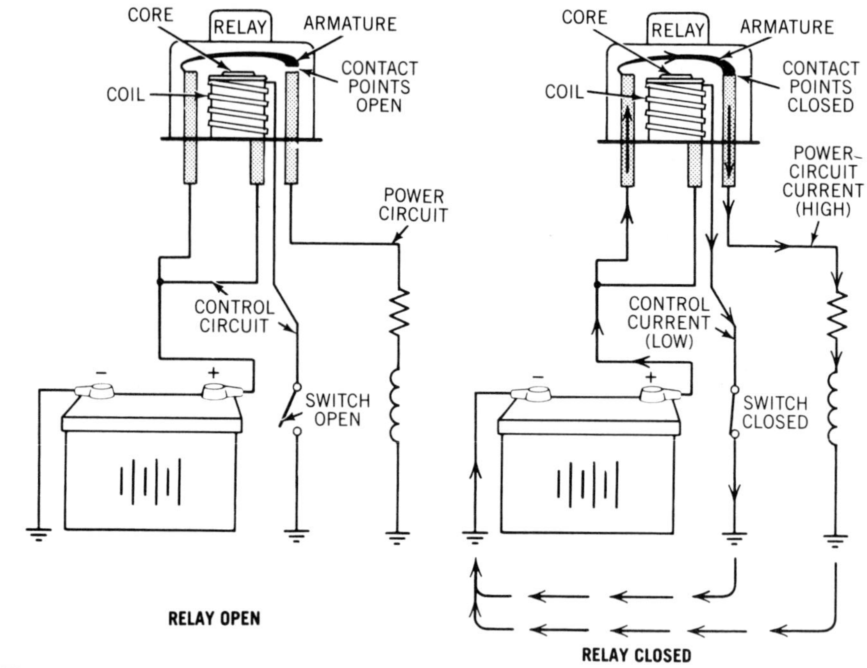


**Figure 13-1.** A small amount of current through the relay coil closes the contacts and allows the relay power circuit to carry a larger amount of current.

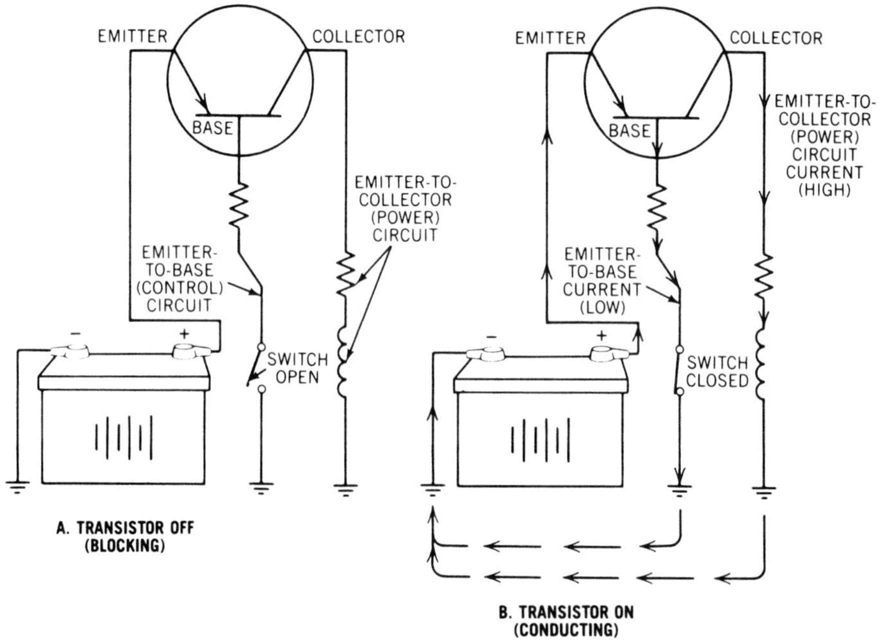


**Figure 13-2.** In this circuit, a transistor has replaced the relay for the same kind of current control.

## Ignition Transistor Operation

Primary current control in an electronic ignition is based on the ability of a transistor to control a high current flow in response to a very small current. The small current is created by a small voltage applied to its base. Chapter 8 explains basic transistor operation. An important feature of a transistor for ignition control is its ability to act as a solid-state relay.

Figure 13-1 shows an electromagnetic relay of the kind used to control a starter motor, a horn, or a headlamp circuit. When the control switch is closed, a small amount of current flows through the relay coil. This sets up a magnetic field that attracts the relay armature, closes the relay contacts, and completes the power circuit. When the relay control switch is open, the electromagnetic field collapses, and the armature spring opens the contacts of the power circuit. Current through the relay coil and control circuit may be only 1 or 2 amperes. Current through the power circuit may be 10 or 20 amperes, or more. (A starter motor can draw several hundred amperes through a starter relay.)

If we replace the relay with a power transistor, we can achieve the same kind of switching. In figure 13-2a, the switch to the transistor base is open. No voltage is applied to the base, and no current flows through the emitter-to-base circuit. Also, the transistor emitter-to-base junction acts as a diode, and no current flows through the emitter-to-collector circuit. If we close the switch and apply a small voltage to the base, we cause a small amount of current to flow through the emitter-to-base circuit. This creates a forward bias on the emitter-to-base junction and allows a larger amount of current to flow through the emitter-to-collector power circuit, figure 13-2b.

This is basically how an electronic ignition system uses transistors to turn primary current on and off. The early transistorized ignitions of 20 or 25 years ago used distributor breaker points as a mechanical switch to control voltage applied to a transistor base, figure 13-3. Full primary current did not flow through the points; it flowed through the transistor. The breaker points carried less than 1 ampere of current. Pitting and burning of the points due to arcing were greatly reduced. But, the points still bounced at high speed and suffered mechanical wear.

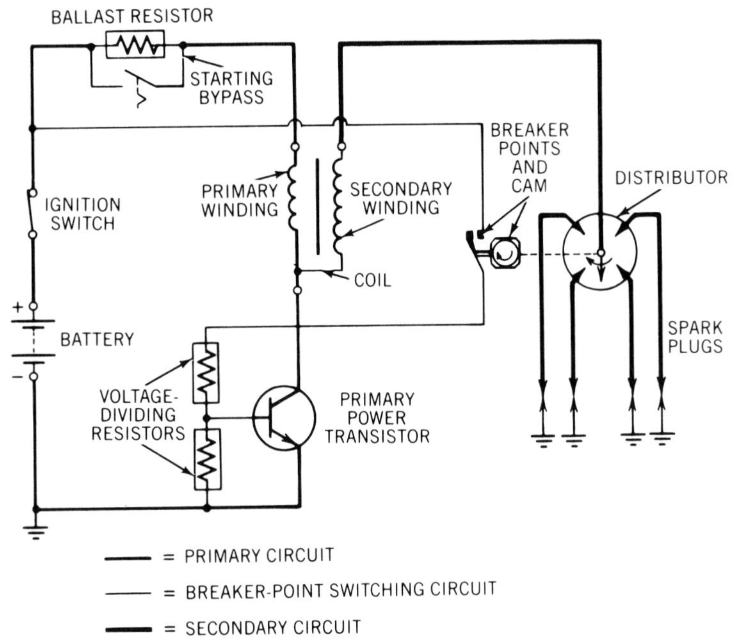


**Figure 13-3.** In this transistorized ignition, breaker points are the trigger for the power transistor, but they do not carry full primary current.

The electronic ignitions that became standard replaced the points with an electronic signal generator that has no rubbing parts. Modern electronic ignitions provide uniform timing, dwell, and secondary voltage at all engine speeds and loads for tens of thousands of miles. Before we go on to look at the signal generators in electronic distributors, you should understand a couple more things about transistor control of primary current.

In an electronic ignition, *no primary current passes through the distributor.* All primary current passes through an **ignition control module**, which is an assembly of transistors, resistors, ca-

### BURNED TO A CRISP

Usually when we think of something that's burned to a crisp, we visualize it as blackened like charcoal. So it is with burned toast, a well done steak, or the ashes of a bonfire. It's not that way with spark plugs, however, it's just the opposite.

The ceramic insulator tip of a spark plug that's badly overheated, "fried," or "burned up" isn't black; it's almost pure white. This condition reflects overly high operating temperatures of the plug that completely burn away any residue or "ash" that might collect on the tip. The plug is literally operating "white hot." A blackened plug, on the other hand, isn't burned to a crisp. It's operating too cold. Black deposits on a plug tip may be gasoline residue from an overly rich air-fuel mixture. You can think of it as "gasoline ash." Black deposits also can be from oil that has gotten into the combustion chamber past leaky piston rings or valve guides and hasn't been completely burned. In either case, the plug wasn't operating hot enough to burn itself clean.

You can't solve engine air-fuel ratio problems or oiling problems by changing spark plugs, but you can tell a lot about engine conditions by learning to recognize a plug that's burned to a crisp and one that isn't.

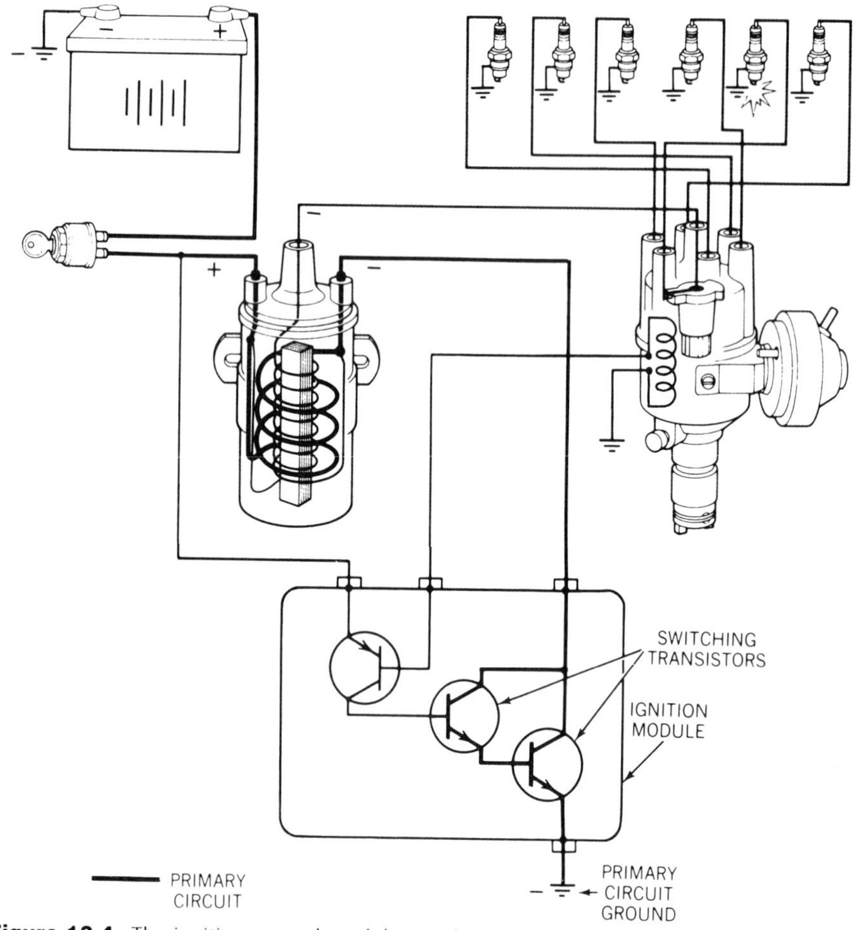


**Figure 13-4.** The ignition control module switches primary current on and off on the ground side of the coil, just as breaker points do in a breaker-point system.

**IS IT SILICON OR SILICONE?**
Silicon is one of the earth's most plentiful elements, but we know it best in its forms as various compounds. Common sand is silicon dioxide $SiO_2$; so is glass. Pure silicon wafers are the foundation for semiconductor integrated circuits. Silicon combines readily with many other elements in various compounds.

We use many rubber and plastic materials, greases, other lubricants, and fluids in our cars, all of which contain silicon. When silicon joins with other elements we call the resulting compounds and mixtures silicone materials, such as silicone rubber and silicone grease. Sometimes we refer to these materials simply as "silicones."

pacitors, and other electronic devices. The ignition coil positive (+) terminal is connected to battery voltage through the ignition switch and receives primary current similarly to a coil in a breaker-point system, figure 13-4. The coil negative (−) terminal is connected to ground through the ignition module. Primary current is thus switched on and off on the ground side of the coil by the transistors in the module.

Most ignition modules use one or more large power transistors to switch primary current. The **power transistors** can reliably handle current up to 10 amperes, much more than breaker points can handle. These are the **switching transistors** of the primary circuit. The switching transistor, or transistors, are controlled by another **driver transistor** that receives timing voltage signals from the signal generator in the distributor. Figure 13-5 is a simplified ignition module circuit. Here is how it works.

When the ignition switch is turned on, a low base current flows through the driver transistor, T1. The driver transistor turns on, and its emitter-to-collector current flows to the base of transistor T2. Transistor T1 requires only very low base current and voltage (a few millivolts and milliamperes), but it produces enough collector current to turn on T2. Voltage applied to the base of T2, turns it on. Emitter-to-collector current from T2 is applied to the base of transistor T3 to switch it on. Transistors T2 and T3 are connected in a current-amplifying arrangement, called a **Darlington circuit**. Ignition primary current flows through the ignition coil and builds up the magnetic field as long as all three transistors are switched on.

As the engine starts and runs, the signal generator in the distributor changes the voltage on the base of transistor T1 each time a piston nears tdc on the compression stroke. The voltage change on the base of T1 turns this transistor off. Its collector current no longer flows to the base of T2. T2 similarly turns off and causes T3 to turn off. This action turns off coil primary current. The coil magnetic field collapses, and secondary voltage is induced and distributed to the appropriate spark plug.

Electronic ignition control modules have several forms, but all work on these principles. Some modules are large assemblies, figure 13-6. Other modules are small integrated circuit assemblies, mounted in the distributor, figure 13-7. Some modules, like the original Chrysler design, use a single large power transistor to switch primary current. Others use several power transistors in a Darlington arrangement.

## ELECTRONIC DISTRIBUTOR TRIGGER DEVICES

You have seen how primary current is switched on and off by transistors in the ignition control module. Timing of

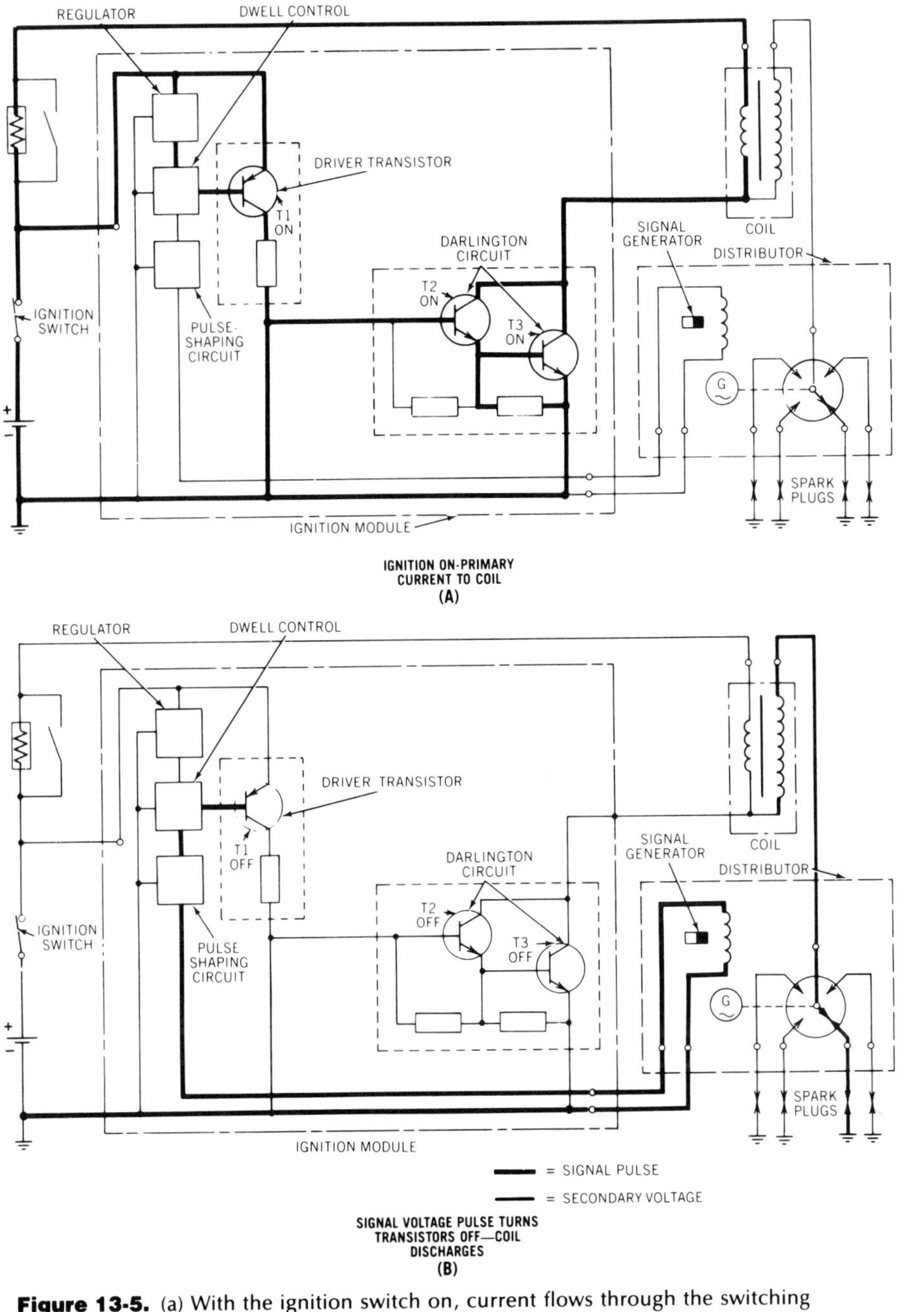


**Figure 13-5.** (a) With the ignition switch on, current flows through the switching transistors to charge the coil. (b) A trigger signal from the distributor turns off the driver and switching transistors to discharge the coil.

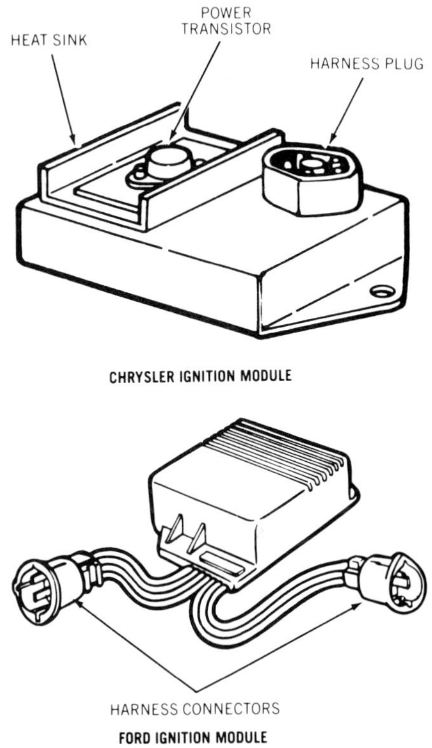


**Figure 13-6.** These Ford and Chrysler ignition modules are mounted in the engine compartment, separately from the distributor. (Champion Spark Plug)

the switching is controlled by a trigger device in the distributor. We have called this device a "signal generator" because it generates a voltage signal that changes the bias on the base of the driver transistor.

The signal generator is a device that triggers ignition much as the cam and breaker points do in a breaker-point ignition. Remember, though, that the signal generator, or trigger device, does not carry ignition primary current. All electronic distributor trigger devices create or alter a voltage signal on the base of the driver transistor. Depending on the driver transistor design, this voltage may be only a few millivolts (thousandths of a volt), and its current may be only a few milliamperes. There are several ways to create such a voltage signal in an electronic distributor:

1. Magnetic-pulse generator
2. Hall-effect switch
3. Metal-detection circuit
4. Optical (light-detection) circuit.

All trigger devices consist of a trigger wheel that rotates with the distributor shaft and a stationary sensor, figure 13-8. Figure 13-9 lists the names that manufacturers have given to these devices. We will generally use the terms "trigger wheel" and "sensor" or "pickup coil" to describe their operation.

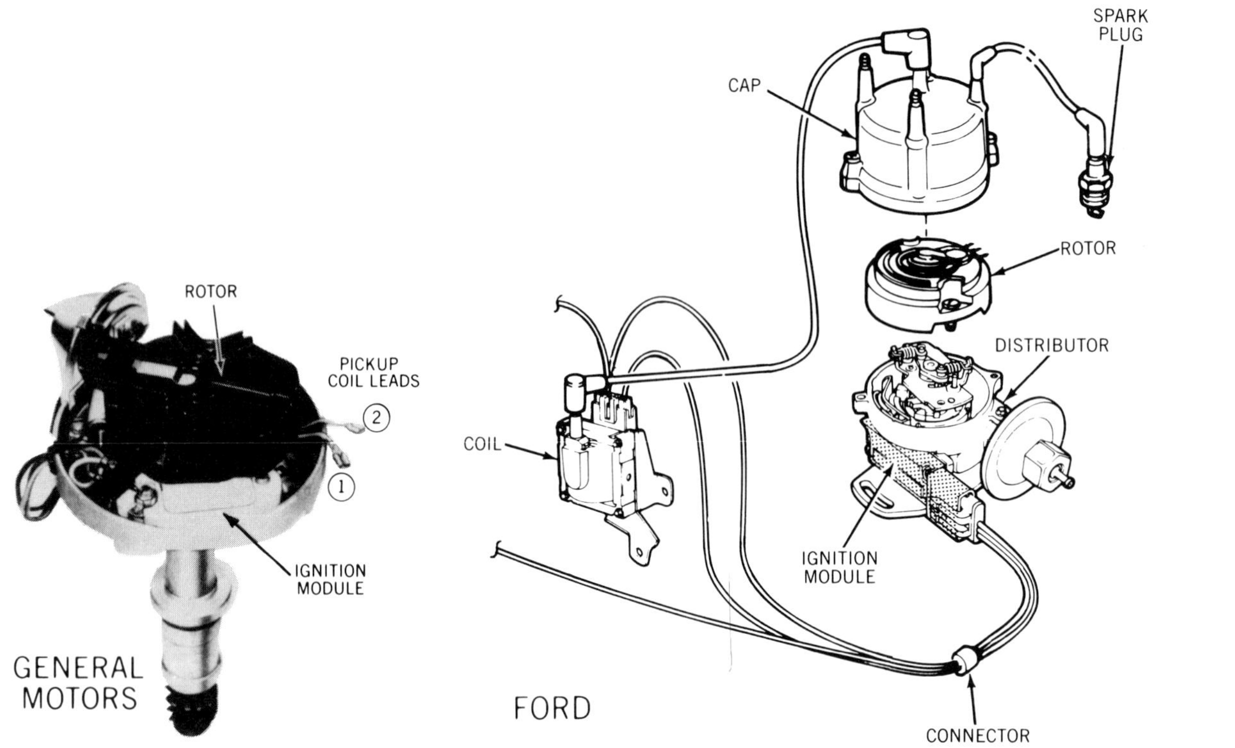


**Figure 13-7.** These ignition modules are mounted in or on the distributor. They work the same as larger modules. (Ford) (Champion Spark Plug)

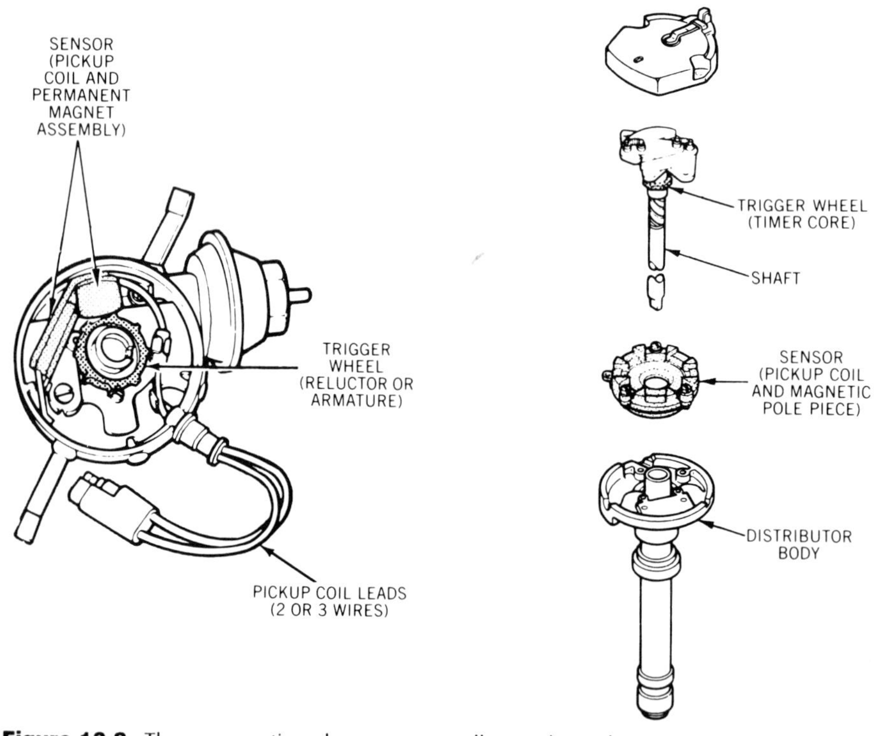


**Figure 13-8.** These magnetic-pulse generators all use a low-reluctance trigger wheel to induce a voltage signal in a pickup coil. (Chrysler) (Chevrolet)

| Manufac-turer | Sensor | Trigger Wheel |
|---|---|---|
| AMC | Sensor | Trigger wheel |
| Bosch | Pickup coil & pole piece | Trigger wheel |
| Chrysler | Pickup coil | Reluctor |
| Ford | Stator | Armature |
| General Motors | Pole piece & magnetic pickup | Timer core |

**Figure 13-9.** Manufacturers have different names for the sensors and trigger wheels in electronic distributors, but all devices serve the same purpose.

## Magnetic-Pulse Generator

Magnetic-pulse generators are the most common trigger devices in electronic distributors. The original Chrysler, Ford, and GM electronic ignitions used them, and still do today. So do most European and Japanese ignitions. These devices have a stationary sensor with a permanent magnet and a pole piece. Fine wire is wrapped around the pole piece to form a pickup coil, figure 13-10. The permanent magnet establishes a field around the pickup coil. If the strength (magnetic flux) of this field changes, it will induce a small voltage in the pickup coil.

Whether the pickup coil has one pole piece or several, it works the same way. If the pickup has several pole pieces, the coil winding is wound around each in series so that the voltages at each pole piece add together for a single signal voltage, figure 13-10.

The trigger wheel for a magnetic-pulse generator is attached to the distributor shaft and usually has as many teeth, or spokes, as the engine has cylinders. The trigger wheel is made of steel that is nonmagnetic but that has low magnetic reluctance (high permeability). As a tooth of the trigger wheel enters the magnetic field around the pole piece and pickup coil, magnetic flux concentrates in the tooth. This changes the magnetic field strength and induces a voltage in the pickup coil. This voltage is proportional to the *rate of change of the magnetic flux*.

Pickup coil voltage is low when a trigger wheel tooth is farthest from the pole piece in degrees of rotation. As a tooth rotates closer to the pole piece, the magnetic field starts to increase, and voltage starts to increase in the pickup coil. Voltage in the pickup coil is an ac voltage, figure 13-11, that develops as follows:

**1.** Voltage increases with positive polarity as a tooth approaches the pole piece.

**2.** Voltage falls to zero when the tooth is directly aligned with the pole piece.

**3.** The magnetic field reverses, and voltage then increases with opposite polarity as the tooth moves away from the pole piece.

Even though magnetic field strength is strongest when the tooth is aligned with the pole piece, the *rate of change* of the flux density is zero at this point. That is why the voltage drops to zero. This point corresponds with the ignition timing point for one cylinder. It is the same as the point where breaker

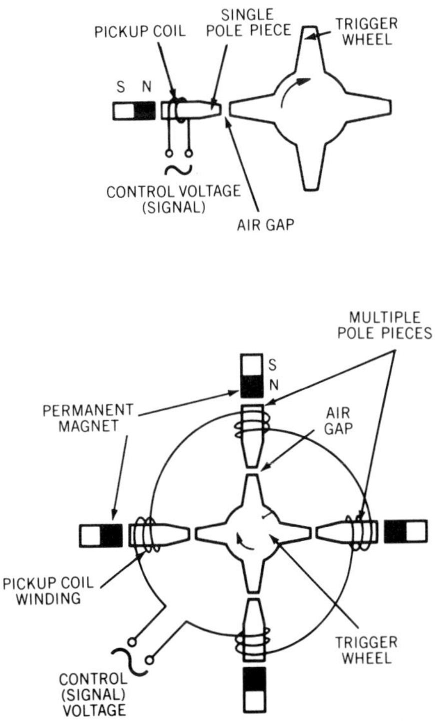


**Figure 13-10.** Whether the pickup coil has a single pole piece or multiple pole pieces, it generates a signal voltage the same way. (Bosch)

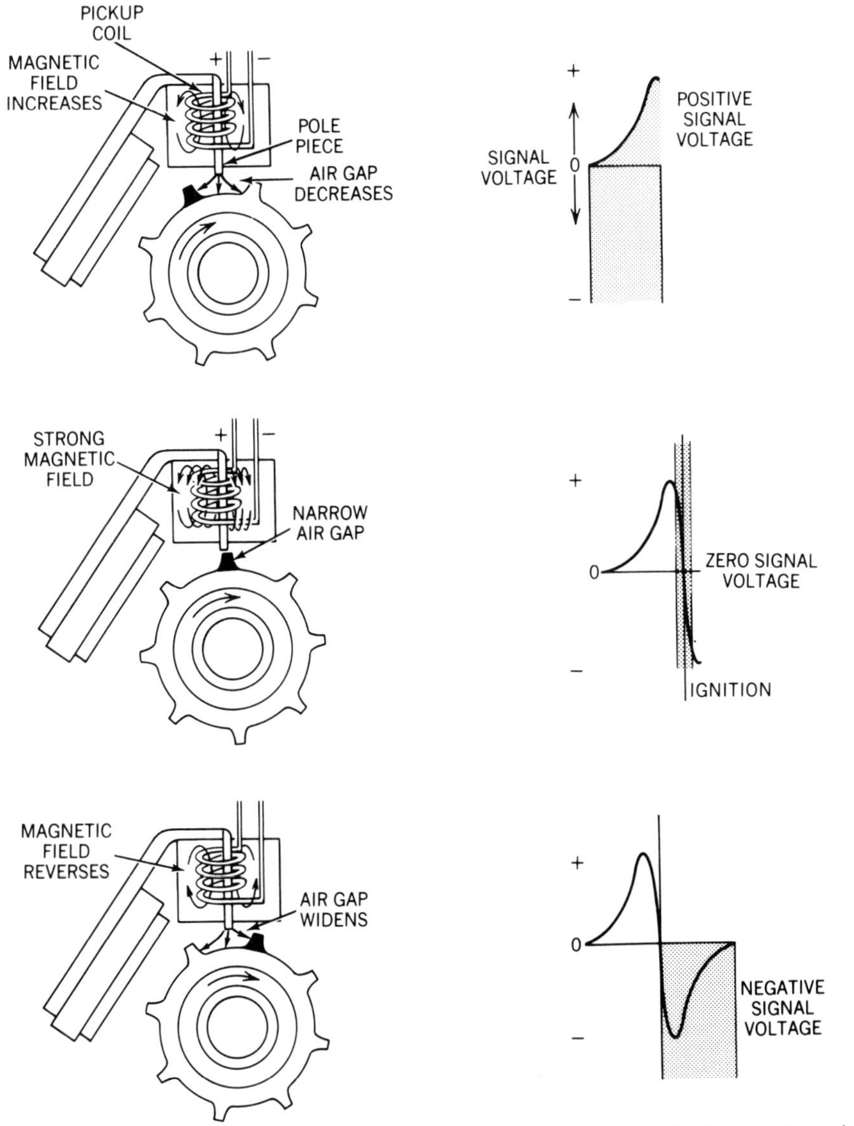


**Figure 13-11.** The pickup coil generates an ac voltage as the trigger wheel moves in and out of the magnetic field.

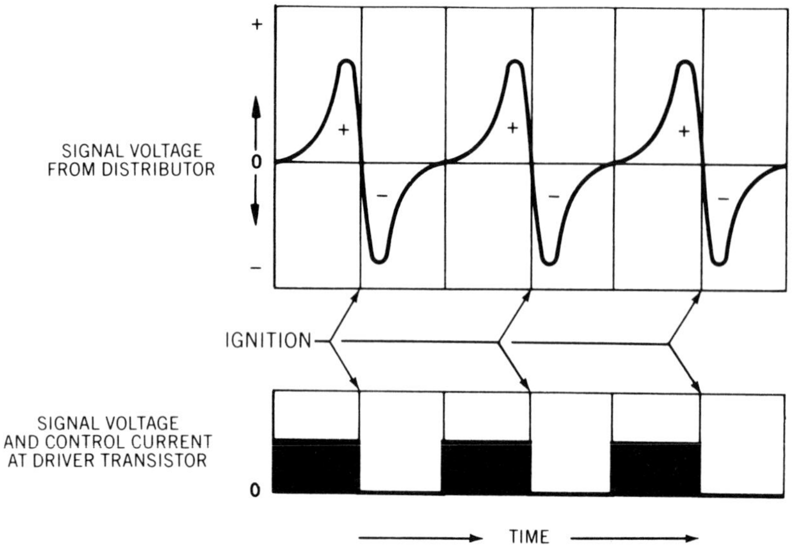


**Figure 13-12.** A circuit in the ignition module converts continuous ac voltage pulses from the pickup coil to dc square-wave pulses that signal the driver transistor. (Bosch)

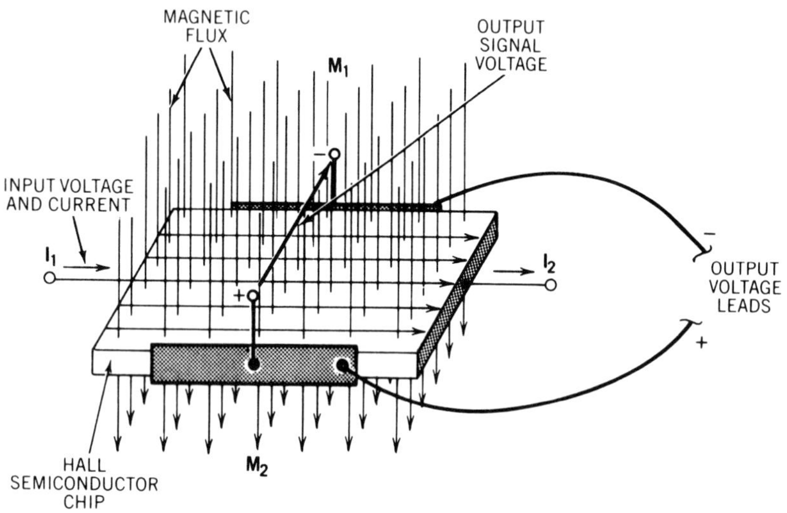


**Figure 13-13.** The Hall-effect output voltage changes as the strength of the magnetic field changes. Input current must stay constant.

points open. The distributor pickup coil is connected to the base of the driver transistor in the ignition module. The zero-voltage signal on the driver transistor changes the transistor bias and turns it off. The switching transistors then turn off primary current, and the ignition coil discharges secondary voltage to the plug.

As the pickup coil voltage induced by one trigger wheel tooth changes to negative polarity, another tooth approaches the pole piece. This starts another positive voltage pulse in the pickup coil. The ignition module receives continuous sine wave ac voltage from the pickup coil. A pulse-shaping circuit in the module converts these signals to dc voltage pulses so that the driver transistor responds to zero-voltage signals, figure 13-12.

## Hall-Effect Switch

Hall-effect switches began to appear in production ignition systems in the 1970's. Bosch introduced such systems in the mid-1970's. One of the first major domestic uses was on Chrysler's Omni and Horizon models in 1978. Their use has increased with other carmakers since.

Like a magnetic-pulse generator, a **Hall-effect switch** has a stationary sensor and a rotating trigger wheel. However, it does not generate a signal voltage in the same way. In fact, it requires a small input voltage to generate an output voltage.

The Hall effect is the ability to generate a small voltage in semiconductor material by passing current through it in one direction and applying a magnetic field to it at a right angle to its surface, figure 13-13. When current flows through a semiconductor chip from $I_1$ to $I_2$ and magnetic flux crosses the chip at a right angle from M1 to M2, voltage develops across the chip at a right angle to the input current. If we hold the input current steady and vary the magnetic field, the output voltage will change in proportion to field strength.

A typical Hall-effect switch in a distributor has a Hall element, a permanent magnet, and a ring of metal blades, or shutters, similar to a trigger wheel. The blades may hang down from the rotor, as on Bosch and Chrysler distributors, figure 13-14. Or, they may be on a separate ring on the shaft, as on the Ford and GM distributors, figure 13-15.

When a shutter blade enters the gap between the magnet and the Hall element, it creates a magnetic shunt that changes the field strength through the Hall element. This causes the Hall output voltage to change, which changes the bias on the ignition driver transistor just as the signal from a magnetic-pulse generator does.

The Hall element is a complex integrated circuit with the Hall-effect semiconductor and an output voltage generator. The generator receives the signal voltage from the Hall chip, processes it, and sends a square-wave voltage pulse to the ignition module. Figure 13-16 summarizes the relationships of the Hall-effect switch and the complete ignition circuit. A key point to remember is that ignition occurs when the rotating vane *leaves* the window between the Hall-effect switch and the magnet.

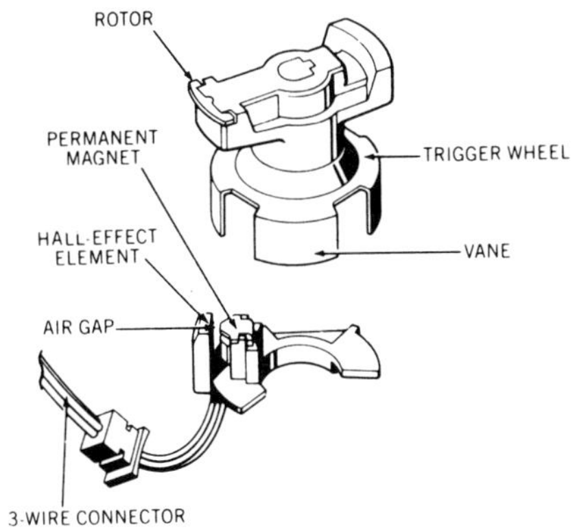


**Figure 13-14.** This Hall-effect distributor has the trigger wheel vanes on the bottom of the rotor. (Bosch)

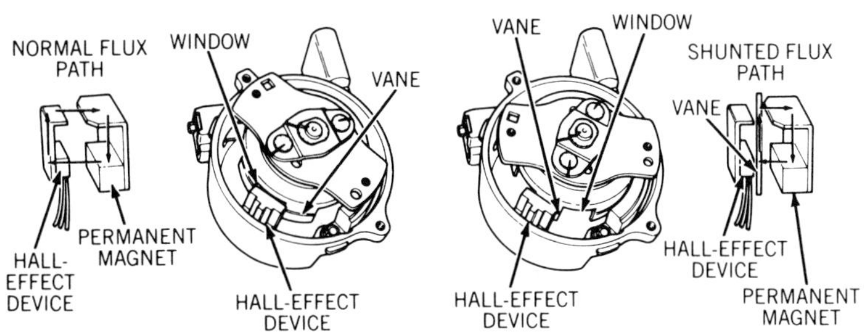


**Figure 13-15.** The Hall-effect wheel of this distributor is on the shaft rotor base. These views show how the vanes and windows change the magnetic field around the Hall-effect device. (Ford)

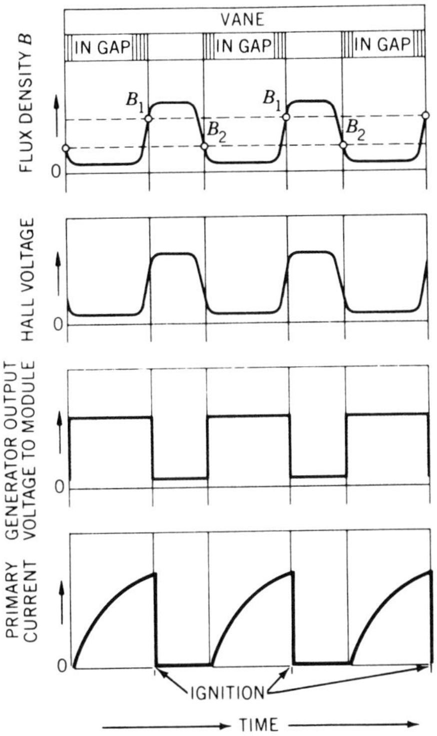


| Vane | Magnetic field ($B$) | Hall voltage | Generator Output Signal Voltage to Module | Ignition Module Transistors |
|---|---|---|---|---|
| not in the air gap | permeates the Hall layer | maximum | minimum | switched off |
| enters the air gap | is deflected away from the Hall layer | drops | increases abruptly | switch *on* |
| in the air gap | very weak at the Hall layer | minimum | maximum | switched *on* (energy storage) |
| leaves the air gap | permeates the Hall layer increasingly | increases | drops abruptly | switch *off* **(ignition point)** |

**Figure 13-16.** The table and graphs explain the relationships of Hall-effect signal voltage and ignition discharge. (Bosch)

Although a Hall-effect switch requires connections for input voltage, its output voltage does not depend on the speed of the rotating trigger wheel. Therefore, it generates a full-strength output voltage even at slow cranking speeds.

## Metal-Detection Circuit

The metal-detection circuit creates an output signal voltage similarly to a magnetic-pulse generator and a Hall-effect switch. This kind of signal generator does not have a permanent magnet, however. It receives an input voltage from the ignition control module to an electromagnet in the distributor sensor. The rotating metal teeth of the trigger wheel affect the strength of the electromagnetic field and cause voltage pulses in the sensor. The ignition control module senses these voltage pulses as signals to open the ignition primary circuit. The module provides primary switching action similar to the action of other ignition modules. Like a Hall-effect switch, a metal-detection circuit provides full-strength voltage signals at slow speeds.

Prestolite was the major maker of metal-detection electronic ignition systems. American Motors cars from 1975 through 1977 used a Prestolite electronic ignition called "Breakerless Inductive Discharge (BID)" ignition.

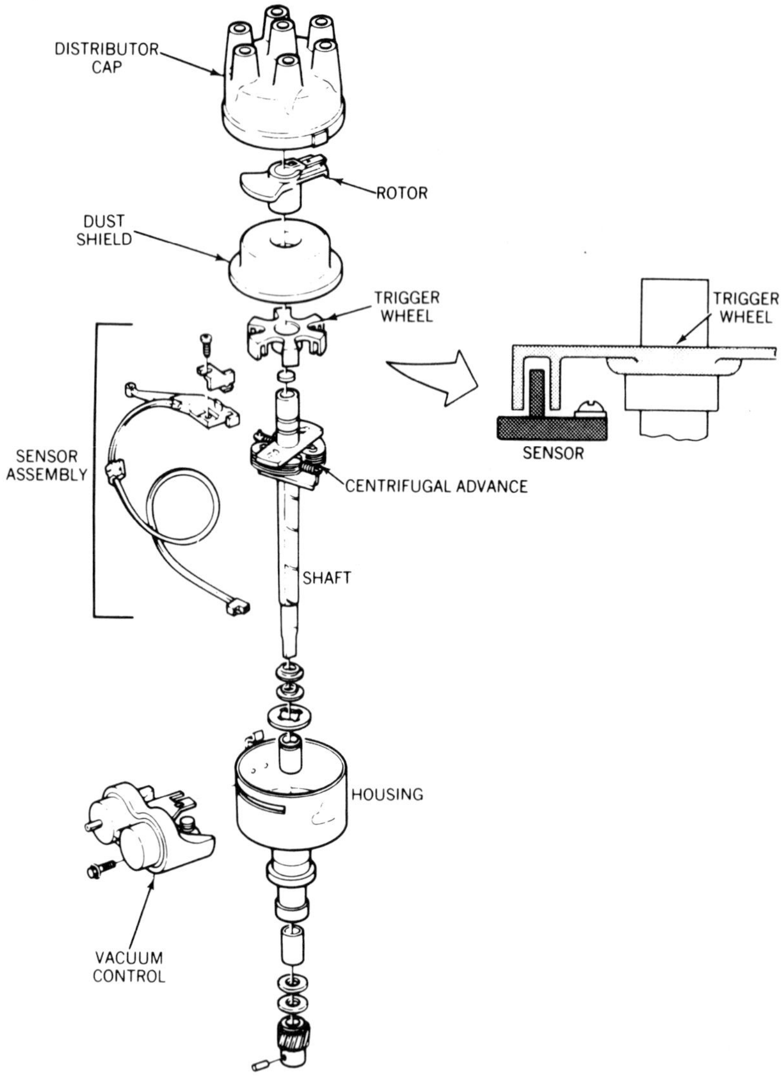


**Figure 13-17.** This Prestolite distributor used by American Motors has a metal-detection triggering device. (AMC)

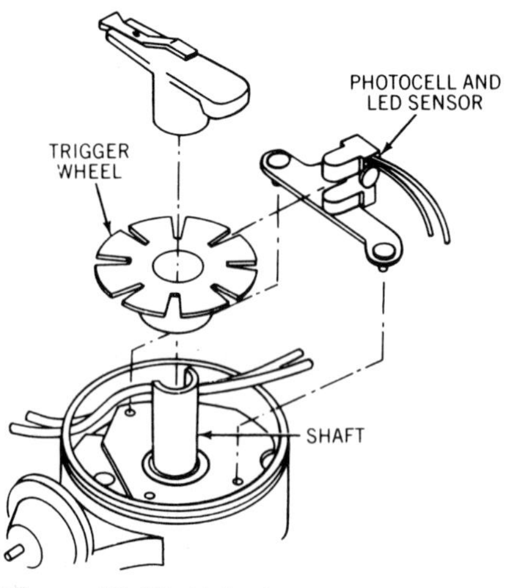


**Figure 13-18.** Light-detection triggering devices have been used in some aftermarket ignition conversions.

International Harvester light-duty gasoline engines of the late 1970's have a variation of the same system. In the AMC-Prestolite distributor, the trigger wheel has teeth with an inverted U shape, figure 13-17, that pass over the sensor. Metal-detection circuits have not been used as original equipment by any major carmaker since the early 1980's, however.

### Optical (Light-Detection) Circuit

An optical signal generator uses the light from a light-emitting diode (LED) to strike a phototransistor and generate a voltage signal. The trigger wheel is a slotted disc that passes between the LED and the phototransistor, figure 13-18. When a slot is between the LED and the transistor, the light beam creates a voltage signal that triggers the ignition module.

Several aftermarket systems have been made to convert breaker-point distributors to optically triggered electronic distributors. Chrysler and Mitsubishi were the first major carmakers to use an original-equipment optical distributor, as explained later in this chapter.

## DWELL IN ELECTRONIC IGNITIONS

In chapter 12, you learned that dwell is the period when the primary circuit is on (closed) and the coil magnetic field is increasing. In a breaker-point ignition, dwell is always measured in degrees of distributor rotation, but it really represents coil charging time. Because electronic ignitions are inductive-discharge systems, they also have a dwell period to charge the coil. However, the start of the dwell period is not controlled by a signal from the distributor as it is when the points close in a breaker-point system.

The ignition control module controls the start of the dwell period through a timing or a current-sensing circuit. When you service ignition systems, you will learn that you can measure dwell in degrees of distributor rotation, using an oscilloscope, figure 13-19. This is an important service measurement for a breaker-point system, and you can use the dwell measurement to troubleshoot several parts of the distributor.

You can use an oscilloscope to observe the dwell period for an electronic ignition, and you can identify some mechanical problems in electronic distributors from the dwell portion of the pattern. However, dwell is not adjustable in most electronic distributors, and you normally do not use the dwell measurement as a service specification. There are two kinds of electronic ignitions in terms of dwell control:

**1.** Fixed-dwell systems
**2.** Variable-dwell systems.

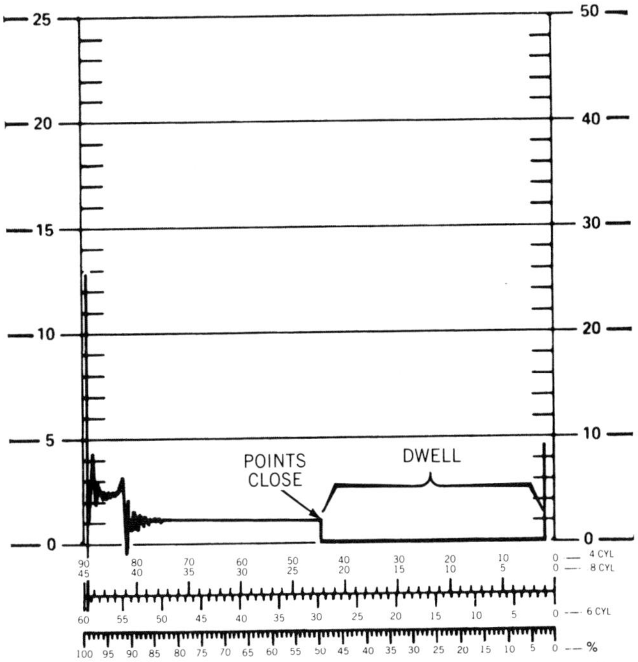


**Figure 13-19.** This primary circuit scope pattern for a breaker-point ignition shows that dwell starts when the points close and stays relatively constant at all speeds.

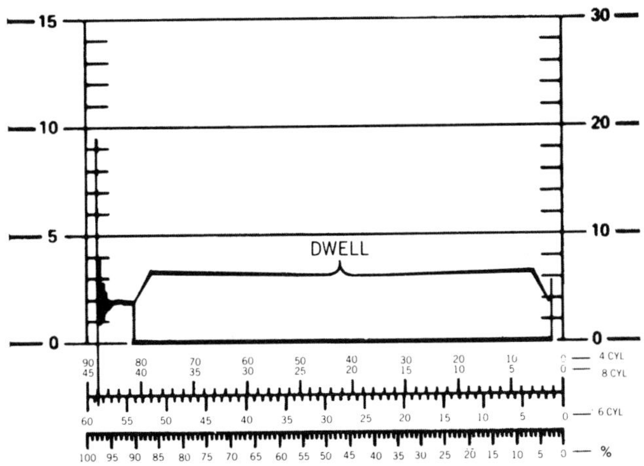


**Figure 13-20.** In a fixed-dwell electronic ignition, dwell starts when the transistor turns on and stays relatively constant at all speeds.

## Fixed-Dwell Electronic Ignitions

A **fixed-dwell electronic ignition** has a ballast resistor in the primary circuit to limit current and voltage. Dwell starts when secondary voltage and current fall below specific levels. The ballast resistor controls primary voltage and current as it does in a breaker-point system. Dwell is the same in distributor degrees at all engine speeds. Figure 13-20 shows a typical fixed-dwell electronic ignition primary circuit oscilloscope pattern.

The original Chrysler electronic ignition is a fixed-dwell system and continues in use on some engines today. The original Ford solid-state ignitions and Dura-Spark II systems are fixed-dwell systems with ballast resistors, as are many Bosch and Japanese electronic ignitions.

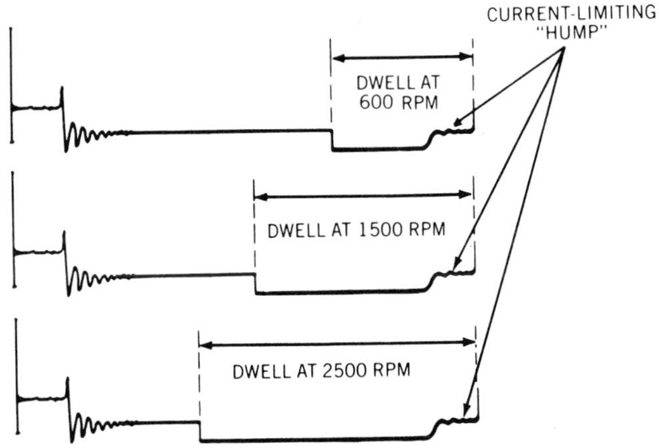


**Figure 13-21.** In a variable-dwell ignition, dwell measured in degrees changes with speed, but dwell *time* stays relatively constant. (Delco-Remy)

## Variable-Dwell Electronic Ignitions

In a **variable-dwell electronic ignition**, the dwell period varies in distributor degrees at different engine speeds but stays relatively constant in actual time. Figure 13-21 shows the primary circuit oscilloscope pattern for a variable-dwell electronic ignition. Notice that the dwell period in distributor degrees is relatively short at idle and increases at higher rpm.

A circuit in the ignition control module senses primary current to the coil and reduces the current when the magnetic field is fully saturated. Notice the small hump toward the end of the dwell periods in figure 13-21. This is the point where current is reduced. Primary current is not shut off, however, and voltage remains applied to the primary until the module receives a trigger signal from the distributor.

Variable-dwell systems do not use a ballast resistor as fixed-dwell and breaker-point ignitions do. The coil and the ignition module receive full battery or charging system voltage at all times when the engine is cranking or running. The variable-dwell feature allows the coil to receive the right amount of current for full magnetic saturation because the dwell *time* re-

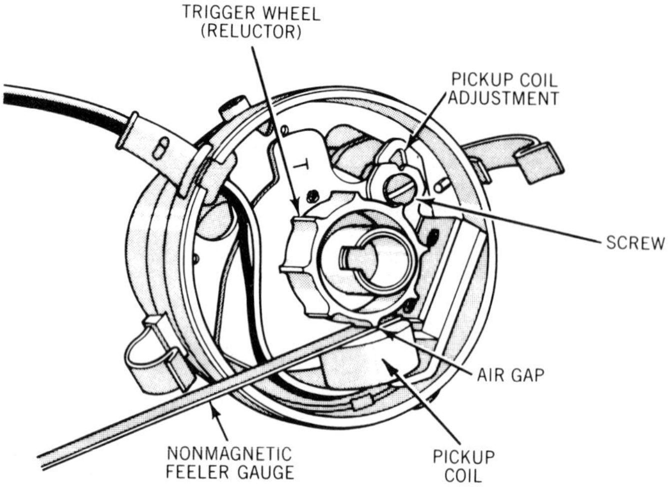


**Figure 13-22.** The air gap between the trigger wheel (reluctor) and pickup coil is adjustable in some distributors. (Chrysler)

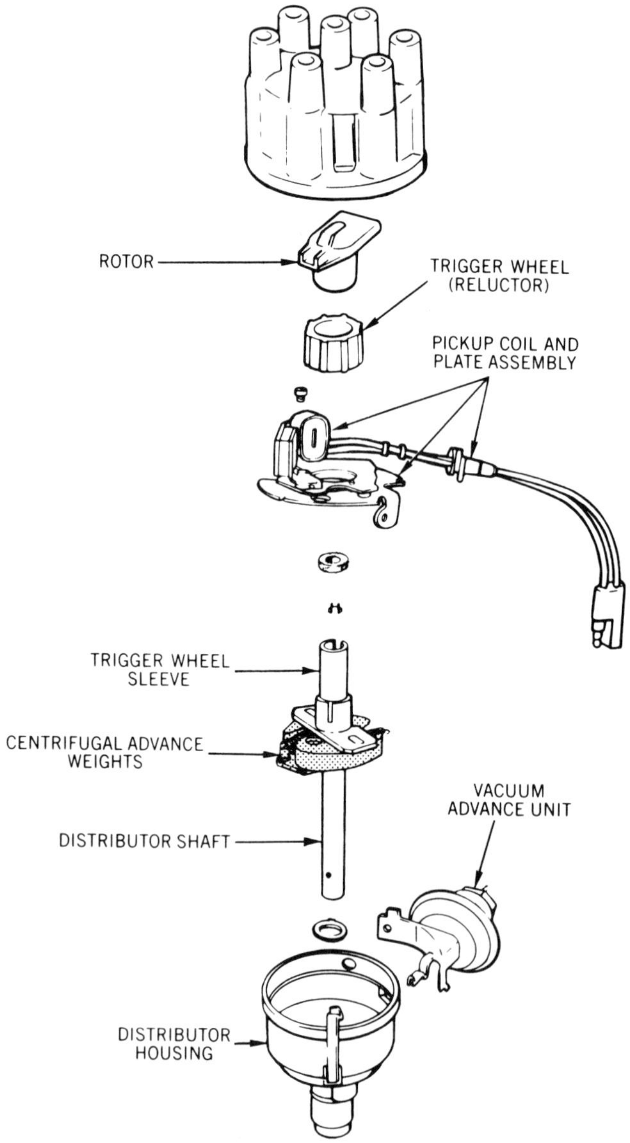


**Figure 13-23.** Early electronic distributors have centrifugal advance mechanisms similar to those in breaker-point distributors. (Chrysler)

mains relatively constant. This ensures enough coil secondary voltage for ignition under the most demanding operating conditions.

Ignition coils for variable-dwell systems generally have higher available voltage capabilities than the coils for fixed-dwell systems. These coils usually have lower primary resistance and a higher turns ratio.

The GM Delco-Remy high-energy ignition (HEI) systems are all variable-dwell systems, as are Ford Motorcraft Dura-Spark I and thick-film integrated (TFI) systems. Other variable-dwell systems include the AMC and International Prestolite ignitions, most Chrysler Hall-effect ignitions, some VW Bosch systems, Marelli systems, and several Japanese ignitions.

### Dwell And Trigger Wheel Gap

The clearance, or gap, between the rotating trigger wheel and the sensor or pickup coil is adjustable on some electronic distributors, figure 13-22. This gap affects the strength of the magnetic field in the pickup coil and thus the strength of the trigger signal. In most distributors, however, it does not affect dwell or timing as the point gap does in a breaker-point distributor. Only in the International Prestolite distributor does the trigger wheel and sensor gap affect dwell.

## SPARK ADVANCE IN ELECTRONIC IGNITIONS

Spark advance in a basic electronic ignition system (on an engine without electronic engine controls) is controlled in the same way it is in a breaker-point system. The trigger wheel is mounted on a sleeve that rotates on the distributor drive shaft, figure 13-23. Centrifugal advance weights rotate the trigger wheel position a few degrees ahead of distributor shaft rotation as engine speed increases. This advances the trigger signal to advance the timing. In a Delco HEI distributor, the advance weights are above the trigger wheel and pickup coil, figure 13-24, as in a Delco V-8 breaker-point

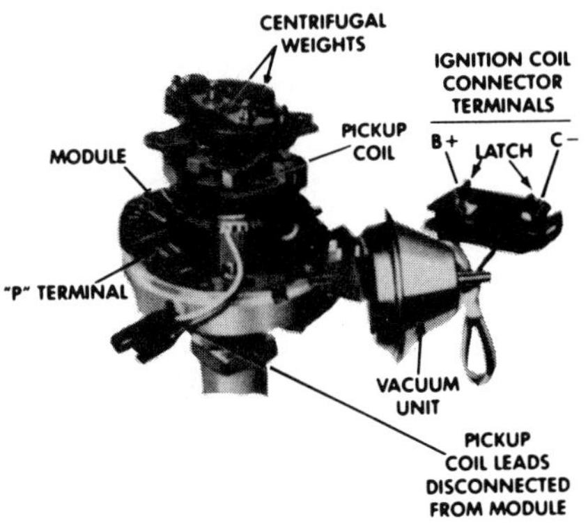


**Figure 13-24.** Centrifugal advance weights in Delco HEI distributors are above the trigger wheel. (Delco-Remy)

distributor. In most other distributors, the weights are below the "breaker plate."

Basic electronic distributors also have vacuum advance mechanisms to change timing in response to engine load. The vacuum advance works as it does in a breaker-point distributor. The vacuum diaphragm linkage is attached to the plate that holds the sensor or pickup coil, figure 13-25. (We can call it the "breaker plate" or the "baseplate.") High vacuum at light load rotates the plate opposite to distributor rotation. This causes the pickup coil or sensor to align with the trigger wheel tooth sooner and advance the timing.

The first generation of electronic ignitions in the early and mid-1970's used the simple centrifugal and vacuum advance mechanisms that breaker-point distributors have. In the late 1970's, carmakers began to equip engines with control systems that perform spark timing and fuel-metering functions electronically. A central computer receives signals from sensors that measure engine temperature, speed, manifold pressure (vacuum), airflow, exhaust oxygen content, and other operating factors. The computer then changes ignition timing more accurately than simple centrifugal and vacuum controls can. Electronic engine controls are commonplace today, and you will learn more about them in a later chapter.

The ignition systems used with electronic engine controls have trigger devices and primary circuit controls of the kind you have studied here. The

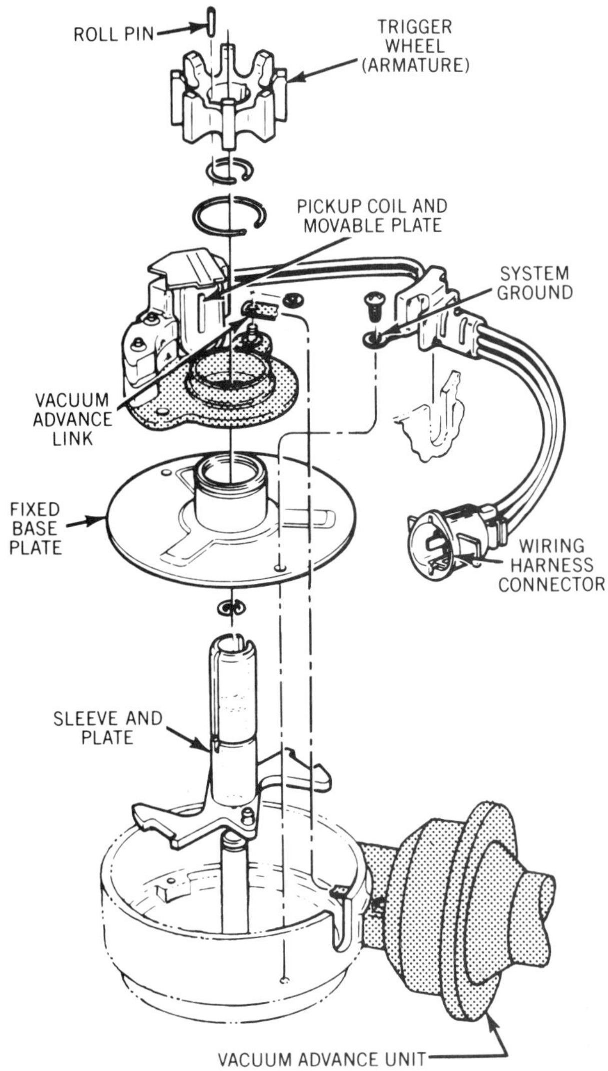


**Figure 13-25.** Early electronic distributors also have vacuum advance mechanisms similar to those in breaker-point distributors. (Ford)

distributors do not have centrifugal and vacuum advance mechanisms. They are principally secondary voltage distributors. Most have magnetic-pulse or Hall-effect triggering devices for basic timing. The triggering devices also act as engine speed sensors or crankshaft position sensors in electronic engine control systems.

## CAPACITIVE-DISCHARGE ELECTRONIC IGNITIONS

Almost all original equipment ignition systems—breaker point and electronic—are inductive-discharge systems. There is another kind of ignition system, however, that you may find on a few engines. This is a **capacitive-discharge (CD) ignition**.

In a capacitive-discharge (CD) ignition, primary current and voltage charge a large capacitor in the ignition module. The time when the ignition module circuit is on and the capacitor is charging corresponds to dwell in an inductive-discharge system. In a CD system, current does not flow through the coil during this time. Instead, it flows to the storage capacitor. A charging circuit in the module has a transformer that increases battery voltage to charge the capacitor to about 300 to 400 volts. While the capacitor is charging, a silicon-controlled rectifier (SCR) in the module acts as an open switch to prevent capacitor discharge, figure 13-26.

When a triggering device—electronic signal generator or breaker

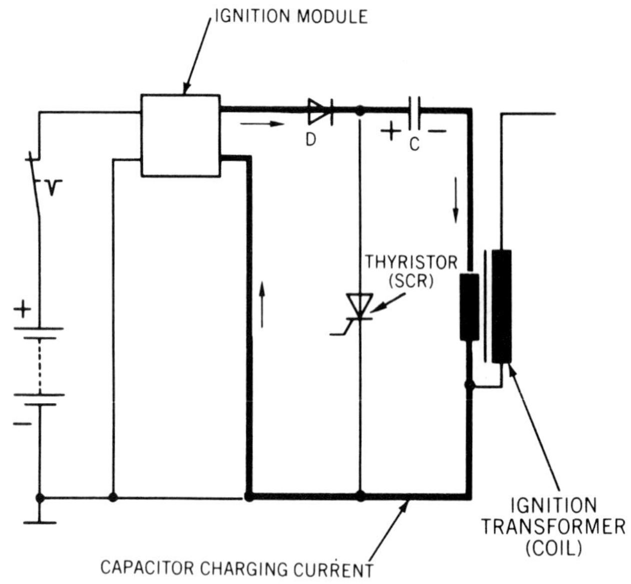


**Figure 13-26.** A CD ignition stores energy by charging the capacitor. The thyristor, or SCR, blocks the discharge circuit. (Bosch)

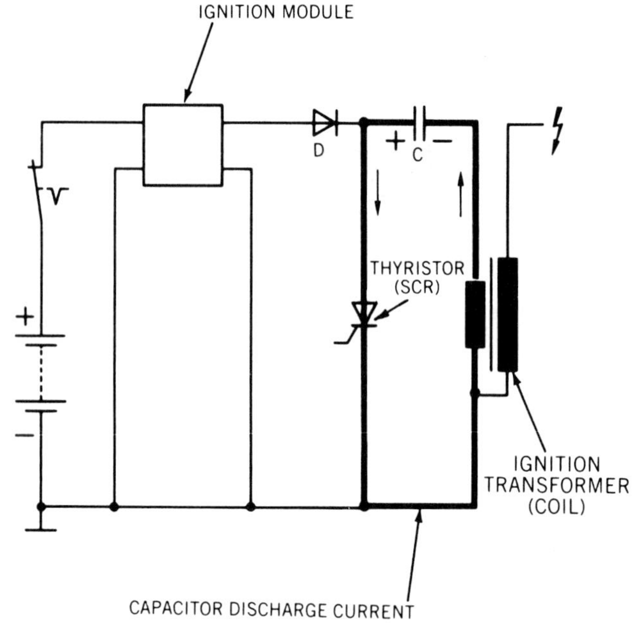


**Figure 13-27.** When the thyristor, or SCR, receives a trigger voltage from the distributor, it allows the capacitor to discharge energy to the coil primary winding. (Bosch)

points—sends a signal pulse to the gate of the SCR, it closes the capacitor discharge circuit. The capacitor then discharges its voltage to the coil primary winding, figure 13-27. The sudden increase in primary voltage and current flow induces high secondary voltage in the coil. Secondary voltage is then distributed to a spark plug as in any other ignition system.

In an inductive-discharge ignition, primary current flows through the coil during a dwell period and is turned *off* to induce high secondary voltage. In a capacitive-discharge ignition, primary current and voltage charge a capacitor. The capacitor is then switched *on* to the coil to induce high secondary voltage.

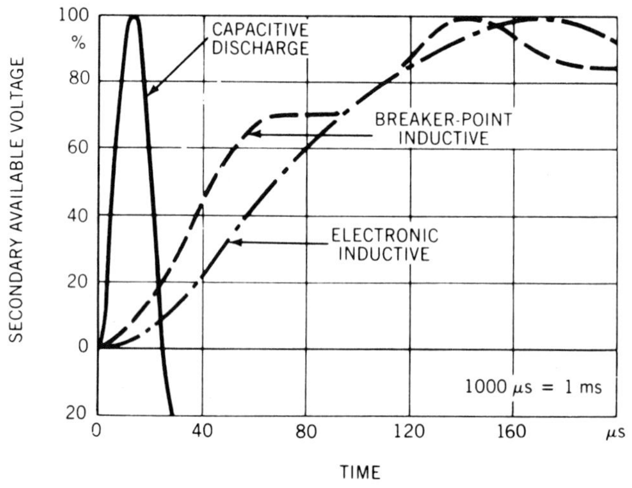


**Figure 13-28.** Secondary voltage can be higher and rise faster in a CD ignition, but ignition duration may not be long enough for many engines. (Bosch)

In either kind of system, secondary voltage is induced by the relative motion of a collapsing or expanding magnetic field. The speed at which the capacitor discharges and the coil field expands is faster than the speed at which the field collapses in an inductive-discharge system. A CD ignition, therefore, can produce a higher and faster firing voltage, figure 13-28. This can be an advantage in high-performance engines or in engines with lean mixtures and other high ignition resistance. The CD spark duration may be too short, however, for reliable ignition in many passenger car engines.

## ORIGINAL EQUIPMENT ELECTRONIC IGNITIONS

The following sections briefly describe the major electronic ignitions used by domestic carmakers, as well as some imported systems. While all are inductive-discharge systems, trigger devices and ignition module circuits vary slightly from one manufacturer to another. The secondary components of these systems do the same job of high-voltage distribution as the secondary components of breaker-point systems. Because they often handle higher voltages, however, many have unique shapes.

All carmakers have modified their ignition systems slightly from year to year and model to model. We can't give you all the details, but we will summarize the basic changes. When you service an electronic ignition on a specific car, refer to the carmaker's

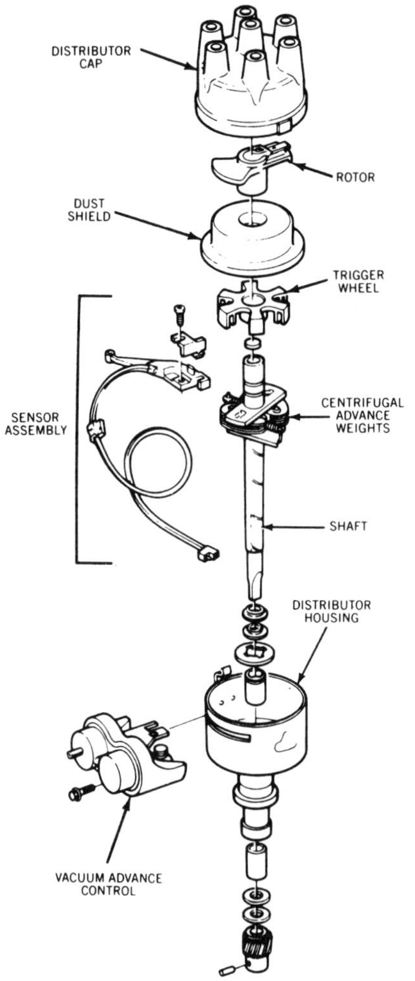


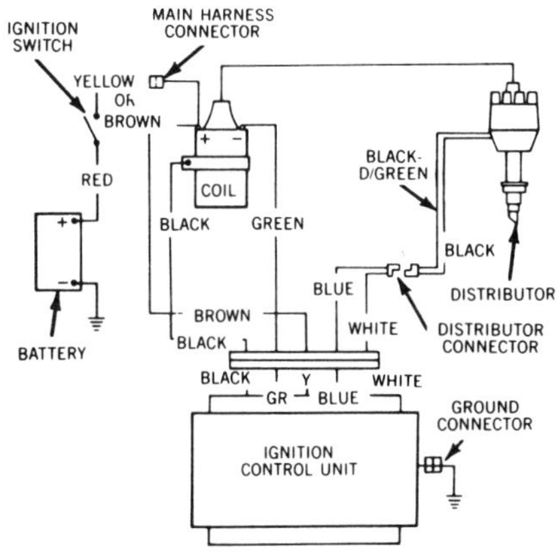


**Figure 13-29.** This Prestolite ignition system and BID distributor were used on 1975–77 AMC engines. (AMC)

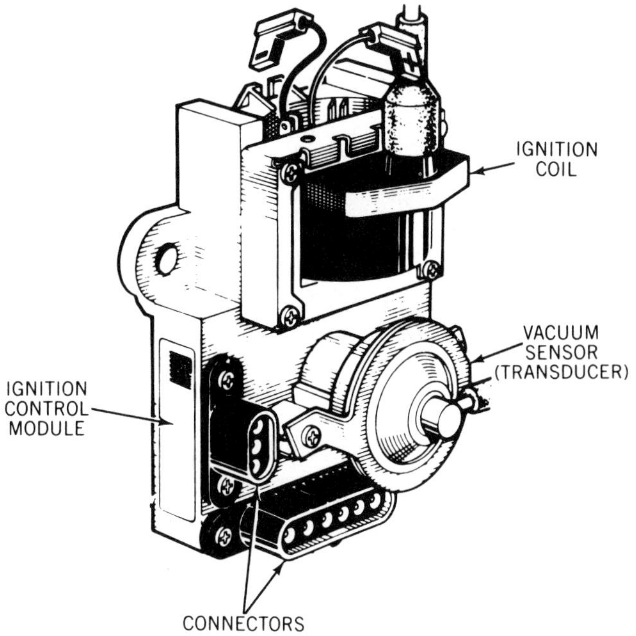


**Figure 13-30.** The coil for the Renix ignition system is mounted above the module. (AMC-Renault)

shop manual or an independent repair manual for exact specifications.

## American Motors Corporation (1975–1987)

The first AMC electronic ignition was the Prestolite breakerless inductive-discharge (BID) system. AMC used the Prestolite system on 6- and 8-cylinder engines from 1975 through 1977. This is a variable-dwell system with a metal-detection triggering device. The ignition control module is mounted remotely from the distributor. Figure 13-29 shows a typical AMC-Prestolite distributor with its unique trigger wheel, as well as its centrifugal and vacuum advance mechanisms. This figure also includes a basic ignition circuit diagram. The BID system uses a conventional-appearing distributor cap, rotor, and ignition coil and 7-mm spark plug cables. Because no ballast resistor is used, full battery voltage is present at the coil and module whenever the system is on. The module provides a steady input voltage for the electromagnet in the distributor sensor.

In 1978, AMC discontinued the Prestolite ignition. AMC uses the Ford Motorcraft solid-state ignition (or Dura-Spark II) on 1978 and later AMC-built 4-, 6-, and 8-cylinder engines. AMC uses the Delco-Remy HEI system in 1980 and later AMC vehicles with GM 4- cylinder and V-6 engines. These systems are described below in the Ford and General Motors sections.

AMC-Renault vehicles use a Renix electronic ignition, built as a joint venture between Renault and Bendix. These inductive-discharge systems are used with electronic engine controls. The distributor simply distributes secondary voltage. The engine computer controls spark timing. The Renix system has a unique-looking coil, mounted on the ignition control module, figure 13-30.

## Chrysler Corporation

Chrysler introduced its electronic ignition on some engines in 1971, and made it standard on all 6- and 8-cylinder engines in 1973. Chrysler's basic electronic ignition uses a distributor with a magnetic pulse generator, figure 13-31. Chrysler calls the rotating trigger wheel a "reluctor" and the stationary sensor a "pickup coil." The basic Chrysler distributor has simple

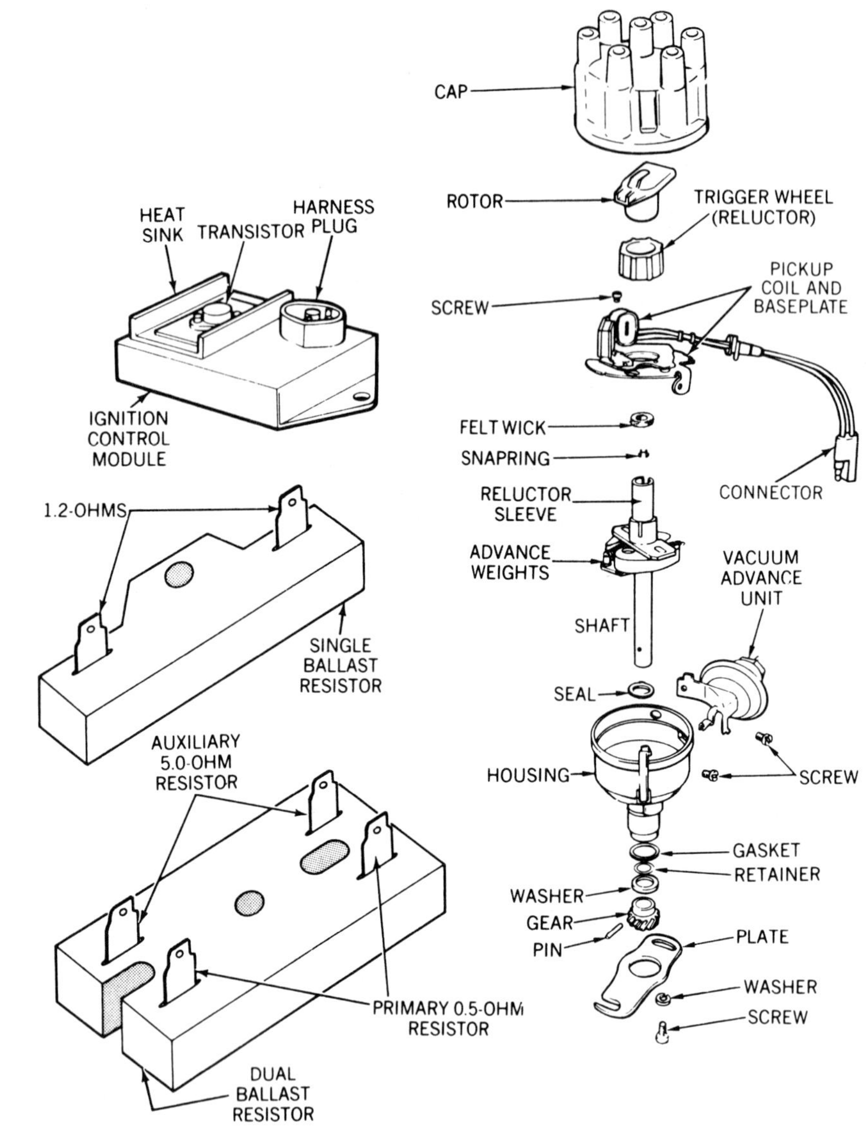


**Figure 13-31.** Typical distributor, ignition module, and ballast resistors for Chrysler's electronic ignition system. (Chrysler)

centrifugal and vacuum advance mechanisms to control spark advance.

The rest of the system consists of a conventional coil, distributor cap, rotor, and 7-mm spark plug cables. The ignition control module is mounted on the firewall or fender panel and has a single power switching transistor to control the primary circuit, figure 13-31. The transistor is on top of the module and has enough voltage across it when the system is on to produce a shock. Figure 13-32 is a diagram of the basic Chrysler ignition.

The basic Chrysler electronic ignition is a fixed-dwell system with a ballast resistor to limit primary voltage and current. Most Chrysler systems built through the late 1970's have a dual-ballast resistor. The 0.5-ohm resistor is the conventional ballast resistor that regulates primary voltage and current. It is bypassed during starting. The 5.0-ohm resistor protects the electronic circuitry against high-voltage surges and current flow. Most late-model versions of this ignition used with electronic engine controls have a single ballast resistor. Some late-model Chrysler electronic spark control (ESC) systems with the magnetic-pulse distributor do not have a ballast resistor. The spark control computer regulates primary current to the coil.

Chrysler has modified its original electronic ignition several times during the past decade to adapt it to various electronic engine control systems. The distributors for some electronic lean-burn (ELB) and electronic spark control (ESC) systems have two pickup coils, figure 13-33. One provides a fixed timing signal for engine starting only. The other provides a timing signal for all conditions when the engine is running. Distributors for early ELB systems had centrifugal advance but no vacuum advance mechanisms. Distributors for later ESC systems have neither centrifugal nor vacuum advance. All spark advance is controlled by the system computer. The air gap between the reluctor and pickup coil, or coils, is adjustable in Chrysler distributors. Specifications vary for different years.

In 1978, Chrysler introduced its first front-wheel drive (FWD) cars, the Dodge Omni and Plymouth Horizon. The electronic ignitions for these cars have a Hall-effect distributor switch, figure 13-34. The original Omni and Horizon ignitions were fixed-dwell types and used a ballast resistor to control primary current and voltage. The original Chrysler Hall-effect distributors also had centrifugal and vacuum advance mechanisms

Late-model Chrysler 4-cylinder engines use Hall-effect distributors in ignition systems with no ballast resistor. The ESC spark control computer controls dwell and spark advance. The distributors have no centrifugal or vacuum advance mechanisms.

The Chrysler optical distributor, figure 13-35, was introduced with the 3.0-liter V-6 engine in some 1987 models. A thin timing disc in the distributor contains two sets of chemically etched slots, figure 13-36, to provide crankshaft angle and tdc signals.

- *Crankshaft angle slots*—Also called the high-data-rate slots, these are located on the outer edge of the plate and occur at crankshaft rotation intervals of 2 degrees. They generate a

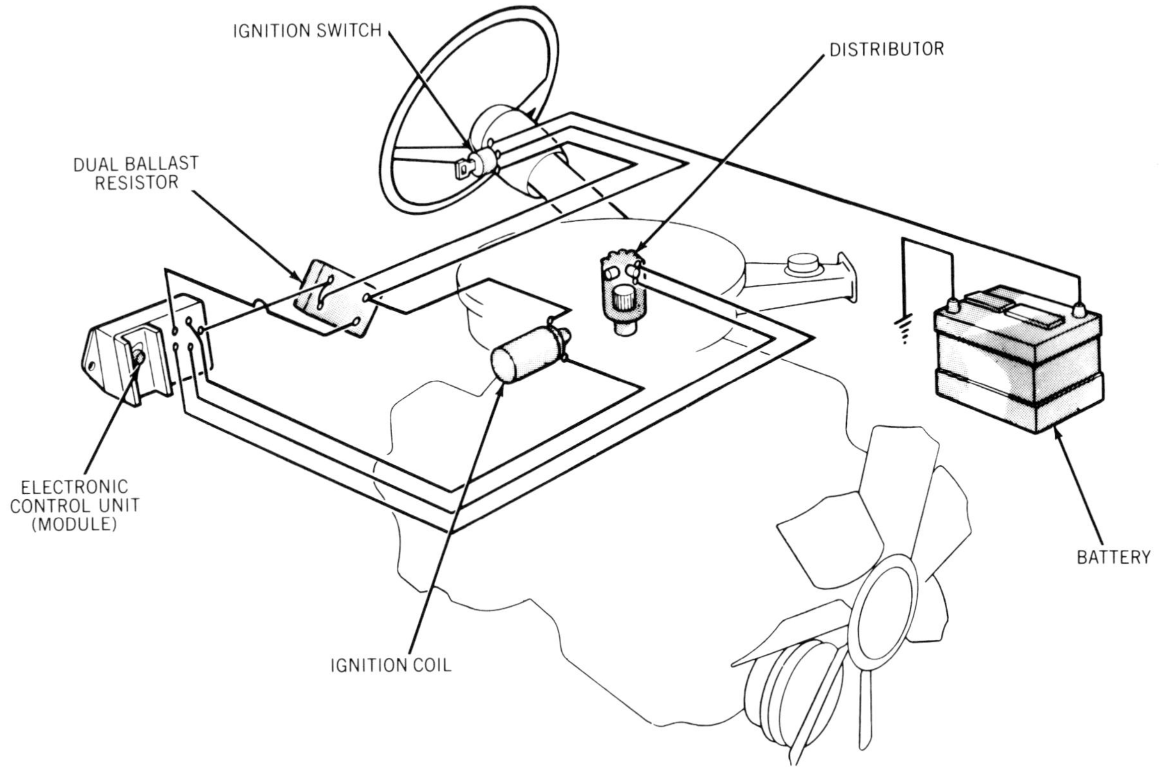


**Figure 13-32.** Typical Chrysler electronic ignition installation. (Chrysler)

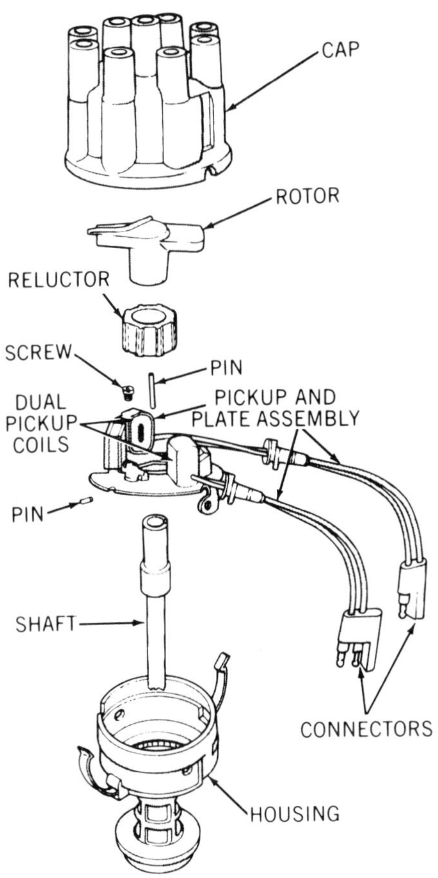


**Figure 13-33.** Many late-model Chrysler distributors have dual START and RUN pickup coils. (Chrysler)

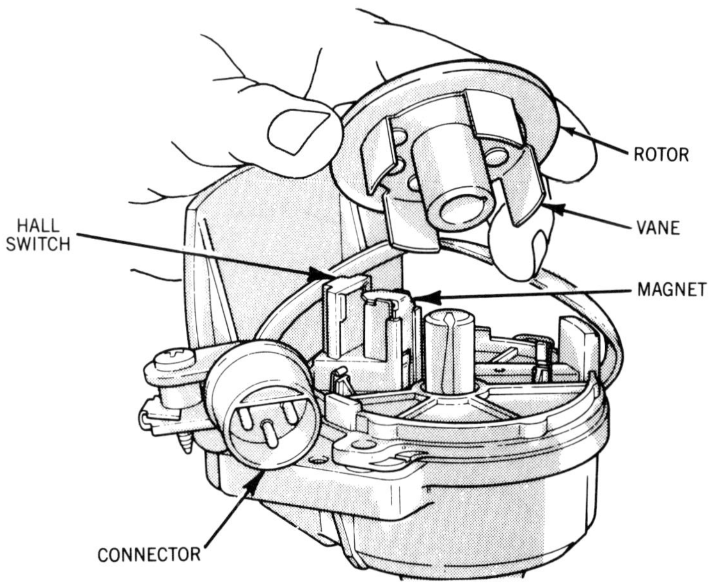


**Figure 13-34.** Chrysler uses a Hall-effect distributor with electronic spark control systems. Early models have centrifugal and vacuum advance. (Chrysler)

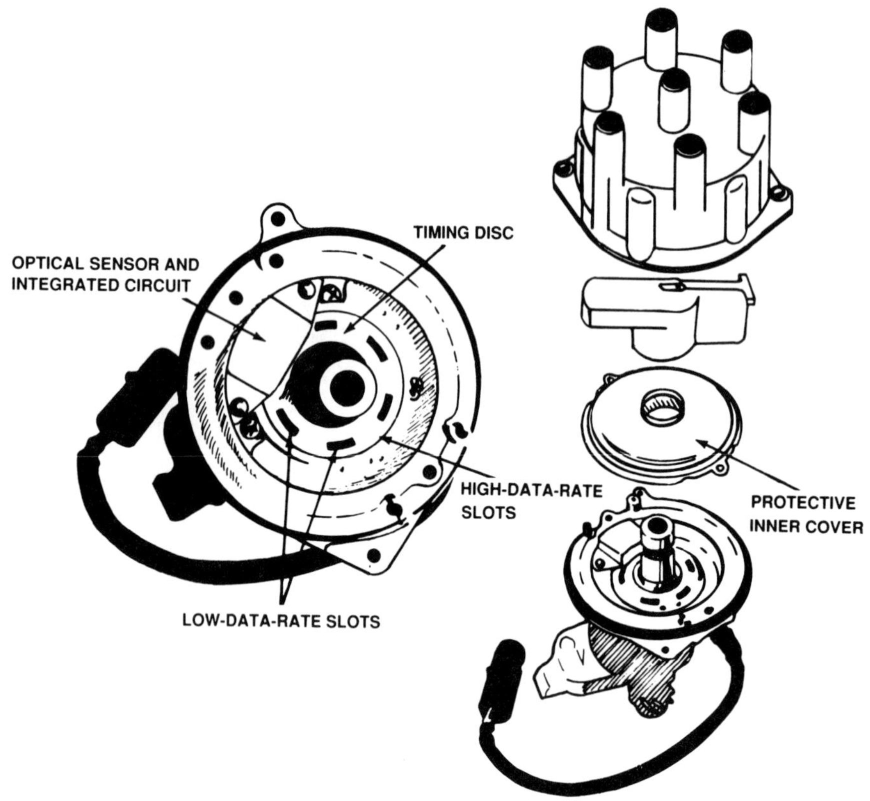


**Figure 13-35.** Chrysler's optical distributor is the first use of a light-detection triggering device to be used in production automobiles. (Chrysler)

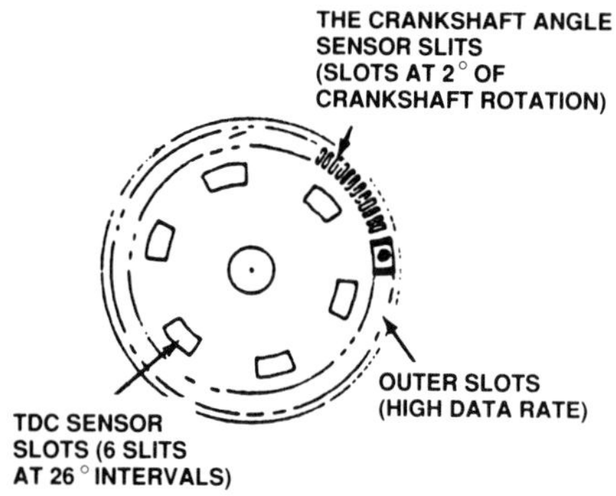


**Figure 13-36.** The outer slots in this timing disc provide crankshaft angle data; the inner slots identify top-dead-center position. (Chrysler)

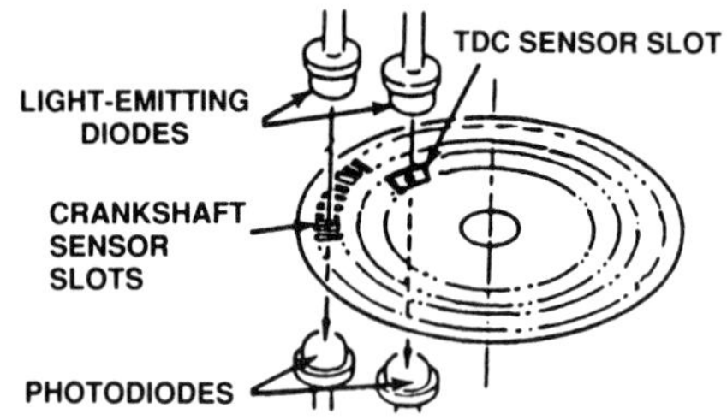


**Figure 13-37.** The timing disc slots interrupt light beams sent to photodiodes by light-emitting diodes (LED'S), creating ac voltage in each diode that is converted into on-off pulses and sent to the logic module. (Chrysler)

signal used to increase ignition timing accuracy at engine speeds up to 1,200 rpm. Engine speed oscillates with the firing pulses of each cylinder during cranking and idle. The high-data-rate signal triggers ignition at the proper crankshaft position instantaneously without regard for the speed oscillations.

• *TDC sensor slots*—The inner slots on the plate correlate to the tdc angle of the crankshaft for each cylinder and provide a low-data-rate signal. The signal triggers the fuel injection system, controls idle speed, and provides ignition timing signals at engine speeds above 1,200 rpm.

The disc revolves between a set of LED's and photodiodes positioned in line with each series of slots, figure 13-37. One set picks up the high-data-rate signal; the other set picks up the low-data-rate signal. The light beam from each LED is focused on its respective photodiode by a mask. As the disc rotates, its two series of slots interrupt the light beams, creating an ac voltage in each photodiode. A hybrid integrated circuit inside the distributor converts the ac voltage into on-off pulses, which are transmitted to the logic module or the single module engine controller (SMEC).

A protective cover is installed between the timing disc (optical sensor) and the rotor, figure 13-35. This cover protects the optical system from contamination and prevents actuation errors that could be caused by electrical noise or EMI.

## Ford Motor Company

Ford introduced its first fully electronic ignition on production cars in mid-1973. It was available on all 6- and 8-cylinder domestic engines in 1974 and became standard in 1975. Looking back on more than a decade of Ford electronic ignitions, it seems that there have been several different systems. The changes, however, have been steady improvements in the original system. Most changes have been in the shape of distributor parts. We can identify Ford ignitions as:

1. Original solid-state ignition (SSI)—1973–76 (cars) and 1974–79 (trucks)
2. Dura-Spark II and III—electrically similar to SSI but with high-dielectric distributor cap, rotor, and cables; 1977 and later cars and trucks
   a. Dual-mode timing—timing control subsystem of Dura-Spark II, 1978 and later
   b. Cranking retard—timing control subsystem of Durak-Spark II, 1979 and later
3. Dura-Spark I—high secondary voltage system; 1977–79 passenger cars
4. Thick-film integrated (TFI) ignition—electrically similar to Dura-Spark I, but with miniature ignition control module installed in the distributor; 1982 and later cars and trucks

**Ford Solid-State Ignition.** Figure 13-38 is an installation drawing of the basic Ford SSI system. Figure 13-39 is a wiring diagram for the primary circuit. The original Ford electronic distributor used a magnetic-pulse generator, as have all Ford distributors except some TFI units. Ford calls the trigger wheel an "armature" and the pickup coil a "stator."

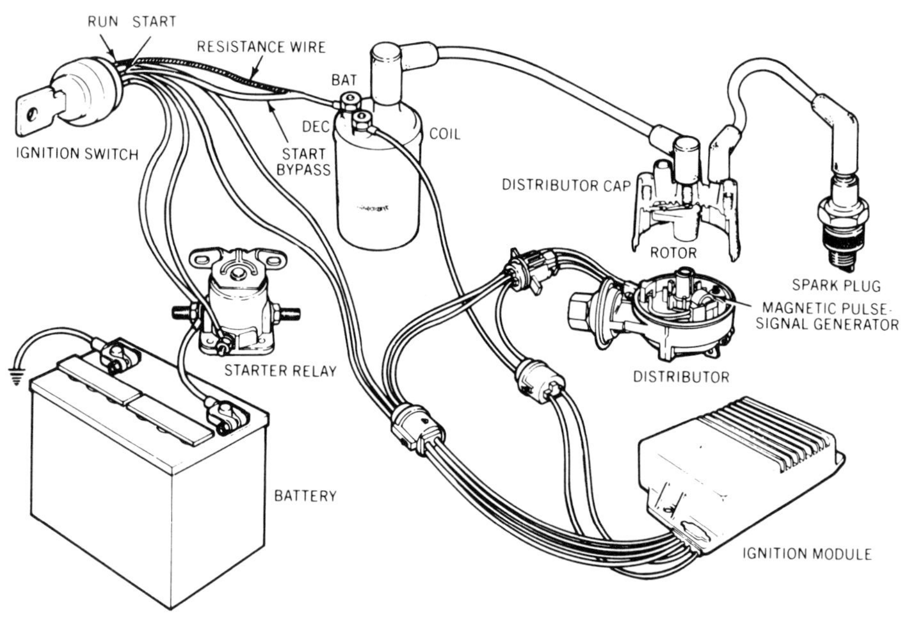


**Figure 13-38.** Ford has modified the original SSI system over the years, but it still has these basic parts. (Ford)

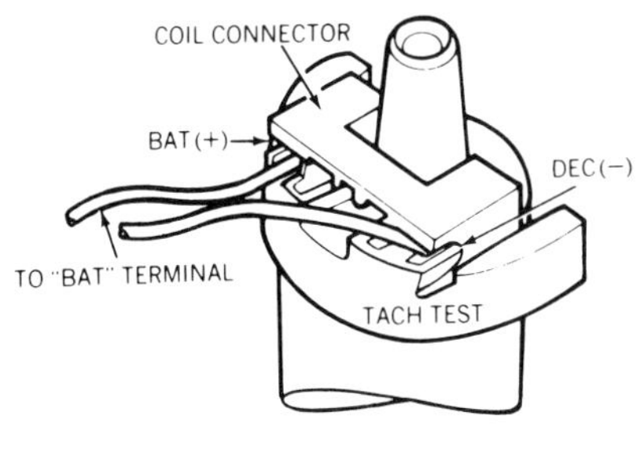


**Figure 13-40.** This connector for the primary terminals of a Ford coil prevents reversed connections. (Ford)

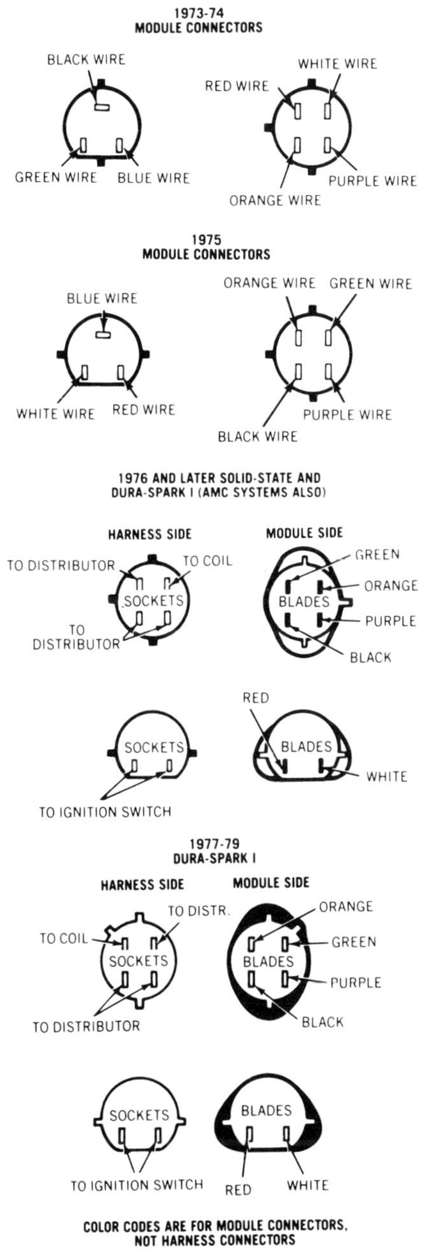


**Figure 13-41.** Ford has changed the ignition module connectors from system to system to prevent misconnection. Wire color codes are the same. (Ford)

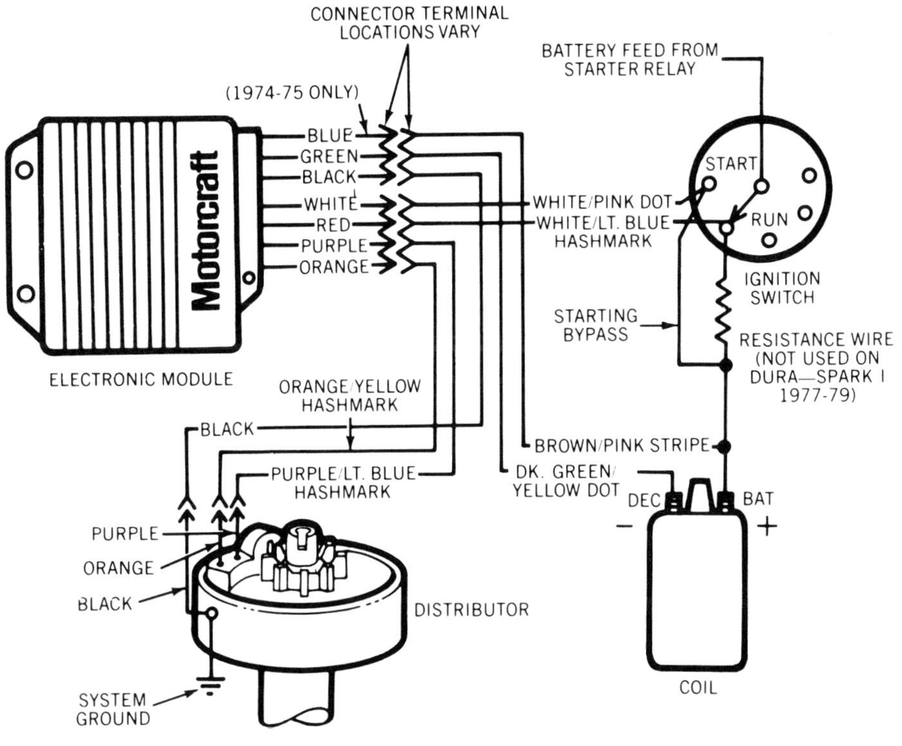


**Figure 13-39.** This basic wiring arrangement is used on all Ford SSI and Dura-Spark systems. (Ford)

The SSI system has a distributor with centrifugal and vaccum advance mechanisms, a conventional-appearing ignition coil, and 7-mm spark plug cables. The SSI system also is a fixed-dwell system with a ballast resistor, as is the later Dura-Spark II system. In 1975, Ford changed the ignition coil connector to a design that prevents reversed primary connections at the coil, figure 13-40.

As Ford refined all of its electronic ignitions over the years, engineers changed the shape and wire arrangement of ignition module connectors so that modules can't be interchanged from one system to another, figure 13-41. In 1976, Ford eliminated the

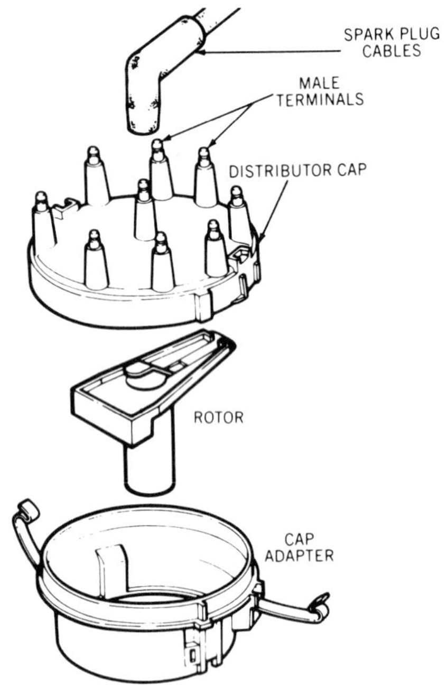


**Figure 13-42.** Typical Dura-Spark distributor cap, rotor, and adapter. (Ford)

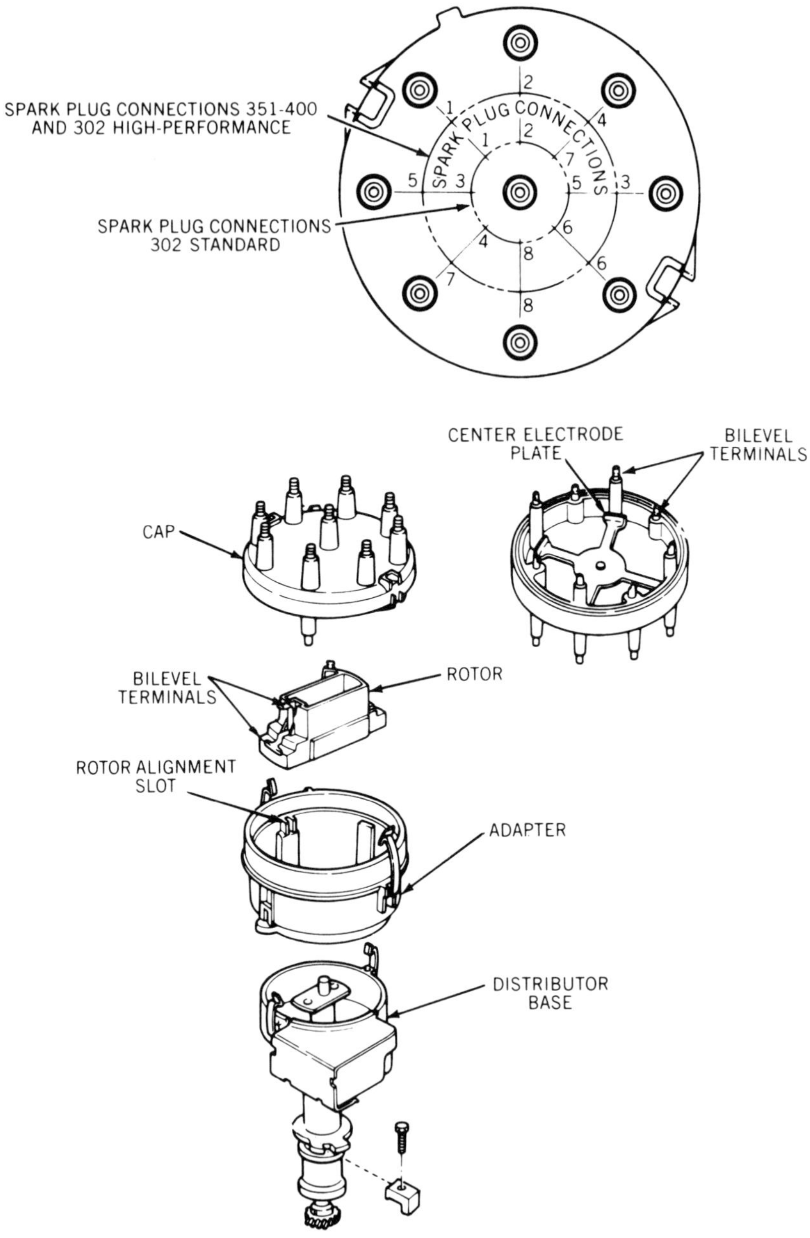


**Figure 13-43.** Dura-Spark I and III distributors have bilevel caps and rotors. Spark plug cables are not in firing order. (Ford)

blue overload shunt wire to the ignition module. All other wire connections and color codes are basically the same for all SSI and Dura-Spark systems, figure 13-39.

**Dura-Spark II and III Ignition.** The primary circuits of Dura-Spark II and III are almost identical to Ford's original SSI system. The resistance of the coil primary and the ballast resistor were lowered slightly to provide higher secondary available voltage. For this reason, the distributor cap and rotor were redesigned to provide more space between the secondary terminals to prevent arcing, figure 13-42. Ford has used a variety of cap and rotor styles on its Dura-Spark systems. All of these designs, along with the plastic materials of the cap and rotor, provide high-dielectric resistance to prevent crossfiring, secondary voltage leakage, and radio interference. These systems also use 8-mm spark plug cables.

Dura-Spark II and III are fixed-dwell systems and use the original magnetic-pulse distributor. Dura-Spark II distributors have centrifugal and vacuum advance mechanisms. Dura-Spark III ignitions are used with electronic engine control systems, and all spark advance is controlled by the system computer. Therefore, the Dura-Spark II and III ignition modules are not identical. Dura-Spark III distributors have caps and rotors with the secondary terminals on two levels, figure 13-43. The greater distance between the terminals prevents secondary voltage arcing. Spark plug cables do not connect to the caps in firing-order sequence, but the cap terminals are numbered by cylinder, figure 13-43.

The dual-mode timing and cranking-retard systems are subsystems of Dura-Spark II. Dual-mode systems have ignition modules with three, rather than two connectors, figure 13-44. The "altitude" version uses a signal from a barometric pressure switch to retard timing at low altitudes. The "economy" version uses a signal from a manifold vacuum switch to retard timing under heavy load. The cranking-retard ignition module senses slow engine speed from the distributor signals during cranking.

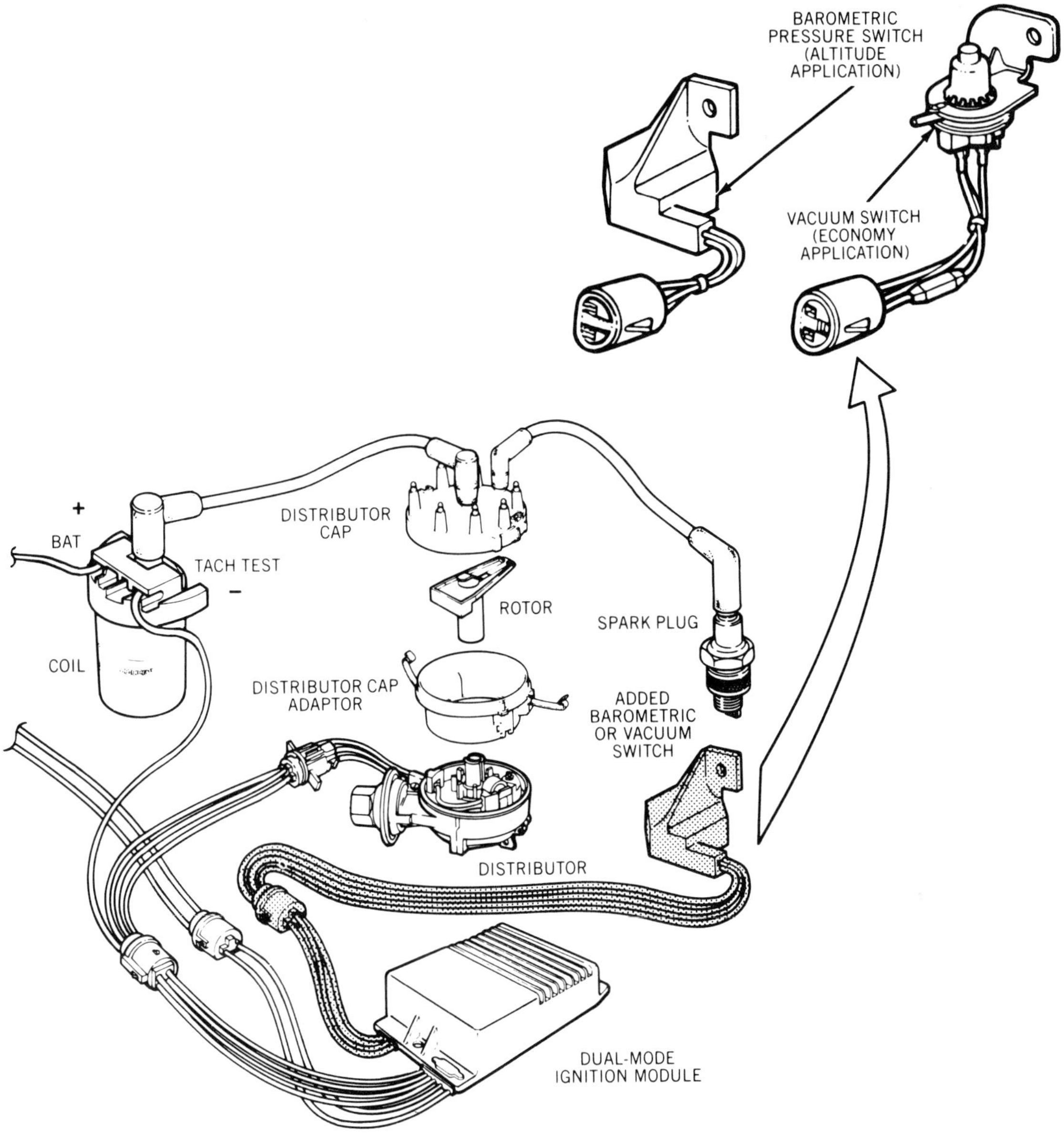


**Figure 13-44.** Typical Dura-Spark II with dual-mode timing. (Ford)

The module electronically retards timing about 18 degrees for easier starting.

**Dura-Spark I Ignition.** Ford's Dura-Spark I system is a high secondary voltage system used with electronic engine controls on some 1977–79 V-8 engines. It uses a magnetic-pulse distributor with centrifugal and vacuum advance. It is a variable-dwell system, however, with no ballast resistor. The ignition module has a current-limiting circuit to control coil saturation.

**Thick-Film Integrated (TFI) Ignition.** Ford introduced its TFI ignition on 1982 1.6-liter 4-cylinder engines in Escort and Lynx models and has expanded its use since. The system is named for the semiconductor manufacturing process used to make the ignition module. Figure 13-45 shows a typical TFI ignition installation, and figure 13-46 is a simplified wiring diagram. The TFI module is a miniature design, mounted on the distributor body, figure 13-47. Like Dura-Spark I, the TFI ignition is a high-voltage, variable-dwell system with no ballast resistor.

The first thick-film ignition ignitions used a magnetic-pulse generator in the distributor and had centrifugal and vacuum advance mechanisms. Later TFI systems for electronic engine controls have Hall-effect switching devices that supply crankshaft position signals. Spark advance is controlled electronically by the EEC-IV engine control computer.

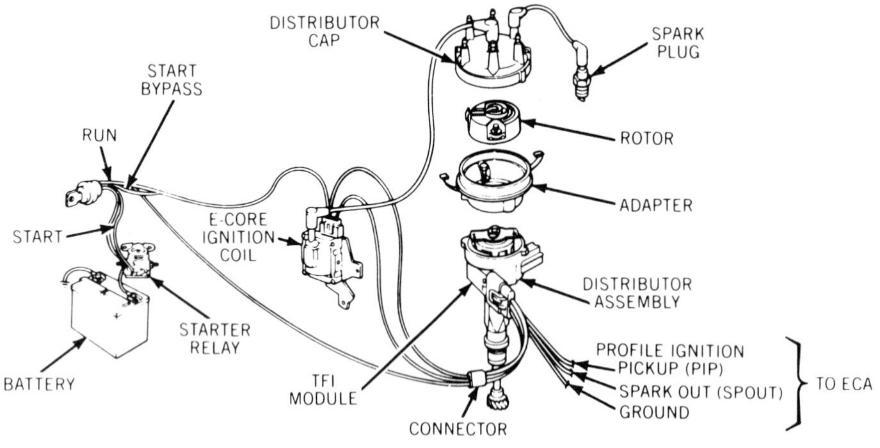


**Figure 13-45.** Typical thick-film integrated (TFI) ignition system installation. (Ford)

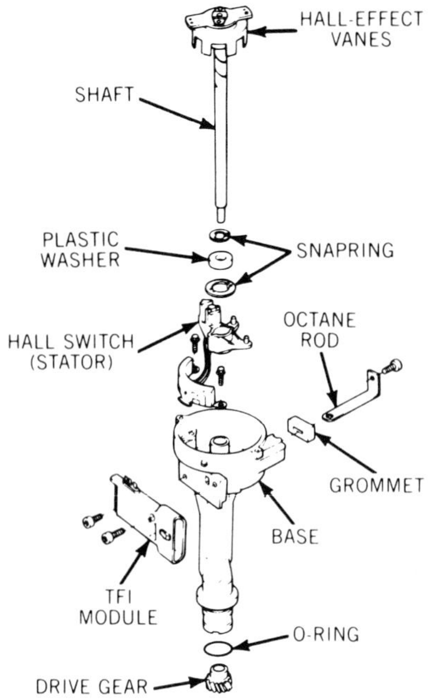


**Figure 13-47.** Typical Ford TFI distributor. (Ford)

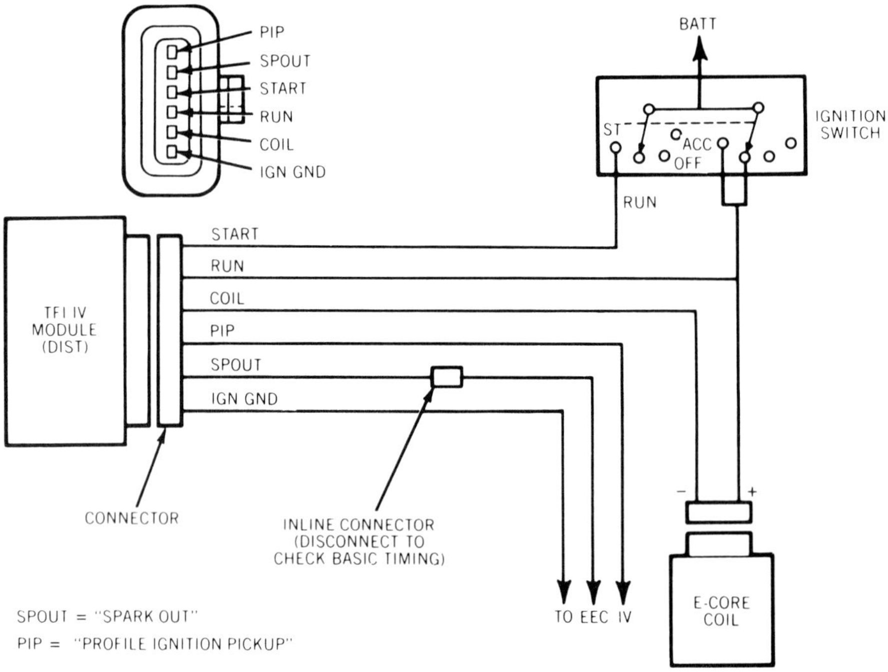


**Figure 13-46.** Typical Ford TFI ignition wiring diagram. (Ford)

## General Motors

The Delco-Remy high-energy ignition (HEI) system was introduced on some GM V-8 engines in 1974. It became standard on all GM engines in 1975. The HEI system was the first major electronic ignition to use a variable-dwell primary circuit with no ballast resistor. In its first decade of use, the HEI system appeared in several versions. But, as with Ford ignitions, the changes were steady improvements and adaptations of the basic HEI design.

**Basic HEI.** The HEI system was the first major system to have all components built into the distributor. A miniature ignition control module is installed in the distributor, and the most common HEI design has the coil mounted in the top of the distributor cap. Figure 13-48 shows a complete HEI distributor. Some systems for in-line 4- and 6-cylinder engines have a coil separate from the distributor for clearance on some engines, figure 13-49. All HEI systems operate alike, however.

The HEI distributor has a magnetic pulse generator, with a trigger wheel called a "timer core." The trigger wheel and the pickup coil pole piece have as many equally spaced teeth as the engine has cylinders, figure 13-50, with one exception. Uneven-firing Buick-built V-6's have three teeth on the trigger wheel and six unevenly spaced teeth on the pole piece. The basic HEI distributor has centrifugal and vacuum advance mechanisms similar to a Delco V-8 breaker-point distributor. HEI distributors contain an RFI suppression capacitor that is not part of the primary circuit. All HEI systems use 8-mm spark plug cables, and many engines use wide-gap spark plugs to take advantage of the system's high-voltage capabilities.

**HEI with C-3.** The basic HEI system has an ignition module with four terminals: two for pickup coil signals, one for battery voltage (B+), and one for the coil ground, figure 13-51. Figure 13-52 is a circuit diagram of the basic HEI system. GM has used six other HEI versions over the years with electronic engine controls and for certain spark timing requirements.

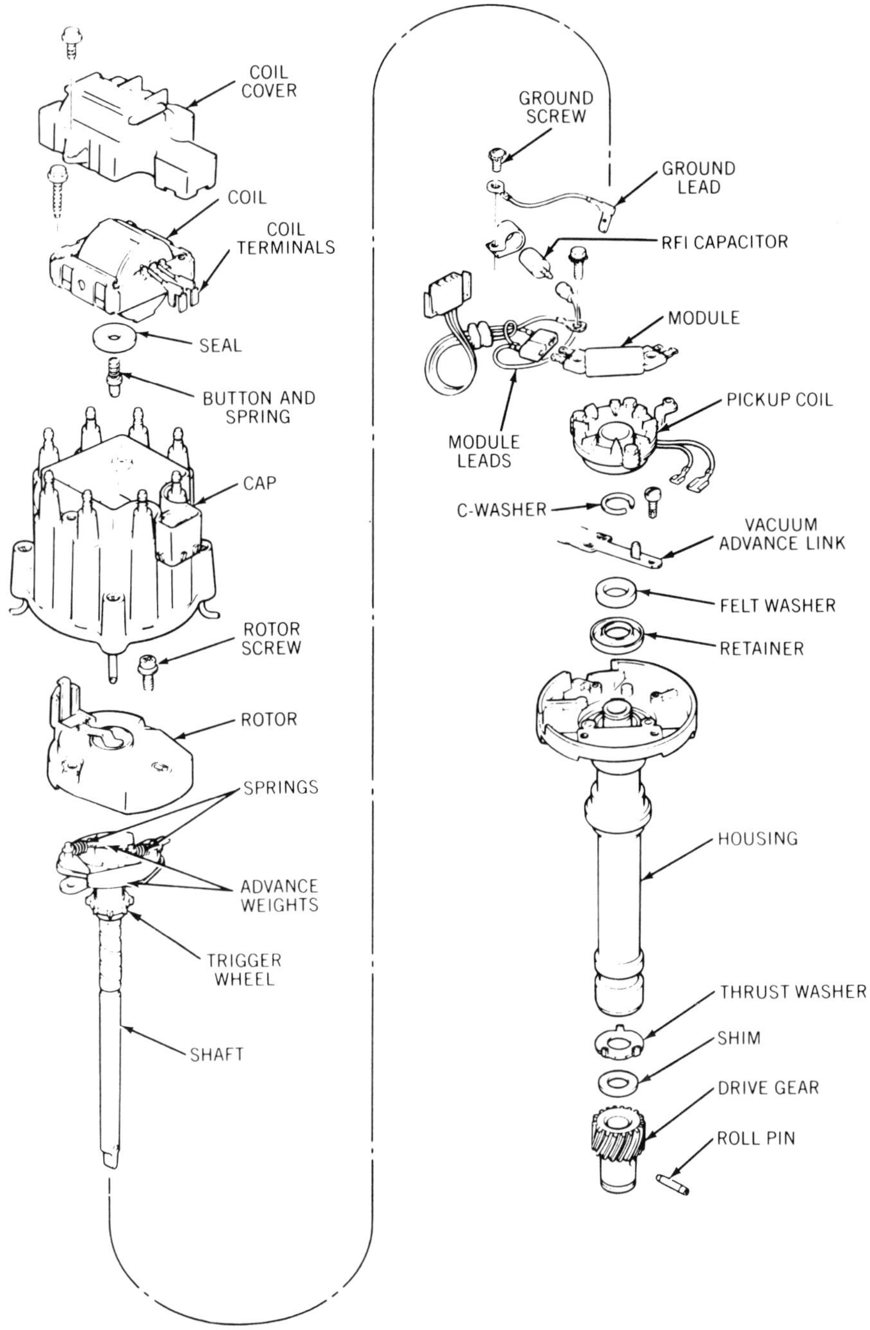


**Figure 13-48.** The complete HEI distributor contains the coil and the ignition module. (Chevrolet)

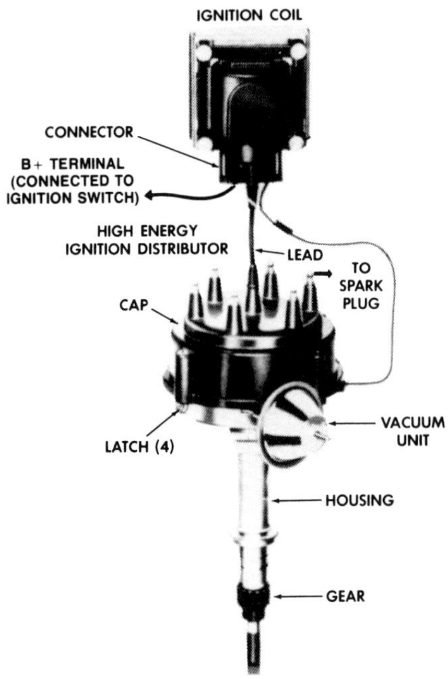


**Figure 13-49.** HEI system with separate coil. (Delco-Remy)

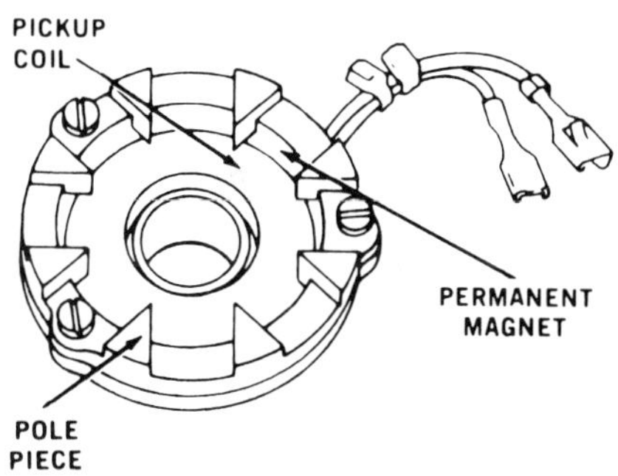


**Figure 13-50.** The HEI pickup coil has multiple pole pieces. (Delco-Remy)

- HEI with electronic spark control (ESC)—This system uses a 5-terminal module and a detonation sensor in the manifold or block. When the sensor detects detonation, it signals the module to retard timing until the detonation stops.
- HEI with electronic spark selection (ESS)—This system uses a 5-terminal module, along with a signal decoder. The decoder monitors signals from the distributor pickup coil, a manifold vacuum switch, and a coolant temperature switch. The decoder signals the ignition module to retard timing for low-temperature starting and to prevent too much advance at cruising speeds.
- HEI with electronic module retard (EMR)—This system uses a 5-terminal ignition module and a manifold vacuum switch or another electronic

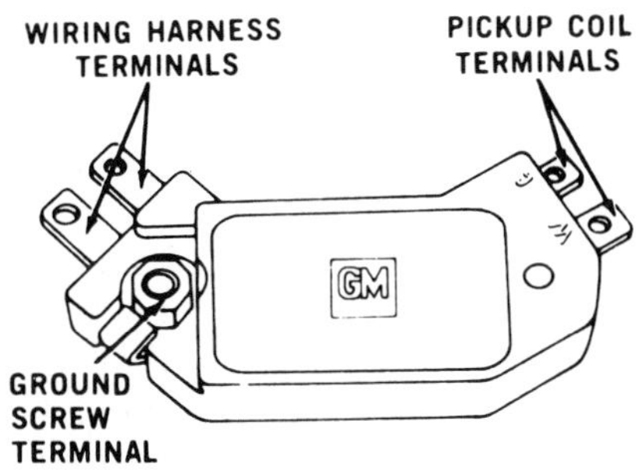


**Figure 13-51.** Basic Delco-Remy HEI control module. (Delco-Remy)

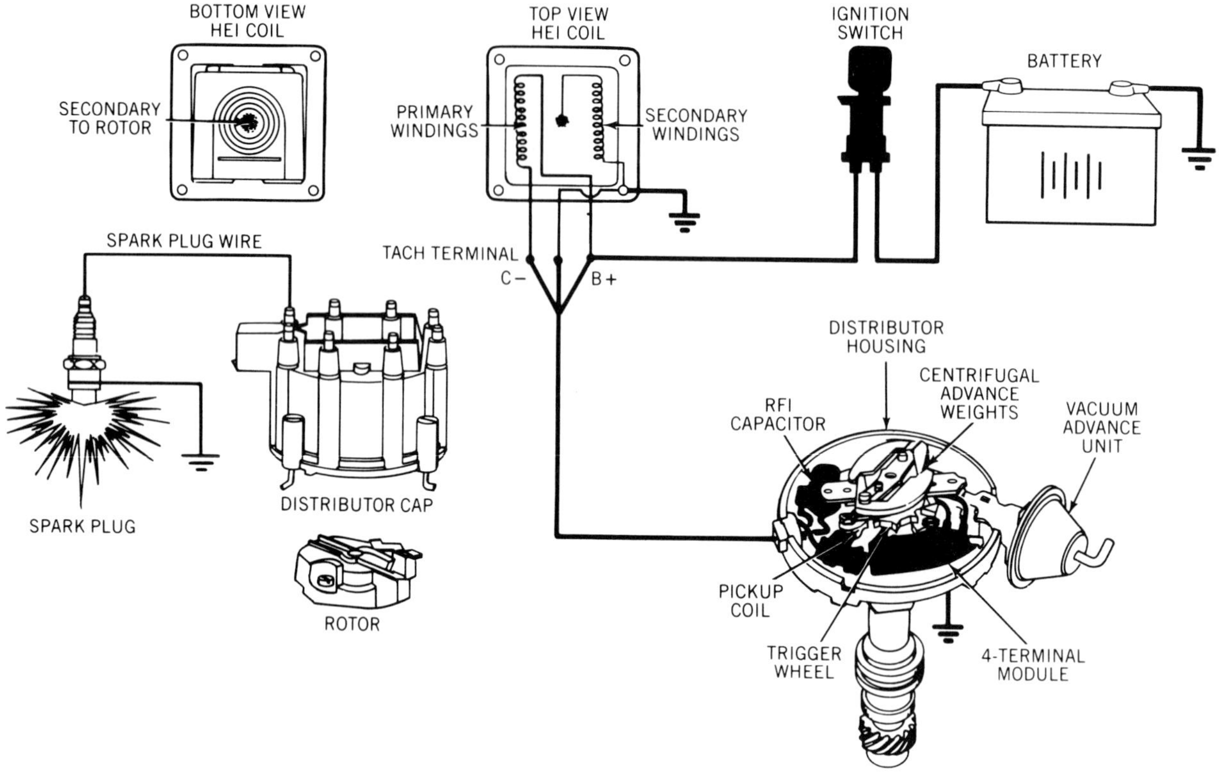


**Figure 13-52.** This is the basic circuit for the Delco-Remy HEI system. (Sun Electric)

module. Under certain conditions of vacuum, temperature, and engine speed, the vacuum switch or EMR module grounds the EMR lead to the HEI module to retard timing about 10 degrees.

- HEI with electronic spark timing (EST)—This system uses a 7-terminal ignition module. EST is used with electronic engine controls, such as computer-controlled catalytic converter (C-4) and computer command control (C-3) systems. The engine control computer controls all spark timing.
- HEI with EST and ESC—This system combines the fully electronic spark control of EST with the detonation sensor of ESC. It is used with some C-3 engine control systems.
- HEI with EST and a Hall-effect switch—This system also is used with C-3 engine controls. It has a Hall-effect switch added to the HEI distributor, figure 13-53, but the distributor also has the magnetic-pulse generator. During cranking, the pickup coil sends timing signals to the HEI module. When the engine starts, the Hall-effect switch takes over and sends crankshaft position signals to the C-3 computer, which then controls all timing electronically.

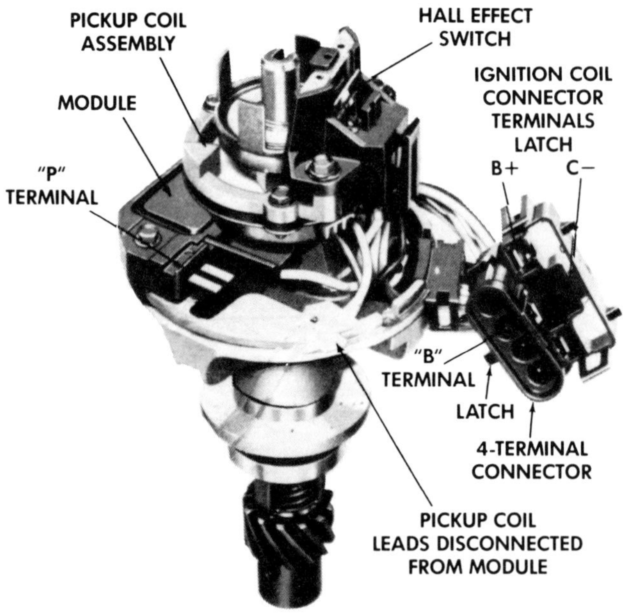


**Figure 13-53.** The Hall-effect switch is a crankshaft position sensor in HEI-EST distributors. (Delco-Remy)

HEI distributors for ESC, ESS, and EMR have centrifugal and vacuum advance. All EST distributors do not.

## DIRECT (DISTRIBUTORLESS) IGNITION SYSTEMS (DIS)

Since the early 1980's, both domestic and foreign carmakers have been developing ignition systems that have no distributors. The first of many direct ignitions appeared in mass production on some fuel-injected, turbocharged, 3.8-liter Buick V-6 engines in 1984. Ford introduced its version on the 1989 Thunderbird supercharged V-6, Taurus V-6, and 2.3-liter ohc engine in the Ranger pickup truck. Chrysler's first DIS installation was in 1990 on the 3.3-liter V-6 engine in minivans and full-size cars. Carmakers have steadily increased DIS use so that by the late-1990's, DIS will be the "conventional" ignition on most engines.

### DIS Types

Direct, or distributorless, ignition systems use multiple coils connected directly to one or two spark plugs. The coil primary circuits are controlled by switching transistors in an ignition module, much like the modules used with electronic ignition distributors. The DIS module still needs a timing signal; but with the distributor gone, the signal must come from another source. Basically, the pickup coils and Hall-effect switches have been moved to the engine block or timing cover and operate with a trigger wheel or ring mounted on the crankshaft.

Direct ignition systems can be classified in two basic categories: individual-coil systems, and 2-cylinder waste-spark systems.

**Individual-Coil DIS.** Several foreign carmakers use individual coils for each spark plug. The coil secondary terminal is attached directly to the plug terminal. Each coil has a wiring harness connector attached to the two primary circuit terminals. The ignition module or engine computer switches each primary circuit separately to charge and discharge the coil. All plugs fire normally, from the center electrode to the ground electrode. Nissan and Saab are two carmakers who have used individual-coil DIS installations.

**Waste-Spark DIS.** Up to the early 1990's, domestic carmakers used the **waste-spark** type of DIS. We will concentrate on this kind of system as we outline the direct ignition systems currently used by GM, Ford, and Chrysler.

A waste-spark direct ignition system is similar to the ignitions used for years in many motorcycle and outboard marine engines. Each end of a coil secondary winding is connected to a spark plug. These two spark plugs are in cylinders that are both at top dead center at the same time: one at the top of the compression stroke, the other at the top of the exhaust stroke. When the coil fires, the spark going to the cylinder on compression ignites the mixture. The spark to the other cylinder is "wasted" at the top of the exhaust stroke.

With distributor-type ignition systems, all spark plugs are connected to the coil and fired in a forward direction; that is, from the center to the side electrode, figure 13-54. In a DIS ignition, one spark plug connected to a coil is fired forward; the other is fired backwards, or from the side to the center electrode, figure 13-55. In other words, one plug in each pair always fires with reverse polarity.

Since it requires about 30 percent more energy to fire a spark plug backwards, the coils used in a DIS system have a different saturation time and primary current flow than a conventional coil. This results in more than 40 kilovolts of available energy, or between 10 and 20 percent more energy than a conventional coil produces.

The coil firing order is determined

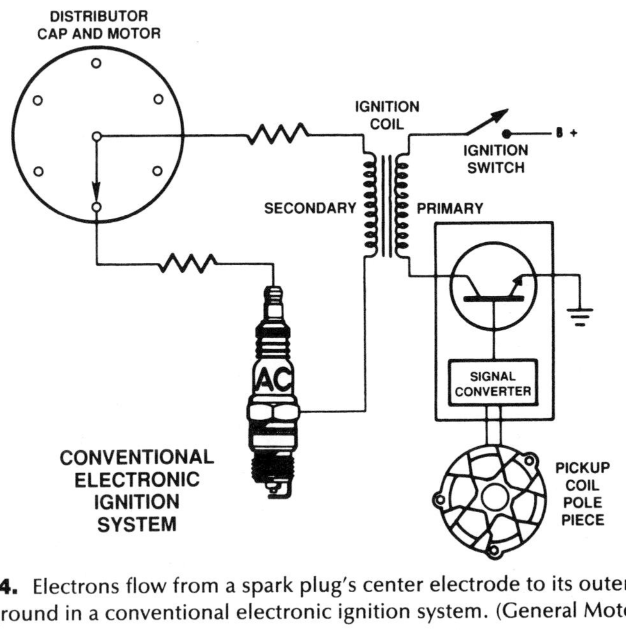


**Figure 13-54.** Electrons flow from a spark plug's center electrode to its outer electrode and then to ground in a conventional electronic ignition system. (General Motors)

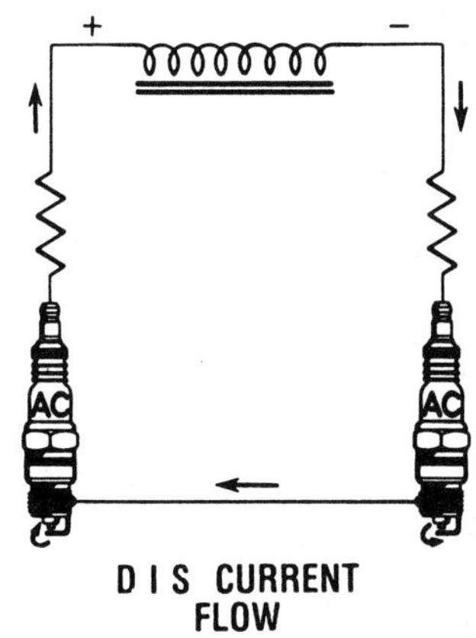


**Figure 13-55.** In a distributorless ignition system, the coil is connected to two spark plugs. It fires one plug forward and one plug backwards. The forward-firing plug ignites the mixture, while the spark of the backward-firing plug is "wasted." (General Motors)

and maintained by the ignition module. When a coil fires, one spark plug fires forward and the other fires backwards at the same time. Polarity and cylinder pressure determines the voltage dropped across each plug. Obviously, the plug in the cylinder on compression will need more voltage to create an arc between the electrodes than will the cylinder on exhaust.

A control circuit inside the module handles primary current flow and limits dwell time. The primary coil winding has very low resistance (under 1 ohm). When combined with an applied voltage of 14 volts, this low resistance produces a theoretical current flow in excess of 14 amperes, which helps to decrease saturation time. However, to prevent damage to system components, the maximum current flow must be kept between 8.5 and 10 amperes. To achieve this, the circuit current is monitored by the module, which modifies the control transistor's base current to limit the collector-emitter current within a range of 8.5 to 10 amperes, figure 13-56.

A form of **closed loop dwell control**, figure 13-57, is used by the module. The previous coil buildup is monitored for maximum current. If maximum current was not achieved, the module increases dwell time to permit full saturation of the coil. If maximum current was achieved, the module decreases dwell time to reduce the wattage consumed by the system. If current limiting takes place before discharge, the module increases dwell for the next firing sequence.

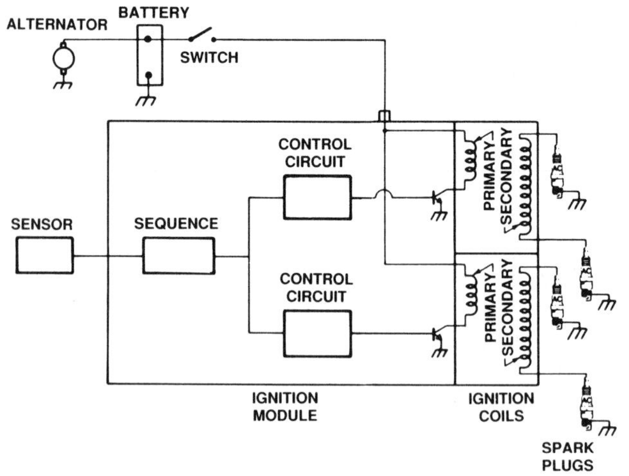


**Figure 13-56.** The ignition module in a direct ignition system contains circuits to control primary current and dwell time limiting. (General Motors)

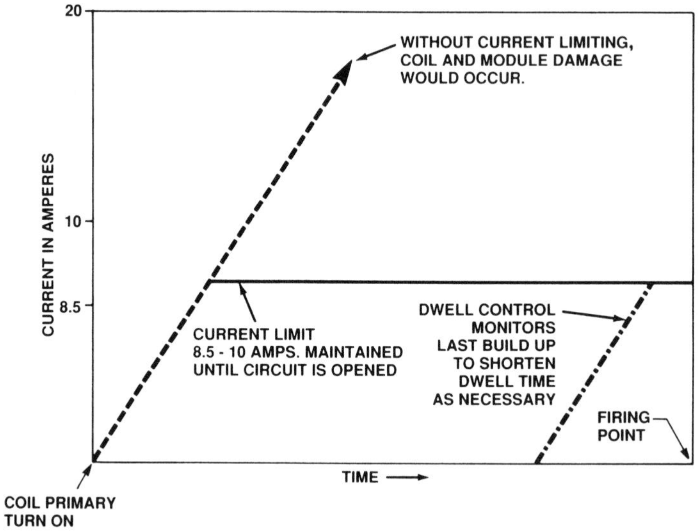


**Figure 13-57.** The closed-loop dwell control feature of a DIS ignition module increases or decreases dwell time to permit full saturation of the ignition coil. (General Motors)

## General Motors DIS Types

General Motors uses the acronym DIS to identify the distributorless systems on Chevrolet- and Pontiac-built 4-cylinder, V-6, and V-8 engines. The systems on Buick-built V-6 engines are called computer-controlled coil (C31) systems. The system on the Oldsmobile-built Quad 4 is called integrated direct ignition (IDI).

Regardless of the marketing names, all these GM systems are waste-spark DIS installations. The major difference between the Buick systems on one hand and the Chevrolet, Pontiac, and Oldsmobile systems on the other is the kind of sensor used for tdc signals to the ignition module. Buick C3I systems use Hall-effect switches. The other GM systems use pickup coil reluctance sensors.

**GM Computer-controlled Coil Ignition (C3I).** The C3I system is used on V-6 engines built by Buick. The basic system contains a coil pack, an ignition module, and Hall-effect sensors to determine crankshaft and camshaft position, figure 13-58. There are three different versions of C3I, each of which can be identified visually.

1. *Type 1*—This version has all three coils molded into a single pack, figure 13-59. If one coil malfunctions, the entire coil pack must be replaced. This system is used with all 1984 and later turbocharged 3.8-liter V-6 engines, and 1985 and later 3.0-liter V-6 engines. Type 1 coil modules are built by Magnavox.

2. *Type 2*—This is the same as type 1, except that the coils can be replaced individually, figure 13-59. The system

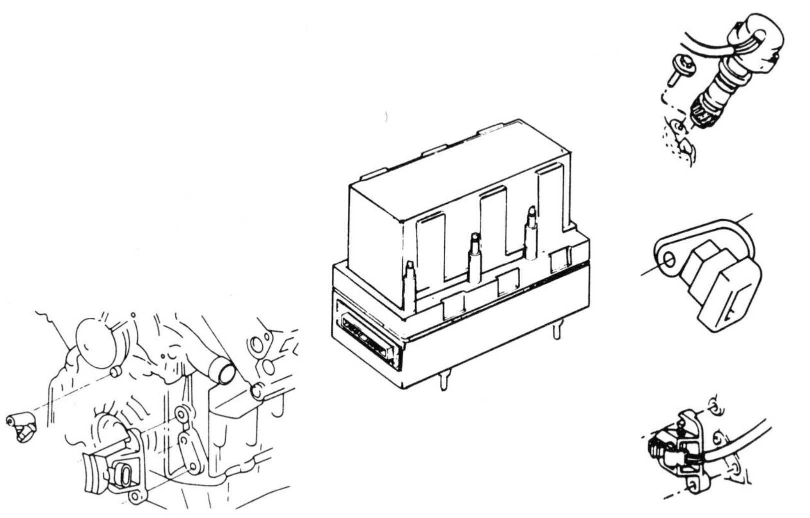

**Figure 13-58.** A computer-controlled coil ignition system uses a module, coil pack, and crankshaft and camshaft sensors. (General Motors)

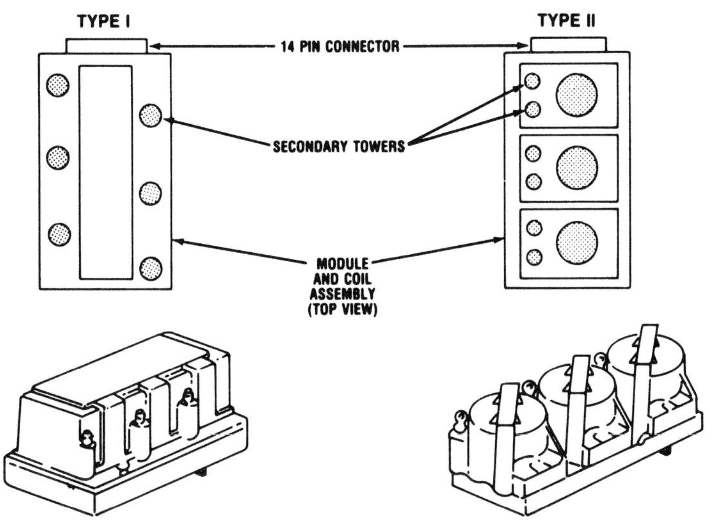


**Figure 13-59.** Type I and type II C3I systems can be identified by the type of coil pack used. (General Motors)

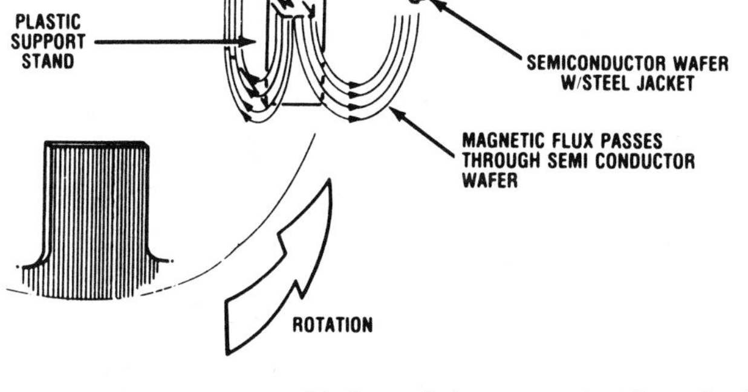


**Figure 13-60.** Alternating interrupter blades and air gaps passing through a Hall-effect switch create a voltage signal to indicate crankshaft position. (General Motors)

is used with 1985–87 3.8-liter V-6 engines with sequential fuel injection (SFI).

**3.** *Type 3*—The coil pack is similar in appearance to, and can be interchanged with, a type 1 pack. The module, however, differs electronically and the connector plugs are not compatible with components from other systems. The type 3 system is used on 1988 and later Buick 3800 V-6 engines.

Hall-effect switch sensors are used to synchronize and fire the coils at the correct time. The ignition module supplies a reference voltage, which is pulled down or dropped by the Hall-effect switch whenever a ferrous metal comes between the semiconductor material and permanent magnet in the switch, figure 13-60. The change in reference voltage tells the module the exact position of the device being monitored. A variation of this is used with the 3.8-liter V-6 camshaft sensor, wherein a moving magnet is rotated past the Hall-effect switch to turn it on and pull down the voltage, figure 13-61.

Three equally positioned metal interrupters pass between the permanent magnet and Hall-effect switch as the crankshaft rotates, figure 13-62. The on-off action of the transistor switches the module reference voltage from high (6 to 8 volts) to low (0 to 0.5 volt). Because each signal is identical in time and amplitude, the module cannot determine which signal it should send to which coil.

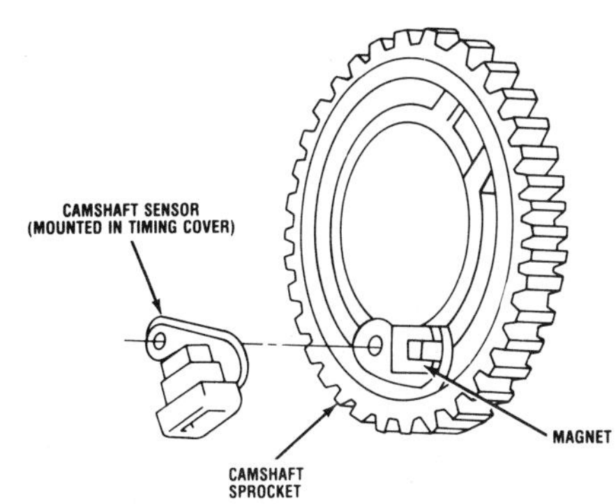


**Figure 13-61.** The camshaft sensor on Buick's 3.8-liter SFI V-6 engine rotates a magnet past the Hall-effect switch to turn it on and pull down the reference voltage. (General Motors)

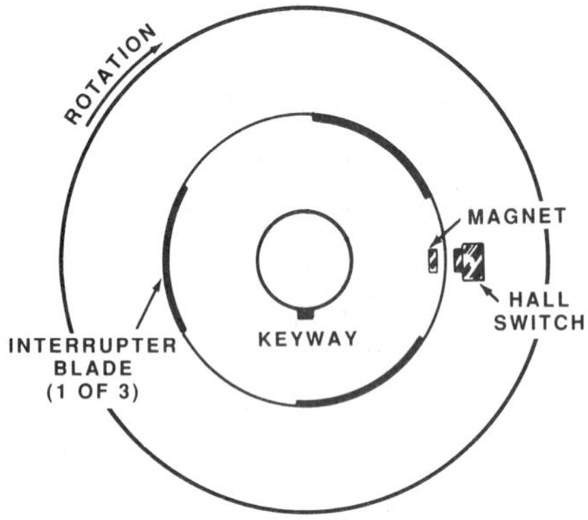


**Figure 13-62.** The Hall-effect transistor switches the ignition module's reference voltage from high to low as the interrupter blades and air gaps pass through the switch. (General Motors)

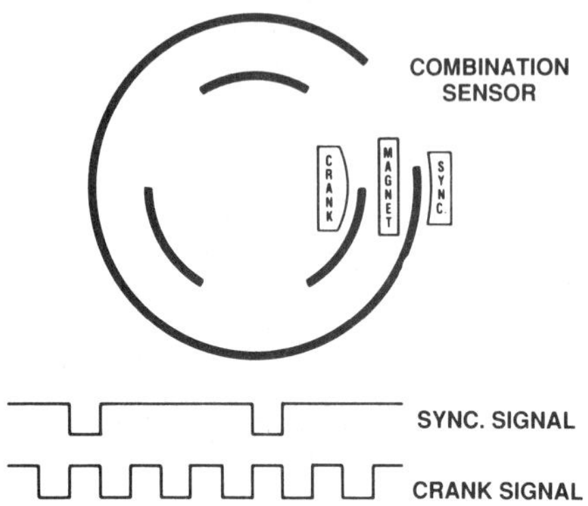


**Figure 13-63.** Dual Hall-effect switches using one magnet provide two sets of signals in the Buick 3.0-liter V-6 design. (General Motors)

The 3.8-liter V-6 camshaft sensor signal is used by the module to identify the correct crankshaft signal and its corresponding coil. Like the crank sensor, the cam sensor also switches the reference voltage from the module. The initial cam sensor signal is synchronized with one of the crankshaft sensor signals during cranking. This synchronization is maintained by the module, which remembers the sequence of the crankshaft sensor until the ignition is turned off. The cam sensor serves no other purpose once the initial ignition setup routine is established.

The 3.0-liter V-6 differs from the 3.8-liter somewhat in signal generation and processing. The combination crankshaft sensor contains a separate Hall-effect switch on each side of the permanent magnet. Two concentric rings on the rear of the harmonic balancer pass between the magnet and the Hall-effect switches. The three inner interrupter rings provide three sensor switches to indicate crankshaft position. The outer ring with a single open area provides one signal per crankshaft revolution, which acts as the synchronization signal to the module, figure 13-63. As with the 3.8-liter V-6, the synchronization signal is required only during engine cranking and coil sequence setup.

Both engines will run without a camshaft or synchronization signal if the engine is running before the sensor failure. However, neither engine will start without the signal. Some 3.0-liter engines may start if the synchronization signal is pulled low during cranking; but if the circuit is pulled low with the engine running, the module resynchronizes the crankshaft signal, often resulting in a stall.

The type 3 system used with the 1988 and later 3800 V-6 engine is slightly different and more sophisticated electronically. It is sometimes referred to as "fast start 1" because it provides a faster startup. The system combines a dual crank sensor (much like that on the 3.0-liter V-6) with a separate camshaft sensor in the front cover (like the 3.8-liter SFI V-6). The crank 1 signal is provided by 18 interrupter rings evenly spaced around the outside of the harmonic balancer, figure 13-64. The signal is called the "18× crank signal" because it delivers 18 pulses per crankshaft revolution. A crank 2 signal comes from the inside ring, which contains three rings with air gaps or windows of 10 degrees, 20 degrees, and 30 degrees

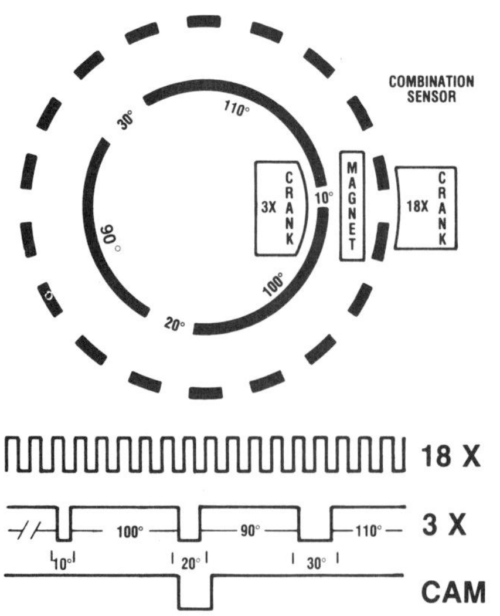


**Figure 13-64.** The type 3 or "fast start 1" C3I system uses dual Hall-effect switches and 18 interrupter blades to provide a more accurate measurement of the crankshaft sensor signals. (General Motors)

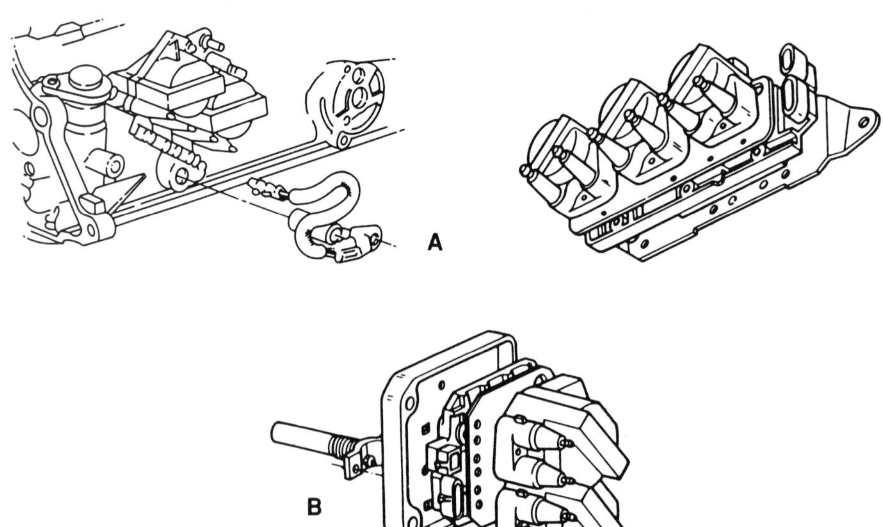


**Figure 13-65.** The direct ignition system (DIS) used with 1987 and later GM 4-cylinder and V-6 engines uses a block-mounted crankshaft sensor instead of Hall-effect switches to provide a crankshaft position signal. (General Motors)

spaced 110 degrees, 100 degrees, and 90 degrees apart, figure 13-64. This is called the "3× crank signal."

The module can energize the proper coil without waiting for the cam signal or the synchronization signal because of the variations in the 3× signal. It can identify the correct coil within 120 degrees of crank rotation and will start firing on whatever coil it identifies first. The 18× pulse is used by the module as a "clock" pulse to determine the length of each 3× pulse. The 18× signal changes once during the 3× 10-degree air gap, twice during the 20-degree gap, and three times during the 3× 30-degree window. The module uses this information to identify which 3× pulse it is reading and then energizes the proper coil.

The module equalizes and buffers the 3× crank signal, sending it to the electronic control module (ECM) on one circuit. This is followed by a buffering of the 18× signal, which is then sent to the ECM on a separate circuit as a more accurate engine speed signal. The camshaft signal is buffered and sent to the ECM on a third circuit to control fuel injection. Since the module requires no camshaft signal to set up or synchronize the coil firing, the engine will start and run even with a malfunctioning cam sensor.

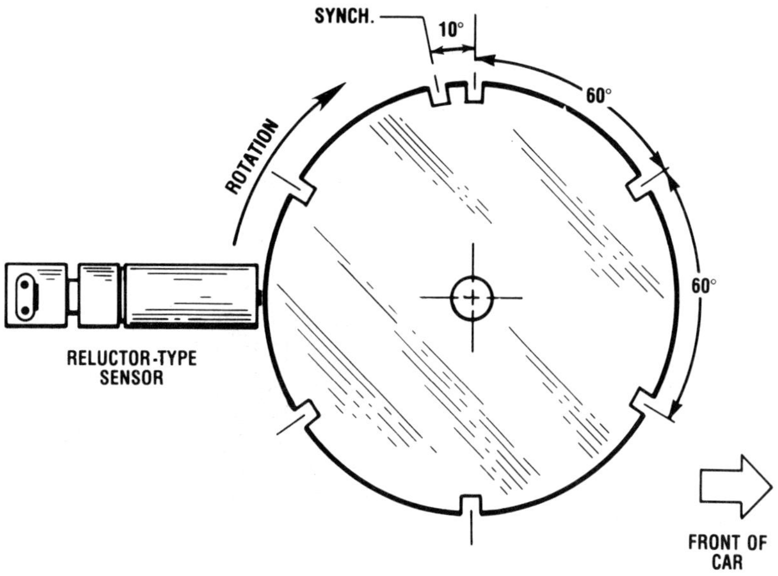


**Figure 13-66.** The crankshaft sensor in a DIS is mounted 0.050 ± 0.020 inch from a rotating reluctor machined in the crankshaft. Notches in the reluctor modulate the sensor signal, with one notch providing the signal used to synchronize the coil firing sequence to crankshaft position. (General Motors)

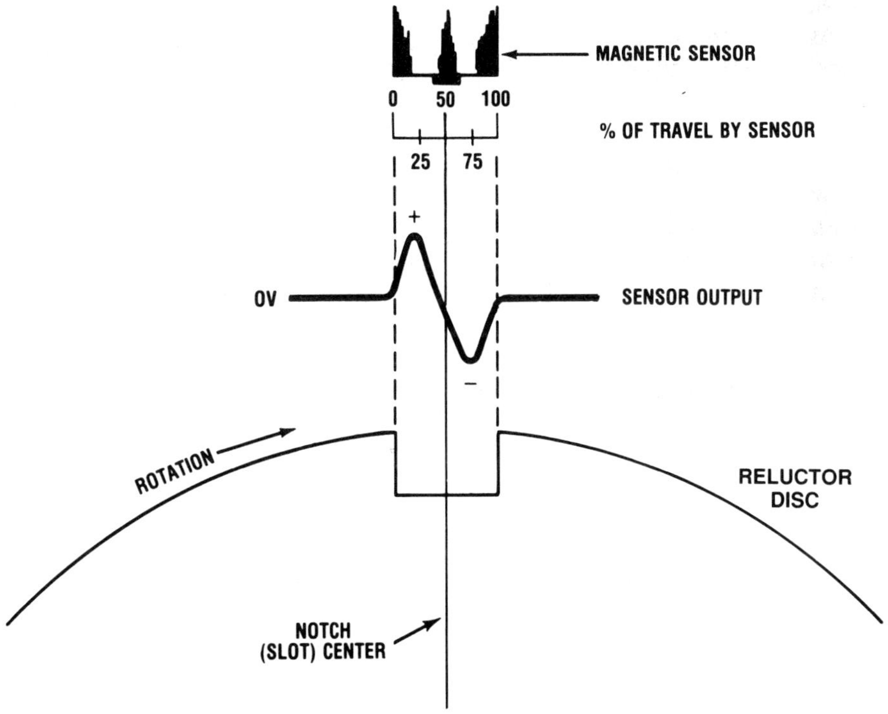


**Figure 13-67.** Interruption of the sensor's magnetic field induces a small ac voltage in the wire wound around the permanent magnet sensor. This voltage change tells the ignition module when to fire each coil. (General Motors)

**GM Direct Ignition System.** The direct ignition system (DIS) is used on Chevrolet 2.0-liter 4-cylinder and 60-degree V-6 engines, as well as the Pontiac-built 2.5-liter 4-cylinder engine. System operation is similar to that of the C3I system, with the major difference in the method of sensing the position of the crankshaft. A magnetic sensor installed in the side of the engine block performs the function of the Hall-effect switches. In the 2.0-liter and V-6 applications, the magnetic sensor is mounted below the module and connects to it externally, figure 13-65A. In the 2.5-liter application, the sensor is mounted on the rear of the module, figure 13-65B.

A reluctor or special wheel cast into the crankshaft has seven machined slots or notches and performs the interrupter function. The sensor contains a permanent magnet with a wire winding and is installed 0.050 ± 0.020 inch from the crankshaft reluctor, figure 13-66. When the crankshaft rotates, the reluctor notches modulate the intensity of the sensor's magnetic field, inducing a small ac, or push-pull, voltage in the wire winding around the sensor, figure 13-67. Since the reluctor is a machined part of the crankshaft and the sensor is installed in a fixed position on the engine block, ignition timing adjustment is not possible, nor required, for DIS.

Six of the seven notches on the crankshaft are spaced at 60-degree intervals around the reluctor surface. The seventh notch is 10 degrees away from one of the evenly spaced notches and provides the signal used by the module to synchronize the coil sequence to the crankshaft position, figure 13-68.

Since one coil fires two spark plugs at a time, the DIS is mainly concerned with coil order and the position of crankshaft rotation in which to fire the coils. The 4-cylinder engines make this determination in a slightly different way from the V-6 engine.

On a 4-cylinder engine, the ignition module monitors the seventh notch to start the firing sequence. Below a specified engine speed, the module controls ignition timing by firing each coil at a preset interval based only on engine speed. This is called **bypass** or **module timing**. The synchronization notch notifies the module to omit the first notch and use the second notch to establish 10° btdc for cylinders 2 and 3. The module then omits the third and fourth notches, using the fifth notch to establish the same timing setting for cylinders 1 and 4. After the sixth and seventh notches pass, the sequence begins again, figure 13-69. This means that the 2/3 coil fires first on initial setup. (Except for the type 3 "fast start 1" system, all DIS and C3I systems start up the second cylinder in the firing order first.) Once engine speed exceeds the specified rpm, the ECM takes over ignition timing.

The V-6 engine system functions much the same, except that the synchronization notch notifies the module to omit the first notch and use the second notch to establish base timing for cylinders 2 and 5. The module then omits the third notch, using the fourth notch to establish the same timing setting for cylinders 3 and 6. Finally, the module omits the fifth notch and uses the sixth notch for cylinders 1 and 4. After the seventh notch passes, the sequence begins again. The firing order in the first crankshaft revolution is 1-2-3 and 4-5-6 in the second crank revolution, figure 13-70.

In both the 4- and 6-cylinder DIS applications, the reference pulse is pulled low by the notch that is 60 degrees ahead of the notch that will fire the cylinder, and then returns to its high state when the cylinder firing notch passes. This change in the reference voltage provides the signal for electronic spark timing (EST) and fuel injection.

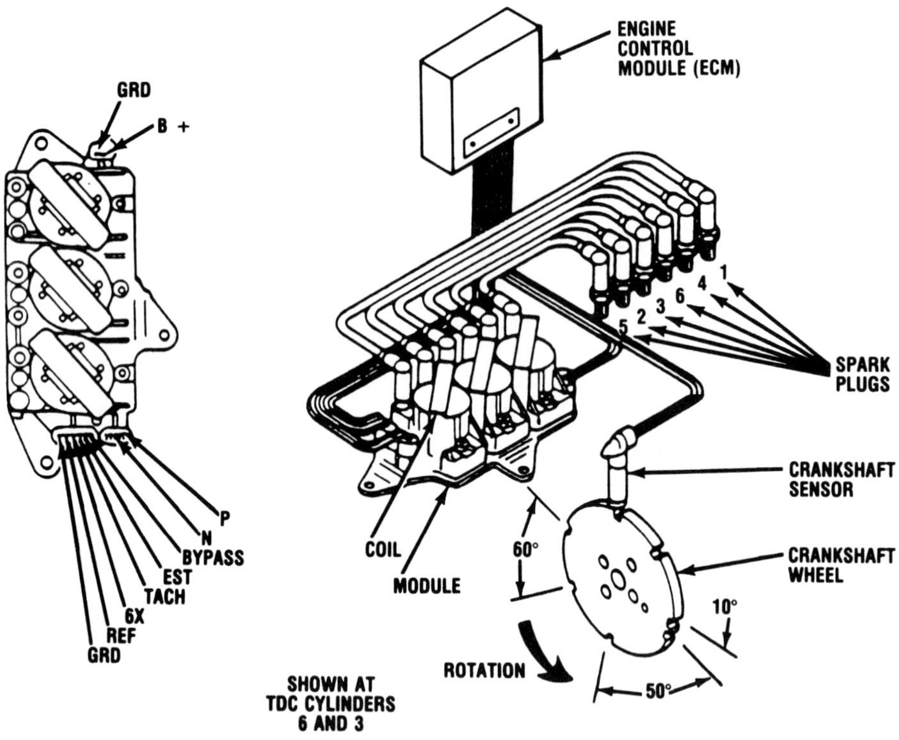


**Figure 13-68.** Operation of the V-6 DIS. (General Motors)

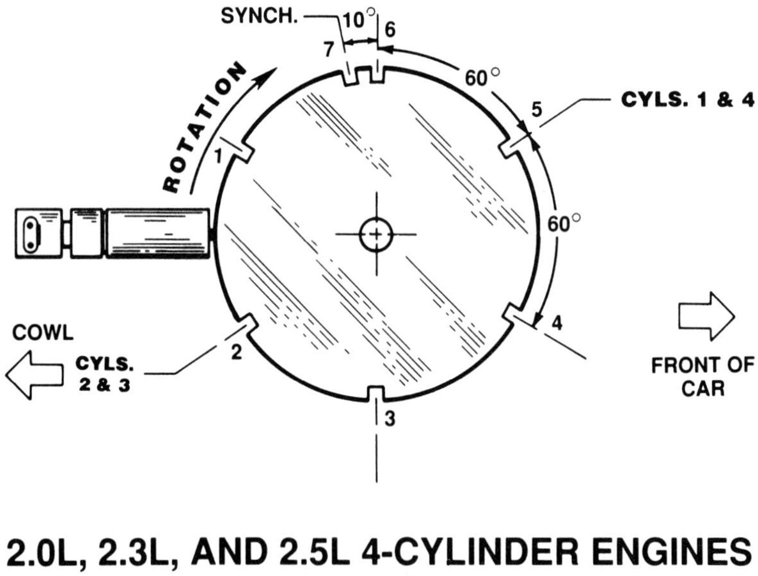


**Figure 13-69.** In the 4-cylinder DIS, reluctor notch 2 fires the ignition coil for cylinders 2 and 3. Reluctor notch 5 fires the cylinder 1 and 4 coil. (General Motors)

**GM Integrated Direct Ignition (IDI).** Used on the Oldsmobile-built Quad 4 engine, figure 13-71, this system functions essentially the same as the 4-cylinder DIS system. However, the coils are housed in an assembly that connects directly to the spark plugs, figure 13-72, eliminating spark plug cables. The 1-piece conductors permit precise routing control, reduce secondary capacitance, and eliminate terminal corrosion.

## Ford DIS Types

Ford's first direct ignition systems were used on some 1989 4-cylinder and V-6 engines. Their use has increased to the point that many current 4-, 6-, and 8-cylinder Ford engines have DIS installations.

The 1989 2.3-liter 4-cylinder engine in Ranger trucks and the 3.0-liter SHO and 3.8-liter supercharged V-6

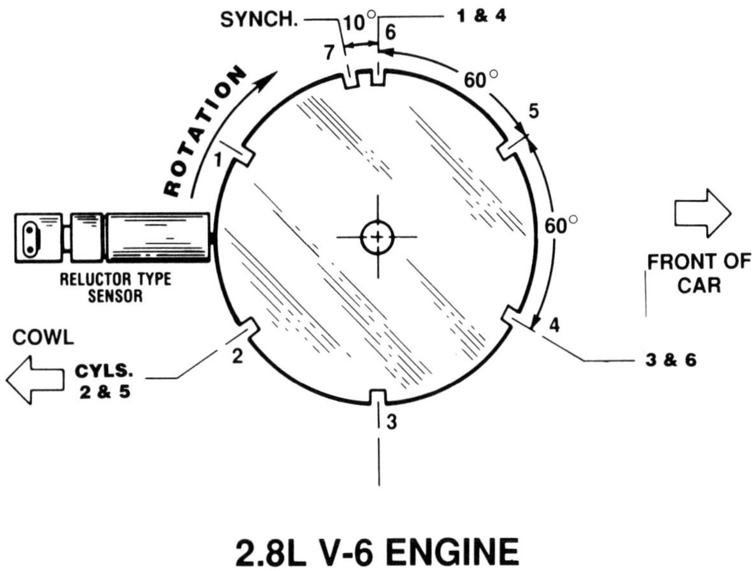


**Figure 13-70.** The V-6 DIS uses reluctor notches 2, 4, and 6 to fire cylinders 2 and 5, 3 and 6, and 1 and 4 respectively. (General Motors)

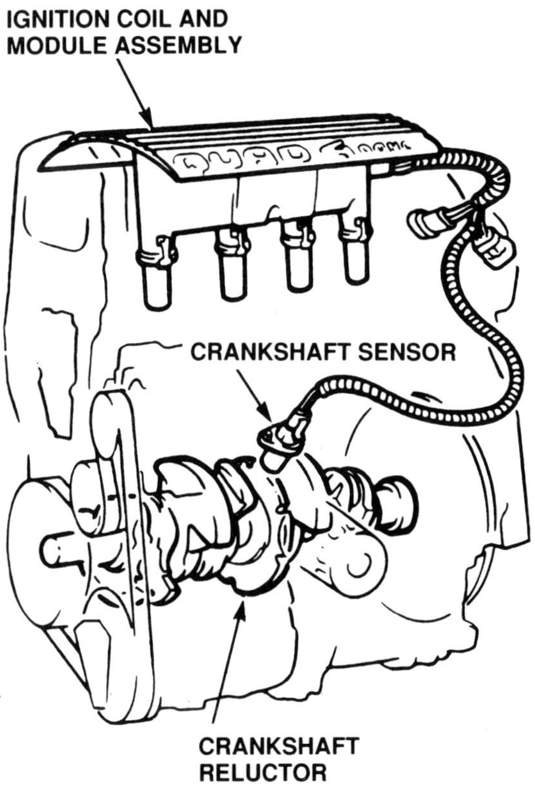


**Figure 13-71.** The integrated direct ignition (IDI) system uses an integral coil and module assembly that connects directly to the spark plugs without secondary wiring. (Oldsmobile)

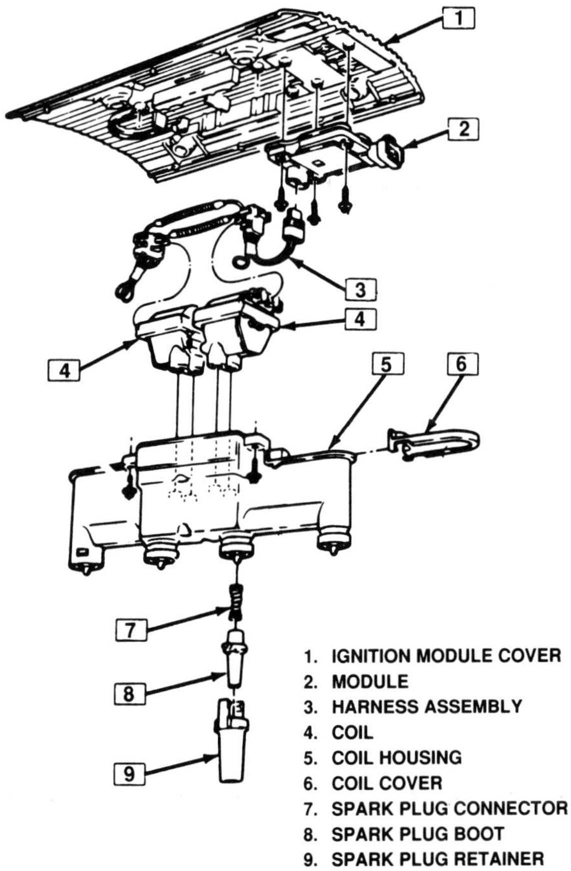


**Figure 13-72.** An exploded view showing placement of the components in the IDI system. (Oldsmobile)

engines used Ford's first-generation DIS systems. These systems use Hall-effect sensors for timing (crankshaft) and cylinder identification (camshaft) signals to the ignition module. The 2.3-liter engine features a unique dual spark plug installation. This system was added to the 4-cylinder engines in Mustangs in 1991 and continued basically unchanged from the 1989 Ranger system.

Ford's second-generation electronic distributorless ignition system (EDIS) replaced the earlier system on 1990 and later V-6 engines and 4-cylinder Escort engines. In 1991 EDIS was added to the 4.6-liter V-8 for LTD's, Grand Marquis models, and Lincoln Town Cars.

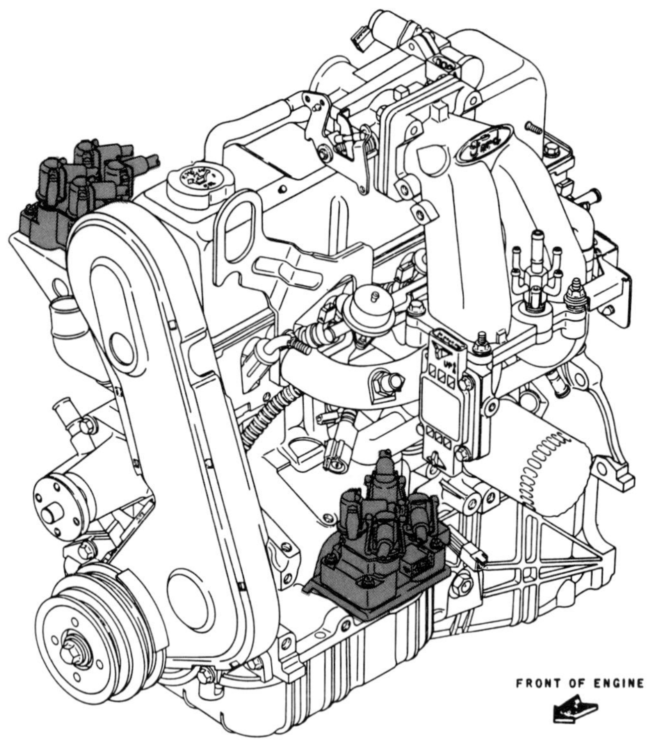


**Figure 13-73.** The Ford 4-2 DIS uses one coil pack for each set of spark plugs. The Ranger truck engine is shown here. Mustang installations have both coil packs at the top right front of the engine. (Ford)

**Ford 4-2 Distributorless Ignition System (DIS).** This DIS is unique in that the 2.3-liter 4-cylinder engine uses two spark plugs per cylinder, with one installed on each side of the combustion chamber. The engine operates at all times on one set of plugs called the primary system. These are installed on the right side of the engine. The second set of plugs (secondary system) is installed on the left side of the engine and switched on and off by the EEC-IV computer. A dual plug inhibit (DPI) circuit allows the computer to switch the ignition system from single- to dual-plug operation according to engine load and speed requirements. During the cranking mode, only the primary plugs fire. When the engine is running, the EEC-IV computer signals the DIS module to switch to the dual-plug mode. The EEC-IV system also controls spark timing and dwell for the ignition system.

The DIS 4-2 ignition also contains two 4-tower DIS coil packs, figure 13-73, and a crankshaft-mounted dual Hall-effect switch sensor. Two coil packs are required because of the dual set of plugs. One coil pack fires the primary set during normal engine operation; the other coil pack fires the secondary set when directed to do so by the EEC-IV computer and DIS module. Figure 13-74 is a block diagram of the Ford DIS 4-2 ignition.

A bracket-mounted, dual-digital-output Hall-effect switch acts as the crankshaft sensor and responds to two rotating metal shutters, or rotary vane cups, mounted on the crankshaft damper in the same way as other Hall-effect switches you have studied. The dual-digital-sensor operation is similar to that in the GM C3I ignition. The profile ignition pickup (PIP) vane cup and Hall-effect switch provide crankshaft position signals (basic spark timing data) to the EEC-IV system. The cylinder identification (CID) vane cup and Hall-effect switch provide a signal used by the DIS module in determining which coil to fire.

The computer-controlled dwell (CCD) feature uses a spark output, or SPOUT, signal sent from the EEC-IV computer to the DIS module. The leading edge fires the coil, and the trailing edge controls the dwell time. A buffered tach signal from the DIS module to the EEC-IV system provides diagnostic information about the ignition system for self-test purposes.

If the CID sensor or circuit fails, the DIS module will randomly select and fire one of the two coils. The engine still can be started by turning the ignition key off and on until the DIS selects the correct firing sequence.

A failure effects management (FEM) program allows the vehicle to run when an ignition failure prevents a SPOUT signal. The EEC-IV computer opens the SPOUT line, and the DIS module fires the coils directly from the PIP output, resulting in a fixed spark timing of 10° btdc with fixed dwell.

**Ford 1989 V-6 Distributorless Ignition System.** This system operates in the same way as the Ford 4-2 DIS system just discussed but the components differ, figure 13-75.

- Three coils are mounted together to form a 6-tower coil pack with an individual tach wire for each coil.
- The camshaft sensor on the 3.0-liter V-6 is mounted at the end of one camshaft.

- The camshaft sensor on the 3.8-liter supercharged V-6 is mounted in the normal distributor location.

Ford's DIS installations on V-6 engines operated with only one spark plug per cylinder. All Ford distributorless ignition systems work on the same principles as the GM DIS versions. In all systems, each coil fires a pair of spark plugs. One spark ignites the air-fuel mixture at the top of the compression stroke. The other spark of the pair is wasted at the top of the exhaust stroke. The secondary circuit for each coil is a series circuit; one plug of each pair has positive (+) polarity, the other has negative (−) polarity.

**Ford Second-Generation Electronic Distributorless Ignition System (EDIS).** Introduced in 1990, EDIS replaces the Hall-effect crankshaft and camshaft sensors with a pickup coil reluctance sensor for crankshaft position and ignition timing signals, figure 13-76. EDIS still works in conjunction with the EEC-IV engine control assembly, but the EDIS module (which has its own microprocessor) has taken over the spark timing control function.

The variable reluctance (VR) crankshaft sensor generates a signal voltage in relation to the rotation of a 36-tooth trigger wheel on the crankshaft. The trigger wheel has one missing tooth, or irregular gap, figure 13-77. As each tooth passes the sensor, it generates an ac voltage signal. When the area of the missing tooth passes, the sensor does *not* produce a voltage change; and this indicates crankshaft position as follows:

- 4-cylinder engine, 90° btdc
- 6-cylinder engine, 60° btdc
- 8-cylinder engine, 45° btdc

When the EDIS module identifies the missing tooth, it then can calculate crankshaft position within 10 degrees at any point in its rotation.

Because the EEC-IV computer was designed to require what Ford calls a "profile ignition pickup" (PIP) signal for ignition timing and engine speed input, a program in the EDIS microprocessor creates a PIP signal from the high-data-rate VR sensor signal, figure 13-78. The rising edge of the square-wave PIP signal is 10° btdc.

The EEC-IV computer sends a "spark angle word" (SAW) signal back to the EDIS module. This is a short square-wave voltage pulse that tells the EDIS module what spark angle is required. The EDIS module then calculates when the appropriate coil should be fired. In this way, the EDIS module has final control over ignition timing advance or retard.

By using coils with revised windings and faster saturation time, the Ford EDIS can operate with shorter dwell

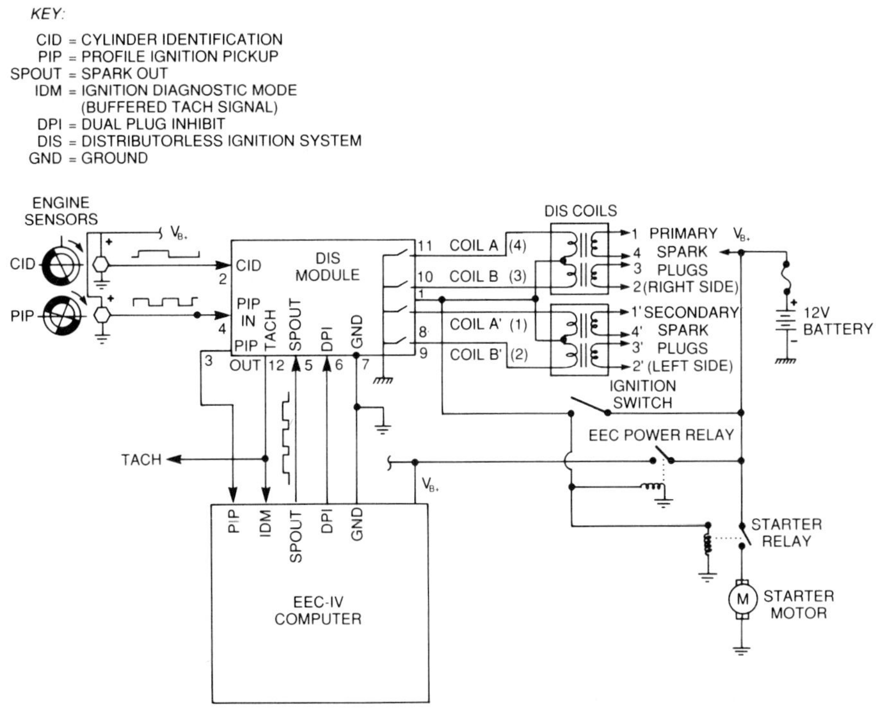


**Figure 13-74.** A block diagram of the Ford 4-2 DIS.

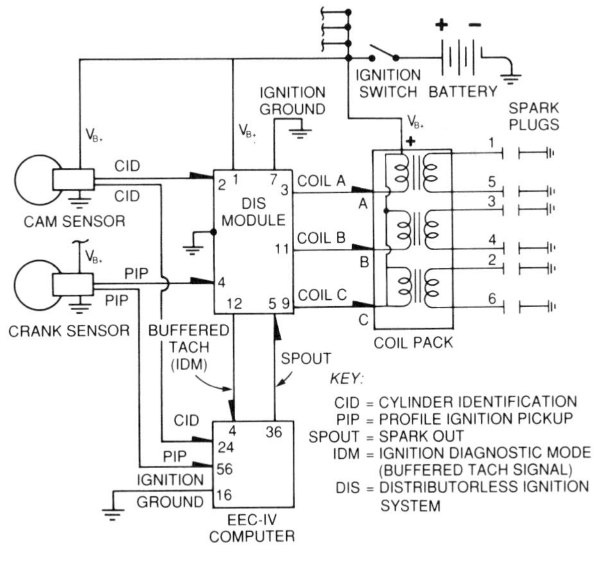


**Figure 13-75.** A block diagram of the Ford V-6 DIS.

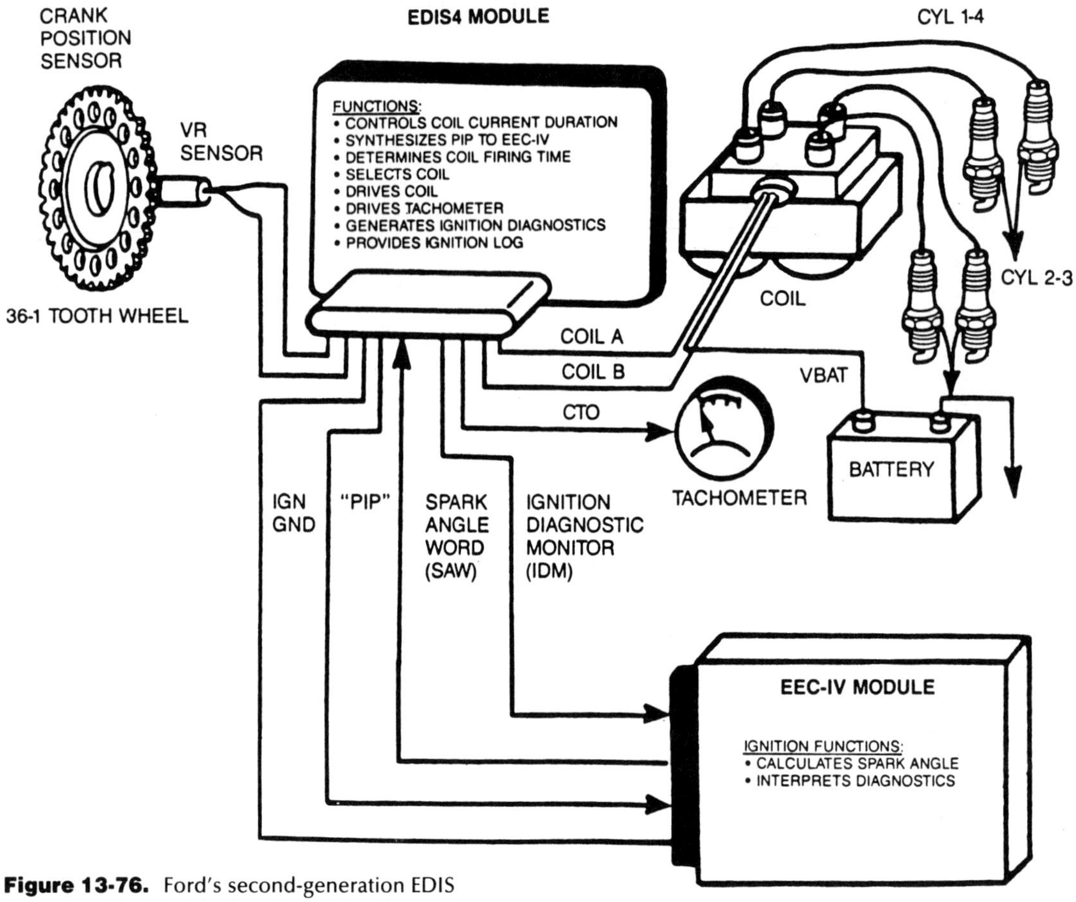


**Figure 13-76.** Ford's second-generation EDIS installation for a 4-cylinder engine. (Ford)

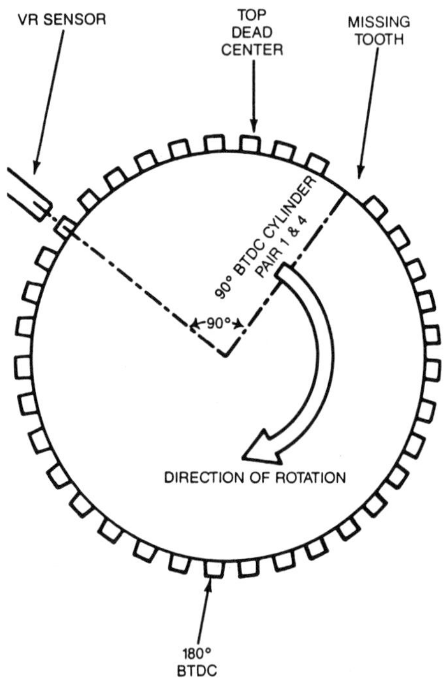


**Figure 13-77.** The variable-reluctance (VR) sensor generates a signal every 10 degrees of crankshaft revolution. The missing tooth is the basis for recognizing cylinder firing order. (Ford)

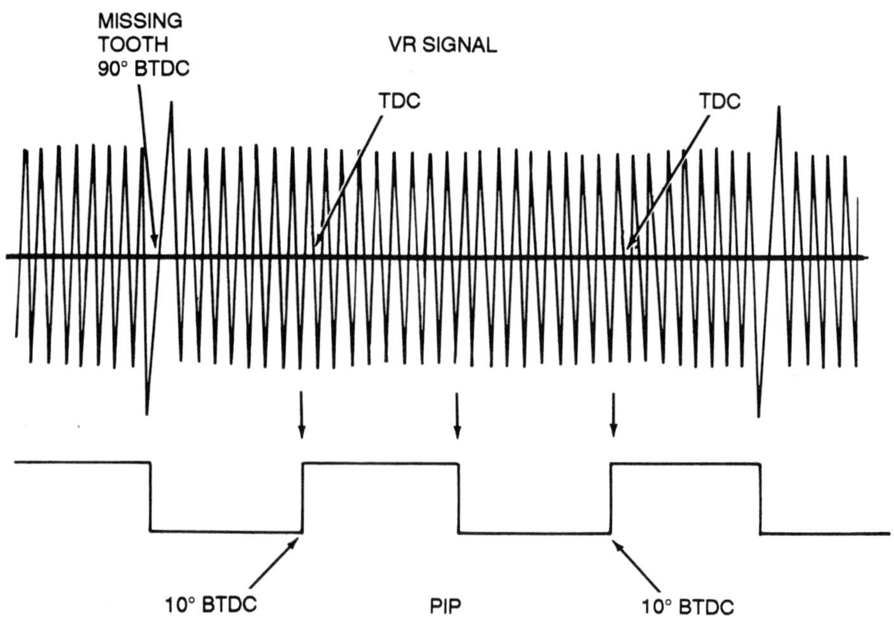


**Figure 13-78.** The EDIS module converts the VR sensor signal to a PIP signal for the EEC-IV computer. (Ford)

time and consequently less heat is developed in the windings. The combination of the high-data-rate crankshaft signal, the SAW signal from the EEC-IV computer, and EDIS timing control makes this short dwell time possible.

The EDIS module also sends an ignition diagnostic monitor (IDM) signal to the EEC-IV computer to indicate the module operating condition. The IDM signal lets the EEC-IV system produce several fault codes for ignition problems.

Figure 13-79 is a diagram of a typical EDIS installation on a 1.9-liter, 4-cylinder engine. Dotted lines indicate a test harness used for troubleshooting. In 1991, Ford added the EDIS to the 4.6-liter, V-8 used in LTD's, Grand Marquis models, and Lincoln Town Cars, figure 13-80.

### Chrysler DIS

Chrysler introduced its first direct ignition system on the 1990 3.3-liter, V-6 used in minivans and full-size cars, figure 13-81. The ignition module functions are integrated into the single-board engine computer (SBEC) for the complete engine control system.

The Chrysler DIS uses input signals from a crankshaft sensor mounted in the transaxle bellhousing and a camshaft sensor mounted in the top of the timing chain cover. The SBEC sends an 8-volt reference voltage to each sensor and receives a return signal voltage that alternates from about 0 to 5 volts.

The crankshaft sensor sends a steady square-wave signal every 20 degrees of crankshaft rotation to the SBEC to indicate engine speed and

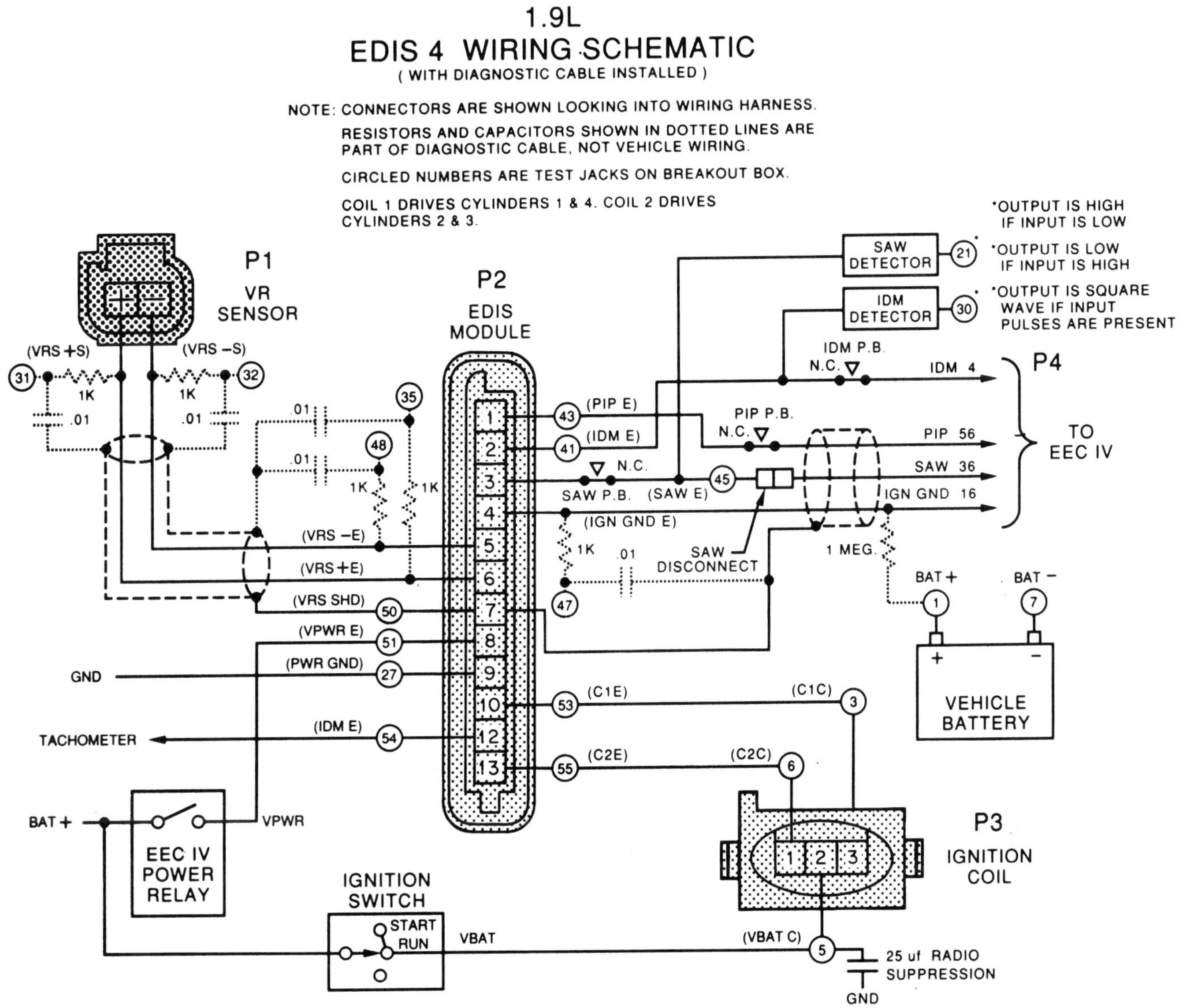


**Figure 13-79.** The wiring diagram for an EDIS installation on a 4-cylinder engine. (Ford)

**Figure 13-80.** Ford adapted EDIS to V-8 engines in 1991. (Ford)

crankshaft position. The SBEC then calculates ignition timing and fuel delivery.

The camshaft sensor operates with a toothed trigger wheel that has five areas with notched slots. Two areas have a single slot; two areas have two slots; and one area has three slots. The trigger wheel also has one long and four short unnotched areas. The arrangement of solid and notched areas produces a signal that lets the SBEC determine piston position and firing order. From this, the SBEC calculates which ignition coil and fuel injectors to fire. The arrangement of crankshaft and camshaft sensor signals produces an ignition and fuel control sequence much like the Buick C3I fast-start system.

The coil pack for Chrysler's V-6 DIS is mounted on the upper right side of the intake manifold. The SBEC controls coil voltage for a waste-spark operation similar to other domestic DIS installations. In 1991, Chrysler adapted this basic system to the 4-cylinder, 24-valve, turbocharged engine in the Dodge Spirit R/T, figure 13-82.

## Imported-Car Electronic Ignitions

Imported-car manufacturers began to use electronic ignitions about the same time as domestic carmakers, and there is nothing "foreign" about their systems. All major imported-car ignitions are inductive-discharge systems. Most have magnetic-pulse generators in the distributor to send timing signals to the ignition module. Bosch, however, makes a Hall-effect distributor used on many Volkswagen engines. Most Japanese ignitions are variable-dwell systems with no ballast resistor and full battery voltage applied to the system at all times.

## SUMMARY

Almost all original equipment electronic ignitions are inductive-discharge systems. They produce high secondary voltage in the same way a breaker-point ignition does. Electronic ignitions, however, use electronic switching circuits rather than breaker points.

Primary current control in an electronic ignition is based on the ability of a transistor to act as a solid-state relay. Transistors turn primary current on and off in response to voltage signals from a triggering device in the distributor. All primary current flows through the switching transistors, not through the distributor.

Distributor triggering devices can be magnetic-pulse generators, Hall-effect switches, metal-detection circuits, or optical circuits. Magnetic-pulse generators and Hall-effect switches are most common. A magnetic-pulse generator uses a low-reluctance trigger wheel to change the field strength of a magnet. Changing magnetic flux induces a voltage in a pickup coil, which switches the driver transistor off. A Hall-effect switch uses a rotating series of blades to change magnetic flux through a semiconductor chip. This varies the output voltage, which switches the driver transistor. A Hall-effect switch requires an input voltage; a magnetic-pulse generator does not.

Basic electronic ignition systems use centrifugal and vacuum advance mechanisms in the distributors to advance and retard timing. These work just like the advance mechanisms in breaker-point distributors. Late-model electronic ignitions used with electronic engine control systems do not have centrifugal and vacuum advance. Spark advance is controlled electronically by the system computer.

All carmakers have changed their electronic ignitions from year to year. These have not been radical changes, however, but steady improvements in basic designs. Electronic ignitions eliminate the problems of mechanical wear and varying timing and dwell that exist in breaker-point ignitions.

The introduction of distributorless ignition systems (DIS) points the way

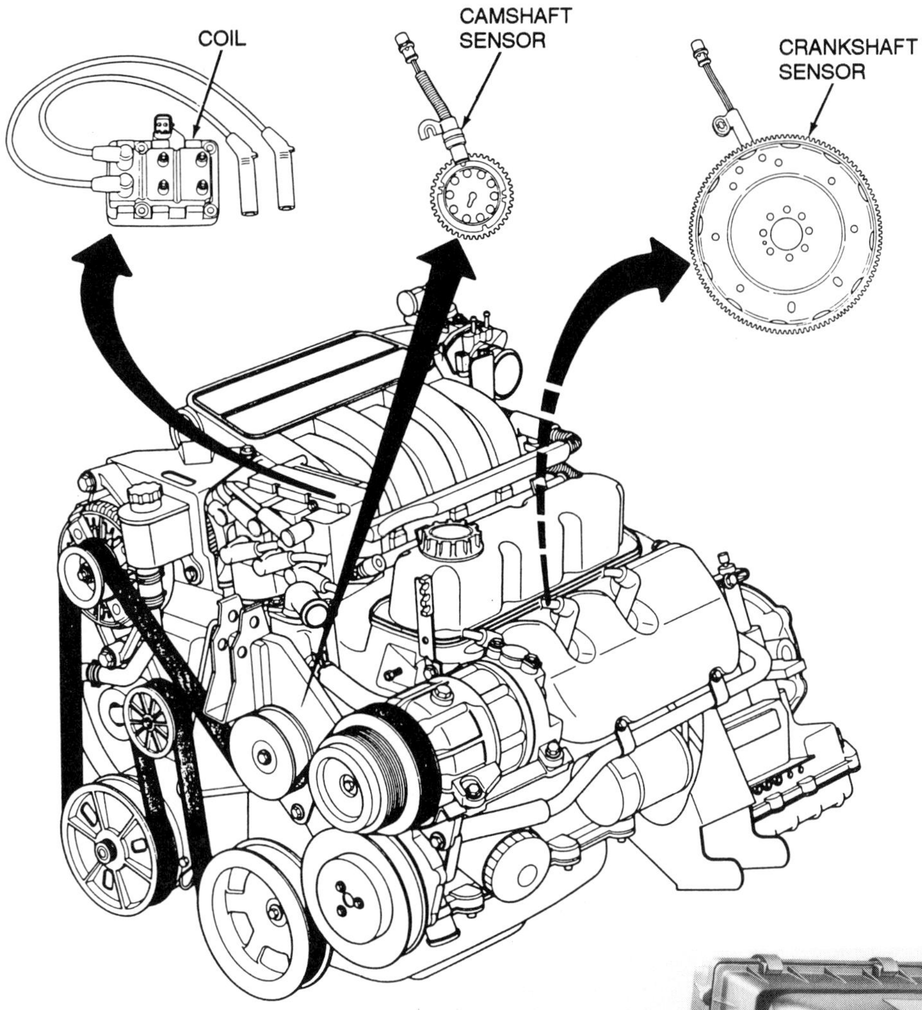


**Figure 13-81.** Chrysler's first DIS installation was on 1990 3.3-liter, V-6 engines. (Chrysler)

to the future. The mechanical distributor will soon go the way of the carburetor. Increased reliance on electronics and elimination of moving mechanical parts results in more precise ignition control with fewer potential problems.

**Figure 13-82.** Chrysler added DIS to the dohc, 24-valve, high-performance engine of the Spirit R/T in 1991. (Chrysler)

## REVIEW QUESTIONS

### Multiple Choice

**1.** Most late-model electronic ignitions are:

**a.** point-controlled systems
**b.** capacitive-discharge systems
**c.** optically triggered systems
**d.** inductive-discharge systems

**2.** Mechanic A says that ignition power transistors generate the basic ignition timing signal. Mechanic B says that ignition primary current flows through the distributor generator to control timing. Who is right?

**a.** A only
**b.** B only
**c.** both A and B
**d.** neither A nor B

**3.** Mechanic A says that ignition transistors can act as solid-state relays. Mechanic B says that ignition transistors can act as current amplifiers. Who is right?

**a.** A only
**b.** B only
**c.** both A and B
**d.** neither A nor B

**4.** All the following are kinds of distributor signal generators, EXCEPT:

**a.** a Hall-effect switch
**b.** an acoustic coupler
**c.** a magnetic-pulse generator
**d.** a metal-detection circuit

**5.** The pickup coil of a magnetic-pulse generator contains:

**a.** a ferromagnetic armature
**b.** a permanent magnet and a pole piece
**c.** a nonmagnetic sensor
**d.** a magnetic reluctor

**6.** "Armature" and "reluctor" are names for the _________ in a magnetic-pulse distributor.

**a.** stator
**b.** pickup coil
**c.** sensor
**d.** trigger wheel

**7.** To generate an output voltage signal, a Hall-effect switch requires:

**a.** an input voltage
**b.** no magnetic field
**c.** a shunt transistor
**d.** a crankshaft position sensor

**8.** Mechanic A says that a Hall-effect switch is an optical signal generator. Mechanic B says that a Hall-effect switch generates full-strength signal voltage at all engine speeds. Who is right?

**a.** A only
**b.** B only
**c.** both A and B
**d.** neither A nor B

**9.** Mechanic A says that electronic ignitions can have a fixed dwell angle or a variable dwell angle. Mechanic B says that coil saturation varies directly with a change in the dwell angle in electronic ignitions. Who is right?

**a.** A only
**b.** B only
**c.** both A and B
**d.** neither A nor B

**10.** Mechanic A says that a variable-dwell ignition requires a larger ballast resistor than a fixed dwell ignition. Mechanic B says that the dwell angle usually increases at high speed in a variable-dwell system. Who is right?

**a.** A only
**b.** B only
**c.** both A and B
**d.** neither A nor B

**11.** Centrifugal advance weights in an electronic distributor move the:

**a.** baseplate (breaker plate)
**b.** pickup coil (sensor)
**c.** trigger wheel
**d.** stator

**12.** The only original-equipment ignition with metal-detection triggering used by a domestic carmaker was used by AMC and made by:

**a.** Delco-Remy
**b.** Motorcraft
**c.** Robert Bosch
**d.** Prestolite

**13.** The first domestic carmaker to use electronic ignition as standard equipment was:

**a.** AMC
**b.** Chrysler
**c.** Ford
**d.** General Motors

**14.** Mechanic A says that the Delco-Remy HEI system is a variable-dwell ignition. Mechanic B says that the Ford Solid-State and Dura-Spark II systems are fixed-dwell systems. Who is right?

**a.** A only
**b.** B only
**c.** both A and B
**d.** neither A nor B

**15.** Dual-mode timing is a feature of:

**a.** HEI systems
**b.** Chrysler ignitions
**c.** Bosch ignitions
**d.** Ford Dura-Spark systems

**16.** Thick-film integrated (TFI) ignition systems are used by:

**a.** AMC
**b.** Chrysler
**c.** Ford
**d.** General Motors

**17.** Which of the following is NOT used in a direct ignition system?

**a.** ignition coil
**b.** ignition module
**c.** distributor
**d.** crankshaft and camshaft sensors

**18.** Student A says that the ignition coil in a direct ignition system is connected to two spark plugs. Student B says that the ignition module controls primary current flow and dwell time limiting. Who is right?

**a.** A only
**b.** B only
**c.** both A and B
**d.** neither A nor B

**19.** Student A says that one spark plug in a direct ignition system fires forward and the other fires backwards. Student B says that all spark plugs in a DIS system fire forward. Who is right?

**a.** A only
**b.** B only
**c.** both A and B
**d.** neither A nor B

**20.** Student A says that a spark plug fired backwards in a direct ignition system requires 30 percent less voltage to create the arc. Student B says that a direct ignition system may use a Hall-effect sensor or a magnetic sensor. Who is right?

**a.** A only
**b.** B only
**c.** both A and B
**d.** neither A nor B

**21.** Student A says that a C3I system can be identified by the appearance of the coil pack. Student B says that the coil packs in all C3I systems are interchangeable. Who is right?

**a.** A only
**b.** B only
**c.** both A and B
**d.** neither A nor B

## Fill in the Blank

**22.** In modern electronic ignitions, breaker points have been replaced by an __________ .

**23.** In an electronic ignition, all primary current passes through an ignition control __________ .

**24.** The most common trigger devices in electronic distributors are __________ generators.

**25.** The Hall effect works by passing current through semiconductor material and then applying a __________ to it at a right angle to its surface.

**26.** In a metal-detection circuit, an input voltage is transmitted to an __________ in the distributor sensor.

**27.** For either a breaker-point ignition or an electronic ignition, the best test instrument for determining problems with dwell is an __________ .

**28.** The type of electronic ignition in which dwell is the same in distributor degrees at all engine speeds is called a __________ ignition.

**29.** In a variable-dwell electronic ignition, the dwell period varies in distributor degrees at different engine speeds but dwell __________ remains relatively constant.

**30.** If an ignition system uses primary current and voltage to charge a capacitor, it is called a __________ system.

**31.** In many ignition modules, two power transistors are arranged in a __________ circuit.

**32.** Ignition switching transistors control primary current. The switching transistors are usually controlled by another __________ transistor.

**33.** A direct ignition system uses the __________ spark theory.

**34.** When an engine with a direct ignition system is first cranked, ignition timing signals are provided by the __________ .

**35.** The direct ignition system requires a __________ pulse to set up the crankshaft signal and ignition coil sequence.

**36.** The ignition module in a direct ignition system allows full coil saturation by a process called __________ .

**37.** In some direct ignition systems, crankshaft position is sensed by the use of a magnetic sensor and crankshaft __________ .

# PART FIVE
# FUEL SYSTEMS AND EMISSION CONTROLS

# INTRODUCTION

All automobile internal combustion engines—gasoline and diesel, alike—burn petroleum fuel to develop power. The scientific principles that you reviewed in Part One are the starting point for understanding engine combustion and combustion control.

Chapter 14 outlines the production of gasoline and diesel fuel. It also explains the important features of these fuels that affect engine performance. Knowing how fuel is refined and the purposes of various additives often can help you identify an engine performance problem.

Certainly, an engine fuel system consists of more than just the carburetor or the fuel injection parts. Chapter 15 explains how fuel delivery and storage components (tanks, pumps, filters, lines, and other parts) affect engine operation. Air induction and evaporative emission control also are important parts of air-fuel control, and this chapter outlines the operation of air intake and EEC systems on late-model vehicles.

Understanding carburetion is basic to understanding gasoline engine operation. Chapter 16 explains the carburetor systems and operation, as well as assist devices and modifications that aid emission control and driveability on late-model vehicles.

# 14 INTRODUCTION TO FUELS AND EMISSION CONTROLS

## WHAT

Both gasoline and diesel fuel come from the same crude oil as motor oil. In chapter 5, you learned about motor oil and how it is formulated. This chapter explains how gasoline and diesel fuel are refined to meet the needs of different engines and to deliver the most heat energy from their chemical energy.

As combustion turns chemical energy into heat energy, petroleum hydrocarbon compounds change into different compounds. Some of these combustion byproducts are harmful emissions that pollute our air. Emission control has been a major part of automobile design and service for over 20 years, and it is directly related to the combustion process. This chapter goes from a discussion of fuels to an explanation of the major automotive emissions and the principles of their control.

## WHY

Many engine performance problems relate directly to the kind and quality of fuel used. If you know how fuels are refined for best engine performance, you will understand how to diagnose and service fuel-related problems. Also, if you know how combustion produces air pollutants and the principles used to control them, you will better understand how the emission devices work on modern engines and how to service them. Before studying the systems that control combustion, then, we will examine the fuels of combustion and their byproducts.

## GOALS

After studying this chapter, you should be able to:

1. Explain the composition, characteristics, and classifications of gasoline and diesel fuel.
2. List the major automotive air pollutants and the parts of the vehicle from which they come.
3. Explain the basic photochemical smog reaction and the effects of temperature and other climatic conditions on air pollution.
4. Explain the basic reasons for crankcase, fuel evaporation, and exhaust emission controls and which pollutants are controlled by each.

## FUEL PRODUCTION

Fuels are combustible materials whose chemical energy is easily converted to heat energy. Gasoline and diesel fuel are extracted from crude oil by **distillation**, **cracking**, and other refining processes. The processes begin by heating crude oil in the absence of air to boil off different hydrocarbon compounds. During distillation, the hydrocarbons of gasoline vaporize at 25° to 200° C (77° to 392° F). The heavier hydrocarbon compounds that make up diesel fuel vaporize at 150° to 360° C (302° to 680° F).

After the different hydrocarbons are boiled out of crude oil, they are cooled to condense them back to liquid. Then heat is applied again to break the large hydrocarbon molecules into smaller ones. This is the **thermal cracking** process. Often a catalyst is added to help break and reshape the molecules. This is called **catalytic cracking**. After cracking, gasoline and diesel fuel are further refined by heating again and blending with hydrogen in a process called **hydrogenation**.

Further refining steps include more catalytic reactions, as well as breaking up and recombining complex hydrocarbon molecules. The important things to remember are that gasoline is lighter, more volatile, and more easily vaporized than diesel fuel. Additionally, many additives are blended into both fuels for efficient combustion. Some of the characteristics and rating methods for both fuels are the same. Others are different. One quality of any fuel is its heat value.

### Heat Value

All engines produce power by converting fuel chemical energy to heat energy. The more heat produced from a fuel, the more chemical energy is converted to mechanical energy and power. The heat value, or potential heat energy, of any fuel can be measured in any units customarily used to measure energy: calories, joules, or British thermal units. One **British thermal unit (Btu)** is the heat energy needed to raise the temperature of one pound of water one degree Fahrenheit. The Btu is the unit used to measure the potential heat energy of gasoline and diesel fuel. Figure 14-1 shows the average Btu's and weights per gallon of gasoline and diesel fuel. The following sections explain some of the char-

| Fuel | Average Pounds per Gallon | Average Btu per Gallon |
|---|---|---|
| Gasoline | 6.0 | 124,000 |
| Diesel | 7.1 | 138,000 |

**Figure 14-1.** Diesel fuel is heavier and contains slightly more Btu's per gallon than gasoline.

acteristics and rating methods for gasoline and diesel fuel.

## GASOLINE

Gasoline is about 86 percent carbon and 14 percent hydrogen by weight. It vaporizes easily and produces tremendous power. Gasoline consists of several hydrocarbon compounds blended with other additives that improve its performance. All gasoline is basically the same, but no two blends are identical.

Refiners blend gasolines for different temperature and altitude conditions in different parts of the country at different times of the year. The gasoline you buy in the summer is not the same as the gasoline you buy in the winter. Refiners use various blending methods and additives to get the desired **volatility**, remove impurities, and improve gasoline octane rating.

### Volatility

Volatility is a measurement of how easily a liquid vaporizes. The more volatile a gasoline is, the more easily it vaporizes. Volatility of all liquids is affected by temperature and altitude. High temperatures or high altitudes, or both, increase volatility. Efficient vaporization is necessary for complete combustion. If gasoline doesn't vaporize completely, the air-fuel mixture will be rich, liquid gasoline may enter the cylinders, and performance, economy, and emission control suffer. If gasoline vaporizes too quickly, the air-fuel mixture can be lean, the engine can be hard to start and misfire, and again performance, economy, and emission control suffer.

Oil companies control volatility by blending different gasoline hydrocarbons that have different boiling points. They produce gasolines with:

- Low volatility—warm weather and high altitudes
- High volatility—cold weather and low altitudes.

Refiners blend gasoline for expected temperature and climate conditions in different parts of the country. However, unexpected temperature changes or long-distance driving between different areas can cause problems related to volatility. The operating conditions affected by gasoline volatility include:

- *Starting and warmup*—Part of the gasoline must vaporize easily to start a cold engine. This percentage is high for low temperatures and low altitudes and low for high temperatures and high altitudes. The same factors affect the speed of warmup, but volatility for warmup does not need to be as high as it does for starting.
- *Smooth acceleration*—As the throttle opens for acceleration, more air enters the engine, and the carburetor delivers extra gasoline. If the extra fuel doesn't vaporize quickly, the air-fuel mixture becomes momentarily lean. This makes the engine stumble. Then as the extra fuel does vaporize, the mixture becomes momentarily rich. Again the engine hesitates. Without proper vaporization during acceleration, combustion is not correct.
- *Vapor lock prevention*—**Vapor lock** occurs when engine heat vaporizes the gasoline in the fuel pump, the lines, and the carburetor fuel bowl. The fuel pump can't pump vapor, so the fuel system "locks up" for lack of liquid gasoline. Highly volatile gasoline increases the chances of vapor lock.
- *Economy*—Gasoline volatility for good fuel economy is a compromise. Low-volatility fuels have a higher heat content and burn more efficiently. But, they increase starting difficulty and warmup time and reduce acceleration slightly. High volatility fuels improve starting, warmup, and acceleration but have a lower heat content. They can produce an overly rich mixture and reduce economy.

### Chemical Impurities

The chemical impurities in gasoline depend on the kind of crude oil from which it was refined. Oil companies try to remove all impurities, but some small amounts remain. Sulfur is an impurity in most crude oil, and it does not burn readily. The sulfur content in gasoline is less than 0.1 percent, but if enough sulfur reaches the crankcase, it combines with water into sulfuric acid ($H_2SO_4$), which corrodes engine parts. Sulfur is a greater problem in diesel fuel and heavier oils than it is in gasoline.

Gasoline impurities cause gum and varnish deposits that clog carburetor and fuel injection passages and cause piston rings and valves to stick. Oxidation inhibitors added to gasoline help prevent gum and varnish formation.

### Gasoline Octane Rating and Knock Control

As the air-fuel mixture is compressed in a cylinder, temperature increases along with pressure. If the temperature rises above the **autoignition** temperature of a particular gasoline, the mixture will ignite prematurely and burn unevenly. Remember that combustion is not an explosion, but controlled burning that must occur smoothly at a precise time to develop full power. Uncontrolled burning—a partial or complete explosion—reduces combustion efficiency, causes engine knocking, and can damage the engine. Several factors affect combustion efficiency, including:

- Compression ratio and compression pressure
- Combustion chamber design and spark plug location
- Air-fuel ratio
- Air-fuel mixture turbulence in the combustion chamber
- Ignition timing
- Engine temperature
- Combustion chamber deposits and general engine condition
- Gasoline antiknock quality (octane rating).

These factors relate to each other in

**Figure 14-2.** Normal combustion occurs when the air-fuel mixture ignites and burns evenly. (Champion)

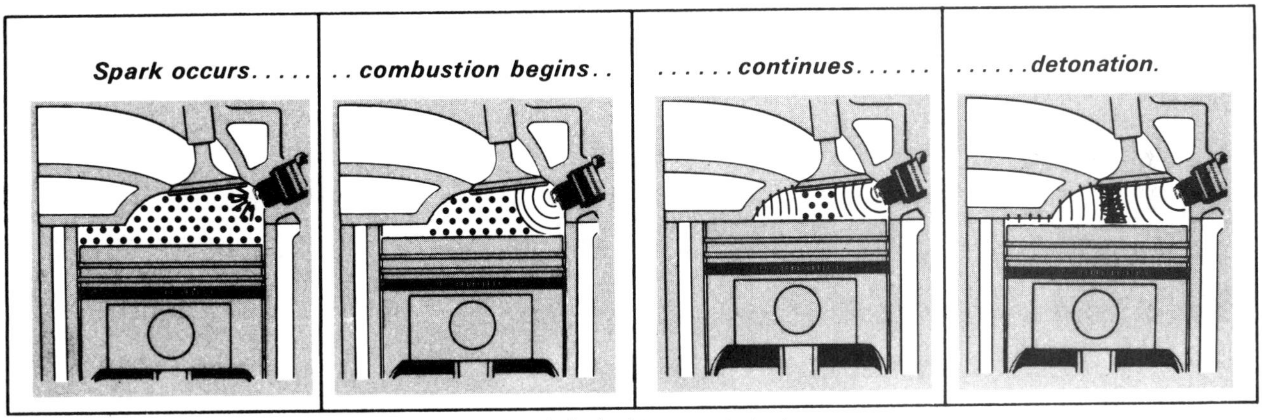


**Figure 14-3.** Detonation occurs when the last part of the air-fuel mixture explodes from high cylinder temperatures. (Champion)

complex and variable ways. You will learn more about them as you study combustion controls and engine design. For now, we will concentrate on gasoline antiknock quality and touch upon the others as they relate to this factor.

**Engine Knocking.** Knocking or pinging that we sometimes hear from an engine is the sound of uncontrolled combustion. These sounds indicate explosions rather than smooth burning. If the explosions are severe enough, or occur too early on the compression stroke, they can damage the engine. Two major forms of knocking can occur in an engine. **Detonation** occurs *after* the time of ignition. **Preignition** occurs *before* the desired ignition time.

When the spark plug ignites the air-fuel mixture, the flame moves rapidly and evenly outward until the mixture burns completely, figure 14-2. This **flame front propagation**, or burn time, takes only about 3 milliseconds. In some cases, the last part of the air-fuel mixture explodes before the flame front reaches it, figure 14-3. This unwanted explosion is due to the combination of combustion temperature and pressure rise, combustion chamber design, and gasoline antiknock quality. This uncontrolled explosion works against normal combustion and creates the knocking that we call detonation.

Preignition occurs when the air-fuel mixture ignites before the spark plug fires. This usually is caused by deposits in the combustion chamber that get hotter than the ignition temperature of the mixture, figure 14-4. Then, the mixture ignites before the spark plug fires. Preignition also can happen when combustion chamber temperature and pressure exceed the knock resistance of the fuel, even when deposits don't form localized hot spots.

A similar problem can occur with poor antiknock gasoline when a hot engine is shut off. Cylinder temperatures remain high and actually rise as hot coolant stops circulating. The engine continues to turn over, and one or two cylinders may draw in a fresh air-fuel charge. If the engine coasts for only two revolutions, this fresh mixture can be compressed and ignited by high cylinder temperatures. A gasoline engine then starts to "diesel" and runs on due to compression ignition.

Gasoline must be blended to resist both detonation and preignition while remaining volatile enough for good vaporization. High antiknock quality ensures stable combustion.

**Octane Rating.** A gasoline's ability to burn stably and resist knocking is expressed as its **octane rating**. The higher the octane number, the better the knock resistance under high temperature and pressure. The octane number up to 100 indicates the per-

**Figure 14-4.** Preignition occurs when a hot spot in the combustion chamber ignites the air-fuel mixture before the spark plug fires. (Champion)

centage of isooctane ($C_8H_{18}$) mixed with *n*-heptane ($C_7H_{16}$) that has the same knock behavior as the gasoline being tested in a test engine. Isooctane has excellent knock resistance and has the octane value of 100. The *n*-heptane has poor knock resistance and has the octane value of 0. Thus, 100 octane gasoline has the antiknock quality of 100-percent isooctane; 87 octane gasoline has the antiknock quality of 87-percent isooctane and 13-percent *n*-heptane.

Using lead, alcohol, and other additives, it is possible to blend gasoline with an octane rating higher than 100. These ratings indicate the theoretical percentage of knock resistance better than that of pure isooctane. Thus, 103 octane gasoline has 3-percent better antiknock qualities than 100-percent isooctane.

Gasoline octane rating is measured in two ways: as **research octane** and as **motor octane**. Both methods use the same variable-compression test engine. The research method measures gasoline knocking tendencies under low load, low temperature, full throttle, and low to medium engine speed. The motor method measures knocking under heavy load, high speed, high temperature, and full throttle. Because of the two different test methods, a single gasoline can have two different octane ratings. Generally, there is about an 8-point difference between the research octane and motor octane for the same gasoline.

Historically, gasolines were marketed as "regular" or "premium" (sometimes called "ethyl") blends. Oil companies used the research octane number when advertising "high octane" or "100 octane" gasoline. In the 1960's, gasoline with an octane rating of 100 or higher was common, and engines had high compression ratios to take advantage of the high-octane antiknock quality. In the early 1970's, tightening emission controls and concern about lead pollution from exhaust caused carmakers to start building engines that would run well on 91 research octane gasoline. At about the same time, another method of expressing gasoline octane was adopted by the oil and automotive industries.

This was not a new octane test method, but rather a way to express the antiknock performance of gasoline more realistically for average use. Known as the gasoline performance information system, or antiknock index, this method uses the average of the research and motor octane ratings for a particular gasoline. It is often called the (R + M)/2 octane number. Since there is approximately an 8-point spread between research and motor

### IS IT METER OR METRE, LITER OR LITRE?

In the reference manuals you use, you will see the spellings *meter* and *metre* or *liter* and *litre*. You might ask, "Which is correct, -er or -re?" The answer is, both are, but neither style is universally "official." The -re spelling is really a British style, similar to the British spellings, *theatre* and *centre*. So, any British books you read will use the spellings *metre* and *litre*. Similarly, the -re spellings is more common in Canada, though not as universal as it is in England.

In the United States, opinions are about equally divided, but -er spellings are more preferred. Lining up on the side of -er are the U.S. Government Printing Office, The American National Standards Institute (ANSI), the Institute of Electrical and Electronics Engineers (IEEE), the American Society for Testing and Materials (ASTM), and the Society of Automotive Engineers (SAE). General Motors formerly used only -re spellings, but in recent years has used both -re and -er. Ford and Chrysler seem to use both spellings equally.

No single spelling is universal in all languages for scientific usage. In some European languages, meter is spelled *metr* or *metro*. Regardless of how you spell it or say it, though, it always means the same thing.

Another interesting point is that when you add any metric prefix to a stem unit, you always write the resulting word as one word. Thus, *kilovolt* and *milliampere* are correct, not *kilo-volt* and *milli-ampere*. Always pronounce any word with a metric prefix with the accent on the first syllable of the prefix. So, *KILL-o-mee-tur* is correct, not *kil-AH-mit-er*.

**Figure 14-5.** Advertised gasoline octane ratings are an average of the motor octane and research octane ratings.

octane numbers, 91 research octane gasoline would have a motor octane number of about 83. The (R + M)/2 octane rating then would be 87. These are the octane ratings you see most often on gasoline pumps, figure 14-5, and in vehicle owner's manuals.

**Tetraethyl Lead.** It is possible to blend 100 octane gasoline without any additives. In fact, some unleaded racing gasoline has been made with an octane rating of 101. The amount of high-octane gasoline that can be extracted from a barrel of crude oil simply by distillation and cracking, however, is expensive and does not yield the most possible gallons of fuel from a barrel of oil.

Refiners blend additives with gasoline to improve antiknock performance and get more high-octane fuel from a barrel of oil. One of the most effective additives is **tetraethyl lead**. Tetraethyl lead, raises the octane rating of gasoline and helps prevent detonation. Lead is a soft metal and also acts as a lubricant for valve faces and seats, much as other soft metals—sodium, lithium, and molybdenum—act as lubricants in greases. Unfortunately, lead also has some harmful effects:

- Besides coating valve faces and seats, lead also forms deposits on combustion chamber surfaces and spark plugs. These deposits shorten plug life and can form hot spots that cause preignition.
- Lead also gets into the crankcase where it dilutes the oil and can cause wear on lower engine parts.
- Lead in exhaust coats the surfaces of catalytic converter substrates, reducing catalytic action and eventually ruining the converter.
- Lead is poisonous and is given off in engine exhaust as a **particulate**. Lead particulate emissions from vehicle exhaust in crowded urban areas are a health hazard.

These last two harmful effects of lead are the major reasons for reducing, and eventually removing, the tetraethyl lead in gasoline. Catalytic converters, introduced on 1975 model cars in the U.S., require that the engines run on unleaded gasoline. Years ago, leaded premium gasoline had up to 4.5 grams of lead per gallon (about 1.2 grams per liter). The highest lead content in gasoline sold in the early 1980's ranged from 0.5 to 1.0 gram per gallon. Low-lead "regular" gasoline had 0.5 gram per gallon or less. As newer cars designed for unleaded gasoline replaced older cars on the road, the demand for high-lead fuel decreased. By July, 1985, the maximum lead content was reduced to 0.5 gram per gallon. By 1986, lead content was no more than 0.1 gram per gallon. All lead is to be removed from gasoline by 1988.

As oil companies removed lead from gasoline, they lowered the octane rating. Most unleaded gasoline has an (R + M)/2 octane rating of about 87 or 88. Low-lead regular had an (R + M)/2 rating of 89 or 90, and premium gasoline (leaded or unleaded) averages about 91 or 92 octane. As octane ratings declined from those of 20 or 30 years ago, carmakers lowered engine compression ratios to burn the lower octane fuel efficiently. Engines of the 1960's had compression ratios as high as 10:1 or 11:1. Today, compression ratios are between 8.5:1 and 9:1. Lower compression also reduces $NO_x$ emissions.

### Other Gasoline Additives

Besides lead to improve octane rating, refiners blend other additives with gasoline. Some do the jobs of similar additives in motor oil. Others are similar to additives in diesel fuel. The major additives are:

1. *Antioxidants* to reduce oxidation and gum formation
2. *Metal deactivators* to deactivate traces of copper and other metal ions that cause gasoline to oxidize in storage
3. *Rust and corrosion inhibitors* to reduce rust and corrosion of fuel system parts
4. *Dyes* to identify specific gasoline blends
5. *Combustion deposit modifiers* to reduce preignition and spark plug fouling
6. *Anti-icing additives*, special alcohols to prevent ice formation on carburetor throttles from a combination of low temperature and low pressure as gasoline vaporizes in cold weather.
7. *Detergents* to remove gum and varnish deposits from carburetor and fuel injection passages and throttles.
8. *Dispersants* to extend PCV valve life, reduce sludge formation, and minimize other engine deposits.

## DIESEL FUEL

Because a diesel relies on compression ignition, almost anything that can be injected into the diesel combustion chamber will burn if it has an ignition point below the compressed air temperature. Dr. Rudolph Diesel built one of his first engines to run on injected coal dust. It blew up, however, so he turned to petroleum-based liquid fuel.

Diesel fuel, like gasoline and motor oil, is a complex blend of hydrocarbons and has many of the same kinds of additives. These include ignition quality improvers, oxidation inhibitors, rust and corrosion preventives, metal deactivators, biocides, pour point depressants, detergents, dispersants, and smoke suppressants.

There are three grades of diesel fuel: 1-D, 2-D, and 4-D. Number 4-D fuel is only for large, low-speed engines used in generators, ships, and locomotives. Number 1 and number 2 diesel fuel are the kinds found at service

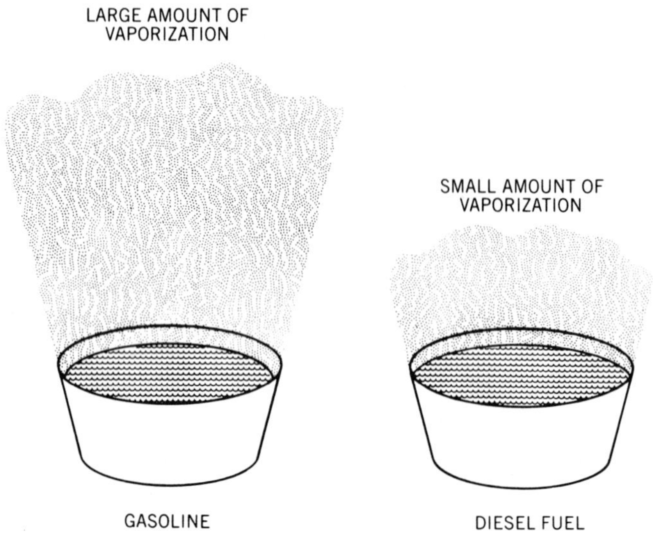


**Figure 14-6.** Gasoline vaporizes readily at atmospheric temperature and pressure. Diesel fuel hardly vaporizes at all.

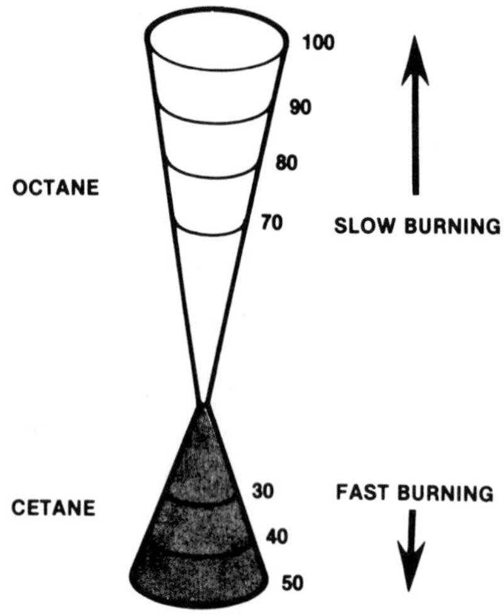


**Figure 14-7.** A high gasoline octane rating indicates a resistance to combustion. A high diesel cetane rating indicates easy combustion. (Ford)

stations, with number 2 a little more common.

Number 2 diesel fuel is blended for high loads and steady speeds because a diesel is basically a constant-speed engine. Large diesel trucks operate with a limited engine speed range and run best on number 2 fuel. Number 1 diesel fuel is more volatile and is designed for engines that operate with frequent load and speed changes. Number 1 and number 2 fuel can be used interchangeably in most automotive diesel engines. They will mix and burn with little or no problem.

Because it is less volatile, number 2 fuel may cause starting problems in cold weather. Number 1 fuel, on the other hand, is more volatile, and may cause smoking problems because it has less time to mix with the compressed air before ignition. Some diesel carmakers recommend just one fuel grade. Others recommend either number 1 or number 2 fuel, depending on the climate and operating conditions.

## Volatility

Volatility of diesel fuel is similar to volatility of gasoline and affects engine performance in the same ways. However, diesel fuel vaporizes at a much higher temperature than gasoline because it is heavier. In fact, diesel fuel hardly vaporizes at all at atmospheric temperature and pressure, figure 14-6. Vaporization occurs *after* the fuel is injected into the combustion chamber. Diesel vehicles don't suffer vapor lock problems or need fuel evaporation emission controls as gasoline-powered cars do. Nevertheless, oil companies carefully control diesel fuel volatility to provide maximum power and economy with minimum soot, engine deposits, and smoke. Diesel fuel volatility is directly related to the fuel cetane number.

## Cetane Number

The cetane rating, or **cetane number**, of diesel fuel indicates the ignition quality or ability of the fuel. The higher the cetane number, the easier the fuel ignites. Remember that a couple milliseconds of ignition lag occur in a diesel engine after injection while the fuel vaporizes and mixes with the air. Ignition lag is less with high-cetane fuel than with low-cetane fuel.

Diesel cetane rating, like gasoline octane rating, is measured in a laboratory test engine. Cetane numbers and octane numbers are both ways to rate fuel performance, but they indicate opposite qualities, figure 14-7. Gasoline with a high octane number resists knocking and does not ignite easily. Diesel fuel with a high cetane number ignites quickly and easily.

Cetane numbers for 1-D and 2-D fuel range from 35 to 65. Most number 1 fuels have a 50 cetane number, and most number 2 fuels have a 45 cetane number. These are generally satisfactory for automobile engines. Normally, the fuel cetane number should be a few units above the cetane requirement for a given engine when new. This is because the engine cetane requirement increases with use. If the cetane number is too high for a given engine, it causes smoking because the fuel doesn't mix completely with the air before it ignites. If the cetane number is too low, the engine can be hard to start and sluggish.

## Specific Gravity and Viscosity

Specific gravity is a measurement of the weight of the fuel compared to water. Diesel fuel is heavier than gasoline and, therefore, has a higher specific gravity. Specific gravity is controlled during refining so that the diesel fuel is heavy enough for good spray penetration into the combustion chamber. If specific gravity is too low, the fuel will ignite too soon and put all the combustion force on a small area of the piston.

### HYDROSTATIC DRIVE

Hydraulic systems made it possible to eliminate the clutch from a car's transmission and gave us fully automatic transmissions that change gear ratios without mechanically "shifting" gears. We can take hydraulic systems a step farther and eliminate gears, drive shafts, chains, and U-joints almost entirely in drive systems called hydrostatic drives.

Hydrostatic drives are used on many farm tractors and other items of agricultural and construction equipment. As the illustration shows, a basic hydrostatic drive consists of an engine to power a hydraulic pump, a control system, high-pressure oil lines, and a hydraulic motor at each driving wheel. Each motor usually is coupled to a planetary gear speed reducer at each wheel for high-torque, low-speed operation. The hydraulic motors provide smooth, infinitely variable power in forward and reverse at each wheel. Moreover, by reversing fluid flow through the motors while the vehicle is moving, the system can provide braking.

The advantages of hydrostatic drive are that power can be transmitted long distances and in directions where it would be difficult to install gears or chain drives. They also provide high torque at low speed and infinitely variable speed control. Hydrostatic drives also allow a central hydraulic system to power other devices on the vehicle besides the driving wheels, such as harvesting, plowing, and digging implements on farm tractors.

The disadvantages of these systems are that the hydraulic pumps and motors are more expensive than gear drives. Also, they transmit power at only 70- to 80-percent efficiency, whereas the efficiency of gear and chain drives is in the 95- and 98-percent range. However, agricultural and construction equipment operating needs often outweigh these cost and efficiency disadvantages.

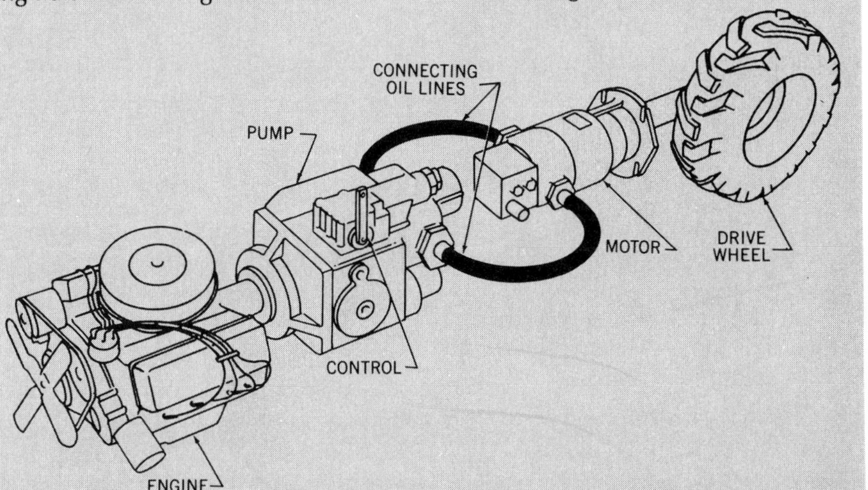


Because diesel fuel is heavier than gasoline, it has a higher viscosity. Viscosity is not important with gasoline, but it is with diesel fuel because the fuel lubricates the close-tolerance parts of the injection pump and injectors. Viscosity must be high enough to form a lubricating film on these parts, but not so high that the fuel clogs the pump and injectors.

Fuel viscosity also affects injector spray pattern. High-viscosity fuel forms large droplets that are hard to vaporize and burn, figure 14-8. Low-viscosity fuel forms a fine, easily burned mist. But, if viscosity is too low, injection pump and injector lubrication suffers.

### Pour Point and Cloud Point

Because diesel fuel contains heavier hydrocarbons and more paraffin than gasoline, it may get cloudy at low temperature. The temperature at which cloudiness forms is the **cloud point**. It indicates that wax crystals are forming; these may clog fuel pumps, filters, and injectors. About 10° F (6° C) below the cloud point, the fuel will barely pour. This is the **pour point**. Refiners use additives to control the cloud point and the pour point and avoid fuel system problems.

Number 1 fuel has lower cloud and pour points than number 2 fuel. Therefore, some carmakers recommend using 1-D fuel in the winter or blending 1-D and 2-D fuels. Kerosene is similar to number 1 diesel fuel and can be added to 2-D fuel to relieve hard starting and fuel system clogging in cold weather.

### Sulfur Content

While sulfur impurities are seldom a problem in gasoline, they can be in diesel fuel. Diesel fuel contains about 0.5-percent sulfur. That's not much, but it is five times more than the 0.1 percent in most gasoline. Sulfur increases engine deposits, corrosion, and wear. Motor oil for diesel engines has additives to minimize these problems caused by fuel.

### Carbon and Ash

Carbon residue and ash deposits are greater problems with diesel fuel than with gasoline. Diesel fuel and motor oil contain additives to reduce carbon

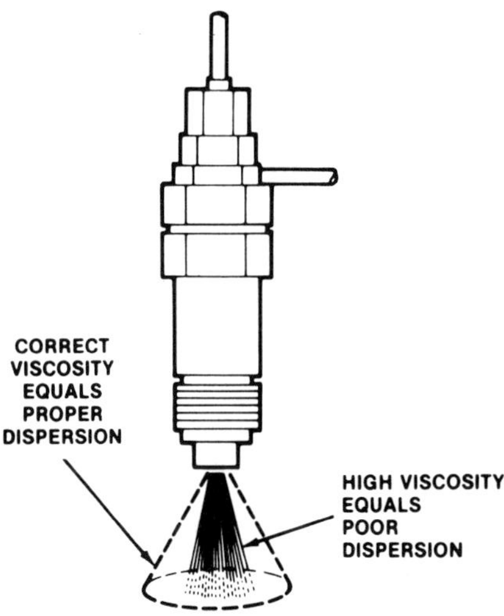


**Figure 14-8.** High-viscosity diesel fuel does not vaporize well when injected into the cylinders. (Ford)

and ash problems such as combustion chamber deposits and clogged injectors.

### Water and Sediment

Diesel fuel can be contaminated by water and sediment during transportation and storage. If enough water, or water sediment, collects in a vehicle fuel tank, it will clog fuel filters and pumps and cause injector wear and rust by preventing lubrication. Many diesel vehicles have water separators and large-capacity filters in the fuel system to eliminate water problems. Also, if enough water collects in the fuel tank, bacteria will grow where the fuel floats on the water. These bacteria, too, form a sediment that clogs fuel system parts. Therefore, some diesel fuel additives contain a biocide to kill the bacteria.

**Figure 14-9.** The major pollutants emitted by automobiles are hydrocarbons (HC), carbon monoxide (CO), and oxides of nitrogen ($NO_x$).

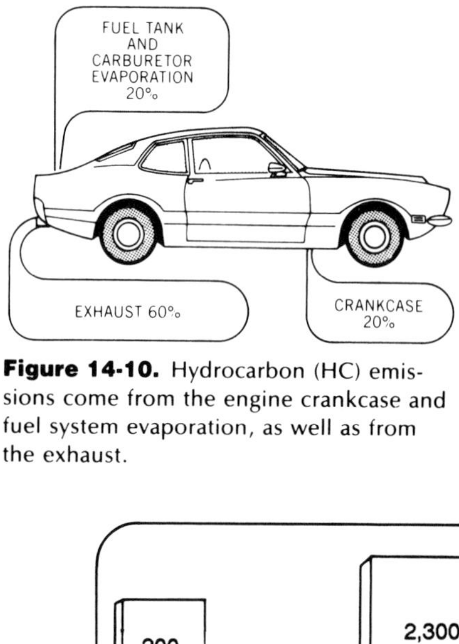


**Figure 14-10.** Hydrocarbon (HC) emissions come from the engine crankcase and fuel system evaporation, as well as from the exhaust.

## INTRODUCTION TO EMISSION CONTROL

Twenty years ago, automobiles were a major source of pollution. Today, automobile emissions are greatly reduced because of government regulations and the engineering efforts of carmakers.

Air pollution is the presence of enough contamination in the atmosphere to injure human, animal, or plant life. Pollution is either natural or manmade. Natural air pollution comes from dust storms, volcanic eruptions, forest fires, and the cycles of plant and animal life. Manmade pollution comes from industrial plants, synthetic chemicals, transportation systems—and automobiles.

An automobile and its internal combustion engine emit three major pollutants: hydrocarbons (HC), carbon monoxide (CO), and oxides of nitrogen ($NO_x$), figure 14-9. These are all gaseous compounds. Additionally, an engine produces some solid emissions of lead, sulfur, carbon, and other particulates, as well as some sulfur oxides. Diesel engines emit less HC and CO but more particulates and sulfur oxides than gasoline engines. $NO_x$ emissions are slightly less from a diesel but still enough to require controls on late-model automobiles.

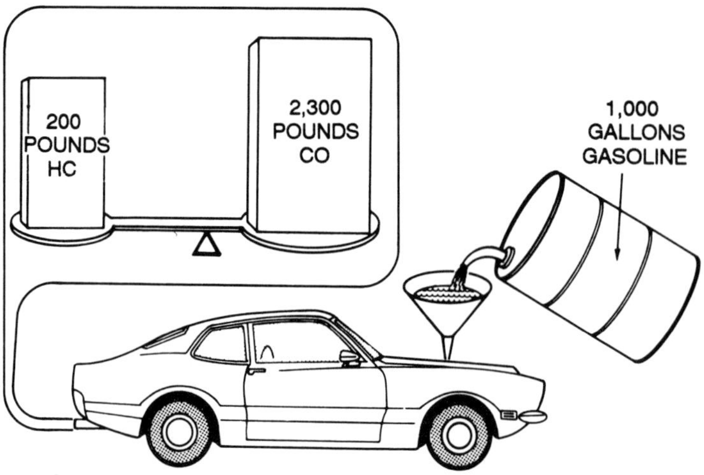


**Figure 14-11.** A car without emission controls emits 200 pounds of HC and 2,300 pounds of CO for every 1,000 gallons of gasoline that it burns.

### Hydrocarbons (HC)

Gasoline, diesel fuel, and motor oil are all hydrocarbons. HC emissions from an automobile are largely unburned fuel in the exhaust, but hydrocarbons are the only major pollutant that come from areas of a car *other than* the exhaust, figure 14-10. The three major sources of HC emissions from a car are:

**1.** Engine exhaust—60 percent

**2.** Crankcase vapors (blowby and oil fumes)—20 percent

**3.** Gasoline fuel system evaporation—20 percent.

Even paint and tires emit traces of hydrocarbons. The film that often clouds the insides of car windows comes from hydrocarbons emitted by vinyl upholstery.

There are more than 200 kinds of hydrocarbons emitted by automobile sources, but hydrocarbons of all kinds can be broken down by complete combustion. If an engine burned its fuel completely, there would be no HC in the exhaust, only water vapor and carbon dioxide ($CO_2$). But combustion is seldom *complete*. Fuel near the edges of the combustion chamber often cools before it can burn and passes out with the exhaust. If the air-fuel mixture is too rich, not all of the fuel burns. If a cylinder misfires, the complete air-fuel charge passes out the exhaust unburned. A car without emission controls produces approximately 200 pounds of HC emissions for every 1,000 gallons of gasoline it uses, figure 14-11.

### Carbon Monoxide (CO)

Carbon monoxide (CO) also results from incomplete combustion. The amount of CO produced depends on how hydrocarbon fuel burns. If the air-fuel mixture is rich, there is not enough oxygen to combine with carbon to form harmless $CO_2$. The air-fuel mixture must be extremely lean to have enough oxygen to form $CO_2$ with no CO. In fact, CO formation can't be eliminated completely from the combustion process in an engine. The same car without emission controls that produces 200 pounds of HC, produces 2,300 pounds of CO from every 1,000 gallons of gasoline, figure 14-11.

CO is an incomplete compound that lacks oxygen. CO is a poisonous gas because it is heavier than oxygen or air and displaces oxygen from the atmosphere. In small amounts, CO causes headaches, dizziness, and nausea. In large amounts, it kills because it deprives the body of oxygen.

### Oxides of Nitrogen ($NO_x$)

High temperatures and pressures of combustion are what produce good performance and fuel economy in an engine. They also produce oxides of nitrogen ($NO_x$). Air is about 21-percent oxygen and 78-percent nitrogen. When combustion temperature exceeds about 2,500° F (1370° C), oxygen and nitrogen combine in large quantities to form $NO_x$. $NO_x$ is not a single compound, but a family of compounds that includes nitrous oxide (NO), nitrogen dioxide ($NO_2$), and nitrogen trioxide ($NO_3$). By themselves, $NO_x$ emissions are no great hazard. When the amounts of $NO_x$ and HC in the air reach the right ratio, however, they combine in the presence of sunlight to form smog.

Automobile $NO_x$ emissions can be reduced by lowering combustion temperatures and pressures. This, however, lowers combustion efficiency, which increases HC and CO emissions. Modern electronic engine controls and 3-way catalysts are the most effective devices for the opposing requirements of HC, CO, and $NO_x$ control.

### Particulates

Solid particulate exhaust emissions are so small that they stay in the atmosphere for a long time before settling. Lead particulates in the air can cause lead poisoning if a person is constantly exposed to them for several years. Carbon and soot particulates can damage people's lungs and respiratory systems. Automobile particulate emissions are controlled mostly by eliminating lead and other additives from gasoline and changing the formulation of fuels, rather than by systems on the car.

### Sulfur Oxides

Some of the sulfur in automobile fuels, especially diesel fuel, combines with oxygen to form **sulfur oxides** in the exhaust. Some of these oxides combine with water vapor in the air to form sulfuric acid. Large amounts of sulfuric acid can be a health hazard in polluted urban air, and it can mix with water and fall as "acid rain." Sulfur oxide pollution from industrial sources is a larger problem than it is from automobiles. But, the large number of diesel vehicles on the road has increased the interest in sulfur oxide automobile emission controls.

### Smog

Not all air pollution is smog. CO, for example, pollutes the air, but it does not form smog. Smog forms when HC and $NO_x$ combine in the presence of sunlight and quiet air. Scientists still don't understand the nature of smog completely, but they call its formation a **photochemical reaction**.

Smog is worse when a layer of warm air forms in the upper atmosphere and holds the smog close to the ground, figure 14-12. This **temperature inversion** keeps cooler air from rising from

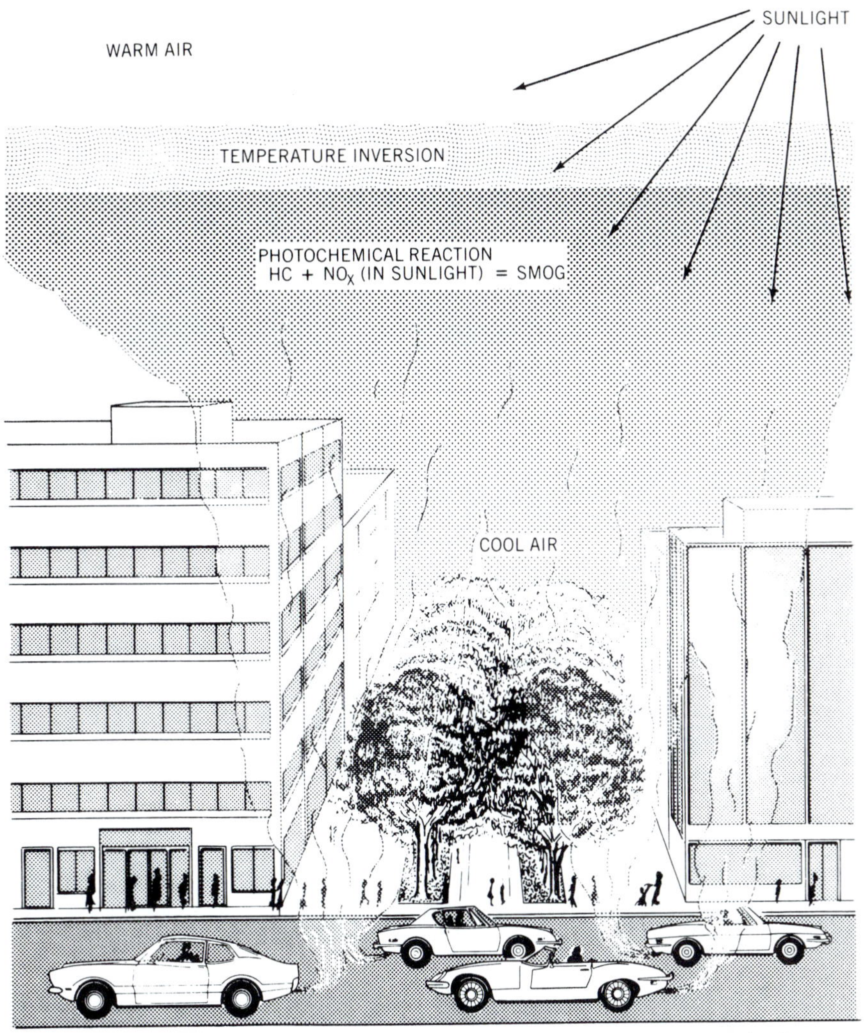


**Figure 14-12.** Smog forms from a photochemical reaction between HC and $NO_x$. A temperature inversion traps the smog at ground level.

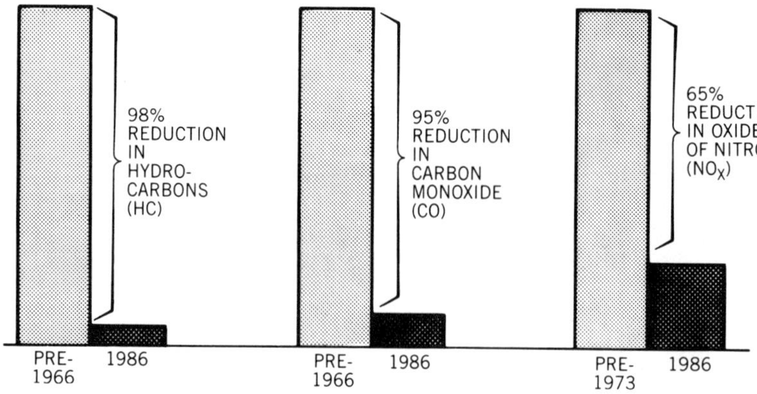


**Figure 14-13.** Automobile $NO_x$ emissions have been reduced more than half, and HC and CO emissions almost eliminated in less than 20 years.

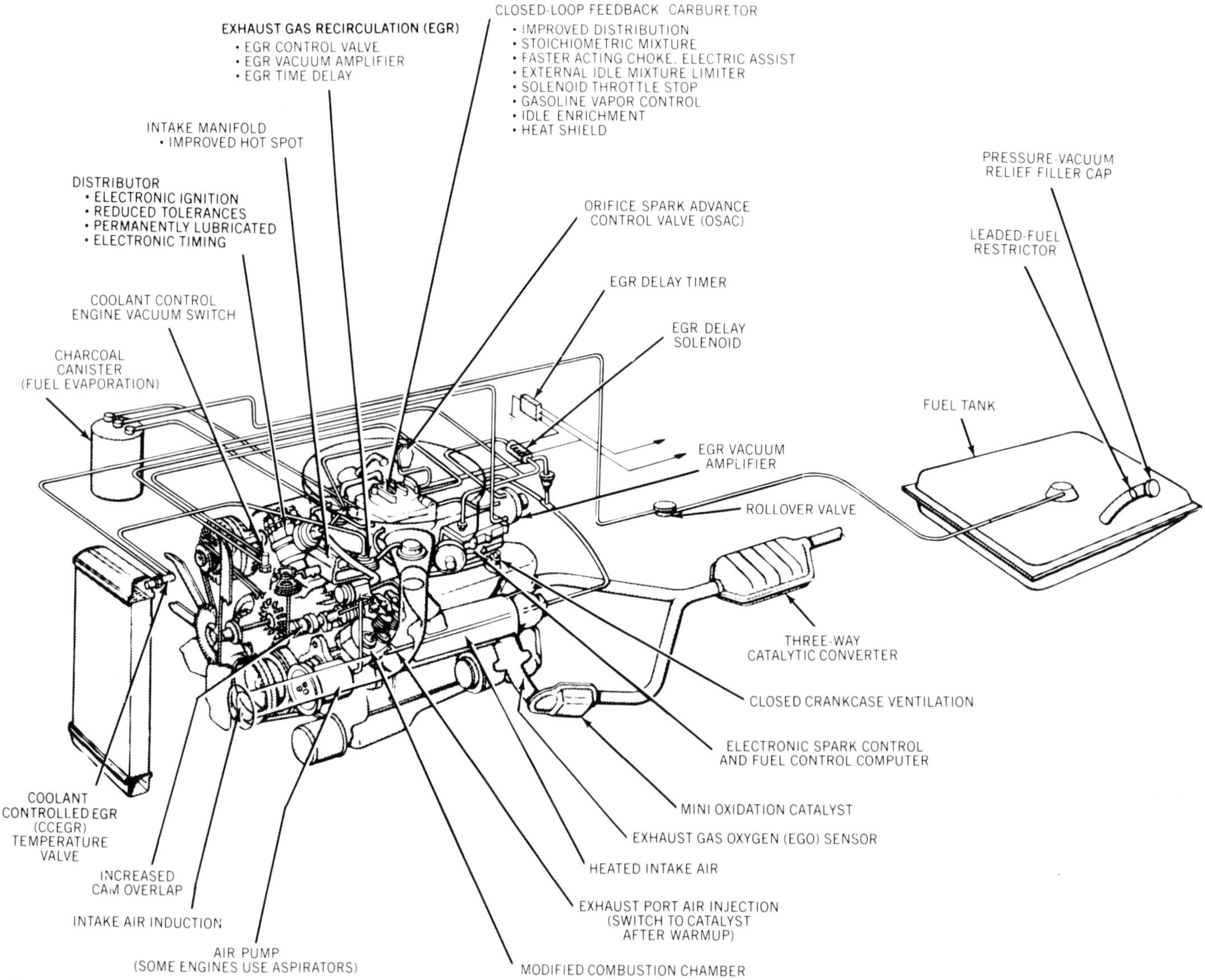


**Figure 14-14.** Emission controls on a late-model automobile are an integral part of the combustion control systems. (Chrysler)

the ground and carrying smog with it. Certain geographic and climatic conditions can create a temperature inversion within a thousand feet of the ground and act as a lid to trap smog. Temperature inversions were first noted and studied in Los Angeles, which suffered some of the worst smog pollution in the country.

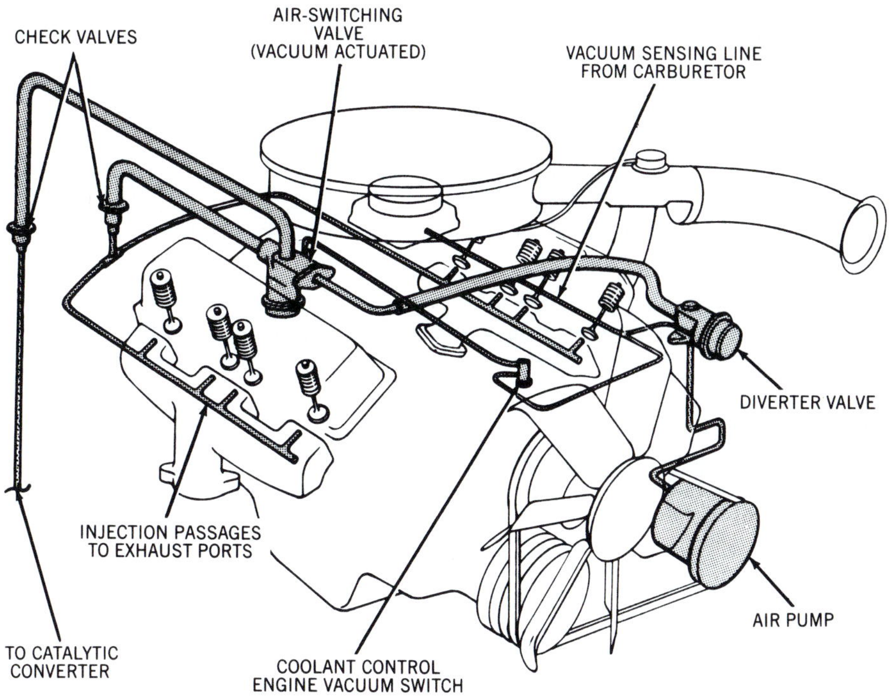


**Figure 14-15.** Air injection systems were among the first HC and CO emission controls. (Chrysler)

## AUTOMOBILE EMISSION CONTROLS

A single car emits only a small amount of pollution in daily use, but there are almost 200 million motor vehicles in the United States. Without emission controls, all of these vehicles would produce almost half of the air pollution in the country.

The automobile industry has done much to reduce motor vehicle air pollution since 1966 when exhaust emission controls were introduced. HC emissions from the engine crankcase and from fuel evaporation (about 40 percent of total HC emissions) have almost been eliminated. HC emissions from the exhaust of late-model cars are less than 1/10 the HC emissions from pre-1966 cars. Total HC emissions from all vehicle sources have been reduced more than 98 percent, and CO emissions have been reduced about 95 percent. Since 1973, $NO_x$ emissions have been reduced about 65 percent, figure 14-13.

Emission controls on modern automobiles are not a separate system, or systems. They are an integral part of an engine's combustion controls—the fuel, ignition, exhaust, lubrication, and cooling systems, figure 14-14. Emission controls, however, can be grouped into families:

**1.** *Positive crankcase ventilation* (PCV) systems that you studied in chapter 5 control HC emissions from the crankcase.

**2.** *Evaporative emission control* (EEC) systems control HC vapor emission from the fuel tank, pump, and carburetor.

**3.** *Exhaust emission controls* are parts of the fuel, ignition, and exhaust systems and other devices that control HC, CO, and $NO_x$. The principal methods used to control exhaust emissions are:

**a.** *Air injection*—Injecting extra air into the exhaust helps oxidize HC and CO to $H_2O$ and $CO_2$, figure 14-15.

**b.** *Engine modifications*—Redesigned combustion chambers and valve timing have helped reduce all three major pollutants.

**c.** *Spark timing*—Early emission systems used retarded timing to reduce HC and $NO_x$. Late-model cars have electronically controlled timing that provides the best ignition timing for emission control, economy, and performance.

**d.** *Exhaust gas recirculation*—One effective way to reduce $NO_x$ is to recirculate some exhaust to the intake air-fuel mixture, figure 14-16. Because the recirculated exhaust doesn't aid combustion, it lowers the combustion temperature, which reduces $NO_x$.

**e.** *Catalytic converters*—The first catalytic converters on 1975–76 cars helped to oxidize HC and CO. Second-generation converters, intro-

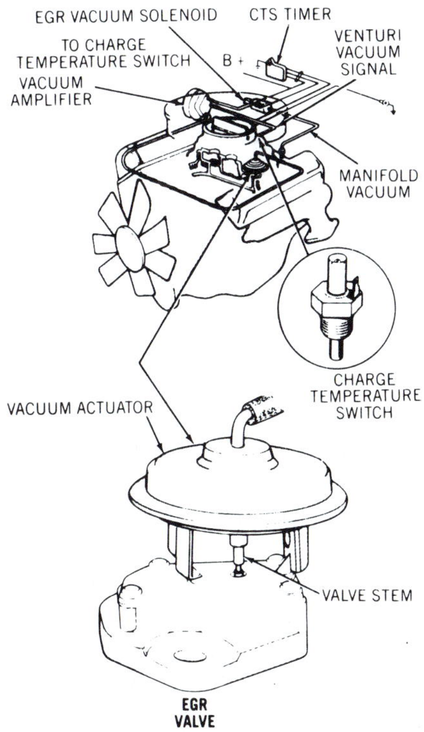


**Figure 14-16.** Exhaust gas recirculation (EGR) is one of the most effective ways to reduce $NO_x$ emissions. (Chrysler)

duced in 1977-78, chemically reduce $NO_x$ emissions.

**f.** *Electronic engine controls and closed-loop fuel systems*—On late-model cars, controls for ignition timing, air injection, exhaust gas recirculation, and fuel delivery are integrated electronically. Sensors measure oxygen content in exhaust, and a computer controls the air-fuel ratio for efficient combustion.

## EMISSION AND FUEL ECONOMY REGULATIONS

The engineering advances that have almost eliminated automobile air pollution did not happen without a lot of scientific research. While air pollution and smog have existed in many parts of the world for generations, they were first studied extensively in Southern California in the late forties and early fifties. By the early 1950's scientists proved that automobile emissions and photochemical reactions were major causes of smog.

In 1961, California required control of crankcase emissions on new cars sold in that state. Carmakers began installing PCV systems on all 1963 cars nationwide. California required the first exhaust emission controls on new 1966 cars, and these systems appeared nationwide on 1967–68 models.

Congress passed the Clean Air Act in 1963, which gave money to states for air pollution control programs. The amended Clean Air Act of 1965 required the Federal government to set standards for automobile emissions, figure 14-17. This law was supplemented by the Air Quality Act of 1967 and its amendments of 1970, 1974, and 1977. The Canadian Ministry of Transport established vehicle emission requirements, beginning with 1971 models.

The U.S. Federal Environmental Protection Agency (EPA) sets standards for all air quality control, including vehicle emissions. The California Air Resources Board (ARB) operates under authority of the EPA to set standards for that state. For many

**Fifty-Thousand-Mile Emission Standards, Gasoline-Powered Automobiles**

| *Year* | *Regulations* | *HC* | *CO* | *$NO_x$* |
|---|---|---|---|---|
| Before Controls[a] | | 850 ppm[b] (16.8 g/mi)[c] | 3.4% (125.0 g/mi) | 1000 ppm (4.0 g/mi) |
| 1966-69 | California | 275 ppm | 1.5% | none |
| 1968-69 | U.S. Federal | | | |
| 1970 | California and U.S. Federal | 4.6 g/mi | 46.0 g/mi | none |
| 1971 | California | 4.6 g/mi | 46.0 g/mi | 4.0 g/mi |
| | U.S. Federal | 4.6 g/mi | 46.0 g/mi | none |
| | Canada | 2.2 g/mi[d] | 23.0 g/mi[d] | none |
| 1972 | California | 3.2 g/mi | 39.0 g/mi | 3.2 g/mi |
| | U.S. Federal and Canada | 3.4 g/mi | 39.0 g/mi | none |
| 1973 | California | 3.2 g/mi | 39.0 g/mi | 3.0 g/mi |
| | U.S. Federal and Canada | 3.4 g/mi | 39.0 g/mi | 3.0 g/mi |
| 1974 | California | 3.2 g/mi | 39.0 g/mi | 2.0 g/mi |
| | U.S. Federal and Canada | 3.4 g/mi | 39.0 g/mi | 3.0 g/mi |
| 1975-76 | California | 0.9 g/mi | 9.0 g/mi | 2.0 g/mi |
| | U.S. Federal | 1.5 g/mi | 15.0 g/mi | 3.1 g/mi |
| | Canada | 2.0 g/mi | 25.0 g/mi | 3.0 g/mi |
| 1977-79 | California | 0.41 g/mi | 9.0 g/mi | 1.5 g/mi |
| | U.S. Federal | 1.5 g/mi | 15.0 g/mi | 2.0 g/mi |
| | Canada | 2.0 g/mi | 25.0 g/mi | 3.0 g/mi |
| 1980 | California | 0.39 g/mi | 9.0 g/mi | 1.0 g/mi |
| | U.S. Federal | 0.41 g/mi | 7.0 g/mi | 2.0 g/mi |
| 1981 | California | 0.41 g/mi | 3.4 g/mi | 1.0 g/mi |
| 1981 and later | U.S. Federal | 0.41 g/mi | 3.4 g/mi | 1.0 g/mi |
| 1982 and later | California | 0.39 g/mi | 7.0 g/mi | 0.7 g/mi |

[a]approximate estimates
[b]ppm = parts per million
[c]g/mi = grams per mile
[d]theoretical flow rate

**Figure 14-17.** Exhaust emission limits for new gasoline-powered automobiles from 1966 to the present.

model years, California emission limits have been more stringent than Federal requirements.

The oil shortages of 1973 and 1979 focused government and industry attention on improving vehicle fuel economy, as well as emission control. The Federal Energy Act of 1975 established the Corporate Average Fuel Economy (CAFE) standards for fuel mileage. These standards are administered by the EPA and the Department of Transportation (DOT).

Engineers then had two, often conflicting, goals to reach: reduced emissions and increased economy. The

**FUEL ECONOMY STANDARDS**

| *Model Year* | *MPG* | *Improvement Over 1974 (%)* |
|---|---|---|
| 1978 | 18.0 | 50 |
| 1979 | 19.0 | 58 |
| 1980 | 20.0 | 67 |
| 1981 | 22.0 | 83 |
| 1982 | 24.0 | 100 |
| 1983 | 26.0 | 116 |
| 1984 | 27.0 | 125 |
| 1985 | 27.5 | 129 |

**Figure 14-18.** Carmakers improved fleet average fuel economy 129 percent in 10 years.

1978 vehicles sold by each carmaker had to average 18 miles per gallon. The mileage standards increased steadily to a 27.5-mpg average for 1985 models, figure 14-18.

## SUMMARY

Gasoline and diesel fuel are petroleum hydrocarbons, refined and blended for performance, economy, and emission control. Additives are used to improve fuel volatility, gasoline antiknock quality (octane rating), and diesel ignition quality (cetane rating). Other additives reduce engine deposits and fuel system contamination.

Motor vehicles have been a major cause of air pollution, largely from the fuel they burn. The major automotive air pollutants are HC, CO, and $NO_x$. Since automobile emission controls were introduced over 20 years ago, automobile pollutants have been reduced 65 to 98 percent.

Emission controls on modern automobiles are not separate "add-on" devices, but integral parts of the combustion control systems. Major emission control methods include PCV systems, air injection, spark timing controls, exhaust gas recirculation (EGR), ignition timing controls, catalytic converters, redesigned engines, and electronic engine controls with closed-loop fuel systems.

Fuel economy regulations of the mid-1970's created the doubly difficult goals of increasing mileage while reducing emissions.

## REVIEW QUESTIONS

### Multiple Choice

**1.** During distillation the hydrocarbons of gasoline and diesel fuel:

- **a.** turn to liquid
- **b.** vaporize
- **c.** stabilize into solids
- **d.** recombine into oxides

**2.** Mechanic A says that the hydrocarbons of gasoline vaporize at lower temperatures than the hydrocarbons of diesel fuel. Mechanic B says that the hydrocarbons of gasoline vaporize at higher temperatures than those of diesel fuel. Who is right?

- **a.** A only
- **b.** B only
- **c.** both A and B
- **d.** neither A nor B

**3.** Thermal cracking is a process in which:

- **a.** distillation of fuels takes place
- **b.** a catalyst is introduced
- **c.** heat is used to break large hydrocarbon molecules into smaller ones
- **d.** hydrogen is used to break large hydrocarbon molecules

**4.** One British thermal unit is:

- **a.** the heat energy needed to raise the temperature of one pound of water two degrees Fahrenheit
- **b.** the unit of rating applied to the potential heat stored in gasoline
- **c.** the heat energy needed to raise the temperature of one pound of water one degree Fahrenheit
- **d.** the heat energy needed to raise the temperature of one pound of water five degrees Fahrenheit

**5.** Refiners use different fuel blending methods to:

- **a.** achieve the desired volatility
- **b.** remove impurities
- **c.** improve a gasoline's octane rating
- **d.** all of the above

**6.** Mechanic A says that volatility is a measurement of how easily a liquid vaporizes. Mechanic B says that the volatility of all liquids is affected by temperature and altitude. Who is right?

- **a.** Mechanic A only
- **b.** Mechanic B only
- **c.** Both A and B
- **d.** Neither A nor B

**7.** Mechanics A and B live in a cold climate. Mechanic A says that a gasoline blended to have low volatility should be used. Mechanic B says that the gasoline must have high volatility. Who is right?

- **a.** Mechanic A only
- **b.** Mechanic B only
- **c.** Both A and B
- **d.** Neither A nor B

**8.** Water and sulfur can combine to produce:

- **a.** hydrochloric acid
- **b.** sulfates
- **c.** peroxide
- **d.** sulfuric acid

**9.** A vehicle carburetor shows excessive gum and varnish buildup. Mechanic A says that the fuel filter should be changed more often. Mechanic B says that the customer could use an oxidation inhibiting additive to reduce the buildup. Who is right?

- **a.** Mechanic A only
- **b.** Mechanic B only
- **c.** Both A and B
- **d.** Neither A nor B

**10.** If engine cylinder temperatures rise above the autoignition temperature:

- **a.** a controlled explosion will occur
- **b.** the air-fuel mixture will ignite too early
- **c.** the fuel will not mix with the air
- **d.** combustion efficiency will be improved

**11.** Mechanic A says that detonation occurs after the time of ignition. Mechanic B says that preignition occurs before the time of ignition. Who is right?

- **a.** Mechanic A only
- **b.** Mechanic B only
- **c.** Both A and B
- **d.** Neither A nor B

### Fill in the Blank

**12.** High ________ quality ensures stable combustion.

**13.** Gasoline octane rating is measured as ________ and ________.

**14.** The octane rating most commonly seen on gasoline pumps is often called the ________ rating.

**15.** Lead used as a fuel additive is given off in engine exhaust as a ________.

**16.** Low engine compression ratios help reduce ________.

**17.** Dispersants are used as fuel additives to extend PCV valve life and ________ ________.

**18.** The least volatile diesel fuel is grade __________.

**19.** The __________ is a measure of the ignition quality of diesel fuel.

**20.** The point at which wax crystals start to form in diesel fuel is called __________.

**21.** An internal combustion engine emits three major pollutants: __________, __________, and __________.

**22.** Diesel engines emit more __________ and *sulfer oxides* than do automobile engines.

**23.** Engine exhaust, crankcase vapors, and __________ are the three major sources of hydrocarbon emission.

**24.** Carbon monoxide is a result of __________.

**25.** An automobile without emission controls can produce up to __________ pounds of hydrocarbons and 2300 pounds of __________ per 1,000 gallons of gasoline.

**26.** Smog is the result of the combination of hydrocarbons, sunlight, and __________.

**27.** The three major families of emission controls are the __________, the __________, and the __________.

**28.** The __________ sets standards for vehicle emissions.

**29.** Gasoline is approximately 86 percent __________ and 14 percent __________ by weight.

**30.** __________ is the blending of hydrogen with gasoline or diesel in order to further refine it.

# 15 FUEL STORAGE, DELIVERY, AND EVAPORATION CONTROL

## WHAT

From the time that crude oil leaves the ground, petroleum products are pumped, stored, transported, and handled. The fuel storage and delivery components on an automobile are important parts of the fuel system. This chapter explains the design, operation, and importance of the:

- Fuel tanks and fillers
- Fuel lines
- Fuel pumps
- Fuel filters
- Fuel gauges
- Air cleaners, filters, and temperature controls
- Evaporative emission controls.

## WHY

All of the fuel storage and delivery components work together as a system to provide a steady supply of clean fuel and air for combustion. Some parts of the fuel system, such as tanks and lines, are simple and need little service. Other parts, such as pumps, filters, and evaporation controls, are more complicated and require periodic maintenance. Whether these parts are simple or complex, each one can affect engine performance. By understanding the purpose and operation of each part, and how all parts work together, you can do a better job of servicing the complete fuel system.

## GOALS

After studying this chapter, you should be able to:

**1.** Explain how the fuel storage and delivery system operates and relates to other vehicle systems.

**2.** Explain the construction, installation, and operation of fuel tanks, lines, pumps, filters, evaporation controls, and air filters.

**3.** Explain the reasons and methods for air intake temperature control.

**4.** Explain the purpose and operation of an evaporative emission control (EEC) system.

## FUEL TANKS AND FILLERS

Most automobile fuel tanks are made of corrosion-resistant steel. The two halves of a tank are formed with ribs for strength and welded together, figure 15-1. Most automobiles have a single tank under the rear of the body, just in front of, or behind, the rear axle. Tanks are secured to vehicle bodies or frames with bolts, nuts, and steel straps, figure 15-2. Some station wagons have a tank on one side of the vehicle, between the inner and outer rear fender panels. Many trucks have fuel tanks under the middle of the frame on either side. Some trucks have two interconnected tanks, figure 15-3, and a few have a second or third auxiliary tank.

Tank design and location are always a compromise with available space, filler location, and the design of the rest of the vehicle. Tank shape, capacity, and location can cause fuel to slosh in the tank as the vehicle moves. Therefore, many tanks have internal baffles to keep fuel from sloshing away from the pickup line, figure 15-1. All fuel tanks include:

- A fuel inlet, or filler, and a filler cap
- A pickup line, or outlet to the fuel lines
- A vent system
- Rollover leakage protection (since 1976).

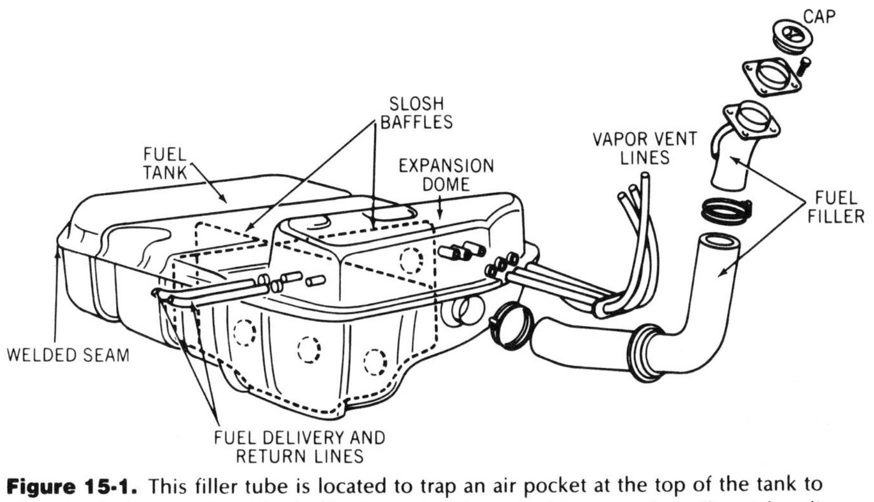


**Figure 15-1.** This filler tube is located to trap an air pocket at the top of the tank to allow for fuel expansion. The baffles keep fuel from sloshing away from the pickup line.

### Fuel Fillers

The fuel filler is a large tube that runs from the tank to an opening on the outside of the vehicle. Fillers on some

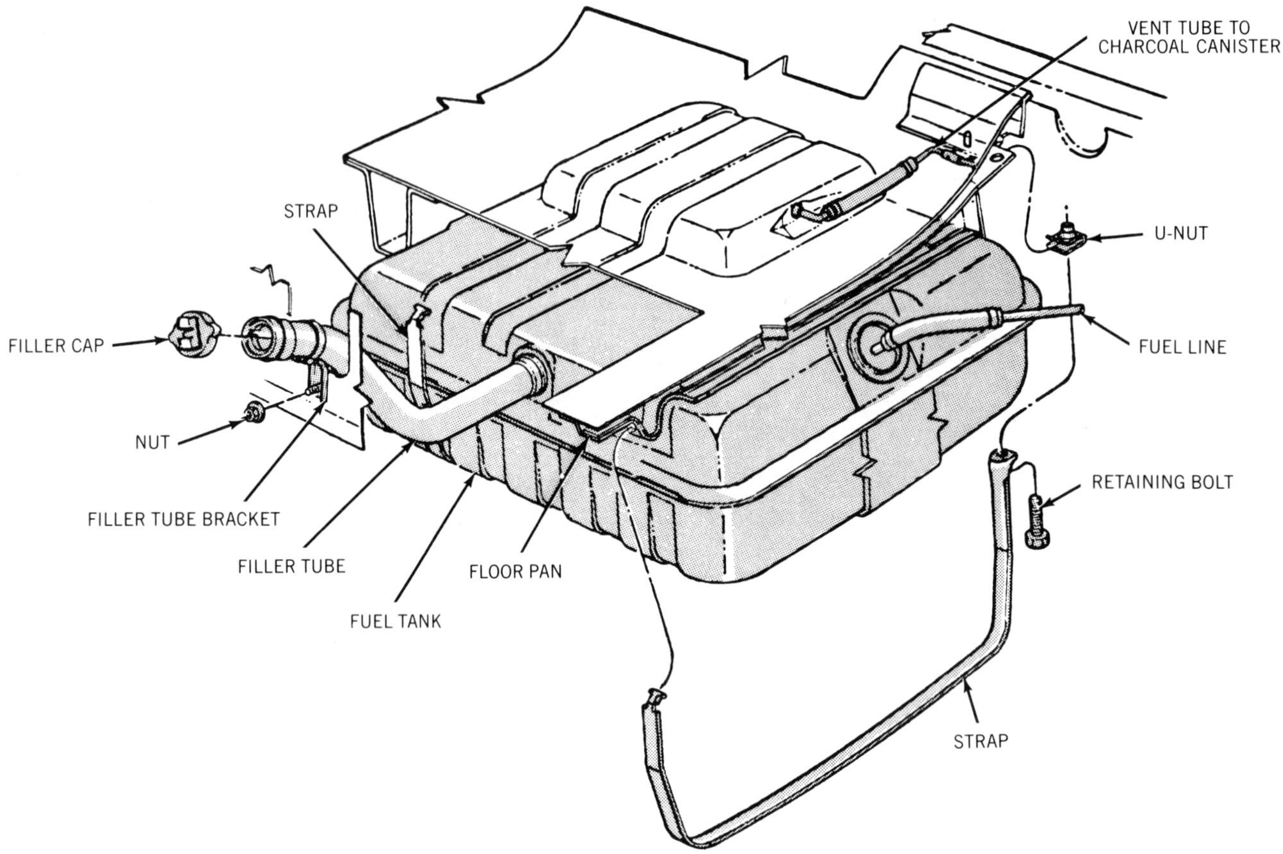


**Figure 15-2.** Fuel tanks on most automobiles are mounted on retaining straps at the rear of the car. (Chrysler)

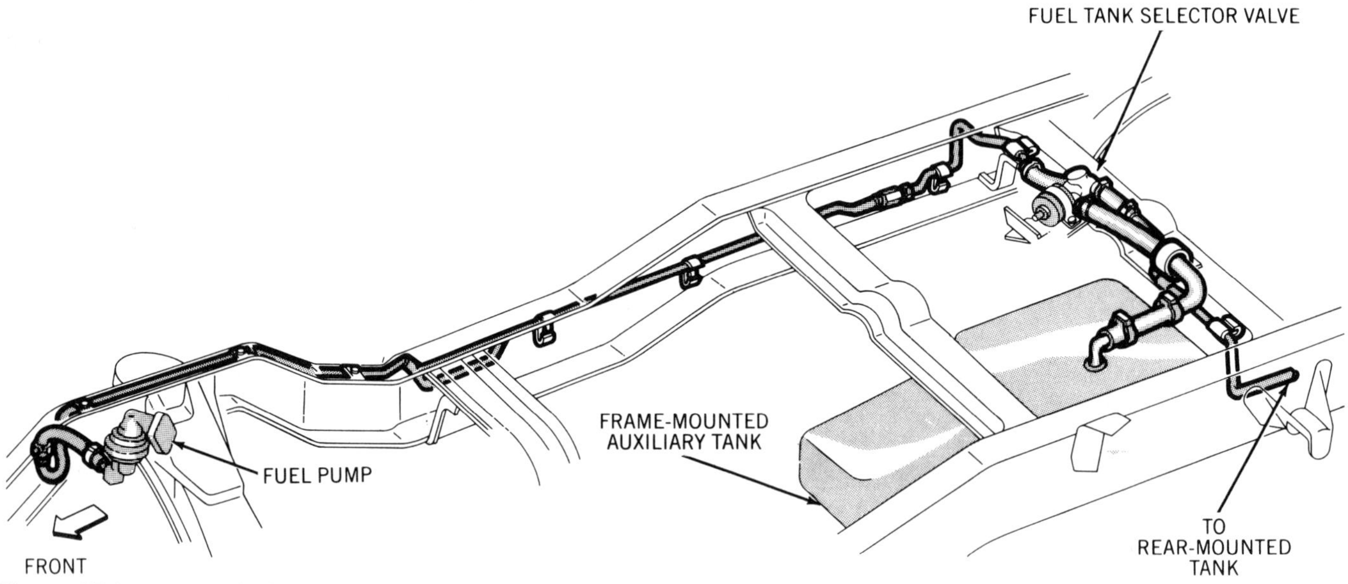


**Figure 15-3.** Some trucks have auxiliary fuel tanks near the midpoint of the vehicle. (Chrysler)

tanks trap a pocket of air in the top of the tank to allow for fuel expansion.

Federal law requires that all vehicles with catalytic converters must use unleaded fuel. All 1975 and later vehicles that need unleaded gasoline have a filler tube with a smaller inlet opening than the opening on cars that can use leaded gasoline, figure 15-4. The small filler opening matches the small diameter of the nozzles on unleaded gasoline pumps. Most fillers for unleaded gasoline also have a spring-loaded flapper to help prevent pouring leaded gasoline into the tank.

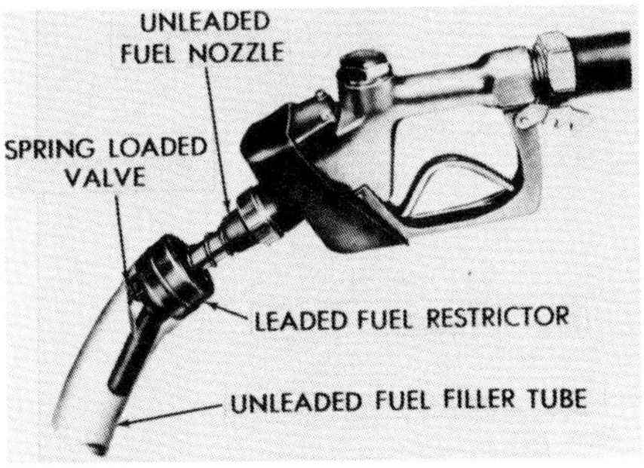


**Figure 15-4.** Vehicles requiring unleaded gasoline have restrictors in the fuel filler tubes to allow only small, unleaded fuel nozzles to enter. (Chrysler)

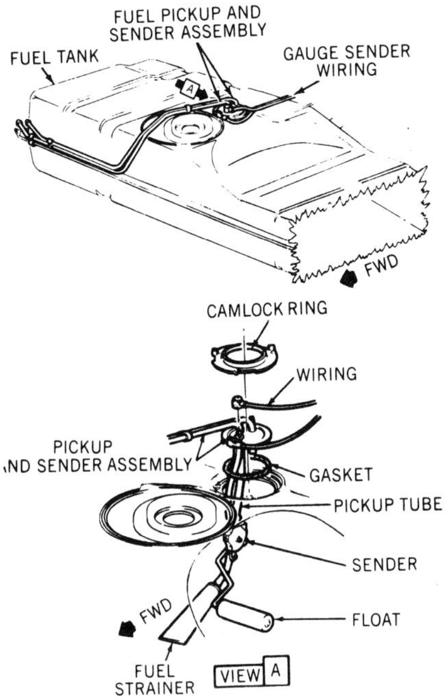


**Figure 15-5.** The fuel gauge sender unit usually is mounted on or near the pickup tube in the fuel tank.

## Fuel Pickup Line

In most fuel tanks, the inlet to the fuel pickup line is slightly above the bottom of the tank to avoid drawing dirt and sediment into the rest of the fuel system. Fuel pickup lines have strainers or filters to keep coarse dirt and sediment out of the lines, figure 15-5. If the tank collects a lot of dirt, the strainer can get clogged and restrict fuel flow. Some fuel systems have electric fuel pumps as part of the fuel pickup assembly in the tank. You will learn more about them later in this chapter.

The fuel gauge sender unit usually is mounted on the pickup tube, figure 15-5. It is a float that rises and falls with the fuel level. Its operation is explained in the fuel gauge section of this chapter.

## Fuel Tank Vent Requirements

As the fuel level drops in the tank, the available air expands to fill the space. This lowers the air pressure and can create a vacuum as tank air pressure drops below atmospheric pressure. If tank air pressure gets low enough, it can create a **vacuum lock** that will keep fuel from being pumped out. In severe cases, this vacuum lock could cause atmospheric pressure to collapse the tank. Therefore, all tanks must be vented to allow more air in as the fuel level drops.

Before 1970, gasoline tanks were vented with a length of tubing open to the atmosphere, or through a vented filler cap. These methods, however, allowed HC vapors to enter the atmosphere. On cars built since 1970 or 1971 with evaporative emission control (EEC) systems, the tanks are vented to the EEC vapor storage canister, figure 15-6. Caps for these tanks may be labeled, "unvented" or "nonvented."

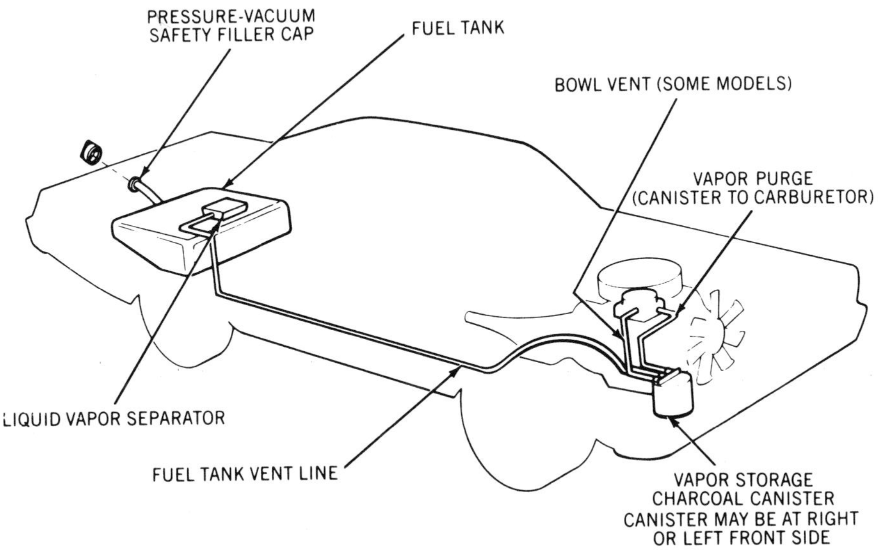


**Figure 15-6.** Gasoline tanks on cars built since 1970 or 1971 are vented to the EEC system vapor canister. (Chrysler)

### DIAGNOSTIC COMPUTERS THEN AND NOW

Much of our modern automobile service and test equipment is based on computer-controlled electronic systems that display and print test results and store vehicle specifications for comparison with test readings. Electronic engine analyzers, wheel alignment equipment, and other devices have made our jobs easier and more accurate. But, the development of these systems didn't happen overnight. It was a long, hard job for engineers to get them to work the way mechanics needed them to work.

The first diagnostic computers were tried in the service business over a decade ago. Many of them didn't last, but they laid the foundation for the systems we use today. In 1972, Pontiac introduced a computer-assisted diagnostic program for its dealer mechanics. Called the PAR-1 program, it allowed mechanics to make a specified series of tests on a car's engine and then use a teletype terminal to send the results to a central computer. If any of the test steps produced a "no" result, the computer told the mechanic several possible cures for the problem.

The idea was good, but it didn't catch on. Because a mechanic had to go to a computer room and communicate with a central computer by teletype lines, it took him out of the service bay and away from the job. When engineers were able to build small powerful computers into a piece of test equipment right in the service bay, computers became practical automobile service tools.

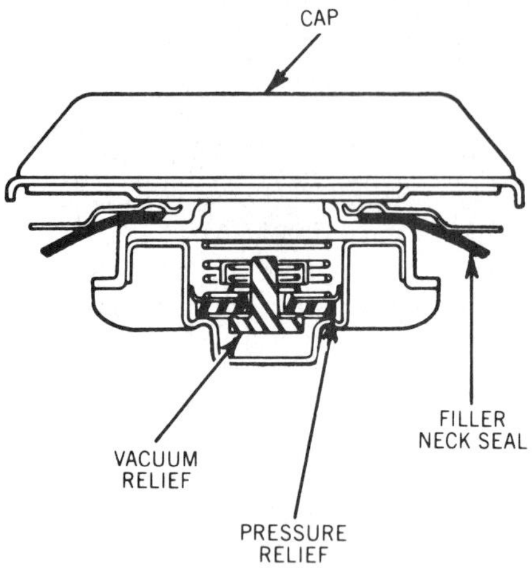


**Figure 15-7.** Fuel tank caps for most cars with EEC systems have pressure and vacuum relief valves. (Ford)

Many, however, have pressure and vacuum relief valves that open to relieve pressures above 0.5 to 1.0 psi and vacuum greater than 0.5 to 1.0 inch of mercury, figure 15-7. Other caps are completely sealed, particularly on early Ford and some GM systems. These systems have separate pressure and vacuum relief valves for tank venting.

### Rollover Leakage Protection

U.S. Federal law requires that all 1976 and later cars have devices to prevent fuel leakage in case the vehicle rolls over. Most of these devices are check valves in fuel lines and vapor lines and in filler caps, figure 15-8. Additionally, GM uses spring-loaded check valves in carburetor inlet fuel filters. Several carmakers use return lines from fuel pumps to tanks to prevent leakage. Other manufacturers have redesigned fuel pumps and carburetor vents for leak prevention.

## FUEL LINES

Rigid lines and flexible hoses carry fuel from tank to pump to carburetor or injection system. Other lines and hoses return excess fuel to the tank and carry gasoline vapors through the EEC system, figure 15-9.

Rigid fuel lines are made of seamless steel tubing to resist vibration and other damage. Whenever a metal fuel line is replaced, only steel tubing must be used. Copper and aluminum tubing must never be substituted. These softer metals will work-harden and leak with excessive vibration.

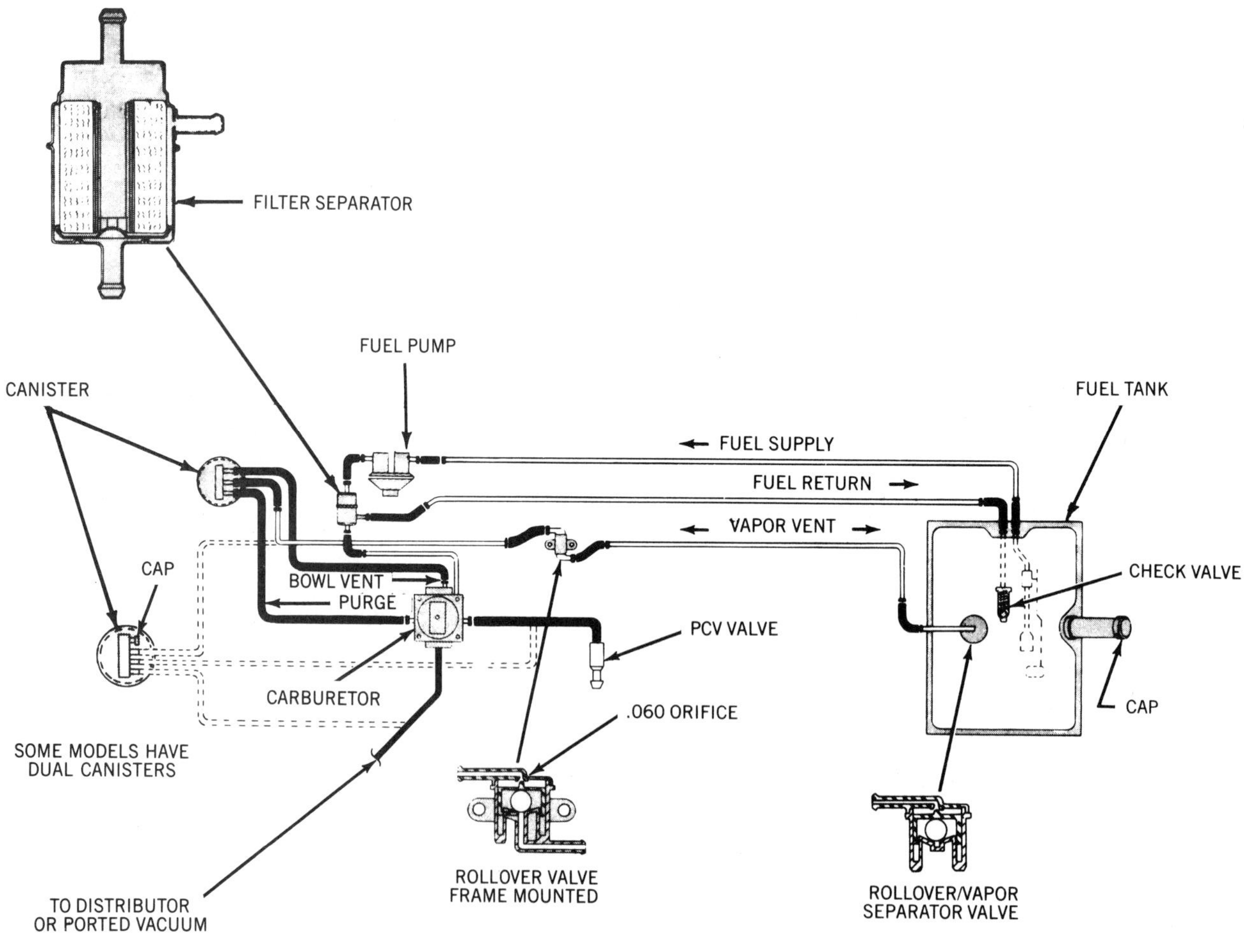


**Figure 15-8.** This fuel system has several check valves to prevent fuel leakage in case the car rolls over. (Chrysler)

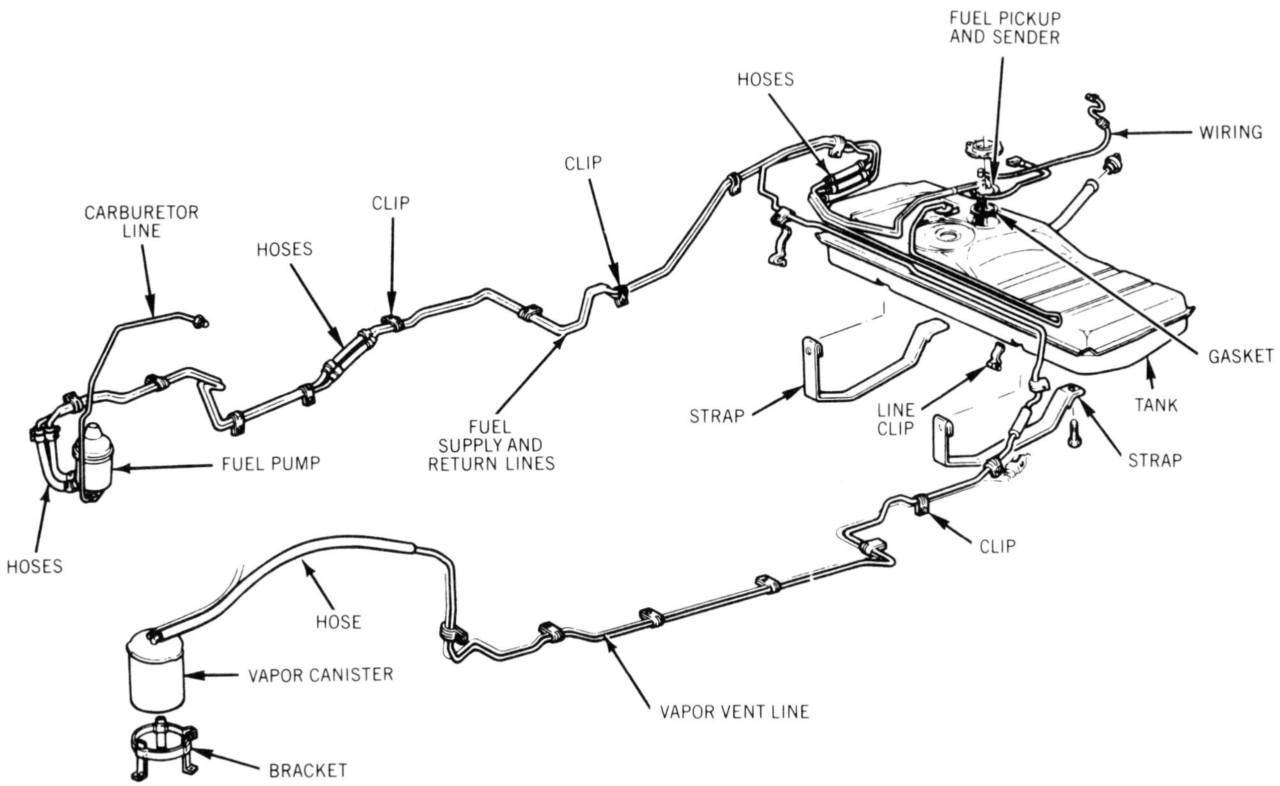


**Figure 15-9.** Steel tubing and rubber hoses carry liquid fuel and fuel vapors from the tank to the engine and the EEC vapor canister. (Chevrolet)

Special rubber hose is used for fuel lines where flexibility is needed, such as between rigid lines on the car body and the engine. Fuel system hoses are made of fuel-resistant material. Ordinary rubber, such as used for vacuum and water hoses, softens and deteriorates when exposed to gasoline or diesel fuel. Vapor vent lines and PCV hoses are made of similar fuel-resistant material. Vapor hoses often are marked EVAP to show their intended use.

Most fuel delivery tubing and hoses are 5/16- or 3/8-inch sizes. Fuel return lines are usually 1/4-inch-diameter tubing and hose. Fuel lines from the tank to the engine are mounted under the car body or inside frame rails and secured by screws and clamps or clips, figure 15-9. Fuel lines are routed away from moving drive train and suspension parts and protected from possible road damage or damage from hoists and jacks. They also are routed away from exhaust systems and other heat sources to prevent vapor lock.

## FUEL PUMPS

All fuel pumps deliver fuel by mechanical pumping action that creates low pressure (suction) at the pump inlet and high pressure at the outlet. Fuel pumps are operated either mechanically or electrically.

### Mechanical Fuel Pumps

Mechanical fuel pumps are diaphragm pumps, mounted on the engine and operated by an **eccentric** lobe on the camshaft or an auxiliary engine shaft. Each revolution of the eccentric moves the pump through one operating stroke. The pump rocker arm can be driven directly by the eccentric; or, as on the Chevrolet small-block V-8 and V-6, by a pushrod between the eccentric and the pump rocker arm, figure 15-10.

A diaphragm pump has a fuel chamber with a movable diaphragm, a diaphragm spring, and inlet and outlet check valves. It delivers fuel in intermittent pressure pulses. As the camshaft eccentric rotates, it pushes one end of the rocker arm, which causes the other end of the arm to pull the diaphragm rod. As the diaphragm moves, it creates low pressure (suction) in the chamber. Atmospheric pressure on the fuel in the tank forces fuel through the lines and into the pump chamber past the inlet check valve, figure 15-11.

As the camshaft eccentric continues to rotate, it releases pressure on the rocker arm and diaphragm rod. The diaphragm spring forces the diaphragm toward the fuel chamber, which increases pressure in the chamber. This pressure closes the inlet check valve, opens the outlet check valve, and forces fuel out of the pump. The outlet check valve maintains pressure in the outlet line and keeps fuel from flowing back to the pump.

A diaphragm pump operates steadily whenever the engine is running, but its delivery rate is controlled by the carburetor or injection system. We

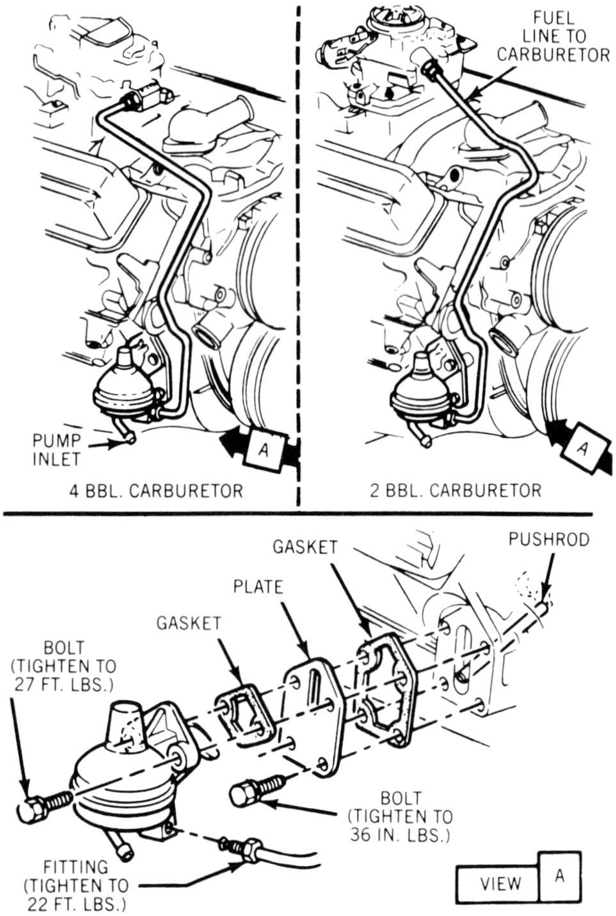


**Figure 15-10.** This Chevrolet fuel pump installation is typical of pump locations on V-type overhead-valve engines. Notice the pushrod between the pump rocker arm and the camshaft. (Chevrolet)

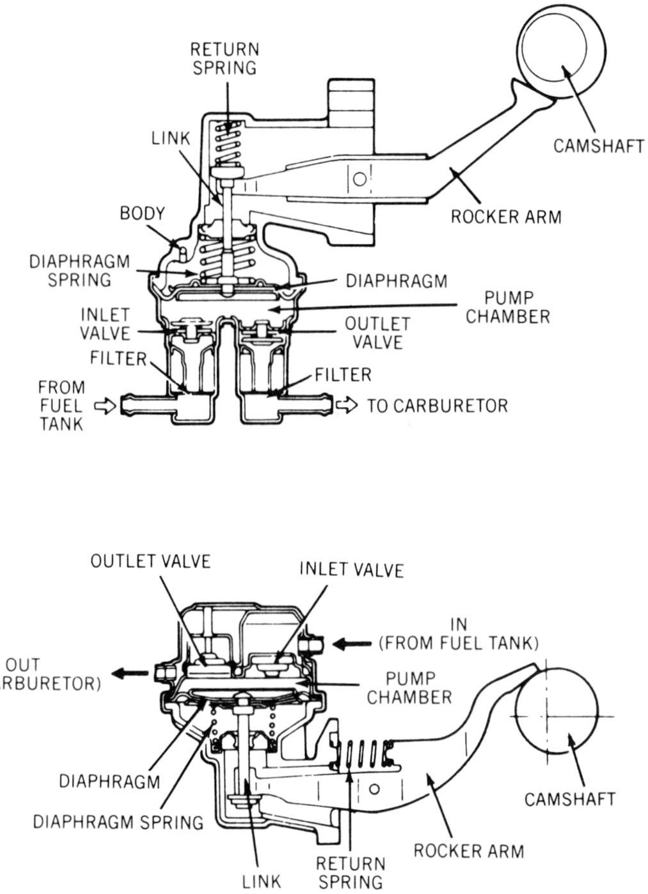


**Figure 15-11.** Typical mechanical, diaphragm-type fuel pump operation. (Chrysler) (Mitsubishi)

can explain how this works by using a carburetor as an example.

A carburetor must maintain a regulated supply of gasoline in its fuel bowl. This is controlled by the carburetor inlet needle valve and float. When the needle valve is open, fuel flows from the pump into the carburetor bowl. As fuel fills the bowl, it raises the float and closes the needle valve. This causes pressure to increase in the line from the pump to the carburetor. When pressure is equal in the line and the pump chamber, it keeps the pump diaphragm spring from moving the diaphragm, figure 15-12. The diaphragm stays in the retracted, "up," position to which it was moved by the rocker arm. A slot in the diaphragm rod or the rocker arm allows the rocker arm to move back and forth without any force on the diaphragm.

As the carburetor uses gasoline from the fuel bowl, the float drops and the needle valve opens. This reduces pressure in the pump chamber and lines and lets the diaphragm pump more fuel to the carburetor. The fuel level in the carburetor varies for different operating conditions, and the needle valve position varies between fully open and closed. Therefore, fuel flow from the pump to the carburetor is always proportional to fuel level in the carburetor bowl and to the rate of flow from the carburetor to the engine.

When the engine is off, the pump diaphragm spring and the outlet check valve maintain fuel pressure in the pump chamber and the lines to the carburetor. If engine heat expands the fuel in the lines, fuel pressure can force the carburetor inlet open and overfill the carburetor bowl, flood the intake manifold, and make the engine hard to start. Gasoline in the lines also may vaporize and cause a vapor lock between the pump and carburetor. Fuel systems use several devices to eliminate the problems of flooding and vapor lock while still maintaining fuel line pressure:

- Some fuel pumps have small holes in the inlet and outlet check valves, to allow pressure to bleed back to the pump inlet line.
- Some systems have a vapor separator between the pump and the carburetor. The carburetor draws liquid from the bottom of the separator. Vapors at the top of the separator return to the tank through a return line, figure 15-13.
- Some inline fuel filters (which you will study in the next section) have a vapor bypass outlet with a return line

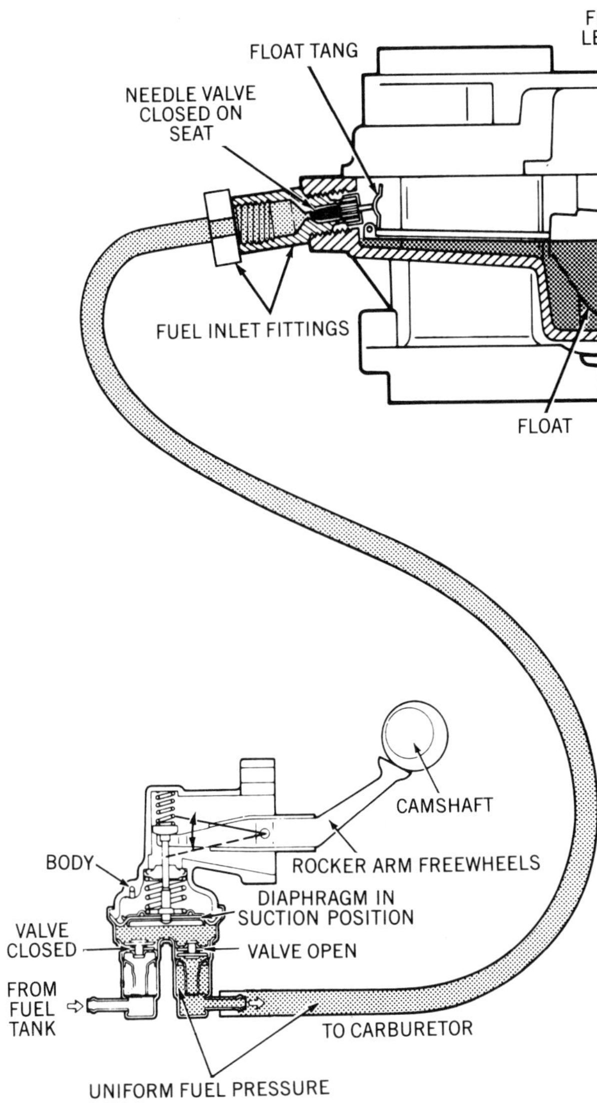


**Figure 15-12.** Pump delivery rate is controlled by the carburetor inlet needle valve. When the valve is closed, pressure in the lines and pump chamber overcomes diaphragm force. (Chrysler)

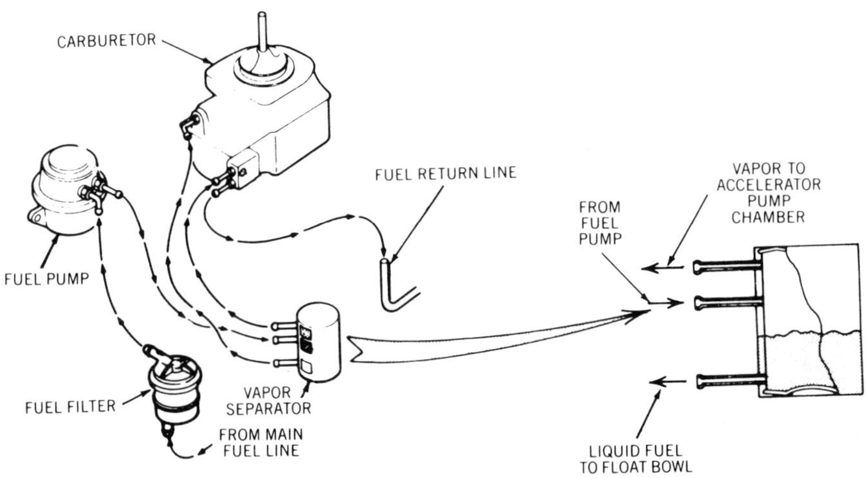


**Figure 15-13.** This vapor separator relieves excessive pressure in the line between the fuel pump and carburetor. It returns gasoline vapors to the tank. (Chrysler)

### THE UNIVERSAL ENGINE

We don't know if anyone keeps score on which automobile engine has been the most widely used for the longest time. But, there are several basic engines built by different carmakers that deserve a place in history for their long production runs of millions of units. Some votes for the alltime universal engine might go to Ford's 4-cylinder Model T engine, built from 1909 through 1927. Another candidate for universal engine would be Ford's venerable flathead V-8, produced from 1932 through 1953. Ford built millions, and they powered everything from passenger cars to trucks to racecars and even a few airplanes. Or, how about Volkswagen's flat 4-cylinder engine? Designed in the late 1930's, it was built in volume from the late 1940's on into the 1980's in several different countries.

One candidate for the world's most universal engine that you have all seen and probably worked on is the famous Chevrolet small-block V-8. It's important because it continues in volume production today, more than 30 years after it was introduced in 1955. Once called "Ed Cole's hotrod" after the man who was Chevrolet General manager when it was designed and later GM President, the small-block Chevy has powered tens of millions of cars, trucks, and championship racecars during its history.

Over the years, the Chevy small-block has been built in displacements ranging from 262 to 400 cubic inches. Here in chronological order by year of introduction are the various sizes of small Chevy V-8's:

| | |
|---|---|
| 265-cid | 1955–57 |
| 283-cid | 1957–67 |
| 327-cid | 1962–69 |
| 302-cid | 1967–69 |
| 350-cid | 1967 to the present |
| 307-cid | 1968–73 |
| 400-cid | 1970–75 |
| 262-cid | 1975–76 |
| 305-cid | 1976 to the present |

If you count up horsepower and engine code variations, you'll wind up with 60 or 70 different versions of this universal engine.

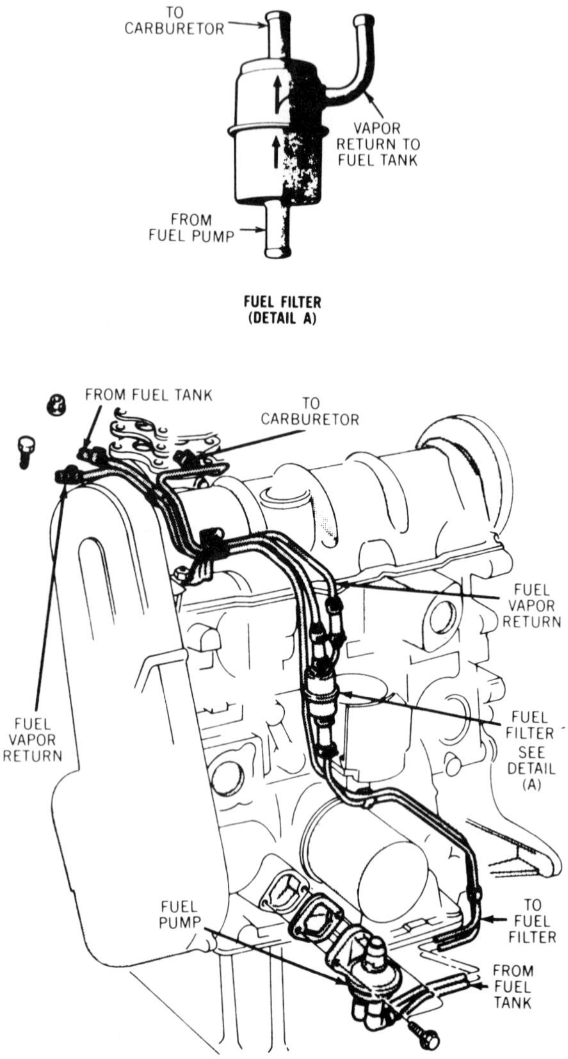


**Figure 15-14.** This inline fuel filter has a vapor bypass nipple that returns fuel vapors through a line to the tank. (Chrysler)

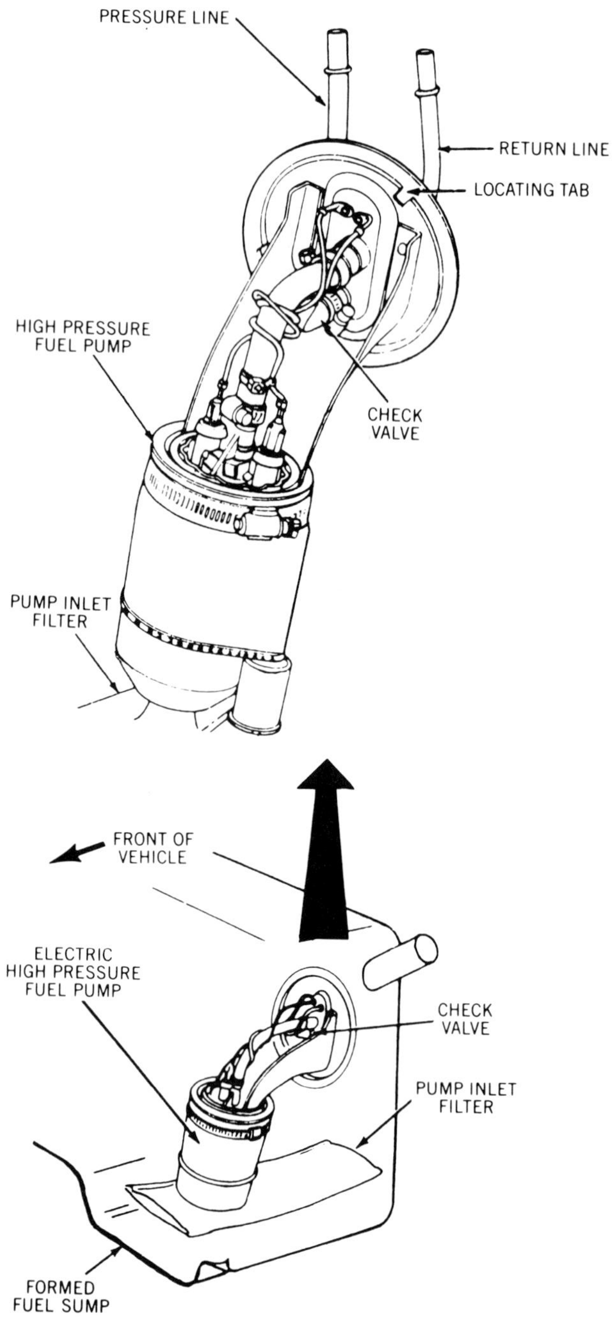


**Figure 15-15.** Many electric fuel pumps are mounted inside the fuel tanks. (Ford)

to the tank, figure 15-14.

• Some fuel pumps have an outlet air chamber that separates vapors from liquid fuel and returns the vapors to the tank.

## Electric Fuel Pumps

Electric fuel pumps are more expensive than mechanical pumps and depend on both correct electrical operation and precise mechanical pumping to deliver fuel. Even though more complicated, electric pumps have two major advantages:

**1.** Because they don't depend on a camshaft eccentric for operation, they can be located anywhere in the fuel system from the tank to the engine. In fact, they work best when located close to the tank because they maintain a constant positive pressure on all of the fuel in the lines. Some fuel injection systems have two electric pumps in series to deliver a constant volume of gasoline at 30- to 40-psi pressure.

**2.** An electric pump delivers fuel as soon as current is supplied to the armature or motor. This allows the pump to pressurize the fuel lines and overcome vapor lock before the engine cranks.

There are two basic kinds of elec-

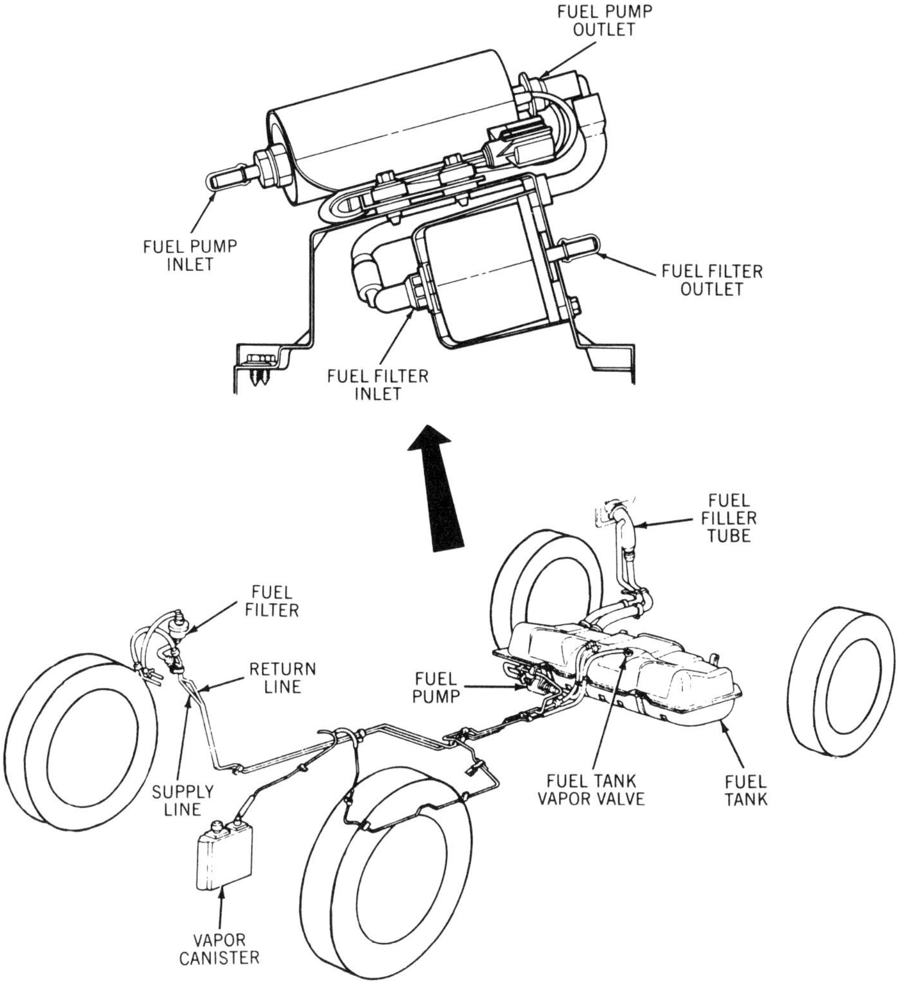


**Figure 15-16.** Some electric fuel pumps are mounted under the rear of the car body or on the frame near the tank. (Ford)

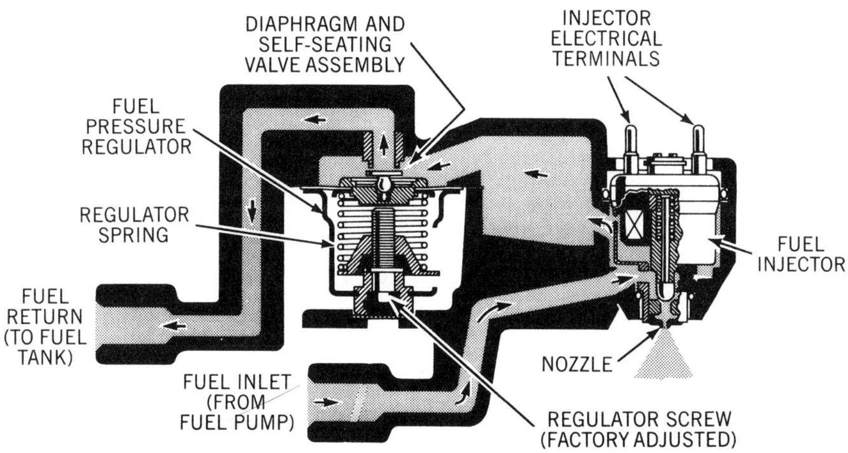


**Figure 15-17.** Gasoline fuel injection systems have pressure regulators that return excess fuel to the tank when pressure rises above the desired maximum. (Rochester)

tric fuel pumps: rotary electric pumps and vibrating-armature pumps. Rotary pumps are more common as original equipment.

**Rotary Electric Pumps.** Almost all original equipment electric fuel pumps are positive-displacement pumps that use an electric motor to turn a roller-vane or impeller pumping mechanism. Many such electric pumps are inside the tank as part of the fuel pickup assembly, figure 15-15. Others are mounted on the outside of the tank or on the rear of the frame near the tank, figure 15-16. Located close to the fuel supply, these pumps provide positive pressure to all parts of the fuel system, which is essential for fuel injection systems.

Because a rotary pump delivers a constant volume of fuel with each revolution, pressure can build up in the lines if the pump delivers more fuel than the carburetor or injection system needs. Some pumps have internal relief valves that return excess fuel to the tank if pressure exceeds a preset limit. Most gasoline injection systems have pressure regulators that limit pressure and return excess fuel to the tank through a return line, figure 15-17.

Rotary electric pumps receive current through two parallel circuits, depending on whether the engine is cranking or running. When the engine is cranking, the pump receives current through the starting contacts of the ignition switch or the starter relay, figure 15-18. As the engine starts, an oil pressure switch opens the starting circuit to the pump and closes a set of "run" contacts, figure 15-18. As long as the engine is running, the pump receives current through the run contacts of the ignition switch and the oil pressure switch. When the engine stops or oil pressure drops below about 2 psi, the pressure switch opens the run contacts in the pump circuit and the pump stops. This kind of control circuit keeps the electric pump from flooding the fuel system when the engine stops or develops low oil pressure for any reason.

Many late-model electric fuel pump installations are controlled by relays,

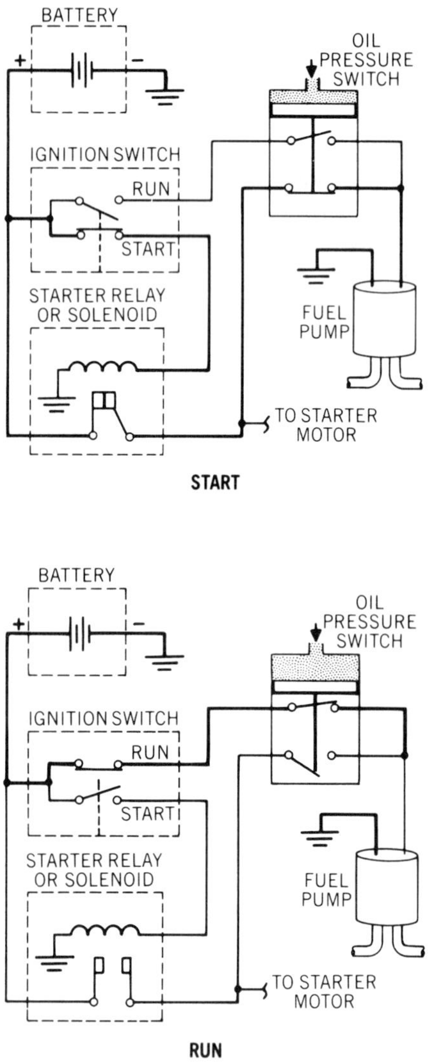


**Figure 15-18.** The fuel pump receives current during cranking through the ignition switch, the starter solenoid, and the normally closed contacts of the oil pressure switch. When the engine is running, it receives current through the ignition switch and the oil pressure switch as long as oil pressure is above a minimum level.

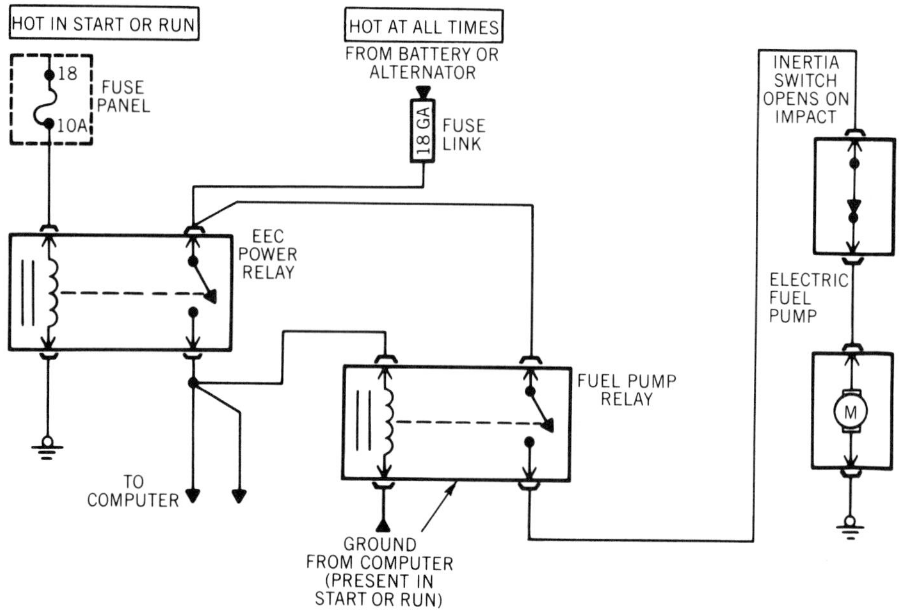


**Figure 15-19.** Many late-model electric fuel pumps are controlled through a pump relay. (Ford)

figure 15-19. When the engine starts, relay control current comes from the start contacts of the ignition switch or through an EEC computer. When the engine is running, relay control current comes from the alternator output current. If the alternator stops turning for any reason, the relay opens, and the fuel pump stops. The system also can be wired so that the computer deenergizes the fuel pump relay when the engine is off. Many systems have an inertia switch or other safety switch that opens the fuel pump circuit in case of an accident, figure 15-19.

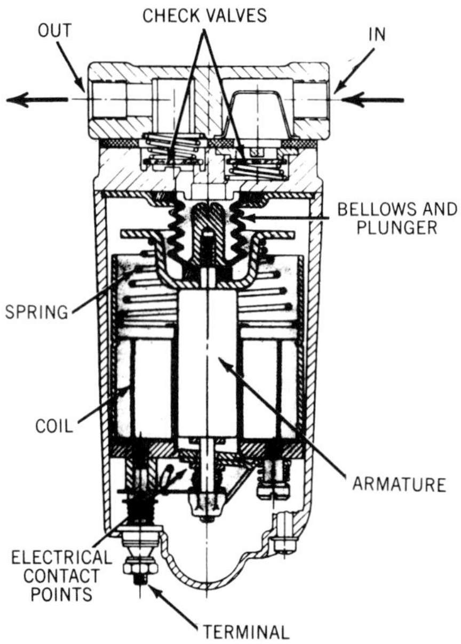


**Figure 15-20.** A vibrating armature electric fuel pump works similarly to a mechanical diaphragm pump.

**Vibrating Armature Pumps.** A vibrating armature pump operates in pulses, similarly to a mechanical diaphragm pump. This kind of pump has an electromagnetic coil with a movable armature attached to a diaphragm, a plunger, or a metal bellows. When current is applied to the coil, it moves the armature downward. The plunger, bellows, or diaphragm develops suction and opens an inlet check valve, figure 15-20. Fuel then enters the pump. As the armature reaches the bottom of its stroke, it opens electrical contacts that deenergize the coil. A return spring forces the plunger, bellows, or diaphragm upward to pump fuel through an outlet check valve.

Fuel output from this kind of pump is regulated as it is from a mechanical pump. As pressure rises in the pump outlet line and chamber, it overcomes the return spring and keeps the armature from moving on an output stroke. The armature contacts stay open. When pressure in the output line drops, the return spring moves the armature to resume pumping. Most vibrating armature pumps are aftermarket accessory or high-performance pumps.

## Fuel Pump Pressure and Volume

Fuel pumps are rated by output pressure and volume. You have seen how output volume is regulated by carburetor inlet valves or pump relief valves. Fuel pump specifications often include an output specification such as 100 liters per hour, 70 gallons per hour, or 2 pints per minute. All of these are specifications for unrestricted output at steady pumping speed. Gasoline pumps for carbureted engines develop 5- to 8-psi output pressure. Some electronic fuel injection systems require pump pressures of 30 to 40 psi. These pressures are achieved with high-volume pumps, relief valves, and regulators that maintain high pressures.

### Diesel Fuel Pumps

The fuel injection pump of a diesel engine is a complex mechanical pump that delivers a precise amount of fuel to each injector depending on engine speed and load. The injection pump also controls ignition timing. Diesel injection pumps are explained in detail in chapter 17.

Besides the injection pump, diesel fuel systems have other pumps that move fuel from the tank to the injection system. These are often called "transfer" or "lift" pumps because they lift and transfer fuel from the tank to the engine-mounted injection pump. Some diesel engines use a mechanical diaphragm pump, identical to a mechanical gasoline pump, for this purpose, figure 15-21. A few diesel automobiles have electric fuel transfer pumps, usually in or near the fuel tank. Bosch and most Japanese distributor injection pumps have built-in, rotary-vane transfer pumps, figure 15-22, that eliminate the need for a separate lift or transfer pump on many vehicles.

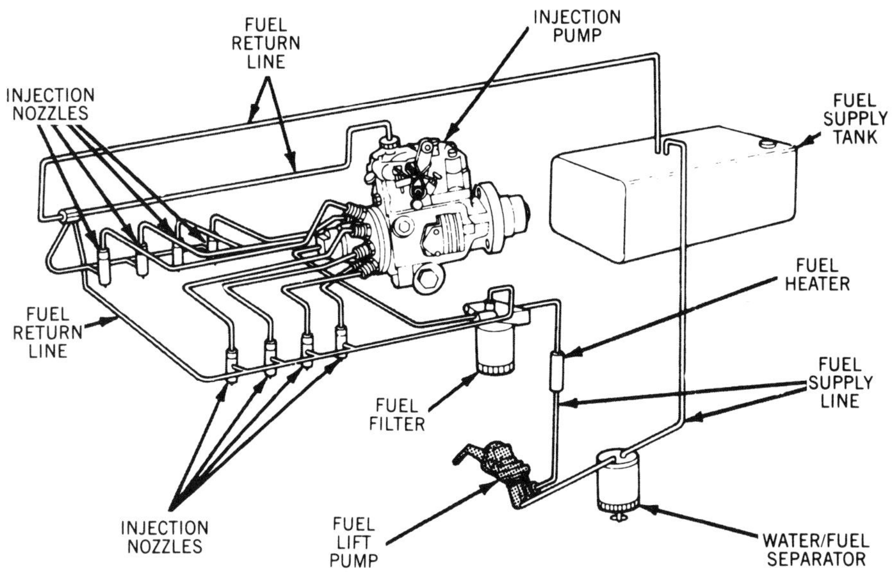


**Figure 15-21.** Many diesel fuel systems have separate "lift" or "transfer" pumps to move fuel from the tank to the injection pump. (Ford)

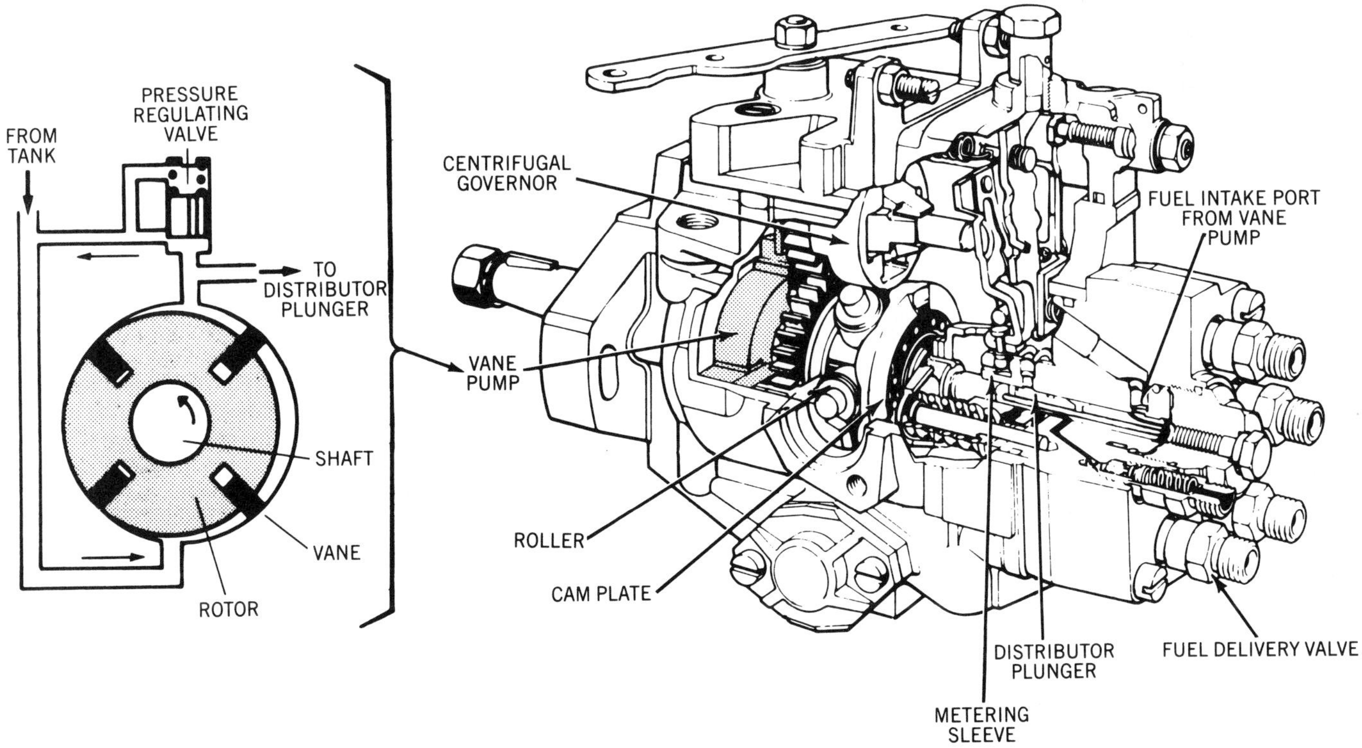


**Figure 15-22.** Bosch and Japanese distributor injection pumps have built-in transfer pumps. (Volkswagen)

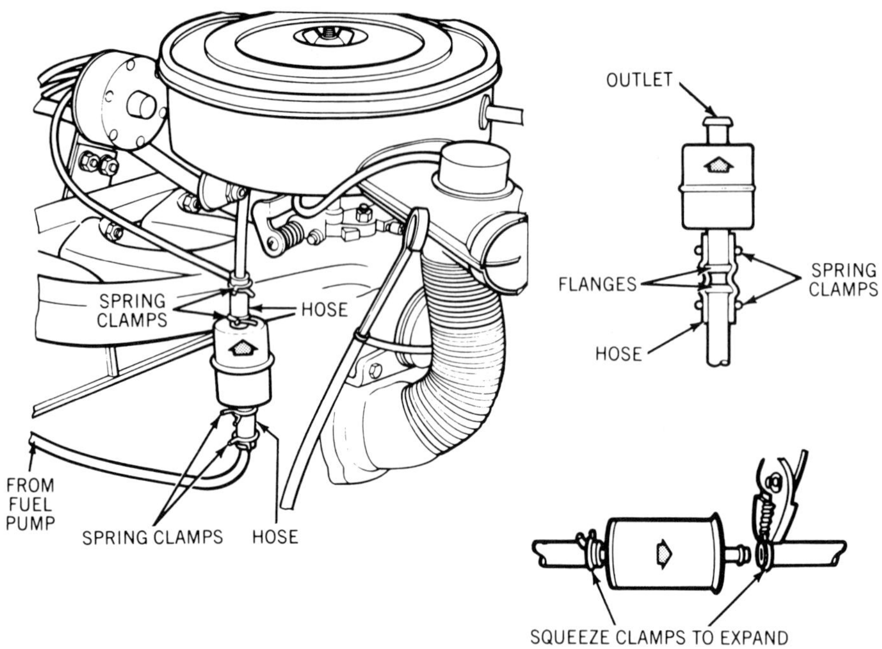


**Figure 15-23.** Inline fuel filters are installed in fuel lines so that gasoline flows in the direction of the arrow. (Chrysler)

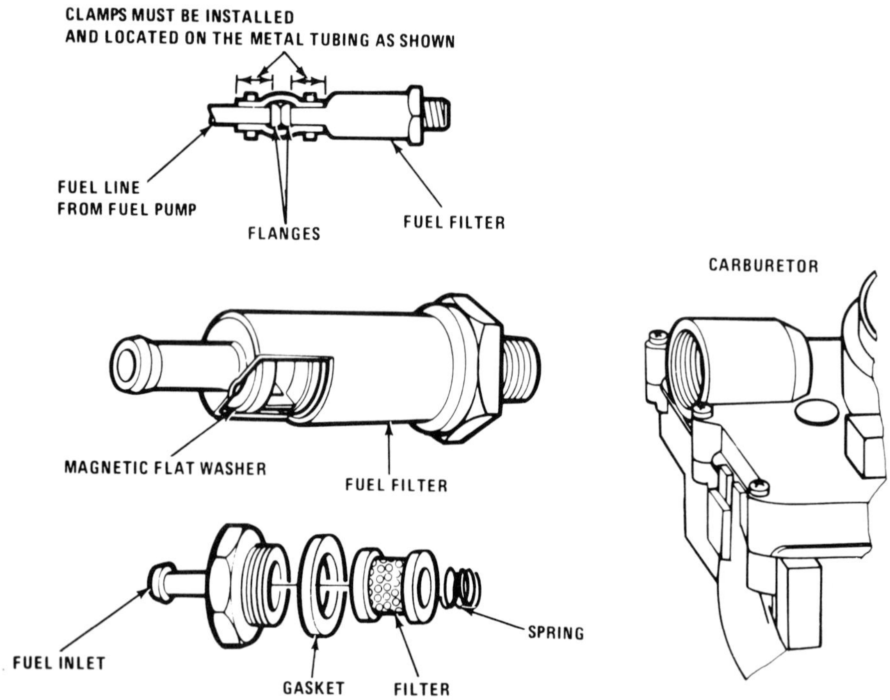


**Figure 15-24.** Some Ford carburetors have fuel filters that screw into the outside of the carburetor inlet. (Chrysler)

## GASOLINE FILTERS

Impurities in gasoline are a smaller problem than they are in diesel fuel. Nevertheless, every gasoline fuel system has one or more filters to remove dirt, rust, water, and other harmful materials from the fuel before it reaches the carburetor or injection system. Gasoline filtration is more critical with fuel injection systems than with carburetors because dirt particles smaller than one **micron** can damage close-tolerance injector parts. Gasoline injection systems often have several filters to remove as much dirt as possible.

You have already seen the strainers and filters used on pickup tubes inside fuel tanks. These remove heavy dirt and sediment that could clog lines to the pump. If the fuel system has a tank-mounted electric pump, the intake filter on the pump is large and has a fine-filtering ability. This is necessary because the filter is not easily replaceable, and the pump must be protected against dirt.

Other filters are installed in the lines before the gasoline reaches the carburetor or injectors. Most filters are replaceable cartridges that contain filter material made of pleated paper. The simplest inline filters are plastic or metal cartridges installed in the fuel lines with clamps and short hoses, figure 15-23. They must be installed so that fuel flows in the direction of the arrow. If installed backwards, the filter can restrict fuel flow. Some of these filters have a bypass nipple to return fuel vapors to the tank. Chrysler, AMC, and many imported cars use these filters.

Many Ford and GM carbureted engines have a fuel filter at the carburetor inlet fitting. These, too, have a pleated paper element. Some Ford filters screw into the outside of the carburetor inlet. The fuel line attaches to the filter with clamps and a short hose, figure 15-24. Most GM filters are inside the carburetor body. Older GM Rochester carburetors had sintered metal filters, but newer ones have paper filters. On late-model GM cars, the filter contains a check valve for rollover leakage protection, figure 15-25.

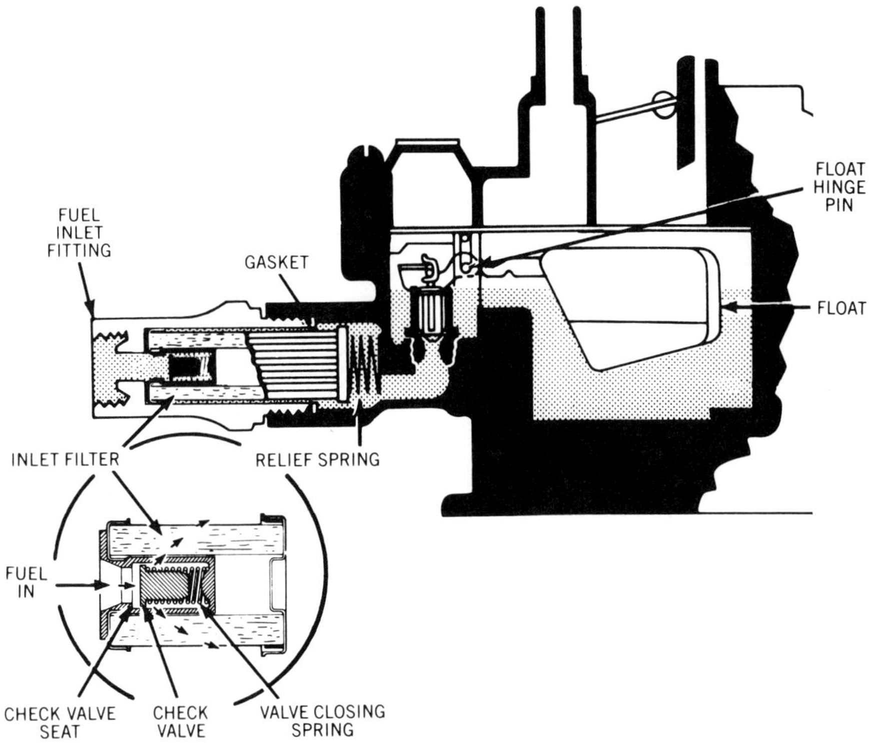


**Figure 15-25.** GM Rochester fuel filters are installed in the carburetor inlet and contain a rollover check valve. (Chevrolet)

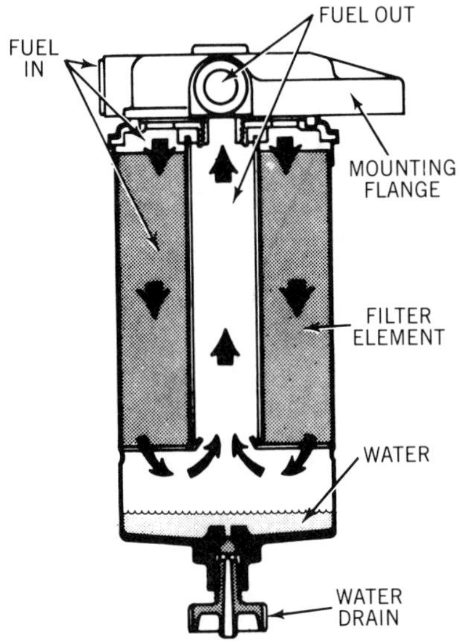


**Figure 15-27.** This Bosch diesel fuel filter has a drain at the bottom of the filter housing to remove water and sediment. (Volkswagen)

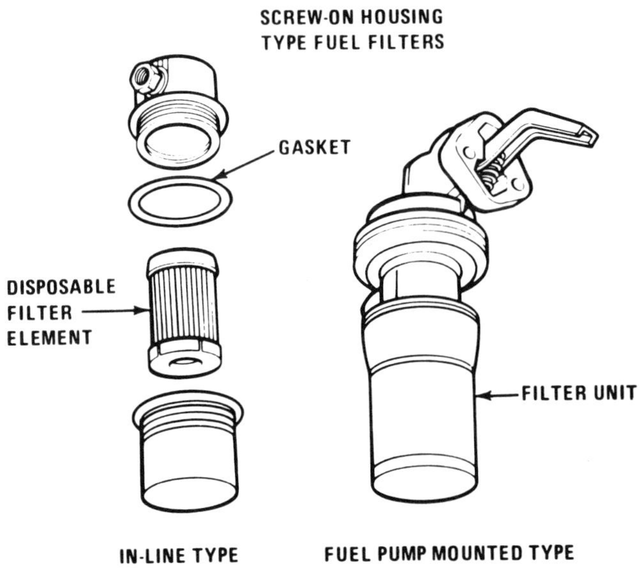


**Figure 15-26.** Large capacity canister-type fuel filters are used with many gasoline fuel injection systems. (Chrysler)

Inline filters usually are in fuel lines on the engine where they are accessible for replacement. Some fuel systems have filters on the outlet side of the pump. Chrysler 6-cylinder engines in the early 1970's had a disposable filter in the pump discharge tower. Older Cadillacs had a similar replaceable filter in the pump outlet. Some big Ford V-8's and Cadillacs with fuel injection have screw-on fuel filters that look like small oil filters, figure 15-26. On Ford engines, they are attached to the fuel pump. On Cadillacs, they are on a bracket near the engine or under the frame near the left rear wheel.

Gasoline fuel injection systems have large capacity, replaceable filters. They may be mounted on the firewall, on an engine compartment bracket, or on the engine.

## DIESEL FUEL FILTERS

Diesel fuel is heavier than gasoline, contains more impurities, and collects water during storage. Diesel injection systems rely on the fuel to lubricate close-tolerance parts. All of these facts make filtration and water separation critical factors in a diesel fuel system.

Every diesel fuel system has at least one large-capacity, replaceable filter. These filters usually have paper filter elements. On some Bosch fuel systems, the filter has a drain at the bottom of the filter housing to remove water that separates from the fuel, figure 15-27. General Motors diesel-powered cars have a fuel-and-water separator that keeps water from reaching the injection pump and nozzles. The separator also senses excessive water in the fuel and lights an indicator lamp to warn the driver.

Ford diesel cars have a unique fuel tank pickup and diesel fuel conditioner, figure 15-28. The operation of these parts illustrates the general requirements for diesel fuel storage, filtering, and conditioning. The Ford fuel tank pickup rests on the bottom of the tank and has nibs around its edge that create low pressure in the incoming fuel. This draws water, along with the fuel, out of the tank and prevents water accumulation in the tank.

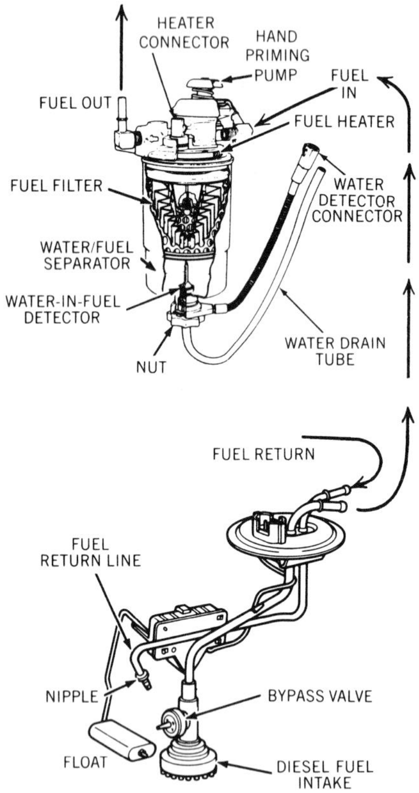


**Figure 15-28.** Ford diesel cars and light trucks have this unique fuel pickup and fuel conditioner. (Ford)

The fuel then goes to a fuel conditioner that:

**1.** Filters the fuel through a large, spin-on, replaceable filter element

**2.** Separates water from the fuel and collects water at the bottom of the filter element

**3.** Has a sensor that lights an indicator lamp to warn the driver to drain excessive water from the filter

**4.** Provides a drain valve to remove water from the system

**5.** Heats the fuel to prevent wax formation in cold weather. The fuel conditioner has a sensor that turns on an electric heater when fuel temperature is between 30° and 55° F (−1° and 13° C) or lower.

## FUEL GAUGES

Every fuel system has an electric gauge that tells the driver how much fuel is in the tank. A float in the tank moves a variable resistor that controls current flow to one of two basic kinds of fuel gauges.

### Balancing-Coil Gauge

A balancing-coil gauge has two electromagnetic coils on either side of the gauge needle. Current flows from the battery and ignition switch through both coils to ground. The wire to the tank sending unit is in parallel with the "full" coil, figure 15-29. When fuel level is high, the tank float moves the variable resistor for maximum resistance. This causes all of the current to flow through both gauge coils to ground. The stronger magnetic field of the "full" coil draws the needle to the "full" position.

As the tank fuel level drops, the float moves the variable resistor for less resistance. More current then flows through the wire to the tank sender unit than through the "full" coil of the gauge. The field of the "full" coil weakens, and the "empty" coil with higher current flow draws the needle toward "empty."

### Thermostatic Gauge

A thermostatic fuel gauge has a tank sender similar to the sender used with a balancing-coil gauge. Current flows through a heater wire and bimetal thermostat in the fuel gauge and then through a wire to the variable resistor of the tank sender, figure 15-30. When fuel level is high, the float moves the resistor for minimum resistance. This allows current through the circuit, which heats the wire and thermostat in the gauge. The expanding thermostat moves the needle to "full." As fuel level drops, resistance increases in the sender circuit. This reduces current in the heater. The thermostat cools and allows the needle to move toward "empty."

### Other Fuel Indicators

The circuit shown in figure 15-30 also has a low-fuel warning lamp, controlled by a thermistor in the tank. The

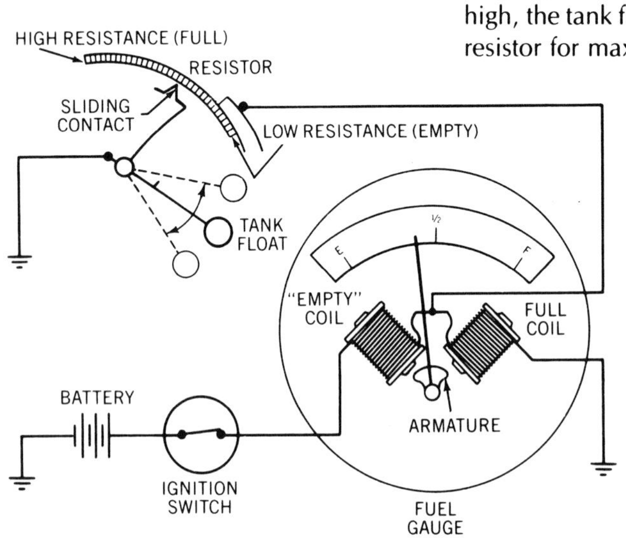


**Figure 15-29.** Typical balancing-coil fuel gauge circuit.

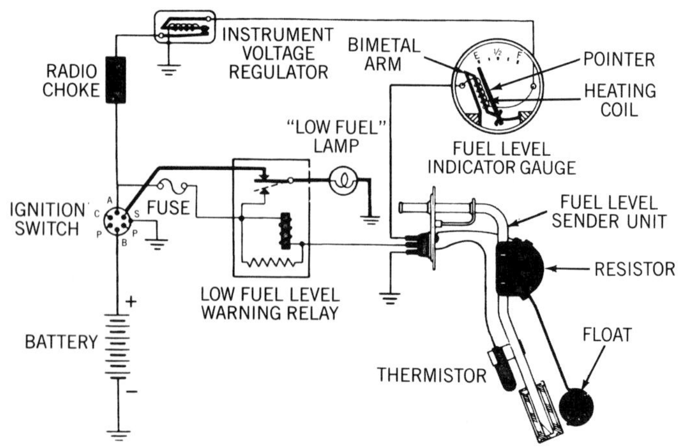


**Figure 15-30.** Typical thermostatic fuel gauge circuit, including a low-fuel warning lamp. (Ford)

thermistor loses resistance as it warms up. Normally, it is submerged in fuel in the tank, which keeps it cool. When fuel level drops, it uncovers the thermistor. The thermistor warms up, loses resistance, and allows a relay to close, which lights the low-fuel warning lamp.

Several late-model cars have electronic trip computers or fuel level and mileage indicators. These indicate fuel level, fuel consumption rate, and (in some cases) miles to empty. These systems use microprocessors to compute fuel level and rate of use from signals received from several sensors in the fuel system.

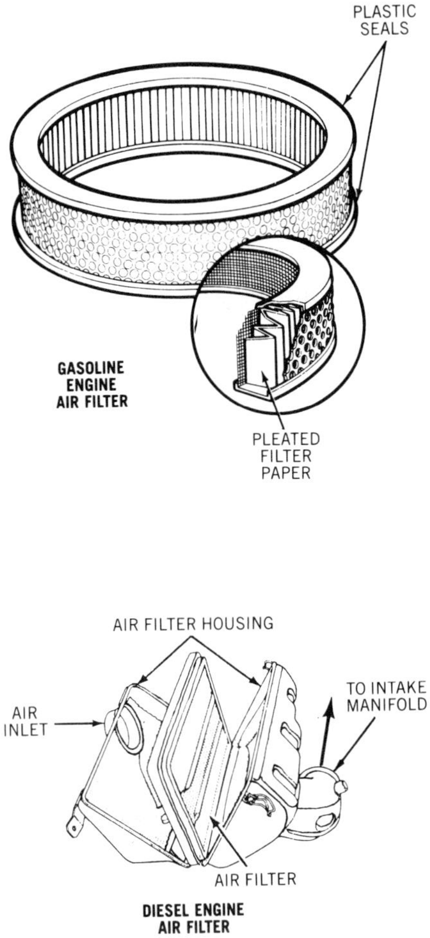


**Figure 15-31.** Most late-model gasoline and light-duty diesel engines use large-capacity paper air filters, like these. (Chrysler) (Ford)

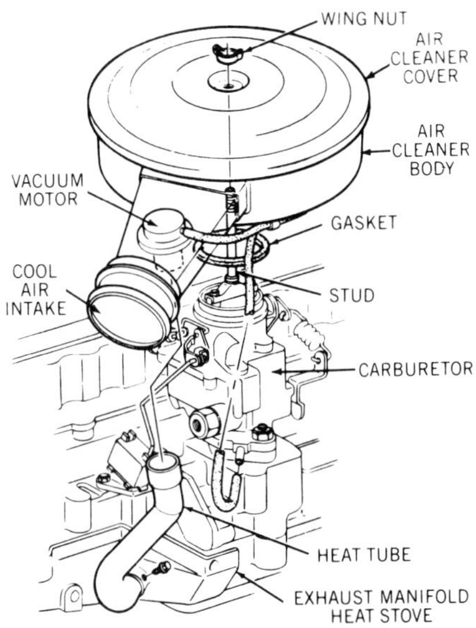


**Figure 15-32.** A basic thermostatically controlled air cleaner mounts on top of the carburetor and has a damper valve to control intake airflow from hot- and cold-air inlet ducts.

## AIR CLEANERS AND FILTERS

For every pound of fuel that it burns, a gasoline engine consumes about 15 pounds of air. Measured another way, it takes about 150,000 gallons of air to mix with 15 gallons of gasoline. A diesel uses even more air. All of this air must be filtered and delivered to the intake manifold at a controlled temperature for efficient fuel vaporization. These are the two main jobs of the air cleaner and filter: filtration and air temperature control. The air cleaner also silences intake air noise and acts as a flame arrester in case of a backfire.

Older cars had oil bath air cleaners or air cleaners with filters made of steel mesh that could be oiled. The oil and mesh trapped large dust and dirt particles and allowed clean air into the engine. Oil bath and wire mesh air cleaners were heavy, messy, and didn't filter as efficiently as modern paper air filters.

The air filters on modern gasoline and diesel engines are made of chemically treated filter paper. The paper is pleated to provide as much filter surface area as possible in a compact air cleaner housing. The edges of the filter are sealed with heat-resistant plastic that seals the filter in the air cleaner and keeps unfiltered air from bypassing the filter, figure 15-31. Air filters for diesel engines often are larger than for comparable displacement gasoline engines, but they are the same kind of replaceable, paper-element filter.

The metal air cleaner housing is a 2-piece metal or plastic housing that holds the filter and has an air intake duct, or snorkel. The simplest air cleaners sit on top of the carburetor or fuel injection air intake, figure 15-32. Many cars, however, have the air cleaner located away from the engine air intake and connected to the intake manifold with rigid or flexible ducts, figure 15-33.

Engineers design air cleaners, fil-

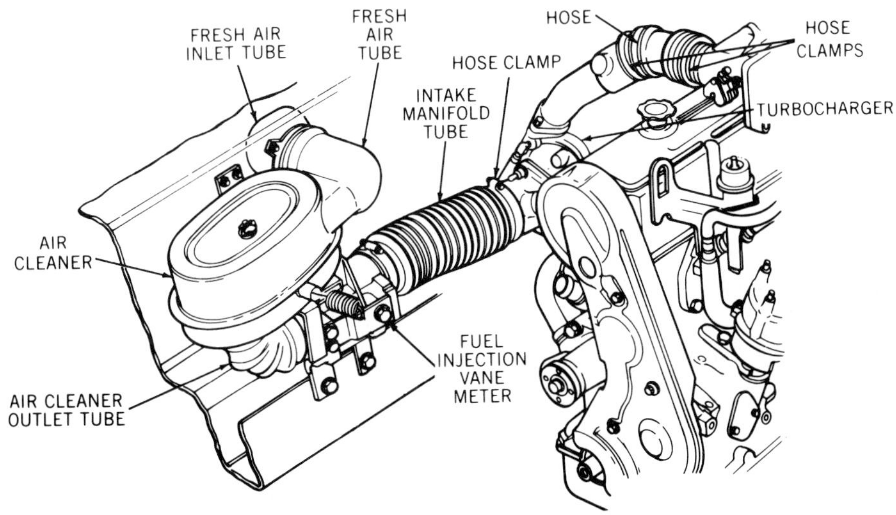


**Figure 15-33.** On this car, the air cleaner is mounted away from the engine and connected to the engine by ducting. (Ford)

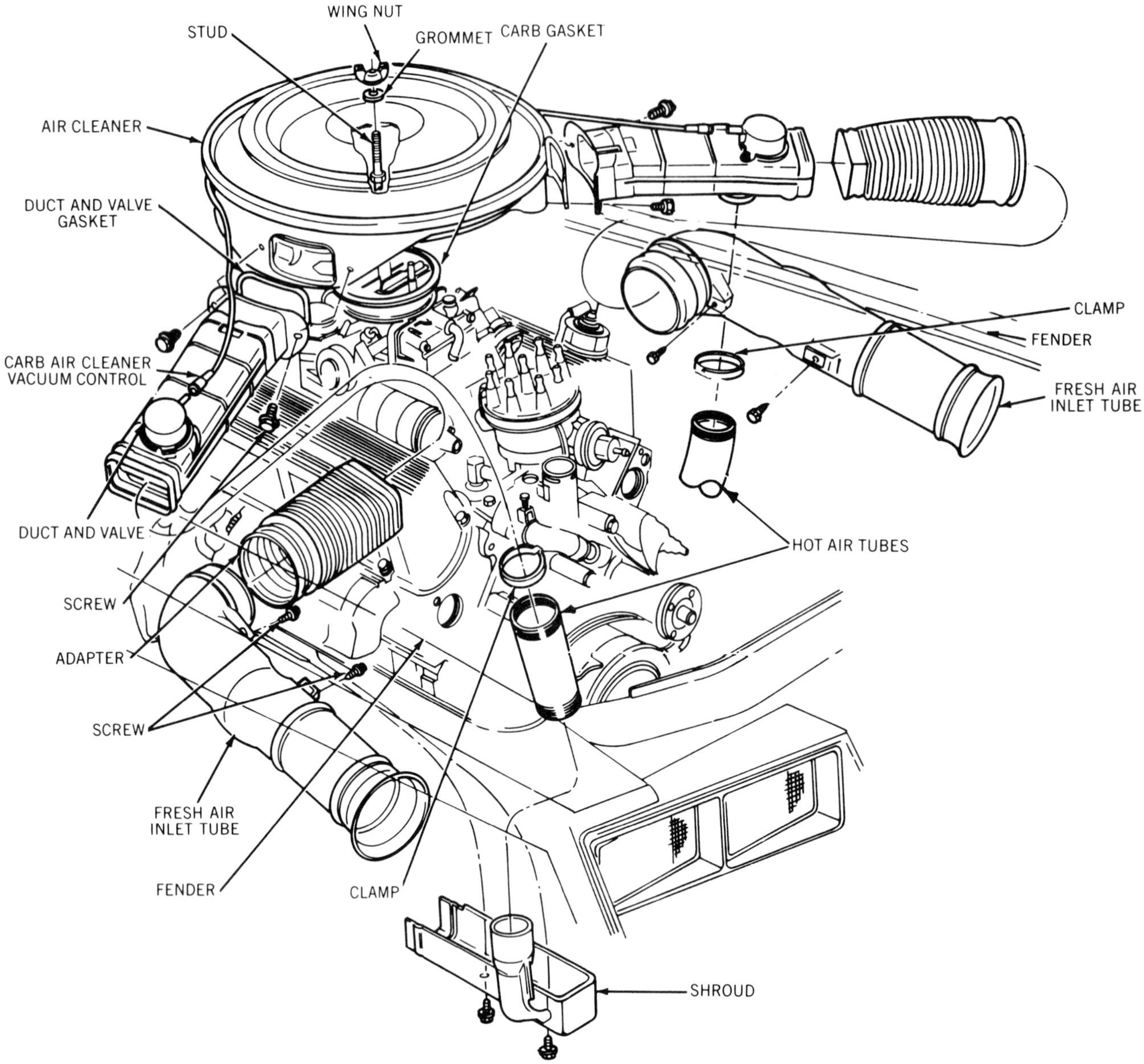


**Figure 15-34.** The air cleaner for this high-output engine has dual air intake ducts. (Ford)

ters, inlet tubes, and ducts to deliver more than the maximum airflow requirements of an engine. The main source of intake air is the air cleaner snorkel or inlet duct. Some air cleaners have two inlet openings, figure 15-34. One provides most airflow for average operation. The other inlet provides cooler outside air when engine compartment temperatures exceed a certain level. Or, the second air inlet may be controlled to open and provide maximum airflow when the engine is operating at wide-open throttle. Air cleaner and duct design often are dictated by the size, location, and shape of other components in the engine compartment and by the structure of the car body.

## THERMOSTATICALLY CONTROLLED AIR CLEANERS

Almost all gasoline-engined cars and light trucks built since the late 1960's have thermostatically controlled air cleaners. These have a combination of ducts and airflow control valves to provide warm air to the engine during warmup and cooler air during operation at normal engine temperatures, figure 15-32.

The major purpose of these air cleaners is to reduce exhaust emissions during warmup without hurting engine performance. They also provide these other benefits:

- Faster warmup and better driveability
- Control of inlet air temperature within a constant range and thus finer control of air-fuel ratios
- Improved fuel economy under all driving conditions

• Reduced throttle valve icing in cold weather.

A thermostatic air cleaner has a conventional metal or plastic air cleaner housing. The inlet air duct, or snorkel, has two openings, figure 15-32. One draws heated air through a duct from a sheet metal shroud, or heat stove, around the exhaust manifold. The other opening draws cooler air from the engine compartment or through a duct from a fresh air inlet at the front of the car. A flapper valve, or damper, in the air cleaner switches intake airflow between the two openings. By measuring intake air temperature and using this measurement to modulate the air inlet source between hot and cold air, the air cleaner regulates air temperature within a range that provides the best emission control, driveability, and economy. Air valve (damper) movement is controlled in two ways:

1. By a vacuum diaphragm, or vacuum motor
2. By a thermostatic bulb.

Air cleaners controlled by a vacuum diaphragm are more common. After the engine is started when cold, intake manifold vacuum is routed to the vacuum motor diaphragm on the air cleaner snorkel, figure 15-35. Vacuum moves the diaphragm to pull the air damper to close the cold air inlet and open the hot air inlet. This draws air from the heat shroud around the exhaust manifold.

The vacuum line to the diaphragm has an air bleed opening operated by a bimetal temperature sensor. The sensor and bleed are in the air cleaner housing so that the sensor can respond to intake air temperature. When air temperature is below 90° to 100° F (32° to 38° C), the sensor holds the air bleed fully closed, and full vacuum is applied to the diaphragm. The engine then draws only hot intake air. As air temperature rises above about

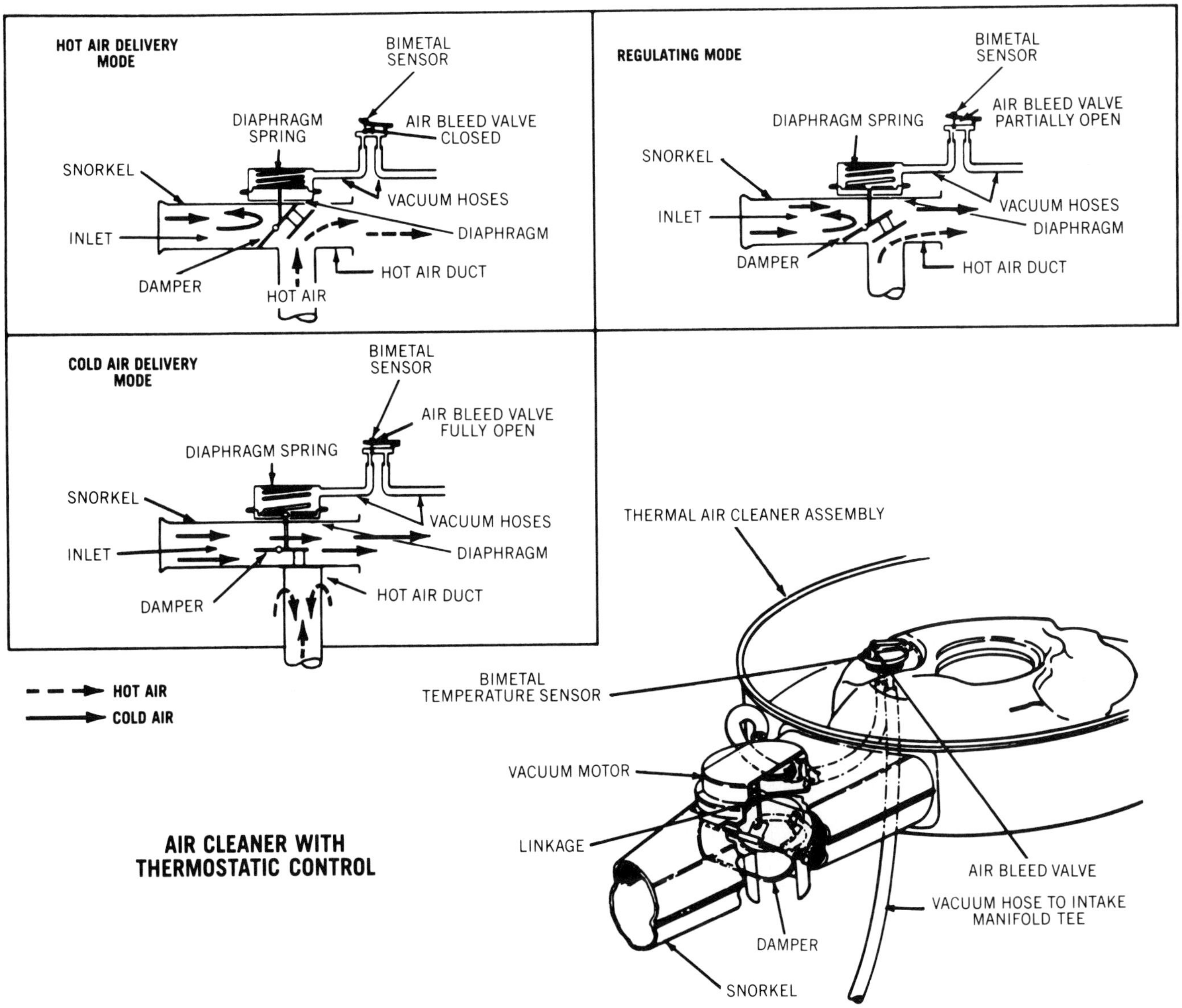


**Figure 15-35.** The vacuum motor moves the air intake damper between the hot and cold positions in this vacuum-operated air cleaner. (GM)

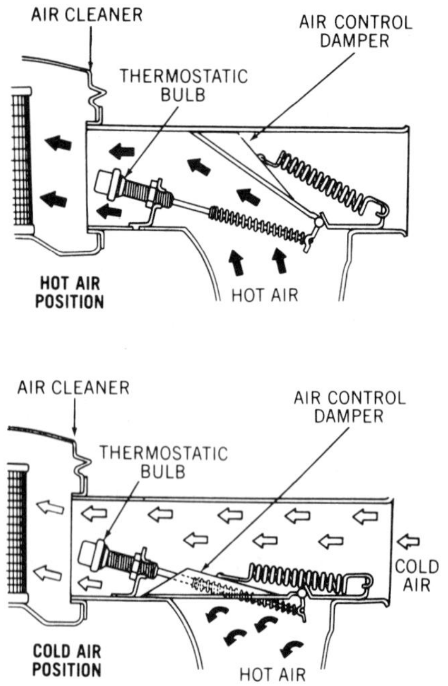


**Figure 15-36.** A thermostatic bulb air cleaner relies on expansion and contraction of the bulb to move the air intake damper. (Ford)

100° F (38° C), the sensor starts to open the air bleed and reduce vacuum at the diaphragm. A spring in the diaphragm housing then starts to move the diaphragm so that the damper begins to open the cold air intake. The engine then draws a combination of heated air and cooler air through the fresh air inlet.

As intake air temperature continues to rise, the temperature sensor opens the vacuum bleed more. At high temperatures, the bleed is fully open, and vacuum to the diaphragm is bled off completely. The spring then holds the damper in the full-cold-air position, and the engine draws air only through the fresh air inlet. Under full-throttle acceleration at any temperature, intake manifold vacuum drops. This allows the diaphragm spring to move the damper to the full-cold-air position to provide maximum airflow when the engine needs it most.

Thermostatic-bulb air cleaners are less common than vacuum-operated ones, but work on the same principles. They have the same kind of air inlet ducts and an air control valve, or damper. The thermostatic bulb works like the wax pellet in a coolant thermostat. The bulb is inside the air cleaner snorkel and exposed to intake air, figure 15-36. When the engine is cold, a spring holds the air control damper in the hot-air position. As intake air temperature rises, the bulb expands and moves linkage to force the damper toward the cold-air position.

All Chrysler and most GM, Ford, and imported car engines have vacuum-controlled air cleaners. Most pre-1972 Ford engines, except large V-8's had bulb-operated air cleaners, as did some pre-1977 Ford 4-cylinder engines. AMC used bulb-operated air cleaners on 6-cylinder and older V-8 engines.

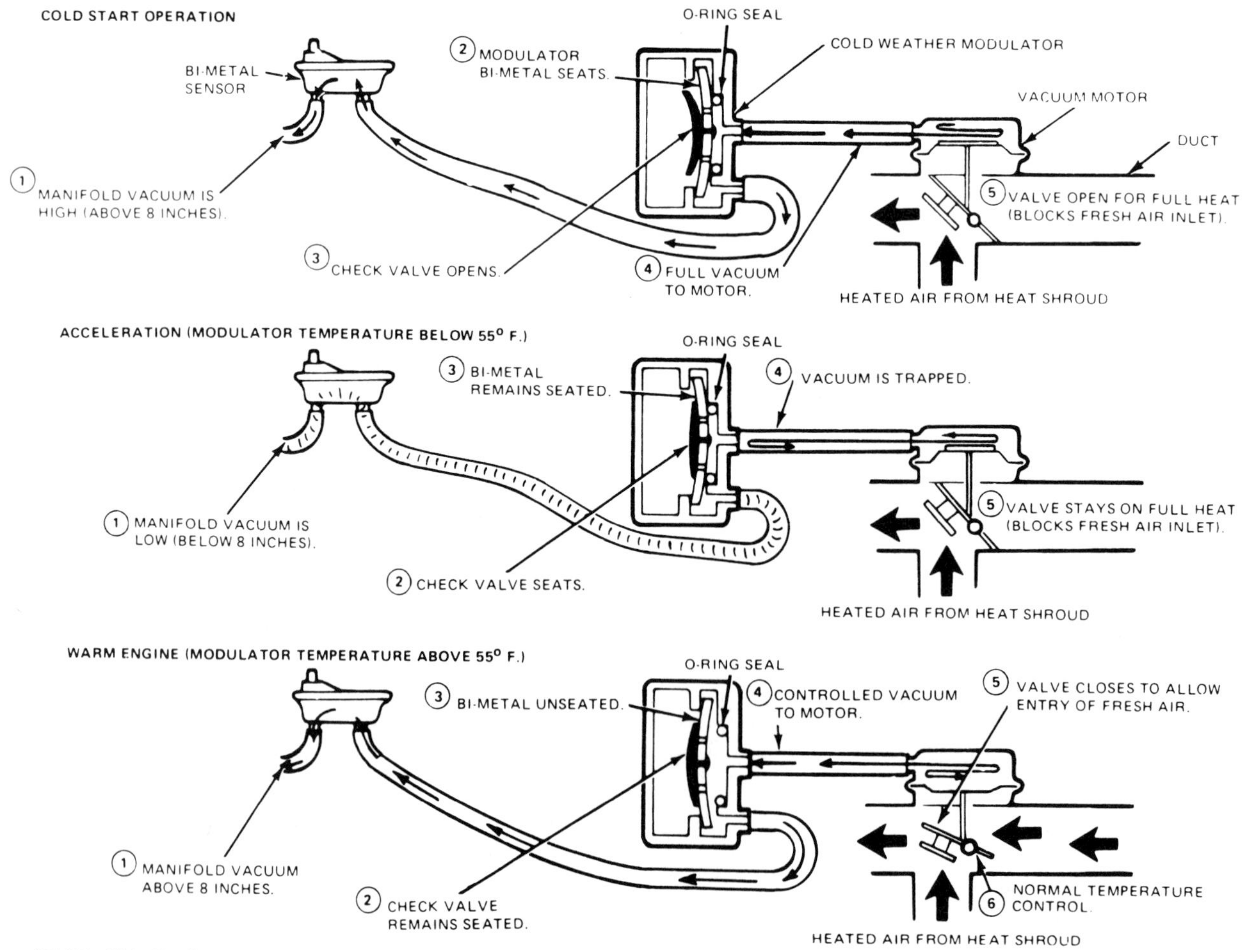


**Figure 15-37.** This Ford system is an example of vacuum modulators used to "fine tune" thermostatic air cleaner operation. (Ford)

Different engines and carburetors or fuel injection systems need air temperature regulation in different ranges. Therefore, carmakers use temperature sensors that start to open the vacuum bleed at temperatures as low as 50° F (10° C) or as high as 120° F (49° C). Some carmakers use vacuum modulators, such as the Ford system shown in figure 15-37. This is a second bimetal temperature sensor and check valve that traps vacuum in the diaphragm housing to hold the damper in the hot-air position at very low temperatures, regardless of manifold vacuum.

Intake air temperature is not as important for diesel engines as for gasoline engines because the air does not mix with, and help vaporize, the fuel in the intake manifold. All fuel vaporization takes place inside the combustion chamber. Most diesel engines use electric glow plugs to heat the combustion chamber of a cold engine. This helps to raise the temperature of the compressed air and ensure complete fuel vaporization and combustion.

Ford has an optional cold weather package for some small diesels. This system circulates engine coolant through passages in the air cleaner housing to keep ice from forming in cold weather.

## EVAPORATIVE EMISSION CONTROL SYSTEMS

As part of its history of tough pollution regulations, California required the first evaporative emission control (EEC) systems on 1970-model cars sold in that state. EEC systems appeared nationwide on all gasoline-powered cars in 1971. These systems trap and store gasoline vapors that otherwise would escape from fuel tanks and carburetor fuel bowls. The systems then return the vapors to the engine intake airflow. EEC systems are not required on diesel-powered vehicles because air pollution from diesel fuel evaporation is not the problem that it is from gasoline.

Carmakers use a variety of parts, controls, and plumbing in EEC systems, but all systems have these parts in common, figure 15-38:

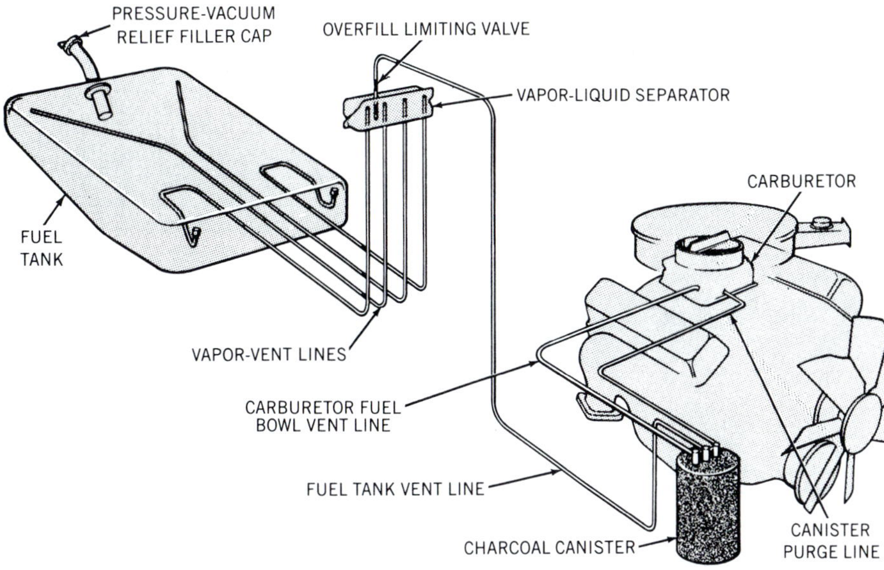


**Figure 15-38.** A typical EEC system has these major parts. (Chrysler)

- Vapor vent lines to collect gasoline vapors from the tank and carburetor fuel bowl
- A way to separate vapors from liquid fuel so that liquid is not drawn into the vapor storage canister
- Vacuum and pressure relief valves for the fuel tank to relieve vacuum that could prevent fuel flow to the pump or high pressure that could damage system parts
- A place to store fuel vapors when the engine is off (usually a canister filled with charcoal)
- A way to purge vapors from the storage canister and route them to the intake airflow or intake manifold when the engine is running.

### Vacuum and Pressure Relief

Earlier in this chapter, you learned why there must be atmospheric pressure above the fuel in the tank to prevent a vacuum lock. EEC systems have vacuum relief valves in the fuel tank caps or elsewhere in the system to let air into the tank when vacuum of about 0.5 inch of mercury develops. If the vacuum relief valve is not in the cap, the cap is sealed. Use of a sealed cap on a system without a separate vacuum relief valve, however, will create a vacuum lock.

A pressure relief valve—in the cap or elsewhere—opens if fuel evaporation raises the air pressure in the tank by 1 or 2 psi above atmospheric pressure. Figure 15-7 shows the construction of vacuum and pressure relief valves in a fuel tank cap. Normally, the vapor venting and storage devices can handle a pressure increase from fuel evaporation. However, on a hot day, a tank that is less than half full may develop more pressure than the vent system can handle. Then, the pressure relief valve opens to release excess pressure. The infrequent operation of this valve does not significantly add to fuel vapor emissions to the air.

### Tank Vent Lines and Overfill Protection

Fuel tanks of EEC systems have a series of vent lines to collect fuel vapors. Some tanks have several lines that enter the tank at its upper corners. With this arrangement, the car can be parked on a steep incline at any angle, and one vent line will always be above the fuel level in the tank. This is necessary to help keep liquid fuel from being forced out of the tank.

To protect against pressure or overflow from an overly full tank, the tanks of EEC systems are designed to trap some air and allow for fuel expansion. Many tanks have a dome that

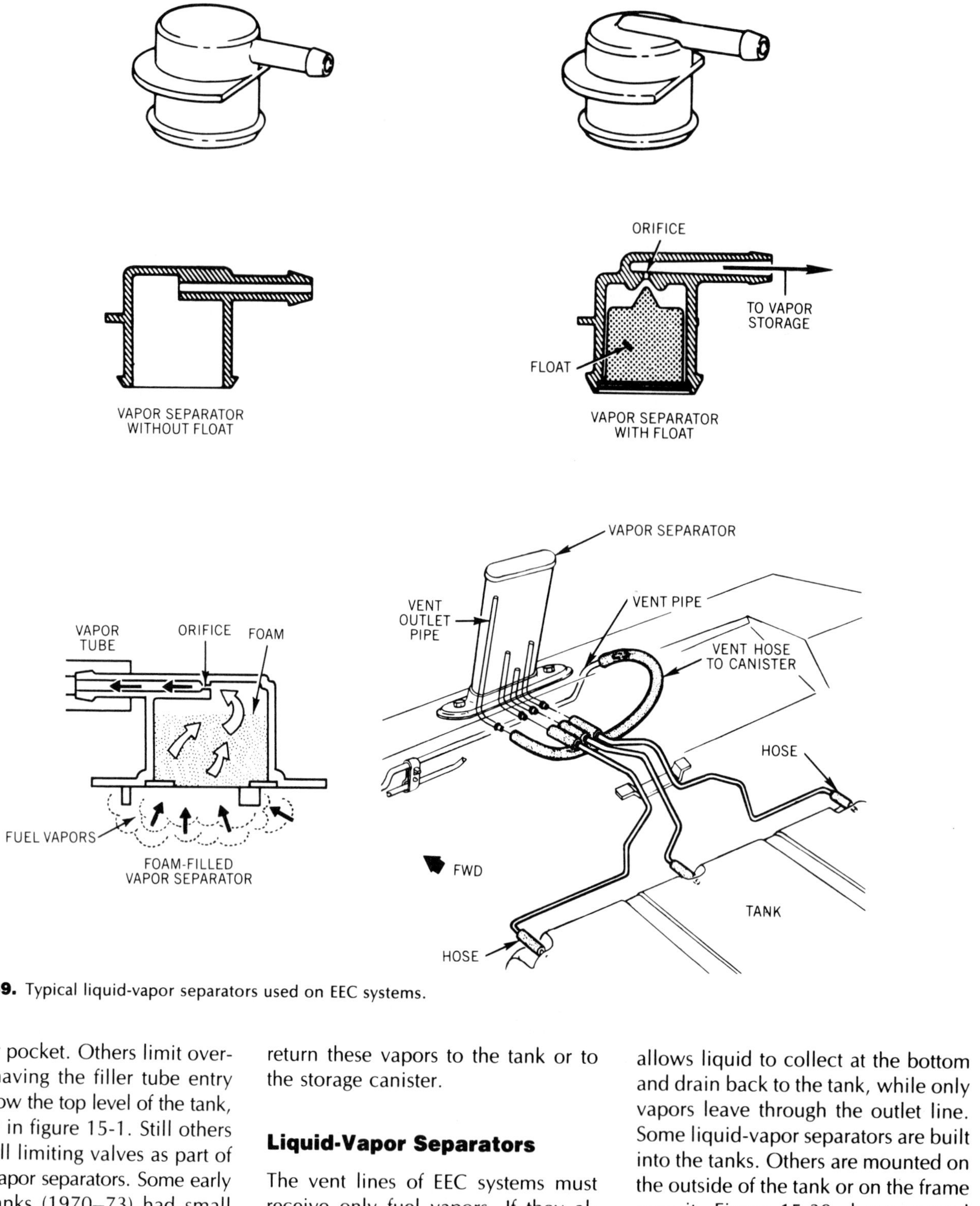


**Figure 15-39.** Typical liquid-vapor separators used on EEC systems.

traps an air pocket. Others limit overfilling by having the filler tube entry located below the top level of the tank, as you saw in figure 15-1. Still others have overfill limiting valves as part of the liquid-vapor separators. Some early EEC fuel tanks (1970–73) had small expansion tanks inside the main tank. These trapped air and limited the initial filling of the tank, but allowed room for fuel expansion.

As you learned earlier in this chapter, heat can cause gasoline vapors to form in fuel pumps, filters, and lines to the carburetor. Pumps and filters with vapor outlets have vent lines that return these vapors to the tank or to the storage canister.

### Liquid-Vapor Separators

The vent lines of EEC systems must receive only fuel vapors. If they allowed liquid gasoline to escape from the tank, the vent system could create a fire hazard. All systems have some way to separate vapors from liquid fuel before the vapors reach the storage canister. One method is to run the tank vent lines to a separator canister that has a single vapor outlet pipe at its highest point. This kind of separator allows liquid to collect at the bottom and drain back to the tank, while only vapors leave through the outlet line. Some liquid-vapor separators are built into the tanks. Others are mounted on the outside of the tank or on the frame near it. Figure 15-39 shows several kinds of liquid-vapor separators.

### Carburetor Venting

A carburetor holds a small amount of gasoline in its fuel bowl, or float bowl, so that it has a steady supply to meter into the intake airflow. As you will learn when you study carburetors, the

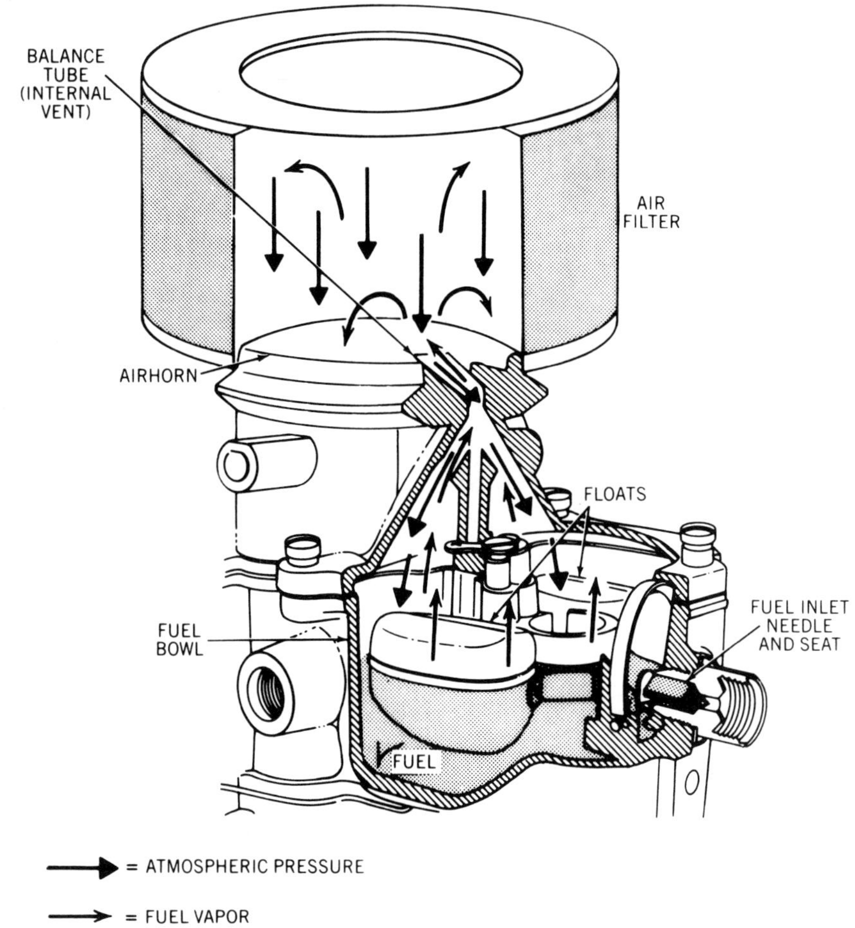


**Figure 15-40.** The balance tube maintains atmospheric pressure on gasoline in the carburetor fuel bowl and acts as a vapor vent to the carburetor airhorn. (Chrysler)

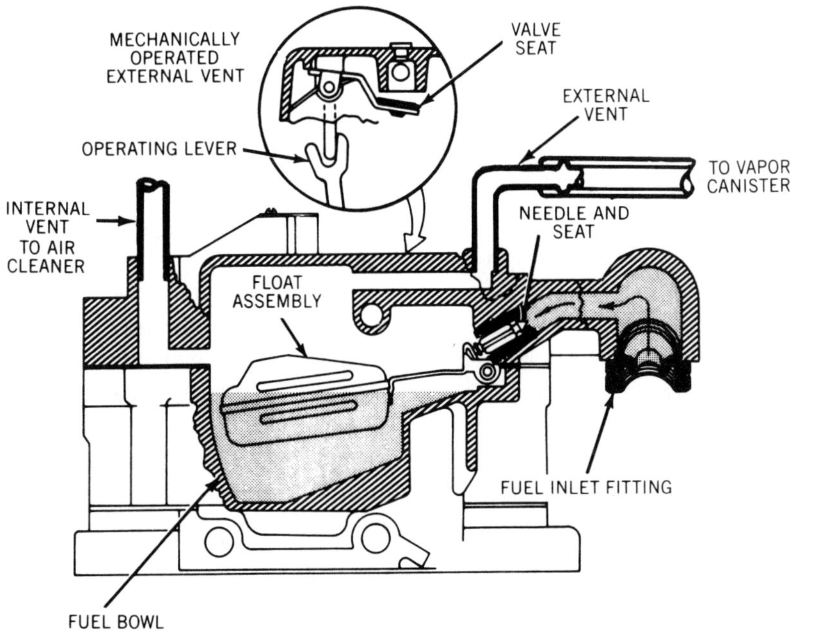


**Figure 15-41.** This external carburetor bowl vent allows vapors to go from the fuel bowl to the EEC vapor storage canister. (Chrysler)

fuel bowl must be vented to atmospheric pressure. This creates the pressure differential needed for precise fuel metering at all times. The fuel bowl vent, however, can allow gasoline vapors to escape when the engine is off or idling. Carburetors for EEC systems are designed to trap these vapors. Basically, carburetors have two kinds of vents: internal and external.

**Internal Vents.** All carburetors have an internal vent, or balance tube, that connects the fuel bowl to the intake air inlet, or airhorn, figure 15-40. This balance tube keeps atmospheric pressure on the fuel in the bowl, which causes fuel to flow through metering passages to the lower pressure area in the carburetor venturi. The balance tube also compensates for the slight air pressure drop that develops through the air filter. When the engine is off, the balance tube allows some vapors from the fuel bowl to collect in the air filter instead of escaping to the atmosphere.

**External Vents.** If the balance tube were the only fuel bowl vent, excess vapors might escape into the carburetor air inlet at high temperatures. This could richen the air-fuel mixture or flood the engine. Fuel evaporation is greatest when a hot engine is idling or first shut off and engine heat vaporizes gasoline in the carburetor. This **percolation** can force gasoline liquid and vapor out of the balance tube and fuel-metering passages and flood the engine. Therefore, most carburetors have an external bowl vent, figure 15-41. On some carburetors, the vent is operated by linkage so that it is open only when the throttle is closed. On cars without EEC systems, the vent is open to the atmosphere. On cars with EEC systems, the carburetor bowl vent is connected to the vapor storage canister. Figure 15-41 shows both the internal and external vents of a carburetor. Linkage connected to the throttle operates the external vent valve.

Many late-model carburetors have solenoid-operated vent valves. Figure 15-42 shows a solenoid that switches the vent passage from an internal vent

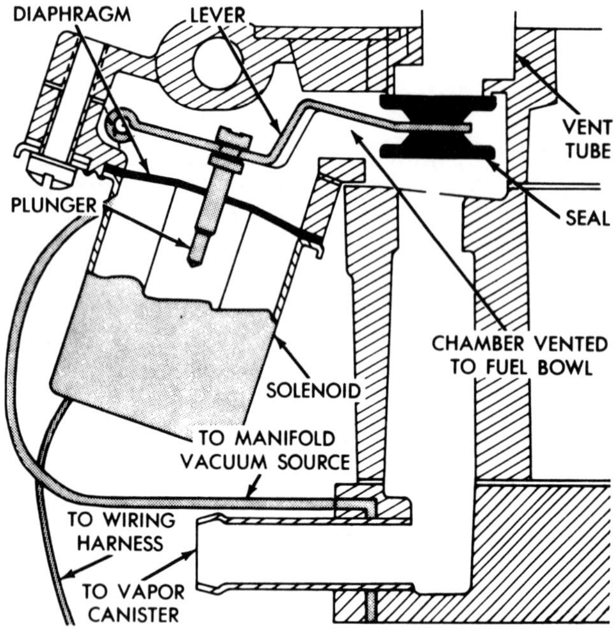


**Figure 15-42.** This bowl vent solenoid switches the vent passages from an internal vent tube to the charcoal canister. (Chrysler)

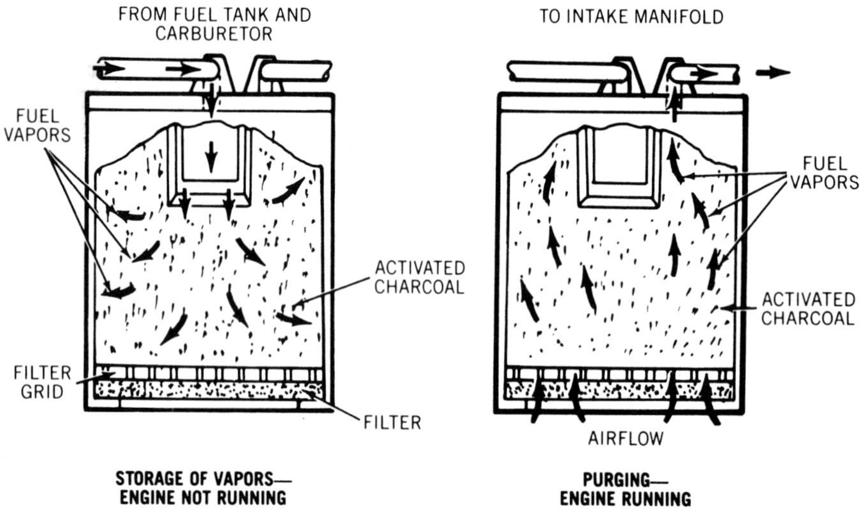


**Figure 15-43.** The vapor storage canister collects vapors from the fuel tank and carburetor and returns them through the purge line to the engine when it is running. (Chrysler)

tube at the top (engine running) to an external vent at the bottom (engine off). The solenoid can be operated through the ignition switch or through an engine control computer.

## Vapor Storage

On all cars and light trucks built since 1972, gasoline vapors from the tank and carburetor are collected and stored in a vapor storage canister. The canister, figure 15-43, is filled with chemically treated charcoal that traps fuel vapors on its surface through the process of **adsorption**. The surface area of the charcoal granules is quite large and will store vapors for a long time. When the engine is running, air flows through the canister to remove the stored vapors. To draw fuel vapors into the intake airflow, the canister purge port is connected to the intake manifold through the PCV line or to the air cleaner inlet.

Canisters receive fresh air in two ways. Most canisters are open to the atmosphere at the bottom, and air enters through an inlet filter, figure 15-43. Some canisters, as on many Chrysler cars, are closed to the atmosphere and receive air from the air injection pump. In these designs, airflow can be controlled by a solenoid to purge the canister only during certain engine operating conditions.

## Vapor Purging

The canister purge line is connected to the air cleaner inlet or to the manifold at the base of the carburetor, usually through the PCV system. Canister purging adds to the intake airflow and can richen the air-fuel mixture. Carburetors used with EEC systems are calibrated to allow for these effects, and the purge rate must be controlled to maintain correct air-fuel ratios at all times. Incorrect canister purging can cause engine surging and hesitation. Purge methods can be either constant or variable, depending on the engine's air and fuel needs. Additionally, a canister can be purged directly through one or more orifices or through a vacuum-controlled purge valve.

A simple constant-purge system without a purge valve has a calibrated orifice in the purge line to the carburetor base, figure 15-44. The orifice is a restriction that creates a constant-purge flow regardless of intake vacuum or airflow. In a simple variable purge system, airflow through the canister varies in proportion to engine intake airflow. This can be done simply by connecting the canister purge line to the air cleaner inlet, also shown in figure 15-44. Airflow through the air cleaner then varies airflow through the canister.

Most late-model purge systems have a vacuum-operated purge valve that opens and closes the purge line in response to intake manifold vacuum, figure 15-45. This figure also shows 2-stage purging with a constant airflow to the manifold and a second purge line that opens in response to carburetor vacuum for extra purging through the valve.

Late-model engines with electronic controls and closed-loop fuel systems are particularly sensitive to canister purging. The extra fuel vapors drawn through the purge line affect the air-fuel ratio control of a feedback carburetor or fuel injection system. Many of these systems have canister purge valves controlled by an electric signal from the engine control computer, figure 15-46.

Because canister purging is critical for electronically controlled engines with fuel-metering feedback systems, most late-model engines use a vacuum solenoid and valves to control canister purge flow. On most systems, purging occurs when the engine is fully warmed and operating in the closed-loop mode at cruising speed. Purging is shut off when the engine is cold, idling, decelerating, or operating at wide open throttle.

The GM system in figure 15-46 uses a solenoid to control vacuum to two different kinds of purge valves.

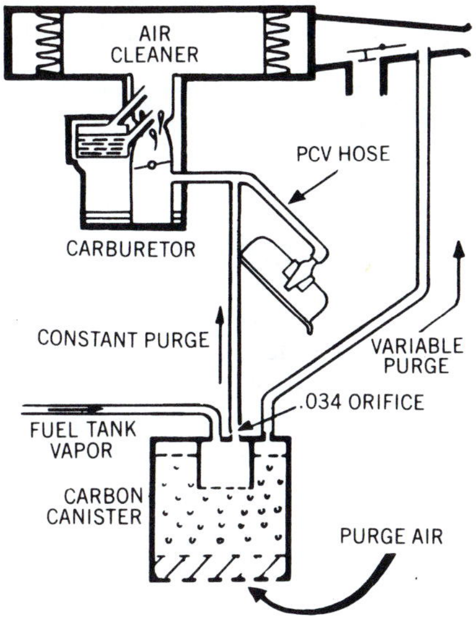


**Figure 15-44.** This EEC system uses a constant-purge orifice to meter fuel vapors to the intake manifold and a variable purge to the air cleaner. (Chevrolet)

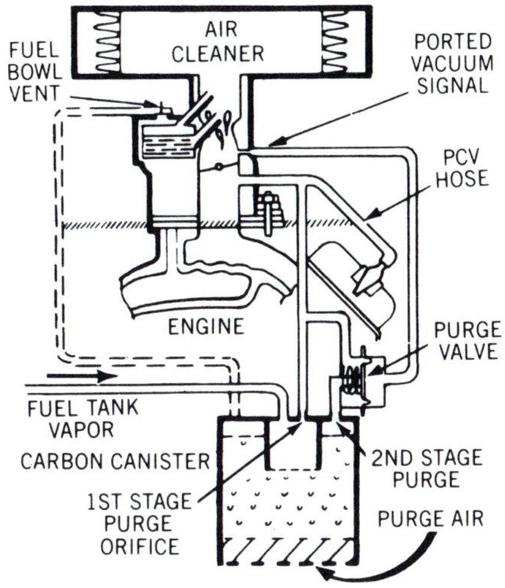


**Figure 15-45.** In this 2-stage purge system, one line provides a constant purge flow to the manifold, while the other is controlled by a vacuum-operated purge valve. (Chevrolet)

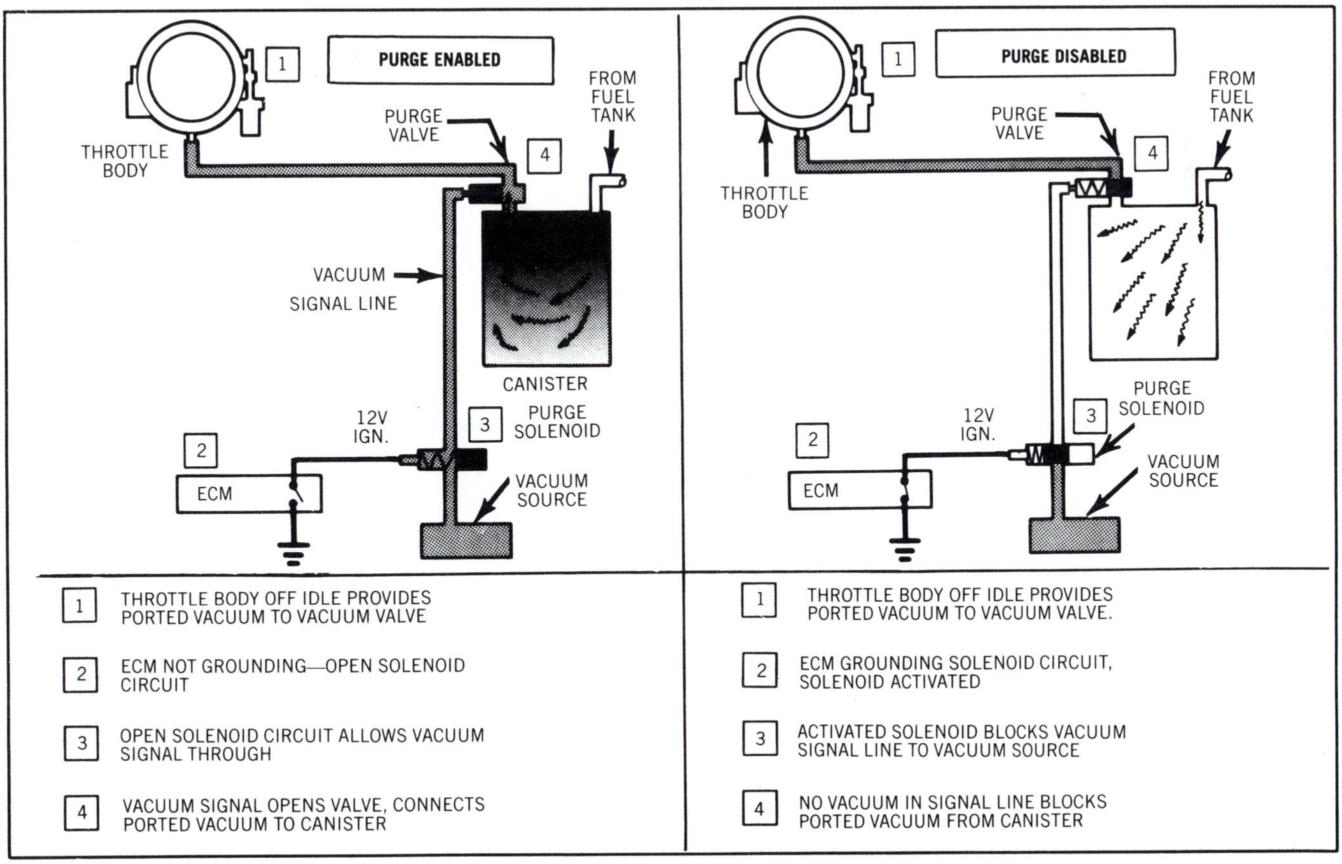


**Figure 15-46.** The solenoid in this electronically controlled system opens and closes the vacuum line to the purge valve in response to signals from the computer. (Cadillac)

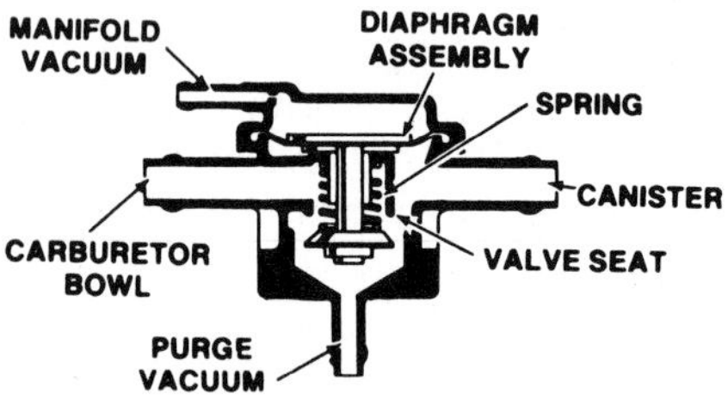


**Figure 15-47.** General Motors type 1 canister purge valve. (GM)

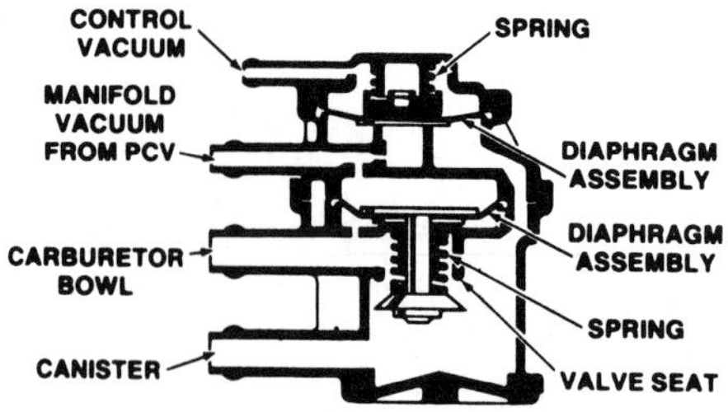


**Figure 15-48.** General Motors type 2 canister purge valve. (GM)

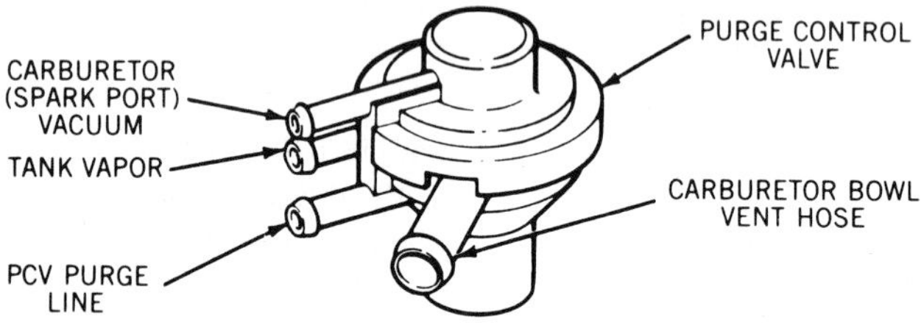


**Figure 15-49.** Ford canister-mounted purge valve. (Ford)

1. In a type 1 valve, figure 15-47, manifold vacuum (through the solenoid) closes the valve to prevent carburetor venting when the engine is running. This allows the canister to vent through the purge vacuum port to the manifold.
2. In a type 2 valve, figure 15-48, manifold vacuum from the PCV system moves the lower diaphragm to close the carburetor bowl vent. When the engine is running above idle, control vacuum moves the top diaphragm to open the purge valve to the PCV system. Control vacuum may be regulated by a solenoid on some engines.

Ford electronic engine control (EEC) systems use the EGR vacuum solenoid or a separate purge solenoid to control vacuum to the canister purge valve. Some late-model Ford canisters purge directly to the air cleaner with no vacuum control. Others purge through a valve to the PCV system, the air cleaner, or both. Ford uses some purge valves that are similar to the GM valves described above. Some are mounted on the canister, figure 15-49, others are installed in the vacuum hoses.

As with other combustion control systems, evaporative emission systems from different carmakers have many variations for specific engines and fuel systems. All, however, have the same purpose and work on the principles outlined here.

## SUMMARY

An automobile must carry enough fuel so that it can run for several hundred miles before refueling. This gasoline or diesel fuel must be stored in a safe tank and delivered to the fuel induction components on the engine. Fuel pumps and lines deliver a steady supply of fuel at uniform pressure and volume to the carburetor or fuel injection system. Fuel filters remove dirt and other contamination from the fuel.

A large volume of clean air at a controlled temperature is equally important for combustion. Air filters remove dirt from the air that would otherwise damage engine parts. Thermostatically controlled air cleaners on gasoline engines control intake air temperature for efficient combustion in a cold or warm engine.

Gasoline carried in an automobile would add to air pollution if hydrocarbon vapors were allowed to escape to the atmosphere. Since 1970 or 1971, all gasoline-powered cars and light trucks have had evaporative emission control (EEC) systems that collect HC vapors and return them to the air-fuel induction system when the engine is running.

## REVIEW QUESTIONS

### Multiple Choice

**1.** All automobile gasoline tanks are vented:

- **a.** through the fuel delivery system
- **b.** internally
- **c.** to prevent vapor lock
- **d.** through vented gasoline caps

**2.** A gasoline supply line is to be replaced. Mechanic A says that copper line will harden and may leak because of excessive vibration. Mechanic B says that a steel line is the only acceptable replacement. Who is right?

- **a.** Mechanic A only
- **b.** Mechanic B only
- **c.** Both A and B
- **d.** Neither A nor B

**3.** All fuel pumps deliver fuel by:

- **a.** electrical pumping action
- **b.** lobe pumping action
- **c.** cam pumping action
- **d.** mechanical pumping action

**4.** A mechanical fuel pump has a movable diaphragm, a diaphragm spring, and:

- **a.** inlet check valves
- **b.** outlet check valves
- **c.** both inlet and outlet check valves
- **d.** an eccentric lobe

**5.** Fuel moves towards a fuel pump from the tank because:

- **a.** gravity forces it
- **b.** of lower pressure in the fuel pump chamber
- **c.** the diaphragm spring pulls it forward
- **d.** the fuel system is a closed system

**6.** Mechanic A says that vapor lock can cause fuel to flow into a vehicle's intake manifold and cause hard starting. Mechanic B says that heat is the cause of the problem. Who is right?

**a.** Mechanic A only
**b.** Mechanic B only
**c.** Both A and B
**d.** Neither A nor B

**7.** An advantage that an electrical fuel pump has over a mechanical fuel pump is:

**a.** that it is less expensive
**b.** that it is lighter
**c.** that it only needs minor maintenance
**d.** that it can deliver fuel immediately to overcome vapor lock

**8.** Mechanic A says that all fuel systems with electric fuel pumps use an oil pressure switch to control the output of the fuel pump. Mechanic B says that they only use relays and in some cases the EEC computer. Who is right?

**a.** Mechanic A only
**b.** Mechanic B only
**c.** Both A and B
**d.** Neither A nor B

**9.** Fuel pumps are rated by:

**a.** unrestricted output
**b.** liters per hour
**c.** output pressure and volume
**d.** gallons per hour

**10.** Mechanic A says that engines with carburetors require fuel pump pressures of 5 to 8 psi. Mechanic B says that fuel injected engines may require four times that amount. Who is right?

**a.** Mechanic A only
**b.** Mechanic B only
**c.** Both A and B
**d.** Neither A nor B

**11.** Which of the following contaminants is more likely found in a diesel fuel system than in a gasoline fuel system?

**a.** water
**b.** dirt
**c.** rust
**d.** dust

**12.** Mechanic A says that a balancing-coil fuel gauge has one coil that must be in balance to move a gauge needle to the "full" indicator on the gauge. Mechanic B says that a balancing-coil fuel gauge has two electromagnetic coils on either side of the gauge needle. Who is right?

**a.** Mechanic A only
**b.** Mechanic B only
**c.** Both A and B
**d.** Neither A nor B

**13.** The major purpose of thermostatically controlled air cleaners is:

**a.** to provide maximum airflow at full throttle
**b.** to provide cooler air when outside temperatures are high
**c.** to provide the most airflow for average performance
**d.** to reduce exhaust emissions during warmup

## Fill in the Blank

**14.** After an engine with a vacuum-diaphragm-controlled air cleaner has started, __________ is routed to the vacuum motor diaphragm.

**15.** EEC systems are used to trap and store __________.

**16.** Fuel tanks in EEC systems are designed to trap some air and allow for __________.

**17.** __________ are used to trap and store fuel vapors.

**18.** On most late-model engines, vapor purging is shut off when the engine is __________, or __________ __________.

**19.** Federal law has required rollover leakage protection in vehicles since __________.

**20.** Mechanical fuel pumps are often operated by an __________ lobe on the camshaft of the engine.

**21.** Electric fuel pumps work well when close to the fuel tank because they maintain a constant __________ on all of the fuel in the lines.

**22.** Dirt particles smaller than one __________ can damage close-tolerance injector parts.

**23.** It takes about 150,000 gallons of air to mix with __________ gallons of gasoline.

**24.** The thermostatic bulb in a thermostatic-bulb air cleaner works like the wax pellet in a __________ thermostat.

**25.** Unlike a gasoline engine, in a diesel engine all fuel __________ takes place inside the combustion chamber.

# 16 GASOLINE CARBURETION

## WHAT

In your study of engine operation and combustion controls, you have learned that gasoline is vaporized and mixed with air before it enters the cylinders. But, you haven't yet seen how the fuel and airflow are metered and mixed for different driving conditions. This chapter explains how the carburetor controls airflow and creates a pressure differential that meters the fuel needed for various conditions from idle to full-throttle, full-load operation.

In this chapter, you will study the air-fuel metering needs of different operating conditions and the carburetor systems, or circuits, that meet these needs. You also will learn about various carburetor designs, control linkages, and devices that help the carburetor reduce exhaust emissions.

## WHY

If an automobile engine operated at a steady speed with an unchanging load, it wouldn't need a carburetor. It would only need two things to mix air and fuel:

- An intake manifold restriction to create the desired amount of airflow for a specific speed and load
- A fuel metering orifice, or jet, to deliver a constant volume of fuel for the required air-fuel ratio.

Automobile operation is not that simple, however. Engine speeds and loads change constantly as a car stops, starts, accelerates, cruises, and decelerates. Air-fuel metering must change with these conditions. Since the first gasoline-engined cars were built, carburetors have met these needs.

Even the simplest carburetor has several fuel metering circuits. Moreover, the carburetor throttle valve controls airflow into the engine and thus engine speed. If any one, or a combination, of the carburetor systems malfunctions, engine operation suffers. As a professional mechanic, you must understand carburetor operation thoroughly to diagnose and correct air-fuel metering problems.

## GOALS

After studying this chapter, you should be able to:

1. Explain the effects of atmospheric pressure and engine vacuum on fuel flow and airflow. Explain venturi action in a carburetor.

2. Identify the major systems of a carburetor and explain what engine operating requirements each satisfies.

3. Identify various types of carburetors and explain the fuel flow and airflow through each.

4. Explain the purposes and operation of various carburetor linkage devices.

5. Explain the operation of various carburetor assist and emission control devices.

## AIR PRESSURE DIFFERENTIAL

Simply put, a carburetor is an air-fuel mixer that operates because of the effects that air pressure has on liquid in a container. You know from Part One of this book that air is gaseous matter and that its weight exerts pressure on everything it touches. The weight of the earth's atmosphere exerts atmospheric pressure of about 14.7 psi (101 kPa or 760 mm Hg) at sea level. If we apply this pressure to liquid in a container and provide an opening in the container to an area of lower pressure, atmospheric pressure will force the liquid to the low-pressure area.

You can see how this works by placing a bucket of water on a bench and then filling a length of hose with more water. Put one end of the hose in the bucket on the bench and the other end in an empty bucket on the floor. As long as there is water in the first bucket and the hose, atmospheric pressure forces water through the hose into the bucket on the floor by **siphon** action, figure 16-1. In this case, atmospheric pressure works with gravity to move liquid from a high point to a low point.

You also can see how an air pressure differential between two points forces liquid from one container to another in the setup shown in figure 16-2. Here, the container filled with liquid is open to atmospheric pressure. A tube connects this container to a second one that has a suction pump. As the pump plunger moves upward, it creates low pressure in the second container. Air pressure in the first container forces liquid through the tube to fill the space in the second container. This is exactly the way a car-

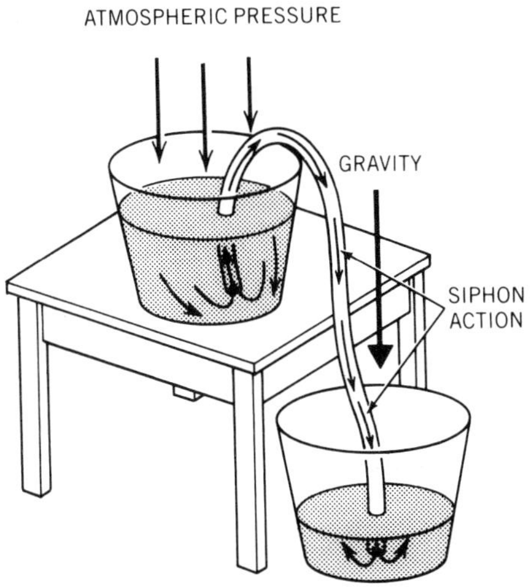


**Figure 16-1.** Atmospheric pressure works with gravity to siphon water from one bucket to another.

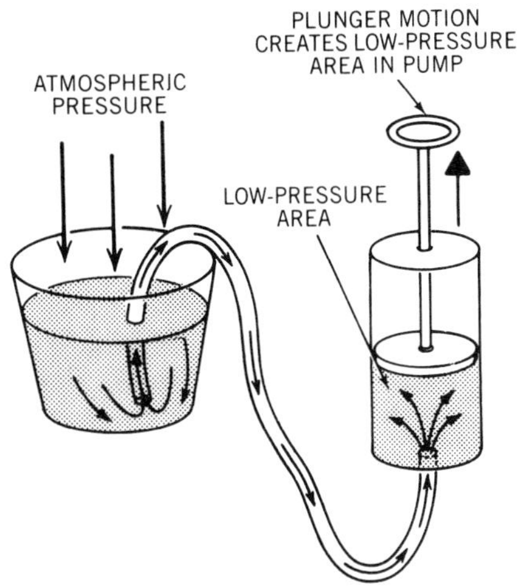


**Figure 16-2.** Atmospheric pressure forces liquid from the container to the pump because the pump plunger creates low pressure in the pump chamber.

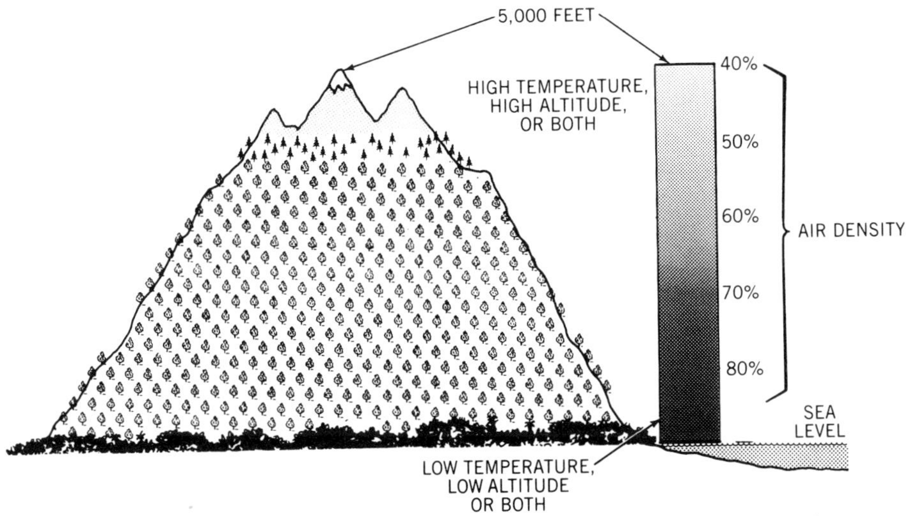


**Figure 16-3.** Air pressure and density decrease as altitude or temperature, or both, increase.

buretor meters gasoline but in a more sophisticated way.

Before we get into carburetor circuits in detail, we should review a few more facts about air pressure that you learned earlier in this book.

## Temperature and Altitude Effects on Air Pressure

The weight of air varies with temperature and height above sea level. Air expands and becomes lighter as temperature rises. This not only reduces pressure, but air density as well. Hot air is less dense than cold air because fewer air molecules occupy every cubic foot or cubic meter of space. Changing air density directly affects engine air-fuel mixtures.

As you rise above sea level, air pressure decreases because the volume of air pressing down on the earth is less. As air pressure drops with increasing altitude, air density also decreases because lower pressure allows the air molecules to expand and fill more space. On a hot summer day in the Sierras or the Rockies, the combined effects of high temperature and high altitude can reduce air pressure and density to a little more than half of what they are at sea level on a cool day, figure 16-3.

## Manifold Vacuum (Low Pressure)

In chapter 2, you learned that air pressure below atmospheric pressure is called vacuum. In chapter 4 you learned that every time a piston moves down on an intake stroke, it enlarges the air space in the cylinder, which lowers the pressure (creates a vacuum). The low-pressure area extends from the cylinder, through the open intake valve, into the intake manifold. This is called manifold vacuum, figure 16-4. The low pressure (vacuum) in the manifold allows atmospheric pressure to force more air through the carburetor, into the engine. The carburetor throttle is a movable restriction in the air intake that regulates vacuum, airflow, and consequently engine speed.

The pressure differential between atmospheric pressure and manifold vacuum controls the amount of fuel drawn, or siphoned, through various carburetor circuits. Atmospheric pressure is applied to gasoline in the carburetor fuel bowl. Vacuum (lower pressure) exists at the discharge nozzles of circuits that draw fuel from the bowl. Atmospheric pressure remains relatively constant. It doesn't change much during several minutes of driving. Manifold vacuum changes constantly depending on engine speed, load, and throttle valve position. Vacuum varies from as high as 21 or 22 inches of mercury (low pressure) to as low as 0 inches of mercury (nearly atmospheric pressure). Because fuel metering is controlled by the difference between atmospheric pressure (which doesn't change much) and vacuum (which does), fuel flow is proportional to the amount of vacuum at the discharge end of a fuel circuit. You will see how this works as you study each circuit, or metering system, in detail.

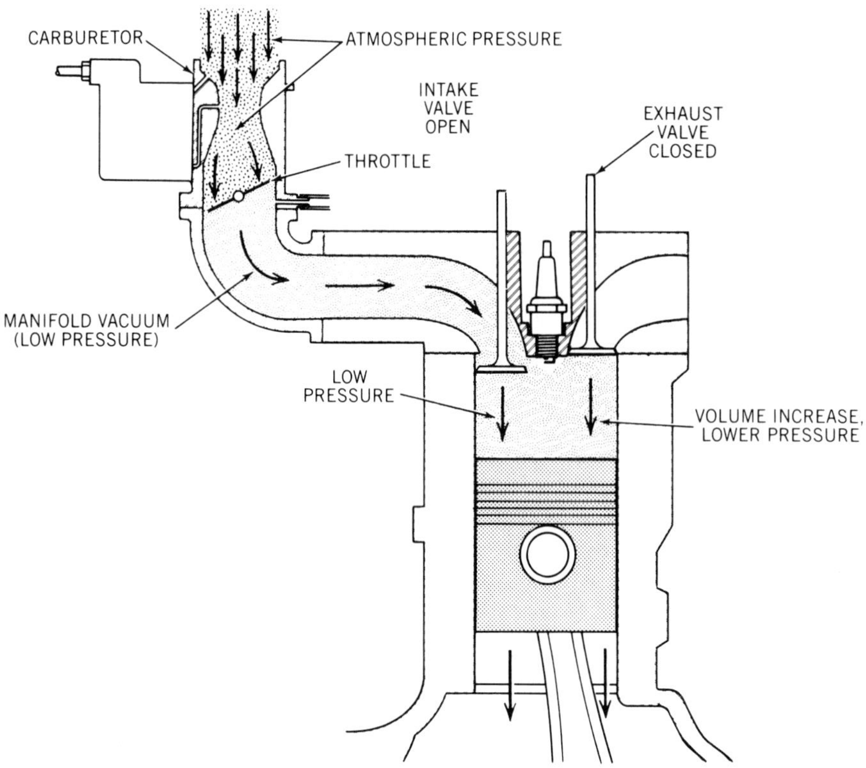


**Figure 16-4.** As a piston moves down, it creates low pressure (vacuum) in the cylinder and manifold. Atmospheric pressure forces more air into the engine through the carburetor.

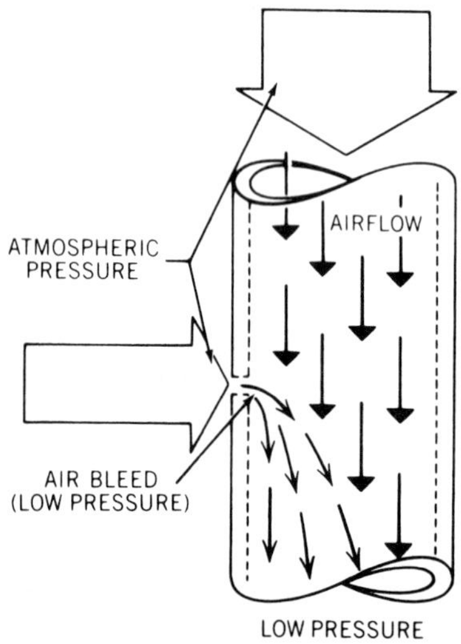


**Figure 16-5.** Airflow through a barrel or cylinder creates low pressure along the sides. This draws more air through a hole into the airstream.

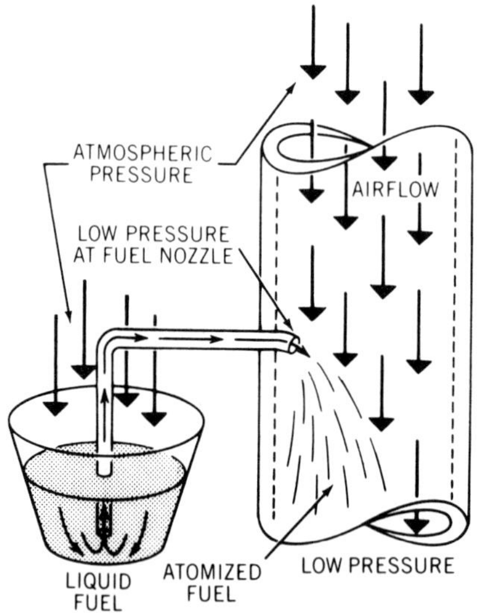


**Figure 16-6.** Atmospheric pressure forces liquid through the tube into the low-pressure area caused by the moving airflow.

## AIRFLOW AND VENTURI ACTION

As the carburetor throttle opens, air moves from the high-pressure area outside the engine to the low-pressure area in the manifold. The throttle opening determines the amount and speed of airflow.

Pressure of a moving airstream is always less at a right angle (perpendicular) to the airflow than it is in the center of the airstream in the direction of airflow. Therefore, air rushing through a carburetor barrel creates low pressure at the sides of the barrel, perpendicular to airflow. A small hole in the side of the barrel causes more air to be drawn into the low-pressure edges of the moving airstream, figure 16-5.

If that hole is connected by a tube to a bowl with fuel in it, the fuel is forced through the tube by the difference between atmospheric pressure on the fuel in the bowl and lower pressure along the sides of the carburetor barrel, figure 16-6. This works on the same principles as the simple siphon and suction pump in figures 16-1 and 16-2. The amount of fuel drawn through the tube depends on the speed of airflow through the barrel. As airflow moves faster, it causes lower pressure at the metering hole, and more fuel flows.

To make a carburetor work efficiently, we need enough airflow velocity through the barrel at all times to draw the right amount of fuel from the bowl. If we put a restriction in the barrel, airflow speeds up as it moves through the restriction. The increased velocity creates more vacuum at the restriction because the air moves faster to get through the restriction. Also, pressure is lower (more vacuum) just below the restriction than above it because of the higher air velocity. Engineers call this a "pressure drop" across the restriction. The restriction is called a **venturi**, and its effect on airflow and pressure is called the **venturi effect**, figure 16-7.

The venturi effect and the pressure drop across the venturi are similar to the actions of voltage and resistance in an electric circuit. The air pressure

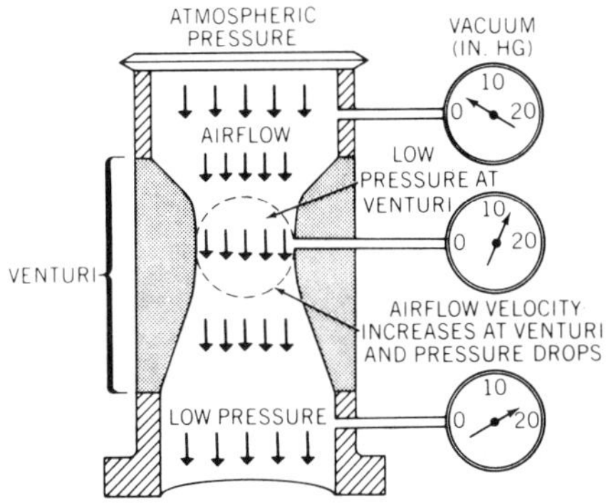


**Figure 16-7.** Airflow velocity increases through the venturi and lowers the air pressure.

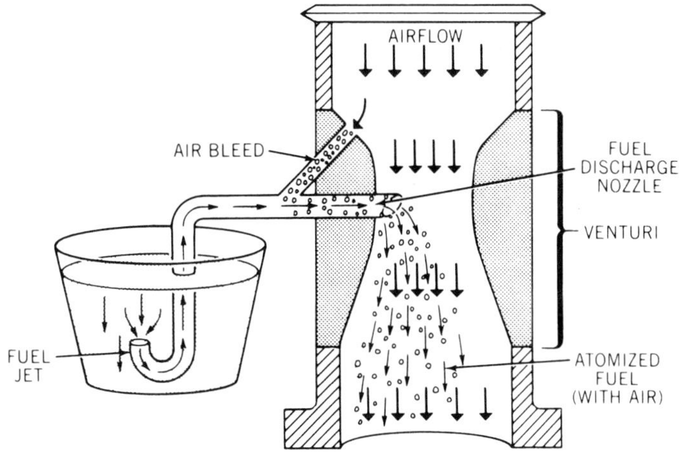


**Figure 16-8.** The main fuel discharge nozzle is at the venturi in the carburetor. Air bleeds help atomize the fuel as it flows through carburetor passages.

differential that moves air molecules is similar to voltage that moves electrons. The venturi resists airflow as a resistor resists current flow, and the venturi creates an air pressure drop as a resistor creates a voltage drop.

If we place a fuel discharge nozzle at, or just below, the venturi, the increased vacuum created by the venturi causes more fuel to flow, figure 16-8. The sizes and locations of venturis and fuel nozzles are critical parts of carburetor design. Their size and placement determine the amount of fuel mixed with varying airflow.

Besides mixing gasoline and air, a carburetor must help to atomize and vaporize the liquid fuel. One way to do this is by placing **air bleeds** in the barrel above the venturi, figure 16-8. An air bleed is an air passage that leads to a fuel passage in the carburetor. This aids atomization and vaporization by mixing air with liquid fuel before it reaches the discharge nozzle in the barrel.

## Carburetor and Manifold Vacuum Levels

There are four air pressure measurements that are important for studying carburetor operation:

1. *Atmospheric pressure*—air pressure outside the carburetor and on the fuel in the fuel bowl
2. *Manifold vacuum*—low pressure

### KNOW YOUR ABC'S

You are probably familiar with X-cars, K-cars, and others in the alphabet soup of corporate car model identities. You also probably know that these letter designations are a handy way that many carmakers have identified basic body shells shared by several divisions of a corporation.

Back in the 1930's, '40's, and '50's, it was simple. General Motors Fisher Body Division, for example, built three basic bodies for all five GM divisions. A-bodies were Chevrolets, Pontiacs, and small Oldsmobiles. B-bodies were big Oldsmobiles and all Buicks except Roadmasters. C-bodies were Buick Roadmasters and Cadillacs. At Chrysler similarly A-bodies were Plymouths and Dodges, B-bodies were DeSotos, and C-bodies were Chryslers.

In the late 1950's and early '60's, the alphabet started getting complicated as carmakers introduced new compact and intermediate models. GM A-bodies then became "intermediates," and B-bodies became "full-size" models sold by all divisions except Cadillac. During the next 20 years, the alphabet of several carmakers grew to include all sorts of H, K, L, T, F, X, J, and other model identifications. Ford never did care for letter identification of different models. The Ford engineering department instead chose to use animal code names and thus developed the "fox platform" and the "panther platform."

Finally, in the early 1980's General Motors had enough alphabet soup and started using numbers to identify new models under development. The corporation also stopped publicly advertising the common model identities. There is a lot of interesting automobile history in the development and changes in basic carlines over 40 or so years. And there also is practical value in knowing these ABC's for professional mechanics. If you can remember that a carmaker may have used a dozen or more nameplates on a single basic body and chassis for several years, it will help you sort out the various model specifications and keep track of the similarities. It's as simple as ABC.

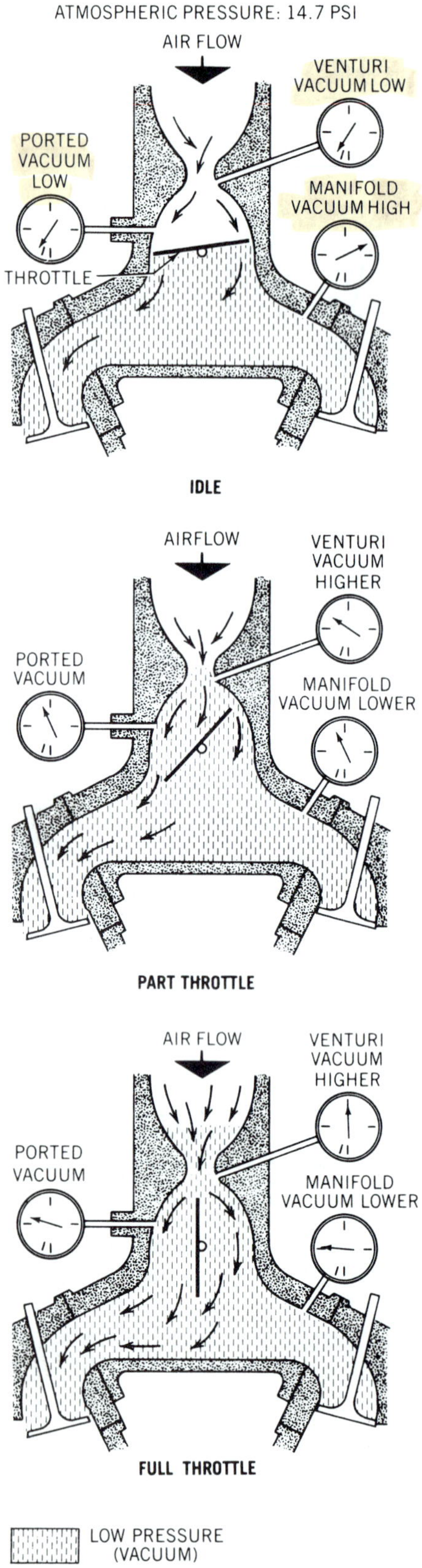


**Figure 16-9.** Atmospheric pressure is constant, but vacuum in tne manifold, above the throttle, and at the venturi changes in relation to throttle position and airflow velocity.

below the carburetor throttle, created by piston action and regulated by the position of the throttle

**3.** *Venturi vacuum*—low pressure in the carburetor barrel, created by airflow through the venturi restriction

**4.** *Ported vacuum*—the low-pressure area just above the throttle in the carburetor barrel.

Figure 16-9 shows where these pressures exist in a carburetor and manifold. Atmospheric pressure always exists above the carburetor, and it stays constant for any temperature and altitude. Manifold vacuum is high when the throttle is closed and lower when it is open. Venturi vacuum increases with the speed of airflow through the carburetor. It is low when the throttle is closed and high when the throttle is open to allow more airflow. Ported vacuum is low when the throttle is closed and increases as the throttle opens.

Late-model engines use ported vacuum to operate distributor diaphragms, EGR valves, and other emission and driveability devices. These devices require vacuum as the throttle is opened, but no vacuum when it is closed. Ports in the carburetor barrel lead to tubes on the carburetor base, figure 16-10. Hoses connect the tubes to the vacuum-operated devices.

Because various vacuum-operated actuators require different vacuum levels at different times, the design of carburetor vacuum ports is a precise science. The size of the port and the location above the throttle determine the timing and amount of vacuum applied to an actuator. EGR vacuum ports, for example, often are slightly higher in the throttle bore than distributor vacuum ports. This delays EGR vacuum until the engine (and carburetor) are in the cruising mode. Some carburetor vacuum ports are 2-stage openings to modulate the vacuum applied to an actuator.

You know that vacuum is simply air pressure lower than atmospheric pressure. But, as you study figure 16-9, you can visualize a "vacuum column" in the manifold and carburetor. The height of the vacuum column, or the location of low pressure, changes as the throttle opens and closes. With a closed throttle, manifold vacuum is high. Almost all of the low-pressure area is below the throttle valve. As the throttle opens, vacuum starts to rise in the carburetor barrel. Airflow also increases as the throttle opens, and this starts to increase venturi vacuum. As the throttle opens farther, vacuum continues to rise in the barrel, and airflow speed increases. More vacuum reaches vacuum ports and fuel discharge nozzles. At wide-open throttle, the vacuum column reaches the top of the carburetor. Actually, manifold vacuum decreases at maximum throttle, but airflow velocity increases and draws the right amount of fuel into the air-fuel mixture.

## ENGINE OPERATION AND CARBURETION REQUIREMENTS

A carburetor has seven fuel-metering systems to serve different air-fuel needs. Before studying these systems, you should know something about the engine's changing air-fuel requirements.

• *Starting*—When an engine starts, intake vacuum and airflow velocity are low because the engine turns slowly. Slow speed, combined with a cold engine, reduces fuel vaporization. Less gasoline reaches the combustion chamber. Therefore, an engine needs a rich air-fuel mixture for starting. The carburetor must have a system to provide this mixture to a cold engine at low speed with low vacuum.

• *Idle*—An engine also needs a rich mixture at idle. Manifold vacuum is high, which promotes vaporization, but low airflow velocity reduces fuel vaporization and some exhaust remains in the cylinders to dilute the incoming mixture. The carburetor must provide this richer mixture at low speed with high manifold vacuum, low airflow, and low vacuum in the carburetor barrel.

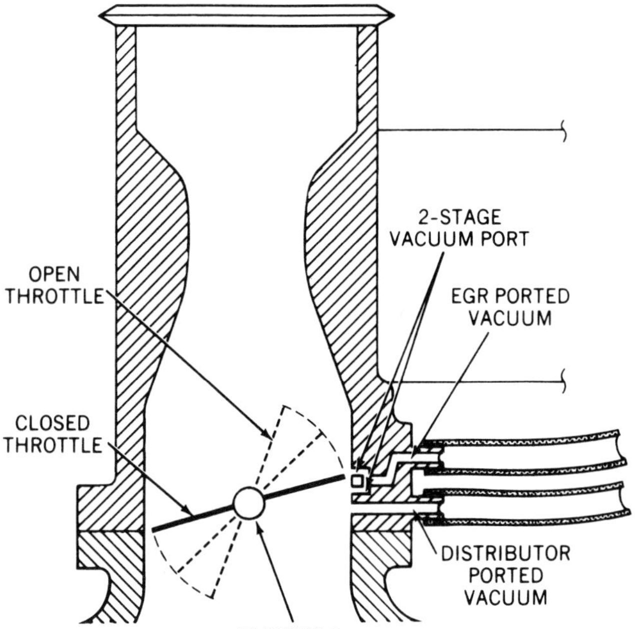


**Figure 16-10.** Ported vacuum operates EGR valves and distributor advance units that require different amounts of vacuum in relation to throttle position.

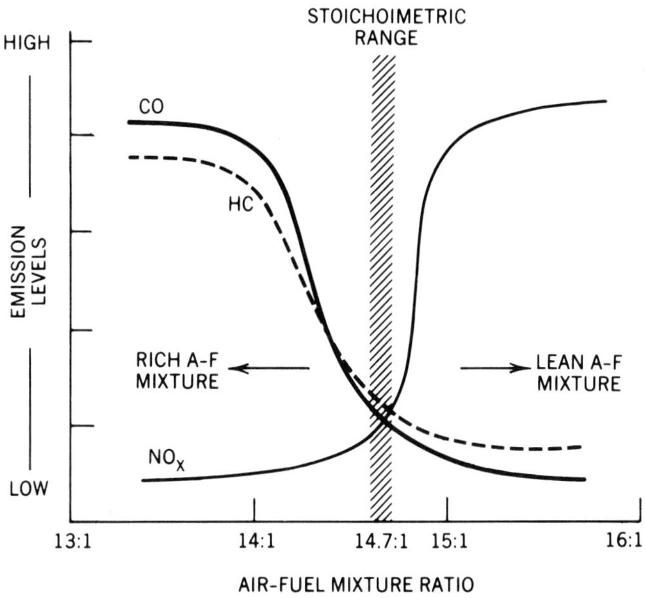


**Figure 16-11.** A stoichiometric air-fuel ratio of about 14.7:1 provides the best combination of performance, economy, and emission control.

• *Low speed*—As a car accelerates gradually, engine speed increases, vacuum rises in the carburetor, and airflow increases. The engine starts to need a leaner mixture, and the carburetor must provide a smooth transition to low-speed cruising requirements.

• *Cruising economy*—At cruising speed, with light engine loads, the best fuel economy is obtained with air-fuel ratios of about 15:1 to 16:1. A carburetor must provide steady fuel delivery at these ratios with high, relatively constant vacuum and airflow.

• *Extra power requirements*—An engine develops maximum power with air-fuel ratios of 12.5:1 to 13.5:1. Extra power demands can occur under several conditions: sudden acceleration, low-speed heavy-load operation, and high-speed full-throttle operation. Low manifold vacuum accompanies all of these conditions. A carburetor must provide extra fuel with low vacuum and low airflow for acceleration and low-speed heavy-load operation, and with low vacuum and high airflow for high-speed operation. Air-fuel ratios needed for maximum power do not change much except at low speed with slower airflow into the engine. Reduced airflow causes poor air-fuel mixing and poor distribution to the cylinders. A slightly richer mixture is necessary to make up for this.

The leaner air-fuel ratios that provide best economy also stay almost the same throughout an engine's operating range. Some enrichment is needed, however, at idle and high speed to prevent a lean misfire and maintain power.

Between the air-fuel ratios of 15:1 or 16:1 for best economy and the ratios of 12.5:1 or 13.5:1 for maximum power is the ratio that provides the best combination of performance, economy, and emission control. This is called the stoichiometric ratio. For most gasoline engines it is about 14.7:1, figure 16-11. At the stoichiometric ratio, most of the fuel HC molecules are oxidized to water vapor and $CO_2$, and the engine produces minimum HC and CO emissions. $NO_x$ formation also is low during stoichiometric combustion but increases with leaner air-fuel ratios. Electronic engine controls and closed-loop fuel systems provide stoichiometric air-fuel ratios of 14.7:1 for most engine operating conditions.

## FUEL ATOMIZATION, VAPORIZATION, AND DISTRIBUTION

All carburetor systems must atomize liquid fuel and help it to vaporize under all operating conditions. The air-fuel mixture then must be distributed evenly to the cylinders. Correct distribution depends on:

• Fuel volatility
• Correct atomization
• Complete vaporization
• Throttle valve position
• Carburetor location on the intake manifold
• Intake manifold design.

Gasoline is atomized as it passes from

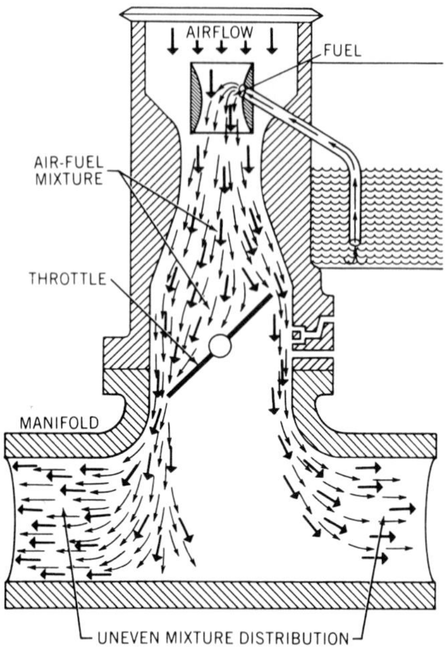


**Figure 16-12.** The throttle valve angle can deflect most of the air-fuel mixture to one side of the manifold.

carburetor discharge nozzles into the moving airstream. It begins to vaporize as it passes the throttle and enters the manifold. Complete vaporization occurs when the fuel is hot enough to boil. These factors affect vaporization:

- The more volatile a fuel is, the lower temperature at which it vaporizes.
- Fuel vaporizes at a lower temperature as pressure decreases. Because manifold vacuum creates a low-pressure area, gasoline vaporizes more easily in the manifold than at atmospheric pressure.
- Heating the fuel increases vaporization.

Poor vaporization can be caused by low fuel volatility, cold intake air, or a cold manifold. You have learned that oil companies blend gasolines of different volatilities for different climates and regions of the country. You also have learned that thermostatic air cleaners and heated intake manifolds overcome the problems of a cold air-fuel mixture. Poor manifold design and incorrect carburetor airflow also can cause poor vaporization. Engineers, however, consider the fuel distribution needs of an engine when they design these parts.

The throttle valve affects distribution because the angle of the throttle directs the mixture against one side of the manifold, figure 16-12. This could cause a rich mixture to move toward some cylinders and a lean mixture to move toward others. Also, cylinders closest to the carburetor may receive more of the mixture than those farther away. Manifolds are designed to minimize these possible problems and distribute the mixture evenly to all cylinders.

The shape, surface, and size of manifold passages determine manifold efficiency. Smooth surfaces speed the mixture flow, but rough surfaces can aid vaporization by breaking up the mixture. Passages must be large enough to supply the needed air-fuel volume and an equal mixture to all cylinders. Passages that are too large or too small will slow down or restrict the mixture flow at different engine speeds. Passages should be free of sharp corners that abruptly change flow direction and cause liquid fuel to drop out of the mixture.

## BASIC CARBURETOR SYSTEMS

All carburetors are not identical. Even a single model can come in dozens of part numbers with internal and external differences. Although there are thousands of variations in designs, all carburetors must meet the same basic air-fuel needs of an engine, and almost all share the following seven systems:

1. Float and fuel bowl system
2. Idle system
3. Low-speed system
4. High-speed, or main metering, system
5. Power system
6. Accelerator system
7. Choke system.

We can study these systems separately, but they are not independent of each other. All work together even though one or another may be more significant for a particular engine operating need.

### The Float and Fuel Bowl System

The fuel bowl is a small reservoir that holds a few cubic centimeters of gasoline to supply all other carburetor systems. Fuel level in the bowl is critical because it determines the fuel level and fuel flow in all other passages of the carburetor. For example, the main fuel metering jet for the high-speed, or main metering, system is in the bottom of the bowl and leads to the main fuel passages of the carburetor. A liquid seeks its own level in any container, so the fuel level in the metering passages is the same as the fuel level in the bowl, figure 16-13. If the level is too high, venturi vacuum at the discharge nozzle will draw too much fuel into the airflow and the mixture will be too rich. If fuel level is too low, too little fuel enters the airflow and the mixture is too lean.

When you studied fuel pumps, you learned that fuel level in the bowl and fuel delivery from the pump are controlled by a float and an inlet needle valve. As gasoline is drawn out of the bowl, the float drops and moves the inlet needle off its seat. This opens the valve and allows more fuel to enter the bowl. As the bowl fills, the float rises and closes the inlet valve, figure 16-14. This shuts off fuel flow and pressure to the carburetor until the fuel level drops again. In some conditions, fuel flow out of the carburetor to the engine and into the bowl from the pump are about equal. In these cases, the needle valve stays partly open to maintain the needed fuel level and flow rate.

When fuel in the bowl is at its highest and the float is holding the inlet needle closed, an air space remains above the fuel. A **balance tube**, or bowl vent, connects this air space to atmospheric pressure at the top of the carburetor airhorn, figure 16-14. This arrangement maintains the differential between atmospheric pressure and engine vacuum needed to move fuel through the carburetor and into the engine. Because the balance tube is inside the air filter, the carburetor is internally balanced. This location also

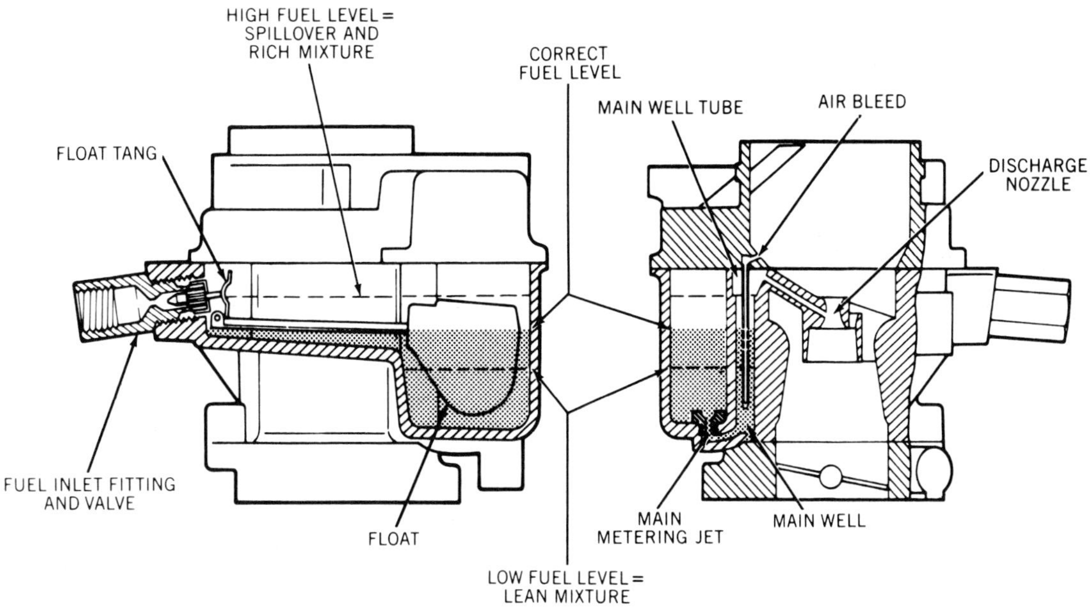


**Figure 16-13.** Fuel level in the bowl affects the amount of gasoline metered into the carburetor airflow. (Chrysler)

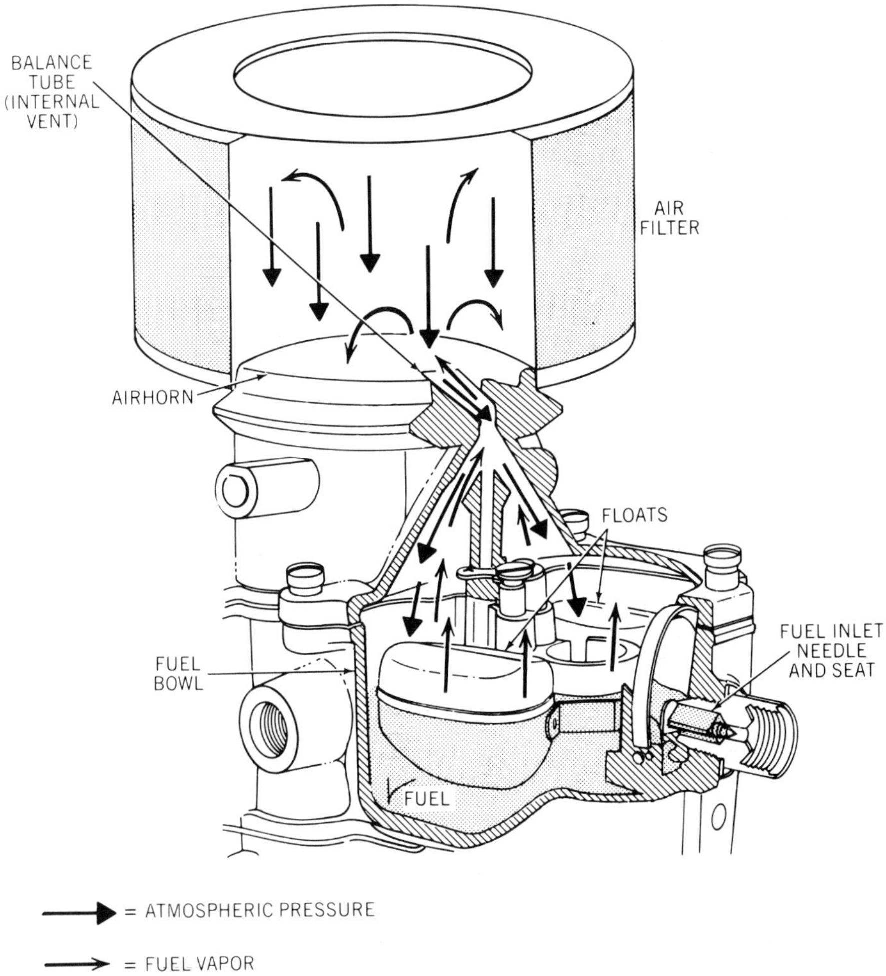


**Figure 16-14.** The float maintains the correct fuel level in the bowl. The balance tube maintains atmospheric pressure above the fuel. (Chrysler)

compensates for any air pressure drop through the air filter.

If the balance tube were not there, or if it were plugged, a vacuum could develop above the fuel as fuel level dropped. This would hold the float up and restrict fuel delivery to the carburetor. Then the engine would be starved for fuel and would stall or run roughly.

The designs of fuel bowls, floats, needle valves, and seats vary, but all work the same way, figure 16-15. Some floats have small springs to keep them from bouncing, and some fuel bowls have baffles to keep fuel from sloshing on turns and on rough roads. Most inlet needles have rubber tips, and the valve seats are made of brass. These nonmagnetic materials minimize the collection of metallic particles on the valve and allow the needle to close the valve if rough spots develop on the seat. If the inlet valve leaks, too much fuel will enter the bowl and the carburetor will flood.

When you studied EEC systems, you learned that fuel bowls are vented to vapor storage canisters when the engine is off to keep fuel vapors out of the air. Many late-model carburetors have plastic fuel bowls to insulate the fuel and reduce evaporation. Some also

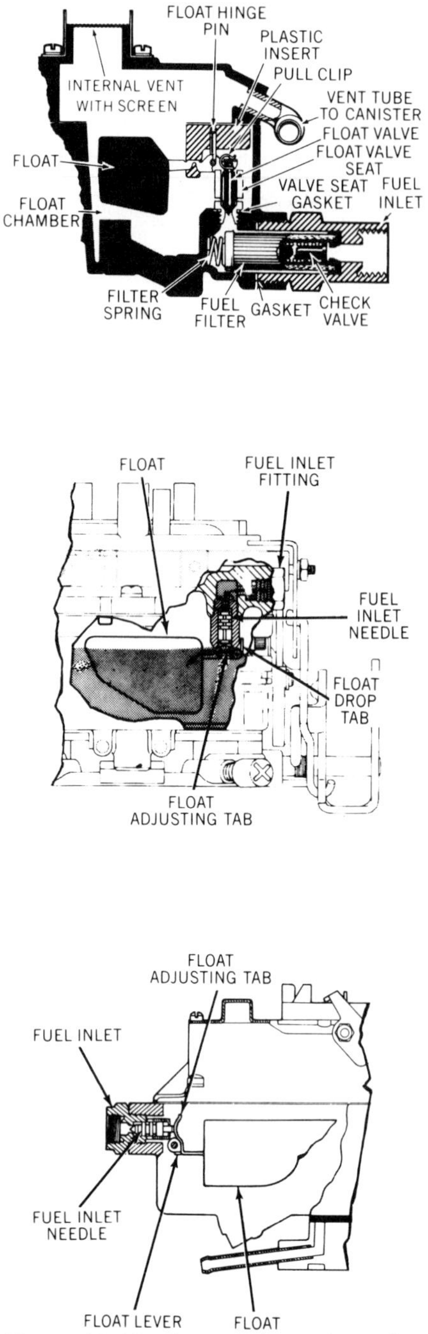


**Figure 16-15.** Float, needle, and seat designs vary, but all work the same way. (Rochester) (Chrysler)

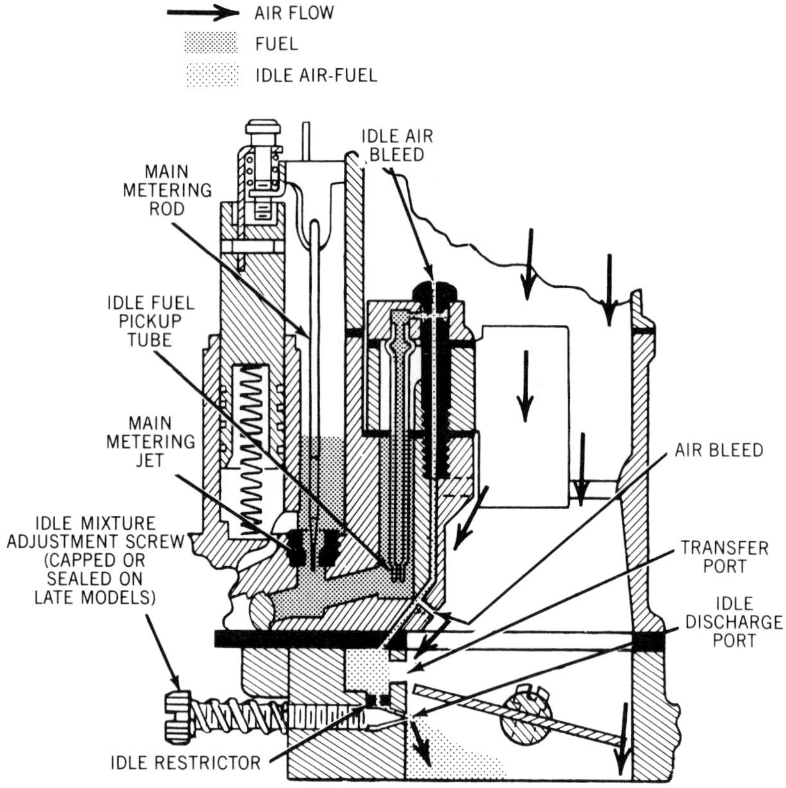


**Figure 16-16.** At idle, fuel is discharged below the throttle. Air bleeds allow the fuel to atomize in the idle passages. (Chrysler)

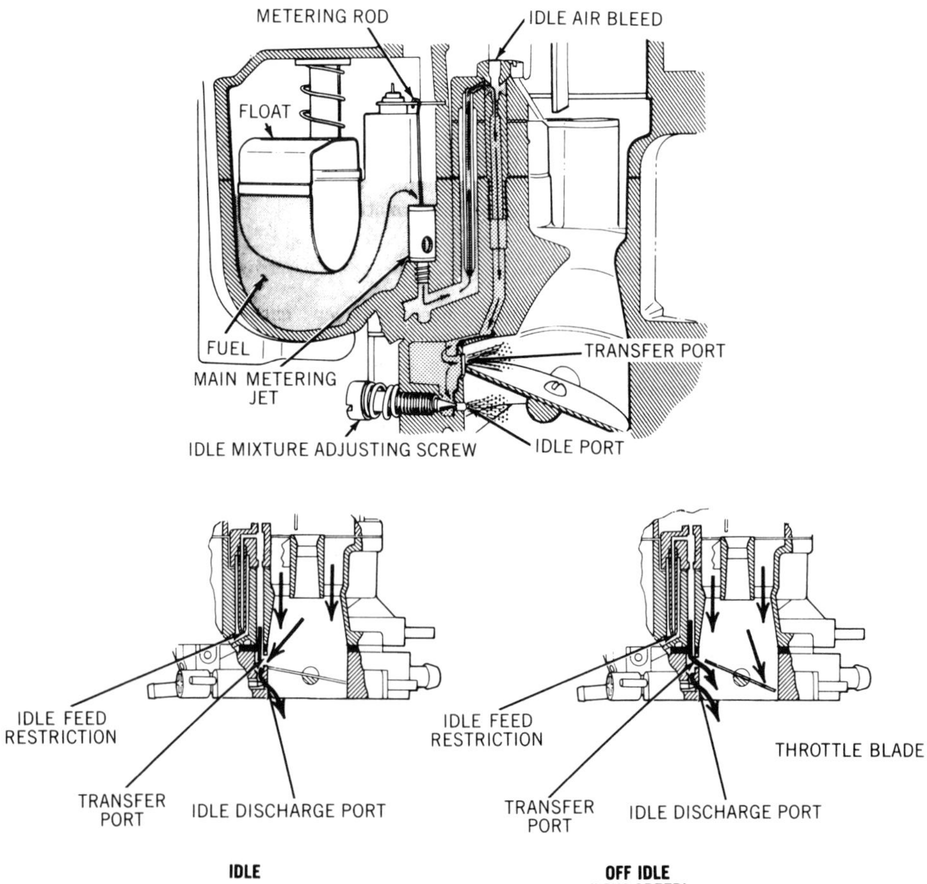


**Figure 16-17.** At idle, air enters the transfer port and mixes with the idle fuel. As the throttle moves off idle, fuel flows out of the transfer port for low-speed operation. (Chrysler)

have an insulator between the carburetor and manifold to protect the fuel bowl from heat. Fuel bowls of late-model carburetors tend to be smaller than the bowls of pre-emission carburetors to reduce the amount of gasoline that can be evaporated by engine heat.

### The Idle System

When an engine is idling, the carburetor allows only minimum airflow to keep engine speed at the minimum. There is little or no venturi vacuum, and no fuel flows from the main metering nozzle at the venturi. The idle system supplies just enough air and fuel to keep the engine running.

Idle discharge ports are located just below the closed throttle because manifold vacuum is high at idle, figure 16-16. The pressure differential between the air in the fuel bowl and manifold vacuum draws fuel through the idle system. Gasoline enters the idle system through the main jet at the bottom of the fuel bowl. It then flows to the idle tube where air bleeds let in air to start breaking up the liquid fuel. The air bleeds also keep fuel from being siphoned out of the idle tube at high speed or when the engine is stopped. Additional air bleeds may be located just above the throttle. Most fuel atomization occurs in the idle passages because of the slight airflow at idle.

Even at idle, the amount of air passing through the carburetor determines engine speed. In most carburetors, the throttle does not close completely, which allows a small airflow at idle, figure 16-16. Other carburetors have an idle air bypass that draws air through a separate passage, past a fully closed throttle.

Small ports, called **transfer ports**, or slots, just above the throttle also allow air into the idle fuel passage, figure 16-17. When the throttle starts to open from idle, it exposes the transfer ports to vacuum. Flow then reverses through the ports. Instead of allowing extra air into the idle fuel passage, the transfer ports (now exposed to vacuum) allow more fuel to flow into the airstream. The transfer ports help to transfer carburetor operation from the idle system to the low-speed system.

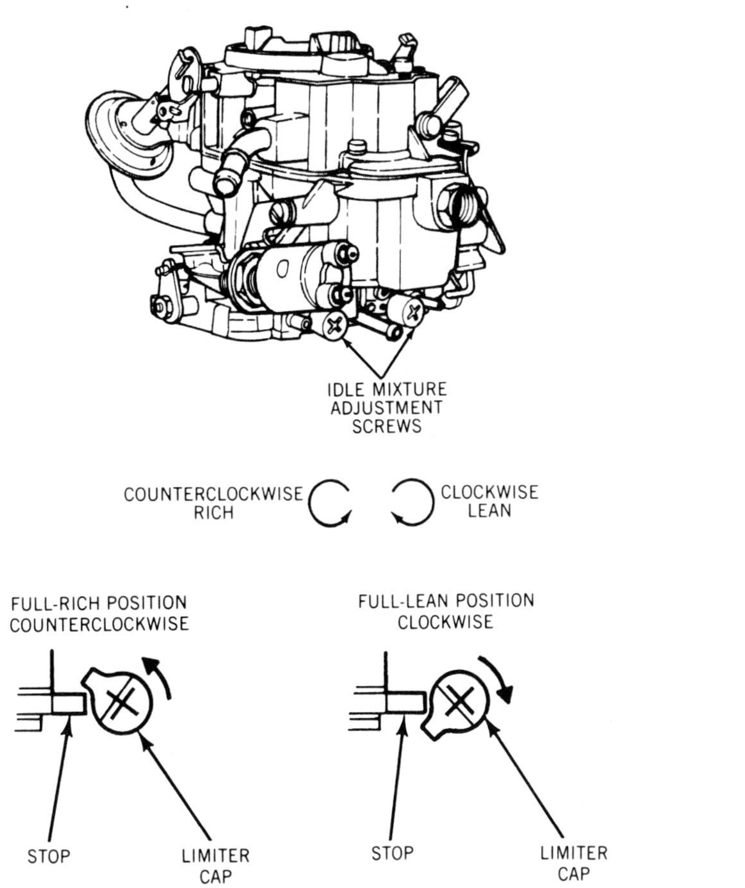


**Figure 16-18.** Adjustment screws control the amount of fuel in the idle air-fuel mixture. Most carburetors from the late 1960's to late 1970's had plastic caps to limit mixture adjustment. (Chrysler)

Idle mixture screws, which are adjustable needle valves, control the amount of gasoline in the idle air-fuel mixture. Each primary barrel of the carburetor has a mixture screw. Turning the screw inward (clockwise) moves the needle tip into the idle passage to restrict fuel, figure 16-18. Turning the screw outward (counterclockwise) moves the needle out to allow more fuel flow.

Pre-emission carburetors had mixture screws that could be adjusted from fully open to fully closed. Most carburetors built from the late 1960's to the late 1970's have plastic limiter caps on the mixture screws that limit adjustment to one turn or less, in or out, figure 16-18. This prevents an overly rich or lean mixture that would increase HC or CO emissions. On most carburetors built since 1979 or 1980, idle air-fuel ratio is calibrated at the factory, and the mixture screws are sealed with steel caps or plugs, figure 16-19. This prevents any tampering with the air-fuel ratio that provides the best idle emission control.

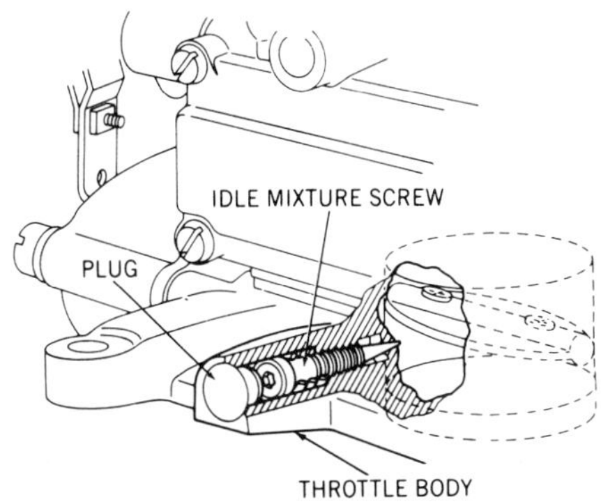


**Figure 16-19.** The idle mixture screws are sealed on most late-model carburetors. (Ford)

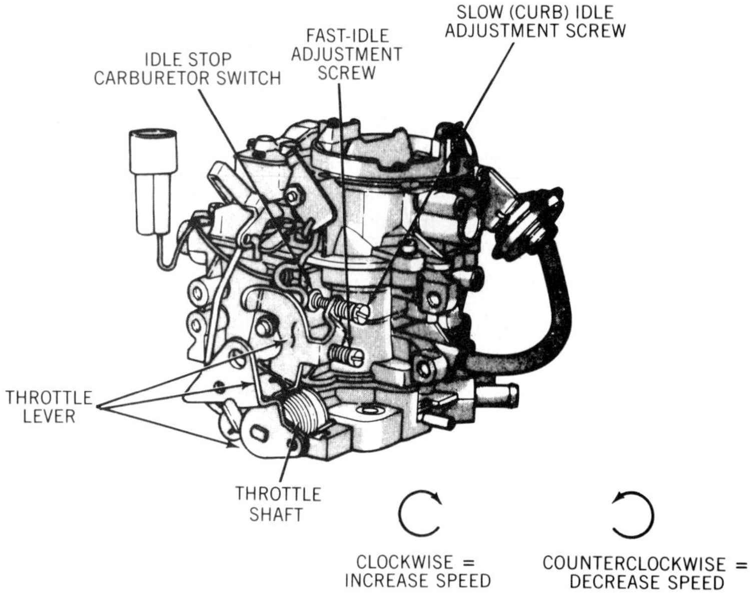


**Figure 16-20.** This basic idle speed screw adjusts the throttle position to regulate idle airflow and engine speed. (Chrysler)

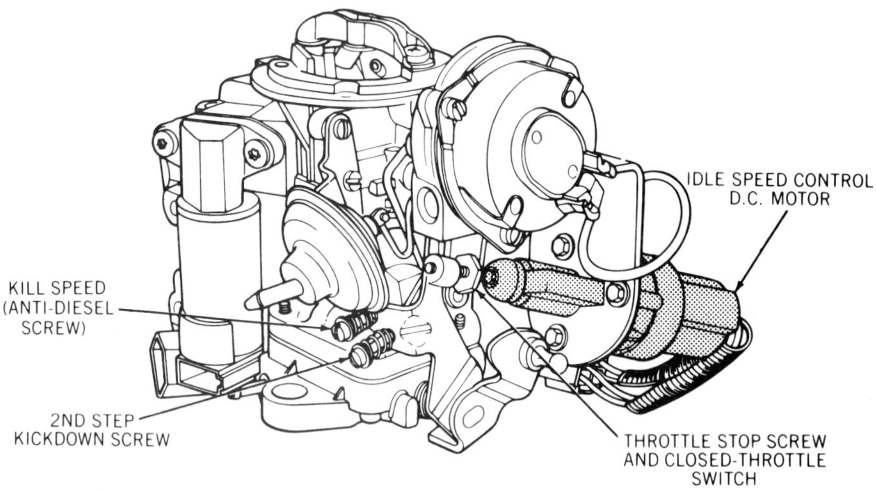


**Figure 16-21.** This idle speed control motor changes idle speed in response to signals from the engine computer. (Ford)

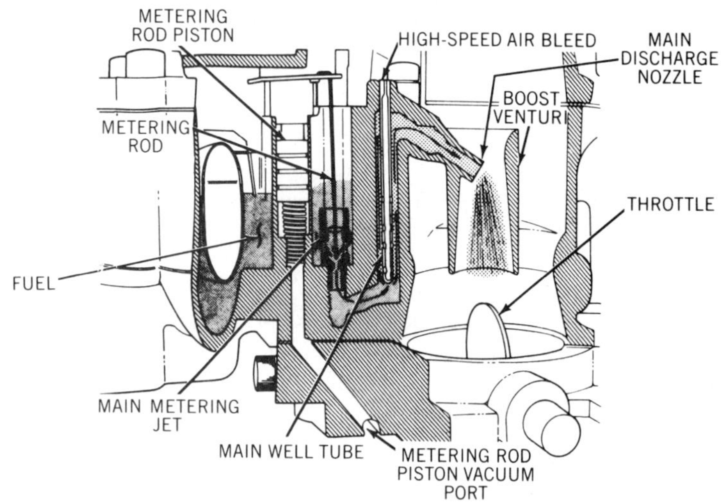


**Figure 16-22.** High-speed (main metering) fuel flow is from the main discharge nozzle at the venturi. (Chrysler)

Idle speed is adjusted by changing the airflow through the carburetor. Most carburetors have an idle speed screw that is an adjustable stop for the closed throttle, figure 16-20. Turning the screw clockwise holds the throttle open farther and allows more airflow. Turning it counterclockwise closes the throttle farther and reduces airflow. A carburetor with an idle air bypass passage has a large screw that varies the opening in the air passage. Turning it clockwise reduces airflow; turning it counterclockwise increases airflow. On many late-model feedback carburetors for electronically controlled engines, idle speed and airflow are regulated by an electric motor or solenoid controlled by the system computer, figure 16-21.

## The Low-Speed System

As the throttle opens for off-idle, low-speed operation, the engine needs more fuel than the idle system can provide. However, airflow through the venturi is not strong enough to draw fuel through the main metering discharge nozzle. As you have seen, flow reverses through the transfer ports. Instead of adding air to the idle system, they now allow more fuel to flow for low-speed acceleration, figure 16-17. Fuel continues to flow from the idle discharge port during transfer, but flow is less than at idle. This operation of the transfer and idle ports provides a steady air-fuel mixture as vacuum increases in the venturi to bring the main metering system into operation.

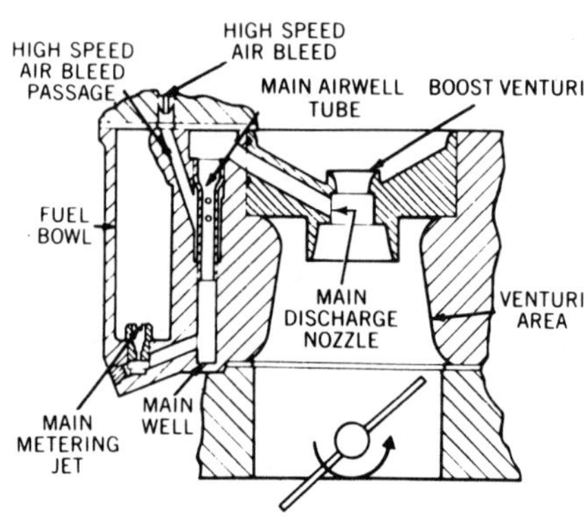


**Figure 16-23.** Multiple boost venturis improve air and fuel mixing. (Ford)

Some engineers consider the low-speed system to be a continuation of the idle system because the same fuel passages are used for idle and low-speed operation. The important points to remember are how fuel and airflow change gradually as the engine comes off idle. The low-speed system (or low-speed operation of the idle system) provides the smooth transition needed from idle to high-speed operation.

## The High-Speed (Main Metering) System

As the throttle opens farther, airflow increases and vacuum increases in the carburetor barrel. Increasing airflow and vacuum increase the pressure drop (vacuum) at the venturi. This causes fuel to flow from the main discharge nozzle, figure 16-22. As the throttle opens wider, fuel flow from the low-speed system decreases, and flow from the high-speed system increases. The main metering system then provides all fuel for high-speed, light-load cruising.

To improve air-fuel mixing efficiency, most carburetors have several "boost" venturis inside of each other. The main nozzle is inside the smallest venturi to increase vacuum action, figure 16-23. The upper venturi creates vacuum that draws the fuel from the bowl. The lower venturis create an airstream that holds the fuel away from the barrel walls. The resulting turbulence improves atomization and air-fuel mixing. A high-speed air bleed adds air to the fuel before it leaves the discharge nozzle to further improve mixing.

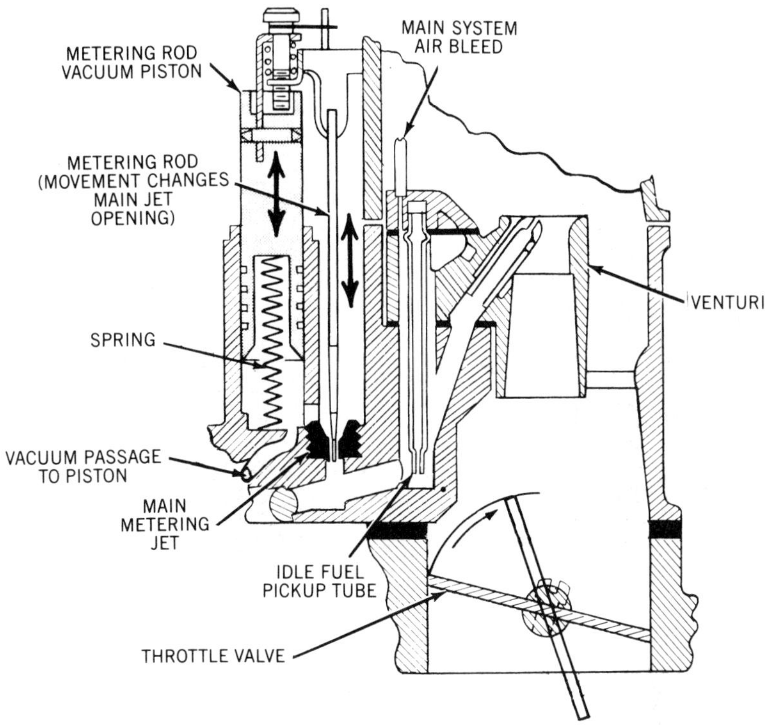


**Figure 16-24.** This power system uses metering rods that are moved out of the main jets to increase fuel flow at full throttle. (Chrysler)

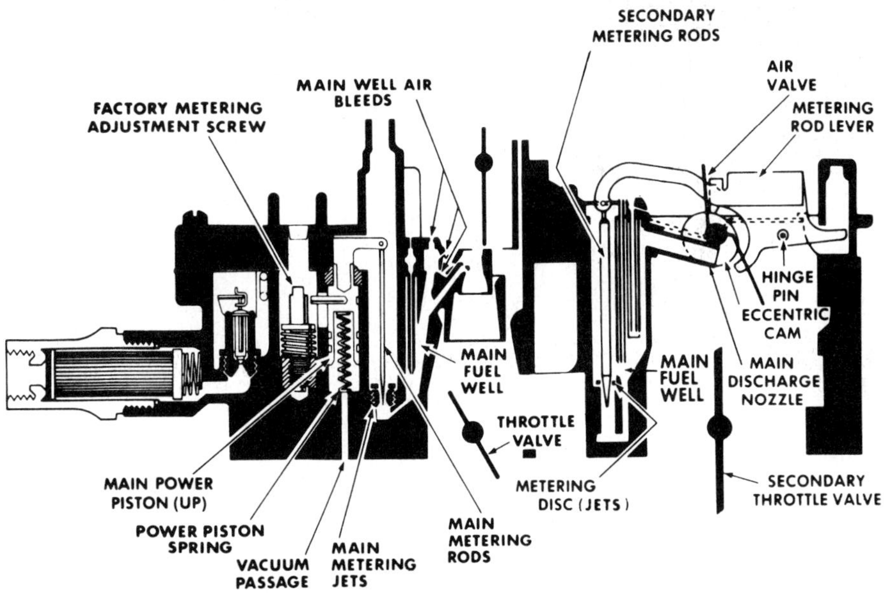


**Figure 16-25.** This cutaway of a Rochester 4-barrel shows metering rods operated by vacuum and by mechanical linkage. (Chevrolet)

## The Power System

The main metering system provides the leanest air-fuel mixture and best economy of all systems. However, when engine load increases and the engine must deliver more power, this mixture is too lean. Extra fuel for extra power comes from the power system, or power valve.

Increased engine loads that require more power always are accompanied by decreased vacuum. Therefore, many power valves are vacuum operated and open when manifold vacuum drops. Increased power demands also are accompanied by wider throttle openings, so many power systems are operated by linkage to the throttle.

One common power system, used on many Carter and Rochester carburetors, consists of a metering rod, or rods. The end of the rod is tapered or stepped and placed in the main jet opening in the fuel bowl. The rod tip restricts the amount of fuel that flows through the jet for light-load, main metering operation. When more power is needed, the rod is moved out of the jet to increase fuel flow, figure 16-24. Metering rods can be operated by a vacuum piston or by mechanical linkage, figure 16-25.

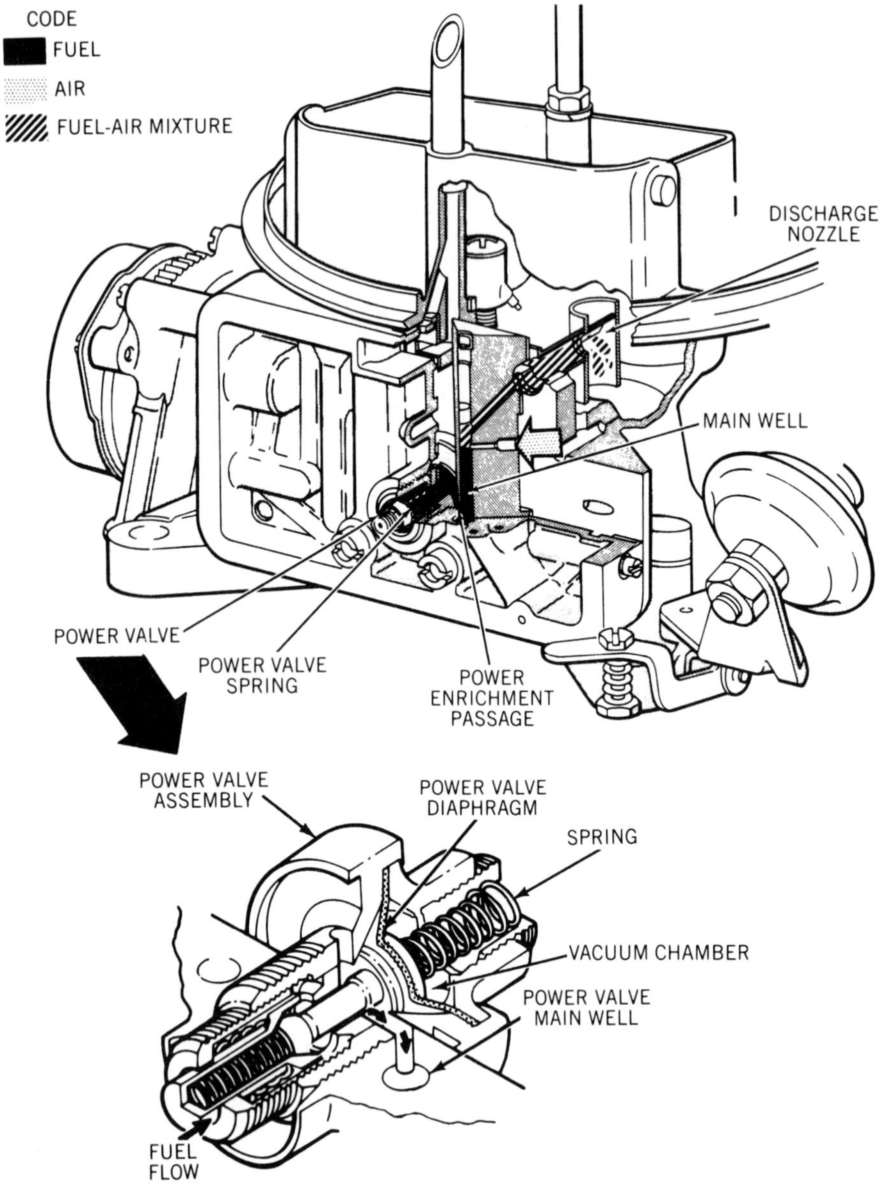


**Figure 16-26.** Manifold vacuum applied to the diaphragm holds this power valve closed. When vacuum drops at full throttle, the valve opens. (Ford)

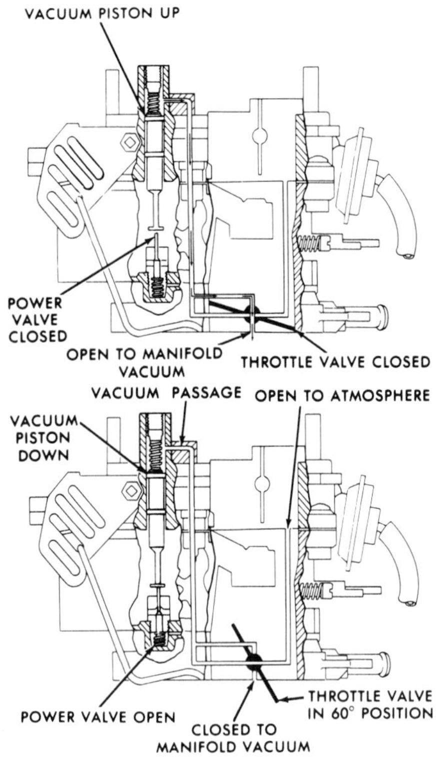


**Figure 16-27.** Vacuum holds the plunger up at part throttle. When vacuum drops at full throttle, the spring moves the plunger down to open the power valve. (Chrysler)

Some power valves are operated by a diaphragm exposed to manifold vacuum, figure 16-26. Under light load, high manifold vacuum moves the diaphragm to close the power valve. When vacuum drops as engine load increases, a spring opens the valve to allow more fuel to flow. You will see these valves on some Holley and Motorcraft carburetors.

Another kind of vacuum-operated power system has a small valve in the bottom of the fuel bowl. A spring holds the valve closed for light-load operation. Directly above the valve is a vacuum piston and a plunger, figure 16-27. Manifold vacuum holds the piston up when engine load is light. As engine load increases and vacuum drops, a spring moves the plunger down to open the power valve.

Most feedback carburetors for electronically controlled engines do not have separate power systems. The mixture control solenoid operates metering rods or air bleeds to control the air-fuel ratio. When the engine needs a rich mixture for extra power, the system computer drives the solenoid to the fully rich position.

## The Accelerator System

When a closed, or nearly closed, throttle is opened suddenly for acceleration, airflow increases faster than fuel flow from the main metering nozzle. This also reduces manifold vacuum and causes a momentarily lean mixture. The lean mixture would cause a momentary engine hesitation, or "flat spot," so the mixture is given an extra shot of fuel by the accelerator system.

The accelerator system consists of an accelerator pump, passages in the carburetor, and a pump discharge nozzle at the venturi. The accelerator pump is a plunger or diaphragm, operated by linkage from the throttle. The pump is in a separate carburetor chamber, or well, with inlet and outlet check valves, figure 16-28. As the throttle closes, the pump draws gasoline into the chamber past the inlet check valve. When the throttle opens quickly, the pump moves to apply pressure to the fuel in the chamber. This closes the inlet and opens the out-

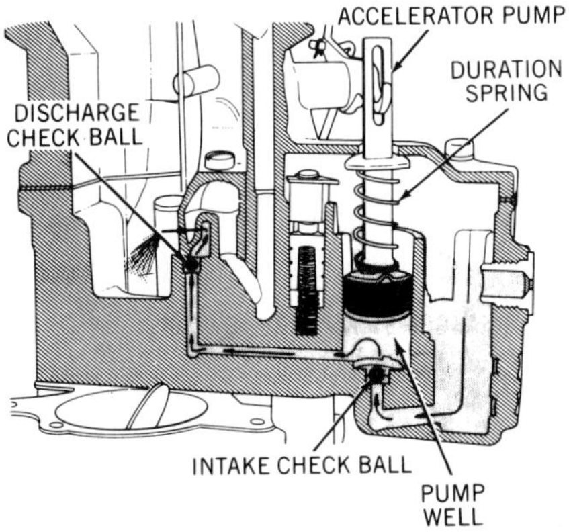


**Figure 16-28.** This simple plunger-type pump illustrates typical accelerator pump operation. (Chrysler)

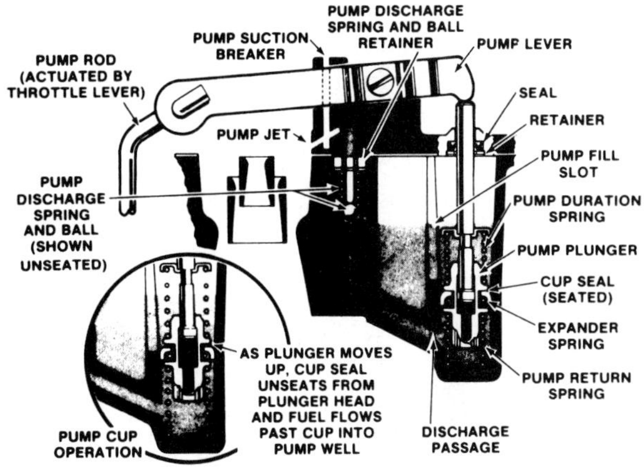


**Figure 16-29.** Upward motion of the plunger creates an opening in the cup for fuel intake. (Rochester)

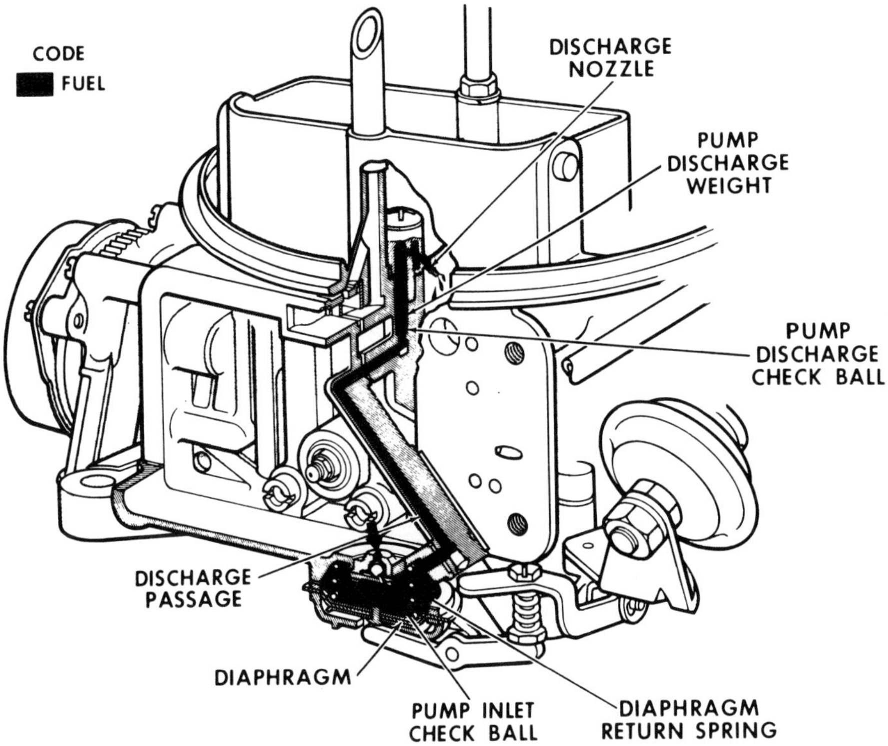


**Figure 16-30.** Diaphragm-type accelerator pumps are used on many Holley carburetors. (Ford)

let check valves and forces fuel to the discharge nozzle at the venturi.

Both check valves can be steel balls, and the outlet ball often is larger. The inlet check valve also can be a diaphragm or part of the pump plunger. Some pumps have an inlet slot in the well or an opening in the top of the pump plunger, figure 16-29. The slot or opening is closed by plunger motion instead of by an inlet check valve.

Many Holley and Motorcraft carburetors have a diaphragm-type accelerator pump, figure 16-30. The diaphragm is operated by a lever attached to the throttle linkage.

All accelerator pumps operate during the first half of throttle movement from closed to open. Most accelerator pumps are operated by a **duration spring**, figures 16-28 and 16-29, so that the time of the pump stroke is always the same, regardless of throttle movement speed. Throttle linkage holds the pump in the returned position and releases the pump as the throttle opens. The spring then moves the plunger or diaphragm to deliver fuel steadily.

At high speed, vacuum at the pump nozzle may unseat the outlet check valve and siphon fuel into the carburetor. In some older carburetors, the extra fuel is included in high-speed fuel metering. Most carburetors have air bleeds in the pump passages to break up the siphon action. Some carburetors also have weights on the outlet check valves or antisiphon valves in the pump plungers.

### The Choke System

The choke system operates only when a cold engine is started, but it is one of the most important systems for driveability and emission control. The choke provides a rich air-fuel mixture to a cold engine at slow cranking speed with low airflow.

To create enough vacuum to draw gasoline from the fuel bowl all the way to the cylinders, the choke valve in the top of the carburetor is closed to restrict airflow, figure 16-31. The choke does not block airflow completely. Its closing angle allows enough air to enter for the engine to start while still creating maximum cranking vacuum.

On older cars (and a few late models), the choke is operated by a cable that runs to the driver's compartment. Most carburetors built in the

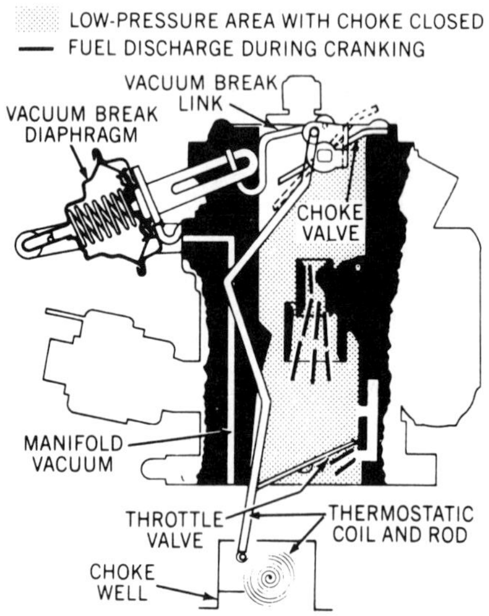


**Figure 16-31.** The closed choke valve creates a low-pressure area throughout the carburetor to draw fuel from the idle and main systems during cranking. (Chevrolet)

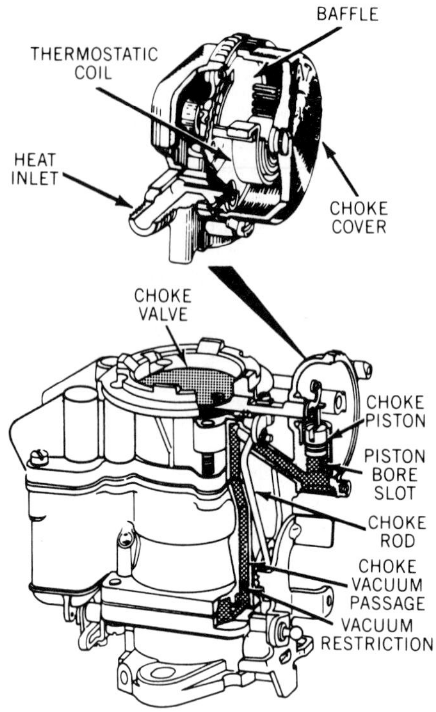


**Figure 16-32.** This typical integral (piston-type) choke has the bimetallic thermostatic spring in a housing on the carburetor airhorn. (AMC)

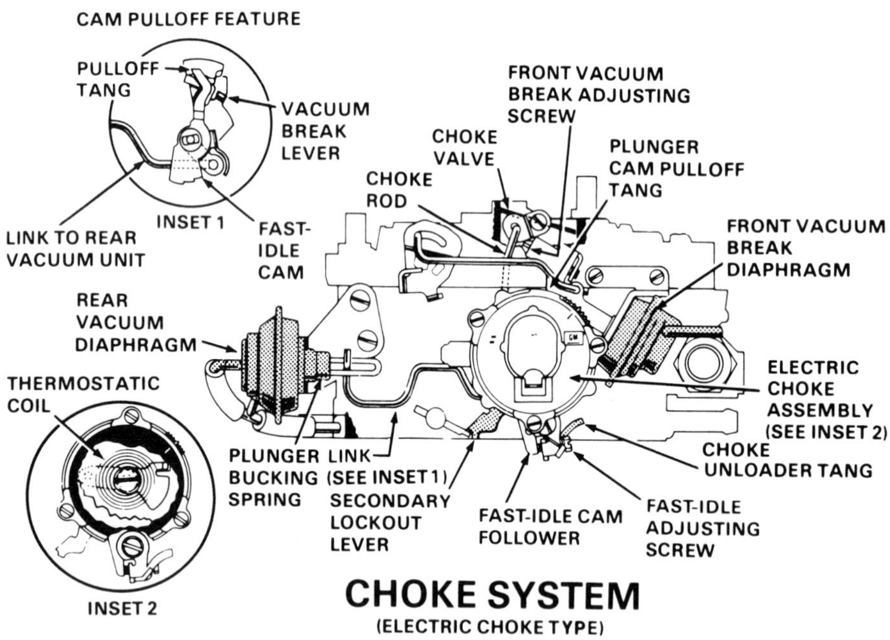


**Figure 16-33.** This Rochester 4-barrel has two vacuum breaks. The front one controls the secondary air valves, as well as the choke. (Chevrolet)

past 35 or 40 years, though, have automatic chokes controlled by a bimetallic thermostat spring, connected to the choke valve. When the spring is cold, its tension closes the choke. As it warms up, the spring loses tension, and lets a vacuum piston or diaphragm open the choke. There are two basic choke designs: integral and remote.

An integral choke has the thermostat spring in a round housing on the carburetor airhorn, figure 16-32. A remote choke has the spring in a well on the intake manifold, figure 16-31, and remote chokes often are called "well" chokes. Linkage connects the spring to the choke valve. To heat the spring of an integral choke, hot air or hot coolant is routed through passages in the choke housing. A remote choke spring is heated by exhaust passing through the heat passages of the intake manifold.

As soon as the engine starts, the choke must open slightly to allow more airflow for a running engine. The choke valve shaft is offset in the carburetor airhorn so that airflow tends to push the choke open. Also, manifold vacuum is applied to a vacuum piston or diaphragm to pull the choke open several degrees. Some integral chokes have a small vacuum piston in the choke housing to do this, and these are often called "piston chokes," figure 16-32. All remote, and many integral, chokes have a vacuum-break diaphragm to pull the choke open as the engine starts, figures 16-31 and 16-33. Either design often is called a "choke pulloff" or "pulldown."

Every choke has a choke unloader, which is a link that pulls the choke open about halfway when the throttle opens completely, figure 16-33. If the engine floods while starting, this provides extra airflow to clear out the extra gasoline. Also, if an engine must accelerate quickly while the choke is closed, it will not get enough air unless the choke is opened.

A cold engine must idle faster than a warm engine to get enough air and fuel to run. Therefore, every choke system includes fast-idle linkage to raise

## THE WANKEL ROTARY ENGINE

A reciprocating engine is a complicated device that must change linear motion to rotating power. It would seem much simpler to build an engine that could develop its power directly through rotary motion. That's what Dr. Felix Wankel did. Dr. Wankel's rotary engine is best known in the Mazda automobiles built in Japan, but Wankel engines have been built for automobiles by NSU in Germany and for aircraft by Curtiss-Wright in the United States. At one time, General Motors had plans to build Wankel-engined automobiles, as well.

The same sequence of intake, compression, power, and exhaust events takes place in a Wankel engine as in a reciprocating engine. But, they occur as stages in one revolution of the engine's 3-sided rotor, figure 1, rather than as strokes of a piston. The Wankel engine has a curved triangular rotor that turns on an eccentric section of a shaft in a chamber. As the rotor turns, its tips follow the housing shape and form separate chambers. Because of the shape of the housing and eccentric rotation of the rotor, the size and shape of the chambers change continuously.

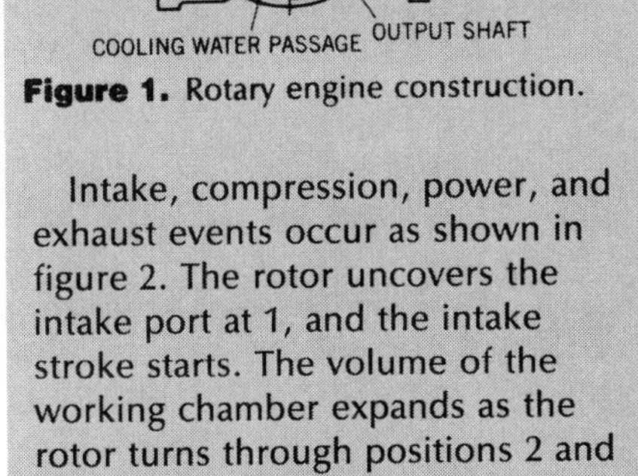


**Figure 1.** Rotary engine construction.

Intake, compression, power, and exhaust events occur as shown in figure 2. The rotor uncovers the intake port at 1, and the intake stroke starts. The volume of the working chamber expands as the rotor turns through positions 2 and 3. Maximum volume exists at 4, and the rotor closes the intake port at 5. The rotor compresses the air-fuel mixture through positions 6, 7, and 8. Ignition occurs at tdc (9), and the power stroke continues through 10, 11, and 12. The rotor uncovers the exhaust port at 13, and the exhaust stroke proceeds from positions 14 through 18, where the rotor closes the exhaust port. From there, another intake stroke starts again at 1.

One revolution of the rotor causes three power pulses, one for each face of the rotor. You can think of each rotor face as a piston, so a single rotor engine has ignition intervals about the same as a 3-cylinder engine has. Wankel engines can be built with two or more rotors, as Mazda engines are. Each power pulse lasts for about

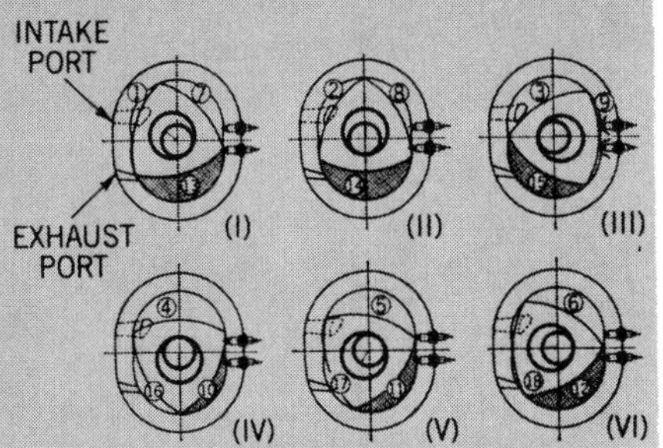


**Figure 2.** Rotary engine operating principles. (Mazda)

3/4 of a revolution. The combination of long power pulses and positioning of multiple rotors can produce a powerful and smooth-running engine.

Although a 2-rotor Wankel engine can produce power equivalent to a 6-cylinder piston engine and has no valves, rods, or other reciprocating parts, it does have some disadvantages. The first major problem with Wankel engines was satisfactory sealing of the rotor tips. Mazda and other manufacturers overcame this problem with their late-model engines. Although it is more difficult to get good fuel economy and emission control from a rotary engine than it is from a reciprocating engine, Mazda has largely overcome these problems in its current engines.

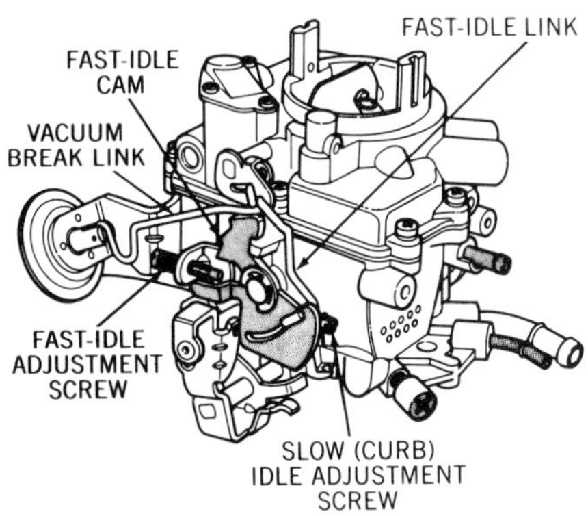


**Figure 16-34.** The fast-idle cam contacts the idle speed screw or a separate fast-idle screw to open the throttle for a faster idle speed when the choke is closed. (Chrysler)

idle speed 400 to 800 rpm above the idle speed of a warm engine. This linkage has a cam operated by a link from the choke valve. The fast-idle cam contacts the idle speed screw or a separate fast-idle screw on the throttle linkage to hold the throttle open more than normal, figure 16-34. Stepping on the accelerator before starting a cold engine allows the choke to close and move the fast-idle cam to the fast-idle position. When the engine warms and choke spring tension decreases, a weight on the cam pulls it back to allow the engine to return to normal idle speed. Fast-idle speed on electronically controlled engines often is regulated by a solenoid or motor, controlled by the system computer.

Because the choke creates a rich air-fuel mixture, it increases HC and CO emissions. Current emission regulations require that late-model engines get off the choke as quickly as possible. The most common way to do this is by warming the choke spring with an electric heater. Most late-model integral and remote chokes have a small heater in the choke housing that receives current from the alternator, figure 16-35. The heater circuit includes a temperature-sensing switch that closes when underhood air temperatures are above 60° to 65° F (17° to 20°C). This allows the heater to heat the spring and open the choke as soon as possible. Some carburetors have 2-stage or 3-stage choke heaters that apply different amounts of heat, depending on air temperature.

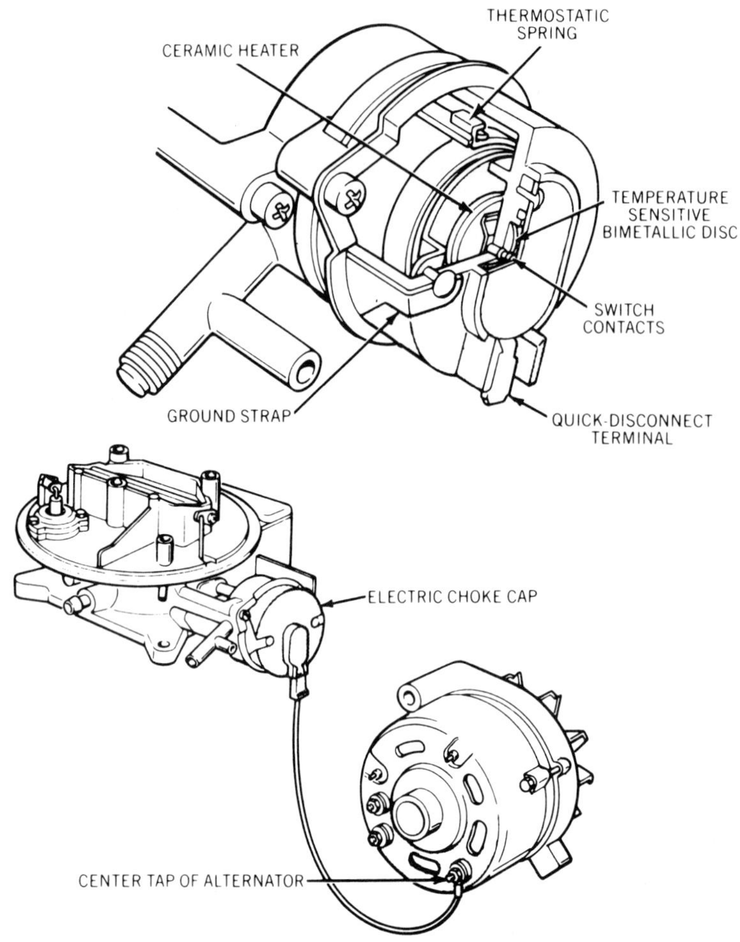


**Figure 16-35.** Electric heaters warm the thermostatic spring to open the choke quickly on late-model cars. (Ford)

Thermostatic spring tension on older chokes is adjustable to change the amount that the choke closes and the speed at which it opens. The choke housings on most carburetors built since 1979 or '80 are sealed to prevent tampering with choke adjustment and upsetting emission controls, figure 16-36.

## CARBURETOR TYPES

All of the systems you have just studied are built into carburetors of different sizes for different engines. Carburetors are commonly identified by the number of barrels and the direction of airflow.

### Carburetor Barrels

Almost all carburetors are 1-barrel, 2-barrel, or 4-barrel types.

**One-Barrel Carburetors.** A 1-barrel carburetor has a single barrel, a single throttle, and a single set of metering systems to supply all air-fuel needs, figure 16-37. Most have airflow capacities of 150 to 300 cubic feet per minute (cfm) and are used on smaller 4- and 6-cylinder engines.

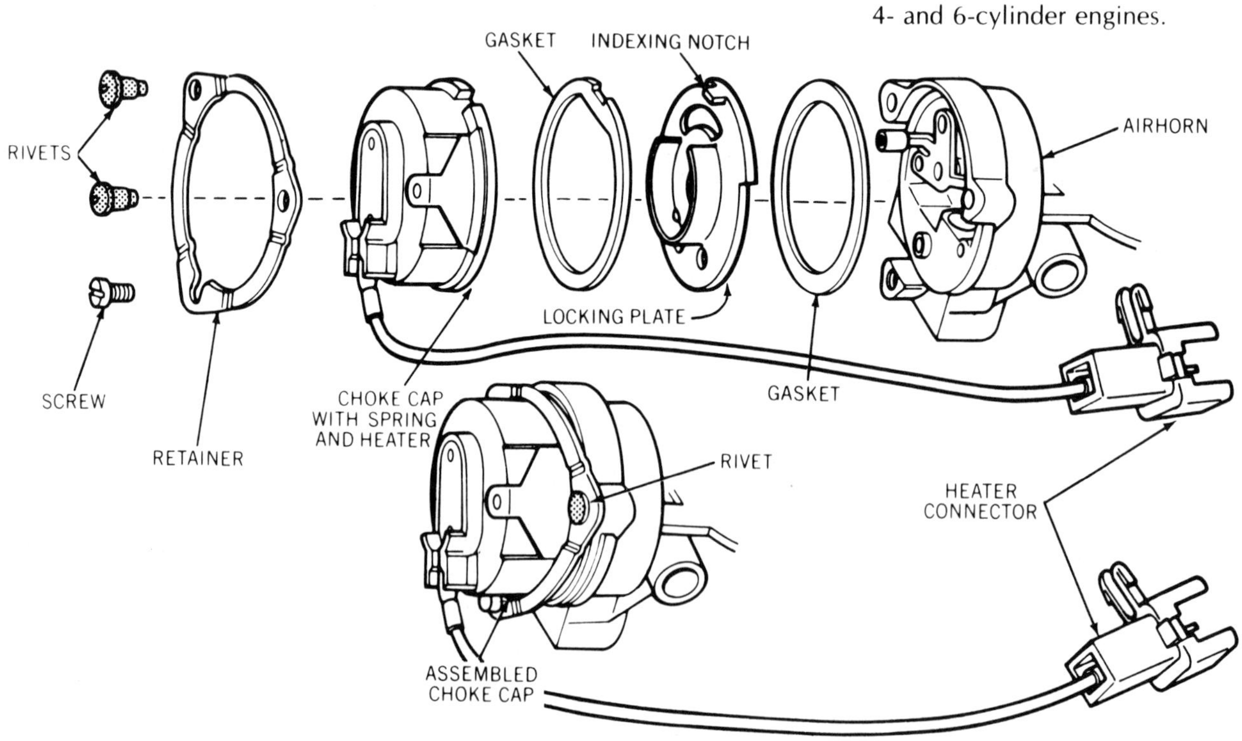


**Figure 16-36.** The choke spring housings on late-model carburetors are riveted closed to prevent tampering with the factory adjustment. (Ford)

**Single-Stage 2-Barrel Carburetors.** A single-stage 2-barrel carburetor is basically two 1-barrel carburetors sharing a common body, figure 16-38. It has a single fuel bowl, float, choke, and accelerator pump. The metering systems for both barrels work simultaneously, but the idle mixture for each barrel is adjusted separately. Single-stage 2-barrel carburetors have two throttle valves on a common shaft that open and close together. These carburetors have airflow capacities of 200 to 550 cfm and are used on many 6- and 8-cylinder engines.

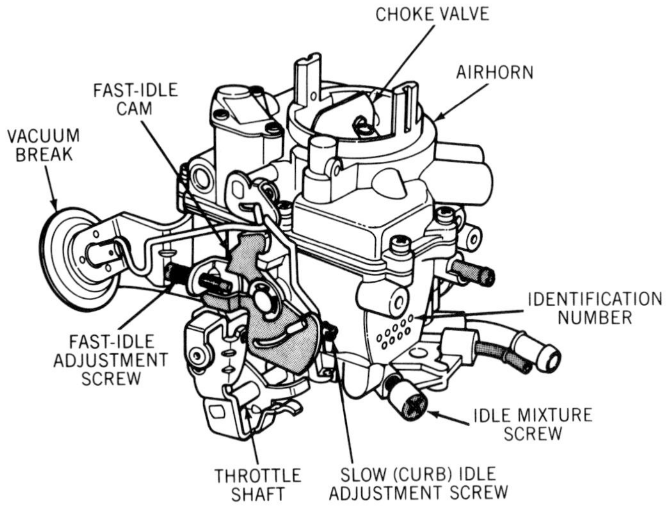


**Figure 16-37.** A 1-barrel carburetor has a single throttle and venturi. (Chrysler)

**Two-Stage 2-Barrel Carburetors.** These carburetors became common in the mid-1970's to improve economy, performance, and emission control for 4-cylinder and small V-6 engines. A 2-stage 2-barrel carburetor has a single float and fuel bowl but two barrels and throttles that operate semi-independently, figure 16-39. The primary barrel is smaller than the secondary. It contains the idle and low-speed systems and supplies the air-fuel mixture at low and moderate loads and speeds. The primary barrel also usually includes the main metering and power systems and the accelerator pump.

The throttle of the secondary barrel opens at higher speeds and loads when the primary throttle is about half open. The secondary barrel supplies additional fuel and air to the engine. It includes transfer, main metering, and power systems. The secondary throttle can be operated by vacuum or mechanical linkage. Some of these carburetors have a choke only for the primary barrel; others have chokes for both barrels. Because they are usually used on small engines, most 2-stage 2-barrel carburetors have airflow capacities of 150 to 300 cfm.

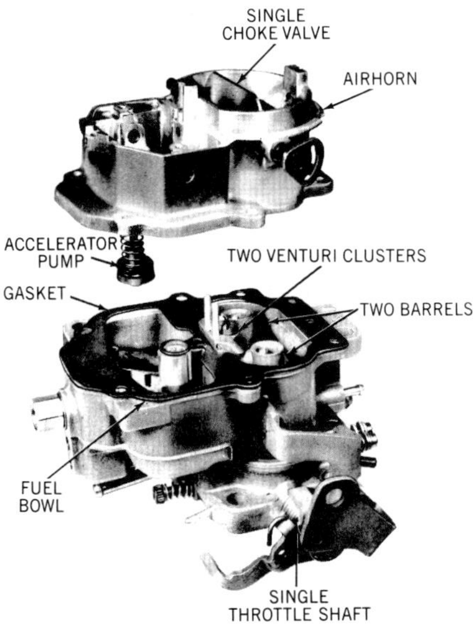


**Figure 16-38.** A single-stage 2-barrel carburetor has two barrels and venturis, and two throttles that operate together. (Chrysler)

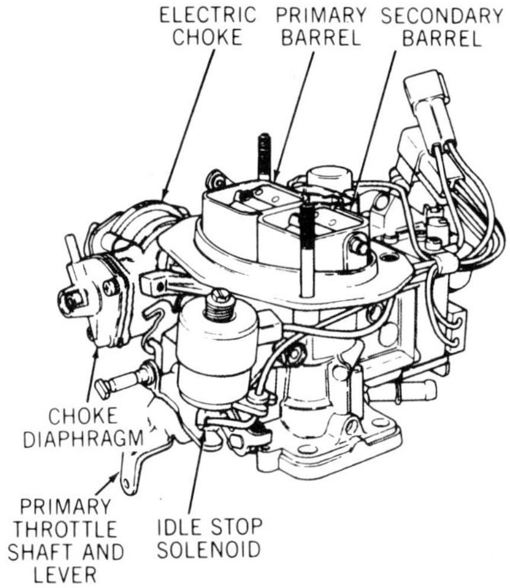


**Figure 16-39.** The throttles of a 2-stage, 2-barrel carburetor operate progressively. (Chrysler)

**Four-Barrel Carburetors.** Used mostly on V-8 engines, 4-barrel carburetors have airflow capacities of 400 to 800 cfm. Some of the largest, however, provide more than 1,000-cfm airflow. A 4-barrel carburetor has two primary and two secondary barrels in one body, figure 16-40. The primary

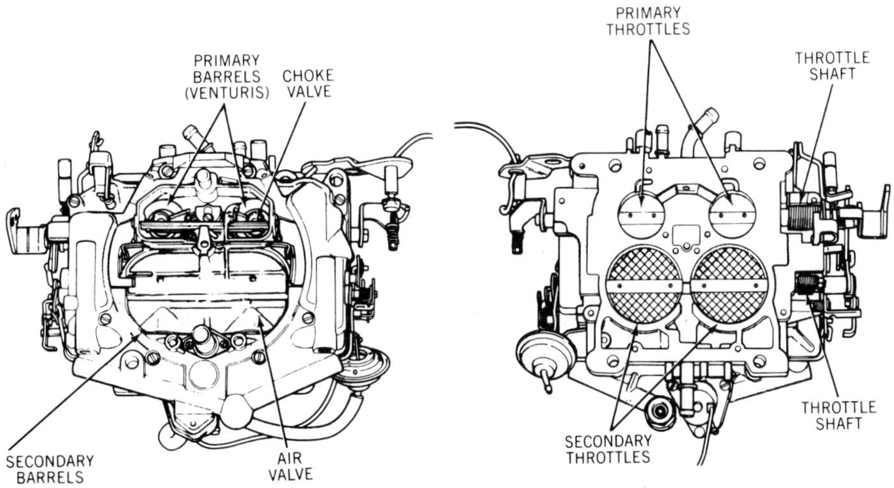


**Figure 16-40.** The primary and secondary systems of most 4-barrel carburetors operate progressively. Only the primary barrels have a choke. (Chrysler)

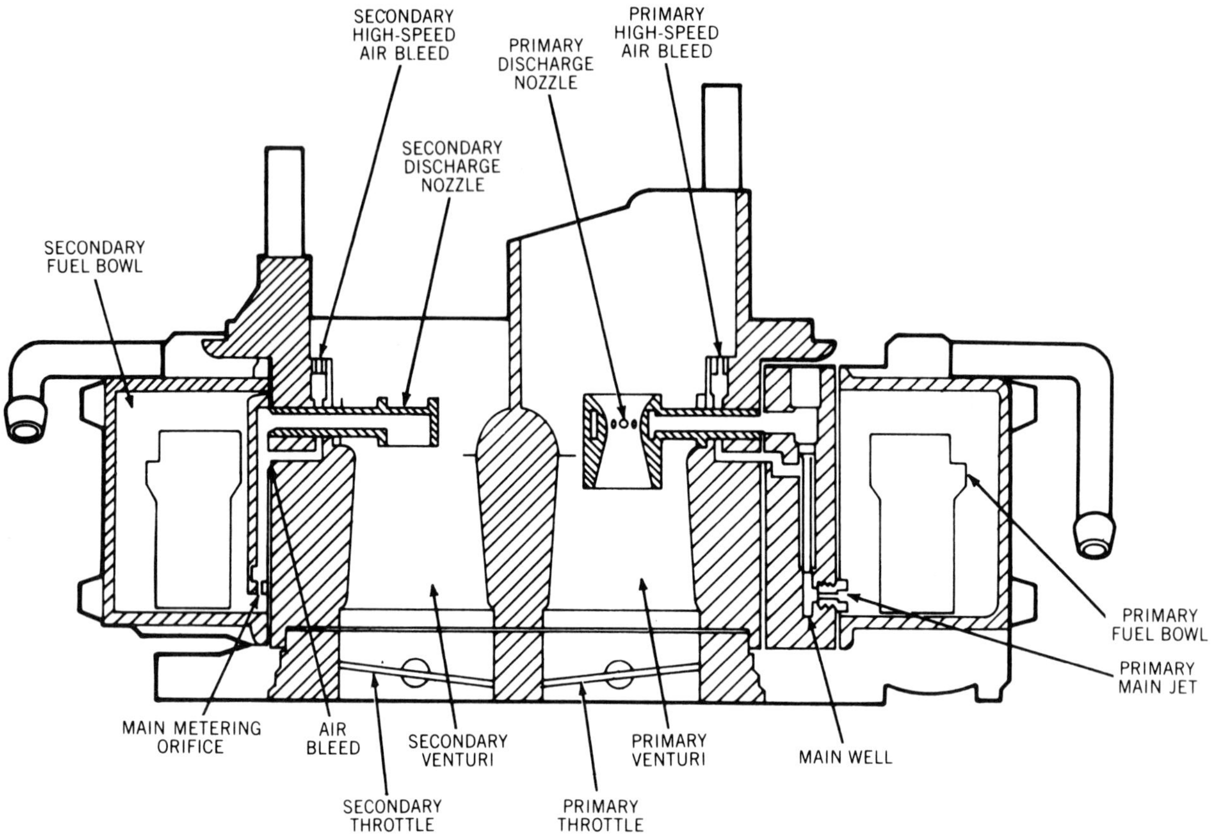


**Figure 16-41.** Airflow and fuel discharge through the secondary barrels of this Holley carburetor are governed by venturi action. (Ford)

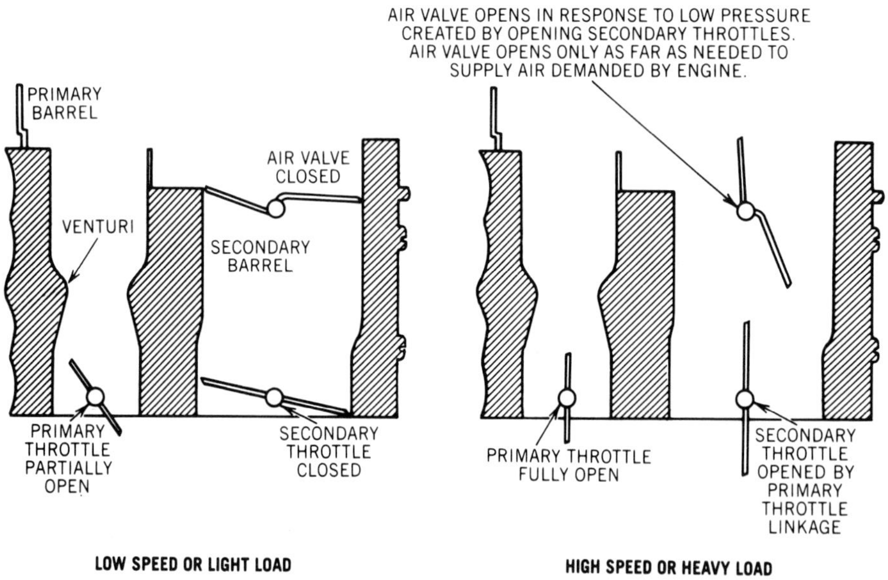


**Figure 16-42.** Airflow through the secondary barrels of many Rochester and Carter 4-barrel carburetors is controlled by air valves.

barrels work like a single-stage 2-barrel carburetor at low and moderate loads and speeds. The secondaries open together like the secondary barrel of a 2-stage 2-barrel carburetor to provide more air and fuel for higher speeds and loads.

The primary barrels have idle, low-speed, high-speed, and power systems. They also contain the choke and accelerator pump. The secondary barrels have separate high-speed and power systems and may have a separate idle system and accelerator pump. Some 4-barrel carburetors have a single fuel bowl and fuel supply for all four barrels. Others have separate bowls and fuel supplies for the primary and secondary barrels.

Airflow through the secondary barrels is provided two ways: by venturi action, figure 16-41, or by **air velocity valves**. Some carburetors have both. Air velocity valves are plates, similar to choke valves, in the secondary barrels, figure 16-42. Low pressure and airflow created when the secondary

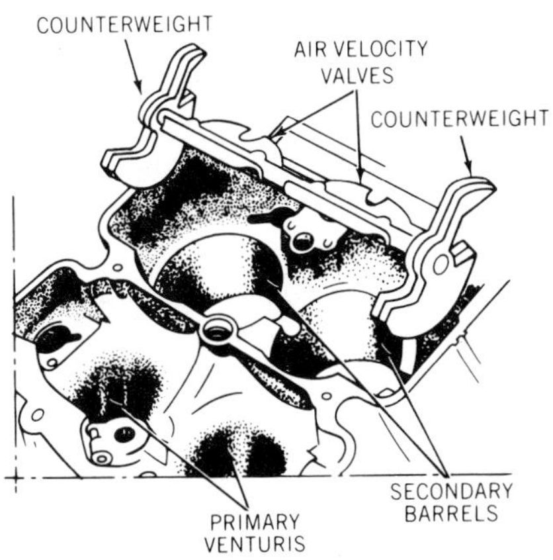


**Figure 16-43.** These auxiliary air velocity valves are opened by airflow through the secondary barrels. Counterweights hold them closed.

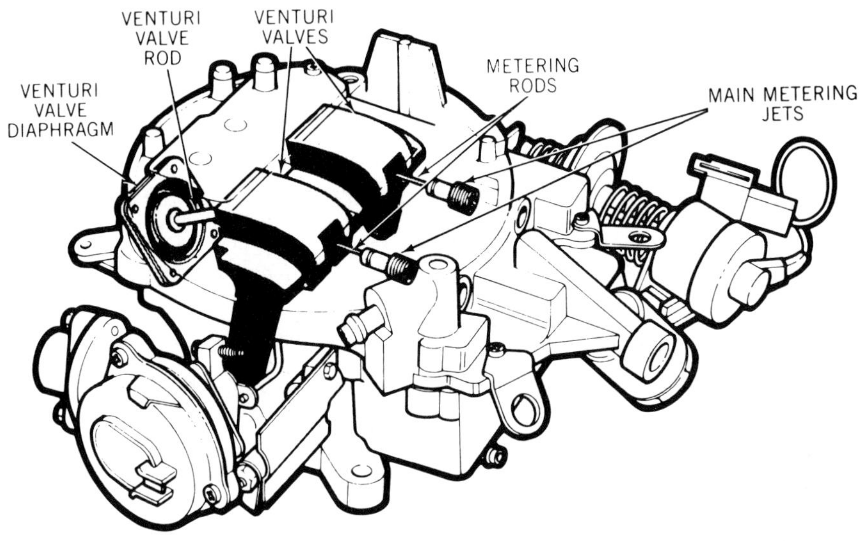


**Figure 16-44.** Ford's variable-venturi carburetor was introduced in 1977 and has been used on many late-model engines. (Ford)

throttles open cause the air velocity valves to open. Late-model Rochester Quadrajet carburetors have vacuum diaphragms to modulate the air valve movement.

Some older Carter and Rochester 4-barrel carburetors have auxiliary air velocity valves inside the secondary barrels, figure 16-43. These valves open in response to airflow through the barrels. When the throttles are closed, counterweights hold the velocity valves closed.

## Carburetor Airflow Direction

Sixty years ago, most automobiles had inline engines and updraft carburetors. The carburetor was below the intake manifold on the side of the engine and delivered the air-fuel mixture upward to the cylinders. Most carburetors had small airflow capacities and restrictive venturis. The updraft design promoted good fuel vaporization for low-speed, low-compression engines. You can still find updraft carburetors on some trucks and small industrial engines.

Today, almost all cars and trucks have downdraft carburetors mounted on top of the intake manifold. A few vehicles, mostly some imported cars with 4-cylinder engines, have sidedraft carburetors. Air flows horizontally through the carburetor and intake manifold. The intake systems of high-performance engines with multiple sidedraft carburetors eliminate the 90-degree bend for mixture distribution that exists below a downdraft carburetor. Air flows almost directly through the carburetors to the intake valves, which aids high-speed induction. Weber and Mikuni are major builders of sidedraft carburetors.

## Variable-Venturi Carburetors

Every carburetor you have seen so far has had one or more fixed venturis in the barrels. A fixed venturi creates a partial vacuum in the barrel that changes with airflow velocity. High velocity increases vacuum, or pressure drop, at the venturi. Since the venturi and barrel size are fixed, but airflow velocity changes with throttle opening, the vacuum at the venturi also changes. Airflow and vacuum at the venturi can draw fuel from the main discharge nozzle for some operating conditions, but not for others. Carburetors have the various idle, power, choke, and accelerator systems to control air-fuel mixture when it can't be regulated by venturi airflow and vacuum.

A carburetor with a variable venturi provides a constant amount of vacuum, or pressure drop, at the venturi because the venturi opening changes as airflow and vacuum change. That is why such a carburetor is called a **constant-depression** carburetor. Variable-venturi carburetors have auxiliary fuel-metering systems, but not the same kinds of choke, idle, power, and accelerator systems that fixed-venturi carburetors have.

Variable-venturi carburetors made by Hitachi, Stromberg, Solex, and SU have been used on many imported cars. The only domestic variable-venturi carburetors made in recent years are Ford's model 2700 VV and 7200 VV 2-barrel Motorcraft carburetors, figure 16-44. The 2700 VV was introduced on some 1977 engines, and the 7200 VV appeared a couple of years later as an electronically controlled, feedback version of the same design. Both continued in use through the mid-1980's. The variable venturis in Ford's carburetor are formed by two plates that slide back and forth across the tops of the barrels. A spring-loaded vacuum diaphragm controls venturi opening. The diaphragm receives a

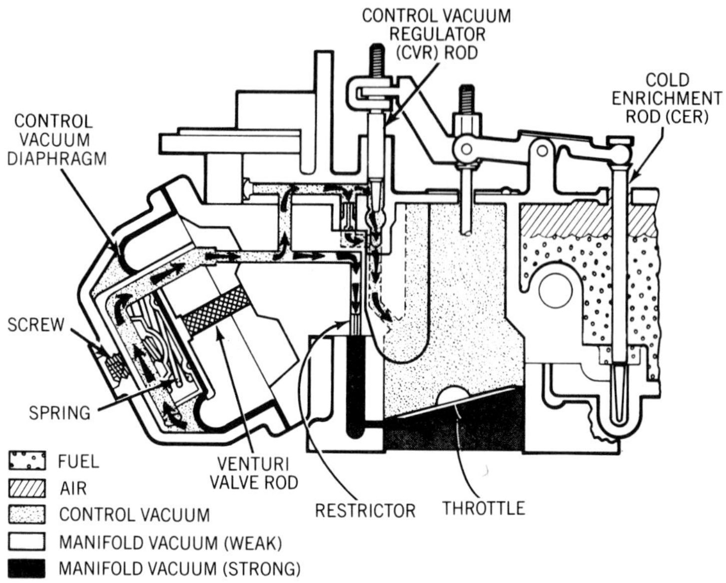


**Figure 16-45.** Control vacuum comes from the carburetor barrel and is modulated by manifold vacuum. (Ford)

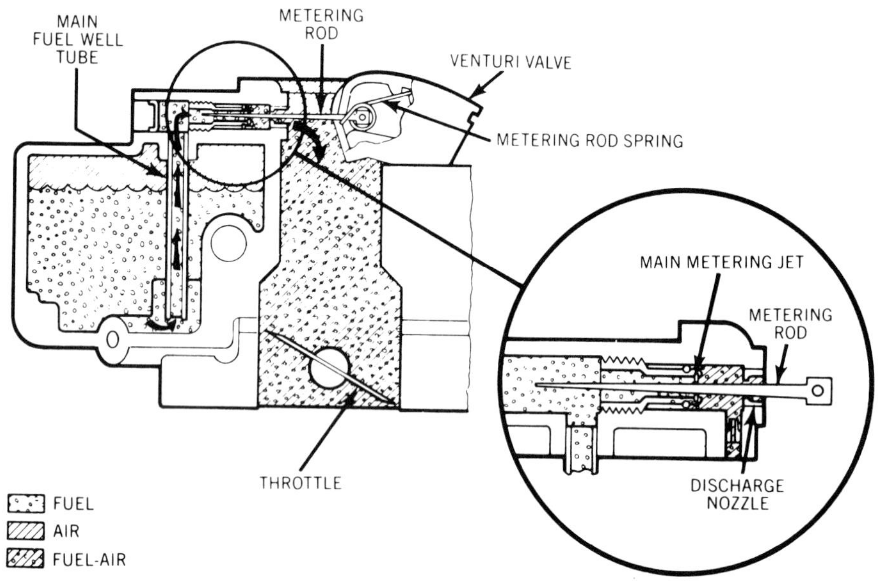


**Figure 16-46.** As the venturi valves open, they pull metering rods out of the main jets to increase fuel flow. (Ford)

vacuum signal from above the throttle in the carburetor barrel, figure 16-45. As the throttle opens, vacuum increases, and the venturis open for more airflow, figure 16-46. Each venturi valve holds a horizontal metering rod that moves in and out of a fixed jet on the opposite side of the barrel as the venturi valve moves. This meters fuel in proportion to airflow and vacuum at the venturi for almost all speeds and loads. Even at idle, venturi vacuum is sufficient to draw fuel from the main jets.

The main metering system of Ford's variable-venturi carburetor supplies the correct air-fuel mixture for most operating conditions, but it does need help from some auxiliary systems. Among these are:

- *An idle trim system*, figure 16-47, that provides a small amount of fuel through discharge ports below the throttle at idle.
- *An accelerator pump* that provides extra fuel for acceleration
- *A full-throttle limiter* that holds the venturis open at full throttle when vacuum is not strong enough to do so.
- *A cold enrichment system* that uses a thermostatic choke spring to move a regulator rod that opens the venturi diaphragm to manifold vacuum. This holds the venturis open wider during cranking to provide more fuel. The linkage also opens a cold enrichment rod (CER) that admits additional fuel during cranking.
- *A cranking enrichment system* on 1977–79 models that uses a solenoid to open a fuel valve and provide more gasoline to the engine while cranking. The solenoid was discontinued on 1980 and later models, and the cold enrichment system modified to provide more fuel.

You will learn about the electronic fuel metering systems of the Ford 7200 VV carburetor when you study electronic engine controls later in this book.

## CARBURETOR LINKAGES

All carburetors have various pieces of mechanical linkage to operate some or all of their systems. The most apparent (but no more important than others) is the throttle linkage from the accelerator pedal to the carburetor throttles. Other linkage pieces operate the automatic choke, the accelerator pump, the fast-idle cam, secondary throttles, and automatic transmission downshifts on some cars. Because linkage design and operation vary widely, we can't cover all kinds here. The following sections, however, explain the purposes of the major kinds of carburetor linkage.

### Throttle Linkage

The throttle linkage provides the driver with control of engine speed, figure 16-48. On some cars, the throttle linkage is a combination of rods and levers. On others, it is a cable that connects the accelerator pedal to the carburetor throttle. All throttle linkage

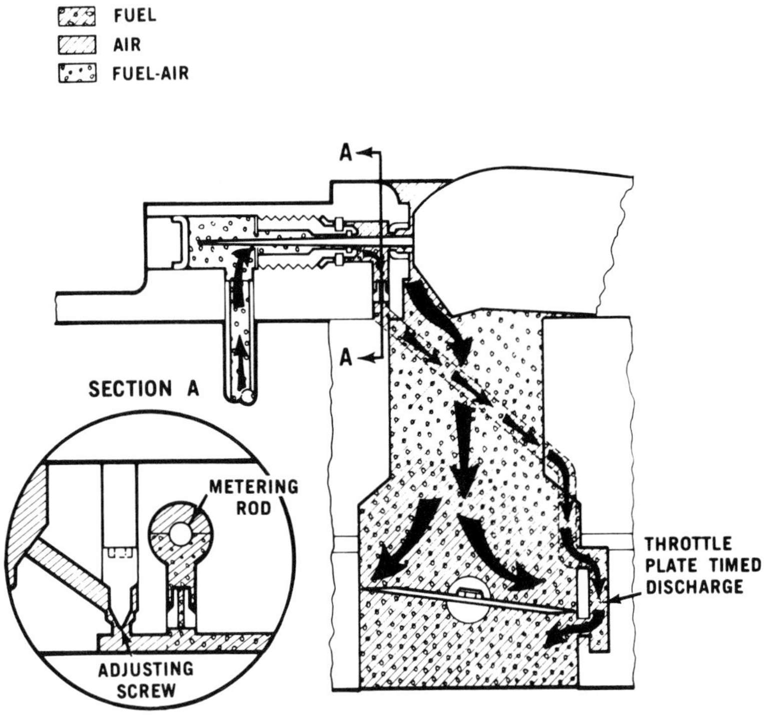


**Figure 16-47.** The idle trim system adds extra fuel to the idle mixture. (Ford)

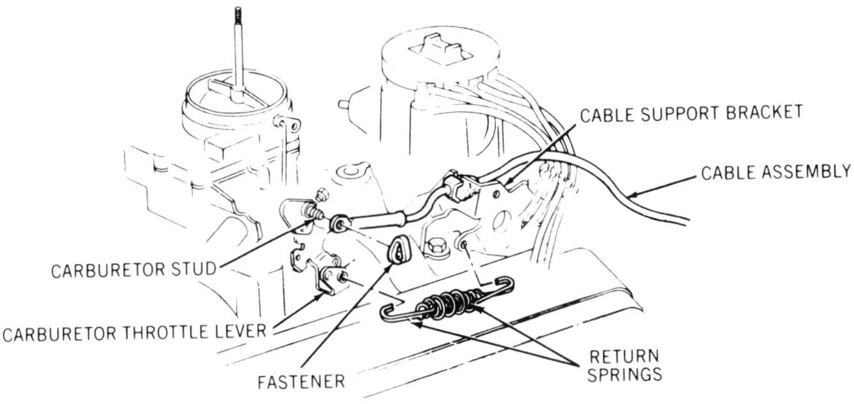


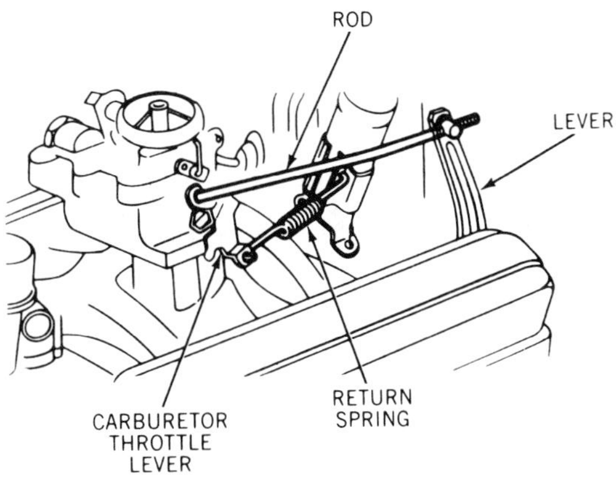


**Figure 16-48.** Throttle linkage can be operated by a cable or by rods. (Chevrolet)

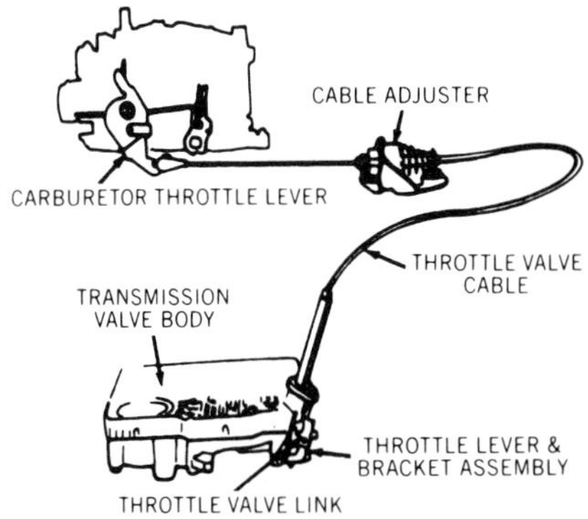


**Figure 16-49.** The cable from the throttle linkage to the automatic transmission controls transmission shift points on this late-model GM car. (Chevrolet)

has springs to close the throttle when the driver releases the accelerator.

## Transmission Linkage

Some cars with automatic transmissions have mechanical linkage connected to the throttle to control transmission shifting in relation to engine speed and load. Rear-wheel-drive Chrysler products have an adjustable throttle rod connected to the throttle valve of Torqueflite transmissions. GM cars with Turbo Hydra-matic 200, 325, and 700R4 transmissions use cable linkage to control transmission shifting in relation to engine throttle position, figure 16-49. Other cars have mechanical linkage to force transmission downshifts as the throttle opens.

## Choke Linkage

All chokes have linkage between the choke valve and the spring that closes, as well as the vacuum break that opens it, figures 16-31 and 16-33. Linkage designs vary depending on where the spring and vacuum break are mounted in relation to each other and to the choke valve.

## Fast-Idle Linkage

The fast-idle linkage is connected to the choke linkage and works with the choke to open the throttle for faster idle when the engine is cold, figure 16-34. When the choke closes, linkage pulls the fast-idle cam to the fast-idle position. As the choke opens, the linkage allows the cam to reduce the fast-idle speed step by step.

### Secondary Throttle Linkage

Secondary throttles of 2-stage 2-barrel carburetors and 4-barrel carburetors are operated by vacuum or by mechanical linkage, or a combination of both. Mechanical secondaries have a rod that connects the secondary throttle shaft to the primary shaft. When the primary throttle is about half open, the linkage starts to open the secondary throttle, figure 16-50. Both reach the fully open position at the same time.

Vacuum-operated secondaries are operated by a vacuum diaphragm on the side of the carburetor. The diaphragm responds to increasing vacuum in the primary barrels as airflow increases. When vacuum is high enough, the diaphragm opens the secondary throttle through linkage, figure 16-51. Diaphragm action is modulated by an air bleed in the secondary barrels. The secondary throttles thus respond to engine vacuum to supply more fuel and air for high-speed, heavy-load operation.

On 4-barrel carburetors with secondary air velocity valves, the velocity valves open and close in response to airflow through the secondary barrels. They are not connected to mechanical or vacuum throttle linkage. All 4-barrel and 2-stage 2-barrel carburetors have secondary lockout linkage that keeps the secondary throttle from opening when the choke is closed.

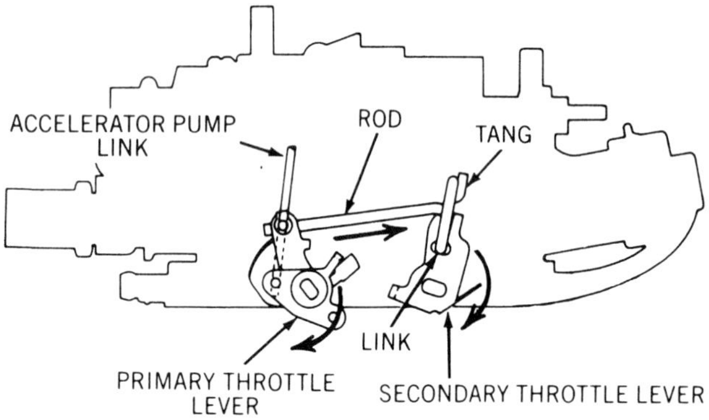


**Figure 16-50.** The secondary throttles on this 4-barrel carburetor are operated by mechanical linkage. (Chevrolet)

## CARBURETOR ASSIST DEVICES AND SYSTEM VARIATIONS

The carburetor is a major emission control device, and engineers have added a number of assist devices and system variations to control emissions related to air-fuel metering. Moreover, accessories such as air conditioning, power steering, and high-output alternators put extra power demands on an engine that can cause driveability problems. Again, engineers have modified carburetors to prevent problems caused by accessory loads and emission controls.

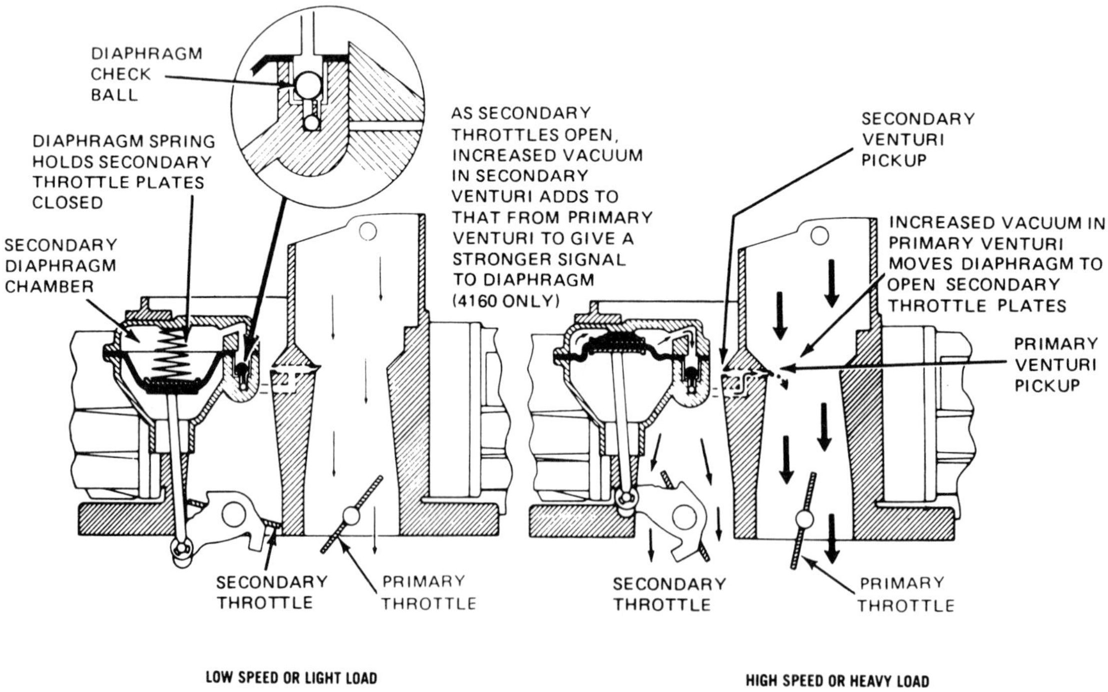


**Figure 16-51.** The diaphragm that controls the secondary throttles responds to vacuum from ports in the carburetor barrels. (Ford)

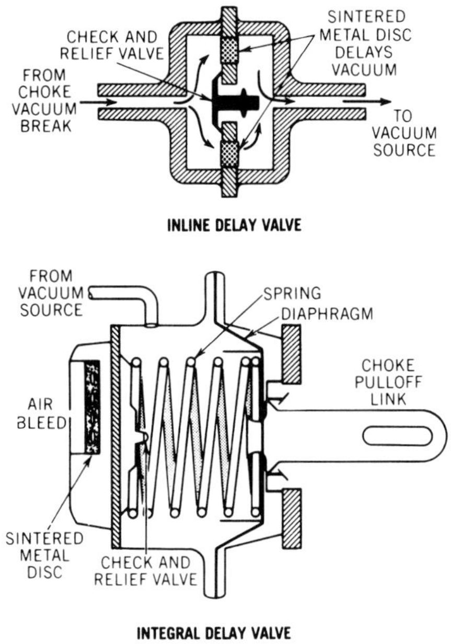


**Figure 16-52.** A choke delay valve slows the application of vacuum to the vacuum break diaphragm.

Carburetor variations and assist devices are so numerous that a complete list and explanation would fill a 1,000-page reference manual. Such manuals exist, and you should have one for major carburetor service. The following sections simply outline some of the major *kinds* of carburetor modifications on late-model engines.

## Choke Controls

You have already learned about electric heaters that cause chokes to open quickly to reduce emissions from rich mixtures. Other devices vary the amount or rate of choke closing or opening.

- *Choke delay valves* are restrictions in the vacuum hose to the vacuum-break diaphragm to slow the rate at which the choke opens when the engine starts, figure 16-52. This prevents stalling and improves driveability for the first minute of operation.
- *Temperature-controlled vacuum breaks* do basically the same thing as vacuum delay valves. A thermostatic vacuum switch in the vacuum hose blocks or restricts vacuum to the vacuum-break diaphragm when the engine is cold. The switch allows more vacuum for wider choke opening when the engine is warm.
- *Delayed, or modulated, choke pulldowns* on some Ford carburetors, figure 16-53, open the choke a few more degrees 6 to 18 seconds after the engine starts.
- *Fast-idle pulloffs* on some GM cars use a solenoid or a second vacuum diaphragm, figure 16-54, to open the choke more and reduce fast-idle speed after several seconds if the engine is warmed up while the car is parked. This prevents overloading a catalytic converter with a rich air-fuel mixture.

## Idle Mixture Controls

Other idle assist devices affect idle mixture and speed.

- *A hot-idle compensator valve* is a bimetallic thermostatic valve that admits more air to the idle system when the engine is hot. This compensates

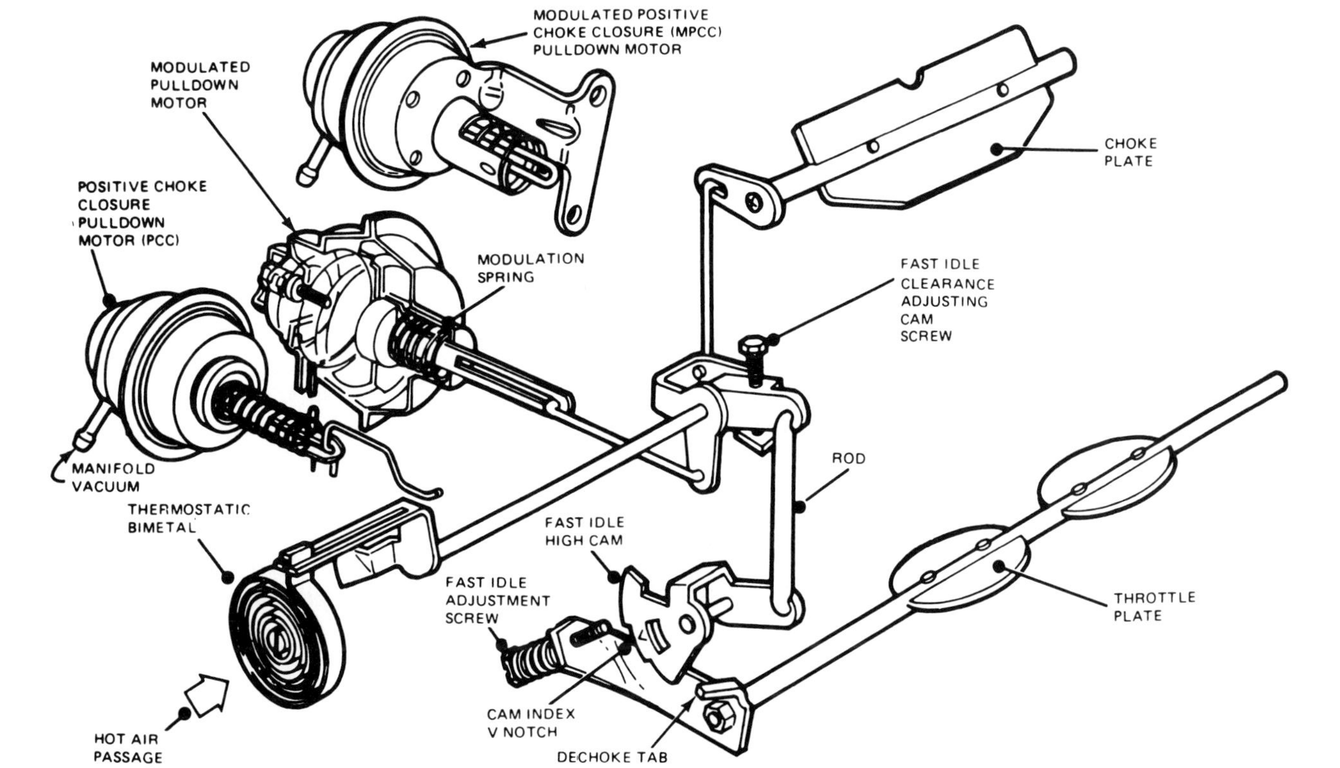


**Figure 16-53.** These Ford vacuum modulators illustrate the different kinds of devices used to control choke pulloff on late-model engines. (Ford)

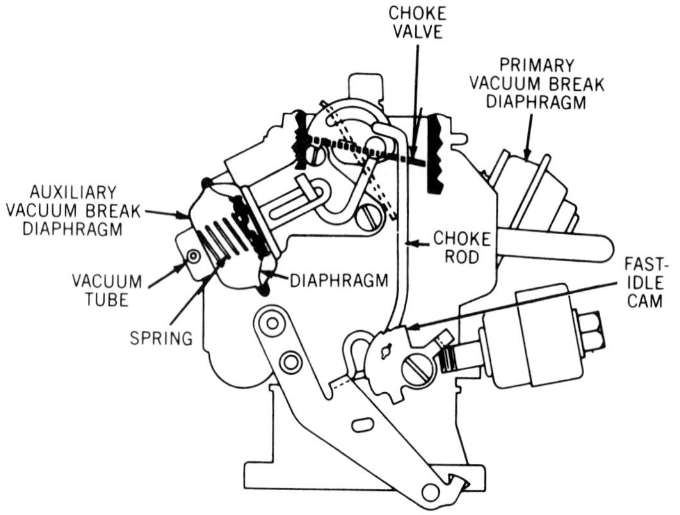


**Figure 16-54.** This auxiliary fast-idle-pulloff diaphragm opens the choke and reduces idle speed after a few minutes of engine operation. (Chevrolet)

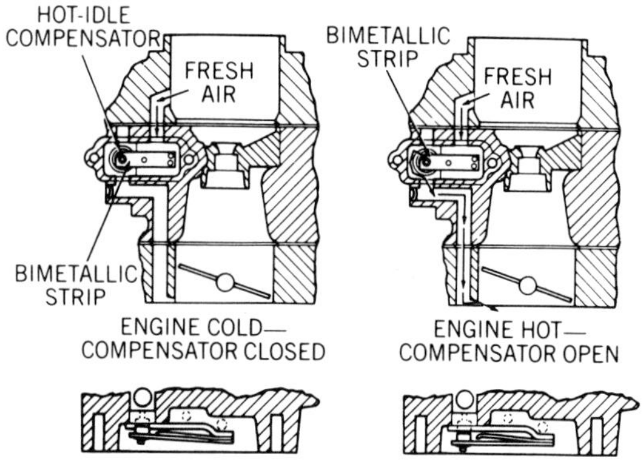


**Figure 16-55.** The hot-idle compensator allows more air into the idle air bleeds of a hot engine. (Ford)

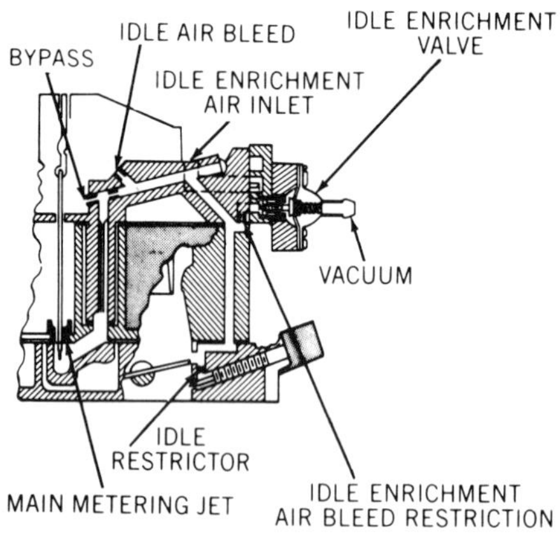


**Figure 16-56.** Chrysler's idle enrichment system closes the idle air bypass when the engine is cold. (Chrysler)

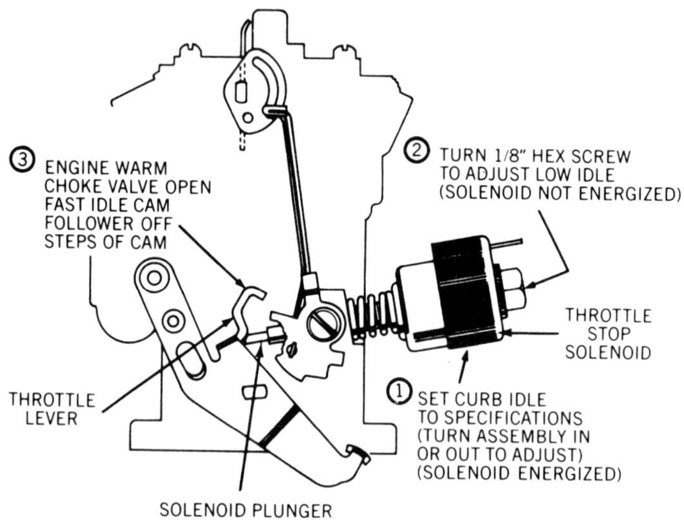


**Figure 16-57.** The throttle stop solenoid plunger extends to hold the throttle at the normal slow-idle position and retracts to let the throttle close farther when the engine is off. (Chevrolet)

for extra fuel vaporized into the idle system of a hot engine. The valve is usually at the top of the carburetor, figure 16-55. Spring tension closes the valve over an air passage when the engine is cold. If the engine gets hot, the bimetal thermostatic spring expands and opens the valve to allow more air through a passage to a port below the throttle. The air leans out a rich mixture that would increase emissions and make the engine run roughly.

• *An idle enrichment valve*, figure 16-56, used on some Chrysler carburetors is a vacuum-operated valve that closes an idle air passage when the engine is cold. It works oppositely from a hot-idle compensator and richens the idle mixture of a cold engine. Vacuum to the valve passes through a coolant temperature switch in the radiator. As coolant warms, vacuum is shut off, and the enrichment valve admits more air to the idle system.

## Idle Speed Controls

A fast-idle pulloff works with both the choke and the idle system to minimize a problem that could affect a major emission device: the catalytic converter. Other assist devices change the idle speed to minimize emission and driveability problems.

1. *A throttle stop solenoid* eliminates dieseling or after-run. High temperatures, lean mixtures, faster idle speeds, and retarded ignition timing of many late-model engines cause combustion chambers to stay hot enough to ignite an idle mixture when the engine is shut off. If the throttle can be closed farther than it is for normal slow idle speed, dieseling can be eliminated by shutting off all airflow.

The solenoid is a movable stop for the idle speed screw. When the ignition is on, the solenoid is energized to extend its plunger, figure 16-57. The idle speed screw or the throttle lever hits the plunger when the throttle is closed, and this maintains normal slow-idle speed. When the ignition is off, the plunger retracts, and the throttle closes farther.

Over the past two decades, carmakers have used many different kinds of throttle solenoids or positioners, figure 16-58, to control:

**a.** Slow (curb) idle

**b.** Shutdown idle

**c.** Idle speed under high engine load

**d.** Throttle deceleration position.

Because of the many variations, you must check the manufacturer's description and adjustment instructions before servicing any throttle solenoid.

**2.** *An air conditioning throttle solenoid* looks like a throttle stop solenoid but is energized by the control circuit for the air conditioning compressor. The compressor creates an extra load that could stall the engine at normal idle speed. When the compressor is turned on, the solenoid plunger extends to hold the throttle open more and maintain the correct idle speed.

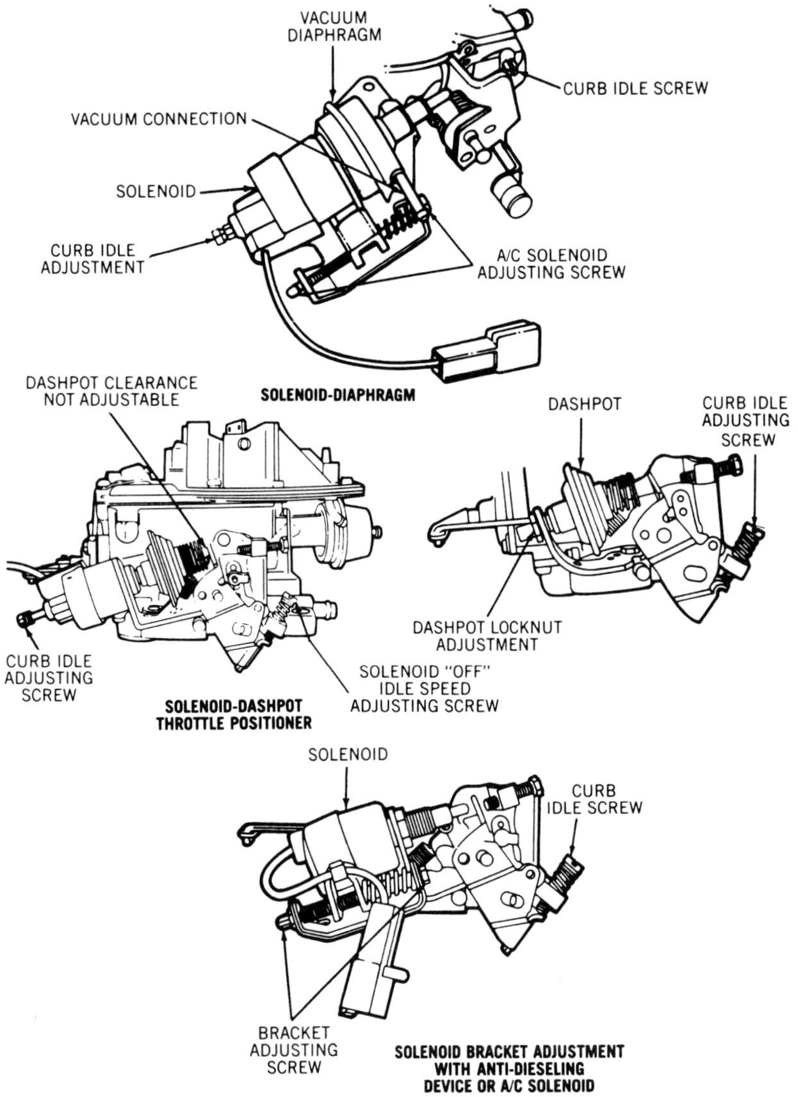


**Figure 16-58.** These throttle positioners illustrate the various devices used to control idle speed on late-model engines. (Ford)

## Deceleration Controls

As a throttle closes abruptly on deceleration, it causes a momentarily rich air-fuel mixture. Airflow stops before fuel stops flowing. The mixture is richer with a manual transmission than with an automatic transmission because the transmission causes the engine to run faster against a closed throttle and pull in more fuel with higher vacuum. The rich mixture causes a sharp rise in exhaust HC emissions. Engineers have used several devices to delay complete throttle closing for a few seconds or to add air to the deceleration air-fuel mixture and reduce HC emissions.

- *A dashpot* is a spring-loaded diaphragm with a plunger. When the throttle closes, a link hits the diaphragm plunger, figure 16-59. Air

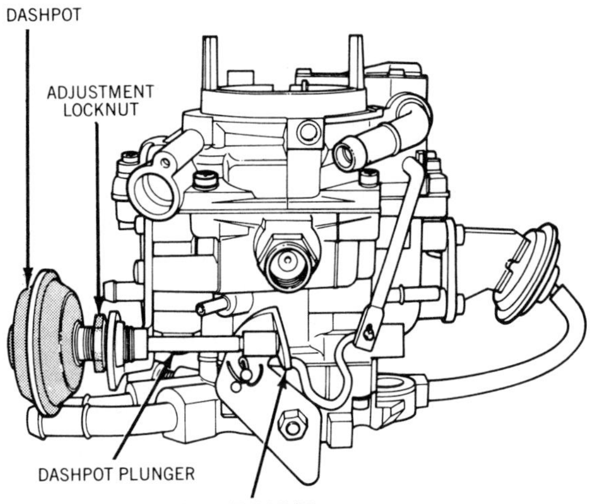


**Figure 16-59.** A dashpot is one of the simplest devices used to delay throttle closing. (Chrysler)

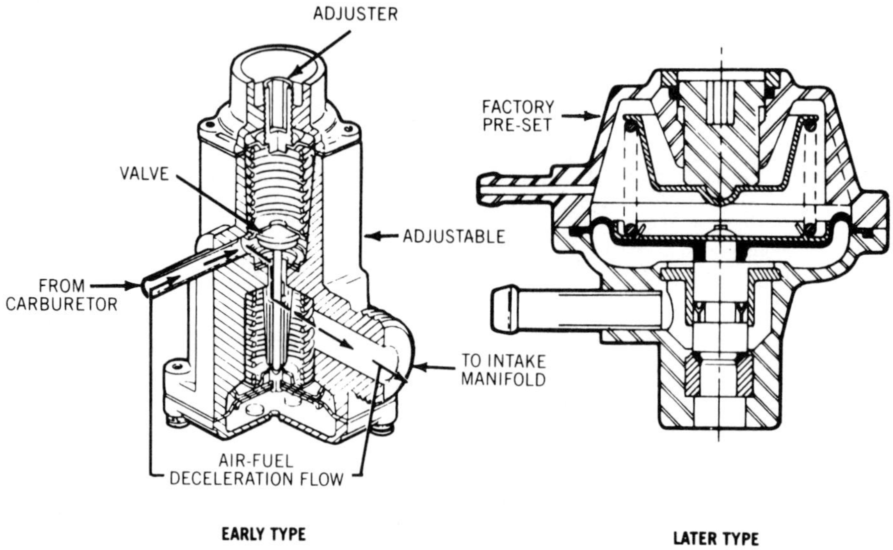


**Figure 16-60.** Ford's decel valves, used during the 1970's, added an extra air-fuel charge to the intake manifold during deceleration. (Ford)

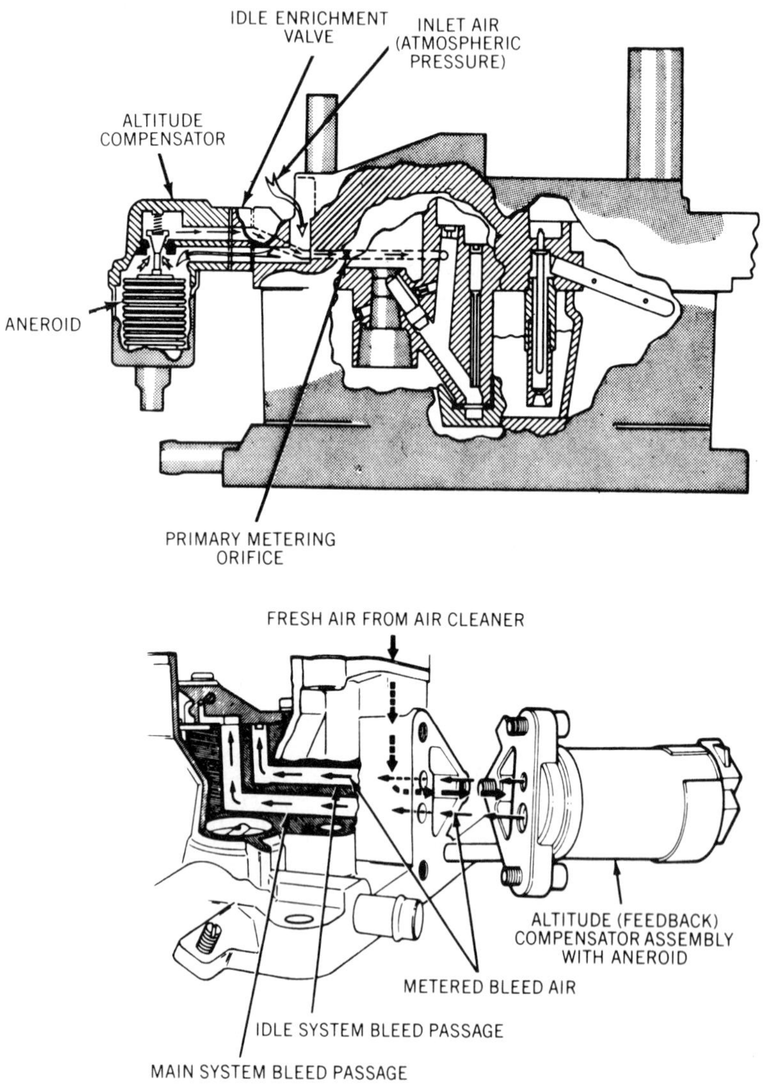


**Figure 16-61.** The aneroid bellows expands at high altitude (low atmospheric pressure) to open additional air bleeds. (Chrysler)

bleeds out of a hole in the diaphragm chamber, and the diaphragm retracts slowly to let the throttle close fully.

- *A deceleration throttle positioner* is a solenoid or diaphragm that momentarily holds the throttle open slightly on deceleration to reduce HC emissions.
- *A deceleration (decel) valve*, figure 16-60, admits an extra air-fuel mixture to the intake manifold for a few seconds on deceleration. These were most common on Ford 4- and 6-cylinder engines during the 1970's, but were used by other carmakers, as well. Increased manifold vacuum with a closed throttle opens the valve diaphragm, which opens a passage from the carburetor to the manifold. This allows an extra air-fuel charge to enter the intake system to slow deceleration speed and reduce HC emissions. Spring action and an air bleed to the diaphragm close the valve after a few seconds.

## Altitude Compensation

At high altitude, atmospheric pressure decreases so that a cubic foot of air weighs less and contains fewer oxygen molecules than at sea level. The same volume of air passes through the carburetor as at sea level, but there is less oxygen to mix with the fuel. The resulting rich mixtures increase CO emissions, reduce mileage, and can make the engine run poorly. These problems are most noticeable above 4,000 feet. Many carburetors have auxiliary devices to provide more air or less fuel at high altitude. Most work automatically. A few need manual adjustment.

An aneroid bellows is a sealed, accordion-shaped chamber that expands as air pressure decreases and contracts as air pressure increases. Many Carter, Rochester, and Motorcraft carburetors have aneroid bellows that automatically open air passages or close fuel passages at high altitude, figure 16-61.

The feedback carburetors of electronic engine control systems automatically compensate for atmospheric pressure changes as they continually adjust air-fuel ratios. This closed-loop fuel control is based on signals from a sensor that measures exhaust oxygen content. The system computer receives the sensor signal and sends an adjustment signal to a solenoid or motor on the carburetor. You will learn how these systems operate in a later chapter.

### Mixture Control Solenoids

Mixture control (MC) solenoids are used on most feedback carburetors for electronic engine control systems. The earliest such systems of the late 1970's used a computer-controlled solenoid to modulate a vacuum valve and thus regulate the amount of vacuum applied to a diaphragm in the carburetor. The diaphragm was connected to metering rods or valves that opened and closed air bleeds and fuel jets.

The most common MC solenoids are those used in Rochester Quadrajet, Dualjet, and Varajet carburetors for GM computer command control (CCC) systems since the early 1980's, figure 16-62. Chrysler also used these Quadrajet carburetors on some V-8 engines with electronic controls.

You will learn more about Rochester MC solenoids in the context of electronic control systems later in this volume. As an introduction, however, remember these principles.

- When the solenoid is energized, it moves a plunger and metering rod down to restrict the main jet and open air bleeds. This leans out the air-fuel mixture.

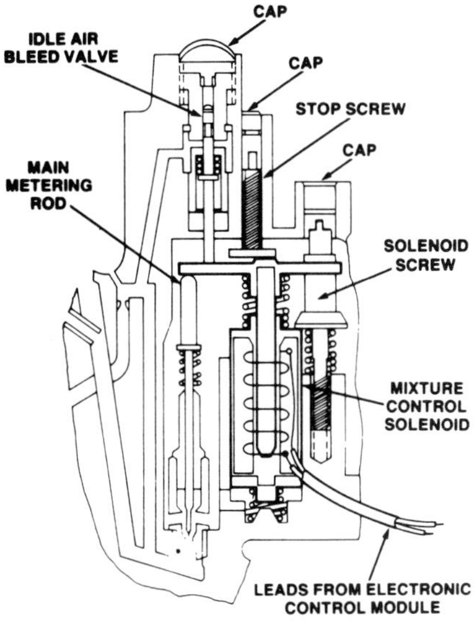


**Figure 16-62.** Cutaway of a mixture control (MC) solenoid installation in a Quadrajet or Dualjet carburetor. (GM)

- When the solenoid is deenergized, it moves the plunger and metering rod up to open the main jet more and close the air bleeds. This richens the air-fuel mixture.

The computer pulses the solenoid at a fixed rate, or cycle time. For GM systems, this is 10 cycles per second, or 10 hertz (Hz). The computer varies the rate of time that the solenoid is energized for each cycle. This is called pulse width modulation, or duty cycle modulation, and you will learn more about it in a later chapter. The result of this modulation has an averaging effect on the air-fuel ratio.

## SUMMARY

A carburetor controls engine airflow by throttling action and fuel metering by the action of an air pressure differential. Atmospheric pressure forces gasoline from the fuel bowl through discharge nozzles in low-pressure areas of the carburetor. A venturi is a restriction in a carburetor barrel that increases airflow velocity and vacuum. Four air pressure or vacuum areas affect carburetor operation: atmospheric pressure, manifold vacuum, venturi vacuum, and ported vacuum.

Carburetors have seven systems to provide proper air-fuel mixtures for all driving conditions: the fuel bowl and float system, the idle system, the low-speed system, the high-speed (main metering) system, the power system, the accelerator system, and the choke system. Carburetors are identified by the number of barrels and the direction of airflow. Variable-venturi carburetors provide a constant pressure drop across the venturi at all airflow rates.

Carburetors have various kinds of linkages and many assist devices and variations to help emission control and driveability.

## REVIEW QUESTIONS

### Multiple Choice

1. An automobile engine would not need a carburetor if it was operated:
   - **a.** at a varying speed with an unchanging load
   - **b.** at a steady speed with an unchanging load
   - **c.** at a varying speed with a changing load
   - **d.** at a steady speed with a changing load.

2. As temperatures rises, air expands and:
   - **a.** becomes lighter
   - **b.** becomes lighter and more dense
   - **c.** becomes more dense
   - **d.** becomes lighter and less dense.

3. Air pressure below atmospheric pressure is called:
   - **a.** vacuum
   - **b.** air pressure drop
   - **c.** depression
   - **d.** differential.

4. The throttle opening in a carburetor controls:
   - **a.** the amount of fuel delivered to the carburetor
   - **b.** the speed and amount of airflow
   - **c.** fuel pressure
   - **d.** atmospheric pressure.

5. A venturi in a carburetor causes:
   - **a.** airflow to slow down as it passes
   - **b.** airflow to stabilize
   - **c.** airflow to speed up
   - **d.** less vacuum.

**6.** The part of a carburetor that primarily aids fuel atomization and vaporization is the:

**a.** throttle
**b.** venturi
**c.** fuel nozzle
**d.** air bleed.

**7.** Ported vacuum is:

**a.** the high-pressure area near the venturi
**b.** the low-pressure area above the throttle
**c.** air pressure just outside the carburetor barrel
**d.** the low-pressure area in the intake manifold.

**8.** As a carburetor throttle starts to open:

**a.** vacuum starts to rise in the barrel
**b.** vacuum starts to increase in the manifold
**c.** vacuum starts to decrease in the barrel
**d.** ported vacuum starts to decrease.

**9.** Mechanic A says that fuel vaporization is poor when an engine is cold. Mechanic B says that since the engine is cold it doesn't require as rich a gasoline mixture as it does when it is warm. Who is right?

**a.** Mechanic A only
**b.** Mechanic B only
**c.** Both A and B
**d.** Neither A nor B.

**10.** At cruising speed the best fuel economy is obtained with an air-fuel ratio of:

**a.** 17:1 to 18:1
**b.** 18:1 to 19:1
**c.** 15:1 to 16:1
**d.** 16:1 to 17:1.

**11.** The best air-fuel ratio for maximum power is:

**a.** 12:1 to 13:1
**b.** 12.5:1 to 13.5:1
**c.** 13:1 to 14:1
**d.** 14.5:1 to 15.5:1.

**12.** The balance tube in a carburetor is plugged. Mechanic A says that this condition could cause the engine to run rich. Mechanic B says that a vacuum could develop in the fuel bowl. Who is right?

**a.** Mechanic A only
**b.** Mechanic B only
**c.** Both A and B
**d.** Neither A nor B.

### Fill in the Blank

**13.** Idle mixture screws control the amount of ________ in the idle air-fuel mixture.

**14.** Limiter caps on carburetor idle mixture screws help control ________ and ________ emissions.

**15.** The ________ provides all fuel for high-speed, light-load cruising.

**16.** The power system, or power valve, provides extra fuel as ________ increases.

**17.** The accelerator system consists of an accelerator pump, passages in the carburetor, and a ________ at the venturi of the carburetor.

**18.** The ________ pulls the choke open about halfway when the throttle opens completely.

**19.** Most cars have ________ carburetors mounted on top of the intake manifold.

**20.** Many carburetors employ a ________ to eliminate dieseling.

**21.** A ________ is a spring loaded diphragm with a plunger.

**22.** A ________ admits an extra air-fuel mixture upon deceleration.

**23.** Idle ________ is adjusted by changing the airflow through a carburetor.

**24.** Most accelerator pumps are operated by a ________ spring.

**25.** In a 2-stage 2-barrel carburetor, the ________ operate semi-independently.

# 17 DIESEL FUEL INJECTION

## WHAT

The last chapter explained gasoline carburetion and how each carburetor system meets specific engine needs. This chapter explains automobile diesel fuel injection systems and how these systems meet the same engine needs. This chapter also explains how diesel combustion chamber design affects engine operation and how the fuel injection system performs the two major jobs of fuel delivery and ignition timing. Finally, you will learn about the types and operations of injection nozzles and diesel auxiliary systems, such as glow plugs, block heaters, and vacuum pumps.

## WHY

Mercedes-Benz built the first production diesel passenger car in 1936. Diesel automobiles have sold well in Europe and other parts of the world for decades, but it was not until the late 1970's that they became commonplace in the United States and Canada. The reasons for their increased popularity were the oil shortages of 1973 and 1979 and the requirement that carmakers improve the mileage of their automobiles over 100 percent by the early 1980's.

In the late 1970's and early 1980's, almost every major carmaker built one or more diesel-powered cars or light trucks. Diesel automobile sales go up and down with the price and availability of fuel. However, light-duty diesel engine service has become an important part of the automobile service business. Millions of diesel-powered vehicles are on the road, and all require regular maintenance. As a professional service technician, you must know the principles of diesel fuel systems and how to service them.

## GOALS

After studying this chapter, you should be able to:

**1.** Explain the fuel and ignition needs of a diesel engine and how the fuel injection system meets them.

**2.** Explain how airflow in a diesel engine is different from airflow in a gasoline engine.

**3.** Identify the principal parts of an automobile diesel fuel injection system.

**4.** Identify and explain the operation of inline and distributor injection pumps, including governing and advance mechanisms.

**5.** Identify different kinds of injection nozzles.

**6.** Explain how diesel combustion chamber design affects ignition and combustion.

**7.** Explain the reasons for glow plugs, block heaters, and vacuum pumps on diesel engines.

## DIESEL ENGINE OPERATION— A REVIEW

You learned in chapter 4 that gasoline and diesel automobile engines both operate on the 4-stroke cycle of intake, compression, power, and exhaust, figure 17-1. You also learned that diesel engines have compression ratios of 16:1 to 22:1. These ratios develop compression pressures of 450 to 800 psi (3,000 to 5,600 kPa) and temperatures of 1300° to 1650° F (700° to 900°C). These temperatures will ignite any fuel present in the combustion chambers. Diesel ignition and combustion are controlled by metering a precise amount of fuel through a pump and injecting it at high pressure into the compressed air in the cylinders. Ignition timing is controlled by timing the fuel injection from the pump, figure 17-2. The fuel injection system, then, is the major combustion control system of a diesel engine. It takes the place of both the carburetor and the spark ignition system of a gasoline engine.

Remember that diesel engines run with an unthrottled air intake and an almost constant volume of air for all operating conditions. Engine speed and power are controlled by varying the *amount* of fuel for air-fuel ratios that range from as lean as 85:1, or more, at idle to as rich as 20:1 at full power.

### Gasoline and Diesel Comparison

Both gasoline and diesel engines are efficient automobile powerplants, and each kind has its advantages and disadvantages. The following paragraphs compare some of the features of gasoline and diesel engines.

• *Fuel power output*—Diesel fuel has somewhat more available heat energy, measured in Btu's per gallon or liter, than gasoline. Therefore, a diesel produces slightly more power from the same volume of fuel than a gasoline engine does. The higher compression of a diesel also ensures that the maximum amount of energy is extracted from the fuel.

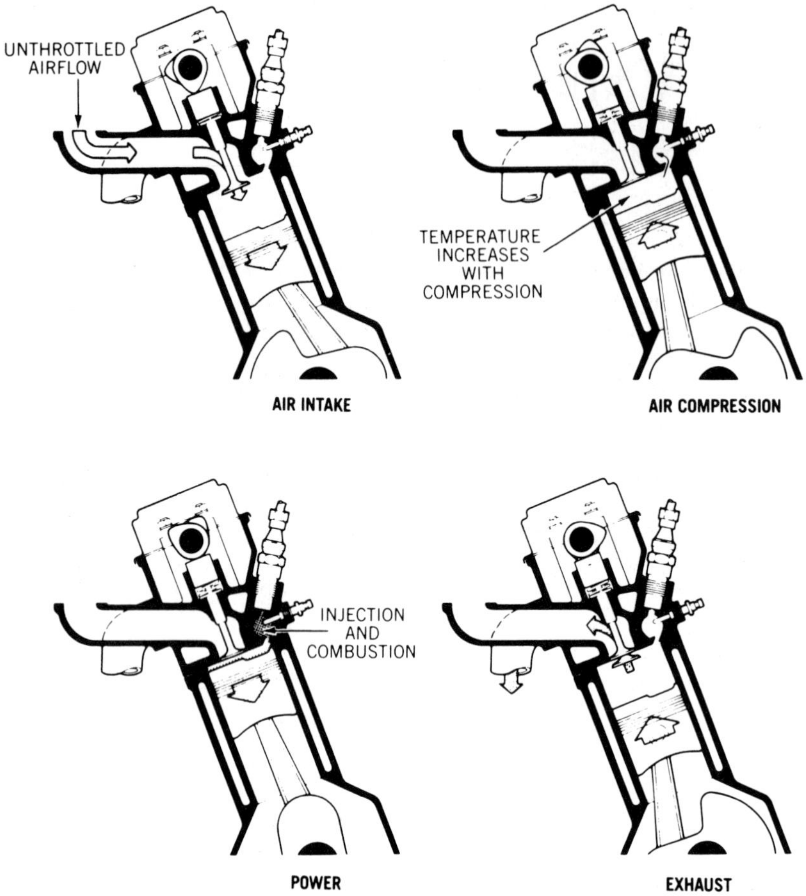


**Figure 17-1.** An automobile diesel engine operates on the same 4-stroke cycle as a gasoline engine. (Volkswagen)

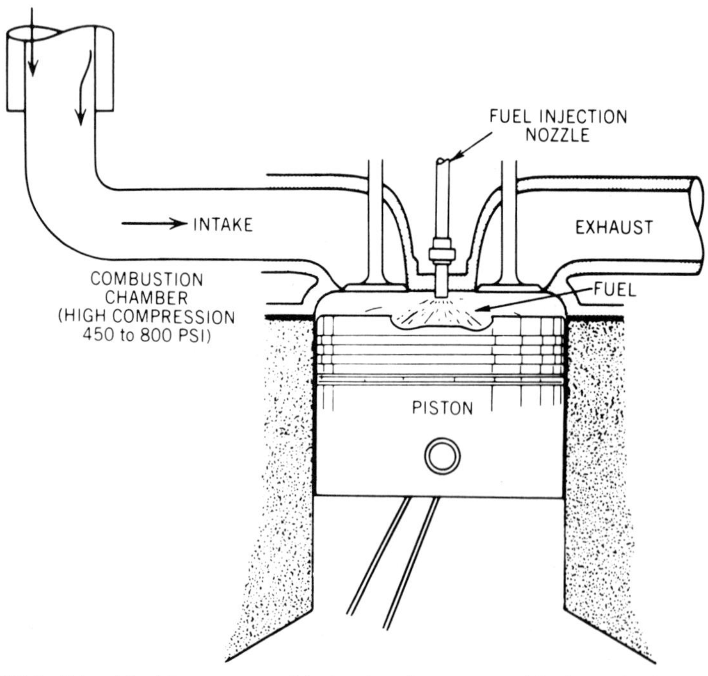


**Figure 17-2.** Diesel ignition occurs with the timed injection of fuel into highly compressed air in the cylinders.

• *Fuel economy*—Because a diesel runs with an unthrottled air intake, it never needs (or has) an air-fuel mixture with excess fuel. The high compression and high air-fuel ratios of a diesel ensure that all fuel is burned. Additionally, a diesel engine has no **pumping losses**.

• *Pumping losses*—This is a term that refers to the power that an engine loses when it works against a restricted throttle to take in air. All gasoline engines use some energy to pull air past a closed throttle valve. A diesel does not. This makes a diesel more efficient, particularly in stop-and-go driving. During acceleration, deceleration, and idling, a diesel can be 40 to 50 percent more fuel efficient than a gasoline engine.

• *Power and torque*—A diesel operates at lower speed, and torque output increases faster with load than in a gasoline engine. Diesel engine speed is limited by fuel ignition lag at high rpm. A high-speed diesel is one that can operate at more than 3,000 rpm. Peak horsepower and torque occur at higher engine speeds in a gasoline engine. Passenger car gasoline engines commonly run at 4,000 to 5,000 rpm or more. Racing engines reach 10,000 rpm.

• *Engine power-to-weight ratio*—A diesel engine is heavier than a gasoline engine to withstand the higher compression pressures and temperatures of its operation. Therefore, it produces less power per pound or kilogram of engine weight.

• *Emissions*—Because of the excess air in a diesel's air-fuel mixture, it emits little HC or CO in the exhaust. Almost all the fuel combines with oxygen to produce useful power. However, high-temperature diesel combustion promotes $NO_x$ formation. $NO_x$ emissions from a diesel are only slightly lower than from a gasoline engine. Particulate emissions from late-model diesels have been reduced, but they are still higher than from a gasoline engine. Additionally, a diesel emits more sulfur oxides than a gasoline engine.

• *Cold starting*—Diesel engines usually are harder to start in very cold weather than gasoline engines are for two major reasons. First, a diesel relies on the temperature of compressed air for ignition. It is harder to develop ignition temperature in cold weather. Most automobile diesel engines have glow plugs to preheat the combustion chambers. Second, diesel fuel can develop contamination at low temperature. The fuel also is more viscous and harder to pump and ignite when cold.

- *Fuel contamination*—You learned in chapter 14 that contamination by water and other impurities can be a problem with diesel fuel. Contamination removal is important because the fuel must lubricate the close-tolerance moving parts of the injection pump and injectors. Therefore, diesel engines require expensive fuel filtration devices.

### 2 STROKES VERSUS 4

At first glance, it might seem that a 2-stroke engine would be twice as efficient as a 4-stroke engine. After all, it produces a power stroke for every single revolution of the crankshaft instead of once for every two revolutions as a 4-stroke engine does. But it's not that simple.

Other advantages of a 2-stroke engine are its simple design, low manufacturing cost, and fewer moving parts. Because 2-stroke gasoline engines have no poppet valves and the related inertia of the valve train, they can run at higher engine speeds than 4-stroke engines. It's not uncommon for a 2-stroke to develop maximum power at 7,000 to 10,000 rpm. Compare that to 3,000 to 6,000 rpm as the maximum power range for most 4-stroke engines.

The disadvantages of 2-stroke engines are that they run roughly at low speeds and have poor heat dissipation because there are fewer no-load strokes between ignition and power strokes. Because intake and exhaust occur at almost the same time, a 2-stroke engine doesn't breathe as efficiently as a 4-stroke engine. Mixing a fresh air-fuel mixture with exhaust preheats and expands the air-fuel charge. Because this increases charge volume, it reduces efficiency and reduces the effective piston pressure that the engine develops. In a 4-stroke engine the intake charge varies from 70 to 90 percent of the maximum cylinder volume. In a 2-stroke engine, it varies from 50 to 70 percent of maximum volume because of exhaust scavenging and charge dilution. The reduced charge volume of a 2-stroke engine reduces engine efficiency and power output.

Because its air-fuel mixture is diluted, a 2-stroke engine requires more fuel than a 4-stroke engine for an equivalent power output. Also, the oil mixed with the fuel to lubricate a 2-stroke gasoline engine creates an overly rich mixture and increases the exhaust emissions. Poor fuel economy and high emissions were the two factors that eliminated 2-stroke engines as desirable automobile powerplants. Two-stroke gasoline engines used to be common as motorcycle powerplants, but as emission regulations were extended to motorcycles in the late 1970's, 2-strokes disappeared from all but off-road cycles. Today, 2-stroke engines are used in outboard marine powerplants, snowmobiles, off-road cycles, chainsaws, and other small industrial equipment. Detroit Diesel 2-stroke diesel engines are still popular, however, as truck powerplants.

## Fuel Requirements for Diesel Operation

A 4-stroke diesel engine in an automobile operates under the same driving conditions as a gasoline engine, and the fuel injection system supplies varying amounts of fuel for these operating needs. Many functions of a fuel injection system are similar to those of a carburetor, but a diesel injection system also has functions that a carburetor does not.

- *Starting*—When a diesel engine starts, it is cold and cranks slowly. As with a gasoline engine, a diesel requires extra fuel for starting to ensure that enough fuel vaporizes for combustion. Injection pumps have devices to meter extra fuel for starting, as a carburetor has a choke.
- *Idling*—Unlike a gasoline engine, a diesel does not need a rich mixture for idling. It needs only enough fuel to keep the engine running. Diesel idle speed is limited by the minimum amount of air a particular engine can take in and still keep running. Most automobile diesels idle at 700 to 800 rpm. The injection system provides just enough fuel to ensure combustion with minimum engine load. The idle air-fuel ratio of 85:1 or more is the highest for any diesel operating condition.
- *Light-load and heavy-load operation*—An automobile diesel engine operates with the same conditions of light-load cruising and deceleration and heavy-load hill climbing, pulling, and acceleration as a gasoline engine. A carburetor has high-speed (main metering), power, and accelerator pump systems to vary fuel metering. A diesel injection pump has devices to vary the amount of fuel metered for changing load conditions, but it does not throttle the engine airflow as a carburetor does. Fuel metering from an injection pump responds directly to engine load and is controlled by a governor to provide more fuel for heavy loads and less for light loads. At light-load, steady-speed operation, a diesel air-fuel ratio is almost as lean as at idle. The engine gets just enough fuel to keep running at a given speed.
- *Ignition timing and advance*—In both gasoline and diesel engines, ignition must occur at a precise time to develop full combustion power in relation to the crankshaft position. Moreover, ignition timing must be advanced at high speeds to allow enough combustion burn time. In a gasoline engine, ignition is timed and advanced by an electrical distributor. In a diesel, it is timed and advanced by the injection pump. This is the major function of a diesel injection system that a gasoline fuel system does not have.
- *Fuel atomization and vaporization*—Diesel fuel, like gasoline, must be atomized into small drops and then

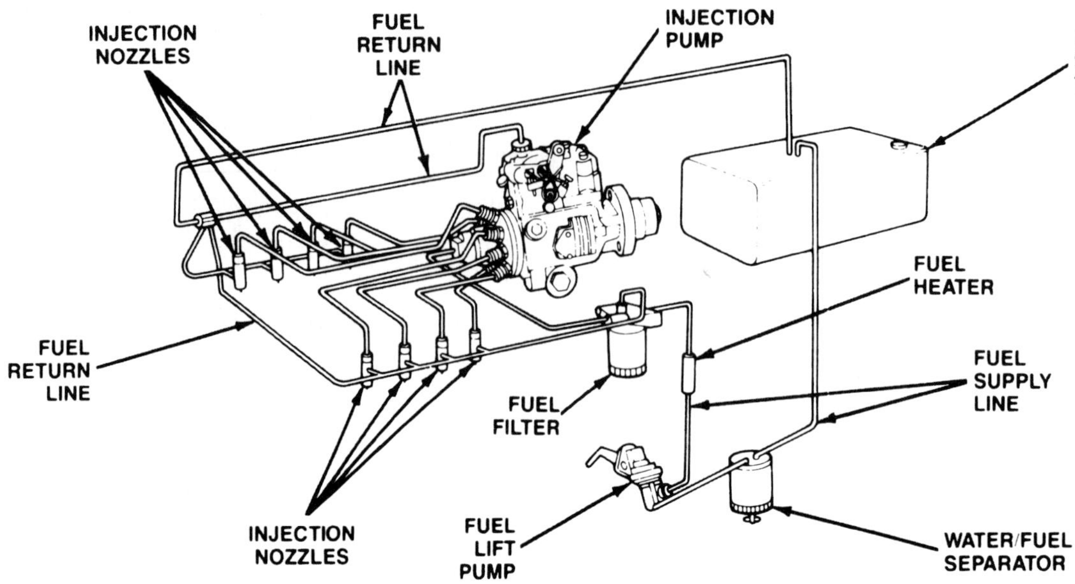


**Figure 17-3.** All diesel fuel injection systems have these basic parts. (Ford)

vaporized before it will mix with air and burn. In a gasoline engine, the fuel system atomizes gasoline as it leaves carburetor discharge nozzles and vaporizes it in the intake manifold. In a diesel, the injection system atomizes fuel as it leaves the injection nozzles, and the fuel vaporizes in the combustion chamber.

As you study diesel fuel injection systems in detail, remember these engine control functions that all systems have:

- The injection pump must develop fuel pressure.
- The pump must control fuel volume for different speeds and loads.
- The pump governor must control minimum and maximum engine speed.
- The pump must control fuel distribution and ignition timing.
- The injectors must atomize the fuel.
- The pump must provide a positive fuel shutoff to stop the engine.

## TYPES OF FUEL INJECTION SYSTEMS

All diesel fuel injection systems have these parts, figure 17-3:

- A fuel tank, delivery lines, and filters or water separators
- A transfer pump, or lift pump, to draw fuel from the tank and supply the high-pressure injection pump
- An injection pump that raises fuel pressure to more than 1,000 psi and that times and meters fuel delivery to each cylinder in firing-order sequence
- Fuel lines to each cylinder injector
- A fuel injection nozzle, or injector, for each cylinder to deliver fuel to the combustion chamber
- Fuel return lines to return excess fuel from the pump and injectors to the tank or the filter.

You already have studied fuel tanks, diesel fuel filters, and some transfer pumps in chapter 15. You now will learn about the kinds of injection pumps and nozzles commonly used on automobile diesel engines.

There are three basic types of diesel injection systems:

1. Common rail
2. Unit injector
3. Pump controlled.

The first two are used on the large diesel engines of trucks, heavy equipment, locomotives, and stationary powerplants that operate in a narrow speed and load range. We will not examine the common-rail and unit-injector systems in this book. The pump-controlled system is used on automobile engines that operate with greater variations in speed and load.

All automobile diesels have an engine-driven pump that develops high fuel pressure and uses that pressure to open the injectors hydraulically. The pump controls all fuel metering, distribution, and timing. Because injection must be timed with the crankshaft and valve positions of the 4-stroke cycle, the same drive belts or gears that drive the engine camshaft drive the injection pump. The pump rotates at camshaft speed (one-half engine speed) and times injection for the top of the power stroke. Automobile injection pumps are either the inline type or the distributor type.

## INLINE INJECTION PUMP

The Robert Bosch pumps, figure 17-4, used on Mercedes-Benz diesels are the most common inline pumps. An inline injection pump has one plunger for each engine cylinder. A camshaft in the pump body operates the plungers in engine firing-order sequence. When a pump cam lobe raises a plunger, the plunger force develops pressure on the fuel and pumps it through the delivery valve and line to the injector.

A fuel supply pump (transfer pump), attached to the injection pump, draws

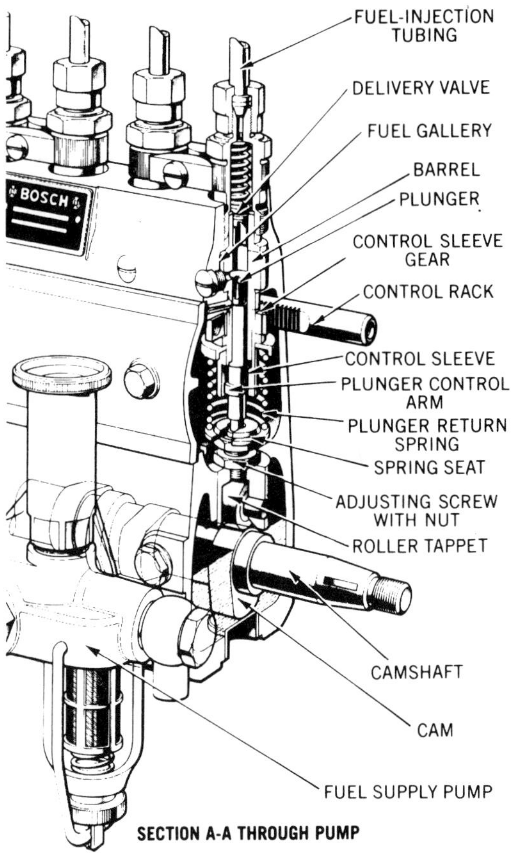


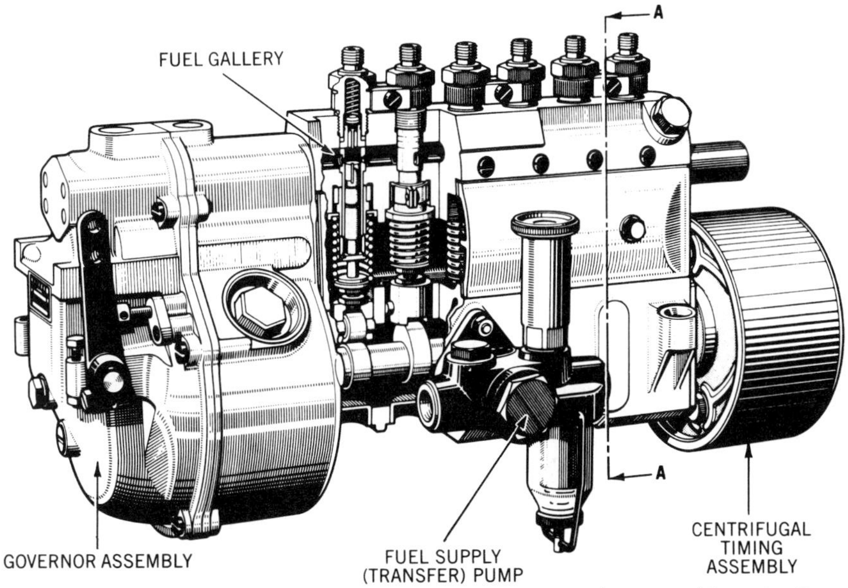


**Figure 17-4.** The Robert Bosch inline injection pump is used on most Mercedes-Benz diesels. (Bosch)

**IS IT A MOTOR OR AN ENGINE?**

You've probably heard people say that if it's powered by electric current, it's a motor; and if it's powered by burning fuel, it's an engine. In some circles it may be technically incorrect to refer to the powerplant of an automobile as a "motor" when it's properly an internal-combustion "engine."

If that's true, someone forgot to tell a lot of the leaders of the automobile industry. Would you like to be the one to tell Henry Ford that he was wrong to call his company Ford *Motor* Company? Maybe Billy Durant was wrong to call his company General *Motors*, but he founded one of history's most successful corporations. Then there is the industry association of vehicle makers called the *Motor* Vehicle Manufacturers Association. And don't forget all of those magazines that have been around for generations, such as *Motor, Motor/Age, and Motor Trend*. In any case, whether you call it a motor or an engine is less important than the fact that you know how each works and how to repair them.

fuel from the tank and fills the injection pump fuel gallery. The gallery fills the barrels of all plungers. Each plunger fits in its barrel with only 0.00008 inch (0.002 mm) clearance to retain pressure during pumping.

### Inline Pump Operation

Each revolution of the pump camshaft moves each plunger on a fixed stroke of the same length, regardless of engine and pump speed. But, the pump must meter varying amounts of fuel for changing speeds and loads. The port-and-helix design of the plungers and barrels accomplishes this. At the bottom of its stroke, the plunger opens one or two fill ports from the fuel gallery into the barrel, figure 17-5. As the cam moves the plunger up, it closes the fill ports and traps fuel in the barrel. The plunger applies pressure to the fuel and forces it past the delivery valve to the injection line.

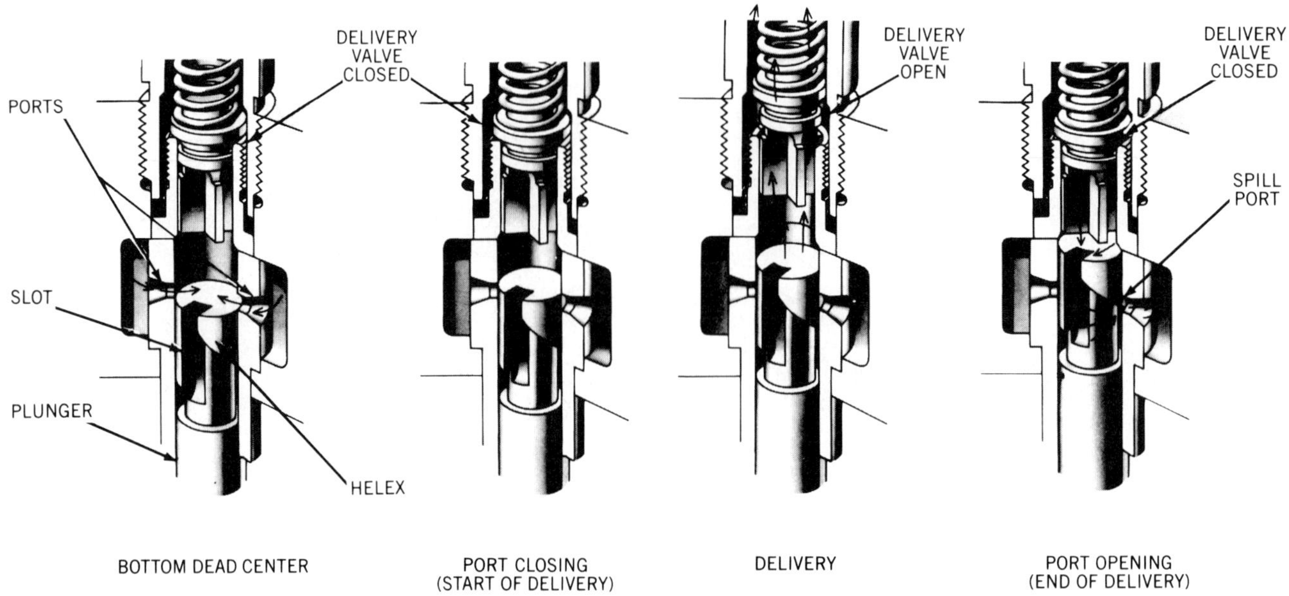


**Figure 17-5.** The plunger always travels the same distance on its stroke, but the position of the helix controls the amount of fuel delivered. (Bosch)

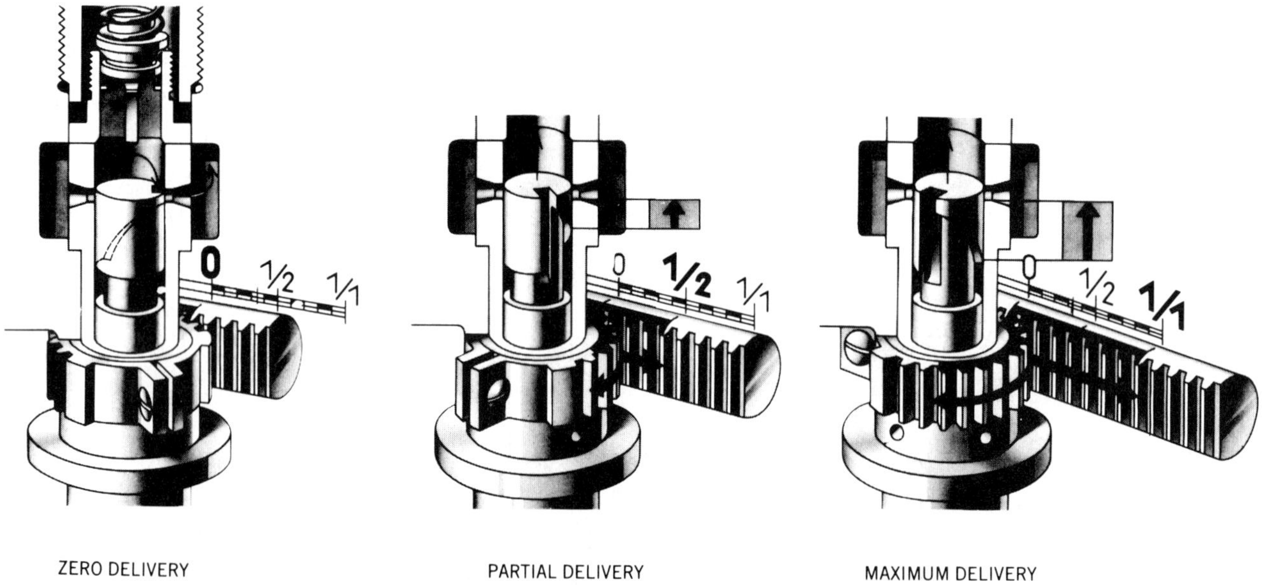


**Figure 17-6.** The control rack and plunger gears turn the plungers to govern fuel delivery. (Bosch)

The side of the plunger has two grooves, or slots. One is vertical and allows fuel to fill the lower area of the plunger and barrel. The other groove is a **helix**, or spiral. As the plunger closes the inlet port, pressure is equal on the fuel above the plunger and in the lower area opened by the vertical slot. As the plunger continues upward, the lower control edge of the helix reopens one port and relieves pressure on the fuel, figure 17-5. Fuel stops flowing through the delivery valve and spills back into the fuel gallery.

Injection occurs during the part of the stroke when the fill ports are closed. Injection stops when the helix control edge reopens the spill port. The engine accelerator linkage and pump governor rotate the plunger in the barrel so that the helix opens the spill port earlier or later to vary the amount of fuel pumped to the injector. Each plunger has a gear segment that meshes with a control rack gear connected to the governor.

For maximum fuel delivery, the control rack rotates each plunger so that the lowest point of the helix control edge aligns with the spill port, figure 17-6. This allows the plunger to pump the maximum fuel during its stroke before the helix uncovers the spill port. For no fuel delivery, the rack rotates the plunger so that the vertical groove aligns with the fill port. In this

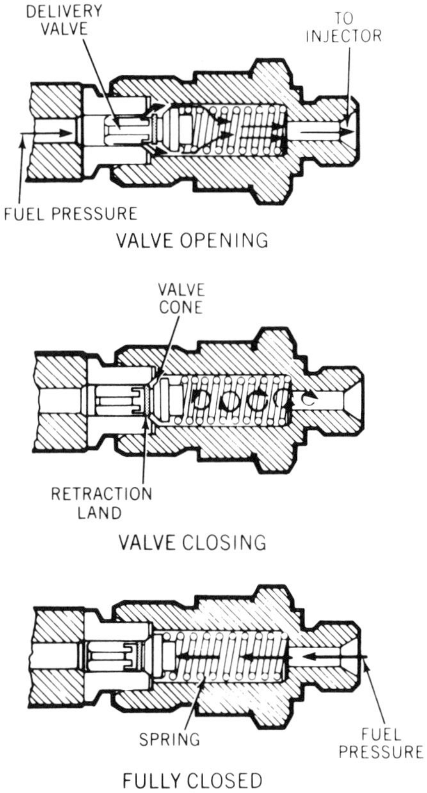


**Figure 17-7.** The fuel delivery valve allows the plunger to build up fuel pressure. The valve closes quickly to prevent injector dribble. (Volkswagen)

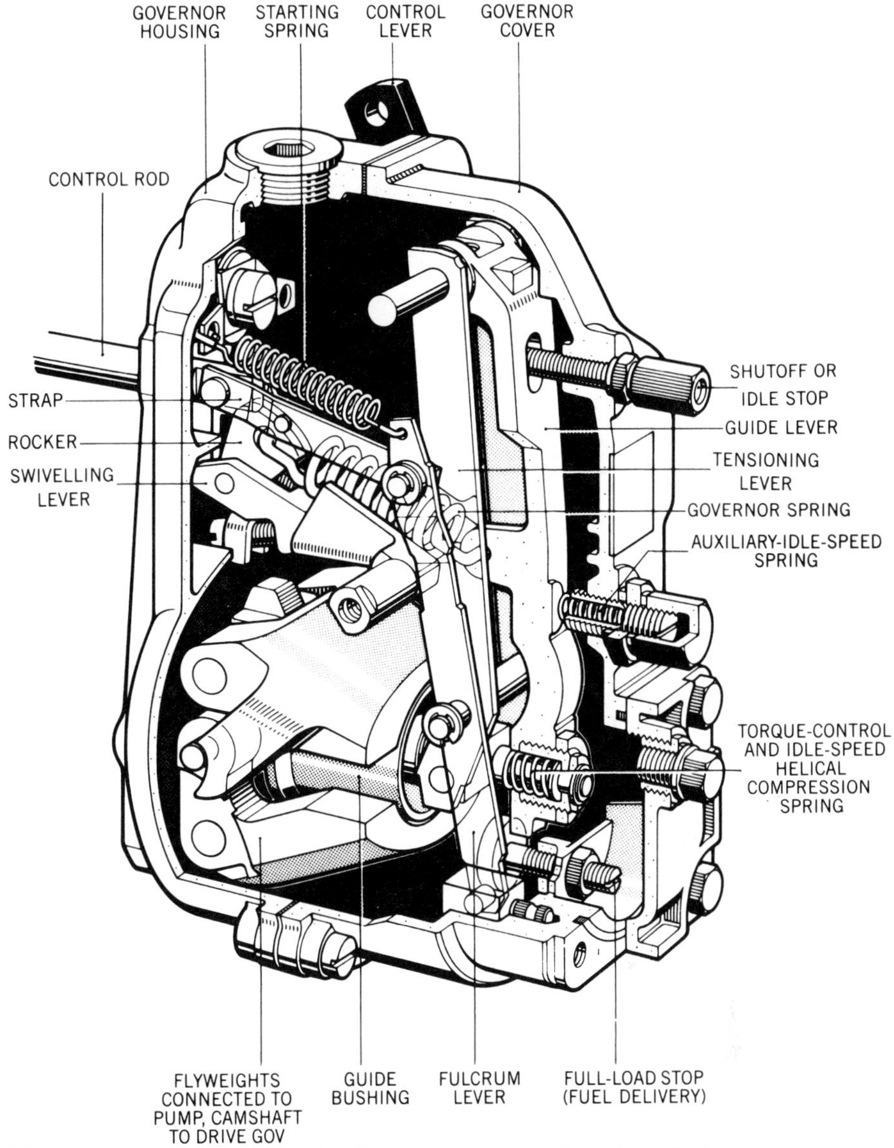


**Figure 17-8.** This Bosch centrifugal governor is mounted on the end of an inline pump. The control rod moves the control rack to regulate injection. (Bosch)

case, the plunger develops no pressure during its entire stroke. Between these two extremes, the governor rotates the plunger so that the helix uncovers the spill port earlier (less fuel) or later (more fuel) according to engine speed and load.

**Fuel Delivery Valve.** The fuel delivery valve at the top of each plunger barrel is held closed by a spring. As the fuel pressure created by plunger movement exceeds spring pressure, the fuel forces the valve off its seat and moves through the injection line to the injector, figure 17-7.

The fuel delivery valve does more than close the plunger barrel so pressure can build up. As the plunger helix uncovers the spill port and fuel pressure drops, the delivery valve snaps closed. This increases the space in the injection line from the delivery valve to the injector, which causes a fast drop in pressure on the fuel in the line, figure 17-7. The fuel expands to fill the available space, and the injector spring closes the injector nozzle quickly. Because the fuel actually moves backward in the line, the delivery valve often is called a **retraction valve**, and the part of the valve that fits into the seat is called the retraction land.

The injection nozzle must close as quickly as it opens. If pressure decreases slowly at the nozzle, the nozzle will dribble fuel after the optimum point of injection. Fuel dribbling from the nozzle will increase smoke and exhaust emissions. Fuel economy will decrease, and unburned fuel will dilute the crankcase oil.

**Governor Operation.** When the engine is started, a control lever moves the injector rack to the maximum fuel position. After the engine starts, the starting lever moves the rack back to the idle position for minimum fuel. As the vehicle accelerates, the accelerator linkage moves the control rack to provide more fuel for the increased engine load. The accelerator controls the injection pump rack through a centrifugal governor on the end of the pump, figure 17-8. Under load at low speed, the governor flyweights that rotate with the pump shaft allow the linkage to move the injector rack for more fuel.

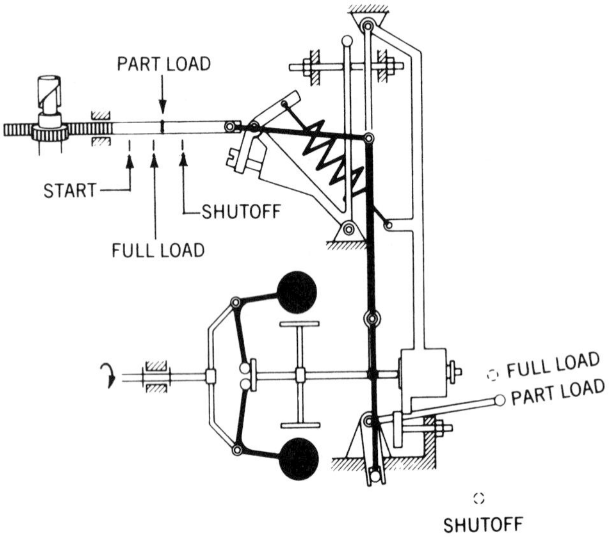


**Figure 17-9.** At part throttle and a light load, the governor moves the control rack to reduce fuel delivery. (Bosch)

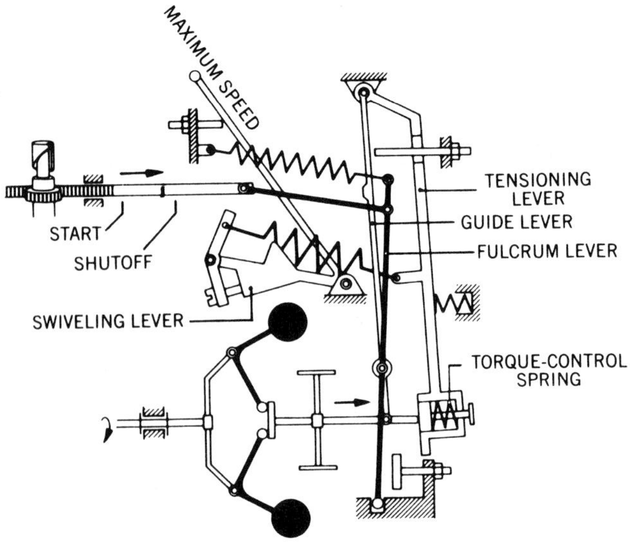


**Figure 17-10.** At maximum governed speed with little or no load, the governor reduces fuel delivery to the minimum. (Bosch)

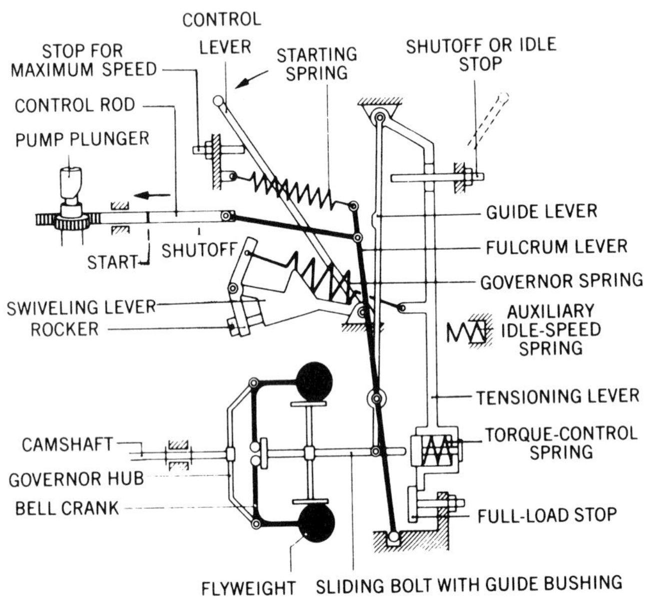


**Figure 17-11.** When engine load increases, the governor slows down and moves the linkage to increase fuel delivery. (Bosch)

As engine speed increases and load decreases, the governor weights swing out and move other linkage to move the rack for less fuel, figure 17-9. At steady cruising, fuel metering is balanced by the position of the accelerator and the governor linkage. If the vehicle accelerates again, load increases, and the engine and governor start to slow down. The governor linkage then moves the rack to inject more fuel for the increased load. Maximum speed is adjusted by a stop screw in the governor that limits linkage and injector rack travel. At maximum speed, the governor again balances the linkage to limit fuel metering, figure 17-10. If engine load increases at top speed (climbing a hill, for example), the governor slows down and moves the linkage to meter more fuel for the heavier load, figure 17-11.

This is a simplified explanation of governor operation. As with other vehicle systems, governor designs vary for specific applications, but all work on these principles. Most automobile diesels have speed-limiting governors that limit the idle and maximum engine speeds. The driver controls intermediate speeds through the accelerator linkage. Besides metering fuel in response to changing load, the governor keeps the engine from stalling at idle and from exceeding a safe maximum speed. A few automobile diesel engines have pneumatic governors that do the same jobs. Industrial engines usually have variable-speed or constant-speed governors.

**Ignition Timing.** At high engine speeds, injection timing must be advanced to allow for ignition lag. The Bosch inline pump has another set of centrifugal weights on the front of the camshaft, figure 17-12. The hub of the centrifugal advance mechanism is attached to the pump drive gear. The advance camplate is attached to the pump camshaft. As speed increases, the weights swing outward and move the camshaft to advance pump timing up to 8 degrees, figure 17-13. Since the pump rotates at one-half engine speed, ignition timing will advance a maximum of 16 degrees. This centrifugal advance mechanism works in the same way that centrifugal advance weights work in an ignition distributor, as you learned in Part Four of this book.

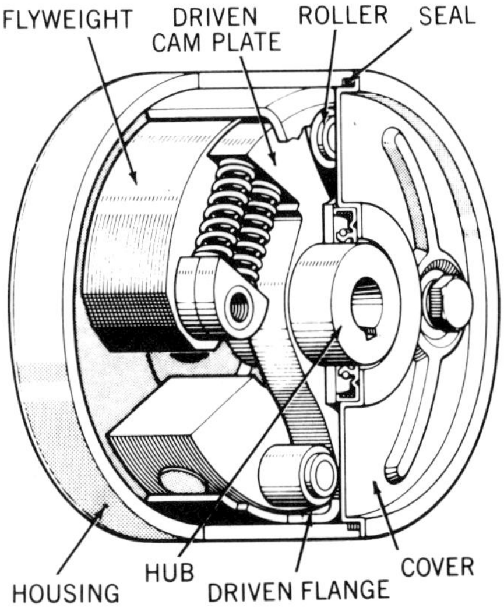


**Figure 17-12.** The Bosch centrifugal advance assembly for an inline pump connects the pump drive gear to the pump camshaft. (Bosch)

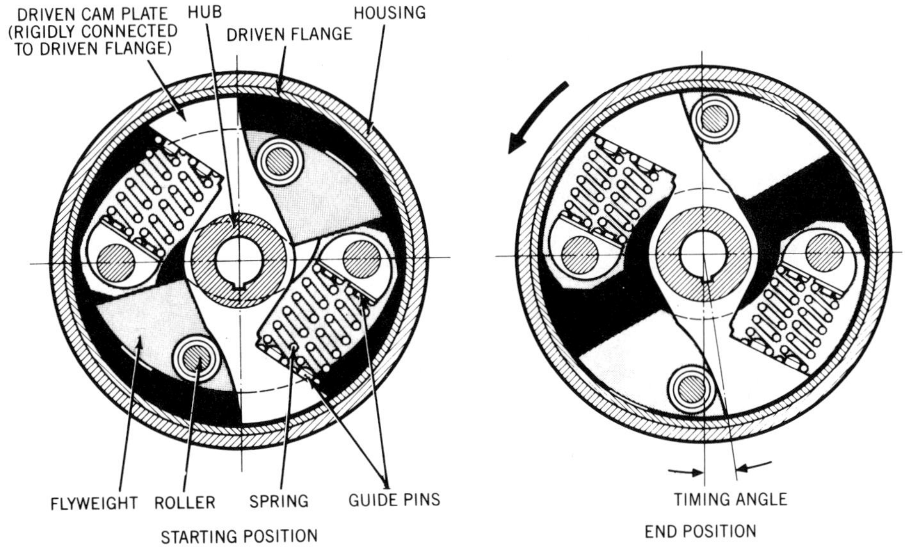


**Figure 17-13.** The centrifugal advance mechanism rotates the pump camshaft to advance injection timing. (Bosch)

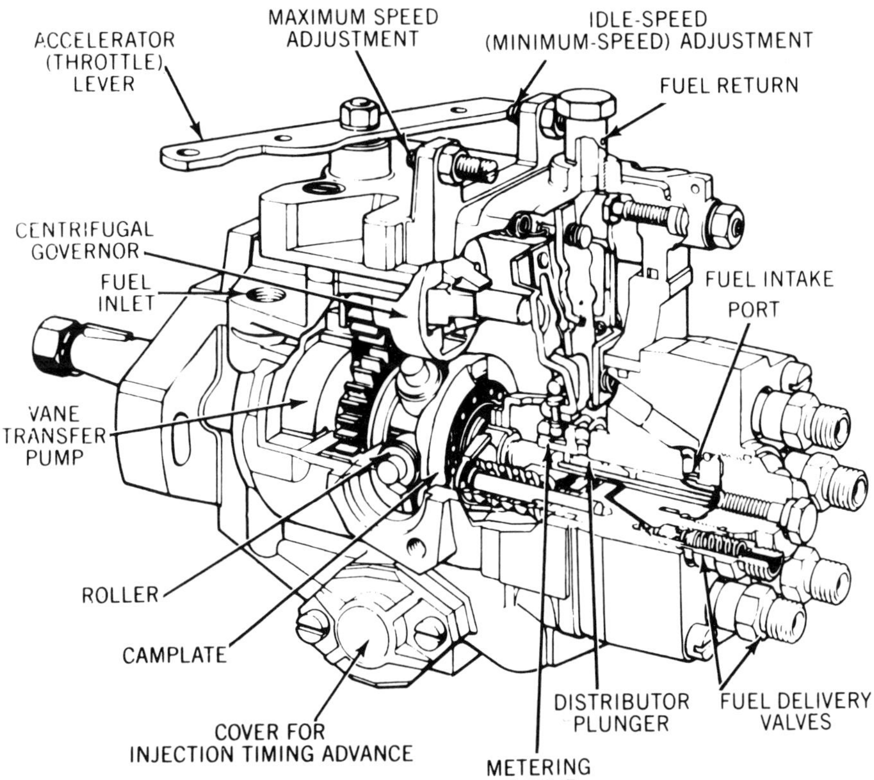


**Figure 17-14.** This Bosch distributor pump has one pumping plunger for all cylinders and distributes fuel as an ignition distributor distributes high voltage. (Volkswagen)

## DISTRIBUTOR INJECTION PUMP

A distributor injection pump has less reciprocating inertia than an inline pump and responds somewhat faster to changing speeds and loads. Therefore, most automobile diesel engines use distributor pumps, figure 17-14. This kind of pump has one pumping element for all cylinders and distributes fuel to each cylinder much as an ignition distributor distributes electrical high voltage to the spark plugs of a gasoline engine. At first glance, a distributor pump may look more complicated than an inline pump, but it works on the same principles of fuel metering, timing, and delivery.

The most common distributor pumps for automobile engines are the Robert Bosch VE or VP pumps and the Standyne Roosa Master DB pump. The Japanese Nippondenso and Diesel Kiki pumps are similar to Bosch pumps. CAV-Lucas or CAV-Simms pumps, used on some GM engines are similar to Standyne pumps.

Like an inline pump, a distributor pump is driven by gears or belts from the engine camshaft and rotates at one-half engine speed, figure 17-15. The entire pump housing is full of diesel fuel at all times for lubrication. Fuel

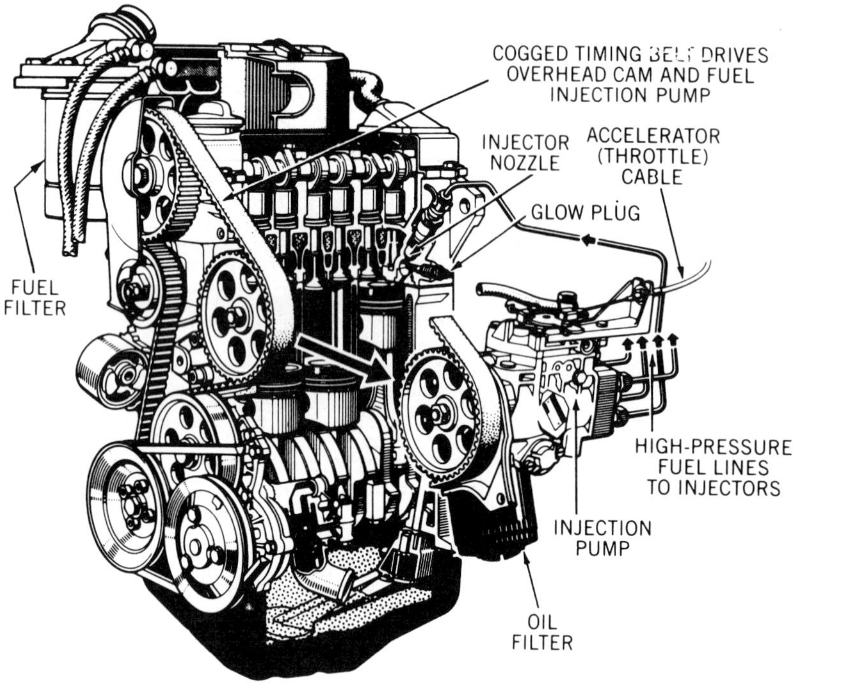


**Figure 17-15.** This distributor injection pump is driven by the same timing belt that drives the engine camshaft. (Volkswagen)

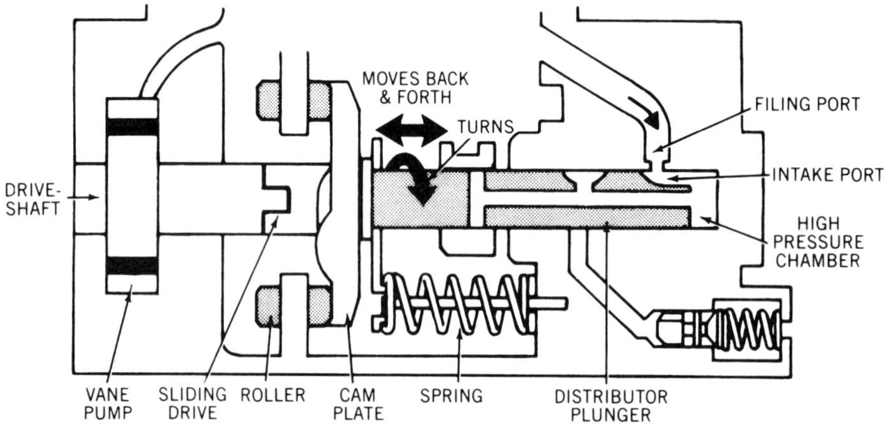


**Figure 17-16.** The Bosch distributor pump plunger moves back and forth and rotates to pressurize and distribute fuel. (Volkswagen)

enters the pump through an inlet and passes through a strainer to the transfer pump. The vane-type transfer pump develops 20 to 80 psi (140 to 550 kPa) of pressure to charge the high-pressure pump. Some Standyne Roosa Master transfer pumps operate with even higher pressures. Bosch pumps use an axial piston that moves back and forth in its bore to pressurize the fuel and rotates to deliver the fuel to the correct engine cylinder. Roosa Master pumps have a pair of small radial pistons in a common bore and a separate rotor to distribute the fuel.

### Bosch Distributor Pump Operation

The pump driveshaft turns the transfer pump, a camplate, and the pump plunger as a unit, figure 17-16. The camplate rotates against fixed rollers, which cause the camplate to move the plunger back and forth as it rotates, figure 17-17. When an intake port in the plunger aligns with the filling port in the housing, fuel flows into the plunger high-pressure chamber. As the plunger turns, it closes the intake port. The camplate pushes the plunger forward to pressurize the trapped fuel to about 1,800 to 2,500 psi (12,400 to 17,225 kPa). As the plunger continues to turn, it opens an outlet port to one of the injection lines in the pump head. Fuel pressure forces open a delivery (retraction) valve in the pump head, and fuel flows to the injector, figure 17-18. Injection lines are connected to the pump head in firing-order sequence, and each line has a separate delivery valve.

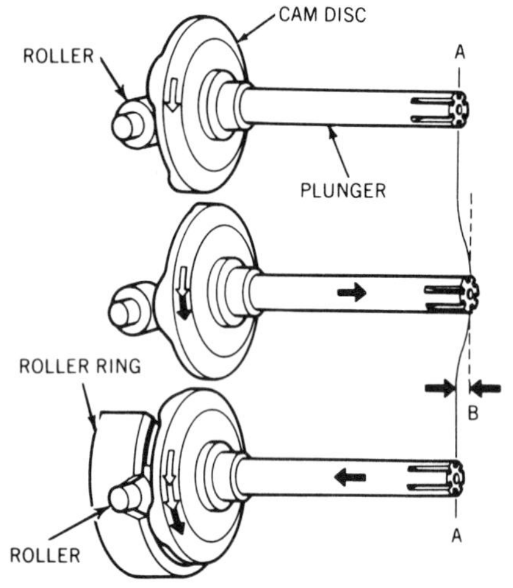


**Figure 17-17.** As the camplate rotates, it moves the plunger back and forth. (Ford)

**Governor Operation.** The Bosch VE pump has a centrifugal governor that moves a metering sleeve on the pump plunger, figure 17-14. As the plunger strokes back and forth, a relief port from the pressure chamber moves in and out of the metering sleeve. The governor can move the sleeve in relation to the relief port to control fuel metering. When the relief port opens, pressure is relieved to stop injection.

The position of the metering sleeve determines the length of the plunger stroke needed to uncover the relief port. The longer the stroke, the more fuel is injected. This is the same principle of outlet metering used in an inline pump.

During starting, figure 17-19, the starting spring and lever move the sleeve to allow the maximum plunger stroke. At idle, figure 17-20, the centrifugal governor weights move a sliding sleeve and lever to uncover the relief port sooner on the pump stroke. This reduces fuel to a minimum. Under acceleration, figure 17-21, the accelerator linkage moves the sleeve control lever to allow more fuel delivery. Governor rotation also slows to allow the sleeve to cover the relief port longer. At maximum speed, figure 17-22, the governor weights move outward and force the metering sleeve to uncover the relief port at the start of the plunger stroke. This limits fuel delivery and maximum speed.

**Ignition Timing.** A hydraulic advance mechanism controls timing and advance in a Bosch distributor pump. Fuel pressure from the transfer pump moves a piston in the bottom of the injection pump housing. The piston is connected to the rollers that cause the camplate to move the injection plunger back and forth. As transfer pump pressure increases, it moves the piston to shift the roller housing so that the camplate strokes the injection plunger sooner, figure 17-23. This advances

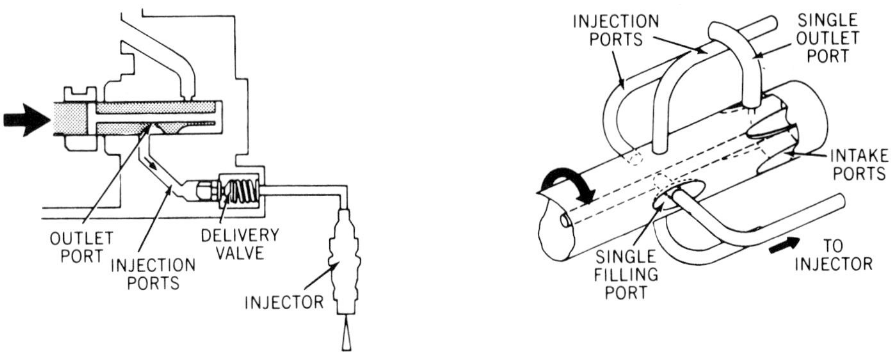


**Figure 17-18.** Plunger rotation opens fuel outlet ports in firing-order sequence. (Volkswagen)

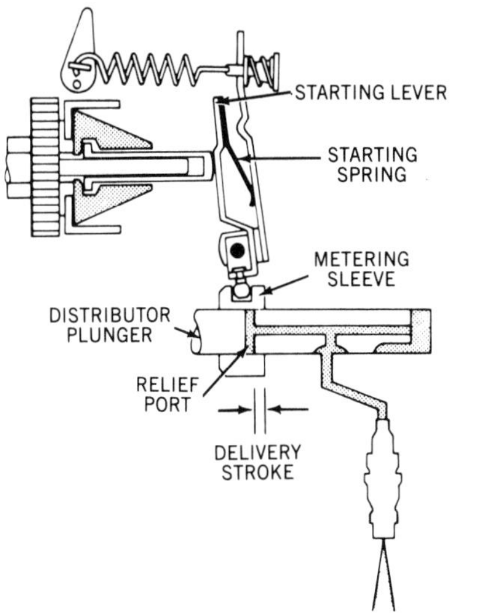


**Figure 17-19.** The starting spring and lever move the metering sleeve to allow maximum fuel delivery for starting. (Volkswagen)

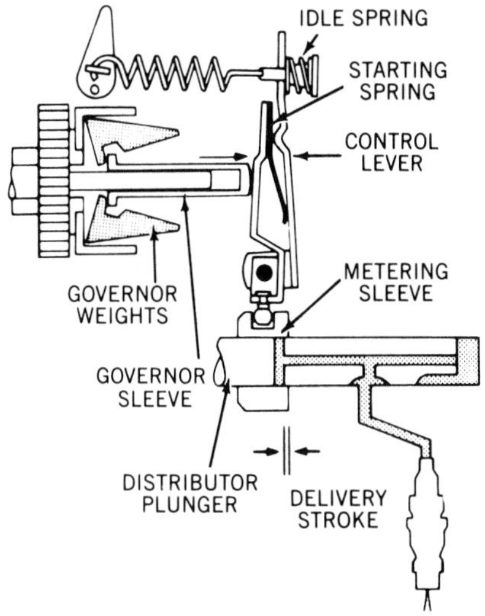


**Figure 17-20.** At idle, the governor weights begin to move out and shift the metering sleeve for less fuel. (Volkswagen)

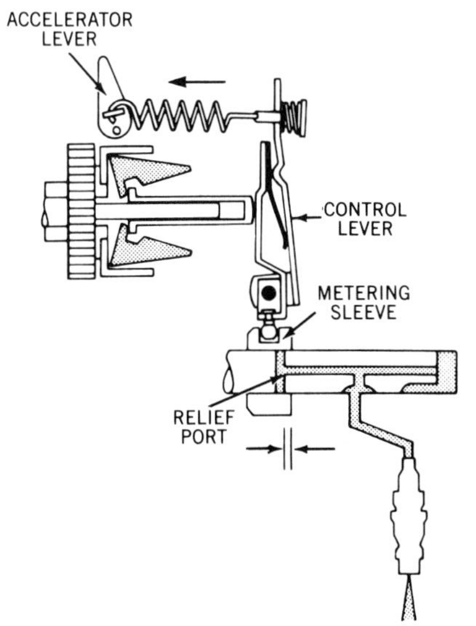


**Figure 17-21.** During acceleration, the accelerator lever moves the control lever and metering sleeve to allow more fuel delivery. (Volkswagen)

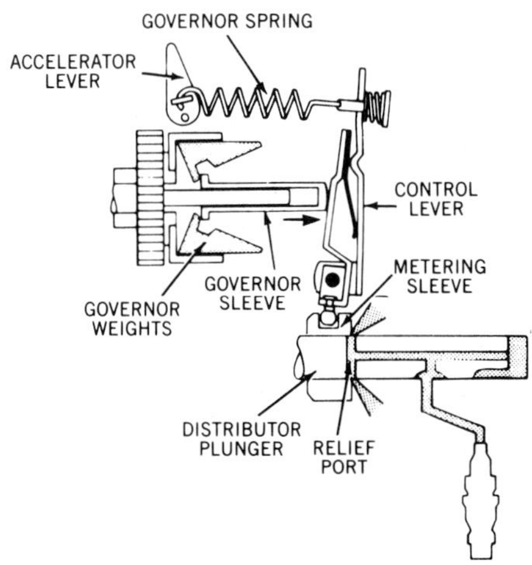


**Figure 17-22.** At maximum speed with light load, the governor moves the metering sleeve so that no fuel is injected. (Volkswagen)

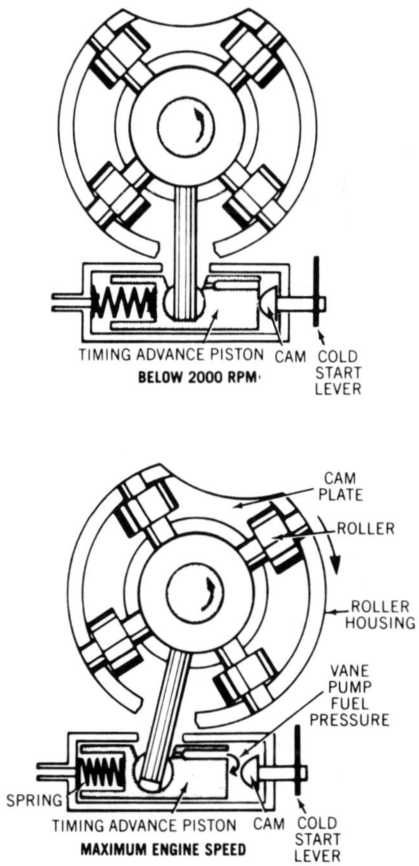


**Figure 17-23.** The timing advance piston moves the roller ring so that the camplate contacts the rollers sooner to advance injection timing. (Volkswagen)

the injection timing. The Bosch VE pump provides a maximum of about 20 crankshaft degrees of injection advance at about 5,000 engine rpm. To start a cold engine, a cold-start lever moves a cam that contacts the advance piston to advance the timing about 5 degrees. The hydraulic advance mechanism of a Bosch distributor pump can provide up to 20 crankshaft degrees of ignition advance at high speed, figure 17-24.

**Fuel Shutoff.** The Bosch distributor pump has a shutoff solenoid controlled by the vehicle "ignition" key. When the ignition switch is on, current is supplied to the solenoid to retract a plunger in the intake port to the injection pump plunger. When the ignition switch is turned off, the solenoid is deenergized and its plunger closes off the fuel intake port to stop the engine, figure 17-25.

## Roosa Master Pump Operation

The Roosa Master pump receives fuel through a central inlet and fuel strainer in the pump head, figure 17-26. A vane-type transfer pump fills the pump housing and delivers fuel to the high-pressure injection pump at 60 to 120 psi (413 to 828 kPa), depending on pump model. Transfer pressure is controlled by a pressure regulator at the inlet end of the pump. Unlike the Bosch pump, the Roosa Master pump varies fuel delivery by inlet metering, rather than outlet metering.

A metering valve varies the size of the inlet opening from the hydraulic head to the pump rotor. A centrifugal governor controls the metering valve inlet opening. Fuel from the transfer pump flows through the inlet opening and rotor to the high-pressure pump. This consists of two radial pistons, or plungers, forced toward each other by rollers that turn with the pump rotor inside a cam ring. As the cam ring and rollers force the pistons toward the center of the pumping chamber, they develop pressure on the fuel and force it past a single delivery (retraction) valve in the rotor, figure 17-27. As the rotor turns, it uncovers outlet ports to the injection lines in firing-order sequence. Ignition advance is controlled hydraulically by a piston that turns the cam ring in its housing to contact the rollers sooner, figure 17-28.

Like Bosch pumps, Roosa Master pumps have solenoids to shut off fuel

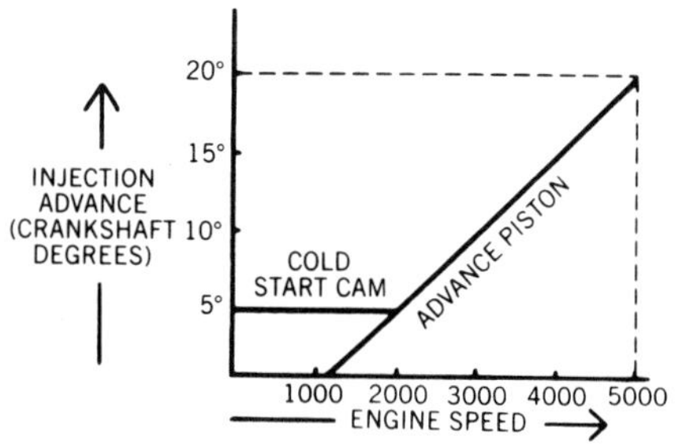


**Figure 17-24.** The Bosch distributor pump can provide up to 20 degrees of injection advance at 5,000 rpm. (Volkswagen)

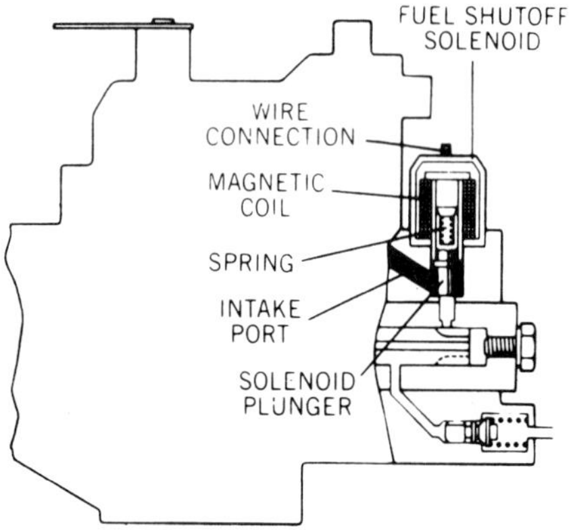


**Figure 17-25.** This Bosch fuel shutoff solenoid closes the fuel intake port to stop the engine. (Volkswagen)

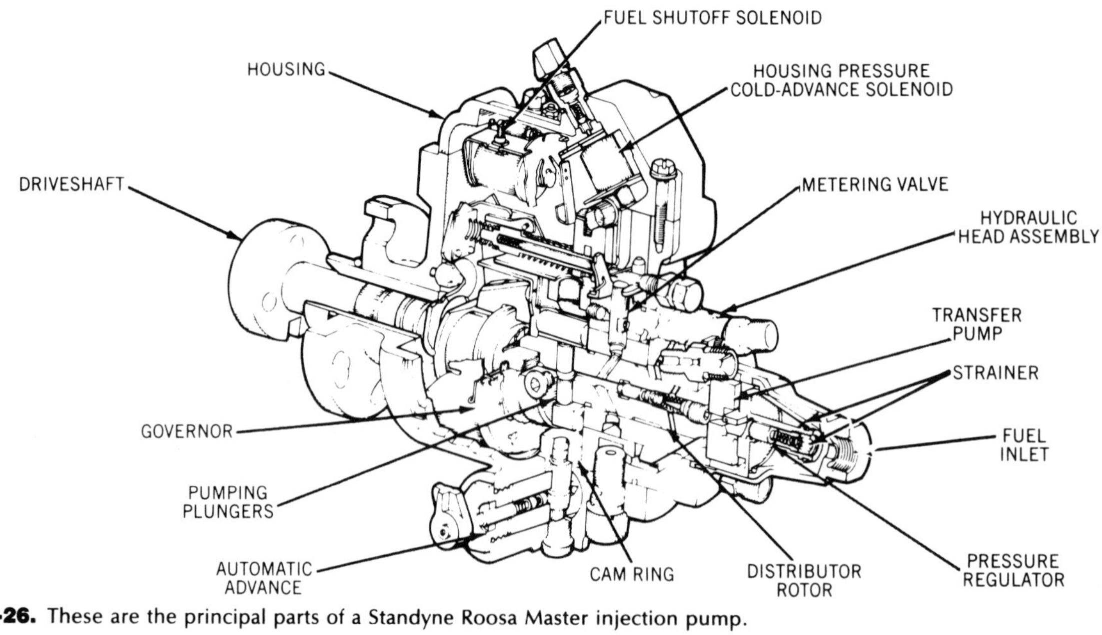


**Figure 17-26.** These are the principal parts of a Standyne Roosa Master injection pump. (Ford)

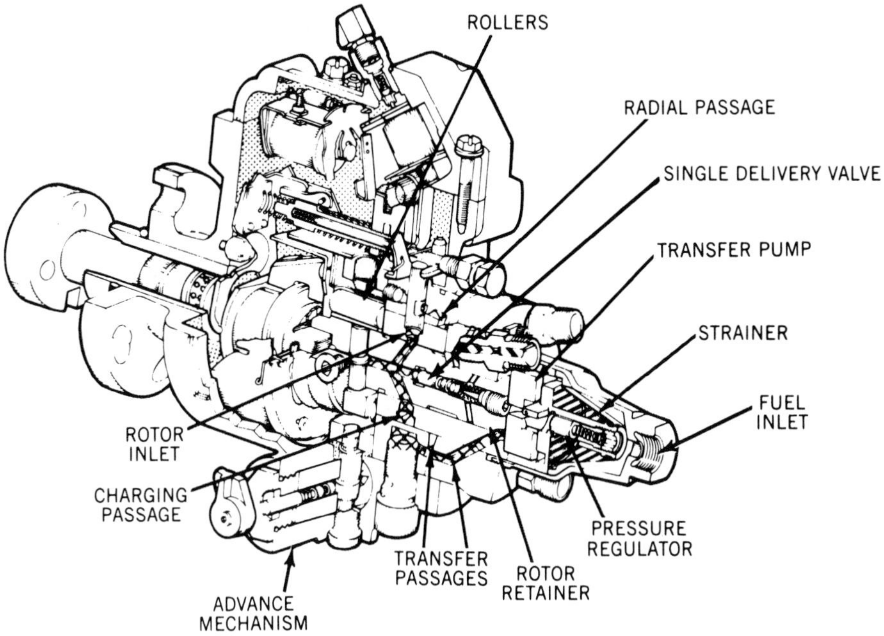


**Figure 17-27.** This drawing shows fuel charging and delivery from the Roosa Master distributor pump. (Ford)

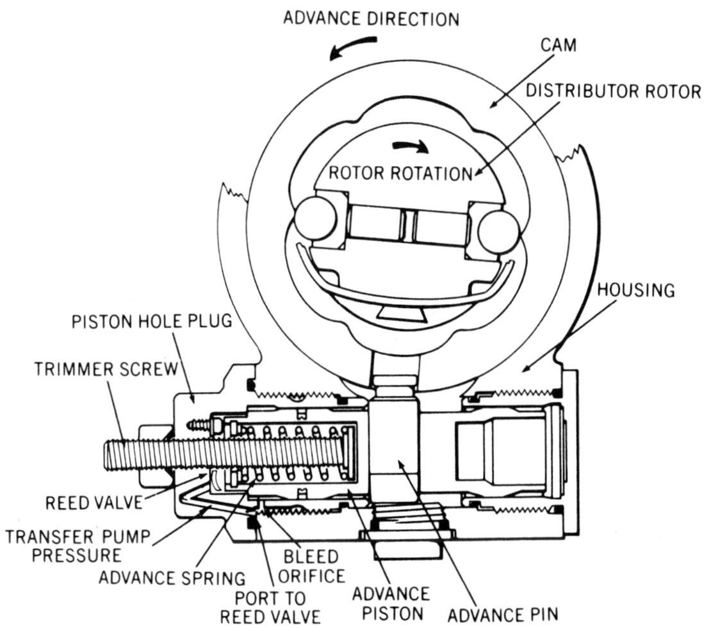


**Figure 17-28.** The Roosa Master hydraulic advance mechanism rotates the cam ring to advance injection timing.

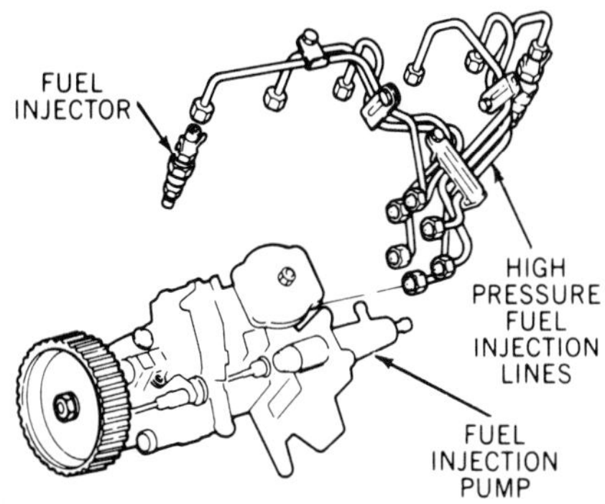


**Figure 17-29.** The high-pressure lines from the pump to the injectors are all exactly the same length. (Ford)

supply and stop the engine. The CAV pumps used on some GM diesels are built on Standyne Roosa Master designs and operate on similar inlet-metering principles.

## FUEL INJECTION AND RETURN LINES

Fuel injection systems use seamless steel tubing for injection lines because of injection pressures of 1,500 psi (10,335 kPa) or more. All lines from the pump to the injectors must be the same length, figure 17-29, because line length determines the time that it takes fuel to travel from the pump to the injector. Therefore injection line length affects ignition timing. Because of these critical pressure and line length requirements, replacement injection lines should always be factory-made items.

Injection pumps and injectors rely on diesel fuel to lubricate and cool close-tolerance moving parts. Excess fuel circulates through pumps and injectors, and return lines carry this fuel back to the tank or to the filter.

## INJECTOR NOZZLES

Injector nozzles are small, but critical, parts of a diesel fuel system. An injection nozzle must open and close with precision to spray an exact amount of fuel into the combustion chamber for an exact amount of time. The design of the nozzle tip controls fuel atomization and has a major effect on combustion.

Injection nozzles for automobile

diesels are hydraulic valves, held closed by spring pressure and opened by fuel pressure. Fuel from the injection pump enters the nozzle holder inlet and moves down an inlet passage to a pressure chamber at the nozzle tip, figure 17-30. As fuel pressure increases, it forces the nozzle needle valve inward against spring pressure. The same fuel pressure then forces fuel out of the nozzle tip, into the combustion chamber. When injection pressure drops, the valve spring closes the injector. Excess fuel circulates through the spring chamber of the injector and out the leakoff connection. A return line then carries the fuel back to the fuel tank.

Each nozzle also has a small inlet filter or strainer to protect the nozzle against dirt. These inlet filters are important service items because a single microscopic dirt particle can cause an injector nozzle leak or an uneven spray pattern. Also, a restricted filter can upset injection and cause poor performance.

Sizes and shapes of injector assemblies vary for different engines, figure 17-31. However, the most common injection nozzles for automobile diesels are the hole type and the pintle type, figure 17-32. Combustion chamber design (which you will learn about in the next section) determines the kind of nozzle used on a particular engine. Hole-type, or orifice, nozzles are used most often with direct injection systems. Pintle nozzles are used most often with indirect injection systems and precombustion chambers. Since most automobile diesels have precombustion chambers, pintle nozzles are most common on automobile diesel engines.

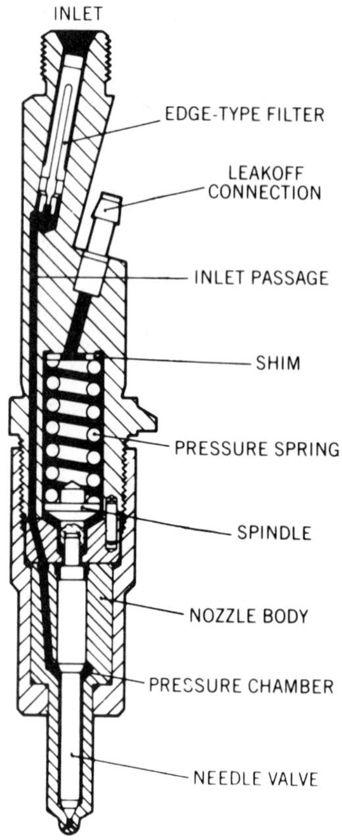


**Figure 17-30.** Fuel from the pump enters the injection nozzle holder and travels to the nozzle tip where fuel pressure forces the needle valve off its seat. (Bosch)

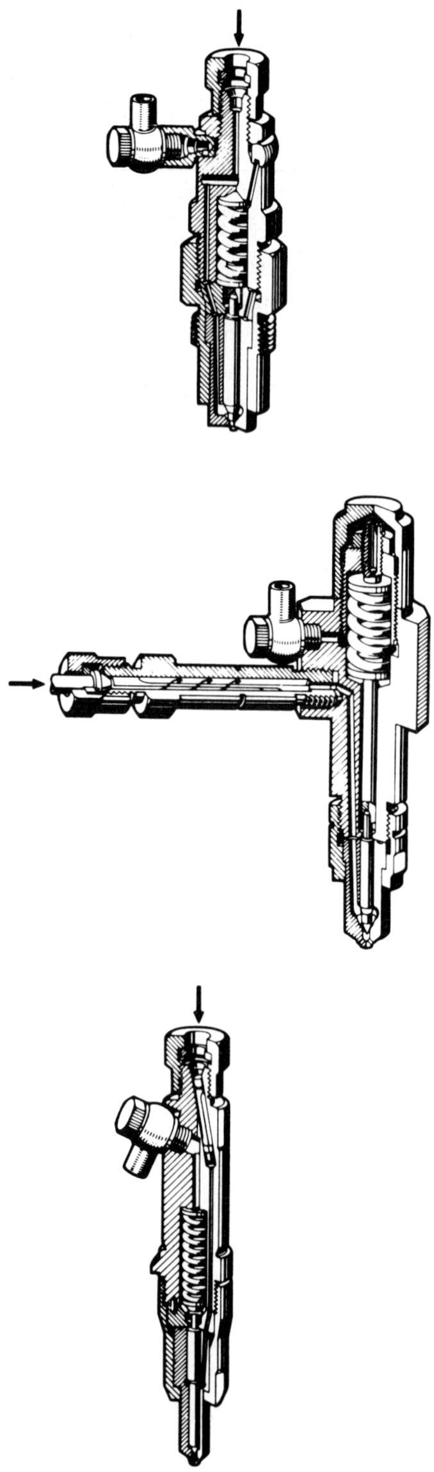

**Figure 17-31.** Typical injector nozzle and holder assemblies. (Bosch)

## Pintle Nozzle Operation

In a pintle nozzle, the valve needle extends through the valve tip when the injector is closed. A tapered section of the needle behind the tip bears against the valve seat to seal the injector. Fuel injection pressure moves the injector needle inward to open the **pintle valve**. Fuel flows around the pintle and out the nozzle tip. This action is similar to the action of a metering rod in a carburetor jet. Pintle nozzles operate with outlet pressures of 1,500 to 2,100 psi (10,335 to 14,470 kPa).

The shape of the pintle determines the shape of the fuel spray cone from the nozzle and the amount of atomization, figure 17-33. Pintle nozzles deliver a small spray cone and incomplete atomization. This is fine for a precombustion chamber because turbulence in the chamber continues to atomize the fuel before combustion in the cylinder.

## Hole-Type (Orifice) Nozzle Operation

Fuel pressure opens the needle valve of a hole-type injector the same way that it opens a pintle injector. The tip of the needle, however, does not extend through the nozzle opening. Fuel moves past the needle and through several small orifices in the nozzle tip, figure 17-34. The size, number, and arrangement of the orifices control fuel atomization and the shape of the spray cone.

Hole-type nozzles are used with direct-injection systems and have outlet pressures of 2,000 to 4,000 psi (13,780 to 27,560 kPa).

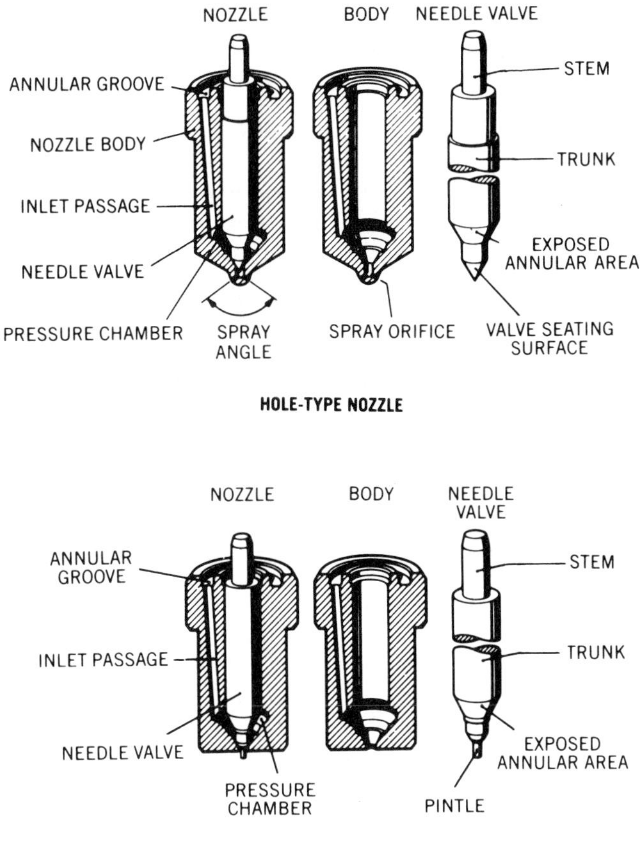


**Figure 17-32.** Comparison drawing of pintle- and hole-type injection nozzles. (Bosch)

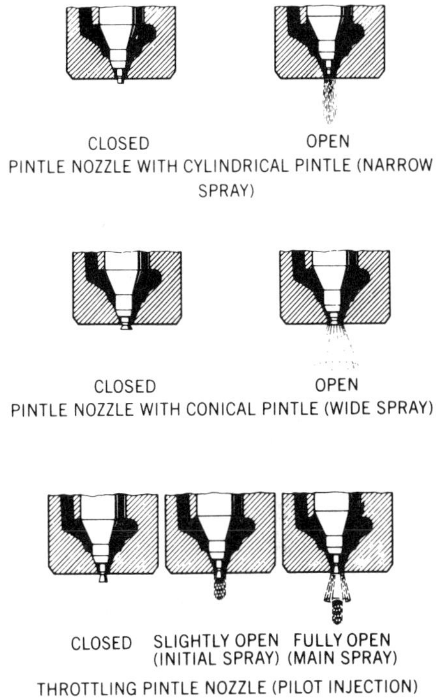


**Figure 17-33.** Typical spray patterns for different pintle nozzles. (Bosch)

## DIESEL COMBUSTION CHAMBER DESIGN

Because diesel fuel does not atomize, vaporize, and mix with air until it enters the combustion chamber, the chamber design (along with the injection nozzle) controls air and fuel mixing. Diesel injection systems are classified as either **direct-injection** or **indirect-injection** systems. In a direct-injection system, the injector sprays fuel directly into the cylinder, on top of the piston. In an indirect-injection system, the injector sprays fuel into a precombustion chamber, separate from the cylinder. An energy cell combustion chamber has features of both the direct and indirect systems. Energy cell chambers, however, are not used on current automobile diesel engines.

Direct- and indirect-injection systems have various advantages and disadvantages. Direct-injection systems provide slightly better fuel economy and specific power for a given amount of fuel than indirect systems do. But, direct-injection engines are noisier and harder to start when cold. Engine parts also must be stronger to withstand greater combustion forces. Fuel economy is slightly lower with indirect injection, but these engines are much quieter and smoother. They also operate with lower injection pressures and combustion forces. With the addition of glow plugs, indirect-injection engines are easier to start in cold weather. For automobile engines, the advantages of indirect-injection systems outweigh their disadvantages. Almost all automobile diesels use indirect injection.

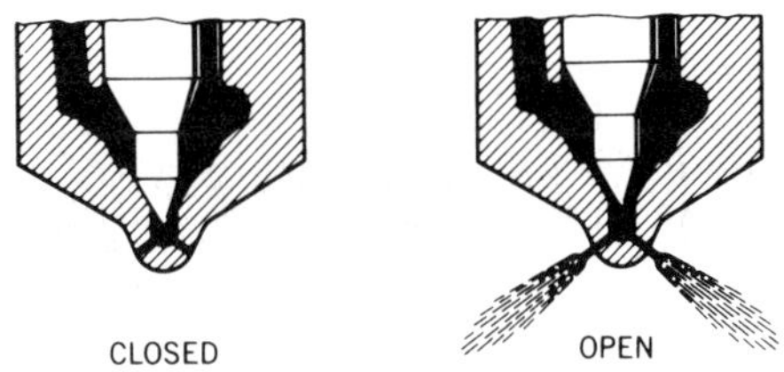


**Figure 17-34.** Typical hole-type nozzle spray pattern. (Bosch)

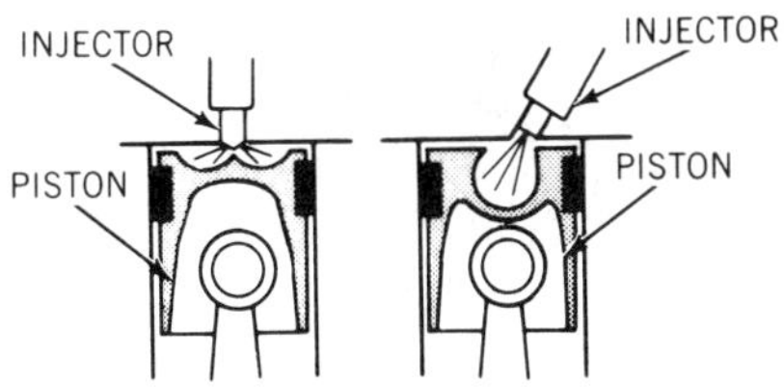


**Figure 17-35.** Typical direct-injection chambers used in heavy-duty diesels.

### Direct Injection

You will find direct-injection systems most often in truck and heavy-equipment engines. The combustion chamber is formed in the top of the piston and designed to create air turbulence that vaporizes the fuel and distributes it throughout the cylinder. Figure 17-35 shows two basic kinds of direct-injection combustion chambers.

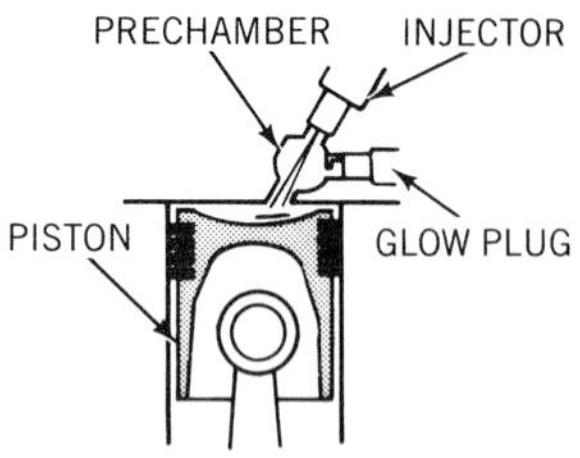


**Figure 17-36.** Typical precombustion chamber used with indirect injection.

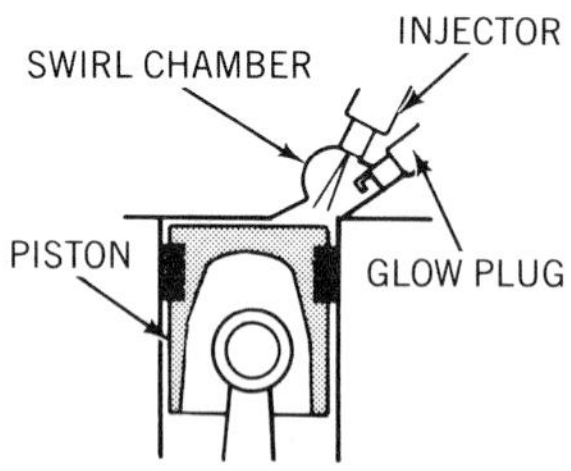


**Figure 17-37.** Typical turbulence, or swirl, chamber used with indirect injection.

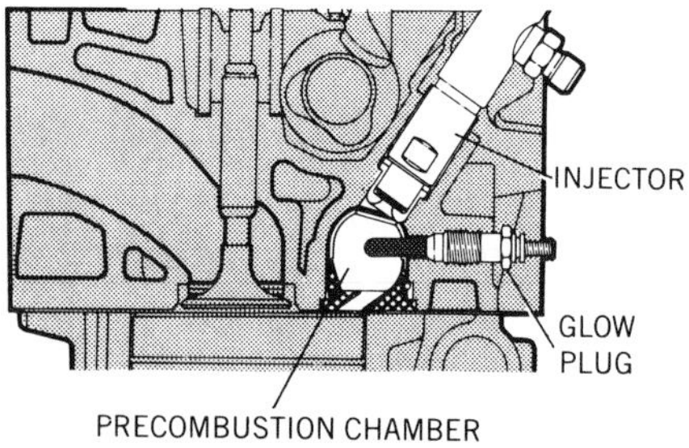


**Figure 17-38.** Glow plugs are installed in the precombustion chambers.

### Indirect Injection

In an indirect-injection system, the injector is in a small chamber, separate from the cylinder and main chamber. A short passage connects the small chamber with the main chamber. There are two kinds of indirect-injection chambers: precombustion and turbulence (or swirl) chambers. They look similar but operate slightly differently.

A precombustion chamber contains about 20 to 35 percent of the cylinder air volume at tdc, figure 17-36. When fuel is injected into this chamber, it creates a rich mixture that ignites rapidly. The burning mixture creates turbulence that mixes the air and fuel. The burning mixture expands through the passage to the main chamber. Final combustion takes place evenly over the top of the piston. The passage from the precombustion chamber is close to perpendicular to the piston top (in line with the connecting rod).

A turbulence chamber contains about 50 to 75 percent of the cylinder air volume at tdc. Fuel is injected into the chamber and ignition occurs similarly to the way it occurs in a precombustion chamber. Because the air volume is greater, turbulence and air-fuel mixing are slightly greater than in a precombustion chamber. The passage that connects the turbulence chamber with the main chamber is closer to parallel with the piston crown (perpendicular to the cylinder and connecting rod), figure 17-37.

Precombustion and turbulence chambers use pintle injection nozzles because they require less fuel atomization from the nozzle than a direct-injection system does. Engine starting is harder with an indirect-injection system because air temperature in the separate chambers does not rise as high with a cold engine as it does with a direct-injection system. Therefore, almost all indirect-injection engines have electric glow plugs to preheat the chambers of a cold engine.

## DIESEL AUXILIARY SYSTEMS

You have now traced the operation of an automobile diesel fuel injection system from the injection pump to the combustion chamber. Diesel engines, however, have other auxiliary systems that you haven't yet studied.

### Glow Plugs

We have mentioned that glow plugs are used as starting aids on most indirect-injection engines. A glow plug is a small plug that screws into the prechamber of each cylinder, figure 17-38, as a spark plug screws into the combustion chamber of a gasoline engine. The glow plug, however, does not ignite the air-fuel mixture as a spark plug does. It simply heats the chamber to help raise the compressed air temperature.

A glow plug is an electrical resistance heater. It receives current from a circuit controlled by the car "ignition" switch. As current flows through the plug, it heats a wire resistance heating element. Figure 17-39 is a cutaway drawing of a typical glow plug. When current is supplied to the glow plug, the heating element can reach temperatures of approximately 1,800° F

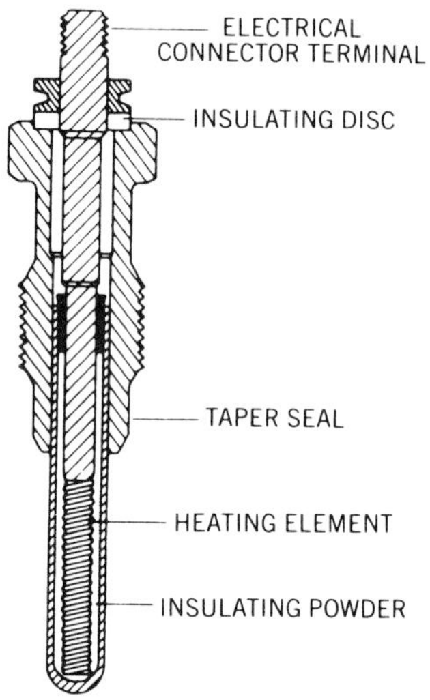


**Figure 17-39.** The tip of a glow plug can reach temperatures above 1,000° C to preheat the intake air charge. (Volkswagen)

(1,000° C). Current for the glow plug circuit varies among different vehicles, but it is often about 300 amperes. This high current is applied for only one minute or less, however.

Although all glow plugs work on the same principles, designs and circuit installations vary. Glow plug circuits can be wired in series or in parallel. A series circuit can carry more current for better heating. If one plug fails, however, the entire circuit will be open. To avoid this problem, most late-model diesel automobiles have the glow plugs connected in parallel or series-parallel circuits. Figures 17-40 and 17-41 show typical glow plug circuits on late-model vehicles.

Almost all glow plug circuits apply full battery voltage (12 volts) to the plugs, but carmakers use two different types of glow plugs. Some circuits have

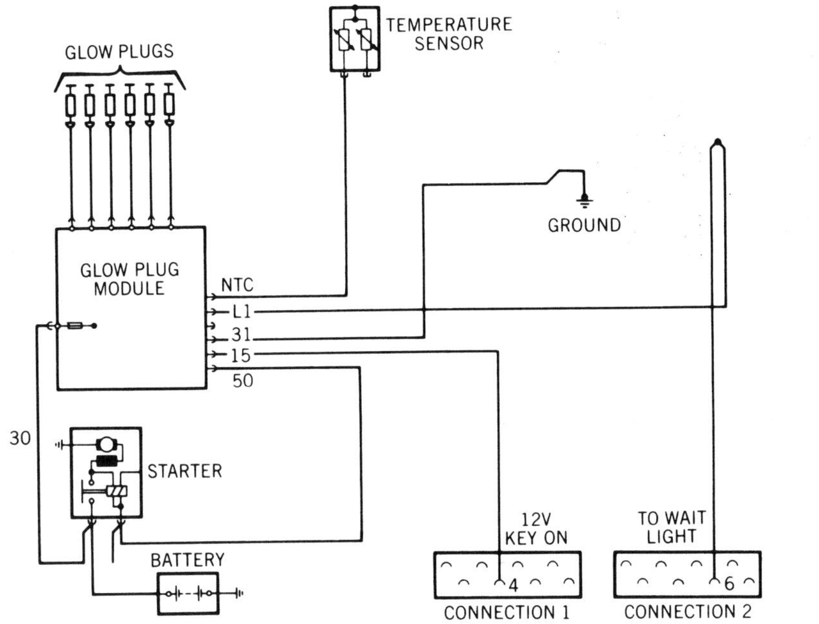


**Figure 17-40.** These six glow plugs are in parallel with the control module. (Ford)

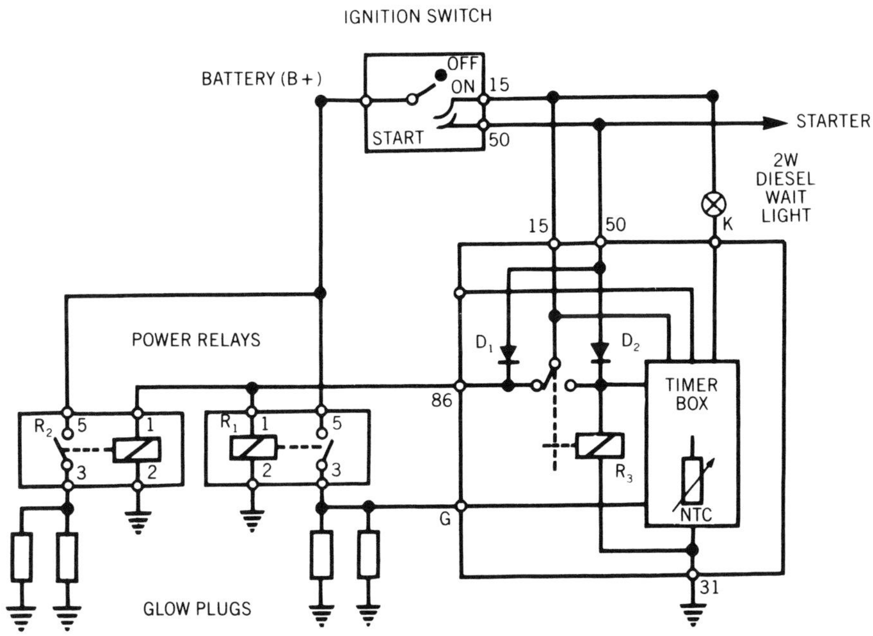


**Figure 17-41.** These glow plugs are wired in a series-parallel arrangement with their power relays.

plugs rated at 12 volts. These are usually called slow-start systems and take 60 seconds or more to warm up. Other circuits have plugs rated at 6 volts. When 12 volts are applied to a 6-volt glow plug, it heats quickly. These fast-start systems can warm up in 10 to 15 seconds.

Operating time and temperature are critical for 6-volt plugs in a 12-volt circuit. If high voltage and current are applied for too long, the plugs will burn out. Operating temperature also is important for 12-volt glow plugs. Although a 12-volt plug will not be damaged by high voltage, it can overheat if it operates too long. To ensure that glow plugs heat properly but do not overheat, carmakers usually design systems with two operating stages:

**1.** *Preglow*—The preglow stage is the initial heating stage for a cold engine. When the driver turns the key on, the glow plug circuit is energized continuously. An indicator lamp lights to tell the driver that the plugs are operating. Preglow time varies for a fast-start or a slow-start system. A timer and a temperature sensor in the circuit will stop the continuous current flow when a specific combination of time and temperature is reached. The system then goes into the afterglow stage.

**2.** *Afterglow*—To avoid overheating the plugs, the circuit cycles on and off in the afterglow stage. This maintains combustion chamber heat until the engine reaches normal operating temperature. Depending on ambient temperature and to aid cold-engine operation, the afterglow stage may continue during one or all of the following conditions:

**a.** Before the driver cranks the engine

**b.** During cranking

**c.** For a minute or two of cold-engine operation

Besides the glow plugs and dashboard indicator lamp, a typical glow plug system usually has a timer, a temperature sensor, and one or two relays to apply preglow and afterglow voltage.

## Block Heaters

In very cold climates in the northern U.S. and Canada, air temperature can get so low that starting a cold diesel is difficult even with glow plugs. A block heater consists of electric heating coils installed in the engine water jackets. When the vehicle is parked overnight or for several hours, the block heater can be plugged into a 110-volt ac electrical power source. The heater keeps coolant temperature warm enough to allow the engine to start easily with the aid of glow plugs.

A block heater also keeps the engine oil warm and reduces moisture condensation and fuel dilution in the crankcase of a cold engine. This im-

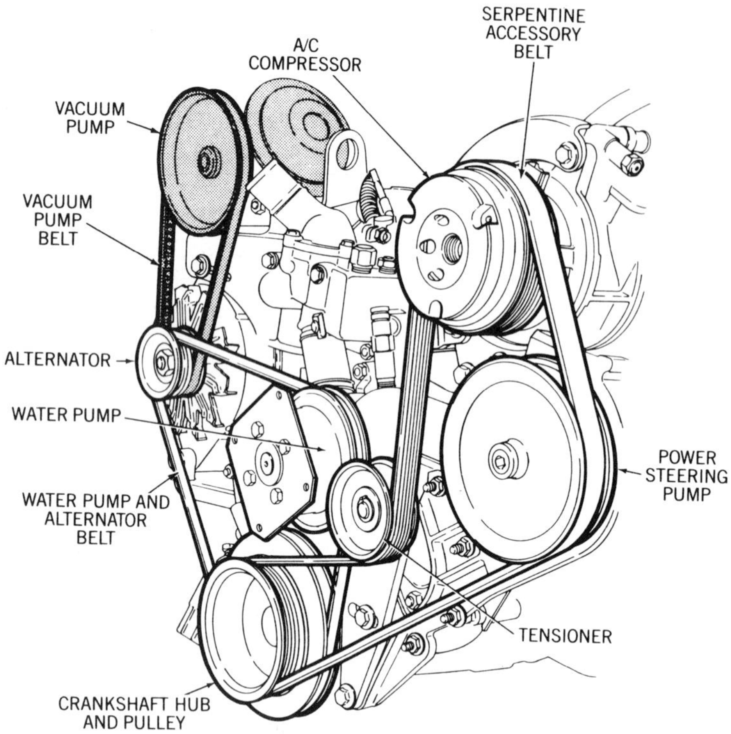


**Figure 17-42.** This vacuum pump is driven by a belt from the engine crankshaft. (Ford)

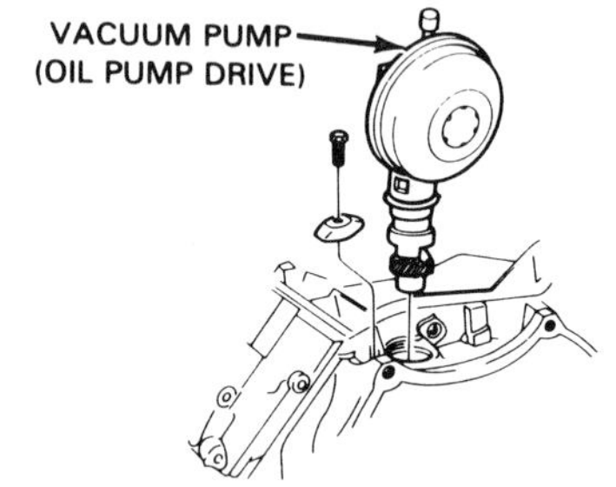


**Figure 17-43.** This GM diesel has a vacuum pump installed where the ignition distributor would be on a gasoline engine. (Cadillac)

proves lubrication and fuel economy with a cold engine and minimizes engine wear. In extremely cold climates, block heaters also are often used in gasoline engines.

## Vacuum Pumps

Many accessories on modern automobiles are operated by engine vacuum. These include power brake boosters, air conditioning controls, and automatic transmission valves. A diesel engine produces no manifold vacuum because of its unthrottled air intake. Therefore, most automobile diesel engines have an auxiliary vacuum pump for vacuum-operated accessories. The vacuum pump is usually a diaphragm pump, driven by an engine drive belt, figure 17-42. On GM diesels developed from gasoline engines, the vacuum pump is driven by the camshaft from the drive gears for an ignition distributor on a gasoline engine, figure 17-43.

## SUMMARY

The number of diesel-powered automobiles has increased greatly since the mid-1970's because of oil shortages and the need to improve fuel economy. Both gasoline and diesel engines are efficient automobile powerplants, and each has its advantages and disadvantages.

A diesel fuel injection system meets many of the same fuel-metering needs that a gasoline carburetor does, but a fuel injection pump also controls diesel ignition timing. A diesel engine atomizes and vaporizes fuel and mixes it with air inside the combustion chambers, not in the carburetor and manifold as a gasoline engine does.

Automobile diesel engines all have pump-controlled injection systems to meet rapidly changing fuel-metering needs. Automobile pumps are either inline pumps or distributor pumps. Injection pumps meter and pressurize fuel and deliver it to injection nozzles in firing-order sequence. Inline pumps and some distributor pumps control fuel amounts by outlet metering. Other distributor pumps control fuel by inlet metering. All injection pumps have governors to control fuel metering in response to changes in engine load and speed. Most pumps on diesel automobiles also have injection advance mechanisms to advance injection timing at high engine speeds.

Pintle-type or hole-type injection nozzles are hydraulically opened valves that control fuel vaporization through precise spray patterns. Combustion chamber design, combined with nozzle design, controls fuel vaporization and air-fuel mixing. Automobile diesel engines have indirect injection systems and precombustion or turbulence chambers.

Many diesel engines have glow plugs and block heaters as starting aids. Also, most automobile diesel engines have vacuum pumps to operate accessories such as brake boosters and air conditioning controls.

## REVIEW QUESTIONS

### Multiple Choice

**1.** Typical diesel engines have compression ratios of:

**a.** 11:1 to 13.5:1
**b.** 16:1 to 22:1
**c.** 25:1 to 45:1
**d.** 60:1 to 80:1.

**2.** Diesel compression pressures typically range from:

**a.** 200 to 400 psi
**b.** 650 to 1050 psi
**c.** 450 to 800 psi
**d.** 150 to 300 psi.

**3.** Mechanic A says that automotive 4-stroke diesels run with an unthrottled air intake. Mechanic B says that diesel air-fuel ratios range from 20:1 to 85:1, or more. Who is right?

**a.** A only
**b.** B only
**c.** both A and B
**d.** neither A nor B.

**4.** A high-speed diesel is one that operates at more than:

**a.** 3,000 rpm
**b.** 5,000 rpm
**c.** 7,500 rpm
**d.** 10,000 rpm.

**5.** A diesel naturally produces *low* exhaust emissions of all of the following, EXCEPT:

**a.** CO
**b.** HC
**c.** $NO_x$
**d.** $SiH_4$.

**6.** Ignition timing and advance for a diesel engine are controlled by the:

**a.** throttle
**b.** governor
**c.** transfer pump
**d.** injection pump.

**7.** Minimum and maximum speeds of a diesel engine are controlled by the:

**a.** injection nozzles
**b.** transfer pump
**c.** altitude compensator
**d.** governor.

**8.** All of the following are parts of a typical diesel fuel system, EXCEPT:

**a.** a pressure reducer
**b.** fuel return lines
**c.** a lift, or transfer, pump
**d.** an injection pump with a governor.

**9.** Mechanic A says that most automobile diesel engines have unit injector systems. Mechanic B says that most automobile diesel engines have inline injection pumps. Who is right?

**a.** A only
**b.** B only
**c.** both A and B
**d.** neither A nor B.

**10.** An inline injection pump controls fuel metering with:

**a.** a retraction valve
**b.** a port-and-helix system
**c.** a distributor port
**d.** inlet metering.

**11.** Mechanic A says that Bosch distributor pumps control fuel volume with outlet metering. Mechanic B says that Standyne Roosa Master pumps control fuel metering differently, with inlet metering. Who is right?

**a.** A only
**b.** B only
**c.** both A and B
**d.** neither A nor B.

**12.** A Bosch VE distributor pump provides maximum ignition advance (in crankshaft degrees and speed) of about:

**a.** 10 degrees at 2,500 rpm
**b.** 12 degrees at 2,400 rpm
**c.** 20 degrees at 5,000 rpm
**d.** 36 degrees at 4,500 rpm.

**13.** Mechanic A says that most automobile diesel injection nozzles are solenoid operated. Mechanic B says that most automobile diesel injection nozzles are pintle nozzles. Who is right?

**a.** A only
**b.** B only
**c.** both A and B
**d.** neither A nor B.

**14.** Automobile diesel engine combustion chambers usually have:

**a.** energy cell chambers
**b.** MAN piston crown chambers
**c.** direct-injection chambers
**d.** precombustion chambers.

**15.** Mechanic A says that some glow plug systems use 6-volt plugs in a 12-volt circuit. Mechanic B says that glow plugs ignite the air-fuel mixture in a cold engine. who is right?

**a.** A only
**b.** B only
**c.** both A and B
**d.** neither A nor B.

### Fill In The Blank

**16.** The term that refers to power that an engine loses when it works against a restricted throttle to take in air is __________.

**17.** A diesel engine is heavier than a gasoline engine to withstand the higher temperatures and __________ of its operation.

**18.** To prevent fuel contamination, diesel engines require expensive __________ devices.

**19.** Diesel idle speed is limited by the minimum amount of __________ a particular engine can take in and still keep running.

**20.** A diesel injection system atomizes fuel as it leaves the injection nozzles, and the fuel vaporizes in the __________.

**21.** A diesel injection pump controls fuel pressure, volume, and timing. It rotates at the same speed as the engine __________.

**22.** A diesel injection nozzle must close quickly to prevent smoking, increased emissions, and a loss of __________.

**23.** Because of high injection pressure, fuel injection systems use seamless __________ tubing for injection lines.

**24.** Injection pumps and injectors rely on __________ to lubricate and cool close-tolerance moving parts.

**25.** If the injector nozzle delivers an uneven spray pattern, a common cause might be a clogged __________.

**26.** A pintle valve is opened by __________ pressure.

**27.** If a diesel engine has a precombustion chamber, it has an __________ -injection system.

**28.** Most diesel truck and heavy equipment engines have __________ -injection systems.

**29.** When a glow plug is in the initial heating stage for a cold engine, it is in a __________ stage.

# 18 COMBUSTION MODIFICATIONS—Spark Timing Controls, Air Injection, Exhaust Gas Recirculation, Catalytic Converters

## WHAT

You have learned that the fuel and ignition systems are the major combustion controls on a gasoline engine. Simple fuel metering and ignition are not enough, however, for efficient engine operation. Even the simplest carburetor has modifications, such as a choke and an accelerator pump, to richen the air-fuel mixture when needed. An engine can't produce full power and good economy if it runs at all times on its basic timing setting. Therefore, centrifugal and vacuum advance mechanisms, or electronic controls, modify the ignition and alter the timing in relation to engine speed and load.

Even before you studied fuel and ignition systems, you knew that combustion produces more than just power in an engine. It also produces combustion byproducts in the exhaust, some of which pollute the air. Since the mid-1960's, when emission control laws took effect, engineers have further modified combustion control systems to reduce emissions. Some of the modifications, such as spark timing controls and exhaust gas recirculation (EGR), change the combustion process while it is taking place. Others, such as air injection and catalytic converters, affect combustion after exhaust byproducts have formed. This chapter outlines the principles of these four major combustion control modifications and how they affect performance, economy, and emissions.

## WHY

To understand how spark timing controls, EGR, air injection, and catalytic converters work, and to service them professionally, you can't think of them as independent systems. They all modify the combustion process and its byproducts. Now that you have studied basic engine operation, as well as fuel and ignition systems, you can look at these other systems and see how they fit into overall engine operation. On most late-model engines, fuel delivery, ignition, and emission controls are all regulated by a central computer. When you service an engine, late model or older, you should think like a service computer. You must understand how all parts work together and how each affects overall engine operation.

## GOALS

After studying this chapter, you should be able to:

**1.** Explain the purposes and operation of various spark-timing emission control devices.

**2.** Explain the purpose and principles of exhaust gas recirculation and identify the principal parts and controls of an EGR system.

**3.** Explain EGR operation under various vehicle operating conditions.

**4.** Explain the purpose and principles of air injection and identify the principal parts of an air injection system.

**5.** Explain the controls used on an air injection system. Explain pulse air injection.

**6.** Explain oxidation and reduction reactions in catalytic converters and how they remove HC, CO, and $NO_x$ from the exhaust.

**7.** Explain why unleaded gasoline is necessary for converter-equipped cars.

## SPARK TIMING CONTROLS

Some engineers say that the ignition system is the major emission control system on an engine. It is certainly one of the most important. Chapter 12 explained the principles of ignition timing and introduced the relationship of timing to emission control. For a better understanding of why ignition timing must be modified to control emissions, we should review the basic reasons for spark advance. Then we will look at spark timing modifications in the larger context of emission controls.

Timing must be advanced as engine speed increases to allow enough burn time for combustion pressure to develop full power. Timing can be advanced more at high speed under light load to allow the engine to burn a lean mixture. Lean mixtures take longer to burn, but they improve fuel economy. As engine load increases, speed decreases slightly and mixtures become richer. Timing should be retarded slightly from a highly advanced position. Centrifugal and vacuum advance mechanisms can adjust timing as speed and load change so that combustion occurs with the greatest heat and pressure. Maximum timing advance for a given speed and load allows the engine to get the most energy possible from its fuel. Unfortunately, it also has harmful side effects.

With maximum advance, the fuel is completely burned as the piston starts its exhaust stroke. Although peak combustion temperatures are hotter, final exhaust gases are cooler and do not heat the exhaust ports and manifold as much as they would if timing were later. Any unburned fuel left in the cylinder is carried out as hydrocarbon (HC) emissions. Exhaust temperatures are not high enough to cause this fuel to combine with any remaining oxygen. Also, maximum spark advance and high temperatures and pressures increase the formation of oxides of nitrogen ($NO_x$) early in the combustion process.

If timing is retarded, combustion is less efficient, and performance and economy will suffer. However, $NO_x$ emissions decrease because combustion temperatures are lower. Also with retarded timing, high temperatures occur later in the combustion process. This increases exhaust gas temperature and helps to oxidize any unburned fuel in the exhaust system. Originally, the principal reasons for spark timing control modifications were to reduce HC and $NO_x$ emissions.

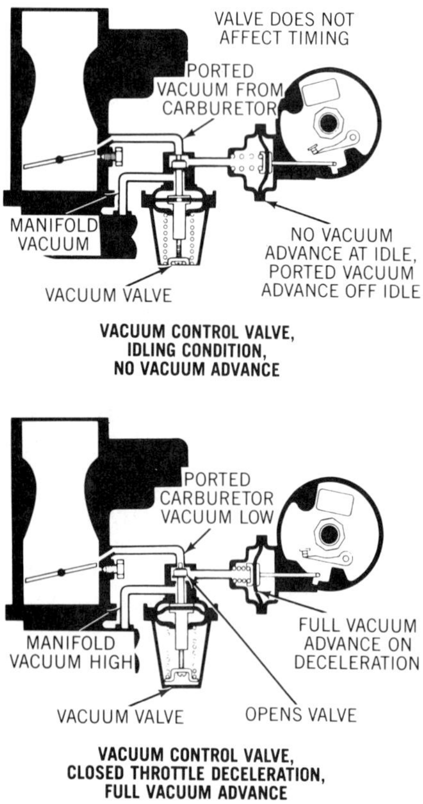


**Figure 18-1.** Used on older cars, this vacuum valve advanced timing during deceleration to control HC and CO. (AC-Delco)

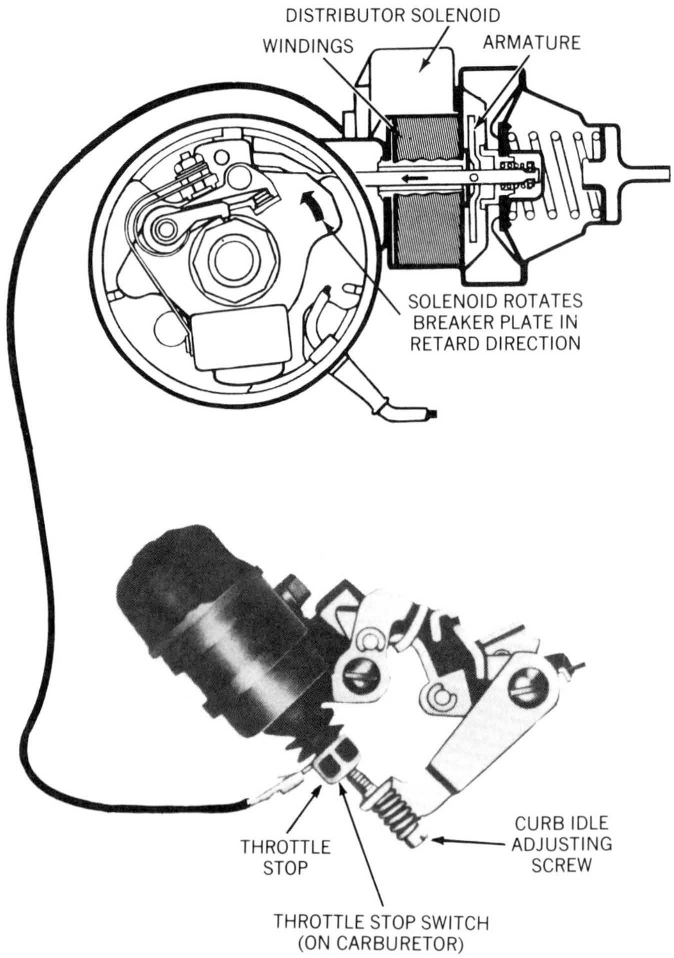


**Figure 18-2.** This distributor retard solenoid was used on some Chrysler cars of 1970–71. (AC-Delco)

You might ask, "Why not just run the ignition system with later initial timing and less overall advance?" If engineers did this, performance and economy would suffer at all times. This isn't necessary because HC and $NO_x$ emissions are highest under certain conditions and not under others. HC emissions are high when a spark plug misfires and fuel does not burn. They also are high with a rich mixture and during deceleration and gear shifting with a manual transmission when high manifold vacuum draws extra fuel through the carburetor. $NO_x$ emissions are high when combustion temperatures and pressures are high and mixtures are lean. By modifying timing only when necessary, engineers were able to maintain reasonable performance and economy while reducing emissions.

## Early Spark Timing Controls

Spark timing controls appeared in the late 1960's and early 1970's, first to help control HC emissions, later for $NO_x$ control. A few early systems also worked to control carbon monoxide (CO) emissions. All of these systems modified the vacuum advance. Early systems changed the timing during starting, deceleration, and idle. Systems introduced in the early 1970's delayed vacuum advance at low and intermediate speeds and allowed it at cruising speeds.

Some Chrysler, Ford, AMC, and Pontiac cars of the late 1960's had deceleration vacuum advance valves, figure 18-1. These were used mostly on cars with manual transmissions because the air-fuel ratio becomes rich during deceleration and gear shifting. This valve momentarily switched vacuum advance control from low **ported vacuum** at the carburetor to high manifold vacuum when the throttle closed. This increased vacuum advance or prevented timing delay during deceleration and shifting. The deceleration vacuum advance valve prevented backfiring on deceleration. It also helped to reduce CO emissions, but it was only marginally effective against HC emissions. It was discontinued in the early 1970's.

Some Chrysler V-8 engines of 1970–71 had distributor retard solenoids to retard vacuum advance during deceleration, figure 18-2. This device worked well with the moderate HC emission requirements of the early 1970's. The solenoid is part of the vacuum advance mechanism and is energized by contacts on the throttle stop solenoid. When the throttle closes, the idle speed screw completes the ground circuit for the distributor solenoid. The distributor solenoid is connected to the vacuum advance linkage and pulls the linkage to a retarded position when energized. This retards timing during deceleration and idle to lower HC emissions. When the throttle opens, the distributor solenoid is deenergized, and the vacuum mechanism can advance timing.

Some other Chrysler engines of 1972–73 had a solenoid on the opposite side of the vacuum advance linkage, figure 18-3. This solenoid was energized during starting to advance timing from a low initial setting. This advance solenoid was not an emission control device, itself, but it allowed lower basic timing to reduce HC emissions.

These early spark timing modifications were effective when emission limits were higher than they are today. In the early 1970's carmakers had to start controlling $NO_x$ emissions, as well as HC and CO.

## Spark Timing Controls of the 1970's

As emission limits became more strict, carmakers adopted three general methods to modify vacuum advance:

- Distributor vacuum retard diaphragms
- Vacuum delay valves
- Speed- and transmission-controlled timing.

Most automobiles of the 1970's used one, or a combination, of these spark advance modifications. You also can find some of these devices (and their principles) in later model electronic engine control systems.

**Distributor Vacuum Retard.** Vacuum retard diaphragms are probably best known on Ford products, figure 18-4, but other carmakers also have used them. The idea is to use engine vacuum to advance timing under some conditions and to retard it under others. Ford did this with a dual-diaphragm mechanism. AMC, Bosch (on some Volkswagens), and some Japanese carmakers have used this device as well. In a dual-diaphragm distributor, one diaphragm is connected to ported vacuum to advance timing for light-load cruising, figure 18-5. The second diaphragm is connected to manifold vacuum, which is high at idle and deceleration. This also is when HC and $NO_x$ formation would be high with advanced timing. The second diaphragm shifts the distributor breaker plate to retard timing under these conditions. The retard diaphragm can override the advance diaphragm to retard timing even when the advance diaphragm is fully advanced.

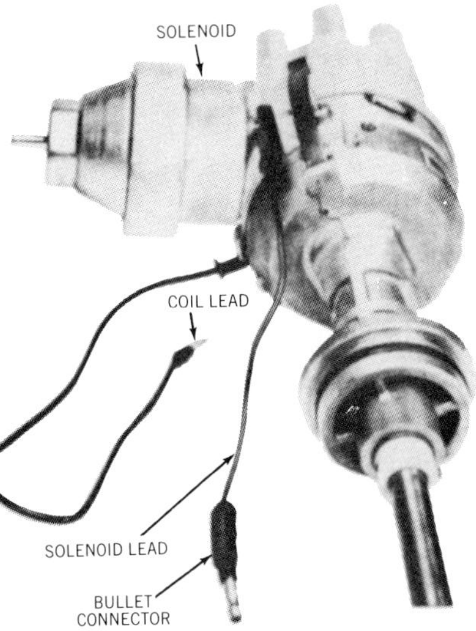

**Figure 18-3.** This distributor solenoid is wired to the start contacts of the ignition switch and advances timing during starting. (Chrysler)

Ford has used dual-diaphragm vacuum units on both breaker-point and electronic distributors, as have other carmakers. Some GM engines of the early 1970's had Delco distributors with single-diaphragm vacuum units. These distributors had two vacuum connections: one to ported vacuum for advance, the other to manifold vacuum for retard. They accomplished the same purpose as dual-diaphragm units.

Carmakers have used vacuum advance and retard distributors along with

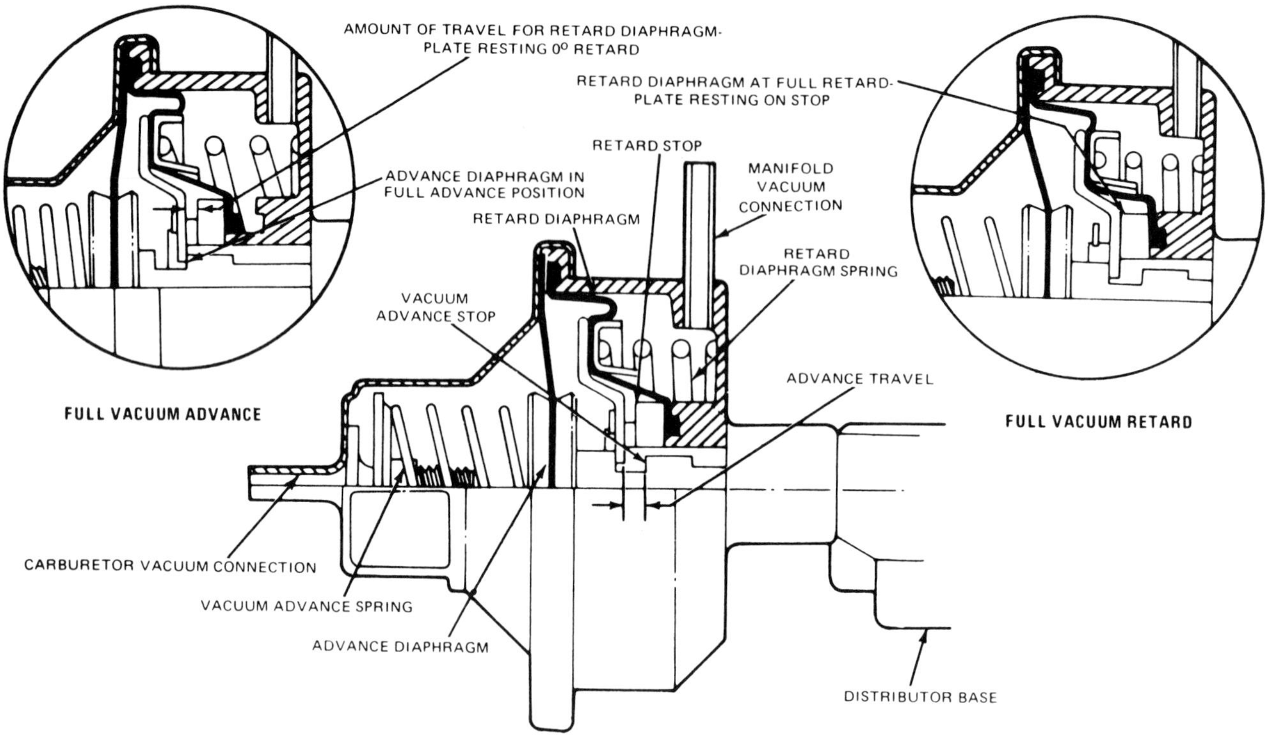


**Figure 18-4.** Dual-diaphragm distributors are best known on Ford products but also have been used by other carmakers. (Ford)

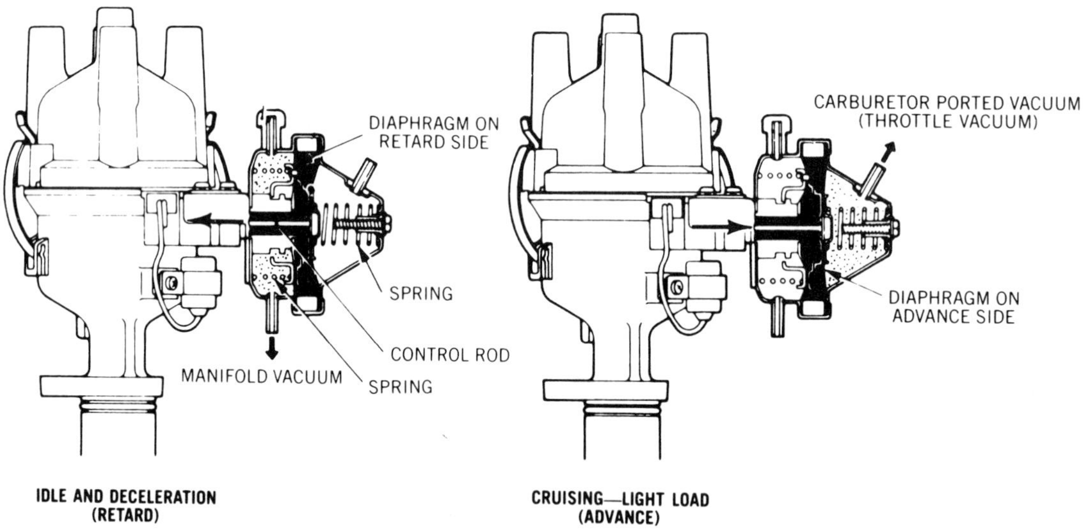


**Figure 18-5.** Ported vacuum advances timing for light-load cruising. Manifold vacuum retards it for deceleration and heavy load. (Chrysler)

vacuum delay valves and speed- or transmission-controlled timing to "fine tune" spark advance for emission control.

**Vacuum-Delay Valves.** A vacuum-delay valve delays the application of engine vacuum to the distributor advance unit for about 15 to 30 seconds to help decrease HC and $NO_x$ emissions. This holds timing in a less advanced condition at low and intermediate speeds than it would be with full vacuum advance. The results are lower overall combustion temperatures, with peak temperature occurring later in the cycle.

The spark delay valve (SDV) used by Ford, GM, and some other carmakers is a small plastic housing with a sintered metal disc, figure 18-6. One side of the valve is connected to ported vacuum; the other is connected to the distributor advance unit. Vacuum must "filter" through the sintered metal disc to reach the distributor. This delays spark advance. When ported vacuum

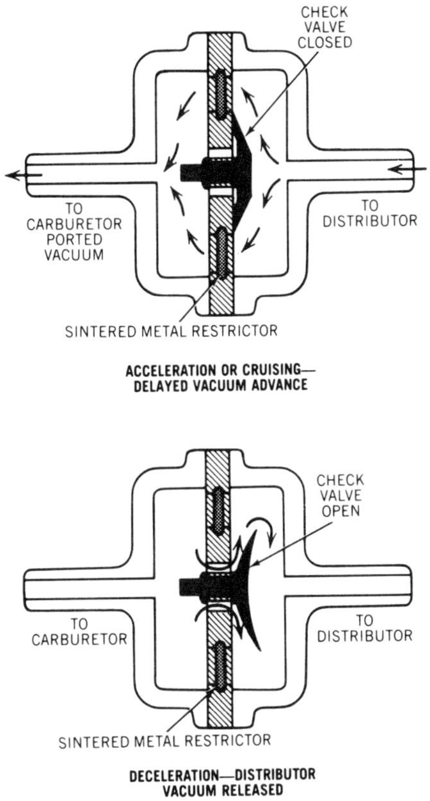


**Figure 18-6.** A spark delay valve "filters" vacuum to the distributor to delay advance. The check valve releases trapped vacuum on deceleration.

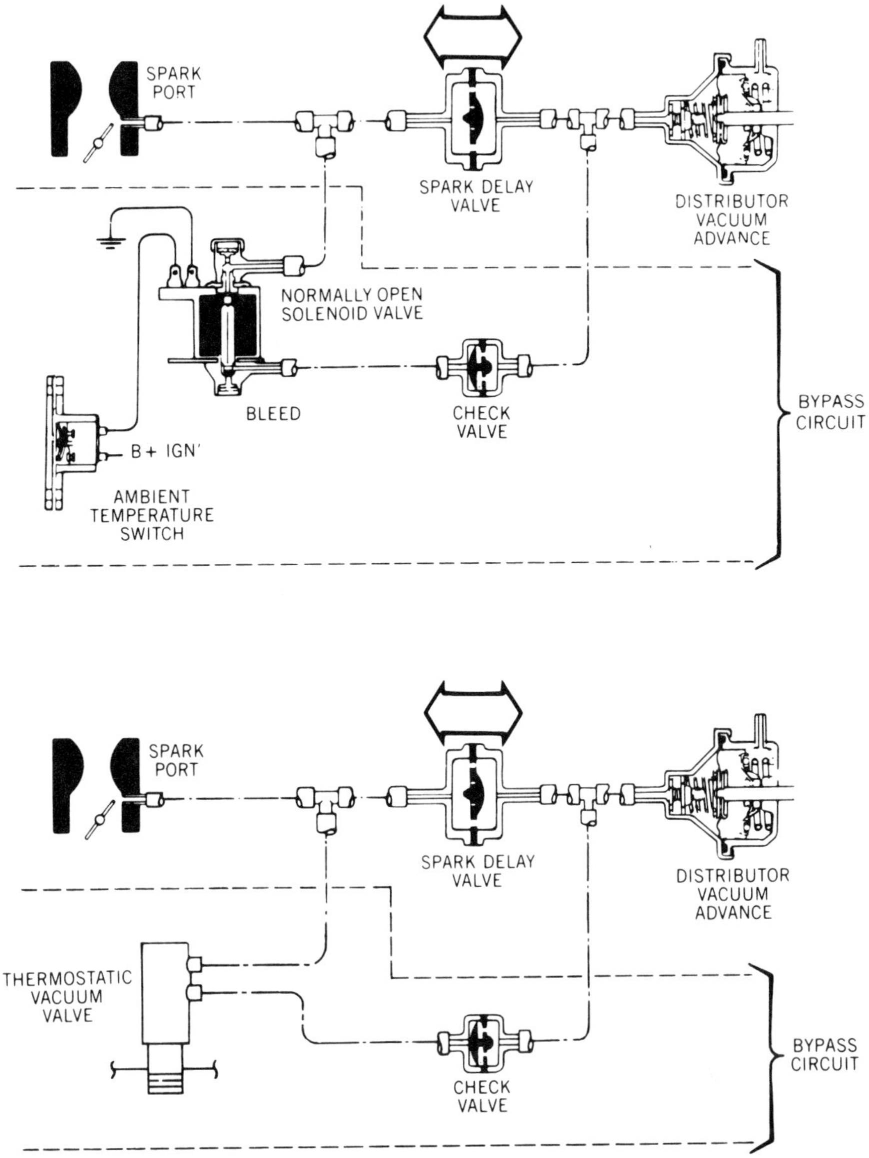


**Figure 18-7.** A vacuum-delay system can have either a thermostatic valve or a solenoid-controlled bypass. The vacuum bypass can occur at high or low temperature, depending on engine requirements. (Ford)

drops under load, deceleration, or idle, a check valve in the SDV opens to release vacuum from the distributor. This prevents too much advance when it isn't needed. Spark delay valves usually are color coded to indicate the amount of delay the valve provides.

In chapter 16, you learned about choke delay valves on carburetors. These valves delay vacuum to the choke vacuum break to prevent stalling when a cold engine starts. The choke delay valves are the same devices as spark delay valves. On carburetors, they improve driveability. On distributors, they control emissions.

Before mid-1973, some Ford engines had delay valve bypass systems that allowed immediate vacuum advance when the engine was cold. These systems had a solenoid vacuum valve in a bypass line around the SDV, figure 18-7. At low temperature, an ambient temperature switch energized the solenoid to open the vacuum bypass valve. Since combustion temperatures are low with a cold engine, full advance could be allowed to aid warmup and driveability. Some post-1973 systems from Ford and other carmakers have high-temperature bypass features. Typically, a thermostatic vacuum valve, controlled by coolant temperature, is in a manifold vacuum line to the distributor. If engine temperature rises above normal, the valve opens and applies manifold vacuum to the distributor. This advances timing to prevent overheating. Similar systems have been built with an ambient air temperature switch that energizes a solenoid to allow vacuum advance at high temperature.

Chrysler's orifice spark advance control (OSAC) valve is another form

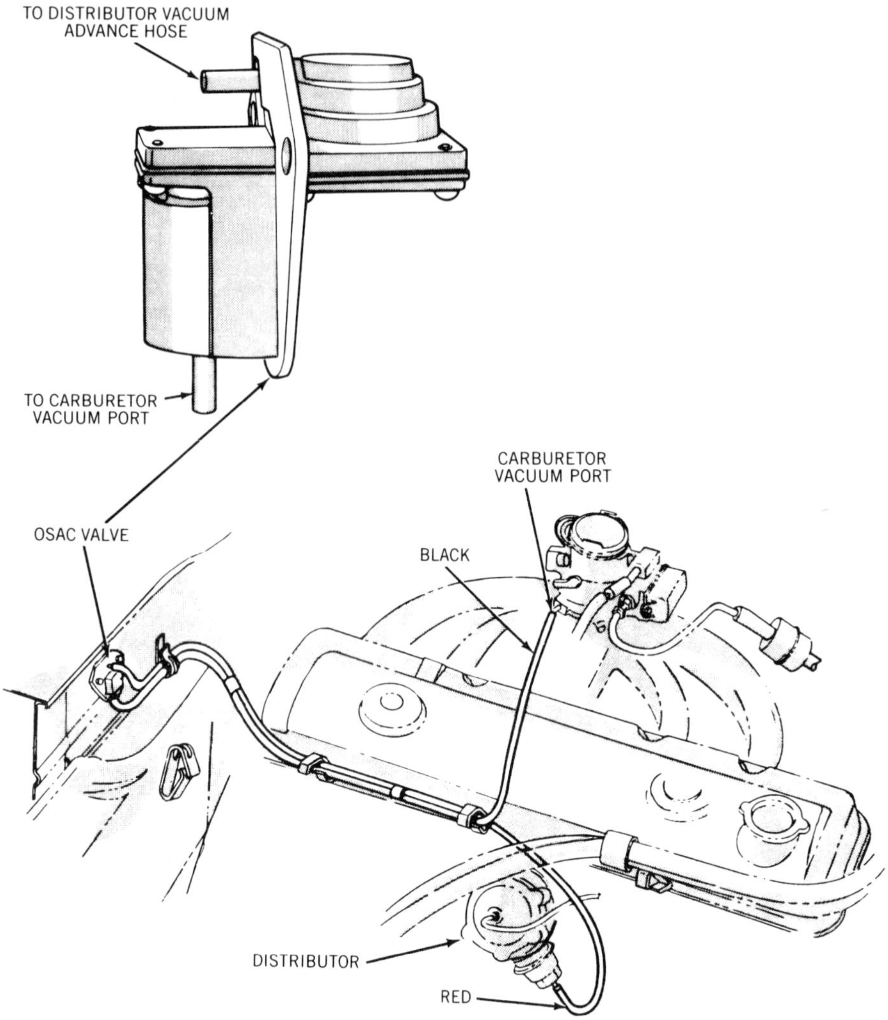


**Figure 18-8.** Chrysler's OSAC valve is another kind of vacuum advance delay valve. (Chrysler)

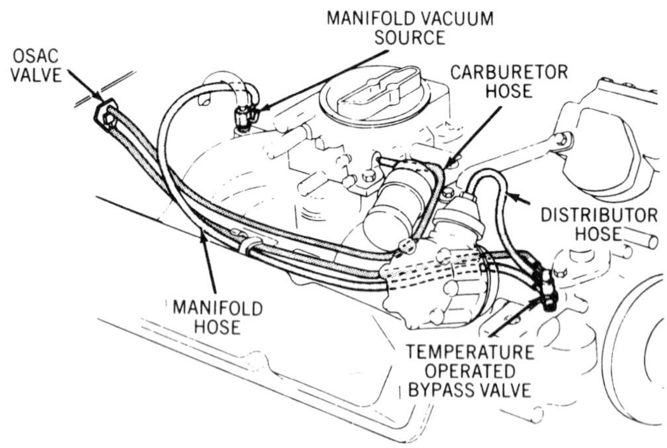


**Figure 18-9.** Chrysler's high-temperature bypass for the OSAC valve uses a coolant-controlled vacuum valve. (Chrysler)

of spark delay valve. Instead of a sintered metal disc, the OSAC valve has an orifice in the vacuum line to the distributor, figure 18-8. This delays vacuum advance just as an SDV does. Chrysler used the OSAC valve from 1973 through the late 1970's. Some early installations had a low-temperature bypass system that allowed full vacuum at low temperatures. Later installations have a thermostatic vacuum valve in the vacuum line to the distributor. This valve is installed in the engine water jacket and opens the vacuum line at high coolant temperature. The valve allows manifold vacuum to the distributor to advance timing and prevent overheating, figure 18-9.

Although the needs of different engines vary, and engineers have used different methods to reach similar goals, all temperature bypass systems have common purposes. They override the vacuum advance control systems to provide vacuum advance at high or low temperatures, or both. A bypass system can provide smoother warmup or prevent engine overheating.

When catalytic converters appeared on 1975 cars, most carmakers reduced the use of vacuum delay valves. They still appeared, however, on many cars through the late 1970's and early 1980's.

**Speed- and Transmission-Controlled Timing.** Many cars of the late 1960's and early 1970's had systems that sensed vehicle speed or transmission gear range and cut off vacuum advance at low and intermediate speeds. Carmakers used various names for these systems, such as:

- Transmission-controlled spark (TCS)
- Speed-controlled spark (SCS)
- $NO_x$ spark control
- Electronic distributor modulator (EDM)
- Electronic spark control (ESC)
- Transmission-regulated spark (TRS).

Regardless of the name, all of these systems work on the same principles and have similar parts. All systems have a solenoid-operated vacuum valve that shuts off vacuum advance at low

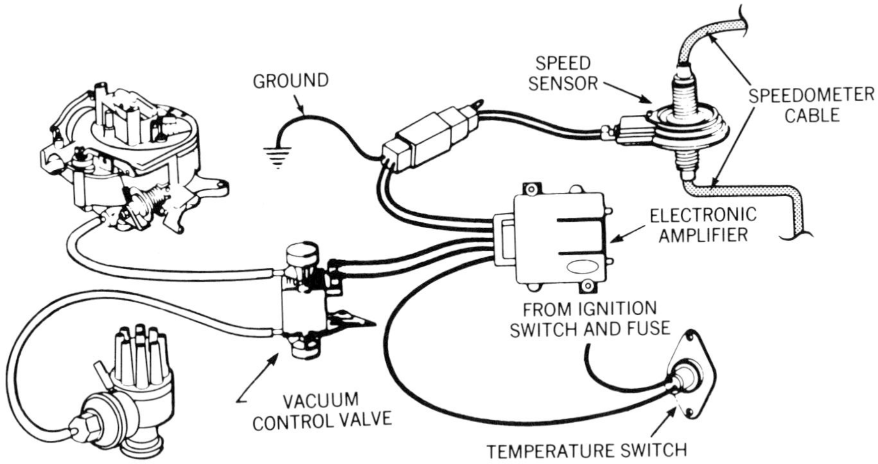


**Figure 18-10.** Speed-controlled spark systems have speed-sensing switches in the speedometer cable. (AC-Delco)

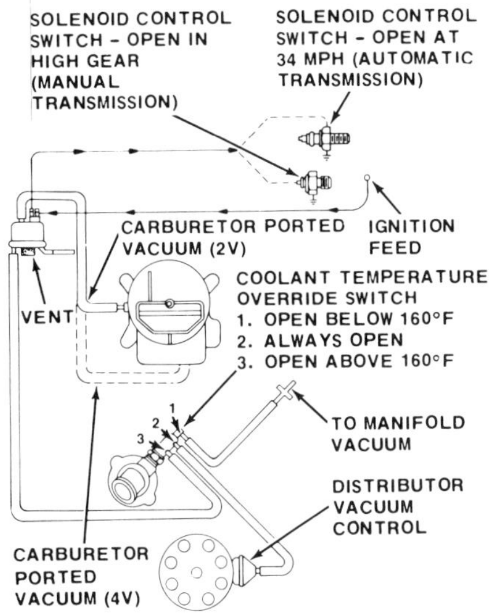


**Figure 18-11.** Transmission-controlled spark systems have switches activated by gear lever position or by automatic transmission hydraulic pressure. (AMC)

speeds. All systems also have a speed-sensing switch or a transmission switch to energize or deenergize the solenoid. All systems also have one or more switches that sense **ambient air temperature** or engine coolant temperature. These are override switches that allow full vacuum advance at high or low engine or air temperatures. Full advance is allowable at low temperatures because combustion temperatures are not at their maximum, and $NO_x$ formation is low. Full advance is necessary at high temperatures to prevent overheating.

Speed-controlled spark systems, such as Cadillac's SCS and Ford's EDM and ESC systems, have speed-sensing switches in the speedometer cables, figure 18-10. These are rotating magnets that turn inside a stationary winding. Above a certain speed, the magnet induces enough voltage in the winding to activate an electronic module or energize the solenoid.

Transmission-controlled spark systems were more common than speed-controlled systems. They use switches activated by shift linkage position on manual transmissions or by hydraulic pressure on automatic transmissions, figure 18-11. The switches are wired to deny vacuum advance at low speeds and allow it at high speeds.

### DISTRIBUTOR ROTOR QUICK TEST

It's possible for a distributor rotor to develop a short-circuit path to ground in the form of a crack or a carbon track that allows high voltage from the coil to go directly to ground in the distributor shaft. Sometimes the short is a small hole that often is called "punch through." Whatever the cause of the short, it keeps the coil voltage from ever reaching the plug cable terminals in the cap, and usually the car won't start.

If you're trying to cure a no-start condition, you might put in a new rotor just to see if that solves the problem. That's really shooting in the dark; it's not the kind of logical troubleshooting a professional mechanic does. If the car still doesn't start, you don't know if the first rotor or the second, or both, was bad. There is, however, a quick and accurate way to check for a shorted rotor on a breaker-point distributor.

Disconnect the coil center cable from the cap but leave it connected to the coil. Remove the cap and move it out of your way. Then, put a small insulator, such as a cardboard match, between the points to block them open. Connect one end of a jumper wire to the insulated breaker-point arm (that's the hot connection to the ignition primary circuit) and turn the ignition switch on. Using insulated pliers, hold the disconnected end of the coil cable about 1/8 to 1/4 inch above the center terminal of the rotor on the distributor shaft. Finally, touch the other end of your jumper wire to ground several times to open and close the primary circuit.

Each time you open the primary circuit, the coil will discharge. If the rotor is shorted, a healthy spark will jump from the coil cable to the rotor and straight through to ground. If the rotor is properly insulated, the secondary circuit from the coil will remain open. No spark will jump to the rotor, but you may see a yellowish-blue corona around the end of the coil cable.

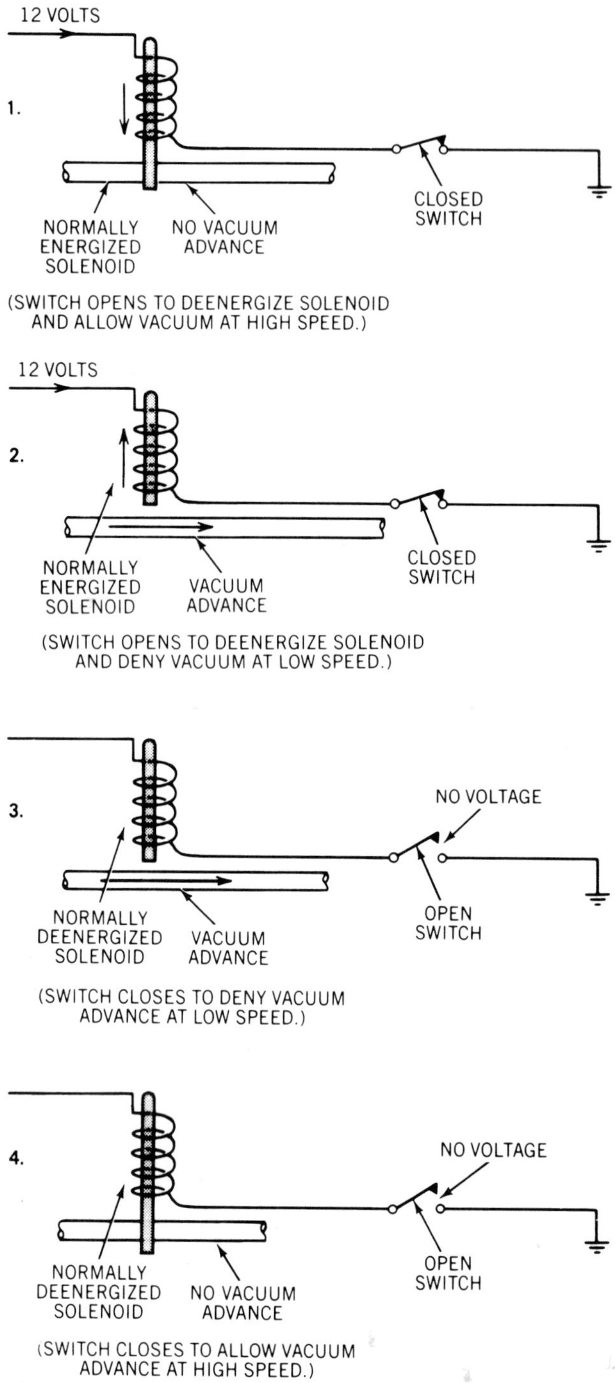


**Figure 18-12.** Vacuum advance solenoids and switches have been built and connected in four different ways. The fourth method shown is the most common.

A solenoid vacuum valve can be **normally energized**, which means that when the ignition is on, voltage is applied to the solenoid windings. Or, the valve can be **normally deenergized**, which means that when the ignition is on, voltage is *not* applied to the windings until a switch closes. Further, the valve can be built to open or close a vacuum line in either the energized or deenergized condition. Therefore, a solenoid-operated vacuum valve can be wired in one of four ways, as shown in figure 18-12.

All of this may sound confusing, and it can be. We haven't the room in this book to detail all of the combinations of speed and transmission switches and solenoids used by carmakers over a 10- or 12-year period. When you service or troubleshoot a spark control system, you will need the electrical and vacuum diagrams for *that specific year and model of car*. There are some principles, however, that you can use to help sort out different systems:

- A **normally open switch** will energize a solenoid when it closes.
- A **normally closed switch** will deenergize a solenoid when it opens.

GM violated this rule of thumb, however, when it used **reversing relays** on some TCS and combination emission control (CEC) systems in 1971 and '72 to energize a solenoid when a normally closed switch opens, figure 18-13.

Many early systems were built so that if a switch, relay, or solenoid failed, vacuum advance would be available at all times. This was fine for driveability and fuel economy, but not for emission control. In mid-1973, the Environmental Protection Agency (EPA) required all carmakers to build their speed- and transmission-controlled spark systems so that *if a part failed, vacuum advance would be shut off at all times*. Therefore, most systems built after March 1973 follow the model shown in the fourth diagram of figure 18-12. A normally open transmission or speed-sensing switch closes at high speed or in high gear. This energizes a normally deenergized solenoid to open a normally closed vacuum line and allow vacuum spark advance. High- and low-temperature override switches were still used in most systems.

While this became the most common speed- or transmission-controlled spark system after 1973, it was not universal. We went through the preceding summary of switch, solenoid, and valve combinations to emphasize the point that you must check vacuum and electrical diagrams to accurately service any specific system.

## The Demise of Spark-Timing Controls

All of the spark-timing control systems of the 1970's used vacuum and electromechanical devices to delay or re-

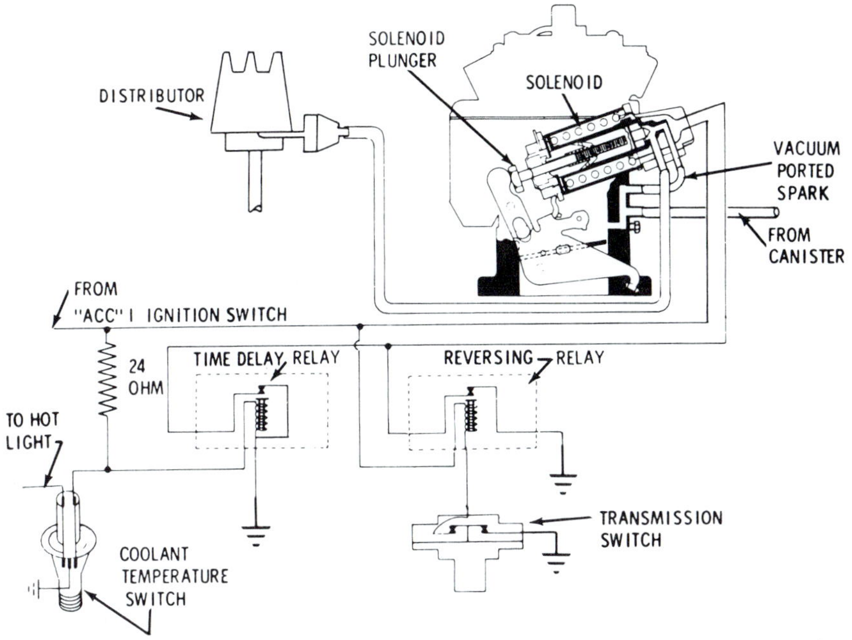


**Figure 18-13.** This diagram of a GM TCS-CEC system shows that you can't take switch and solenoid positions for granted in any system. (Oldsmobile)

tard vacuum advance. As carmakers developed exhaust gas recirculation (EGR) in the mid-1970's, EGR became more effective for reducing $NO_x$ emissions. Many carmakers, therefore, dropped or simplified the spark control systems. Vacuum delay valve and vacuum-retard distributors continued into the late 1970's to assist HC and $NO_x$ control.

By the late 1970's, most carmakers had introduced the first electronic engine control systems, and the earlier timing control systems disappeared. Electronic engine controls use the same principles of delaying timing under certain speed and load conditions to control emissions, but they do it more precisely. Electronic systems use vacuum, speed, and temperature sensors to send information to a central computer, which calculates the best timing for performance, economy, and emission control. You will learn more about these systems in later chapters.

## EXHAUST GAS RECIRCULATION

The first EGR systems appeared on some 1971–72 cars in California when that state had the first $NO_x$ emission regulations. When $NO_x$ regulations became effective nationwide in 1973, almost all carmakers added EGR systems to their vehicles. Today, EGR is an integral part of almost every electronic engine control system.

### Principles of EGR

In Part One of this book, you learned that oxygen will combine with many other elements at high temperatures in an oxidation reaction. These reactions form compounds called oxides. Combustion in an engine is an oxidation reaction. Oxygen combines with carbon and hydrogen to form $H_2O$ vapor, $CO_2$, and CO. These are all oxides. At high temperature, oxygen also combines with nitrogen in the air to form NO, $NO_2$, and other oxides of nitrogen. These are the $NO_x$ compounds that contribute to smog.

$NO_x$ production increases greatly at combustion temperatures above 2,500° F (1,370° C). Engineers found that if they could lower combustion temperatures, they could limit $NO_x$ formation. Retarded ignition timing helped to control $NO_x$, but it was not as effective as EGR. Recirculating a small amount of exhaust back to the intake air-fuel mixture dilutes the mixture. Exhaust gases are inert; they contain neither enough oxygen nor fuel to aid combustion. When exhaust mixes with the intake charge, it displaces fuel and air that *do* contribute to combustion. The diluted mixture contains less chemical energy and burns at a lower temperature. While EGR reduces $NO_x$ formation, it also reduces power and economy. The challenge for engineers was to control EGR so that just enough exhaust is recirculated under the conditions when $NO_x$ formation is highest.

### When EGR Is Needed

$NO_x$ formation is highest when combustion temperatures are high and air-fuel ratios are lean. A rich air-fuel ratio has more carbon and hydrogen to combine with oxygen, which leaves less oxygen to combine with nitrogen. This is why $NO_x$ increases greatly with air-fuel ratios leaner than about 14.7:1. All of these facts mean that EGR is needed only with high temperatures and lean mixtures. These conditions occur during moderate acceleration and cruising speeds from 30 to 70 mph. Under all other conditions, EGR is unnecessary and undesirable. Late-model EGR systems on gasoline engines *turn off* exhaust recirculation at:

- Low temperature (cold-engine warm-up)
- Idle (rich mixture, low combustion temperature)
- Low speed (low combustion temperature)
- High-speed, wide-open throttle (rich mixture)
- Deceleration (rich mixture, low combustion temperature).

EGR systems on diesel engines work slightly differently, as you will learn later in this section. However, diesel EGR systems work on the same principle of diluting the air-fuel mixture under the leanest conditions.

### Early EGR Systems

Chrysler used the first EGR systems on 1972 California cars. Exhaust was recirculated through jets in the intake

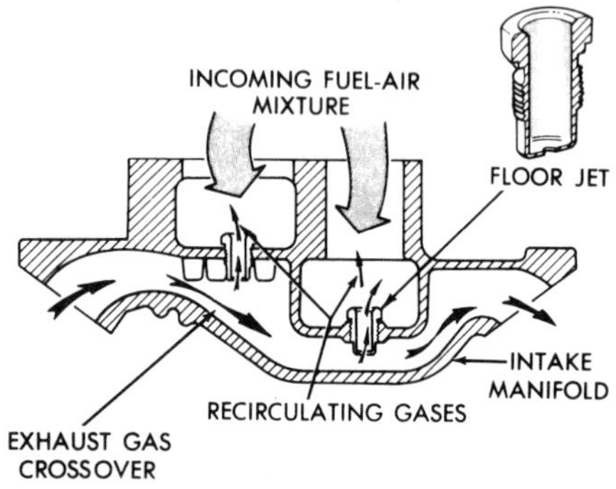

**Figure 18-14.** Chrysler used this floor jet EGR system on some 1972–73 engines, but it did not provide the precise control that an EGR valve does. (Chrysler)

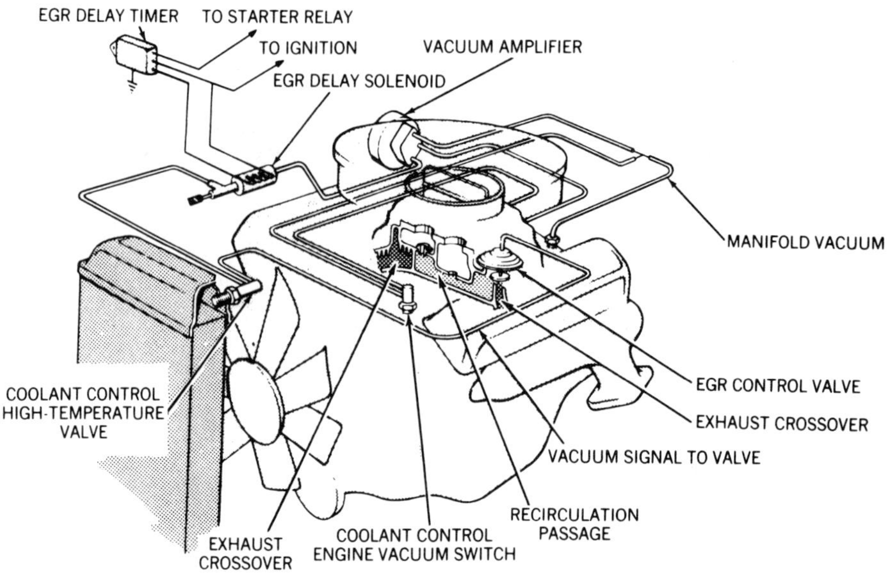

**Figure 18-15.** An EGR valve on a V-type engine is installed in the intake manifold exhaust crossover passage. (Chrysler)

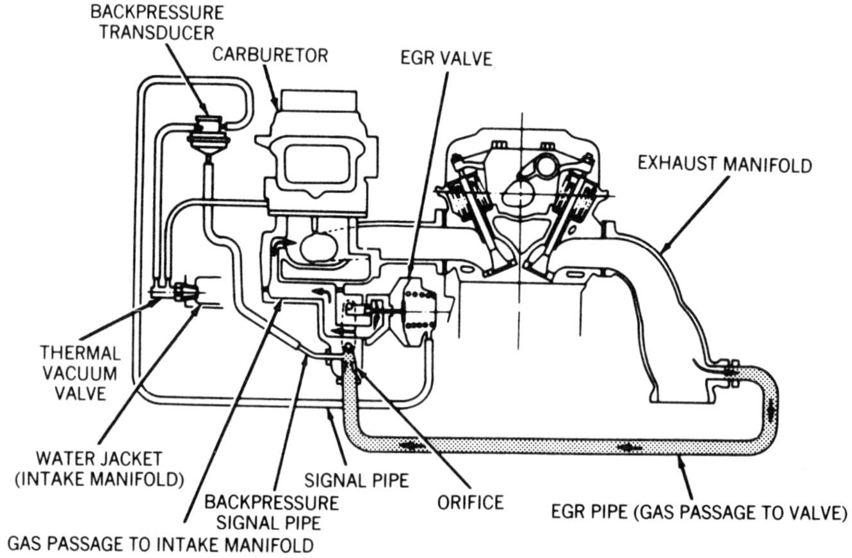

**Figure 18-16.** On an inline engine, the EGR valve can be connected to the manifolds by stainless steel tubing. (Isuzu)

manifold below the carburetor, figure 18-14. The jets created an opening to the exhaust crossover passage that provided carburetor heat for engine warmup. These "floor jets" relied on the combined effects of manifold vacuum and exhaust pressure to force exhaust through a calibrated orifice. The floor jet system was simple, but it allowed exhaust to enter the intake manifold when it wasn't needed, such as when the engine was cold or at idle. This caused poor warmup, rough engine operation, and other problems. Chrysler abandoned the floor jet EGR system after 1973.

Also in 1972, Buick began using a vacuum-operated poppet valve to control EGR. Other carmakers adopted the vacuum EGR valve, and it became standard by the mid-1970's.

In a basic EGR valve installation, the valve is installed in an exhaust crossover passage in the intake manifold, figure 18-15, or connected to the exhaust manifold by stainless steel tubing, figure 18-16. A spring in the vacuum diaphragm holds the EGR valve closed when vacuum is not applied. The vacuum diaphragm is connected to a carburetor vacuum port above the throttle. Ported vacuum is shut off at idle and deceleration and weak at wide-open throttle when EGR is not needed. EGR valves used with ported vacuum are made to close when vacuum drops at wide-open throttle. Ported vacuum is strongest at moderate acceleration and cruising when EGR is needed. Thus a simple vacuum-operated valve can turn EGR on and off for basic $NO_x$ control, figure 18-17.

## Advanced EGR Systems

The simple valve described above does not provide the precise EGR control needed to maintain performance and economy. For example, ported vacuum is the same with a cold or a hot engine. So, the valve will allow EGR during warmup, which hurts driveability. Even on the earliest systems, carmakers installed a temperature valve in the vacuum line to the EGR valve. This usually was a thermostatic vacuum valve in the radiator or engine water jacket, figure 18-18. At low coolant temperatures, the valve closes to block vacuum to the EGR valve. As the engine warms up, the valve opens

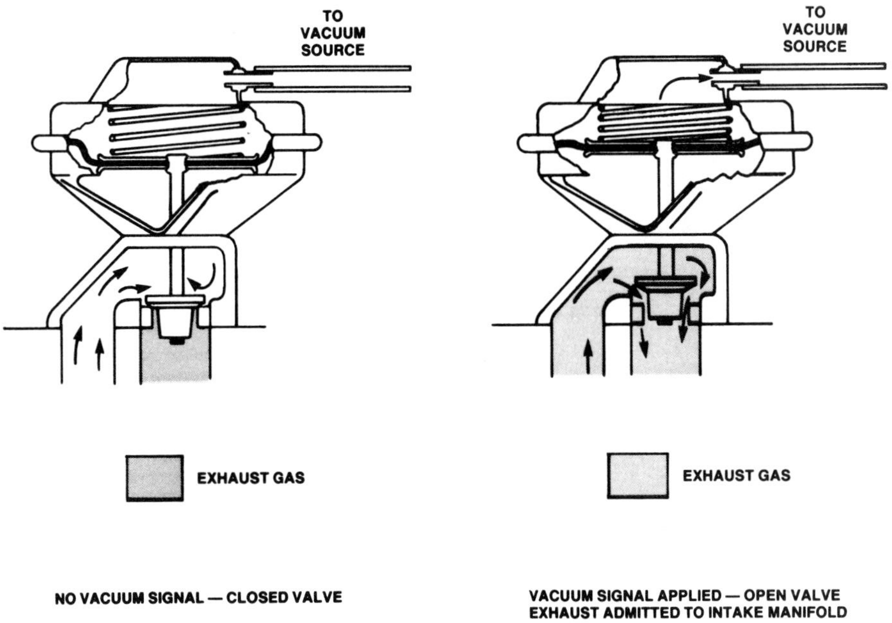


**Figure 18-17.** Ported vacuum opens the EGR valve at moderate acceleration and cruising when $NO_x$ formation is highest. (GM)

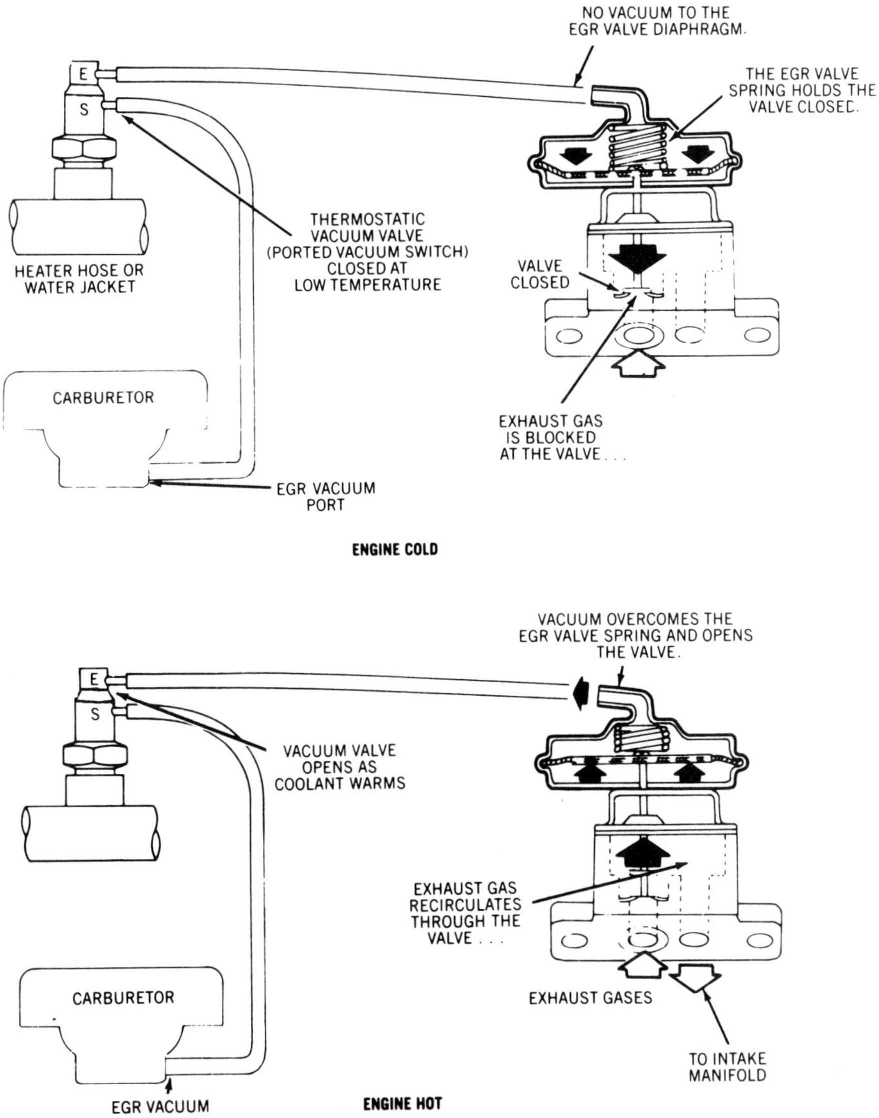


**Figure 18-18.** A coolant temperature vacuum valve blocks vacuum to the EGR valve at low engine temperatures. (Ford)

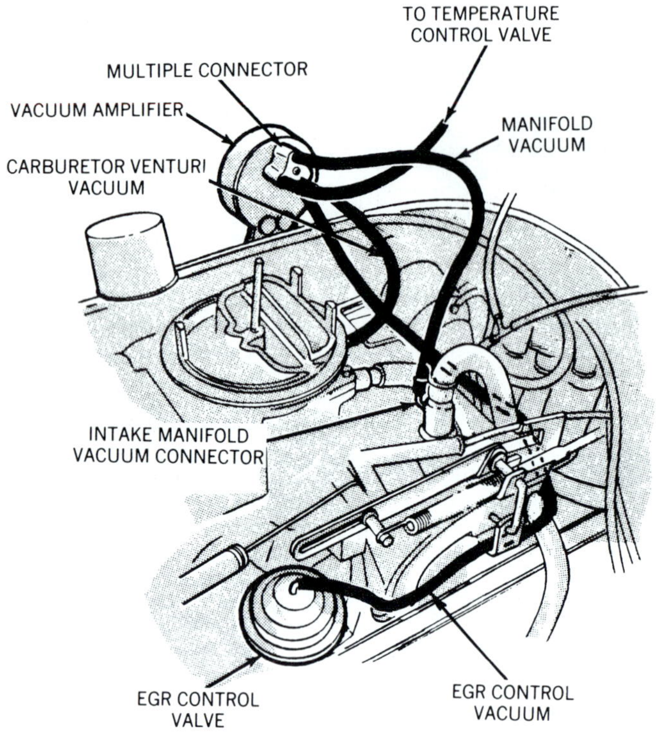


**Figure 18-19.** In this system, manifold vacuum operates the EGR valve, but it is controlled by venturi vacuum. (Chrysler)

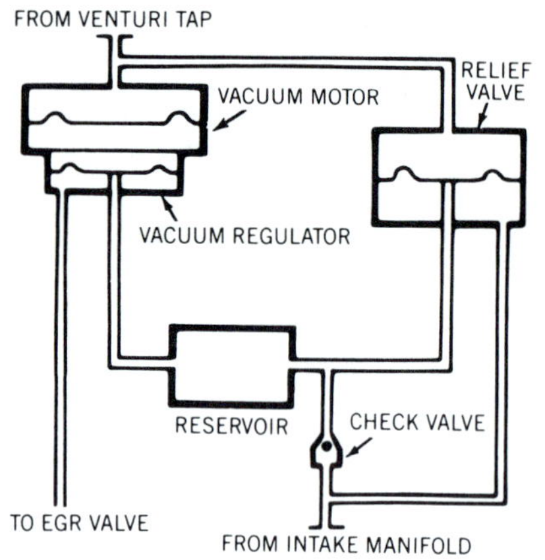


**Figure 18-20.** Venturi vacuum on the large diaphragm opens the smaller diaphragm to allow manifold vacuum to operate the EGR valve. (Ford)

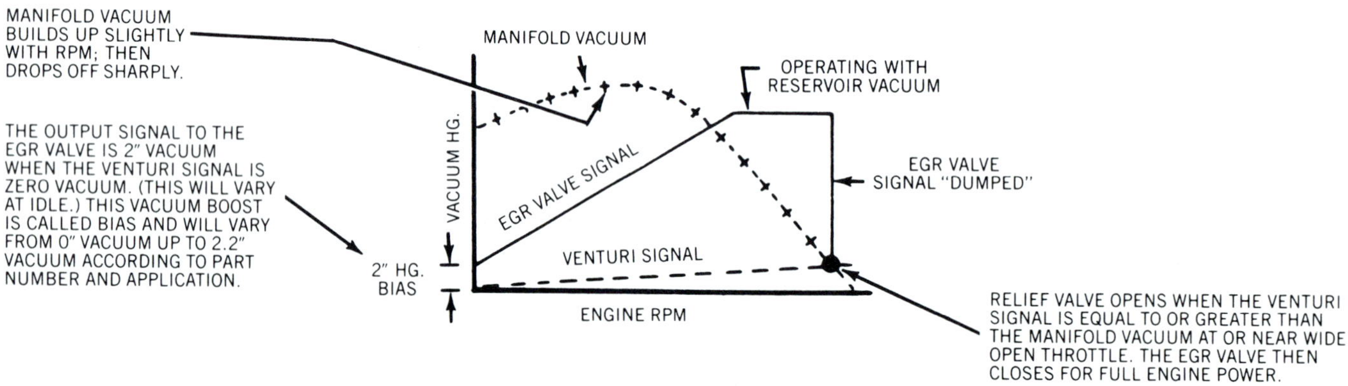


**Figure 18-21.** This graph shows the relationships of different vacuum signals for a venturi-vacuum EGR system. (Ford)

the EGR vacuum line. Carmakers call this valve a ported vacuum switch (PVS), a coolant temperature override (CTO) valve, a temperature vacuum switch (TVS), a thermal vacuum valve (TVV), and several other things. Regardless of the name, they all work the same.

**Vacuum Sources.** The EGR valves we have seen thus far operate on ported vacuum from the carburetor. Ported vacuum changes with engine load and throttle position and is a good vacuum source for EGR control. Carmakers have used other vacuum sources for EGR operation, however. Ford, Chrysler, and AMC all used venturi vacuum sources on EGR systems from the middle 1970's through the early 1980's. These systems use a vacuum port at the carburetor venturi to provide a weak control vacuum signal that is proportional to airflow. The venturi vacuum is not strong enough to open the EGR valve, so these systems use a vacuum amplifier and vacuum reservoir to open the EGR valve with manifold vacuum, figure 18-19.

The reservoir stores manifold vacuum. Weak venturi vacuum operating on a large diaphragm of a 2-stage valve opens a smaller diaphragm to connect the reservoir to the EGR valve, figure 18-20. At wide-open throttle, a relief valve compares venturi and manifold

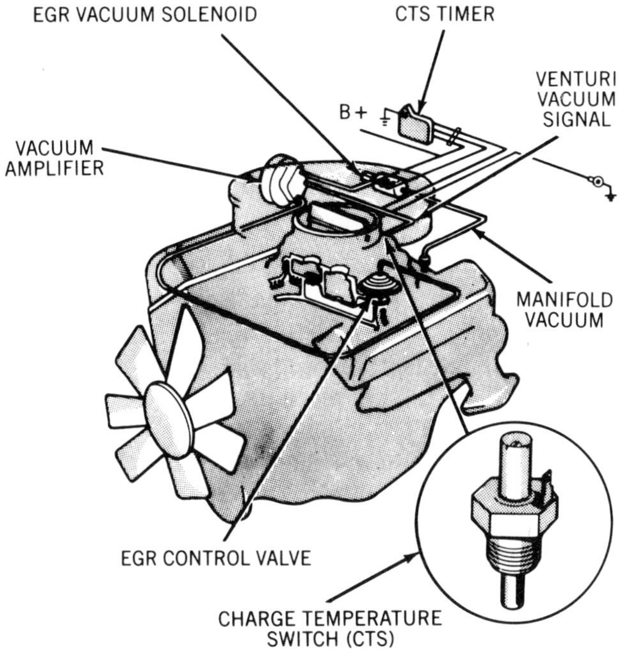


**Figure 18-22.** This EGR system has both a timer and a charge temperature switch to control EGR during warmup. (Chrysler)

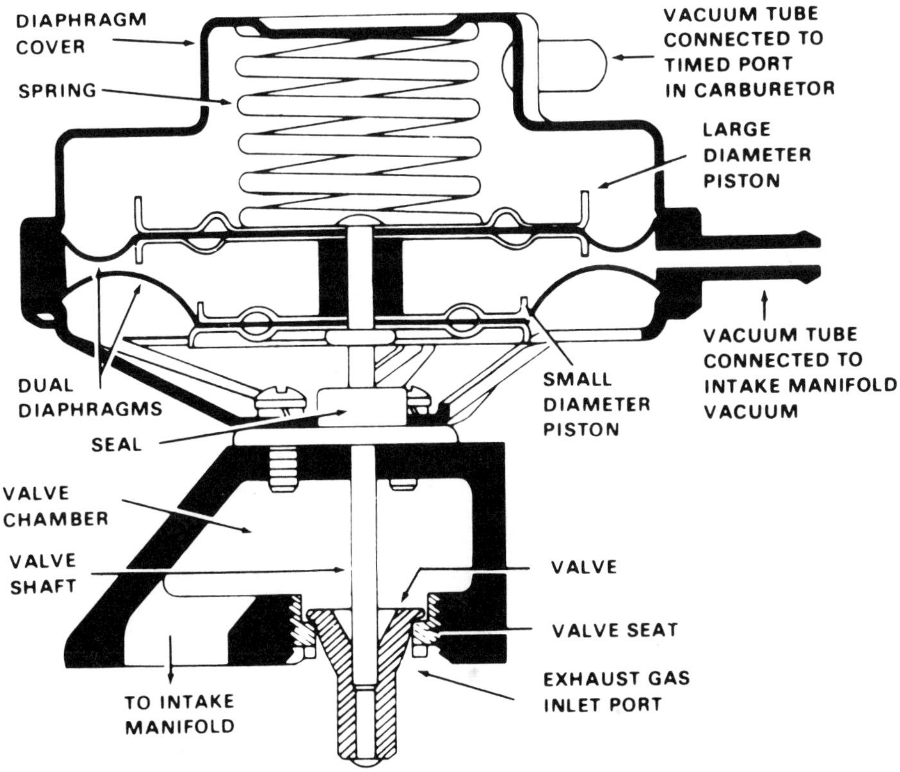


**Figure 18-23.** This dual-diaphragm valve modulates EGR in relation to engine load. (Buick)

vacuum. When venturi vacuum equals manifold vacuum, the relief valve dumps the EGR vacuum to the atmosphere and shuts off EGR. Figure 18-21 is a graph that shows the relationships of manifold vacuum, venturi vacuum, and the vacuum signal needed at an EGR valve. Venturi vacuum systems provide good EGR modulation proportional to engine load for some engines, but reservoirs and amplifiers can develop leaks.

As EGR systems developed, engineers used many devices to modulate EGR flow. Some cars have ambient temperature switches that bleed off part of the EGR vacuum at high and low temperatures. Some cars have time-delay switches and solenoids that block EGR when an engine is started. Figure 18-22 shows such a system, which also has an intake charge temperature switch to turn off the vacuum solenoid when intake temperature is cold. Other cars have vacuum bias valves or dual-diaphragm EGR valves, figure 18-23, that use manifold vacuum to modulate ported vacuum at cruising speed to eliminate surge. Figure 18-24 shows an electronic speed-controlled system that Ford used in the mid-1970's to turn off EGR at high speed. Most late-model EGR systems use backpressure **transducers** and solenoids controlled by an engine control computer to modulate EGR.

## Late-Model EGR Controls

American Motors was the first domestic carmaker to use an EGR backpressure transducer on some 1973 cars. Most carmakers now use them. A backpressure transducer can be separate from the EGR valve, figure 18-25, or built into the valve, itself, figure 18-26. The transducer consists of a tube or a hollow EGR valve stem, open to exhaust backpressure in the EGR passage. Backpressure operates on a diaphragm to close an air bleed in the EGR vacuum line. When backpressure is low at light-load cruising, a spring moves the transducer diaphragm to open the air bleed. This reduces vacuum at the EGR valve and

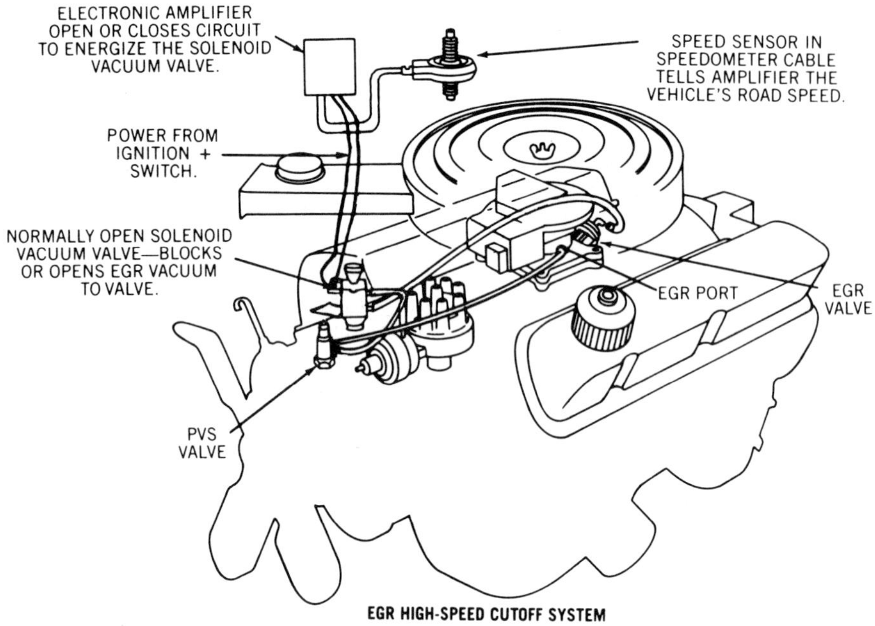


**Figure 18-24.** This system uses a speed sensor and electronic amplifier to shut off EGR at high speed. (Ford)

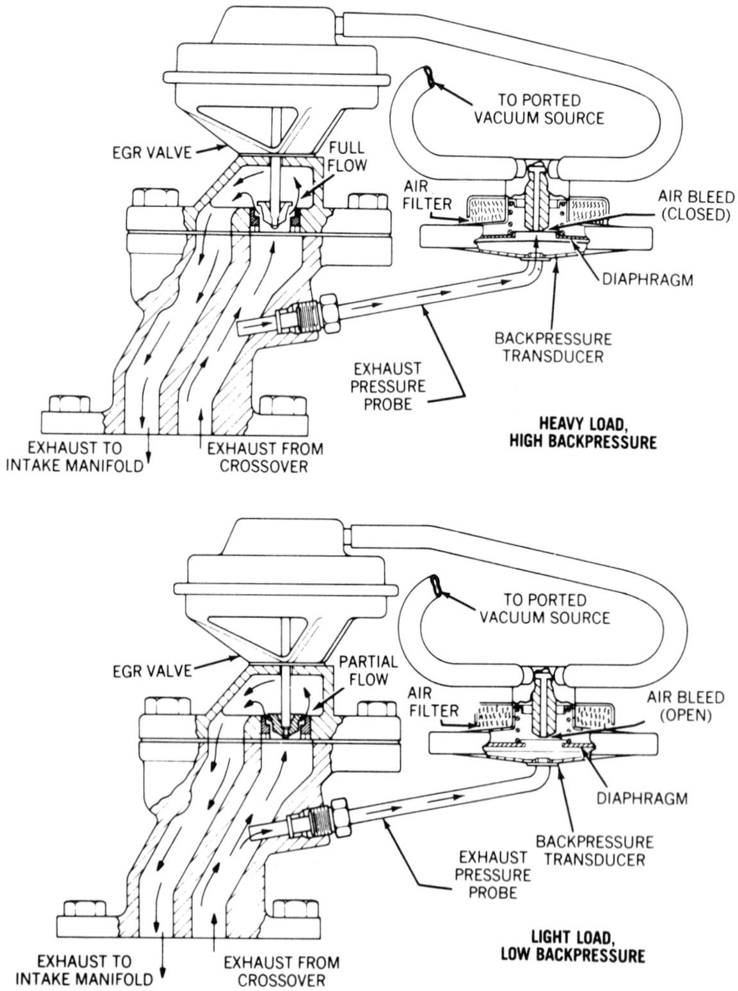


**Figure 18-25.** This EGR backpressure transducer is separate from the EGR valve. (Oldsmobile)

reduces the amount of exhaust recirculated. Exhaust backpressure increases as cruising load increases. This is when combustion temperatures and $NO_x$ formation are highest. Backpressure in the transducer forces the diaphragm to close the air bleed and increase vacuum to the EGR valve.

**Positive- and Negative-Backpressure EGR Valves.** The basic backpressure transducer just described has come to be called a positive-backpressure EGR valve because exhaust backpressure helps the valve to open. General Motors introduced the positive-backpressure EGR valve shown in figure 18-26 in 1977. This positive backpressure valve will not open with just vacuum applied to the diaphragm port. With enough vacuum applied to the diaphragm, however, and enough backpressure through the stem of the valve to close the internal control valve (air bleed), the valve will open. When it opens, backpressure drops, and the spring again opens the air bleed. The valve moves back toward the closed position. All of this happens quite quickly, and the result is that valve opening is modulated in relation to vacuum and exhaust pressure. The EGR valve pintle and the air bleed cycle back and forth about 30 times per second in normal operation.

The General Motors negative-backpressure EGR valve, figure 18-27, was introduced in 1979. It too relies on a combination of vacuum and exhaust backpressure to control the valve opening, but exhaust pressure has an opposite effect.

A small spring below the diaphragm closes an air bleed. When vacuum is applied to the valve diaphragm, the valve starts to open quickly. Exhaust backpressure in the hollow valve stem is then exposed to intake manifold vacuum on the other side of the valve pintle. These two pressures interact to create low pressure (vacuum), or "negative backpressure" that travels through the valve stem to move a lower diaphragm against the small

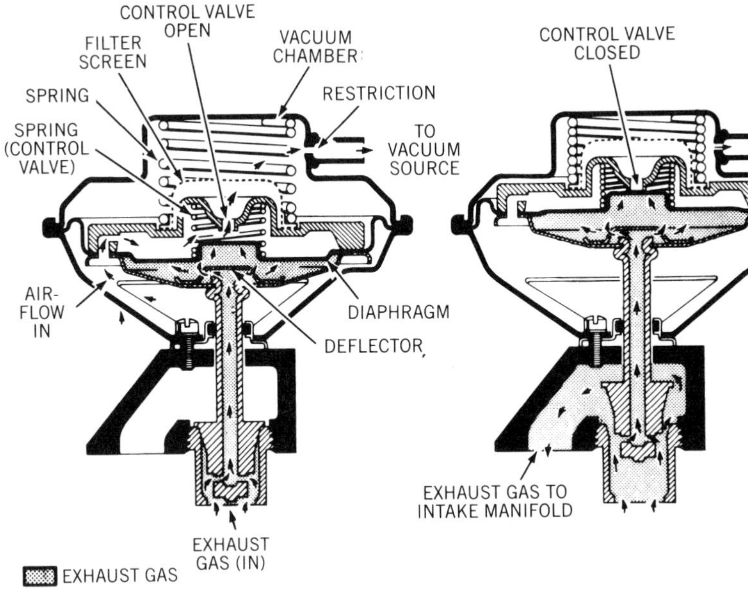


**Figure 18-26.** The backpressure transducer is built into this EGR valve. (GM)

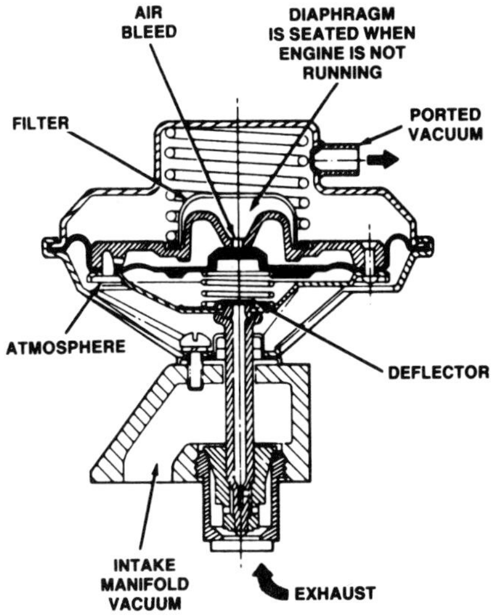


**Figure 18-27.** This negative backpressure valve controls EGR as a constant percentage of airflow. (GM)

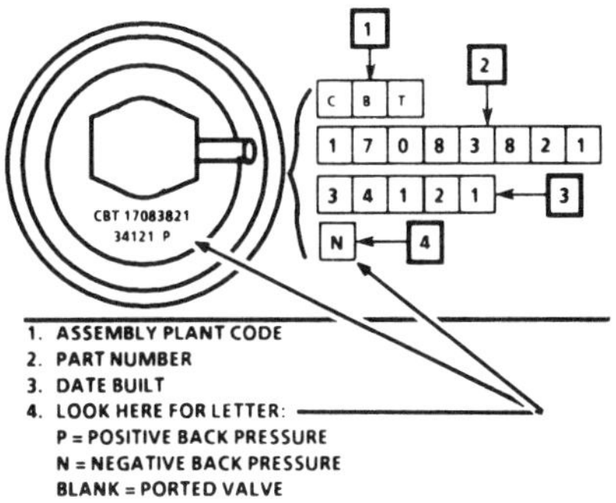


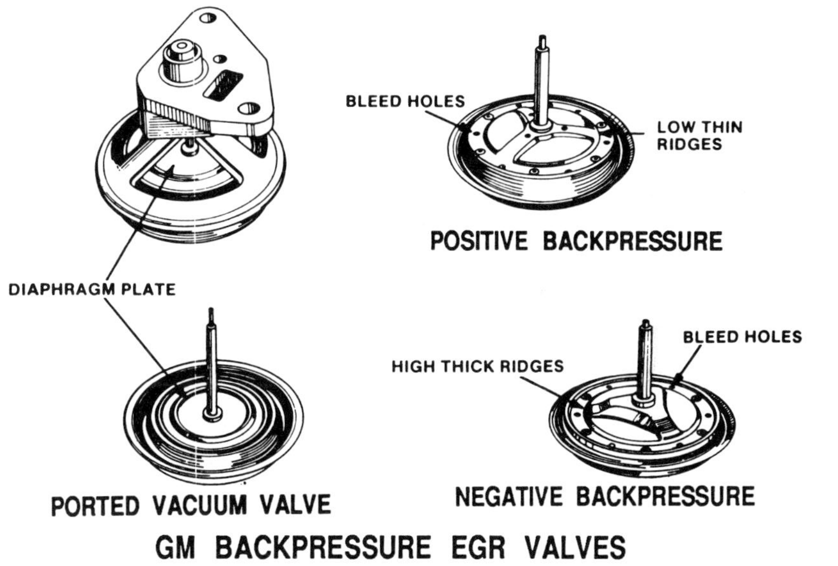


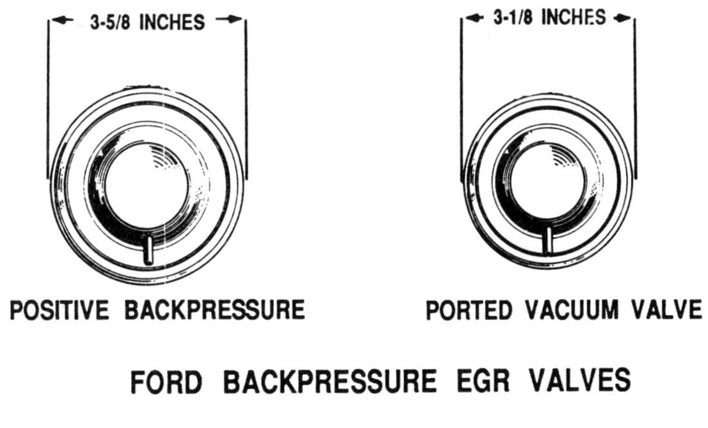


**Figure 18-28.** General Motors and Ford EGR valve identification.

spring pressure. This opens the air bleed to reduce vacuum on the EGR valve. The valve then moves toward the closed position. Again, this happens quickly, and the valve opening is modulated in relation to vacuum and exhaust pressure. The EGR valve pintle and the air bleed cycle back and forth about 30 times per second in normal operation.

The negative-backpressure valve will open with only vacuum applied to the diaphragm port. It opens slightly faster than the positive-backpressure type and is often used where detonation control under light acceleration is critical.

You can identify GM EGR valves made since 1984 by the letter P or N as part of the identification numbers on the top of the valve, figure 18-28. If no letter is present, the valve is a simple ported vacuum valve with no backpressure transducer. You also can identify a positive- or negative-backpressure or ported vacuum GM valve by the shape of the diaphragm plate, figure 18-28.

Ford also uses positive-backpressure EGR valves, as well as simple ported vacuum valves. You can distinguish these from each other by the diameter of the vacuum diaphragm, figure 18-28. Some late-model Ford 4-cylinder engines have backpressure variable transducers that modulate EGR vacuum in relation to two different backpressure sources, figure 18-29. This transducer works as a downstream regulator to bleed off EGR vacuum until closed by exhaust backpressure. This Ford system also uses thermal and ported vacuum valves on the upstream side of the EGR valve to control the application of vacuum.

Chrysler uses positive-backpressure transducers on the EGR valves of some late-model V-6 and V-8 truck engines. These are similar to the GM transducer shown in figure 18-25.

Additional vacuum control solenoids for EGR valves are discussed in the next section.

**EGR Control Solenoids.** Most EGR systems on electronically controlled engines have one or two solenoids in the vacuum line to the EGR valve. Figure 18-30 shows three simple examples. The computer may energize a solenoid to cut off vacuum to the EGR valve at low and high temperatures and at idle or wide-open throttle. On some systems, the solenoid controls an air bleed in the EGR vacuum line to reduce exhaust recirculation under light load.

Many Ford EGR systems use a pair of solenoids to control EGR vacuum, figure 18-31. One solenoid applies vacuum to the EGR valve; the other vents vacuum when EGR is not wanted. The solenoids are controlled by the electronic engine control system computer.

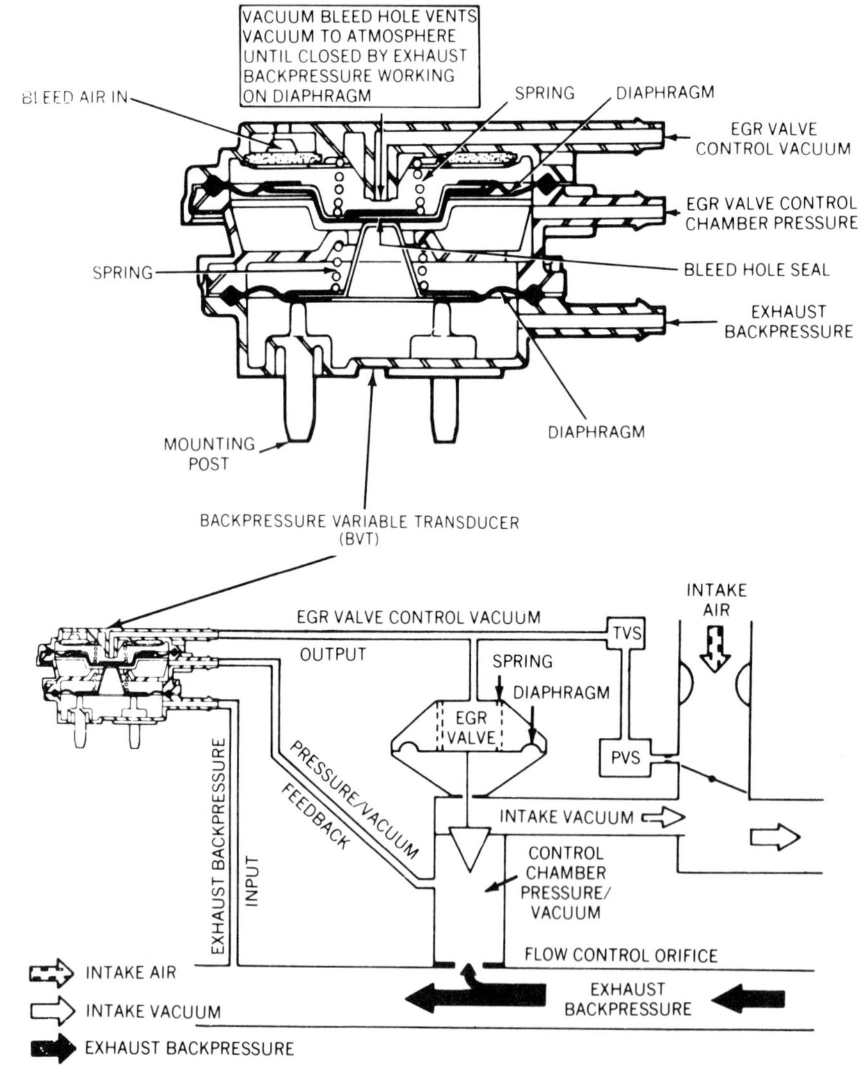


**Figure 18-29.** Several late-model Ford EGR systems use the remote backpressure variable transducer. (Ford)

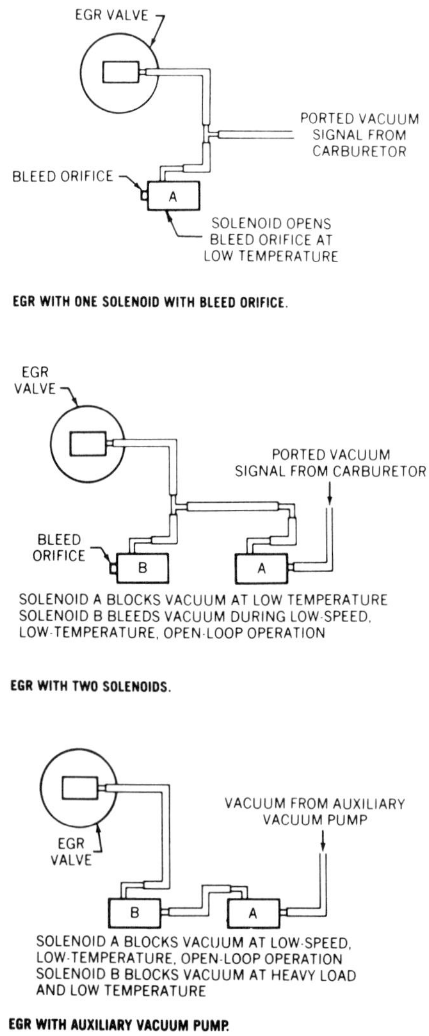


**Figure 18-30.** The solenoids that control EGR vacuum are operated by the electronic engine control computer. (GM)

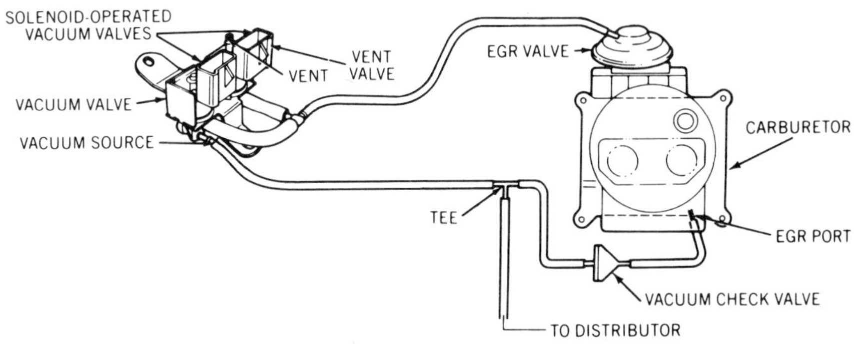


**Figure 18-31.** Typical Ford EGR solenoid installation. (Ford)

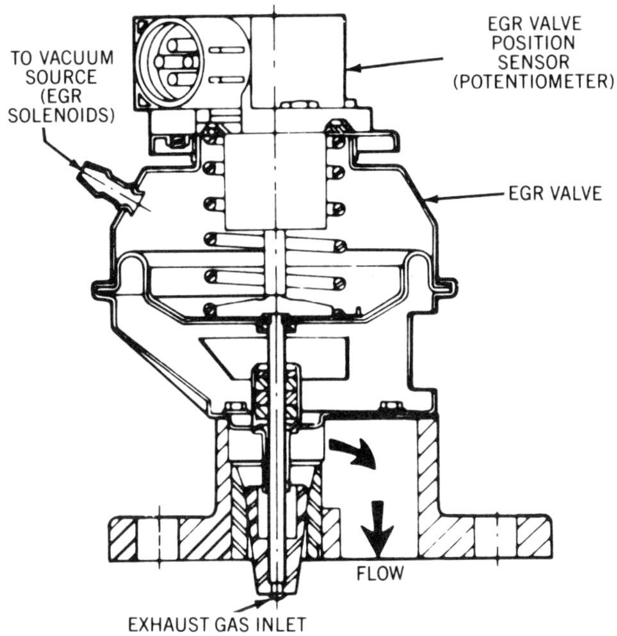


**Figure 18-32.** Some EGR valves have potentiometers as valve position sensors. (Ford)

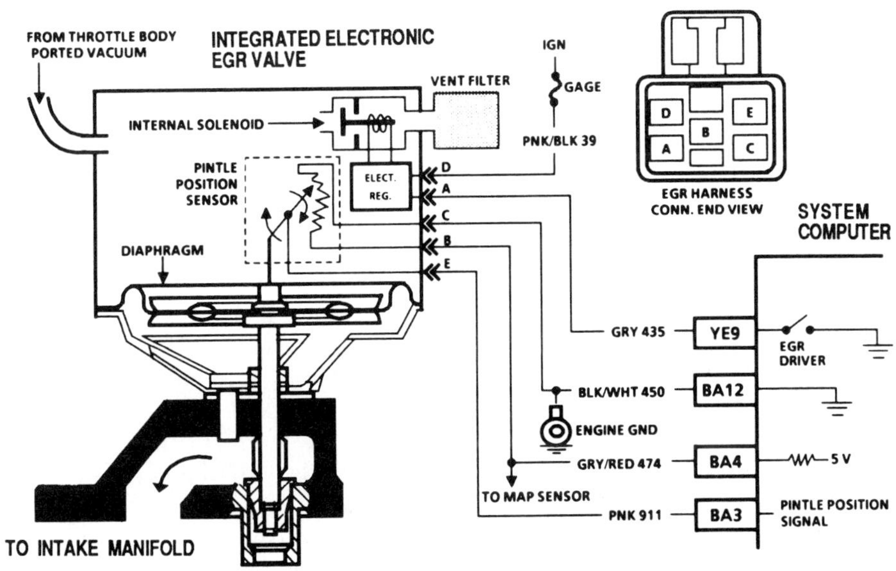


**Figure 18-33.** The General Motors integrated electronic EGR valve contains a pulse-width-modulated vent solenoid and a potentiometer used as a pintle position sensor.

Volume Two, the *Practice Manual*, contains more information on EGR control solenoids and basic test methods.

**EGR Valve Position Sensors.** Many EGR systems on electronically controlled engines provide a feedback signal to the system computer that indicates EGR valve position. For example, Ford has used EGR valves with position sensors since the late 1970's, figure 18-32. These sensors are potentiometers, usually mounted on top of the EGR valve stem. As the valve opens and closes, the sensor sends a varying return voltage signal to the computer that indicates valve position.

General Motors used a similar sensor on the integrated electronic EGR valve on some 1987–88 engines, figure 18-33. This overly complicated, expensive valve was not successful in production and lasted only about two model years.

**Solenoid EGR Valves.** In the late 1980's GM introduced a kind of EGR valve that is not vacuum operated. The so-called "digital" EGR valves have two or three solenoids controlled by the engine computer, figure 18-34. The solenoids receive battery voltage through a common connector. The computer controls the ground side of each solenoid individually. When the computer closes the solenoid ground, the solenoid moves a valve plunger to open an EGR orifice. Depending on EGR requirements, the computer energizes the solenoids individually or in combinations to provide varying amounts of exhaust recirculation.

Exhaust enters the valve chamber through an open port. When a solenoid opens an orifice, exhaust exits to the intake manifold, figure 18-35. GM claims improved EGR flow because the flow rate depends only on orifice size, not plunger position. The swivel action of the plunger also improves valve sealing.

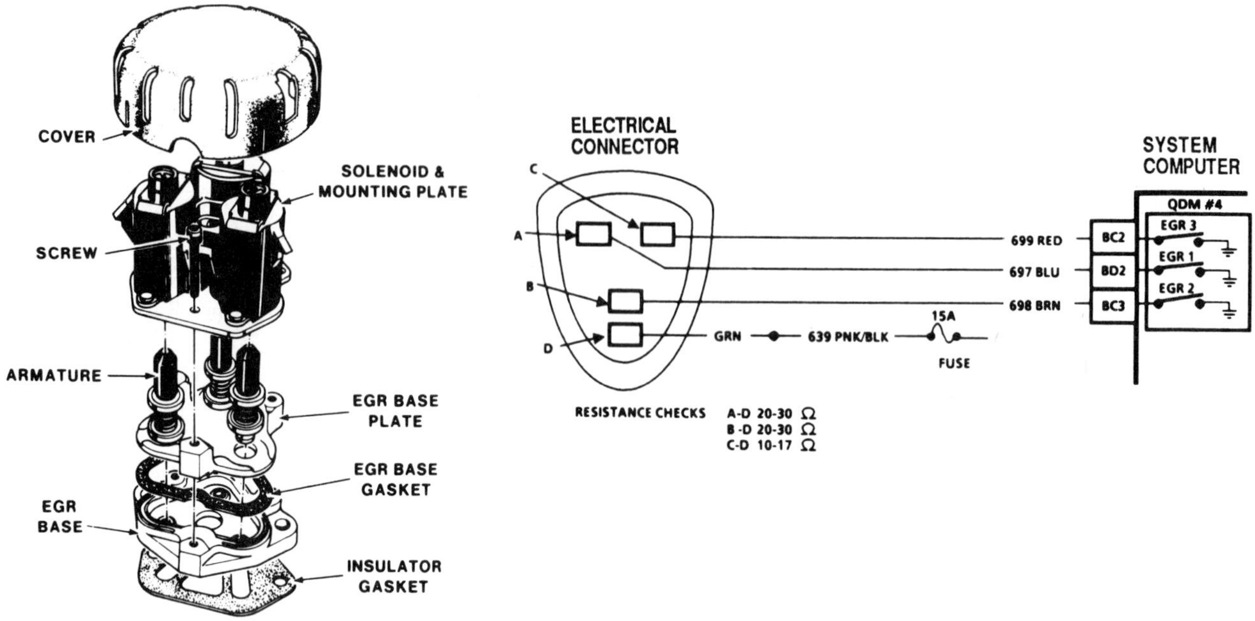


**Figure 18-34.** The General Motors digital EGR valve has two or three solenoids that open different sized orifices for EGR flow. The solenoids receive battery voltage on a common wire but are switched individually by the system computer.

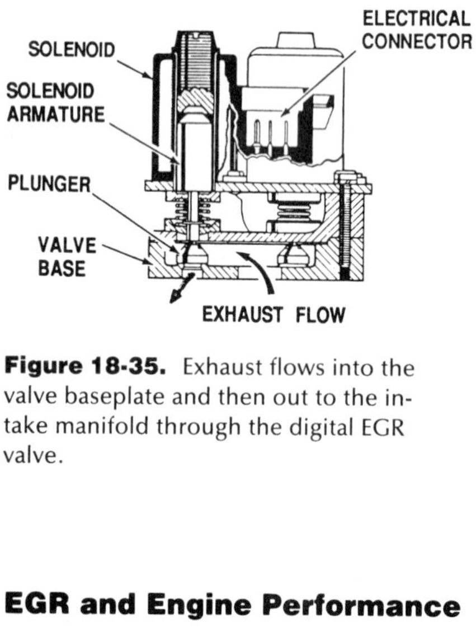


**Figure 18-35.** Exhaust flows into the valve baseplate and then out to the intake manifold through the digital EGR valve.

## EGR and Engine Performance

Because EGR dilutes the intake air-fuel mixture, it reduces the chemical energy of the intake charge volume. This, in turn, reduces the heat energy available for combustion and reduces peak combustion pressure. All of these facts led to the belief that EGR reduces engine power. Some engines of the early 1970's with the first EGR systems did, in fact, develop less power under some conditions than equivalent engines without EGR. Many car owners got the idea that disconnecting the EGR system would increase engine power and efficiency. That idea persists in some minds even today. Not true.

Although EGR *can* reduce thermal efficiency, engineers soon discovered a secondary benefit. EGR also can control engine detonation. Detonation is the uneven burning (an explosion) of the air-fuel mixture. It usually occurs toward the end of the combustion burn time due to high temperature and pressure. You know that detonation reduces engine performance and efficiency. Severe detonation can damage the engine.

Tetraethyl lead in gasoline helps to control detonation, but the unleaded fuel required by cars with catalytic converters has made detonation control a critical problem since the mid-1970's.

Diluting the air-fuel mixture with exhaust reduces the temperatures and pressures that cause detonation. Another basic way to control detonation is to reduce ignition advance. EGR on late-model engines allows ignition timing to be designed for better efficiency while still minimizing detonation. The engineering equations get very complicated, but the effects are easy to see. An inoperative or disconnected EGR system on a late-model engine usually causes detonation, which you can hear as engine pinging. The pinging reduces performance and can lead to engine damage.

## Diesel EGR Systems

At first glance, a diesel EGR system seems to work backwards from a gasoline engine EGR system. In terms of applying EGR for different operating conditions, it does. In terms of controlling $NO_x$ formation, it does not. $NO_x$ formation is highest with excess oxygen (lean air-fuel ratio) and high temperatures. These conditions simply exist during different operating modes for a diesel than for a gasoline engine.

Diesel air-fuel ratios are leanest at idle. High combustion temperatures also exist at idle because of the excess oxygen. Therefore, a diesel requires maximum EGR at idle (when fully warmed up) and decreasing EGR as speed and load increase. This is what seems backwards about a diesel EGR system, but the principles of $NO_x$ control are unchanged.

Almost all diesel automobile engines have vacuum pumps. Among other jobs, the pump supplies the vacuum to operate a conventional EGR valve. At idle with a fully warmed up engine, the EGR valve opens to meter exhaust into the intake manifold where it dilutes the intake air charge. Figure 18-36 shows a typical diesel EGR system with these parts:

- *A vacuum pump* provides vacuum for the EGR valve, power brakes, air conditioning, and other vehicle systems.
- *A vacuum regulator valve* attached to the fuel injection pump controls EGR vacuum in relation to the amount of fuel being injected. As fuel injection increases, vacuum decreases.
- *A solenoid* to block vacuum when the engine is cold or the transmission is in high gear, or both.
- *The EGR valve* operates as a simple ported-vacuum EGR valve on a gasoline engine.

Figure 18-36 is a simplified diagram that shows a basic diesel EGR system. Different carmakers also include vacuum reservoirs, vacuum-delay valves, vacuum-reducing valves (to close the EGR valve quickly on acceleration), engine speed sensors, idle switches, electronic control modules, and transmission or torque converter switches. As with gasoline engines, all of these devices modulate EGR for best driveability and emission control.

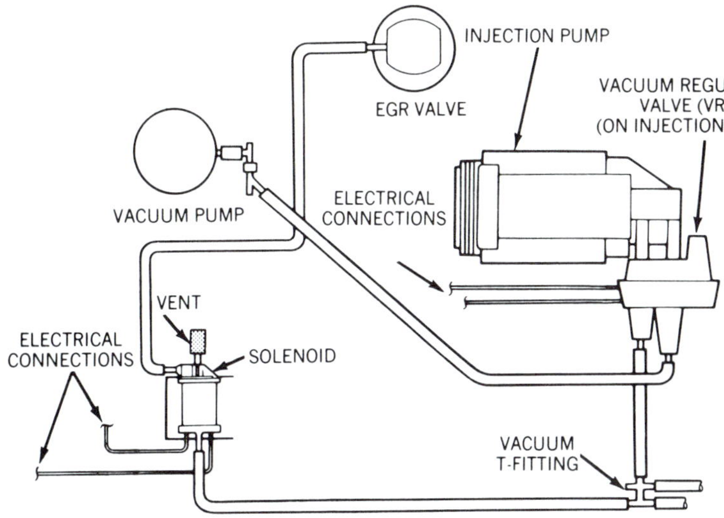


**Figure 18-36.** Basic diesel EGR system.

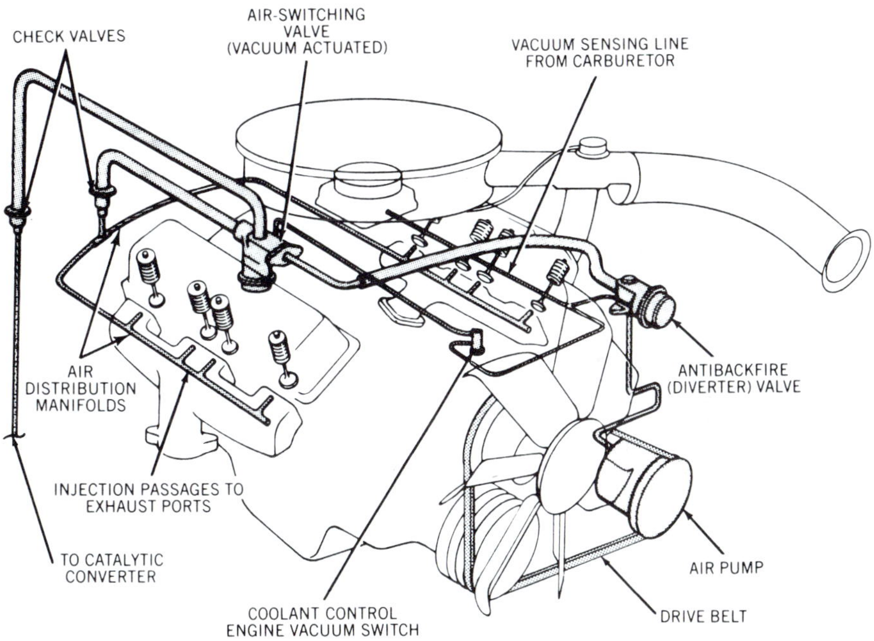


**Figure 18-37.** All basic air injection systems have these common parts. (Chrysler)

## AIR INJECTION SYSTEMS

Installed on some 1965 California cars, air injection systems were among the first emission devices used to control HC and CO. Air systems also are modifications to the basic engine combustion process. By pumping air into the exhaust system immediately after the hot gases leave the cylinders, these systems promote continued oxidation of any HC and CO in the exhaust. Remember that combustion is an oxidation reaction, but the results are usually incomplete. An air injection system continues the oxidation so that HC and CO combine with $O_2$ to form $H_2O$ vapor and $CO_2$.

Carmakers have given their air injection systems different names, such as GM's Air Injection Reactor (AIR) and Ford's Thermactor. They all work the same way, however, and have the same basic parts, figure 18-37.

**1.** A belt-driven air pump with an inlet air filter

**2.** Air distribution manifolds and injection nozzles

**3.** One or more exhaust check valves

**4.** An antibackfire valve

**5.** Hoses for air distribution and vacuum control.

Late-model systems used with 3-way catalytic converters also have various air-switching valves and solenoids to control air delivery for different operating conditions.

### Air Injection Pumps

The air pump, itself, is one of the most common parts of all air systems. All domestic carmakers use centrifugal pumps made by GM's Saginaw Division, figure 18-38. The pump is driven

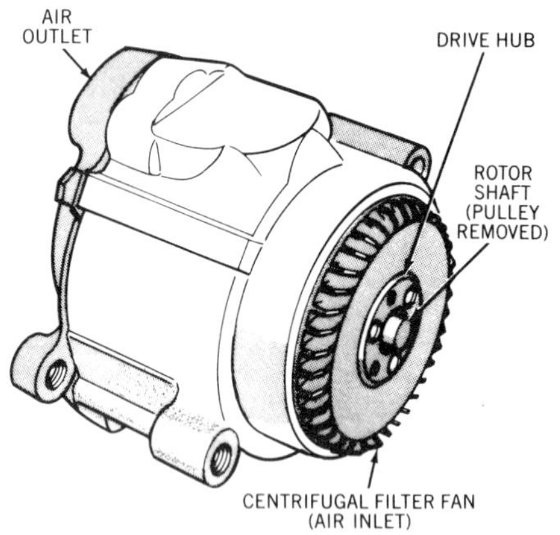


**Figure 18-38.** All domestic carmakers use air pumps made by GM's Saginaw Division. (Chrysler)

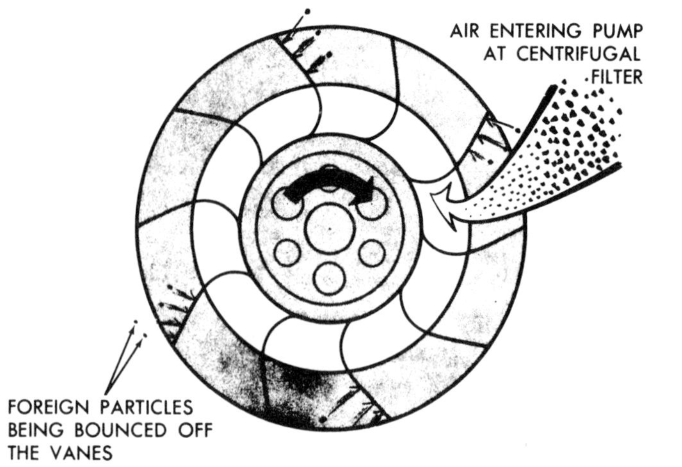


**Figure 18-39.** The air pump filter fan removes dust and dirt from incoming air by centrifugal force. (Chevrolet)

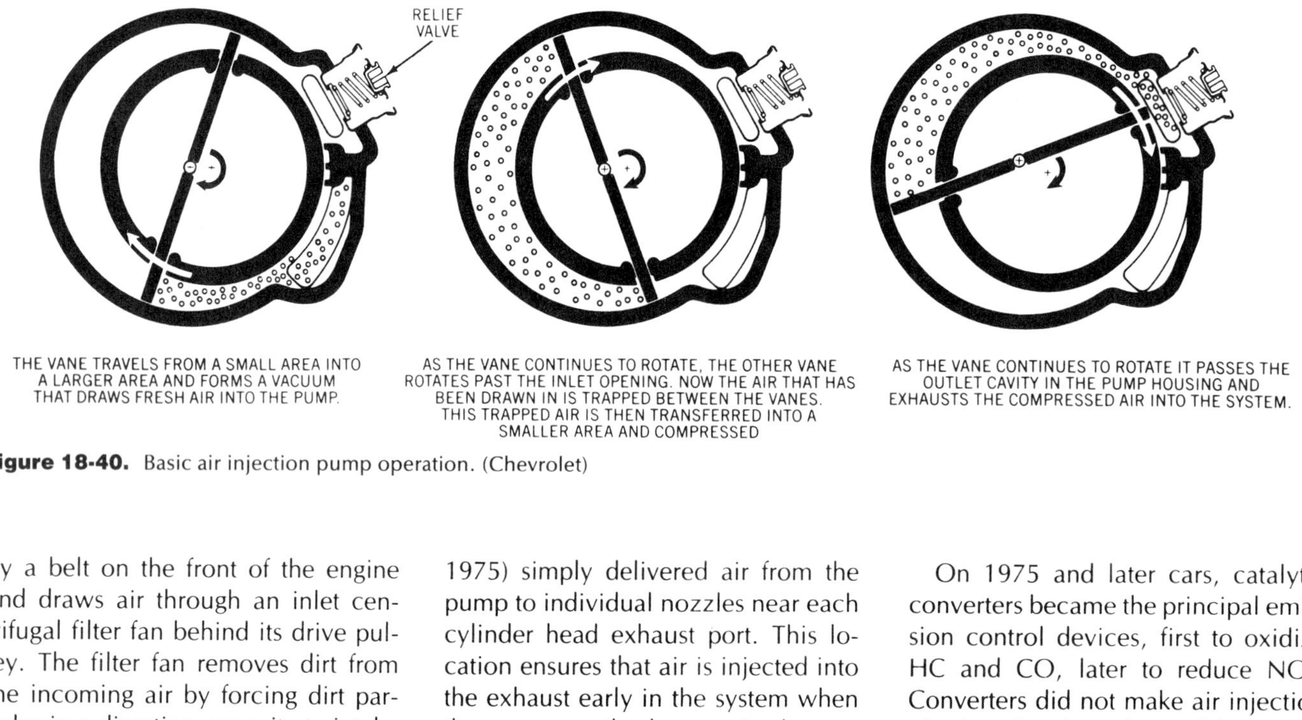


**Figure 18-40.** Basic air injection pump operation. (Chevrolet)

by a belt on the front of the engine and draws air through an inlet centrifugal filter fan behind its drive pulley. The filter fan removes dirt from the incoming air by forcing dirt particles in a direction opposite to intake airflow, figure 18-39. Most pumps have a built-in relief valve to relieve excess air pressure at high engine speed. Some systems have the relief valve in the antibackfire valve, however. Figure 18-40 shows air injection pump operation. This figure also shows a built-in relief valve.

## Air Distribution Manifolds and Injection Nozzles

Early air injection systems on cars without catalytic converters (before 1975) simply delivered air from the pump to individual nozzles near each cylinder head exhaust port. This location ensures that air is injected into the exhaust early in the system when the gases are the hottest. It also ensures equal air injection for the exhaust from each cylinder.

Air is delivered to the exhaust system in one of two ways. Many engines have an external air manifold, or manifolds, to distribute air through tubes and stainless steel nozzles, figure 18-41. The nozzles are screwed into the exhaust manifolds or cylinder heads near each exhaust valve. Some engines have air distribution passages cast into the cylinder heads, figure 18-42. The passages have openings to inject air near each exhaust valve.

On 1975 and later cars, catalytic converters became the principal emission control devices, first to oxidize HC and CO, later to reduce $NO_x$. Converters did not make air injection obsolete, but they changed the air distribution requirements. Most late-model cars use air injection to improve converter efficiency and to speed catalyst warmup. Air need not always be injected near each exhaust port. Rather, it can be injected into the exhaust manifold outlet or into the exhaust pipe upstream from the converter. Air mixes with the exhaust and aids complete HC and CO oxidation in the converter. It also raises the exhaust temperature so that the converter reaches operating temperature faster.

**RACERS' INGENUITY**

Racecar builders historically have been known for their ingenuity, their ability to invent parts when none were available, and to adapt existing parts to new uses. Many years ago, some racing engine builders found a new use for some common emission control parts.

High crankcase pressure can be a critical problem in a racing engine. A 400- or 500-cid V-8 drag racing engine may run at high speed, full throttle, and full load for only 6 to 10 seconds. Even in that brief time, crankcase pressure can get high enough to blow oil seals and force oil past the piston rings into the cylinders. The same engine in a lakes racer or a Bonneville car can develop crankcase pressure that will literally blow the oil pan off the engine if it is not released.

The simple solution that some engine builders found for this problem is to install an oil breather cap with large-diameter tubing on the valve cover (both valve covers of a V-type engine). They connect the other end of the tubing to an air injection check valve and tap the check valve into the exhaust header at the front of the exhaust collector, at a right angle to the exhaust flow. This is the point of maximum exhaust flow and lowest pressure in the exhaust system. The exhaust flow acts like airflow in a carburetor barrel. It creates low pressure that opens the check valve. (This also is the same principle on which a pulse air injection valve works.) This applies low pressure to the tubing connected to the engine crankcase. Crankcase pressure is thus relieved directly into the exhaust flow. If the engine backfires on deceleration or high pressure otherwise occurs in the exhaust, the check valve closes to protect the crankcase.

This is an example of racers' ingenuity taking a part designed for one purpose and finding another, equally beneficial, use for it.

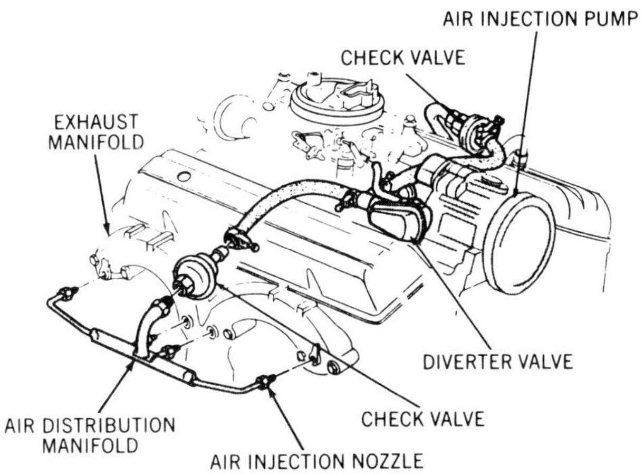


**Figure 18-41.** Many air injection systems have external air distribution manifolds and injection nozzles. (GM)

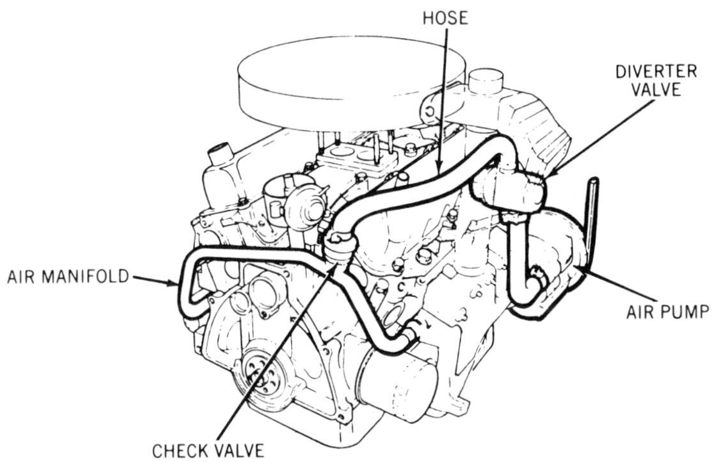


**Figure 18-42.** Some engines have air injection passages cast into the cylinder heads. (Ford)

Air distribution got more complicated when $NO_x$ reduction converters were introduced on some 1977 cars. While oxidation adds $O_2$ to HC and CO, reduction removes $O_2$ from $NO_x$ compounds. A $NO_x$ converter needs extra air in the exhaust to help it warm up. After it is warm, however, it works best with less air in the exhaust. All of these factors led to the development of air-switching systems, which you will learn about later in this chapter.

## Exhaust Check Valves

All air injection systems have one or more check valves to protect the pump, hoses, and other valves from reverse exhaust flow. The check valve is usually near the air distribution manifold inlet, downstream from all other components, figures 18-41 and 18-42. It allows air to flow from the pump to the air manifold and nozzles, but it keeps exhaust backpressure out of the air injection system, figure 18-43. If an engine backfired and had no check valve, the backfire could blow hoses off the air injection system and damage the air pump. The check valve has a spring-type metal disc or reed that closes the air line under exhaust backpressure.

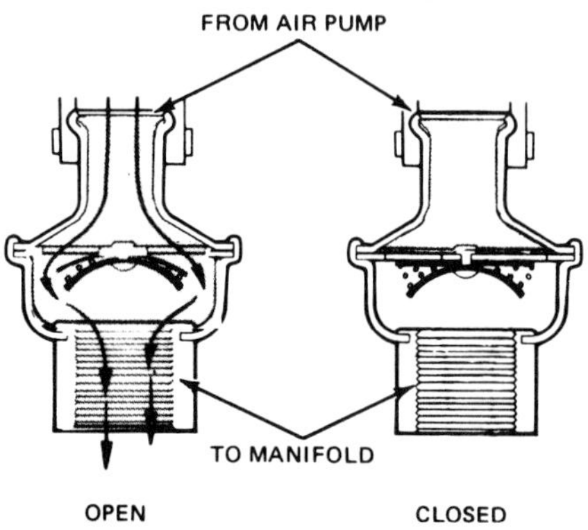


**Figure 18-43.** The air injection check valve closes to protect the air system when high pressure occurs in the exhaust. (Ford)

## Antibackfire Valves

During deceleration, high manifold vacuum creates a rich air-fuel mixture. Part of this mixture passes through the cylinders unburned and richens the exhaust. If air were injected into the exhaust during deceleration, it would combine with the excess unburned gasoline and cause a backfire. To prevent this, air injection is shut off during deceleration by an antibackfire valve. These valves are diaphragm-operated valves, controlled by manifold vacuum. There are two basic kinds, identified by the direction in which they redirect airflow: a **gulp valve** and a **diverter valve**.

**Gulp Valves.** Gulp valves were used on many early air injection systems but gradually gave way to diverter valves on most later systems. You still will find them on some late-model cars, however. A gulp valve has an inlet air connection to the air pump and an outlet connection to the intake manifold, figure 18-44. This arrangement puts it in parallel with the main air delivery from the pump to the exhaust system. A spring under the valve diaphragm normally holds the valve closed. The diaphragm housing has a vacuum port connected to the intake manifold. When manifold vacuum is high during deceleration, it moves the diaphragm against spring pressure. This opens the gulp valve and allows air to flow through the air outlet to the intake manifold. Air injection is redirected into the intake manifold to lean the mixture and prevent backfiring.

Air bleed holes, or balance restrictions, in the gulp valve diaphragm bleed off vacuum after a few seconds and allow the valve to close. Gulp valves can open when manifold vacuum is high other than during deceleration. If the valve opens at idle, for example, it can cause hard starting and rough idle. At high speed, manifold vacuum may not be high enough, even with a closed throttle, to open the valve quickly. This can allow backfiring.

**Diverter Valves.** A diverter valve also redirects air pump output away from the exhaust during deceleration. A diverter valve is connected in series with the air pump output and the air distribution manifolds. Air passes through it at all times on the way to the exhaust system, figure 18-45. Like a gulp valve, a diverter valve uses a diaphragm controlled by a spring and manifold vacuum to redirect airflow. The spring normally holds the valve so that airflow passes through the outlet to the exhaust system. High vacuum during deceleration moves the diaphragm and valve to close the outlet to the exhaust and open an outlet to the atmosphere. Most diverter valves have an outlet muffler to reduce air noise during deceleration. Or, the valve diverts air to the engine air cleaner. Some diverter valves have the pressure relief valve

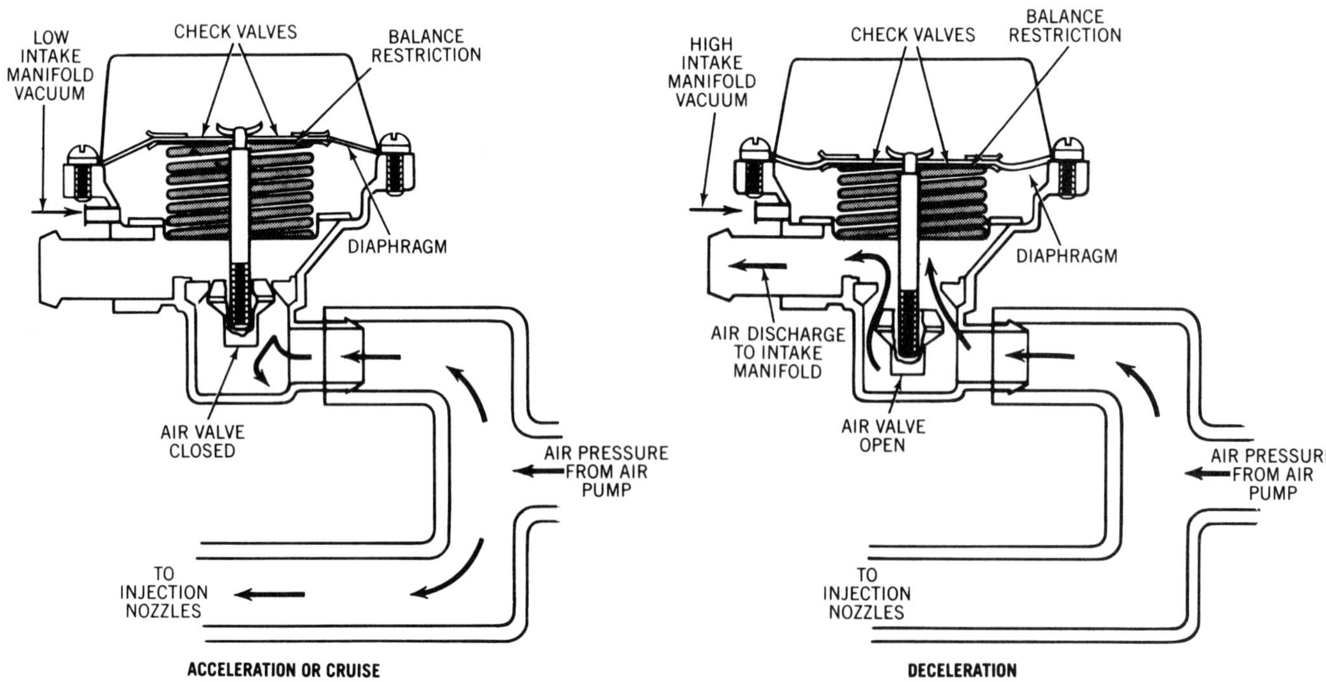


**Figure 18-44.** A gulp valve redirects air injection to the intake manifold during deceleration.

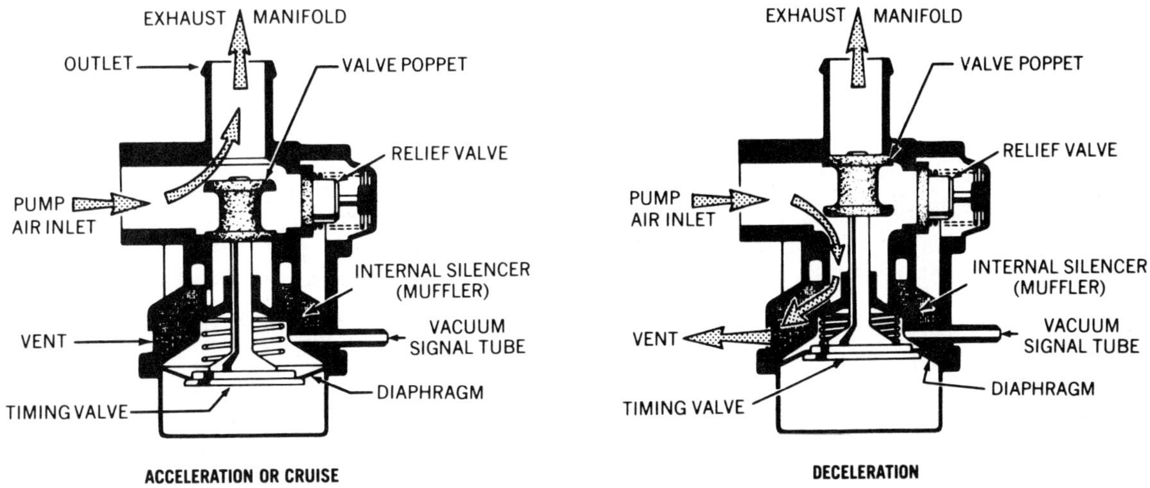


**Figure 18-45.** A diverter valve diverts air injection from the exhaust system to the atmosphere during deceleration. (GM)

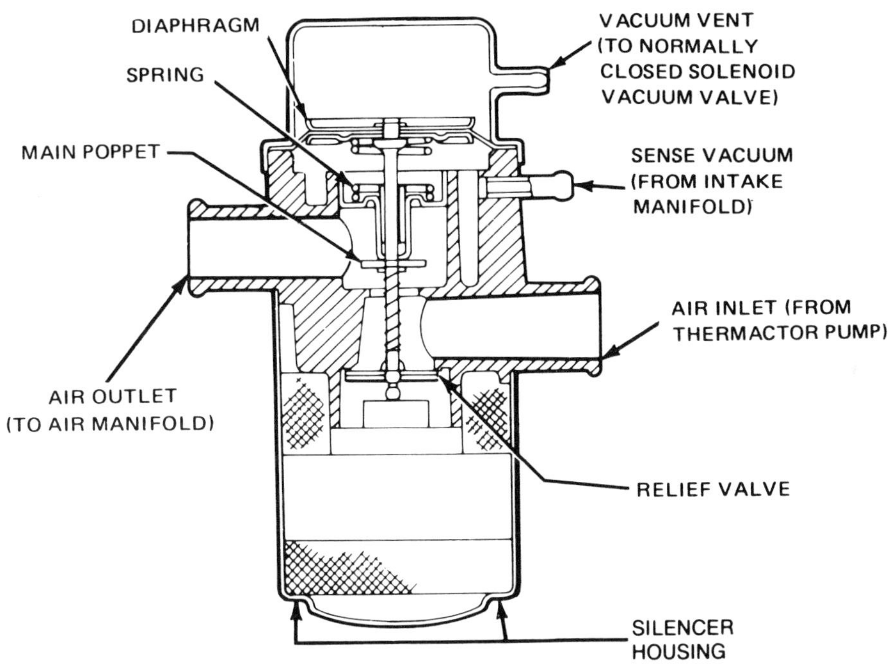


**Figure 18-46.** Ford calls this diverter valve a timed air bypass valve with a vacuum vent. (Ford)

for the air injection system. A diverter valve does not affect the intake air-fuel mixture and is more trouble free than a gulp valve.

## Air Control Variations

As with other emission control systems, carmakers fine-tuned air injection systems to provide the best combination of driveability and emission control for different engines. This led to many variations on the simple antibackfire valves described above. With the introduction of oxidation catalytic converters in 1975, carmakers had to divert air injection for conditions other than deceleration. During cold engine warmup, the rich air-fuel mixture creates high HC in the exhaust. Air injection for some engines had to be diverted during warmup to prevent catalyst overheating. Similarly, prolonged idling can overheat the catalyst. This, too, requires diverting air injection on some engines.

Ford Motor Company has used several different diverter valves, along with combinations of supplementary vacuum differential valves, vacuum solenoids, idle valves, and vacuum vent valves. These provide brief examples of system variations used to fine tune air injection.

Ford calls many of its diverter valves, "bypass valves." Figure 18-46 shows a timed air bypass valve with a second chamber above the diaphragm. This can be connected to a solenoid valve that opens the chamber to atmospheric air during warmup or create a timed vacuum bleed for diverter action during deceleration. The pressure relief portion of this bypass valve is a separate large disc at the bottom of the valve stem. Differential air pressure on the large disc relieves excess air to the atmosphere. Other systems with a timed air bypass valve use an idle vacuum vent valve, figure 18-47. This releases vacuum from the upper chamber of the bypass valve during idle, which causes air injection to be diverted.

Figure 18-48 shows a system with a bypass valve that uses vacuum to *hold the valve open* for air injection. Vacuum passes through the air cleaner thermostatic vacuum switch (TVS). At cold temperature, the TVS blocks vacuum, spring pressure closes the bypass valve, and air injection is diverted. During prolonged idle and deceleration, ported vacuum is blocked by the throttle. Vacuum bleeds down through the delay valve, and the bypass valve diverts air injection.

The bypass valve in figure 18-48 is used in some systems with a vacuum differential valve (VDV), figure 18-49. The VDV dumps vacuum quickly during deceleration and allows the bypass valve to divert air pump output to the atmosphere.

These air injection control variations used during the mid-1970's were an introduction to the more sophisticated air-switching and air control systems used with oxidation and reduction catalysts.

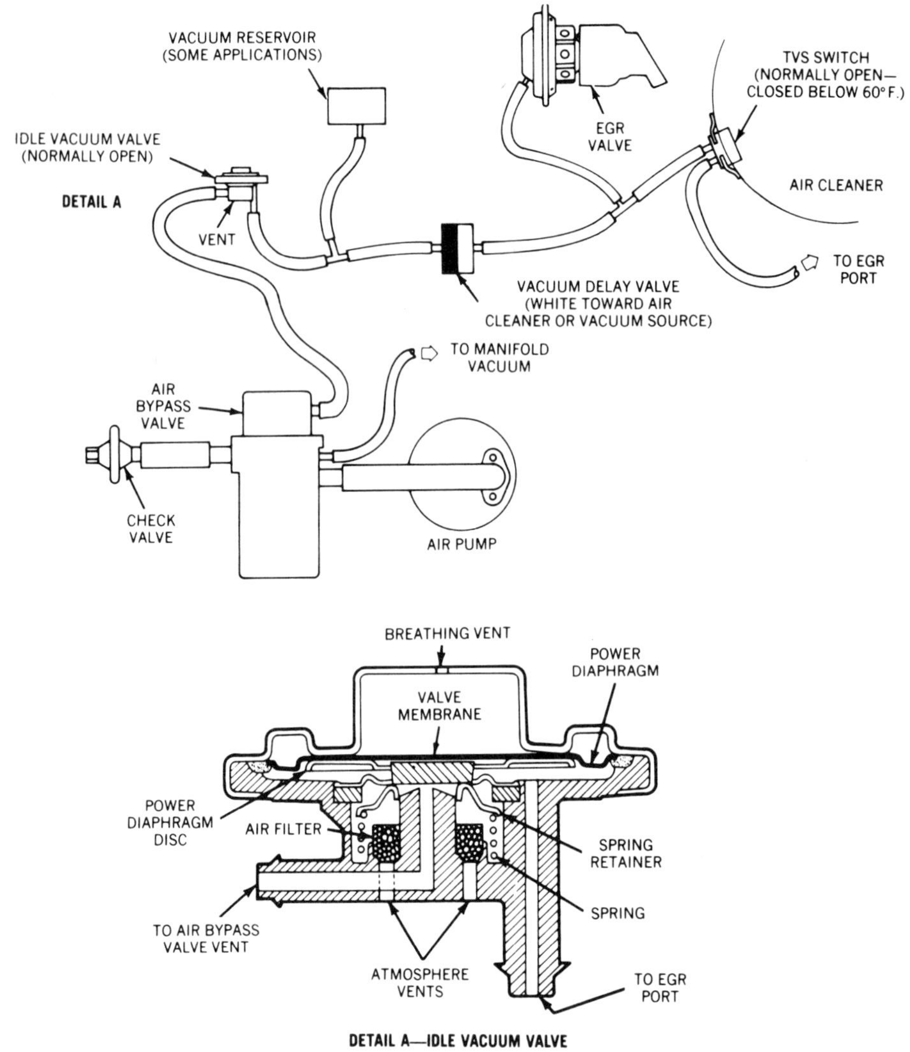


**Figure 18-47.** The idle vacuum valve allows the air bypass valve to divert air pump output at idle. (Ford)

## Air Injection with Catalytic Converters

The introduction of oxidation (HC and CO) converters in 1975 and reduction converters ($NO_x$) in 1977 complicated air injection systems. As with spark timing controls and EGR, we haven't the room in this book to detail all of the various valves that carmakers have

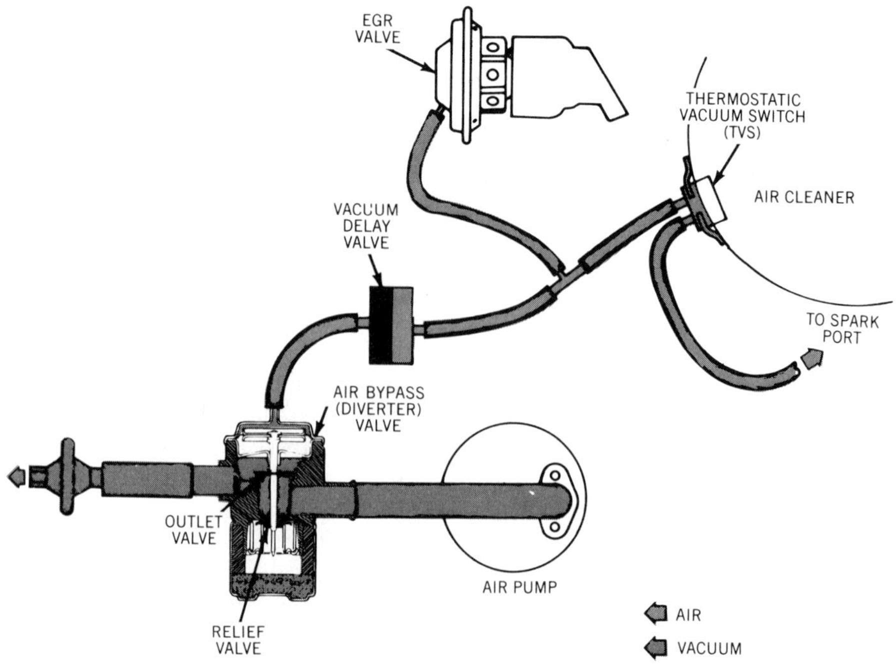


**Figure 18-48.** The TVS and delay valve control vacuum to this diverter (bypass) valve. (Ford)

used to control air injection on late-model cars. But, if you understand the common goals of all systems, you can figure out each carmaker's version. Many air injection systems on late-model cars have the control variations described in the preceding section. Additionally, air injection assists the catalysts and emission control in general for the following reasons:

1. Catalysts—either oxidation or reduction—work best when heated to 400° to 500° F (204° to 260° C) and higher. Extra air in the exhaust during engine warmup raises the temperature and heats the converter faster.

2. Catalysts can be damaged by overly rich HC and CO concentrations in the exhaust. Large amounts of HC and CO are present in the exhaust of a cold engine (with the choke closed). Extra air in the exhaust dilutes the HC and CO.

3. Exhaust oxygen sensors, used with electronic engine controls, must heat up like catalysts in order to work. Extra air in the exhaust during warmup speeds this process.

These are the three reasons why late-model air injection systems deliver air to an upstream point in the exhaust during warmup. Duration of warmup air injection must be controlled carefully, however, to prevent catalyst overheating. The computers of electronic engine control systems monitor both time and engine temperature to do this.

During warmup, air usually enters the exhaust near the manifold outlet, ahead of the exhaust oxygen sensor, figure 18-50. When the engine reaches operating temperature, air injection requirements change:

1. The exhaust oxygen sensor will send false signals to the engine computer if extra air is added to the exhaust leaving the cylinders.

2. The $NO_x$ reduction converter must separate oxygen from the $NO_x$ compounds. Extra air will work against this reaction.

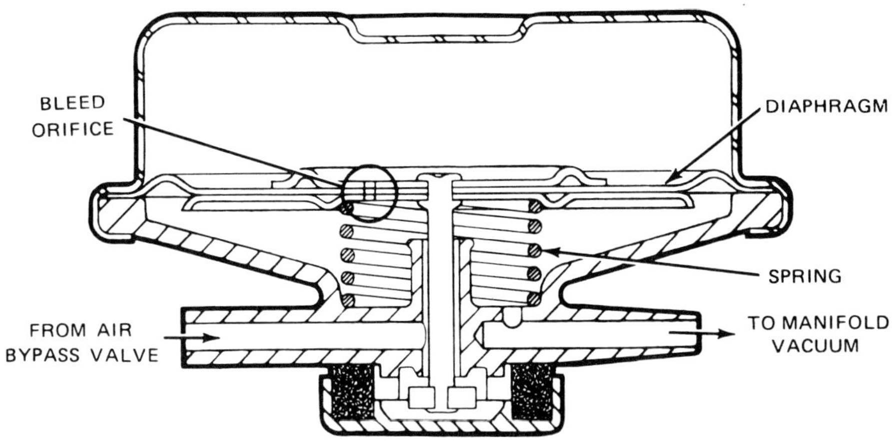


**Figure 18-49.** This vacuum differential valve dumps vacuum to the bypass valve during deceleration. (Ford)

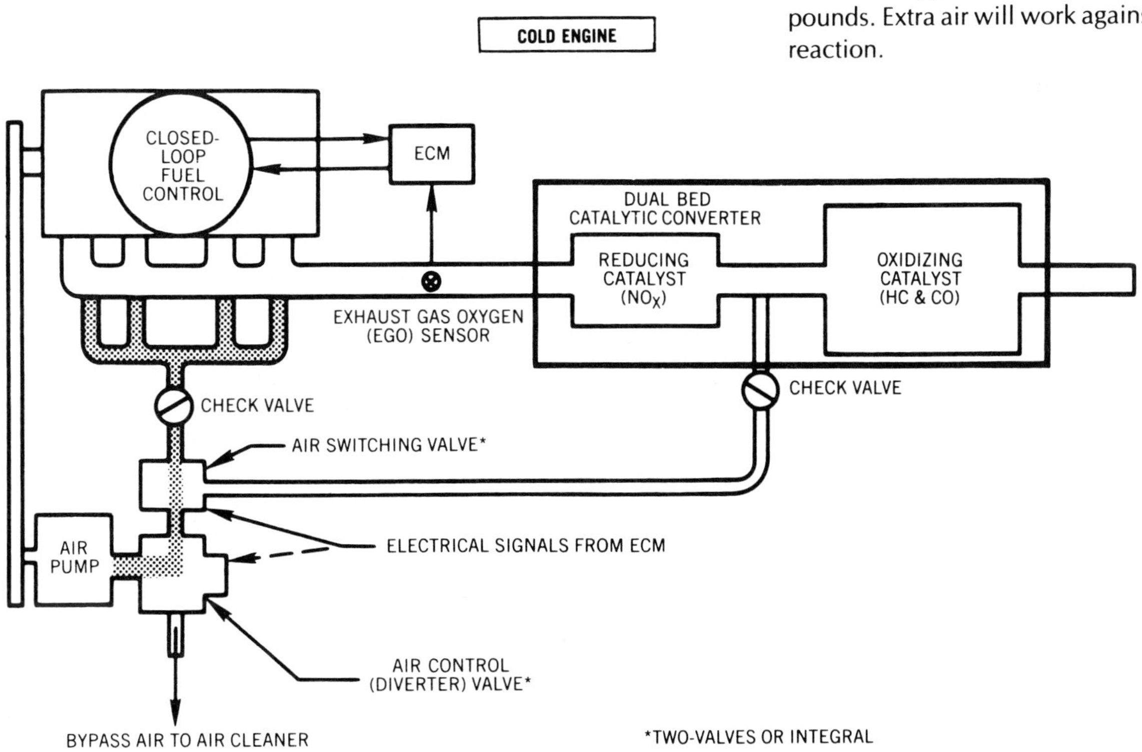


**Figure 18-50.** Air is injected ahead (upstream) of the exhaust oxygen sensor and both catalysts when the engine is cold. (Cadillac)

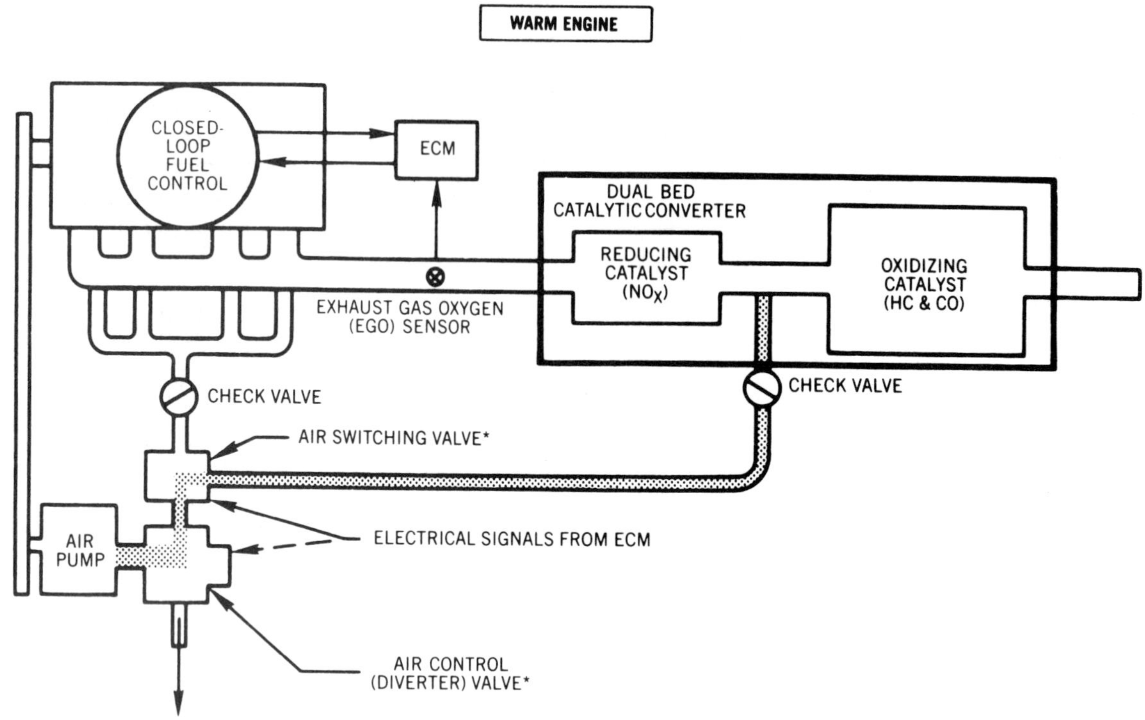


**Figure 18-51.** Air is injected after (downstream) the exhaust oxygen sensor and the reduction catalyst when the engine is warm. (Cadillac)

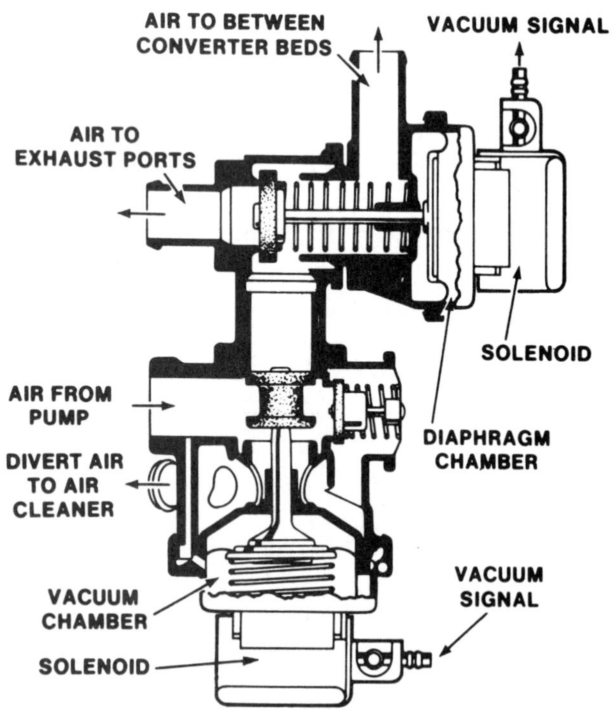


**Figure 18-52.** This EDES valve does several air switching and diverting jobs on a late-model air injection system. (GM)

**3.** The HC and CO oxidation converter will continue to be most efficient with extra air in the exhaust.

For these reasons, air injection must be switched to a downstream point in the exhaust when the engine is warm. Air enters the exhaust *after* the $O_2$ sensor and $NO_x$ reduction catalyst and *before* the HC and CO oxidation catalyst, figure 18-51. Some cars have separate reduction and oxidation converters with an air inlet tube in the pipe between them. Other cars have both catalysts in one housing with an air inlet between the two catalysts.

Additionally, air must be diverted from the exhaust entirely during deceleration and extended high-speed or heavy-load operation. This is necessary to keep the catalysts from overheating. Carmakers use a wide variety of control valves on late-model air injection systems, figure 18-52. GM, alone, has 16 different air control valves for different applications of the computer command control (CCC) engine control system. Air control valves are regulated by:

- Vacuum diaphragms connected to manifold vacuum or ported vacuum
- Air pump output pressure, working against vacuum or a spring in a diaphragm chamber
- Solenoids in the vacuum line to open or close the vacuum supply to a diaphragm.

All air control valves include the function of a simple diverter valve or gulp valve to shut off air injection during deceleration. Air control valves also can switch air injection from upstream to downstream in the exhaust system, or shut air off entirely at sustained high speeds or heavy loads.

### Pulse Air Injection

Since 1975, several carmakers have used **pulse air injection** systems that have no air pump. These systems use pulse air valves that are similar to check valves. The valve is a spring-loaded diaphragm or reed valve, connected by tubing to the exhaust port for each cylinder, figure 18-53. When an engine exhaust valve closes, the inertia of the moving exhaust in the manifold creates a brief low-pressure pulse (slight vacuum). This vacuum allows atmospheric air pressure to open the pulse air valve so that air can enter the exhaust, figure 18-54. When exhaust pressure rises again, it forces the pulse air valve closed, as a check valve. This action provides "free" air injection, with no power-consuming air pump.

Pulse air valves must be connected to the exhaust system upstream, where negative pressure pulses are strongest. Therefore, the air delivery can't be switched downstream between reduction and oxidation catalysts. Pulse air systems are most effective on cars with *only* an HC-CO oxidation catalyst. A few cars with 3-way catalysts, however, use pulse air systems for extra air in the exhaust of a cold engine. These systems have valves in the air inlet supply to the pulse air valves to close the air supply when the engine is warm.

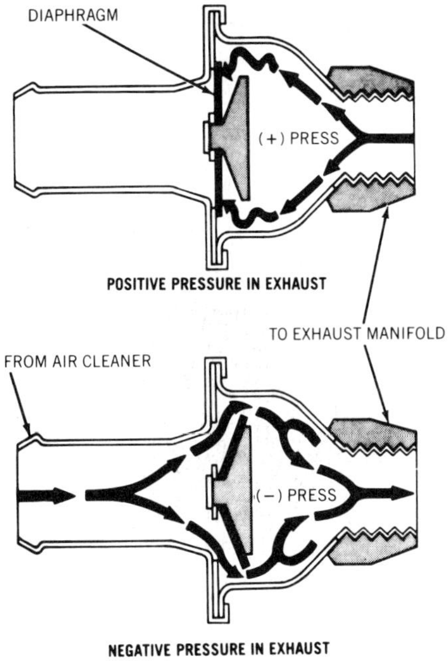


**Figure 18-54.** Pulse air systems draw air through a valve that opens when a negative pressure pulse (slight vacuum) occurs in the exhaust. (Chrysler)

## CATALYTIC CONVERTERS

You have read about catalytic converters as they relate to other emission control devices, particularly air injection. You also have read about catalytic converters as they relate to unleaded gasoline and fuel systems in general, and you have learned about basic oxidation and reduction reactions elsewhere in this book. Catalytic converters, like other emission controls, modify the basic combustion process. Since 1975, they have been the major emission control devices on most cars. As such, they affect all other combustion control systems. Now, we can take a closer look at converter construction and operation.

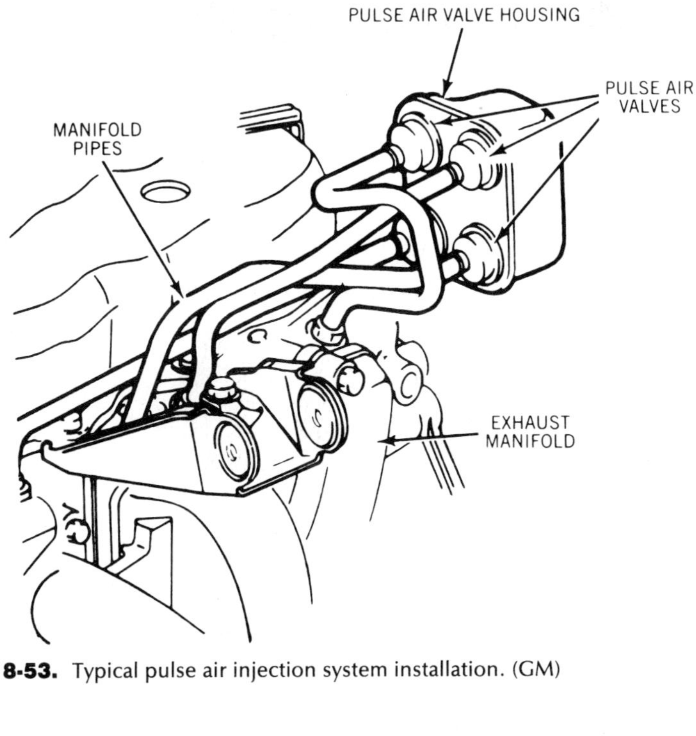


**Figure 18-53.** Typical pulse air injection system installation. (GM)

### Catalytic Reactions

A catalyst is an element or compound that promotes a chemical reaction without being affected by the reaction. An oxidation catalyst promotes oxidation; a reduction catalyst promotes reduction. Oxidation catalysts are **platinum** and **palladium**. The most common reduction catalysts are platinum and **rhodium**. These elements are called **noble metals**.

All exhaust catalysts work best when hot. In fact, they don't work at all when cold. You learned earlier that a catalytic converter must heat up to 400°

to 500° F (204° to 260° C) to start working. Engineers call this the **light-off temperature**, at which the catalyst is 50-percent effective. A fully hot converter reaches temperatures of 900° to 1,600° F (480° to 870° C) for complete efficiency. These temperatures are almost as hot as the cylinder combustion temperatures. They must be to modify the combustion process. Moreover, a converter remains hot long after the engine is shut down. This is why converters on automobiles have heat shields to protect the car body and anything underneath from the high temperatures. In spite of the heat, the catalytic reactions do not produce the open flame of a simple burning reaction.

The converters do not contain pure platinum, palladium, and rhodium. Rather these materials are deposited on ceramic **substrates** to create a large surface area of catalyst for contact with the exhaust. Exhaust flows through the converter housing and over the catalyst-coated substrate.

### Converter Installations

Most cars with only an oxidation catalyst have one converter in the exhaust system, below the passenger compartment, figure 18-55. If the converter were farther back in the exhaust system, gases would cool too much for effective oxidation. Some cars (notably a few Ford and Chrysler products of the late 1970's) have a small "miniconverter" attached to the outlet of the exhaust manifold, figure 18-56. This oxidation converter has a small catalyst surface area and is closer to the engine. It heats up fast and starts the HC and CO oxidation sooner with a cold engine. A large main converter under the floorboards completes the reaction.

Converter systems with both oxidation and reduction catalysts are called **2-stage or 3-way catalytic converter systems**. The $NO_x$ reduction catalyst is always ahead of the oxidation catalyst in the system because it works best with less air and a richer HC and CO content in the exhaust. Air injection switching, which you studied earlier, allows air to be di-

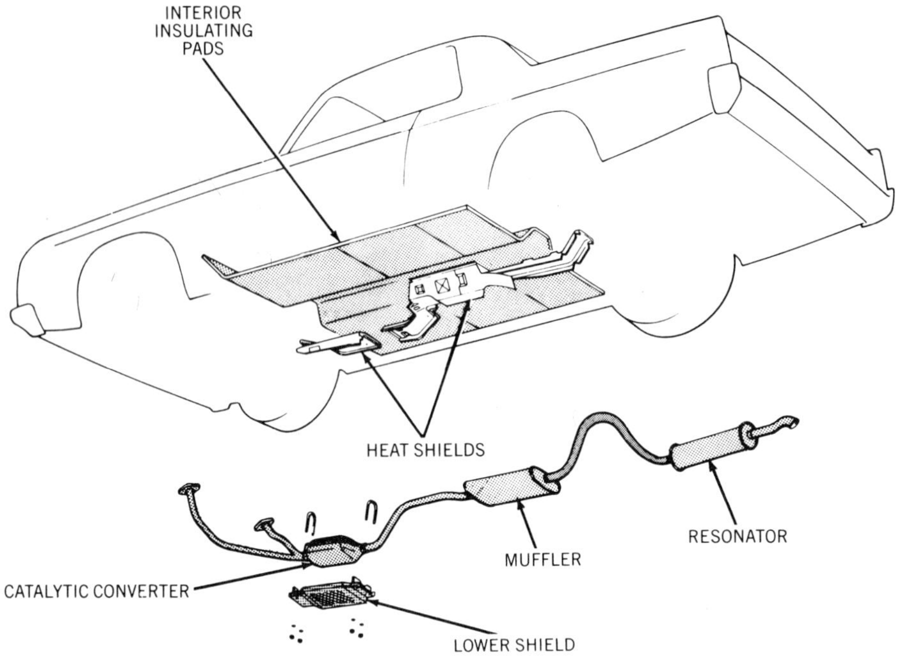


**Figure 18-55.** Most cars with an oxidation converter have one main converter under the floor. (Chrysler)

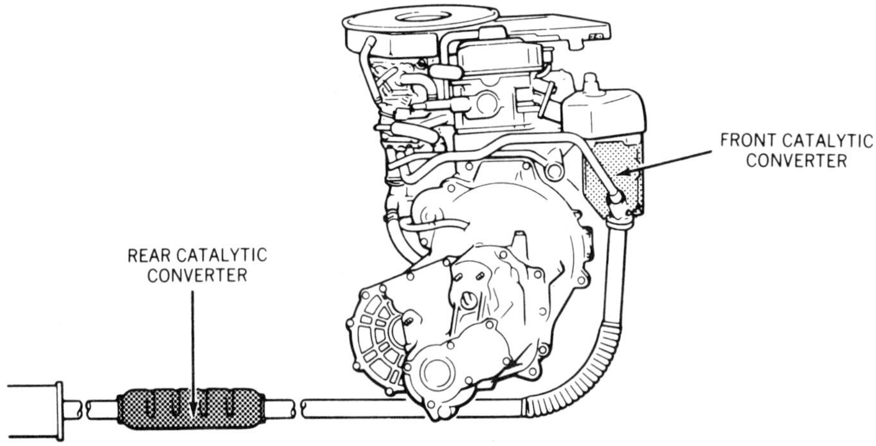


**Figure 18-56.** Some cars have a small "miniconverter" ahead of the main catalyst to start the reaction sooner. (Chrysler)

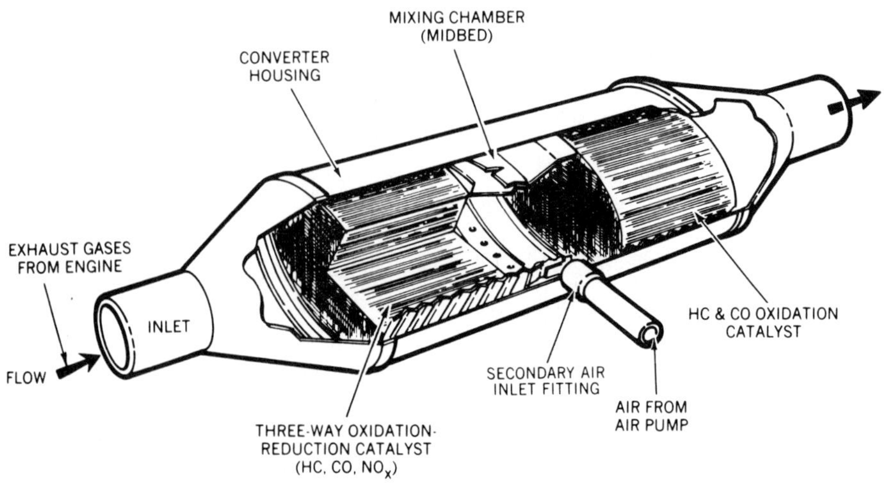


**Figure 18-57.** The reduction catalyst is always upstream from the oxidation catalyst, whether in a separate housing or the same housing. (Ford)

verted around the reduction catalyst to the oxidation catalyst after the converters warm up. The two catalysts can be in separate converter housings, or they can be in separate chambers of a single housing, figure 18-57.

On many 3-way (2-stage) systems, the first catalyst in the system is a **hybrid**, reduction and oxidation catalyst. It works best to reduce $NO_x$ when the CO content in the exhaust is between 0.8 and 1.5 percent. The second converter is strictly an oxidation catalyst to complete the HC and CO conversion. Figure 18-58 shows the 2-stage reaction.

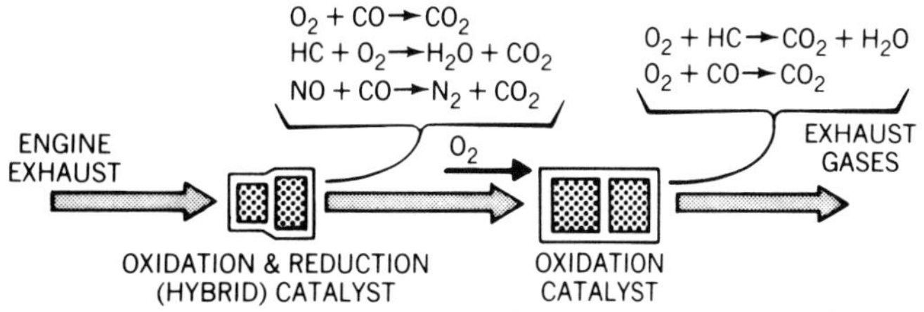


**Figure 18-58.** Catalytic reduction and oxidation reactions occur in this order.

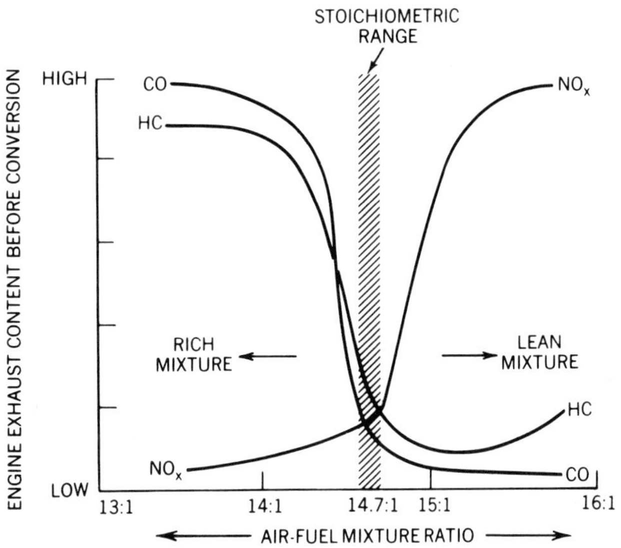


**Figure 18-59.** Three-way (2-stage) converters control emissions most effectively with a 14.7 : 1 air-fuel ratio.

## Stoichiometric Air-Fuel Ratio

Oxidation and reduction converter systems work best when the engine air-fuel ratio is regulated at 14.7:1. This stoichiometric ratio provides the most complete combustion of gasoline and air and the fewest harmful byproducts. Figure 18-59 shows that HC and CO emissions are high at air-fuel ratios richer than 14.7:1. $NO_x$ emissions increase rapidly with ratios leaner than 14.7:1. Therefore, a 3-way converter system has an exhaust oxygen sensor to measure oxygen in the exhaust. The voltage signal from this sensor goes to an engine control computer. The computer then controls the air-fuel mixture for the needed 14.7:1 ratio. All 3-way converter systems are used with an air-fuel metering feedback system or electronic engine controls, which you will learn about in a later chapter.

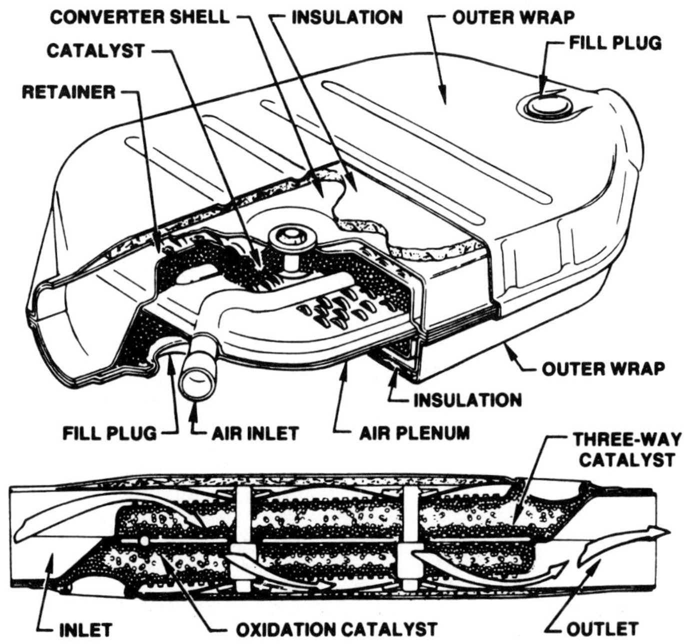


**Figure 18-60.** The pellet-type converter is used on many GM cars. (GM)

## Converter Design

There are two kinds of converter design: pellet type and monolithic. In either kind, the oxidation and reduction catalysts are on ceramic substrates in a steel housing, or shell. The AC Division of General Motors makes **pellet-type converters** used on many GM and on AMC cars of the 1970's. The catalyst metals are deposited on ceramic pellets, which are placed in beds in the converter housing, figure 18-60. All GM early oxidation converters were the pellet type. Late-model GM cars with 3-way catalysts have

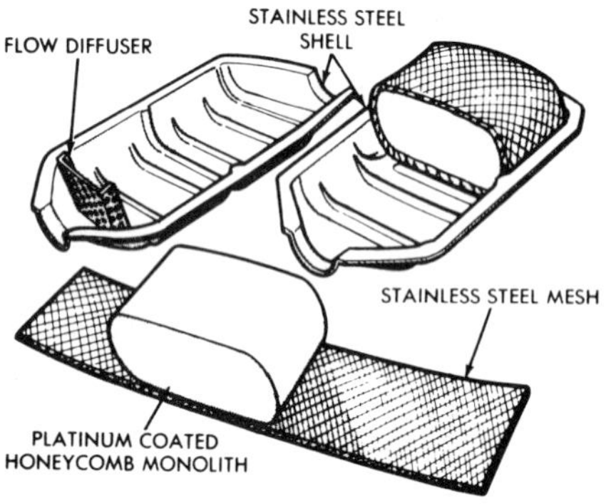


**Figure 18-61.** In a monolithic converter, the catalyst is on a honeycomb core. (Chrysler)

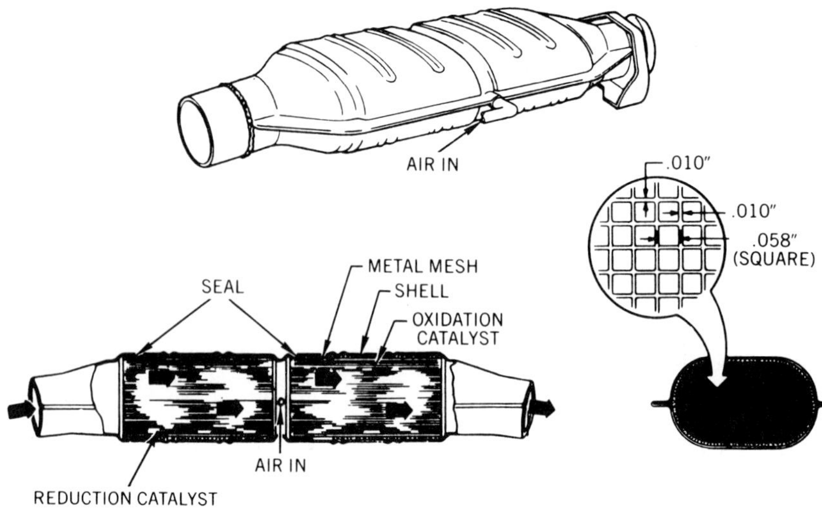


**Figure 18-62.** Exhaust flow through a monolithic converter with reduction and oxidation catalysts in the same housing. (GM)

either pellet-type or monolithic converters.

In a **monolithic converter**, the catalysts are deposited on a large ceramic honeycomb block, or "biscuit," which is installed in the converter shell, figure 18-61. Exhaust flows through the pores of the ceramic monolith, which provides a large surface area for the catalytic reaction. If a monolithic converter fails or wears out, the entire converter must be replaced. The pellets in GM pellet-type converters can be removed and replaced in the shell.

Most late-model converters, pellet type and monolithic, have oxidation and reduction catalysts in a single housing, because it is the most economical and efficient design. Exhaust first passes through the reduction catalyst pellets or monolith, then through the oxidation catalyst, figures 18-60 and 18-62. Space exists between the two catalysts for downstream air injection to the reduction catalyst after the system reaches operating temperature.

### Converter Operation

Catalytic converters have no moving parts and require no periodic service. Carmakers warrant catalyst effectiveness for 50,000 miles or 5 years, as required by law. Eventually, catalysts can wear out, however. When they do, the pellets, or the entire converter, must be replaced for proper emission control.

Converters can be damaged, however, before the catalyst wears out. Lead emissions from leaded gasoline can coat the catalysts and reduce their effect. Continued use of leaded fuel can destroy a converter, but a small amount of lead in the exhaust will eventually burn off of the catalyst. The converter then resumes almost normal operation.

Even though a converter operates at high temperature, excessive temperatures can damage it. The most common cause of catalyst overheating is too much unburned fuel reaching the converter. This happens when a cylinder misfires. It also happens if the air-fuel mixture is continually too rich. Long periods of idling (with a rich mixture) also raise converter temperatures.

Here are basic precautions that will help avoid catalyst damage or overheating:

**1.** Use only unleaded gasoline in a converter-equipped vehicle.

**2.** Do not try to start an engine with a dead battery by pushing the car for a compression start. Crank the engine with a booster battery and jumper cables.

**3.** Don't turn off the ignition while a vehicle is moving.

**4.** Don't crank an engine for more than 60 seconds when it is flooded or firing irregularly.

**5.** Do not disconnect (ground or short circuit) spark plugs to test an ignition system for more than 30 seconds.

**6.** Fix engine performance problems, such as dieseling and misfiring, as soon as possible.

## SUMMARY

Exhaust emission control requires changes in engine combustion. Some emission systems, such as EGR and spark timing controls, modify combustion as it takes place. Others, such as air injection and catalytic converters, affect the exhaust after combustion. All have the same goals: to reduce HC, CO, and $NO_x$.

Spark timing controls retard timing under acceleration, low speeds, and moderate loads—when combustion temperatures are high—to cut down HC and $NO_x$ emissions. EGR recirculates exhaust to the intake manifold to dilute the air-fuel mixture. This lowers temperatures and reduces $NO_x$ formation. EGR systems have controls so that exhaust is recirculated only when $NO_x$ formation is highest.

Early air injection systems injected air into exhaust manifolds or exhaust ports to continue HC and CO oxidation at a point where exhaust gases are hottest. Late-model air injection systems aid the catalytic converters. They are not the principal HC and CO control systems. On a converter-equipped car, air is injected into the manifold outlet or the exhaust pipe to assist the catalytic reaction and help heat the converter. On a car with a reduction catalyst, air is switched downstream when the engine is warm.

Catalytic converters use reduction catalysts to reduce $NO_x$ emissions and oxidation catalysts to control HC and CO. The reduction catalyst is always installed ahead of the oxidation catalyst. All converters must warm up above 400° F (204° C) to be effective. Cars with 2-stage, 3-way converters all have fuel-metering feedback control systems to maintain a 14.7:1 air-fuel ratio.

Early emission controls used electromechanical switches and solenoids and vacuum valves to control spark timing, EGR, and air injection. Late-model combustion control systems employ the same principles and have the same goals as earlier systems, but they use different devices. Electronic engine control computers regulate these functions more precisely than electromechanical switches, relays, solenoids, and vacuum valves ever could. You will see how in later chapters.

## REVIEW QUESTIONS

### Multiple Choice

**1.** Mechanic A says that EGR and spark timing modifications change the combustion process while it is taking place. Mechanic B says that air injection and catalytic converters affect combustion after exhaust byproducts have formed. Who is right?

**a.** A only
**b.** B only
**c.** both A and B
**d.** neither A nor B

**2.** Originally, the principal reasons for spark timing control modifications were to control:

**a.** CO emissions
**b.** CO and HC emissions
**c.** $NO_x$ and HC emissions
**d.** particulate and CO emissions

**3.** Mechanic A says that high HC emissions can be caused by a misfiring spark plug. Mechanic B says that HC emissions are high during deceleration. Who is right?

**a.** A only
**b.** B only
**c.** both A and B
**d.** neither A nor B

**4.** Mechanic A says that CO emissions are high with a lean mixture. Mechanic B says that $NO_x$ emissions are high with a lean mixture and high temperature. Who is right?

**a.** A only
**b.** B only
**c.** both A and B
**d.** neither A nor B

**5.** Mechanic A says that early spark timing control systems modified vacuum advance. Mechanic B says that early spark timing controls worked to control CO emissions. Who is right?

**a.** A only
**b.** B only
**c.** both A and B
**d.** neither A nor B

**6.** Mechanic A says that a distributor retard solenoid retards spark advance during deceleration. Mechanic B says that a distributor advance solenoid advances timing during starting. Who is right?

**a.** A only
**b.** B only
**c.** both A and B
**d.** neither A nor B

**7.** All of the following are general kinds of vacuum-advance modifications EXCEPT:

**a.** speed- and transmission-controlled timing
**b.** "loadamatic" distributor control
**c.** vacuum-delay valves
**d.** distributor vacuum-retard diaphragms

**8.** Mechanic A says that spark-delay valves work the same way as choke-delay valves. Mechanic B says that an orifice spark advance control (OSAC) valve is a kind of spark-delay valve. Who is right?

**a.** A only
**b.** B only
**c.** both A and B
**d.** neither A nor B

**9.** Exhaust gas recirculation (EGR) systems control emissions of:

**a.** CO
**b.** particulates
**c.** HC
**d.** $NO_x$

**10.** In a gasoline engine, EGR should be off at:

**a.** low speed and low temperature
**b.** wide-open throttle
**c.** idle
**d.** all of the above

**11.** Mechanic A says that EGR should be turned on at full engine load for a diesel. Mechanic B says that EGR should be on at cruising speed for a gasoline engine. Who is right?

**a.** A only
**b.** B only
**c.** both A and B
**d.** neither A nor B

**12.** Late-model air injection systems may have:

**a.** gulp valves
**b.** diverter valves
**c.** air-switching valves
**d.** all of the above

**13.** Mechanic A says that air-switching systems aid catalytic converter warmup. Mechanic B says that air-switching systems aid HC and CO control in the oxidation converter. Who is right?

**a.** A only
**b.** B only
**c.** both A and B
**d.** neither A nor B

**14.** All of the following are reasons for air-switching systems on late-model cars EXCEPT:

- **a.** to cool the exhaust gas oxygen (EGO) sensor
- **b.** to dilute the HC and CO in the exhaust of a cold engine
- **c.** to aid warmup of oxidation and reduction catalysts
- **d.** to aid HC and CO oxidation in the catalytic converter

**15.** All of the following are catalyst elements used in catalytic converters EXCEPT:

- **a.** rhodium
- **b.** germanium
- **c.** platinum
- **d.** palladium

**16.** Mechanic A says that exhaust catalysts work best when cooled by air injection. Mechanic B says that catalysts work best below 400°F (204° C). Who is right?

- **a.** A only
- **b.** B only
- **c.** both A and B
- **d.** neither A nor B

**17.** Mechanic A says that the HC-CO oxidation converter is always ahead of the $NO_x$ reduction converter in the exhaust system. Mechanic B says that the oxidation converter removes oxygen from the HC and CO compounds. Who is right?

- **a.** A only
- **b.** B only
- **c.** both A and B
- **d.** neither A nor B

**18.** The stoichiometric air-fuel ratio for a gasoline engine is approximately:

- **a.** 12.5:1
- **b.** 13.6:1
- **c.** 14.7:1
- **d.** 16:1

**19.** Exhaust catalysts can be damaged by:

- **a.** overheating
- **b.** an overly rich air-fuel mixture
- **c.** lead additives in gasoline
- **d.** all of the above

### Fill in the Blank

**20.** Timing can be advanced at high speed under ________ load to allow an engine to burn a lean mixture.

**21.** Lean mixtures take ________ to burn than rich mixtures.

**22.** Maximum spark advance and high combustion temperatures and pressures increase the formation of ________ emissions.

**23.** With retarded, or delayed, ignition timing, high temperatures occur ________ in the combustion process.

**24.** Spark timing controls of the early 1970's delayed vacuum advance at ________ and ________ speeds and allowed vacuum advance at ________ speeds.

**25.** Most transmission-controlled spark systems use a normally ________ transmission switch and a normally ________ vacuum solenoid.

**26.** If a part fails in a transmission-controlled spark system built after mid-1973, vacuum advance is usually ________ at all times.

**27.** Many late-model engines have a ________ transducer to control EGR operation.

**28.** Besides reducing $NO_x$ formation, EGR systems also control ________ in a gasoline engine.

**29.** Air injection systems promote ________ of HC and CO in the exhaust.

**30.** A gulp valve diverts air injection to the ________ manifold. A diverter valve diverts air injection to the ________.

# 19 SUPERCHARGING AND TURBOCHARGING

## WHAT

Turbocharging is a form of supercharging, which is a way to increase the air charge that enters an engine. Supercharging and exhaust-driven turbochargers have been used on trucks, aircraft engines, and some passenger cars for decades. However, modern turbocharger installations have been refined greatly compared to earlier systems. This chapter outlines the principles of turbocharging used on many late-model automobiles. As you study this chapter, remember that these systems simply modify, or supplement, the basic internal combustion process.

## WHY

For many years, supercharging and turbocharging were used on high-performance and industrial engines. In the 1970's, however, engineers took a fresh look at these technologies as ways to reach the conflicting goals of good performance, fuel economy, and emission control. Today, every major carmaker offers turbocharging on one or more cars.

As a professional mechanic, you will have to service these systems as a regular part of your job. Turbochargers are commonplace, not exotic. They serve the same engine operating needs that you have already studied. You will find all of the principles of airflow, pressure, and fuel metering in these systems that you find in a normally aspirated carbureted engine.

## GOALS

After studying this chapter, you should be able to:

**1.** Explain the effects of supercharging on an internal combustion engine.

**2.** Explain turbocharger operation and identify the principal parts of a turbocharger.

**3.** Explain turbocharger boost control, mixture control and cooling, spark timing, and detonation control.

## SUPERCHARGING PRINCIPLES

All of the engines you have studied so far—gasoline and diesel—have been normally aspirated, or **naturally aspirated** engines. "Aspirated" describes how something breathes: human beings, animals, or engines. A naturally aspirated engine draws in air by the natural process of atmospheric pressure forcing air into the low-pressure area of the intake manifold. Low pressure (vacuum) is created in the manifold by the pumping action of the pistons. It is controlled by the valves and, in a gasoline engine, by the throttle.

Atmospheric pressure and the pumping action of the pistons do a fine job of allowing a basic engine to breathe. Remember, however, that one important measure of an engine's efficiency is its volumetric efficiency. This is a percentage figure that indicates the volume of air an engine draws in compared to the maximum it *could* draw in, which is its total displacement volume. A passenger car engine averages about 70 to 80 percent volumetric efficiency through all operating conditions. For example, an average 300-cid (5-liter) V-8 engine will draw in 210 to 240 cubic inches of air for every 8 intake strokes in a typical stop and go and cruising cycle.

A racing engine can be tuned for 100-percent volumetric efficiency at the peak of its power curve. Actually, intake and exhaust tuning can achieve better than 100-percent volumetric efficiency in some cases. Overall volumetric efficiency of any naturally aspirated engine, however, is limited by the density of the air that atmospheric pressure forces into the engine. Remember that atmospheric pressure and air density are greatest at sea level for any given temperature. At high altitude, atmospheric pressure and air density decrease. At sea level, atmospheric pressure is approximately 14.7 psi; at high altitude, it may be as low as 8 or 9 psi. Volumetric efficiency changes directly as air density changes. Atmospheric pressure also happens to change in the same general relationship.

If we could increase the pressure on the incoming air above atmospheric pressure, we could force more air into each cylinder during each intake stroke. More air would allow more fuel to be mixed with the air while still maintaining the desired air-to-fuel ratio. More fuel and air together equal more potential chemical energy to be released through combustion. All of this adds up to more power, and that is exactly what supercharging provides. **Supercharging** is the process of increasing the density of air entering an engine above that which can be drawn in by the natural pumping action. Supercharging allows a proportionally increased fuel volume and a denser air-fuel charge. Overall, the weight of the air-fuel charge is greater than for a naturally aspirated engine. This produces more power because power varies directly with the weight of the air-fuel charge consumed per unit of time.

### Boost Pressure

A supercharger increases intake air pressure above atmospheric pressure. The amount of increase is the **boost pressure**, or simply boost. If a supercharger provides 10 psi of boost, the engine operates with atmospheric pressure plus 10 psi. At sea level, that's 14.7 + 10 psi, or 24.7 psi total intake pressure. If atmospheric pressure is 10 psi at high altitude, the additional 10 psi of boost pressure provides 20-psi total intake pressure. Superchargers can be built with controls that allow increased boost at higher altitudes to make up for lower atmospheric pressure. That's one reason why a lot of supercharger development was done by the aircraft industry over 50 years ago.

### Supercharging and Compression

Supercharging the air-fuel mixture raises the engine compression pressure and temperature. High compression produces more power from a given amount of air and fuel than low compression can. A high compression ratio affects power and volumetric efficiency similarly to the way supercharging does. In the 1960's, compression ratios of some engines were as high as 10:1 or 11:1. But those were the days before $NO_x$ controls and unleaded gasoline.

A high compression ratio on a naturally aspirated engine has two benefits:

**1.** It improves volumetric efficiency because the piston displaces a larger percentage of total cylinder volume on each stroke.

**2.** It improves thermal efficiency because it raises compression temperatures and provides more nearly complete combustion.

High compression ratios also have disadvantages:

**1.** The compression ratio is a fixed part of engine design. Therefore, temperatures and pressures are high in all operating modes and can increase exhaust emissions during idle, deceleration, and choked or part-throttle operation.

**2.** High compression increases $NO_x$ production.

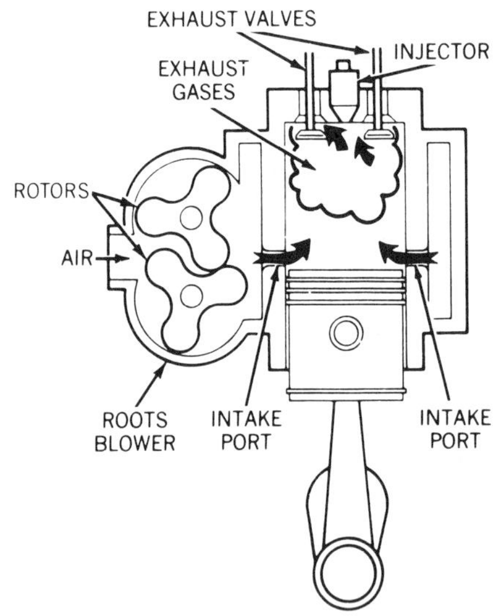


**Figure 19-1.** Roots blowers are positive-displacement air pumps used on 2-stroke diesels and as superchargers on some racing engines. (Ford)

**Figure 19-2.** This Roots blower installation on a late-model Z28 is typical of high-performance aftermarket systems. (B&M Automotive)

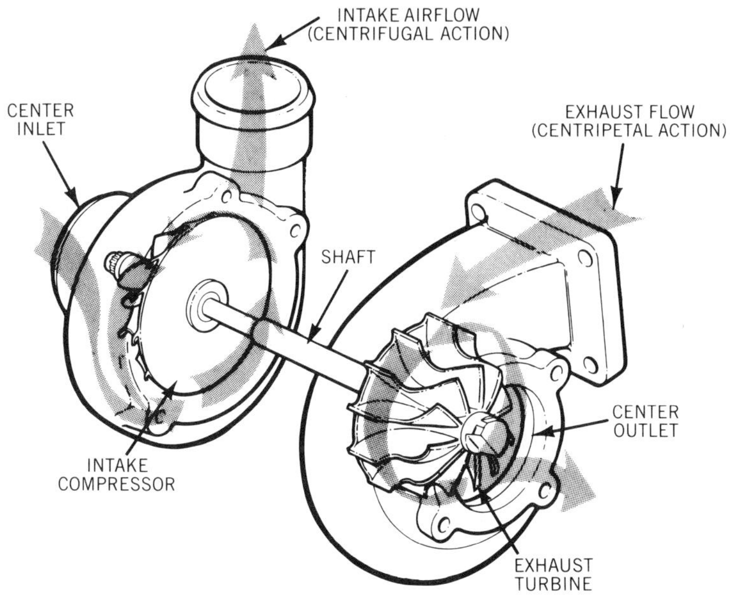


**Figure 19-3.** A centrifugal supercharger draws air through a central inlet and uses centrifugal force to accelerate the air and pump it through the outlet. Exhaust applies centripetal force to the turbine and flows inward to the center outlet.

**Figure 19-4.** This belt-driven centrifugal supercharger was a Ford factory option in the late 1950's.

**3.** High compression requires high-octane gasoline with large amounts of antiknock additives. Lead is the most efficient antiknock additive, but it damages catalytic converters and is a harmful emission, itself.

Since the early 1970's, emission control requirements have lowered compression ratios to the range of 8:1 to 9:1. An advantage of supercharging is that it works best with lower compression ratios. Supercharging can increase the air-fuel charge for high compression pressure when power is needed, and it allows the engine to operate with lower pressures when full power is not needed.

### Supercharging Methods

There are several supercharger types and drive methods. The supercharger can be:

- A positive-displacement pump that moves the same volume of air on each pumping cycle regardless of speed, or
- A variable-displacement pump that moves a greater volume of air as speed increases.

The most common positive-displacement supercharger is the lobe-type, or **Roots blower**, figure 19-1. They are used on 2-stroke diesel engines where they are not primary superchargers, but assist air intake and exhaust scavenging. Roots blowers also are used as superchargers on many high-performance gasoline engines, figure 19-2. Vane-type and axial-flow positive-displacement superchargers also have been built, but they are not very common. A positive-displacement supercharger draws in a large volume of air, compresses it into a small space, and then forces it at higher pressure through an outlet.

A centrifugal pump is the most common and efficient variable-displacement supercharger. Like any centrifugal pump, it draws the material to be pumped (air) through a central inlet. It then uses centrifugal force to move the air to the perimeter of the pump housing and force it through an outlet, figure 19-3. Because centrifugal force increases as speed increases, a centrifugal supercharger takes in more air and creates more boost pressure as speed increases.

Any supercharger—positive- or variable-displacement—can be driven by gears or belts from the engine crankshaft, figure 19-4. Many superchargers have been built—and still are built—with these mechanical drives. They have limitations, however. A mechanically driven supercharger can use up to 20 percent of an engine's power, even when supercharger boost is not needed. The efficiency of mechanically driven centrifugal superchargers is limited by engine speed. They require high overdrive ratios to get enough speed for adequate boost. This causes high gear or belt speeds and mechanical wear. There is another way to drive a supercharger that provides almost "free" power. That is the heat energy in an engine's exhaust.

Remember that almost 50 percent of combustion heat energy is lost in the engine exhaust. If we install a turbine drive wheel in the hot exhaust flow and connect it by a shaft to a centrifugal supercharger, we can put the exhaust heat energy to useful work. That is the principle of a **turbocharger**.

## TURBOCHARGER CONSTRUCTION AND OPERATION

A turbocharger is a simple and compact device with few moving parts. But, it is built to precise, close tolerances and operates under extreme heat and wear conditions. A turbocharger has two chambers with two rotating wheels connected by a shaft, figure 19-5. Exhaust passes from the engine manifold into the drive turbine housing. The turbocharger must be as close

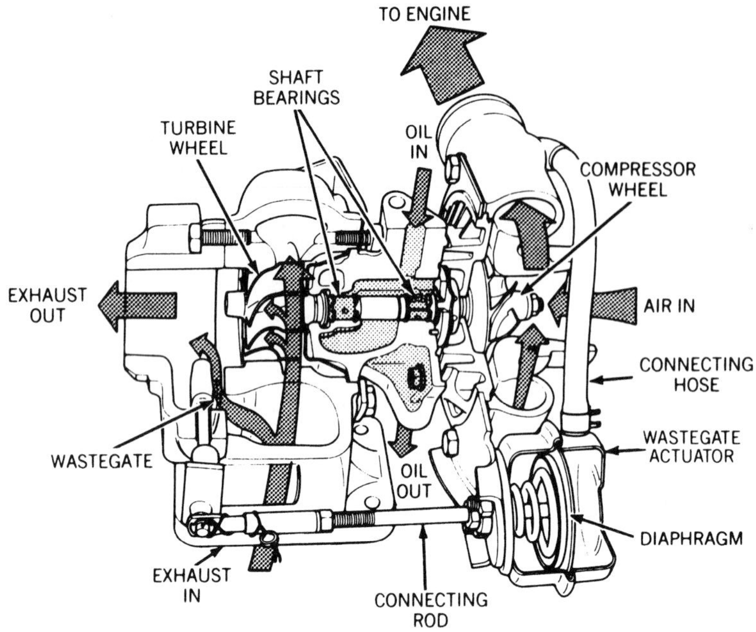


**Figure 19-5.** Basic turbocharger construction. The turbo must be as close as possible to the exhaust manifold for maximum heat energy. (Ford)

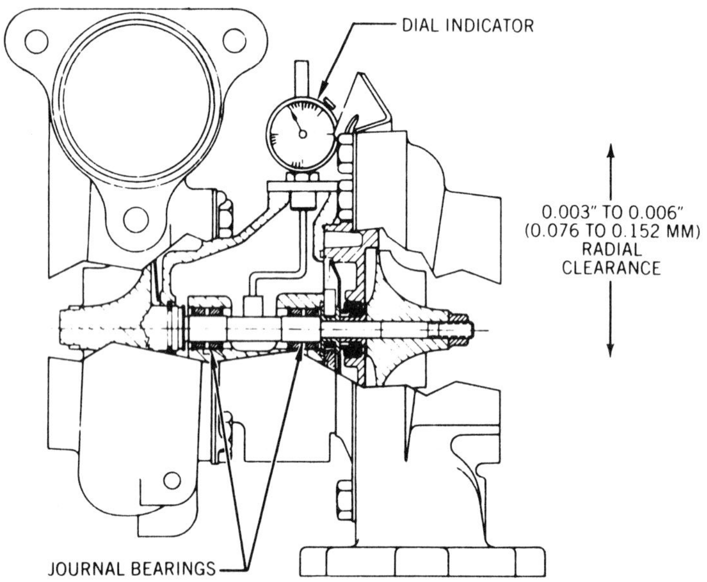


**Figure 19-6.** The radial clearance of shaft journal bearings is 0.003" to 0.006" (Garrett)

as possible to the engine to receive the hottest exhaust. It is the heat energy and expansion of the exhaust, more than the speed or pressure, that deliver rotating force to the turbine.

The exhaust-driven turbine operates by the **centripetal force** of the exhaust gas applied to the tips of the turbine blades. Exhaust flows inward, exits through the center outlet, and flows to the exhaust system, figure 19-3. A shaft connects the rotating exhaust turbine to the compressor wheel. Compressor rotation draws intake air through a center inlet and uses centrifugal force to pump it to an outlet at the housing perimeter. The turbocharger center housing supports the turbine and compressor shaft in a pair of bearings. The center housing and bearings are lubricated by engine oil, figure 19-5.

The turbine and compressor wheels operate with close clearances in their chambers to minimize leakage around the blades. If exhaust leaks around the turbine blades, much of the heat energy for compressor rotation is lost. Similarly, if intake air leaks around the compressor blades, the turbocharger will not develop full boost pressure.

When a turbocharged engine starts and runs at low speed, exhaust heat and pressure are low. The turbine idles at about 1,000 rpm. The compressor doesn't turn fast enough to provide boost. Intake air simply passes through the compressor, and the engine runs naturally aspirated. When the engine runs faster or under heavy load, exhaust heat and flow increase. The turbine and compressor then turn faster. There is no brake and very little rotating resistance on a turbocharger shaft. When a turbo starts to accelerate, it will keep accelerating to the maximum speed it can develop from the heat energy of the exhaust. When the engine is running at full power, turbocharger speeds are as high as 100,000 to 150,000 rpm.

When an engine decelerates from top speed, it drops from 5,000 rpm or more to idle in one or two seconds. Its internal friction, pumping resistance, and the load of the drive train cause fast deceleration. A turbocharger has none of these loads on it, and it turns 10 to 20 times faster than the engine. The turbo can take a minute or more to decelerate to its idle speed after the engine returns to idle

You probably have the idea by now that the close clearances of the turbine and compressor wheels in their housings and the high rotating speeds place critical requirements on the turbo bearings. The bearings typically allow 0.003 to 0.006 inch (0.076 to 0.152 mm) radial clearance, figure 19-6, and 0.001 to 0.003 inch (0.025 to 0.075 mm) axial clearance (endplay) on the shaft, figure 19-7. Because the bearings must maintain these clearances at extreme speeds, they need a constant supply of clean oil. The close turbine and compressor clearances and high rotating speeds also mean that if a small dirt particle enters the exhaust or intake housings, it can damage the turbocharger. As we said, a turbocharger is a simple device, and it can be trou-

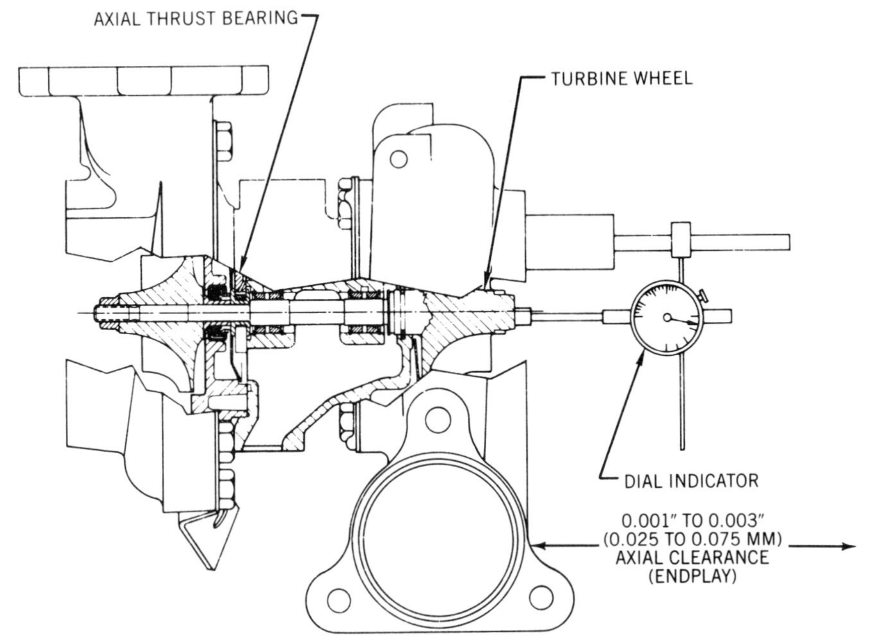


**Figure 19-7.** The axial clearance (endplay) of the shaft thrust bearing is 0.001" to 0.003". (Garrett)

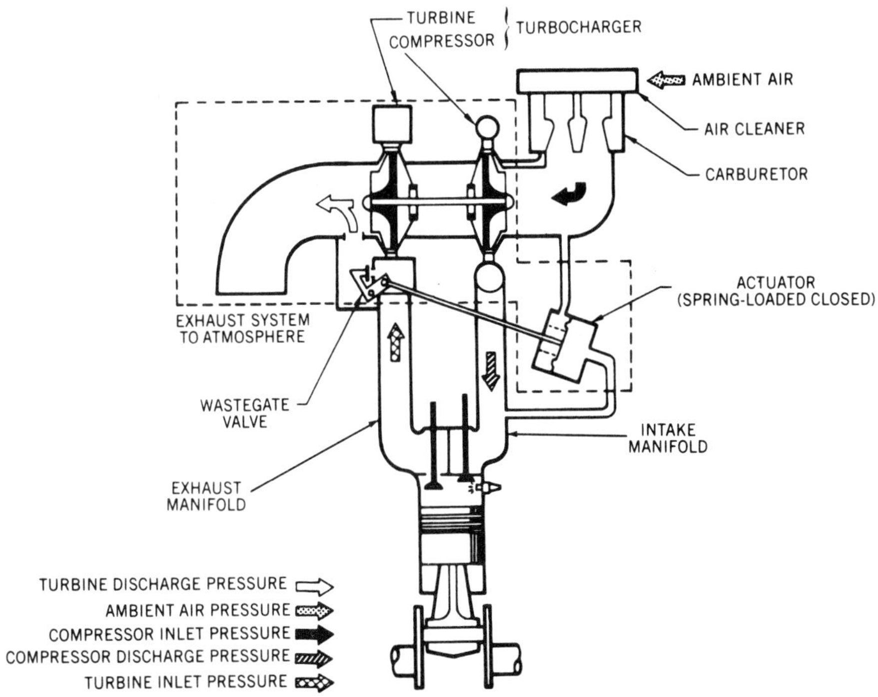


**Figure 19-8.** Locating the turbine and the compressor close to the exhaust and intake manifolds, respectively, reduces turbo lag. (Garrett)

ble-free as long as two conditions are maintained:

**1.** The turbo shaft and thrust bearings must constantly receive proper lubrication.

**2.** The intake and exhaust housings must be kept free of even the smallest dirt particles.

Because of the close clearances of the shaft bearings and the high exhaust temperatures, heat is a critical factor in turbocharger operation. To minimize heat damage and prolong turbo life, Chrysler and several other carmakers use water-cooled turbochargers. Engine coolant circulates through separate passages in the turbo center housing to carry off excess heat.

### Turbocharger Size and Response Time

A turbocharger uses exhaust flow for its power, and moving exhaust gas (like any material) has inertia. When an engine accelerates or has a load applied, one or two seconds pass before the exhaust flow develops enough energy to accelerate the turbine. The turbine, itself, must overcome its own inertia as it accelerates. The compressor, too, must overcome its inertia and the inertia of the intake airflow. Because it is a variable-displacement pump, it doesn't supply useful boost at low speed. All of these inertia factors cause **turbo lag**, the delay during acceleration when a turbo doesn't supply boost. While turbo lag may occur for only a second or less, it has created challenges for engineers to overcome.

The size of the turbine and compressor wheels has a large effect on turbo response time. Large turbine and compressor wheels accelerate slowly; small ones accelerate faster. A small turbine and compressor, however, may not have the airflow capacity for a large engine. They may actually restrict the intake and exhaust at high speed. The exhaust flow and intake airflow capabilities of a turbocharger must match the intake and exhaust breathing capacities of a specific engine to minimize turbo lag. The location of the turbocharger also affects response time. If the turbine is close to the exhaust manifold and the compressor outlet is close to the intake manifold, turbo lag is reduced, figure 19-8.

## TURBOCHARGER INSTALLATION

A turbocharger—any supercharger—can be installed either upstream or downstream from the engine's fuel-metering source (carburetor or fuel injectors). If the turbocharger is upstream (on the air intake side) from the fuel source, it compresses only air and delivers the denser air charge to the carburetor or injectors. If the turbo-

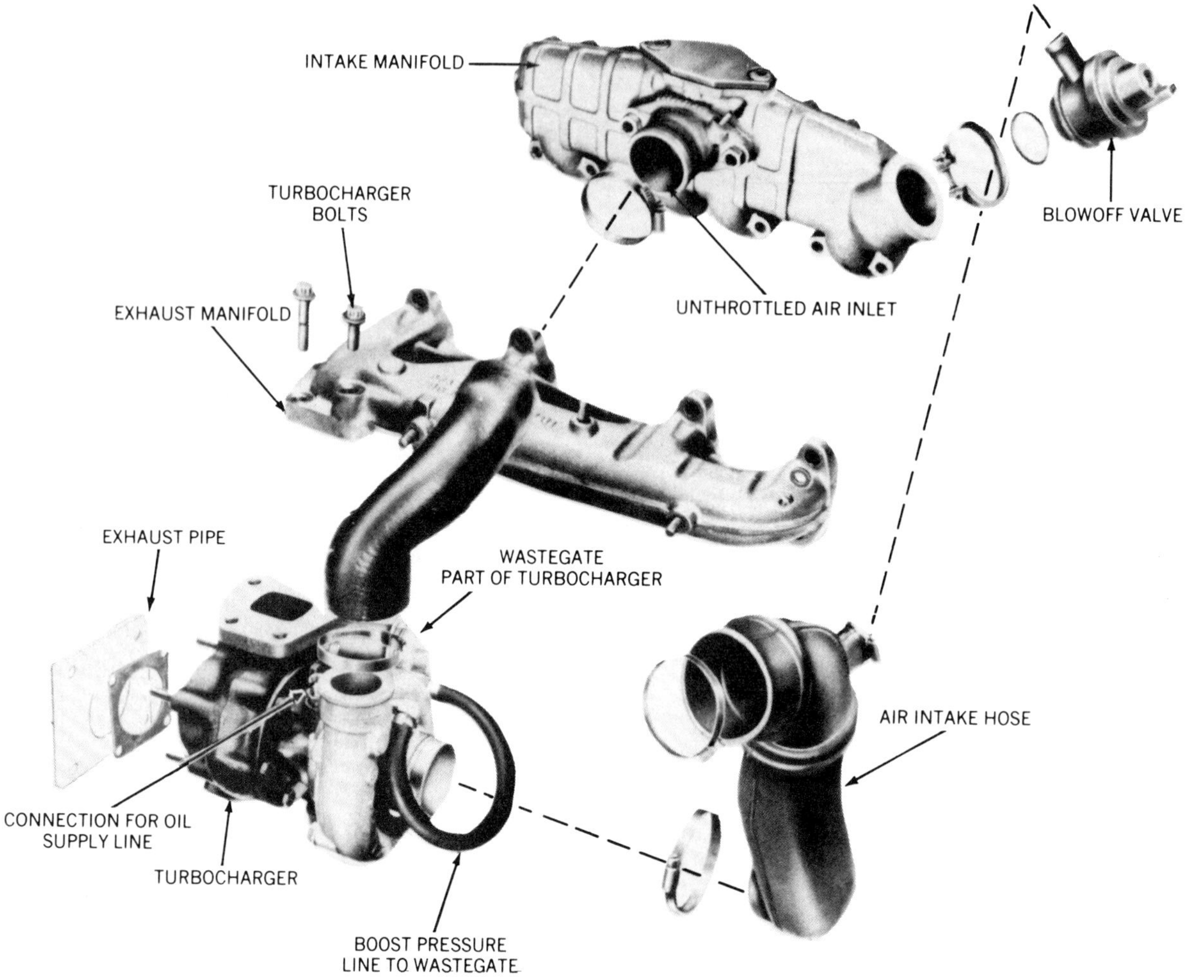


**Figure 19-9.** On a diesel, the turbo is at the inlet of the intake manifold and delivers compressed air to the cylinders. (Volkswagen)

charger is downstream from the fuel source, it compresses the air-fuel mixture and delivers the denser mixture to the cylinders. Let's look at some common turbo installations.

## Diesel Turbocharging

Turbochargers were common on diesel trucks long before they were common on automobiles. A diesel turbo installation is about the simplest possible arrangement. A diesel runs with an unthrottled air intake. The injection pump is calibrated and governed to inject the right amount of fuel in relation to engine speed, load, and intake air volume.

A diesel turbo is mounted at the inlet of the intake manifold, figure 19-9. It compresses intake air and increases the air volume (density) in the manifold. The fuel injection pump of a turbo diesel has a manifold pressure compensator that varies the amount of fuel injected in relation to intake air pressure. This is usually a pressure-operated diaphragm that moves the governor linkage, figure 19-10.

On most diesel automobiles, the pressure compensator responds to manifold pressure changes both from turbocharger boost and from altitude. Atmospheric pressure decreases at high altitudes and increases at low altitudes. Turbo boost, however, changes independently of atmospheric pressure. The manifold pressure compensator responds to the combined effects of atmospheric pressure *and* turbo pressure and varies fuel metering accordingly.

## Turbocharging with Gasoline Fuel Injection

Turbocharger installations on most gasoline engines with fuel injection are similar to installations on diesel engines. The turbo is located at the manifold air intake and compresses only air, figure 19-11. Most turbocharged, fuel-injected gasoline engines have individual port or manifold injection nozzles for each cylinder. Fuel is injected into the compressed air charge just before it enters the cylinder, figure 19-12. The throttle is between the turbocharger and the injectors to regulate airflow. Some gasoline turbo engines, such as Mitsubishi engines, have just one or two injection nozzles in a throttle body on the intake manifold. The turbo is still upstream from the injec-

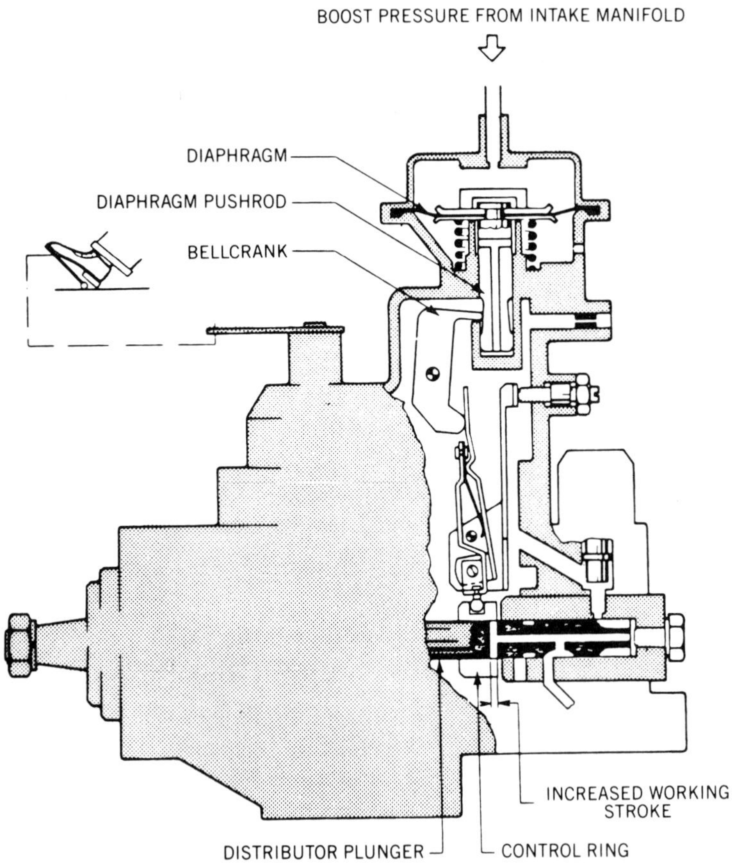


**Figure 19-10.** The boost diaphragm moves a plunger that operates a bellcrank. The bellcrank operates linkage that moves the governor control ring to close the spill port and increase fuel metering. (Volkswagen)

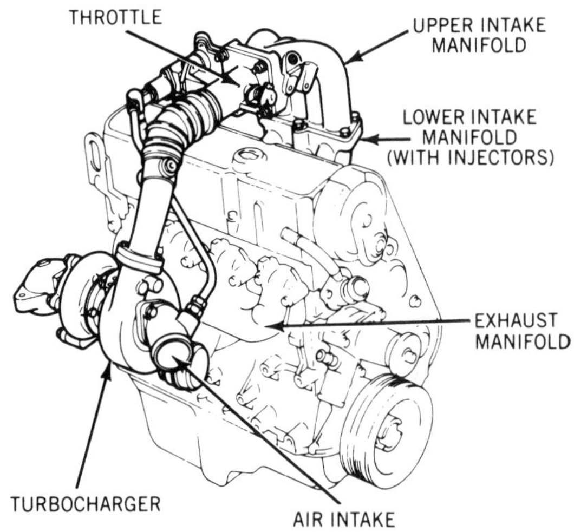


**Figure 19-11.** With port fuel injection, the turbo is usually at the intake manifold inlet, with the throttle between the turbo and the injectors. (Ford)

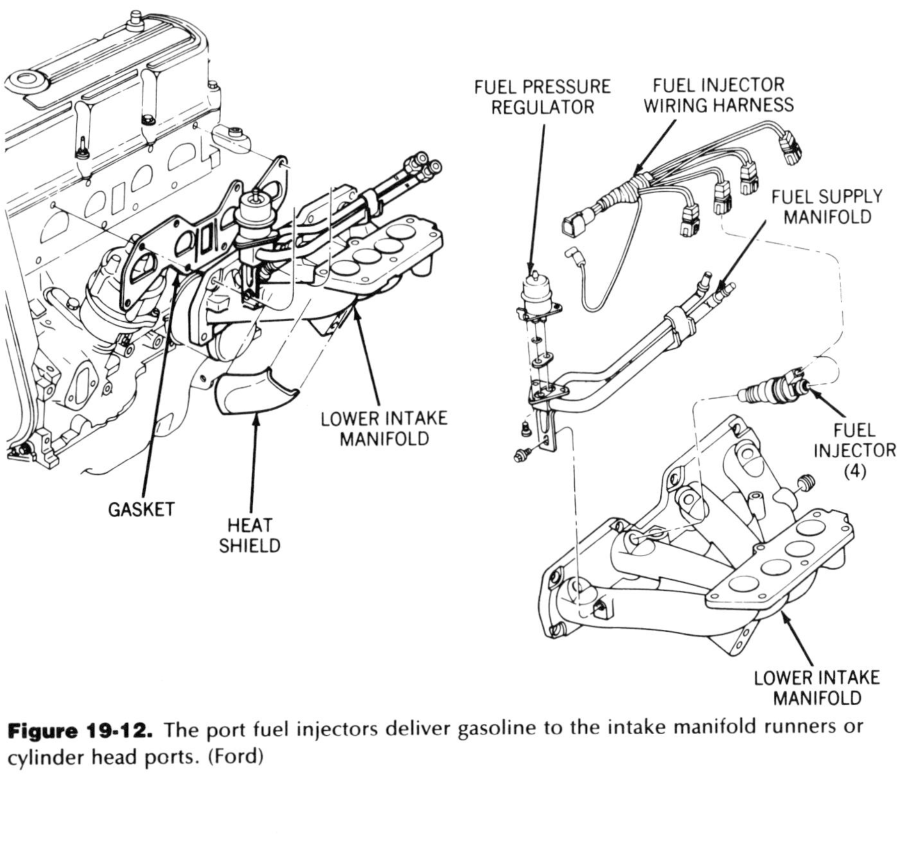


**Figure 19-12.** The port fuel injectors deliver gasoline to the intake manifold runners or cylinder head ports. (Ford)

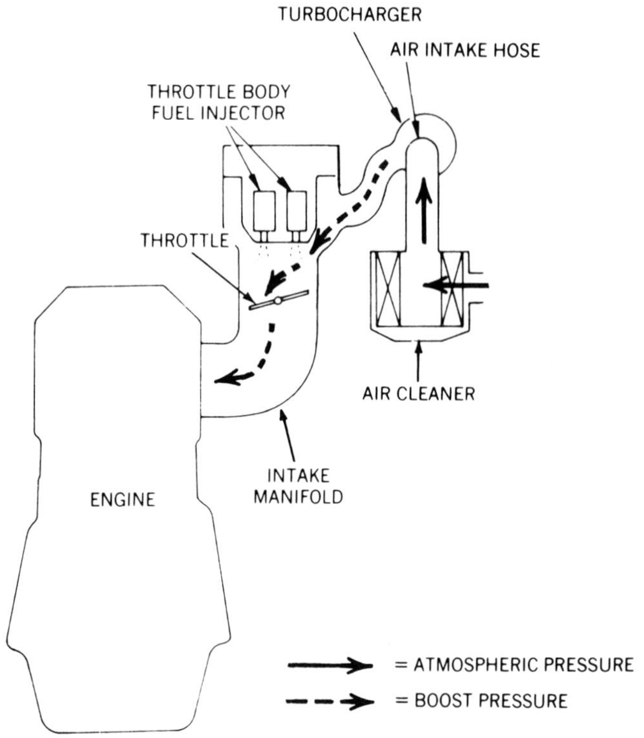


**Figure 19-13.** This Mitsubishi turbo installation with throttle body injection has the throttle downstream from the turbo and the injectors. (Mitsubishi)

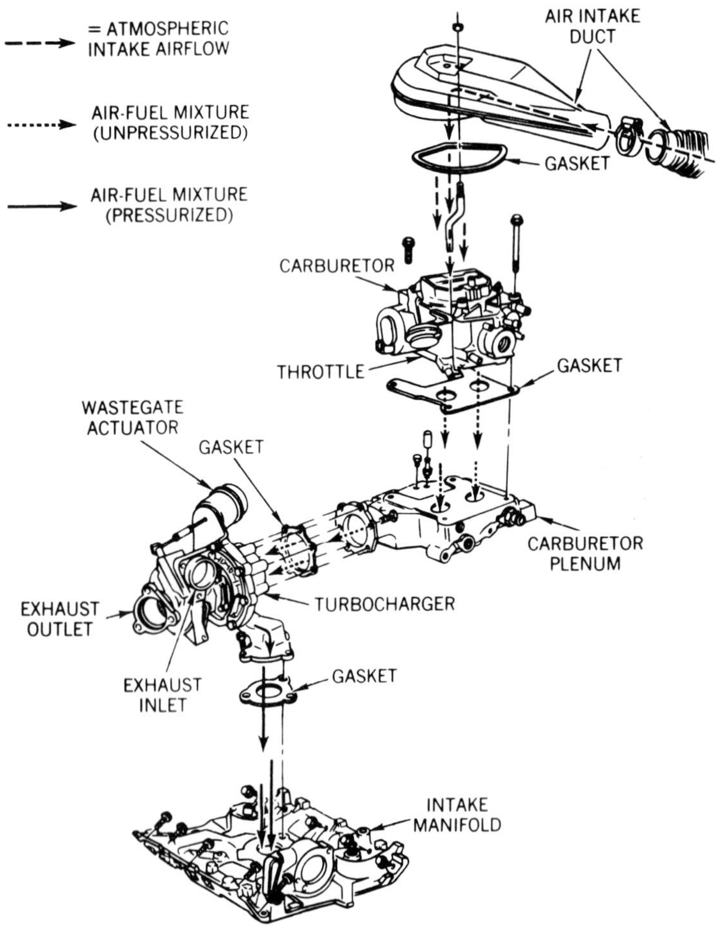


**Figure 19-14.** With the carburetor at the compressor inlet, the turbo compresses the air-fuel mixture. (Buick)

tors and delivers compressed air to them. The throttle, however, is downstream from the injectors and regulates the intake volume of the compressed air-fuel mixture, figure 19-13.

The injection system has a manifold pressure sensor or an airflow sensor that tells the injection control unit the air pressure or volume in the manifold. This regulates fuel metering to correspond to the intake air volume, whether the engine is running with no boost or full boost. The pressure or airflow sensor does the same job as the manifold pressure compensator on a turbo diesel. You will learn how it works when you study gasoline injection.

## Turbocharging with Gasoline Carburetion

A carburetor can be on the intake or the outlet side of a turbocharger. Each design has advantages and disadvantages, but an installation with the carburetor at the turbo inlet is more common.

With the carburetor on the compressor inlet, the turbo increases the airflow and pressure drop through the carburetor. Atmospheric pressure on the fuel bowl forces more gasoline into the increased airflow to provide the needed air-fuel ratio. The turbo compresses the air-fuel *mixture* and delivers it through the manifold to the cylinders, figure 19-14.

Locating the carburetor at the turbo inlet allows the use of an essentially standard carburetor with no modifications to withstand boost pressure. But, it also places the carburetor farther from the intake manifold. This can cause slight engine hesitation when the throttle opens quickly. Remote carburetor location also can complicate throttle and choke linkages and the preheating of the air-fuel mixture of a cold engine. The turbo also requires a special seal between the compressor and the center housing to keep the air-fuel mixture out of the engine oil system. Additionally, fuel may tend to separate from the compressed mixture as it passes through the turbo and manifold. Careful design of manifolds,

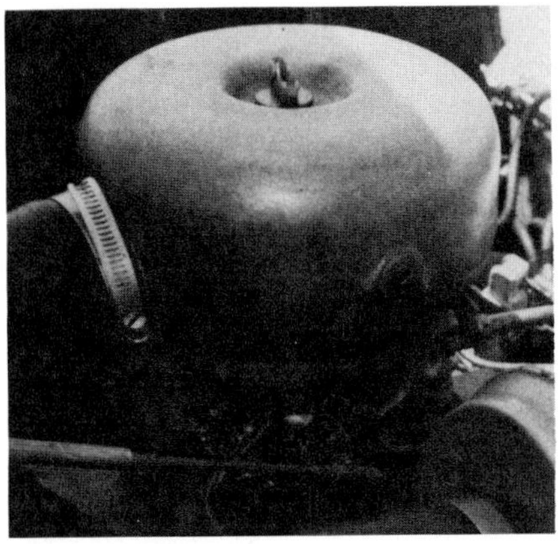

**Figure 19-15.** Locating the turbo so that it pressurizes the carburetor makes fuel metering more difficult, especially at boost pressures above 8 to 10 psi.

**MORE LAWS**

Today, we take it for granted as a scientific fact that heat is a form of energy. But, it was approximately 150 years ago (about 1850) that the Englishman, James Prescott Joule, proved that heat was a form of energy that could be converted to mechanical energy. Actually, engineers had been doing it for a long time, but they didn't fully understand the principles. Joule proved that one calorie (a standard unit for measuring heat energy) was equal to 4.2 joules of mechanical energy. (He named the new unit after himself, and it is still the name used in the modern metric system for the universal unit of energy measurement.)

Joule's work led to a new branch of physics that studies the relationship between heat and work—thermodynamics. There are three fundamental laws of thermodynamics, all based on the theory that heat is energy resulting from the motion of molecules in a substance. The greater the amount of motion, the greater the heat energy.

***The first law of thermodynamics*** is that a specific amount of heat can be changed into a specific amount of mechanical energy, or work, and no more. Conversely, a specific amount of mechanical energy can be changed into a specific amount of heat. This is another application of the principle of conservation of energy. You can't increase it from one form to another; you can only change it.

***The second law of thermodynamics*** is that heat can't move from a cold body to a hot body; it can only move from hot to cold. This may seem obvious, but without a full understanding of the principle, engineers couldn't have built your car's engine to operate as it does, nor could air conditioning have been invented.

***The third law of thermodynamics*** says that as the temperature of any substance comes closer to absolute zero, its usable energy approaches absolute zero.

turbocharger size, and carburetors, however, has overcome most of these problems and made carburetor location at the turbo inlet the more common arrangement.

With the carburetor on the compressor outlet, the turbocharger delivers high pressure to the carburetor airhorn, figure 19-15. This means that the carburetor must be calibrated for correct fuel metering with more than atmospheric pressure on the fuel bowl. But, when the turbo is not providing boost, the carburetor must meter fuel at atmospheric pressure. These factors make carburetor calibration very difficult when the carburetor is at the turbo outlet. Moreover, the entire carburetor must be modified to withstand boost pressure without fuel leakage. The fuel system, too, must deliver gasoline at higher pressure to overcome boost pressure at the carburetor. Air-fuel ratio control is much more difficult with a pressurized carburetor.

## TURBOCHARGER CONTROLS

Turbocharger installation is not a simple matter of bolting a turbo to an engine and gaining free power. A turbocharger raises compression pressures and loads on an engine's moving parts. It also raises engine and underhood temperatures. Most late-model engines are strong enough to stand the higher compression pressures of turbocharging without major structural changes. Since compression ratios of most current naturally aspirated engines range from 8:1 to 9:1, compression ratios usually aren't lowered when a turbo is installed. A few carmakers, however, lower compression one-quarter to one-half point (from 9:1 to 8.75:1 or 8.5:1, for example) when installing a turbo. Because of higher engine and underhood temperatures, many turbocharged cars have oil coolers and larger capacity radiators. Turbocharged engines also require shorter oil change intervals than naturally aspirated engines.

High pressures and temperatures created by a turbo require controls for three factors:

1. Boost pressure
2. Air-fuel mixture temperature
3. Detonation.

These are not three independent factors; they are interrelated. Higher intake and compression pressures raise intake mixture temperatures and combustion temperatures. Higher pressures and temperatures can combine to cause detonation.

### Boost Pressure Control

As a turbo accelerates to 100,000 or 150,000 rpm, boost increases steadily. Boost pressure of any supercharger must be limited to the maximum that an engine can stand without detonation or serious damage. On a turbocharger, this is done most efficiently with an exhaust **wastegate**, figure 19-16, which controls exhaust flow into the turbine. The wastegate is a bypass valve at the exhaust inlet to the turbine. It allows all of the exhaust to flow into the turbine, or it diverts part of the exhaust past the turbine to the exhaust pipe. The wastegate is operated by a diaphragm actuator, exposed to compressor outlet pressure. When pressure rises to the desired maximum, it forces the diaphragm to move a rod that opens the wastegate,

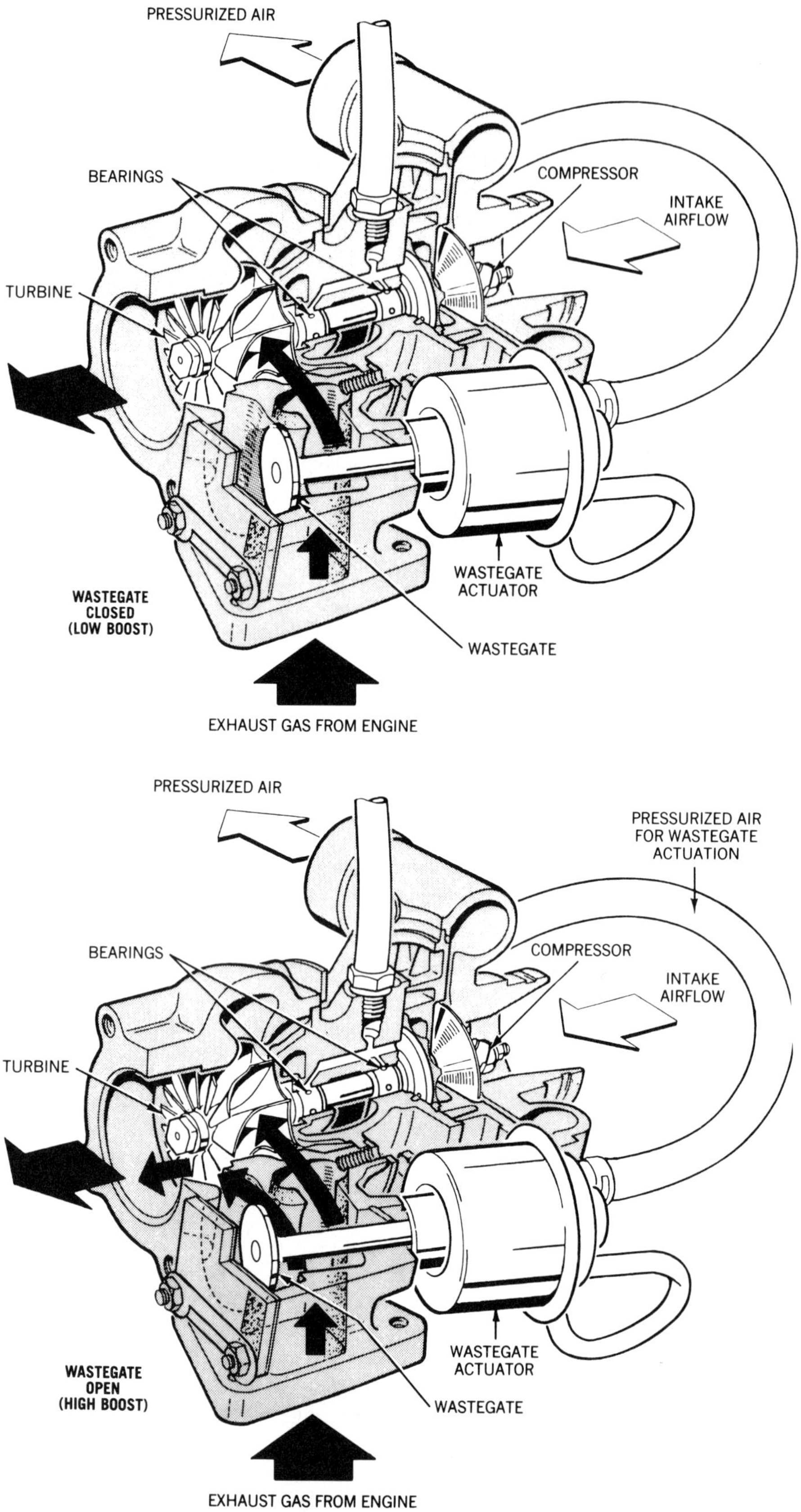


**Figure 19-16.** High boost pressure at the compressor outlet forces the actuator to open the wastegate and divert part of the exhaust past the turbine. (Volkswagen)

figure 19-16. This diverts part of the exhaust (in some cases, most of the exhaust) to the turbine outlet and limits its maximum turbine and compressor speed.

The pressure line to the actuator is at the compressor outlet. During deceleration, pressure at this point rises sharply because the compressor is working against the closed throttle and an engine that needs little airflow. The actuator opens the wastegate fully and eliminates overboost at closed throttle.

Boost pressure on most early OEM turbocharger installations was limited to 6 or 8 psi. Since the early 1980's turbocharging has been refined so that many engines now operate consistently with 10 to 14 psi of boost pressure. All turbochargers with wastegate boost control operate as described above. However, wastegate operation can be modulated to allow higher boost with high-octane fuel and lower boost with low-octane fuel, figure 19-17.

A wastegate can be used on a turbocharger that is either upstream or downstream from the carburetor or injectors. It does not affect, and is not affected by, intake air-fuel mixture. Boost pressure also can be controlled by a **blowoff valve**. This is a large, spring-loaded relief valve that opens when pressure exceeds the desired maximum. If the turbo is upstream from the carburetor or injectors and compresses only air, the blowoff valve can relieve excess air pressure directly to the atmosphere or back to the air intake, figures 19-9 and 19-18. If the turbo is downstream from the carburetor and compresses air and fuel, the valve *must* relieve pressure back to the compressor inlet.

A blowoff valve can be operated by manifold vacuum to relieve boost during idle, deceleration, and choked operation. This lowers combustion temperature and pressure and $NO_x$ emissions under these conditions. A blowoff valve also can be operated by vacuum on one side of a diaphragm and compressor pressure on the other. Volkswagen, BMW, and some other turbo diesels have both a wastegate and a blowoff valve for overboost protection, figure 19-18.

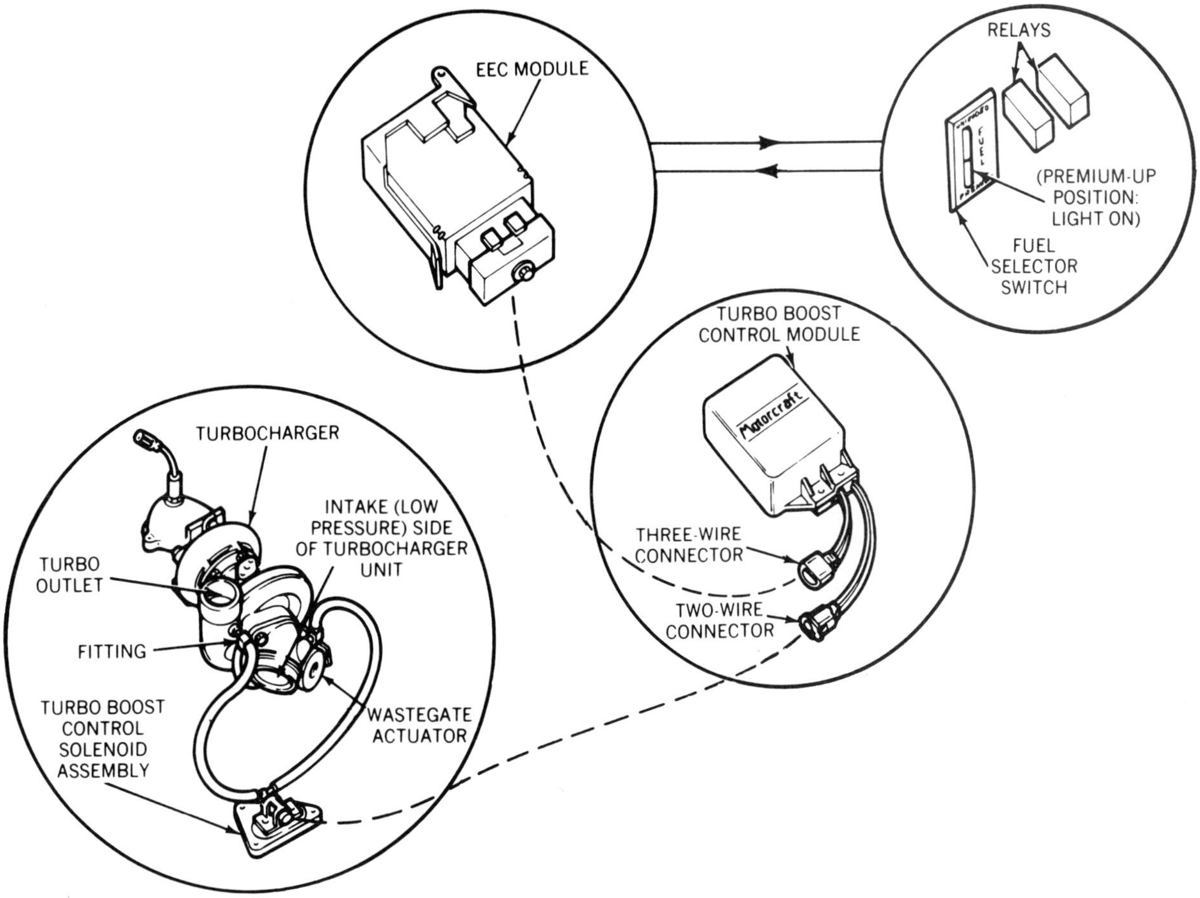


**Figure 19-17.** When the driver turns the boost control switch to PREMIUM, the solenoid opens a bypass to divert pressure from the actuator line back to the compressor inlet. This allows the wastegate to stay closed until boost reaches 14 psi. (Ford)

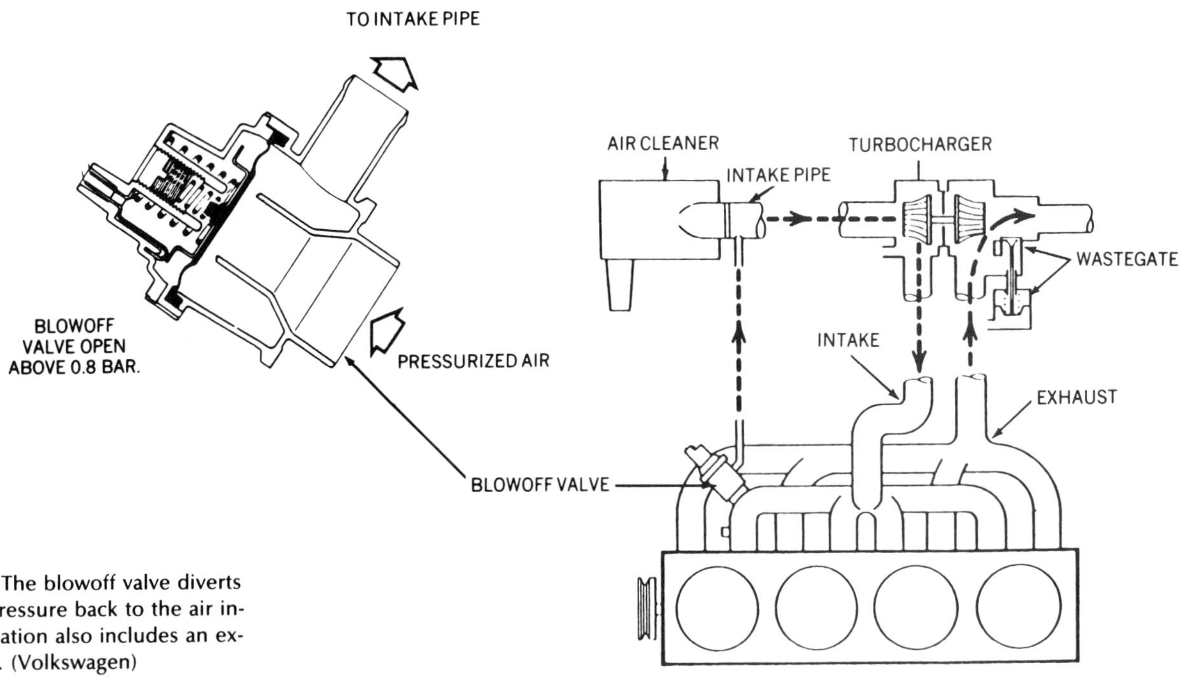


**Figure 19-18.** The blowoff valve diverts high manifold pressure back to the air intake. This installation also includes an exhaust wastegate. (Volkswagen)

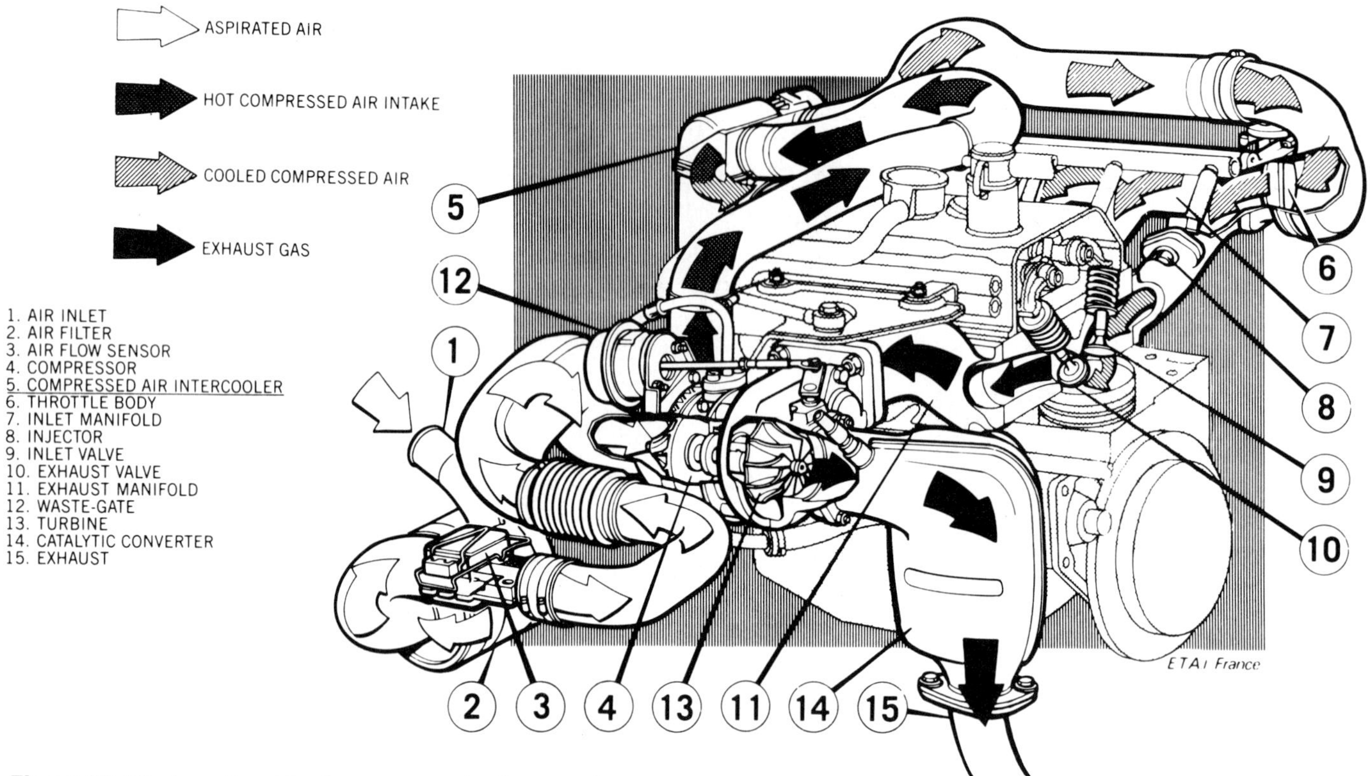


**Figure 19-19.** Renault and Ford were among the first carmakers to use air-to-air intercoolers to lower the supercharged air temperature.

A turbo diesel also may have an overboost protection system that stops fuel injection if the wastegate fails and pressure gets too high. This is usually a diaphragm-operated device on the injection pump.

### Intake Mixture Cooling

As turbocharging raises the pressure of the intake air, it also raises the temperature of the air-fuel mixture. This has two undesirable effects. Higher temperature reduces the charge density. Thus, it works against the boost pressure that increased the charge density and the temperature to begin with. Secondly, a high-temperature air-fuel mixture entering the cylinders tends to detonate more easily. Detonation in a turbo engine with high compression pressures can burn a piston, bend a rod, or otherwise damage the engine very quickly. There are two common ways to cool the intake mixture of a turbocharged engine: water injection and intercooling.

Water injection is not often used on OEM turbo installations, but it is common on aftermarket turbo systems. It is installed most easily on a carbureted engine where the turbo is downstream from the carburetor. An electric or vacuum-operated pump injects a fine spray of water into the carburetor inlet at maximum boost. The water adds nothing to the combustion process, but it vaporizes along with the gasoline and cools the intake charge. It is expelled as water vapor in the exhaust. Effective water injection depends on precise operation of pumps and valves, and it requires the driver to keep the water supply tank filled. Because of these variables, most carmakers have chosen intercooling to cool the intake mixture.

An **intercooler** is a heat exchanger, through which the compressed air passes from the turbo to the manifold. Most intercoolers are the air-to-air design, figure 19-19. Compressed air from the turbo passes through the radiator-like intercooler and transfers heat to ambient airflow around the intercooler. Intercoolers work best on systems with fuel injection or a carburetor downstream from the turbo. Only air passes through the intercooler, and there is no chance of fuel separation from the mixture as it cools. Intercoolers also can be built as air-to-water heat exchangers, but they are more complicated and heavier than air-to-air intercoolers.

### Detonation and Spark Timing Control

High temperature and pressure are not the only causes of detonation in a turbocharged engine. If spark timing is overly advanced, the dense air-fuel mixture will ignite too soon, burn unevenly, and develop peak cylinder pressure too early on the power stroke. All of these can cause knocking and engine damage. Ignition timing must be retarded from maximum advance when a turbo engine is at peak boost.

One easy way to do this on a gasoline engine is with manifold pressure retard on the distributor vacuum advance diaphragm. This method was used on several turbocharged engines of the 1960's and 1970's. The distributor vacuum unit is built so that manifold pressure acts on a diaphragm to retard timing at high boost. The retard diaphragm moves the breaker plate

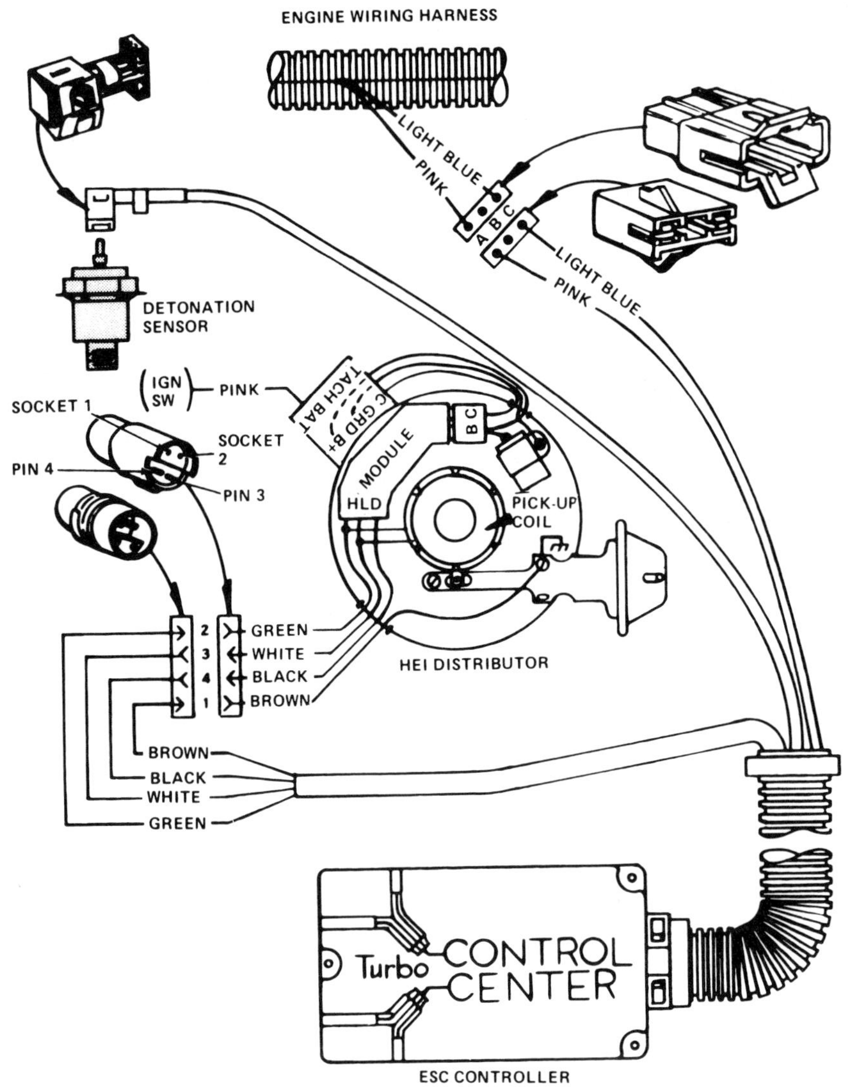


**Figure 19-20.** This detonation sensor used with electronic ignition signals an electronic module to retard timing when knocking occurs. (Buick)

> **BAR EXAMINATION**
>
> In California, the BAR is the Bureau of Automotive Repair that licenses repair shops and regulates the automobile service business. There is another bar, however, that you will read about as you study turbochargers and fuel injection systems. It is a unit of pressure measurement.
>
> One bar is basically one unit of atmospheric pressure. This is barometric pressure, from which the word, "bar" comes. One standard bar is one kilogram of force applied to one square centimeter. This equals 14.2 pounds per square inch, which is close to 14.7-psi atmospheric pressure at sea level at a temperature of 0° C (32° F). Using these standards corrects the pressure measurements for altitude and temperature. One bar, therefore, is equivalent to 14.7 psi or 760 millimeters of mercury (mm Hg) or 101.28 kPa. The bar is an older metric unit, used primarily by the Robert Bosch company and European carmakers for turbocharger and fuel injection pressure specifications.
>
> If you look at European service literature, you will see turbocharger boost specifications such as 0.6 bar or 1.2 bar. These specifications mean that the turbo *adds* that much pressure to atmospheric pressure. For example, boost of 0.6 bar equals atmospheric pressure (14.7 psi) + 0.6 bar (8.82 psi), or 23.52 psi total intake pressure.
>
> Similarly, injector fuel pressure might be specified as 2 bar or 2.5 bar. These specifications equal 29.4 psi and 36.75 psi, respectively.

opposite to the way it moves for advance. This works similarly to the way a dual-diaphragm distributor retards timing with high manifold vacuum at idle and deceleration.

A turbocharged diesel can be built with similar boost pressure retard on the injection pump. A diaphragm, actuated by boost pressure, moves the injection timing advance linkage to retard injection slightly at high boost pressures.

Boost pressure retard of ignition timing works well enough on a simple turbo installation, but it doesn't offer the precise timing control of electronic detonation sensing. Many late-model turbo engines with electronic controls use a sensor that signals the computer to retard timing when detonation occurs. The sensor is a piezoelectric crystal that generates a voltage or varies the return signal of a reference voltage when engine knock occurs, figure 19-20. The sensor detects detonation as physical hammering within the engine and signals the computer to retard timing instantly. Timing is usually retarded in 2- to 4-degree increments until knocking stops. The computer then gradually restores the maximum spark advance that the engine needs for efficient operation.

## TURBOCHARGER DEVELOPMENT IN THE LATE 1980'S AND EARLY 1990'S

As turbocharging became commonplace during the 1980's, engineers continued to work on improving turbo performance and durability. Among their goals were to:

- Improve turbo life and reliability through better cooling
- Improve low-speed turbo response
- Integrate turbocharger control with electronic engine control systems.

### Liquid-Cooled Turbochargers

Since the mid-1980's most carmakers with turbocharged engines have adopted turbos with liquid cooling passages surrounding the center bearings. Figure 19-21 shows engine coolant lines routed to a turbo center housing. Engine coolant flows through "water jackets" around the oil passages in the turbo to lower the lubricant temperature.

## Improved Turbocharger Response

As explained earlier in this chapter, turbo lag has always been a challenge for engineers. One way to improve turbo response at low speed is to make the rotating turbine and compressor wheels lighter. In 1985, Nissan introduced a silicon-ceramic compressor wheel that weighs only half as much as a steel wheel. Since then, other manufacturers also have adopted lightweight ceramic rotating parts for their turbochargers.

**Figure 19-21.** Tubing connected to this turbocharger lets engine coolant circulate through the center housing to cool the shaft bearings.

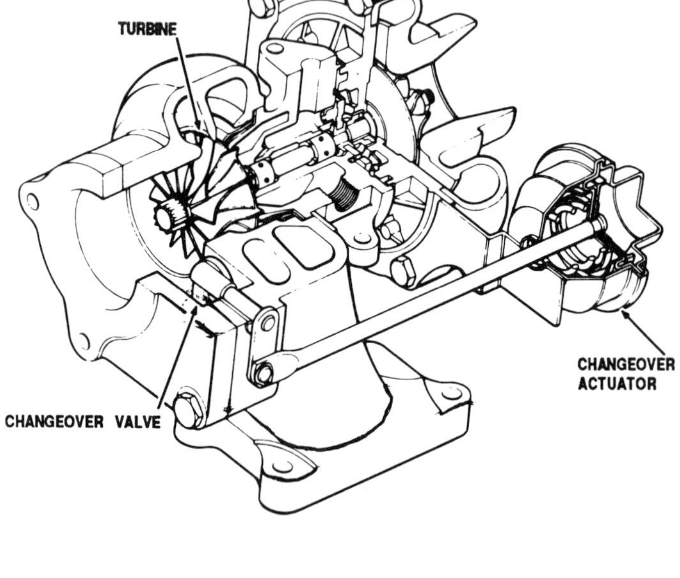


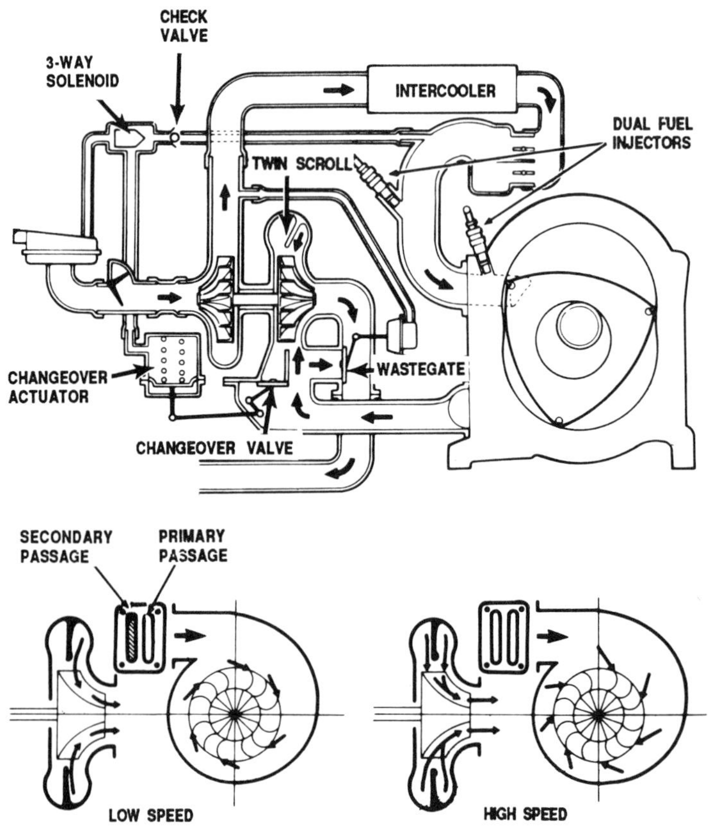


**Figure 19-22.** Mazda's dual-inlet turbo uses a twin-scroll turbine housing and a changeover valve to improve exhaust speed and flow at low speed.

Another approach to improving turbo response has been to control the inlet exhaust gas flow to the turbine. If exhaust flow can be accelerated and made to strike the turbine blades as close to a 90-degree angle as possible, the turbine will accelerate faster. Let's look at a couple of examples of how this can be done.

The variable-nozzle turbine (VNT) was tried by Chrysler (for one year) and Nissan. A VNT turbo has a movable curved flap at the turbine inlet. With the flap closed, exhaust flow entering the turbine speeds up and strikes the turbine closer to a 90-degree angle. The turbocharger operates in this condition at low speed for quick turbo response. With the flap fully open at high engine speed, exhaust backpressure is reduced and increased exhaust volume keeps the turbine spinning at high speed. The VNT flap is usually controlled by a vacuum-operated diaphragm. The engine control computer regulates a pulse-width-modulated vacuum solenoid to control vacuum applied to the diaphragm and thus the flap position. Nissan used a VNT turbo on some home-market cars in the mid-1980's, and Chrysler used it for one year: 1990. The complexity and cost of VNT controls have prevented this technology from becoming commonplace in the 1990's.

Another approach to improving turbo response is Mazda's dual-inlet turbine, figure 19-22. A changeover actuator, operated by intake boost pressure, controls the openings to the two exhaust passages. At low engine speed, exhaust through the primary passage flows at high speed and strikes the turbine at a 90-degree angle. This is much like the operation of a VNT turbo. At higher speed, the changeover actuator opens the secondary exhaust passage. This slows down gas flow, but increases volume and reduces backpressure. Boost pressure to the changeover actuator is controlled by a solenoid, which is operated by the engine computer.

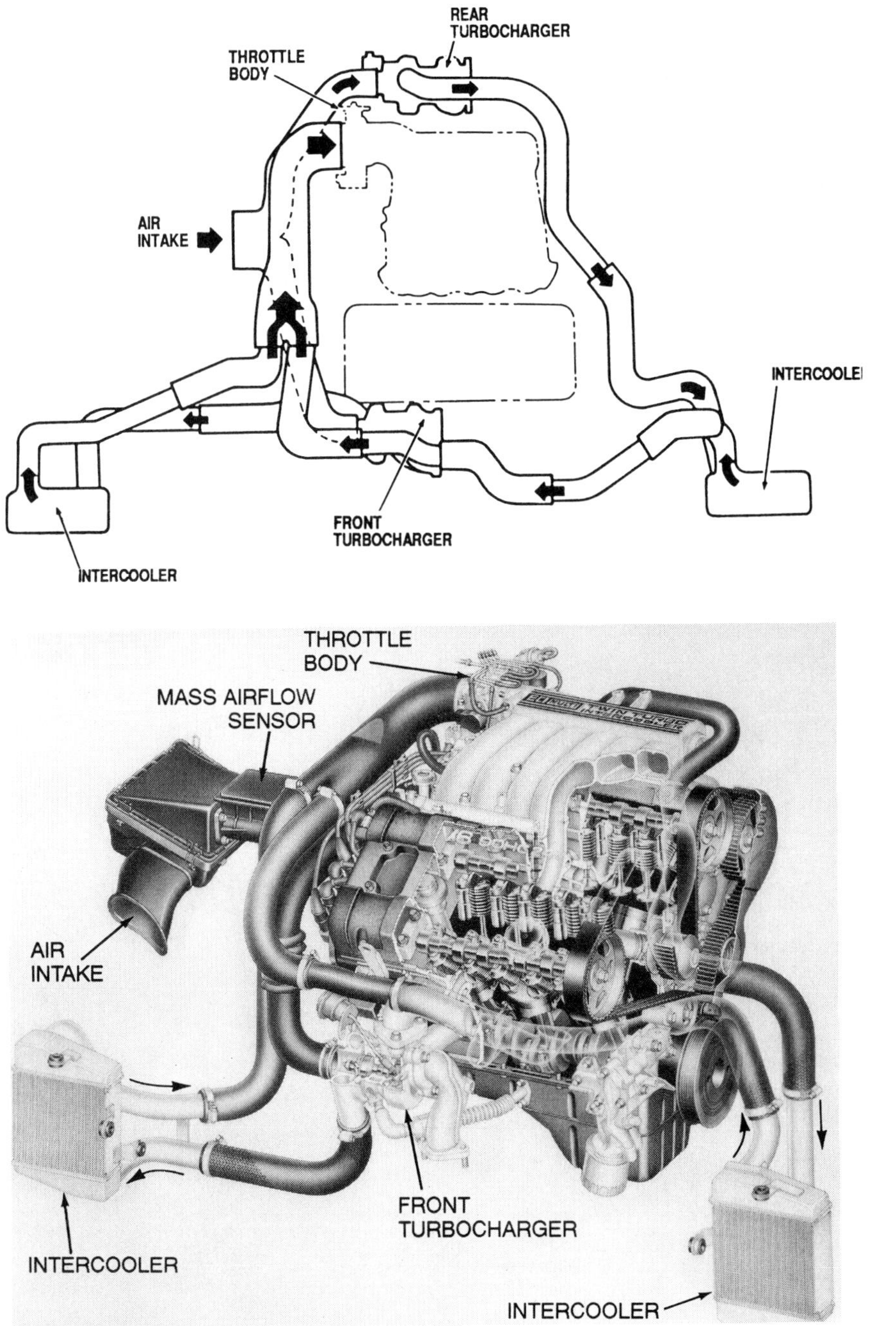


**Figure 19-23.** This Mitsubishi-Chrysler twin turbo installation with two intercoolers surrounds the engine with a maze of intake air plumbing. (Chrysler)

## Twin Turbos and Electronic Controls

On high-priced, high-performance cars, manufacturers can afford to install more exotic systems, such as twin turbos with twin intercoolers and more advanced electronic controls. Figure 19-23 shows a complex twin-turbo installation used by Chrysler and Mitsubishi since 1991. Two small, lightweight turbos are driven independently by the exhaust from each bank of a V-type engine. The smaller turbos respond fast at low engine speed, and their combined volume provides efficient boost throughout the engine's operating range. The wastegate actuator for each turbo is regulated by the engine computer through a pulse-width-modulated solenoid, figure 19-24. The solenoid bleeds off boost pressure in the actuator line.

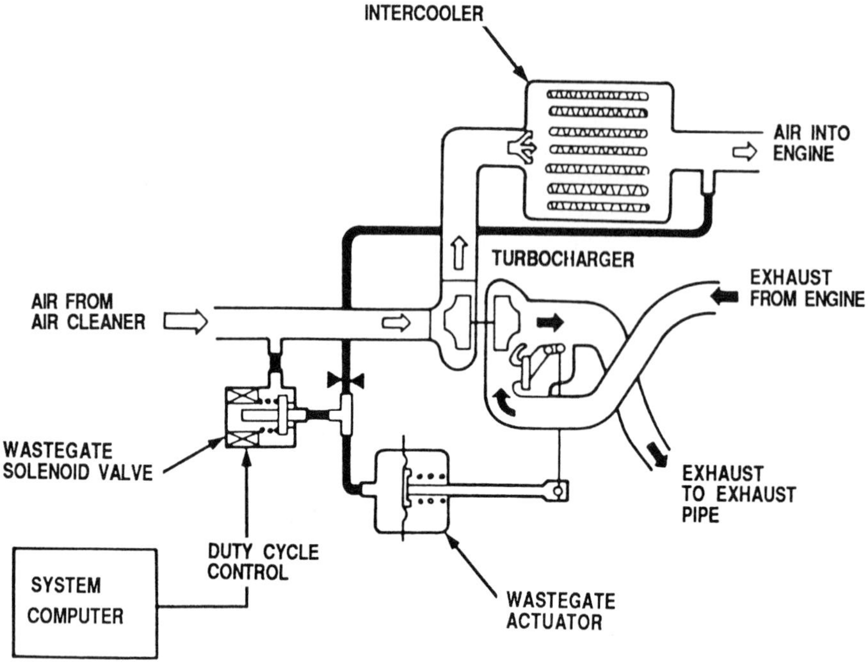


**Figure 19-24.** The Mitsubishi-Chrysler turbo wastegate is regulated by a computer-controlled solenoid valve that bleeds off boost to the wastegate actuator. (Chrysler)

The 1993 Mazda RX-7 has a unique twin-turbo system with a small high-speed turbo to provide low-speed boost and a larger turbo, which is engaged at higher engine speed for boost throughout the engine's power range. Turbo engagement and wastegate control are regulated by the engine control computer.

## SUPERCHARGING FOR THE 1990'S

Mechanical supercharging is actually an older technology than turbocharging, but from the mid-1960's through the 1980's turbocharging was considered a path to free horsepower. In the 1980's, however, engineers took another look at the positive aspects of mechanical supercharging. Several designs were tried on limited produc-

### THE BIG WHITE CARS

Ken Purdy, the famous motorsports journalist, gave the name "The Big White Cars" to the Mercedes-Benz grand prix cars of the 1930's. The prodigious acceleration of the Mercedes W125 racecar came from a supercharged 5.6-liter (344 cubic inch) straight-8 engine that developed more than 640 horsepower. It was widely reported that the W125 could accelerate from a standstill and *overtake* another car traveling at 100 mph in less than a mile. Not bad for 1937.

The heritage of the W125 was Daimler-Benz longtime interest in supercharger technology. In 1921, two supercharged Mercedes models were introduced at the Berlin motor show, and in 1924, supercharged Mercedes racecars were entered in the Indianapolis 500 (with Dr. Ferdinand Porsche as chief engineer).

Certainly the most famous road-going Big White Cars were the Mercedes K, S, SS, SSK, and SSKL models built from 1926 through 1934. The SSK of the early 1930's had a 7.1-liter (436 cubic inch) 6-cylinder engine that produced 170 horsepower with the Roots blower disengaged and 225 with the blower engaged.

The supercharger, shown here, was a Roots blower mounted along the front side of the engine. The drive gears were engaged by rods and levers when the driver mashed the pedal to the metal. The resulting shriek of the blower and its gears and the roar of the engine were deafening. Top speed was reported to be 117 mph for cars weighing close to two tons.

The racing special SSKL models were lighter and developed more power—about 300 horsepower with the blower engaged. Top speed was about 145 mph for those brave enough to try.

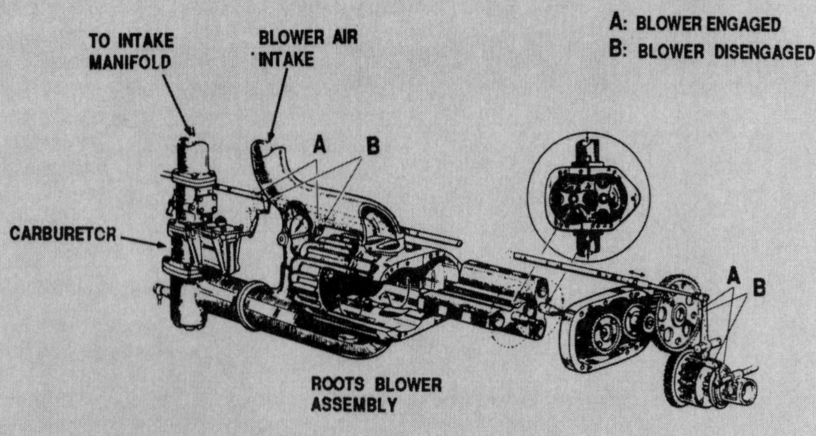


The whine of the blower rotors and the shriek of the drive gears created the banshee scream of Mercedes' Big White Cars in the 1930's.

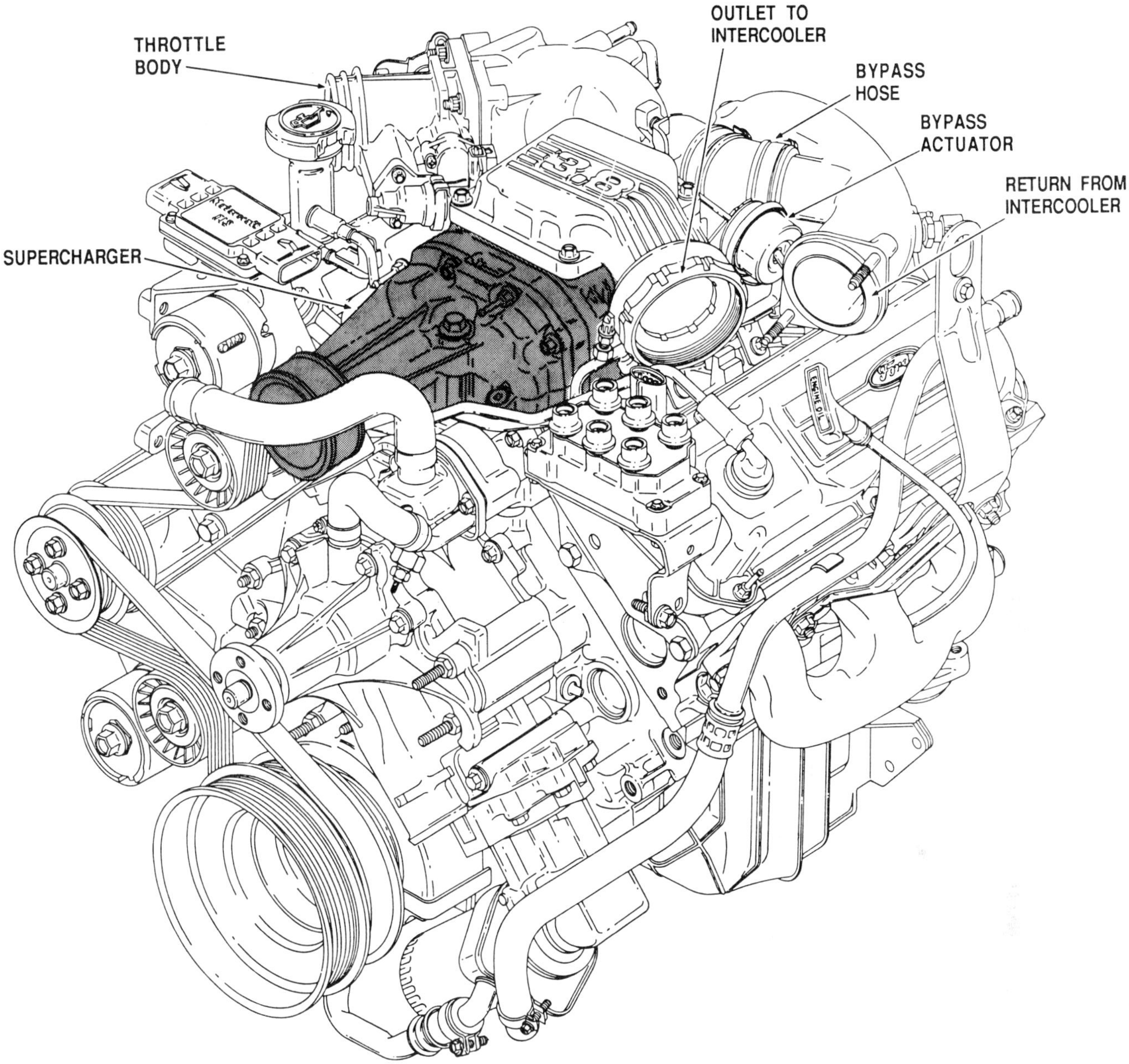


**Figure 19-25.** Ford's supercharged 3.8-liter V-6 is available in 1989 and later Thunderbird Super Coupes and some Cougar models. (Ford)

tion cars, such as the G-lader blower on Volkswagen's Corrado, the twin-screw Sprintex blower, and the pressure-wave Comprex blower used on some Opel diesels in Europe. In the late 1980's the Roots blower, described earlier in this chapter, emerged as the most common supercharger on production passenger car engines.

Toyota, Ford, and General Motors have installed Roots blowers on some late-model high-performance cars. Toyota used a Roots blower on the 1.6-liter, 4-cylinder engine for the MR2 for just two years: 1988–89. This installation was interesting for the computer-controlled electromagnetic clutch used to disengage the blower when boost was not required.

## Ford Roots Blower Installations

In 1989, Ford introduced the Thunderbird Super Coupe with a Roots blower on the 3.8-liter V-6, figure 19-25. This engine also has been installed in some Mercury Cougar models. The Ford blower is built by Eaton Corporation and features a unique airflow design. Intake air enters through the rear of the blower housing, figure 19-26. Compressed air exits through the top to an air-to-air intercooler, from which it returns to the inlet at the rear of the intake manifold.

The Ford blower delivers 12 psi maximum boost at low manifold vac-

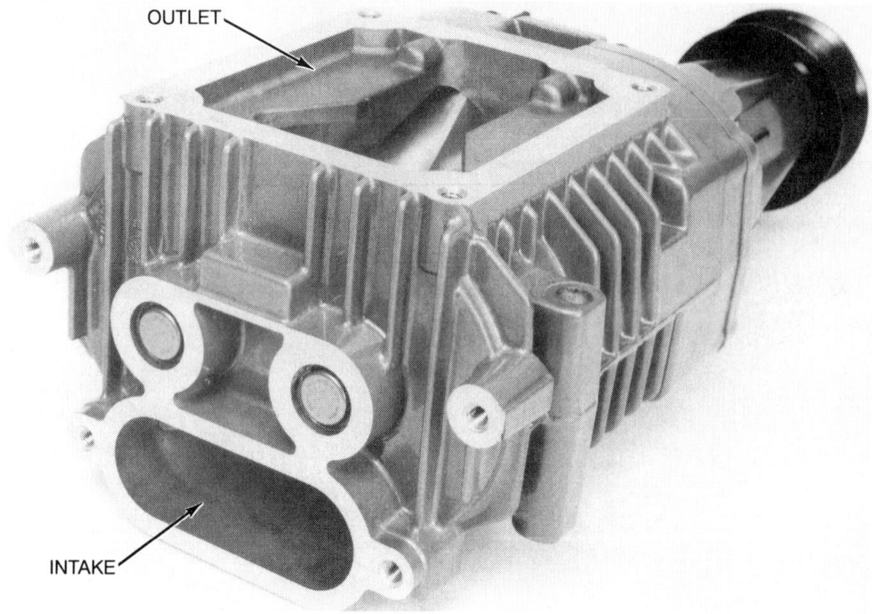


**Figure 19-26.** The Eaton blower used by Ford has the air inlet at the lower rear and the compressed air outlet at the top. (Ford)

**Figure 19-27.** The GM supercharged 3800 V-6 was introduced in full-size Buicks, Oldsmobiles, and Pontiacs in 1992. (GM)

uum. A vacuum-operated bypass actuator bleeds boost pressure back to the blower inlet during cruising, deceleration, and idle when vacuum is high. The 3-lobe rotors of the Ford blower are formed at a 60-degree spiral to reduce pressure pulsations and blower noise.

## GM Roots Blower Installations

In 1992, General Motors introduced a Roots-blown version of the "3800" 3.8-liter V-6, figure 19-27. This engine is used in high-line models of full-size sedans from Buick, Oldsmobile, and Pontiac. Intake air enters through a throttle body on the rear of the blower (left side of the car with a transversely mounted engine), figure 19-28. Compressed air exits from the bottom of the blower directly to the intake manifold plenum. To make the blower fit under the low hoods of the cars in which it is used, GM designed the intake manifold as the lower half of the blower housing. The engines do not have charge air intercoolers as Ford supercharged engines do.

Like Ford, GM uses a vacuum-operated bypass actuator to bleed boost pressure back to the blower inlet during cruising, deceleration, and idle when vacuum is high, figure 19-29. The bypass valve also limits maximum boost to about 8 psi because of the torque limitations of the 4T60E automatic transmission. The GM blower also has spiral rotors to reduce noise and pressure pulsations.

## Roots Blower Advantages and Disadvantages

Compared to a turbocharger, a Roots supercharger has some distinct advantages and disadvantages. A Roots blower can provide boost pressure immediately off idle with no lag. Because it is driven mechanically by the engine, however, it takes back some of the extra power it provides. Ford estimates that its Roots blower consumes about 60 horsepower during full-load operation. This drops to only about 3/4 horsepower at light-load, 55-mph cruising speed, however.

To provide fast boost response, most Roots blowers are overdriven to turn faster than crankshaft speed. Typical blower drive ratios are about 1.5 to 2.5 times faster than engine speed. The rotors of a Roots blower may turn at 10,000 to 15,000 rpm, but this is nowhere nearly as fast as the 100,000- to 150,000-rpm speeds of turbocharger turbine shafts. Consequently, high-temperature lubrication is not as critical for a Roots blower as it is for a turbo. Ford's Roots blower has its own oil sump and requires a special 90-weight synthetic lubricant (Ford part number E95Z19577A). There is no filter or specified change interval, but fluid level should be checked every 30,000 miles.

Because of the high temperatures of turbochargers, special heat shields are required on most turbo engines. Heat shields are not required for Roots blowers.

## SUMMARY

Turbocharging is a proven technology that has achieved new popularity since the 1970's. A turbocharger is an exhaust-turbine-driven centrifugal supercharger that increases intake air pressure above atmospheric pressure. Boost pressure from a turbo increases the intake charge density. This forces more air and fuel into the cylinders for increased combustion. Turbocharging increases an engine's volumetric efficiency and works like a high compression ratio to develop more power from the air-fuel mixture. Turbocharging, however, increases engine power only on demand, when extra power is needed. In this way, it allows a small engine to provide good fuel economy and emission control, as well as extra power.

Both turbochargers and gasoline injection systems are used with many late-model electronic engine control systems that you will learn about in a later chapter.

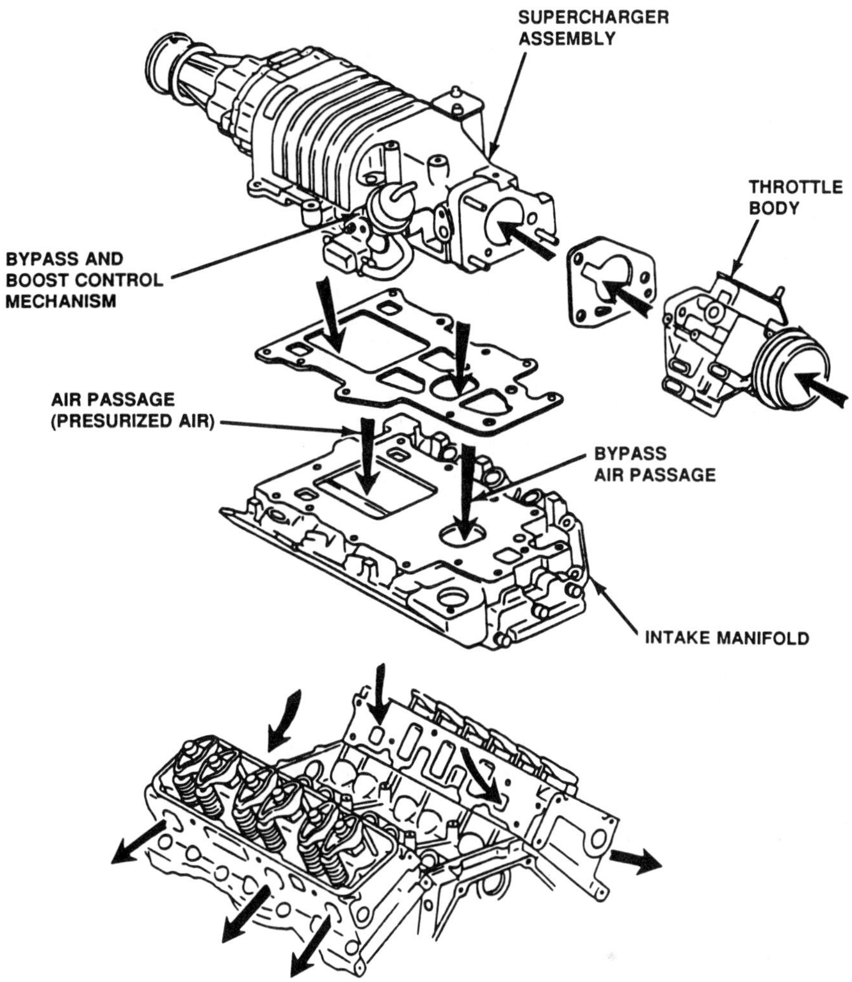


**Figure 19-28.** The GM blower draws intake air through a throttle body at the rear of the blower and delivers compressed air directly to the intake manifold. (GM)

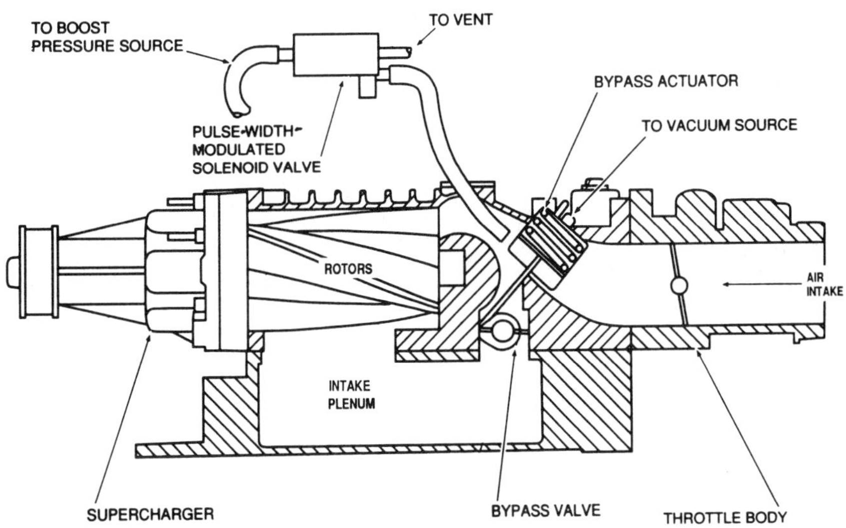


**CROSS SECTION OF SUPERCHARGER**

**Figure 19-29.** The bypass actuator and bypass valve redirect air to the blower inlet to limit boost pressure on the GM blower. (GM)

## TECHNOLOGY BLOWING IN THE WIND

Although the Roots blower has emerged as the supercharger of choice for the early 1990's, several other mechanical supercharger designs have been tried by several carmakers. Most successful in production is Volkswagen's G-lader blower used on Corrado models. "Lader" means "charger" in German, and the blower housing is shaped roughly like the letter "G."

The G-lader has a spiral shaped displacer that oscillates inside the G-shaped chamber of the housing. An eccentric shaft moves the displacer vertically and horizontally. It does not rotate. Air is drawn into the blower as the displacer moves away from the outer chamber wall. Continued oscillation of the displacer compresses the air and forces it out through ports in the center of the housing. Volkswagen says the G-lader is lighter and quieter than a Roots blower and has minimal wear between the displacer and the housing. Boost pressure is almost 12 psi, and the G-lader consumes only about 17 horsepower at maximum throttle and rpm.

The Sprintex blower has male and female rotors that mesh inside a figure-8 housing. As shown in the diagram, air enters at the front of the rotors as the lobes mesh. Air is compressed as the rotor clearance decreases and then exhausted through ports in the side of the case. Airflow is axial rather than radial as in a Roots blower, which results in lower air outlet temperature. Intercooling is said to be unnecessary with the Sprintex blower.

The Comprex supercharger from the Swiss firm, Asea Brown Boveri, has been used on some Opel and Mazda diesel vehicles sold in their home markets. In the Comprex, exhaust pressure waves act directly on intake air through slotted ports in a rotating drum.

Will these or any other mechanical supercharging technologies overtake the Roots blower during the 1990's? We'll have to wait for the 21st century to say for sure.

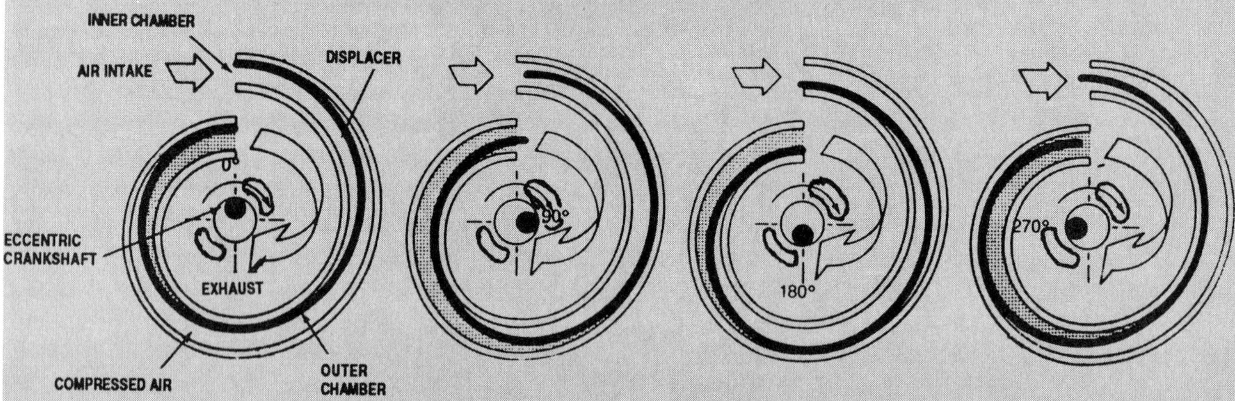


Axial airflow through the Sprintex blower enters at the front of the rotors and exits at the rear through ports in the side of the blower housing.

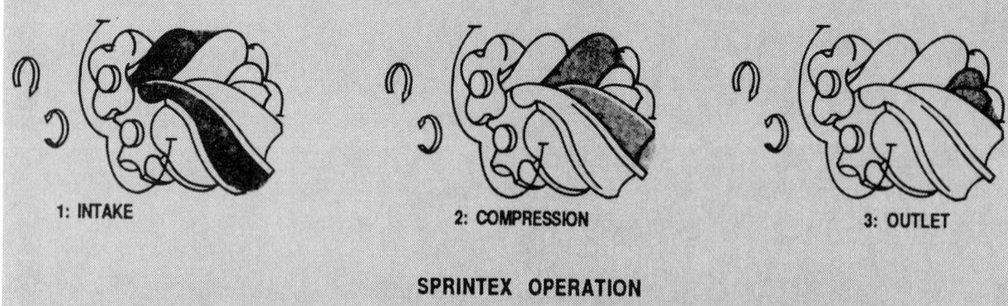


The G-lader supercharger compresses air between an oscillating displacer and the G-shaped passages in the blower housing.

# REVIEW QUESTIONS

## Multiple Choice

1. High compression ratios do all of the following EXCEPT:
   a. decrease $NO_x$ formation
   b. improve volumetric efficiency
   c. improve thermal efficiency
   d. raise compression temperature and pressure

2. Mechanic A says that turbochargers are variable-displacement pumps. Mechanic B says that Roots blowers are also variable-displacement pumps. Who is right?
   a. A only
   b. B only
   c. both A and B
   d. neither A nor B

3. In a turbocharger, engine exhaust drives the:
   a. compressor
   b. wastegate
   c. turbine
   d. blowoff valve

4. Turbocharger maximum speeds can be as high as:
   a. 5,000 to 10,000 rpm
   b. 50,000 to 80,000 rpm
   c. 100,000 to 150,000 rpm
   d. 20,000 to 40,000 rpm

5. Mechanic A says that most of the energy that drives a turbocharger comes from exhaust pressure. Mechanic B says that most energy to drive a turbocharger comes from exhaust heat. Who is right?
   a. A only
   b. B only
   c. both A and B
   d. neither A nor B

6. Typical radial clearance of turbocharger shaft bearings is:
   a. 0.03 to 0.06 inch
   b. 0.3 to 0.6 inch
   c. 0.0010 to 0.0025 inch
   d. 0.003 to 0.006 inch

7. Typical axial clearance (endplay) of turbocharger shaft bearings is:
   a. 0.003 to 0.006 inch
   b. 0.001 to 0.003 inch
   c. 0.010 to 0.030 inch
   d. 0.006 to 0.010 inch

8. The brief hesitation from the time the throttle opens until the turbocharger starts delivering boost is called:
   a. overboost control
   b. ignition lag
   c. wastegate delay
   d. turbo lag

9. Mechanic A says that most carbureted, turbocharged engines have the carburetor downstream from (on the pressure side of) the turbo. Mechanic B says that with port fuel injection, the injectors are on the downstream side of the turbo. Who is right?
   a. A only
   b. B only
   c. both A and B
   d. neither A nor B

10. High pressures and temperatures created by a turbocharger require controls for:
    a. detonation
    b. air-fuel mixture temperature
    c. boost pressure
    d. all of the above

11. To limit boost pressure, most turbochargers have:
    a. water injection
    b. detonation sensors
    c. a wastegate
    d. a priority valve

12. A large, spring-loaded relief valve that limits boost pressure is called a:
    a. wastegate
    b. blowoff valve
    c. priority valve
    d. altitude compensator

13. A heat exchanger that lowers the temperature of the compressed intake air charge is called:
    a. a blowoff valve
    b. an intercooler
    c. a wastegate
    d. a detonation control module

14. Mechanic A says that a detonation sensor controls spark knock, or pinging. Mechanic B says that distributor manifold pressure retard controls spark knock. Who is right?
    a. A only
    b. B only
    c. both A and B
    d. neither A nor B

15. Most detonation sensors contain:
    a. an aneroid capsule, or bellows
    b. a vacuum diaphragm
    c. a piezoresistive crystal
    d. none of the above

## Fill in the Blank

16. The amount of air pressure increase above atmospheric pressure that a supercharger provides is called ________.

17. An engine that draws in air only by allowing atmospheric pressure to force air into the low-pressure area of the intake manifold is called a ________.

18. The percentage figure that indicates the volume of air an engine draws compared to the maximum it could draw in is called its ________.

19. The process of increasing the density of air entering an engine above that which can be drawn in by the natural pumping action is called ________.

20. Increasing an engine's boost pressure raises the engine temperature and ________ pressure.

21. A turbocharger uses ________ energy to boost engine performance.

22. In a turbocharger, the turbine operates by ________ force.

23. The delay during acceleration when a turbo doesn't supply boost is called ________.

24. The ________ is a bypass valve that controls exhaust flow into the turbine to prevent the boost pressure from damaging the engine.

25. Some turbocharged engines have an ________, which acts as a heat exchanger to cool the intake mixture.

# PART SIX
# ELECTRONIC ENGINE CONTROL SYSTEMS

# INTRODUCTION

The final part of this volume brings together all of the engine systems and subsystems that you have studied in the earlier chapters. These last chapters are both a *summary* of what you have studied previously and an *introduction* to the future of automotive technology.

Electronic engine control systems on late-model cars integrate the operation of fuel, ignition, emission control, and other vehicle systems. Fully integrated electronic control systems have been state-of-the-art engineering for over a decade. All the principles of chemistry, physics, electricity, and electronics have their most relevant applications in these systems. Your ability to understand and service them will determine the rest of your automotive service career.

Part Six begins with two chapters on the principles of microcomputer systems and an introduction to sensors and actuators. These chapters provide an overview of computer system principles and the common sensors (input signal devices) and actuators (output devices) used by most carmakers.

The last chapter in this part traces the development of specific carmakers' control systems and the integration of electronic fuel injection as the predominant, "conventional," fuel system on late-model cars. After studying these chapters, you will realize that there are far more similarities than differences among the engine control systems used by various manufacturers.

# 20 MICROCOMPUTER ELECTRONIC SYSTEMS

## WHAT

A modern automobile may have a dozen or more separate "computers" on board to control engine operation, air conditioning, instrument displays, suspension operation, power steering, braking, and other vehicle functions. Engineers use computers to design automobiles and to test their operation. Automobile factories have computers to control automatic manufacturing equipment. Computers do many jobs for different people in different applications. But all computers, whatever their size and purposes, operate on similar principles.

## WHY

You probably will never design or build an automotive computer or modify its operation—at least not as a service technician. If you know the basic parts of a computer system and understand its operation, however, you can recognize when it is not operating correctly and what parts are not working properly. This understanding also will help you to recognize how and why a system might malfunction, which lets you apply basic knowledge to specific troubleshooting. If you know the principles of all computers, you will see more similarities than differences among various systems, and you will understand the operation of new systems as carmakers introduce them each year.

## GOALS

After studying this chapter, you should be able to:

**1.** Explain the four basic functions of all computers.

**2.** Explain the difference between analog and digital computers, as well as analog and digital voltage signals.

**3.** Understand basic digital logic and how it is similar to operation of simple relays.

**4.** Explain the principal parts of a computer.

**5.** Explain the difference between open-loop and closed-loop control system operation.

## WHAT IS A COMPUTER?

Actually, "computer" is probably not the best name for these devices. The first computers built 50 or more years ago were designed by scientists and mathematicians to do complex mathematical calculations. Their principal job was to compute. The name stuck.

The earliest computers and today's modern versions use the same principles of converting information to combinations of voltage signals that represent combinations of numbers. The number combinations might represent other kinds of information: speed, temperature, or even letters and words, for example. The computer processes the signals by computing the numbers they represent and then delivering information in computed, or processed, form.

### Computer Functions

The operation of any computer is divided into the four basic functions of input, processing, storage, and output. These functions aren't really unique to computers, however. You can identify them in other systems, such as the mechanical drives and hydraulic systems of an automobile. Here is how the four functions operate in a computer.

**1.** *Input*—The computer receives a voltage signal from an input device. The device might be a pushbutton on an instrument panel or a **sensor** on an engine component. Keyboard terminals for office computers are also input devices.

Automobiles have many mechanical, electrical, and magnetic sensors that measure things like car speed, engine speed, temperature, air pressure, and airflow. The sensors convert the measurements to voltage input signals for the computer. An input sensor can be something as simple as a switch that opens and closes a circuit to the computer.

**2.** *Processing*—The computer takes the input signals and switches them through many series of electronic logic circuits in its programmed instructions. These logic circuits pro-

Figure 20-1. Most automobile computers are contained in metal boxes and mounted in the passenger compartment.

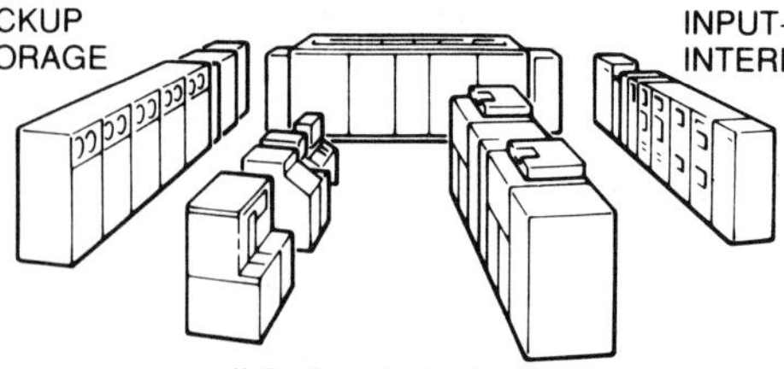


Figure 20-2. Mainframe computers occupy large spaces and are used in commercial, industrial, or scientific applications. (Toyota Motor Company)

cess the input signals into output commands, which are another series of voltage signals that the computer sends to other devices in the system.

3. *Storage*—Every computer has an electronic memory to store its program instructions. Also, many programs require that some input signals be stored and not processed immediately, or be stored for later reference. Some output signals also may be stored momentarily before sending them to other parts of the system. The information storage ability of a computer is its "memory," and we will look at different kinds of memory later.

4. *Output*—All of the information processing and memory ability of a computer would not be worth much if it did not give back some other information or instructions. After processing the input signals, the computer returns output voltage signals to various output devices. On an office computer, the output device might be a printer or a cathode ray tube (CRT) display screen. On an automobile, the output device might be an instrument panel display or a system **actuator**. Actuators are devices that adjust engine idle speed, control air conditioning temperature, regulate fuel metering, adjust suspension height, and keep brakes from locking.

Another feature of many computer systems is that an output signal from one system can also be an input signal to another. Thus, a temperature-control computer can send an output signal that engages an air conditioning compressor clutch. That same signal can also be an input signal to an engine-control computer. The engine computer then sends an output signal to another actuator to control idle speed with the increased engine load. Not all air conditioning and engine computers work exactly this way, but this example shows how some systems are interconnected through their output and input functions.

Regardless of its purpose, complexity, or size, every computer system operates with the four basic functions of input, processing, storage, and output. Remembering these four functions will help you to organize your troubleshooting. When you test a system, you will be trying to isolate a problem to one of these areas. You can adjust or repair some input and output devices. You can only replace a processing or storage component, as well as some other kinds of input and output devices.

## Computer Sizes

Automotive computers are installed in small metal boxes, figure 20-1, and are generally mounted in the passenger compartment under the instrument panel or behind a kick panel. Because of their small size, it's often difficult to visualize the similarities between these small computers and large computer systems. Historically, computer engineers recognize three general sizes, or classes, of computers: mainframe computers, minicomputers, and microcomputers.

Mainframe computers are the large, complex systems built by IBM, Cray Research, and other major manufacturers, figure 20-2. The processing and storage equipment often occupies a large room or an entire building built just to house the computer. A mainframe computer can have thousands of separate input devices, including keyboard terminals, video input devices, system sensors, and connections to output signals from other computers. A large mainframe computer can do several different jobs at once and can send signals to thousands of different output devices, including printers, display screens, system actuators, and other computer systems. Many of the computers used

**Figure 20-3.** The silicon chip containing the memory circuits can be seen through the window in this GM PROM. The entire circuit package is about 1 1/2 inches long.

to design automobiles and to control automatic factory equipment are mainframe computers. Scientists at research laboratories have used large mainframe supercomputers to do calculations that earlier mathematicians only dreamed about but never lived long enough to do.

Modern minicomputers have the computing power of earlier mainframe computers, but they are much smaller. Many businesses use a single minicomputer system for their accounting records, sales transactions, and word processing. Schools use them to process student and class records. A minicomputer can have many different input and output devices and can do several jobs simultaneously. It does not fill a large room or an entire building, and it usually does not have the processing capacity or speed of a large, modern mainframe computer.

A microcomputer is a computer in a box. Apples, IBM PC's, and other personal computers are all microcomputers. Most are smaller than a large television set but have the computing power of mainframe computers built 30 or 40 years ago. Most personal microcomputers are designed for use by one person at a time, although they can be connected to larger mini or mainframe computers and serve as input terminals or auxiliary computers for a larger system. Most microcomputers can only do one job or "run one program" at a time, but these are often complex processing operations. A microcomputer does not have the computing power, speed, and accuracy of a modern mainframe. It can, however, do many different jobs very well, including the operation of vehicle systems.

## Onboard Computers

This brief summary of computer types and sizes brings us to the microprocessor, or "computer on a chip." Through the electronic technology of very large scale integration (VLSI), engineers can pack tens of thousands of transistor logic circuits on a chip of silicon only one-quarter or three-eighths of an inch square. Installed in its circuit package with all connecting leads, a microprocessor is about the size of a small drawing eraser, figure 20-3. The microprocessor receives the computer input signals and sends out

### SILICON DIP

If it weren't for silicon DIP's, computers would still be far too expensive to use on automobiles. PROM's could not be replaced to cure driveability problems, and personal computers (IBM PC's and Apples) would cost tens of thousands of dollars and be reserved for very large business offices.

So what is this marvelous DIP? Is it some kind of special chemical? a magic elixir? what?

A DIP is a dual inline package, a standardized way to package IC devices so that they can be plugged into circuit boards on different kinds of computers. When you look at a computer circuit board, you see many small rectangular devices with rows of small metal connector legs projecting downward on each side. These are dual inline packages (DIP's).

Microprocessors, coprocessors, memory chips, input-output circuits, and other integrated circuit chips are installed in the small plastic packages. Their circuit ends are connected to the leg connectors of the dual inline package. In this way, one standard microprocessor or memory chip can be used in a wide variety of computer circuits. Dozens of memory chips can be installed in parallel on a circuit board to give a computer an enormous amount of electronic storage.

Dual inline packages range in size from about 1/4-inch long to 2 or 3 inches long. Widths vary, as does the spacing of the connectors. All sizes, however, conform to a variety of standardized shapes that allow computer engineers to design circuit boards in the most economical way.

When you change a PROM in an engine control or body control computer, you are handling a dual inline package device. Without dual inline package standards—silicon DIP—we wouldn't have the powerful, economical computers that we use in our homes, shops—and cars.

the output signals. Its thousands of circuits are the computer for the entire system.

It is interesting to note that the microprocessors used in onboard automobile computers are the same ones used in personal microcomputers. Because of their versatility, microprocessors can be programmed for many different functions. When the microprocessor and other IC memory components are programmed for specific purposes, they can be installed on a circuit board in a computer to control engine operation or to do financial calculations at a businessman's desk.

The early automobile onboard computers handled one function and made 10 engine adjustments per second. A decade later, today's onboard computer can adjust 27 functions while processing up to 2 million instructions per second. When the Cadillac Allanté was introduced, its computer had 90.7 kilobytes of memory, or more than the best personal computer available in 1986.

We have mentioned computer programs, memory, voltage signals, and processing in this introduction. The remaining sections of this chapter explain how these and other computer functions operate in automobile systems.

## ANALOG AND DIGITAL PRINCIPLES

Every computer needs instructions to do any job, just as you need written instructions to do a repair job you have never done before. The instructions must be organized into a given sequence to perform a particular task. The instructions for a computer are called its **program**, but a computer cannot read words or even numbers. A computer can only read voltage signals.

Because a computer is based on IC chips with thousands of transistors that can respond to thousands of combinations of voltage signals, all we have to do is "translate" words, numbers, and other information into various combinations or amounts of voltage that the computer can understand. That's what the programming languages do.

### Analog and Digital Computers

In chapter 8, we introduced the terms **analog** and **digital** to describe transistor operation as an amplifier or as a relay. Computers and their input and output signals also can be either analog or digital. Understanding these terms will help you understand computer system operation.

Analog means that a voltage signal or a processing function is infinitely variable in relation to something being measured or an adjustment to be made, figure 20-4. Digital means that an input or output signal or a processing decision is either yes-no, high-low, or on-off, figure 20-5. Most automobile system operating conditions are analog variables. Temperature, for example, doesn't change abruptly from zero to 100 or 200 degrees. It varies in infinite steps from high to low. The same is true of engine speeds, vehicle speed, airflow, fuel consumption, and other factors.

A computer may need to measure temperature changes through a range from 0° to 250° F. This can be "translated" for the computer as an analog voltage that varies from 0 volts at 0 degrees to 5 volts at 250°. Any analog signal between 0 and 5 volts represents a proportional temperature between 0 and 250° F.

A digital signal is voltage that is either high or low, or current that is either on or off, with no changing range of voltage or current in between. Suppose that a computer needs to know that engine temperature is either above or below a specific level, say 100° F. It doesn't need to know the exact temperature, only whether it's above or below a certain point. This can be translated to a digital signal of no voltage below 100° F and any arbitrary voltage when temperature exceeds 100° F. In this way, a digital signal is similar to a simple switch that opens and closes a circuit.

This kind of signal is called a digital signal because the on and off signals are processed by the computer as the digits, or numbers, 0 and 1. Digital computers use a mathematical system called the **binary system**, which uses only these two digits. Any number from our conventional decimal number system can be translated into a combination of binary 0's and 1's. So can any word from any language if we use specific combinations of 0's and 1's to represent letters of the alphabet.

A transistor working as an amplifier is the basis for an analog computer. The output current or voltage varies in direct proportion to the input current or voltage. The amplified output is thus analogous to the input. The ratio of input change to output change remains constant.

A transistor working as a relay is the basis for a digital computer. The transistor output switches from cutoff to saturation as the input signal switches from off to on. Output is either high or low, on or off. The on and off output signals represent the binary digits 1 and 0 for a digital computer.

Some early engine computers were analog devices that received and delivered analog input and output signals. Analog computers, however, are susceptible to changing temperature and supply voltage and to signal interference. They also are expensive to build. Digital computers are not as susceptible to changing temperature and supply voltage or signal interference. Digital computers also are less expensive to build and faster than analog computers. Almost all current automobile computers are based on digital microprocessors.

Digital computers change analog input signals to digital **bits** (binary digits) of information through analog-to-digital (AD) converter circuits. The microprocessor processes data as high-low, on-off voltage signals through digital logic networks made of thousands of **MOSFET** transistors. Based on its computations, the computer sends an output voltage to an instrument panel display or to a system actuator to change spark timing,

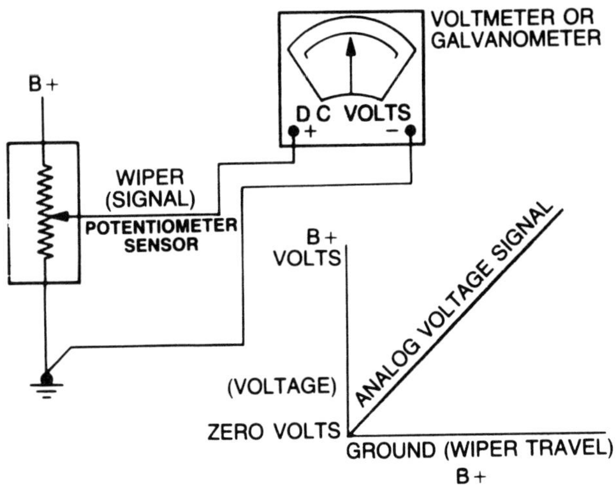


**Figure 20-4.** An analog sensor circuit. (Ford)

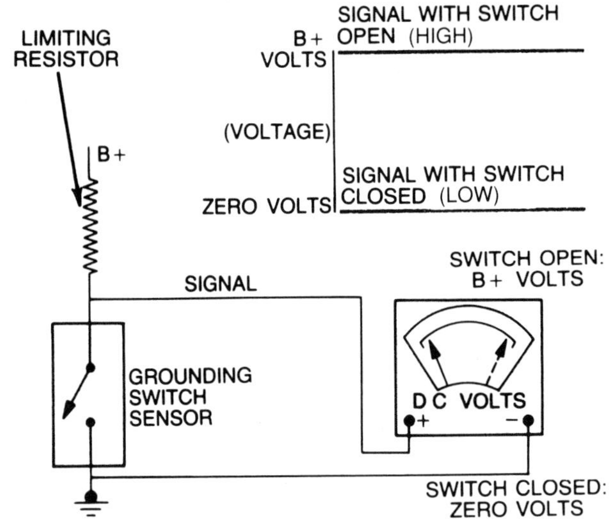


**Figure 20-5.** A digital sensor circuit. (Ford)

fuel metering, cruise control speed, air conditioning temperature, or another condition.

These onboard computers can process thousands of digital signals per second. They do it faster than humans can respond to corresponding words and numbers because their circuits can switch the voltage signals on and off in billionths of a second.

## Binary Numbers

On and off voltage signals can represent the digits 1 and 0 of the binary number system. You may have worked with the binary system in mathematics classes, but let's review how it works in a digital computer.

Our customary decimal system uses the numbers 0 through 9, and we call it a base 10 system. For single numbers from 0 through 9, we write one column of numbers, or numbers in the ones position. When we reach numbers above 9, we add another column or position to the left: the tens position. Thus the number 10 equals one ten and zero ones; 29 equals two tens and nine ones. Each successive position to the left multiplies the number by a power of ten:

$$1 = 10^0$$
$$10 = 10^1$$
$$100 = 10^2$$
$$1000 = 10^3$$

The base 10 decimal system is not the only possible system of numerical notation. Mathematicians do, in fact, use number systems that work on base 8, base 12, base 16, and base 2—the binary system. In the binary system, whole numbers are grouped from right to left as they are in the decimal system. Because the system uses only two digits, the first position must equal 0 or 1. To write the value of 2, we must use the second position. We write it 10, but it equals one two and zero ones. Each successive position to the left multiplies the number by a power of two:

$1 = 2^0$ = decimal number 1
$10 = 2^1$ = decimal number 2
$100 = 2^2$ = decimal number 4
$1000 = 2^3$ = decimal number 8

Similarly, we can change other decimal numbers to binary numbers. For example:

| *Decimal Number* | *Binary Equivalent* |
|---|---|
| 3 | 11 |
| 5 | 101 |
| 7 | 111 |
| 9 | 1001 |

We could go on like this indefinitely, translating any decimal number you can think of into a binary equivalent. It may seem like an academic exercise, but it has a practical use in a digital computer. If we arrange several thousand transistor circuits in different series and parallel combinations in a microprocessor, we can make them switch on and off in combinations to equal any binary number faster than you can write the decimal equivalent.

## Analog-to-Digital Conversion

Let's go back to our earlier examples of temperature signals for a computer. If a computer only needs to know that temperature is below or above a certain point—100° F, for example—we can connect a digital sensor that acts as a simple switch. Below 100° F, the switch is open and the computer receives no input voltage from that sensor circuit. Above 100° F, the switch closes and the computer receives an input voltage signal. That's a simple digital, on-off circuit: on = 1, off = 0.

Suppose, however, that the computer needs to know the exact temperature (or any temperature in a range) within one degree. Earlier, we used the example of a sensor that measures temperature from 0° F to 250° F. The sensor may send an analog voltage signal that varies from 0 to 5 volts. Each 1-volt change in the signal equals a 50-degree change in temperature. But this computer needs more precise temperature measurements than 50-degree increments. Fortunately, automotive computers and many sensors are precise enough to operate on signals that change by fractions of a volt. The fractions can be as small as 0.001-volt, or 1-millivolt, increments.

If our computer needs to know the temperature within one degree, it must respond to sensor voltage changes of 0.020 volt (20 millivolts). Most automotive computers can do this. If 0 voltage equals 0-degree temperature and 5 volts equals 250 degrees, then:

1.0 volt = 50 degrees
0.2 volt = 10 degrees
0.1 volt = 5 degrees
0.02 volt = 1 degree

If the temperature is 150 degrees, the sensor signal is 3.00 volts. If temperature rises to 151 degrees, sensor voltage rises to 3.02 volts.

Of course, the temperature passes through fractions of a degree as it changes from 150 degrees to 151 degrees, and the sensor signal passes through smaller fractions of a volt as it changes from 3.00 to 3.02 volts. These are true analog changes. Figure 20-6 shows a changing curve for any analog variable. It could be temperature; it could be voltage. It doesn't matter. The analog curve traces a continuous line from low to high and back to low. You can mark an infinite number of points along such an analog curve.

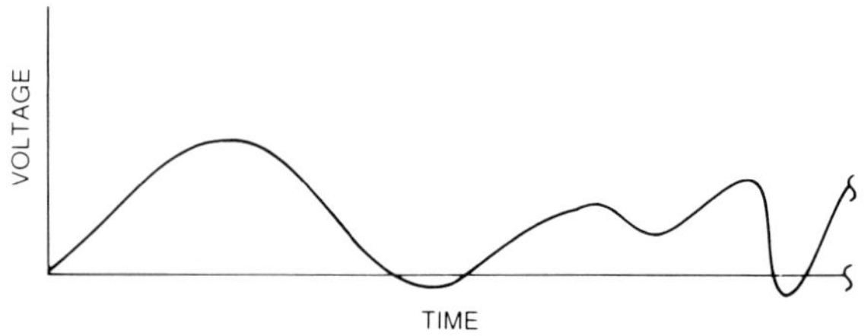


**Figure 20-6.** A continuously variable analog signal.

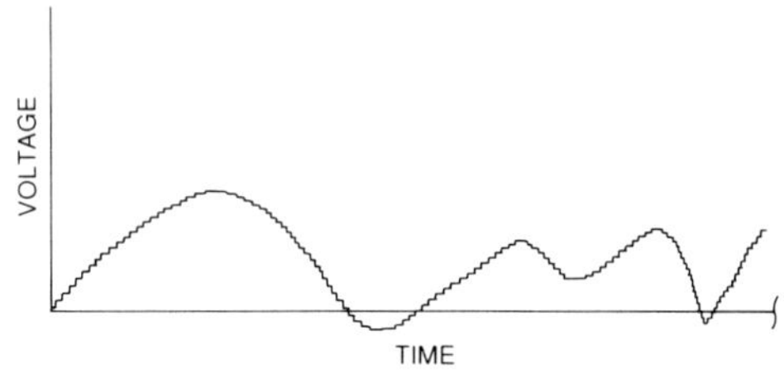


**Figure 20-7.** A digitized representation of the analog signal in figure 20-6.

Our computer, however, is a digital computer. It only needs to process signals that equal one degree temperature changes. The computer converts the analog voltage from the sensor to a series of 0.020-volt (20-millivolt) changes for each degree. The computer does this through analog-to-digital (AD) conversion circuits. Engineers refer to this process as "digitizing" an analog signal. The result is a stairstep graph of digital signals that approximates the true analog signal, figure 20-7. The smaller and closer together the digital increments, the more closely the digital inputs resemble the actual analog measurement.

The process of analog-to-digital conversion takes us beyond the scope of an automobile service text, but it does bring us back to the binary number system. Transistors can be built to switch on and off at different voltage levels or with different combinations of voltage signals. In our sample computer, groups of transistors must switch from no conduction (off) to full conduction (on, or saturation) at 20-millivolt increments. Remember also that we can convert any decimal number to a binary number. We can display that binary number, or create an input signal, by varying combinations of transistors that are on or off. Our 150-degree temperature in binary numbers is:

10010110

When the temperature changes to 151 degrees, the binary number changes to:

10010111

We can represent these binary combinations of 1's and 0's by switching different groups of transistors on and off. The computer reads the resulting combinations of voltage signals as the binary numbers and does its calculations. Because electrons of current travel at the speed of light through miniature computer circuits, all of these actions happen faster than you can read the decimal number 150.

## Number System Conversion

Digital computers do all their processing in binary digits (bits) of 1's and 0's. Binary equivalents of decimal numbers can get quite awkward for people to work with, however. The binary number 10010110, which equals decimal 150, is quite long enough. Consider the binary equivalent of an engine speed of 2,500 rpm: 100111000100. To simplify these awkward binary numbers, computer engineers use other number systems: the octal system (base 8), the hexadecimal system (base 16), and the binary coded decimal (BCD) system.

These other number systems allow engineers to load programming information into computers with simpler number combinations than they could with binary numbers. The computer then changes the octal, hexadecimal, or BCD numbers into binary numbers. The octal and hexadecimal systems work well for this because they are based on numbers that are powers of 2: $8 = 2^3$, $16 = 2^4$. Conversions between binary numbers and base 8 or base 16 numbers are relatively easy for a digital computer.

Conversions between decimal (base 10) and binary numbers are not as easy for the computer because decimal 10 is not a power of 2. Extra circuits are needed and the calculations are more complicated. Engineers do use binary coded decimal (BCD) numbers for some operations, however. BCD numbers use groups of four binary numbers to represent each digit of a decimal number.

Again, octal, hexadecimal, and BCD mathematics takes us beyond the scope of an automotive service text. We won't go into these number systems because they are subjects for several complete texts in themselves. Different number systems are important, however, as examples of various ways in which information can be coded and processed. Any number system can be used to represent any measurable value. Similarly, any number can be converted to binary 1's and 0's for digital processing as on-off voltage signals. And this brings us to the subject of digital logic.

## DIGITAL LOGIC

How does a computer compute or decide to turn an output signal on and off? You can spend years studying digital logic, but all digital computers manipulate data bits with three basic logic circuits, or **logic gates**: the NOT, AND, and OR gates. These terms describe the switching functions of the circuits, not the electronic construction. We call them "gates" because the circuits act as gates for output voltage signals in response to different combinations of input signals. Each gate accepts a voltage signal at its input terminals and transforms it according to the switching arrangements of its internal circuits. The transformed signal appears at the output terminal of the logic gate. Inside each logic gate is one or more active switching elements. These are controlled by the input voltage and, in turn, control the output voltage.

Just as atoms are the fundamental building blocks of matter, logic gates are the fundamental building blocks of all digital computer operations. Logic gates are the thousands of field-effect transistors (FET's) in a microprocessor. The symbols in figures 20-8 through 20-12 represent logic functions, not electronic construction. The tables next to the symbols are called "truth tables." They indicate the logical (truthful) combinations of input and output signals for each logic gate.

A digital microprocessor computes by switching output voltage on and off in response to one or more input voltage signals. The absence or presence of input voltage changes the transistors of the logic gate from cutoff to full saturation. This equals an *off* (low) or an *on* (high) signal (a binary 0 or 1). Input and output signals can be combined in logical combinations to equal binary numbers.

Just as there are two current flow theories (conventional and electron flow) and two rules for determining electromagnetic current direction (right-hand and left-hand), there are two rules for explaining digital logic.

**1.** Positive logic defines the most positive (highest) voltage as a logical 1 (on). Therefore, negative (low) voltage is a logical 0 (off).

**2.** Negative logic defines the most negative (lowest) voltage as a logical 1 (on). Therefore, positive (high) voltage is a logical 0 (off).

Additionally, high or low voltage can result from closing a circuit and allowing current to flow to a switching point. It also can result from leaving a circuit open with high voltage applied to a switching point. When the circuit is closed, voltage drops low at that point as current flows through the complete circuit. The following examples are based on positive logic and the simple application of voltage to logic gate inputs.

### Basic Logic Gates

The simplest logic gate is a NOT gate, figure 20-8, which inverts the signal. If voltage to its single input terminal is high, or on (logic 1), output voltage is low, or off (logic 0). If input logic is logic 0, output logic is logic 1. A NOT gate simply changes binary 1's to 0's and vice versa. Figure 20-8 also shows the truth table for a NOT gate and the logic symbol: an A with a bar over the top. (The overbar represents an inverted output signal in any logic symbol.)

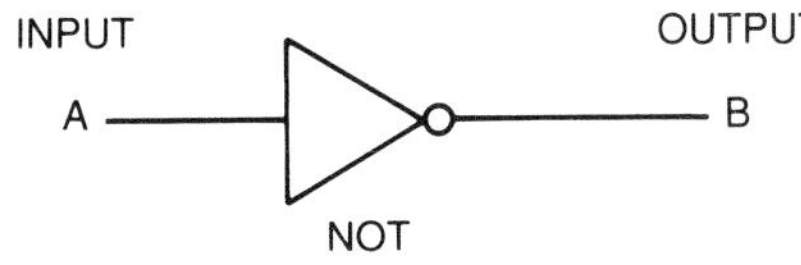


| A | B |
|---|---|
| 0 | 1 |
| 1 | 0 |

$\bar{A}$
(OVERBAR)

**Figure 20-8.** A NOT gate, logic symbol, and truth table.

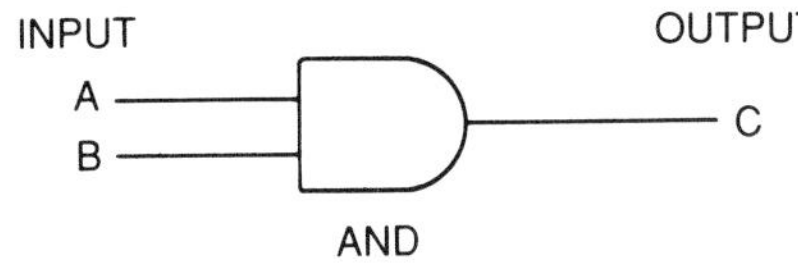


| A | B | C |
|---|---|---|
| 0 | 0 | 0 |
| 0 | 1 | 0 |
| 1 | 0 | 0 |
| 1 | 1 | 1 |

A • B = C
(LARGE DOT)

**Figure 20-9.** An AND gate, logic symbol, and truth table.

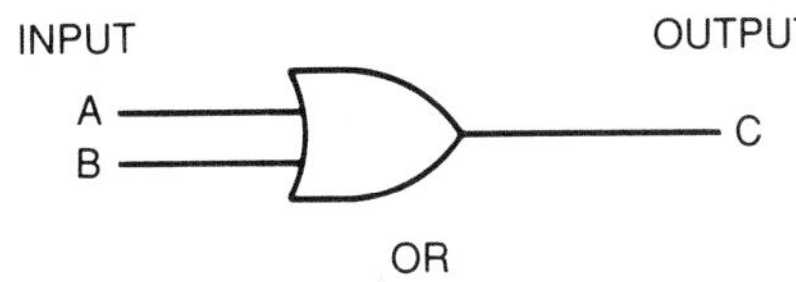


| A | B | C |
|---|---|---|
| 0 | 0 | 0 |
| 0 | 1 | 1 |
| 1 | 0 | 1 |
| 1 | 1 | 1 |

A + B = C
(PLUS SIGN)

**Figure 20-10.** An OR gate, logic symbol, and truth table.

An AND gate has at least two inputs; it may have more. Figure 20-9 shows a simple AND gate with two inputs. Output is high (1) only when both inputs are high (1). If either or both of the inputs are low (0), output is low (0).

An OR gate, figure 20-10, also has two or more inputs and one output. The output is high (1) when any one or more of the inputs are high (1). Output

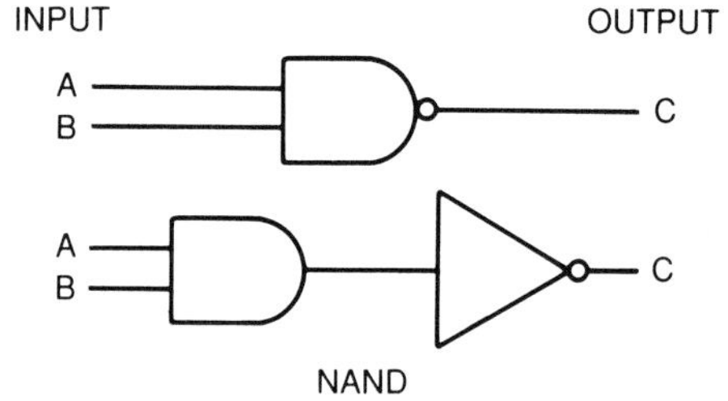


| A | B | C |
|---|---|---|
| 0 | 0 | 1 |
| 0 | 1 | 1 |
| 1 | 0 | 1 |
| 1 | 1 | 0 |

$\overline{A \cdot B} = C$

**Figure 20-11.** A NAND gate, logic symbol, and truth table.

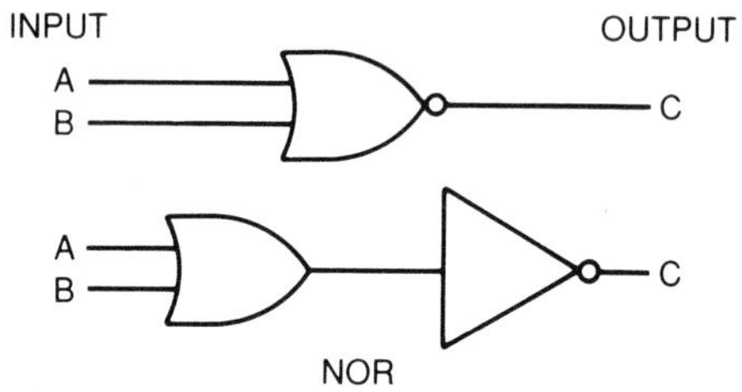


| A | B | C |
|---|---|---|
| 0 | 0 | 1 |
| 0 | 1 | 0 |
| 1 | 0 | 0 |
| 1 | 1 | 0 |

$\overline{A + B} = C$

**Figure 20-12.** A NOR gate, logic symbol, and truth table.

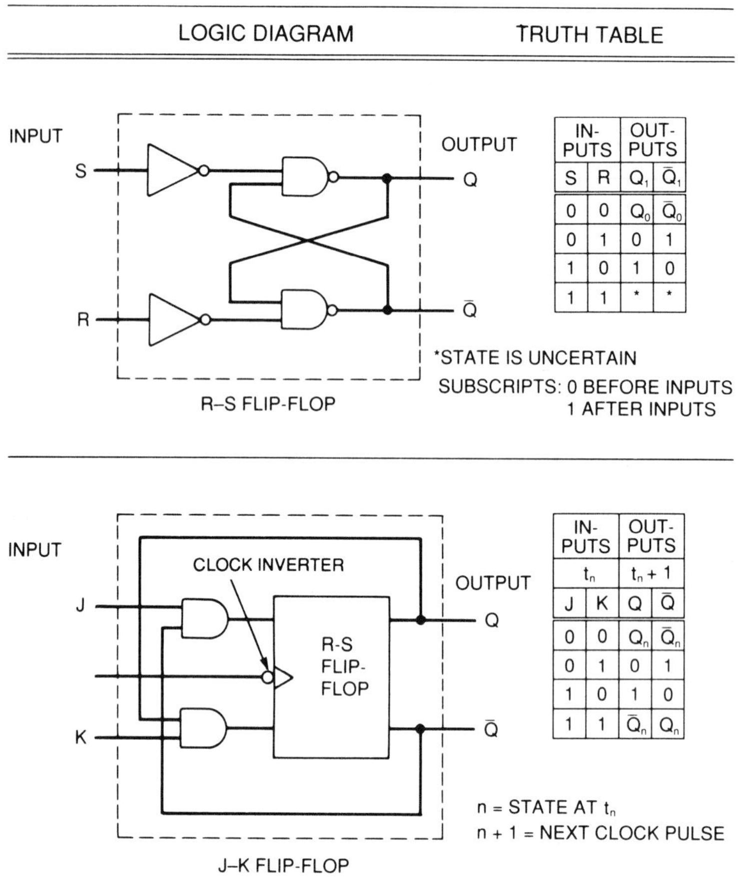


| IN-PUTS | | OUT-PUTS | |
|---|---|---|---|
| S | R | $Q_1$ | $\bar{Q}_1$ |
| 0 | 0 | $Q_0$ | $\bar{Q}_0$ |
| 0 | 1 | 0 | 1 |
| 1 | 0 | 1 | 0 |
| 1 | 1 | * | * |

| IN-PUTS | | OUT-PUTS | |
|---|---|---|---|
| $t_n$ | | $t_n + 1$ | |
| J | K | Q | $\bar{Q}$ |
| 0 | 0 | $Q_n$ | $\bar{Q}_n$ |
| 0 | 1 | 0 | 1 |
| 1 | 0 | 1 | 0 |
| 1 | 1 | $\bar{Q}_n$ | $Q_n$ |

**Figure 20-13.** Flip-flop circuits (R-S and J-K types) are the basis of computer memory. They "remember" their output settings until they receive new input signals.

is low (0) only when all inputs are low (0). Figures 20-9 and 20-10 also show the logic gate truth tables and symbols.

Other logic functions can occur when these basic gates are combined. An inverter (NOT gate) placed after an AND or an OR gate inverts the output signal and creates a NAND (not AND), figure 20-11, or a NOR (not OR) gate, figure 20-12. Notice that the NAND and NOR gates can be drawn as an AND or an OR gate followed by a NOT gate, or as the basic AND and OR gates with a small circle on the output. The circle represents an inverted output on any logic gate symbol.

**Other Logic Circuits.** A combination of OR, AND, and NOR gates can create an "exclusive OR," or XOR circuit in which the output is high only when one or the other input is high, but not both. XOR and AND gates can be combined into adder circuits, which are the basis of electronic calculators.

All of these logic gates are combinational logic circuits whose outputs are determined only by the present inputs. They have no memory. Other logic circuits can be built from these basic gates that remember previous inputs and do not change their outputs until they receive new input signals. These are called sequential logic circuits, and they are the basis of computer memory circuits. Flip-flop circuits, such as those in figure 20-13, hold their output signals when input signals are removed. They generate new outputs only when they receive new combinations of inputs.

Additionally, a computer engineer can build any arrangement of logic gates for any processing function by combining transistors in various circuit combinations. Fortunately, the engineer does not have to collect thousands of separate resistors, diodes, capacitors, and transistors and then wire them together on a circuit board to build his or her computer. Thirty years ago, before integrated circuits were invented, engineers did just that. Even earlier, engineers had to wire together thousands of vacuum tube amplifiers to perform digital logic operations. Today, all of the transistors needed for different logic gates are built into IC chips. IC devices are available with different combinations of gates per circuit and different input and output connections for various gates.

Also fortunately, you don't have to be a computer engineer to troubleshoot an automobile onboard com-

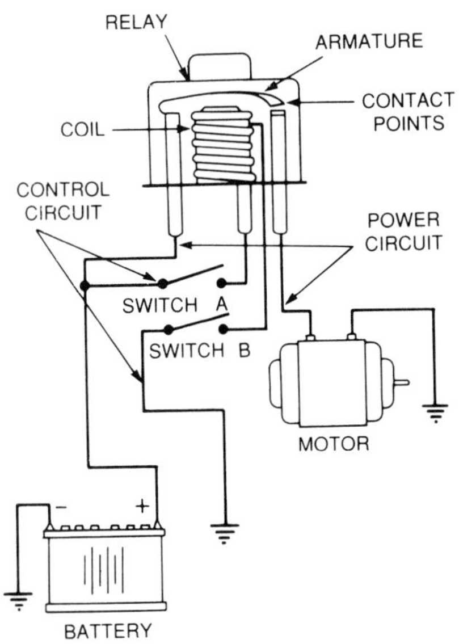


**Figure 20-14.** Switches A and B work like an AND gate in the control circuit to energize (close) the power circuit.

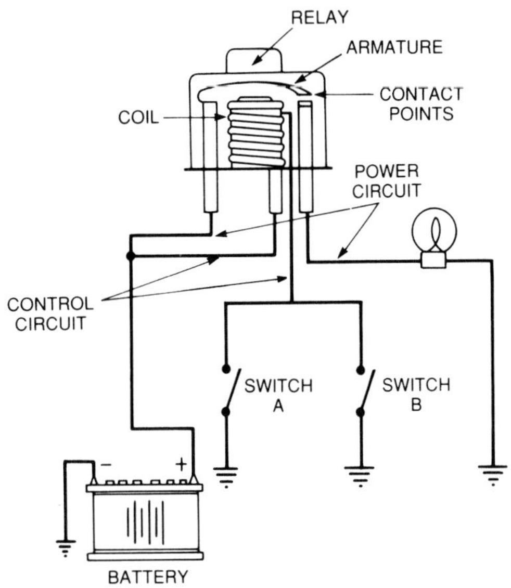


**Figure 20-15.** Switches A and B work like an OR gate in the control circuit to energize (close) the power circuit.

puter. Understanding these fundamentals of computer logic circuits and the way they operate on the electrical principles of voltage and current will help you understand computer operation as you study it in this and later chapters.

### Relay Logic

It's often difficult to visualize computer logic operations because the processing takes place so fast and on circuits so small that you can't see the parts without a microscope. We can, however, build a simple computer with some relays and switches.

Suppose that we take one relay and connect the voltage supply terminal of its control circuit to the battery. Then we put a switch in that wire, figure 20-14. Next, we connect the other control circuit terminal to ground, also through a switch. The relay will be actuated and complete the power circuit only when both switches are closed. We have just built an AND gate. We have also built a common starter relay circuit to switch voltage for a starter motor. Even though we drop voltage across the relay coil and the voltage level is different at each switch, both switches must be closed (high) to produce an output voltage through the relay. This makes it a basic logic gate with the AND function.

Suppose we take another relay and connect two switches in parallel to one terminal of the control circuit. Each switch can complete a circuit to ground. The other control circuit terminal is connected to the battery, figure 20-15. We can actuate the relay and provide an output signal in the power circuit by closing either or both of the control switches. This is a basic OR gate. We can take this relay and these switches and install them in a dome lamp circuit to turn on the interior lights when either or both doors of a car are open.

These are only two examples of "relay logic" circuits that have been used on automobiles for decades. Individual computer logic gates are also this simple. In fact, you may hear computer technicians talk about computer "relay logic."

If you look at the input switch connections in figures 20-14 and 20-15, you can see that different combinations of series and parallel connections can produce different logic functions. Semiconductor engineers do the same thing on IC chips by arranging thousands of transistors in various series and parallel combinations to produce thousands of logic gates.

## COMPUTER PROGRAMS

Earlier in this chapter, we mentioned that any computer needs instructions to do its job. These instructions are the computer program. Computer designers use programming languages to translate instructions into binary numbers (or octal or hexadecimal or BCD numbers). The numbers are then translated into voltage signals that a computer can "read" to understand its program instructions.

The program instructions that a computer uses to understand its input data, do its calculations, and send its output commands consist of:

- The mathematical instructions, in binary form, for processing data.
- Information on engine and vehicle constant values, such as number of cylinders, displacement, compression ratio, gear ratios, type of fuel system, emission devices, vehicle weight, and accessories.
- Information on variable values, such as engine and vehicle speed combinations, temperature, airflow, fuel flow, timing, EGR flow, air injec-

tion operation, transmission gear position, vehicle load, braking effort, and other factors.

The first two elements are fixed values and can be loaded into computer memory rather easily. To load the variable values into memory, engineers must simulate the vehicle, or a vehicle system, in operation. They load programs with all of the possible variable conditions for any system into a large mainframe computer. Then they let the mainframe computer calculate all the combinations of those variables and determine the control program for the vehicle onboard computer.

One example of this system simulation is the process called **engine mapping**. Engineers operate a vehicle on a dynamometer and manually adjust speed, load, spark timing, fuel metering, airflow, and other variables. Using a larger computer with the dynamometer, they determine the optimum fuel, ignition, exhaust gas recirculation (EGR), and other output settings for best performance, economy, and emission control. This information becomes part of the program for the onboard computer. It is stored in the programmable read-only memory (PROM) chip for the computer on that particular model of car.

Engine mapping creates a 3-dimensional performance graph for a particular engine, vehicle, and accessory combination, figure 20-16. When the PROM is installed in a computer, it "remembers" the settings for timing, EGR, fuel metering, and other adjustable variables. Then, based on speed, temperature, airflow, and other information from its sensors, the computer adjusts timing, fuel, EGR, and other conditions for best performance, economy, and emission control.

By storing this information on one PROM, carmakers can use one basic computer for many vehicle models. To make the engine control system work for a specific model, they install the correct PROM. Some carmakers use a single PROM chip plugged into the computer, figure 20-17. Others use a larger calibration module, fig-

## COMPUTERS PLUS

When you get into troubleshooting, replacing parts, and otherwise servicing onboard computer systems, you will probably feel overwhelmed with thousands of part numbers and other variations. Don't worry about it. Don't try to memorize all the part numbers. You're not alone. Here are some statistics to show you what the world's largest corporation has to deal with in terms of electronic variations.

In 1986, General Motors introduced its Computerized Automotive Maintenance System (CAMS). CAMS is a computer-based diagnostic system that lets a technician plug into a GM car and troubleshoot electronic problems. CAMS provides online test procedures, specifications, and service bulletins for all GM vehicles from about 1980 to the present.

As of the 1988 model year (the last year for which complete figures are available), CAMS represented the following volume of data "crunched" into a computer database:

- Over 7,000 part numbers for ECM PROM's from 1980 to 1988
- 60,000 unique part numbers for the CAMS system, its parts, and its software
- 3,000 part numbers for different electrical connectors on all GM vehicles
- 6.2 million lines of computer code to make the CAMS system operate
- 37 dedicated service bays at the GM technical center to test the CAMS application software

If these numbers seem overwhelming, you're probably right. By 1992, GM quietly suspended CAMS development, because of high cost and lack of dealer support.

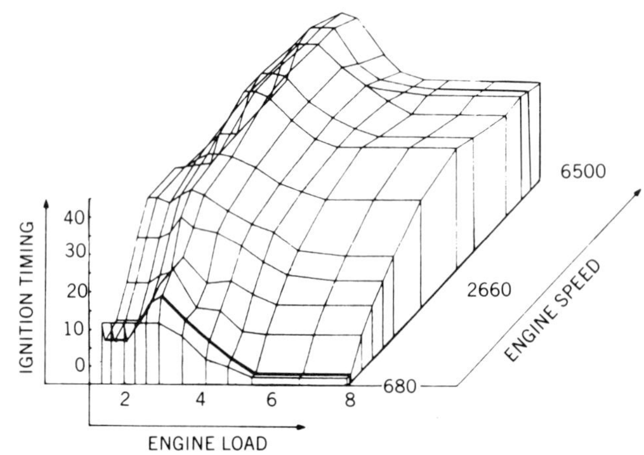


**Figure 20-16.** This engine map calculates the optimum ignition point for each combination of speed and load. Similar maps are computed for fuel metering and other operating variables. (Porsche)

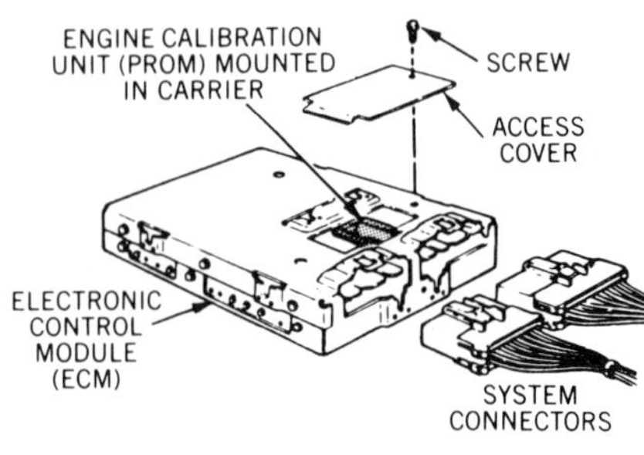


**Figure 20-17.** The General Motors CCC computer uses a single removable PROM. (Chevrolet)

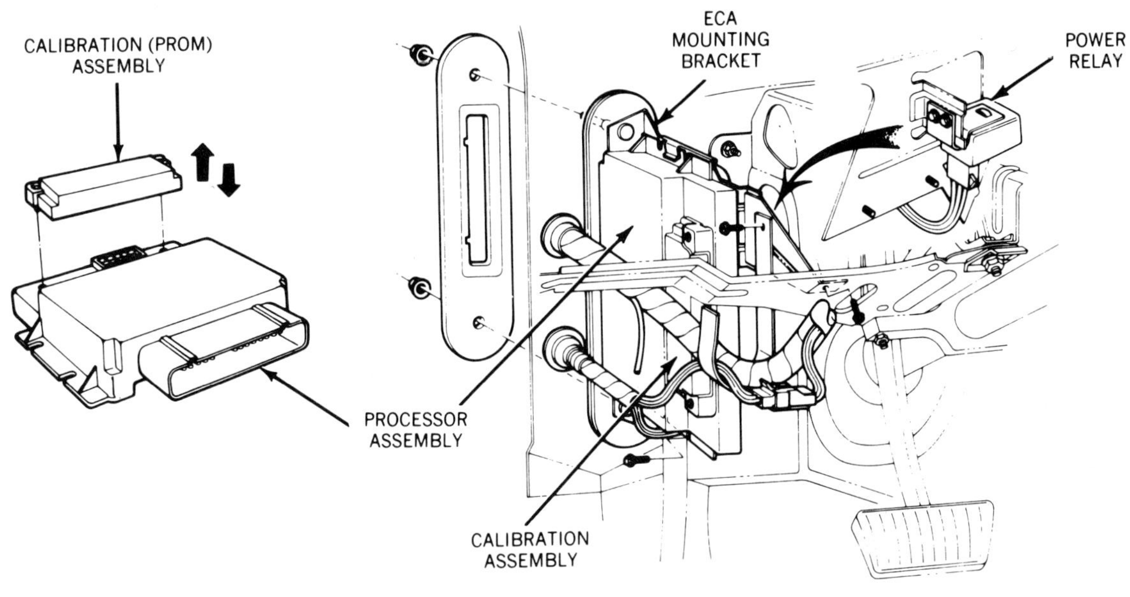


**Figure 20-18.** The calibration assembly (module) of this Ford EEC computer contains the system PROM. (Ford)

ure 20-18. Engine computers have become economical and versatile because engineers can use a single **computer hardware** assembly with interchangeable PROM's to change the **software**. When you service engine control systems, you will learn that checking for the correct PROM installation is an important service step.

## PARTS OF A COMPUTER

Up to this point, our discussion of computers has concentrated on computer functions, computer logic, and software. The software is the programs and logic functions stored in the electronic circuits of the computer. You can't see it and touch it; you can only see the results of its operation. You can see and touch the computer hardware, however. The hardware is the mechanical, magnetic, electrical, and electronic parts that make up the physical structure of a computer. Although the sizes, arrangements, and quantity of computer hardware vary from mainframes to microcomputers, you can understand the basic structure of a digital computer from the block diagram in figure 20-19.

### Central Processing Unit (CPU)

The microprocessor, which we discussed earlier, is the **central processing unit (CPU)** of the computer. All of the mathematical operations and logical decisions occur in its integrated circuits. (Some microcomputers have more than one microprocessor.) You can consider the microprocessor as the heart of the computer because it performs the essential processing function (along with the microcomputer memory). It is only part of the complete hardware structure, however.

### Computer Memory

Other electronic IC devices provide the computer storage, or memory, function. These are not processing circuits; they simply store data for the CPU. Memory circuits can store three basic kinds of data:

1. The computer operating program
2. Input data from system sensors
3. Output data to system actuators

Often a computer will receive an input signal or send an output signal, but it also must remember the signal data for later reference. Because an automobile computer operates with complex programs and large amounts of input and output data, the memory circuits often are as large and sophisticated as the microprocessor circuits.

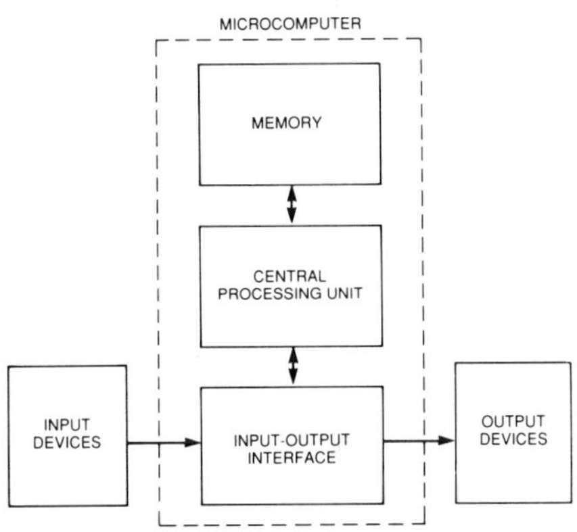


**Figure 20-19.** The basic structure of a digital computer.

**Read-Only Memory (ROM).** Computers use different memories for different purposes. Memory can be divided into two different types: memory that can be changed, and memory that can't be changed. In an automotive onboard computer, both kinds are stored in the computer's IC chips. For any computer to do its job properly, its program instructions must be stored in memory and remain unchanged. Also, the program can't be lost when power is turned off. This kind of stored, permanent memory is **read-only memory (ROM)**. Engineers sometimes call the programs stored in ROM "firmware" instead of software because they cannot be changed.

The operating program is a series of instructions organized into a given sequence to perform a given task. The operating program includes system strategy and lookup tables. The system strategy is a blueprint used by the computer to control engine operation under a wide variety of operating conditions and includes all of the decision-making logic and equations necessary to accomplish its job. The lookup tables contain specifications and calibrations required to make the engine perform as it should. For example, when a cold engine is started and the temperature sensor sends its signal to the computer, the microprocessor determines the engine operating condition and selects the most appropriate system strategy for it. The exact duration of engine warmup time is then determined by comparing the sensor temperature reading to the warmup time code in the lookup tables.

For most automotive computers, the operating program is loaded into an IC chip called a **programmable read-only memory (PROM)**. Computer engineers do this by applying a series of voltage pulses to a PROM chip that sets the switching positions of the transistors in prescribed on and off conditions. The small amount of current that flows with each voltage pulse blows a fuse on each circuit path after the transistor is set. The switching positions, therefore, become permanent memory and cannot be changed. Input signal voltages to be processed by the program are applied across parallel circuits and don't affect the program memory. The programming process is called "burning a PROM." A PROM differs from ROM because it can be removed from the computer and replaced with a different PROM containing new design information. ROM is permanently installed in the computer.

A variation of programmable read-only memory (PROM) is called **erasable, programmable, read-only memory (EPROM)**. The EPROM differs from PROM in two ways: the computer can write information into it for permanent storage, and it can be erased and reprogrammed by the carmaker (other PROM's are programmed by the chipmaker and have a 1-time use).

This first feature makes the EPROM ideal for handling important automotive functions such as storage of odometer information. General Motors uses EPROM's to store the option content of a vehicle equipped with a body computer module (BCM).

The erasable feature allows EPROM memory to be changed either by electrically altering the memory one byte at a time **(an electrically erasable programmable read-only memory—EEPROM)** or by exposing the memory to ultraviolet light through a small window in the top of the chip for 15 to 20 minutes. Intel, a leading chipmaker, has developed a "flash" memory that can be electrically erased in less than one second. The blank chip memory then can be reprogrammed with new information in less than four seconds. Chrysler uses EEPROM's to record engine fault codes on late-model cars. This type of chip memory allows a dealership to electrically update or change a module memory instead of replacing the entire module.

**Random-Access Memory (RAM).** Input and output signals that are stored temporarily are held in computer memory circuits called **random-access memory (RAM)**. Actually, "read-write memory" would be a better name because information can be written into and read from RAM circuits. They can be changed, while ROM circuits cannot (except for EEPROM).

Two types of RAM memory are used in automobile computers: volatile and nonvolatile. Information in the volatile RAM is erased when the ignition key is turned off. However, a volatile RAM can be wired directly to the battery so that its information is not erased by shutting off the ignition. This type of RAM is often called **keep-alive memory (KAM)**. However, RAM and KAM circuits lose their memory when power is disconnected from them. That's the reason why disconnecting the battery in a car with a "programmable" radio results in the loss of all the station settings, which have to be reset when the battery is reconnected. All those settings were stored in RAM, not ROM.

The nonvolatile RAM memory retains its information even when power is disconnected. This type of RAM is often used for storing odometer information in an electronic speedometer. The memory chip retains the accumulated mileage of the vehicle. If the speedometer is replaced, the odometer chip must be removed and installed in the new speedometer.

**Adaptive Learning Strategy.** Computer programs now in use contain many backup or fail-safe modes and are sophisticated enough to compensate for such things as sensor failure. For example, turning on the fuel pump in an electronically controlled fuel injection system is done through a relay. If the relay fails, however, the computer will take its signal from the oil pressure switch. In most cases, the driver will never realize that the fuel pump relay has failed.

A variety of other factors (such as variations in production, fuel quality, and engine wear) may gradually move the real needs of the engine away from the ideal values designed into the computer program. In addition to failure, sensors may gradually shift their response range, providing a signal that the computer questions. To compensate for such problems, an **adaptive learning strategy** is incorporated in many computer programs. If a controlled value falls outside the original design parameters, the adaptive learning strategy modifies the original program to restore proper operation.

Program modifications made by the adaptive learning strategy are stored in RAM that is connected directly to the battery. The modifications remain intact when the ignition is shut off. When the battery is disconnected, however, the modifications are lost and vehicle driveability becomes unsatisfactory.

Vehicle driveability will also be unsatisfactory if a malfunctioning sensor is replaced, since the program has compensated for signals from the unreliable sensor and is now receiving signals that conflict with its adapted program. If you are unaware of the presence of an adaptive learning strategy, you may question your diagnosis, because the vehicle will not perform any better immediately after replacing the sensor.

After the vehicle in question has been driven about 15 to 20 miles under varying driving conditions, the adaptive learning strategy will correct the program, and vehicle driveability will return to normal.

General Motors uses a pair of functions within its electronic control module (ECM) on fuel-injected engines called integrator (short-term) and block learn (long-term). They control fuel delivery by changing the injector duty cycle, just as the mixture control solenoid dwell (duty cycle) does on carbureted engines. The two functions interact with a base fuel calculation stored in the ECM's memory to make minor adjustments in the air-fuel mixture of the injection system. The ECM adds or subtracts fuel from the base calculation according to the feedback signal received from the exhaust gas oxygen (EGO) sensor.

The integrator function is a short-term, temporary change in fuel delivery that functions only in closed-loop mode. It monitors the EGO sensor voltage and adds or subtracts fuel according to the sensor data. Integrator uses a value of 128 as neutral, figure 20-20. When integrator reads 128, the EGO sensor is reporting the results of a 14.7:1 air-fuel mixture burning in the engine cylinder. Block learn continually watches the integrator and makes the same correction in its own values.

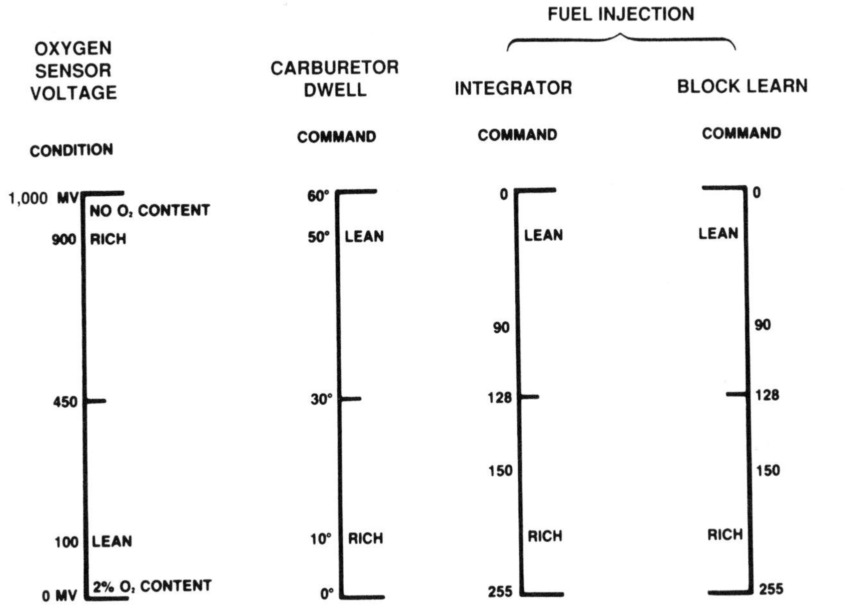


**Figure 20-20.** The relationship between oxygen sensor voltage, carburetor dwell, integrator, and block learn. (General Motors)

A (LOAD vs. RPM)

| | | | |
|---|---|---|---|
| 128 | 128 | 128 | 128 |
| 128 | 128 | 128 | 128 |
| 128 | 128 | 128 | 128 |
| 128 | 128 | 128 | 128 |

B (LOAD vs. RPM)

| | | | |
|---|---|---|---|
| 128 | 130 | 128 | 129 |
| 129 | 130 | 130 | 131 |
| 133 | 136 | 129 | 128 |
| 135 | 132 | 129 | 130 |

C (LOAD vs. RPM)

| | | | |
|---|---|---|---|
| 121 | 130 | 128 | 128 |
| 119 | 120 | 119 | 121 |
| 115 | 117 | 116 | 121 |
| 122 | 119 | 126 | 120 |

**Figure 20-21.** Computer memory dealing with block learn divides engine operating range for a given combination of load and speed conditions into 16 blocks. In **A**, all blocks are operating at a 14.7:1 air-fuel ratio under all engine load and speed conditions. **B** shows examples of block learn readings when the oxygen sensor reads a slightly lean exhaust. Similar examples are shown in **C** for a slightly rich exhaust. (General Motors)

Block learn cannot make as great a correction in fuel delivery, but its correction remains for a longer time. The name comes from the concept that divides the operating range of the engine for any given combination of load and speed into 16 cells or blocks, figure 20-21. (Block learn also uses the number 128 as representing a 14.7:1 air-fuel ratio.) A given fuel delivery is stored in each block in the computer memory. As the operating range moves into a particular block, the fuel delivery is based on the value stored in that memory block. Each time that block learn makes a correction, the integrator makes the same correction in its values.

A scan tool is used to obtain integrator and block learn data from the ECM. Properly connected to the car's test connector, the scan tool taps into the ECM serial data transmission link and reads the inputs and outputs that the ECM receives and sends. The base fuel calculation appears to the scan tool as the number 128. If a higher number is read, fuel is being added; a number less than 128 indicates that the ECM is subtracting fuel from the mixture. When the integrator function notes a condition that remains too high or too low for a predetermined length of time, block learn takes over and makes a long-term correction.

Integrator and block learn are forms of adaptive memory but with some limitations. An example is the result of a vehicle in stop-and-go traffic for a long period of time. Much of that time is spent at idle, with occasional minor accelerations and stops. The integrator function compensates for this operating condition, and after a period of time, block learn takes over. By the time that the traffic condition suddenly clears and the vehicle is ready to resume cruise speeds, block learn has revised the air-fuel mixture as a long-term correction. This can cause a driveability problem. To solve the problem, GM has resorted to a technique of imposing predetermined limits on the integrator and block learn function called "clamping."

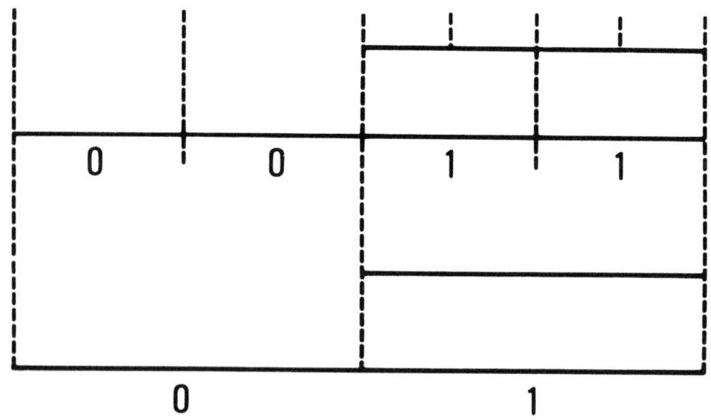


**Figure 20-22.** Computers use clock pulses to distinguish between these two signals. (General Motors)

### Input and Output Circuits

Computer processing and memory circuits can't do anything without receiving input signals and sending output signals. The microprocessor, however, is not connected directly to every input and output device in the system. Other IC devices receive and send the signals. Most of these provide parallel connections for the microprocessor so that it can read several input signals and send several output signals simultaneously.

### Converter or Conditioning Circuits

A microprocessor is limited in the types of voltage signals with which it can work. In many cases, voltage signals provided by sensors must be converted or conditioned before the microprocessor can use them. Earlier in this chapter, we said that many sensors in vehicle computer systems send analog voltage signals to the computer. The analog signals must be changed to digital data, or "digitized," for the microprocessor. The conversion is done by analog-to-digital (AD) converter circuits. These are generally housed on separate IC devices, connected to the microprocessor. A few sensors, however, contain their own input conditioning circuitry.

Other sensor voltage signals may be too small or not have the correct voltage range for use by the microprocessor circuits. Such voltage signals must be amplified before they are digitized. Amplification is performed; then the signals are sent through the AD converter circuitry.

Although most vehicle system actuators and instrument displays (the output devices) are digital components, some are analog devices. This means that the digital output signals from the microprocessor must be converted to analog values. Digital-to-analog (DA) converter circuits do this. Suppose a computer must control radio volume at a preselected level as the radio signal strength changes. This requires a changing analog voltage to be applied to a radio amplifier transistor. The DA converter circuits produce the analog signal from the microprocessor digital signals. Actually, the analog output is still a "digitized" signal that changes in discrete stair-steps. The digitized curve, however, is close enough to the analog curve that it produces the desired response.

### Computer Clock Rates and Timing

To this point, we know that a computer receives input data in the form of voltage signals from various sensors. It processes the input data and sends output voltage signals to various actuators. The processing of the data is done by a microprocessor, which acts as the "brain" by communicating with various memories.

All of this communication, from input to output, is accomplished using the binary code, consisting of lengthy strings of 0's and 1's that pulse through the computer from input to output. Have you wondered yet how the computer components are able to tell when one pulse ends and another starts? Or, put another way, how does a computer tell the difference between a 01 and a 0011, figure 20-22?

**Clock Pulses.** Computers contain a crystal oscillator or clock generator, figure 20-23. This clock delivers a constant stream of steady pulses one bit in length. The microprocessor and memories both watch the clock pulses while they are interfacing, figure 20-24. In this way, they are able to distinguish between a 01 and a 0011 since they know how long each voltage pulse is supposed to last. The process is extended to include the input and output circuits, figure 20-25.

**Computer Speeds (MHz).** Some computers operate faster than others. Computer speed is measured in the

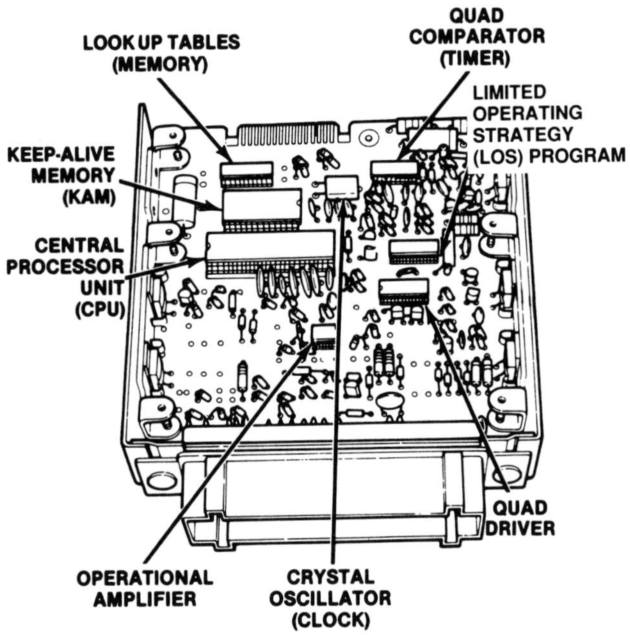


**Figure 20-23.** A computer contains a clock generator, which provides constant voltage pulses the length of one bit. (Ford)

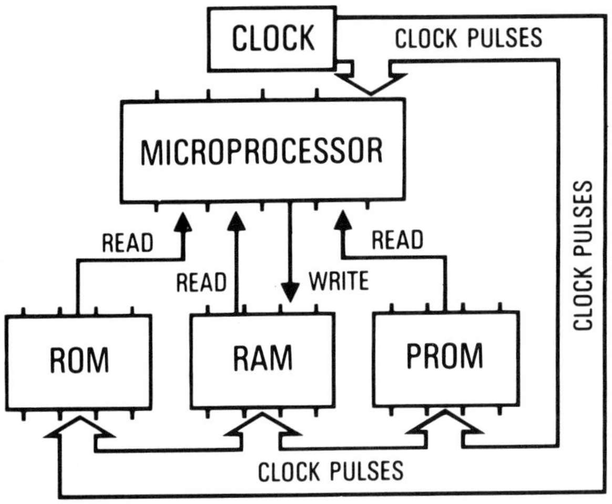


**Figure 20-24.** The microprocessor and memories watch the clock pulses. (General Motors)

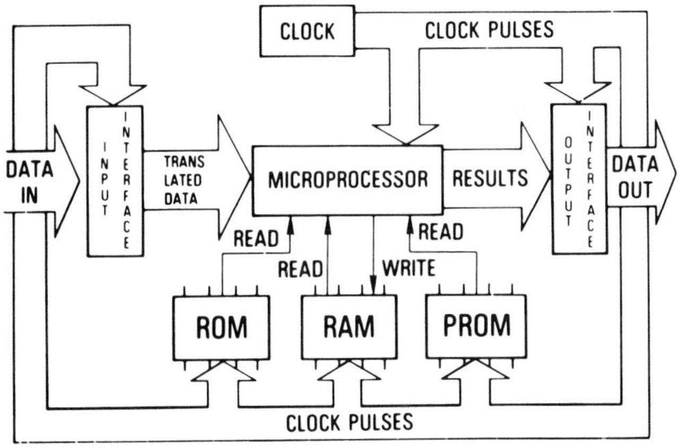


**Figure 20-25.** Input and output circuits also watch the clock pulses. (General Motors)

cycle time, or clock speed, in hertz (4.7 MHz, 8.0 MHz, etc.) required to perform certain measurements such as adding two 10-digit numbers.

The clock speed in hertz is difficult to relate to the real world in terms of actual speed, since many other factors besides processor speed enter into the rate at which a computer functions. A computer that operates at 8.0 MHz may actually work more slowly than one that operates at 4.7 MHz because of an inefficient program, a slower data transfer rate, or an inferior operating system.

**Baud Rate.** All manufacturers use the **baud rate**, or bits per second, as a standard measure of how quickly a computer transmits data. Baud rate is most useful in estimating the length of time one computer requires to transmit a specific amount of data to another computer. The standard storage of a single alphabetic or numeric character uses 8 bits per byte. Adding one start bit and one stop bit brings the transmission requirement of one character to 10 bits. To determine the maximum number of words per minute that can be transmitted electronically, the baud rate is divided by 10. Thus, a baud rate of 300 means that about 30 words can be sent or received per minute. A baud rate of 1200 will transmit about 120 words per minute.

When you study electronic engine control systems later in this text, you will study examples of vehicles with high-baud and low-baud computers. As Chrysler, GM, Ford, and foreign carmakers have developed their computer systems, they have gone from low baud rates on computers of the early 1980's to high baud rates on computers of the later 1980's and 1990's. Baud rate, or data transmission speed, is important in computer system operation and troubleshooting. Generally, higher baud rates are necessary for advanced, more precise control operations and communication on late-model vehicles.

**Processor Bit Size.** Microprocessor bit size is another factor that determines the speed of a computer. The early microprocessors were capable of transmitting 4 bits at a time and thus were called 4-bit processors. Since a byte contains 8 bits, the 4-bit processor had to make two passes to process or transmit a byte. The 4-bit processor was superseded by an 8-bit processor, which could process or transmit a byte in a single pass. However, since some data requires more than a single byte, even 8-bit processors must make more than one pass to process or transmit such data. Processors may be designed to handle 16 bits, 32 bits, or more. Automotive processors generally use 4-, 8- or 16-bit processors, depending upon their application.

### Computer Installations

Automotive onboard computers are called electronic control units, modules, and assemblies. Some are even called computers. All have the basic features of input, processing, storage, and output. All of the computer hardware is mounted on one, or a few, small circuit boards and installed in a metal case. The metal case helps to shield it from radiofrequency interference (RFI) and other sources of signal interference. Multipin connectors or edge connectors on the circuit boards join the computer to wiring harnesses. The harnesses connect the computer to the input and output devices (sensors and actuators) of the system.

An onboard computer is usually installed under the instrument panel or behind a kick panel in the passenger compartment, although a few engine computers are located in the engine compartment. Carmakers usually favor locations inside the passenger compartment to keep the computer as far as possible from the extremes of temperature, dirt, and vibration found elsewhere on the vehicle.

Some onboard computers are single-function units that control just one operation. Others are multifunction units that regulate all engine systems, for example. Even a simple ignition module can "compute" timing based on an input signal from a distributor. A more complex computer can use other input signals from speed, temperature, and vacuum sensors (besides distributor signals) to vary the timing for different temperatures, speeds, and loads.

## CONTROL SYSTEM OPERATING MODES

Any computer control system can have different operating modes. The computer does not respond to information from its sensors in exactly the same way all the time. It can ignore sensors under some conditions, or it can respond to one signal in several ways, depending on what several other sensors tell it. Because a computer can be selective, engineers can design control systems with two basic operating conditions, or modes: **open loop** and **closed loop**.

### Open-Loop Control

Open-loop control means that a computer—or any control device—adjusts a system to operate in a certain way. If there is an error or a deviation in the results of the operation, a signal of the error or deviation does not get sent back to the computer or controller. Figure 20-26 shows a simple block diagram of open-loop system control.

Open-loop control is not unique to computer systems. You exercise open-loop control over your car radio when you set the volume for a certain level. The volume control switch signals the amplifier for a certain power output, and that's where it stays. If you drive through a tunnel, the volume weakens because the input signal gets weaker. But the amplifier power output doesn't change until you readjust the volume control. The radio system itself didn't receive an error **feedback** signal that the volume was too low. It relied on an outside force for adjustment. Similarly, a manually operated car heater works in an open-loop control mode. You turn it on and set it for maximum heating. The system stays in that operating mode until you readjust it or turn it off.

### Closed-Loop Control

Closed loop means that the computer, or controller, is reading and responding to an output error signal, or feedback signal, figure 20-27. Again, closed-loop control is not limited to computer systems. We use closed-loop control systems every day. A temperature-control thermostat is a good example.

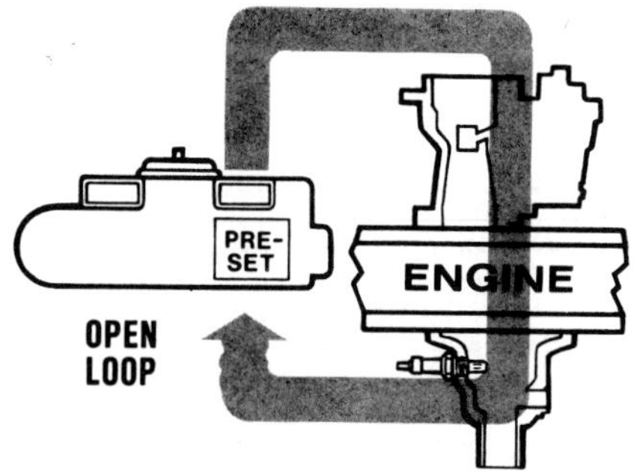


**Figure 20-26.** When a system is in open loop, it operates according to preset values and ignores any feedback messages sent by a sensor. (Chrysler)

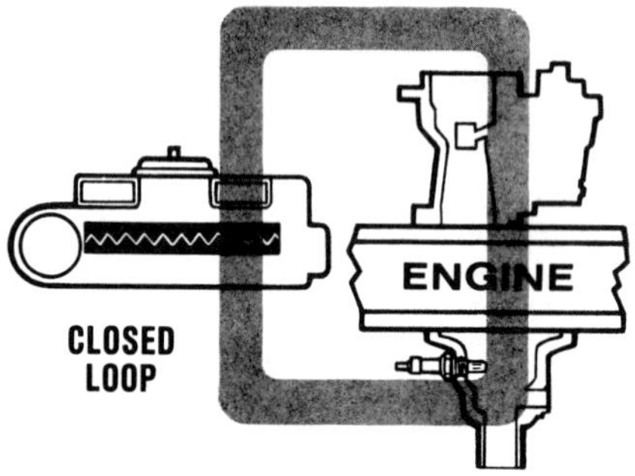


**Figure 20-27.** Once a system moves into closed loop, it adjusts its operation according to feedback messages from a sensor. (Chrysler)

You can set an oven thermostat for a cooking temperature of 350° F or a furnace thermostat for 75° F. If the oven or the room temperature is below the setting, the controller turns on the oven or furnace. When temperature rises above the setting, the controller turns off the heat. This is a simple kind of closed-loop control, and it is an example of **limit-cycle control**. Limit-cycle control is a control system that monitors the system output and responds only when output goes above or below selected limits. The system tries to maintain an average operating condition. For a simple temperature-control system, the limits may be ±5° or ±25° or whatever the system designer chooses. A furnace controller may turn the heat on at 70° F and off at 80° F. The result is an average room temperature of 75° F.

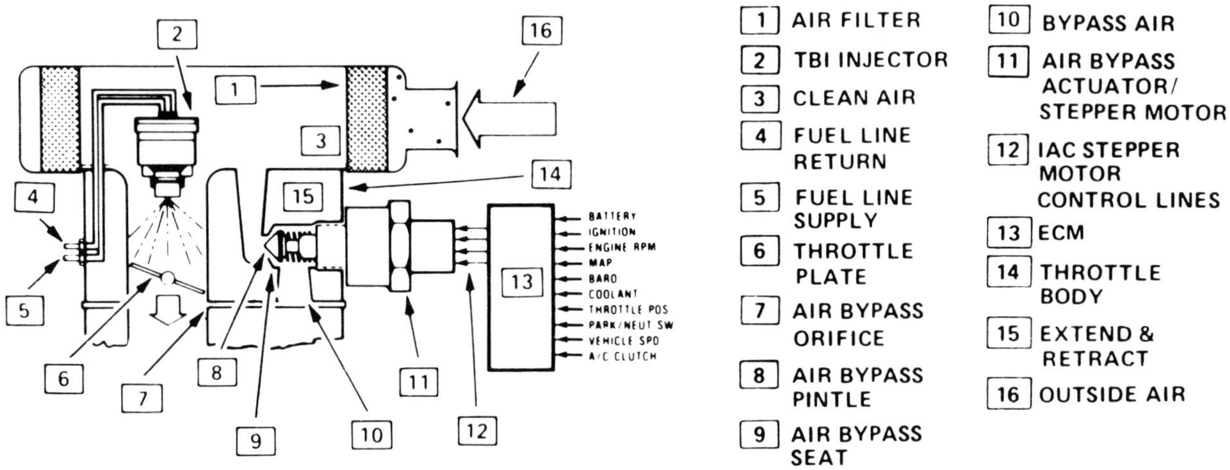


**Figure 20-28.** The computer (ECM) adjusts the idle air control (IAC) stepper motor to control airflow and idle speed. (Chevrolet)

A limit-cycle controller can be a simple thermostatic switch, or it can be a computer. A simple automotive air conditioner thermostat is a closed-loop, limit-cycle controller. It turns the air conditioning compressor on and off, depending on evaporator cooling temperature. A speed governor for an industrial engine is another kind of limit-cycle controller. You can set the throttle for a steady 2,000 rpm with no engine load. If engine load increases, speed will decrease. Centrifugal weights in the governor respond to the speed change and move the throttle to regain the 2,000-rpm setting.

The same principle works in an engine control system that uses a computer as a controller. The computer PROM can be programmed with the desired idle speed (plus or minus upper and lower limits) for a particular vehicle and engine. If idle speed rises or falls from the upper and lower limits of the setpoint for any reason, input from a speed sensor tells the computer to send an output signal to an idle speed control motor or other actuator. Figure 20-28 shows a computer-controlled idle air control (IAC) valve in a fuel injection throttle body. In response to engine speed signals, the computer controls the motor that moves the IAC valve in and out of the air passage. This changes the idle airflow and thus regulates idle speed.

**Proportional control** is another kind of closed-loop control. A proportional control system uses a feedback signal just as a limit-cycle system does. The control electronic devices subtract the feedback signal from the output signal to get an error signal. The controller, or computer, then changes its output signal in proportion to the error signal. If the error signal shows that output is 10 percent higher than it should be, the computer reduces the output by 10 percent.

Engine control computers use proportional closed-loop control for many output signals. For closed-loop fuel control, the error sensor is the exhaust gas oxygen sensor. The sensor signal tells the computer whether the air-fuel mixture passing through the engine is too rich or too lean. (You will learn how it does this later.) The computer can use the signal to readjust fuel metering, or it can ignore the signal. When it ignores the signal, the computer is operating the fuel system in open loop. When the computer responds to the signal, it is exercising proportional closed-loop control. If the oxygen sensor says the mixture is too lean, the computer adjusts the variable fuel actuator to enrich the mixture. If the computer calculates that the mixture is 10 percent too lean, it enriches the mixture by 10 percent, figure 20-29.

Fuel metering is not the only proportional closed-loop function performed by an engine computer. Many systems have knock sensors to detect detonation (pinging). Spark timing is a major factor that affects detonation. When the knock sensor detects detonation, it sends a signal to the computer, which then retards timing until detonation stops. That's another kind of proportional output control.

## Modified-Loop Control

On some engines, the exhaust oxygen sensor cools off at idle and cannot send a reliable feedback signal. Under this condition, the computer may return to a modified form of open-loop control in which it regulates fuel metering according to programmed values but still monitors and controls spark timing and idle speed. This kind of modified-loop control also may be required if the catalyst cools off at idle.

Some late-model vehicles also go to a modified-loop form of operation at light-throttle cruising speed. Under certain conditions, the computer may provide air-fuel ratios slightly leaner than 14.7:1 to improve fuel economy

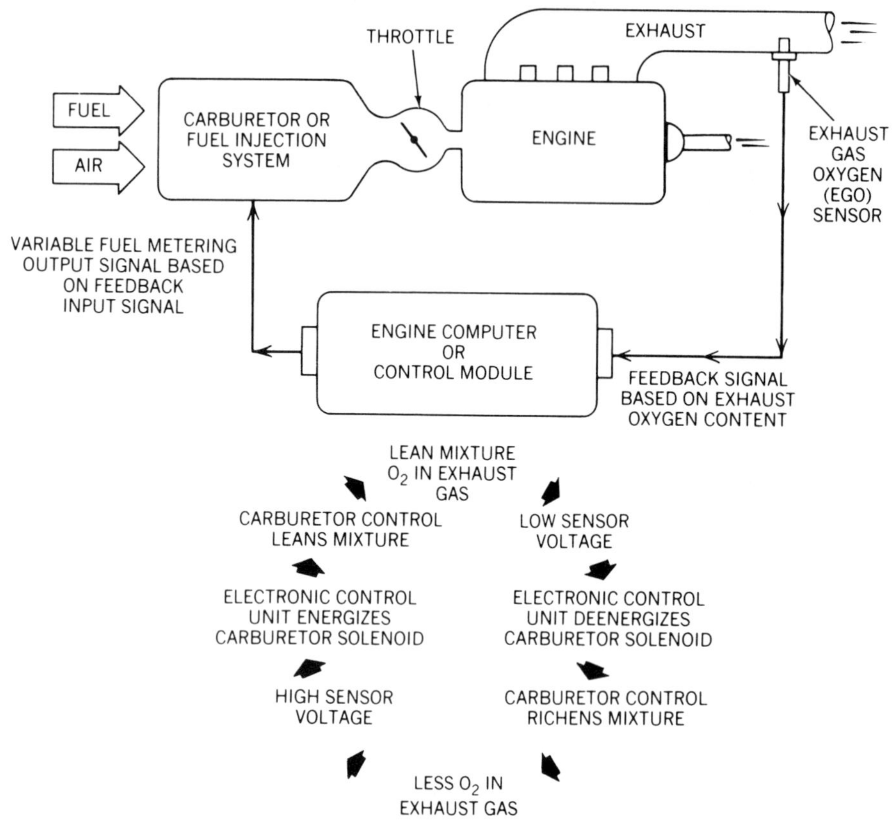


**Figure 20-29.** In closed-loop operation, the computer responds to the output error feedback signal and readjusts the input.

without adversely affecting emissions or driveability.

Complete and accurate diagnosis of any control system requires that you know the kinds of operating modes—open, closed, or modified—used on a specific vehicle and the conditions of each mode.

## SUMMARY

Every computer—large mainframe or small microcomputer—works on the four principal functions of input, processing, storage, and output. Personal microcomputers and automotive onboard computers use the same digital microprocessors for their operation.

Computers, as well as input and output devices, can operate on analog or digital signals. An analog signal is infinitely variable. A digital signal, or processing operation, varies in discrete on-off or high-low increments. Most variable measurements on an automobile produce analog signals, but computers change these to digital signals for processing.

Most automotive computer systems are digital systems. These systems convert measurements to binary numbers (combinations of 0's and 1's) that are represented by on-off or high-low voltage signals. These voltage signals are switching pulses for thousands of transistor logic gates. The basic NOT, AND, and OR logic gates are the building blocks of all digital computers.

Computer programs are the instructions, in numerical form, that a computer follows to do its job. Programs are part of computer software. The physical structure of any computer is the hardware. Computer hardware consists of the central processing unit (CPU) and other circuit devices for computer memory, input and output signals, and signal conversion.

Computer programs are stored in ROM and RAM. ROM is permanent stored memory; RAM can be changed and is not permanent. Adaptive learning strategies allow the computer program to alter its program and accommodate small deviations from the norm caused by engine wear, poor fuel quality, sensor values that have shifted, and other factors.

When more than one onboard computer is used on a vehicle, a body computer module (BCM) may be used as a master computer. It receives input data from the other computers and sends output data back.

Automotive computers are used in instrumentation and control systems. An instrumentation system measures some variable quantities and displays its output information to the driver. A control system regulates the operation of another system.

A control system can operate in open-loop or closed-loop modes. In an open-loop mode, the system does not respond to an output feedback, or error, signal. In a closed-loop mode, the computer responds to a feedback error signal and readjusts the output to a selected value. Closed-loop operation can provide limit-cycle control or proportional control.

Although the computer and its microprocessor are the heart of an onboard computer system, the system cannot operate without its sensors and actuators (input and output devices). These are the subjects of the next chapter.

## REVIEW QUESTIONS

### Multiple Choice

**1.** The operating program for a specific engine and vehicle is stored in the computer:

**a.** random access memory (RAM)
**b.** programmable read-only memory (PROM)
**c.** keep-alive memory (KAM)
**d.** relay logic

**2.** Student A says that an engine computer receives input information, processes data, stores data, and sends output information. Student B says that late-model engine computers are based on digital microprocessors. Who is right?

**a.** A only
**b.** B only
**c.** both A and B
**d.** neither A nor B

**3.** Student A says that the computer can read but not change the information stored in RAM. Student B says that the computer can read and change the information stored in ROM. Who is right?

**a.** A only
**b.** B only
**c.** both A and B
**d.** neither A nor B

**4.** A digital computer uses the binary number system. This consists of:

**a.** 2 numbers
**b.** 4 numbers
**c.** 6 numbers
**d.** 8 numbers

**5.** Student A says that digital computer switching circuits are called NOT, AND, and OR gates. Student B says that NOT gates can be added to AND gates and OR gates to create NAND and NOR gates. Who is right?

**a.** A only
**b.** B only
**c.** both A and B
**d.** neither A nor B

**6.** Engine mapping is used by automotive engineers to:

**a.** increase horsepower and torque
**b.** produce a 3-dimensional performance graph
**c.** provide more space in the engine compartment
**d.** all of these

**7.** Student A says that onboard computers work with binary numbers. Student B says that onboard computers work with voltage signals. Who is right?

**a.** A only
**b.** B only
**c.** both A and B
**d.** neither A nor B

**8.** The microprocessor can do all of the following except:

**a.** ignore its system strategy under specified conditions
**b.** respond in two or more ways to the same input signal
**c.** receive and accept an input signal from other computers
**d.** ignore input from its sensors under specified conditions

**9.** Student A says that a digital signal is either on or off. Student B says that an analog signal changes proportionally to the quantity measured. Who is right?

**a.** A only
**b.** B only
**c.** both A and B
**d.** neither A nor B

**10.** Student A says that a volatile RAM will retain its information even when power is removed. Student B says that a nonvolatile RAM will be erased when power is removed. Who is right?

**a.** A only
**b.** B only
**c.** both A and B
**d.** neither A nor B

**11.** Student A says that the microprocessors used in onboard automotive computers are the same as those used in personal minicomputers. Student B says that a minicomputer is "a computer in a box." Who is right?

**a.** A only
**b.** B only
**c.** both A and B
**d.** neither A nor B

**12.** Student A says that a transistor working as a relay is the basis for a digital computer. Student B says that a transistor working as an amplifier is the basis for an analog computer. Who is right?

**a.** A only
**b.** B only
**c.** both A and B
**d.** neither A nor B

**13.** Digital computers convert analog input signals from sensors to digital bits through:

**a.** a DA circuit
**b.** an AD circuit
**c.** a MOSFET transistor
**d.** binary numbers

**14.** Student A says that an onboard computer can act as an instrumentation system. Student B says that an onboard computer can act as a control system. Who is right?

**a.** A only
**b.** B only
**c.** both A and B
**d.** neither A nor B

**15.** Student A says that when the computer ignores the EGO sensor signal, it is exercising limit-cycle control. Student B says that when the computer responds to the EGO sensor, it is exercising proportional closed-loop control. Who is right?

**a.** A only
**b.** B only
**c.** both A and B
**d.** neither A nor B

**16.** Student A says that computers send and receive digital signals through data links. Student B says that computer speed is a function of MHz, baud rate, and processor bit size. Who is right?

**a.** A only
**b.** B only
**c.** both A and B
**d.** neither A nor B

**17.** Computers distinguish the length of each voltage pulse representing a bit by watching a:

**a.** clock
**b.** data link
**c.** serial bus
**d.** nonvolatile memory

**18.** Student A says that RAM is used for permanent storage because it can be written into. Student B says that a volatile RAM retains its information when power is removed. Who is right?

a. A only
b. B only
c. both A and B
d. neither A nor B

### Fill in the Blank

19. A digital computer processes information in binary digits called ________ .

20. An AD converter circuit changes ________ signals into ________ signals.

21. An input or output signal that is either high-low or on-off is a ________ signal.

22. The engineering process of measuring the best combinations of timing, air-fuel ratio, and other variables for various operating conditions is called ________ .

23. The two basic operating modes for all late-model engine control systems are ________ and ________ .

24. In closed-loop operation, the computer receives an output error, or ________ signal from the EGO sensor.

25. The ________ in a computer keeps all the components going at the same pace and tells the microprocessor the length of one bit.

26. A signal whose value constantly varies is called (a/an) ________ signal.

27. The instructions used by a computer are called its ________ .

28. Sensor voltage signals may have to be ________ before being digitized by the converter circuit.

# 21 INTRODUCTION TO SENSORS AND ACTUATORS

## WHAT

Input and output are two of the four basic computer system functions. The computer processing and storage circuits can do nothing without an input signal from a sensor or without sending an output to an actuator. Input and output signals connect the computer system to another system being controlled or monitored. Chapter 20 summarized the operation of computer control systems. This chapter examines the design and function of common sensors and actuators.

## WHY

A single computer system can have a dozen or more input and output devices. If any one does not send a correct input signal or respond properly to an output signal, the entire system can malfunction. Most of your electronic service work will deal with the system sensors and actuators.

The operating principles and design technology of input and output devices are based on the electrical and electronic principles you have already learned. Sensors used in different systems have many common features. The same is true of actuators. If you understand the general way in which a specific sensor or actuator works and the operating condition that it measures or controls, you can test it quickly and accurately. Therefore, it is a good idea to examine the principles of sensors and actuators before working with them in individual systems.

## GOALS

After studying this chapter, you should be able to:

1. Explain what a transducer does.
2. List five of the seven variable conditions and quantities sensed by an automotive sensor and give an example of a correct sensor to be used for each condition or quantity.
3. List and explain the four major characteristics required of an automotive sensor.
4. Explain the difference between a generating and nongenerating sensor.
5. Identify five different types of resistive sensors and explain their use in an automotive control system.
6. Explain how solenoids, relays, and stepper motors are used as actuators.

## TRANSDUCERS

All sensors and actuators are **transducers**. A transducer is any device that converts (transduces) one form of energy to another. Sensors transduce motion, temperature, light, and other kinds of energy to electrical energy in the form of voltage signals. Actuators transduce electrical energy (voltage and current) into mechanical work. Some carmakers refer to specific sensors as transducers, such as a throttle position transducer. This particular kind of transducer senses the position of an engine throttle and produces a corresponding input signal to a computer.

We can also classify display devices in the broad category of transducers. An instrument panel gauge converts an electrical signal to mechanical work as the analog movement of a needle across a dial. A digital instrument transduces electrical energy to light energy in an illuminated display. Whether a sensor, an actuator, or a display device is called a transducer or not, each one performs the function of a transducer—which allows them to produce computer input signals or put output signals to work.

Transducer functions are not limited to electrical and electronic devices. Automotive systems also use many mechanical transducers. A simple vacuum diaphragm on an ignition distributor, figure 21-1, is a mechanical transducer. It transduces air-pressure fluctuations on a movable diaphragm into mechanical work to move a breaker plate and advance or retard ignition timing. Similar vacuum transducers use air-pressure signals to

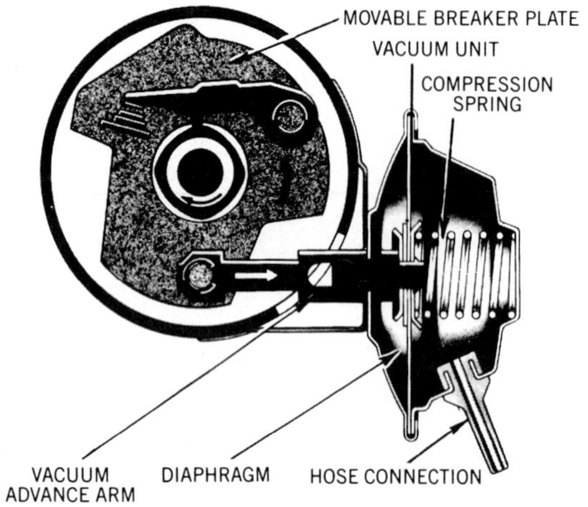


**Figure 21-1.** A distributor vacuum advance diaphragm changes air pressure fluctuations into mechanical work, moving a breaker plate to advance or retard ignition timing. (Chrysler) (Bosch)

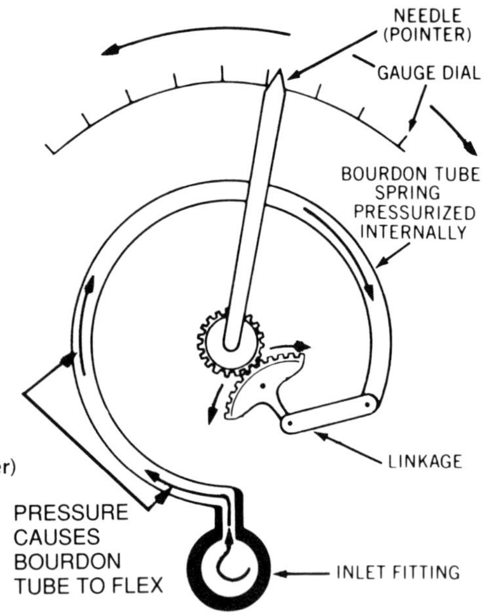


**Figure 21-2.** A direct reading pressure gauge changes pressure to move a needle around a dial.

open and close air conditioning doors and to increase the force applied to a brake master cylinder. Many exhaust gas recirculation (EGR) systems have similar diaphragm-type transducers to control EGR flow. A direct reading pressure gauge, figure 21-2, transduces pneumatic or hydraulic pressure to the movement of a needle around a dial.

It's helpful to compare the operation of mechanical and electrical or electronic transducers to recognize common purposes. Many sensors and actuators on modern automotive computer systems do the same jobs that mechanical transducers did on earlier vehicles. If you recognize the input and output signals or functions with which a system operates and the system's overall purpose, you can understand the similarities between electrical, electronic, and mechanical sensors and actuators.

Also, many automotive computer systems use combinations of electrical and electronic sensors and actuators and mechanical transducers to achieve their goals. Electrical solenoids and vacuum diaphragms often are combined to produce a required analog output action. Some sensors are solid-state electronic devices, but many are electromechanical transducers that use a mechanical transducer to produce an electrical signal.

## SENSORS

Because a computer can only read voltage signals, an automotive sensor must convert motion, pressure, temperature, light, and other energy to voltage. Automotive sensors are switches, timers, resistors, transformers, and generators, figure 21-3. They can send either analog or digital signals. The computer converts analog signals to digital form for processing. It may also have to amplify certain voltage signals before converting them. Automotive sensors send their input signals to the computer in one of the following ways:

1. *By generating a voltage*—Only a generator sensor can produce a signal voltage on its own. Various means are used to create the voltage signal, depending upon the type of sensor involved. Some sensors use a specific type of quartz crystal (piezoelectric sensor). Others use electrically conductive materials, such as zirconium dioxide, or operate on electromagnetic principles. Generator sensors are explained in more detail later in this chapter.

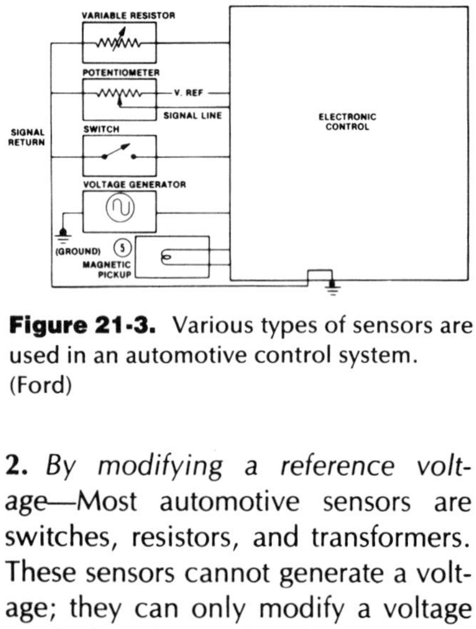


**Figure 21-3.** Various types of sensors are used in an automotive control system. (Ford)

2. *By modifying a reference voltage*—Most automotive sensors are switches, resistors, and transformers. These sensors cannot generate a voltage; they can only modify a voltage applied to them. Therefore, switches and resistive sensors must operate with a reference voltage ($V_{ref}$) from the computer. This is a fixed voltage supplied to the sensor by a voltage regulator within the computer. Most computer control systems operate with a 5-volt reference voltage. Some use a 9-volt reference voltage. In any case, reference voltage must be less than minimum battery voltage to prevent inaccurate sensor signals. The com-

puter sends the reference voltage to the sensor. As the sensor changes, the return voltage is altered and relayed to the input conditioners for conversion, then transmitted to the microprocessor for use.

### Physical Qualities and Operating Conditions

The best way to start examining sensors is to look at the physical qualities or operating conditions that they sense. Whether a computer system controls engine operation, instrument panel displays, antilock braking, or any other vehicle operation, its sensors measure some or all of the following variable conditions and qualities:

1. Motion and position
   a. linear (straight line)
   b. angular (rotation)
2. Speed
   a. engine rpm (angular)
   b. vehicle speed (linear)
   c. transmission and driveline speed (angular)
   d. rate of change (acceleration and deceleration of the engine, the vehicle, or the wheels)
3. Physical mass or volume
   a. airflow
   b. fuel flow
   c. exhaust oxygen content
4. Temperature
   a. air
   b. liquid (coolant, oil, and other fluids)
5. Pressure
   a. hydraulic (fuel, oil, refrigerant, and transmission and brake fluid)
   b. pneumatic (barometric pressure, boost pressure, and engine vacuum)
   c. physical pressure (engine detonation or knock)
6. Light
7. Sound or noise

Most of these conditions and quantities are analog variables, but current vehicle computers use digital microprocessors. Some computer programs require only a digital, high-or-low, signal for some of the conditions listed above. Other programs require an analog input signal. Therefore, a system can use either a digital or an analog sensor to monitor many of these variables. The choice of a sensor depends upon the system program and the control or instrumentation requirements.

### Sensor Characteristics

Whether a sensor sends a digital or an analog signal, the following characteristics, or operating features, affect the selection of a sensor for a particular system. These features also establish the specifications for troubleshooting and service.

**Repeatability.** A simple digital sensor must switch from a low to a high signal at a specific point in the range of values that it senses. A temperature switch may close at 160° F. It may open again at 160° or 180° F. The opening and closing switching points are design considerations for the system engineer. Once they are determined, the switch must close and open consistently at the required points. This is the necessary repeatability required of a sensor.

An analog sensor produces a signal that varies in proportion to the quantity or condition being measured. It, too, must deliver a repeatable signal for each increment of change. If an analog temperature sensor must provide a 20-millivolt signal change for each degree of temperature change, it must do so throughout its operating range. Moreover, the signal increments must be repeatable both upwards and downwards.

**Accuracy.** We expect sensors to be accurate, but accuracy is not a simple matter of right or wrong. All sensor signals have tolerances, or limits, on their accuracy. The degree of accuracy also is a design consideration. Once the accuracy requirements are determined, however, the sensor must operate consistently within those tolerances.

A digital temperature switch may close at 160° F, but it need not always switch at exactly 160° F. Its required accuracy may be 160° ± 1° F or 160° ± 10° F. These design tolerances establish the test specifications that are used to check the sensor operation.

Similarly, an analog sensor may provide a signal change of 20 ± 2 millivolts for each 1 degree ± 0.1 degree of temperature change. These tolerances are based on the accuracy requirements of the computer system and also help to determine the test specifications for the sensor.

**Operating Range.** A digital temperature switch closes at a specific temperature: 160° ± 5° F, for example. It may open again at the same point, or at another point: 180 degrees ± 5° F, for example. Simple throttle position switches that sense idle and wide-open positions operate similarly. So do simple pressure switches. The operating range of a digital sensor is just one or two switching points.

The operating range of an analog sensor is broader. An analog temperature sensor may be required to provide accurate, proportional signals through a range from 0° to 250° F, or from 100° to 300° F. An analog pressure sensor's range may be from 5 to 50 psi, or from 0 to 30 inches of mercury. An analog throttle position sensor may deliver signals that are proportional to throttle positions from 0 degrees to 90 degrees. The operating range is called the **dynamic range** of the sensor, and it is also reflected in test specifications. Outside of the upper and lower limits of the dynamic range, the sensor signal is not proportional to the variable value being measured. Most computers are programmed to ignore a signal outside of the dynamic range.

**Linearity.** Within the dynamic range, an analog sensor signal must be as consistently proportional to the measured value as possible. This is an expression of signal **linearity**. Linearity is an expression of sensor accuracy throughout its dynamic range. If a temperature sensor signal varies in 20-millivolt increments for each 1 degree of temperature change, it should do so ideally from the lower to the upper limits of its dynamic range. A perfect linearity signal curve is a straight diagonal line, figure 21-4.

In practice, no sensor has perfect linearity. The signal curve of a typical sensor is S shaped, figure 21-5, varying more or less from a perfectly linear straight line. Linearity is usually most accurate near the center of the dynamic range. Computer programs often contain memory data to compensate for nonlinear signals in a sensor's range. Signal processing circuits adjust the input signal to give the computer a true indication of the variable measured by the sensor.

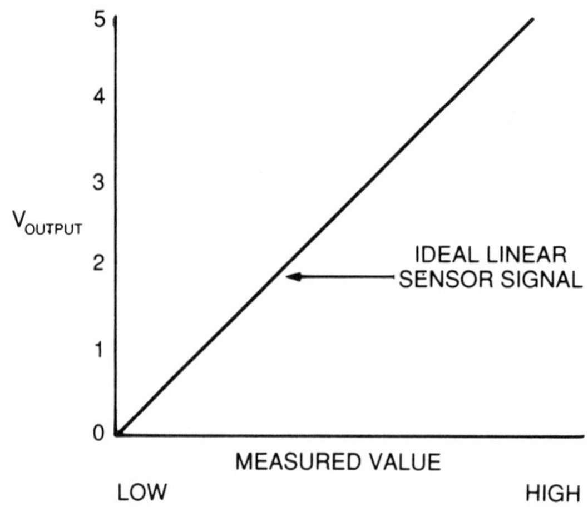


**Figure 21-4.** A perfectly linear sensor signal represents a straight line, with output voltage proportional to the changes in the measured value.

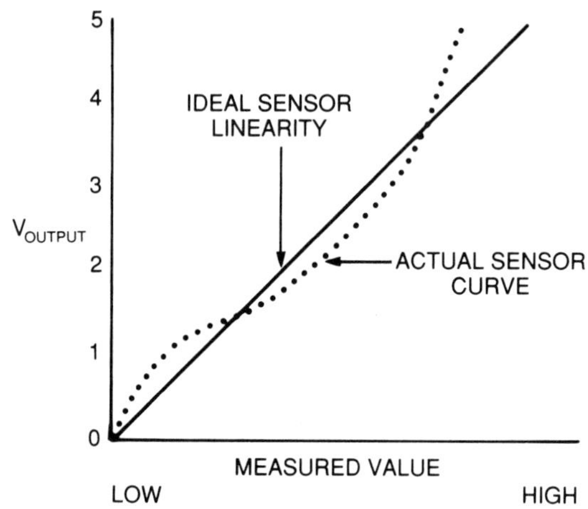


**Figure 21-5.** Actual linearity of a sensor varies from the ideal proportional signal.

**Other Sensor Features.** Repeatability, accuracy, operating range, and linearity are the requirements that define sensor design. They also are features that will help you recognize different kinds of sensors and test their operation.

All automotive sensors are built for long-term, reliable operation in a harsh environment. They must provide reliable signals in extremes of temperature, vibration, and dirt. Simplicity aids reliability, and thus many sensors are simple, general-purpose items. The same pressure sensor, for example, may be used in a throttle-body fuel injection system and in a multiport injection system. A single common temperature sensor may also have different uses in different systems.

Sensors provide low-voltage, low-current signals. Low-power operation means that they do not put a large load on the vehicle's electrical system, but it also makes them susceptible to electromagnetic interference (EMI). Therefore, sensor design and installation often include EMI suppression and shielding devices. Wiring location and low-resistance connections are critical for proper sensor signal transmission. All of these sensor characteristics are important points for correct testing and service. With these facts in mind, we can continue with an examination of specific sensor types.

## Switches and Timers

A switch is the simplest digital sensor. A switch opens or closes a circuit and delivers a high or a low voltage signal, figure 21-6. One switch terminal receives input voltage from the electrical system, often through the computer. The other terminal returns a high-voltage signal when the switch closes and a low-voltage signal when it opens. The computer can respond to either a high or low signal, or both, depending upon its program.

**Switches.** All switches are mechanical devices, but they can be operated by mechanical linkage or contacts, by pressure, or by temperature. A simple throttle-position switch, figure 21-7, is closed by linkage or by a contact stop when the throttle is closed at idle or wide open, or both.

Pressure switches, figure 21-8, sense hydraulic or pneumatic pressure. Engine oil pressure switches and

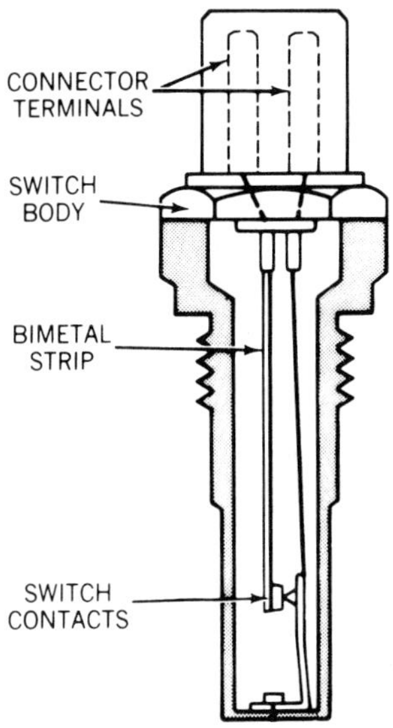


**Figure 21-6.** When this temperature switch closes a ground connection, signal voltage is high. When it opens the circuit, voltage is low.

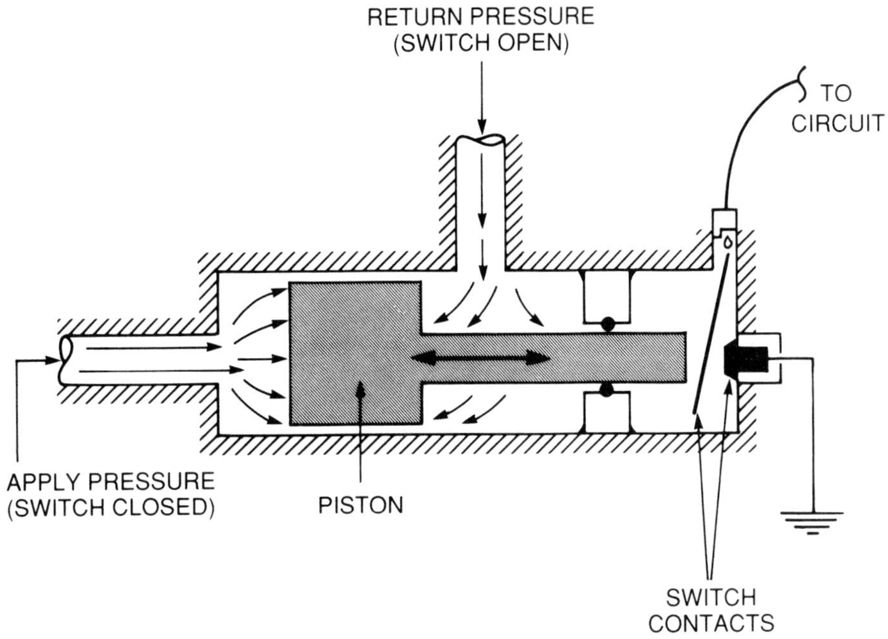


**Figure 21-8.** This hydraulic pressure switch is similar to one that you might find in a transmission. Applying pressure from fluid in one direction moves the piston to close the switch contacts. Return pressure moves the piston in the other direction, and the spring-loaded switch contacts open.

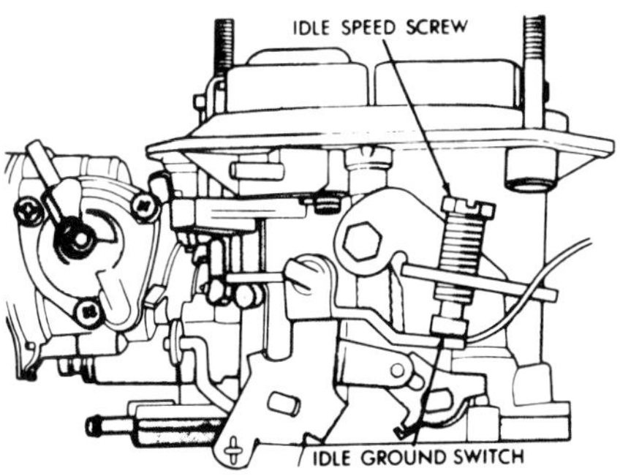


**Figure 21-7.** This throttle position (idle ground) switch sends a simple on-off signal to the computer. (Chrysler)

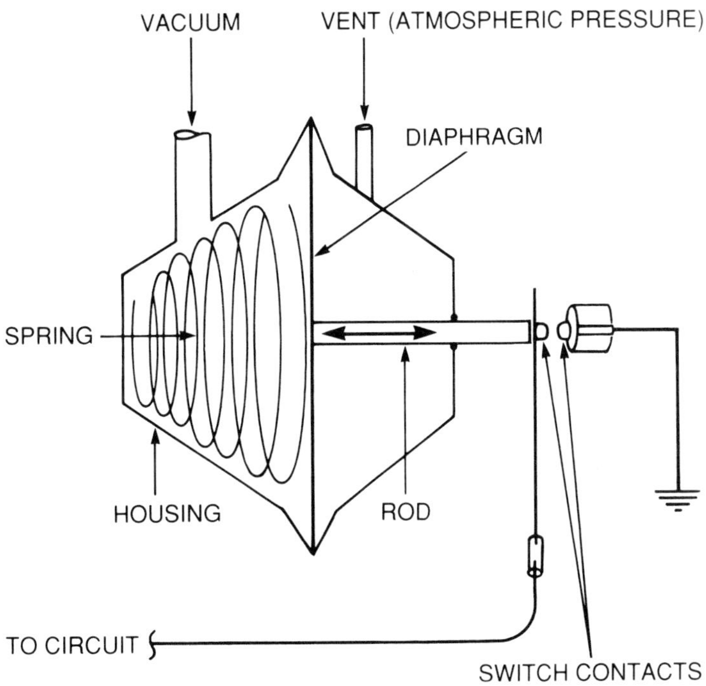


**Figure 21-9.** The combination of atmospheric pressure and vacuum (low pressure) move the diaphragm against spring pressure to open the switch contacts. When vacuum is cut off, spring pressure moves the diaphragm to close the switch.

transmission pressure switches are older than computers, but many are used with modern electronic control systems. Fluid pressure acting on a movable piston or diaphragm causes the switch contacts to open and close. Most air conditioning systems have similar switches to sense high and low refrigerant pressure. Simple pneumatic pressure switches also have movable diaphragms that open and close the switch contacts, figure 21-9. The diaphragm can be actuated by positive air pressure or by negative pressure (vacuum).

Pressure switches also can be actuated by an **aneroid** bellows, or capsule. The aneroid capsule is an evacuated metal bellows with low pressure (vacuum) inside. Pressure variations outside the bellows cause it to expand and contract. This movement can open and close switch contacts just as a diaphragm can. Because aneroid capsules are more expensive than diaphragms and their movements more precise, their use is generally reserved for analog sensors rather than simple switches.

Temperature switches open and close as temperature rises and falls past a specified switching point. The simplest temperature switch is a **bimetal** element that forms part of the circuit conductor. The element is made of two strips of metal joined together, figure 21-6. Each metal strip expands and contracts at a different rate when exposed to changing temperature. These different expansion rates cause the bimetal element to bend as it heats and cools. Most bimetal switches are closed when cold. As surrounding temperature heats the element, it bends to open the switch contacts. The switch closes again when the element cools and bends back.

Bimetal switches also can be heated by current flowing through the element. Because most computer system sensors are low-power devices, however, current for an electrically heated bimetal switch usually is provided by a separate circuit. This leads to the design of a common timer, or timed switch.

**Timers.** A timer combined with a switch can delay a signal. The timer can be electronic circuitry in the computer, or it can be part of the switch. Many switch and timer combinations are thermo-time switches, as used with Bosch gasoline fuel injection systems, figure 21-10. The switch delivers the signal that controls the cold-start injector. Closing the switch depends on temperature from the engine or from current passing through the switch. When the engine is cold, current passes through the heating winding. After a specified time, the current heats the bimetal thermostat and opens the circuit to turn off the cold-start injector.

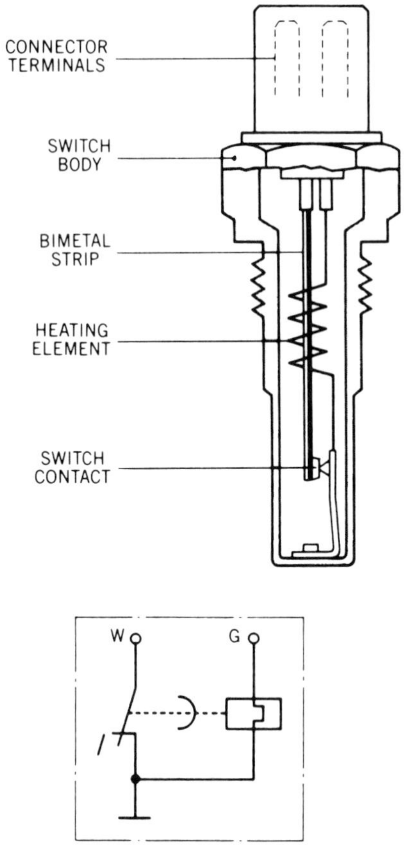


**Figure 21-10.** This thermo-time switch limits the operating time of the cold-start injector according to engine temperature. (Bosch)

## Resistors

A resistor can send an analog voltage signal that is proportional to temperature, pressure, motion, or other variables. A resistor, like a switch, cannot generate a voltage. It can only modify a voltage applied to it. Therefore, automotive resistive sensors operate with a reference voltage from the computer.

Two kinds of variable resistors are commonly used to regulate voltage or current: rheostats and potentiometers. Of the two, potentiometers are more common as computer input sensors. A potentiometer changes mechanical motion to a voltage signal and often is used to indicate the position of a specific component, such as a valve. Thermistors, piezoresistive strain gauges, and hot-wire or hot-film sensors are other resistive sensors used to measure temperature and pressure.

**Potentiometers.** A potentiometer is a variable resistor with three terminals. Reference voltage is applied to one end of the resistor, and the other end is grounded, figure 21-11. The third terminal is connected to a movable wiper or contact that slides across the resistor. Depending on whether the contact is near the supply end or the ground end of the resistor, return voltage will be high or low. Constant current through the resistor maintains constant temperature. Therefore, resistance does not change due to temperature variations. This ensures a constant voltage drop across the resistance so that return voltage changes only in relation to sliding contact movement.

The movable contact of a potentiometer can move in an arc or in a straight line. Thus these devices can be used to sense linear motion and position or rotary motion and angular position. Potentiometers are commonly used as throttle position sensors (rotary) and as EGR valve position sensors (linear).

**Rheostats.** A rheostat is a 2-terminal variable resistor, more often used as a current-control device than as a computer system sensor. Voltage is applied to a fixed terminal at one end of the resistor. The other terminal is connected to a movable contact and provides a return current path, figure 21-12. Voltage drop across the resistor varies with the position of the movable contact. Current flows through varying amounts of resistance. As the contact moves to increase resistance, current decreases and vice versa. Resistor

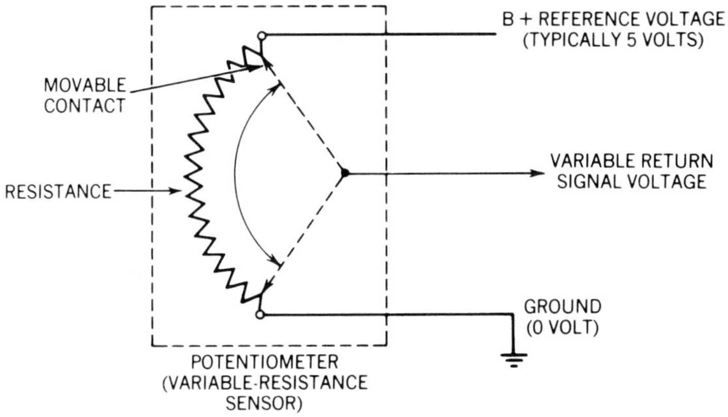


**Figure 21-11.** A potentiometer is a variable resistor with three terminals. Signal voltage is taken from the movable contact.

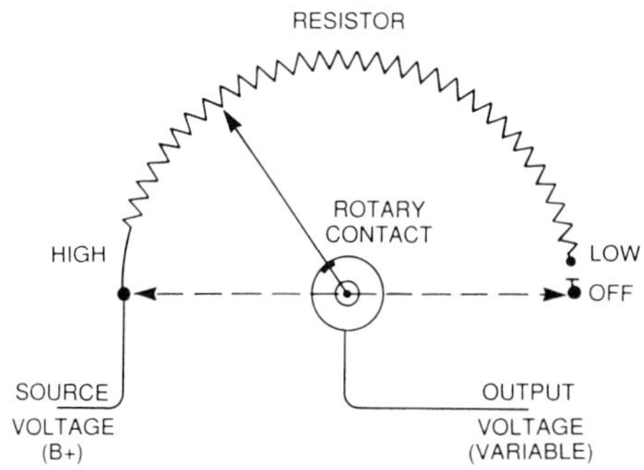


**Figure 21-12.** A rheostat is a 2-wire variable resistor. Rheostats were often used in older heater motor controls.

temperature also varies as current increases and decreases.

Rheostats are excellent control devices for circuits that require variable current. These include instrument panel lamp circuits (dimmer control) and fuel tank sender circuits (variable current for a magnetic-coil gauge). Although rheostats and potentiometers appear similar and both work with variable resistance, a rheostat is not a good computer input sensor.

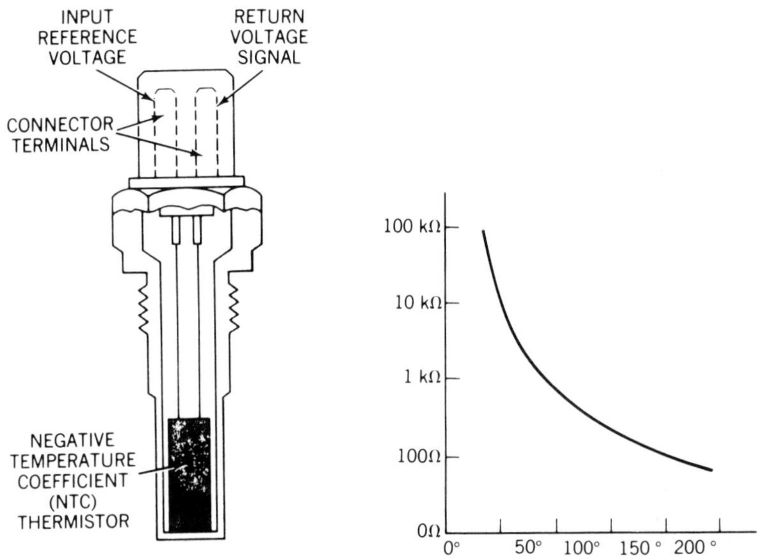


**Figure 21-13.** The graph shows that the resistance of a thermistor decreases as temperature increases. (Bosch)

**Thermistors.** A **thermistor** is a solid-state variable resistor with two connector terminals, figure 21-13. The resistance of any resistor changes as temperature changes, but the resistance variations throughout the operating range of a thermistor make it very accurate as an analog temperature sensor.

Thermistors are divided into two groups: **positive temperature coefficient (PTC) resistors** and **negative temperature coefficient (NTC) resistors**. These names simply mean that:

- The resistance of a PTC thermistor increases as temperature increases.
- The resistance of an NTC thermistor decreases as temperature increases.

Figure 21-14 shows the operating characteristics of PTC and NTC thermistors. Both kinds are used in automotive systems, but the NTC thermistor is more common. Heat can come from an external source or from current through the resistor. Self-heated PTC thermistors can be used as current limiters and timers in electronic systems. Externally heated NTC thermistors are the most common analog sensors for engine coolant temperature and intake air temperature.

NTC thermistors can also be used in indicator circuits, figure 21-15. In this case, the thermistor controls voltage applied to a low-fuel warning lamp. When fuel covers the thermistor in the tank, its temperature is low and its

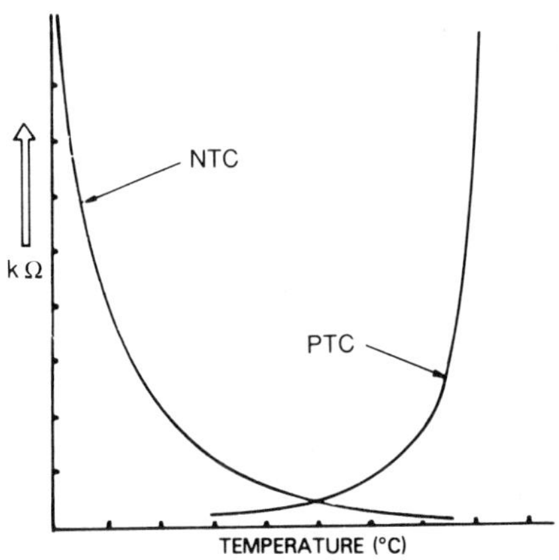


**TEMPERATURE CHARACTERISTIC OF THE THERMISTOR**

**Figure 21-14.** Temperature characteristics of PTC and NTC thermistors. (Toyota)

resistance is high. Voltage applied to the lamp is low, and little current flows through the circuit. When the fuel level drops, the thermistor is uncovered. It warms up and its resistance decreases. Higher voltage is applied to the lamp, more current flows, and the lamp lights.

NTC thermistor sensors are made of semiconductor material with high resistance at low temperature and low resistance at high temperature. The computer applies the reference voltage to one thermistor terminal and takes the return voltage from the other, figure 21-13. As the thermistor warms up, return voltage increases.

PTC thermistors are not commonly used as temperature sensors, but they are used for heating elements, such as in electric chokes and heated exhaust gas oxygen sensors.

**Piezoresistive Sensors.** In chapter 7 you learned that a piezoelectric crystal develops voltage across its surfaces when pressure is applied to it. Similar crystals change their resistance when pressure is applied to them. This feature makes piezoresistive sensors excellent devices for analog pressure measurement.

A typical piezoresistive sensor is made of a small silicon diaphragm sealed to a quartz plate, figure 21-16. The chamber in the middle is evacuated to create a vacuum. Four resistors are formed around the edge of the diaphragm by doping the silicon. Pressure applied to the diaphragm causes it to deflect, which changes the resistance of the resistors. The resistors are connected to an external circuit in a **Wheatstone bridge** arrangement. A constant reference voltage is applied to the sensor input terminal.

This bridge circuit is a series-parallel arrangement between the input terminal and ground. Resistors $R_1$ and $R_2$ are in series, and they are in parallel with series resistors $R_3$ and $R_4$. When no pressure is applied to the sensor diaphragm, the resistances are equal. The voltage drops across $R_1$, $R_2$, $R_3$, and $R_4$ are equal; and there is no voltage difference between terminals

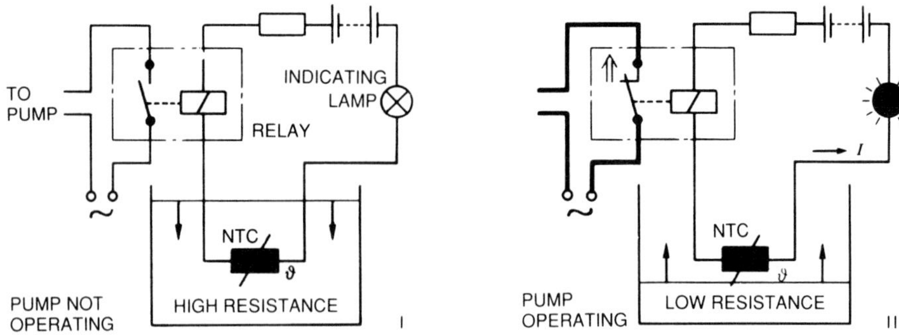


**Figure 21-15.** The application of an NTC thermistor in an indicator circuit. (Bosch)

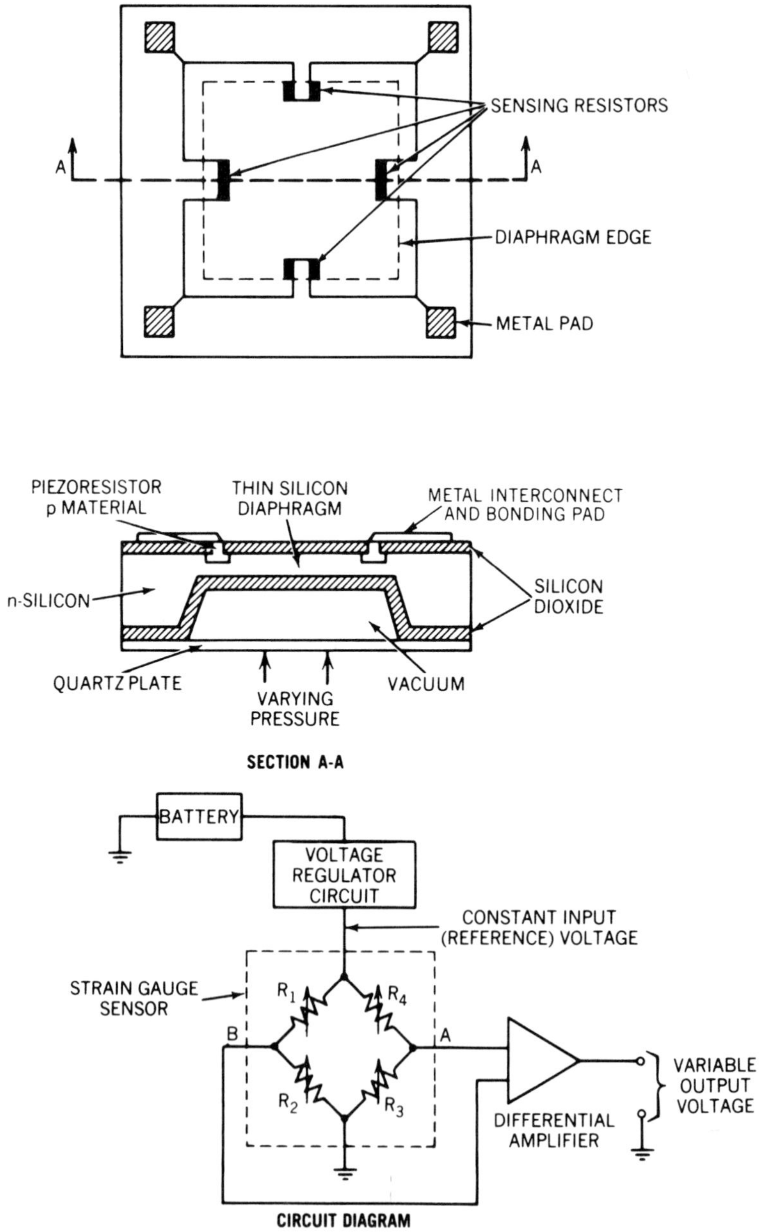


**Figure 21-16.** This piezoresistive MAP sensor varies signal voltage as it flexes.

A and B, which are the output connections. The bridge circuit is balanced.

When pressure on the sensor changes, it causes the resistances to change proportionally to the pressure on the diaphragm. $R_1$ and $R_3$ increase, and $R_2$ and $R_4$ decrease. This unbalances the bridge circuit and creates a voltage difference between terminals A and B. This is the sensor output voltage, which is also proportional to the pressure applied to the silicon diaphragm.

Piezoresistive sensors are commonly used to measure engine intake manifold absolute pressure (MAP). They can also be used as mechanical strain gauges to measure physical force.

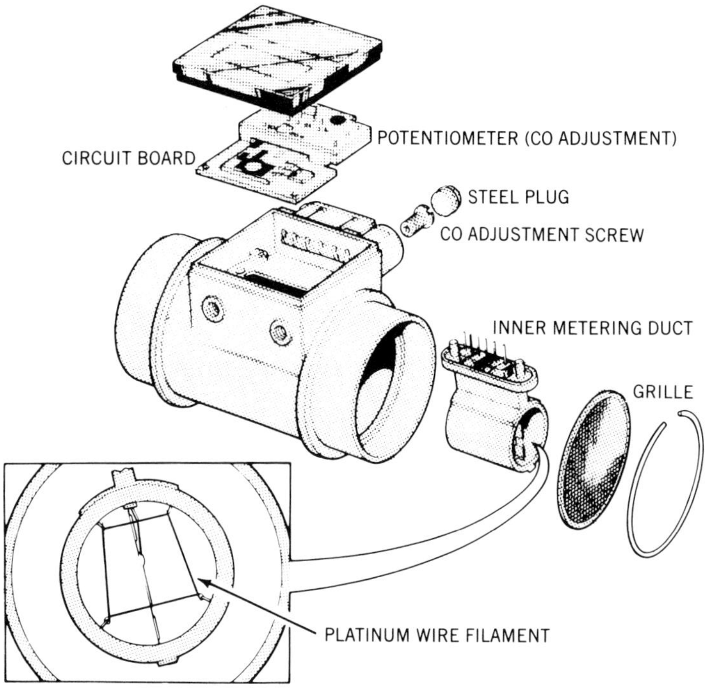


**Figure 21-17.** This hot-wire air mass sensor is used on Bosch LH-Jetronic EFI systems. (Volvo)

**Magnetoresistive Sensors.** Some magnetic materials change their resistance in a rotating magnetic field. Magnetoresistors can be connected in a bridge circuit similar to the piezoresistors in figure 21-16. The computer supplies the constant reference voltage. A rotating magnetic field changes the resistance of the resistors and unbalances the bridge circuit. This produces a variable output voltage proportional to the rate of change of the magnetic field.

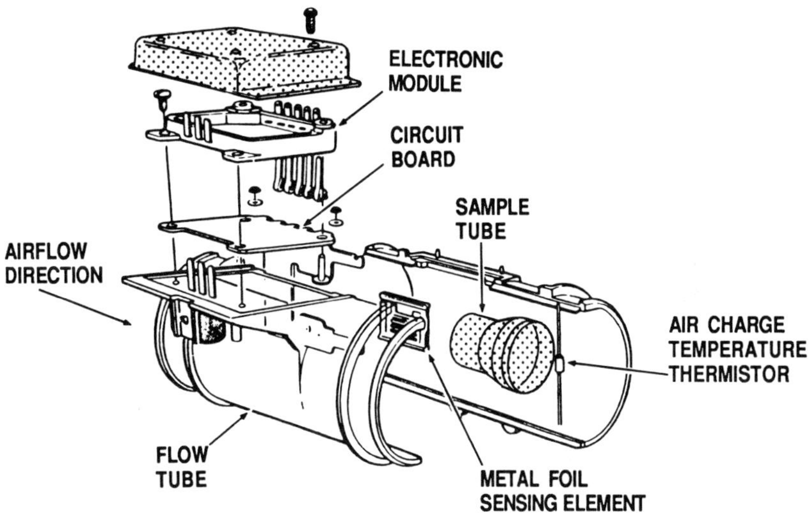


**Figure 21-18.** The GM heated-film mass airflow sensor sends a variable-frequency signal to the computer.

**Heated Resistive Sensors.** The application of electric current to a platinum wire or metal-foil sensing screen as part of a bridge circuit can be controlled by the computer to maintain the wire at a constant temperature regardless of the ambient air temperature. The current required to maintain the wire or screen temperature is then changed into a voltage signal at a resistor.

This type of sensor is most useful in measuring mass airflow and is found in some fuel injection systems, figure 21-17. Air flowing past the hot wire or screen tends to cool it, according to the temperature and velocity of the airflow. As this occurs, the computer applies more current to maintain the wire at a constant temperature. The amount of current required to maintain the wire temperature is interpreted by the computer as the amount of air mass taken into the engine.

**Other Heated Airflow Sensors.** Both GM and Ford have used heated mass airflow sensors on port fuel injected engines since the mid-1980's. Chevrolet-designed 305- and 350-cid (5.0- and 5.7-liter) V-8's use a Bosch-type heated-wire sensor similar to the one shown in figure 21-17. This kind of mass airflow sensor sends a variable voltage signal to the computer that is proportional to the mass (weight) of air entering the engine.

The mass airflow sensors on GM V-6 engines (both Chevrolet- and Buick-designed) use a heated-film sensing element, figure 21-18. Some of these GM sensor assemblies also contain a thermistor to measure the temperature of the intake air charge, figure 21-19.

The computer sends a 5-volt reference voltage to the sensor, and the sensor sends back a variable frequency signal that is proportional to the mass of air entering the engine. The amplitude, or level, of the signal voltage produces a constant dc square-wave signal of 0 to 5 volts. The frequency of the signal varies, however, from 30 to 150 hertz (Hz). Low frequency equals low airflow; high frequency equals high airflow. Ford uses similar frequency-modulated mass airflow sensors on several late-model engines.

GM variable-frequency MAF sensors are made by Delco Electronics and by Hitachi. Reliability problems with the Delco sensors on 1987–88 2.8-liter V-6 engines caused GM to discontinue this sensor on many engines. In the early 1990's the use of heated, variable-frequency MAF sensors declined at GM, while their use increased on Ford engines.

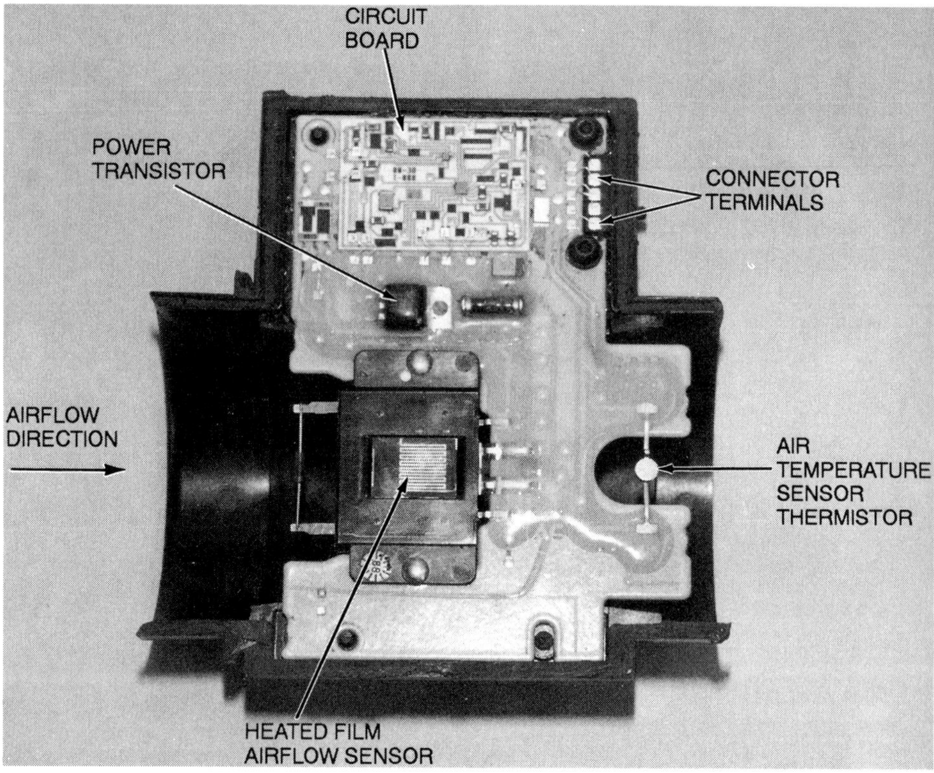


**Figure 21-19.** This cutaway photo of a GM mass airflow sensor shows the heated-film sensing element and the air temperature thermistor sensor.

**Karman Vortex Mass Airflow Sensors.** Mitsubishi and Chrysler (on Mitsubishi-built engines) have used a MAF sensor known as a Karman vortex sensor since the mid-1980's. This sensor operates with a 5-volt reference voltage, as resistive sensors do, and sends a variable-frequency signal to the computer, as GM and Ford frequency-modulated MAF sensors do. Its operating principles are different, however.

The sensor has a triangular Karman vortex generating column in the middle of the intake airflow, figure 21-20. The column disturbs airflow and generates turbulence called Karman vortices (after the 19th century German physicist who discovered the phenomenon).

The 5-volt reference voltage is applied to an amplifier in the sensor that drives a transmitter. The transmitter sends ultrasonic sound waves across the sensor opening, at a right angle to the airflow. A receiver on the other side of the sensor opening receives the sound waves.

When there is no airflow through the sensor, the sound waves take a fixed time to reach the receiver. This becomes a reference time signal. As air flows through the sensor, past the column, vortices develop in the airflow. As the sound waves strike the vortices, they speed up or slow down and create a variable signal at the receiver that alternately rises and falls in pitch. The frequency of the signal is directly proportional to the number of vortices, which are directly proportional to the amount of airflow. A modulator circuit converts the ultrasonic sound signal to a square wave signal of varying frequency. This is the sensor input signal to the computer. Once again, low frequency equals low airflow; high frequency equals high airflow.

**Transformers.** Like resistors, transformer sensors send an analog return signal to the computer in response to a reference voltage. The transformer is a set of induction coils with a movable core, figure 21-21. This kind of device is called a **linear variable differential transformer (LVDT) sensor**. The electromagnetic coupling between the input and output windings varies with core position. An oscillating reference voltage (usually 10 kHz) is applied to the input winding. Two output windings are balanced so that output voltage of each is the same with the core in the center position. When the core moves, the output voltage of one winding exceeds that of the other winding. The two output winding voltages are processed to produce a single sensor signal voltage that is proportional to amount and direction of movement.

This kind of transformer sensor is used as one type of manifold absolute pressure (MAP) sensor. In this design, core movement is controlled by an aneroid bellows or capsule. Pressure variations on the outside of the aneroid cause it to expand and contract and move the core. LVDT sensors also have been used to measure linear movement, such as throttle position.

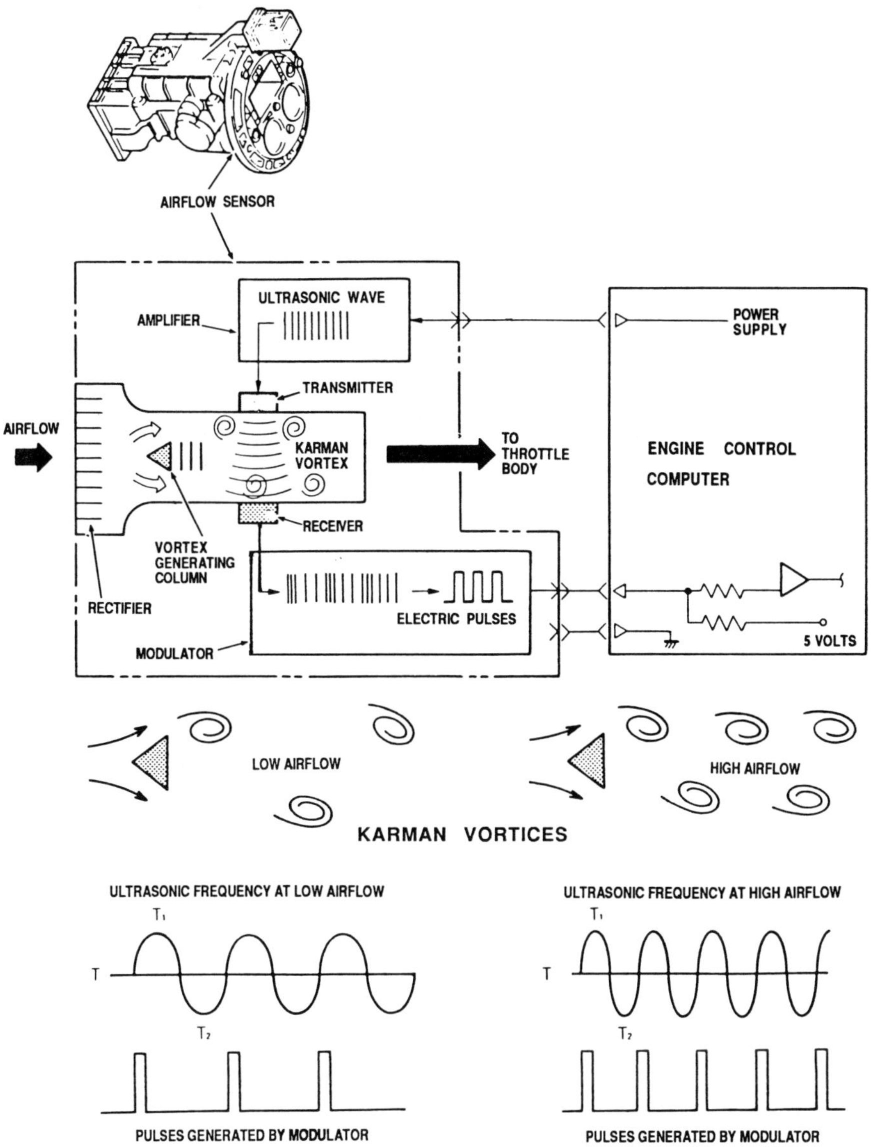


**Figure 21-20.** The Mitsubishi-Chrysler Karman vortex mass airflow sensor. (Chrysler)

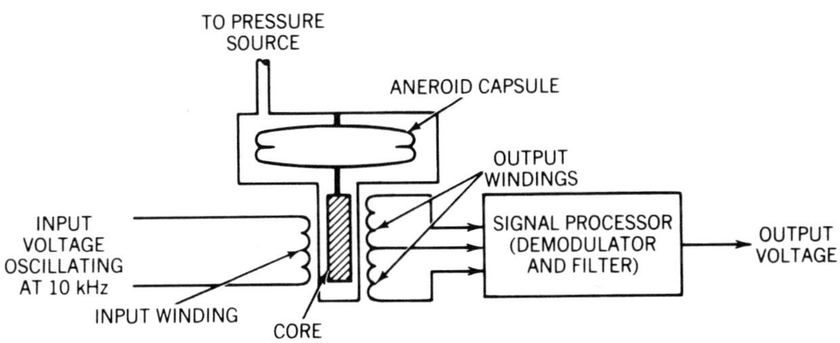


**Figure 21-21.** This MAP sensor is a transformer that creates a voltage differential between two output windings.

In these designs, the core is moved by mechanical linkage.

As pressure and position sensors, LVDT sensors do not have the accuracy and repeatability of resistive-bridge sensors and potentiometers, respectively. Therefore, their use in engine control systems declined in the early 1980's. Their durability and other positive features have let them make a comeback as height sensors in electronic ride control systems, however.

## Voltage Generators

Every sensor we have seen thus far sends an input signal in response to a reference voltage. A few sensors, however, can generate a signal voltage. When you studied electronic ignitions in chapter 13, you learned about distributor pulse generators. These use a rotating trigger wheel to vary the reluctance around a magnetic pole piece and pickup coil, figure 21-22. Some carmakers call them **reluctance sensors**. Engine computers can use these signals as a reference for basic timing. Signal frequency also indicates engine speed.

Pulse generators also can be used to indicate crankshaft position, engine speed, and firing order. Some systems use a pulse generator or reluctance sensor in the engine block to respond to a tooth or a cutout on the harmonic balancer or the flywheel, figure 21-23. Bosch, Ford EEC-I & II, and GM distributorless ignition systems use these sensors. Some engines have two separate sensors of this kind to indicate crankshaft speed and position. Other systems use a Hall-effect switch in the distributor, figure 21-24, to indicate crankshaft position and basic timing.

## Hall-Effect Switches

The operation of a Hall-effect switch was explained as an ignition timing sensor in chapter 13. A Hall-effect switch receives an input voltage and returns a square-wave signal voltage. The amplitude of the signal voltage may be 0 to 5, 7, 9, or 12 volts. The

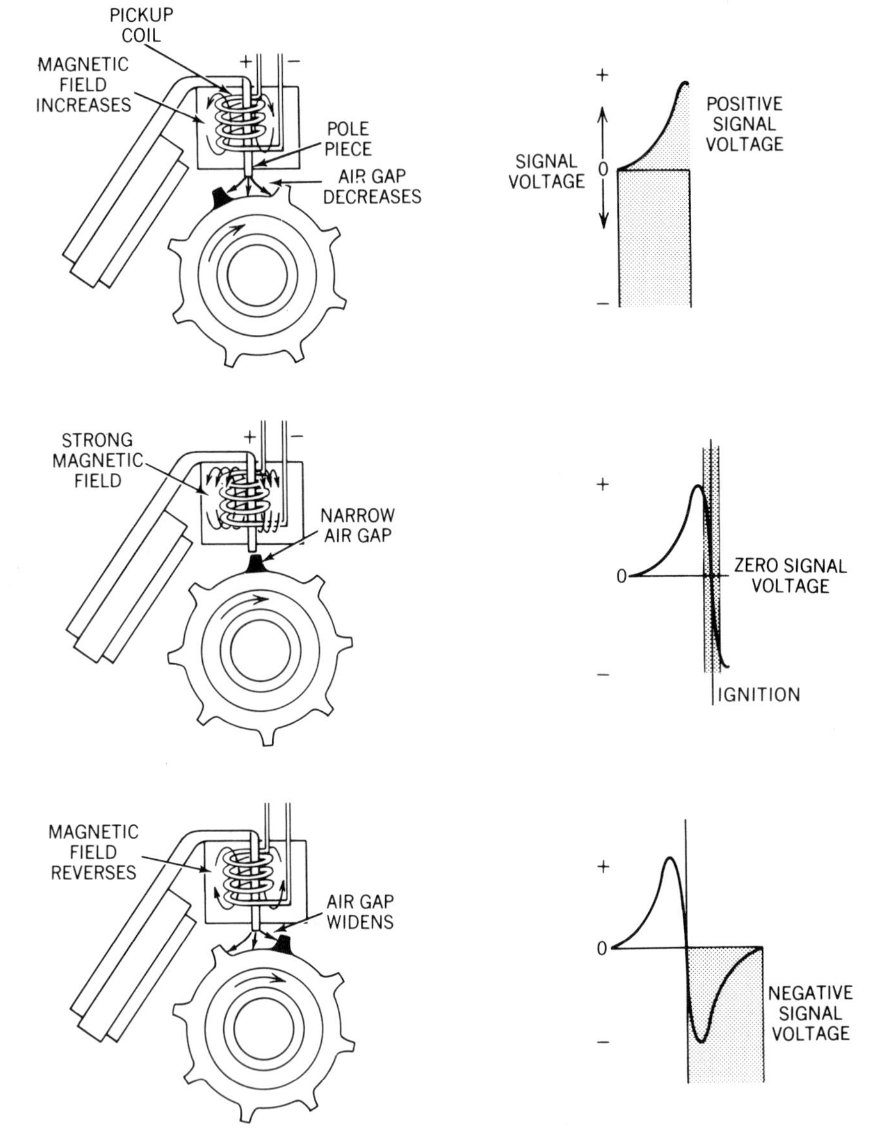


**Figure 21-22.** The rotating trigger wheel varies the reluctance around a pole piece and pickup coil to generate voltage.

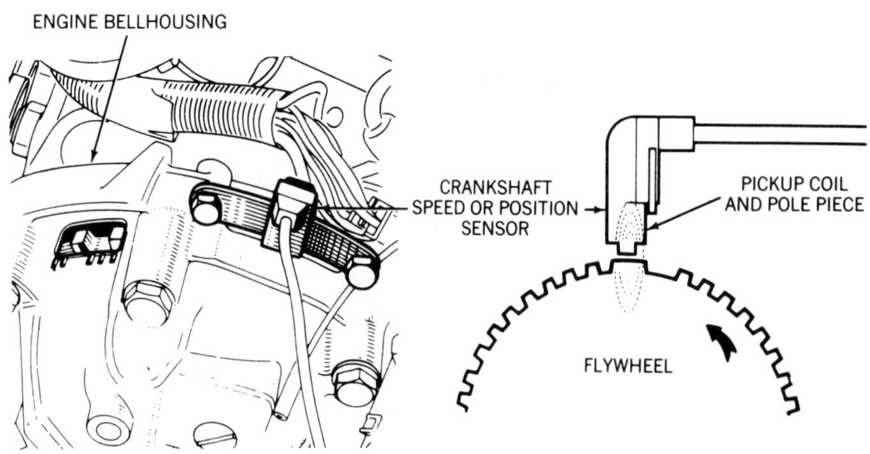


**Figure 21-23.** Some GM, Ford, Renault, and other Bosch systems use crankshaft position sensors in the engine block. (AMC-Renault)

voltage level does not matter as much as the shape of the signal.

The square wave is a perfect digital signal that is either high or low. The computer can be programmed to react to either the rising or falling edge of the signal. The important feature of a Hall-effect switch is that the high-low signal can precisely indicate both speed and position. They can indicate crankshaft speed and position from a position inside a distributor, figure 21-24, or from a location on the engine crankshaft damper. Because the high or low condition of the signal voltage depends on the position of a shutter blade in relation to a magnet and the Hall sensor element, a Hall-effect switch can be an excellent position sensor for electronic ride height control systems.

## Exhaust Gas Oxygen Sensor

The exhaust gas oxygen (EGO) sensor is a unique voltage generator of zirconium dioxide that measures exhaust oxygen content. It generates analog signals from 0 to 1 volt by comparing the difference between oxygen in the exhaust and oxygen in the ambient air.

The EGO sensor is based on the **lambda** concept. Lambda is the Greek symbol engineers use to indicate the ratio of one number to another. For air-fuel control, lambda indicates the ratio of excess air to stoichiometric air quantity. That's why Bosch calls EGO sensors lambda sensors.

At an air-fuel ratio of 14.7:1, as much air as possible combines with fuel. There is no excess air, and there is no shortage of air. Lambda, therefore, equals 1. With a lean mixture ratio of 15, 16, or 17:1, there is excess air left after combustion. The lambda ratio of excess air to desired air is greater than 1. It may be 1.03, 1.07, 1.15, or some other number. With a rich mixture of 12, 13, or 14:1, there is a shortage of air and the lambda ratio is less than 1. It may be 0.97, 0.93, 0.89, etc. With lambda (excess air) ratios less than 0.8 or greater than

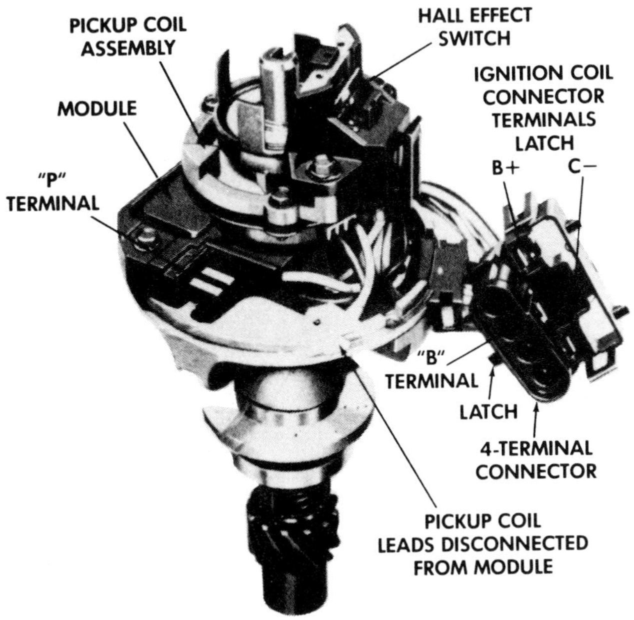


**Figure 21-24.** This Hall-effect switch in a Delco HEI distributor indicates crankshaft position and basic timing. (Delco-Remy)

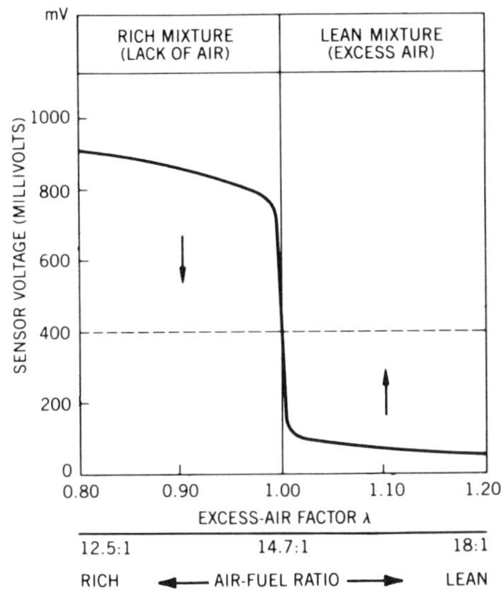


**Figure 21-25.** Lambda (EGO) sensor voltage at 800° C (1,472° F). (Bosch)

1.20, the engine will not run. These roughly equal air-fuel ratios of 11.7:1 and 18:1.

The EGO sensor works as a **galvanic battery** to generate voltage up to about 1 volt. Its effective signal range is from about 0.1 to 0.9 volt (100 to 900 millivolts). When exhaust oxygen content is low (rich mixture), sensor voltage is high (450 to 900 millivolts). When exhaust oxygen content is high (lean mixture), sensor voltage is low (100 to 450 millivolts). Figure 21-25 shows the operating range of the sensor at a temperature of about 800° C (1,472° F). Notice that sensor voltage changes most rapidly near a lambda ratio of 1 (air-fuel ratio of 14.7:1), which makes it ideal for maintaining a stoichiometric ratio. The sensor must warm up to at least 300° C (572° F) before it will generate an accurate signal, and it provides the fastest switching times at about 800° C. That is the major reason for open-loop fuel control with a cold engine.

The EGO sensor is installed in the exhaust manifold or headpipe, figure 21-26, and consists of two platinum electrodes separated by a zirconium dioxide ($ZrO_2$) ceramic electrolyte. $ZrO_2$ attracts free oxygen ions, which are negatively charged. One electrode is exposed to ambient (outside) air through vents in the sensor shell, figure 21-27, and collects many $O_2$ ions. It becomes a more negative electrode. The other electrode is exposed to exhaust gas and also collects $O_2$ ions. However, it collects fewer ions and becomes more positive. When there is a large difference between oxygen in the exhaust and oxygen in the air (rich mixture), the negative oxygen ions on the outer electrode

### HOW SENSITIVE ARE PRESSURE SENSORS?

One of the first electronic pressure sensors for automotive use was designed by Delco Electronics in 1976. Delco made more than 32 million manifold absolute pressure (MAP) and barometric or BMAP sensors for automobiles in the decade following that first one. Today's MAP and BMAP sensors continue to shrink in size while increasing in sensitivity. Today's sensor uses a silicon membrane only 0.001 inch thick, (see photo). That's pretty thin for a piece of silicon strong enough to survive both the corrosive gases and extreme heat found in an engine compartment, yet sensitive enough to tell whether it's mounted right side up or upside down. The next generation of MAP sensors is right around the corner and will combine the sensor with its signal circuitry on a single silicon membrane.

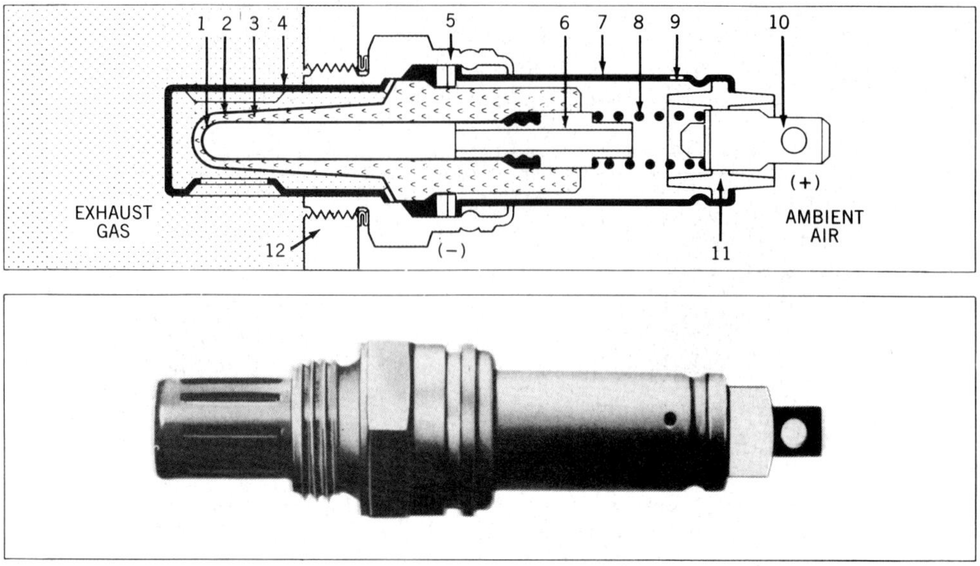


**Figure 21-26.** The EGO sensor is installed in the exhaust manifold and is exposed to ambient air as well as to exhaust. (Bosch)

move to the positive inner electrode. This is simple electric current (electron flow). The sensor then develops a voltage between the two electrodes. When there is more oxygen in the exhaust (lean mixture), there is less difference between $O_2$ ions on the electrodes and a lower voltage. Figure 21-28 summarizes the relationships of engine air-fuel mixture, exhaust oxygen content, sensor voltage, and computer response.

An important point to remember about an EGO sensor is that it measures oxygen; it does not measure air-fuel ratio. If the engine misfires, no oxygen is consumed in combustion. There is a large amount of oxygen in the unburned exhaust mixture, and the sensor will deliver a false "lean mixture" signal. This is one reason why computer control of ignition timing and EGR is essential for effective fuel metering control.

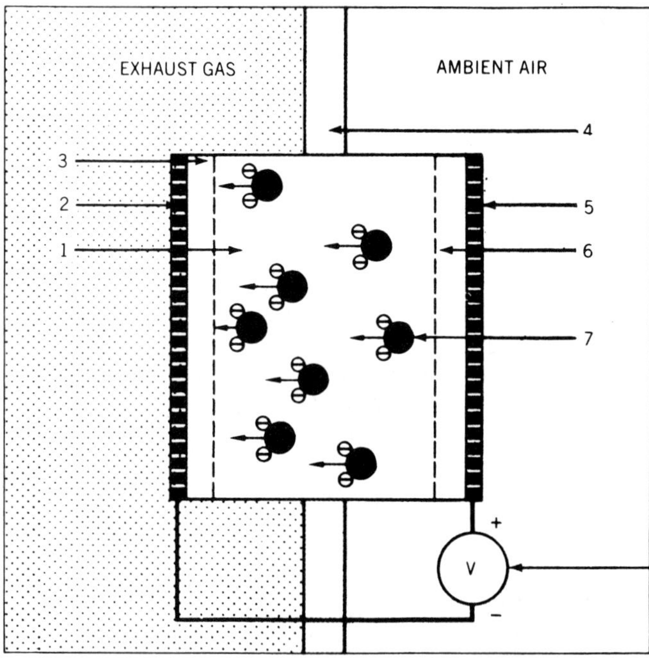


**Figure 21-27.** The $ZrO_2$ ceramic electrolyte attracts $O_2$ ions to the electrodes and thus generates a voltage. (Bosch)

All EGO sensors work on the principles explained above, but construction details differ. Some sensors have a single-wire connector for the output signal. The ground connection exists between the sensor shell and the exhaust manifold or pipe, figure 21-26. Other sensors have a 2-wire connector that provides a ground connection through the computer, figure 21-29. Single-wire and 2-wire sensors are not interchangeable. Figure 21-29 also shows a silicone boot that protects the sensor and provides a vent opening for ambient air circulation. The boot position is important for such a sensor. If it is pushed down too far on the body, it will block the air vent and create an inaccurate signal voltage.

There is a growing trend toward the use of heated EGO sensors. These sensors have a third wire that delivers battery current of 1 ampere or less to the sensor electrodes whenever the ignition switch is on. This current warms the sensor faster with a cold engine and helps to maintain sensor temperature for accurate voltage sig-

| *Air-Fuel Mixture* | *Exhaust Oxygen Content* | *Oxygen Sensor Voltage* | *Computer Output Signal* | *Fuel Metering Change To* | *Idle Airflow* |
|---|---|---|---|---|---|
| Rich | Low | High | Decrease Fuel Metering | To Lean | Increase |
| Lean | High | Low | Increase Fuel Metering | To Rich | Decrease |

**Figure 21-28.** This table shows the relationships of air-fuel mixture, exhaust $O_2$, sensor voltage, and computer response.

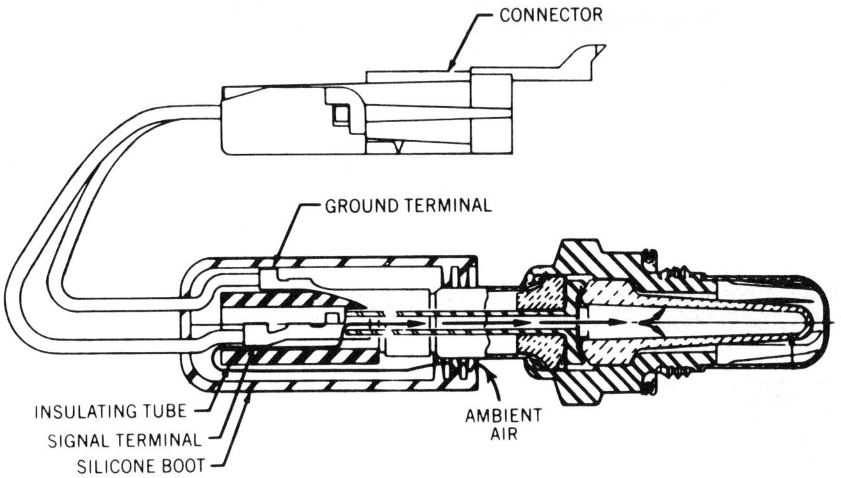


**Figure 21-29.** This EGO sensor has a 2-wire connector and a protective silicone boot. (AC-Delco)

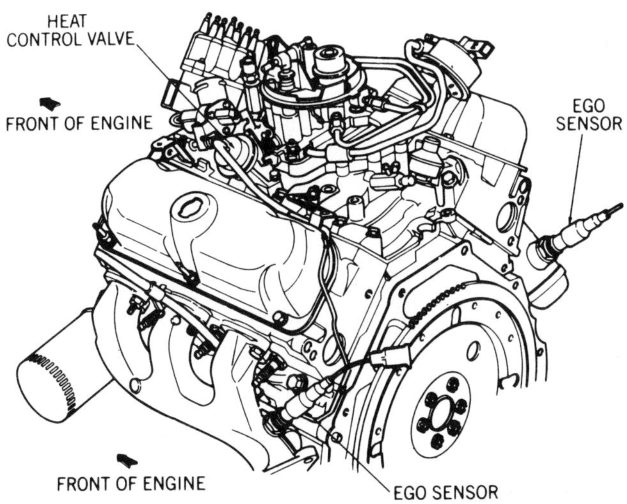


**Figure 21-30.** Some V-type engines have two EGO sensors, one in each exhaust manifold. (Ford)

nals at all times. Heated EGO sensors often are used on turbocharged engines where the sensor is installed downstream from the turbo. The turbo absorbs much of the heat energy in the exhaust, which delays sensor warmup on a cold engine and can cause the sensor to cool during long periods of idling and low-speed operation. A heated EGO sensor overcomes these problems. Some heated EGO sensors have an additional pair of wires to supply heating current to a PTC thermistor in the sensor.

EGO sensors are always installed in the exhaust manifolds or headpipes, but exact locations vary for different engines. Some V-type engines have separate sensors in each manifold, figure 21-30. Other V-type engines have only a single sensor. Sensors are often located low in the engine compartment and access can be difficult. Other sensors are installed in the center of the exhaust manifold and access is easy, figure 21-31. Some Toyota engines have two EGO sensors: one before and one after the catalytic converter. Engineers determine the best sensor location for accurate $O_2$ measurement during laboratory testing of each engine design.

**Titania Exhaust Gas Oxygen Sensors.** The zirconia EGO sensors described in the previous section are by far the most common and have been in use for well over a decade. Another kind of EGO sensor has been used by

**Figure 21-31.** This EGO sensor is easily accessible in the center of the manifold, at the front of the engine.

some carmakers, such as Jeep, Nissan, and Toyota. The sensing element is titanium dioxide, or titania.

The titania sensor provides the same kind of feedback signal as a zirconia sensor to indicate the relative rich or lean condition of the exhaust. The titania EGO sensor *does not* generate a signal voltage as a zirconia sensor does, however. Rather, it is a variable resistor that requires a constant reference voltage from the system computer. The resistance of the titania element varies in relation to the oxygen concentration in the exhaust. The resistance of the titania element also changes in relation to temperature, so the sensor has a heating element to maintain a constant temperature.

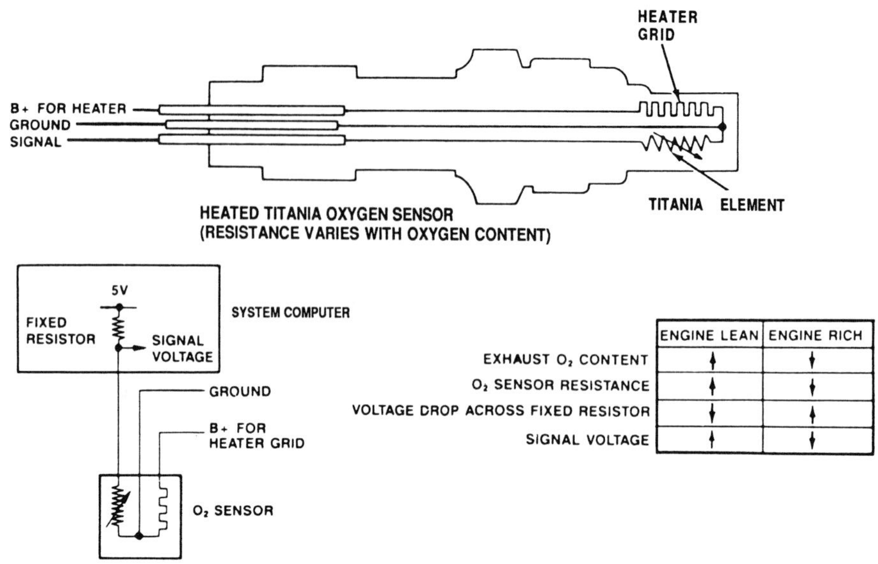


**Figure 21-32.** The titania EGO sensor used in 1987–90 Jeeps operates on a 5-volt reference from the computer. (Chrysler)

The titania EGO sensor used on 1987–90 244-cid (4.0-liter) Jeep engines receives a 5-volt reference voltage from the computer, figure 21-32. A fixed resistor is in series with the sensor, and the signal voltage is measured between the sensor and the fixed resistor. As sensor resistance changes, it drops more or less voltage; and the signal voltage changes. This is similar to the circuits for most coolant temperature sensors.

When exhaust oxygen content is low (rich mixture), sensor signal voltage is low—below 2.5 volts. When exhaust oxygen content is high (lean mixture), sensor signal voltage is high—above 2.5 volts. Notice that the direction of the signal voltage in relation to high or low oxygen content is opposite to the signal from a zirconia sensor. Because the computer is programmed for this kind of signal, it can read it correctly and control fuel metering as any other feedback system does.

The titania EGO sensor used on some Toyota Corollas and V-6 trucks since 1989 receives a 1-volt reference voltage from the computer and returns a signal that varies from 0.1 to 0.9 volt. Because of the way the circuit is built, figure 21-33, signal direction in relation to exhaust oxygen is similar to a zirconia sensor. High signal voltage above 0.5 volt indicates low oxygen content, or a rich mixture. Low signal voltage below 0.5 volt indicates a lean mixture.

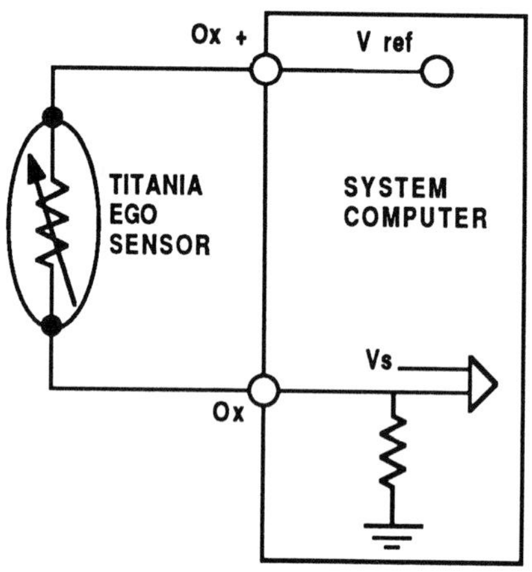


**Figure 21-33.** Simplified circuit diagram of the Toyota titania EGO sensor.

## Other Sensors

Here is a brief summary of other sensors used in electronic engine control systems.

**1.** *Airflow sensors*—Airflow sensors are usually analog devices that use a potentiometer to vary a return voltage signal in proportion to airflow volume, figure 21-34. They provide information to regulate fuel flow in relation to airflow as a venturi does in a carburetor.

**2.** *Manifold pressure, vacuum, and barometric pressure sensors*—These can be load sensors, air volume sensors, or both. Most are analog devices that use a peizoresistive sensor, figure 21-35, or a transformer or potentiometer operated by a vacuum diaphragm or aneroid bellows, figure 21-21. These sensors allow the computer to adjust fuel metering in relation to air volume and engine load. They also allow the computer to simultaneously adjust timing and EGR in relation to load. Remember that vacuum is engine negative pressure; barometric pressure is atmospheric pressure; and manifold absolute pressure is barometric pressure minus vacuum. A combination of this information allows more precise engine control than manifold vacuum can provide directly.

**3.** *Temperature sensors*—If a computer only needs to know whether coolant or air temperature is above or below a single point, a simple bimetal switch will do the job, figure 21-6. If a computer needs to know any temperature within a range, a thermistor is

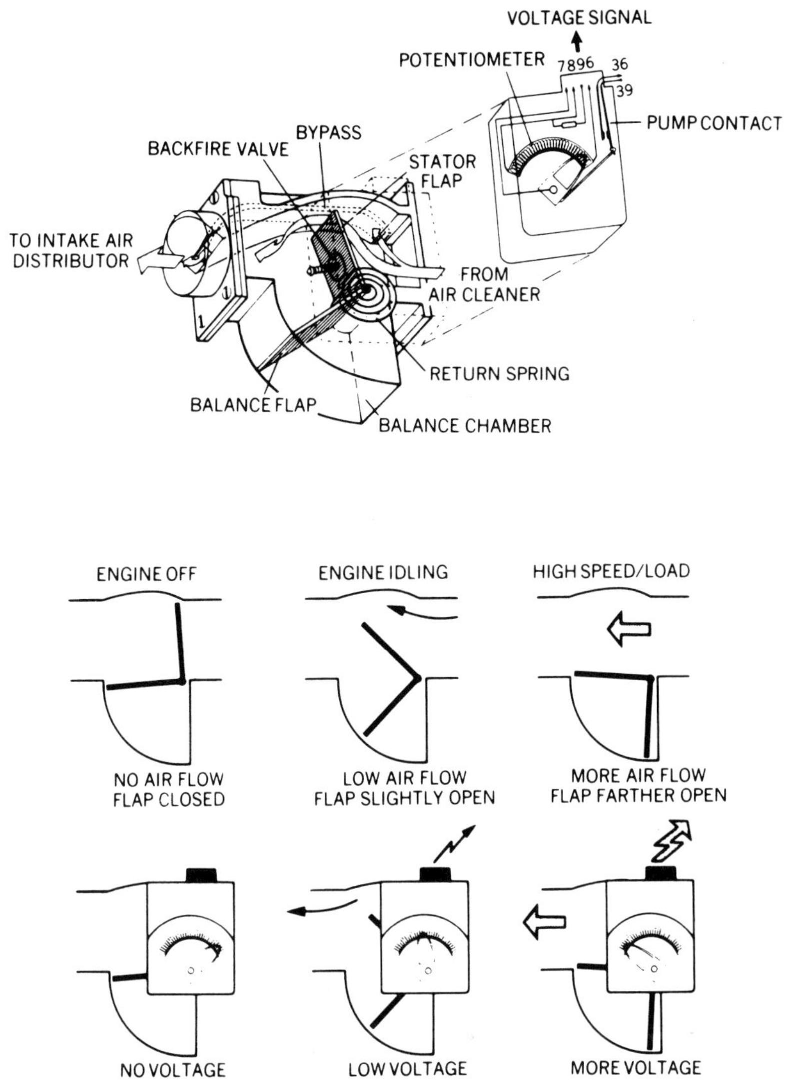


**Figure 21-34.** This airflow sensor uses a potentiometer to change analog movement into a variable voltage signal. (Volkswagen)

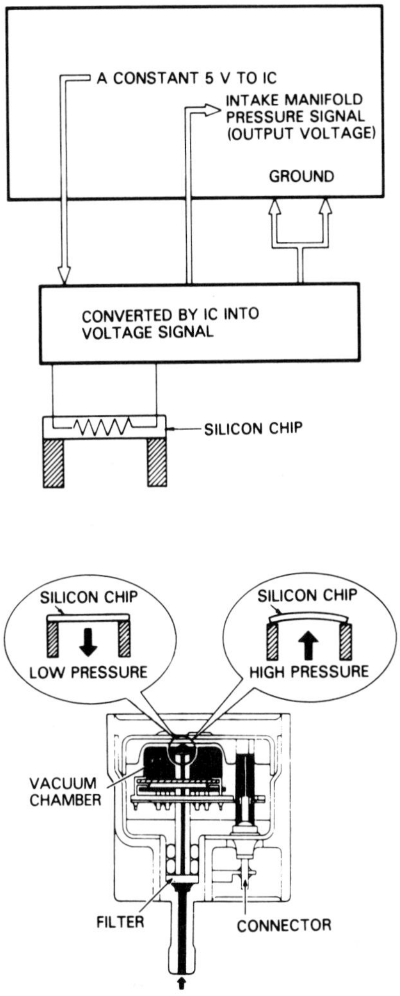


**Figure 21-35.** Operation of a piezoresistive MAP sensor. (Toyota)

necessary, figure 21-13. For fuel metering, coolant and air temperature sensors do the jobs of a carburetor choke coil and supplement the control of a thermostatic air cleaner. They also allow the computer to fine tune ignition timing and EGR flow.

**4.** *Throttle position sensors*—A throttle position sensor indicates whether an engine is at idle, wide open throttle, or a point in between. It is both a load and a speed sensor. A throttle position sensor can be a simple wide-open throttle switch or idle position switch, figure 21-36, to indicate the extremes of throttle travel with a high or low voltage. Many systems use potentiometers as throttle sensors, figure 21-37. The throttle moves a sliding contact across a resistor to send an analog signal that indicates the position and speed of throttle movement.

**5.** *Ignition timing, crankshaft position, and engine speed sensors*—We looked at these sensors at the beginning of this section. They send the computer both analog and digital signals to control timing, fuel metering, and EGR.

**6.** *Vehicle speed sensors*—Some control systems need to know vehicle speed to control torque converter lockup and transmission gear selection. Speed sensors are usually pulse

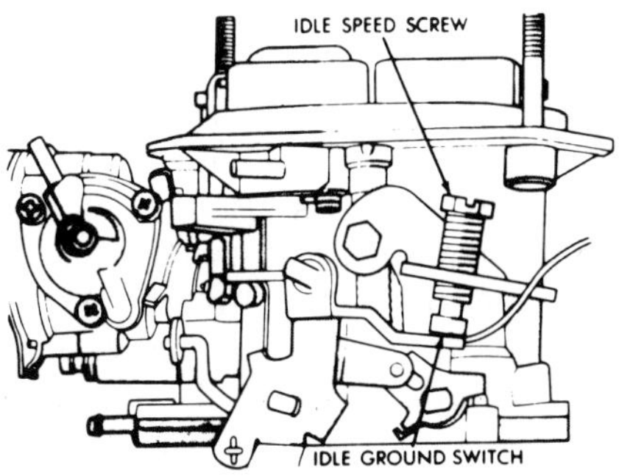


**Figure 21-36.** A simple wide-open throttle switch or idle position switch indicates the extremes of throttle travel with a high or low voltage. (Chrysler)

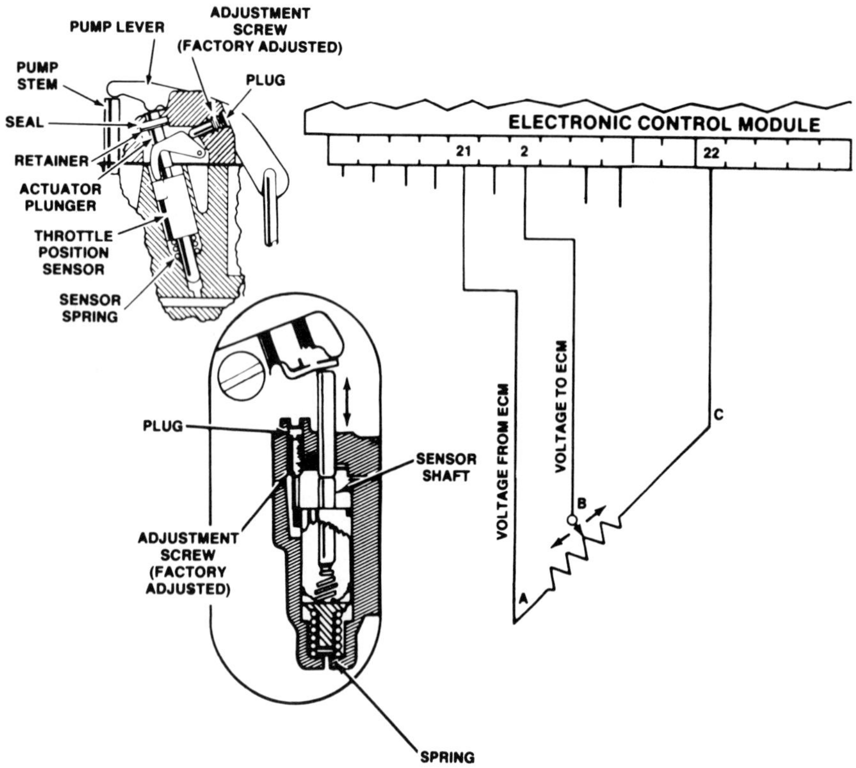


**Figure 21-37.** This GM throttle position sensor is a potentiometer, mounted in the carburetor fuel bowl. (GM)

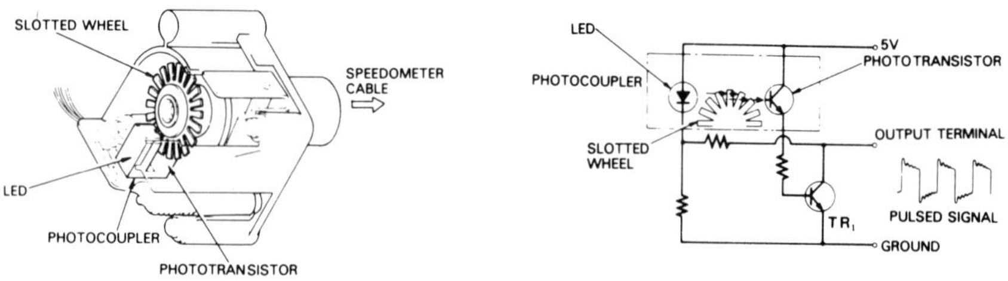


**Figure 21-38.** This speed sensor uses an LED and a photocell in the speedometer. (Toyota)

generators or optical sensors in the speedometer, figure 21-38.

7. *EGR sensors*—Knowing the EGR flow rate at any time is important for controlling timing and fuel metering, as well as EGR valve operation. Many EGR valves have a transducer that consists of a sliding-contact potentiometer on the top of the valve stem, figure 21-39. Through engine mapping, the computer knows that a certain return voltage signal equals a certain valve position, which equals a specific EGR flow rate. When combined with speed, load, and temperature signals, the computer uses this information to adjust timing, EGR, and fuel metering.

8. *Air conditioning sensors*—Most air conditioning sensors are switches that indicate when the compressor clutch is engaged. The A/C compressor can use 5 to 10 extra horsepower, which increases engine load. When the computer knows that the compressor is engaged, it can adjust timing, fuel flow, idle speed, and other variables to compensate for the extra load.

9. *Detonation sensors*—Most detonation sensors contain a piezoresistive crystal that changes resistance when pressure is applied to it. When detonation does not exist, pressure is uniform across the crystal, figure 21-40. Reference voltage is applied to the sensor and the return signal voltage remains at its programmed value. When detonation occurs, it changes the pressure applied to the sensor crystal. This changes the sensor resistance and, thus, the return signal voltage.

## ACTUATORS (CONTROL SYSTEMS)

An actuator is an output device that changes voltage signals to mechanical action. Most engine control actuators are solenoids or relays. Some, however, are stepper motors. As you know:

- A solenoid is an electromechanical device that uses magnetism to move a metal core, thus changing an electrical voltage into mechanical movement.
- A relay uses one electric current to control another electric current.
- A stepper motor is a dc motor that moves in incremental steps from deenergized (no voltage) to fully energized (full voltage).

### Solenoids

A computer controls voltage to a solenoid, but it usually does not switch the system voltage. Battery voltage is applied to one solenoid terminal, and the computer opens and closes a ground switch connected to the other terminal, figure 21-41. Thus most actuators are grounded through the computer, and a secure computer ground connection is essential for proper system operation. The solenoid can be energized for a random length of time, or it can vibrate on and off at a certain number of cycles per second.

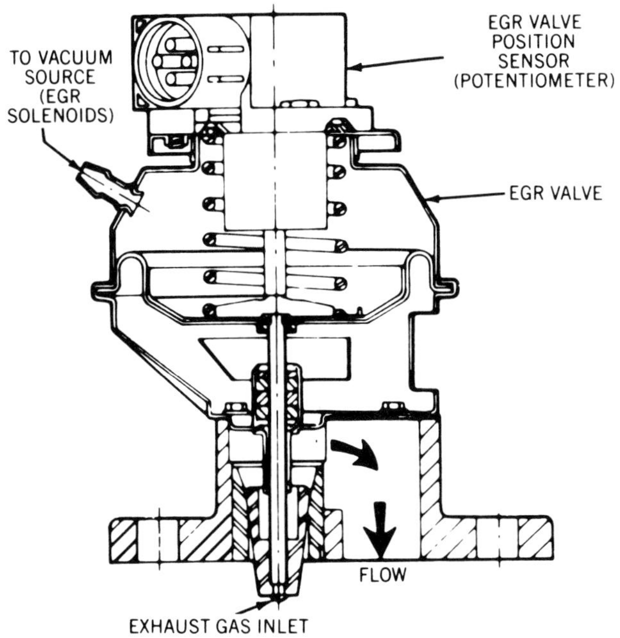


**Figure 21-39.** This potentiometer senses EGR valve position. (Ford)

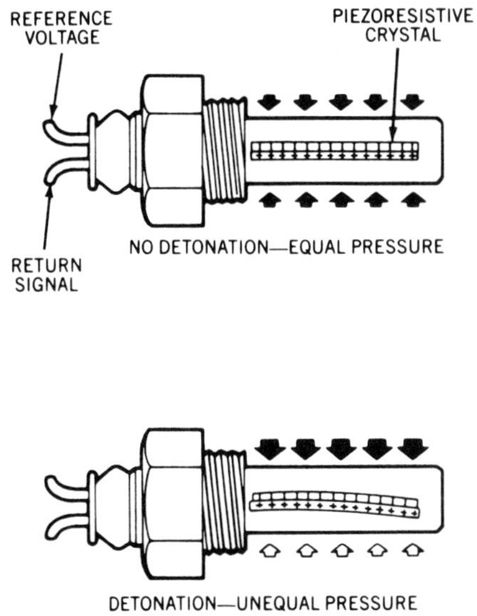


**Figure 21-40.** Detonation (knock) sensors have a piezoresistive crystal that flexes when detonation occurs. (Audi)

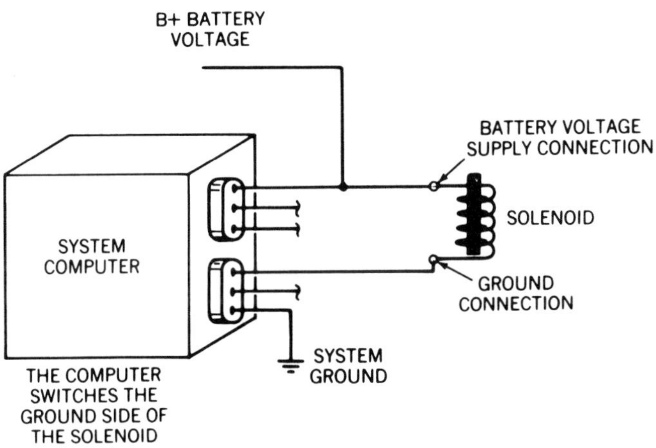


**Figure 21-41.** Most engine control system solenoids are grounded through the computer so the computer does not have to switch system voltage.

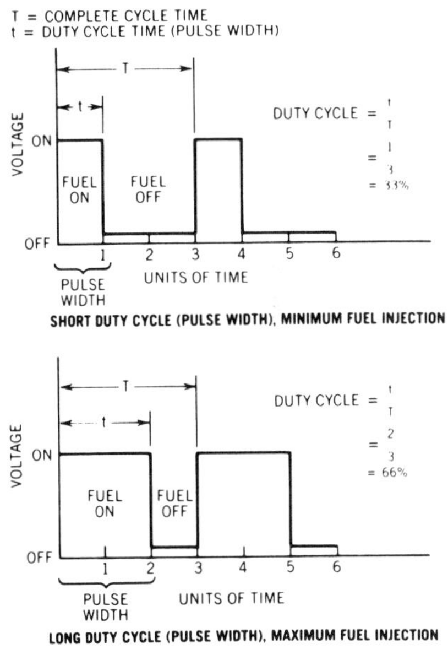


**Figure 21-42.** Pulse width is the duration of solenoid on-time. Duty cycle is the percentage of on-time during one complete cycle.

**Pulse Width and Pulse Width Modulation.** Computer control of solenoid-operated metering devices uses a concept called **pulse width modulation (PWM)**. This is a term you will hear in relation to many solenoid-operated metering devices. If a solenoid, such as a canister purge solenoid or an EGR vacuum solenoid, stays energized indefinitely, we don't concern ourselves with pulse width modulation. The complete operating cycle of any other solenoid-operated metering device is the sequence from on to off to back on again. A device can operate at any number of cycles per second, depending upon its design: 10, 20, 60, and so on.

For a fast-acting, on-off solenoid, however, **pulse width** is the amount of time that the device is energized, measured in milliseconds. An electronic injection nozzle used in a fuel injection system provides a good example of pulse width. The computer varies injector pulse width in relation to the amount of fuel a particular engine requires at any given time. A short pulse width delivers little fuel; a longer pulse width delivers more fuel, figure 21-42. The computer modulates, or varies, the pulse width to establish the duty cycle, which will achieve the required solenoid output.

## MULTIFUNCTION SENSORS

Wouldn't it be nice if one sensor could do everything? Nice but not likely. How about one type of sensor that can do several different things? Also nice, and quite probable. A piezoelectric level switch has been developed that allows a given sensor to function differently in different fluids. An oscillator drives the piezo element, transferring the mechanical vibrations it creates to the sensor diaphragm, which has an inherent frequency. Submerging the diaphragm in a fluid dampens the sensor, shifting the resonating frequency to a different range. The resonating frequency is transmitted to a control module that uses a sweep generator and resonance discriminator for signal processing. The piezo oscillator frequency is swept between a lower and upper frequency limit. The module circuitry processes the frequency range and amplitude transmitted by the sensor and compares it to a value in memory to determine the level of the fluid being measured. If the fluid level is low, the module circuit activates a display. The illustration shows how several sensors of this type could function with different fluids and a single control unit.

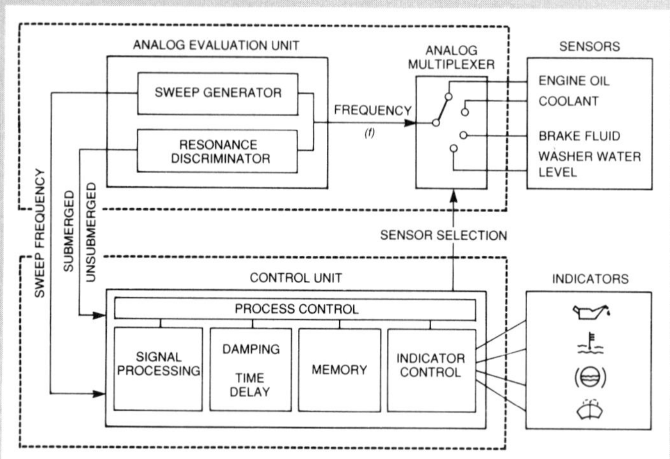


**Duty Cycle.** The **duty cycle** is the percentage of the complete cycle during which the solenoid is on (energized). Another way to say it is that the duty cycle is the ratio of pulse width to complete cycle width. Pulse width is measured in absolute time; duty cycle is measured as a percentage. A timed voltage pulse from the computer determines the duty cycle. Figure 21-43 shows the relationship of pulse width, variable duty cycle, and fixed cycle time for a feedback carburetor mixture-control solenoid.

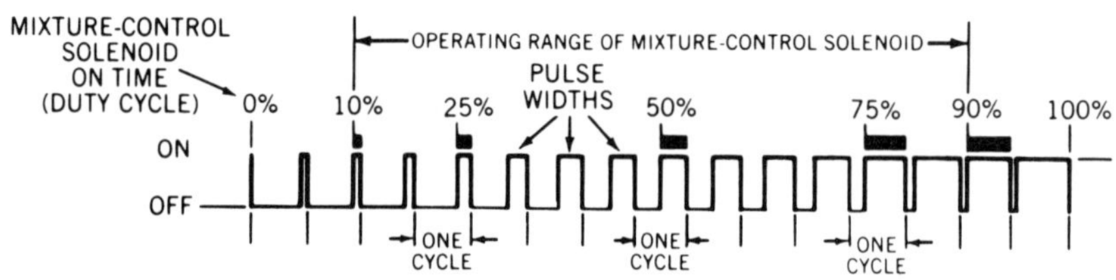


**Figure 21-43.** Duty cycle is the percentage of solenoid on-time and can be changed to control fuel metering. Total cycle time remains constant, but duty cycle and pulse width can vary. (GM)

**Relays.** The computer output driver opens and closes the control circuit of a relay on command from the microprocessor. When the control circuit is open, the power circuit stays open and no power is delivered through the relay to the device being controlled, figure 21-44A. When the control circuit is closed, the energized coil generates a magnetic field that causes the relay switch to close. This allows current to flow through the relay, operating the control device, figure 21-44B.

### Stepper Motors

Stepper motors are also digital actuators. A typical stepper motor can have 100 to 120 steps of movement from deenergized to fully energized, depending on voltage. Stepper motors are commonly used as idle speed controls. The dc motor generally acts directly on the throttle linkage of carbureted engines, figure 21-45. In fuel injection systems, it usually controls an idle air bypass inside the throttle body.

## SUMMARY

All sensors and actuators are transducers. That is, they convert or transduce one form of energy to another. Sensors are input devices used to convert light, motion, temperature, and other forms of energy into voltage signals (electrical energy) for processing by the computer. Actuators are output devices that change the voltage signals of the computer into mechanical work.

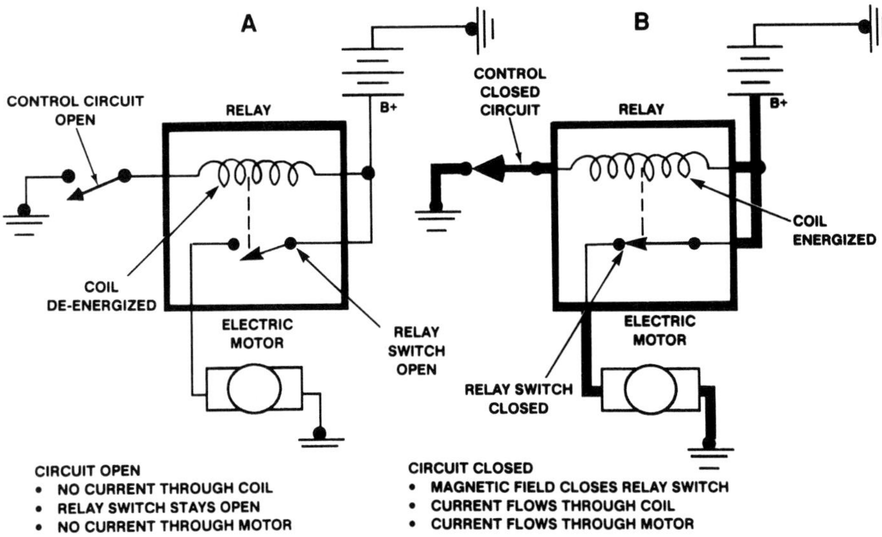


**Figure 21-44.** A relay used as an actuator is controlled by the computer output driver on command from the microprocessor. (Ford)

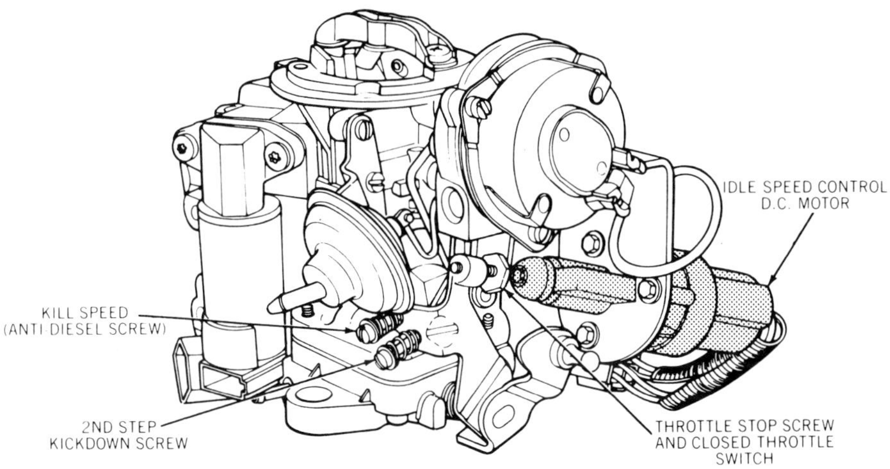


**Figure 21-45.** Idle speed is controlled by the engine computer through a stepper motor called an idle speed control motor. (Ford)

Automotive sensors are switches, timers, resistors, transformers, and voltage generators. Their input to the computer may take the form of analog or digital signals. Analog signals are converted into digital signals by special circuits in the computer for use by the microprocessor. Sensors must have certain characteristics. These are repeatability, accuracy, a specific operating range, and linearity.

Sensors relay voltage signals to the computer either by controlling a reference voltage or by generating voltage. Sensors that operate with a reference voltage receive a 5-volt or a 9-volt signal from the computer. This voltage is modified by the sensor and returned to the computer, which is able to interpret the returned voltage to indicate the engine condition being sensed. Sensors that generate their own voltage convert the condition they are sensing into a specific voltage according to the degree of the condition.

Automotive actuators are generally solenoids or relays, but may also be stepper motors. They translate voltage signals from the computer into mechanical motion to change an engine operating condition according to the input from the sensors.

## REVIEW QUESTIONS

### Multiple Choice

1. Digital computer switching circuits are known as:
   - **a.** logic gates
   - **b.** logic relays
   - **c.** the binary system
   - **d.** AD converters

2. An exhaust gas oxygen sensor is an example of a:
   - **a.** thermistor
   - **b.** potentiometer
   - **c.** generator
   - **d.** voltage divider pickup

3. An engine detonation sensor uses a:
   - **a.** reference voltage
   - **b.** piezoresistive crystal
   - **c.** bimetal element
   - **d.** Wheatstone bridge

4. A stepper motor:
   - **a.** is a type of relay
   - **b.** is either on or off
   - **c.** is a less complicated solenoid
   - **d.** moves in discrete steps

5. The reference voltage sent to a sensor by the computer:
   - **a.** must exceed battery voltage
   - **b.** must equal battery voltage
   - **c.** must be less than minimum battery voltage
   - **d.** has no relation to battery voltage

6. The simplest of all digital sensors is a:
   - **a.** switch

b. potentiometer
c. relay
d. none of these

7. Student A says that a variable resistor with three terminals is called a resistor. Student B says that a variable resistor with two terminals is called a potentiometer. Who is right?

a. A only
b. B only
c. both A and B
d. neither A nor B

8. The percentage of time a solenoid is energized in relation to its total cycle time is called the:

a. frequency
b. duty cycle
c. pulse width modulation
d. linear variation factor

9. Student A says that the resistance of a PTC thermistor increases as temperature increases. Student B says that the resistance of an NTC thermistor decreases as temperature decreases. Who is right?

a. A only
b. B only
c. both A and B
d. neither A nor B

10. Student A says that piezoresistive sensors are excellent devices for digital pressure measurement. Student B says that piezoresistive sensors are used to measure engine intake manifold absolute pressure (MAP). Who is right?

a. A only
b. B only
c. both A and B
d. neither A nor B

11. Student A says that an exhaust gas oxygen (EGO) sensor measures oxygen, not air-fuel ratio. Student B says that the EGO sensor is based on the lambda concept. Who is right?

a. A only
b. B only
c. both A and B
d. neither A nor B

12. Student A says that pulse width is the amount of time that a solenoid is energized. Student B says that duty cycle is the percentage of the complete cycle during which the solenoid is off (deenergized). Who is right?

a. A only
b. B only
c. both A and B
d. neither A nor B

13. Which of the following does not act as an actuator?

a. solenoid
b. relay
c. stepper motor
d. linear variable differential transformer (LVDT)

### Fill in the Blank

14. The time of solenoid pulse width is measured in __________.

15. A thermistor is a variable resistor whose resistance __________ as it is heated.

16. A distributor pulse generator is sometimes called a __________ sensor.

17. An exhaust gas oxygen (EGO) sensor functions as a __________ battery.

18. The duty cycle of a solenoid is the __________ of pulse width to complete cycle width.

19. A semiconductor diode that will transmit light whenever electrical current passes through it is called an __________.

20. A sensor that converts the condition it is sensing into a specific voltage according to the degree of the condition is called a __________.

21. The four characteristics required of all sensors are repeatability, accuracy, operating range, and __________.

# 22 ELECTRONIC ENGINE CONTROLS (AND GASOLINE FUEL INJECTION)

## WHAT

Chapter 8 in Part One of this book introduced the principles of solid-state electronics. Chapters 20 and 21 explained the operation of computer systems and the common sensors (input signal devices) and actuators (output devices) used by most carmakers. Part Five explained the principles of fuel delivery, carburetion, and emission controls.

Chapter 22 now brings these previous chapters together and traces the development of specific carmakers' control systems and the integration of electronic fuel injection as the predominant, "conventional," fuel system on late-model cars.

You may want to review the earlier chapters before you begin this one. As you study chapter 22, you will see that there are more similarities than differences among the engine control systems used by various carmakers.

## WHY

Chapters 1 and 8 introduced the terms "system integration" and "system engineering" to describe how parts and subsystems are organized to operate as complementary parts of a larger system. For almost two decades, electronic engine control systems have integrated the operation of fuel, ignition, emission control, and other vehicle systems.

You will practice "service integration" when you work on these systems. You know that when you troubleshoot and repair a late-model car, you must look at the complete combustion control system. You can't service *just* the ignition, the fuel system, or the emission controls. A single temperature sensor signal to a central computer can affect ignition timing, fuel metering, EGR, and air injection. To service such a system professionally, you must know what the input and output signals mean, what they should be, and how to recognize them when they are wrong. This chapter will help you to understand the principles of engine control systems so you can service them accurately.

## GOALS

After studying this chapter, you should be able to:

1. Identify the engine control systems used by various carmakers and list and explain the functions they monitor and control.
2. Identify and explain the differences between the following fuel injection features:
   a. port injection and throttle body injection
   b. continuous and intermittent injection
   c. mechanical and electronic injection
3. Identify and explain the operation of engine sensors, actuators, and auxiliary systems in gasoline injection systems.
4. Identify various analog and digital inputs from engine sensors and outputs to different control devices (actuators). Explain how each is used to monitor or control ignition timing, fuel metering (air-fuel ratio), idle speed, and other engine functions.
5. Identify various air-fuel ratio control devices in different carburetors and fuel injection systems and explain how they work.

## HISTORY OF ENGINE CONTROL SYSTEMS

Electronic engine controls first appeared in the 1977 model year. Early versions were partial-function systems that regulated ignition timing *or* fuel metering, but not both. Some were introduced with 3-way catalysts to provide a constant stoichiometric ratio. Carmakers quickly expanded their systems to control many engine functions. All systems have one or more of the following features. Most late-model systems have all:

- Electronic timing control instead of vacuum and centrifugal advance
- A 3-way catalytic converter and an exhaust gas oxygen (EGO) sensor for stoichiometric air-fuel ratio control
- Electronic control of fuel metering in the carburetor or fuel injection system
- Electronic control of EGR, air injection switching, vapor canister purging
- Open- and closed-loop operating modes for exhaust sensor feedback, fuel metering, and other control functions
- Electronic control of automatic transmission shifting and torque converter lockup

The exhaust gas oxygen (EGO) sensor is a key part of fully integrated engine control systems. Some carmakers call it simply an oxygen ($O_2$) sensor.

**REBIRTH OF AN OLD IDEA**

In the late 1970's Nissan (Datsun) introduced a twin-ignition system on some 4-cylinder engines. Called the NAPS-Z system, it features two spark plugs per cylinder. The electronic distributor and single ignition coil fire both plugs in each cylinder simultaneously. This provides more uniform flame front propagation to burn the mixture completely in a high-swirl combustion chamber.

Nissan adopted the twin-ignition system on selected engines for more efficient performance, economy, and emission control. It was not, however, a revolutionary new idea. Rather, it was a modern rebirth of an old idea.

In the late 1930's and early 1940's, the Electric Auto-Lite Company built several kinds of twin ignition systems that had two sets of breaker points, two ignition coils, and two spark plugs per cylinder. Each set of points, coil, and plugs was a completely separate ignition system that provided two sparks per cylinder simultaneously. Nash automobiles of the late 1930's used 6- and 8-cylinder versions of this system, and some versions were adapted for use on Ford flathead V-8 racing engines.

The early twin-ignition systems were complicated and expensive to build and service. Improved combustion chamber design, higher octane gasoline, and other ignition advancements made the early systems obsolete after World War II. Nissan, however, saw that an old idea can have a modern application in the NAPS-Z system.

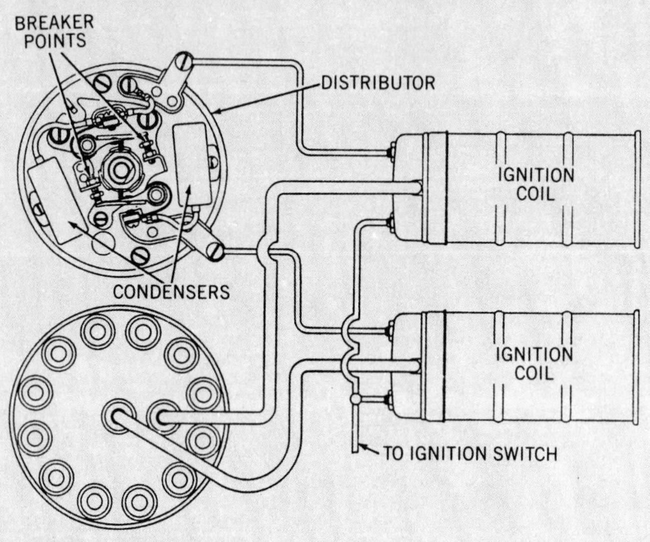


This breaker-point twin-ignition system was used on some engines of the late 1930's and early 1940's.

### Chrysler Corporation

Introduced in late 1976, Chrysler's electronic lean-burn (ELB) system was the first system from a domestic carmaker to replace mechanical spark advance controls with electronic devices. The ELB system, figure 22-1, used sensors for vacuum, throttle position, coolant temperature, and air temperature to send input signals to the spark control computer. Engine speed and basic timing signals came from the electronic distributor pickup coil. Chrysler systems have used distributors with one or two pickup coils. The two pickup coils send different timing signals for starting and running conditions. Chrysler 4-cylinder engines have Hall-effect distributors.

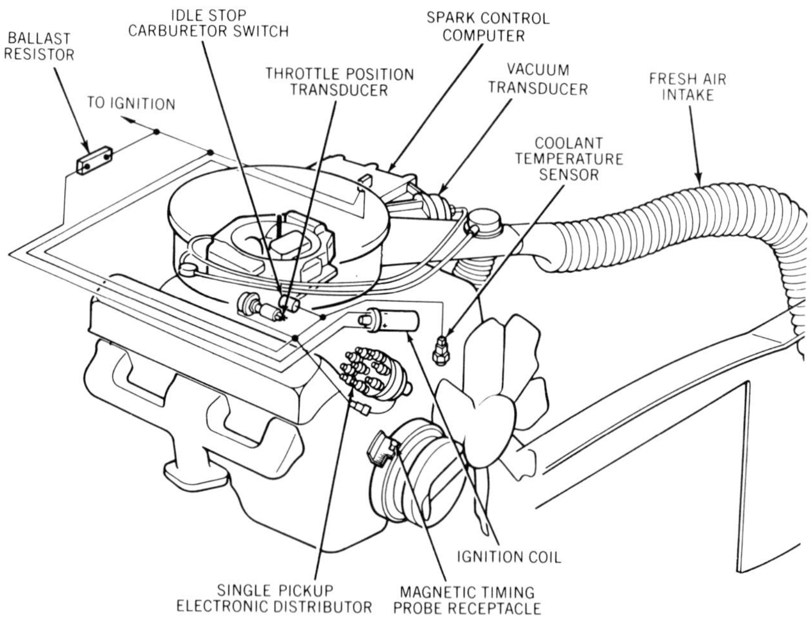


**Figure 22-1.** Chrysler's ELB system was the first system from a domestic carmaker to replace centrifugal and vacuum advance mechanisms with electronic timing control. (Chrysler)

The sensors simply convert the information on engine speed and vacuum to voltage signals for the computer. The computer adjusts timing fast and precisely by processing the input signals and regulating the output signal to the ignition module. Chrysler distributors for electronic control systems have never had vacuum advance devices. The 1976 distributors had centrifugal advance, but this was eliminated in 1977. The computer also controls timing in relation to temperature because it gets input signals from the coolant and air temperature sensors.

The ELB system had a standard carburetor, calibrated for air-fuel ratios as lean as 16:1 to 18:1. The lean

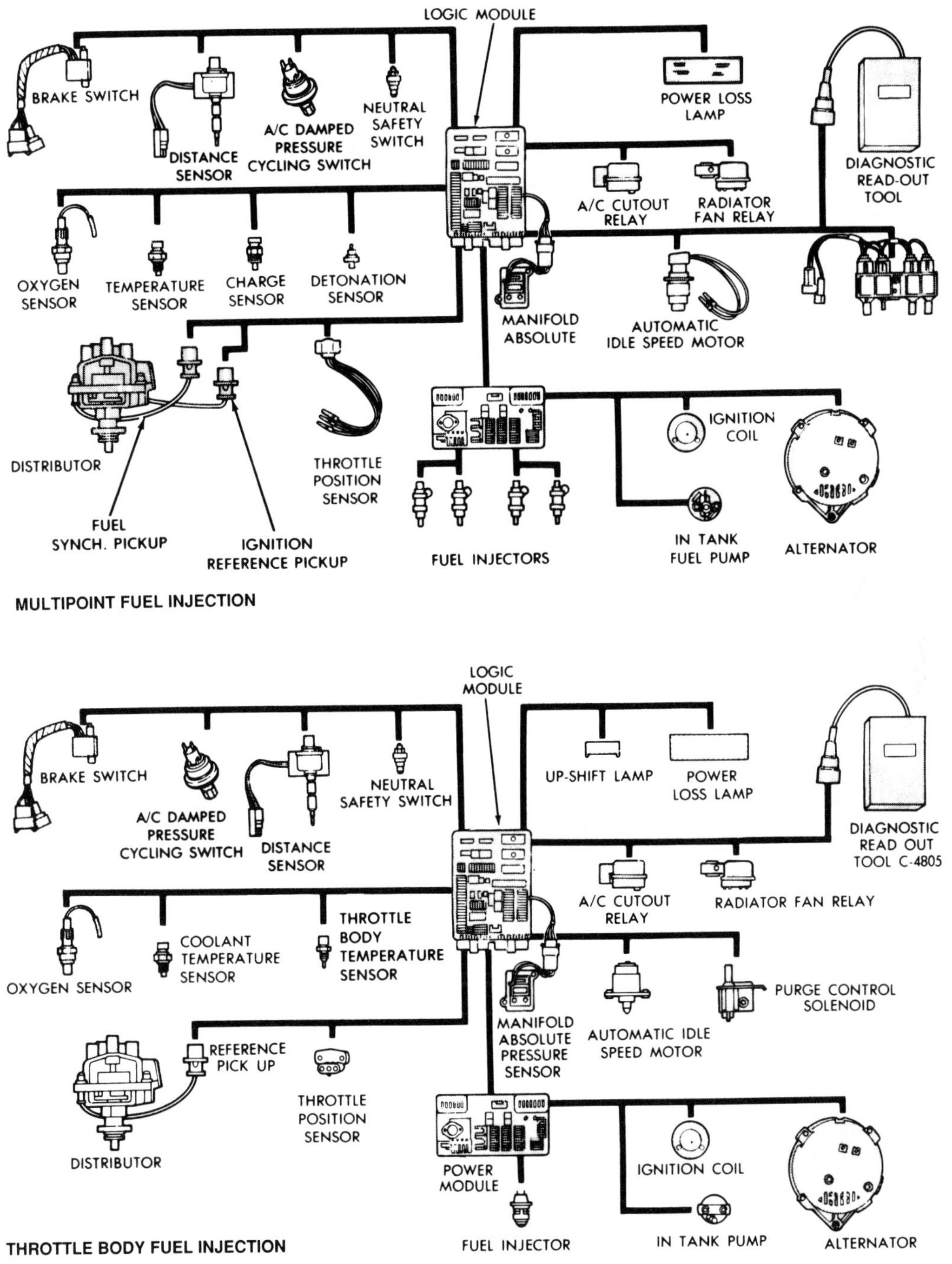


**Figure 22-2.** Chrysler's late-model engine control systems regulate fuel metering, ignition timing, and emission control with port and throttle body injection, as well as with feedback carburetors. (Chrysler)

mixtures could be ignited and burned efficiently with electronically controlled timing, while they could not be with traditional mechanical controls. This system improved mileage and was fine for emission requirements of 1976–78. It did not include a 3-way catalyst, an EGO sensor, or feedback-controlled fuel metering.

In 1979, Chrysler renamed the ELB system, electronic spark control (ESC) and recalibrated carburetors for air-fuel ratios closer to 14.7 : 1. In the same year, Chrysler introduced its first 3-way catalysts, EGO sensors, and feedback-controlled carburetors. In later years, Chrysler added a detonation sensor and other devices, and put more functions under electronic control. Chrysler calls its systems electronic spark advance (ESA) and electronic feedback carburetor or electronic fuel control (EFC), in addition to ESC. Since the early 1980's Chrysler has used fully integrated systems on 4-, 6-, and 8-cylinder engines with injection, carburetors, and turbocharging, figure 22-2. In 1983, Chrysler added self-diagnostic capabilities to its fuel-injected ESC systems.

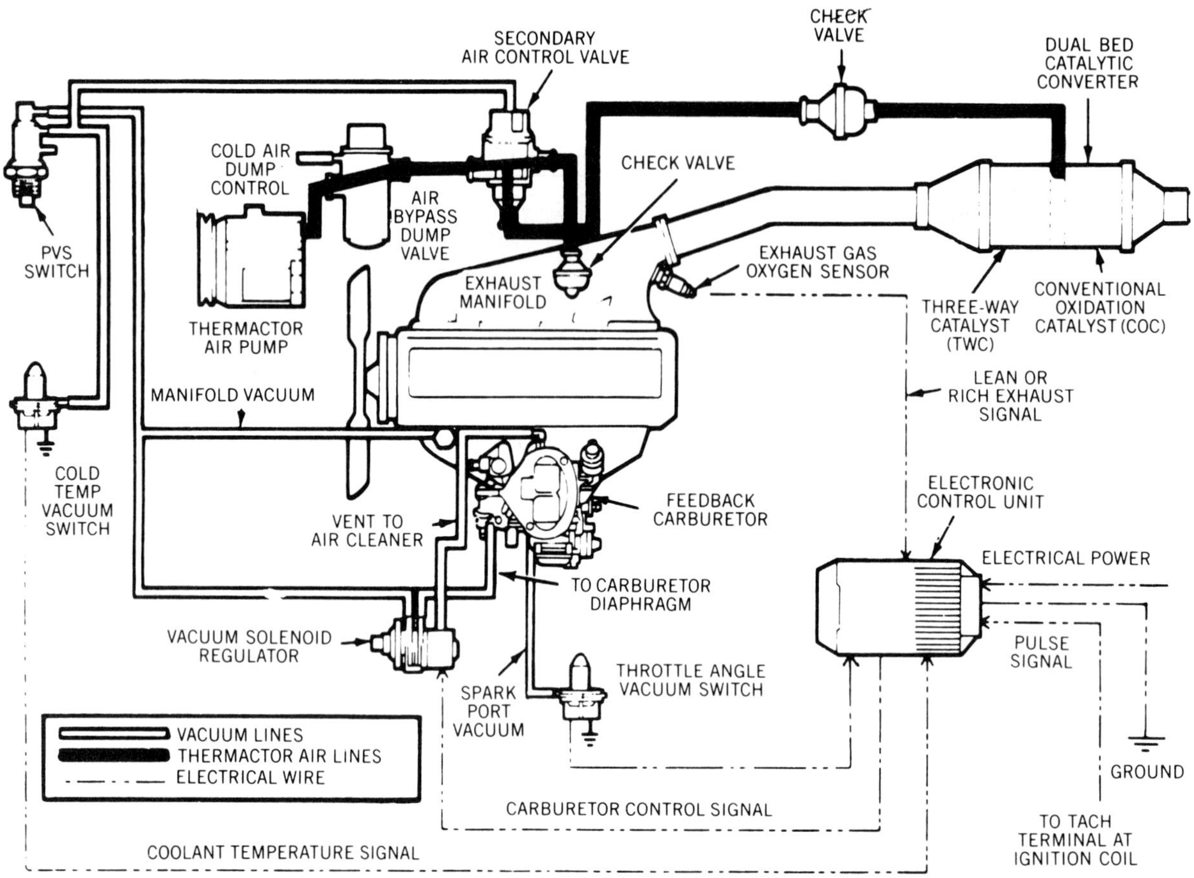


**Figure 22-3.** Ford's first electronic control, feedback carburetor system regulated only fuel metering, not ignition timing. (Ford)

## Ford Motor Company

In 1978, Ford introduced an electronically controlled feedback carburetor system with a "three-way catalyst and conventional oxidation catalyst" (TWC-COC). Used only on 2.3-liter 4-cylinder engines, figure 22-3, this system had an EGO sensor, an analog computer, and a vacuum control solenoid. It controlled only fuel metering for a 14.7 : 1 air-fuel ratio as needed by the 3-way catalyst. Ignition was controlled by the standard ignition module, and distributors had centrifugal and vacuum advance. The fuel electronic control unit, however, received an engine speed signal from the ignition module. In 1980, Ford renamed this system the microprocessor control unit (MCU) system, figure 22-4, to reflect a change from an analog to a digital computer. System function remained the same on many engines through the 1980's.

Also in 1978, Ford introduced the first of its electronic engine control (EEC) systems. EEC-I, figure 22-5, was used on 302-cid (5-liter) V-8's in Lincoln Versailles to control spark timing, EGR, and air injection. It did *not* have a 3-way catalyst, an EGO sensor, or feedback fuel control.

EEC-I has Ford's 2700 VV variable-venturi carburetor and Dura-Spark II ignition with no centrifugal or vacuum advance. The computer controls timing signals to the ignition module and receives input signals from manifold pressure and barometric sensors, coolant and air temperature sensors, a crankshaft position sensor, an EGR valve position sensor, and a throttle position sensor. The computer also controls a solenoid that regulates vacuum to the air injection diverter valve and two solenoids that control air pressure to the EGR valve. The EGR valve works on air pressure from the air injection pump rather than on vacuum, figure 22-6. The EGR system also includes an EGR cooler, through which engine coolant circulates to lower recirculated exhaust temperature. Dura-Spark distributors used with EEC systems have unique bilevel rotors and cap terminals, figure 22-7, to prevent crossfiring within the distributor.

In 1979, Ford added feedback fuel control for the 3-way catalyst, vapor canister purging, and air injection switching to the EEC system and called it EEC-II. Used on 351-cid (5.8-liter) V-8's, figure 22-8, EEC-II has an EGO sensor and a 7200 VV carburetor with a fuel control stepper motor. The EGR valve is vacuum operated, but vacuum is regulated by two solenoids similar to those of EEC-I. In 1980, Ford added self-diagnostic capabilities to the computer and modified the system to operate with throttle body injec-

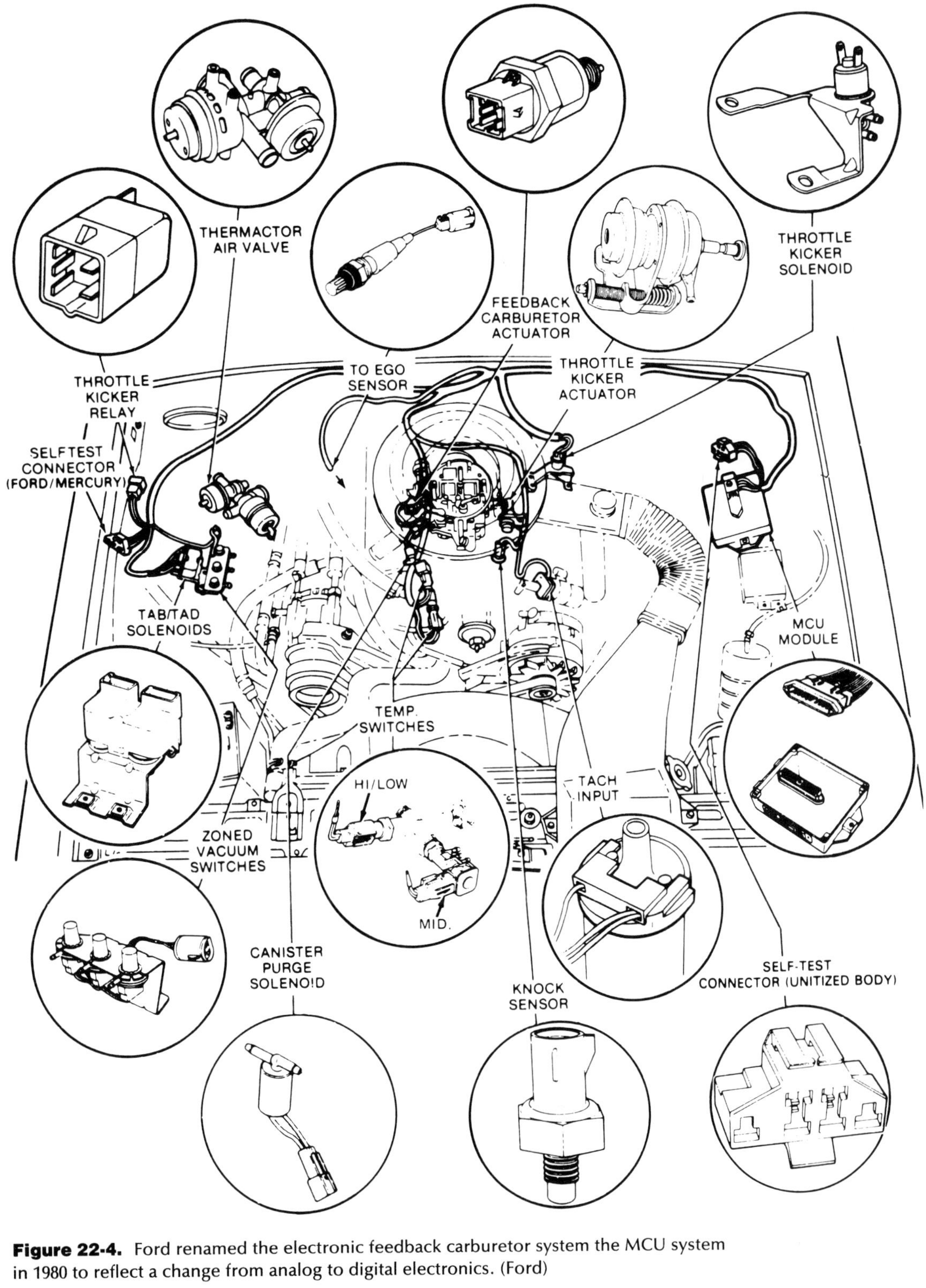


**Figure 22-4.** Ford renamed the electronic feedback carburetor system the MCU system in 1980 to reflect a change from analog to digital electronics. (Ford)

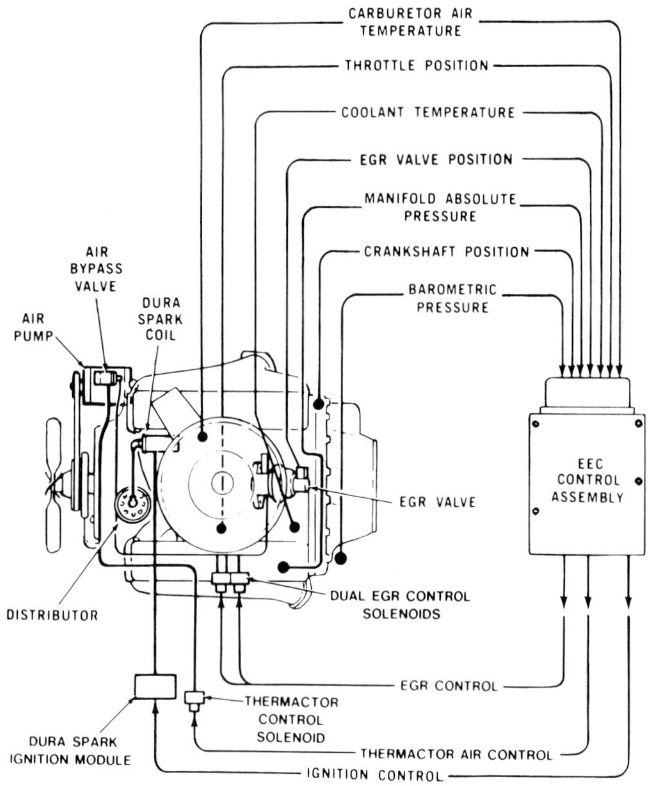


**Figure 22-5.** Ford's EEC-I system was used only on 1978 Lincoln Versailles models to control ignition timing, EGR, and air injection. (Ford)

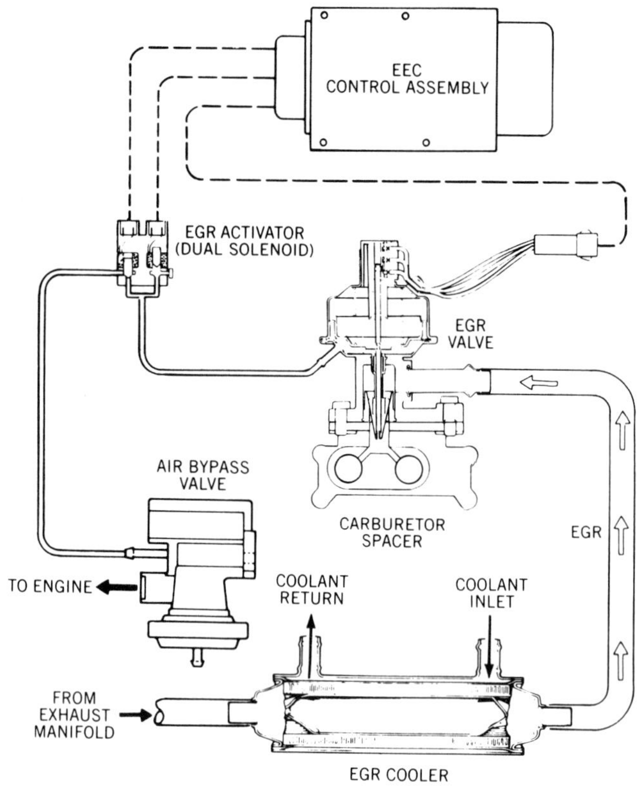


**Figure 22-6.** Ford's EEC-I system had an air-pressure-operated EGR valve. Ford used an EGR cooler on this and some later EGR systems. (Ford)

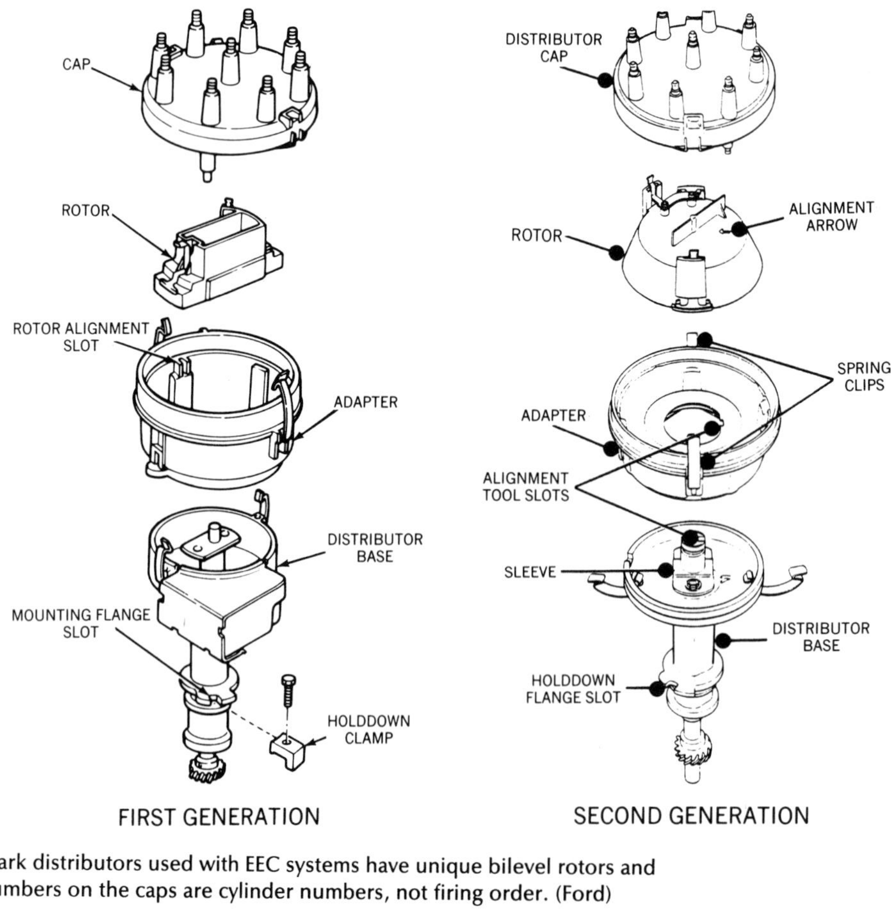


**Figure 22-7.** Dura-Spark distributors used with EEC systems have unique bilevel rotors and terminals in the cap. Numbers on the caps are cylinder numbers, not firing order. (Ford)

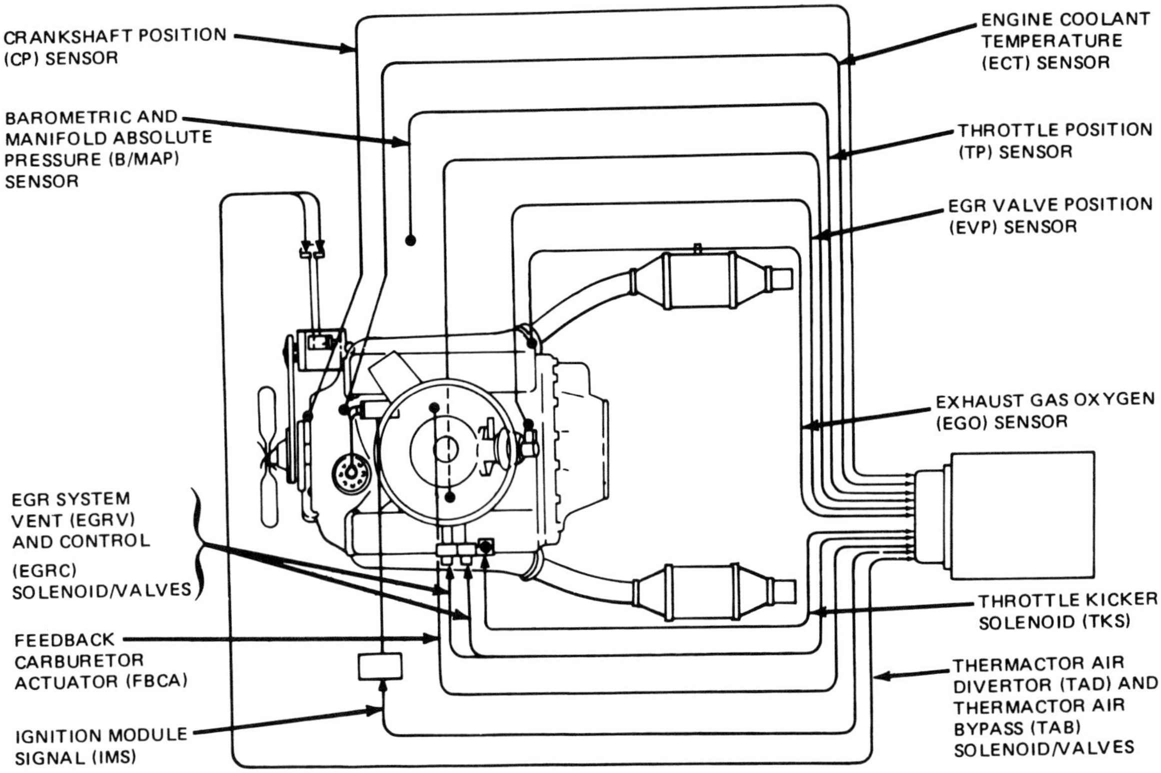


**Figure 22-8.** EEC-II was Ford's first fully integrated system to control timing, fuel metering, and emission devices. (Ford)

tion, figure 22-9, as well as the 7200 VV carburetor. This system is called EEC-III and is used on many 302- and 351-cid (5.0- and 5.8-liter) V-8's.

Ford continued EEC-III through 1984 and MCU systems through the late 1980's. Also in 1983, Ford introduced EEC-IV on 1.6-liter 4-cylinder engines. This system, figure 22-10, features a redesigned computer, Ford's thick-film-integrated (TFI) ignition, and port fuel injection. EEC-IV regulates all timing, fuel, and emission control functions. In 1984 and later models, Ford extended EEC-IV to most 4-, 6-, and 8-cylinder engines with both carburetors and injection, including Festiva, Tracer, and other captive-import models. After 1985, the only carbureted engines in Ford automobile EEC systems were some 1.9- and 2.3-liter engines and some police vehicles.

## General Motors

Like other carmakers, GM has developed its systems since the late 1970's. In 1977–78, Oldsmobile Toronados had a system called microprocessor

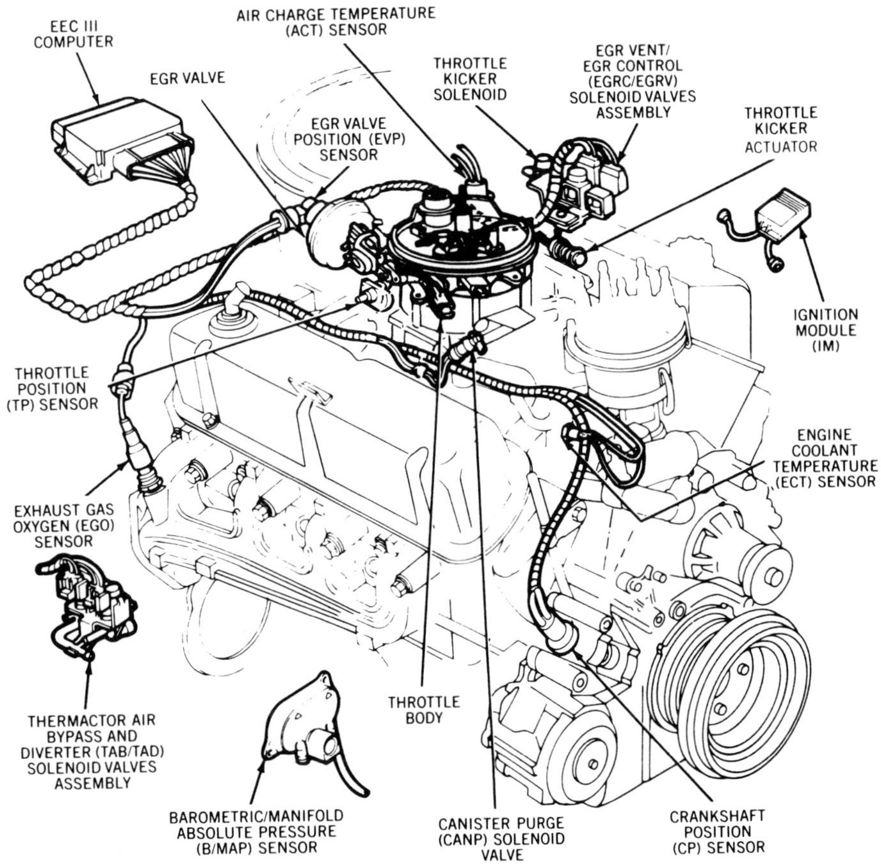


**Figure 22-9.** EEC-III is similar to EEC-II but used with throttle body injection on some engines. (Ford)

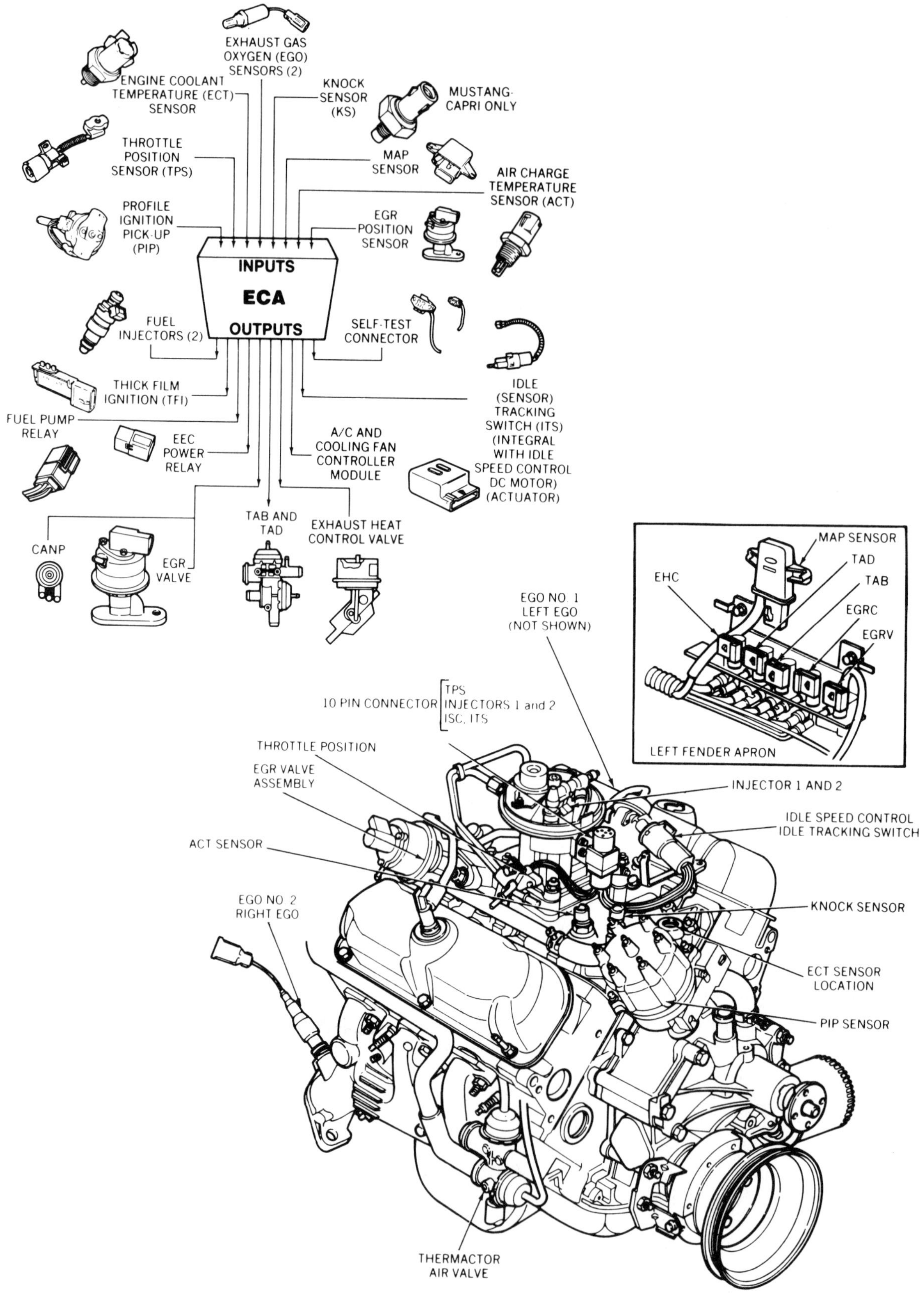


**Figure 22-10.** Introduced in 1983, EEC-IV became Ford's standard engine control system in the mid-1980's. (Ford)

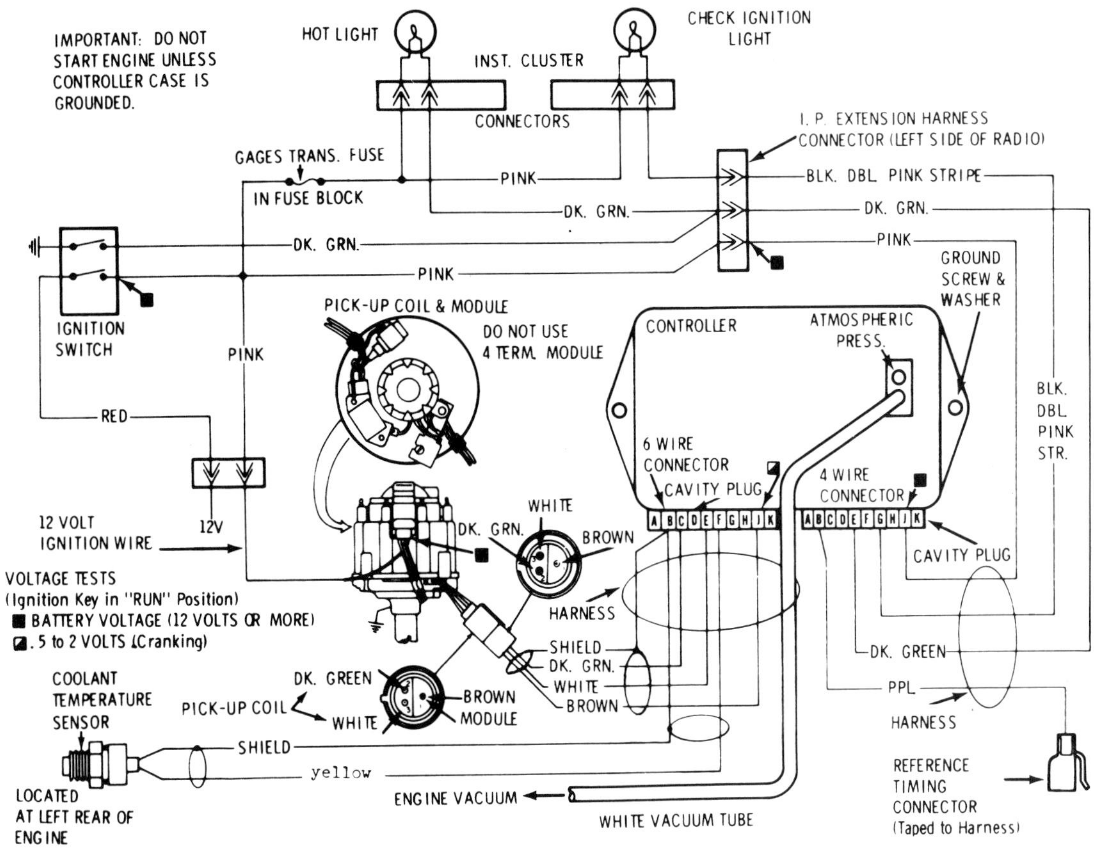


**Figure 22-11.** Oldsmobile's MISAR system of 1977–78 was GM's first electronically controlled timing system. (Oldsmobile)

sensing and automatic regulation (MISAR) that controlled only timing, figure 22-11. It was discontinued after 1978. In 1978, Cadillac began using electronic spark selection (ESS) that retards timing during cranking and cold operation. The electronic ignition chapter of this text summarizes the electronic timing modifications used with GM high-energy ignition (HEI) systems. Some of these systems are used with independent electronic injection systems, such as the early Seville EFI, and some are used with fully integrated engine systems.

Also in 1978, GM introduced an electronic fuel control (EFC) system on some 4-cylinder engines, figure 22-12. Like Ford's TWC-COC and MCU systems, it had a 3-way catalyst, an EGO sensor, and a feedback carburetor. While this 1978–79 system provided electronic fuel control, it did not have electronic ignition timing. In 1979, GM installed its first fully integrated computer-controlled catalytic

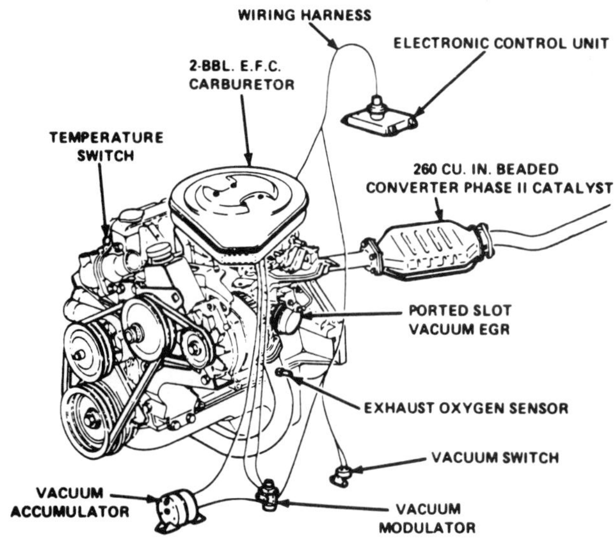


**Figure 22-12.** GM's first electronic fuel control system was used on 1978–79 4-cylinder California engines. (GM)

### MODULATED DISPLACEMENT

Ford experimented with it on 6-cylinder truck engines in the late 1970's. Cadillac marketed it as the V8-6-4 system in 1981. The modulated-displacement engine uses solenoid-operated actuators to change the valve rocker arm pivot points. The solenoid moves a blocking plate that slides over the top of the spring-loaded valve selector. This releases spring force on the normal rocker arm pivot and allows the rocker to pivot on the tip of the valve, figure 1. The valve springs hold the valves closed. The pistons continue up and down, but the cylinders receive no intake charge and produce no exhaust. They are effectively shut off. Air in the cylinder is compressed and expanded. No power is produced, and no fuel is consumed.

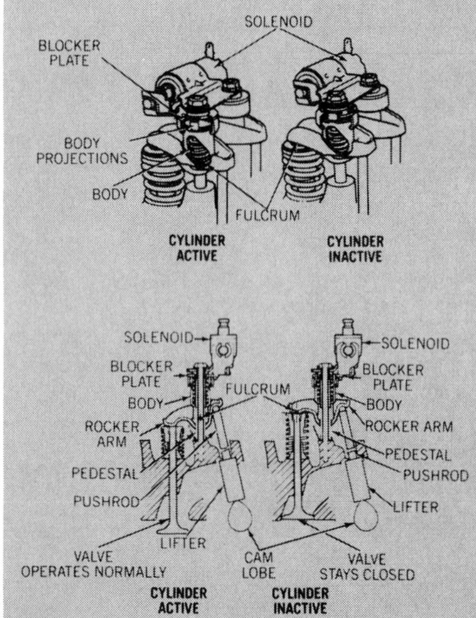


**Figure 1**

In Cadillac's V8-6-4 system, selectors are on cylinders 1, 4, 6, and 7, figure 2. In 6-cylinder operation, the engine computer shuts off cylinders 1 and 4. For 4-cylinder operation, the computer shuts off cylinders 1, 4, 6, and 7. Before the engine is allowed to run on 4 or 6 cylinders, it must be at normal temperature, the transmission in high gear, and the fuel injection system in closed loop. The computer deactivates and reactivates cylinders according to engine load.

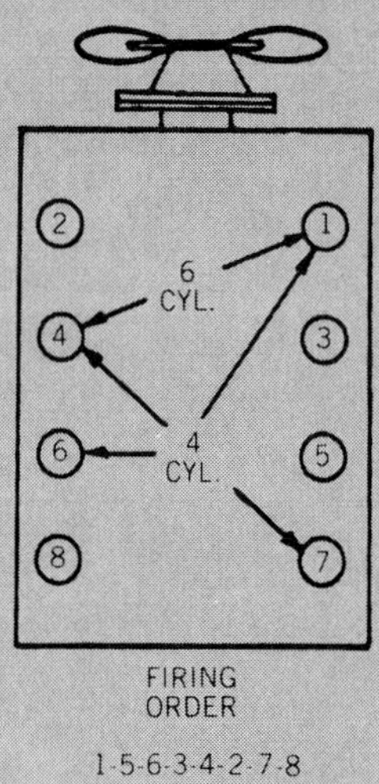


**Figure 2**

Cadillac's V8-6-4 system did not provide the long-term reliability and drivability the manufacturer wanted. It was discontinued after 1982, but the modulated-displacement concept may return on future engine systems.

converter (C-4) system on 1980-model front-wheel-drive X-cars, figure 22-13.

Used with feedback carburetors, the C-4 system receives input signals from engine speed and crankshaft position sensors, manifold and barometric pressure sensors, coolant and air temperature sensors, throttle position sensors, and an EGO sensor. The computer controls fuel metering, EGR, air injection switching, and ignition timing. The C-4 system also has self-diagnostic capabilities. In 1981, GM modified the C-4 system and renamed it computer command control (CCC or C-3). The CCC system, figure 22-14, can monitor up to 15 vehicle functions and can control the following operations:

- Electronic spark timing (EST)
- Fuel metering—feedback carburetor or fuel injection
- Idle speed
- Exhaust gas recirculation
- Air injection
- Intake manifold heat
- Fuel vapor canister purging
- Torque converter lockup

The CCC system is now used by all GM divisions on most cars and light trucks. The CCC self-diagnostic capabilities are greater than those of C-4, and the CCC system is used on carbureted and fuel-injected engines, with and without turbochargers.

Throughout the 1980's, the CCC control system and GM onboard computers have developed steadily and acquired more capabilities. In 1981 and 1982, most CCC systems were used with carbureted engines and were known as "minimum function" and "full-function" systems. Minimum-function systems were used on T-cars (Chevettes and Pontiac 1000 models) and some Oldsmobile V-8's. They had fewer sensors and did not provide the control sophistication of full-function systems. Today, the terms "minimum-function" and "full-function" are often used to refer to carbureted CCC systems. In fact, carbureted engines sometimes are distinguished from fuel-injected engines by calling them CCC, or C-3, engines.

All late-model GM fuel injection systems use the basic CCC system introduced in the early 1980's but often referred to by the names given the various injection systems. Since 1986, systems with high-speed computers are often called "P-4" systems. Regardless of the popular names for GM engine control systems, all are evolutionary developments of the original systems introduced on 1980–81 models. Later sections of this chapter explain GM injection systems in detail and outline the continuing development of the onboard computers.

## Bosch Lambda and Motronic Systems

Historically, many European carmakers have used ignition, fuel injection, and electronic systems made by the Robert Bosch company. Bosch was a pioneer in diesel and gasoline fuel in-

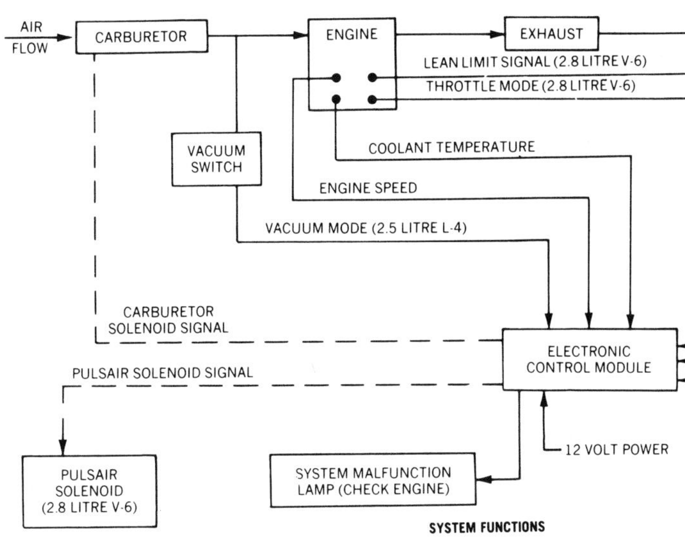


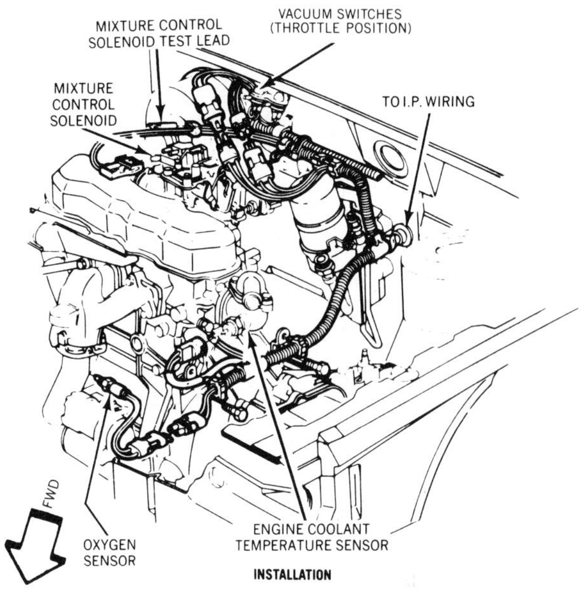


**Figure 22-13.** GM's first fully integrated engine control system was the C-4 system used on 1980 X-cars. (Oldsmobile)

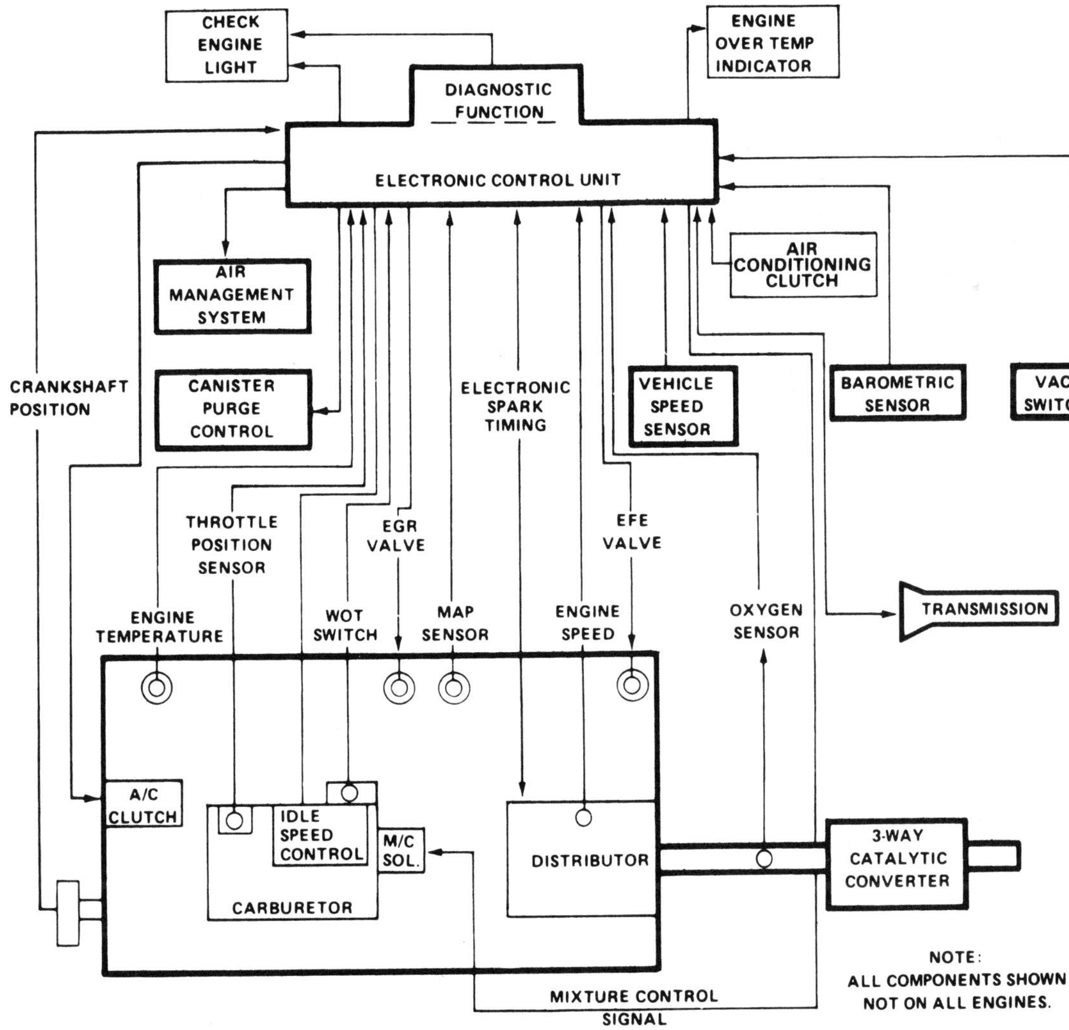


**Figure 22-14.** The CCC system became the standard GM engine control system in the early 1980's. (Chevrolet)

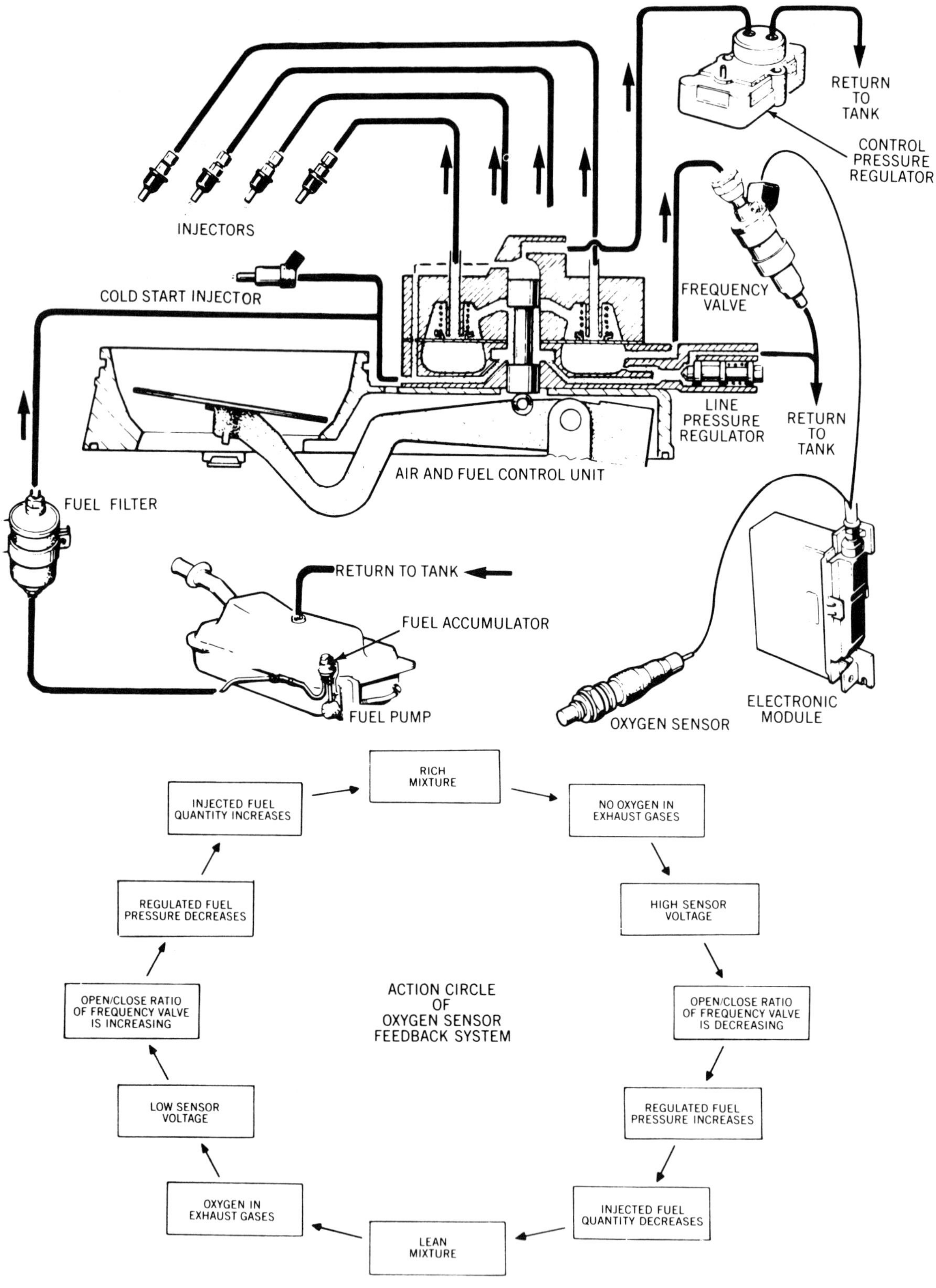


**Figure 22-15.** One of the first Bosch lambda fuel feedback control systems was used with K-Jetronic injection on 1977 Volvos. (Volvo)

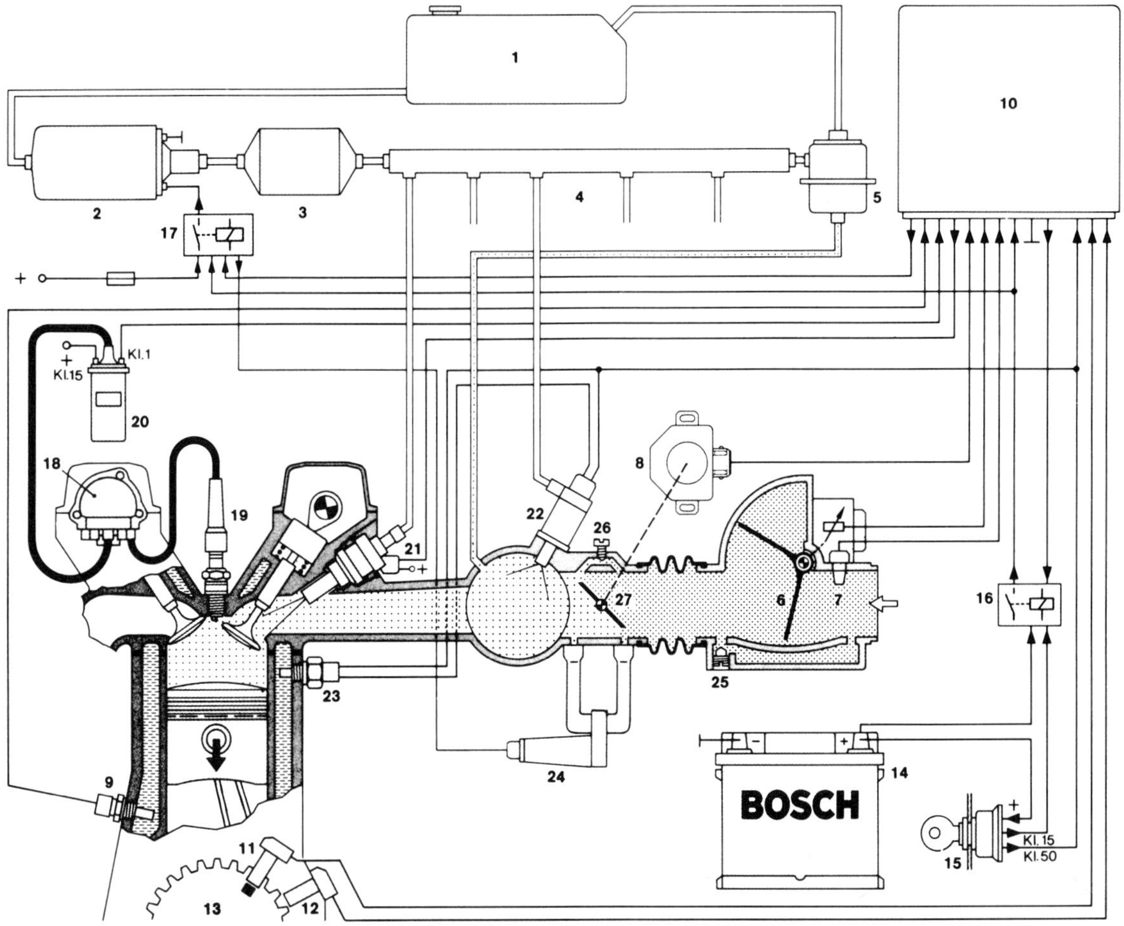


1 FUEL TANK, 2 FUEL PUMP, 3 FUEL FILTER, 4 DISTRIBUTOR TUBE, 5 PRESSURE REGULATOR, 6 AIR-FLOW SENSOR, 7 INTAKE-AIR TEMPERATURE-SENSOR, 8 THROTTLE-VALVE SWITCH. 9 ENGINE-TEMPERATURE SENSOR, 10 CONTROL UNIT, 11 REFERENCE-MARK SENSOR, 12 SPEED SENSOR, 13 FLYWHEEL, 14 BATTERY, 15 IGNITION AND STARTER SWITCH, 16 MAIN RELAY. 17 PUMP RELAY, 18 HIGH-VOLTAGE DISTRIBUTOR, 19 SPARK PLUG, 20 IGNITION COIL, 21 INJECTION VALVE, 22 START VALVE, 23 THERMO-TIME SWITCH, 24 AUXILIARY-AIR DEVICE. 25 IDLE-MIXTURE ADJUSTMENT SCREW, 26 IDLE-SPEED ADJUSTMENT SCREW, 27 THROTTLE VALVE.

**Figure 22-16.** The Bosch Motronic system combines lambda fuel control and electronic timing with L-Jetronic injection. (Bosch)

jection and holds many basic patents for injection and electronic systems.

In 1977, Volvo installed an electronically controlled fuel-metering system on cars sold in the U.S. Volvo's lambda-sond system was made by Bosch and used with K-Jetronic injection, figure 22-15. Later adopted by other carmakers, the lambda-sond system was the first with a 3-way catalyst and EGO sensor. The electronic unit receives an input signal from the EGO sensor and sends an output signal to a timing valve in the K-Jetronic system. The valve varies injection control pressure and regulates the fuel delivered by the continuous injection nozzles. The original lambda-sond system does not regulate ignition timing or other emission functions. These came later with the Motronic system.

The Bosch Motronic system, or digital motor electronics (DME), combines L-Jetronic injection with lambda feedback fuel control and electronic spark timing, figure 22-16. Besides a 3-way catalyst and EGO sensor, the Motronic system uses input signals from crankshaft speed and position sensors, the pulse generator in the ignition distributor, and the L-Jetronic sensors. The computer regulates the injector pulse width for air-fuel ratio control and adjusts ignition timing for combined speed and load conditions. The system also shuts injection off completely during closed-throttle deceleration.

The Motronic system is a full-function control system based on the Bosch L-Jetronic injection system, but late-model versions of the mechanical K-Jetronic injection system have some of the same control capabilities. The KE-Jetronic system includes fuel cutoff on deceleration, altitude compensation, and engine speed limiting functions. All of these are regulated by the system computer, or electronic control unit. The KE-Jetronic system, however, does not include ignition timing control and other management functions in its computer. These are handled by other electronic control units. A later section of this chapter explains Bosch gasoline injection systems in more detail.

### Control System Development

From these summaries of engine control systems used by various carmakers, you probably have the idea that there has been a lot of experimentation and continuous development. You're right. We haven't the room in this book to give you all the specifications for each manufacturer for every year and model. Consider these trends during the 1980's, however:

• The earliest control systems used analog computers with limited control functions and slow processing speed. In the early 1980's digital computers replaced earlier analog units, and processing speed increased.

• The earliest systems were designed primarily to control feedback fuel metering. A few were designed as spark timing control systems. Carmakers quickly combined these functions in fully integrated systems.

• Computer size has decreased as speed and processing capability has increased. The earliest onboard computers were relatively large units with many discrete components soldered to several circuit boards. Chrysler's first computers had separate circuit boards for power components and for computer logic devices. The latest Chrysler systems have all devices integrated into a single-module engine controller (SMEC) or single-board engine controller (SBEC) computer. GM's first CCC computers communicated with sensors and actuators at a 160-baud data transmission speed. The latest CCC computers communicate at 8192 baud.

• Computer networks have appeared as the latest generation of control systems. On GM cars, the engine control module (ECM) shares sensor signals with the body computer module (BCM). Late-model Chrysler systems and some imported systems work similarly. Ford and other manufacturers introduced similar computer networks in the early 1990's.

All of these trends illustrate the fact that when you service a particular system, you will need the manufacturer's system description, procedures, and specifications from a shop manual or an independent service manual. Control systems will continue to evolve and grow during the next decade, and you must always work with the latest service procedures for a specific year and model of vehicle.

Remember, though, that the laws of electricity, physics, and chemistry don't change from one engine to another. All systems operate on the same principles, and all engineers have the same goals of best performance, driveability, emission control, and economy for the cars they design. The rest of this chapter outlines engine control principles and will help you recognize more similarities than differences among systems.

## GASOLINE FUEL INJECTION REQUIREMENTS

Air volume, air pressure, and airflow into the engine are the factors that control fuel metering with a carburetor *and* with an injection system. A fuel injection system must provide the same kinds of air-fuel ratios that a carburetor does for these basic operating conditions:

• *Starting*—Low intake vacuum, low airflow, a cold engine, and poor fuel vaporization require a slightly rich air-fuel ratio.

• *Idle*—Low airflow, high manifold vacuum, low carburetor vacuum, and poor vaporization still require a slightly rich mixture.

• *Low speed*—Increasing engine speed, airflow, and carburetor vacuum allow the air-fuel ratio to become leaner.

• *Cruising economy*—With light engine loads and high, relatively constant, vacuum and airflow, the engine provides best economy with lean air-fuel ratios.

• *Extra power requirements*—An engine needs slightly rich air-fuel ratios for heavy loads and acceleration (low vacuum and low airflow) and for full-throttle operation (low vacuum and high airflow).

A carburetor provides rich air-fuel ratios of 12.5:1 to 13.5:1 and lean ratios of 15:1 to 16:1. A fuel injection system does exactly the same things for exactly the same engine needs. Late-model injection systems and carburetors used with electronic engine controls have fuel-metering devices to maintain the stoichiometric air-fuel ratio of 14.7:1 for most operating conditions. The rich and lean limits stay much closer to the ideal 14.7:1 ratio than they did on earlier engines.

Fuel injection and carburetors have a lot in common in their operating requirements. The parts that carburetors and injection systems use to meet these needs look different, but every system of a carburetor has a fuel injection counterpart. Whether mechanical or electronic, port or throttle body, all systems have these basic groups of parts:

• Fuel delivery system

• Air control system

• Engine sensors and auxiliary systems

## GASOLINE FUEL INJECTION SYSTEMS

For many years, gasoline fuel injection was used on high-performance engines. In the 1970's, however, engineers took a fresh look at gasoline injection, combined with electronic engine control, as a way to reach the conflicting goals of good performance, fuel economy, and emission control. Today, all major carmakers offer fuel injection on most of their vehicles.

Fuel injection is steadily forcing the carburetor into obsolescence. Today, fuel injection is a major feature of most fully integrated engine control

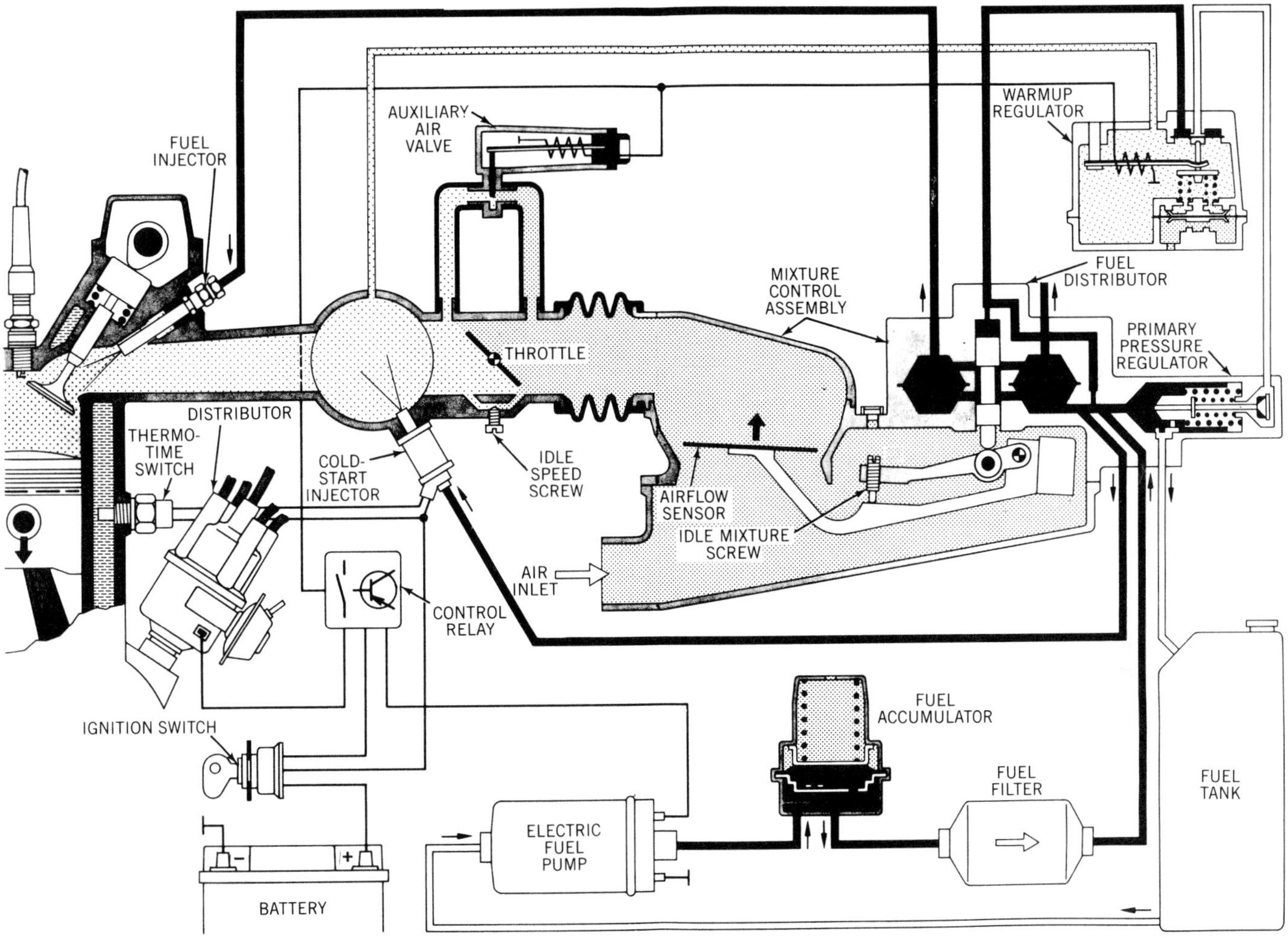


**Figure 22-17.** Bosch K-Jetronic fuel injection is a mechanical continuous injection system. (Bosch)

systems. The following sections explain the various kinds of injection systems and their common principles. Gasoline injection can be classified as:

- Mechanical or electronic injection
- Port (or manifold) injection or throttle body injection
- Continuous or intermittent injection

Mechanical injection systems inject fuel by using the pressure of gasoline in the injector lines to open the injector valves. They are usually continuous, which means that they inject fuel constantly while the engine is running. Most electronic systems use solenoid-operated injectors to spray fuel intermittently (in timed pulses) into the manifold or intake ports. Port, or manifold, systems have individual injectors for each cylinder. Throttle body systems have one or two injectors to spray fuel into the intake manifold inlet, above the throttle. These distinctions are not clear-cut. All electronic systems are not exclusively port or throttle body systems. All mechanical systems have some electronic parts. When you study a specific injection system, you must ask yourself how it is controlled, how the injection nozzles are located, and how it operates.

Before we examine the principles of all systems, we will look at some of the most common specific systems. This introduction will help you recognize the different devices that do similar jobs.

## Bosch K-Jetronic

Introduced in the early 1970's on many European cars, the Bosch K-Jetronic is a mechanical continuous injection system (CIS), figure 22-17. The individual port injectors inject fuel by gasoline pressure at the nozzles. The system regulates this pressure in relation to air volume, which is measured by sensing airflow.

## Bosch KE-Jetronic

The Bosch KE-Jetronic system is a development of the basic K-Jetronic system with increased electronic control capabilities. KE-Jetronic systems have additional sensors for throttle position, airflow sensor position, and en-

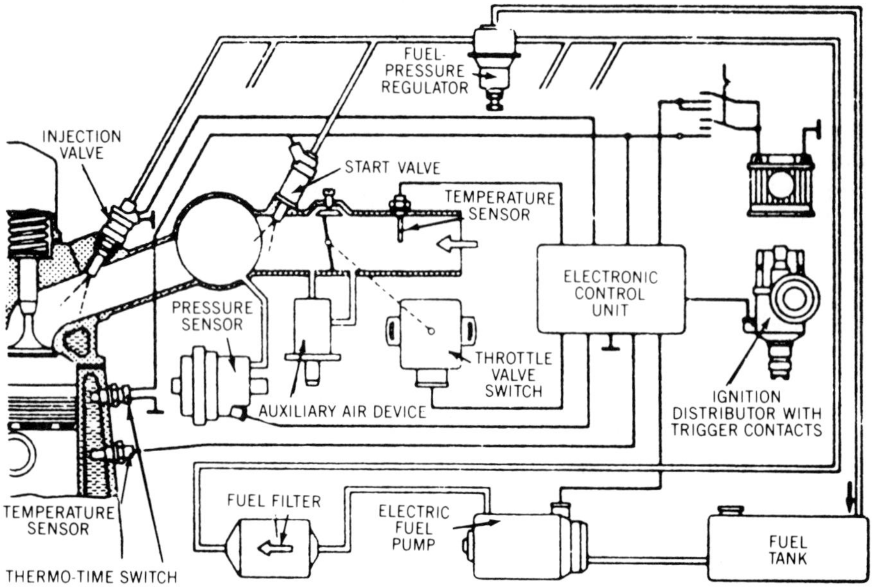


**Figure 22-18.** Bosch D-Jetronic fuel injection is an electronic intermittent injection system, controlled by manifold pressure. (Bosch)

gine coolant temperature. KE systems have a more powerful computer than basic K-Jetronic systems and use input signals from the ignition distributor and a heated EGO sensor (in most installations). The electrohydraulic pressure actuator built into the KE-Jetronic fuel distributor controls fuel pressure, and thus fuel quantity, in response to signals from the system computer. Idle speed is programmed into the computer and cannot be adjusted in service.

## Bosch D-Jetronic

This electronic fuel injection (EFI) system uses individual solenoid-operated port injectors, figure 22-18. Fuel is metered in relation to manifold air

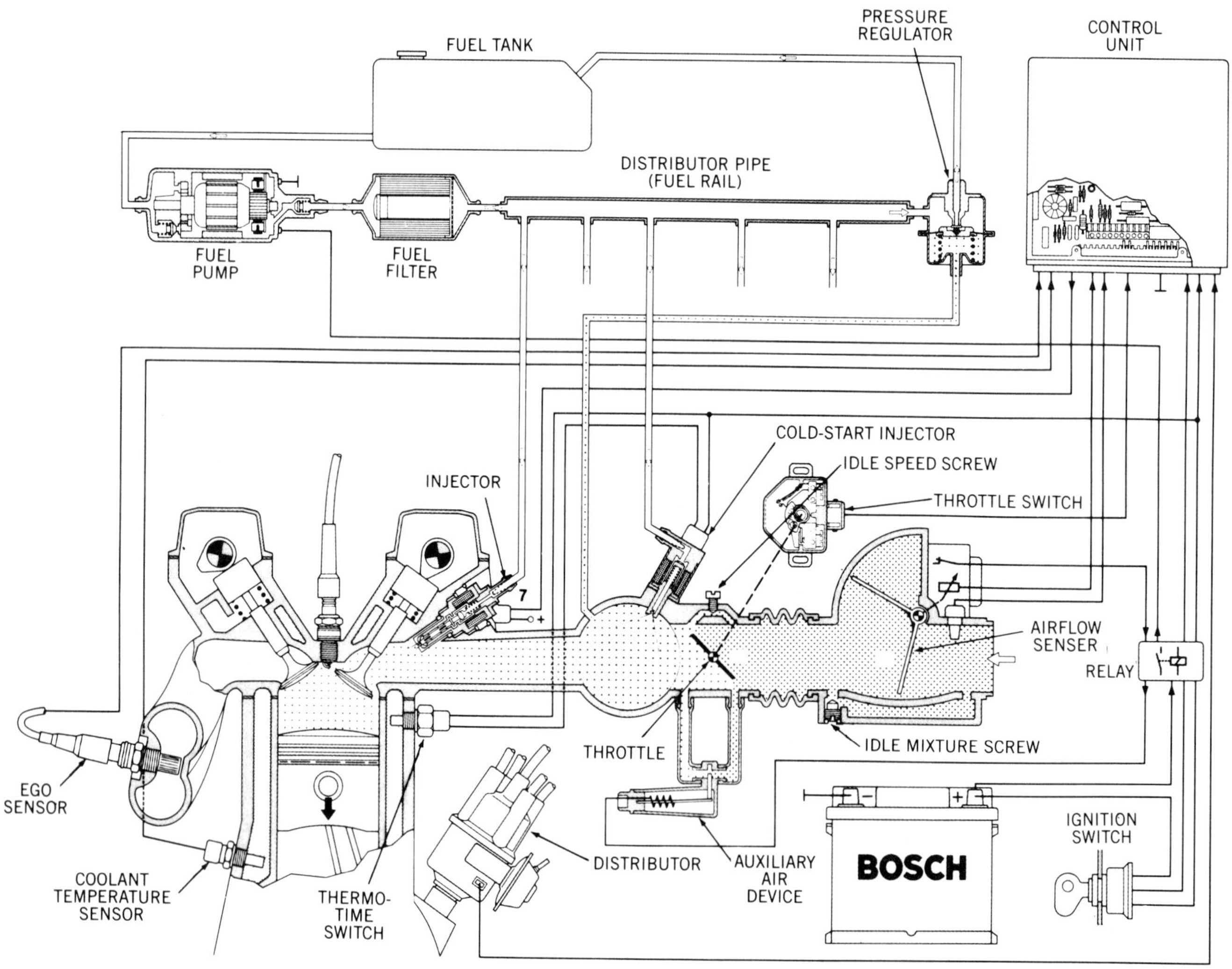


**Figure 22-19.** Bosch L-Jetronic fuel injection is an electronic intermittent injection system, controlled by airflow measurement and engine speed. (Bosch)

pressure and engine speed. Injection is in intermittent pulses. D-Jetronic injection was used on 1968–73 Volkswagens and other European cars.

## Bosch L-Jetronic

Introduced in 1974, the Bosch L-Jetronic system uses solenoid-operated intermittent injectors for each port. As with the mechanical K-Jetronic system, fuel metering is controlled in relation to airflow. The L-Jetronic system does this, however, with an electronic module that responds to a voltage signal from an airflow sensor, figure 22-19. The module also regulates fuel metering in relation to engine speed and temperature. This system often is called airflow-controlled (AFC) injection.

## Bosch LH-Jetronic

Introduced in the 1980's, the LH-Jetronic system, figure 22-20, is similar to the L-Jetronic system. Air volume, however, is measured by a hot-wire air mass sensor rather than an airflow sensor. This provides more precise measurement of the molecular volume of combustible air entering the engine.

## Bendix Electronic Port Injection System

The Bendix EFI system used on 1976–79 Cadillac Sevilles, figure 22-21, was the first modern port injection system in widespread use on domestic cars. This system has individual port injectors that deliver fuel intermittently. The electronic module controls injection in relation to manifold pressure, throttle position, engine speed, and air and coolant temperature. The last versions of this system included an exhaust gas oxygen (EGO) sensor and feedback control of fuel metering.

The Bosch D-Jetronic and Cadillac-Bendix systems were very similar in design and operation. In fact, Bosch D-Jetronic design was based on earlier engineering work by Bendix.

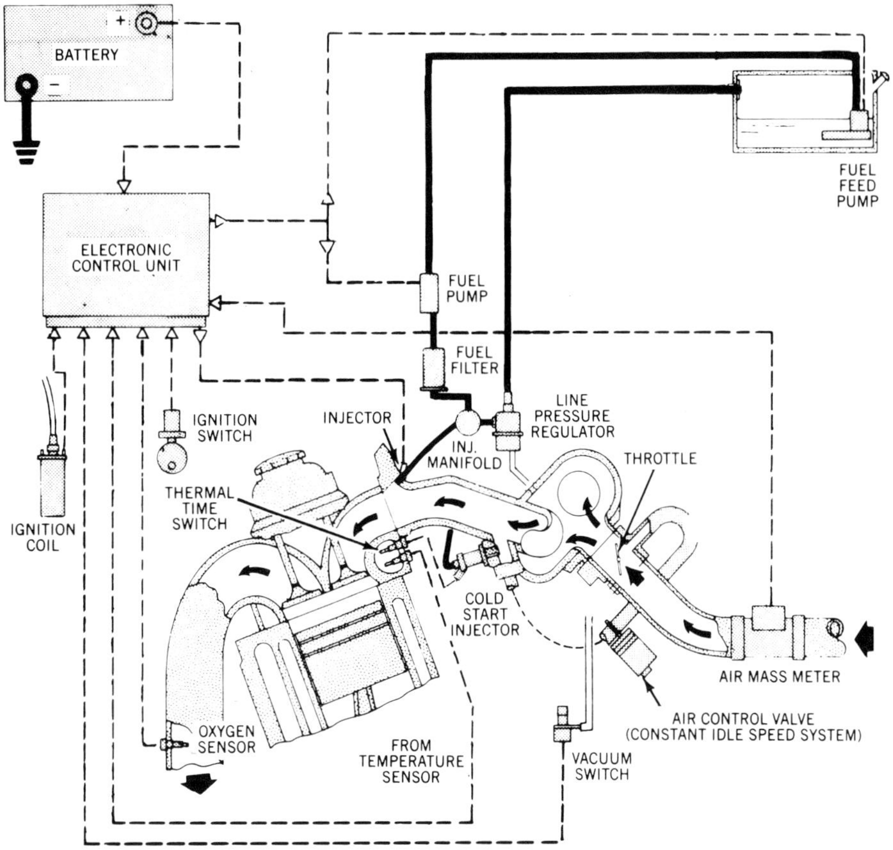


**Figure 22-20.** Bosch LH-Jetronic fuel injection is similar to L-Jetronic but controlled by an air mass sensor. (Volvo)

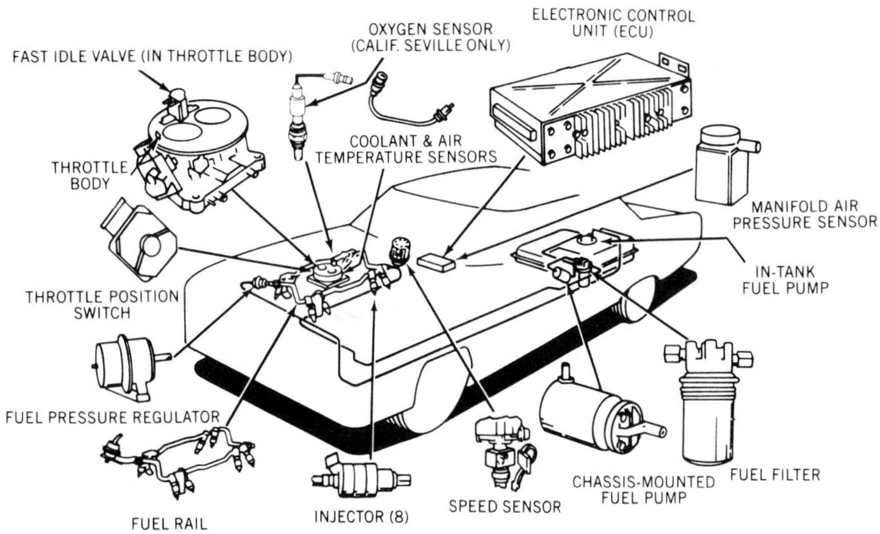


**Figure 22-21.** The Bendix EFI system with port injectors was used on 1976–79 Cadillac Sevilles. (Cadillac)

## Late-Model Port Injection Systems

Many EFI systems with individual port injectors used since the early 1980's are based on Bosch D-Jetronic, L-Jetronic, and LH-Jetronic designs. In fact, many of the components are made under patent license from the Robert Bosch Corporation.

**General Motors Systems.** General Motors introduced port injection systems on its 1984 models and now uses three different types:

1. Port fuel injection (PFI) or multiport (multipoint) fuel injection (MFI or MPFI)
2. Tuned port injection (TPI)
3. Sequential fuel injection (SFI)

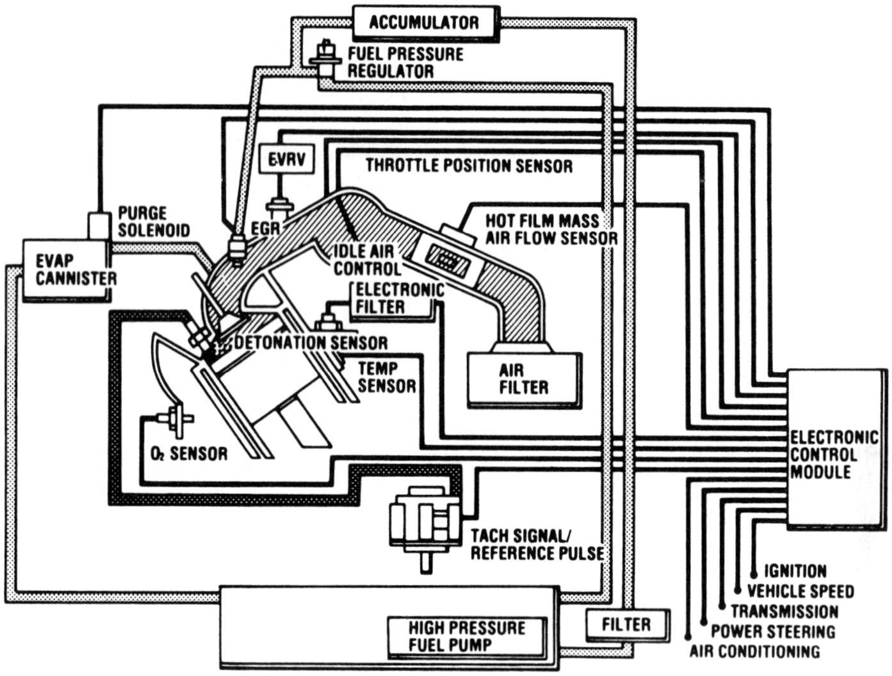


**Figure 22-22.** The GM multiport fuel injection (MFI) system uses a hot film mass airflow sensor and fires the injectors once every crankshaft revolution. (General Motors)

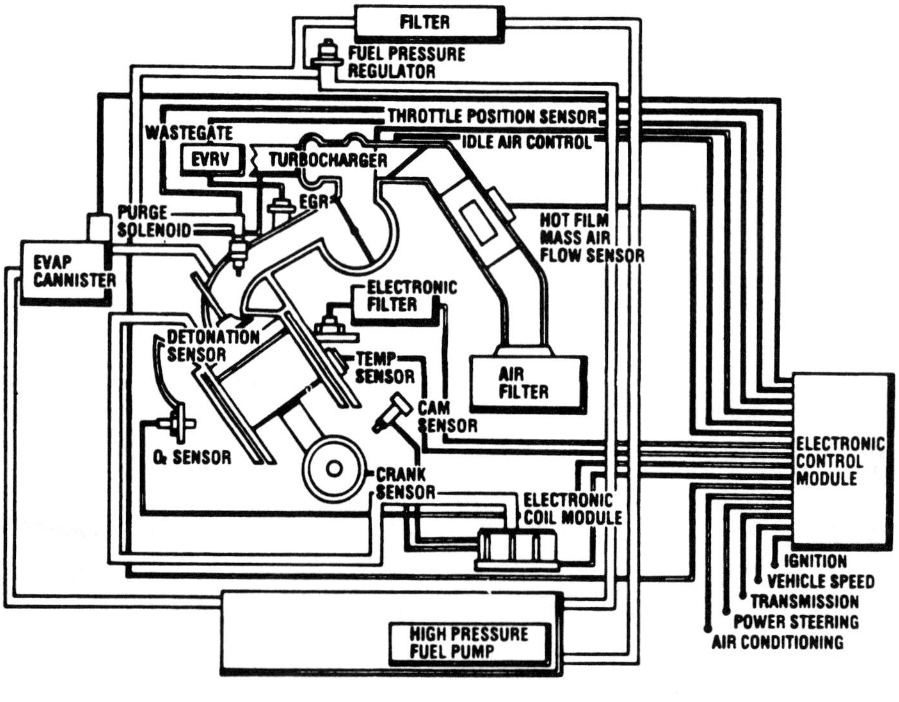


**Figure 22-23.** The GM sequential fuel injection (SFI) system also uses a hot film mass airflow sensor, but fires the injectors one at a time according to engine cylinder firing order. (General Motors)

PFI and MFI systems, figure 22-22, are both simultaneous double-fire injection systems. All injectors fire once during each revolution of the engine, mixing two injections of fuel with intake air for each combustion cycle. MFI systems were used on 1984–85 engines. PFI systems are used on 4-cylinder and small V-6 engines with conventional HEI distributors.

TPI operates essentially the same as PFI and MFI, but uses an intake air plenum that has individually tuned air intake runners to provide optimum airflow for each cylinder. TPI systems are found on the high-performance V-8 engines used in 1985 and later Corvettes, as well as Camaro and Firebird models.

SFI design and system components, figure 22-23, are similar to those of PFI, MFI, and TPI, but the injectors are triggered in firing order sequence. SFI injection systems are found on GM engines with a distributorless ignition system (DIS). The SFI 3.8-liter V-6, figure 22-24, is typical of late-model GM port injection systems.

In 1992, GM introduced a unique central port injection (CPI) system on V-6 truck engines, figure 22-25. The CPI system has one central injection valve. Its solenoid is controlled by the system computer as other solenoid-operated EFI injectors are. CPI is really a cross between port injection and throttle body injection (TBI) systems. In fact, you can think of it as "distributed TBI." The computer pulses the injector continuously, similar to a TBI injector. The distribution tubes and nozzles equalize fuel delivery to the ports, and the intake manifold can be tuned for better airflow than a TBI system.

All GM port injection systems operate with advanced versions of the basic CCC system introduced in 1981.

**Figure 22-24.** This 1991 Buick 3800 V-6 is typical of late-model GM port injection systems. (Buick)

**Ford Systems.** Ford's port injection systems are called electronic fuel injection (EFI). The 4-cylinder EFI systems are based on the Bosch L-Jetronic design with simultaneous double-fire injector triggering. The first Ford EFI system appeared on the 1.6-liter 4-cylinder engine in 1983 Escort and Lynx vehicles and is used essentially unchanged on certain 1.6- and 1.9-liter engines in other models. A variation of the system is found on 1984 and later 1.6- and 2.3-liter turbocharged engines. Figure 22-26 shows a port injection system installation on a Ford 4-cylinder engine.

A simultaneous injection system also is used on the 183-cid (3.0-liter)

Ford also uses a simultaneous injection system on the 183-cid (3.0-liter) V-6 introduced in the 1986 Taurus and Sable models. The system is similar to those used on 4-cylinder engines.

Ford introduced sequential fuel injection (SFI) systems on 1986 302-cid (5.0-liter) V-8's, and on 1988 232-cid (3.8-liter) V-6's, figure 22-27. Injectors are triggered in firing order sequence by the EEC-IV computer. Each injector

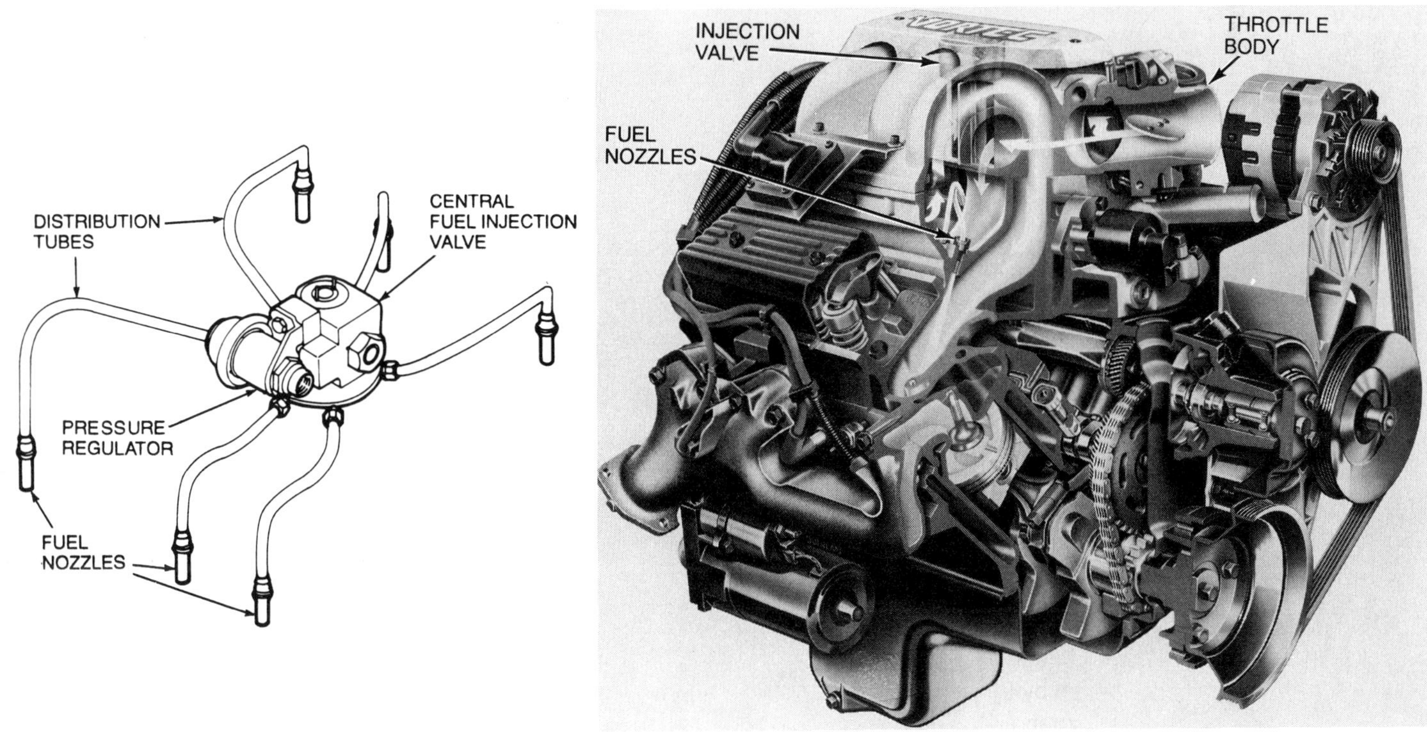


**Figure 22-25.** GM introduced the central port injection (CPI) system on some 1992 truck engines. You can think of it as "distributed TBI." (Chevrolet)

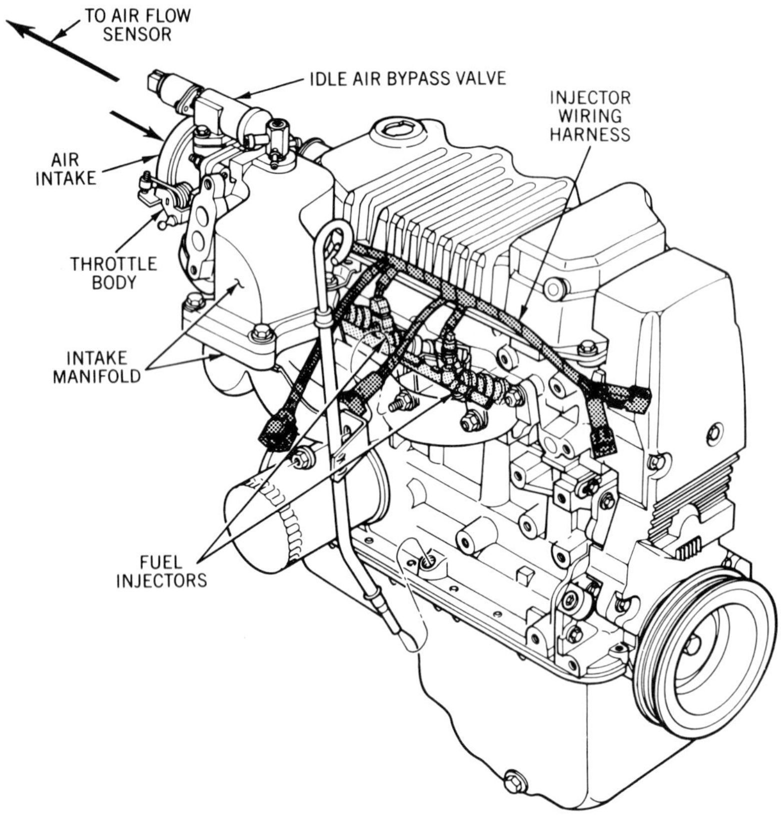


**Figure 22-26.** Ford port EFI systems are similar to Bosch L-Jetronic systems. (Ford)

**Figure 22-27.** Sequential fuel injection on Ford's 1990 3.8-liter V-6. (Ford)

fires once every two crankshaft revolutions.

The early Ford EFI port injection systems used vane airflow sensors based on the Bosch design and patents for L-Jetronic systems. Ford port injection systems also use manifold absolute pressure (MAP) and barometric pressure sensors for air measurement. Several Ford port injection systems of the late 1980's and early 1990's use mass airflow (MAF) sensors, figure 22-28. The sequential injection on the Ford "modular" 281-cid (4.6-liter) V-8, figure 22-29, is typical of Ford's current direction for MAF-controlled port injection systems.

**Chrysler Systems.** The Chrysler grouped double-fire port injection system, figure 22-30, is similar to a Bosch D-Jetronic system. It uses a MAP sensor for air measurement, Bosch injectors, and Chrysler electronics. Chrysler uses this basic port injection system on almost all 1984 and later 2.2- and 2.5-liter 4-cylinder engines.

Chrysler's basic fuel injection systems are examples of so-called **speed density** systems. Fuel metering in a speed density system is based on engine speed (rpm) and air density, or pressure, as measured by a MAP sensor. (More properly, they should be called speed *and* density systems.)

Chrysler's port injection systems on V-6 and V-8 engines are either speed density or mass-airflow-controlled systems. MAF systems are found mostly on Mitsubishi-built engines. In 1992, Chrysler introduced port injection systems on V-6 and V-8 truck engines, figure 22-31.

**Japanese Systems.** Most port injection systems used by major Japanese carmakers were derived from the Bosch L-Jetronic system. Nissan was the first to introduce EFI in 1975; Toyota's system first appeared in 1979. These applications controlled only fuel injector duration. However, the systems used by Nissan since 1981 and Toyota since 1983 are completely integrated with the engine

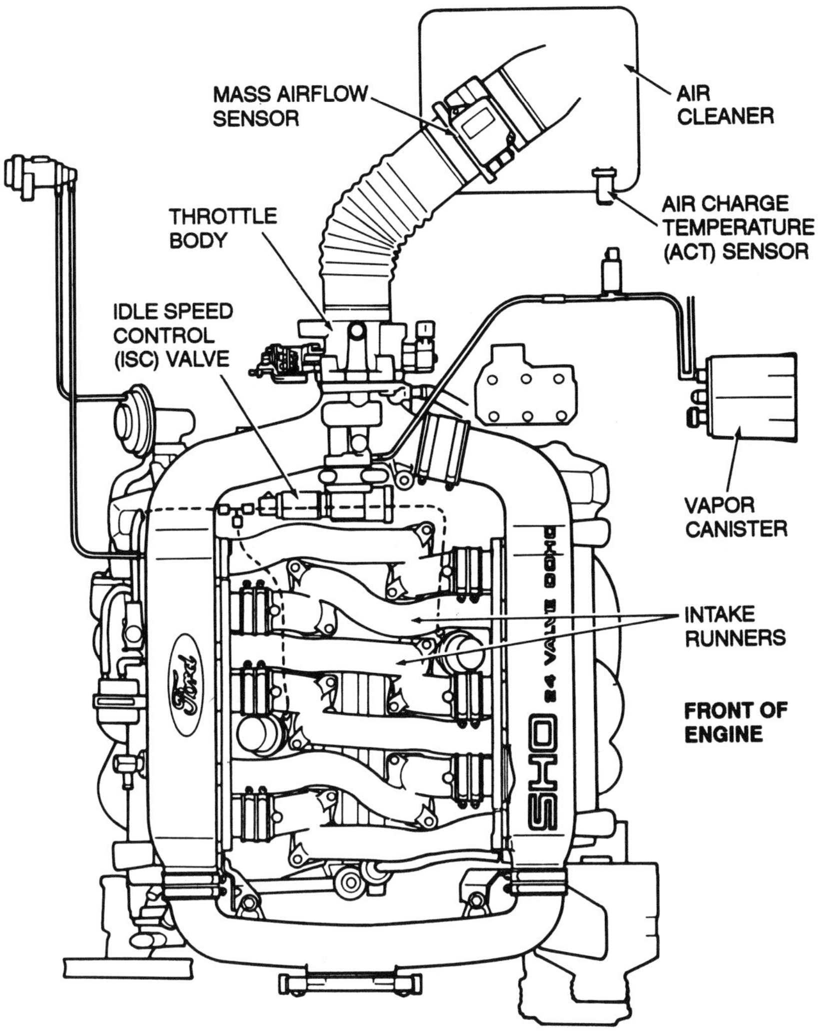


**Figure 22-28.** The SHO V-6 and several other late-model Ford engines have MAF-controlled fuel injection. (Ford)

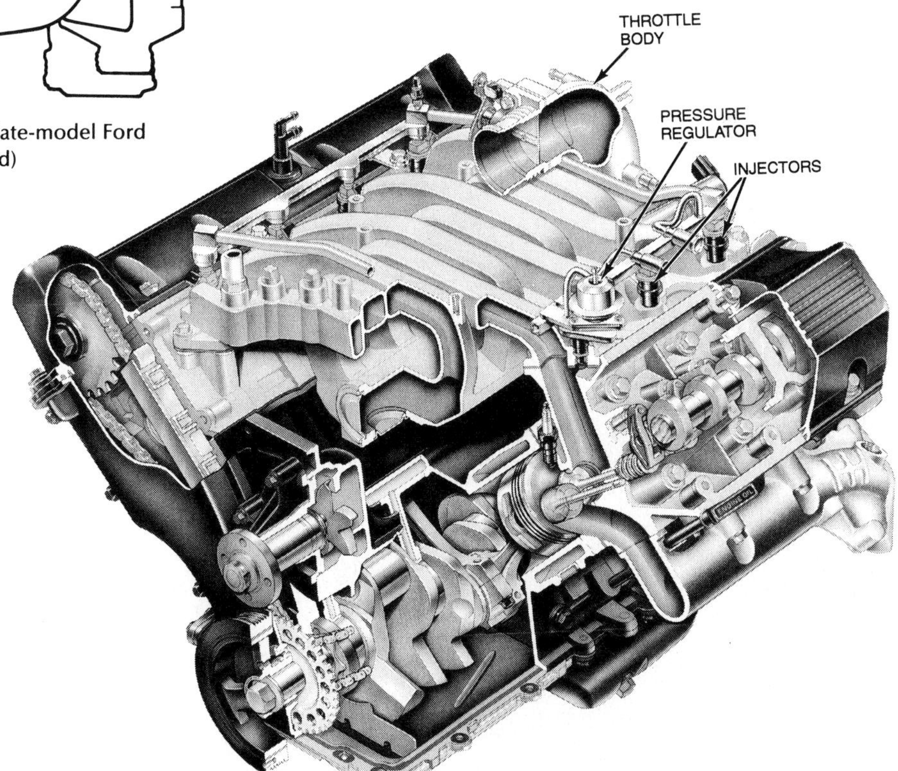


**Figure 22-29.** Ford's "modular" V-8 with sequential injection and mass airflow control is used in several carlines for the 1990's. (Ford)

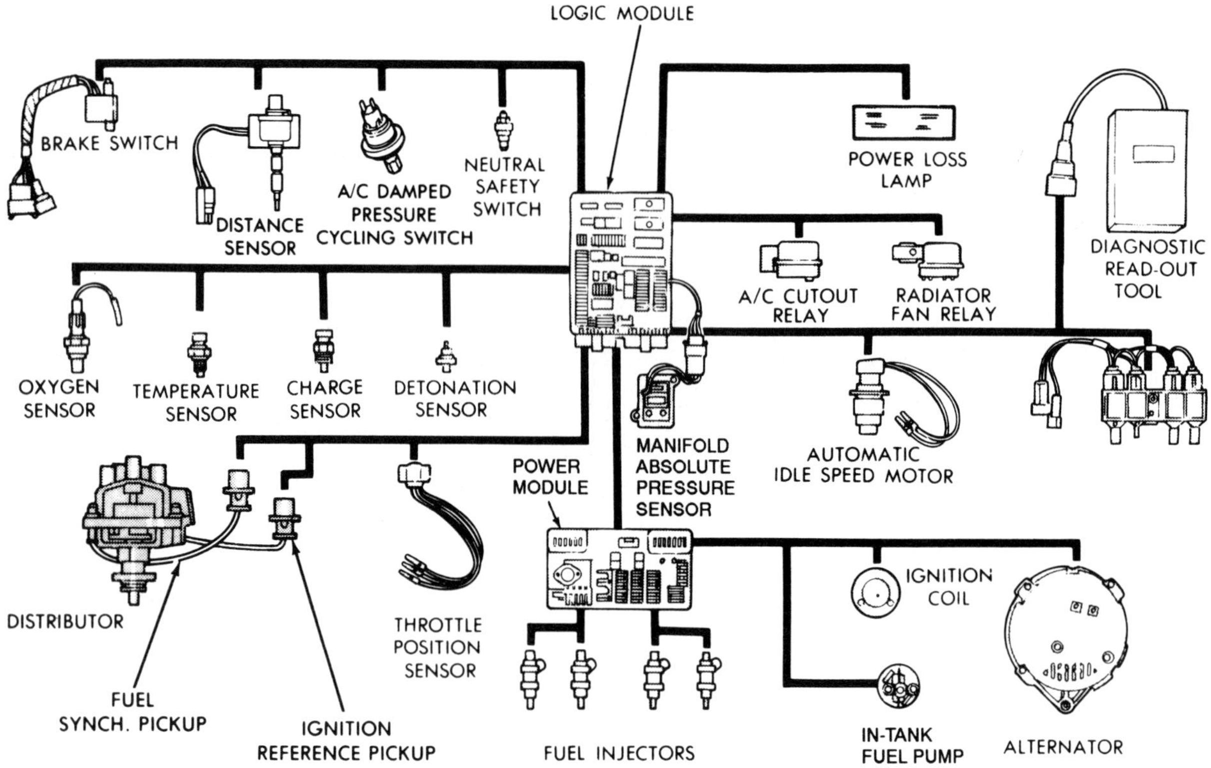


**Figure 22-30.** Chrysler uses a grouped, double-fire port injection system based on Bosch D-Jetronic system principles.

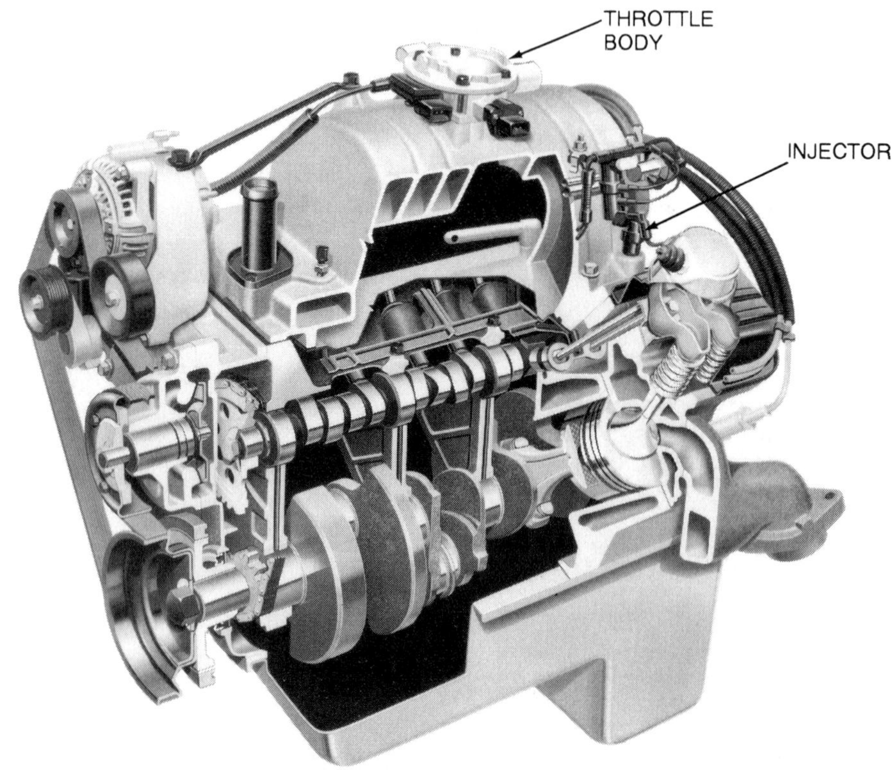


**Figure 22-31.** Chrysler introduced port-injected V-6 and V-8 truck engines in 1992. The unique high-rise intake manifold has tuned intake runners. (Chrysler)

control system and incorporate idle speed, EGR, and ignition timing control according to injection system requirements. Figure 22-32 shows a late-model Toyota injection system, based on Bosch designs.

## Throttle Body Injection

Throttle body injection (TBI) systems first appeared on domestic engines in the early 1980's. Today, gasoline injection systems are about equally divided between individual port injection and TBI systems.

Except for the Chrysler V-8 system, all TBI systems have similar parts. One or two solenoid-operated injectors are located in a throttle body assembly that resembles the base of a carburetor. The injectors deliver fuel intermittently to the intake air charge. A manifold pressure sensor measures intake air volume, and the electronic module controls fuel injection volume proportionally.

**GM-Rochester Systems.** The first GM TBI system appeared on 1980 Cadillacs and was called digital fuel injection (DFI). All GM TBI systems are manufactured by the Rochester Division of General Motors and use either one or two injectors in the throttle body housing. There are seven basic models currently in use:

- Models 100, 200, and 220 are 2-barrel, dual-injector assemblies. The injectors in some model 220 units are calibrated to flow fuel at different rates from each other.
- Models 300, 500, and 700 are 1-barrel, single-injector assemblies. Except for minor differences in fuel metering and airflow, models 300 and 500 are identical. The model 700 appeared in 1987 as a replacement for the other two models on 4-cylinder engines. The model 700 uses a low-pressure Multec injector.
- Model 400 is an assembly of two single-barrel, single-injector TBI units on a single intake manifold. It is also known as crossfire injection, because the electronic control module (ECM) is programmed to fire the TBI units in an alternating, or crossfire, sequence. The model 400 is found on some 1982–83 305-cid (5.0-liter) V-8 engines used in Camaros and Firebirds.

Rochester TBI systems are now used on many 4-, 6-, and 8-cylinder GM engines. Figure 22-33 shows typical Rochester model 300 and model 400 TBI assemblies. Because these systems are used with electronic engine controls, the TBI units include throttle position sensors and idle air control valves or idle speed motors.

**Ford Systems.** Ford also introduced a TBI system on some 1980 V-8 engines. Now called central fuel injection (CFI), the high-pressure Ford TBI system has been used mostly on V-6 and V-8 engines through the early 1980's. A single-injector low-pressure TBI unit is used on the 1985 and later 2.3-liter HSC and some 2.5-liter HSC 4-cylinder engines.

Figure 22-34 shows a typical Ford 2-barrel, dual-injector TBI assembly,

**Figure 22-32.** The Nippondenso injection system on this Toyota is similar to the Bosch L-Jetronic system.

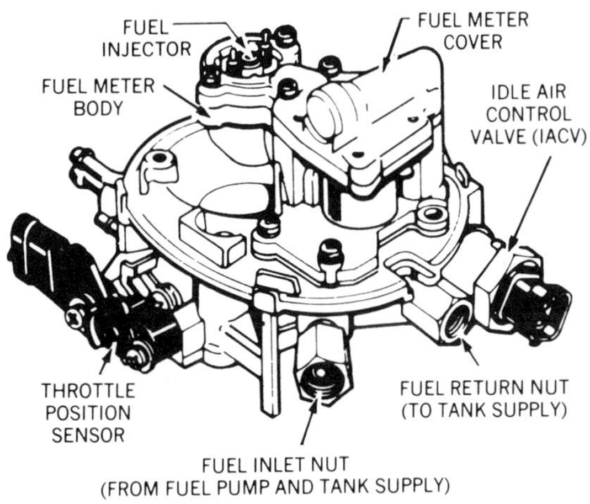


MODEL 300 TBI ASSEMBLY

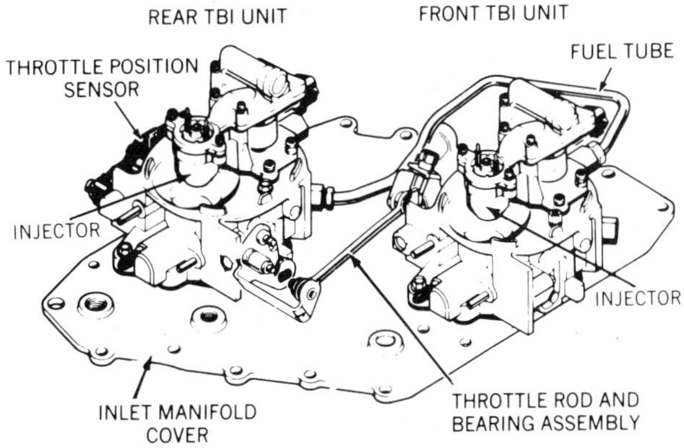


MODEL 400 TBI ASSEMBLY

**Figure 22-33.** First used on 1980 Cadillacs, Rochester TBI systems are now used by all GM car divisions. (Rochester)

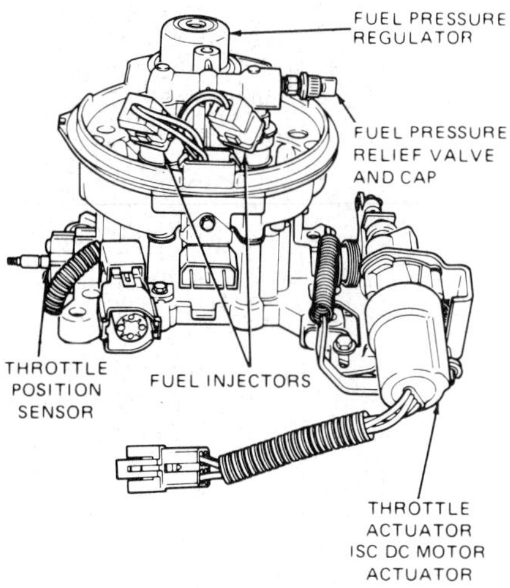


**Figure 22-34.** Ford calls its throttle body injection "central fuel injection (CFI)," but it works similarly to other systems. (Ford)

which Ford calls a fuel charging assembly. Because Ford's TBI systems are used with electronic engine controls, they also include throttle position sensors and idle speed control motors.

**Chrysler Systems.** Chrysler introduced its first TBI system on 1981 V-8 engines in Imperials and now uses other TBI systems on 4-cylinder engines. Like GM's crossfire injection, the Chrysler continuous flow system proved to be a technological dead end. The Chrysler V-8 system, figure 22-35, uses dual injectors to deliver fuel continuously in response to varying pressure from a control pump. This system measures air volume with a unique airflow sensor in the air cleaner inlet.

Chrysler's 4-cylinder TBI systems are similar to other TBI systems for small engines. From 1981 through 1985, Chrysler used a single-injector, high-pressure system with a Bendix fuel injector, figure 22-36. The system was redesigned for 1986 and later models to operate at low pressure using a Bosch fuel injector, figure 22-37. All late-model Chrysler TBI systems are part of fully integrated electronic engine control systems.

**Imported Car TBI Systems.** Several imported cars use TBI systems that are similar to those on domestic vehicles. Figure 22-38 shows the major components of an AMC-Renault TBI system. Although built in North America, these vehicles use a typical example of one type of imported TBI system.

An atypical example is the TBI system used on turbocharged Japanese vehicles built by Mitsubishi and sold under both the Mitsubishi and Chrysler nameplates in the United States. This system uses a Karman vortex airflow sensor in the air cleaner, figure 22-39, to measure intake airflow

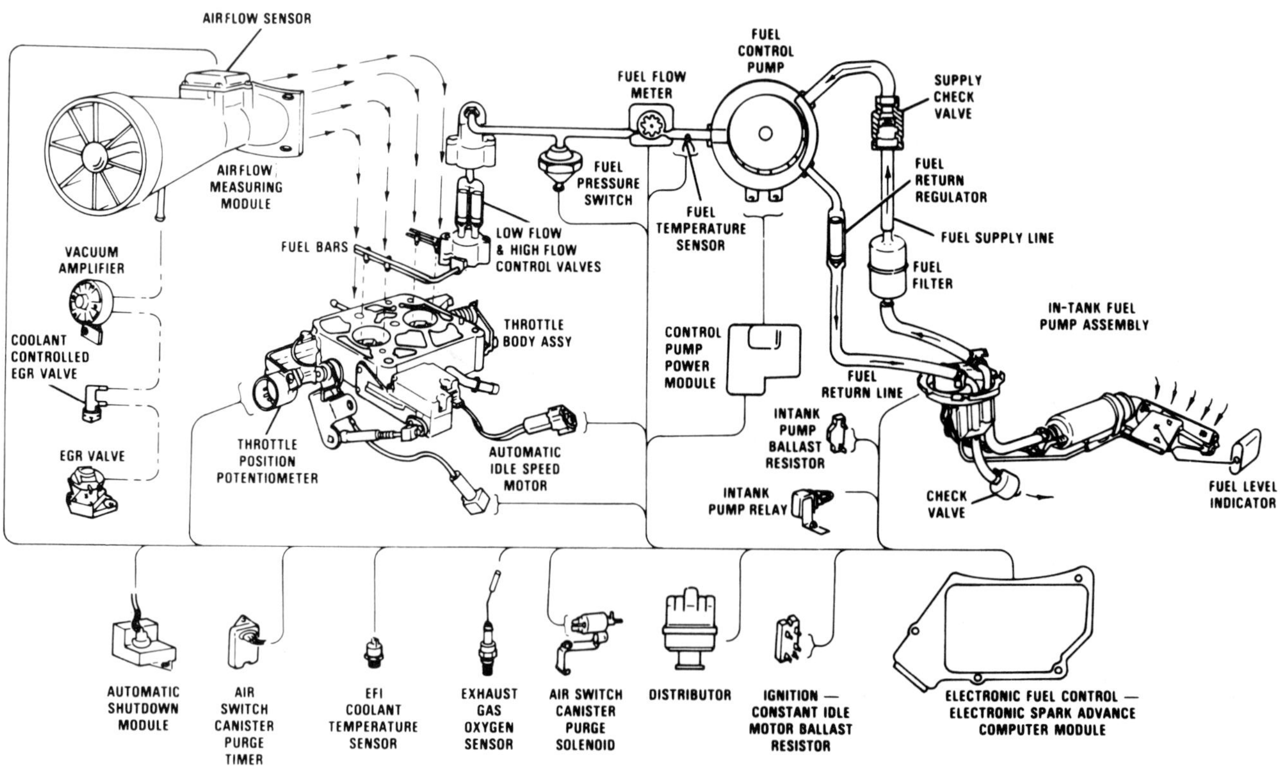


**Figure 22-35.** Chrysler's V-8 TBI system used continuously operating injectors, regulated by control pump pressure. Chrysler later replaced these systems with Quadrajet carburetors under extended warranty. (Chrysler)

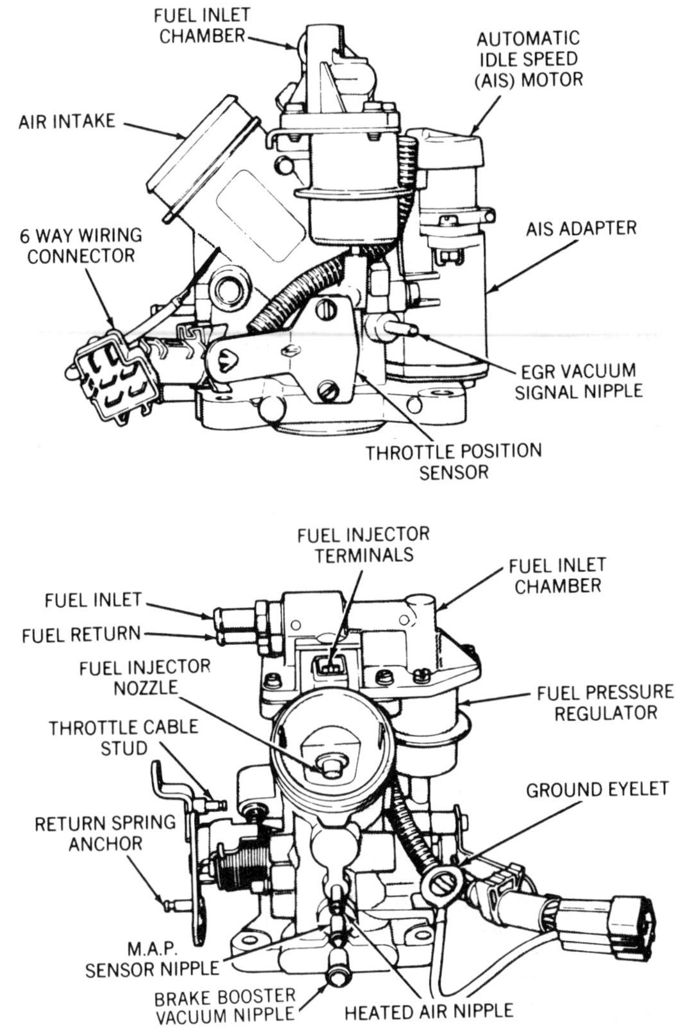


**Figure 22-36.** Chrysler's high-pressure throttle body injection assembly for 4-cylinder engines. (Chrysler)

rate. The dual injectors mounted in the injection mixer or throttle body inlet port, figure 22-40, work like other solenoid-operated injectors and are fired in an alternating sequence. The electronics of this system, however, are considerably more complex than the other TBI systems discussed.

## FUEL INJECTION SUBSYSTEMS

The major subsystems of any fuel injection system are:

- The fuel delivery system
- The air control system
- The engine computer, sensors, and auxiliary systems

Chapters 20 and 21 explain computer operation and the sensors commonly used with any engine control system. The following sections explain the fuel delivery system, including the injectors, and review some key points about air control and system sensors.

### The Fuel Delivery System

The fuel delivery system consists of a pump, a filter, a pressure regulator, the injectors, and fuel distribution lines.

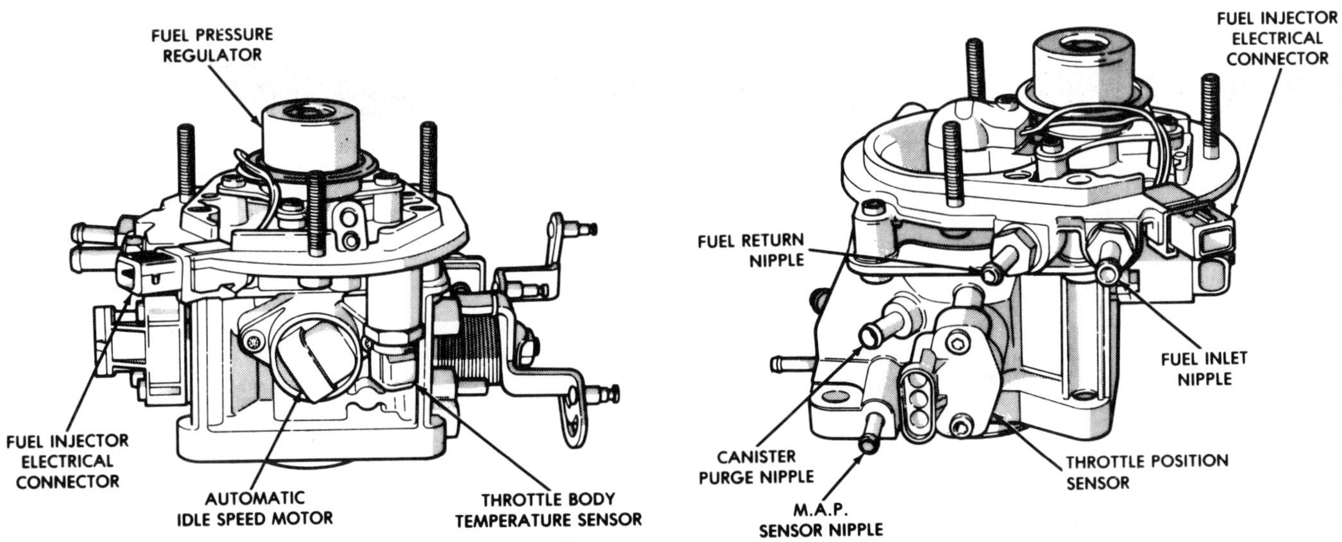


**Figure 22-37.** Chrysler's low-pressure throttle body assembly for 4-cylinder engines. (Chrysler)

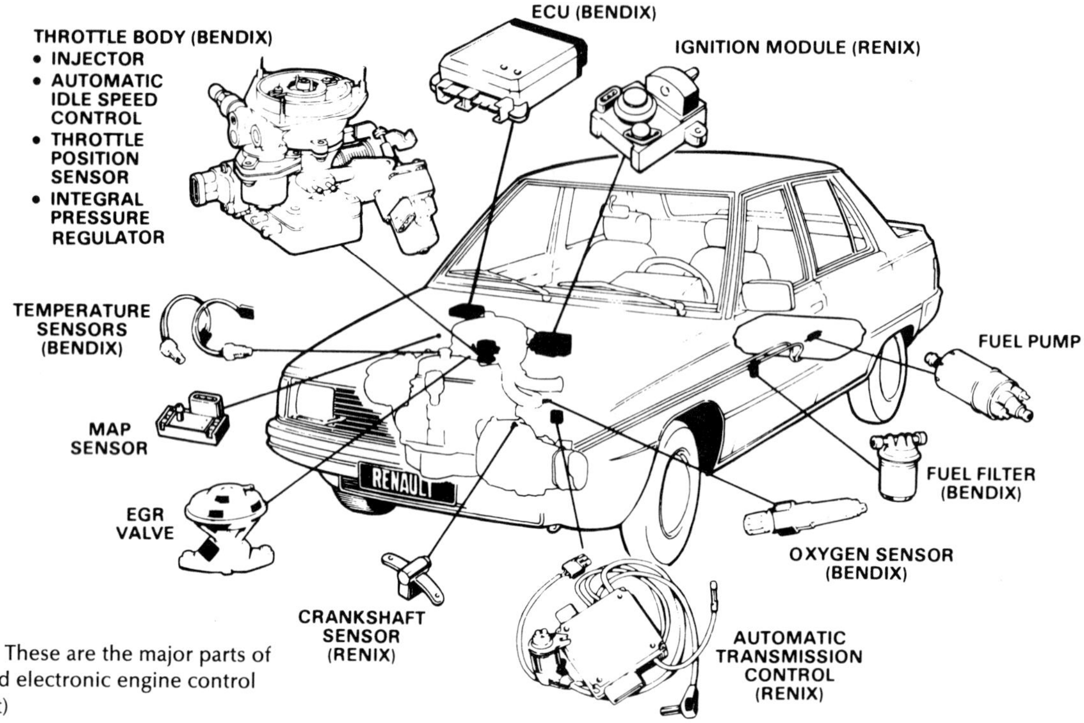


**Figure 22-38.** These are the major parts of Renault's TBI and electronic engine control system. (Renault)

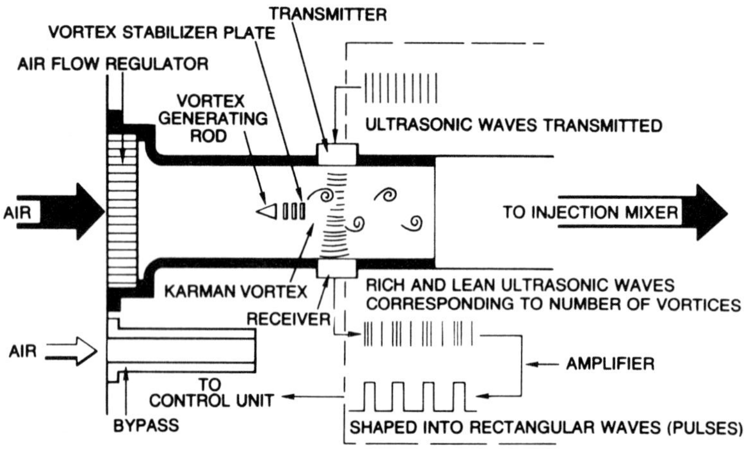


**Figure 22-39.** The Mitsubishi TBI system measures intake airflow by passing it through a vortex generator to create ultrasonic waves. The ultrasonic waves are amplified, shaped into rectangular waves, and sent to the control unit. (Chrysler)

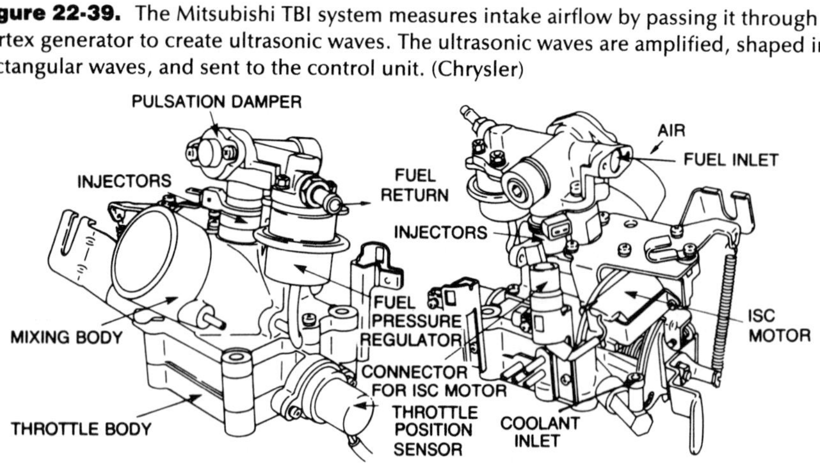


**Figure 22-40.** The Mitsubishi ECI throttle body assembly used on turbocharged 4-cylinder engines. (Chrysler)

### The Fuel Pump

An injection system needs constant, uniform fuel pressure to the injectors. An electric pump can provide this much better than a mechanical pump can. Most systems have a positive-displacement, vane or roller pump, figure 22-41, in or near the fuel tank. Some cars have two pumps, one inside and one outside the tank, figure 22-42. The pump always supplies more fuel than the injectors need and works with the pressure regulator to control fuel volume and pressure.

### The Fuel Filter

Clean fuel is as important for gasoline injection as it is for diesel systems. The fuel delivery system has one or more large filters, figure 22-43, that remove dirt particles larger than 10 to 20 microns (0.00025 to 0.00050 inch). Most systems also have an inlet strainer or filter on the fuel pickup tube in the tank. The injectors, themselves, have inlet filter screens to remove any final dirt before fuel enters the nozzles.

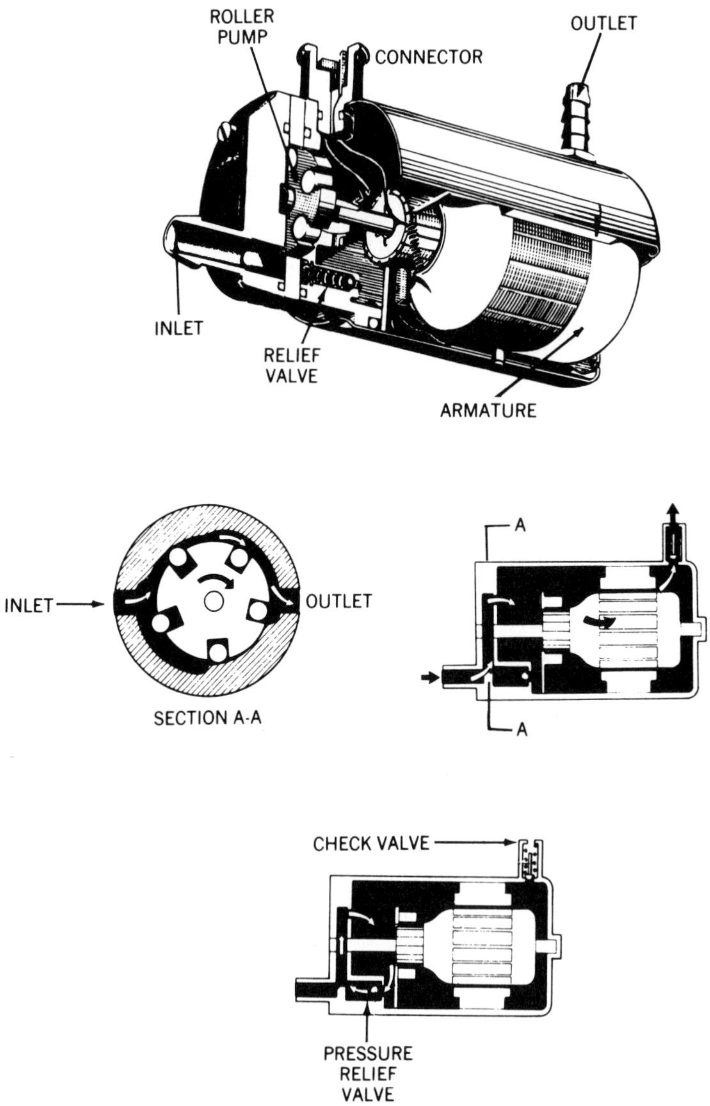


**Figure 22-41.** An electric pump provides the continuous steady fuel pressure needed by an injection system. (Volvo)

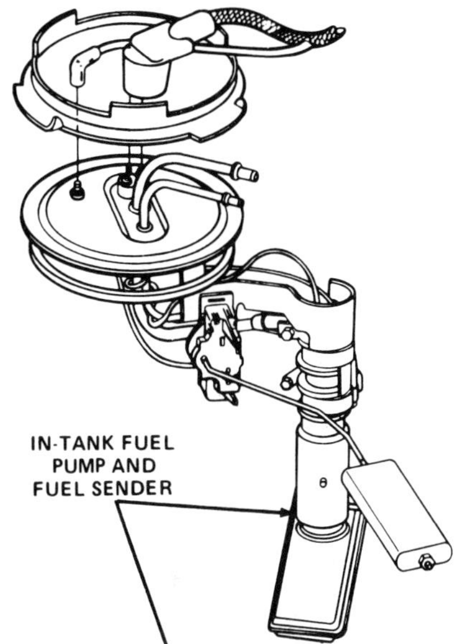


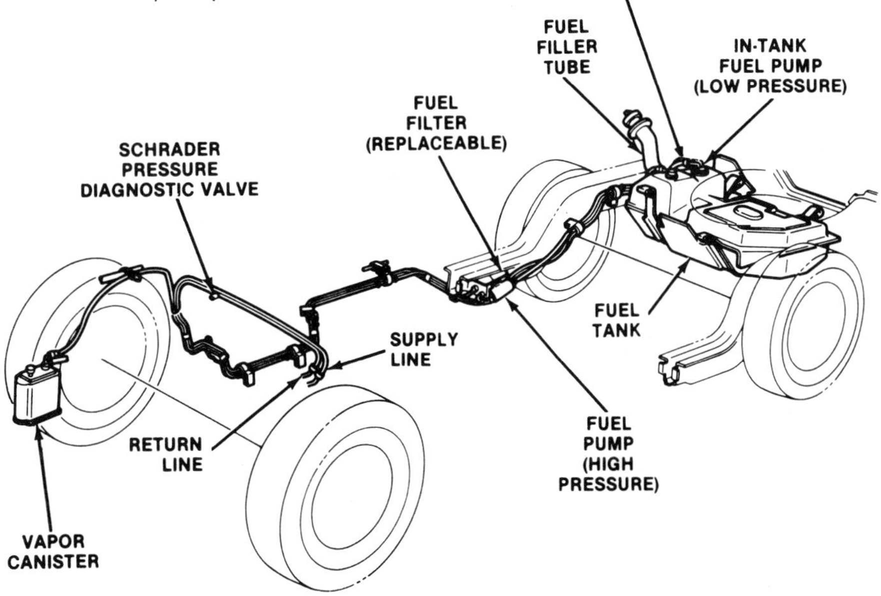


**Figure 22-42.** Many gasoline injection systems have one pump in the tank and another outside the tank. (Ford)

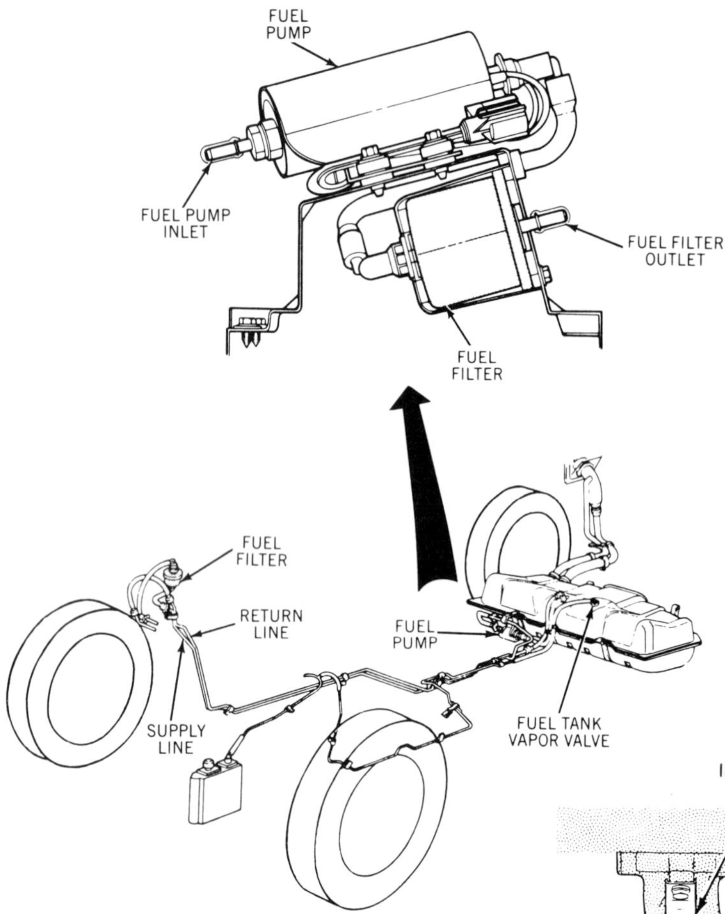


**Figure 22-43.** Gasoline injection systems use high-capacity fuel filters. Some systems have two or more filters. (Ford)

## The Fuel Lines

The fuel lines are arranged to circulate fuel continuously from the tank to the injectors and back to the tank. Like diesel systems, gasoline injection systems all have return lines. This ensures constant fuel volume and pressure and helps to prevent fuel heating and vapor lock. Gasoline injection lines are made of steel tubing, but they do not have to stand the high pressures of diesel injection lines. Also, they do not have to be of equal lengths. Flexible hose is often used in low-pressure return lines.

## The Pressure Regulator

The pressure regulator is a critical part of every injection system. The regulator and the pump maintain the necessary constant pressure at the injectors. In electronic injection systems, the regulator is a diaphragm-operated

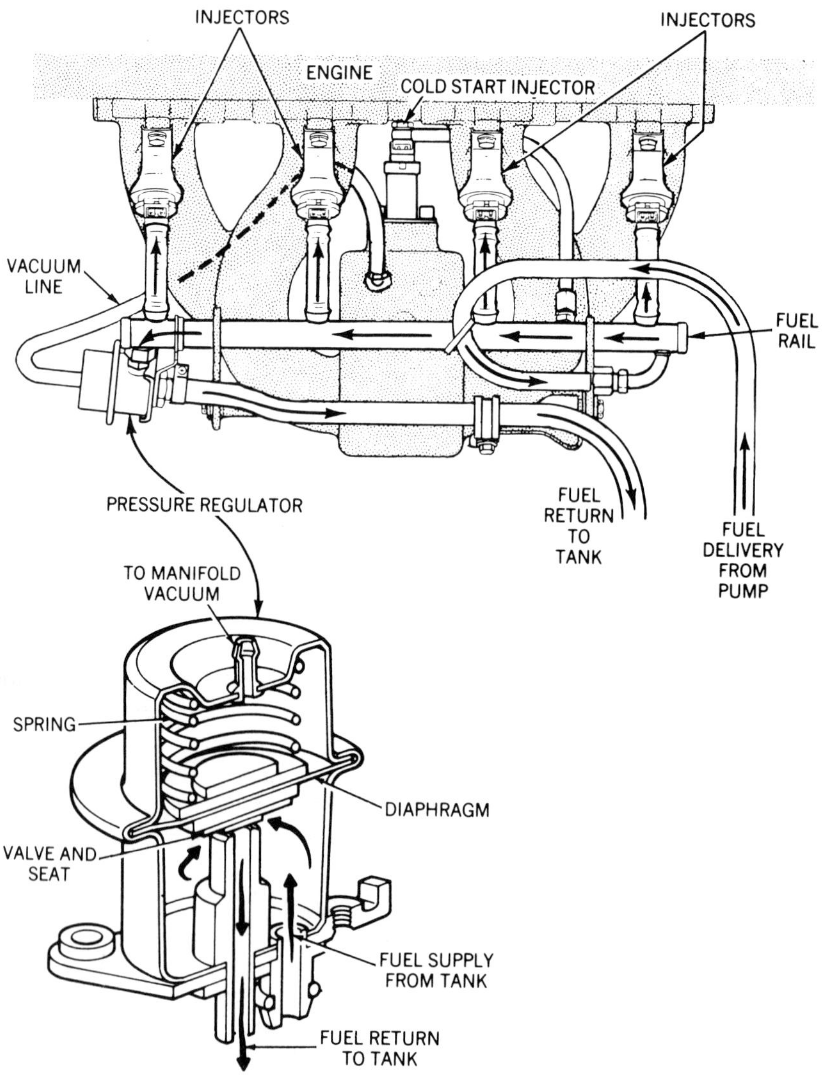


**Figure 22-44.** The pressure regulator on most EFI systems controls fuel pressure in relation to manifold pressure. (Ford) (Volvo)

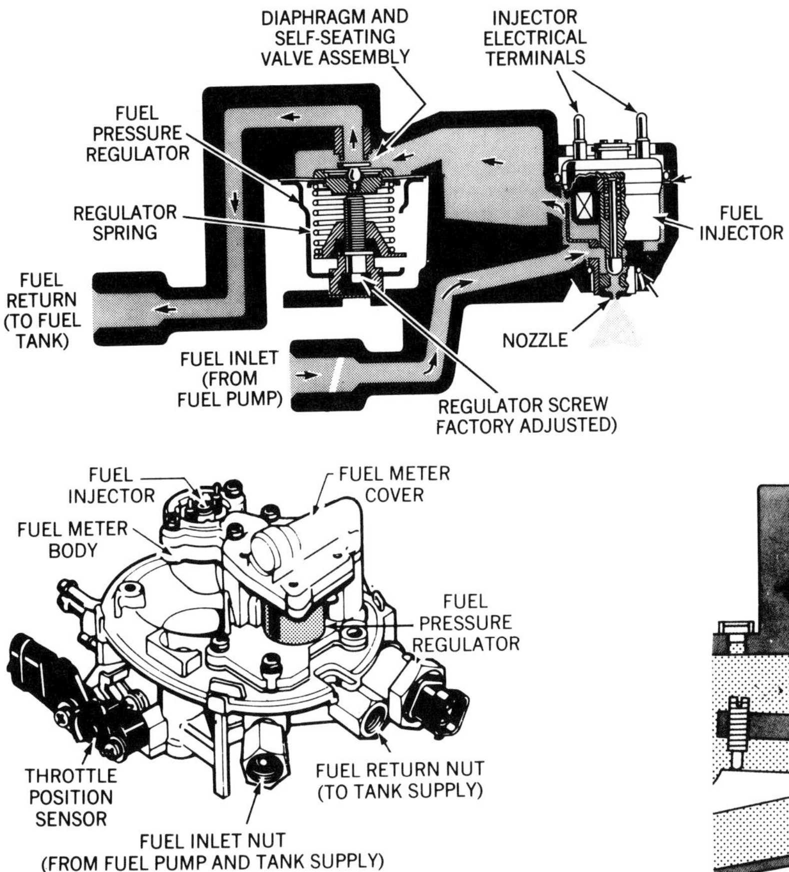


**Figure 22-45.** This TBI pressure regulator is on the return side on the injector and works on the opposing forces of fuel pressure and spring and air pressure. (Rochester)

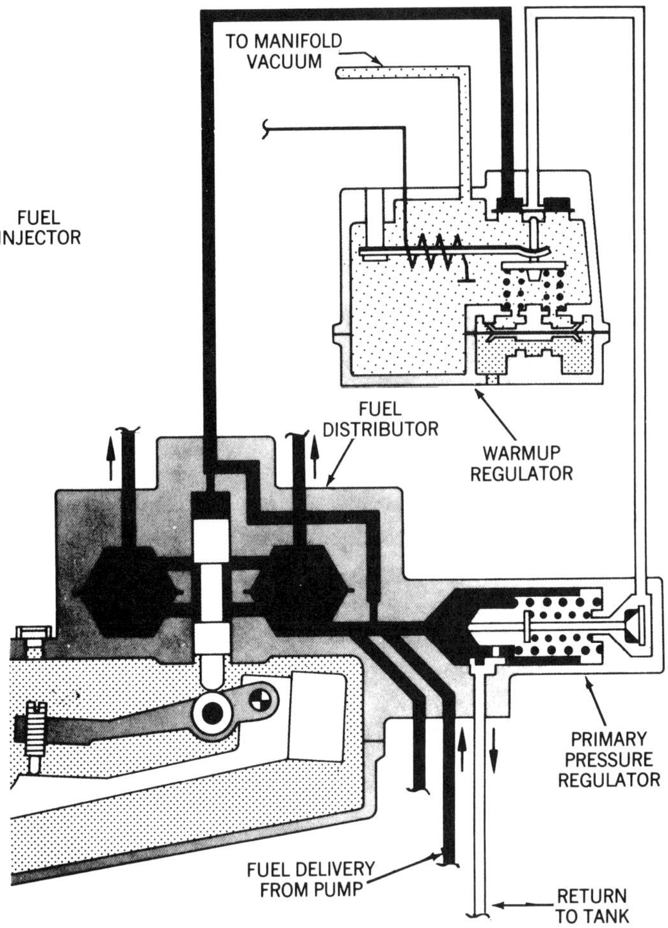


**Figure 22-46.** Bosch K-Jetronic injection uses a main pressure regulator and a separate warmup regulator to control injection pressure. (Bosch)

valve, downstream (on the return side) from the injectors. The fuel chamber has an inlet from the distribution pipe, or fuel rail, and an outlet to the return line. Most regulators are controlled by a combination of spring pressure and manifold vacuum applied to the diaphragm. The vacuum chamber of the regulator is connected to the intake manifold so that fuel pressure is regulated in relation to manifold pressure or vacuum, figure 22-44. The combination of spring force and diaphragm action controls the amount of fuel returned to the tank, and thus fuel pressure. When the fuel pump is off, the regulator closes the return line to maintain pressure at the injectors.

The regulator in a TBI system is mounted on the throttle body, figure 22-45. Although it may look different, it works the same as a regulator on a fuel rail. Functionally, it controls fuel pressure on the return side of the injector. The regulator in figure 22-45 works on the opposing forces of fuel pressure on one side and spring pressure and air pressure on the other. It is not connected to manifold vacuum, but the spring side of the diaphragm is exposed to inlet air pressure inside the throttle body. The air inlet is usually near the injector tip so that the regulator senses the same air pressure that the injector is exposed to.

Mechanical injection systems work on a difference between fuel pump pressure and a control pressure developed by the fuel distributor. The pressure regulator is on the control-pressure side of the fuel distributor and connected to the fuel return line, figure 22-46. Control pressure and spring force work on one side of the regulator and pump pressure on the other to vary the return line opening. Bosch K-Jetronic systems have a primary regulator and a warmup regulator to vary control pressure for different operating needs.

Fuel pressure in a gasoline injection system is less than one-tenth the pressure in a diesel system because it does not have to inject fuel against cylinder pressures. Most port injection systems operate with pressures at the injectors of 30 to 50 psi (207 to 345 kPa). Most throttle body systems operate with injector pressures of 10 to 20 psi. Fuel pressure must only be high enough to get the necessary fuel through the nozzle and to aid atomization. Because throttle body injectors mix fuel with air at the manifold inlet, there is more time for atomization in the manifold. Throttle body injectors, therefore can work with lower pressures than port injectors. Ford and Chrysler, however, have used high-pressure TBI systems.

## Mechanical Injection Nozzles

Bosch K-Jetronic is the major mechanical gasoline injection system. The individual injectors, figure 22-47, have spring-loaded tips that seal the nozzles against fuel pressure. When pressure exceeds about 45 psi (310 kPa), it forces the nozzle open. Fuel pressure thus controls the amount of nozzle opening and the amount of fuel that enters the manifold. Injection pressure is controlled by the mixture control unit through a combination of pump pressure and control pressure. Under full-load operation, injection pressure may go as high as 70 psi (490 kPa). Bosch K-Jetronic nozzles inject fuel continuously while the engine is running. Fuel collects, or waits, at the intake port to mix with the air charge. Spring force and injection pressure cause the nozzle to vibrate and ensure good atomization. When the engine is off, the spring closes the nozzle to hold reserve pressure in the lines.

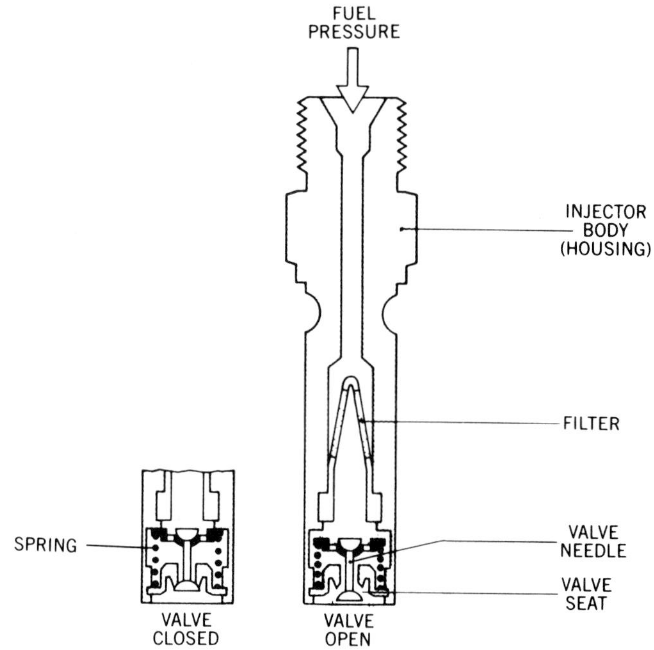


**Figure 22-47.** Fuel pressure and spring force work against each other to control the opening of a mechanical injection nozzle. (Bosch)

## Electronic Injection Nozzles

A solenoid-operated electronic injector, figure 22-48, has an armature and a needle valve or ball valve assembly. A spring holds the needle or ball against the valve seat when the solenoid is deenergized. When the solenoid is energized, it pulls the armature against spring force to unseat the valve and inject fuel. The valve moves only about 1 mm. The design of the injector tip ensures good fuel atomization and spray pattern.

The injector always opens the same distance, but the amount of fuel injected is controlled by injector **pulse width**. Pulse width is a term you will hear in relation to many solenoid-operated metering devices. It is the amount of time that the device is energized, measured in milliseconds. The electronic module varies injector pulse width in relation to the amount of fuel a particular engine needs at any time. A short pulse width delivers little fuel; a longer pulse width delivers more fuel. Figure 22-49 shows the relationships of engine operating cycles, ignition frequency, and injection timing

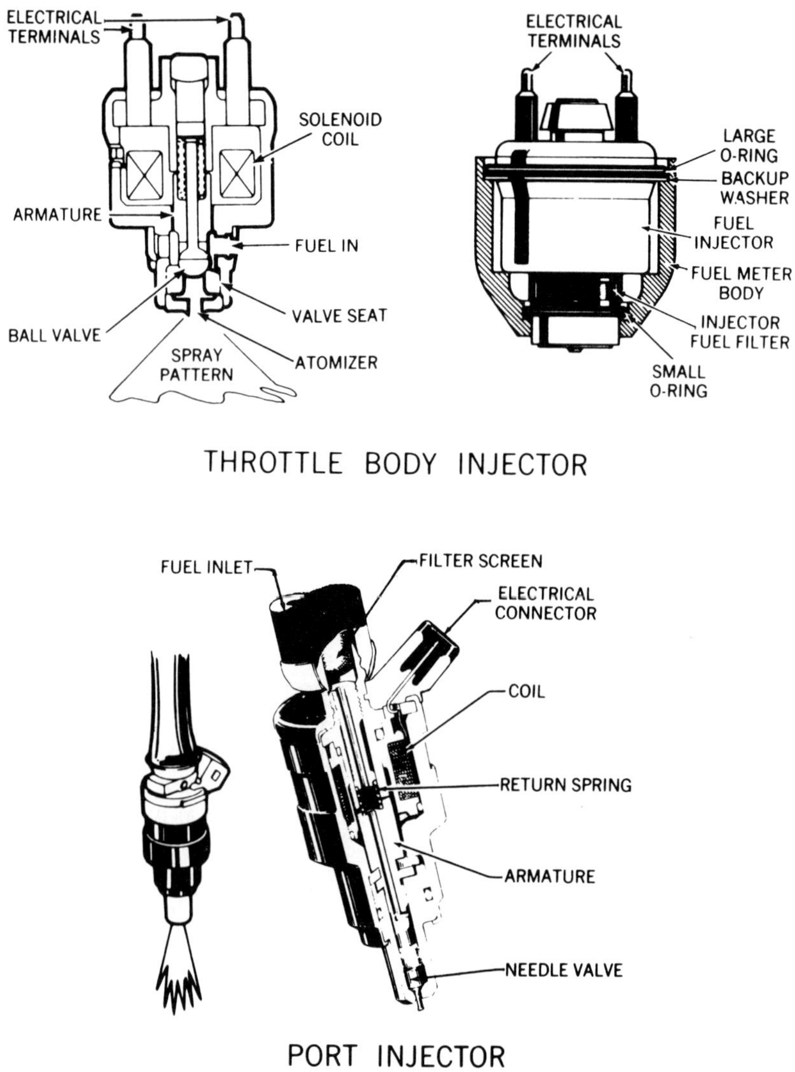


**Figure 22-48.** EFI nozzles are opened intermittently by timed solenoid action. (Rochester) (Volvo)

for a 4-cylinder engine. Time "Tj" in the bottom section is the injector pulse width.

You may hear pulse width discussed in relation to the duty cycle of other solenoid-operated devices. Pulse width and duty cycle are related, but they are not the same thing. The complete operating cycle of any solenoid device is the sequence from off to on to back off again. Duty cycle is the ratio of on-time to total cycle time. Duty cycle is a useful measurement for solenoids or circuits that operate on a fixed number of cycles per second (hertz). Because the cycle time of a port fuel injector varies with engine speed, duty cycle is not a useful measurement of port fuel injector operation.

The system computer calculates injector pulse width from information provided by sensors that measure airflow, engine speed, load, temperature, and throttle position. Throttle body injectors supply fuel for all cylinders from one or two nozzles, so their operating cycles and pulse widths must be matched to overall airflow and engine operation.

Electronic port fuel injectors can be energized, or fired, in several ways. The most common are:

- *Simultaneous*—All injectors fired at the same time, once every crankshaft revolution, figure 22-49. This is the simplest injector operating sequence, especially for 4-cylinder engines.
- *Grouped, single fire*—The injectors are divided into two equal groups that fire alternately. Only one group fires once each crankshaft revolution. One injector pulse is used for each combustion stroke. Because only two injectors can be fired close to the point

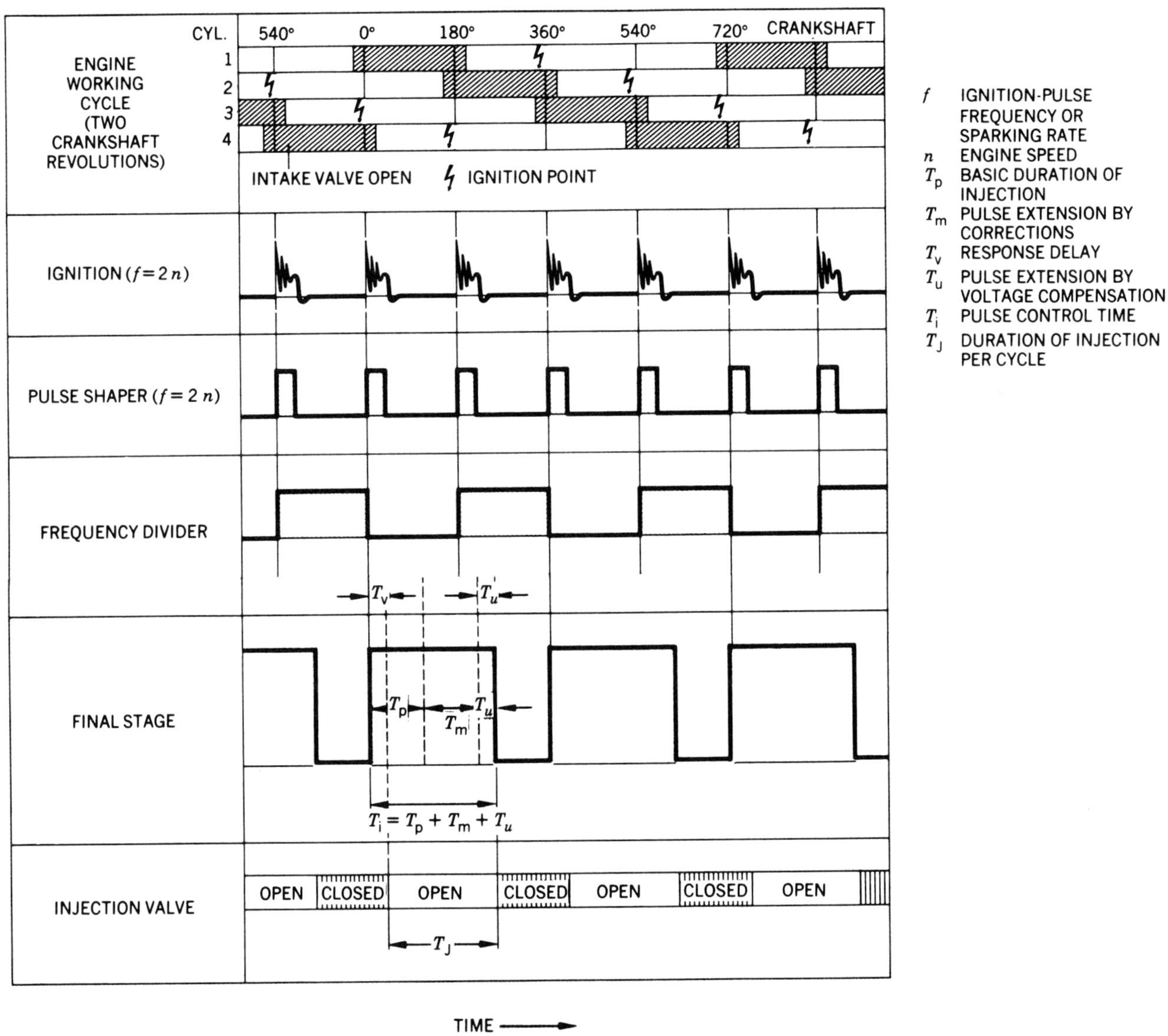


**Figure 22-49.** These graphs show the relationships of engine working cycles, ignition, and injection timing for a 4-cylinder engine. All four injectors operate simultaneously in this system. (Bosch)

where the intake valves open, the fuel charge for the other cylinders must wait in the manifold for varying periods of time. A completely new fuel charge is delivered only once for every two crankshaft revolutions, so the computer must wait for a complete engine firing cycle before it can alter the air-fuel mixture.

• *Grouped, double fire*—The injectors are divided into two equal groups, but both groups fire once each crankshaft revolution. Two injector pulses are used for each combustion stroke. The air-fuel mixture remains in the manifold for a shorter time before the intake valve opens, and the computer can adjust the mixture twice as often as for a single-fire system.

• *Sequential*—The injectors are energized in engine firing order sequence. Each fires before the intake valve for its cylinder opens. Sequential injection provides the most precise air-fuel control because the computer can adjust the mixture between firings of individual cylinders. It also is the most expensive and complex port injection system and requires a number 1 cylinder synchronizing signal from the engine camshaft to fire the injectors in the correct order.

## AIR CONTROL SYSTEM

Fuel metering through a carburetor is controlled directly by the difference between air pressure in the fuel bowl and air pressure in the carburetor barrel. Intake air-fuel volume is controlled by the pumping capacity of the engine, the airflow size of the carburetor, and the throttle. A fuel injection system must provide the same kind of mixture control, but the fuel is isolated from the airflow, behind the injection nozzles. Therefore, all systems must have a way to measure intake air volume and then meter fuel accordingly. All air control systems have a throttle to control air volume and one of the following devices to measure it:

- An airflow meter or sensor
- A manifold pressure sensor
- An air mass sensor.

### Throttle

The throttle in a fuel injection system works exactly like the throttle in a carburetor. Connected to the accelerator linkage, the throttle regulates the amount of air allowed into the engine. On port injection systems, the throttle is on the air inlet side of the injectors and regulates only airflow. On TBI systems, the throttle is between the injectors and the manifold and controls the amount of air-fuel mixture entering the engine. While the throttle regulates airflow, a sensor measures the speed of the airflow, the pressure in the manifold, or the molecular mass of the air entering the engine. All of these indicate the intake air volume and relate directly to the fuel volume the injectors must provide.

### Air Sensors

You probably will read some carmakers' references to techniques used for measuring air intake and other operating conditions to regulate injection fuel metering. Two common terms used to describe injection control are **speed density** and **mass airflow**. Speed density systems use a manifold absolute pressure (MAP) sensor and engine rpm as the principal sensor input signals to determine the amount of fuel to be injected, which is controlled by regulating injector pulse width. Speed density systems also use input signals from temperature sensors and maps of engine variables such as EGR flow and volumetric efficiency.

A mass airflow sensor uses a sensor that measures the molecular mass, or weight, of intake air. Simple airflow-controlled fuel injection systems use a vane-type sensor that measures the velocity of intake air.

The principles of vane airflow sensors, MAP sensors, barometric pressure sensors, and MAF sensors are covered in chapter 21. We will review them briefly in the following sections. Bosch K-Jetronic mechanical injection systems use a unique airflow sensor that regulates injection pressure.

**Bosch K-Jetronic Airflow Sensor.** The K-Jetronic airflow sensor is a movable plate deflected by intake airflow, figure 22-50. The plate moves a counterweighted pair of levers that operate on the control plunger in the fuel distributor. The airflow sensor and the fuel distributor are the complete mixture control unit for the system. The sensor deflects the control

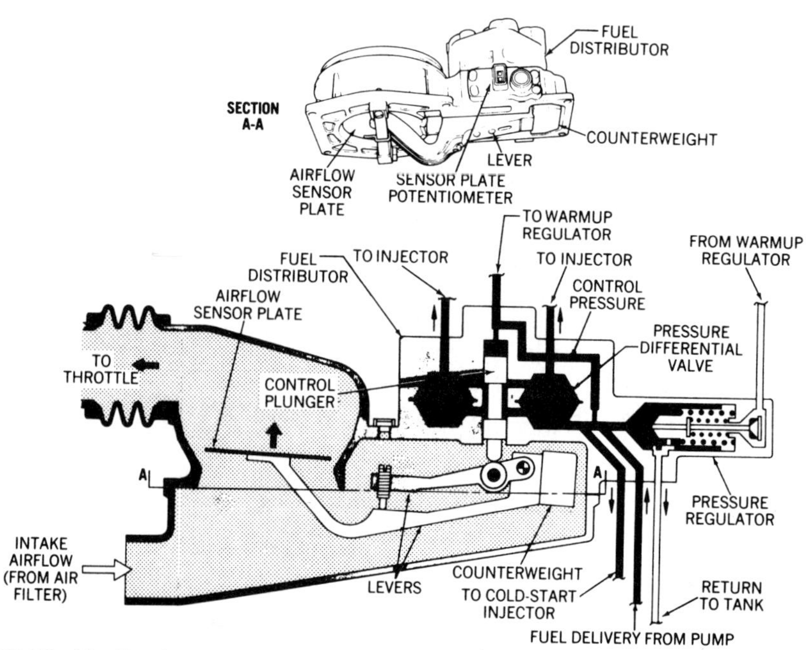


**Figure 22-50.** The Bosch K-Jetronic airflow sensor controls fuel pressure to the continuous injectors. (Bosch) (Audi)

plunger in proportion to airflow. The control plunger regulates fuel pump pressure to the differential pressure valve for each cylinder. Fuel control pressure on the other end of the control plunger balances airflow sensor force. The combination of these pressures at the pressure differential valves regulates injection in relation to airflow.

**Vane Airflow Sensors.** Vane airflow sensors are based on the original Bosch L-Jetronic system. They also are used on some Ford systems and by many Japanese carmakers. The movable vane is attached to a potentiometer, figure 22-51. The potentiometer receives a 5-volt reference voltage from the computer. Intake airflow deflects the vane, and the potentiometer sends a return signal voltage to the computer that is proportional to airflow velocity and thus to the amount of air being pumped in by the engine.

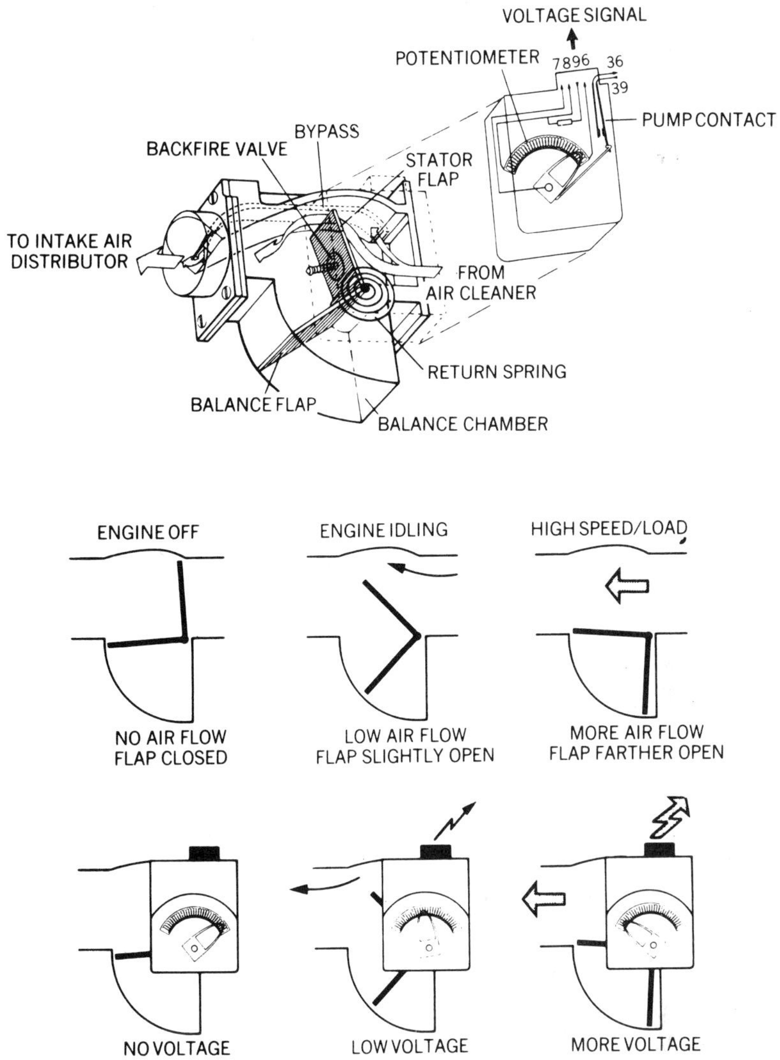


**Figure 22-51.** The vane airflow sensor used on Bosch, Ford, and other EFI systems sends a variable voltage signal to the electronic control module. (Volkswagen)

**Manifold Absolute Pressure (MAP) and Barometric Pressure Sensors.** MAP sensors measure the absolute pressure in the intake manifold, which is the difference between atmospheric (barometric) air pressure and manifold vacuum. Barometric pressure sensors measure the atmospheric pressure. These pressure measurements are the principal indications of engine load. They are major input signals used by the computer to control the air-fuel ratio and ignition timing. Figure 22-52 is a table that compares manifold pressures in absolute values and gauge values. Gauge pressure is what you see on a test gauge, whether in pounds per square inch (psi) kilopascals (kPa), or inches of mercury (in. Hg). It is pressure above or below atmospheric pressure. Absolute pressure is a measurement that includes atmospheric pressure. Thus, 0 **psig** equals 14.697 **psia**; a turbocharger boost pressure of 12 psig equals 26.70 psia (12 + 14.697).

Three pressure measurements or calculations are necessary. The computer usually measures two directly and calculates the third. The three principal air pressure conditions that affect engine operation are:

- *Barometric pressure* is the ambient atmospheric air pressure. Barometric pressure measures the weight of the atmosphere on the earth. It also is the outside pressure of the air entering the engine. Barometric pressure changes with altitude and temperature. Standard barometric pressure at sea level is 14.7 pounds per square inch (psi), 101.3 kilopascals (kPa), or 29.9 inches of mercury (in. Hg). It decreases with altitude or with increasing temperature.

- *Manifold vacuum* is the pressure in the intake manifold *below* atmospheric pressure when the engine is running. This is the traditional measure of manifold pressure that has been used in the automotive service industry for generations. Manifold vacuum is measured in relation to atmospheric pressure. High vacuum is low pressure. Manifold vacuum is traditionally expressed in inches of mercury (in. Hg).

- *Manifold absolute pressure (MAP)* is the algebraic sum of atmospheric (barometric) pressure and vacuum. It is the relative difference between air pressure outside the manifold and vacuum inside. MAP is measured in relation to zero pressure (high vacuum) and is inversely proportional to vacuum. When vacuum is high, MAP

| *Pounds Per Square Inch Absolute* | *Pounds Per Square Inch Gauge and Inches of Mercury Gauge* | *Kilopascals Gauge and Centimeters of Mercury Vacuum* | *Kilopascals Absolute* | *Inches of Mercury Absolute* |
|---|---|---|---|---|
| 29.70 psia | 15 psig | 104.43 kPag | 204.76 kPaa | 60.45 in. Hg |
| 26.70 psia | 12 psig | 82.74 kPag | 184.07 kPaa | 54.35 in. Hg |
| 24.70 psia | 10 psig | 68.95 kPag | 170.28 kPaa | 50.28 in. Hg |
| 21.70 psia | 7 psig | 48.27 kPag | 149.60 kPaa | 44.17 in. Hg |
| 19.70 psia | 5 psig | 34.48 kPag | 135.81 kPaa | 40.10 in. Hg |
| 16.70 psia | 2 psig | 13.79 kPag | 115.12 kPaa | 33.99 in. Hg |
| 14.697 psia | 0 psig/in. Hg | 0.00 kPag/cm Hg | 101.33 kPaa | 29.92 in. Hg |
| 13.71 psia | 2 in. Hg | 5.08 cm Hg | 94.53 kPaa | 27.92 in. Hg |
| 12.24 psia | 5 in. Hg | 12.70 cm Hg | 84.39 kPaa | 24.92 in. Hg |
| 11.26 psia | 7 in. Hg | 17.78 cm Hg | 77.63 kPaa | 22.92 in. Hg |
| 9.78 psia | 10 in. Hg | 25.40 cm Hg | 67.43 kPaa | 19.92 in. Hg |
| 8.80 psia | 12 in. Hg | 30.48 cm Hg | 60.68 kPaa | 17.92 in. Hg |
| 7.33 psia | 15 in. Hg | 38.10 cm Hg | 50.54 kPaa | 14.92 in. Hg |
| 5.85 psia | 18 in. Hg | 45.72 cm Hg | 40.34 kPaa | 11.92 in. Hg |
| 4.87 psia | 20 in. Hg | 50.80 cm Hg | 33.58 kPaa | 9.92 in. Hg |
| 3.89 psia | 22 in. Hg | 55.88 cm Hg | 26.82 kPaa | 7.92 in. Hg |
| 2.42 psia | 25 in. Hg | 63.50 cm Hg | 16.68 kPaa | 4.92 in. Hg |
| 0.00 psia | 29.92 in. Hg | 75.996 cm Hg | 0.00 kPaa | 0.00 in. Hg |

Note: All values are taken at sea level.

**Figure 22-52.** Pressure relationship chart.

is low—well below barometric pressure. When vacuum is low, MAP is high—closer to barometric pressure. MAP can be expressed in kPa, in. Hg, or psi.

Barometric pressure (BARO), manifold vacuum, and MAP are related to each other as follows:

- BARO = MAP + vacuum
- Vacuum = BARO − MAP
- MAP = BARO − vacuum

Figure 22-53 illustrates these relationships.

Supercharger or turbocharger boost operation also affects manifold pressure. When a supercharger or turbocharger is providing boost pressure, manifold absolute pressure rises *above* atmospheric pressure.

Depending on the control system and sensors used on an engine, the computer uses a combination of the barometric pressure, MAP, or vacuum sensor input signals to control fuel metering and spark advance.

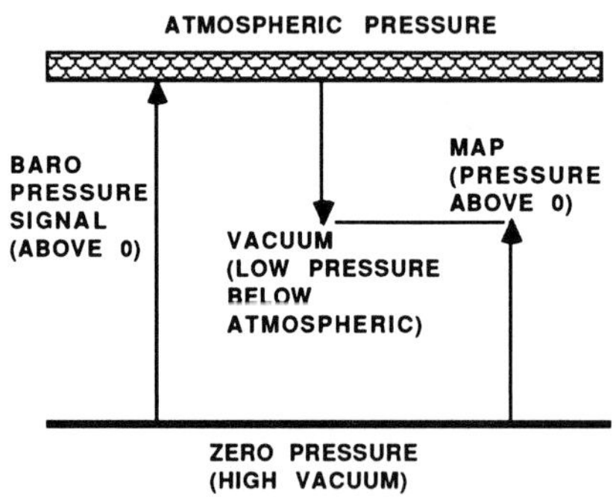


**Figure 22-53.** The sum of vacuum and MAP readings always equals barometric pressure.

Most pressure sensors are mounted remotely from the engine and connected to the intake manifold by a length of tubing. Some systems have separate MAP and BARO sensors. Others take a barometric pressure reading from the MAP sensor when the ignition is turned on, just before cranking the engine. At that time, manifold pressure equals ambient barometric pressure. Some systems check and update the barometric reading with the engine at wide-open throttle when manifold pressure again equals barometric pressure. Other systems, such as many Chrysler installations, use a solenoid to switch the MAP sensor back and forth between manifold pressure and barometric pressure.

MAP sensors are the basis for speed density fuel injection systems. They originated on Bosch D-Jetronic systems in the late 1960's and are used on most TBI and many port injection systems today.

**Mass Airflow (MAF) Sensors.** MAF sensors measure the molecular mass, or weight, of the intake air. They provide the most accurate measurement of intake air in relation to fuel metering, because air-fuel ratios are calculated by weight.

In most MAF sensors, a heated wire or conductive film sensor receives current from the computer. Current heats the sensor; airflow cools it. The computer tries to maintain the sensor at a constant temperature. As airflow increases, the current requirement increases. The MAF sensor sends a variable voltage or variable frequency signal to the computer that is proportional to mass airflow.

Some Mitsubishi and Chrysler engines use a unique Karman vortex airflow sensor that generates turbulence in the intake airflow and uses ultrasonic sound waves to measure mass airflow.

## Fuel Injection Auxiliary Systems

In our examination of fuel injection, you haven't seen any chokes, power valves, accelerator pumps, or idle speed and mixture controls. Remember that a fuel injection system must provide enrichment and mixture control for the same engine conditions that a carburetor does. Fuel injection does it with a combination of engine sensors and auxiliary metering devices. Enrichment is provided by varying the pulse width of an electronic injector or the pressure on a mechanical injector. Cold starting is aided by coolant and air temperature sensors that tell the system to enrich the mixture. Injection pulse timing in an EFI system is governed by engine speed and crankshaft position sensors. The following sections summarize some of the common engine sensors and auxiliary metering systems.

**Cold Starting.** Extra air is provided by an auxiliary air bypass that works like an air bypass circuit in a carburetor. A valve opens an air passage in the intake system to allow more air into the engine, figure 22-54. The opening is controlled by a thermostatic switch or a timer, or both.

Bosch K-, D-, and L-Jetronic systems and similar port injection systems have a start valve, or cold-start injector, to add fuel to the system for starting, figure 22-55. The cold-start injector is a separate electronic injector that sprays fuel into a central area of the manifold when engine temperature, speed, and airflow are low. A thermo-time switch senses coolant temperature to close the injector. Current applied to the switch heating coil opens the circuit after a prescribed amount of time.

**Idle Enrichment.** Idle enrichment in an electronically controlled fuel injection system results from changing the pulse width of an electronic injector or the pressure on a mechanical injector. However, the input signals from the MAP or MAF, BARO, throttle position, engine speed, and vehicle speed sensors are required to allow the computer to make the proper decisions.

**Idle Speed Control.** Fuel injection idle speed is usually controlled by one of two devices. An idle air control (IAC) valve, controlled by the computer, has a movable pintle that opens

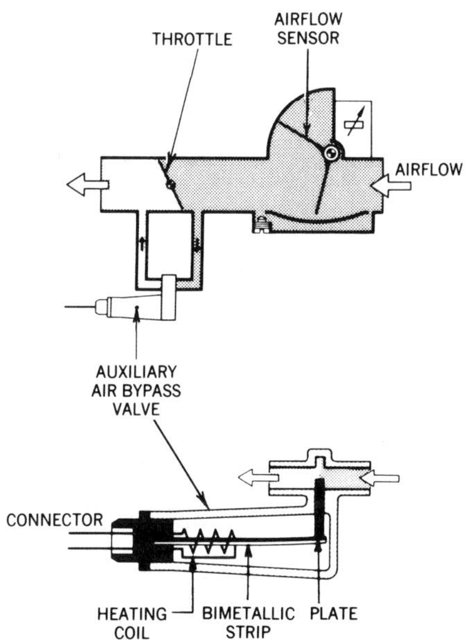


**Figure 22-54.** Most port injection systems have air bypass valves to aid cold starting. (Bosch)

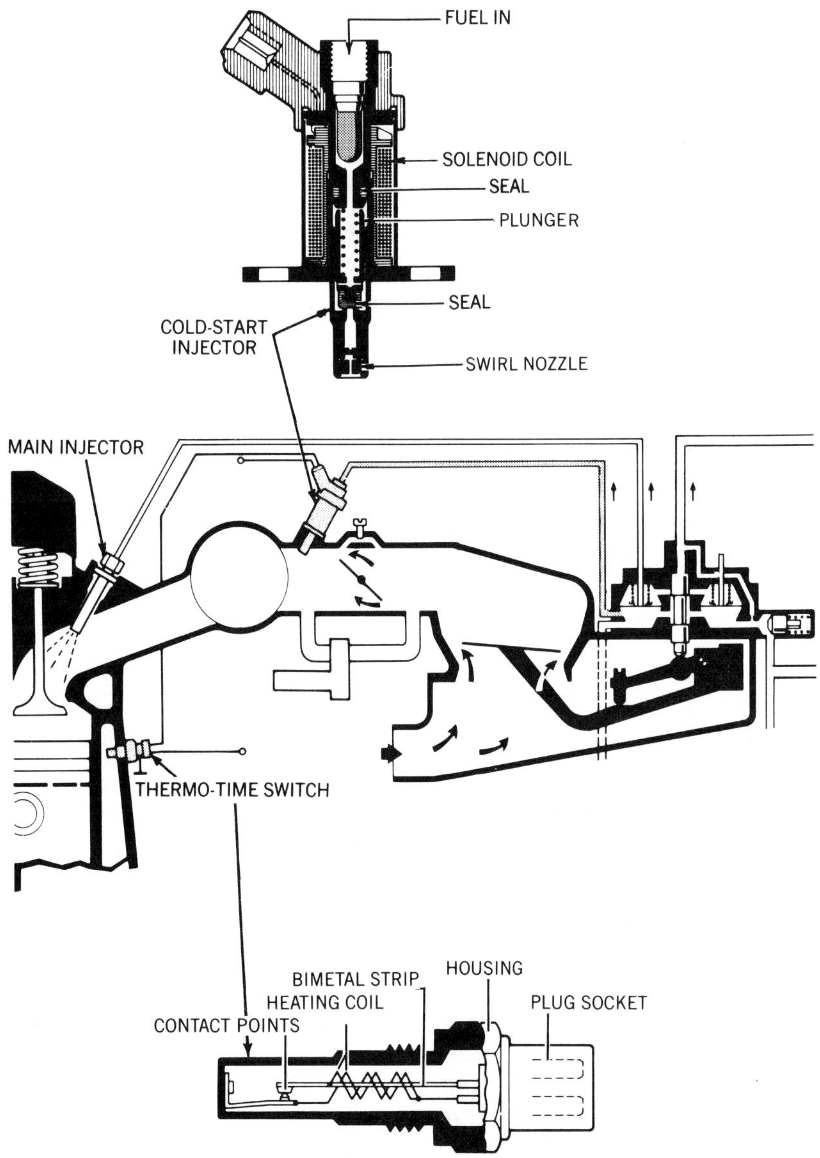


**Figure 22-55.** Bosch and several other port injection systems have a cold-start valve, or auxiliary injector, to deliver fuel to the manifold of a cold engine. (Volkswagen)

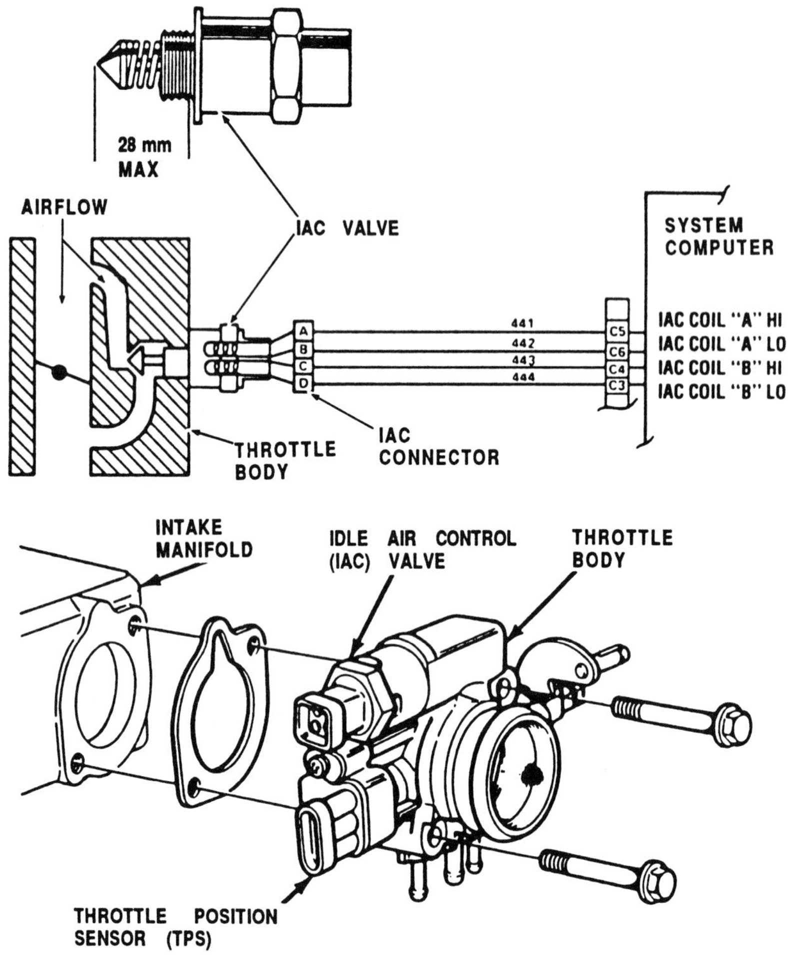


**Figure 22-56.** Typical idle air control (IAC) valve installation.

and closes a bypass passage in the throttle body, figure 22-56. Most port injection systems use this kind of idle speed control device, which works similarly to the cold-start air bypass shown in figure 22-54.

Some TBI systems also use an IAC valve. Others use an idle speed control (ISC) motor, figure 22-57. The ISC motor operates against the linkage on the throttle body to control idle speed. Some ISC motors contain an idle-stop, or idle tracking, switch that closes to signal the computer when the throttle is in the idle position.

The motors of both IAC valves and ISC motors are reversible dc motors. The computer reverses current through the motor windings or alternately energizes different sets of windings to move the motor in both directions.

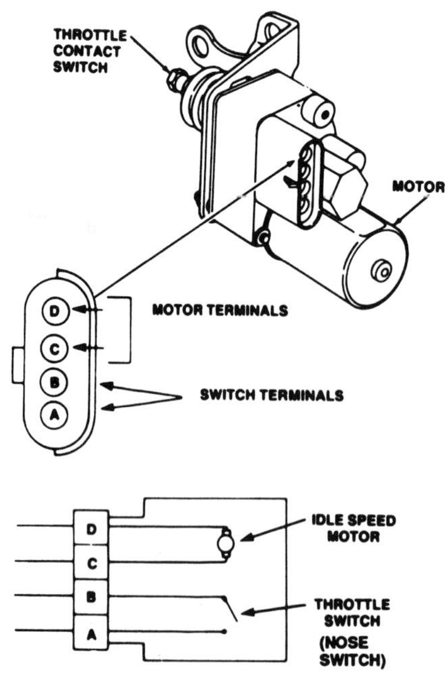


**Figure 22-57.** Typical idle speed control (ISC) motor installation.

**Acceleration and Load Enrichment.** When an engine needs extra fuel for extra power, airflow and manifold pressure change suddenly. The airflow or pressure sensors signal the control module of an EFI system to increase the injector pulse width. Thus, enrichment is provided by the same systems that control light-load metering. In a mechanical system, the same airflow and pressure changes affect fuel pressure control on the metering assembly to increase fuel volume.

## CARBURETOR FUEL CONTROL ACTUATORS

Although fuel injection has become the conventional fuel system of the 1990's, many engine control systems built in the 1980's used carburetors with computer-controlled actuators. The actuators are usually stepper motors of solenoids.

A fuel control solenoid or stepper motor normally operates so that the midpoint of its range or the 50-percent duty cycle gives a 14.7 : 1 ratio during closed loop. Most such actuators allow a ±2 air-fuel ratio range. That is, the system provides ratios from 12.7 : 1 to 16.7 : 1. If the system is not working correctly, however, the computer may drive the actuator to one side or the other of the midpoint to maintain a 14.7 : 1 closed-loop ratio. The system then may not provide the proper rich or lean mixtures for other conditions.

### Stepper Motors

Stepper motor fuel control is used in some Motorcraft and Carter carburetors. The motor moves metering pins in main circuit air bleeds, figure 22-58, to control the air-fuel ratio. Moving the pins into the bleeds enriches the mixture (less air); moving them out leans the mixture (more air). The stepper motor in a Ford 7200 VV carburetor moves a valve to control the air-fuel ratio in one of two ways:

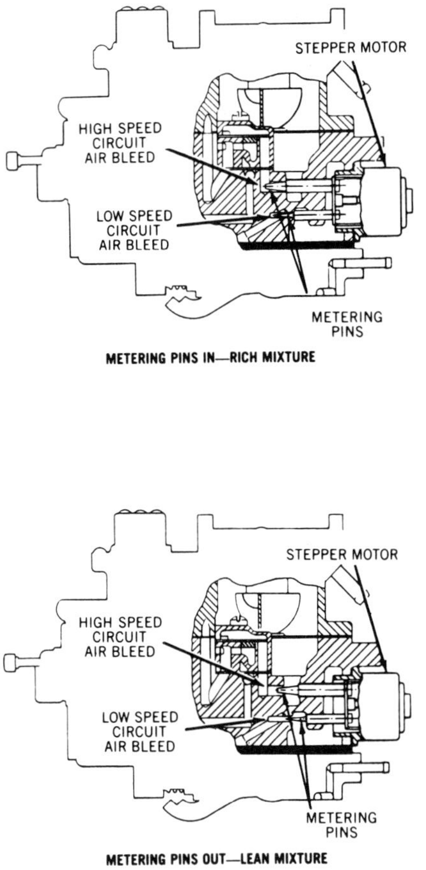


**Figure 22-58.** This stepper motor controls air bleeds in the high-speed and low-speed circuits to vary the air-fuel ratio. (AMC)

1. In early models, the valve opens a passage to allow control vacuum to enter the fuel bowl, figure 22-59. This lowers the air pressure on the fuel and creates a leaner air-fuel ratio.
2. In later models, the valve opens air bleeds that allow additional air into the air-fuel mixture at the discharge jets.

## Mixture Control Solenoids

Mixture control solenoids can control air-fuel ratios in several ways. All, however, work on the variable duty cycle principle. Some solenoids control vacuum to diaphragms in a carburetor, figure 22-60. The vacuum diaphragms move metering rods in and out of carburetor jets to regulate fuel flow, and they open and close air bleeds for idle and main-metering circuits. Some Carter carburetors have

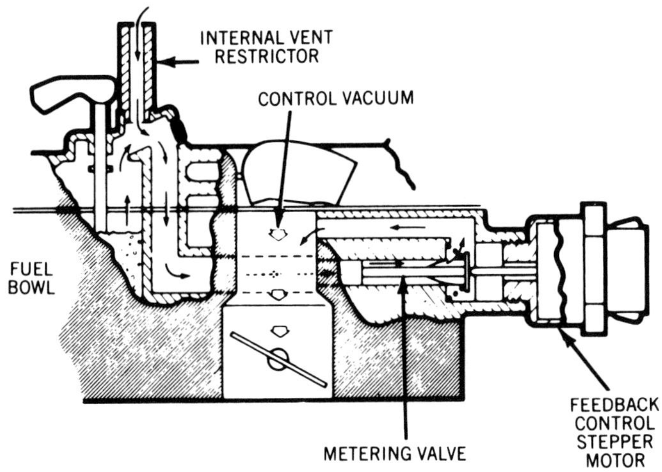


**Figure 22-59.** The stepper motor in Ford's early 7200 VV feedback carburetor varies air pressure on the gasoline in the fuel bowl. (Ford)

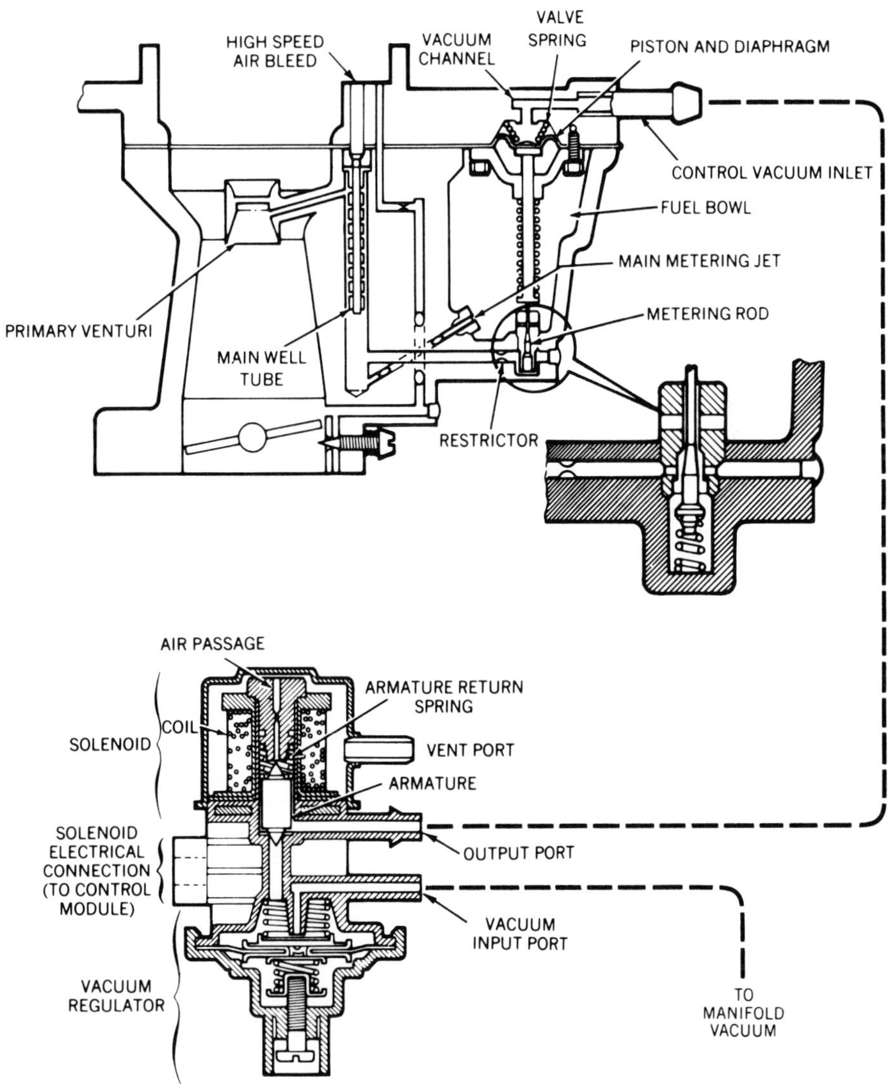


**Figure 22-60.** This feedback solenoid controls vacuum on carburetor diaphragms that move metering rods and open or close air bleeds. (GM)

solenoids that work on a variable duty cycle to open and close air bleeds in the carburetor, figure 22-61.

Rochester carburetors used in GM C-4 and CCC systems have mixture solenoids that regulate fuel flow directly, rather than by regulating a vacuum signal. Mounted in the carburetor, the mixture solenoid controls a metering rod in the main jet, figure 22-62. It also controls a rod that opens and closes an idle air bleed, figure 22-63.

## AIR-FUEL RATIO, TIMING, AND EGR EFFECTS ON ENGINE OPERATION

You have learned how air-fuel ratios, timing, and EGR individually affect overall engine operation. By reviewing these effects, you can see how computer control can integrate the operation of all these factors and obtain the best engine efficiency.

### Air-Fuel Ratio

Figure 22-64 shows the effects of a variable air-fuel ratio with fixed timing and engine speed. With air-fuel ratios richer than 14.7:1, fuel consumption and HC and CO emissions are high, but $NO_x$ emissions are relatively low. Torque is highest at ratios of 12:1 to 16:1. Above 16:1, torque starts to drop and HC and $NO_x$ increase. HC emissions and fuel consumption increase with lean mixtures because the engine can misfire and pass fuel through the cylinders unburned. This also reduces power and torque.

**The Stoichiometric Air-Fuel Ratio.** The stoichiometric air-fuel ratio of 14.7:1 provides the most complete combination of fuel and air during combustion. It also provides the best combination of performance, economy, and emission control. It is somewhat of a compromise between maximum power and maximum economy, but both of these factors are good at 14.7:1. For successful emission control, the stoichiometric ratio depends on the use of a 3-way, oxidation-reduction catalyst.

Figure 22-65 shows that a 3-way catalyst has the best combined conversion efficiency for all three pollutants at a 14.7:1 air-fuel ratio. With richer mixtures, HC and CO conversion efficiency is lower. With leaner mixtures, $NO_x$ conversion efficiency is lower. The range of conversion efficiency is quite narrow, only about ±0.05 air-fuel ratio, (that is, 14.65:1 to 14.75:1). A carburetor or injection system without feedback control cannot maintain this narrow range, but a computer-controlled system can.

### Ignition Timing

Figure 22-66 shows how spark timing affects fuel consumption and HC and $NO_x$ emissions with fixed engine speed and air-fuel ratio. With timing near tdc, or slightly retarded, emissions are low, but fuel consumption is

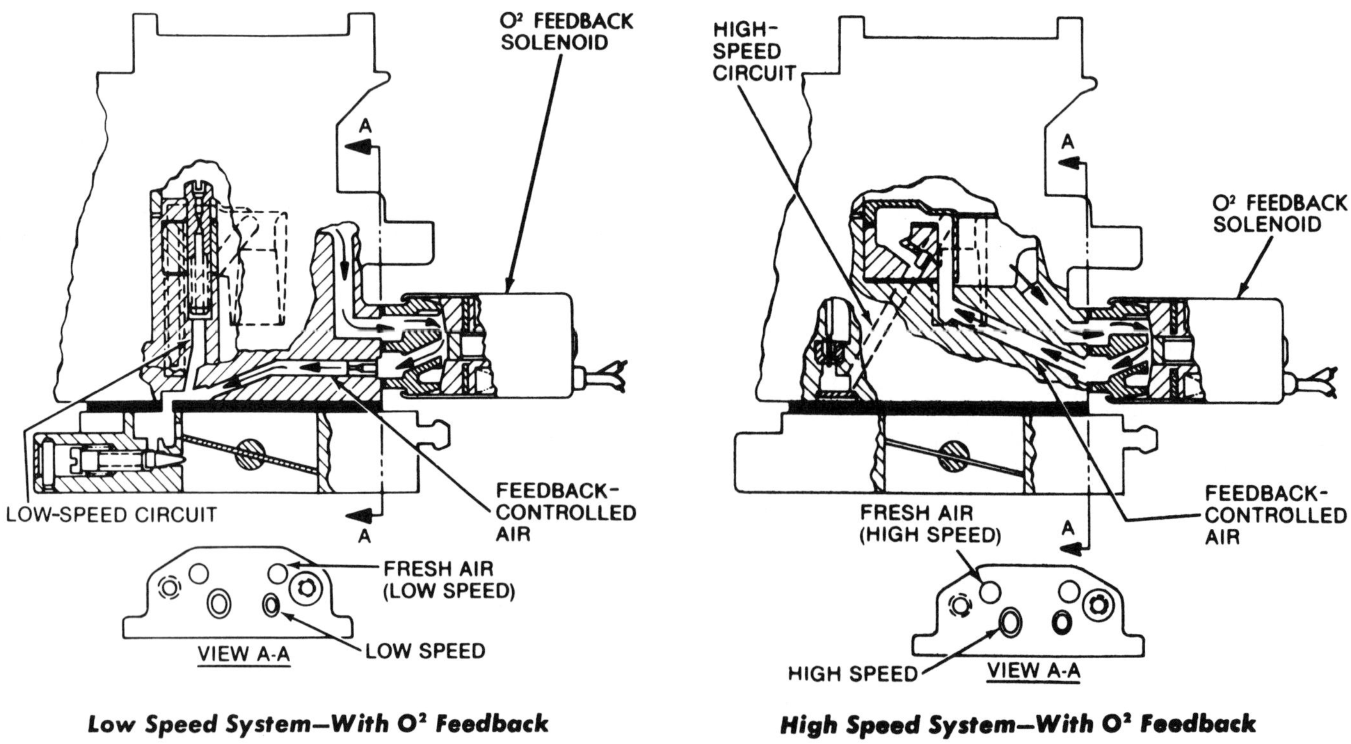


**Figure 22-61.** This solenoid opens and closes high- and low-speed air bleeds with a variable duty cycle. (Chrysler)

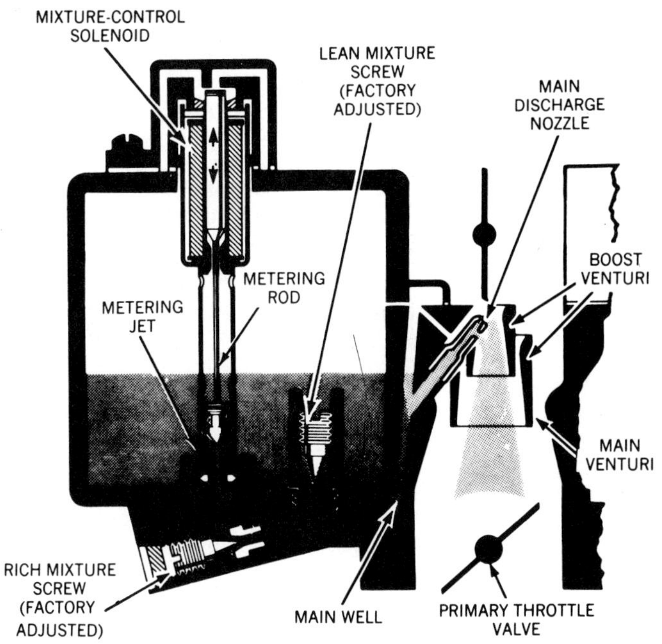


**Figure 22-62.** This mixture solenoid in a Rochester carburetor controls metering rod position. (GM)

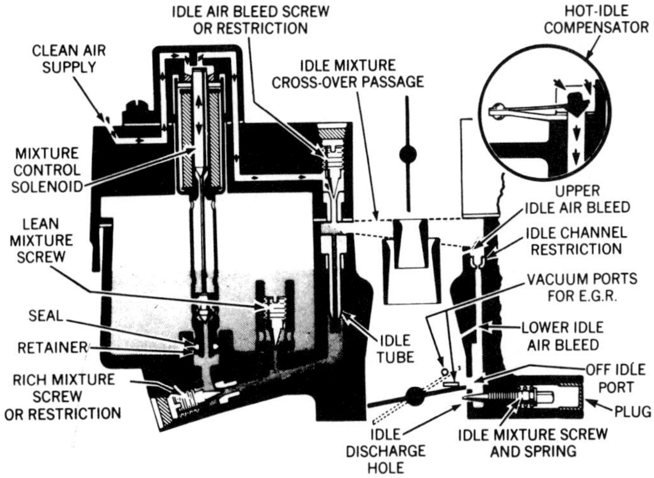


**Figure 22-63.** The same mixture solenoid shown in figure 22-62 also controls an idle air bleed passage. (GM)

high. As timing is advanced, fuel consumption decreases, but emissions increase. During engine mapping, engineers calculate the best timing for any combination of speed and air-fuel ratio. This is generally the point where the HC emission and fuel consumption curves cross, which also corresponds to peak torque.

### Exhaust Gas Recirculation

Figure 22-67 shows that $NO_x$ emissions decrease sharply as EGR is added to the air-fuel mixture. But, as EGR increases, so do HC emissions and fuel consumption. Again, during engine mapping, engineers calculate the EGR percentage that yields the best compromise between $NO_x$ control, HC emissions, and economy. Remember that both EGR and ignition timing are important variables for detonation control on late-model engines, so engineers also consider detonation control when calculating EGR percentage.

### Computer Control

Centrifugal and vacuum advance devices control spark timing in relation

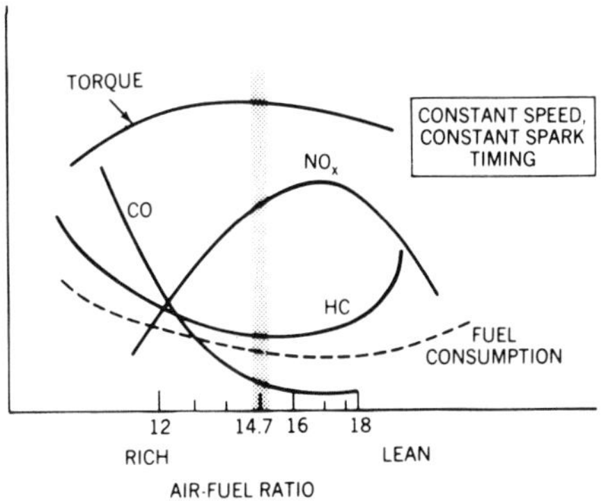


**Figure 22-64.** Typical fuel consumption, emission, and torque curves with changes in air-fuel ratio.

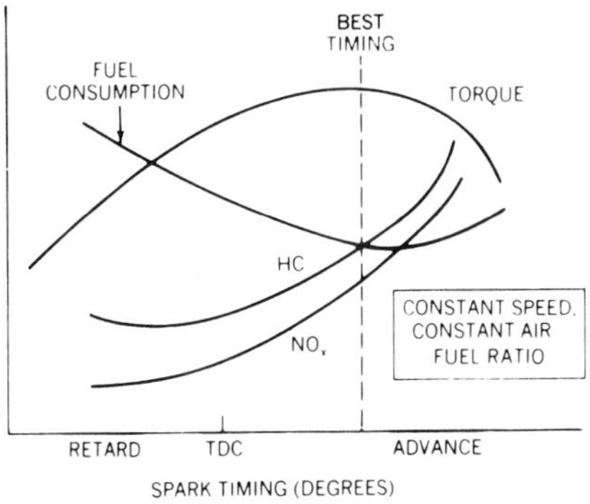


**Figure 22-66.** Typical fuel consumption, emission, and torque curves with changes in spark timing.

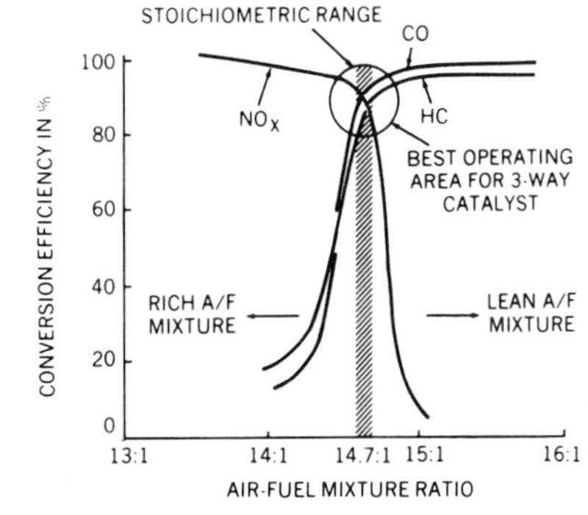


**Figure 22-65.** Three-way catalyst efficiency is at its best with a 14.7:1 air-fuel ratio. (GM)

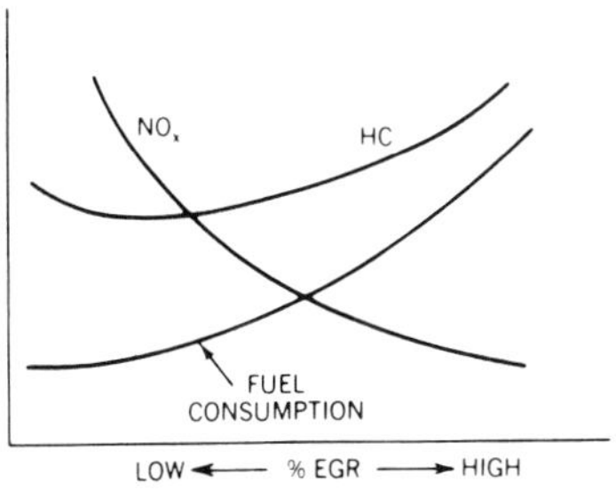


**Figure 22-67.** As EGR increases, $NO_x$ decreases; but HC and fuel consumption increase.

to engine speed and load. Airflow through a carburetor controls fuel metering similarly, and manifold vacuum can provide basic control for EGR. All of these simple controls operate more or less independently, however. A computer doesn't do anything to engine variables that mechanical controls don't do. But, a computer regulates them faster and more precisely, and can "compute" the combined effects of changing several variables simultaneously.

## SUMMARY

Electronic engine control systems began as single-function systems by metering fuel or controlling ignition timing. They were expanded to control a variety of engine functions, such as EGR, air injection switching, vapor canister purging, idle speed, torque converter lockup, and transmission shift points. Such fully integrated control systems act as engine management systems.

Gasoline fuel injection is a proven technology that established itself in the 1970's and 1980's. It has become the standard fuel system of the 1990's as part of fully integrated control systems. Gasoline injection systems use pressure-operated or solenoid-actuated nozzles to spray fuel into the intake air charge. Injection systems meter fuel in relation to intake air mass, just as carburetors do; but air sensors are needed to measure the incoming air. Gasoline injection systems can be classified as port or throttle body systems, electronic or mechanical systems, and continuous or intermittent systems.

### Sensor Summary

All engine control systems measure most, or all, of the following engine and vehicle operations. The following list summarizes typical sensor functions.

*Exhaust gas oxygen*—The exhaust oxygen sensor is usually a voltage-generating galvanic battery that produces a voltage from 0 to 1 volt (0 to 1000 millivolts) depending on the ratio of oxygen in the exhaust compared to oxygen in the outside air. (A few carmakers use a titanium thermistor-type exhaust sensor that modifies a reference voltage rather than generating its own voltage.)

*Engine coolant temperature*—Coolant temperature can be measured by two kinds of sensors:

- A simple bimetal switch for a high or low temperature indication
- A thermistor sensor for an analog measurement of changing temperature in degrees

*Intake air temperature or ambient air temperature*—A thermistor sensor normally is used for an analog measurement of changing temperature in degrees.

*Engine speed (rpm)*—The computer uses a tachometer signal produced by any of the following:

- A distributor pickup coil
- A Hall-effect switch or an optical sensor in a distributor
- A pickup coil or Hall-effect switch used as a crankshaft position sensor.

*Crankshaft position*—This is usually the same sensor that indicates engine speed.

*Camshaft position*—Particularly for sequential injection systems, the computer uses a cylinder identification signal to synchronize fuel injection and ignition timing. Usually:

- A Hall-effect switch or an optical sensor in a distributor, or
- A pickup coil or Hall-effect switch in the engine block or on the crankshaft damper.

*Throttle position*—A potentiometer sends an analog variable voltage signal that is proportional to throttle position.

*Manifold vacuum or manifold absolute pressure (MAP)*—A variable-resistance or variable-capacitance pressure sensor sends an analog voltage or frequency signal that is proportional to air pressure (or vacuum) in the intake manifold.

*Barometric pressure (ambient air pressure)*—A variable-resistance or variable-capacitance pressure sensor sends an analog voltage or frequency signal that is proportional to barometric (atmospheric) air pressure.

*Intake mass airflow (MAF)*—A heated wire or heated film sensor sends an analog voltage or frequency signal that is proportional to the weight (molecular mass) of air entering the engine.

*Detonation (spark knock or ping)*—A piezoelectric or piezoresistive sensor acts as a microphone to send an analog voltage when detonation occurs.

*EGR valve position*—A potentiometer sends an analog variable voltage signal that is proportional to EGR valve stem position.

*Battery or charging system voltage*—The computer takes a direct analog voltage reading from the charging system or a power relay.

*Air conditioning controls*—The computer receives on-off (high-low) switch signals from the A/C controls.

*Power steering pressure switch*—The computer receives an on-off (high-low) switch signal from a high-pressure switch in the power steering system.

*Brake switch*—The computer receives an on-off (high-low) switch signal from a normally closed brake switch used for cruise control and torque converter lockup operation.

*Park-neutral selection for automatic transmissions*—The computer receives an on-off (high-low) signal from the park-neutral switch.

*Transmission gear position switches*—The computer receives on-off (high-low) switch signals from automatic or manual transmission switches

*Vehicle speed sensor (VSS)*—An analog speed signal is produced by any of the following:

• A pickup coil (permanent magnet) sensor

• An optical sensor in the speedometer head

• A magnetic reed switch

*Engine running time or time since startup*—This is calculated by the engine computer.

Some systems may not regulate all of the operations listed previously or may not get input signals from *all* of the sensors listed above.

## REVIEW QUESTIONS

### Multiple Choice

**1.** The first electronic engine control system used by a domestic carmaker was the:

**a.** GM Oldsmobilė MISAR system
**b.** Ford EEC I system
**c.** Chrysler electronic lean-burn system
**d.** AMC Motronic system

**2.** The electronic engine control system used on most late-model GM cars is called:

**a.** electronic spark and fuel control (ESFC)
**b.** computer-controlled catalytic converter (C-4)
**c.** computerized emission control (CEC)
**d.** computer command control (CCC or C-3)

**3.** Mechanic A says that an engine computer receives input information, processes data, stores data, and sends output information. Mechanic B says that most late-model engine computers are based on digital microprocessors. Who is right?

**a.** A only
**b.** B only
**c.** both A and B
**d.** neither A nor B

**4.** Most late-model engine control systems control all of the following EXCEPT:

**a.** spark timing
**b.** air-fuel ratio
**c.** EGR
**d.** intake airflow

**5.** The simplest digital sensor is a:

**a.** timer
**b.** switch
**c.** transducer
**d.** solenoid

**6.** Most automobile engine sensors are:

**a.** solenoids
**b.** transformers
**c.** magnetic-pulse generators
**d.** resistors

**7.** Mechanic A says that a resistive sensor must receive a steady reference voltage. Mechanic B says that the output (return) voltage of a resistive sensor changes as resistance changes. Who is right?

**a.** A only
**b.** B only
**c.** both A and B
**d.** neither A nor B

**8.** Mechanic A says that an exhaust gas oxygen (EGO) sensor is a thermistor. Mechanic B says that an EGO sensor measures air-fuel ratio. Who is right?

**a.** A only
**b.** B only
**c.** both A and B
**d.** neither A nor B

**9.** A negative temperature coefficient (NTC) sensor is a:

**a.** transducer
**b.** thermistor
**c.** potentiometer
**d.** all of the above

**10.** A sensor that produces a variable output voltage in relation to force or pressure is a:

**a.** potentiometer
**b.** thermistor
**c.** piezoresistive sensor
**d.** none of the above

**11.** Mechanic A says that a reluctance sensor can indicate engine temperature. Mechanic B says that a reluctance sensor can indicate engine speed. Who is right?

**a.** A only
**b.** B only
**c.** both A and B
**d.** neither A nor B

**12.** The EGO sensor is a:

**a.** thermistor
**b.** potentiometer
**c.** bimetal sensor
**d.** galvanic battery

**13.** Mechanic A says that heated EGO sensors are often used on turbocharged engines. Mechanic B says that heated EGO sensors eliminate the need for air switching. Who is right?

**a.** A only
**b.** B only
**c.** both A and B
**d.** neither A nor B

**14.** Mechanic A says that airflow sensors are analog sensors. Mechanic B says that manifold absolute pressure (MAP) sensors are analog devices. Who is right?

**a.** A only
**b.** B only
**c.** both A and B
**d.** neither A nor B

**15.** Coolant and air temperature sensors do the jobs of a:

**a.** carburetor choke coil
**b.** carburetor power circuit
**c.** vacuum advance diaphragm
**d.** centrifugal advance weights

**16.** Mechanic A says that throttle position sensors are both load and speed sensors. Mechanic B says that a throttle position sensor can be a simple on-off switch. Who is right?

**a.** A only
**b.** B only
**c.** both A and B
**d.** neither A nor B

**17.** Most engine control system actuators are:

**a.** transducers
**b.** solenoids
**c.** potentiometers
**d.** thermistors

**18.** All of the following are kinds of gasoline fuel injection system EXCEPT:

**a.** port injection
**b.** precombustion chamber injection
**c.** throttle body injection
**d.** continuous injection

**19.** Mechanic A says that Bosch K-Jetronic systems are continuous injection systems. Mechanic B says that K-Jetronic systems are mechanical systems that work on differential fuel pressure. Who is right?

- **a.** A only
- **b.** B only
- **c.** both A and B
- **d.** neither A nor B

**20.** Bosch L-Jetronic and similar systems measure airflow with a:

- **a.** manifold absolute pressure sensor
- **b.** hot-wire air mass sensor
- **c.** vane-type airflow sensor
- **d.** barometric pressure sensor

**21.** Bosch LH-Jetronic and similar systems measure airflow with a:

- **a.** manifold absolute pressure sensor
- **b.** hot-wire air mass sensor
- **c.** vane-type airflow sensor
- **d.** barometric pressure sensor

**22.** Mechanic A says that most throttle body injection (TBI) systems measure airflow and air volume with manifold absolute pressure (MAP) or barometric pressure sensors, or both. Mechanic B says that most TBI systems also have hot-wire air mass sensors. Who is right?

- **a.** A only
- **b.** B only
- **c.** both A and B
- **d.** neither A nor B

**23.** Most port-type electronic fuel injection (EFI) systems are based on designs by:

- **a.** Rochester Products Division of GM
- **b.** Bendix Division of Allied Corporation
- **c.** Nippondenso
- **d.** Robert Bosch GmbH

**24.** The first modern port injection system on domestic cars was made by Bendix and used on:

- **a.** 1983 and later Chevrolet Corvettes
- **b.** 1980 and later Chrysler K-cars
- **c.** 1973 Ford Mustang SVO models
- **d.** 1976–79 Cadillac Sevilles

**25.** Mechanic A says that TBI systems are mechanical, continuous injection systems. Mechanic B says that TBI systems do not require fuel pressure regulators. Who is right?

- **a.** A only
- **b.** B only
- **c.** both A and B
- **d.** neither A nor B

**26.** Mechanic A says that the duty cycle of a mechanical (continuous) injection nozzle is controlled by an electronic module. Mechanic B says that the duty cycle of a solenoid-operated injector is the percentage of on-time to total cycle time. Who is right?

- **a.** A only
- **b.** B only
- **c.** both A and B
- **d.** neither A nor B

**27.** Electronic port injectors are usually energized:

- **a.** continuously
- **b.** individually in firing-order sequence
- **c.** in groups or all together
- **d.** at a fixed duty cycle

**28.** Mechanic A says that the throttle in a TBI system is between the injectors and the manifold. Mechanic B says that the throttle in an EFI system works the same as a throttle on a carbureted engine. Who is right?

- **a.** A only
- **b.** B only
- **c.** both A and B
- **d.** neither A nor B

**29.** All of the following are kinds of airflow sensors EXCEPT:

- **a.** a throttle position sensor
- **b.** a movable plate that operates a plunger in the fuel distributor
- **c.** a platinum hot-wire sensor
- **d.** a movable-vane potentiometer sensor

**30.** Manifold pressure sensors may be based on:

- **a.** an aneroid capsule or bellows
- **b.** a diaphragm
- **c.** a piezoresistive crystal
- **d.** all of the above

**31.** A hot-wire air mass sensor measures:

- **a.** airflow velocity entering the engine
- **b.** manifold absolute pressure
- **c.** molecular mass of air entering the engine
- **d.** all of the above

**32.** A cold-start injector is a:

- **a.** mechanical, continuous injector controlled by intake airflow
- **b.** bypass airflow injector
- **c.** coolant-controlled detonation regulator
- **d.** solenoid-operated auxiliary injector

**33.** Mechanic A says that a throttle position sensor is a variable potentiometer. Mechanic B says that some EFI systems use an on-off throttle position switch. Who is right?

- **a.** A only
- **b.** B only
- **c.** both A and B
- **d.** neither A nor B

## Fill in the Blank

**34.** In a solenoid-operated electronic fuel injection system, the amount of fuel injected is controlled by injector ________, or duty cycle.

**35.** Fuel metering through a carburetor is controlled directly by the difference between air pressure in the ________ and air pressure in the carburetor barrel.

**36.** The throttle in a fuel injection system regulates the amount of ________ allowed into the engine.

**37.** Manifold absolute pressure is a combination of atmospheric pressure and ________.

**38.** An EFI system must know engine speed and ________ position to time injection pulses and calculate pulse width.

**39.** A thermistor is a variable resistor whose resistance ________ as it is heated.

**40.** A digital computer processes information in binary digits called ________.

**41.** The computer's converter circuits change ________ input signals into digital information.

**42.** An input or output signal that is either on-off or high-low is a ________ signal.

**43.** The two basic operating modes for all late-model engine control systems are ________ and ________.

**44.** At the stoichiometric air-fuel ratio of 14.7 : 1, HC, CO, and $NO_x$ emissions are ________.

**45.** An exhaust gas oxygen (EGO) sensor measures ________.

**46.** The ratio of excess air to stoichiometric air quantity is called the ________ ratio.

**47.** Air and coolant temperature sensors often are a type of resistor that is called a ________ .

**48.** To time injection pulses and calculate injector pulse width, an EFI system must have sensors for engine ________ and crankshaft ________ .

**49.** In most port injection systems, idle air flow passes through an ________ passage.

**50.** As vacuum decreases, manifold absolute pressure ________ .

**51.** Bosch mechanical injection systems work on a difference between fuel pump pressure and a control pressure developed by the ________ .

# APPENDIX METRICS*

*Reprinted through the courtesy of Ford Motor Company

## INTRODUCTION

The specifications for threaded fasteners define mechanical properties such as tensile strength, yield strength, proof load, and hardness. For vehicle safety, replacement fasteners must have the same specifications as original-equipment fasteners.

Most customary inch-size and metric fasteners have markings that indicate the strength of the fastener. The illustrations on the following pages describe and explain these markings.

The following pages also contain tables of English and metric equivalencies and conversion factors that are useful in the automotive service profession.

## NOMENCLATURE FOR BOLTS

### (ENGLISH) INCH SYSTEM
**Bolt, 1/2-13 × 1**

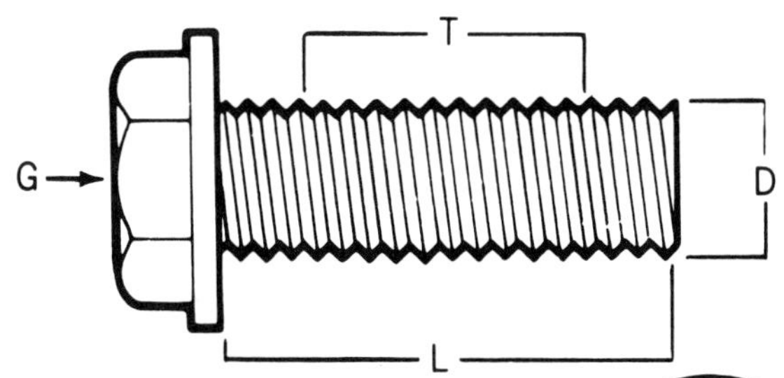


G—GRADE MARKING (BOLT STRENGTH)
L—LENGTH, (INCHES)**
T—THREAD PITCH (THREAD/INCH)
D—NOMINAL DIAMETER (INCHES)

### METRIC SYSTEM
**Bolt M12-1.75 × 25**

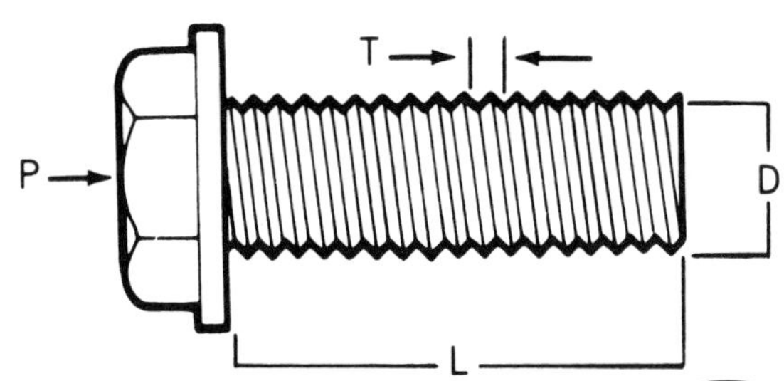


P—PROPERTY CLASS* (BOLT STRENGTH)
L—LENGTH (MILLIMETERS)**
T—THREAD PITCH (THREAD WIDTH CREST TO CREST MM)
D—NOMINAL DIAMETER (MILLIMETERS)

*THE PROPERTY CLASS IS AN ARABIC NUMERAL DISTINGUISHABLE FROM THE SLASH SAE ENGLISH GRADE SYSTEM.
**THE LENGTH OF ALL BOLTS IS MEASURED FROM THE UNDERSIDE OF THE HEAD TO THE END.

# BOLT STRENGTH IDENTIFICATION

## (ENGLISH) INCH SYSTEM

English (Inch) bolts—Identification marks correspond to bolt strength—increasing number of slashes represent increasing strength.

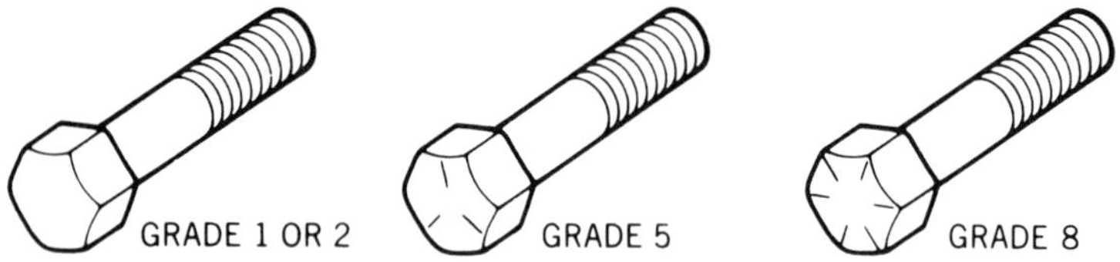


## METRIC SYSTEM

Metric bolts—Identification class numbers correspond to bolt strength—increasing numbers represent increasing strength. Common metric fastener bolt strength properties are 9.8 and 10.9 with the class identification embossed on the bolt head.

# HEX NUT STRENGTH IDENTIFICATION

## (ENGLISH) INCH SYSTEM

| GRADE | HEX NUT GRADE 5 | HEX NUT GRADE 8 |
|---|---|---|
| IDENTIFICATION |  3 DOTS |  6 DOTS |

INCREASING DOTS REPRESENT INCREASING STRENGTH.

## METRIC SYSTEM

| CLASS | HEX NUT PROPERTY CLASS 9 | HEX NUT PROPERTY CLASS 10 |
|---|---|---|
| IDENTIFICATION |  ARABIC 9 |  ARABIC 10 |

MAY ALSO HAVE BLUE FINISH OR PAINT DAUB ON HEX FLAT. INCREASING NUMBERS REPRESENT INCREASING STRENGTH.

# OTHER TYPES OF PARTS

Metric identification schemes vary by type of part, most often a variation of that used of bolts and nuts. Note that many types of English and metric fasteners carry no special identification if they are otherwise unique.

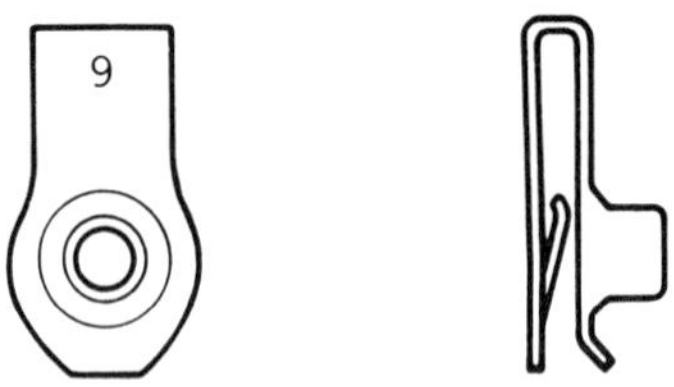


—Stamped U-Nuts

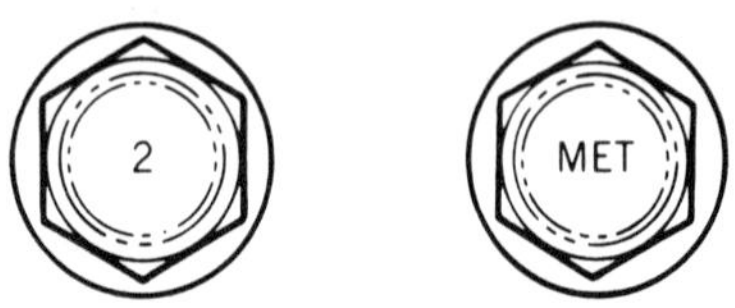


—Tapping, thread forming and certain other case hardened screws

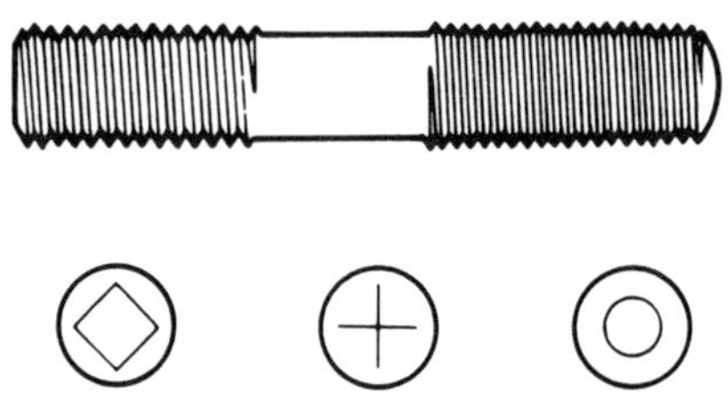


—Studs, Large studs may carry the property class number. Smaller studs use a geometric code on the end.

# ENGLISH METRIC CONVERSION

| DESCRIPTION | MULTIPLY | BY | FOR METRIC EQUIVALENT |
|---|---|---|---|
| ACCELERATION | FOOT/SEC$^2$ | 0.304 8 | METER SEC$^2$ (M/S$^2$) |
| | INCH/SEC$^2$ | 0.025 4 | METER SEC$^2$ |
| TORQUE | INCH-POUND | 0.112 98 | NEWTON-METERS (N·M) |
| | FOOT-POUND | 1.355 8 | NEWTON-METERS |
| POWER | HORSEPOWER | 0.746 | KILOWATTS (KW) |
| PRESSURE OR STRESS | INCHES OF WATER | 0.2488 | KILOPASCALS (KPA) |
| | POUNDS/SQ.IN. (PSI) | 6.895 | KILOPASCALS (KPA) |
| ENERGY OR WORK | BTU | 1.055. | JOULES (J) |
| | FOOT-POUND | 1.355 8 | JOULES (J) |
| | KILOWATT-HOUR | 3 600 000. OR $3.6 \times 10^6$ | JOULES (J = ONE W'S) |
| LIGHT | FOOT CANDLE | 10.76 | LUMENS/METER$^2$ (LM/M$^2$) |
| FUEL PERFORMANCE | MILES/GAL | 0.425 1 | KILOMETERS/LITER (KM/L) |
| | GAL/MILE | 2.352 7 | LITERS/KILOMETER (L/KM) |
| VELOCITY | MILES/HOUR | 1.609 3 | KILOMETERS/HR. (KM/H) |
| LENGTH | INCH | 25.4 | MILLIMETERS (MM) |
| | FOOT | 0.304 8 | METERS (M) |
| | YARD | 0.914 4 | METER (M) |
| | MILE | 1.609 | KILOMETERS (KM) |
| AREA | INCH$^2$ | 645.2 | MILLIMETERS$^2$ (MM$^2$) |
| | | 6.45 | CENTIMETERS (CM$^2$) |
| | FOOT$^2$ | 0.092 9 | METERS$^2$ (M$^2$) |
| | YARD$^2$ | 0.836 1 | METERS$^2$ |
| VOLUME | INCH$^3$ | 16 387. | MM$^3$ |
| | INCH$^3$ | 16.387 | CM$^3$ |
| | QUART | 0.016 4 | LITERS (1) |
| | QUART | 0.946 4 | LITERS |
| | GALLON | 3.785 4 | LITERS |
| | YARD$^3$ | 0.764 6 | METERS$^3$ (M$^3$) |
| MASS | POUND | 0.453 6 | KILOGRAMS (KG) |
| | TON | 907.18 | KILOGRAMS (KG) |
| | TON | 0.90718 | TONNE |
| FORCE | KILOGRAM | 9.807 | NEWTONS (N) |
| | OUNCE | 0.278 0 | NEWTONS |
| | POUND | 4.448 | NEWTONS |
| TEMPERATURE | DEGREE FAHRENHEIT | 0.556 (°F −32) | DEGREE CELSIUS (°C) |

# DECIMAL AND METRIC EQUIVALENTS

| FRACTIONS | DECIMAL INCH | METRIC MM | FRACTIONS | DECIMAL INCH | METRIC MM |
|---|---|---|---|---|---|
| 1/64 | .015625 | .397 | 33/64 | .515625 | 13.097 |
| 1/32 | .03125 | .794 | 17/32 | .53125 | 13.494 |
| 3/64 | .046875 | 1.191 | 35/64 | .546875 | 13.891 |
| 1/16 | .0625 | 1.588 | 9/16 | .5625 | 14.288 |
| 5/64 | .078125 | 1.984 | 37/64 | .578125 | 14.684 |
| 3/32 | .09375 | 2.381 | 19/32 | .59375 | 15.081 |
| 7/64 | .109375 | 2.778 | 39/64 | .609375 | 15.478 |
| 1/8 | .125 | 3.175 | 5/8 | .625 | 15.875 |
| 9/64 | .140625 | 3.572 | 41/64 | .640625 | 16.272 |
| 5/32 | .15625 | 3.969 | 21/32 | .65625 | 16.669 |
| 11/64 | .171875 | 4.366 | 43/64 | .671875 | 17.066 |
| 3/16 | .1875 | 4.763 | 11/16 | .6875 | 17.463 |
| 13/64 | .203125 | 5.159 | 45/64 | .703125 | 17.859 |
| 7/32 | .21875 | 5.556 | 23/32 | .71875 | 18.256 |
| 15/64 | .234375 | 5.953 | 47/64 | .734375 | 18.653 |
| 1/4 | .250 | 6.35 | 3/4 | .750 | 19.05 |
| 17/64 | .265625 | 6.747 | 49/64 | .765625 | 19.447 |
| 9/32 | .28125 | 7.144 | 25/32 | .78125 | 19.844 |
| 19/64 | .296875 | 7.54 | 51/64 | .796875 | 20.241 |
| 5/16 | .3125 | 7.938 | 13/16 | .8125 | 20.638 |
| 21/64 | .328125 | 8.334 | 53/64 | .828125 | 21.034 |
| 11/32 | .34375 | 8.731 | 27/32 | .84375 | 21.431 |
| 23/64 | .359375 | 9.128 | 55/64 | .859375 | 21.828 |
| 3/8 | .375 | 9.525 | 7/8 | .875 | 22.225 |
| 25/64 | .390625 | 9.922 | 57/64 | .890625 | 22.622 |
| 13/32 | .40625 | 10.319 | 29/32 | .90625 | 23.019 |
| 27/64 | .421875 | 10.716 | 59/64 | .921875 | 23.416 |
| 7/16 | .4375 | 11.113 | 15/16 | .9375 | 23.813 |
| 29/64 | .453125 | 11.509 | 61/64 | .953125 | 24.209 |
| 15/32 | .46875 | 11.906 | 31/32 | .96875 | 24.606 |
| 31/64 | .484375 | 12.303 | 63/64 | .984375 | 25.003 |
| 1/2 | .500 | 12.7 | 1 | 1.00 | 25.4 |

# TORQUE CONVERSION

| NEWTON METERS (N·M) | FOOT-POUNDS |
|---|---|
| 1 | 0.7376 |
| 2 | 1.5 |
| 3 | 2.2 |
| 4 | 3.0 |
| 5 | 3.7 |
| 6 | 4.4 |
| 7 | 5.2 |
| 8 | 5.9 |
| 9 | 6.6 |
| 10 | 7.4 |
| 15 | 11.1 |
| 20 | 14.8 |
| 25 | 18.4 |
| 30 | 22.1 |
| 35 | 25.8 |
| 40 | 29.5 |
| 50 | 36.9 |
| 60 | 44.3 |
| 70 | 51.6 |
| 80 | 59.0 |
| 90 | 66.4 |
| 100 | 73.8 |
| 110 | 81.1 |
| 120 | 88.5 |
| 130 | 95.9 |
| 140 | 103.3 |
| 150 | 110.6 |
| 160 | 118.0 |
| 170 | 125.4 |
| 180 | 132.8 |
| 190 | 140.1 |
| 200 | 147.5 |
| 225 | 166.0 |
| 250 | 184.4 |

| FOOT-POUNDS | NEWTON METERS (N·M) |
|---|---|
| 1 | 1.356 |
| 2 | 2.7 |
| 3 | 4.0 |
| 4 | 5.4 |
| 5 | 6.8 |
| 6 | 8.1 |
| 7 | 9.5 |
| 8 | 10.8 |
| 9 | 12.2 |
| 10 | 13.6 |
| 15 | 20.3 |
| 20 | 27.1 |
| 25 | 33.9 |
| 30 | 40.7 |
| 35 | 47.5 |
| 40 | 54.2 |
| 45 | 61.0 |
| 50 | 67.8 |
| 55 | 74.6 |
| 60 | 81.4 |
| 65 | 88.1 |
| 70 | 94.9 |
| 75 | 101.7 |
| 80 | 108.5 |
| 90 | 122.0 |
| 100 | 135.6 |
| 110 | 149.1 |
| 120 | 162.7 |
| 130 | 176.3 |
| 140 | 189.8 |
| 150 | 203.4 |
| 160 | 216.9 |
| 170 | 230.5 |
| 180 | 244.0 |

# GLOSSARY

**absolute zero.** The temperature at which molecular motion stops and all substances possess minimal electron energy. Equal to −273° C or −459.67° F.

**absorption muffler.** A kind of muffler that reduces noise by changing sound energy to friction energy through sound-absorbing material.

**ac ripple.** The small waves at the top of a rectified dc voltage oscilloscope pattern. They are the peaks of the ac voltage sine waves. Any dc voltage from a rectified ac source always has a slight ac ripple.

**actuator.** Any device that receives an output signal from a computer and does something in response to the signal.

**adaptive learning strategy.** A strategy that modifies the original computer program to compensate for changes in system parameters caused by various factors such as engine wear, driving habits, poor fuel quality, or component malfunctioning

**adsorption.** The collection of gas, vapor, or dissolved solid on the *surface* of a solid material.

**aeration.** The process of exposing liquid to air circulation and trapping air bubbles in the liquid. Mixing air or another gas with liquid.

**aerobic.** Curing in the presence of oxygen.

**after top dead center (atdc).** The position of a piston after it has passed top dead center of a stroke, measured in degrees of crankshaft rotation.

**air bleed.** A small opening that allows air to enter (bleed into) a fuel or vacuum passage in a carburetor.

**alternating current (ac).** Current flow that continually changes direction between two points through a conductor in a given amount of time.

**ambient air temperature.** Temperature of the air surrounding an area or device.

**American Wire Gauge (AWG).** A system for specifying wire size (conductor cross section area) by a series of gauge numbers.

**ampere (A).** The unit for measuring current flow. One ampere equals a current flow of $6.28 \times 10^{18}$ electrons per second.

**ampere-hour rating.** A battery current-capacity rating that indicates the amperes of current a battery can supply steadily for 20 hours without voltage dropping below 1.75 volts per cell. It also is called the "20-hour discharge rating."

**anaerobic.** Curing in the absence of oxygen.

**analog.** A signal that varies proportionally with the information that it measures. In a computer, an analog signal is voltage that fluctuates over a range from high to low.

**analogy.** A likeness. A similarity in some respects between two things that are otherwise dissimilar.

**aneroid.** An evacuated metal bellows containing a vacuum that changes in length as a response to changes in atmospheric pressure.

**anode.** The positively charged electrode in a voltage cell. The positive-polarity terminal of a component.

**antioxidant.** A kind of oil additive that reduces oil oxidation and the formation of varnish and carbon deposits.

**API service classifications.** The system for classifying motor oil resistance to engine wear, oxidation, corrosion, and dirt and carbon formation.

**area test.** A general test made to a complete vehicle system, or systems, to determine overall system operation.

**armature.** The movable part in a relay or the rotating part of a generator or motor consisting of a conductor wound around a laminated iron core.

**asynchronous.** Data transmission that incorporates a constant tone between sending and receiving. A start pulse precedes the transmission of new data, and a stop pulse tells the computer that the transmission has been completed.

**atmospheric pressure.** Pressure on the surface of the earth created by the weight of air in the atmosphere. Atmospheric pressure changes with altitude and temperature. At sea level it is 14.7 psi at 68° F (760 mm Hg at 20° C).

**atom.** The smallest part of a chemical element that has all the properties and characteristics of the element.

**atomize.** The action of breaking a liquid into a fine mist or small droplets.

**autoignition.** The spontaneous ignition of a fuel due to high temperature. The autoignition temperature of a fuel is the temperature at which the fuel will ignite by itself.

**available voltage.** The maximum high-voltage capability of an ignition coil. Also called open-circuit voltage, peak voltage, or no-load voltage.

**axis.** The rotation centerline of a wheel, a gear, or a shaft.

**backpressure.** Pressure that slows the flow of exhaust gas from an engine. Backpressure is caused by restrictions in the exhaust system. Also, any similar pressure caused by restrictions that slow gas or liquid flow in a system.

**balance tube.** The vent that opens the carburetor fuel bowl to atmospheric pressure at the airhorn and thus provides balanced pressure for fuel metering.

**base.** The center layer of a bipolar transistor, made of doped material opposite from the collector and the emitter.

**base unit.** One of the seven basic units that are the foundation of SI metric measurements.

**basic (initial) timing.** The ignition timing setting of an engine at idle speed.

**battery cell.** An assembly of a positive plate group, a negative plate group, separators, and connectors in a case with electrolyte. A complete chemically and electrically active unit of a battery.

**baud rate.** Computer data transmission speed in digital bits per second or characters per second.

**BCI group number.** A battery size identification number that indicates the length, width, height, terminal design and location, and other physical features of a battery.

**bearing web (saddle).** The area of an engine crankcase that holds a main bearing for the crankshaft.

**before top dead center (btdc).** The position of a piston before it reaches top dead center of a stroke, measured in degrees of crankshaft rotation.

**bias voltage.** Voltage applied across a semiconductor PN junction, such as the PN junction of a diode.

**bimetal.** Two strips of metal, each of which expands and contracts at a different rate when exposed to changes in temperature.

**bimetal element.** A strip made of different metals that expand at different rates when heated. Commonly used in circuit breakers and temperature switches to open the contacts when high current flows through the circuit.

**bimetal temperature sensor.** A temperature sensor (thermostat) made by joining two metals with different heat expansion rates. As the metals heat, they expand at different rates and bend to open a valve or move clutches, air cleaner vacuum valves, and choke coils.

**binary system.** The mathematical system that uses only the digits 0 and 1 to present information. The mathematical system used by digital computers.

**bipolar.** The general name for NPN and PNP transistors because current flows through semiconductor materials of both polarities. Holes and electrons are both used as current carriers.

**bit.** A binary digit (0 or 1). Bit combinations are used to represent letters and numbers in digital computers. Eight bits equal one byte.

**block learn.** The factor that represents long-term operation and long-term correction of fuel metering on a General Motors fuel injection system.

**blowby.** Combustion gases that leak past piston rings into the crankcase. Blowby can contain unburned fuel, water vapor, acids, and exhaust.

**blowoff valve.** A large pressure relief valve on the outlet side of a turbocharger compressor. It opens to relieve pressure when boost reaches a desired maximum pressure.

**bonding straps.** Additional ground cables installed between the engine and chassis to provide a low-resistance ground path for various electrical circuits.

**boost pressure.** The amount of air pressure increase above atmospheric pressure provided by a supercharger or turbocharger.

**bore.** The diameter of a cylinder. Also the machining process of enlarging or finishing a hole.

**boss.** A machined circular hole in a part, into which another part is installed.

**bottom dead center (bdc).** The exact bottom of a piston stroke.

**bound electrons.** Five or more tightly held electrons in the valence shell of an atom.

**bourdon tube.** A small coiled tube attached to a gear mechanism that turns a gauge needle. When pressure is applied to the tube, it flexes and moves the needle.

**brake horsepower (bhp).** The horsepower developed at an engine's crankshaft.

**brake mean effective pressure.** Mean effective pressure measured as force at the engine crankshaft.

**breakdown voltage.** The prescribed, or designed, voltage at which a zener diode allows reverse current flow.

**British thermal unit (Btu).** A standard unit for measuring heat energy. One Btu is the amount of heat needed to raise the temperature of one pound of water one degree Fahrenheit at atmospheric pressure.

**burn time.** The time in milliseconds during which the fuel ignites and burns to develop cylinder pressure.

**bus.** A common conductor, or transmission path, shared by several circuits, or components, in a computer.

**bus bar.** A solid metal strap or bar in a fuse block that acts as a common conductor for several fuses.

**bypass oil filter.** An oil filter that receives only part of the oil supply on a bypass line from the main oil gallery.

**bypass or module timing.** Ignition timing provided by the module of a direct ignition system during cranking until the synchronization signal is received and ignition coil setup is achieved.

**byte.** A unit of computer data consisting of eight digital bits.

**cam.** An irregularly shaped wheel, or other object, on a rotating shaft or other movable part; used to produce reciprocating motion in a contacting part. In an engine valve train, a cam moves a valve lifter to open a valve.

**camshaft.** A rotating shaft or other movable part with irregularly shaped protrusions or lobes used to produce reciprocating motion in a contacting part. In an engine valve train, camshaft motion moves a valve lifter to open a valve.

**capacitance.** The ability of two conducting surfaces to accumulate a voltage charge and store that voltage when separated by an insulator.

**capacitive-discharge (CD) ignition.** An ignition system that stores its primary energy in a capacitor and then discharges the capacitor to create an expanding magnetic field in the coil.

**case ground.** A ground connection directly from the case of a part to the vehicle engine or chassis.

**catalytic cracking.** The process of cracking hydrocarbon molecules by using a chemical catalyst.

**catalytic reaction.** A chemical reaction caused by a material (a catalyst) that is not, itself, changed by the reaction.

**cathode.** The negatively charged electrode in a voltage cell.

**cathode ray tube (CRT).** An electron-beam tube in which the beam is focused on a small area of a luminescent screen and varied in position and intensity to produce a visible pattern. Used as the display screen in a TV set, an oscilloscope, or a computer monitor.

**central processing unit (CPU).** The calculating part of a computer that makes calculations and logical decisions by comparing input data with data in memory.

**centrifugal (mechanical) advance.** A distributor mechanism that uses centrifugal weights to advance timing in relation to engine speed.

**centrifugal force.** A force that tries to keep moving objects traveling in a straight line. When combined with rotating force, centrifugal force causes an object to move away from the center of rotation and circle the center.

**centripetal force.** Force acting on a rotating body that is applied at the outer edge and directed inward. The opposite of centrifugal force, which moves outward.

**cetane number.** A measurement of diesel fuel's ability to ignite and burn quickly.

**check valve.** A pressure-regulating valve

that allows gas or fluid flow in one direction but prevents flow in the opposite direction.

**choke coil.** A coil wound with fine wire, used to absorb oscillations that occur in a circuit when the circuit is opened or closed.

**circuit.** An unbroken circular path through which electric current can flow. Any path through which force can be applied or matter can move. A hydraulic system also is a circuit.

**claw pole.** The finger, or claw, of an alternator rotor that concentrates the magnetic flux and forms the magnetic field. A claw-pole rotor has interlaced sets of claw poles to form alternating north-south poles.

**clearance volume.** The volume of the combustion chamber and area around the top of the piston with the piston at top dead center.

**closed loop.** The system operating condition when an output error signal is fed back to the system controller, which then readjusts the system to eliminate the error.

**closed-loop dwell control.** A feature of a distributorless ignition at which the ignition control module adjusts dwell time to provide full coil saturation based on the previous coil buildup.

**cloud point.** The low temperature at which wax crystals form in diesel fuel or motor oil.

**code.** A numerical code, generated by a vehicle control system to indicate that a fault has occurred in a particular subsystem, circuit, or part. Carmakers refer to codes as follows: GM, *trouble* codes; Ford, *service* codes; Chrysler, *fault* codes.

**cold-cranking rating.** A battery current-capacity rating that indicates the amperes of current a battery can supply for 30 seconds at 0° F (−18° C) with no cell voltage dropping below 1.2 volts.

**collector.** The outer layer of semiconductor material that receives the majority current carriers in a bipolar transistor.

**commutator.** A segmented ring attached to one end of an armature shaft in a motor (or a dc generator). A commutator acts as a rotary switch to provide a continuous current path from an external circuit through carbon brushes and through the armature windings.

**compound.** A substance, or form of matter, formed by molecules of two or more different elements. Molecules of compounds are chemically bound together by electron sharing.

**compression pressure.** Pressure created by the piston as it moves from bdc to tdc on the compression stroke without having the air-fuel mixture ignited.

**compression ratio.** The ratio of the total volume of a cylinder and combustion chamber to the clearance volume of the chamber alone.

**computer hardware.** The mechanical, magnetic, electrical, and electronic devices that make up the physical structure of a computer.

**condenser.** A capacitor that usually is made of two sheets of metal foil that are separated by an insulator and housed in a small metal can.

**conductor.** A material that has many free electrons and provides easy current flow. Metals are good electrical conductors.

**constant depression.** The constant pressure drop, or vacuum, developed in a variable-venturi carburetor.

**continuity.** Unbroken or continuous. A circuit or electrical device has continuity if it has an unbroken path for current flow.

**conventional theory.** The theory that electric current flows from positive to negative. Also called the positive current flow theory.

**conversion factor.** Any number used to multiply or divide a measurement in one set of units to change it to an equal measurement in another set of units.

**counterelectromotive force (CEMF).** An induced voltage that opposes the source voltage and any increase or decrease in source current flow.

**cracking.** The oil refining process of breaking large hydrocarbon molecules into smaller ones.

**crankshaft throw.** The offset portion of a crankshaft to which a connecting rod and rod bearing are attached. Each throw creates a lever with the crankshaft centerline as the pivot point (fulcrum).

**crossflow head.** A cylinder head with intake ports on one side and exhaust ports on the other side.

**curb idle.** The normal slow-idle speed of a fully warmed up engine.

**cycling.** The action of changing from one condition to an opposite condition and back to the original condition. Battery cycling is the electrochemical action from fully charged to discharged and back to fully charged.

**darlington pair (circuit).** A current-amplifying arrangement whereby one transistor acts as a preamplifier and creates a larger base current for a second transistor. Often described as a "piggyback" power transistor, which requires only three external circuit connections.

**d'Arsonval movement.** A small current-carrying coil in the field of a permanent magnet. Used as the measuring movement in electrical meters.

**data packet.** One complete package, or serial data transmission cycle, from a vehicle that provides control system operating parameters. Also called a data *frame*.

**datum line.** The primary reference dimension for an object, from which all other dimensions are determined.

**delta stator.** A 3-winding alternator stator with the ends of each winding connected to each other; there is no neutral junction. Schematically, the arrangement looks like the Greek letter delta, or a triangle. A delta stator produces high current at low speed.

**demultiplexer (DEMUX).** The solid-state switching device that handles the output data from a body computer module (BCM) to an actuator when the BCM is engaged in the process of sequential sampling.

**derived unit.** Any compound metric unit formed by multiplying two or more base, supplementary, or other derived units.

**detent.** A position of a multiple-pole switch that can be felt or in which the switch will stay until moved to another position.

**detented.** A switch with multiple detent positions.

**detergent-dispersant.** A group of oil and fuel additives that reduce sludge and dirt formation by suspending dirt particles in the oil or fuel.

**detonation.** An unwanted, uncontrolled explosion of the air-gasoline mixture due to high temperature and pressure after spark ignition.

**dielectric.** An insulating material between the plates of a capacitor.

**dieseling.** The tendency of a gasoline engine to keep running after the ignition is off. Also called afterrun. Caused by high combustion chamber temperatures and a throttle that is slightly open.

**digital.** A signal that is either on or off and that is translated into the binary digits 0 and 1. In a computer, a digital signal is voltage that is either low or high, or current flow that is on or off.

**digital microprocessor.** A miniature digital computer on a single silicon chip integrated circuit (IC).

**diode.** An electronic device made by joining P-material to N-material at a junction. It allows current flow in one direction and blocks it in the other.

**direct current.** Current flow that is always in one direction through a conductor.

**direct drive.** A 1 : 1 gear ratio where both drive and driven gears turn at the same speed.

**direct injection.** A diesel fuel injection system that injects fuel directly into a combustion chamber in the cylinder. The chamber often is in the piston crown.

**discrete device.** An individual diode, transistor, or other electronic part that performs only one function and is connected to a circuit by wire leads.

**displacement.** The measurement of the volume of air moved by a piston as it travels from bottom to top of its stroke. Engine displacement is the displacement of one cylinder multiplied by the number of cylinders in the engine.

**distillation.** The process of using heat to vaporize different parts (fractions) of a liquid mixture at different temperatures, then the condensing and collecting of the distilled liquids.

**diverter valve.** An air injection antibackfire valve that diverts airflow to the atmosphere during deceleration.

**dopant.** The other elements added to pure silicon or germanium in the doping process to change the electrical characteristics of the semiconductor material.

**doping.** The addition of a second element to a pure semiconductor. The second element has either three or five valence electrons.

**double-overhead-cam (dohc) engine.** An overhead-cam engine with separate camshafts for intake and exhaust valves.

**drain.** The part of a field-effect transistor that receives the current-carrying holes or electrons. Similar to the collector in a bipolar transistor.

**driver transistor.** The smaller transistor in an ignition control module that receives a timing signal from the distributor and switches the power transistor off.

**dry-sump oil system.** An engine lubrication system in which the oil supply is stored in a tank separate from the engine.

**duplex serial data line.** A data link between two computers that allows each computer to send and receive data from the other. An external clock line between the two computers controls which one sends and which one receives.

**duration spring.** The spring that operates the carburetor accelerator pump plunger or diaphragm.

**duty cycle.** The percentage of time that a solenoid is energized during one complete on-off cycle. Duty cycle is the ratio of on-time to total cycle time.

**dwell angle.** The number of degrees of distributor cam rotation during which the points and the primary circuit are closed and current flows through the coil primary winding. Also called cam angle or simply, dwell.

**dynamic inertia.** The tendency of an object in motion to keep moving.

**dynamic range.** The operating range of a sensor.

**dynamometer.** An instrument used to measure the torque and power output of an engine.

**eccentric.** Off center. A circular shaft lobe that has a center different from the centerline of the shaft.

**eddy current.** Small induced current in the armature of a motor (or a dc generator), created by CEMF. Eddy currents oppose circuit current flow.

**electric current.** The controlled, directed movement of electrons from atom to atom in a conductor. It is caused by electromotive force (voltage).

**electrical load.** The working devices of an electrical circuit, such as lamps and motors, that change electrical energy to heat, light, or motion.

**electrically erasable programmable read-only memory (EEPROM).** Computer memory program circuits that can be erased electrically with a voltage pulse and then reprogrammed.

**electrochemical.** A reaction or condition in which chemical energy is changed to electrical energy, or vice versa. Electrochemical action in a battery produces voltage through the reaction of two dissimilar electrodes in a conductive electrolyte.

**electrochemistry.** Voltage developed by the chemical action of two dissimilar materials in a conductive chemical solution.

**electrode.** A conductor through which electric current enters or leaves a device. A positive or negative electrically charged terminal. A conductive, electrochemically active part of a battery.

**electrolysis.** Chemical decomposition of a material in the presence of a liquid (electrolyte) and electric current.

**electrolyte.** The chemical solution in a battery that provides a medium for plate materials to create voltage and conduct current. Automobile battery electrolyte is sulfuric acid ($H_2SO_4$) plus water ($H_2O$).

**electromagnet.** An iron bar or rod surrounded by a current-carrying coil. Current flow creates a magnetic field that concentrates in the iron and magnetizes it.

**electromagnetic.** The relationship of electrical energy and magnetic energy.

**electromagnetic induction.** The generation of voltage in a conductor by relative motion between the conductor and a magnetic field.

**electromagnetic interference (EMI).** Electrical impulses emitted as high-frequency transmission or radiation from electromagnetic devices. EMI can interfere with computer data transmission, as well as with radio and TV reception.

**electromagnetism.** Magnetism caused by electric current flow.

**electomotive force (emf).** The electrical force that causes current flow (voltage).

**electron.** The smallest individual part of an atom. Electrons have a negative (−) charge and orbit the nucleus of an atom.

**electron theory.** The theory that electric current flows from negative to positive.

**electrostatic field.** The electrically charged area around two points or surfaces that have a voltage difference between them.

**element.** A chemical element is a substance that contains all identical atoms. A fundamental unit of matter. There are 103 known elements, of which 92 exist in nature and 11 are manmade.

**emitter.** The outer layer of semiconductor material that supplies the majority current carriers in a bipolar transistor.

**energy.** The ability, or capacity, to do work. Energy is the work that a system or object *can* do by changing from a given state or position into another state or position.

**engine mapping.** The process of measuring the best combinations of timing, air-fuel ratio EGR, and other controllable variables for various speeds, loads, and temperatures.

**equivalent resistance.** Total series resistance of all branches of a parallel circuit.

**erasable programmable read-only memory (EPROM).** Computer memory program circuits that can be erased and reprogrammed. Erasure is done by exposing the IC chip to ultraviolet light.

**ethylene glycol.** The compound that resists boiling and freezing and that is mixed with water to form engine coolant.

**excitation current.** Field current of an alternator. It magnetically excites the field. During starting, it comes from the battery; when running, it comes from the alternator output.

**excitation voltage.** Battery or charging system voltage that delivers field current to the alternator.

**exponent.** Any number written to the right and above another number to indicate how many times the other number is multiplied or divided by itself. Also called a "power."

**extended-core spark plug.** A spark plug with the insulator and both electrodes made to extend farther into the combustion chamber than they do on a standard spark plug.

**farad (F).** The unit for measuring capacitance.

**feedback.** The process of sending an output error signal back to the system controller, or computer, so that the computer can readjust system output control.

**field-effect transistor (FET).** A transistor through which current flow is controlled by voltage in a capacitive field. A unipolar transistor.

**firing order.** The sequence in which ignition occurs in cylinders of an engine.

**firing (ionization) voltage.** The high voltage needed to ionize the air and fuel molecules at the spark plug gap and create an ignition spark across the gap.

**fixed-dwell electronic ignition.** An electronic ignition with a ballast resistor to control primary voltage and current to the coil. The dwell angle in degrees is fixed and does not vary with engine speed.

**flame front propagation.** The controlled burning process, or spread of the combustion flame front, from the spark plug throughout the combustion chamber.

**flux density.** The number of magnetic flux lines that pass through one square centimeter of any area.

**flux lines.** Magnetic lines of force.

**forward bias.** Voltage applied across a diode that causes current to flow across the junction.

**four-stroke cycle.** The engine operating cycle in which a piston makes four full strokes (intake, compression, power, and exhaust) and 720 degrees of crankshaft revolution to develop power from the combustion process. Also called the Otto cycle.

**frame.** One complete serial data transmission cycle, or package from a vehicle that provides control system operating parameters. Also called a *data packet*.

**free electrons.** Three or fewer loosely held electrons in the valence shell of an atom.

**friction.** The force that resists motion between the surfaces of two objects, or forms of matter.

**friction horsepower.** Engine power lost to internal friction of the engine. Power consumed to overcome internal engine friction.

**full-flow oil filter.** An oil filter through which the entire oil supply passes from the pump before entering the engine oil galleries and lines.

**gain.** The ratio of amplification in an electronic device.

**galvanic battery.** A source of dc voltage. In an exhaust gas oxygen sensor, voltage is generated by a difference in oxygen content near two electrodes.

**gasket creep.** The compression of a gasket that occurs after bolts are tightened.

**gate.** The part of a field-effect transistor that controls the capacitive field and current flow. Similar to the base in a bipolar transistor.

**gear ratio.** The number of revolutions made by a drive gear compared to the number of revolutions made by a driven gear. It is usually expressed as a ratio of drive revolutions to one driven revolution, such as 3 : 1 or 0.75 : 1.

**gear reduction.** A gear ratio where the drive gear rotates faster than the driven gear. The driven gear turns at reduced speed with increased torque. A ratio of 2 : 1 is a gear reduction ratio.

**gross brake horsepower.** Brake horsepower of an engine with only the fuel, oil, and water pumps and built-in emission controls as accessory loads.

**grounded circuit (ground).** An unwanted connection between part or all of a circuit that bypasses current directly to ground. A short circuit to ground.

**gulp valve.** An air injection antibackfire valve that redirects airflow to the intake manifold during deceleration.

**half-wave rectification.** The process of rectifying only one-half of an ac voltage (either positive or negative) and producing pulsating dc.

**Hall-effect switch (sensor).** A semiconductor device that produces a voltage pulse dependent on the presence of a magnetic field.

**hard code.** A trouble code from a vehicle computer system that indicates a problem that is permanently present at the time of testing. A hard code may or may not keep the system from operating.

**hard conversion.** Redesigning a part to convert its design dimensions from one measurement system to another.

**header.** A low-restriction exhaust manifold, usually made of welded steel tubing.

**heat sink.** Any object that absorbs and dissipates heat from another object.

**heat range.** The indication of a spark plug's ability to dissipate, or transfer, heat from its center electrode to the engine.

**helical gear.** An external or internal gear on which the teeth are at an angle to the gear axis, or centerline. All teeth are at the same angle.

**helix.** A spiral or three-dimensional curve. In a fuel injection pump, the spiral groove that opens the spill port and controls fuel metering.

**hertz (Hz).** A unit of frequency equal to 1 cycle per second.

**hole.** The place in a semiconductor valence shell from which an electron is missing. A positive charge carrier.

**horsepower (hp).** One horsepower equals 33,000 foot-pounds of work done in one minute. The amount of power to do an equivalent amount of work in an equivalent amount of time.

**hybrid.** Something of mixed composition. A catalytic converter composed of both oxidation and reduction catalysts.

**hybrid circuit.** An electronic circuit built from discrete devices and IC chips.

**hydraulics.** The science, or study, of using liquids to do work.

**hydrogenation.** The chemical process of adding hydrogen to a compound.

**hydrometer.** An instrument used to measure specific gravity of liquids, such as battery electrolyte or engine coolant.

**idler.** A gear, pulley, or sprocket that turns with drive and driven members but does not change the overall drive-to-driven ratio. An idler gear changes direction of rotation. Idler pulleys or sprockets are used as tensioners in belt and chain drives.

**ignition control module.** The electronic assembly of transistors and other solid-state devices that controls ignition primary current.

**ignition interval.** The degrees of crankshaft rotation between ignition in cylinders in firing order sequence.

**ignition lag.** In a diesel engine, the time from the injection of fuel until the fuel vaporizes and ignites; usually about 2 milliseconds.

**impedance.** The combined opposition to current flow created by resistance, capacitance, and inductance in a circuit. In dc electrical systems, impedance is largely resistance.

**indicated horsepower.** The theoretical maximum power produced inside an engine's cylinders, calculated by measuring combustion pressure.

**indirect injection.** A diesel fuel injection system that injects fuel into a small prechamber separate from the cylinder.

**induced current.** Current in a conductor generated by magnetic flux lines due to relative movement of either the magnetic field or the conductor.

**induced voltage.** Voltage in a conductor generated by magnetic flux lines due to relative movement of either the magnetic field or the conductor.

**induction coil.** Two conductors wound into coils, with one placed inside the other. Transformers and ignition coils are examples.

**inductive-discharge ignition system.** An ignition system based on the induction of high voltage in a coil.

**inertia.** The tendency of an object in motion to keep moving and the tendency of an object at rest to remain at rest.

**inertia engagement.** A general type of starter motor that uses the rotating inertia of the motor armature shaft to engage the drive pinion with the engine flywheel. A Bendix drive is the most common inertia-engagement starter drive.

**insulator.** A material that has many bound electrons and therefore opposes current flow. Many nonmetallic compounds such as glass, rubber, and plastics are good insulators.

**integrated circuit (IC).** A complete electronic circuit of many transistors and other devices, all formed on a single silicon chip.

**integrator.** The factor that represents short-term operation and short-term correction of fuel metering on a General Motors fuel-injection system.

**intercooler.** An air-to-air or air-to-water heat exchanger. In a turbocharger installation, an intercooler lowers the temperature of the intake air charge.

**intermittent code.** An electronic engine control system trouble code that indicates a problem that may or may not exist at the time of testing. An intermittent code occurs on and off, but will be recorded in long-term memory, if the computer has such memory.

**internal combustion engine.** An engine in which the air-fuel mixture is burned inside the engine power chamber.

**internal ring gear.** A gear wheel with an open center and teeth cut on its inner circumference. Also called an *annulus gear* or simply a ring gear.

**ion.** An atom that has become electrically unbalanced by losing or gaining electrons. An ion can be positive (fewer electrons than protons) or negative (more electrons than protons).

**keep-alive memory (KAM).** Random-access memory that is retained by keeping a voltage applied to the circuits when the engine is off.

**kinetic energy.** The specific energy of motion.

**lambda.** The ratio of one number to another. For exhaust gas measurement, lambda is the ratio of excess air to ideal air for complete combustion.

**left-hand rule.** The current flow and flux direction rule based on the electron theory of current flow. Grasping a conductor with the left-hand thumb pointing in the direction of current flow (toward +) causes the fingers to wrap around the conductor in the direction of the flux lines.

**light-emitting diode (LED).** A gallium-arsenide diode that releases energy as light when holes and electrons collide.

**lightoff temperature.** The temperature at which a catalytic converter becomes 50 percent effective, approximately 400° to 500° F (204° to 260° C).

**linear.** In a straight line.

**linearity.** The expression of sensor accuracy throughout its dynamic range.

**liquid crystal display (LCD).** A display that sandwiches electrodes and polarized fluid between two pieces of glass. The application of voltage to the electrodes rearranges the light slots in the fluid and causes light to pass through it.

**logic gate.** Logic circuits used by a computer to manipulate data bits and make variable decisions. The three basic gates are the NOT, AND, and OR gates.

**magnetic field.** The area around a magnet formed by magnetic lines of force. The area that is influenced by the magnet's energy.

**magnetic flux.** A magnetic field.

**magnetic saturation.** The full strength (maximum flux density) of a magnetic field with a steady amount of current.

**magnetism.** The form of energy caused by the alignment of atoms in some materials. The ability of a material to attract iron.

**manifold absolute pressure (MAP).** A combination of atmospheric (barometric) pressure and vacuum. MAP equals barometric pressure minus vacuum.

**mass.** The measure of the inertia of an object or form of matter or its resistance to acceleration.

**mass airflow (MAF).** The measurement of the molecular mass, or weight, of air entering an engine

**matter.** All things in the universe that occupy space, have mass and weight, and are subject to inertia and momentum. Matter may be in gas, liquid, or solid form.

**mean effective pressure.** The average cylinder pressure developed throughout one complete 4-stroke cycle.

**menu.** A list of vehicle tests or programs available from a scan tool or other computerized test equipment. A selection is made from the menu to perform a specific test or to read computer data.

**menu driven.** Referring to a computer program that provides the user with a "menu" or list of choices. Making a choice brings up another page of the menu and requires that another choice be made. In this way, the user tells the computer what he wishes to know and the program displays the desired information.

**micron.** A unit of length equal to one-millionth ($10^{-6}$) of a meter, or 0.00254 inch.

**microprocessor.** A miniature digital computer built as an integrated circuit on a single silicon chip. The microprocessor is the central processing unit (CPU) of an automotive computer.

**minus rule.** The rule that indicates a low-capacity ignition condenser. "Minus metal (pitting) on the minus (−) point indicates a minus (low) capacity condenser."

**mixture.** A substance, or form of matter, consisting of two or more elements or compounds that are not bound together chemically by shared electrons.

**molecule.** Two or more atoms of the same or different elements chemically joined by sharing valence electrons.

**momentary contact.** A kind of switch that operates only when held in position. When released, it returns to its normally open or normally closed position.

**momentum.** The force of continuing motion. The momentum of a moving object equals its mass times its speed.

**monolithic converter.** A catalytic converter that has the catalyst deposited on a single large honeycomb block (a monolith), through which the exhaust flows.

**MOSFET.** A metal-oxide semiconductor, field-effect transistor. A type of integrated-circuit (IC) device used in a microprocessor

**motor octane.** Gasoline antiknock quality tested under conditions of heavy load, high speed, high temperature, and full throttle.

**movie.** A recording of data transmitted from a vehicle to a scan tool. Also called a *snapshot.*

**multigrade (multiviscosity) oil.** An oil whose viscosity is tested and rated at both low temperature and 210° F (−18° and 99° C).

**multimeter.** An electrical test instrument that contains two or more different meters or performs two or more test functions. Often a combined ammeter, voltmeter, and ohmmeter.

**multiple.** A measurement unit that is larger than the stem unit through multiplying by a power of 10.

**multiplexer (MUX).** A solid-state switching device that handles the input data from a given sensor for a body computer module (BCM) when it is engaged in the process of sequential sampling. Multiplexing is the process of transmitting several signals simultaneously over the same data line.

**multiplex wiring system.** An electrical circuit in which signals are transmitted simultaneously by a peripheral serial bus or over optical fiber cables. Several devices share signals on a common conductor.

**mutual induction.** The creation of voltage in one coil by the increase and decrease in the magnetic field around another coil.

**N-material.** Silicon or germanium doped with phosphorus or arsenic so that it has excess free electrons and a negative charge.

**naturally (normally) aspirated.** The natural process of breathing, or taking in air, from the force of atmospheric pressure. A naturally aspirated engine is one without a supercharger to increase intake air pressure.

**negative-ground electrical system.** A dc electrical system in which the battery negative terminal is connected to ground. Most modern automobiles have negative-ground systems.

**negative temperature coefficient (NTC) resistor (thermistor).** A thermistor whose resistance decreases as temperature increases—high resistance at low temperature and low resistance at high temperature.

**net brake horsepower.** Brake horsepower of a fully equipped engine with all accessory loads.

**neutral junction.** The center tap to which the common ends of Y-type stator windings are connected.

**neutron.** A neutral part of an atom with neither positive nor negative charge. Neutrons are located in the nucleus, along with protons.

**noble metals.** The group of metallic elements that includes platinum, paladium, and rhodium. These elements resist corrosion and do not combine easily in chemical reactions, but they serve as chemical catalysts.

**normally closed (NC) switch.** A switch whose contacts are closed until an outside force opens them to open the circuit.

**normally deenergized.** The condition of a solenoid or other electrical load that receives electrical power only after a switch is activated.

**normally energized.** The condition of a solenoid or other electrical load that receives power directly from an electrical system without having a switch activated.

**normally open (NO) switch.** A switch whose contacts are open until an outside force closes them to complete the circuit.

**nucleus.** The center of an atom, made up of protons and neutrons. The center, or core, of any body of matter.

**octane rating.** The measurement of a gasoline's antiknock qualities, compared to 100% isooctane.

**ohm ().** The unit used to measure the amount of electrical resistance in a circuit. One ohm is the amount of resistance present when one volt forces one ampere of current through a circuit.

**open circuit.** An incomplete, or broken, circuit. An open circuit has infinite resistance and allows no current flow.

**open-circuit (no-load) voltage.** Applied voltage measured when a circuit is open and no current is flowing. Usually battery voltage in automobile electrical systems. Also the voltage discharge from a coil secondary winding when not connected to a complete circuit. The same as available voltage for given primary voltage and current.

**open loop.** The system operating condition when an output error signal is not fed back to the system controller.

**orifice.** A restriction in a line (gas or liquid) that regulates pressure by creating a pressure differential across the restriction.

**oscillate.** To move back and forth between two points in regular cycles.

**oscilloscope.** A test instrument that produces a trace of electron motion on a cathode ray tube. The trace is an analog voltage representation that is proportional to amplitude and frequency.

**overdrive.** A gear ratio where the drive gear rotates slower than the driven gear.

The driven gear is overdriven at increased speed with decreased torque. A ratio of 0.75 : 1 is an overdrive ratio.

**overhead-cam (ohc) engine.** An engine with the camshaft mounted in the cylinder head, rather than in the block.

**overhead valve (ohv) engine.** An engine with the intake and exhaust valves in the cylinder head, above the combustion chambers.

**oxidation reaction.** Any chemical reaction in which a compound is formed by adding oxygen to another element or compound.

**P-material.** Silicon or germanium doped with boron or gallium so that it has a shortage of free electrons and a positive charge.

**paladium.** An oxidation catalyst used in catalytic converters.

**parallel circuit.** A circuit with two or more branches or paths through which current can flow.

**parallel data.** Data transmission in which each component of the data is sent at the same time on its own line.

**parameter.** A measured value of control system input or output operation. Parameters include voltage signals, as well as temperature, pressure, speed, and other data.

**particulate.** Microscopic solid particles emitted to the air. A form of pollution.

**pellet-type converter.** A catalytic converter that has the catalyst deposited on small ceramic pellets, through which the exhaust flows.

**percolation.** Boiling and vaporization of a liquid that causes the liquid to move from one area to another.

**peripheral serial bus.** A data link between two computers in which one link sends and the other link receives data transmissions.

**permeability.** The ease with which a material can be penetrated by magnetic flux lines. Iron is more permeable than gases such as air.

**permanent magnet (PM) generator sensor.** A sensor using a pickup coil wound around a permanent magnet polepiece to generate an ac voltage signal as a toothed wheel (reluctor or armature) rotates past it. Also called a reluctance sensor or a pickup coil sensor.

**phase angle.** The angular location of alternator conductors in relation to the rotating field. A phase is a fraction of a complete ac cycle, usually expressed as an angle between sine waves. A 3-phase alternator has three windings 120 degrees apart and produces overlapping sine wave voltage.

**photochemical reaction.** A chemical reaction in the presence of sunlight. HC and $NO_X$ react in the presence of sunlight to form smog.

**photoelectricity.** Voltage developed by light striking the surface of some metals.

**piezoelectric crystal.** A substance that can generate a voltage when subjected to mechanical pressure.

**piezoelectricity.** Voltage developed by pressure applied to certain crystal materials.

**piezoresistive sensor.** A substance whose electrical resistance changes when subjected to mechanical pressure.

**pinpoint test.** A specific test made to a single component or a single function to isolate a malfunction.

**pintle valve.** Any valve formed by a pin that passes through a round opening. A diesel fuel injection valve of this design.

**platinum.** An oxidation catalyst used in catalytic converters. Also can be a reduction catalyst.

**pneumatic.** Systems using air or other gases for mechanical work. The study of the mechanical properties of gases, especially compressed air.

**PN junction.** The point where two opposite kinds of semiconductor material (P and N) join together.

**point gap.** The distance between the breaker points when they are completely open. A wide gap creates a short dwell angle; a narrow gap creates a longer dwell angle.

**polarity.** The possession or existence of two opposing forces or characteristics, such as the north and south poles of a magnet or the positive and negative terminals of an electrical circuit.

**pole.** The opposing ends of a magnet, identified as north (N) and south (S). The areas of a magnetic field where the lines of force are concentrated.

**poppet valve.** A valve that opens and closes by back-and-forth linear motion.

**ported vacuum.** Vacuum taken from a port above the throttle. The amount and timing of vacuum application depends on throttle position. Ported vacuum is shut off at idle.

**positive displacement pump.** A pump that displaces, or delivers, the same amount (volume) of liquid with each revolution regardless of speed.

**positive engagement.** A general type of starter motor that uses the positive mechanical action of a solenoid or a movable pole shoe to engage the drive pinion with the engine flywheel.

**positive-ground electrical system.** A dc electrical system in which the battery positive terminal is connected to ground.

**positive temperature coefficient (PTC) resistor (thermistor).** A thermistor whose resistance increases as temperature increases—low resistance at low temperature and high resistance at high temperature.

**potential energy.** Stored, unreleased energy.

**potentiometer.** A variable resistor that acts as a voltage divider to produce a continuously variable output signal proportional to mechanical position.

**pour point.** The low temperature (lower than the cloud point) at which diesel fuel or motor oil will barely pour or flow.

**power.** The rate, or speed, at which work is done. Power equals the amount of work times the amount of time needed to do it.

**powers of 10.** The number of times that 10 is multiplied or divided by itself to get a larger or smaller number. Powers of 10 are indicated by exponents.

**power transistor.** The high-current transistor, or transistors, that carry primary current and switch it on and off in an ignition control module. Any transistor that carries high current.

**preignition.** An unwanted, uncontrolled ignition of the air-gasoline mixture from a combustion chamber hot spot before spark ignition.

**pressure.** Force exerted on a given unit of surface area. Measured in pounds per square inch (psi) or kilopascals (kPa). Pressure = force/area.

**pressure differential.** A difference in pressure between two points. The pressure differential between 50 psi and 10 psi is 40 psi.

**pressure-regulating valve.** A hydraulic valve that controls pressure developed in a system and the release of that pressure.

**primary battery.** A battery whose electrochemical reaction cannot be reversed. The battery dies when one of the electrodes is destroyed by the electrochemical action; it is not rechargeable.

**primary circuit.** The low-voltage ignition circuit, originating at the battery, that induces high voltage in the coil.

**primary (ballast) resistor.** The resistor in an ignition primary circuit that stabilizes voltage and current to the coil primary winding.

**primary winding.** A coil winding that uses voltage and current flow to create a magnetic field and induce a voltage in a second winding.

**primary wiring.** Low-voltage automobile wiring as distinguished from high-voltage ignition secondary wiring.

**program.** The job instructions for a computer. The program is usually stored in programmable read-only memory (PROM) devices for a vehicle computer.

**programmable read-only memory (PROM).** A computer memory IC chip that can be programmed once to store the computer program.

**proton.** The part of an atom with a positive (+) charge. Protons are located in the center of an atom, along with neutrons.

**pulley.** A grooved, or toothed, wheel that engages a belt in a belt-drive system. Also called a *sheave*.

**pulse air injection.** An air injection system that uses negative pressure (vacuum) pulses in the exhaust to draw air into the exhaust system.

**pulse width.** The operating time, usually measured in milliseconds, of an intermittent device, such as a solenoid.

**pulse width modulation (pwm).** A continuous on-off cycling of a solenoid a specified number of times a second. The pulse width (on-time) can be changed, or modulated, to produce a variable duty cycle. (See *duty cycle*.)

**pumping losses.** Power used and wasted by a gasoline engine to take in air past a restricted throttle.

**radiofrequency interference (RFI).** High-voltage discharges from ignition systems (and other sources) that interfere with radio and television transmission.

**random-access memory (RAM).** Computer read-write memory on which information can be written and from which it can be read.

**read-only memory (ROM).** The permanent program memory of a computer. Instructions can be read from ROM, but nothing can be written into it and it cannot be changed.

**reciprocating engine.** An engine in which pistons move up and down in cylinders. Combustion of an air-fuel mixture at one end of the cylinders develops power and causes a reciprocating motion. Also called a piston engine.

**rectifier bridge.** A single assembly that contains three or all six rectifying diodes for an alternator.

**rectify.** To change alternating current to direct current.

**reduction reaction.** Any chemical reaction in which oxygen is removed from a compound.

**reference voltage.** The constant voltage applied by a computer to a resistive sensor. The sensor sends a varying return voltage related to the quantity being measured.

**relative motion.** A difference in movement between two objects or forces.

**relay.** An electromagnetic switch that uses a small amount of current in one circuit to open or close a circuit with greater current flow. Often used for remote control of a circuit.

**relief valve.** A valve that releases system pressure (gas or liquid) when it exceeds a predetermined value.

**reluctance.** The opposition, or resistance, to magnetic flux lines or a magnetic field.

**reluctance sensor.** A magnetic pulse generator that sends a voltage signal in response to varying reluctance of a magnetic field.

**remote ground.** A ground connection through a wire between an electrical part and the vehicle engine or chassis.

**required voltage.** The specific high voltage required to ionize and fire a spark plug under any given combination of operating conditions.

**research octane.** Gasoline antiknock quality tested under conditions of low load, low temperature, full throttle, and low to medium speed.

**reserve-capacity rating.** A battery current-capacity rating that indicates the number of minutes a battery can supply 25 amperes of current with no cell voltage dropping below 1.75 volts.

**resistor spark plug.** A spark plug with a resistor of about 10,000 ohms built into the center electrode.

**resonator.** A kind of muffler that reduces noise by reflecting sound waves back toward the noise source.

**retraction valve.** A diesel fuel injection pump delivery valve, so called because it causes fuel to retract in an injection when it closes and reduces injection pressure.

**reverse bias.** Voltage applied across a diode that prevents current flow across the junction.

**reverse polarity.** The condition in which the ignition primary circuit has a reversed connection (usually at the coil). This causes the secondary voltage polarity at the spark plug to be reversed and force current from a positive (+) center electrode to a negative (−) ground electrode.

**reversing relay.** A relay that reverses the action of a switch. Closing a switch applies current to the control contacts of the relay to open the normally closed power contacts of the relay.

**revolutions per minute (rpm).** A measurement of engine rotating speed. The number of complete revolutions made by the crankshaft in one minute.

**rheostat.** A variable resistor used to regulate the strength of an electrical current.

**rhodium.** A reduction catalyst used in catalytic converters.

**right-hand rule.** The current flow and flux direction rule based on conventional current flow theory. Grasping a conductor with the right-hand thumb pointing in the direction of current flow (toward −) causes the fingers to wrap around the conductor in the direction of the flux lines.

**rocker arm ratio.** The ratio of the distance from the rocker arm pivot point to the pushrod center on one end and the valve stem center on the other. Rocker arm ratio multiplies cam lift to determine actual valve lift.

**roller chain.** A drive chain in which the cross piece of each link is a small roller that rotates in a pin to reduce friction as it engages the teeth of a sprocket.

**Roots blower.** A positive-displacement blower or supercharger that uses two lobe-shaped rotors to draw in and compress air.

**running change.** A change made by the

manufacturer to a particular car model during the course of a model year. Running changes often cause revised specifications and service procedures.

**running mates.** Cylinders in which ignition occurs 360° from each other.

**scan tool (scanner).** A diagnostic computer that is used to communicate with and test a vehicle onboard computer. Because it "scans the serial datastream" on a GM car, it came to be called a "scan tool, or scanner."

**secondary battery.** A battery whose electrochemical reaction can be reversed. The battery electrodes and electrolyte can be restored by reversing the current flow.

**secondary circuit.** The high-voltage ignition circuit, originating at the coil secondary winding, that delivers high voltage to the spark plugs.

**secondary winding.** A coil winding that receives an induced voltage from the magnetic field created by current flow in a primary winding.

**semiconductor.** An element with four valence electrons. Semiconductors are neither good conductors nor good insulators but will conduct current under controlled conditions.

**sensor.** Any device that sends an input signal to a computer.

**sequential sampling.** A complex switching method involving a multiplexer and demultiplexer, which allows a body computer module (BCM) to deal with sensor input and actuator output in sequence and in a logical manner.

**serial data line.** A data link between two computers that transmits the digital words one after another, or serially, from one computer to the other. The computers are clocked together and synchronized to the data being transmitted.

**serial datastream.** The stream of computer data transmitted serially from a General Motors car to a scan tool.

**series circuit.** A circuit with only one path for all current flow.

**series-parallel circuit.** A circuit with some loads in series with the power source and other loads, and some loads in parallel with each other.

**serpentine drive belt.** A V-ribbed drive belt installation that uses a single belt to drive several engine accessories.

**short circuit (short).** An unwanted connection between two conductors that bypasses part of a circuit.

**shunt.** A low-resistance electrical connection or branch circuit in parallel with another connection or branch circuit.

**silent chain.** A drive chain made from a series of flat metal links pinned together so that the teeth of each link align to form teeth similar to those on an internal ring gear.

**silicon-controlled rectifier (SCR).** A common type of thyristor used in automative circuits.

**SI metric.** The modern international metric system, used by the automobile industry and other industries.

**sine wave voltage.** The changing voltage polarity between positive and negative of alternating current. The voltage pattern is based on the trigonometry sine function.

**single-grade (straight-grade) oil.** An oil whose viscosity is tested and rated at either, but not both, 0° or 210° F (−18° or 99° C).

**single-overhead-cam (sohc) engine.** An overhead cam engine with a single camshaft for intake and exhaust valves.

**single-phase voltage.** The complete sine wave voltage induced in one conductor during one revolution of an alternator rotor.

**siphon.** The use of a tube or hose to allow atmospheric pressure to force a liquid upward, over a barrier, and then downward to a point lower than its original level.

**sludge.** Heavy, black engine deposits formed when oil mixes with blowby and water vapor.

**snapshot.** A recording of data transmitted from a vehicle to a scan tool. Also called a *movie*.

**soft code.** A trouble code from a vehicle computer system that indicates a problem that is not present at the time of testing. A soft code indicates an intermittent problem that occurred sometime before testing, but which is not present now. Soft codes are stored in long-term computer memory. Usually they are automatically erased after 50 or 100 ignition on-off cycles.

**soft conversion.** Restating the dimensions of a part or some measured value in units of a measurement system other than the system originally used. The part is not physically changed.

**software.** The various programs and data in ROM and RAM that provide a computer with memory and operating instructions

**solenoid.** A device similar in operation to a relay, but movement of the armature or movable iron core changes electrical energy into mechanical motion.

**solid-state electronics.** Electronic systems based on solid semiconductor devices such as diodes and transistors.

**spark plug reach.** The dimension from the spark plug seat to the tip of the shell (the end of the threaded portion). Spark plug reach must match the depth of the spark plug hole in the cylinder head.

**spark voltage.** The voltage, lower than firing voltage, needed to sustain a spark across a plug gap after initial ionization and firing.

**specific energy.** Energy that has been released from its potential form and is doing specific work as heat, light, or mechanical energy.

**specific gravity.** The weight of a volume of liquid divided by the weight of an equal volume of water at equal temperature and pressure. The ratio of the weight of any liquid to the weight of water, which has a specific gravity of 1.000.

**speed density.** More correctly, speed *and* density. A fuel injection control method that uses engine speed (rpm) and air density measured by a MAP sensor to regulate fuel metering in relation to engine air intake.

**sprocket.** A toothed wheel that engages a chain in a chain-drive system.

**spur gear.** An external or internal gear on which the teeth are parallel with the gear axis, or centerline.

**starting bypass.** The parallel circuit path that bypasses the ballast resistor when an engine is started.

**state of charge.** The percentage of charge present in a battery at any time. It is indicated by electrolyte specific gravity and battery voltage.

**static electricity.** Opposite voltage charges present on two opposing surfaces. The charges do not move; they are static. Usually caused by the friction of rubbing two materials together. Also the kind of charge present in a capacitive field.

**static inertia.** The tendency of an object at rest to remain at rest.

**stem unit.** Any metric unit to which a prefix can be added to indicate larger or smaller measurements to some power of 10. Kilopascal is formed by adding the prefix, "kilo," to the stem unit, "pascal."

**stepper motor.** A dc motor that moves in incremental steps between a no-voltage (deenergized) position and a full-voltage (energized) position.

**stoichiometric ratio.** An air-fuel ratio of approximately 14.7 : 1, which provides the most complete combustion and recombination of air and fuel.

**stroboscopic.** The effect of a rapidly flashing light that makes moving objects appear to stand still.

**stroke.** The distance that a piston travels from bottom to top in a cylinder. Also, one complete bottom-to-top or top-to-bottom movement of a piston.

**submultiple.** A measurement unit that is smaller than the stem unit through dividing by a power of 10.

**substrate.** A foundation material upon which other materials are deposited. The ceramic core (pellets or monolith) of a catalytic converter, or the basic silicon wafer of a semiconductor device.

**sulfur oxides.** Compounds formed by the combination of sulfur and oxygen. A form of air pollution.

**supercharging.** The process of using a mechanical pump to increase intake air pressure for an internal combustion engine.

**supplementary unit.** The two SI metric units for angular measurement that are based on abstract geometrical concepts rather than on physical measurement of a quantity.

**switching transistor.** The power transistor in an ignition control module. Any transistor used as a switch or relay to turn current on and off. (See *power transistor*.)

**symmetrical.** The same pattern on each side of a centerline. The light beam of a symmetrical high-beam headlamp is positioned the same distance on each side of the center.

**synchronous data.** A constant data flow with regular clock pulses that incorporates a sync pulse at the beginning of each new piece of data to tell the computer that this is a new word.

**synthetic motor oil.** Oil compounded by breaking down and recombining petroleum molecules or by formulating lubricant molecules from nonpetroleum materials.

**television-radio-suppression (TVRS) cables.** High-resistance spark plug cables that have nonmetallic conductors to reduce radiofrequency interference.

**temperature inversion.** A layer of warm air at high altitude that traps cooler air near the ground.

**tensioner.** An idler pulley or sprocket that maintains necessary tension on a drive belt or chain.

**tetraethyl lead.** A gasoline antiknock additive that raises the gasoline octane rating.

**thermal cracking.** The process of cracking hydrocarbon molecules by applying heat.

**thermal energy.** The specific energy of heat.

**thermistor.** A variable resistor whose resistance changes when temperature changes.

**thermocouple.** A temperature-measuring device made of two dissimilar metals joined so that voltage is generated between two points that is directly proportional to the temperature difference between the two points.

**thermoelectricity.** Voltage developed by the flow of heat through two dissimilar metals.

**three-phase alternating current.** Three overlapping, evenly spaced, single-phase ac voltage sine waves that make up the complete ac output of an alternator.

**three-phase voltage.** Three overlapping ac voltage sine waves generated by an alternator with a 3-winding stator. The sine waves are at equally spaced, varying phases.

**threshold voltage.** The minimum conduction voltage for junction diode operation.

**throttling.** The process of regulating the amount of air-fuel mixture entering an engine to control speed and power.

**timing (spark) advance.** Ignition occurring sooner before top dead center than initial timing. Any advance in timing is measured in degrees of crankshaft rotation.

**titania.** Titanium dioxide. A compound used as a temperature-sensing device in one kind of exhaust oxygen sensor.

**top dead center (tdc).** The exact top of a piston stroke.

**torque.** Twisting or turning force. The action of a force that produces rotating mechanical work around an axis. Torque equals the distance to the axis times the force applied.

**transducer.** A device that changes one form of energy to another. For example, a transducer can change motion to voltage, or air pressure to motion. An automotive sensor, actuator, or display device is a transducer.

**transfer port.** The slot-shaped metering port, or ports, just above the throttle in a carburetor barrel that transfers fuel flow for low-speed operation.

**transistor.** A 3-element semiconductor device of NPN or PNP materials that *trans*fers electrical signals across a res*istor*ance.

**turbocharger.** An exhaust-turbine-driven centrifugal (variable-displacement) supercharger.

**turbo lag.** The brief hesitation from the time the throttle opens until a turbocharger starts delivering boost.

**turns ratio.** The ratio of the number of turns in a coil secondary winding to the number of turns in the primary winding. Turns ratio indicates voltage multiplication in a coil.

**two-stage (three-way) catalytic converter system.** A system that has two catalysts, oxidation and reduction, to work against three pollutants, HC, CO, and $NO_X$.

**two-stroke cycle.** The engine operating cycle in which a piston makes two strokes (intake-compression and power-exhaust) and one 360-degree revolution to develop power from the combustion process.

**vacuum.** Air or gas pressure lower than atmospheric pressure. Complete vacuum is the complete absence of gaseous matter.

**vacuum advance.** A distributor advance mechanism that uses a vacuum diaphragm and linkage to advance timing in relation to engine load.

**vacuum lock.** A low-pressure condition in a fuel tank or other reservoir that keeps fluid from flowing, or being pumped, from the tank.

**valence shell.** The outer electron shell in any atom that contains the electrons that join with electrons of other atoms to create molecules. The valence shell contains 1 to 8 electrons.

**valve duration.** The number of crankshaft degrees during which a valve is open.

**valve lift.** The maximum distance that a valve opens, measured in a decimal-inch fraction or millimeters.

**valve overlap.** The period (in crankshaft degrees) when the intake and exhaust valves for one cylinder are open simultaneously. Overlap occurs between the exhaust and the intake strokes.

**valve timing.** The relationship between crankshaft position and the opening and closing positions of the intake and exhaust valves.

**valve train.** All of the parts from camshaft to valves that open and close the valves in an engine.

**vaporize.** To change solid or liquid matter into a gaseous form. Vapor is gaseous matter.

**vapor lock.** A condition in which gasoline vaporizes in the fuel pump, the lines, or the carburetor fuel bowl due to high temperature. Fuel flow stops because of vapor lock.

**variable displacement pump.** A pump such as a centrifugal pump, whose output volume changes with speed.

**variable-dwell electronic ignition.** An ignition that uses full battery voltage on the coil with no ballast resistor. Dwell time and coil saturation are controlled by circuits in the ignition module. Dwell angle varies with engine speed.

**varnish.** Engine deposits formed by high-temperature oxidation of motor oil.

**venturi.** A restriction in an airflow passage, such as a carburetor barrel, that increases airflow velocity and reduces pressure.

**venturi effect.** The relationship of increased airflow velocity and reduced pressure created by a venturi.

**vernier scale.** A fine, auxiliary scale that indicates fractional parts of a larger scale.

**viscosity.** The measurement of a liquid's tendency to resist flowing.

**viscous drive.** A drive mechanism, such as a fan clutch, that connects drive and driven members through the viscosity of a fluid.

**volatility.** A measure of a liquid's ability to change into a vapor.

**volt.** The unit used to measure the amount of electrical force or energy.

**voltage.** The electromotive force that causes current to flow. The potential force that exists between two points when one is positively charged and the other is negatively charged.

**voltage cell.** Two dissimilar materials suspended in a conductive and reactive substance.

**voltage reserve.** The difference between the required voltage needed to fire a spark plug and the maximum available voltage that a coil can produce.

**volume.** The physical size of an object, or form of matter, for a given mass. An object's length, width, and height.

**volumetric efficiency.** Expressed as a percentage, the comparison of the actual volume of air taken into an engine to the theoretical maximum amount that the engine could draw in (its displacement).

**wastegate.** A pressure-operated valve that diverts part of the exhaust around a turbocharger turbine. It opens to limit turbine speed when boost reaches a desired maximum pressure.

**waste spark.** An ignition system in which one coil fires two spark plugs at one time. The spark plug in the cylinder on compression ignites the mixture; the spark fired in the cylinder on its exhaust stroke is wasted.

**watt (W).** The unit of electric power. The amount of electrical work done in a given time. Watts = amperes × voltage.

**weight.** The measure of the earth's gravitational force, or pull, on an object.

**wet-sump oil system.** An engine lubrication system in which the oil supply is stored in the engine oil pan.

**Wheatstone bridge.** An arrangement of resistors that allows resistance to be measured by measuring current and voltage.

**wiring harness.** A collection of wires encased in a plastic covering or conduit and routed to specific areas of the vehicle. Most harnesses terminate in plug-in connectors.

**work.** The transfer of energy from one object or system to another as potential energy becomes specific energy. Mechanical work is the application of energy through force to produce motion.

**work harden.** The effects of vibration and bending that cause a piece of metal to become hard and brittle.

**Y-type stator.** A 3-winding alternator stator in which one end of each winding is connected to a neutral junction. The other end of each winding is connected between a positive and a negative diode. Schematically, the windings are arranged like a letter "Y." A Y-type stator produces high voltage at low speed.

**zener diode.** A diode that allows reverse current flow above a prescribed voltage without being damaged.

**zirconia.** Zirconium dioxide. A compound used as an oxygen-sensing device in one kind of exhaust oxygen sensor. When exposed to oxygen, zirconia acts as a galvanic battery and generates a voltage.

# INDEX

— engine ok at idle – on A/C engine stall ?!

– engine OK at idle – used loop telephone engine stall?